PHYTOREMEDIATION

ENVIRONMENTAL SCIENCE AND TECHNOLOGY

A Wiley-Interscience Series of Texts and Monographs

Edited by JERALD L. SCHNOOR. *University of Iowa*
ALEXANDER ZEHNDER, *Swiss Federal Institute for Water Resources and Water Pollution Control*

A complete list of the titles in this series appears at the end of this volume.

PHYTOREMEDIATION

Transformation and Control of Contaminants

Edited by

Steven C. McCutcheon

U.S. Environmental Protection Agency
National Exposure Research Laboratory
Athens, Georgia

Jerald L. Schnoor

The University of Iowa
Iowa City, Iowa

A JOHN WILEY & SONS, INC., PUBLICATION

Published by John Wiley & Sons, Inc., Hoboken, New Jersey.

Published simultaneously in Canada.

For general information on our other products and services please contact our Customer Care Department within the U.S. at 877-762- 2974, outside the U.S. at 317-572-3993 or fax 317-572-4002.

Wiley also publishes its books in a variety of electronic formats. Some content that appears in print, however, may not be available in electronic format.

Library of Congress Cataloging-in-Publication Data is available.

ISBN: 0-471-39435-1

Printed in the United States of America

10 9 8 7 6 5 4 3 2 1

Dedication

To the U.S. Environmental Protection Agency Office of Research and Development Contaminated Sites Program, the U.S. Strategic Environmental Research and Development Program, and the European COST 837 project for the seminal support and organization of phytoremediation research and development.

In memory of

Rajiv Bhadra (1964–2002), a member of the Editorial Review Board and a rising star in phytoremediation who will be sorely missed
Eugene P. Odum (1913–2002) and **Howard T. Odum** (1924–2002), outstanding and exceptional brothers who pioneered systems ecology and ecological engineering that will be vital to the future of phytoremediation

Contents

Series Preface

Environmental Science and Technology

We are in the third decade of the Wiley-Interscience Series of texts and monographs in Environmental Science and Technology. It has a distinguished record of publishing outstanding reference texts on topics in the environmental sciences and engineering technology. Classic books have been published here, graduate students have benefited from the textbooks in this series, and the series has also provided for monographs on new developments in various environmental areas.

As new editors of this Series, we wish to continue the tradition of excellence and to emphasize the interdisciplinary nature of the field of environmental science. We publish texts and monographs in environmental science and technology as it is broadly defined from basic science (biology, chemistry, physics, toxicology) of the environment (air, water, soil) to engineering technology (water and wastewater treatment, air pollution control, solid, soil, and hazardous wastes). The series is dedicated to a scientific description of environmental processes, the prevention of environmental problems, and preservation and remediation technology.

There is a new clarion for the environment. No longer are our pollution problems only local. Rather, the scale has grown to the global level. There is no such place as "upwind" any longer; we are all "downwind" from somebody else in the global environment. We must take care to preserve our resources as never before and to learn how to internalize the cost to prevent environmental degradation into the product that we make. A new "industrial ecology" is emerging that will lessen the impact our way of life has on our surroundings.

In the next 50 years, our population will come close to doubling, and if the developing countries are to improve their standard of living as is needed, we will require a gross world product several times what we currently have. This will create new pressures on the environment, both locally and globally. But there are new opportunities also. The world's people are recognizing the need for sustainable development and leaving a legacy of resources for future generations at least equal to what we had. The goal of this series is to help understand the environment, its functioning, and how problems can be overcome; the series will also provide new insights and new sustainable technologies that will allow us to preserve and hand down an intact environment to future generations.

JERALD L. SCHNOOR AND ALEXANDER J. B. ZEHNDER

Preface

SCOPE AND COVERAGE

This book covers phytotransformation, phytodegradation, rhizosphere degradation, and phytocontainment of xenobiotic organic pollutants and select inorganic compounds that plant enzymatic processes transform or mineralize. Consistent with this coverage, the term phytoremediation is defined to encompass the use of green plants, fungi, algae, bacteria, and microbial mats when one of three vital plant processes is involved in waste management. These processes include (1) photoautotrophic conversion of sunlight to useful energy and use of atmospheric carbon dioxide to synthesize new biomass, thus fueling plant and rhizosphere microbial control and metabolism of contaminants; (2) green-liver metabolism involving transformation, conjugation, and sequestration of contaminants and the resulting by-products (*e.g.*, plant and fungal glycosylation and lignification); and (3) plant transpiration to control the movement of contaminants in water, soil, and air. Other recent books (cited in Chapter 1 of this book) cover phytoextraction of toxic metals, phytovolatilization of selenium, and phytostabilization of metals and organics. Where necessary for complete coverage of phytoremediation terminology and fundamentals, a few overlaps occur to maintain consistency. In one case, McIntyre (Chapter 30 of this book) introduces two new plant databases—one for rhizodegradation of petroleum hydrocarbons, and one for metals accumulation. In the other case, Rock (Chapter 31 of this book) reviews several field evaluations of phytoremediation; include some disappointing field results for phytoextraction of lead.

To broadly cover the latest advances from fundamental investigation to field testing of concepts, seven sections are the basis of organization for this book *Phytoremediation.* Each section starts with fundamental contributions that define the state-of-the-science and ends with chapters on the applications of fundamental and heuristic concepts in practical settings. The first three chapters overview the state of the science and practice, including review of technical, economic, social, and regulatory issues in translating the research to date into practical cleanup applications. Burken (Chapter 2 of this book) covers the vital green-liver concept of Heinrich Sandermann, Jr. for plant metabolism. Section II covers fundamental and important advances involving enzymatic metabolic processes, proteomic and genomic bases of plant tolerance, phytotoxicity of selected xenobiotic chemicals, and fundamental physiological processes that include rooting and root ecology, and evapotranspiration. Sections III, IV, and V cover the spectrum of fundamental investigation to field testing for aromatic and hydrocarbon contaminants, explosives, and chlorinated solvents, respectively. Section VI covers the latest modeling, design, and field application advances, starting with the latest in phytohydraulic control and modeling that establish the state-of-the-practice, then concentrates on applied management techniques for wastewaters, leachates, and brines. Section VII presents the latest advances in genetic engineering and screening of plants that may be useful in managing atmospheric nitrous oxides and halocarbon pollution, use of plants to control methyl *tertiary*-butyl ether (MTBE), phytodegradation of cyanide in soil, and rhizodegradation and phytodegradation of dissolved perchlorate. The final two chapters of this section, and of the book, introduce the first plant databases and review current field evaluations of important types of phytoremediation. The heuristic plant-based practices of land farming (including sewage spraying), constructing treatment wetlands, and developing riparian buffers that predate coining the term phytoremediation in 1991 are covered to unify recent chemical-specific treatment approaches with the ecological engineering of wetlands, buffers, or tree, grass, and cultivar plantings. The appropriate practices are introduced and evaluated in the scientific context of specific pathways, transformation products, kinetics, and efficiency in achieving cleanup standards or acceptable residual risks. The exceptions involve a few cases where field experience with trees, grasses, and other vegetation establish some beneficial effects, but the scientific basis has not yet been fully explored.

INTENDED AUDIENCE

This book is intended to be a definitive reference for leaders in the research and practice of phytoremediation as well as those students entering the field. Practicing engineers, ecologists, foresters, agronomists, and extension agents; waste site managers; and regulatory experts will find this book to be a definitive reference on the phytoremediation that is possible, feasible, and proven for organic and some inorganic pollutants in water, soil, and air. Each chapter has a summary of practical implications. Where necessary to define fundamental principles for broad audiences, a glossary and definitions in the text are judiciously applied. Système International (SI) units are used (along with English common in most practical chapters) and care taken to avoid confusion over decimal points and numerical expression. (The U.S. practice of using a period as the decimal point and avoiding the use of the comma in favor of a space in denoting thousands is used for numbers greater than 9999, *e.g.*, 99 800.) Attention has also been paid to the consistent use of significant digits and scientific nomenclature in each chapter to facilitate practical applications of the knowledge worldwide.

The coverage of existing and new practices includes wetland construction, land farming, tree and crop plantation, riparian buffer management, and biotechnology-based waste treatment unit processes to treat most of the major organic xenobiotic contaminants (*e.g.*, phenols, hydrocarbons, surfactants, pesticides, explosives, and chlorinated solvents) and a few inorganic contaminants that plants mineralize or volatilize. The underlying theme is the use of *in situ*, sustainable, and renewable biotechnology to protect humans and the environment, but heuristic, short-term, energy-intensive methods are described and evaluated when necessary for complete waste management coverage of this innovative, evolving field. In some cases, design guidance can be distilled from this book.

Experts and students in allied fields will also find this book to be the definitive introduction to the science and practice of phytoremediation. Some of the allied fields include plant, fungal, and bacterial biochemistry, genetics, and proteomics; enzymology and metabolic engineering; biotechnology; ecology and ecological engineering; wetland ecology and hydrobiology; plant biology and other life sciences; plant, crop, and soil sciences and agronomy; forestry and silviculture; botany, plant physiology, and root ecology; plant toxicology; environmental chemistry and science; environmental, biological, bioresource, irrigation, agricultural, chemical, and civil engineering; microbiology and bioremediation; hazardous waste management; groundwater hydrology and hydrogeology; biometeorology; water resource management; alternative biofuel production; biogeochemistry, global change modeling, and risk assessment involving plants as sinks and sources; indoor and outdoor air pollution control; landscaping; land use planning and management; and environmental and ecological economics and management.

Graduate and undergraduate students interested in phytoremediation should find this book to be an indispensable reference to practical case studies as well as definitive process research on why phytoremediation works and where current gaps in knowledge exist that can be filled by enterprising thesis and dissertation research over the next decade or longer. Course instructors and curricula planners in the evolving phytoremediation and ecological engineering programs of study will find this book an adequate text to provide fundamental background and case studies until texts are tailored to this purpose. Assignments should be easily derived from the practical elements and practical implications summarized in each chapter. For course organization and planning, the sections group together all the work on major contaminant problems such as aromatic and hydrocarbon contaminants, explosives, and chlorinated solvents following the overview and the basics of phytoremediation—green-liver metabolism, tissue culturing and enzymology, proteomics and useful plant biochemistry approaches, basis of plant tolerance, root ecology and control, and evapotranspiration. The latest advances available for study include new approaches to treating air pollutants, MTBE, and perchlorate in groundwater and wastewater, and cyanide in soil, especially from the numerous abandoned town gas sites that was the source for street and home lighting in the late nineteenth and early twentieth centuries. Hopefully, an instructors' guide with assignments and engineering design will follow soon.

ORGANIZATION AND PEER REVIEW

The extensive involvement of many research groups was also intended to be exceptional. The authors of the various chapters represent almost all of the leading teams developing phytoremediation of organic contaminants. In some cases, authorship is shared among some of the most productive teams for focused, concise coverage of important topics. Other chapters were specifically directed to full coverage of all known work of importance, especially to involve younger collaborators and future leaders in the field. Despite the outreach involved, a few have surged into prominence in this dynamic field since the writing began 2 years ago. The Editorial Review Board was therefore set up to engage some emerging leaders during the process, some who were constrained by time, but mostly those with review skills and foresight into the coverage necessary.

A remarkable trait of almost all of the research and development teams engaged in phytoremediation became evident early on, as authors were selected. Almost all U.S. teams have dual leadership from a science discipline and from engineering. European research and development is a bit different, where the strength and leadership in the field comes from the marvelous organization and coordination of the COST 837 project led by Jean-Paul Schwitzguébel of the Swiss Federal Institute of Technology in Lausanne, Switzerland, and Tomas Vanek of the Czech Academy of Sciences in Prague. Despite the inadequate funding for research and development noted in Marmiroli and McCutcheon (Chapter 3 of this book), the outlook for phytoremediation based on these productive teams and organizations is very good. Government and industry leaders and the public should look forward to the development of additional cost-saving methods that effectively manage widespread, moderately toxic contamination and some more toxic hot spots using sustainable, natural processes that can be easily engineered for the benefit of humankind and the ecosystem of this planet.

To ensure the best quality coverage, all chapters including McCutcheon and Schnoor (Chapter 1 of this book) were independently peer reviewed and accepted for the book by one of the editors not associated with the authors. In the case of McCutcheon and Schnoor (Chapter 1 of this book), a senior member of the Editorial Review Board, Alan Baker, was empowered as acting editor to assess the independent reviews and determine if the chapter was sufficient. The reviews started with outlines of each chapter to ensure coverage and coordination. Because of these reviews by the Editorial Review Board and authors of other chapters, some gaps in coverage were filled with a second round of invited contributions. Board members and the editors also highlighted the fast-developing work and the important topics. Once the chapters were completed, three to seven reviews were undertaken. One editor, one member of the Editorial Review Board, and one of the better-known experts in the area (if this did not involve the editor or Board member) reviewed each chapter. Authors of other chapters commented on overlaps in coverage, but also provided outstanding technical criticism. Most of the reviews were focused on chapters that defined a consensus on the state-of-the-practice that hopefully will lead to several design guidance documents after this book is published. All revisions were further evaluated editorially. Despite the rigor in review, only one chapter was declined.

EXPRESSIONS OF APPRECIATION

We thank the authors of each chapter for the quality coverage, especially those who teamed with others to cover gaps in knowledge. Christina Negri and Paul Schwab took on the daunting tasks of distilling extensive work on tree and grass root ecology, respectively, into a phytoremediation context. The support of Hector Flores on the Editorial Review Board was important in the editors enlarging the scope of the book to cover this important, but previously neglected area. Victor Medina led the international coverage of plant tolerance and worked with Elena Maestri to translate extensive literature on pesticides and metals toxicity into knowledge that supports the phytoremediation of other classes of xenobiotic compounds. Jerry Schnoor organized coverage of work involving MTBE. Paul Olson covered vital work at the University of Oklahoma and the developing work at Colorado State University with experience from the University of California at Berkeley *via* the work of Elizabeth Pilon-Smits. Hans Harms and his coauthors also put the independent axenic tissue culture investigations of John Fletcher at the University of Oklahoma and the U.S. Environmental Protection Agency laboratory at Corvallis into a fuller and more practical context with his extensive work over the years. Jim Jordahl reached out to Kenth Hasselgren of SWECO VBB VIAK AB on the valuable experience in tree plantation in Sweden and northern Europe. Similar leadership of Sandy Eberts, Larry Davis, Stefan Trapp, and Hiromichi Morikawa was also vital to the quality and coverage of this book and is much appreciated by the editors. In addition, the outstanding reviews of other chapters by Larry Davis and Tomas Vanek were notable and much appreciated.

We also thank the Editorial Review Board members for the outstanding reviews of outlines and final chapter drafts. Rajiv Bhadra, before his untimely death, reviewed all the chapters on explosives and provided insightful criticism and suggestion for improvement and consistency. Alan Baker served as acting editor to avoid a conflict of interest in accepting Chapter 1 by the editors. Sridhar Susarla provided valuable reviews and assistance with the final proofs. Evelyn Drake provided broad insights that led to better coverage of hydrocarbon phytoremediation and root ecology, and enabled the Board to foresee some developments in allied areas. Rashalee Levine was an early advocate in providing precise plant nomenclature and quality assurance information so that many more insights can be distilled after publication of the book.

In addition, the reviews of other experts are very much appreciated for the rigor in examining the technical quality of the work and suggesting better forms of expression when appropriate. These reviewers include Pedro Alvarez of the University of Iowa; Ellie Best of the U.S. Army Waterways Experiment Station; Om Parkash Dhankher of the University of Georgia; Shoeleh Di Julio of California State University; Steve Dwyer of the U.S. Department of Energy Sandia National Laboratory; Andre Gerth of BioPlanta; Leon Kochian of the U.S. Department of Agriculture Plant, Soil and Nutrition Laboratory; Peter Kulakow of Kansas State University; Louis Licht of EcoloTree; Steven Link of Washington State University; Ed Mead of the U.S. Army Corps of Engineers; Anu Ramaswami of the University of Colorado, Denver; Paul Thomas of Thomas Consultants, Inc.; William J. (Jody) Waugh of the U.S. Department of Energy Environmental Sciences Laboratory; Jody Wireman of the U.S. Air Force Human Systems Centre; Keith Halford of the U.S. Geological Survey; Paul Brantley of the U.S. Geological Survey; and John Fletcher of the University of Oklahoma.

We also thank David Tsao of the BP Corp. for the coordination of coverage of the field in another publication. David edited the monograph *Phytoremediation* for the Springer-Verlag series *Advances in Biochemical Engineering/Biotechnology* (vol. 78) during the time this book was being produced.

Bob Esposito of John Wiley was most helpful in organizing and maintaining momentum for this effort along with Danielle Lacourciere in the final stages. Vera Madison provided outstanding editorial and office support in Athens that was much appreciated, especially in final stages. Most importantly, we appreciate the support of our families and colleagues during this immense undertaking.

Finally, we find three remarkable sources of support for the international authorship of this book. The U.S. Environmental Protection Agency Office of Research and Development Contaminated Sites Program supported both editors and several authors, especially through the Hazardous Substance Research Centers. This program funded a 1991 proposal of Ila Raskin in which the term phytoremediation was first defined. Next, the U.S. Strategic Environmental Research and Development Program also has supported both editors and several authors directly and indirectly. This program was conceived in 1991 by former U.S. Senators Sam Nunn of Georgia and Al Gore of Tennessee as an appropriate way to redirect the peace dividend of defense technology development to the international clean up of cold war facilities in the U.S. and Europe. Madelyn Creedon of the Senate Defense Appropriations Committee suggested the appointment of a renowned Science Advisory Board to ensure use of the best science in developing phytoremediation and other innovative waste management techniques. Finally, the European project COST 837 is a wonderful organization of much needed fundamental research on phytoremediation that has supported or aided many authors in the last several years. Other government, industrial, and commercial support for the field of phytoremediation has also been important and is acknowledged in the chapters where appropriate. For example, Marmiroli and McCutcheon (Chapter 3 this book) note the entrepreneurial and government leadership that has occurred in making phytoremediation a useful technology.

DISCLAIMER

This contribution has been reviewed in accordance with the U.S. Environmental Protection Agency peer and administrative review policies and approved for publication but reflects the views of the editors and not necessarily those of the Agency.

Steven C. McCutcheon
Athens, Georgia

Jerald L. Schnoor
Iowa City, Iowa

Contributors

M. Kathy Banks, School of Civil Engineering, Purdue University, West Lafayette, Indiana

Scott W. Beckman, Division Manager, SAIC, Hackensack, New Jersey

Claudia Bock, formerly with the Institute of Plant Nutrition and Soil Science Federal Agricultural Research Centre (FAL), now with the National Reference Laboratory, Federal Office of Consumer Protection and Food Safety (BVL), Berlin, Germany

Maria Bokern, formerly with the Institute of Plant Nutrition and Soil Science Federal Agricultural Research Centre (FAL), Braunschweig, Germany

Bruce G. Bugbee, Crop Physiology Laboratory, Department of Plants, Soils, and Biometeorology, Utah State University, Logan, Utah

Joel G. Burken, Department of Civil Engineering, University of Missouri-Rolla, Rolla, Missouri

Sheau-Yun Chiang, Project Environmental Engineer, Earth Tech Inc., Roswell, Georgia

Helle Christiansen, TeamProtection A/S, Kobenhavn, Denmark

Barton D. Clinton, Research Ecologist, Coweeta Hydrologic Laboratory, U.S. Department of Agriculture, Forest Service Southern Research Station, Otto, North Carolina

Harry R. Compton, Environmental Response Team Center, U.S. Environmental Protection Agency, Edison, New Jersey

Lawrence C. Davis, Department of Biochemistry, Kansas State University, Manhattan, Kansas

Annette C. Dietz, Senior Staff Scientist, GeoSyntec Consultants, Knoxville, Tennessee

William J. Doucette, Utah Water Research Laboratory, Department of Civil and Environmental Engineering, Utah State University, Logan, Utah

Sandra M. Eberts, Department of Defense Environmental Conservation Program, U.S. Geological Survey, Columbus, Ohio

KATHERINE J. ELLIOT, Research Ecologist, Coweeta Hydrologic Laboratory, U.S. Department of Agriculture, Forest Service Southern Research Station, Otto, North Carolina

HENRIETTE M. EMOND, Northwest Regional Technology Leader, Natural Treatment Systems, Phytoremediation Team, CH2M Hill Consulting Engineers and Scientists, Portland, Oregon

LARRY E. ERICKSON, Center for Hazardous Substance Research, Department of Chemical Engineering, Kansas State University, Manhattan, Kansas

EDWARD G. GATLIFF, Applied Natural Sciences, Inc., Hamilton, Ohio

Jon S. Ginn, Select Engineering Service, Inc., Ogden, Utah

MILTON P. GORDON, Department of Biochemistry, University of Washington, Seattle, Washington

LEO F. GUERIGUIAN, Office of Water, U.S. Environmental Protection Agency, Washington, D.C.

HANS HARMS, formerly with the Institute of Plant Nutrition and Soil Science Federal Agricultural Research Centre (FAL), Braunschweig, Germany

GREGORY J. HARVEY, Environmental Safety and Health Division Engineering Directorate, U.S. Air Force Aeronautical Systems Center, Wright-Patterson Air Force Base, Ohio

RAY R. HINCHMAN, formerly with the Energy Systems Division, Argonne National Laboratory, Argonne, Illinois

STEVEN R. HIRSH, Region III, U.S. Environmental Protection Agency, Philadelphia, Pennsylvania

CHRISTOPHER F. HOEHAMER, National Research Council Fellow, c/o National Exposure Research Laboratory, U.S Environmental Protection Agency, Athens, Georgia

KEVIN M. HOWE, Omaha District, U.S. Army Corp of Engineers, Omaha, Nebraska

STACY L. HUTCHINSON, Department of Biological and Agricultural Engineering, Kansas State University, Manhattan, Kansas

MYRNA E. JACOBSON, Wrigley Institute, Department of Biology, University of Southern California, Los Angeles, California

PETER M. JEFFERS, Chemistry Department, State University of New York, Cortland, New York

SONYA A. JONES, U.S. Geological Survey, Austin, Texas

JAMES L. JORDAHL, Senior Technologist, Natural Treatment Systems, Phytoremediation Team, CH2M Hill Consulting Engineers and Scientists, Des Moines, Iowa

YOSHIFUMI KAWAMURA, Department of Mathematical and Life Sciences, Graduate School of Science, Hiroshima University, Higashi-Hiroshima, Japan

Marit Kolb, formerly with the Institute of Plant Nutrition and Soil Science Federal Agricultural Research Centre (FAL), now with the Institute of Ecological Chemistry and Waste Analysis, Technical University, Braunschweig, Germany

Steven L. Larson, Waterways Experiment Station, U.S. Army Corp of Engineers, Vicksburg, Mississippi

Christina D. Liddy, Chemistry Department, State University of New York, Cortland, New York

Mark F. Madison, Principal Technologist, Natural Treatment Systems, Phytoremediation Team, CH2M Hill Consulting Engineers and Scientists, Portland, Oregon

Elena Maestri, Division of Genetics and Environmental Biotechnology, Department of Environmental Sciences, University of Parma, Parma, Italy

Marta Marmiroli, Division of Genetics and Environmental Biotechnology, Department of Environmental Sciences, University of Parma, Parma, Italy

Nelson Marmiroli, Division of Genetics and Environmental Biotechnology, Department of Environmental Sciences, University of Parma, Parma, Italy

Dale H. Matey, formerly with Lockheed Martin, now with the Environmental Response Team Center, U.S. Environmental Protection Agency, Edison, New Jersey

Steven C. McCutcheon, National Exposure Research Laboratory, U.S. Environmental Protection Agency, Athens, Georgia

Terry C. McIntyre, Environmental Biotechnology Program, Environment Canada, Hull, Quebec/Ontario, Canada

Victor F. Medina, Pacific Northwest National Laboratory, Richland, Washington

Hiromichi Morikawa, Graduate Department of Gene Science, Faculty of Science, Hiroshima University, Higashi-Hiroshima, Japan

Donald D. Moses, Omaha District, U.S. Army Corps of Engineers, Omaha, Nebraska

Marie Quitterie Motte, Agricultural and Environmental Engineer, Phytoremediation Team, CH2M Hill Consulting Engineers and Scientists, Portland, Oregon

Muralidharan Narayanan, GlobeSpanVirata, Inc., Red Bank, New Jersey

M. Cristina Negri, Energy Systems Division, Argonne National Laboratory, Argonne, Illinois

Lee A. Newman, Norman J. Arnold School of Public Health, University of South Carolina, Columbia, South Carolina and the Savannah River Ecology Laboratory, Savannah River Site, Aiken, South Carolina

Valentine A. Nzengung, Department of Geology, University of Georgia, Athens, Georgia

Paul E. Olson, formerly with the Department of Chemical and Bioresource Engineering and Biology Department, Colorado State University, Fort Collins, Colorado

Walter L. O'Niell, National Research Council Fellow, c/o National Exposure Research Laboratory, U.S Environmental Protection Agency, Athens, Georgia

Christopher J. Pajak, Utah Water Research Laboratory, Utah State University, Logan, Utah

John Pierson, Senior Research Engineer, Georgia Tech Research Institute, Georgia Institute of Technology, Atlanta, Georgia

Elizabeth A. H. Pilon-Smits, Biology Department, Colorado State University, Fort Collins, Colorado

John J. Quinn, Environmental Assessment Division, Argonne National Laboratory, Argonne, Illinois

Kenneth F. Reardon, Department of Chemical Engineering, Colorado State University, Fort Collins, Colorado

Steve A. Rock, National Risk Management Research Laboratory, U.S. Environmental Protection Agency, Cincinnati, Ohio

F. Michael Saunders, School of Civil and Environmental Engineering, Georgia Institute of Technology, Atlanta, Georgia

William H. Schneider, Weston Solutions, Inc., Albuquerque, New Mexico

Jerald L. Schnoor, Civil and Environmental Engineering, University of Iowa, Iowa City, Iowa

A. Paul Schwab, Department of Agronomy, Purdue University, West Lafayette, Indiana

Jean-Paul SchwitzguÉbel, Laboratory for Environmental Biotechnology, Swiss Federal Institute of Technology Lausanne, Switzerland

Troy L. Settle, Devon Energy Corp., Carthage, Texas

Tanya Q. Shang, Department of Chemistry and Biochemistry, University of Delaware, Newark, Delaware

Jacqueline V. Shanks, Department of Chemical Engineering, Iowa State University, Ames, Iowa

Jason K. Smesrud, Project Technologist, Natural Treatment Systems, Phytoremediation Team, CH2M Hill Consulting Engineers and Scientists, Portland, Oregon

Shannon C. Smith, Environmental Management Directorate, Hill Air Force Base, Utah

Murali Subramanian, Department of Chemical Engineering, Iowa State University, Ames, Iowa

Misa Takahashi, Graduate Department of Gene Science, Hiroshima University, Higashi-Hiroshima, Japan

Phillip L. Thompson, Department of Civil and Environmental Engineering, Seattle University, Seattle, Washington

Stefan A. J. Trapp, Environment and Resources, Technical University of Denmark, Lyngby, Denmark

Tomas Vanek, Department of Plant Cell Cultures, Institute of Organic Chemistry and Biochemistry, Czech Academy of Sciences, Prague, Czech Republic

James M. Vose, Project Leader, Coweeta Hydrologic Laboratory, U.S. Department of Agriculture, Forest Service Southern Research Station, Otto, North Carolina

James W. Weaver, National Exposure Research Laboratory, U.S Environmental Protection Agency, Athens, Georgia

Sara K. Winnike-McMillan, University of North Carolina, Chapel Hill, North Carolina

Larry R. Westholm, formerly with the School of Civil and Environmental Engineering, Georgia Institute of Technology, Atlanta, Georgia

N. Lee Wolfe, National Exposure Research Laboratory, U.S. Environmental Protection Agency, Athens, Georgia

John G. Wrobel, Directorate of Safety, Health and Environment, U.S. Army Garrison Aberdeen Proving Ground, Aberdeen Proving Ground, Maryland

Qizhi Zhang, Technical Development Division, Argonne National Laboratory, Argonne, Illinois

Guangxuan Zhu, School of Civil and Environmental Engineering, Georgia Institute of Technology, Atlanta, Georgia

Editorial Review Board

SECTION I

OVERVIEW OF SCIENCE AND APPLICATIONS

1

OVERVIEW OF PHYTOTRANSFORMATION AND CONTROL OF WASTES

S. C. McCutcheon and J. L. Schnoor

SUMMARY

Phytoremediation is evolving into a cost-effective means of managing wastes, especially excess petroleum hydrocarbons, polycyclic aromatic hydrocarbon, explosives, organic matter, and nutrients. Applications are being tested for cleaning up contaminated soil, water, and air. A number of important botanical processes have been discovered, including phytoextraction and hyperaccumulation from soil, plant-assisted microbial degradation of hydrocarbons in soil, use of specific enzymatic processes involved in created wetland treatment, and several other means of transforming and sequestering organic pollutants. Plant metabolism using sunlight energy and atmospheric carbon dioxide to produce organic matter is fundamentally different from heterotrophic microbial respiration requiring energy, carbon, and nutrients from soil or water. As a result, green plant transformation, conjugation, and sequestration are vital new tools in waste management that are categorized along with methods of vegetative control of pollutants to manage contaminated groundwater plumes and soil. In addition, this chapter highlights the more pressing research and development needs.

GLOSSARY AND ABBREVIATIONS

MTBE: Methyl *tert*-butyl ether, typically used as a gasoline additive.

PAH: Polycyclic aromatic hydrocarbon, a typical cocontaminant in oily wastes due to combustion.

Proof of concept: Laboratory, bench, or field pilot tests that prove or establish that a remediation approach or engineering concept is viable and ready for further testing and optimization of the process for expected field

Phytoremediation: Transformation and Control of Contaminants,
Edited by Steven C. McCutcheon and Jerald L. Schnoor.
ISBN 0-471-39435-1 (cloth)

conditions. Appropriate testing requires rigorous quality control plans and specifies beforehand what criteria will be used to determine whether the concept is viable. Alternatively, testing in independent laboratories or by independently derived plans and oversight will suffice to achieve proof of concept. The degree of testing should be tailored to the extent to which the concept is derived from fundamental principles and scientific evidence in the peer-reviewed literature.

Proof of principle: Laboratory, bench, or field pilot tests that prove or establish that a remediation approach or engineering concept is appropriate for use over a range of field conditions. These rigorous demonstrations ensure that enough testing and optimization has occurred over the range of conditions for most applications. Success or failure of testing must be based on typical site cleanup criteria derived from U.S. state and federal (province and national) standards for water, air, and soil quality or human and ecological risk assessment. These rigorous demonstrations should involve independent testing or oversight to certify results. The next step is the writing of design manuals for widespread applications.

RDX: Hexahydro-1,3,5-trinitro-1,3,5-triazine, a secondary explosive used in munitions.

Rhizosphere: Soil and microbes adjacent to and influenced by plant roots.

Self-engineering or self-design of ecosystems: Capability of ecosystems to change and control surrounding environment and the use of this capability to allow plants or other ecosystems to modify and control local environments to achieve objectives useful for humankind and the ecosystems involved.

TCE: Trichloroethylene, a common solvent.

TNT: 2,4,6-Trinitrotoluene, a secondary explosive used in munitions and commercial blasting.

Xenobiotic: Compound that is foreign to a living organism.

INSIGHTS INTO THE DEVELOPMENT OF PHYTOREMEDIATION

Phytoremediation involves the use of vascular plants, algae, and fungi either to remove and control wastes or to spur waste breakdown by microorganisms in the rhizosphere. Included in this group of organisms are photoautotrophic plants and saprophytic fungi in which glycosylation (glycosyl groups added to a protein to form a glycoprotein) normally occurs in the transformation of organic chemicals. Remediation exclusively based on heterotrophic microorganisms is the basis of the allied field of bioremediation. However, these vital ecological communities do not operate in isolation, and thus an important type of phytoremediation is the use of plants to spur greater heterotrophic degradation or transformation of contaminants by rhizosphere bacteria (rhizodegradation). Because most applications involve photoautotrophic plants, phytoremediation is primarily solar powered and thus more sustainable, especially compared to the typical mechanical approaches to hazardous waste management. An important distinction is that vascular green plants have the marvelous ability to self-engineer or exert limited control over the rhizosphere, local biogeochemistry, availability of water and nutrients, and the local microclimate. As such, phytoremediation is an important part of the new field of ecological engineering.

The wastes that potentially can be managed using phytoremediation are quite diverse, including heavy metals, metalloids, radionuclides, salts, nutrients, xenobiotic organic chemicals, sewage, and air pollutants. In this first decade after realizing that phytoremediation could become a powerful tool in waste management, new applications are being conceived and tested at a rapid pace. Although not yet fully validated, phytoremediation has the great promise of saving money in cleaning up and healing the Earth.

Raskin *et al.* (1994) and Raskin (1996) reported that they first coined the term phytoremediation in a 1991 proposal funded by the U.S. Environmental Protection Agency Superfund Program (grant #R81869), a consistent supporter of this innovative technology. Cunningham and Berti (1993) first used the term in the open technical literature, noting, however, that the use of plants in land farming of wastewaters has been occurring for at least 300 years. The past few decades have also seen wetland applications and the use of plants in air pollution control. In the 1970s, plant-based remediation of metals-contaminated soil and dredge material slurries was proposed (Cunningham and Lee 1995). However, matrix toxicity was thought to limit these applications (Cunningham *et al.* 1996) until seminal work on hyperaccumulation (Brooks *et al.* 1977, Brooks 1998) and other methods to manage phytotoxicity were established in the 1980s and 1990s (Chaney 1983, Cunningham and Berti 1993, U.S. DOE 1994, Raskin *et al.* 1994, Baker *et al.* 1995, Dushenkov *et al.* 1995, Kumar *et al.* 1995, McCutcheon *et al.* 1995, Salt *et al.* 1995, Schnoor *et al.* 1995).

This book focuses on the transformation of xenobiotic organic chemicals, degradation of some inorganic compounds (nitrous oxide gas, perchlorate, and cyanide), and the use of evapotranspiration to control hazardous wastes in soil, groundwater, and air. The use of green plants to accumulate metals and radionuclides and to stabilize metals and organic pollutants in soil is covered by Vangronsveld and Cunningham (1998), Raskin and Ensley (2000), and Terry and Banuelos (2000). Thus, these topics are only mentioned in this chapter and this book for continuity and context, except that McIntyre (this book) presents a new plant database on metals accumulation with an allied database on plants that spur degradation of petroleum hydrocarbons. Rock (this book), also in the context of field evaluations, covers the lack of success so far in proving that lead accumulation by plants is a feasible cleanup technology.

Potential Applications

Within just a few years, phytoremediation has bloomed into a number of interesting, potential applications for treating specific elemental and organic contaminants. As presented in Table 1-1, these chemicals or class-specific treatments provide additional choices beyond the older heuristic plant-dominated practices used to treat mixtures of wastes, for example, sewage.

In some cases, the newer chemical-specific phytoremediation approaches are developing into primary remedies to clean up hazardous waste sites. Already plant-assisted rhizodegradation of petroleum hydrocarbons and polycyclic aromatic hydrocarbons (PAHs) can be a primary method of waste management for some sites and source removal is not necessary under some conditions (see Hutchinson *et al.* this book). According to proof of concept testing both in the laboratory and in the field, created treatment wetlands should be able to buffer aquatic plants from extremely high concentrations of explosives in soils and supersaturated "pink waters" from ammunition plant waste streams (see McCutcheon *et al.* this book). In the case of chlorinated solvents and methyl *tert*-butyl ether (MTBE), plants tolerate much higher concentrations compared to the risk-based cleanup standards used for the protection of humans and wildlife (Medina *et al.* this book, Winnike-McMillan *et al.* this book). As a result, other treatment alternatives may be necessary to overcome some of the limitations of phytoremediation such as shallow to moderate rooting depths of plants (*e.g.*, pumping and irrigating with deeper contaminated groundwater) and the seasonal variability in the rate of treatment by plants (*e.g.*, equalization).

Despite the development of primary cleanup remedies for some situations, many phytoremediation applications are expected to be supplemental approaches in cleaning up areas bordering "hot spots" of contamination or following more extreme treatment (*e.g., in situ* permanganate treatments of free-phase sources of chlorinated solvent). The niches for phytoremediation are primarily defined at the moment by a common sense analysis of what green plants have evolved to do as part of the ecosystem, and what these marvelous organisms cannot do (Cunningham *et al.* 1995a, Cunningham *et al.* 1996). In the developing areas of phytoremediation, pilot testing has not been sufficient to sharply define the engineering limitations or even to confirm the many proofs of concept available from laboratory investigations.

Based on the current common sense approach, phytoremediation seems ideal to clean up moderate to low levels of some contaminants over extensive areas and to restore sites to more pristine conditions by treating residuals that may be below risk levels for a U.S. Comprehensive Environmental Response, Compensation, and Liability Act or U.S. Resource Conservation and Recovery Act driven cleanup. Phytoremediation is also ideal (1) for site maintenance to prevent the widespread and slow buildup of residual wastes; (2) in green-field development (ecological or phytorestoration in urban areas); (3) as a buffer to potential waste releases; (4) in voluntary cleanup efforts; (5) for

TABLE 1-1 Potential Applications of Phytoremediation and Other Waste Management Practices Involving Green Plants

Potential application	How *in situ* and *ex situ* concepts work[a]	Applications or important bench or field testing for various plants [and wastes subject to treatment in square brackets][b]
Cleanup of soils, sediments, and other residuals[c]		
Tree or shrub plantation or other forestry and landscaping applications	Large quantities of soil moisture transpired (includes deep planting, infiltration control, or root training; also see vegetative capping) to transport contaminants to roots and change soil physically, chemically, and microbiologically by root growth, dieback, and osmotic pressure, and may involve: 1. *In situ*: phytoextraction, stabilization, volatilization, and containment of metals, radionuclides, and salts 2. *In situ*: phytotransformation, degradation, volatilization, stabilization, accumulation, rhizodegradation, and hydraulic control of organic and inorganic chemicals 3. *Ex situ*: phytodegradation or stabilization of compost containing explosives, PAHs, and other organic chemical residues	1. (a) Phytoextraction field pilots: willow (*Salix* spp.) [Cd, Zn, and Cu] Sweden (Greger and Landberg 1999), and hybrid prairie cascade willow (*Salix pentandra* × *Salix* ×*blanda* [*Salix babylonica* × *Salix fragilis*]) and weeping willow (*Salix alba* var. *niobe*) [Cr Hg, Se, and Zn] Argonne National Laboratory-West, Idaho (Negri *et al.* 1998). (b) Observed hyperaccumulation: [Ni] (Brooks 1998, Baker and Smith 2000). (c) Observed phytoextraction: [transuranic elements] Savannah River Site, South Carolina (U.S. DOE 2000). (d) Field phytostabilization testing not successful: hybrid poplar (*Populus deltoides* × *Populus nigra* DN-34) [As and Cd] Whitewood Creek, South Dakota, gold mine tailings and Dearing, Kansas, smelter site (Schnoor 2000) 2. (a) Field tests: [PAHs, petroleum hydrocarbons, phenols, and PCBs] several U.S. sites; Axelved, Denmark; other European sites; and subarctic regions (Fletcher and Hegde 1995, Hegde and Fletcher 1996, Olson and Fletcher 2000, Coyle 2000, Fletcher 2001, Landmeyer 2001, Ferro *et al.* 2001, Trapp and Karlson 2001, Leigh *et al.* 2002, Palmroth *et al.* 2002, Olson *et al.* this book and Jordahl *et al.* this book). (b) Greenhouse and bench pilots: hybrid poplar (*Populus deltoides* × *Populus nigra* DN-34) [TNT, HMX, RDX, and atrazine] (Thompson 1997, Thompson *et al.* 1998, Schnoor 2002). (c) Field and bench: Salicaceae [CN] Holte, Denmark (Trapp and Karlson 2001, Trapp and Christiansen this book). (d) Greenhouse: [nitrobenzene] (McFarlane *et al.* 1990). (e) Tissue cultures: Paul's Scarlet rose (*Rosa* spp.) [PCBs and phenols] (Butler *et al.* 1992, Lee and Fletcher 1992) 3. Unproven concept

(continues)

TABLE 1-1 (*continued*)

Potential application	How *in situ* and *ex situ* concepts work[a]	Applications or important bench or field testing for various plants [and wastes subject to treatment in brackets][b]
Grass, forb, herb, and fern planting	1. *In situ*: (a) rhizodegradation of organic compounds and (b) preparation of soil with pioneer plants	1. (a) Field evaluations: grasses [petroleum hydrocarbons and PAHs] at several U.S. and Danish sites (Trapp and Karlson 2001, Olson *et al.* this book, Hutchinson *et al.* this book, Rock this book); see also worldwide plant database (McIntyre this book). Crested wheat grass (*Agropyron desertorum*) [pentachlorophenol and phenanthrene] (Ferro *et al.* 1994). Observed rhizodegradation: wild flowers (*Senecio glaucus*) [crude oil] Kuwait from Gulf War spills (Trapp and Karlson 2001). Microcosm and mesocosm: the leguminous goat rue (*Galega orientalis*) [benzene, toluene, and xylene] (Suominen *et al.* 2000). Greenhouse: Timothy grass (*Phleum pratense*), red clover (*Trifolium pratense*), and sunflower (*Helianthus annuus*) [2,4,5-trichlorophenoxyacetic acid] (Boyle and Shann 1998). (b) Perennial vetiver (*Vetiveria zizanioides*) grass [vegetate petroleum hydrocarbon contaminated soil] (Cunningham *et al.* 1996) and various indigenous plants [PCBs] Czech Republic (Leigh *et al.* 2001)
	2. *In situ* and *ex situ*: phytostabilization and accumulation of metals and organic chemicals	2. See Vangronsveld and Cunningham (1998), Brooks (1998), and Terry and Bañuelos (2000). (a) Phytomining greenhouse pilot and observations: *Streptanthus polygaloides, Alyssum bertolonii*, and *Berkheya coddii* [Ni] Red Hills, California; Italy; South Africa; and Albania (Kelly and Guerin 1995, Nicks and Chambers 1995, 1998, Robinson *et al.* 1997, U.S. DOE 2000). (b) Hyperaccumulation field pilots and observations: [Zn] Palmerton, Pennsylvania (Brown *et al.* 1995, Comis 1995, Cornish *et al.* 1995, Kelly and Guerin 1995, Baker and Smith 2000, Aarts *et al.* 2000, Lasat *et al.* 2001) and [Cd, Cu, and Co] (Baker and Smith 2000). (c) Stabilization of northern Swedish mine spoils: [As, Cd, Cu, and Zn] (Stoltz and Greger 2000). (d) Observed extraction: *Astragalus* spp. [Se] (Negri *et al.* 1998, Brooks 1998, Zayed *et al.* 2000). (e) Stabilization: quack grass (*Agropyron repens* L.) [Pb] Poland (Kutrys and Gawronski 2000). (f) Revegetation [As, Cd, Cu, Pb, and

		Zn] Anaconda Smelter in Montana and Big River Mine Tailings Superfund sites (U.S. DOE 2000). (g) Greenhouse and bench phytoextraction: [Cs] (Negri *et al.* 1998, Negri and Hinchman 2000). (h) Bench phytostabilization: bladder champion {*Silene vulgaris* (Moench.) Garcke L.} [variety of metals] (Brown *et al.* 1995). (i) Bench transformation: kudzu (*Pueraria thunbergiana*) [DDT] (Garrison *et al.* 2000).
	3. *In situ* and *ex situ*: phytotransformation	3. Lab growth media: lacy scorpion weed (*Phacelia tanacetifolia*) [chlorinated phenols] (Smrcek 2000).
	4. *In situ* and *ex situ*: erosion control	4. See standard erosion control manuals for plants and soil conditions for each U.S. state and province of countries worldwide.
	5. *In situ*: infiltration control to prevent soil leaching	5. Has not been tested except for vegetative capping {see "Clean up of wastewater, sludge, leachate, groundwater, and runoff" "Tree, shrub, forb, herb, and grass applications" number (4)}.
Agronomic cultivation and cropping practices	1. *In situ*: phytoextraction, stabilization, and volatilization of metals, radionuclides, and salts	1. (a) Field feasibility testing: [Zn and Cd] and ready for feasibility testing [Co] (Brown *et al.* 1994, 1995, U.S. DOE 2000). (b) 20-acre (8-hectare) pilot and other testing: [Se and B] California (Banuelos *et al.* 1993, Negri *et al.* 1998, U.S. DOE 2000, Bañuelos 2000). (c) Field: Indian mustard (*Brassica juncea*) and sunflower (*Helianthus annuus*) [Pb] at sites in Connecticut, New Jersey, and Massachusetts attempted but unsuccessful (Rock this book).[d] (d) Field and greenhouse pilots: [Cs] Brookhaven National Laboratory, New York, and Idaho National Engineering and Environmental Laboratory (Broadley and Willey 1997, Negri *et al.* 1998, Negri and Hinchman 2000). (e) Parris Island Romaine lettuce (*Lactuca sativa* L. var. *longifolia*), maize (*Zea mays* L.), and inbred FR-37 Merlin red fescue (*Festuca rubra*) [Cd, Zn, and Pb], (f) Field and greenhouse: canola (*Brassica napus*) and *Kochia scoparia* [Cr and Zn] at Argonne National Laboratory-West, Idaho (Negri *et al.* 1998). (g) Greenhouse: [U] Ashtabula and Fernald, Ohio soils (Negri and Hinchman 2000, U.S. DOE 2000). (h) Observed hyperaccumulation: [As, Cs, Sr, and U] (Ma *et al.* 2001)
	2. *In situ*: phytotransformation, degradation, volatilization, stabilization, accumulation, and stimulation for organic chemicals	2. (a) Pilot: carrot (*Daucus carota*) [DDT] (Cunningham *et al.* 1996). (b) Greenhouse transformation: [nitrobenzene] (McFarlane *et al.* 1990). (c) Bench transformation: [nonyl phenol, nonyl-phenoxy-ethoxy-ethanol, pyrene, linear alkyl-benzenesulfonate, *bis*(diethyl-hexyl)phthalate, pesticides] (Kure *et al.*

(*continues*)

TABLE 1-1 (*continued*)

Potential application	How *in situ* and *ex situ* concepts work[a]	Applications or important bench or field testing for various plants [and wastes subject to treatment in brackets][b]
		2000, Chaudhry *et al.* 2002, Coleman *et al.* 2002). (d) Bench rhizodegradation: bean (*Phaseolus vulgaris*) [2,5-dichlorobenzoate] (Crowley *et al.* 1996). (e) Lab growth media transformation: [chlorinated phenols, TNT, and dinitrotoluene] (Smrcek 2000, Snellinx *et al.* 2000). (f) Bench accumulation: [dioxins, benzofurans, and some pesticides] (Campanella and Paul 2000, Trapp and Karlson 2001, Chaudhry *et al.* 2002)
	3. *Ex situ*: phytoextraction, stabilization, and volatilization of metals, radionuclides, and salts from excavated soil or sediment	3. Field (unsuccessful): [Pb] (Cunningham and Lee 1995, Rock this book).[d] Field pilot: sunflower (*Helianthus annuus* L.) [U] (Dushenkov *et al.* 1997). Bench: Indian mustard (*Brassica juncea* L. Czern.), sunflower (*Helianthus annuus* L.), and various grasses [Cu, Cr, Ni, Cd, and Zn] (Dushenkov *et al.* 1995, 1997)
	4. *Ex situ*: enzymatically active plant material harvested and mixed or slurried with solid wastes such as excavated soil placed in landfills	4. Bench: [explosives, phenols, aniline, and mycotoxins] (Dec and Bollag 1994, Todd 1996, Messner *et al.* 2000, Medina *et al.* 2002)
Created, constructed, engineered, or managed treatment wetlands, and aquatic plant lagoons	1. *In situ*: contaminated soils or sediments diked and flooded or excavated to below the water table, seeded with natural wetland sediment or planted with wetland plants	1. Field pilots: indigenous plants at Iowa Army Ammunition Plant [RDX and TNT] (Thompson *et al.* this book)
	2. *Ex situ*: contaminated soils or slurries excavated, spread in a containment area, flooded, and planted with wetland plants	2. Bench: parrot feather (*Myriophyllum aquaticum*) and stonewort (*Nitella* spp.) [TNT] (McCutcheon *et al.* this book)

Cleanup of wastewater, sludge, leachate, runoff, and groundwater		
Tree, shrub, forb, herb, and grass applications	1. *In situ*: phreatophytes transpire large quantities of groundwater or soil moisture to dewater aquifers (using deep planting, planting trees in shallow wells, infiltration control, and root training) and control soil infiltration and leaching of contaminants. See tree plantations under "Clean up of soils, sediments, and other residuals" on how phyto-transformation, stabilization, and accumulation may occur 2. *Ex situ*: phytoirrigation 3. *In situ*: tree and grass buffers filter sediment and other particulate matter and sorb and infiltrate nutrients and other pollutants from runoff 4. Vegetative capping of wastes *ex situ*: soil cap plus vegetation designed to (a) control or limit infiltration and retard leaching of hazardous wastes from piles of solid wastes, landfills, and other containment facilities, and (b) allow waste degradation 5. Land farming (*ex situ*): wastewater, sludge, or slurried sediment sprayed or applied on planted or existing forests, grasslands, or crops, or applied, ploughed, and planted so that rhizodegradation occurs	1. See Jordahl *et al.* this book, for design guidance. (a) Field pilots: Salicaceae [chlorinated solvents, BTEX, nitrate, and tritium] Carswell Air Force Base, Texas (Eberts *et al.* this book), Aberdeen Proving Grounds, Maryland (Hirsch *et al.* this book), Argonne National Laboratory, Illinois (Quinn *et al.* 2001), and Houston, Texas (Hong *et al.* 2001) are demonstrating at least partial aquifer and contaminant control. (b) Field pilots: *Eucalyptus* spp. and salt cedar (*Tamarix* spp.) [As] hydraulic control (Jordahl *et al.* this book) and redwood (*Sequoia sempervirens*) [landfill leachate] Santa Rosa, California (U.S. DOE 2000). (c) Greenhouse and bench rhizodegradation testing: Salicaceae and *Eucalyptus cineria* [perchlorate] (Nzengung and McCutcheon this book). (d) Phytovolatilization in lab: alfalfa (*Medicago sativa*) [MTBE and 1,4-dioxane] (Schnoor 2000, Winnike-McMillan *et al.* this book). (e) Bench hydroponics: poplars (*Populus simonii* and *Populus tremula*) [TNT] (Vanek *et al.* 2000) 2. Field testing: [municipal wastewater, landfill leachate, chlorinated solvents, MTBE, petroleum hydrocarbons, and nutrients] U.S., Canada, Sweden, and New Zealand (Trapp and Karlson 2001, Jordahl *et al.* this book). Lab cell cultures and whole plants: [some PCBs, PAHs, and 4-*n*-nonylphenol] (Harms *et al.* this book) 3. Applications: see Haan *et al.* (1994) and McCutcheon (2000) for list of typical grasses, shrubs, and trees, and contaminant trapping efficiencies for buffers. Field tests: hybrid poplar (*Populus* spp.) buffers [sediment, nutrients, and atrazine] Iowa (Schnoor *et al.* 1995, Burken and Schnoor 1996). 4. Field testing: Salicaceae and understory grasses [landfill leachates of metals and organic chemicals] 13 U.S. landfills (Benson *et al.* 2001, Licht *et al.* 2001, Jordahl *et al.* this book). Rigorous assessment under way at 12 landfills by U.S. Environmental Protection Agency Alternative Cover Assessment Program (Rock this book) 5. Applications: variety of trees, grasses, and crops [domestic sewage low in metals, petroleum hydrocarbon wastes, and pentachlorophenol from wood treating wastes] (Ferro *et al.* 1994, Trapp and Karlson 2001, Jordahl *et al.* this book)

(*continues*)

TABLE 1-1 (*continued*)

Potential application	How *in situ* and *ex situ* concepts work[a]	Applications or important bench or field testing for various plants [and wastes subject to treatment in brackets][b]
Created, constructed, engineered, or managed treatment wetlands, and aquatic plant lagoons	1. *In situ*: excavation to or below water table and seeding with natural wetland sediment or planting	1. Field and bench pilots: wetland plant community [RDX and TNT] Iowa Army Ammunition Plant (Thompson *et al.* this book and McCutcheon *et al.* this book)
	2. *In situ*: at groundwater, landfill, or mine seepage points or in a surface drainage network to capture and treat contaminated water	2. Applications: wetland plants [landfill leachate, acid mine drainage = 1 waste, not 2 and Se] (Reed and Hines 1993, Walski 1993, Horne 2000, Zayed *et al.* 2000) and field pilot: wetland plants [chlorinated solvents] Athens, Georgia (www.appliedphyto.com)
	3. *In situ*: rhizofiltration from surface waters by planting aquatic plants or floating other plants in the contaminated water and harvesting the plants after biosorption and filtration	3. (a) Field, greenhouse, and bench: wetland plants [Cd, Co, Pb, Hg volatilization, Cu, Zn, Ag, Cr, and Cs] Brookhaven National Laboratory, New York, and elsewhere (Hussain and Jamil 1992, Lasat *et al.* 1997, U.S. DOE 2000, La Starza *et al.* 2000, Negri and Hinchman 2000). (b) Field observations: [elements from fly ash] Savannah River Plant, South Carolina (Rodgers *et al.* 1978). (c) Observations: *aufwuchs* [Hg] Wisconsin lakes (Cope and Rada 1992). (d) Rhizofiltration greenhouse pilot: hydroponic hybrid sunflowers (*Helianthus annuus* 'SF-187' and 'Mammoth Giant') [U] Ashtabula, Ohio (Dushenkov *et al.* 1997, Negri and Hinchman 2000). (e) Field and bench hydroponics: [Cs and Sr] in Chernobyl, Ukraine, ponds and wetlands (Vasudev *et al.* 1996, Negri *et al.* 1998, Negri and Hinchman 2000, U.S. DOE 2000). (f) Bench hydroponics: Indian mustard (*Brassica juncea* L. Cern. '426308'), sunflower (*Helianthus annuus* L.), and various grasses [Cd, Pb, Cu, Zn, Ni, and Cr(VI)] (Dushenkov *et al.* 1997). (g) Bench plant screening: pondweed (*Potamogeton natans*) [Cu and Zn] (Fritioff and Greger 2000). (h) Bench transformation: Canadian waterweed (*Elodea canadensis*) [DDT and organophosphorus pesticides] (Gao *et al.* 2000a, 2000b). (i) Bench accumulation: [pesticides] (Chaudhry *et al.* 2002, Coleman *et al.* 2002)
	4. *Ex situ*: pumping of contaminated groundwater or runoff into treatment wetlands	4. Bench: hydroponic aspen (*Populus tremula* × *Populus tremuloides*) [Pb] (Spirochova *et al.* 2000). Bench: stricta (*Hygrophila corymbosa*) [Cd] (La Starza *et al.* 2000), and parrot feather (*Myriophyllum aquaticum*) and hydroponic sunflower (*Helianthus annuus*) [TNT, RDX, HMX, dinitrotoluene, nitrobenzene, and nitrotoluene] (Snellinx *et al.* 2000, McCutcheon *et al.* this book)
	5. Greenhouse rhizofiltration and *ex situ* phytoslurry unit processes, and outdoor treatment lagoons and wetlands for domestic and industrial wastewaters	5. (a) Applications: wetland communities [P, nitrate, particulates, heavy metals, Se, some pesticides and other organics, and some pathogens] (Reed and Hines 1993, Todd and Josephson 1996, Horne 2000, Zayed *et al.* 2000). (b) Bench and pilot testing: [phenols, anilines, and other aromatic compounds] (Dec and Bollag 1994, Roper *et al.* 1996, Buchanan *et al.* 1998, Coyle 2000)

		(c) Bench pilots: [Cr, Mn, Cd, Ni, Cu, Pb, Zn, and U] (Tripathi and Chandra 1991, Garg and Chandra 1994, Salt *et al.* 1995, Vasudev *et al.* 1996). (d) Bench: freshwater aquatic plants and cell cultures [TNT, nitroglycerine, phenols, nitrate, domestic and industrial wastewaters, storm water, and landfill leachates] (Gerth and Bohler 2000, Schnoor 2002, Vanek *et al.* 2000, Xantholis and Dumont 2000)
	6. Unit processes for concentration of oil field brines by halophyte evapotranspiration or salt excretion	6. Field demonstration: halophytes [concentration of oil field brines] Oklahoma (Negri, Hinchman, and Settle this book)
Cleanup of air pollution		
Living walls, roof gardens, and plant biofilters	1. Vegetation is planted on building walls and roofs, slopes, and all open space to remove greenhouse gases and other pollutants from the atmosphere and provide greater surface and microclimates for cooling and particulate deposition 2. Plant terrariums filter indoor air for carbon monoxide and smoke	Living walls were patented in 1970s, but rigorous testing does not seem to have occurred for outdoor or indoor applications. Morikawa *et al.* of this book establishes that genetic engineering is necessary and then only partial control of NO_x by plants can be achieved. Jeffers and Libby of this book report screening of plants that assimilate and transform atmospheric halogenated hydrocarbon pollutants, but no bench or field testing seems to have occurred
Phyto-vapor extraction	Contaminated vadose gases removed by roots *in situ*	Unproven concept (Cunningham *et al.* 1996)

[a]See Table 1-2 for definitions of the different physiological processes that contribute to different applications of phytoremediation.
[b]When numerous species are involved, plants are not listed. See each citation for the plants involved.
[c]Also see land farming for sludges later in the section "Cleanup of wastewater, sludge, leachate, runoff, and groundwater."
[d]Evaluations establish that additional laboratory bench testing for extraction of Pb (and other tightly bound metals, *e.g.*, Cr, U, Cs, and Pu) is necessary to establish proof of concept and then optimize use of chelating agents consistent with plant uptake (Rock this book).

The following abbreviations are used—BTEX: benzene, toluene, ethyl benzene, and xylenes; DDT: dichloro-diphenyl-trichloro-ethane; MTBE: methyl *tert*-butyl ether; PAHs: polycyclic aromatic hydrocarbons; PCBs: polychlorinated biphenyls; RDX: hexahydro-1,3,5-trinitro-1,3,5-triazine; and TNT: trinitrotoluene.

nonpoint source pollution control; and (6) as a more active form of monitored, natural attenuation (that has until now only involved microbial processes).

Segregated by media (soil, water, and air), Table 1-1 organizes the potential applications of vegetation that are known as of 2002. The next section reviews the botanical processes that are involved in the different applications conceived to date. For example, tree plantations are categorized as an application to manage both soil and water (and eventually air) pollution, but Table 1-1 notes the specific processes involved, including phytoextraction, stabilization, degradation, transformation, volatilization, rhizodegradation, and hydraulic control to contain and treat various pollutants in soil, sediment, wastewater, sludge, leachate, runoff, groundwater, and air.

The separation of planting applications from physiological processes allows a clearer integration of past heuristic practices involving green plants in waste management with the more chemical-specific developments of the last decade. Waste and resource management practices that preceded coining the term phytoremediation in 1991 include land farming, created treatment wetlands and plant lagoons, engineered buffer strips, vegetative capping, and revegetation. The primary basis for enlarging the phytoremediation field to include the pre-1990s practices is that these heuristic engineering applications have served as the basis for all of the newer phytoremediation concepts. Newer phytoremediation methods that take advantage of the pre-1990s practical experience (*e.g.*, planting grass for the rhizodegradation of petroleum hydrocarbons and constructing treatment wetlands for the restoration of explosives-contaminated sites) are ready for limited applications. The concepts involving the enzymology of phytotransformation and the biochemistry and genetics of phytoaccumulation, however, are still under development.

Table 1-1 integrates some existing practices with newer chemical or class-specific developments that have the promise of providing a sound basis for waste management. For example, tree plantation, grass planting, and agronomic cultivation are not only used in land farming of petroleum wastes, sewage, and sludge, but may also achieve phytohydraulic containment, degrade chlorinated solvents, and transform explosives and other xenobiotic chemicals. Created wetlands not only treat mine drainage, landfill leachate, urban runoff, and sewage, but also control groundwater seepage contaminated with RDX and TNT. Other existing and developing waste management practices such as land farming, vegetative buffer strips, and vegetative capping of landfills can similarly be related to tree or grass planting (Table 1-1). Other long-standing practices, such as revegetation to prevent erosion and to control nonpoint source pollution, are covered under grass planting.

Table 1-1 does not explicitly cover monitored, natural attenuation involving plants, which is just now being recognized (2002) as a potential application of phytoremediation. See McCutcheon *et al.* (this book), Weaver *et al.* (this book), and Doucette *et al.* (this book), for observations involving natural wetlands formed by beaver dams, seepage wetlands, evapotranspiration of existing vegetation, forest encroachment, and forest and grassland attenuation of explosives and chlorinated solvents. Despite these observations, much continued development is necessary before drafting monitoring protocols. For example, natural revegetation of explosives-contaminated soil may require several decades and may result in ecologically suboptimal communities due to phytotoxicity. Projecting the monitoring period and the costs are not possible at this time (2002). Other limitations involve public policy on how much of the natural assimilative capacity of soils and wetlands should be devoted to long-term hazardous waste remediation. Most natural wetlands, for example, are not very effective at waste treatment (Horne 2000) compared to the capacity for providing habitat and other resources, and thus should be protected from waste disposal.

Other vegetative practices such as bioengineering for slope stability and for stream bank restoration are not included in Table 1-1. Although both prevent excessive sediment and nutrient loading to streams, the rational for exclusion from Table 1-1 is that these methods are not immediately recognized as being important in waste management.

Engineered Treatment Wetlands

Engineered wetlands for waste treatment are one existing practice that is increasingly a vital part of phytoremediation (Horne 2000). In fact, created wetlands for treating contaminated water and sediment is an aquatic analog application to rhizosphere biodegradation of contaminants in soil. In addition, see McCutcheon *et al.* (this book) and Thompson *et al.* (this book) on the use of wetland flooding for the treatment of contaminants in soil by phytotransformation, rather than by enhancement of indigenous microorganisms. Although the biochemistry for the wetland treatment of xenobiotic compounds is not well known, the ecology of plant and microbial communities and the biogeochemical cycling of nutrients and natural elements has been frequently studied. Despite the lack of knowledge of plant and microbial biochemistry, managed treatment wetlands offer one of the best opportunities to maintain anoxic conditions to facilitate some xenobiotic transformation and to stabilize organic pollutants and metals, or to use wet–dry or anaerobic–aerobic cycling to irreversibly bind microbial and phytotransformation products and metals (see McCutcheon *et al.* this book and Nzengung *et al.* this book).

Used for many years (Young 1996), the construction of treatment wetlands has increased greatly in the U.S. in the last two decades as a result of favorable aesthetics, capital costs, operation and maintenance costs, and the positive experience that has been gained. Thousands of constructed wetlands operating in the U.S. are treating domestic, municipal, and industrial wastewater, including sewage and wastewaters from mining, pulp and paper, petrochemical, chemical, textile, and food processing. Current regulations favor using wetlands to treat sewage from towns of less than 5000 people (Horne 2000).

Many more applications of treatment wetlands occur in Europe and other parts of the world.

Created treatment wetlands are a logical solution for waste streams containing excess organic carbon, nutrients, particulate matter, and metals, and in situations where the waste stream is under control (*i.e.*, in a pipe) and land is available and cheap. Performance data for wetland treatments have been summarized for the removal of biochemical oxygen demand, total suspended solids, pathogens, phosphorus, nitrogen, metals, and trace organic compounds by Reed and Hines (1993). Long-term utilization of wetland plants and sulfate-reducing conditions can result in an increase in pH and a decrease of toxic metal concentration in acid mine drainage (Walski 1993, Wieder 1993). Aluminum, cadmium, copper, chromium, lead, iron, and zinc have all been removed by more than 94 percent in recent applications (Glass 1999). Figure 1-1 presents the elements suitable for stabilization and accumulation in wetland sediments. Often, either some storage of waste is advantageous to equalize flow during periods of wet weather or to store wastewater during winter periods of decreased performance.

The free-water-surface wetland and the subsurface-flow wetland are the two primary types of constructed treatment wetlands. The free-water-surface wetland contains a basin or channel with sediments to support the roots of plants and shallow water (usually 8 to 46 centimeters or 3 to 18 inches deep) flowing through the system, underlain by compacted clay or a liner to prevent seepage. Communities may include free-floating, submergent, and emergent plants. In free-water-surface wetlands, Eurasian water milfoil (*Myriophyllum spicatum*), parrot feather (*Myriophyllum aquaticum*), pondweed (*Potamogeton pusillus*), and arrowroot (*Sagittaria* spp.) are used. Introduction of nuisance species that may grow without bound is a significant design challenge. Native wetland species are sometimes required at U.S. federal sites.

The subsurface-flow wetland includes a bed (0.3 to 0.9 meters or 1 to 3 feet deep) that contains porous media (rock or gravel) to support roots and biofilms. Wastewater flow normally remains below the upper surface of the rock or gravel media; therefore all the plants in these treatment wetlands are emergent varieties. In both types of wetlands, microorganisms are thought to be responsible for waste treatment (perhaps because aquatic plant metabolism has not been fully explored). Both aerobic and anaerobic degradation occur in microzones of the gravel bed. Biomass accumulates in the rock and gravel of the subsurface flow system similar to that in a trickling filter. Microorganisms degrade organic chemicals as a primary substrate, and utilize and incorporate nutrients into biomass. Total suspended solids and pathogens are removed by filtration as the water flows through the porous media, roots, and microbial biomass. Toxic metals bind to biosolids in the gravel bed; both the roots and microbial biomass are effective sorbents. Roots require oxygen to survive, but microorganisms in anaerobic microzones of the gravel bed can reduce sulfate to sulfide such that some metals are precipitated. In addition, nitrate can be denitrified resulting in nitrogen removal from the system. The

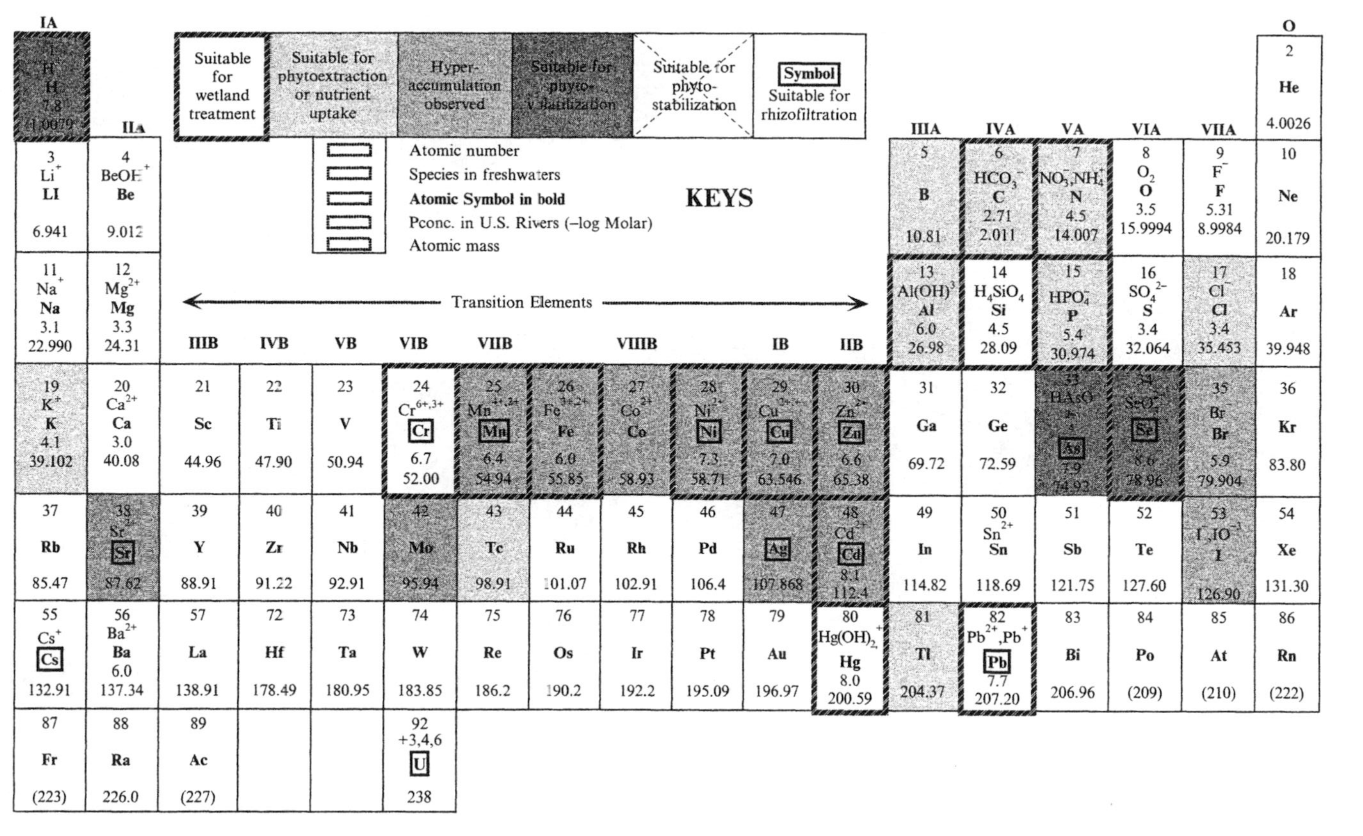

Figure 1-1 Periodic table of elements for which phytoextraction (McIntyre this book), rhizofiltration (Brooks 1998), phytostabilization, and phytovolatilization may be possible cleanup options. Note that some elements have not been evaluated to determine if some type of phytoremediation is a feasible cleanup option. Selenium and arsenic are also subject to phytoextraction and hyperaccumulation by some plants. All elements that can be hyperaccumulated should be subject to cleanup using phytoextraction. Note that tritium is phytovolatilized and hydrogen ions, of which pH is a measure, can be controlled with wetlands specifically designed to treat acid mine drainage. Two field evaluations of lead phytoextraction indicate that chelators mobilize significantly more of the tightly bound elements (Rock this book) such as lead, chromium, cesium, uranium, plutonium, and others than some plants can take up. Some wetlands do not remove and accumulate all the metals listed under all conditions (*e.g.*, lead and nickel, Horne 2000).

role of the plants is to provide (1) a root surface for filtration and sorption of suspended solids, metals, and pathogens; (2) a niche and soluble food source for supporting the growth of indigenous microorganisms; (3) a cover to shade the water in preventing the excessive growth of algae; and (4) an uptake mechanism for nutrients. Cattails (*Typha* spp.), bulrush (*Scirpus* spp.), and common reeds (*Phragmites* spp.) are emergent vegetation that is often utilized.

Potential problems and disadvantages of constructed treatment wetlands involve wet weather, reduced winter operations, loss of the plant community for nutrient control, long-term performance problems associated with plugging of the porous media, maintenance of the plant communities, and mineralization and release of nutrients and metals during and after warm periods. In addition, photosynthetic plants are effective at pumping oxygen to roots and maintaining aerobic conditions near the root surfaces. Wetlands also can be an attraction for both children and wildlife. An attractive nuisance exists if selenium, metals (Horne 2000), and organic pollutants are accumulated, or if organic by-products are slow to degrade.

PLANT PHYSIOLOGICAL PROCESSES

Based on the physiological action of plants, at least ten different processes have been identified that assist in the management of polluted soil, water, and air. Table 1-2 lists these processes.

Table 1-2 also provides a number of alternative names that are used for different physiological processes involving green plants. The derivative of the Greek word for plant, *phyton*, has been used extensively since 1993 (Cunningham and Berti 1993) in formulating new remediation concepts and to emphasize the rapidly expanding view of the usefulness of plants in waste management. Distinguishing the special role of plant metabolism particularly has played a role in the early merging of scientific advances with engineering practice. The early merging of research and development has accelerated field testing of the new concepts (U.S. DOE 1994, Schnoor *et al.* 1995, Cunningham *et al.* 1995a, Cunningham *et al.* 1996), and the reaching out to broader scientific and engineering audiences (Cunningham and Berti 1993, Cunningham and Lee 1995, McCutcheon *et al.* 1995, Cunningham *et al.* 1995b, Cunningham and Ow 1996, McCutcheon 1998).

The best means to understanding the botanical processes that are important in phytoremediation is to consider the following organizing concepts:

- "Green-liver" concept of plant metabolism (Sandermann 1992) that is reviewed by Burken of this book
- Separate processes that completely degrade or mineralize contaminants from the less desirable processes that only transform or partially degrade contaminants (concept pioneered by the bioremediation field)

TABLE 1-2 Types of Phytoremediation Based on Physiological Processes (Schnoor 1997, McCutcheon 1998, Schnoor 2002)

Type	Processes involved	Wastes that may be treated	Media	Status of applications of the process that have been proven	References
Phytoextraction (including chelator-induced), phytoaccumulation, phytoconcentration, phytotransfer, hyperaccumulation, and phytomining	Chemicals taken up with water, by cation pumps, absorption, and other mechanisms, and usually translocated aboveground. Harvested shoots or roots put in hazardous waste landfills or could be smelted after volume reduction by incineration or composting. Hyperaccumulation is *ca.* 100 times normal plant accumulation of elements and is 0.01 percent by dry weight for Cd and other rare elements, 0.1 percent for most heavy metals, and 1 percent for Fe, Mn, and other common elements	Metals, metalloids, radionuclides (see Figure 1-1), perchlorate, BTEX, PCP, short-chained aliphatic and other organic compounds not tightly bound to soils	Soil only	Field proof of concept: failed two evaluations using chelators for Pb; thus questionable for Cr, Cs, and other tightlyboundelements. New lab proof of concept now required. Phytomining for Ni: two U.S. locations and testing in Albania and South Africa. Proof of concept: 1993 to 1995 for Cd, Ni, Zn, Cu, Se, B, and other elements. Field proof of concept: Ni, Zn, Sr, Cs, and Cd from long-term application of sludges using Brassicaceae hyperaccumulators in U.K.; Mariupol and Chernobyl regions, Ukraine; and Pennine region, U.K. (plus Ag, Al, Co, Fe, Mo, and Mn)	Raskin *et al.* (patent 5 364 451 http://patft.uspto.gov), Banuelos *et al.* (1993), Baker *et al.* (1995), Raskin *et al.* (1994), Kumar *et al.* (1995), Salt *et al.* (1995), Brooks (1998), Lasat *et al.* (1997), Raskin and Ensley (2000), Terry and Banuelos (2000), U.S. DOE (2000), Schnoor (2002), Nzengung and McCutcheon (this book), McIntyre (this book), Rock (this book)

(*continues*)

TABLE 1-2 (*continued*)

Type	Processes involved	Wastes that may be treated	Media	Status of applications of the process that have been proven	References
				Pilot field testing at eastern U.S. sites: six unproven with Pb (*Brassica juncea*) and two proven with Zn and Cd (*Thlaspi caerulescens*) Bench testing for arid western U.S. site for Cr, Zn, Hg, Ag, and Se (*Salix spp., Kochia scoparia*, and *Brassica napus*) and perchlorate (wetland halophytes)	
Rhizofiltration (phytofiltration), blastofiltration, phyto or biosorption, biocurtain, biofilter, contaminant uptake, and epuvalisation	Compounds taken up, sorbed, or precipitated by roots (rhizofiltration) and young shoots (blastofiltration) or sorbed to fungi, algae and bacteria (biosorption): 10 percent to 60 percent dry weight of plant may be accumulated metals.	Metals, radionuclides, organic chemicals, nitrate, ammonium, phosphate, and pathogens	Wetlands, wastewater, landfill leachates, surface water, and pumped groundwater	Proof of concept for phytosorption: aquatic plants (Pb, Cu, Fe, Cd, and Hg) Proof of concept for rhizofiltration: sunflower (*Helianthus annuus*) and Indian mustard (*Brassica juncea*) (Pb, Cr, Mn, Cd, Ni, Cu, U(VI), Zn, and Sr) Field proof of concept: sunflower (*Helianthus annuus*) at Chernobyl, Ukraine (Cs and Sr) and field pilot, Ashtabula, Ohio for U	Raskin *et al.* (1994), Dushenkov *et al.* (1995), Dushenkov (1997), Salt *et al.* (1995), Vasudev *et al.* (1996), Cunningham and Ow (1996), McCutcheon (1998), Raskin and Ensley (2000), Dushenkov and Kapulnik (2000), (http://epuvaleau.fsagx.ac.be/EPUVALISATION-fiche1.htm)

Phytostabilization (biogeochemical), biomineralization, phytosequestration, and lignification	1. Revegetation to prevent erosion and sorbed pollutant transport 2. Plants control pH, soil gases, and redox that cause speciation, precipitation, and sorption to form stable mineral deposits 3. Humification, lignification, and covalent or irreversible binding of some organic compounds are expected	Metals (Figure 1-1), phenols, tetrachloromethane, trichloromethane, and other chlorinated solvents	Soil, mine tailings, wetlands, and leachate pond sediments	1. Revegetation grasses established for different metals dominated wastes in U.K. and U.S. 2. Lab proof of concept for Pb and Cr(VI) 3. Bench proof of concept for some pesticides, phenols, and anilines	Smith and Bradshaw (1972), Salt *et al.* (1995), Cunningham *et al.* (1995a), Cunningham *et al.* (1996), Vangronsveld and Cunningham (1998), Raskin and Ensley (2000), Terry and Banuelos (2000), U.S. DOE (2000)
Phytocontainment: 1. Phyto or solar pumping, phytohydraulic control, and phytohydraulic barriers (also biobarriers) 2. Control of soil and landfill leaching 3. "Pump and tree" phytoirrigation, or other plant treatment *ex situ*	Trees and other phreatophytes transpire large quantities to contain shallow groundwater plumes or contaminated soil leaching by reversing horizontal aquifer hydraulic gradient, or vertical soil moisture pressure gradient (infiltration and leaching minimized) both year-round or seasonally to fully or partially capture	Water-soluble contaminants (*e.g.*, chlorinated solvents, MTBE, explosives, other organic contaminants, salts, and some metals)	Groundwater, vadose zone, wetlands, wastewater, and leachate	1. Field proof of principle: *Populus* spp. (TCE, PCE, MTBE, and CCl_4) 2. Concept not proven 3. Proposed and undergoing testing: (a) pine (*Pinus* spp.) (TCE and by-products) and (b) *Salix* spp. (organic solvents, MTBE, petroleum hydrocarbons, and nutrients)	Cunningham and Lee (1995), Cunningham *et al.* (1996), Quinn *et al.* (2001), Hong *et al.* (2001), Trapp and Karlson (2001), Doucette *et al.* (this book), Eberts *et al.* (this book), Hirsch *et al.* (this book), Weaver *et al.* (this book), Jordahl *et al.* (this book), Winnike-McMillan *et al.* (this book)

(*continues*)

TABLE 1-2 (*continued*)

Type	Processes involved	Wastes that may be treated	Media	Status of applications of the process that have been proven	References
	contaminants. Applications normally coupled with rhizo and phytodegradation				
Brine volume reduction	Brines pumped onto halophytes planted in wetlands that accumulate or excrete salt and the smaller volume transported and disposed of more economically	Oil field brines	Deep groundwater	Wetland halophytes pilot tested in Oklahoma oilfield	Negri, Hinchman, and Settle (this book)
Phytovolatilization, biovolatilization, and phytoevaporation	Volatile metals and organic compounds are taken up, re-speciated and transpired. Some recalcitrant organic compounds are more easily degraded in the atmosphere. Multimedia transfer requires a risk assessment	Se, tritium, As, Hg, *m*-xylene, chlororbenzene, tetrachloro-methane, trichloromethane, trichloroethane, and other chlorinated solvents	Soils, sludges, wetlands, and groundwater	Field proof of principle: Se from wastewaters and soil. Field proof of concept: tritium from groundwater Current technical consensus: TCE volatilization is not significant but site risk assessments are required to be certain. The risk of volatilization of other organic pollutants has not been explored. Transgenic plants volatize Hg in the lab but re-deposition from the atmosphere makes field applications unlikely	Cunningham and Lee (1995), Cunningham *et al.* (1996), Brooks (1998), Zayed *et al.* (2000), Banuelos (2000), Raskin and Ensley (2000), Trapp and Karlson (2001), Quinn *et al.* (2001), McCutcheon and Rock (2001)

Phytostimulation, rhizodegradation, rhizosphere bioremediation, and plant-assisted bioremediation	Plant exudation, root necrosis, and other processes provide organic carbon and nutrients to spur soil bacteria growth by two or more orders of magnitude in number; stimulate enzyme induction and cometabolic degradation by mycorrhizal fungi and the rhizo-microbial consortium; provide diverse root zone habitat; and attenuate chemical movements and concentrations. Live roots pump oxygen to aerobes, and dead roots may support anaerobes or leave aeration channels	BTEX, other petroleum hydrocarbons, PAHs, PCP, perchlorate, pesticides, PCBs, and other organic compounds	Soils and wetlands	Field proof of concept: BTEX, other hydrocarbons, PAHs, PCP, and TCE Fungi: (a) field-scale tests of white rot fungus (*Phanerochaete chrysosporium* degradation of BTEX and (b) proof of concept for DDT, dieldrin, endosulfan, pentachloronitro-benzene, and PCP	Ferro *et al.* (1994), Cunningham *et al.* (1995a), Cunningham *et al.* 1996), Olson *et al.* this book, Hutchinson *et al.* (this book)
Phytodegradation, phytoassimilation, phytotransformation, phytoreduction, phytooxidation, phytolignification, and phytodetoxification	Aquatic and terrestrial plants take up, store, and biochemically degrade or transform organic compounds to harmless by-products, products used to create new plant biomass, or by-products that are further broken down by microorganism and other processes to less	Chlorinated solvents (CCl_4, trichloromethane, tetrachloro-methane, HCA, PCE, TCE, DCE, and VC), methyl bromide, tetrabromoethene, tetrachloroethane, dichloroethene, atrazine, DDT,	Soil, sediment, wetlands, wastewater, surface water, groundwater, and air	Proof of principle: (a) field—*Populus* spp. Carswell Air Force Base, Texas; Aberdeen Proving Grounds, Maryland; and using lysimeters at Tacoma, Washington (TCE); and (b) horseradish peroxidase pilot tested in unit process to degrade phenols, aniline, and other aromatic	Groeger and Fletcher (1988), Dec and Bollag (1994), Cunningham *et al.*(1996), Nzengung and Jeffers (2001), Shang *et al.* (this book), Nzengung *et al.* (this book), Doucette *et al.* (this book), Eberts *et al.* (this book),

(*continues*)

TABLE 1-2 (*continued*)

Type	Processes involved	Wastes that may be treated	Media	Status of applications of the process that have been proven	References
	harmful compounds. Growth and senescence enzymes, sometimes in series, are involved in plant metabolism, assimilation or detoxification. Reductive and oxidative enzymes may be serially involved in different parts of the plant	other Cl and P-based pesticides, PCBs, phenols, anilines, nitriles, TNT, DNT, RDX, HMX, NB, picric acid, NT, nitromethane, nitroethane, and nutrients		contaminants in wastewater. Proof of concept: *Rosa* spp. 'Paul's Scarlet' rose (PCBs)	Hirsch *et al.* (this book), Weaver *et al.* (this book), Jordahl *et al.* (this book), Winnike-McMillan *et al.* (this book)
Phytophotolysis	Contaminant translocated from soil or water into leaves and broken down by photolysis	RDX	Soil, wastewater, wetlands, and groundwater	Proof of concept being investigated	
Phytoslurry	Enzymatically active plant material ground and slurried with wastewater contaminated soil, or sediment	DNT and TNT	Wastewater, soil, and sediment	Lab proof of concept	Medina *et al.* (2002)

The following abbreviations are used—BTEX: benzene, toluene, ethyl benzene, and xylene; DCE: dichloroethene; DDT: dichloro-diphenyl-trichloro-ethane; DNT: dinitrotoluene; HCA: hexachloroethane; HMX: octahydro-1,3,5,7-tetranitro-1,3,5,7-tetraazocine; MTBE: methyl *tert*-butyl ether; NB: nitrobenzene; NT: nitrotoluene; PAHs: polycyclic aromatic hydrocarbons; PCBs: polychlorinated biphenyls; PCE: tetrachloroethene; PCP: pentachlorophenol; RDX: hexahydro-1,3,5-trinitro-1,3,5-triazine; TCE: trichloroethene; TNT: 2,4,6-trinitrotoluene; U.K.: United Kingdom; and VC: vinyl chloride.

- Distinguish active plant uptake from sorption and other passive processes that sequester contaminants in different ways (Nzengung and Jeffers 2001)

These fundamental concepts aid in limiting the proliferation of terminology for phytoremediation as noted in Table 1-3.

Phytoextraction, Hyperaccumulation, Phytoaccumulation, and Phytomining

Phytoextraction is the uptake of metals (see Figure 1-1 for examples) and other pollutants from soil and the translocation of those contaminants either into harvestable aboveground plant tissue or into belowground root tissue (Kumar *et al.* 1995). Hyperaccumulation, which is a special case of phytoextraction, is an interesting characteristic of a few plants involving uptake and accumulation of more than 0.1 percent (by dry weight of plant) nickel (Brooks *et al.* 1977), zinc, copper, chromium, and other trace metals (Brooks 1998; see

TABLE 1-3 Other Terminology for Phytoremediation

Term	Definition
Botano-remediation, green remediation, phytotechnology, or phytoremediation technologies	Phytoremediation (Cunningham *et al.* 1995b, CH2M Hill 2001, McCutcheon and Rock 2001, McIntyre this book, Rock this book)
Phytodecontamination	Phytodegradation, transformation, extraction, volatilization, and rhizodegradation; with phytostabilization used to categorize all waste management approaches involving green plants (Cunningham *et al.* 1995b, Cunningham *et al.* 1996). Similar to the use of green plants to decontaminant metals–laden soils (Baker *et al.* 1995)
Phytolignification	Cunningham *et al.* (1996) seem to have coined the term to distinguish vascular and unicellular plant lignification, which seems redundant if the term phyto is taken to mean all plants
Phytodetoxification	Distinguishes plant metabolism to obtain carbon and nutrients from metabolism to alleviate stress (McCutcheon 1998)
Bioconcentration	Phytoextraction, rhizofiltration (Raskin *et al.* 1994), and sorption of metals and other contaminants to algae, fungi, and bacteria (Crist *et al.* 1988)
Phyto–land farming	Redundant use of phyto for renewed emphasis on the role of plants in an established practice (Cunningham and Ow 1996)

Figure 1-1). For cadmium and other rare metals, hyperaccumulator plants accumulate more than 0.01 percent by dry weight of plant material. Hyperaccumulation of iron, manganese, and other common elements equals or exceeds 1 percent of dry plant weight (Baker 1995, Baker and Reeves 1996, Brooks 1998). These criteria typically represent two orders of magnitude greater accumulation than that found in plants growing in uncontaminated soils (Brooks 1998). The tree *Sebertia acuminata* in New Caledonia accumulates the highest measured amount of metal—as much as 10 percent nickel in the sap, giving a bluish color (Reeves *et al.* 1995, Cunningham *et al.* 1995b, Cunningham *et al.* 1996, Brooks 1998). Phytomining follows from biogeochemical prospecting and botanical geology (Rosler and Lange 1972, Brooks 1998). Observations and pilot studies establish that some plants can extract enough metals from soil to be used as an alternate mineral extraction method, which might be practiced on hazardous waste sites in the future (Chaney 1983, Baker *et al.* 1995, Nicks and Chambers 1995, 1998, Reeves *et al.* 1995, Brooks and Robinson 1998). Often semantically misused for phytomining, biomining is the use of microorganisms to oxidize and solubilize metals from heaps of sulfide minerals (*e.g.*, copper leaching using *Thiobacillus ferrooxidants*) (Brooks 1998) and is not considered part of phytoremediation. Phytoaccumulation refers to the removal of inorganic and organic compounds (but primarily the latter) from soil and water to distinguish phytoextraction of metals (McCutcheon 1998).

Phytoextraction was one of the most promising processes for commercialization [the first patent applied for in Japan in 1980 (T. Utsunomiya, application number 55-72959, from Cunningham *et al.* 1995a) and in the U.S. in 1994 (Raskin *et al.*, U.S. Patent 5 364 451, http://patft.uspto.gov)]. Numerous sites in the U.S. and worldwide are contaminated with metals from mining and other industrial activities, some mining sites even predating written history (Reeves *et al.* 1995, Brooks 1998). In the U.S., lead, chromium, cesium, and uranium are particularly troublesome to clean up, and bioavailability for plant extraction has become a major issue. Chelates have been used to overcome the lack of mobility (Brooks 1998), resulting in more leaching into deeper soil and groundwater that undermines regulatory and commercial support for proposals to use phytoextraction of lead, chromium, cesium, and uranium. Thus, commercialization in the short term is currently (2002) in doubt, pending the establishment of the laboratory proof of concept and pilot optimizations of the dose of chelation agents in relation to lead and chromium uptake by plants. Two U.S. Environmental Protection Agency Superfund Innovative Technology Evaluations (Rock this book) have not validated the use of phytoextraction, and two of the very few companies formed to use this approach have failed or reorganized (Marmiroli and McCutcheon this book). Successful bench testing in the arid western U.S. with much deeper aquifers (over 100 meters or several hundred feet deep) may salvage a niche for this treatment technology (Negri *et al.* 1998). Nevertheless, radionuclides already occur as deep as 64 meters (210 feet) at the Idaho National Environmental Engineering Laboratory (Schneider 1988) before the testing of phytoextraction of metals occurred at the site.

Rhizofiltration, Blastofiltration, Phytoadsorption, and Contaminant Uptake

Early work established that aquatic plants, seaweed, algae, bacteria (*e.g., Klebsiella pneumonia*), and fungi are good biosorbents for metals (*i.e.*, lead, copper, cadmium, iron, and mercury) and radionuclides (Crist *et al.* 1988, Xue *et al.* 1988, Wilde and Benemann 1993, Raskin *et al.* 1994, Brooks 1998, Wolverton and Wolverton 2001). This work seemed to be lumped under the term contaminant uptake. Biosorbent aquatic plants include water hyacinth (*Eichhornia crassipes*), reed (*Phragmites communis* and other species), cattail (*Typhus latifolia*), pennywort (*Hydrocotyle umbellata*), duckweed (*Lemna* spp., *Spriodela* spp., and *Wolffia* spp.), water velvet (*Azolla pinnata*), and water milfoil (*Myriophyllum spicatum*). Alga families that involve good biosorbents include *Chlorella, Anabaena, Chlamydomonas, Cladophora, Nostoc, Spirogyra, Vaucheria, Chroococcus*, and several other families. Except for the work of Todd and Josephson (1996) and of Wolverton and Wolverton (2001), the lack of economical culturing and harvesting methods and difficulty in getting concentrated biomass in contact with contaminated waters seem to have prevented pilot and full-scale testing (Dushenkov and Kapulnik 2000). A breakthrough occurred in the early 1990s with the use of terrestrial plants grown hydroponically to achieve much greater biomass in the form of extensive root systems to "filter," adsorb, and take up metals from water (Raskin *et al.* 1994, Dushenkov *et al.* 1995, Dushenkov and Kapulnik 2000). This process is rhizofiltration (Figure 1-1). In addition to adult roots taking up metals from water, young plant seedlings grown in aquacultures are also similarly effective. Using the Greek word *blastos* for budding or embryonic, this concept was termed blastofiltration (Dushenkov and Kapulnik 2000). Field and pilot tests are promising (Tables 1-1 and 1-2), but full-scale cleanup using rhizofiltration and other sorptive processes remains unproven except for the limited heuristic applications of Wolverton and Wolverton (2001) and of Todd and Josephson (1996).

Phytostabilization, Biomineralization, or Phytosequestration

Phytostabilization refers to the holding of contaminated soils in place by vegetation and thereby immobilizing toxic contaminants in soils (Vangronsveld and Cunningham 1998). Establishment of rooted vegetation prevents the transport of windblown dust, an important pathway for human exposure at hazardous waste sites. Biomineralization—enhancement of the natural biogeochemical processes of forming mineral deposits (especially sulfide minerals)—is the key to stabilization of metals (Kalin 2002). Humidification, covalent binding, and long-term sequestration are the key processes to stabilize organic and some inorganic contaminants. Phytohydraulic control to prevent the leaching of contaminants from soils and vegetative capping are similar processes.

Phytostabilization is particularly applicable for elemental contaminants in soils at waste sites for which the best alternative is often to hold these pollutants in place to (1) prevent greater bioavailability during removal, (2) avoid more worker exposure, and (3) minimize disruption of wetlands and other critical ecosystems. In general, phytostabilization can be temporarily effective for metals tightly bound to soils. Biomineralization seems to be the only sustainable type of phytostabilization.

Unlike phytoextraction, the ideal plants for phytostabilization of metals in soils are those that do not take up or mobilize the contaminants (*i.e.*, by producing natural surfactants), or that limit uptake to the roots, thus avoiding translocation to the shoots. Aboveground translocation and, to a lesser degree, belowground root accumulation, expose the food chain to the potential toxicity associated with bioaccumulation of metals (not observed in studies to date, U.S. DOE 2000). Tolerance of the highest contaminant concentrations is another trait of the ideal plant for phytostabilization. The use of metals by plants is very limited, so that stabilizing these elements *in situ* is sometimes the best alternative at sites with low contaminant levels (*i.e.*, near risk thresholds) and at sites where vast areas are contaminated such that large-scale removal action or other *in situ* remediation is not presently feasible. For those radionuclides that have short half-lives, such as [^{90}Sr] with a half-life of 2.8 years, holding the contaminant in place long enough for decay to occur can be feasible.

Phytostabilization of lead, cadmium, zinc, arsenic, chromium, molybdenum, cesium, strontium, uranium, and other elements (see Figure 1-1) is theoretically possible under the limited conditions of biomineralization. Many divalent metals (*e.g.*, lead, cadmium, zinc, nickel, copper, cobalt, and mercury) precipitate as sulfide minerals under anaerobic conditions, so occasionally the strategy may be to lower redox conditions in the soil environment and to reduce sulfate to sulfide. In addition, some relative stability is possible for redox metals that have less mobile valence states in natural settings. Chromium, for example, is much less mobile and less toxic in the reduced state [Cr(III)] than in the oxidized state [Cr(VI)]. A self-sustaining or low-maintenance application of phytostabilization may be possible for elements that are more stable in an oxidized state. Arsenic and molybdenum are less mobile in the oxidized valence states [arsenate As(V) and molybdate Mo(V)], and thus vigorously growing, deep rooting or deep planted trees, shrubs, or wetland plants might keep these wastes oxidized and immobile. Sorbents such as Fe(III)-oxides may be necessary during seasonal lulls in growth to buffer the lack of continuous pumping of oxygen by plants into the soil. Finally, root control will be difficult, but the work of Eberts *et al.* (this book) implies that exudation may sustain reducing conditions below healthy oxidized roots.

Revegetation is rarely sustainable, especially if overfertilization for excessive production of organic material is required. Excessive eutropy and senescence to produce reducing conditions and divalent metals stabilization may not be sustainable because of subsequent rooting and "pumping" of oxygen into the stabilized waste. Plants require oxygen, and the roots of most plants will not grow into anaerobic zones that contain stabilized metals. However, phreatophytes and wetland plants such as cypress (*Taxodium* spp.) can penetrate a few centimeters into anaerobic zones and then tend to "pump" or introduce oxygen into the wastes to mobilize reduced metals. Therefore, combinations of phreatophytes and other vegetation in controlled succession with root training may be necessary for longer-term stabilization of some metals.

Immobilization of toxic elements such as lead, cadmium, zinc, and arsenic by revegetation may require soil amendments to increase soil organic matter, raise the pH (using lime), or bind some constituents with phosphate or carbonate (Schnoor 2000). Without amendments, cadmium readily translocates to the leaves of many plants, which represents a risk of food chain bioaccumulation such that this exposure pathway may limit the consideration of phytostabilization at some waste sites. Nevertheless, the role of plants in sustaining soil organic matter, pH, and binding agents is not well defined.

Phytostabilization of xenobiotic organic pollutants is based on sequestration processes (and thus the term phytosequestration), which may be better classified as phytotransformation according to the green-liver theory (Sandermann 1992). These processes include humification, covalent or irreversible binding, and lignification (Cunningham *et al.* 1996), which also seem to be the basis of composting (Bruns-Nagel *et al.* 2000, Jerger and Woodhull 2000). As a result, composting may be viewed as phytoremediation if potato wastes and other components add peroxidase and other plant enzymatic activity and are established to be a critical component. Alternatively, composting might be classified as phytoremediation if fungal transformation, conjugation, and sequestration (green liver phases) involving lignification, glycosation, and other plant-like processes dominate over microbial processes. The key to the phytostabilization of organic compounds is the benefit from plants and fungi in achieving greater humification, irreversible binding, and lignification. Humification is the incorporation of organic contaminants into soil organic matter. Humification reduces bioavailability, and consistent with covalent binding, occurs from both microbial and plant mediated metabolism or transformation. Specifically, phytostabilization pertains to increased humification and binding of contaminants by soil bacteria whose numbers and activity are enhanced by the presence of plants, and humification and binding by fungi and by external plant enzymes (Schnoor *et al.* 1995). Lignification by fungi, which does not occur within bacteria, would appear to be important in both phytostabilization and phytotransformation. In this chapter, algae and fungi biosorption (and any transformation), with biomass harvest and disposal, exclusive of higher plants, are considered as phytoremediation. Thus, only increased humification and binding by microorganisms enhanced by plants would seem to constitute phytostabilization of organic xenobiotic compounds in the rhizosphere.

Phytocontainment and Brine Volume Reduction

Both phytocontainment and brine volume reduction involve the use of plants to control and manipulate the near-surface-water balance using enhanced evapotranspiration. Phytocontainment involves (1) phyto or solar pumping, solar pump and tree, phytohydraulic control, and phytohydraulic barriers or biobarriers (Strand *et al.* 1995, Cunningham *et al.* 1996, Schnoor 1997, 2002); (2) control of soil and landfill leaching (Weaver *et al.* this book); and (3) "pump and tree" (Trapp and Karlson 2001), phytoirrigation (Jordahl *et al.* this book), or other *ex situ* plant treatments. The design objective of all three approaches is to keep contaminated soils, buried solid wastes, or polluted aquifers as dry as possible.

Phytohydraulic control or solar pump and tree utilizes phreatophytes to transpire or "pump" large quantities of water from the saturated zone and to create a cone of depression through which contaminants at the top of an aquifer cannot migrate. The key to the process is for the trees to be rooted into the capillary zone (*i.e.*, not taking much soil water from recent infiltration of rainfall) and for the aquifer to be "tight" or of limited porosity. If the plume is captured, no contaminants move off-site. If the plume extends over the depth of an aquifer that flows in the general case of a Dupuit–Forchheimer flow (Landmeyer 2001, Weaver *et al.* this book), trees must dewater the aquifer for complete control. However, with downgradient infiltration and dilution, and natural attenuation, complete control is rarely necessary or achievable (without supplementary mechanical pumping). Aquifer dewatering is not sustainable without irrigation in the upper soil to keep the trees alive during the severe water stress that may result.

It is also possible to use trees and grasses with dense fibrous root systems to help transpire large amounts of soil water (from the vadose zone) and to limit vertical percolation of water through waste. Pump and tree or phytoirrigation is the *ex situ* concept of mechanically pumping or supplementing the transpiration of trees or other plants to fully achieve hydraulic control of a contaminated plume and then irrigating trees or other vegetation. The irrigated plants may be growing over the plume (preferred) or off-site under controlled conditions. The irrigated vegetation must take up and transpire all the contaminated water before migration through the soil into the upper groundwater aquifer.

Phytocontainment also should be coupled with phytotransformation, phytodegradation, or rhizodegradation of the organic chemicals that are taken up by vegetation. If the absorbing plants do not at least transform the contaminants to benign products, then the vegetation must undergo proper harvesting and disposal dependent on the residual human and ecological risk of the accumulated contaminants and by-products.

When the goal is to create a hydraulic barrier to capture a plume, the trees must capture and transpire groundwater. Yet, physiologically, trees expend less energy to transpire near-surface soil moisture from infiltration of more recent precipitation. The extent to which trees transpire moisture from unsaturated, near-surface soil reduces the likelihood of achieving aquifer control.

Thus, surface diversion of infiltration from runoff, deep planting into the capillary zone or aquifer, and root training by deep irrigation or chemical means are critical design issues. Ferro *et al.* (2001), Hong *et al.* (2001), Landmeyer (2001) and show excavated poplar (*Populus* spp.) root systems with uncontrolled near-surface rooting to capture infiltrated runoff. Quinn *et al.* (2001) discuss deep planting to force plant transpiration from confined groundwaters. See Negri, Gatliff *et al.* (this book) and Hutchinson *et al.* (this book) for more information on design to control and direct rooting.

Phreatophytic trees such as poplar, cottonwood, and willows (Salicaceae) can send roots down to the groundwater table as much as 3 to 5 meters deep (10 to 16 feet) and pump water for transpiration into the atmosphere. The roots of willows (*Salix* spp.), cypress (*Taxodium* spp.), and a few other plants can persist for a limited time below the water table due to enhanced oxygen pumping to the roots during seasonal fluctuations. Use of an auger for planting unrooted "poles" of poplar (*Populus* spp.) trees, buried up to the crown down to a depth of 2.4 to 3 meters (8 to 10 feet) is a standard practice for getting roots closer to the water table quickly. However, as Ferro *et al.* (2001) demonstrate, deeper roots may be abandoned if infiltration is not controlled. A few phytoremediation companies specialize in developing deep root systems (planting trees in a well boring) that can reach even 15 meters (50 feet) down to confined aquifers or the groundwater table (U.S. DOE 2000, Quinn *et al.* 2001, see U.S. patents 5 829 191 and 5 829 192 by Gatliff involving infiltration control sleeves in the bore hole, www.uspto.gov).

It is important to clearly understand water relations in phytoremediation designs. Evapotranspiration refers to the amount of water that is both transpired by trees and evaporated from the soil (see Jordahl *et al.* this book and Vose *et al.* this book). Once canopy closure occurs, only a negligible amount of water evaporates from the soil.

Very few transpiration measurements actually exist for phytoremediation systems and some literature claims are dubious. When planted densely, poplars (*Populus* spp.) and willows (*Salix* spp.) usually can reach canopy closure (*i.e.*, capturing almost all the sunlight) after 3 to 5 years, depending on maintenance and care of the trees. Transpiration increases each year until a plateau is reached at the time of canopy closure (Ferro *et al.* 2001, Jordahl *et al.* this book). It is the plateau level for each season that must be estimated for phytoremediation designs.

The theoretical maximum amount of transpiration is based on the amount of energy received from the sun and that required to evaporate water. For mid-latitudes (35 to 50 degrees north) during the growing season, the earth receives approximately 30 million Joules per square meter per day of solar insolation. About 2.5×10^6 Joules evaporates 1 liter (0.3 gallons) of water per day. Thus, the thermodynamic limit of evaporation is 12 liters per square meter per day (12 millimeters per day or 0.5 inches per day) into very dry air. However, plants are not perfectly efficient; energy is required to lift the water from the aquifer to the atmosphere and to overcome friction or resistance to flow up the xylem. Typical crops such as corn (*Zea mays*) can transpire about 4 to 5 liters per square meter per day (4 to 5 millimeters per day or approximately 0.2 inches per day) during growth. Poplars (*Populus* spp.) can perform almost 30 percent better than corn (*Zea mays*) if these trees are rooted in the surficial aquifer, but only for about 4 to 6 months of the year, depending on the location. Thus, the best expected phytoremediation performance (where the trees have canopied and are rooted in a shallow groundwater table) is 4.5 liters per square meter per day $\times 1.3 \times$ (6 months per 12 months) $\times$ (365 days per year) $\times$ (1 cubic meter per 1000 liters) $= 1.07$ meters per year which is about 1 million gallons per acre per year. For a good phytoremediation effort, phreatophyte evapotranspiration rates typically range from about 2.6×10^5 to 6.5×10^5 liters per hectare per year (0.4 to 1 million gallons per acre per year).

Vose *et al.* (this book) shows that transpiration increases with the diameter of the tree, increasing solar radiation, decreasing humidity (vapor pressure deficit), and increasing temperature. Generally, hardwood trees transpire less than conifers. Vose *et al.* of this book also shows that transpiration increases with precipitation. In wetter climates with greater soil moisture, taking water from the soil is energy efficient for trees and other vegetation. However, if phreatophytic trees can be planted into shallow groundwater, then very high transpiration rates are possible in dry climates.

In recent years, there have been dozens of applications of phytoremediation for the uptake and capture of plumes containing chlorinated solvents such as trichloroethylene (TCE). Some of these applications are documented successes (Eberts *et al.* this book and Hirsh *et al.* this book).

Plants create hydraulic barriers or achieve hydraulic control, both horizontally and vertically. Evapotranspiration also can be great enough to prevent the downward migration of contaminants related to percolation through the unsaturated zone. Vertical evapotranspiration-driven phytocontainment is also the underlying principle for vegetative capping. Vegetative caps are called "alternative covers" and "evapotranspiration landfill covers." The purpose is to increase evapotranspiration from the soil cover of a landfill to prevent leaching of hazardous waste and enhance bioremediation with the limited oxygen and moisture that does enter the waste.

Landfill caps have been required in the U.S. by the Resource Conservation and Recovery Act since the mid-1980s. Hazardous waste landfills are regulated by Subtitle C in the Federal Register (40 DFR 264 and 265), and by Subtitle D (40 CFR 257 and 258) that includes wastes that are not hazardous such as those in municipal landfills. Superfund sites often have materials that pose a relatively high level of risk and are excavated and disposed in hazardous waste landfills. The caps for these landfills have traditionally employed grasses rooted in a shallow soil layer, underlain by a drainage layer and an impermeable geomembrane [such as high density polyethylene, 1 to 1.5 millimeters or 40 to 60 mils (1 mil is one thousandth of an inch) thick]. Below that is a layer of low hydraulic conductivity material (usually compacted clay of less than 10^{-6} centimeter per second saturated hydraulic conductivity) and then the waste itself. Grasses rooted in the uppermost soil layer reduce erosion, increase evapotranspiration, and protect the geomembrane cover. Because the depth of the soil is shallow, the amount of water that can be stored in the soil layer is minimal, and large rainfall events result in precipitation water running down the sloped geomembrane cover for collection at the base of the landfill.

Vegetative caps provide a completely different approach to landfill capping and closure. The Remedial Technology Demonstration Forum (consortium of U.S. industries and federal agencies) Phytoremediation of Organics Action Team has defined a vegetative cap as "a long-term, self-sustaining cover of plants growing in and over materials that pose environmental risk; a vegetative cap reduces that risk to an acceptable level and requires minimal maintenance." In an alternative vegetative cap, trees and shallow-rooted vegetation are rooted in a mulch layer and organic-rich topsoil, underlain by native soil from the site, and a compacted soil layer at the base of the rooting zone (usually 2 to 3 meters or 7 to 10 feet). Roots may grow through the compacted soil layer directly into the waste. In the absence of a geomembrane and a compacted clay hydraulic barrier, more water and oxygen percolate through the cap and enter the waste. This hastens biodegradation of the waste. In some designs, leachate can be collected and added to the trees and vegetation growing in the cap. The recycled leachate is normally added by drip irrigation through tubes buried in the soil layer so that there is no aboveground exposure to the leachate and volatile organic compounds cannot volatilize into the atmosphere.

Other advantages of the alternative vegetative cap are (1) more rapid "stabilization" of the wastes, (2) decreased gas production after 5 to 20 years, and (3) earlier access to the site for alternative uses (*e.g.*, parkland, municipal building construction, and similar land uses). Disadvantages include (1) the possibility of phytotoxicity, (2) pests or weather destroying the trees and decreasing the efficiency of the alternative cap, (3) lack of a track record to guide selection and design, and (4) strict state regulations that do not allow alternative caps.

The most important problem in the U.S. is that the Resource Conservation Recovery Act does not explicitly countenance the use of vegetative covers. The Act stipulates that a regulatory agency may approve an alternative final cover design that achieves an infiltration reduction equivalent to that of the recommended design. However, with vegetative covers, the idea is to allow limited percolation through the landfill waste. The Act provides no guidance for alternative cover designs that seek to improve overall performance and the rate of biodegradation of the waste.

Since 1998, the U.S. Environmental Protection Agency has sponsored the Alternative Cover Assessment Program to conduct side-by-side evaluations with traditional covers (Benson *et al.* 2001). To date, a dozen alternative landfill caps are under evaluation (Rock this book). These covers are being used on municipal and industrial landfill sites for control of leachate, at petrochemical sites for control of benzene and other organic contaminants, at land farms for degradation of petroleum hydrocarbons, and on construction debris landfills (Glass 1999). Pending these performance evaluations, the rationale is compelling and the potential market for alternative covers is significant. The number of landfills in the U.S. is very large and this could be one of the major uses of phytoremediation in the future, but only if U.S. states amend regulations to allow for alternative covers.

Brine volume reduction in oil field waste management involves pumping brines into created treatment wetlands of halophytes. Two types of halophytes have been field pilot tested—one accumulates salt and the other excretes salt. Preliminary risk assessments indicate that the salt-laden wetland vegetation can be grazed or harvested for cattle feed rich in salt to reduce the cost of supplements (see Negri, Hinchman, and Settle this book).

Phytovolatilization or Biovolatilization

Phytovolatilization (Terry *et al.* 1995) or biovolatilization as used by Cunningham and Ow (1996) is a multimedia transfer of pollutants from soil or water to the atmosphere. This process can only be used if risk-assessment and follow-up monitoring establish that sufficient risk reduction occurs. For example, see the risk assessment conducted for the phytovolatilization of tritium at the Argonne National Laboratory near Chicago (Negri *et al.* 2001). For elemental contaminants (see Figure 1-1) this process involves uptake and transformation or respeciation to volatile forms that are released through the roots, stems, or leaves. Volatile organic compounds are taken up and transpired with water vapor or diffused out of roots, stems, and leaves. The following two discoveries cover the short-term and long-term potential of this process.

First, Terry *et al.* (1992) found selenium volatilization from cultivars and other plants (Terry and Zayed 1994), including wetland plants. Root and rhizosphere transformation of selenium to volatile dimethyl selenide avoids accumulation of toxic levels of selenium in plants and the food chain, especially in birds and fish, as was discovered in 1982 in Kesterson Reservoir located in California. Air transport of dimethyl selenide is not known to represent a health risk; in fact, many soils are selenium poor (Reeves *et al.* 1995, Banuelos 2000), and plants with limited accumulation may represent a good supplement to cattle feed. Other methylated metalloids such as arsenic may also be good candidates for this process (CH2M Hill 2001), depending on local risk assessments of the volatilized material. Documentation of wetlands that volatilize selenium from oil refinery wastes establishes the concept in the field.

The second discovery, which is of long-term importance, is that *Arabidopsis thaliana* can be genetically engineered with bacterial genes to resist mercury poisoning. In addition, the same genetic engineering provides genes to demethylate organic complexes of mercury and transpire elemental mercury (Rugh *et al.* 1996, Rugh *et al.*, 1998). Practically, fast-growing, high-biomass plants are unlikely to be genetically engineered to spread less toxic elemental mercury through the air (already the primary route from source areas). The redeposited mercury can again methylate and bioaccumulate in cleaner areas. Nevertheless, testing the concept of transplanting bacterial genes into plants to increase tolerance, transformation, and accumulation is a remarkable precedent.

Phytovolatilization of organic chemicals is a two-edged sword, and being similar to that of mercury transpiration, will be very site dependent. Even though the half-life of TCE in the atmosphere is on the order of hours to days and atmospheric dispersion significant, the lack of general and local risk assessments have slowed application of trees for phytohydraulic control of TCE and MTBE plumes. However, the scientific consensus is that phytovolatilization of TCE is acceptably small (immeasurable to 9 percent), and that continued monitoring is sufficient until a general risk assessment can be undertaken (McCutcheon and Rock 2001). See Winnike-McMillan *et al.* of this book, on potential risks of phyto and soil volatilization of MTBE. However, further unpublished studies associated with Hong *et al.* (2001) have shown that very little MTBE is volatilized under field conditions, suggesting storage or metabolism by poplar (*Populus* spp.) trees in the field.

Phytostimulation, Rhizo(sphere)degradation, Rhizosphere Bioremediation, or Plant-Assisted Bioremediation

Rhizosphere bioremediation and rhizofiltration require contaminants to be at or near the roots. Root necrosis associated with fluctuating water tables, soil moisture, and exudation adds significant amounts of organic carbon and nutrients to the soil. Plants release 10 to 20 percent of total carbon fixed in photosynthesis through the roots into the soil (Whipps and Lynch 1985, Olson *et al.* this book). Root biomass and exudates are rich in biodegradable organic macromolecules that stimulate microbial growth (Schnoor 2000). Thus, the concentration of heterotrophic microorganisms in the rhizosphere can be up to 100 times greater than in the bulk soil (Rovira and Davey 1974, Paul and Clark 1989, Atlas and Bartha 1993). This zone of high microbial concentration and site of organic releases from plants appears to foster increased rates of removal of a wide variety of environmental pollutants, including chlorinated solvents, PAHs, and pesticides (for reviews, see Schnoor *et al.* 1995, Schnoor 1997, Negri, Gatliff, *et al.* this book, Olson *et al.* this book).

In dry years, trees produce roots that follow the declining groundwater table. In wet years, trees will slough roots (root turnover) and create dissolved organic carbon. This dissolved organic carbon can serve as substrate for microorganisms to consume dissolved oxygen and even create anoxic zones for anaerobic decomposition of chlorinated solvents (Eberts *et al.* this book). Facilitation of microbial decomposition of chlorinated solvents (*i.e.*, TCE → *cis*-dichloroethylene → vinyl chloride → ethene) is very useful unless microbial degradation stops at a metabolic intermediate such as vinyl chloride that is more toxic than the parent compound.

Some aromatic compounds in the abandoned roots and exudates from live roots (*e.g.*, phenols, l-carvone, and *p*-cymene) induce dioxygenase enzymes that degrade polychlorinated biphenyls (Donnelly *et al.* 1994, Fletcher and Hegde 1995, Hegde and Fletcher 1996, Gilbert and Crowley 1997). Thus, such compounds could serve as primary substrates for the co-oxidation of various priority pollutants (Anderson *et al.* 1993) and support the proliferation of specific degraders (Jordahl *et al.* 1997, Miya and Firestone 2000). However, abandoned roots and exudates also contain easily degradable compounds such as sugars and amino acids (Schnoor 2000). Such compounds could hinder PAH biodegradation if the exudates repress enzyme induction, serve as preferential substrates (*e.g.*, diauxy), increase the sorption capacity, and decrease PAH bioavailability. Therefore, the potential beneficial effects of root exudates and detrimental substrate interactions should be evaluated under realistic conditions.

Research examining the fate of PAHs in planted soil strongly indicates that adsorption onto roots may be appreciable, but uptake and translocation of compounds from roots to foliar portions of the plants are negligible if the chemical has four or more rings (Sims and Overcash 1983, Edwards 1998). Apparently, such hydrophobic compounds sorb strongly to soil particles, which precludes vegetative uptake (Bell 1992). Therefore, PAH removal in phytoremediation experiments has been generally attributed to enhanced microbial degradation (Aprill and Sims 1990, Qui *et al.* 1994, Schwab and Banks 1994, Epuri and Sorensen 1997, Banks *et al.* 1998). However, the mechanisms responsible for this beneficial effect, which is not always observed (Banks *et al.* 1998), are not fully understood.

The capability of soil microorganisms to degrade PAHs is well documented, and numerous PAH-degrading bacteria have been isolated from various genera, including *Pseudomonas*, *Mycobacterium*, *Flavobacterium*, *Acinetobacter*, *Arthrobacter*, *Bacillus*, and *Nocardia* (Gibson and Subramanian 1984). These bacteria generally initiate PAH oxidation using dioxygenase enzymes that incorporate both atoms of molecular oxygen (as OH^-) into one of the aromatic rings. Nevertheless, PAH compounds tend to be poor inducers of such catabolic enzymes. Therefore, the presence of PAHs is ineffective in enhancing or exerting selective pressure for degradative microorganisms. This, in addition to the poor bioavailability, is probably a major reason for the recalcitrance of higher-molecular-weight PAHs. Root exudates may improve both the bioavailability of hydrophobic organic compounds such as polychlorinated biphenyls (PCBs) and PAHs as well as the biodegradation of these chemicals.

Microbial PAH degradation is often a cometabolic process that depends on the presence of lower-molecular-weight aromatic compounds that trigger enzyme induction (Bauer and Capone 1988, Heitkamp and Cerniglia 1988, Atlas and Bartha 1998). Thus, it has been hypothesized that PAH biodegradation in soil can be enhanced by introducing plants with roots that deposit or exude phenolic compounds (Aprill and Sims 1990, Schwab and Banks 1994, Fletcher 2001).

Phytodegradation, Phytotransformation, Phytoreduction, Phytooxidation, and Phytolignification

An important distinction in biotreatment occurs between complete biological degradation or mineralization of xenobiotic compounds and the transformation of the same chemicals along other pathways. Typically, plants transform xenobiotic compounds and then conjugate or bind the products to the cell wall or store the conjugates in vacuoles (see Burken this book). Catabolic microorganisms transform xenobiotic compounds during metabolism, usually resulting in mineralization to carbon dioxide and water. As a result, catabolism is preferred over phytotransformation for the most sustainable outcome.

Some plants employ reductive transformation and some oxidative transformation. Because these transformation pathways can be the very same for bacteria, phytoreduction and phytooxidation are used to distinguish causative agents in mixed biotic systems (Nzengung and Jeffers 2001, McCutcheon and Rock 2001). Phytolignification (Cunningham *et al.* 1996) is also a metabolic process but one probably best categorized as phytostabilization, along with humification and irreversible binding. Because prokaryotes are not involved in lignification (only fungi and vascular plants are), the use of the term phyto is probably redundant with the use of the term lignification.

HETEROTROPHIC MICROBIAL AND PHOTOAUTOTROPHIC REMEDIATION PROCESSES

Bioremediation and phytoremediation are both biological treatments, but these technologies are scientifically quite different in several aspects. Bioremediation has had a head start of at least two decades (and longer, considering work with industrial biological wastewater treatment) in developing a science base for hazardous waste management. Foremost in importance to all waste management disciplines is the recognition of the vital difference between sustainable mineralization and transformation (Spain *et al.* 2000). The ultimate achievement of reliable, well-understood, and cost-effective waste management practices that reduce hazardous wastes to benign substances (*i.e.*, water, nutrients, carbon dioxide, methane, and other compounds) has recently been adopted by bioremediation specialists and should be adopted by all as an important part of the multimedia, risk reduction approach of modern environmental and ecological engineering. Under this new paradigm has come the development of biodegradation methods to treat benzene, toluene, ethylbenzene and xylene (BTEX), TCE, and other hazardous xenobiotic chemical wastes. In most cases, these same biodegradation processes are enhanced or better controlled with vegetation in both soils and treatment wetlands. In some cases, however, vegetation has no effect, or worse, may produce so called "dead end" metabolites or mobilize some contaminants. Dead end metabolites are transformation products that are more persistent than the original pollutant and maintain some residual risk or toxicity to living organisms, including humans.

Plants tend to transform and store metabolic products, whereas heterotrophic microorganisms are more likely to mineralize xenobiotic compounds for the energy, carbon, and nutrient content. Plants use sunlight as an energy source and atmospheric carbon dioxide as a source of carbon. However, plants respire, and thus require some catabolic enzymes to break down lignin, cellulose, coumarins, flavonoids, and other complex molecules produced by these photoautotrophs. This means that for every biomolecule that has evolved, a microbial or botanical enzyme process exists somewhere to mineralize the molecule. Evolved biomolecules include natural pesticides, allelopathic chemicals, and growth hormones, which plants, insects, and other forms of life produce. When anthropogenic chemicals have a similar natural analog, these compounds tend to be easily recycled by natural processes. This may include immediate degradation by heterotrophic microorganisms for energy or nutrients, or transformation, conjugation, and storage by plants and then delignification and degradation much later after the plant dies and decays. Dissimilarities between xenobiotic molecules and natural substances usually necessitate longer periods for transformation and degradation to occur. Xenobiotic molecules that are toxic in some form and persist too long [*i.e.*, natural degradation rates are small compared to industrial mass production rates as is the case for dichloro-diphenyl-trichloro-ethane (DDT)] have been banned and are not ideal candidates for phytoremediation except for transforming the recalcitrant molecules already in the environment so that microbial degradation of the products will occur. Some anthropogenic molecules may not be subject to any natural biological process and can only be managed by energy-intensive methods similar to those used in the creation of the molecule. This case might include some dioxins.

Unfortunately, bioremediation lacks many of the cost-saving attributes of the self-engineering capability of plants—the capability to rapidly change soil and water pH, redox conditions, organic content, nutrient availability, and other geochemical conditions. A particularly difficult hurdle has been transplanting purified bacterial cultures to the environment and having these specialized organisms survive (Spain *et al.* 2000). This has tended to limit the potential applications to bioslurry reactors (*e.g.*, bioslurry reactors used to degrade dinitrotoluene). In many attempts, cultures of bacteria and fungi are not able to survive competition with local communities regardless of the adaptation techniques tried to date. By contrast, mature cottonwood trees control soil redox conditions well enough to enhance natural microbial dechlorination and degradation of TCE (Eberts *et al.* this book).

Self-engineering by plant ecosystems almost always occurs symbiotically with microorganisms, not as frequently in the absence of microorganisms (anexic cultures), and rarely in the absence of green plants which power the process with conversion of solar energy to organic molecules. For these reasons, neither phytoremediation nor bioremediation alone will be very sustainable practices but in combination can be very powerful and cost-effective.

Nevertheless, no natural processes except volcanism, earthquakes, tsunamis, hurricanes, tornados, and wild fire seem to rival the concentration of energy that goes into refining elements such as aluminum, cadmium, zinc, lead, and many others, and into creating synthetic organic chemicals. Therefore, the natural *in situ* processes of bioremediation and phytoremediation should be anticipated to require much longer cleanup time compared to that for the energy-intensive, rapid creation of some xenobiotic substances or excavation and treatment. In some cases, the diffuse solar energy and limited mechanical creation, transportation, and addition of supplements cannot be concentrated by natural process to rival the energy-intensive creation or segregation of chemicals, minerals, and the associated wastes. In other cases, neither the natural slow burning that heterotrophic degradation or respiration represents nor the partial fragmentation and reassembly into cell walls and other structures that transformation represents can keep up with the creation and spread of some xenobiotic compounds such as polychlorinated biphenyls (PCBs) that have been banned as a result. At hazardous waste sites, the light engineering enhancements of the self-engineering capabilities of plants will not overcome the concentrated toxicity quickly enough to reduce toxicity and protect the nearby ecosystem and human populations. Only when exposures are limited and the site controlled for months to years will systems of plants and microorganisms be capable of using lower-grade solar energy to degrade concentrated anthropogenic wastes or influence speciation and dispersion of elemental pollutants.

Microorganisms mutate and can sometimes evolve rapidly and thus do not need extensive self-engineering capabilities. Millions of generations of microorganisms have come and gone under the stress of anthropogenic pollution (now widespread) during the centuries-long lifetimes of many trees. Nevertheless, the continual problem with evolving annual weeds that farmers and the pesticide industry face means plants are not genetically static, but the evolution is clearly not as fast as that for microorganisms.

In contrast to the advantages of prokaryotes, plants produce and sustain more biomass and have tissues involving more specialized enzymes in concentrated areas. This is one reason that plants can better self-engineer. A plant represents a multitude of specialized cells that can not only transform, conjugate, and store toxins but also process and store water, energy, and nutrients more readily. Microorganisms, even at the community level, lack this organization and can die off rapidly, but also quickly increase in numbers with the return of the basic elements of water, energy, and nutrients. Plants are also largely stationary, and as a result are evolved with elaborate defenses against pests, especially those in the soil (*e.g.*, the plant *Arabidopsis thaliana* has about 25 000 genes whereas bacteria such as *Escherichia coli* have approximately 3000 genes).

In addition to vascular plants spurring most biodegradation in the rhizosphere (rhizodegradation), some phytodegradation and phytotransformation processes seem faster, or are at least as fast as bioremediation processes (see Table 1-4). For the vast majority of organic chemicals, the pathways and

TABLE 1-4 Transformation Kinetics for Various Xenobiotic Compounds Due to Hydrolysis and Microbial and Phytotransformation

	Half-life [and second order rate constant in liters per gram per day in terms of total microbial or plant mass present]		
Chemical	Hydrolysis[a]	Microbial transformation[b]	Phytotransformation[c]
Carbon tetrachloride	>10 years	39 days [89 000]	38 hours [0.0022]
Hexachloroethane	>10 years	>30 days [1 200 000]	8 hours [0.0104]
Tetrachloroethylene	>10 years	35 days [990 000]	5 days [0.0007]
Trinitrotoluene	NA	3 days [16 000]	20 minutes [0.25]
Triaminotoluent	>10 years	ND	2 minutes [2.5]
Hexahydro-1,3,5-trinitro-1,3,4-triazine (RDX)	NA	>20 days [>1 700 000]	21 hours [0.0040]
Benzonitrile	0.3 years	30 days [1 200 000]	11 hours [0.0076][d]

[a]For pH 7, including both neutral and alkaline hydrolysis contributions (Jeffers *et al.* 1989).
[b]Based on bacterial densities of 10^5 organisms per liter and a typical cell mass of 2×10^{-13} grams (Bowen 1966, Uhlmann 1979, Gaudy and Gaudy 1980).
[c]*Spirogyra* spp. at a density of 200 grams wet weight per liter (Lee Wolfe *et al.*, 1995, unpublished data, U.S. Environmental Protection Agency Laboratory, Athens, Georgia).
[d]Laura Carreira (1995, unpublished data, U.S. Environmental Protection Agency Laboratory, Athens, Georgia).

The following abbreviations are used—NA: not applicable; ND: not determined.
Note: The half-life is 0.69 divided by the pseudo first order transformation rate constant that is the microbial[b] or plant[c] density multiplied by the second order rate constant given in this table.

kinetics of microbial and botanical transformation and degradation are not known well enough to avoid the need for bench and field pilot testing to fashion the optimal design of biological treatments using microorganisms and plants.

FUNDAMENTAL AND APPLIED RESEARCH NEEDS

Ecological Community Effects

A shortcoming of most phytoremediation trials to date has been the reliance on monocultures. Only some rhizodegradation tests and some wetland pilots (see Thompson *et al.* this book) have attempted to master the use of ecosystems and communities. Nevertheless, even these forays have shortcomings. Monocultures of grass are used with native microbial communities for many rhizosphere applications to date. It is not clear if the grass selection process has fully considered native prairie communities from the Great Plains of the U.S. or other steppes, or if the monocultures currently used are optimal. Use of monocultures of wetland plants was definitely a problem at Milan Army Ammunition Plant (see McCutcheon *et al.* this book). Terrestrial monocultures of poplar and cottonwood (*Populus* spp.) have suffered damage from beavers and wind, and are subject to wipeout by disease.

Tackling the use of communities in phytoremediation is daunting. There is rudimentary heuristic design information from each state for revegetation and erosion control, and some information on native and indigenous plants. However, these data do not seem to cover too many sustainable communities, and rarely is the community niche specified or known for various species of plants used in revegetation. There is certainly not enough information to select pioneers to phytoremediate in phases. Jordahl *et al.* (this book) covers weed control for tree plantations, but in many potential applications the hardy pioneers necessarily will be terrestrial and aquatic weeds that must be carefully considered in design with methods such as planting only male plants for eradication after treatment. Each U.S. state and provinces in other countries will normally have local weed control and invasive-species bans that must be taken into account in fashioning an ecosystem approach to phytoremediation. A Presidential Order did require U.S. government facilities to use native species in revegetation.

Proteomics and Enzymology

Schwitzguebel and Vanek (this book), Wolfe and Hoehamer (this book), and Medina *et al.* (this book) indicate that less than 100 metabolic enzymes from plants, which are useful in phytoremediation, have been identified and partially characterized to date. However, recent genomic advances indicate that many vascular plants probably produce 25 000 or more proteins. *Arabidopsis thaliana* has about 25 000 genes, loblolly pine (*Pinus taeda*) at least 110 000, and maize (*Zea mays*), wheat (*Triticum aestivum*), and rice (*Oryza sativa*) from 25 000 to 40 000. Thus, more than 25 000 to 110 000 plant biomolecules may exist per species. Nevertheless, the full set of molecules must cover not only plant enzymes but also the full range of biomolecules involved in all other physiological functions. However, each plant species also will involve different enzymes and some genes express more than one protein. Thus, many natural plant enzymes seem to await discovery for additional uses in hazardous waste management. This seemingly extensive potential of natural enzymes does not preclude the need for genetic engineering, as some exotic anthropogenic compounds do not seem to have natural analogs. Without natural analogs or time to evolve, plants and microorganisms may be unable to eventually break down some xenobiotic compounds.

Several methods have been used in phytoremediation to characterize the enzymology of plant transformations and degradation. Dec and Bollag (1994) selected a specific enzyme, the plant peroxidase, and looked for applications of the enzyme in wastewater treatment. The result was a proof of concept of a unit process using additions of horseradish (*Armoracia rusticana*) peroxidase and peroxide, and energy-intensive process control to remove phenol from industrial wastewater (Roper *et al.* 1996, Buchanan *et al.* 1998). The cost of peroxidase and cofactor additions, plus the control necessary, seems to have precluded widespread application to date (Dec and Bollag 1994). In addition, questions remain regarding the toxicity of precipitated residuals (Buchanan *et al.* 1998). More recently, Coyle *et al.* (1999) have explored uses of peroxidase and cofactor applications to soils for the cleanup of pesticides. Wolfe and Hoehamer (this book) have begun exploring phosphatases to transform and degrade phosphorus-based pesticides and nerve agents.

Next, Strand *et al.* (1995), Gordon *et al.* (1998), and Newman *et al.* (1997, 1999) explored the pathway of TCE oxidation by hybrid poplar (*Populus trichocarpa* × *Populus deltoides*) trees. They noted that the pathway was very similar to the mammalian liver degradation of TCE (see Shang *et al.* this book) and presumed that the plant cytochrome P-450 enzyme must be responsible. Doty *et al.* (2000) later found enhanced TCE degradation from transgenic tobacco (*Nicotiana tabacum*) engineered with the more reactive human P-450 2E1.

Finally, Wolfe and coworkers took a third, more classical approach from biochemistry that is analogous to trial-and-error searches of plant and insect material for medicinal activity that can be developed into medicines (see McCutcheon *et al.* this book). The discovery of fast reactions of various classes of chemicals mixed into anoxic sediments was traced to plant enzymes. Classical biochemical isolation and segregation of the activity of sediment organic matter produced a crude extract using various sizing, metals affinity, ion exchange, and precipitation columns. The extract was placed in the spleen of a mouse and then in the spleen of a goat to get a double antibody that was used to further characterize the isolated activity. The enzyme-linked immunosorbent assay (ELISA) based on the antibodies indicated that plants were the source of these stable proteins in reducing sediments, but unfortunately extractions directly from plants produced a much smaller protein, implying that the original extracts were probably conjugated with fulvic acids in a narrow size range. Surprisingly, in all three isolations to date, the enzymatic activity tentatively was traced to nitroreductase, dehalogenase, and nitrilase activities in plants, and not to microorganisms.

The strategies of Bollag and coworkers (Dec and Bollag 1994), Strand *et al.* (1995), and Wolfe and coworkers fall into two general approaches of biochemistry. The first is a direct approach of selecting a characterized enzyme and finding a role in phytoremediation as pioneered by Dec and Bollag (1994). The second is characterizing a fast, natural process and tracing the process to specific enzymes presumptively (Strand *et al.* 1995) or by classical protein extraction as practiced by Wolfe and coworkers. Neither approach is yet fully scientifically satisfactory, but in practical terms the latter is much preferred for accelerating testing at a field scale. Starting with specific characterized enzymes requires expensive engineering trial and error to find a role in waste management (Roper *et al.* 1996, Buchanan *et al.* 1998). However, starting with uncharacterized natural processes allows early pilot engineering to determine the kinetics and tentative ranges of operating conditions that can be accelerated to the field as an inexpensive *in situ* process on an engineering basis while the science base is being established. See the early successful application of created wetlands to restore TNT- and RDX-contaminated groundwater documented in Thompson *et al.* (this book), and the developing science and engineering base in Subramanian and Shanks (this book), Jacobson *et al.* (this book), and McCutcheon *et al.* (this book). Nevertheless, understanding the proteomics of the peroxidase, laccase, and various other plant enzymes involved in plant pesticide tolerance (Medina *et al.* this book) has been vital to the broad understanding of phytotransformation possibilities.

Genetic Engineering of Plants

For at least two reasons, some genetic engineering of plants is necessary. Firstly, plant metabolism rarely mineralizes xenobiotic contaminants. Therefore, some transgenic plants and other organisms will be necessary to achieve sustainable *in situ* recycling for all anthropogenic chemicals. Secondly, some exotic xenobiotic compounds do not seem to have analogous natural biomolecules. Genetic engineering could be useful if it is possible to produce hybrid or *de novo* enzymes to transform and mineralize exotic molecules.

In genetically engineering *Arabidopsis thaliana* and yellow poplar (*Liriodendron tulipifera*) to tolerate and phytovolatilize mercury, Rugh *et al.* (1996), Rugh *et al.* (1998) established that genetic engineering of plants with prokaryotic genes is feasible for improved phytoremediation. Because microorganisms seem to have a broader degradative range than plants, transplantation should be explored to allow more plants (*e.g.*, phreatophytes) to do more than just transform some organic contaminants. Already genetically modifying enzymatic metabolism (proteomics) of pesticides by crops to tolerate pesticides is well established (Stomp *et al.* 1993, Raskin 1996, Hooker and Skeen 1999, Medina *et al.* this book). These investigations plus classical breeding to achieve better physiological capabilities such as greater transpiration rates should be explored to determine if hybrid plants can clean up hazardous waste sites effectively and control other conditions.

An example is the development of Roundup Ready® (Monsanto Co., http://tess.uspto.gov/bin/showfield?f=doc&state=2h97gg.2.31) (glyphosate-tolerant) soybeans. Glyphosate (*N*-phosphonomethyl glycine) is the active ingredient in Roundup® (Monsanto Co., http://tess.uspto.gov/bin/showfield?f=doc&state=2h97gg.2.18), a widely used herbicide effective against both grasses and broadleaf weeds. The new line of soybeans contains a bacterial 5-enolpyruvylshikimate-3-phosphate synthase that is more resistant to glyphosate inhibition, allowing the modified soybean plants to withstand applications of the herbicide without reduction in yield (Delannay *et al.* 1995, Padgette *et al.* 1995).

Several transgenic plant species have been developed already with phytodegradation applications in mind. Tobacco (*Nicotiana tabacum*) plants containing the human cytochrome P-450 2E1 were able to transform up to 640 times the amount of TCE compared with control plants (Doty *et al.* 2000). These transgenic plants also showed increased uptake and metabolism of ethylene dibromide, another halogenated hydrocarbon found in groundwater.

Transgenic tobacco (*Nicotiana tabacum*) plants expressing a microbial pentaerythritol tetranitrate reductase have higher tolerance to the explosives glycerol trinitrate and TNT (French *et al.* 1999). Denitration of glycerol trinitrate was also more rapid and complete in the transgenic seedlings.

These plant improvements have great potential for field applications, assuming public acceptance of the use of more genetically modified organisms. One major advantage of genetically engineered plants is that specific enzymes for degradation of a contaminant could be transferred to a plant species that is indigenous to an ecosystem or has other desirable remediation properties such as rapid growth, deep root structures, or high water uptake.

Paleo-Ecology

The works of Olson and Fletcher (2000), Fletcher (2001), Leigh *et al.* (2001), Olson *et al.* (2001), and Waugh (2001) strongly imply the need to engage in paleoecological, early digenetic (including contaminant digenesis), prepetrographic, petrologic, stratigraphic, and pedogenic investigations of sites that are cleaning up naturally from vegetative and microbial succession. These forensic investigations are necessary to replace decades-long pilot investigations normally used to establish the proof of principle in the field. Investigations could involve the following:

1. Search of historical records (aerial photos, old land use and character maps, forest resource and mineralogical investigations, soils maps, and industrial plant records)
2. Review of business records for industrial waste production and the identification of technological artifacts associated with waste management
3. Study of natural preservations of pollen and other species-specific plant materials in dateable soil layers and other locations
4. Dendrochronological studies
5. Investigating local plant ecology (species and community effects from climate and other changes) and physiology (roots and organs that leave records in the soil)
6. Study of the ecology of local ants, worms, and burrowers that aerate, disrupt, and mix soil layers
7. Stratigraphic and other geoarcheological analyses
8. Investigation of calcareous remains and other vestiges of animal life associated with specific species or communities of vegetation and soil layer mixing
9. Synthetic investigation of the early digenesis of site contaminants at a molecular scale in terms of site-specific geochemistry and mineralogy of soils leading eventually to mineral and rock formation—specifically investigating the contaminant and soil geochemical digenesis at decadal time scales

and in some cases the interaction of naturally occurring petroleum with soil minerals and natural surfactants originating with plants and animals

10. Dating by standard isotopic analysis
11. Radionuclide stratigraphic dating of the 1950s deposits from atmospheric testing of nuclear bombs and from more recent accidents such as Chernobyl
12. Radiocarbon dating of pre and early industrial effects

See Olson *et al.* (this book) for other insights on what additional research is necessary.

Acknowledgments

Drew Ekman of the University of Georgia kindly provided genome sizes for various plants, animals, and microorganisms. Alan Baker of the advisory panel served as acting editor in accepting this chapter; Nelson Marmiroli and Paul Flathman provided insightful and through peer reviews used in evaluation of the chapter and then to immeasurably improve the writing and expression of ideas. Bob Swank was a supporting and motivating force for this research over the last decade, who also reviewed the chapter. Support for the work of Steve McCutcheon that went into this chapter came from the U.S. Environmental Protection Agency Superfund Program, the Strategic Environmental Research and Development Program (including Project 720 managed by the Army Waterways Experiment Station) of the Departments of Defense and Energy and the U.S. Environmental Protection Agency, an internal grant from the National Exposure Research Laboratory, and cooperative interagency research agreements with the U.S. Air Force Restoration Division at Wright Patterson Air Force Base and Human Systems Centre at Brooks Air Force Base, U.S. Navy Southern Command, and U.S. Army Waterways Experiment Station. Support for Jerry Schnoor came from the Groundwater Remediation Technology Analysis Center and the EPA Hazardous Substances Research Center at Kansas State University. This paper has been reviewed in accordance with the U.S. Environmental Protection Agency peer and administrative review policies and approved for publication. Mention of trade names or commercial products does not constitute endorsement or recommendation for use.

REFERENCES

Aarts, M., A. Assuncao, S. de Folter, P. da Costa Martins, and H. Schat (2000) Metal transporters of *Thlaspi caerulescens*: tools for improved heavy metal accumulation in non-accumulators? In: First Scientific Workshop, Phytoremediation 2000: State of the Art in Europe (an Intercontinental Comparison). COST Action 837, Plant biotechnology for the removal of organic pollutants and toxic metals from wastewaters and contaminated sites, held in Herisonissos, Crete, Greece, April 6–8 (http:// lbewww.epfl.ch/ COST837/).

Anderson, T.A., E.A. Guthrie, and B.T. Walton (1993) Bioremediation in the rhizosphere: plant roots and associated microbes clean contaminated soil. *Environ. Sci. Technol.* **27**: 2630–2636.

Aprill, W. and R.C. Sims (1990) Evaluation of the use of prairie grasses for stimulating polycyclic aromatic hydrocarbon treatment in soils. *Chemosphere* **20**: 253–265.

Atlas, R.M. and R. Bartha (1993) *Microbial Ecology: Fundamentals and Applications*. Benjamin/Cummings Redwood City, California.

Atlas, R.M. and R. Bartha (1998) *Microbial Ecology: Fundamentals and Applications*. 4th ed. Benjamin/Cummings Redwood City, California.

Baker, A.J.M. (1995) Metal hyperaccumulation by plants: our present knowledge of the ecophysiological phenomena. In: Fourteenth Annual Symposium: Current Topics in Plant Biochemistry, Physiology and Molecular Biology – Will Plants Have a Role in Bioremediation. The Interdisciplinary Plant Group, University of Missouri, Columbia, April 19–22, pp. 7–8.

Baker, A.J.M. and R.D. Reeves (1996) Metal accumulating plants: the biological resource and its commercial exploitation in soil clean-up technology. In: Phytoremediation. Proceedings of International Business Communications conference, May 8–10, Arlington, Virginia.

Baker, A.J.M. and A. Smith (2000) Metal hyperaccumulator plants: biological resources for exploitation in the phytoextraction of metal-contaminated soils. In First Scientific Workshop, Phytoremediation 2000: State of the Art in Europe (an Intercontinental Comparison). COST Action 837, Plant biotechnology for the removal of organic pollutants and toxic metals from wastewaters and contaminated sites, held in Herisonissos, Crete, Greece, April 6–8 (http://lbewww.epfl.ch/COST837/).

Baker, A.J.M., S.P. McGrath, C.M.D. Sidoli, and R.D. Reeves (1995) The potential for heavy metal contamination. *Mining Environmental Management*. September: 12–14 [reprinted with permission from (1994) The possibility of *in situ* heavy metal decontamination of polluted soils using crops of metal-accumulating plants. *Resources, Conservation and Recycling*. 11, Elsevier Science B.V., Amsterdam, The Netherlands].

Banks, M.K., A.P. Schwab, R.S. Govindaraju, P. Kulakow, S.L. Lewis, and K.I. Rathbone (1998) Craney Island phytoremediation project. Final report to Advanced Applied Technology Demonstration Facility. Rice University, Houston, Texas.

Banuelos, G. (2000) Factors influencing field phytoremediation of selenium-laden soils. In: *Phytoremediation of Contaminated Soil and Water*. N. Terry, and G. Banuelos, eds. Lewis Publishers, Boca Raton, Florida, pp. 41–59.

Banuelos, G.S., G. Cardon, B. Mackey, J. Ben-Asher, L. Wu, P. Beuselinck, S. Akohoue, and S. Zambrzuski (1993) Boron and selenium removal in boron-laden soils by four sprinkler irrigated plant species. *J. Environ. Qual.* **22**: 786–792.

Bauer, J.E. and D.G. Capone (1988) Effects of co-occurring aromatic hydrocarbons on degradation of individual polycyclic aromatic hydrocarbons in marine sediment slurries. *Appl. Environ. Microbiol.* **54**: 1649–1655.

Bell, R.M. (1992) Higher plant accumulation of organic pollutants from soils. U.S. Environmental Protection Agency Report EPA/ 600/ R92/ 138. Cincinnati, Ohio.

Benson, C., T. Abichou, W. Albright, G. Gee, and A. Roesler (2001) Field evaluation of alternative earthern final covers. *Int. J. Phytoremed.* **3**(1): 105–127.

Bowen, H.J.M. (1966) *Trace Elements in Biochemistry*. Academic Press, New York.

Boyle, J.J. and J.R. Shann (1998) The influence of planting and soil characteristics on mineralization of 2,4,5–T in rhizosphere soil. *J. Environ. Qual.* **27**: 704–709.

Broadley, M.R. and N.J. Willey (1997) Differences in root uptake of radiocaesium by 30 plant taxa. *Environ. Pollut.* **97**(1–2): 11–15.

Brooks, R.R., ed. (1998) *Plants that Hyperaccumulate Heavy Metals*. CAB International, New York.

Brooks, R.R. and B.H. Robinson (1998) The potential use of hyperaccumulators and other plants for phytomining, Chapter 15. In: *Plants that Hyperaccumulate Heavy Metals*. R.R. Brooks, ed. CAB International, New York.

Brooks, R.R., J. Lee, R.D. Reeves, and T. Jaffre (1977) Detection of nickeliferous rocks by analysis of herbarium specimens of indicator plants. *J. Geochem. Explor.* **7**: 49–57.

Brown, S.L., R.L. Chaney, J.S. Angle, and A.J.M. Baker (1994) Phytoremediation potential of *Thlaspi caerulescens* and bladder champion for zinc- and cadmium-contaminated soil. *J. Environ. Qual.* **23**: 1151–1157.

Brown, S.L., R.L. Chaney, J.S. Angle, and A.J.M. Baker (1995) Zinc and cadmium uptake by hyperaccumulator *Thlaspi caerulescens* grown in nutrient solution. *Soil Sci. Soc. Am. J.* **59**: 125–133.

Bruns-Nagel, D., K. Steinbach, D. Germsa, and E. von Low (2000) Composting (humidification) of nitroaromatic compounds, Chapter 13. In: *Biodegradation of Nitroaromatic Compounds and Explosives*. J.C. Spain, J.B. Hughes, and H.J. Knackmuss, eds. Lewis Publishers, Boca Raton, Florida, pp. 357–393.

Buchanan, I.D., J.A. Nicell, and M. Wagner (1998) Reactor models for horseradish peroxidase-catalyzed aromatic removal. *J. Environ. Eng.* **124**(9): 794–802.

Burken, J.G. and J.L. Schnoor (1996) Phytoremediation: plant uptake of atrazine and role of root exudates. *J. Environ. Eng.* **122**: 958–963.

Butler, J.M., A.W. Groeger, and J.S. Fletcher (1992) Characterization of monochlorinated biphenyl products formed by Paul's Scarlet rose cells. *Environ. Contam. Toxicol.* **49**: 821–826.

Campanella, B. and R. Paul (2000) Phytoremediation of persistent pollutants in soil: study of the absorption mechanisms in two cucurbitacean species and future applications. In: First Scientific Workshop, Phytoremediation 2000: State of the Art in Europe (an Intercontinental Comparison). COST Action 837, Plant biotechnology for the removal of organic pollutants and toxic metals from wastewaters and contaminated sites, held in Herisonissos, Crete, Greece, April 6–8 (http://lbewww.epfl.ch/COST837/).

Chaney, R.L. (1983) Plant uptake of inorganic waste constituents. In: *Land Treatment of Hazardous Wastes*. J.E. Parr, P.B. March, and J.M. Kla, eds. Noyes Data Corp., Park Ridge, New Jersey, pp. 50–76.

Chaudhry, Q., P. Schroeder, D. Wreck-Reichhart, W. Grajek, and R. Marecik (2002) Prospects and limitations of phytoremediation for the removal of persistent pesticides in the environment. *Environ. Sci. Pollut. Res.* **9**(1): 4–17.

CH2M Hill (2001) Guidance for successful phytoremediation. Report prepared for the American Institute of Chemical Engineers Center for Waste Reduction Technologies. New York.

Coleman, J.O.D., C. Frova, P. Schroeder, and M. Tissut (2002) Exploiting plant metabolism for the phytoremediation of persistent herbicides. *Environ. Sci. Pollut. Res.* **9**(1): 18–28.

Comis, D. (1995) Metals-scavenging plants to cleanse the soil. *Agricultural Research.* U.S. Department of Agriculture, Agricultural Research Service, November: 4–9.

Cope, W.G. and R.G. Rada (1992) Accumulation of mercury by *aufwuchs* in Wisconsin seepage lakes: implications for monitoring. *Arch. Environ. Contam. Toxicol.* **23**: 172–178.

Cornish, J.E., W.C. Goldberg, R.S. Levine, and J.R. Benemann, (1995) Phytoremediation of soils contaminated with toxic elements and radionuclides. In: *Bioremediation of Inorganics.* R.E. Hinchee, J.L. Means, and D.R. Burris, eds. Battelle Press, Columbus, Ohio.

Coyle, C. (2000) The use of plant peroxidases for the *in situ* phytoremediation of soils and wastewaters contaminated with phenolic compounds. In: First Scientific Workshop, Phytoremediation 2000: State of the Art in Europe (an Intercontinental Comparison). COST Action 837, Plant biotechnology for the removal of organic pollutants and toxic metals from wastewaters and contaminated sites, held in Hersonissos, Crete, Greece, April 6–8 (http:/ / lbewww.epfl.ch/ COST837/).

Coyle, C., P. Duggan, and M. Godinho (1999) The development of a phytoremediation technique for the detoxification of soils contaminated with phenolic compounds using horseradish peroxidase (*Armoracia rusticana*): preliminary results. *Int. J. Phytoremed.* **1**(2): 189–202.

Crist, R.H., K. Oberholser, D. Schwartz, J. Marzoff, and D. Ryder (1988) Interactions of metals and protons with algae. *Environ. Sci. Technol.* **22**(7): 755–760.

Crowley, D.E., M.V. Brennerova, C. Irwin, V. Brenner, and D.D. Focht (1996) Rhizosphere effects on biodegradation of 2,5-dichlorobenzoate by a bioluminescent strain of root-colonizing *Pseudomonas fluorescens. Fed. Eur. Microbiol. Soc.* **20**: 79–89.

Cunningham, S.D. and W.R. Berti (1993) Remediation of contaminated soil with green plants: an overview. *In Vitro Cell. Dev. Biol.* **29**P: 207–212.

Cunningham, S.D. and C.R. Lee (1995) Phytoremediation: plant-based remediation of contaminated soils and sediments. In: *Bioremediation: Science and Applications.* Soil Science Society of America, Special Publication 43, Madison, Wisconsin, pp. 145–156.

Cunningham, S.D. and D.W. Ow (1996) Promises and prospects of phytoremediation. *Plant Physiol.* **110**: 715–719.

Cunningham, S.D., W.R. Berti, and J.W. Huang (1995a) Phytoremediation of contaminated soils. *Trends Biotechnol.* **13**: 393–397.

Cunningham, S.D., W.R. Berti, and J.W. Huang (1995b) Remediation of contaminated soils and sludges by green plants. In: *Bioremediation of Inorganics.* R.E. Hinchee, J.L. Means, and D.R. Burris, eds. Battelle Press, Columbus, Ohio, **3**(10): 33–54.

Cunningham, S.D., T.A. Anderson, A.P. Schwab, and F.C. Hsu, (1996) Phytoremediation of soils contaminated with organic pollutants. In: *Advances in Agron.* Academic Press, New York, **56**: 55–114.

Dec, J. and J.-M. Bollag (1994) Use of plant material for the decontamination of water polluted with phenols. *Biotechnol. Bioeng.* **44**: 1132–1139.

Delannay, X., T.T. Bauman, D.H. Beighley, M.J. Buettner, H.D. Coble, M.S. DeFelice, C.W. Derting, T.J. Diedrick, J.L. Griffin, E.S. Hagood, F.G. Hancock, S.E. Hart, B.J. LaVallee, M.M. Loux, W.E. Lueschen, K.W. Matson, C.K. Moots, E. Murdock, A.D. Nickell, M.D.K. Owen, E.H. Paschall, L.M. Prochaska, P.J. Raymond, D.B. Reynolds, W.K. Rhodes, F.W. Roeth, P.L. Sprankle, L.J. Tarochione, C.N. Tinius, R.H. Walker, L.M. Wax, H.D. Weigelt, and S.R. Padgette (1995) Yield evaluation of a glyphosate-tolerant soybean line after treatment with glyphosate. *Crop Sci.* **35**: 1461–1467.

Donnelly, P.K., R.S. Hegde, and J.S. Fletcher (1994) Growth of PCB-degrading bacteria on compounds from photosynthetic plants. *Chemosphere* **28**(5): 981–988.

Doty, S.L., T.Q. Shang, A.M. Wilson, J. Tangen, A.D. Westergreen, L.A. Newman, S.E. Strand, and M.P. Gordon (2000) Enhanced metabolism of halogenated hydrocarbons in transgenic plants containing mammalian cytochrome P450 2E1. *Proc. Nat. Acad. Sci. USA* **97**: 6287–6291.

Dushenkov, S. and Y. Kapulnik (2000) Phytofiltration of metals. In: *Phytoremediation of Toxic Metals*. I. Raskin and B.D. Ensley, eds. John Wiley, New York.

Dushenkov, V., P.B.A.N. Kumar, H. Motto, and I. Raskin (1995) Rhizofiltration: the use of plants to remove heavy metals from aqueous streams. *Environ. Sci. Technol.* **29**(5): 1239–1245.

Dushenkov, S., D. Vasudev, Y. Kapulnik, D. Gleba, D. Fleisher, K.C. Ting, and B. Ensley (1997) Removal of uranium from water using terrestrial plants. *Environ. Sci. Technol.* **31**: 3468–3474.

Edwards, N.T. (1998) Assimilation and metabolism of polycyclic aromatic hydrocarbons by vegetation: an approach to this controversial issue and suggestions for future research. In *Polycyclic aromatic hydrocarbons: a decade of progress*. M. Cooke and A.J. Dennis, eds. 10th International Symposium. Battelle Press, Columbus, Ohio, pp. 211–229.

Epuri, V. and D.L. Sorensen (1997) Benzo(*a*)pyrene and hexachlorobiphenyl contaminated soil: phytoremediation potential. In: *Phytoremediation of Soil and Water*. E.L. Kruger, T.A. Anderson, and J.R. Coats, eds. American Chemical Society Symposium Series No. 664. Washington, D.C., pp. 200–222.

Ferro, A.M., R.C. Sims, and B. Bugbee (1994) Hycrest crested wheatgrass accelerates the degradation of pentachlorophenol in soil. *J. Environ. Qual.* **23**: 272–281.

Ferro, A., J. Chard, R. Kjelgren, B. Chard, D. Turner, and T. Montague (2001) Groundwater capture using hybrid poplar trees: evaluation of a system in Ogden, Utah. *Int. J. Phytoremed.* **3**(1): 87–104.

Fletcher, J. (2001) Rhizosphere remediation of recalcitrant soil contaminants: an important component of long-term sustained biosystem treatment. In: Phytoremediation: State of the Science Conference. U.S. Environmental Protection Agency Report EPA/625/R-01/011b, held in Boston, Massachusetts, May 1–2, 2000. pp. 139–147 (http://www.epa.gov/ORD/NRMRL/Pubs/625R01011b/625R01011bchap6.pdf).

Fletcher, J.S. and R.S. Hegde (1995) Release of phenols by perennial plant-roots and their potential importance in bioremediation. *Chemosphere* **31**: 3009–3016.

French, C.E., S.J. Rosser, G.J. Davies, S. Nicklin, and N.C. Bruce (1999) Biodegradation of explosives by transgenic plants expressing pentaerythritol tetranitrate reductase. *Nat. Biotechnol.* **17**: 491–494.

Fritioff, A. and M. Greger (2000) Is *Potamogeton natans* a good phytoremediator of metal contaminated storm water. In: First Scientific Workshop, Phytoremediation 2000: State of the Art in Europe (an Intercontinental Comparison). COST Action 837, Plant biotechnology for the removal of organic pollutants and toxic metals from wastewaters and contaminated sites, held in Herisonissos, Crete, Greece, April 6–8 (http://lbewww.epfl.ch/COST837/).

Gao, J., A.W. Garrison, C. Hoehamer, C.S. Mazur, and N.L. Wolfe (2000a) Uptake and phytotransformation of *o,p′*-DDT and *p,p′*-DDT by axenically cultivated plants. *J. Agric. Food Chem.* **48**(12): 6114–6120.

Gao, J., A.W. Garrison, C. Hoehamer, C.S. Mazur, and N.L. Wolfe, (2000b) Uptake and phytotransformation of organophosphorus pesticides by axenically cultivated plants. *J. Agric. Food Chem.* **48**(12): 6121–6127.

Garg, P. and P. Chandra (1994) The duckweed *Wolffia globosa* as an indicator of heavy metal pollution: sensitivity to Cr and Cd. *Environ. Monit. Assess.* **29**: 89–95.

Garrison, A.W., V.A. Nzengung, J.K. Avants, J.J. Ellington, W.J. Jones, D. Rennels, and N.L. Wolfe (2000) Phytodegradation of *p,p′*-DDT and the enantiomers of *o,p′*-DDT. *Environ. Sci. Technol.* **34**: 1663–1670.

Gaudy, A. and E. Gaudy (1980) *Microbiology for Environmental Scientists and Engineers.* McGraw-Hill, New York.

Gerth, A. and A. Bohler (2000) Scale-up of phytoremediation techniques. In: First Scientific Workshop, Phytoremediation 2000: State of the Art in Europe (an Intercontinental Comparison). COST Action 837, Plant biotechnology for the removal of organic pollutants and toxic metals from wastewaters and contaminated sites, held in Herisonissos, Crete, Greece, April 6–8 (http://lbewww.epfl.ch/COST837/).

Gibson, D.T. and V. Subramanian (1984) Microbial degradation of aromatic hydrocarbons. In *Microbial Degradation of Organic Compounds.* D.T. Gibson, ed. Marcel Dekker, New York.

Gilbert, E.S. and D.E. Crowley (1997) Plant compounds that induce polychlorinated biphenyl biodegradation by *Arthrobacter* sp. Strain B1B. *Appl. Environ. Microbiol.* **63**(5): 1933–1938.

Glass, D.J. (1999) Current market trends in phytoremediation. *Int. J. Phytoremed.* **1**: 1–8.

Gordon, M., N. Choe, J. Duffy, G. Ekuan, P. Heilman, I. Muiznieks, M. Ruszaj, B.B. Shurtleff, S. Strand, J. Wilmoth, and L.A. Newman (1998) Phytoremediation of trichloroethylene with hybrid poplars. *Environ. Health Persp.* **106**(4): 1001–1004.

Greger, M. and T. Landberg (1999) Use of willow in phytoextraction. *Int. J. Phytoremed.* **1**(2): 115–123.

Groeger, A.G. and J.S. Fletcher (1988) The influence of increasing chlorine content on the accumulation and metabolism of polychlorinated biphenyls (PCBs) by Paul's Scarlet Rose cells. *Plant Cell Rep.* **7**: 329–332.

Haan, C.T., B.J. Barfield, and J.C. Hayes (1994) *Design Hydrology and Sedimentology for Small Catchments.* Academic Press, New York.

Hegde, R.S. and J.S. Fletcher (1996) Influence of plant growth stage and season on the release of root phenolics by mulberry as related to development of phytoremediation technology. *Chemosphere* **32**(12): 2471–2479.

Heitkamp, M.A. and C.E. Cerniglia (1988) Mineralization of polycyclic aromatic hydrocarbons by a bacterium isolated from sediments below an oil field. *Appl. Environ. Microbiol.* **54**: 1612–1614.

Hong, M.S., W.F. Farmayan, I.J. Dortch, C.Y. Chiang, S.K. McMillan, and J.L. Schnoor (2001) Phytoremediation of MTBE from a groundwater plume. *Environ. Sci. Technol.* **35**(6): 1231–1239.

Hooker, B.S. and R.S. Skeen (1999) Transgenic phytoremediation blasts onto the scene. *Nat. Biotechnol.* **17**: 428.

Horne, A. (2000) Phytoremediation by constructed wetlands. In: *Phytoremediation of Contaminated Soil and Water*. N. Terry and G. Banuelos, eds. Lewis Publishers, Boca Raton, Florida, pp. 13–39.

Hussain, M.S. and K. Jamil (1992) Appearance of new proteins in water hyacinth weevils (*Neochetina eichhornae* Warner), under the influence of metal bioaccumulation. *Arch. Environ. Contam. Toxicol.* **22**: 214–218.

Jeffers, P.M., L.M. Ward, L.M. Woytowitch, and N.L. Wolfe (1989) Homogenous hydrolysis rate constants for selected chlorinated methanes, ethanes, ethenes and propanes. *Environ. Sci. Technol.* **23**: 965–969.

Jerger, D.E. and P.M. Woodhull (2000) Applications and cost of biological treatment of explosives-contaminated soils in the U.S., Chapter 14. In: *Biodegradation of Nitroaromatic Compounds and Explosives*. J.C. Spain, J.B. Hughes, and H.J. Knackmuss, eds. Lewis Publishers, Boca Raton, Florida, pp. 395–423.

Jordahl, J.F., P.J. Alvarez, and J.L. Schnoor (1997) Effect of hybrid poplar trees on microbial populations important to hazardous waste bioremediation. *Environ. Toxicol. Chem.* **16**(6): 1318–1381.

Kalin, M. (2002) Pollution control – wetlands and ecological engineering. In: Ecological Engineering: Implementing the Profession. Proceedings of the American Ecological Engineering Society Annual Meeting, April 28–30, University of Vermont, Burlington.

Kelly, R.J. and T.F. Guerin (1995) Feasibility of using hyperaccumulating plants to bioremediate metal-contaminated soil. In: *Bioremediation of Inorganics*. R.E. Hinchee, J.L. Means, and D.R. Burris, eds. Battelle Press, Columbus, Ohio.

Kumar, P.B.A.N., V. Dushenkov, H. Motto, and I. Raskin (1995) Phytoextraction: the use of plants to remove heavy metals from soils. *Environ. Sci. Technol.* **29**(5): 1232–1238.

Kure, L., G. Mortensen, F. Laturnus, and P. Ambus (2000) Uptake of sewage sludge associated organic contaminants in crop plants. In: First Scientific Workshop, Phytoremediation 2000: State of the Art in Europe (an Intercontinental Comparison). COST Action 837, Plant biotechnology for the removal of organic pollutants and toxic metals from wastewaters and contaminated sites, held in Hersonissos, Crete, Greece, April 6–8 (http://lbewww.epfl.ch/COST837/).

Kutrys, S. and S.W. Gawronski (2000) Quack grass – wild grass species as a candidate for phytoremediation. In: First Scientific Workshop, Phytoremediation 2000: State of the Art in Europe (an Intercontinental Comparison). COST Action 837, Plant biotechnology for the removal of organic pollutants and toxic metals from wastewaters and contaminated sites, held in Herisonissos, Crete, Greece, April 6–8 (http://lbewww.epfl.ch/COST837/).

Landmeyer, J. (2001) Monitoring the effect of poplar trees on petroleum-hydrocarbon and chlorinated-solvent contaminated ground water. *Int. J. Phytorem.* **3**(1): 61–85.

Lasat, M.M., W.A. Norvell, and L.V. Kochian (1997) Potential for phytoextraction of ^{137}Cs from a contaminated soil. *Plant Soil* **195**(1): 99–106.

Lasat, M.M., M. Furhmann, S.D. Ebbs, J.E. Cornish, and L.V. Kochian (1998) Phytoremediation of radiocesium-contaminated soil: evaluation of cesium-137 bioaccumulation in the shoots of three plant species. *J. Environ. Qual.* **27**: 165–169.

Lasat, M.M., N.S. Pence, D.L.D. Letham, and L.V. Kochian (2001) Zinc phytoextraction in *Thlaspi caerulescens. Int. J. Phytorem.* **3**(1): 129–144.

La Starza, S.R., C. Damiano, A. Frattarelli, P. Ferrazza, and A. Figliolia (2000) Cadmium detoxification by aquatic plant *Hygrophilia corymbosa* "stricta", *in vitro* growth. In: First Scientific Workshop, Phytoremediation 2000: State of the Art in Europe (an Intercontinental Comparison). COST Action 837, Plant biotechnology for the removal of organic pollutants and toxic metals from wastewaters and contaminated sites, held in Herisonissos, Crete, Greece, April 6–8 (http://lbewww.epfl.ch/COST837/).

Lee, I. and J.S. Fletcher (1992) Involvement of mixed function oxidase systems in polychlorinated biphenyl metabolism by plant cells. *Plant Cell Reports* **11**: 97–100.

Leigh, M.B., J.S. Fletcher, D.P. Nagle, M. Mackova, and T. Macek, (2001) Vegetation and fungi at Czech PCB-contaminated sites as bioremediation candidates. In: *Proceedings of the Sixth International Symposium on In Situ and On-Site Bioremediation.* A. Lesson, E.A. Foote, M.C. Banks, and V.S. Magar, eds. held in San Diego, California June 4–7. Battelle Press, Columbus, Ohio.

Leigh, M.B., J.S. Fletcher, X. Fu, and F.J. Schmitz (2002) Root turnover: an important source of microbial substrates in rhizosphere remediation of recalcitrant contaminants. *Environ. Sci. Technol.* **36**: 1579–1583.

Licht, L., E. Aitchison, W. Schnabel, M. English, and M. Kaempf, (2001) Landfill Capping with Woodland Ecosystems. *Pract. Periodical Hazard., Toxic Radioact. Waste Manag.* **5**(4): 175–184.

Ma, L.Q., K.M. Komart, C. Tu, W. Zhang, Y. Cai, and E.D. Kennelley (2001) A fern that hyperaccumulates arsenic. *Nature* **409**: 579.

McCutcheon, S.C. (1998) Phytoremediation: applications and limitations. *PBI Bull.* September: 1–4.

McCutcheon, S.C., J.C. Hayes, C.A. Williams, C.P. Weisskopf, and S.J. Klaine (2000). Evaluation of vegetative filter strips to control urban runoff into Charleston Harbor and other coastal waters. Report prepared for the South Carolina Department of Health and Environmental Control. Charleston, South Carolina.

McCutcheon, S.C. and S. Rock (2001) Phytoremediation: state-of-the-science conference and other developments. Special Issue on the 2000 EPA Conference on Phytoremediation. *Int. J. Phytoremed.* **3**(1): 1–11.

McCutcheon, S.C., N.L. Wolfe, L.H. Carreira, and T.Y. Ou (1995) Phytoremediation of hazardous waste. In: *Innovative Technologies for Site Remediation and Hazardous Waste Management: Proceedings of the National Conference.* R.D. Vidic and F.G.

Pohland, eds. Environmental Engineering Division, American Society of Civil Engineers, New york, held in Pittsburgh, Pennsylvania, July 26, pp. 597–604.

McFarlane, C., T. Pfleeger, and J. Fletcher (1990) Effect, uptake and disposition of nitrobenzene in several terrestrial plants. *Environ. Toxicol. Chem.* **9**: 513–520.

Medina, V.F., S.L. Larson, L. Agwaramgbo, and W. Perez (2002) Treatment of munitions in soils using phytoslurries. *Int. J. Phytoremed.* **4**(2): 143–156.

Messner, B., H. Sandermann, A. Schaeffner, O. Thulke, and S. Wegener (2000) Functional genomics of *Arabidopsis* glycosyl transferases. In: First Scientific Workshop, Phytoremediation 2000: State of the Art in Europe (an Intercontinental Comparison). COST Action 837, Plant biotechnology for the removal of organic pollutants and toxic metals from wastewaters and contaminated sites, held at, Herisonissos, Crete, Greece, April 6–8 (http://lbewww.epfl.ch/COST837/).

Miya, R.K. and M.K. Firestone (2000) Phenanthrene-degrader community dynamics in rhizosphere soil from a common annual grass. *J. Environ. Qual.* **29**: 584–592.

Negri, M.C. and R.R. Hinchman (2000) The use of plants for the treatment of radionuclides. Chapter 8. In: *Phytoremediation of Toxic Metals – Using Plants to Clean Up the Environment*. I. Raskin and B.D. Ensley, eds. John Wiley, New York.

Negri, M.C., R.R. Hinchman, and G.T. Ulbricht (1998) Operable Unit 9–04 phytoremediation bench scale testing, A: interceptor canal mound soil, and (with L.R. Skubal) B: main cooling tower blowdown ditch soil. Draft final reports. Argonne National Laboratory, Department of Energy, Chicago, Illinois.

Negri, M.C., R. Hinchman, and J. Wozniak (2001) Capturing a "mixed" contaminant plume: tritium phytoevaporation at Argonne National Laboratory. In: Phytoremediation: State of the Science, Conference. U.S. Environmental Protection Agency Report EPA/625/R-01/011b, held May 1–2, 2000. Boston, Massachusetts, pp. 189–208 (http://www.epa.gov/ORD/NRMRL/Pubs/625R01011b/625R01011bchap9.pdf).

Newman, L.A., S.E. Strand, N. Choe, J. Duffy, G. Ekuan, M. Ruszaj, P.B. Shurtleff, J. Wilmouth, P. Heilman, and M.P. Gordon (1997) Uptake and biotransformation of trichloroethylene by hybrid poplars. *Environ. Sci. Technol.* **31**: 1062–1067.

Newman, L.A., X. Wang, I.A. Muiznieks, G. Ekuan, M. Ruszaj, R. Cortellucci, D. Domroes, G. Karscig, T. Newman, R.S. Crampton, R.A. Hashmonay, M.G. Yost, P.E. Heilman, J. Duffy, M.P. Gordon, and S.E. Strand (1999) Remediation of TCE in an artificial aquifer with trees: a controlled field study. *Environ. Sci. Technol.* **33**(13): 2257–2265.

Nicks, L.J. and M.F. Chambers (1995) Farming for metals. *Mining Environ. Manage.* **3**: 15–18.

Nicks, L.J. and M.F. Chambers (1998) A pioneering study for the potential of phytomining for nickel, Chapter 14. In: *Plants that Hyperaccumulate Heavy Metals*. R.R. Brooks, ed. CAB International, New York.

Nzengung, V. and P. Jeffers (2001) Sequestration, phytoreduction, and phytooxidation of halogenated organic chemicals by aquatic and terrestrial plants. *Int. J. Phytoremed.* **3**(1): 13–40.

Olson, P.E. and J.S. Fletcher (2000) Ecological recovery of vegetation at a former industrial sludge basin and its implication to phytoremediation. *Environ. Sci. Pollut. Res.* **7**: 195–204.

Olson, P.E., J.S. Fletcher, and P.R. Philp (2001) Natural attenuation/phytoremediation in the vadose zone of a former industrial sludge basin. *Environ. Sci. Pollut. Res.* **8**(4): 243–249.

Padgette, S.R., K.H. Kolacz, X. Delannay, D.B. Re, B.J. LaVallee, C.N. Tinius, W.K. Rhodes, Y.I. Otero, G.F. Barry, D.A. Eichholtz, V.M. Peschke, D.L. Nida, N.B. Taylor, and G.M. Kishore (1995) Development, identification, and characterization of a glyphosate-tolerant soybean line. *Crop Sci.* **35**: 1451–1461.

Palmroth, M.R.T., J. Pichtel, and J.A. Puhakka (2002) Phytoremediation of subarctic soil contaminated with diesel fuel. *Bioresource Technol.* **84**(3): 221–228.

Paul, E.A. and F.E. Clark (1989) *Soil Microbiology and Biochemistry*. Academic Press, San Diego, California, pp. 81–84.

Qui, X., S.I. Shah, E.W. Kendall, D.L. Sorenson, R.C. Sims, and M.C. Engelke (1994) Grass-enhanced bioremediation for clay soils contaminated with polynuclear aromatic hydrocarbons. In: *Bioremediation through Rhizosphere Technology*. T.A. Anderson, and J.R. Coats, eds. American Chemical Society Symposium Series No. 563. Washington, D.C., pp. 142–157.

Quinn, J., M.C. Negri, R.R. Hinchman, L.P. Moos, and J.B. Wozinak (2001) Predicting the effect of deep-rooted hybrid poplars on the groundwater flow system at a large-scale phytoremediation. *Int. J. Phytoremed.* **3**(1): 41–60.

Raskin, I. (1996) Phytoremediation, In: *Phytoremediation*. Proceedings of International Business Communications Conference, held in Virginia, Arlington, May 8–10.

Raskin, I. and B.D. Ensley, eds. (2000) *Phytoremediation of Toxic Metals – Using Plants to Clean Up the Environment*. John Wiley, New York.

Raskin, I., P.B.A.N. Kumar, S. Dushenkov, and D.E. Salt (1994) Bioconcentration of heavy metals by plants. *Curr. Opin. Biotechnol.* **5**: 285–290.

Reed, S.C. and M. Hines (1993) Constructed wetlands for industrial wastewaters. In: *Proceedings of the 48th Industrial Waste Conference*. Purdue University, Lewis Publishers, Boca Raton, Florida, pp. 9–17.

Reeves, R.D., A.J.M. Baker, and R.R. Brooks (1995) Abnormal accumulation of trace metals by plants. *Mining Environ. Manage*. September: 4–8.

Robinson, B.H., A. Chiarucci, R.R. Brooks, D. Petit, J.H. Kirkman, P.E.H. Gregg, and V. De Dominicis (1997) The nickel hyperaccumulator plant *Alyssum bertolonii* as a potential agent for phytoremediation and phytomining of nickel. *J. Geochem. Explor*. **59**(2): 75–86.

Rodgers, J.H., Jr., D.S. Cherry, and R.K. Guthrie (1978) Cycling of elements in duckweed (*Lemna perpusilla*) in an ash settling basin and swamp drainage system. *Water Res.* **12**: 765–770.

Roper, J.C., J. Dec, and J.-M. Bollag (1996) Using minced horseradish roots for the treatment of polluted waters. *J. Environ. Qual.* **25**: 1242–1247.

Rosler, H.J. and H. Lange (1972) *Geochemical Tables*. Elsevier Publishing Company, New York, pp. 365–374.

Rovira, A.D. and C.B. Davey (1974) Biology of the rhizosphere. In: *The Plant Root and Its Environment*. E.W. Carson, ed. University Press of Virginia, Charlottesville, pp. 153–204.

Rugh, C.L., H.D. Wilde, N.M. Stack, D.M. Thompson, A.O. Summers, and R.B. Meagher (1996) Mercuric ion reduction and resistance in transgenic *Arabidopsis thaliana* plants expressing a modified bacterial *mer*A gene. *Proc. Nat. Acad. Sci. USA* **93**: 3182–3187.

Rugh, C.L., J.F. Senecoff, R.B. Meagher, and S.A. Merkle (1998) Development of transgenic yellow poplar for mercury phytoremediation. *Nat. Biotechnol.* **16**: 925–928.

Salt, D.E., M. Blaylock, N.P.B.A. Kumar, V. Dushenkov, B.D. Ensley, I. Chet, and I. Raskin (1995) Phytoremediation: a novel strategy for the removal of toxic metals from the environment using plants. *Biotechnol.* **13**: 468–474.

Sandermann, H., Jr. (1992) Plant metabolism of xenobiotics. *TIBS* **17**: 82–84.

Schneider, K. (1988) Plutonium leak in Idaho symptom of atomic ills. *N. Y. Times Nat.* April 17, p. 24.

Schnoor, J.L. (1997) Phytoremediation. Ground-Water Remediation Technology Analysis Center Technology Evaluation Report TE-98-01. Concurrent Technologies Corp., Pittsburgh, Pennsylvania.

Schnoor, J.L. (2000) Phytostabilization of metals using hybrid poplar trees. In: *Phytoremediation of Toxic Metals: Using Plants to Clean up the Environment*. I. Raskin and B.D. Ensley, eds. John Wiley, New York, pp. 133–150.

Schnoor, J.L. (2002) Phytoremediation of soil and groundwater. Ground-Water Remediation Technology Analysis Center Technology Evaluation Report TE-02-01. Concurrent Technologies Corp., Pittsburgh, Pennsylvania (www.gwtac.org/pdf/phyto_e_2002.pdf).

Schnoor, J.L., L.A. Licht, S.C. McCutcheon, N.L. Wolfe, and L.H. Carriera (1995) Phytoremediation of organic and nutrient contaminants. *Environ. Sci. Technol.* **29**: 318A–323A.

Schwab, A.P. and M.K. Banks (1994) Biologically mediated dissipation of polyaromatic hydrocarbons in the root zone. In: *Bioremediation through Rhizosphere Technology*. T.A. Anderson and J.R. Coats, eds. American Chemical Society Symposium Series No. 563. Washington, D.C., pp. 132–141.

Sims, R.C. and M.R. Overcash (1983) Fate of polynuclear aromatic compounds (PNAs) in soil-plant systems. *Resid. Rev.* **88**: 1–68.

Smith, R.A.H. and A.D. Bradshaw (1972) *Min. Ind.* **81**: 230–257.

Smrcek, S. (2000) "*In vitro*" plant biodegradation of pentachlorophenol. In: First Scientific Workshop, Phytoremediation 2000: State of the Art in Europe (an Intercontinental Comparison). COST Action 837, Plant biotechnology for the removal of organic pollutants and toxic metals from wastewaters and contaminated sites, April 6–8, Herisonissos, Crete, Greece (http://lbewww.epfl.ch/COST837/).

Snellinx, Z., J. Vangronsveld, H. Clijsters, and D. van der Lelie (2000) Phytoremediation of nitroaromatic compounds. In: First Scientific Workshop, Phytoremediation 2000: State of the Art in Europe (an Intercontinental Comparison). COST Action 837, Plant biotechnology for the removal of organic pollutants and toxic metals from wastewaters and contaminated sites, held in Herisonissos, Crete, Greece, April 6–8 (http://lbewww.epfl.ch/COST837/).

Spain, J.C., J.B. Hughes, and H.J. Knackmuss, eds. (2000) *Biodegradation of Nitroaromatic Compounds and Explosives*. Lewis Publishers, Boca Raton, Florida.

Spirochova, I., T. Vanek, P. Soudek, and J. Puncocharova (2000) A study of accumulation of heavy metals by *in vitro* culture of plants. In: First Scientific

Workshop, Phytoremediation 2000: State of the Art in Europe (an Intercontinental Comparison). COST Action 837, Plant biotechnology for the removal of organic pollutants and toxic metals from wastewaters and contaminated sites, held in Herisonissos, Crete, Greece, April 6–8 (http://lbewww.epfl.ch/COST837/).

Stoltz, E. and M. Greger (2000) Suitable species for plant establishment on water saturated mine spoil. In: First Scientific Workshop, Phytoremediation 2000: State of the Art in Europe (an Intercontinental Comparison). COST Action 837, Plant biotechnology for the removal of organic pollutants and toxic metals from wastewaters and contaminated sites, held in Herisonissos, Crete, Greece, April 6–8 (http://lbewww.epfl.ch/COST837/).

Stomp, A.M., K.-H. Han, S. Wilbert, and M.P. Gordon (1993) Genetic improvement of tree species for remediation of hazardous wastes. *In Vitro Cell Dev. Biol.: Plant* **29P**: 227–232.

Strand, S.E., L. Newman, M. Ruszaj, J. Wilmoth, B. Shurtleff, M. Brandt, N. Choe, G. Ekuan, J. Duffy, J.W. Massman, P.E. Heilman, and M.P. Gordon (1995) Removal of trichloroethylene from aquifers using trees. In: *Innovative Technologies for Site Remediation and Hazardous Waste Management, Proceedings of the National Conference*. R.D. Vidic, and F.G. Pohland, eds. Environmental Engineering Division, American Society of Civil Engineers, New york, held in, pp. 605–612. Pittsburgh, Pennsylvania, July 26.

Suominen, L., M.M. Jussila, K. Mäkeläinen, M. Romantschuk, and K. Lindström (2000) Evaluation of the *Galega-Rhizobium galegae* system for the bioremediation of oil-contaminated soil. *Environ. Pollut.* **107**(2): 239–244.

Terry, N. and G. Banuelos, ed. (2000) *Phytoremediation of Contaminated Soil and Water*. Lewis Publishers, Boca Raton, Florida.

Terry, N. and A.M. Zayed (1994) Selenium volatilization by plants. In: *Selenium in the Environment*. W.T. Frankenberger, Jr. and S. Benson, eds. Marcel Dekker, New York.

Terry, N., C. Carlson, T. K. Raab, and A. M. Zayed (1992) Rates of selenium volatilization among crop species. *J. Environ. Qual.* **21**: 341–344.

Terry, N., A. Zayed, E. Pilon-Smits, and D. Hansen (1995) Can plants solve the selenium problem? Proceedings of the 14th Annual Symposium, Current Topics in Plant Biochemistry, Physiology and Molecular Biology: Will Plants Have a Role in Bioremediation? University of Missouri, Columbia, April 19–22, pp. 63–64.

Thompson, P.L. (1997) Phytoremediation of munitions (RDX, TNT) wastes at the Iowa Army Ammunition Plant with hybrid poplar trees. Ph.D. dissertation. University of Iowa, Iowa City.

Thompson, P.L., L.A. Ramer, and J.L. Schnoor (1998) Uptake and transformation of TNT by hybrid poplar trees. *Environ. Sci. Technol.* **31**: 975–980.

Todd, J. and B. Josephson (1996) The design of living technologies for waste treatment. *Ecol. Eng.* **6**: 109–136.

Todd, S. (1996) Phytoremediation of dinitrotoluene contaminated soils using parrot feather (*Myriophyllum aquaticum*). M.S. thesis. Department of Civil Engineering, Auburn University, Alabama.

Trapp, S. and U. Karlson (2001) Aspects of phytoremediation of organic pollutants. *J. Soils Sed.* **1**: 1–7.

Tripathi, R.D. and P. Chandra (1991) Chromium uptake by *Spirodela polyrrhiza* (L.) Schleiden in relation to metal chelators and pH. *Bull. Environ. Contam. Toxicol.* **47**: 764–769.

Uhlmann, D. (1979) *Hydrobiology*. John Wiley, New York.

U.S. Department of Energy (DOE) (1994) Summary report of a workshop on phytoremediation research needs, held at Santa Rosa, California, June 24–26. Report DOE/ EM-0224. Oak Ridge, Tennessee.

U.S. Department of Energy (DOE) (2000) Proceedings of the Workshop on Phytoremediation of Inorganic Contaminants. Report INEEL/EXT-2000-00207. Idaho National Engineering and Environmental Laboratory, Idaho Falls.

Vanek, T., A. Nepovim, M. Hubalek, and Z. Jalova (2000) Phytoremediation of explosives from laboratory experiments to field applications. In: First Scientific Workshop, Phytoremediation 2000: State of the Art in Europe (an Intercontinental Comparison). COST Action 837, Plant biotechnology for the removal of organic pollutants and toxic metals from wastewaters and contaminated sites, held in Herisonissos, Crete, Greece, April 6–8 (http://lbewww.epfl.ch/COST837/).

Vangronsveld, J. and S.D. Cunningham, eds. (1998) *Metals-Contaminated Soils*. Springer, New York.

Vasudev, D., S. Dushenkov, A. Epstein, Y. Kapulnik, B. Ensley, G. Huddleston, J. Cornish, I. Raskin, B. Sorochinsky, M. Ruchko, A. Prokhnevsky, A. Mikheev, and D. Grodzinsky (1996) Removal of radionuclide contamination from water by metal-accumulating terrestrial plants. In: *In Situ Soil and Sediment Remediation*, Spring National Meeting of the American Institute of Chemical Engineers, February 25–29, New Orleans, Louisiana.

Walski, T.M. (1993) Long-term effects of wetland treatment of mine drainage. *J. Environ. Eng.* **119**: 1004–1005.

Waugh, J. (2001) Growing a 1,000-year landfill cover. In: Phytoremediation State of the Science Conference. U.S. Environmental Protection Agency Report EPA/625/ R-01/011b, held at Boston, Massachusetts, May 1–2, 2000. (http://www.epa.gov/ ORD/NRMRL/Pubs/625R01011b/625R01011bchap26.pdf).

Whipps, J.M. and J.M. Lynch (1985) Energy losses by the plant in rhizodeposition. In: *Plant Products and the New Technology*. K.W. Fuller and J.R. Gallon, eds. Ann. Proc. Phytochem. Soc. Eur. Oxford Univ. Press, Oxford, United Kingdom. **26**: 59–71.

Wieder, R.K. (1993) Ion input-output budgets for five wetlands constructed for acid coal mine drainage treatment. *Water Air Soil Pollut.* **71**: 231–270.

Wilde, E.W. and J.R. Benemann (1993) Bioremoval of heavy metals by the use of microalgae. *Biotechnol. Adv.* **11**: 781–812.

Wolverton, B.C. and J.D. Wolverton (2001) *Growing Clean Water, Nature's Solution to Water Pollution*. WES Publishers, Picayune, Mississippi.

Xantholis, D. and I.R. Dumont (2000) Epuvalisation: a biological purification technique using plants, framework, experiences and application domains for past, present and future. In: First Scientific Workshop, Phytoremediation 2000: State of the Art in Europe (an Intercontinental Comparison). COST Action 837, Plant biotechnology for the removal of organic pollutants and toxic metals from wastewaters and contaminated sites, held in Herisonissos, Crete, Greece, April 6–8 (http:// lbewww. epfl.ch/COST837/).

Xue, H.B., S. Werner, and L. Sigg (1988) The binding of heavy metals to algal surfaces. *Water Res.* **22**(7): 917–926.

Young, P. (1996) The new science of wetland restoration. *Environ. Sci. Technol.* **30**: 292A–296A.

Zayed, A., E. Pilon-Smits, M. deSouza, Z.-Q. Lin, and N. Terry, (2000) Remediation of selenium-polluted soils and waters phytovolatilization. In: *Phytoremediation of Contaminated Soil and Water*. N. Terry and G. Banuelos, eds. Lewis Publishers, Boca Raton, Florida, pp. 61–83.

2

UPTAKE AND METABOLISM OF ORGANIC COMPOUNDS: GREEN-LIVER MODEL

J. G. Burken

SUMMARY OF PRACTICAL IMPLICATIONS

Plant metabolism of organic compounds is a vital phytoremediation process for sustainable waste management. Metabolism appears to follow detoxification or elimination metabolic processes, which are collectively called the "green-liver" model. For this metabolism to take place, the compounds must enter plant tissues; therefore uptake of organic compounds is also covered, with specific applications for phytoremediation.

Plant metabolism of organic compounds shares many processes with mammalian liver function, including (1) the capability to detoxify contaminants, (2) the specific enzymatic pathways, and (3) removal of the compounds from the susceptible organelles. The similarities extend to the very structure of the enzymes involved. The major difference is in the ultimate fate—storage as opposed to excretion in mammals. By understanding the processes that occur within the plants used in phytoremediation, a better knowledge of the potential ecological impacts can be gained. Furthermore, new applications can be developed with the knowledge of plant metabolism and, through molecular engineering techniques, a new paradigm in the plant metabolism of contaminants and phytoremediation should result.

GLOSSARY

Anabolism: The biochemical process, usually requiring energy, that involves the synthesis of complex cell constituents from simpler molecules.

Apoplasm: The region of the plant which is outside the plasma membrane, *i.e.*, the cell wall and the intercellular (free) space.

Phytoremediation: Transformation and Control of Contaminants,
Edited by Steven C. McCutcheon and Jerald L. Schnoor.
ISBN 0-471-39435-1 (cloth)

Catabolism: Destructive metabolism, often involving organic compounds, resulting in the release of energy and the breakdown of complex materials within the organism.

Endodermis: The layer of cells in root tissues forming a sheath around the stele. The casparian strip, a hydrophobic barrier that links the endodermal cell walls to block the apoplastic path, characterizes this layer of cells.

Phloem: Vascular tissues responsible for transport of plant growth and metabolism regulators (*i.e.*, hormones), and photosynthetic compounds from the points of origin (sources) to the point of use or storage (sinks); cells are alive at maturity.

RCF—Root Concentration Factor: Calculated as the sorbed equilibrium concentration in the root tissues divided by the bulk aqueous concentration at the root surface.

Stele: Primary vascular bundle found in roots and shoots comprised of the xylem, phloem, and other tissues.

Symplasm: The region of the plant bounded by the plasma membrane, linking interconnected cells and the associated cytoplasmic components *via* plasmodesmata.

TSCF—Transpiration Stream Concentration Factor: The equilibrium concentration in the transpiration stream divided by the bulk aqueous concentration at the root surface.

Xylem: Vascular tissue responsible for the transport of water and dissolved minerals from the roots to the aerial plant organs and tissues; cells are dead at maturity.

INTRODUCTION

Interest in the metabolism of organic compounds by plants has existed for decades. Initially, the interest was centered on the metabolism of pesticide compounds in crop species, with a focus on the potential contamination of crops. Herbicide metabolism was also of great interest in understanding selectivity, mode of action, and herbicide resistance. Initial identification of plant detoxification of herbicides can be traced back more than 40 years (Castelfranco *et al.*, 1961). This research produced a mechanistic, molecular model of plant metabolism that has been tested. In recent years, the advent of phytoremediation for organic contaminants has broadened the interest in plant metabolism of organic compounds. A number of researchers in many fields have taken an interest in the fate and metabolism of various environmental contaminants that previously were not considered. Phytoremediation investigations have expanded the research beyond crops to look at trees and wetland plants.

OVERVIEW

The "green-liver" model is based on some of the metabolic processes by which plants can detoxify or store organic compounds that enter tissues. The concept first appeared in 1977, based on enzyme isolation from plant cell cultures and on the metabolism of nonpolar compounds previously believed to persist in vegetation, including 1,1,1-trichloro-2,2-*bis*-(4′-chlorophenyl) ethane (DDT) and benzo(*a*)pyrene (Sandermann 1994, Sandermann *et al.* 1977). Given the similarity to the detoxification mechanisms in mammalian hepatic systems, the term "green liver" was adopted (Sandermann 1994).

Plant metabolism differs in purpose from microbial metabolism for both anthropogenic and natural compounds. Being photoautotrophic, plants do not rely on organic compounds as a source of energy for catabolism or a source of carbon for anabolism; rather, plants use photosynthesis to harness light as the energy source. Through the energy capture of photosynthesis, plants fix carbon dioxide as carbohydrates. These carbohydrates (*i.e.*, photosynthates) are used in respiring plant cells as the source of catabolic energy and anabolic carbon. Therefore, plants do not harbor the extensive array of catabolic and anabolic enzymes possessed by heterotrophic bacteria or fungi, which rely on organic compounds found in the environment as the source of energy and carbon. In essence, plant metabolism of organic carbon other than photosynthates can follow general transformation and isolation processes to avoid build up and potential toxicity to sensitive organelles, rather than specific metabolism to enter catabolic and anabolic cycles.

Plant metabolism of xenobiotic chemicals is similar to liver functions in a number of ways. One similarity is in the metabolic steps that are involved. The individual metabolic steps that occur in sequence are almost homologous to mammalian liver function (see Table 17-4, Shang *et al.* this book). In both plants and mammals, xenobiotic contaminants are essentially eliminated and removed from the active organelles. Perhaps the greatest similarity that supports the green-liver model is in the broad range of substrates and in enzymes that act on these substrates. The transformation and elimination mechanisms of green-liver metabolism have been shown to be active for a diverse and wide range of xenobiotic organic compounds (Komossa *et al.* 1995).

PLANT UPTAKE OF ORGANIC COMPOUNDS

Plant metabolism has been studied for broad classes of contaminants in various laboratory experiments. Compounds studied range from highly polar compounds like methyl *tert*-butyl ether (MTBE) and phenol to highly hydrophobic polychlorinated biphenyls and DDT (Sandermann *et al.* 1984). Experiments range from whole-plant studies to cell culture suspensions. Plant cell cultures are of great importance when considering plant metabolism and the underlying molecular enzymology and biology. However, in the application of phytoremediation, whole-plant systems are more representative of full-scale application and potential.

When considering the whole-plant metabolism of organics, the organic chemical must first enter the plant tissue. Entry can occur by translocation of soil water with the transpiration stream, by transfer from soil gas, or from atmospheric gases or water to the foliar tissues (Figure 2-1). After being transported into the plant cells, an organic chemical can be metabolized if the individual compound serves as a substrate for the enzymes present. Enzymatic activity in specific plant tissues should also be considered, as plants can possess highly specific tissues that serve a variety of functions. Translocation is the most prevalent way for organic compounds to enter plant tissues. However, in wetlands and roots below the water table, passive diffusion also can occur. This chapter focuses on translocation in the transpiration stream because foliar uptake is of lesser consequence when considering phytoremediation of waste streams or contaminated soil and water. However, foliar uptake has been proposed as a major sink of some anthropogenic contaminants in the atmosphere (Simonich and Hites 1994, Jeffers and Liddy this book, Morikawa *et al.* this book).

Plant uptake of organic chemicals has been studied for many years, with the early emphasis placed on agrochemicals. Shone and Wood (1972, 1974) initially evaluated movement of herbicides and fungicides in the translocation stream (or xylem flow). The term transpiration stream concentration factor (TSCF) was used to normalize the compound concentration in the translocation stream with respect to the concentration in the bulk solution in the root zone. Similar work was performed by Briggs *et al.* (1982) and Briggs *et al.* (1983) who developed TSCF relationships as a function of chemical properties. The best correlation related the TSCF to the logarithm of the octanol–water partition coefficient ($\log K_{ow}$), a measure of chemical hydrophobicity (Figure 2-2). Other research has looked at the relationship between uptake or the TSCF and chemical properties, with the focus on agrochemicals and crops (Ryan *et al.* 1988, Hsu *et al.* 1990). A number of recent studies have investigated the uptake of priority pollutants by plant species adaptable to phytoremediation (Aitchison 1998, Burken and Schnoor 1998, Davis *et al.* 1998, Thompson *et al.* 1999, Orchard *et al.* 2000).

Data in Figure 2-2 are from studies that looked at a range of compounds for one plant type. In addition to the data presented in Figure 2-2, a number of studies have determined TSCF values for a certain compound, or for a variety of compounds in different plants and settings (Davis *et al.* 1998, Aitchison *et al.* 2000, Orchard *et al.* 2000, Rubin and Ramaswami 2001). These studies have shown that TSCF can range from 0.1 to 0.9 for trichloroethylene in different treatments and plants (Davis *et al.* 1998). Comparable ranges were also observed for different experimental arrangements looking at

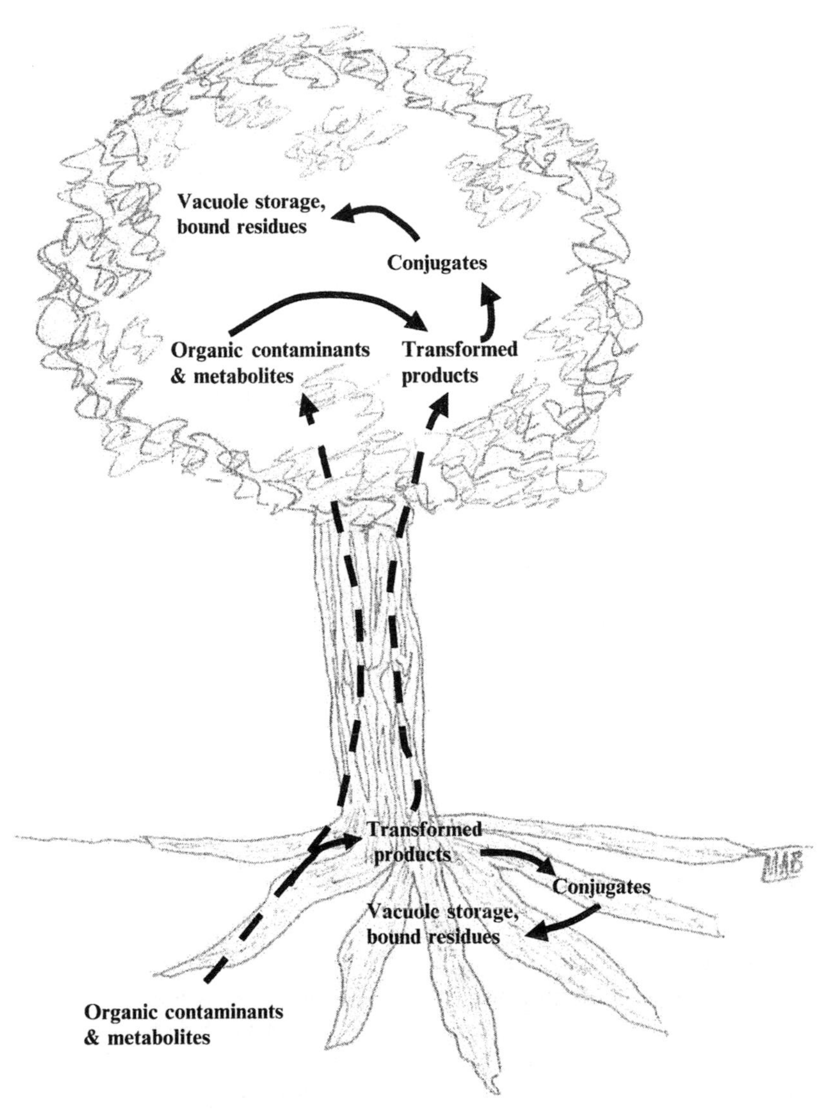

Figure 2-1 Impact of "green-liver metabolism" on organic pollutants in phytoremediation applications. Solid lines represent cellular-level metabolism, as a part of the green-liver concept. Dashed lines represent plant-scale transport into and through different plant tissues.

similar plants and compounds. Variation for trichloroethylene translocation by hybrid poplars (various *Populus* hybrids) covers the range from 0.02 to 0.75 (Burken and Schnoor 1998, Davis *et al.* 1998, Orchard *et al.* 2000). Most TSCF studies mentioned were performed in hydroponic reactors. In considering the data gathered in uptake studies, the methodology and

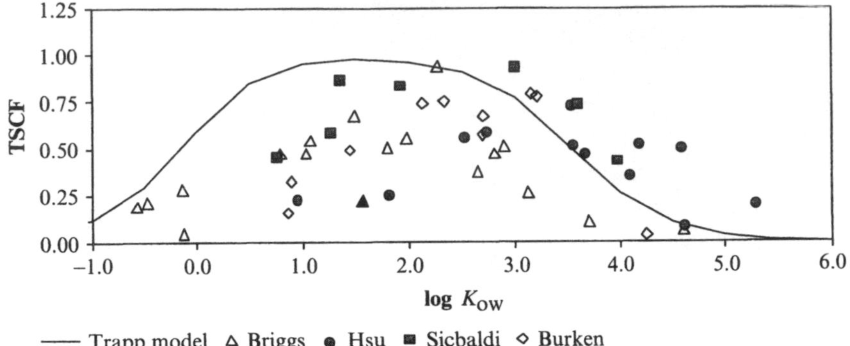

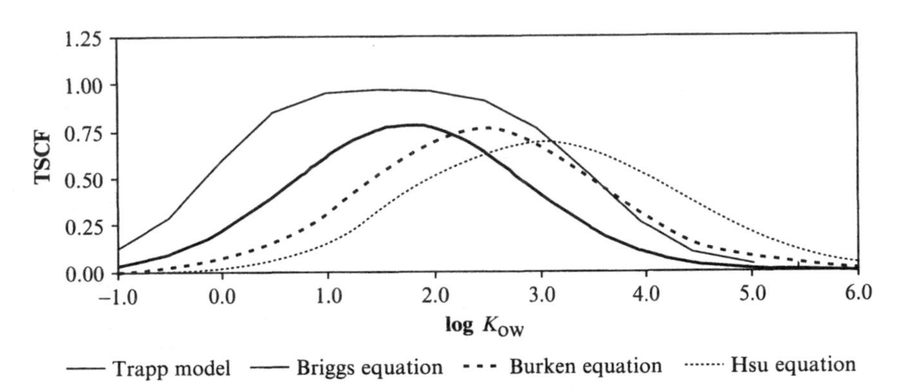

Figure 2-2 Uptake of various organic compounds presented as the transpiration stream concentration factor (TSCF). Adapted from Trapp (2000), original data in Briggs *et al.* (1983), Hsu *et al.* (1990), Sicbaldi *et al.* (1997), Burken and Schnoor (1998).

experimental period should be considered. In addition, the actual compound concentration in the transpiration stream was rarely measured; rather, concentrations were inferred by summation of selected fates or as the removal from the bulk solution. Finally, the experimental period must allow sufficiant time for the compound to reach equilibrium with respect to sorption to root tissues, which by definition should not be considered in the TSCF calculations.

The results of uptake *via* the transpiration stream show a general pattern. Uptake is greatest for compounds with a log K_{ow} in the range of 1 to 4, with the maximum uptake ranging from 1.8 (Briggs *et al.* 1982) to 3.1 (Hsu *et al.* 1990). Compounds with a log K_{ow} greater than 4 become bound to cell membranes either as the compounds move through the apoplast or when reaching the endodermis of the root tissues. To enter the xylem, the chemicals must pass through the cell membranes (*i.e.*, symplast) associated with the endodermis. The observed root concentration factors (RCFs) consistently show high affinity for hydrophobic compounds in the roots (Shone and Wood 1974, Briggs *et al.* 1982, Briggs *et al.* 1983, Topp *et al.* 1986, Burken and Schnoor

1998, Trapp 2000). The affinity of organic compounds for root tissues also has been described as a logarithmic function of K_{ow} as shown by Briggs *et al.* (1982)

$$\log \mathrm{RCF} = 0.77 \ \log K_{ow} - 1.52 \quad (2\text{-}1)$$

Hydrophilic compounds ($\log K_{ow}$ less than 1) can readily move through the apoplast but have a low permeability and limited transport in the symplast of the endodermis (*i.e.*, organic membranes). Thus these compounds do not diffuse into and through these organic membranes. In the different TSCF data sets, this limited transport of low $\log K_{ow}$ compounds is apparent, but exceptions are obvious. There are indications that the $\log K_{ow}$ alone is not an absolute predictor for the membrane permeability of very lipophobic (*i.e.*, hydrophilic) compounds (Trapp 2000). In a number of studies, the observed TSCF for hydrophilic compounds is higher than anticipated. Aitchison *et al.* (2000) determined the TSCF of 1,4-dioxane to be 0.72, as compared to an extrapolated value of 0.14 using the relationship of Briggs *et al.* (1982) and an extrapolation of 0.03 using the relationship developed with the same poplar clone, *Populus deltoides* × *Populus nigra* 'DN-34', using the same laboratory arrangement (Burken and Schnoor 1998). Similarly, the TSCF values measured for methyl-*tert*-butyl ether (MTBE) are higher than expected, with values of 0.6 (Davis *et al.* 1998) to 1 (Rubin and Ramaswami 2001). These two compounds, methyl-*tert*-butyl ether and 1,4-dioxane, both contain an ether bond, but the significance of the chemical structure similarity relative to the transport is not known.

Although the relationships based solely on $\log K_{ow}$ are not strict indicators, there is considerable work validating the concept. In two independent translocation studies utilizing pressure cells and detopped soybean (*Glycine max*) plants, very similar findings were observed (see data from Hsu *et al.* 1990 and Sicbaldi *et al.* 1997 in Figure 2-2). Hsu *et al.* (1990) developed a relationship for the pesticide cinmethylin and analogs. The maximum TSCF value was for a compound with a $\log K_{ow}$ of 3.08. Sicbaldi *et al.* (1997) found that a $\log K_{ow}$ of 3 related to the highest TSCF for the various chemicals tested. In addition, the maximum TSCF values for the relationships were similar (0.70 to 0.78). Topp *et al.* (1986) correlated uptake to another single property of the compounds tested, the molecular weight. These correlations fit the data well for the uptake studies performed in soil. However, there was only one compound tested with a low molecular weight (less than 150 atomic mass units), or low $\log K_{ow}$ (less than 2.5) to evaluate the theory of low TSCF values for small, polar compounds. Comparison of the TSCF values generated in other studies (Briggs *et al.* 1982, Hsu *et al.* 1990, Burken and Schnoor 1998) and related to molecular weights of the compounds tested did not provide better correlation than relationships based on $\log K_{ow}$. The limited correlation of molecular weight to TSCF is very apparent in the work by Hsu *et al.* (1990), as all compounds tested were bicycloalkanes with relatively high molecular weights. The compound with the highest TSCF (0.72) had a molecular weight greater than 240.

The Gaussian relationships of Figure 2-2 are simple correlations based on only one chemical property. These simple, predictive relationships are also based on data from very different laboratory arrangements, usually utilizing only one type of plant in each of the studies [*i.e.*, barley (*Hordeum vulgare*), hybrid poplar (*Populus deltoides* × *Populus nigra*), or soybean (*Glycine max*)]. Comparisons of the uptake of the same organic compounds by different plants have shown obvious differences (Walker 1972, McFarlane *et al.* 1990). Whereas the differences in uptake between species appear to be related to lipid content, a full understanding of plant structure and properties relative to uptake characteristics does not yet exist.

Nearly all of the uptake studies presented here were performed in hydroponic systems, void of the competitive sorption of soil matrices. Simple modeling approaches have suggested that the maximal uptake from organic soils would be for compounds with a $\log K_{ow}$ near 0.5 or 1 (Ryan *et al.* 1988, Hsu *et al.* 1990), *versus* 3 observed for the hydroponic systems. The competitive sorption to soils would limit the plant uptake by keeping the compounds in the soil profile. However, sorption to the soil does not decrease the TSCF value for the uptake of a compound; sorption only decreases the aqueous concentration at the root surface.

The uptake studies represented here also primarily utilized nonionic compounds. Compounds carrying an ionic charge are subject to more regulated uptake, as compared to the passive uptake model outlined above for nonionic organic compounds (Trapp 2000). Active transport of ionized compounds has also been documented. The ion trap effect is shown to be the dominant process in active transport, particularly for weak electrolytes. The effect results in a TSCF greater than 1 for some compounds. Trapp (2000) offers a comprehensive discussion on the uptake of ionizable organic compounds and the ion trap effect.

All relationships discussed herein should be considered only as a starting point to estimate what plant uptake is generally possible or anticipated. In all of these discussions regarding TSCF and RCF values for organic compounds, the simple, single-variable relationships are invalid if metabolism or degradation of the compounds occurs during or prior to uptake. More advanced plant uptake models do exist (Boersma *et al.* 1991, Behrendt and Brüggemann 1993, Trapp *et al.* 1994, Trapp 2000) and go far beyond the single-variable predictive relationships for organic compound uptake. For a more in-depth approach to plant uptake and transport of organics, these models should be investigated. These models take into account diffusive exchange between soil and roots, sorption to soils, metabolism, phloem transport, and other processes. Nonetheless, these models rely on the same data sets to generate the TSCF values reported here, and thus encompass the same uncertainty with respect to entry into the plant tissues. To use a more advanced, multivariant approach, a greater number of input variables is required. However, for many plant-contaminant-soil systems, reliable information concerning these variables is not available. As research continues, more understanding and more data will be produced.

GREEN-LIVER MODEL

The three common steps in plant and mammalian metabolism are transformation, conjugation, and elimination or storage (Figure 2-3). The first two steps are highly similar in mammalian and plant systems, including the enzymes involved. Mammalian and plant metabolic action in the third step diverges in terms of ultimate fate. In mammalian systems, the elimination can include excretion with other metabolic waste products, whereas plants

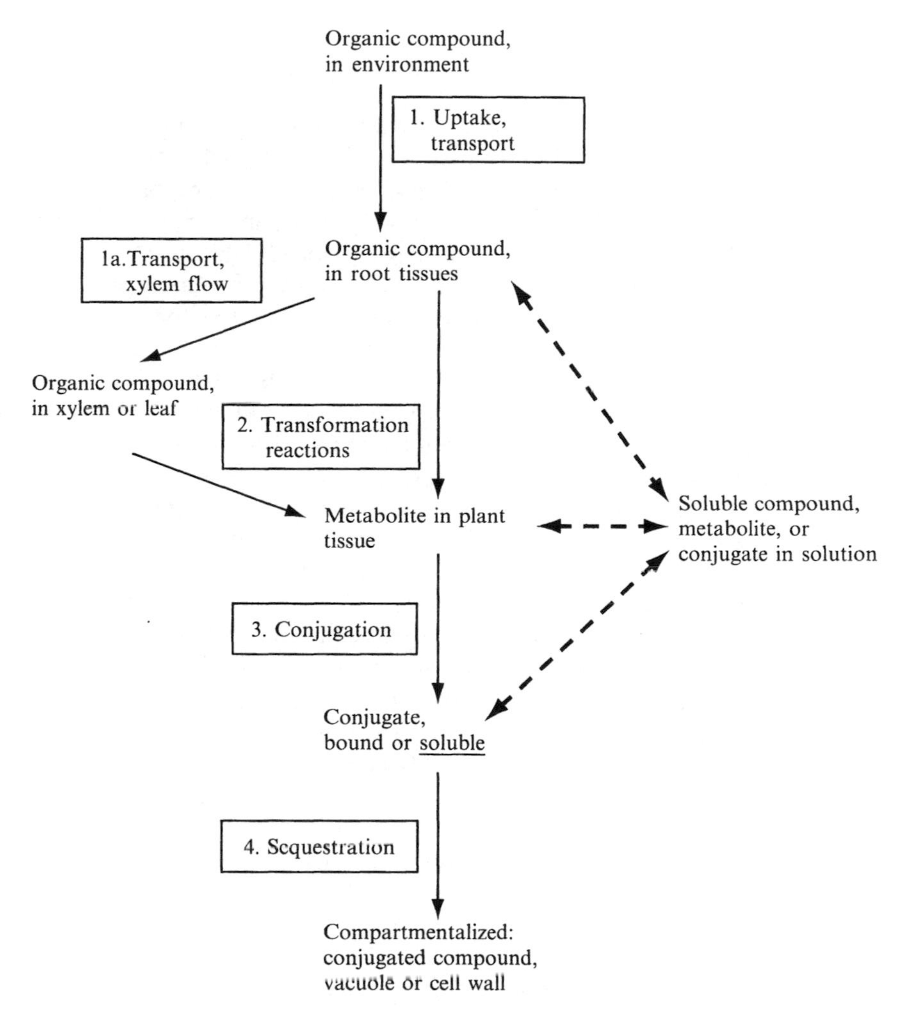

Figure 2-3 Schematic of organic contaminant uptake and metabolism by plants in a phytoremediation application. Metabolic steps in the green-liver model are shown individually. Dashed lines represent pathways that are possible in aquatic phytoremediation. Adapted from Burken *et al.* (2000).

eliminate the conjugates from the metabolic organelles by sequestering the conjugated products in nonmetabolic tissues. The first work to outline these basic steps dates back nearly 100 years. Sandermann (1994) traces the foundation of the concept back to Knoop in 1905, who observed from feeding fatty acids with phenyl substituents to dogs that (1) the fatty acid chains were shortened in β-oxidation (transformation), (2) the resulting products were bound to the amino acid glycine (conjugation), and (3) the conjugates were excreted from the body in urine (an elimination, analogous to compartmentalization in plants). The sequential metabolic steps for plants are illustrated in Figure 2-3 and outlined below.

Transformation

The initial step in the metabolism of organic compounds in plants is the transformation of the initial substrate. The initial transformation can be many different reactions, including a number of oxidations or reductions. Oxidation of lipophilic compounds is very important in increasing solubility and in providing an opportunity for conjugation (Cole 1983). The most prevalent transformation has been identified as hydroxylation (adding an OH^-). The hydroxylation of atrazine was initially viewed as the sole detoxification mechanism before the discovery of conjugates that could act to detoxify hydroxyatrazine further (Lamoureux *et al.* 1970). The addition of the –OH functional group results in a suitable site for conjugation to occur. In one example, uptake of the herbicide bentazone was followed by rapid metabolism to 6-hydroxybentazon in laboratory studies. Appearance of 6-hydroxybentazon was followed by the accumulation of glycosylhydroxybentazon (Leah *et al.* 1991). The hydroxylation step evidently resulted from cytochrome P-450 transformation, as inhibition of P-450 activity resulted in chlorosis and leaf necrosis following benzatone application. The oxidation of many priority pollutants has also been observed in phytoremediation studies. Newman *et al.* (1997, 1999) detected numerous oxidized metabolites of trichloroethylene with –OH functional groups, such as trichloroacetic acid, dichloroacetic acid, and trichloroethanol. Transformation metabolites were detected in a number of studies that included cell cultures, highly controlled laboratory studies, and large pilot-scale studies in the field, where all studies utilized hybrid poplar (*Populus trichocarpa* × *Populus deltoides* 'H11-11') trees (Shang *et al.* this book). Furthermore, the same metabolites of trichloroethylene have been identified in field sites where vegetation was exposed to trichloroethylene-contaminated groundwater (Doucette *et al.* 1998).

Cytochrome P-450 is a key enzyme family in the transformations. Oxidations by P-450 enzyme families are the predominant detoxification mechanism for many plants (Sandermann 1992, Chiapella *et al.* 1995, Komossa *et al.* 1995, Coleman *et al.* 1997, Ohakawa *et al.* 1999). Herbicide detoxification has been addressed by the herbicide industry for years. A number of herbicide additions (herbicide safeners) have been synthesized to target the P-450 function and decrease plant detoxification of herbicides. The function of many P-450 enzymes from plants are summarized and tabulated by Ohakawa *et al.* (1999), who found over 360 P-450 gene sequences. Schwitzguébel and Vanek (this book) discuss some of the specific P-450 enzymes and the activities. Komossa *et al.* (1995) present the activities a schematic of cytochrome P-450–catalyzed hydroxylations of selected herbicides, the plant species involved, and the location of the ring or ring-methyl hydroxylation. In addition, the P-450 activity of many plants has been assayed; of nine plants studied, avocado pear (*Persea americana*) had the highest activity (Chiapella *et al.* 1995).

In numerous plants, a number of transformations take place simultaneously. Whereas P-450 is the prevalent enzyme for transformations, other enzymes are also active. Chiapella *et al.* (1995) assayed the peroxidase activity of selected plants. Another important enzyme family, peroxidases play a significant role in the transformation step of the green-liver model. In many studies, numerous water-soluble metabolites have been found after uptake or direct exposure of xenobiotic compounds. Overall, a number of different enzyme functions have been observed to carry out initial transformations. In addition to the predominant P-450 and peroxidase activities, peroxygenases and carboxylesterases are active in the initial transformation process (Komossa *et al.* 1995). While oxidations comprise the majority of transformations that take place, reduction processes have also been identified (Scheunert 1992). Of these processes, the reduction of nitro groups is of special interest as phytoremediation of nitroaromatic compounds and explosives has been extensively investigated (McCutcheon *et al.* this book, Shanks *et al.* this book, Thompson *et al.* this book). This topic is also covered in this chapter.

Conjugation

Following transformation, conjugation is predominantly the next step in the detoxification or metabolism of xenobiotic compounds. However, some compounds are conjugated without being preceded by transformation. In either case, the resulting conjugates are generally more water-soluble and have reduced toxicity from herbicidal activity. Conjugates can be deposited in vacuoles or incorporated into bound residues through sequestration. Besides soluble conjugate formation, direct conjugation to lignin can occur. This conjugation is covered under sequestration in the following section.

Primary conjugation occurs with malonic acid, D-glucose, glutathione, cysteine, and other amino acids, although many other constituents can act as conjugate molecules in plants (Scheunert 1992, Komossa *et al.* 1995). Examples of detoxification of herbicides by glutathione have been extensively identified in plant, mammalian, and other animal metabolic studies. Glutathione *S*-transferases (GSTs) are enzymes that catalyze the conjugation of glutathione to a variety of substrates. These enzymes and others utilize the functional groups produced from the transformations, as the sites for conjugation. The activity of GSTs is considered to be one of the primary inactivation mechanisms for toxic chemicals. Found in nearly all animals and plants, GSTs were first identified over 30 years ago (Shimabukuro *et al.* 1970).

In addition to inactivation of toxic compounds, GSTs have been shown to be important in (1) the management of secondary plant metabolites (*e.g.*, anthrocyanin pigments) and plant hormones (*e.g.*, auxin), (2) response to oxidative stress, (3) lipid peroxidation, and (4) defense against pathogens (Marrs 1996 and associated references). Reviews by Marrs (1996) and Edwards *et al.* (2000) provide extensive discussion on the specificity of four types of GSTs. Four GST classes have been grouped according to structure and sequence. These classes phi, zeta, tau, and theta represent a standardization of nomenclatures for the previous classes, type 1 to 4, respectively (Edwards *et al.* 2000). The tau and phi classes are specific to plants, whereas the plant GSTs in zeta and theta classes are similar to mammalian GSTs, even sharing the same classification (Edwards *et al.* 2000).

Whereas regulation of secondary plant products and stress response are the primary function of GSTs, these proteins are ubiquitous in plants and active toward a broad range of xenobiotic substrates. Pflugmacher *et al.* (2000) discovered GST activity for five nitrobenzene derivatives, two pesticides (atrazine and fluorodifen), and a natural substrate (cinnamic acid) in 59 different algal and plant species. The high activities observed in marine macroalgae toward substrates that are rarely encountered indicate that GST activity is extremely widespread and active toward many substrates (Pflugmacher *et al.* 2000).

A number of other conjugates can be formed. Glucosidation has also been identified and examined. In initial work, Frear (1968) identified and isolated a glucosyltransferase from soybean (*Glycine max*) plants. Glucosidation results in soluble glucosides that can be incorporated into lignin or stored as a soluble conjugate in a vacuole, discussed in the following section "Sequestration." Specific enzymatic activities for conjugation are also active across wide ranges of plant species, similar to that of P-450. *O*-Glucosyltransferase activities for chlorinated phenols, 2,2-*bis*-(4-chlorophenyl)-acetic acid (a DDT-metabolite), and indole-3-acetic acid were observed in 59 different algal and plant species and 4 plant cell suspension cultures (Pflugmacher and Sandermann 1998).

Plant tissue specificity is also a consideration. Schmidt *et al.* (1995) quantified the activity of 3,4-dichloroanaline *N*-glucosyl- and *N*-malonyltransferase in wheat (*Triticum aestivum*) and soybeans (*Glycine max*), and demonstrated that different plant tissues contained substantially different activity. This advancement (Schmidt *et al.* 1995) is important for phytoremediation in that many types of enzyme activity studies are performed with standardized wheat (*Triticum aestivum* L. 'Heines Koga II') and soybean (*Glycine max* L. Merr. 'Mandarin') cell suspension cultures (Komossa *et al.* 1995). However, extrapolation of cell culture data to whole-plant metabolism can be misleading for some compounds (Harms *et al.* this book, Schwitzguébel and Vanek this book, Shang *et al.* this book).

Sequestration

The third step in the green-liver metabolic model is the sequestration of the conjugated xenobiotic compounds. This step is analogous to mammalian excretion, essentially removing xenobiotic compounds from metabolic tissues. The major difference between plant metabolism and mammalian hepatic function is that plants to do not have an active excretion mechanism to remove the conjugated compound from the organism; rather, the conjugate is isolated through storage or sequestration from the catabolic and anabolic metabolism of the plant. This process may also be termed compartmentalization.

Sequestration can lead to at least three terminal fates within the plant tissues: storage in cell vacuole, storage in the apoplast, or covalent binding to cell walls. Active transport is necessary to store conjugates in a vacuole or the apoplast. To reach a vacuole or the apoplast, the conjugate must be moved through the tonoplast or the plasma membrane (Coleman *et al.* 1997). This movement is facilitated and controlled by a glutathione pump that targets conjugation "tags." The active transport of the pump is necessary as diffusion of the conjugates is not feasible. The pH of the cytosol is approximately 7.4, and therefore the glutathione conjugates are anionic (Coleman *et al.* 1997). The anionic conjugates must then be actively managed in any transmembrane transport. These glutathione pumps are adenosine triphosphate (ATP)-dependant, tonoplast transporters that move the conjugate from the cytosol to the vacuole. Coleman *et al.* (1997) give a more extensive review of the ATP transporter activity and mode of action.

In both plants and animals, glutathione pumps attach to and transfer a number of glutathione conjugates across membranes. In animals, selective excretion eliminates glutathione conjugates from the body altogether. In plants, the transport results in an accumulation of the conjugates, away from the cytosol. The term "storage excretion" has even been coined to describe this process (Lamoureux and Rusness 1989, Marrs 1996). Despite the number of glutathione conjugates involved, neither cysteine conjugates nor nonconjugated glutathione were transported (Martinoia, 1993). Because diffusion of the conjugates is not feasible, movement back into the cytosol following transport *via* the glutathione pump is not possible. This isolates the conjugates permanently from the metabolic functions of the plant cell.

Vacuole storage of glutathione conjugates has been reported in numerous plants but perhaps best illustrated in recent work by Coleman *et al.* (1997). Monochlorobimane was used as a model compound because the bimane forms a blue fluorescent conjugate with glutathione. The conjugate was shown to accumulate rapidly in the vacuole for a number of species. Another significant finding of this work was the tissue specific activity within the leaf. Mesophyll cells had vacuole storage of the bimane conjugate whereas epidermal cells did not (Coleman *et al.* 1997). Tissue-specificity should be considered when extrapolating cellular activity to whole-plant interactions with xenobiotic compounds.

Bound residues are defined by being nonextractable and thus, by definition, elude analysis. Often, bound residues are no more understood than the name implies. However, methods to fractionate the residues have been developed (Langebartels and Harms 1985). This sequential method generates fractions composed of starch, protein, cellulose, hemicellulose, pectin, and lignin. Using this technique or similar methods, the residue of the initial compound that is covalently bound, is clarified. Khan and Dupont (1986) found that a major portion of bound residues are associated with lignin—the phenolic, structurally nonrepeating polymer that makes up much of plant cell walls. Covalent binding into the polymer of lignin is often termed lignification (Cunningham *et al.* 1996). Studies by Bokern and Harms (1997) using radiolabeled 4-nonylphenol investigated inextricable residues and characterized the residues with respect to the cell wall fractions containing radioactivity. In 7 of 12 cell cultures, lignin was the main fraction to which [^{14}C] was associated. In other research, the metabolism of [^{14}C]-phenoxyacetic acid and the formation of associated bound residues were studied in soybean (*Glycine max*) leaves and stems. Lignin was not the only tissue involved in binding. In leaves, hemicellulose and lignin were preferential components for the binding of [^{14}C], whereas in stems, pectin and lignin were the preferential fractions for binding. Lignification has also been directly observed in phytoremediation studies. Castro *et al.* (2001) identified lignification of triazoles in sunflowers (*Helianthus annuus*), and covalently bound residues of DDT have been observed in experiments with Canadian waterweed (*Elodea canadensis*) (Garrison *et al.* 2000).

Overall Process

The rationale for the green-liver model is not the serendipitous degradation of xenobiotic contaminants but the transformation of natural biochemicals. All of the enzymes responsible for green-liver metabolism have the capability to transform or control natural compounds produced by plants and other organisms. Table 2-1 was adapted from Sandermann (1992) and shows the natural substrates (biochemicals) for a number of plant enzyme classes along with the xenobiotic compounds that also serve as substrates for these enzymes. Enzymatic activity towards natural substrates is important in secondary metabolism. The same enzymes inactivate and transform plant regulators after the initial regulation activity and function are served. The extensive number of substrates also with which enzymes react allows plants to transform many xenobiotic compounds. Additional, in-depth discussion of specific enzyme activity towards xenobiotic compounds appears in other chapters (Harms *et al.* this book, Medina *et al.* this book, Schwitzguébel and Vanek this book, Shang *et al.* this book).

The green-liver model is a grouping of many enzymatic functions into one overall concept for the metabolism of organic compounds. In an overview, Komossa *et al.* (1995) presented the relative metabolite breakdown and different fates for 28 compounds in standardized cell cultures. These data

TABLE 2-1 Plant Enzyme Classes with Activity For Natural and Xenobiotic Compounds

Enzyme class	Natural substrate	Xenobiotic substrate
Cytochrome P-450	Cinnamic acid, pterocarpans	4-Chlor*N*-methyl-analine
Glutathione S-transferases	Cinnamic acid	Fluorodifen, alachlor, atrazine
Carboxylesterases	Lipids, acetylcholine	Diethylhexylphthalate
O-Glucosyltransferases	Flavonoids, coniferyl alcohol	Chlorinated phenols
O-Malonyltransferases	β-D-Glucosides of flavonoids and isoflavonoids	β-D-Glucosides of pentachlorophenols and of 4-hydroxy-2,5-dichlorophenoxyacetic acid
N-Glucosyltransferases	Nicotinic acid	Chlorinated anilines, metribuzin
N-Malonyltransferases	1-Aminocyclopropyl carboxylic acid, D-amino acids, anthranilic acid	Chlorinated anilines

Note: Adapted from Sandermann (1992). See original article for more extensive discussion and original references.

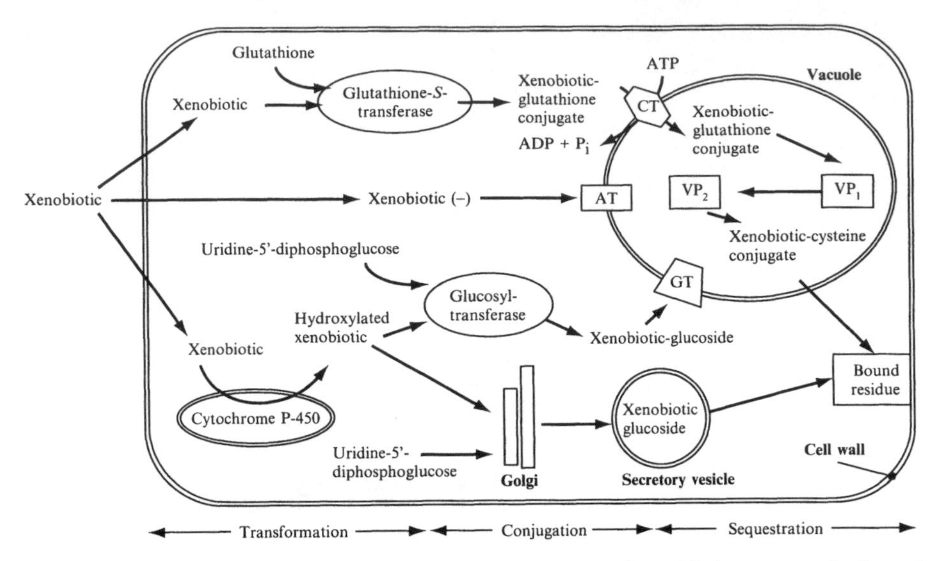

Figure 2-4 Enzymatic functions responsible for the metabolism of xenobiotic compounds. Several metabolic reactions take place in or associated with individual organelles or compartments. Abbreviations — CT: glutathione conjugate transporter; AT: ATP-dependant xenobiotic anion transporter; GT: ATP-dependent glucoside-conjugate transporter; VP: vacuolar protein. Adapted with permission from Coleman *et al.* (1997).

show great variability in the outcome of plant–contaminant interactions. The different possibilities within the model are apparent in Figure 2-4. The figure shows the interrelationship of the many steps, and in fact only a small sampling of the different specific enzymatic activities (shown in Table 2-1) that can be included when considering green-liver plant metabolism.

PESTICIDE EXAMPLES

The majority of the early work on the green-liver model involved herbicides, in particular atrazine and related *s*-triazine compounds. Over 40 years ago, Castelfranco *et al.* (1961) showed that resistant plants were able to transform 2-chloro-4,6-*bis*(ethylamino)-*s*-triazine rapidly to the 2-hydroxy analogue. In subsequent years, cytochrome P-450 was found to be responsible for this transformation.

Other stages of green-liver metabolism were clearly outlined before the responsible molecular reactions were understood or identified. Shimabukuro and Swanson (1969) evaluated atrazine selectivity and metabolism in peas (*Pisum sativarum*) and sorghum (*Sorghum vulgare*). In leaf-tissue studies, the atrazine-resistant sorghum (*Sorghum vulgare*) tissues metabolized atrazine to hydroxyatrazine (highly water-soluble) and an insoluble residue as shown in Table 2-2 (Shimabukuro and Swanson 1969). Subsequent studies showed that

TABLE 2-2 Atrazine Metabolism in Sorghum (Sorghum vulgare) Leaf Disks

	1.5-hour incubation period (percent)	7-hour period (percent)
Water-soluble metabolite	14.5	62.3
Atrazine	77.2	21.9
Unidentified	1.8	2.6
Insoluble residue	6.5	13.2

Note: Data presented as percent of initial moles added as [^{14}C]-labeled atrazine. Adapted from Shimabukuro and Swanson (1969).

atrazine was metabolized *via* hydroxylation of the 2-chloro group and the formation of cysteine and glutathione conjugates (potentially the water-soluble metabolites observed earlier), as well as *N*-dealkylation (Shimabukuro 1968, Shimabukuro and Swanson 1969). Bound metabolites have also been identified as a terminal fate for atrazine and other *s*-triazines. The bound residue increased from 12 percent of the [^{14}C] introduced as atrazine after 48 hours to 52 percent of the [^{14}C] after 336 hours.

Similar metabolism has been observed in phytoremediation studies (Burken and Schnoor 1996, Burken and Schnoor 1997). In these studies with hybrid poplar (*Populus deltoides* × *Populus nigra* 'DN-34') trees, hydroxyatrazine, *N*-dealkylation metabolites, and other water-soluble metabolites were isolated *via* [^{14}C] analysis and liquid chromatography techniques; although the exact chemical structure of the water-soluble metabolites was not established. The temporal metabolite profile is outlined in Table 2-3. Bound residues increased over time as the concentration of atrazine decreased, resulting in up to 15 percent of the total plant uptake after 48 days of growth following an initial 13-day exposure. In longer-term experiments, poplar trees exposed to an initial atrazine dose were analyzed after 52 and 80 days. Although uptake was limited by sorption to the organic matter in the loam soil, atrazine was taken up into plant tissues but did not accumulate as the parent compound. Atrazine comprised only 8 percent

TABLE 2-3 Atrazine and Metabolites in Hybrid Poplar (*Populus deltoides* × *Populus nigra* 'DN-34') Trees

Total incubation time (days)[a]	Atrazine (percent)	Hydroxy atrazine (percent)	*N*-dealkylation products (percent)	Soluble metabolites (unidentified) (percent)	Bound residues (percent)
13	38.5	4.6	37.9	10.6	8.4
48	24.4	4.6	28.9	26.3	15.8

[a]Atrazine and metabolites removed from soil profile at 13 days; thereafter hybrid poplar trees continued to grow in contaminant-free media.

Note: Results presented as percent of total moles present in hybrid poplar trees as determined by extraction or total oxidation followed by [^{14}C] analysis.

of the extractable [^{14}C] in plant tissues exposed for 80 days, whereas the nonsoluble bound residues and soluble metabolites were the dominant forms. The soluble metabolites were hypothesized to be conjugates, similar to those involving glutathione and malonate as seen in numerous other studies.

Sandermann (1994) showed that two herbicides, 2,4-dichlorophenoxyacetic acid (2,4-D) and pentachlorophenol, undergo hydroxylations as the initial transformations. Yet, the ultimate fates were different, even though both are chlorinated, substituted, monoaromatic compounds. Following hydroxylation, 2,4-D is conjugated with D-glucose and malonyl residues. The soluble glucoside conjugate is then stored in the vacuole. In contrast, the pentachlorophenol-hydroxylated product is directly conjugated to lignin and stored in apoplastic cell-wall tissue.

NITROAROMATICS AND EXPLOSIVES EXAMPLES

A number of studies established the occurrence of plant uptake and metabolism of nitroaromatic compounds and explosives with nitro-functional groups. The most commonly studied compounds are 2,4,6-trinitrotoluene (TNT) and hexahydro-1,3,5-trinitro-1,3,5-triazine (RDX). Over 40 plant species have been studied in conjunction with these compounds and a compilation of results has been summarized (Burken *et al.* 2000). The metabolism of TNT has been of primary interest. Initial transformations of TNT can be either oxidations or reductions. Oxidation can occur either at the methyl group or directly on the ring. Reduction of the nitro groups produce hydroxylamino or amino groups. The initial reduction of TNT has been the most commonly reported transformation, often evident from the appearance of the mono-amino derivatives, 2-amino-4,6-dinitrotoluene and 4-amino-2,6-dinitrotoluene (Burken *et al.* 2000 and associated references). In terrestrial and aquatic studies, the mono-amino derivatives were commonly detected at highest levels in the root tissues and were generally not detected at similar levels in foliar tissues (Thompson 1997, Scheidemann *et al.* 1998). Oxidation has recently been observed in TNT transformation, with six individual oxidation metabolites being identified (Bhadra *et al.* 1999a). Oxidation appears to occur following initial transformation that produces hydroxylamino derivatives.

Conjugation of TNT-transformation products also has been reported (Harvey *et al.* 1991, Bhadra *et al.* 1999b). These conjugates have been characterized through acid hydrolysis and mass spectrometry. The formation of soluble, *i.e.*, extractable, conjugates appears to be the gateway to the formation of bound residues of the TNT-derived conjugates. Bhadra *et al.* (1999b) showed that bound residues increased as the conjugates turn over (Table 2-4).

Metabolism of RDX has been investigated but not to the extent of TNT. There is less understanding of the exact transformations that RDX undergoes on entering plant tissues. However, the fate is well documented and is quite different from that of TNT. Translocation of parent RDX to foliar tissues was

TABLE 2-4 Mole Balance of TNT, Mono-Amino Metabolites, Conjugated Metabolites, and Bound Residues in Hairy Root Cultures of Periwinkle (*Catharanthus roseus*)

	Fraction of initial TNT moles		
Time (hours)	Nitroaromatics	Conjugates	Bound
0	1.0 ± 0.00	0 ± 0.00	0 ± 0.00
7	0.45 ± 0.02	0.15 ± 0.04	0.08 ± 0.02
24	0.18 ± 0.01	0.20 ± 0.06	0.14 ± 0.02
31	0.15 ± 0.02	0.22 ± 0.06	0.14 ± 0.07
47	0.12 ± 0.02	0.26 ± 0.01	0.25 ± 0.01
75	0.05 ± 0.01	0.22 ± 0.01	0.29 ± 0.01

Note: Results are based on [^{14}C] analysis in microorganism-free cultures. Adapted from Bhadra *et al.* (1999b).

much more prevalent and overall RDX was more persistent than TNT. Accumulation of RDX has been observed in many species such as lettuce (*Lacuta sativa*), tomato (*Lyopersicum esculentum*), corn (*Zea mays*), soybean (*Glycine max*), sorghum (*Sorghum sudanese*), wheat (*Triticum aestivum*), and hybrid poplar (*Populus deltoides* × *Populus nigra*) (Larson 1997, Price *et al.* 1997, Thompson *et al.* 1999). The different fates of TNT and RDX were clearly shown in the work by Thompson (1997), where [^{14}C]-labeled TNT and RDX were introduced to hydroponically grown hybrid poplar (*Populus deltoides* × *Populus nigra* 'DN-34') trees. Trinitrotoluene was not detected in the foliar tissues and only limited [^{14}C] was translocated in any form. In contrast, RDX was detected as the predominant [^{14}C] form in the leaves, and the leaves contained the majority of the [^{14}C] (Figure 2-5). The great variation observed in the fate of TNT and RDX provides insight into the variability of the different reactions involved in plant metabolism of anthropogenic compounds and the substrate specificity of the enzymes involved. Phytoremediation of nitroaromatic compounds and explosives, and the specifics of the metabolism involved, are covered in much greater detail in other chapters (McCutcheon *et al.* this book, Shanks *et al.* this book, Thompson *et al.* this book) and in other works (Burken *et al.* 2000).

IMPACTS ON THE APPLICABILITY OF PHYTOREMEDIATION

Plant metabolism of compounds in phytoremediation studies has been shown and is in agreement with the green-liver model (Burken and Schnoor 1997, Thompson *et al.* 1998, Bhadra *et al.* 1999a, Bhadra *et al.* 1999b, Newman *et al.* 1999, Shanks *et al.* this book, Thompson *et al.* this book). Where anthropogenic compounds and plants are in direct contact, there is concern for contamination of the food chain. The impacts of bioaccumulation of compounds such as DDT, dioxin, and polychlorinated biphenyls up the food chain are well documented. While metabolized in plants, the impacts of DDT and associated

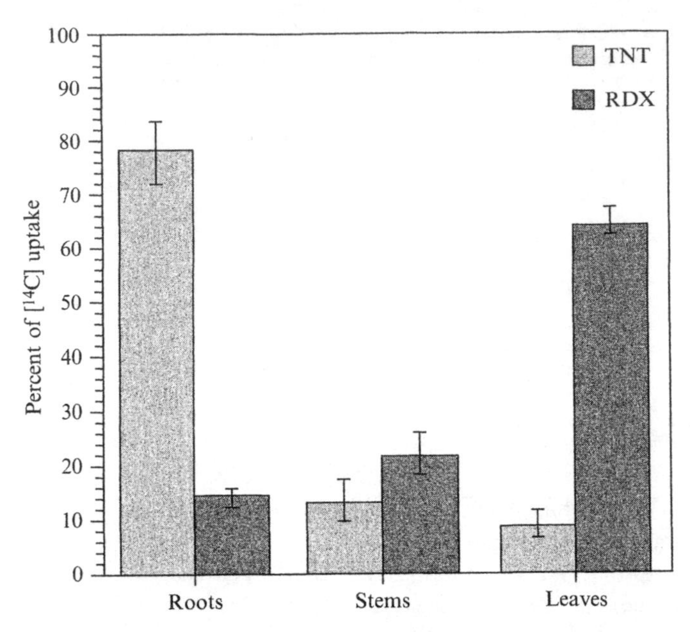

Figure 2-5 Contrasting distributions of [^{14}C] between hybrid poplar (*Populus deltoides* × *Populus nigra* 'DN-34') tissues for TNT and RDX after uptake from hydroponic solution for 2 days. The majority of the [^{14}C] remained in the roots for TNT, whereas the leaves contained the bulk of RDX-related [^{14}C]. Each bar is based on the average of three replicates, and error bars represent ±1 standard deviation. Reprinted with permission from Thompson (1997).

metabolites are quite severe in birds. Because no one wants to be responsible for any recurrence of future ecosystem exposures by anthropogenic compounds, some hesitate to accept phytoremediation as a widely applied treatment for cleaning contaminated sites. These concerns are founded in the lack of knowledge and data relating to the general interactions of organic pollutants and plants.

The storage of organic compounds (the terminal fate in the green-liver model) and the presence of transformed and conjugated derivatives are also causes for concern. In addition, the potential toxicity of metabolites is not well understood. The metabolism of organic compounds appears to reduce toxicity, and this reduction in toxicity serves to protect the plant. However, through translocation and incorporation into the plant biomass, the potential exposure to humans and the ecosystem may be greatly increased compared to confined, subsurface contaminants. Therefore, the possible ingestion of metabolized organics is still a matter of concern, even though the specific toxicity is thought to be reduced.

Bound residues resulting from plant metabolism have been extensively studied, but concerns still exist. Many studies have looked at bound residues and toxicity has not been generally observed. Research has shown that bound residues can be unavailable to animals that are not able to degrade the cellulose or lignin of plant cell walls. Bakke *et al.* (1972) fed sheep and rats methanol-insoluble residues of sorghum plants that had been grown in and had taken up [^{14}C]-labeled atrazine. The radiolabel was almost totally excreted in the feces of both species. In another study, alfalfa containing bound residues of the carbamate herbicide propham was fed to rats and sheep. These bound residues were excreted almost entirely within one day for rats and two days for sheep. Rats were also fed bound residues of radishes that had been treated with dieldrin or carbofuran, and again low bioavailability was observed (Khan *et al.* 1987). In all these cases in which bound residues were fed to rats and sheep, there appeared to be no metabolic interaction with the bound residues. Bound residues under these conditions could be considered as harmless to animals (Klein and Scheunert 1982). However, bound residues cannot be considered as totally non-bioavailable. Lignin-bound 3,4-dichloroanaline was found to be bioavailable to rats to a small degree (Sandermann *et al.* 1990). However, the method used to generate the bound residues can strongly influence bioavailability in such studies (Sandermann *et al.* 1992). Khan *et al.* (1986) also studied deltamethrin residues in bean plants and found that the residues were partially bioavailable. Although these bound residues were bioavailable, no toxic effects were observed. This observation indicates that even though residues can persist in plant tissues for extreme periods of time, bound metabolites may or may not have toxicological impacts of a chronic nature. Whereas bound residues appear to have varying levels of bioavailability and potential toxicological impacts are unknown, conjugation products are vulnerable to cleavage in animal digestive systems (Sandermann *et al.* 1990, Sandermann *et al.* 1992). This may result in the release of the transformed product or potentially even the parent compound. Although there is little direct evidence that plant uptake and metabolism raise even a moderate toxicological concern, many questions remain. Consideration should be given to transformation products, conjugates, and bound residues when contemplating toxicity of xenobiotic chemicals instead of solely focusing on parent compounds.

TRANSGENIC PLANTS: ENHANCING PLANT METABOLISM

Molecular tools have made it possible to enhance certain traits and even to transfer the metabolic capabilities from one species to another. The genetic engineering of crops with genes from other plants and bacteria has resulted in herbicide resistant crops. This resistance in many cases is the expression of certain enzymes to detoxify the herbicides through the processes outlined here as green-liver metabolism. The use of genetic engineering "closes the loop" with regard to plant metabolism of anthropogenic compounds: starting with discovery and proceeding to investigation, to comprehension, and finally to manipulation and engineering of the enzymatic pathways that comprise the green-liver metabolism of anthropogenic compounds.

Acknowledgments

I thank Drs. Stefan Trapp, Philip Thompson, Lee Newman, Marshall Porterfield, and Heinrich Sandermann, Jr. for their contributions and input. I would also like to thank Juel Gibbons, Xingmao Ma, Mary Bordner, and Amanda Gilbertson for their contributions and comments on the manuscript.

REFERENCES

Aitchison, E.W. (1998) Phytoremediation of 1,4-dioxane by hybrid poplar trees. Master's thesis. Civil and Environmental Engineering. University of Iowa, Iowa City, Iowa, 151 pages.

Aitchison, E.W., S.L. Kelley, P.J. Alvarez, and J.L. Schnoor (2000) Phytoremediation of 1,4-dioxane by hybrid poplar trees. *Water Environ. Res.* **72**(3): 313–321.

Bakke, J.E., R.H. Shimabukuro, K.L. Davison, and G.L. Lamoureux (1972) Sheep and rat metabolism of the insoluble ^{14}C residues present in ^{14}C treated sorghum. *Chemosphere* **1**(1): 21.

Behrendt, H. and R. Brüggemann (1993) Modelling the fate of organic chemicals in the soil plant environment: model study of root uptake of pesticides. *Chemosphere* **27**(12): 2325–2332.

Bhadra, R., R.J. Spanggord, J.B. Wayment, and J.V. Shanks (1999a) The characterization of oxidation products of TNT metabolism in aquatic phytoremediation systems of *Myriophyllum aquaticum. Environ. Sci. Technol.* **33**(19): 3354–3361.

Bhadra, R., D.G. Wayment, J.B. Hughes, and J.V. Shanks (1999b) Confirmation of conjugation processes during TNT metabolism by axenic plant roots. *Environ. Sci. Technol.* **33**(3): 446–452.

Boersma, L.M., C. McFarlane, and F.T. Lindstrom (1991) Mathematical model of plant uptake and translocations of organic chemicals: application to experiments. *J. Environ. Qual.* **20**(Jan–Mar): 137–146.

Bokern, M. and H.H. Harms (1997) Toxicity and metabolism of 4-*n*-nonylphenol in cell suspension cultures of different plant species. *Environ. Sci. Technol.* **31**(7): 1849–1854.

Briggs, G.G., R.H. Bromilow, and A.A. Evans (1982) Relationships between lipophilicity and root uptake and translocation of non-ionized chemicals by barley. *Pestic. Sci.* **13**: 495–504.

Briggs, G.G., R.H. Bromilow, A.A. Evans, and M. Williams (1983) Relationships between lipophilicity and root uptake and distribution of non-ionised chemicals in barley shoots following uptake by the roots. *Pestic. Sci.* **14**: 492–500.

Burken, J.G. and J.L. Schnoor (1996) Phytoremediation: plant uptake of atrazine and the role of root exudates. *J. Environ. Eng.* **122**(11): 958–963.

Burken, J.G. and J.L. Schnoor (1997) Uptake and metabolism of atrazine by hybrid poplar trees. *Environ. Sci. Technol.* **31**(5): 1399–1406.

Burken, J.G. and J.L. Schnoor (1998) Predictive relationships for uptake of organic contaminants by hybrid poplar trees. *Environ. Sci. Technol.* **32**(21): 3379–3385.

Burken, J.G., J.V. Shanks, and P.L. Thompson (2000) Phytoremediation and plant metabolism of explosives and nitroaromatic compounds. In: *Biodegradation of Nitroaromatic Compounds and Explosives*. J.C. Spain, J.B. Hughes, and H.J. Knackmuss, eds. CRC Press, Boca Raton, Florida, pp. 239–275.

Castelfranco, P., C.L. Foy, and D.B. Deutsch (1961) Non-enzymatic detoxification of 2-chloro-4,6-bis(ethylamino)-*s*-triazine (simazine) by extracts of *Zea mays*. *Weeds* **9**: 580–591.

Castro, S., L.C. Davis, and L.E. Erickson (2001) Plant enhanced remediation of glycol-based aircraft deicing fluids. *Pract. Periodical Hazard., Toxic, Radioact. Waste Manag.* **5**(3): 141–152.

Chiapella, C., P. Ysern, J. Piera, and M. Llagostera (1995) A plant metabolic activation system for *Persea americana* with cytochrome P-450-dependent and peroxidase activities. *Mutat. Res.* **329**: 11–18.

Cole, D. (1983) Oxidation of xenobiotics in plants. *Prog. Pest. Biochem. Technol.* **3**: 199–253.

Coleman, J.O.D., M.M.A. Blake-Kalff, and T.G.E. Davies (1997) Detoxification of xenobiotics by plants: chemical modification and vacuolar compartmentation. *Trends Plant Sci.* **2**: 144–151.

Cunningham, S.D., T.A. Anderson, A.P. Schwab, and F.C. Hsu (1996) Phytoremediation of soils contaminated with organic pollutants. In: *Advances in Agron.* Academic Press, New York, **56**: 55–114.

Davis, L.C., S. Vanderhoof, J. Dana, K. Selk, K. Smith, B. Golpen, and L.E. Erickson (1998) Movement of chlorinated solvents and other volatile organics through plants monitored by fourier transform infrared (FT-IR) spectrometry. *J. Haz. Waste Res.* **1**(4): 1–26.

Doty, S.L., T.Q. Shang, A.M. Wilson, J. Tangen, A.D. Westergreen, L.A. Newman, S.E. Strand, and M.P. Gordon (2000) Enhanced Metabolism of halogenated hydrocarbons in transgenic plants containing mammalian cytochrome P450 2E1. *Proc. Nat. Acad. Sci. USA* **97**(12): 6287–6291.

Doucette, W., B. Bugbee, S. Hayhurst, W. Plaehn, D. Downey, S. Taffinder, and R. Edwards (1998) Phytoremediation of dissolved-phase trichloroethylene using mature vegetation. In: *Bioremediation and Phytoremediation: Chlorinated and Recalcitrant Compounds*. G.B. Wickramanayake and R.E. Hinchee, eds. Battelle Press, Columbus, Ohio, pp. 251–256.

Edwards, R., D.P. Dixon, and V. Walbot (2000) Plant glutathione *S*-transferases: enzymes with multiple functions in sickness and health. *Trends Plant Sci.* **5**(5): 193–198.

Frear, D.S. (1968) Herbicide metabolism in plants—I. purification and properties of UDP-glucose: arylamine *N*-glucosyl-transferase from soybean. *Phytochemistry* **7**: 913–920.

Garrison, A.W., V.A. Nzengung, J.K. Avants, J.J. Ellington, W.J. Jones, D. Rennels, and N.L. Wolfe (2000) Phytodegradation of *p,p′*-DDT and the enantiomers of *o,p′*-DDT. *Environ. Sci. Technol.* **34**: 1663–1670.

Harvey, S.D., R.J. Fellows, D.A. Cataldo, and R.M. Bean (1991) Fate of the explosive hexahydro-1,3,5-trinitro-1,3,5 triazine (RDX) in soil and bioaccumulation in bush bean hydroponic plants. *Environ. Toxicol. Chem.* **10**: 845–855.

Hsu, F.C., R.L. Marxmiller, and A.Y. Yang (1990) Study of root uptake and xylem translocation of cinmethylin and related compounds in detopped soybean roots using a pressure chamber technique. *Plant Physiol.* **93**: 1573–1578.

Khan, S.U. and S. Dupont (1986) Bound pesticide residues and their bioavailability. In: *Pesticide Science and Biotechnology.* R. Greenhalgh and T.R. Roberts, eds. Sixth International Congress of Pesticide Chemistry, Blackwell Scientific, Ottawa, Canada.

Khan, S.U., S. Kacew, and H. Akhtar (1986) Bioavailability of bound carbon-14 residues in rats from bean plants treated with carbon-14 deltamethrin. *Chemosphere* **15**: 923–928.

Khan, S.U., S. Kacew, S. Dupont, G.D. Stratton, Jr., and W.B. Wheeler (1987) Bioavailability in rats of bound residues from radishes treated with either radiolabeled dieldrin or carbofuran *J. Agric. Food Chem.* **35**(1): 89–93.

Klein, W. and I. Scheunert (1982) Bound pesticide residues in soil, plants and food with particular emphasis on the application for nuclear techniques, *Agrochemical Fate in Food and Environment*, Proc. International Symposium. International Atomic Energy Agency, Vienna, Austria, pp. 177–205.

Komossa, D., C. Langebartels, and H. Sandermann, Jr. (1995) Metabolic processes for organic chemicals in plants. In: *Plant Contamination: Modeling and Simulation of Organic Chemical Processes.* S. Trapp and J.C. McFarlane, eds. CRC Press, Boca Raton, Florida, pp. 69–103.

Lamoureux, G.L. and D.G. Rusness (1989) Propachlor metabolism in soybean plants, excised soybean tissues, and soil. *Pestic. Biochem. Physiol.* **34**: 187–204.

Lamoureux, G.L., R.H. Shimabukuro, H.R. Swanson and D.S. Frear (1970) Metabolism of atrazine in excised sorghum leaf sections. *J. Agric. Food Chem.* **18**(1): 81–87.

Langebartels, C. and H. Harms (1985) Analysis for nonextractable (bound) residues of pentachlorophenol in plant cells using a cell wall fractionation procedure. *Ecotox. Environ. Safe.* **10**: 268.

Larson, S.L. (1997) Fate of explosive contaminants in plants. In: *Bioremediation of Surface and Subsurface Contamination.* R.K. Bajpai and M.E. Zappi, eds. The New York Academy of Sciences. **829**: 195–201.

Leah, J.M., T.L. Worrall, and A.H. Cobb (1991) A study of benzaton uptake and metabolism in the presence and absence of cytochrome P-450 and acetyl-coenzyme A carboxylase inhibitors. *Pestic. Biochem. Physiol.* **39**: 232–239.

Marrs, K.A. (1996) The function and regulation of glutathione *S*-transferases in plants. In: *Annual Review of Plant Physiology and Molecular Biology*, Annual Review, Palo Alto, California. **47**: 127–158.

Martinoia, E. (1993) An ATP-dependent glutathione-*S*-conjugate "export" pump in the vacuolar membrane of plants. *Nature* **364**: 247–249.

McFarlane, J.C., T. Pfleeger, and J.S. Fletcher (1990) Effect, uptake and disposition of nitrobenzene in several terrestrial plants. *J. Environ. Toxicol. Chem.* **9**: 513–520.

Newman, L.A., S.E. Strand, N. Choe, J. Duffy, G. Ekuan, M. Ruszaj, B.B. Shurtleff, J. Wilmoth, P.E. Heilman, and M.P. Gordon (1997) Uptake and transformation of TCE by hybrid poplars. *Environ. Sci. Technol.* **31**(4): 1062–1067.

Newman, L.A., X.P. Wang, I.A. Muiznieks, G. Ekuan, M. Ruszaj, R. Cortellucci, D. Domroes, G. Karscig, T. Newman, R.S. Crampton, R.A. Hashmonay,

M.G. Yost, P.E. Heilman, J. Duffy, M.P. Gordon, and S.E. Strand (1999) Remediation of trichloroethylene in an artificial aquifer with trees: a controlled field study. *Environ. Sci. Technol.* **33**(13): 2257–2265.

Ohakawa, H., H. Tsujii, M. Shimoji, Y. Imajuku, and H. Imaishi (1999) Cytochrome P-450 biodiversity and plant protection. *J. Pestic. Sci.* **24**: 197–203.

Orchard, B.J., W.J. Doucette, J.K. Chard, and B. Bugbee (2000) Uptake of trichloroethylene by hybrid poplar trees grown hydroponically in flow-through plant growth chambers. *Environ. Toxicol. Chem.* **19**(4): 895–903.

Pflugmacher, S. and H. Sandermann, Jr. (1998) Taxonomic distribution of plant glucosyltransferases acting on xenobiotics. *Phytochemistry* **49**(2): 507–511.

Pflugmacher, S., P. Schroder, and H. Sandermann, Jr. (2000) Taxonomic distribution of plant glutathione *S*-transferases acting on xenobiotics. *Phytochemistry* **54**(3): 267–273.

Price, R.A., J.C. Pennington, S.L. Larson, D. Neumann, and C.A. Hayes (1997) Plant uptake of explosives from contaminated soil and irrigation water at the former Nebraska Ordnance Plant, Mead, Nebraska. Tech. Report EL-97-11 U.S. Army Engineer Waterways Experiment Station, Vicksburg, Mississippi.

Rubin, E. and A. Ramaswami (2001) The potential for phytoremediation of MTBE. *Water Res.* **35**(5): 1348–1353.

Ryan, J.A., R.M. Bell, J.M. Davidson, and G.A. O'Connor (1988) Plant uptake of non-ionic organic chemicals from soils. *Chemosphere* **17**: 2299–2323.

Sandermann, H., Jr. (1992) Plant metabolism of xenobiotics. *TIBS* **17**: 82–84.

Sandermann, H., Jr. (1994) Higher plant metabolism of xenobiotics: the "green liver" concept. *Pharmacogenetics* **4**: 225–241.

Sandermann, H., Jr., H. Diesperger, and D. Scheel (1977) Metabolism of xenobiotics by plant cell cultures. In: *Plant Tissue Culture and Its Biotechnical Application.* W. Barz, E. Reinhard, and M.H. Zenk, eds. Springer-Verlag, Berlin, pp.178–196.

Sandermann, H., Jr., D. Sheel, and T. Von der Trenk (1984) Use of plant cell cultures to study the metabolism of environmental chemicals. *Ecotox. Environ. Safe.* **8**: 167.

Sandermann, H., Jr., M. Arjmand, I. Gennity, R. Winkler, C.B. Struble, and P.W. Aschbacher (1990) Animal bioavailability of defined xenobiotic lignin metabolites. *J. Agric. Food Chem.* **38**(9): 1877–1880.

Sandermann, H., Jr., T.J. Musick, and P.W. Aschbacher (1992) Animal bioavailability of 3,4-dichloroaniline metabolite fraction in wheat. *J. Agric. Food Chem.* **40**(10): 2001–2007.

Scheidemann, P., A. Klunk, C. Sens, and D. Werner (1998) Species dependent uptake and tolerance of nitroaromatic compounds by higher plants. *J. Plant Physiol.* **152**: 242–247.

Scheunert, I. (1992) Fate of pesticides in plants and in soil fauna. *Chem. Plant Protect.* **8**: 77–103.

Schmidt, B., C. Rivero, and B. Thiede (1995) 3,4–dichloroanaline *N*-glucosyl- and *N*-malonyltransferase activities in cell cultures and plants of soybean and wheat. *Phytochemistry* **39**(1): 81–84.

Shimabukuro, R.H. (1968) Atrazine metabolism in resistant corn and sorghum. *Plant Physiol.* **43**: 1925–1930.

Shimabukuro, R.H. and H.R. Swanson (1969) Atrazine metabolism, selectivity, and mode of action. *J. Agric. Food Chem.* **17**(2): 199–205.

Shimabukuro, R.H., H.R. Swanson, and W.C. Walsh (1970) Glutathione conjugation: atrazine detoxification mechanism in corn. *Plant Physiol.* **46**(2): 103–107.

Shone, M.G.T. and A.V. Wood (1972) Factors affecting absorption and translocation of simazine by barley. *J. Exp. Bot.* **23**(74): 141–151.

Shone, M.G.T. and A.V. Wood (1974) A comparison of the uptake and translocation of some organic herbicides and a systemic fungicide by barley: I absorption and relationship to physico-chemical properties. *J. Exp. Bot.* **25**(85): 390–400.

Sicbaldi, F., G.S. Sacchi, M. Trevisan, and A.A.M. Del Re (1997) Root uptake and xylem translocation of pesticides from different chemical classes. *Pestic. Sci.* **50**: 111–119.

Simonich, S.L. and R.A. Hites (1994) Importance of vegetation in removing polycyclic aromatic hydrocarbons from the atmosphere. *Nature* **370**(7 July): 49–51.

Thompson, P.L. (1997) Phytoremediation of munitions (RDX, TNT) waste at the Iowa Army Ammunition Plant with hybrid poplar trees. Ph.D. dissertation. Civil and Environmental Engineering. The University of Iowa, Iowa City, Iowa.

Thompson, P., L. Ramer, and J. Schnoor (1998) Uptake and transformation of TNT by hybrid poplar trees. *Environ. Sci. Technol.* **32**(7): 975–980.

Thompson, P.L., L.A. Ramer, and J.L. Schnoor (1999) Hexahydro-1,3,5-trinitro-1,3,5-triazine translocation in poplar trees. *Environ. Toxicol. Chem.* **18**(2): 279–284.

Topp, E., I. Scheunert, A. Attar, and F. Korte (1986) Factors affecting the uptake of ^{14}C-labeled organic chemicals by plants from soil. *Ecotoxicol. Environ. Safe.* **11**: 219–228.

Trapp, S. (2000) Modeling uptake into roots and subsequent translocation of neutral and ionisable organic compounds. *Pest Manag. Sci.* **56**: 767–778.

Trapp, S., C. Mc Farlane, and M. Matthies (1994) Model for uptake of xenobiotics into plants: validation with bromacil experiments. *Environ. Toxicol. Chem.* **13**(3): 413–422.

Walker, A. (1972). Availability of atrazine to plants in different soils. *Pestic. Sci.* **3**: 139–148.

3

MAKING PHYTOREMEDIATION A SUCCESSFUL TECHNOLOGY

N. Marmiroli and S. C. McCutcheon

SUMMARY OF PRACTICAL IMPLICATIONS

Phytoremediation is becoming a proven technique; soon, the method should be inexpensive, esthetic, and effective for large areas such as urban brownfields. This approach will require longer periods, and accumulations must be monitored until the necessary residual management techniques and plant communities are optimized for important contaminants. If the development follows the course of other biotechnologies, the commercialization of phytoremediation will advance through several stages. Already, U.S. start-up companies are accumulating patents from university and government research and are being sold for a return on investment or are doing good consulting business, whereas some have misjudged the opportunities. Europe may not have full protection of plant biotechnology discoveries to support commercialization of recent fundamental advances. In both the U.S. and Europe, excellent opportunities are in peril from lack of public investment in phytoremediation research and development. Environmental remediation has a low return on investment that requires public funding to commercialize the latest biotechnologies in maintaining the quality of life on this planet.

INTRODUCTION

Plants have the marvelous capability to survive in many hostile environments on this planet (Larcher 1995). These autotrophs survive by

1. Absorbing metals from soil and water to make stems and leaves less appealing to grazers

Phytoremediation: Transformation and Control of Contaminants,
Edited by Steven C. McCutcheon and Jerald L. Schnoor.
ISBN 0-471-39435-1 (cloth)

2. Synthesizing natural insecticides
3. Producing allelopathic compounds to prevent competition from other plants for water and nutrients
4. Stabilizing, precipitating, and binding contaminants in the rhizosphere
5. Exuding sugars and other simple food sources to spur the growth of symbiotic microorganisms in the rhizosphere
6. Fertilizing soil annually with abandoned roots containing lignin, flavonoids, and coumarins that spur microbial cometabolism of complex aromatic pollutants
7. Responding to growth hormones in the saliva of bison and other large animal grazers
8. Changing soil redox conditions to favor symbiotic microorganisms
9. Tolerating high concentrations of some contaminants and using others as nutrients in pioneering restrictive environments
10. Removing salt from water and soil to live in brackish to saline marshes (Mohr and Schopfer 1995)
11. Performing many other remarkable functions to self-engineer local environments (McCutcheon and Mitsch 1994)

These natural advantages, most of which were discovered long ago by botanists and ecologists, are now the basis of a new decontamination strategy known as phytoremediation (Cunningham and Berti 1993, U.S. DOE 1994, Raskin *et al.* 1995, Schnoor *et al.* 1995, Flathman and Lanza 1998).

Different kinds of plants contribute in different ways to phytoremediation (see McCutcheon and Schnoor this book for definitions of the types of applications and the botanical processes involved). A clear relationship exists between plant type and the advantages and disadvantages for application in any specific remediation. In applying phytoremediation, particular attention must also be given to the final disposition of the plants and other residues at the end of a phytoremediation project (Table 3-1).

NATIVE AND INDIGENOUS, CULTIVATED, AND GENETICALLY ENGINEERED PLANTS

There are advantages and disadvantages in using native or indigenous *versus* cultivated or genetically engineered plants for phytoremediation. Figure 3-1 puts many of the trade-offs into a broader context of changes in ecosystem and human risk, genetic manipulation effort, and increased residual disposal problems.

Generally, sustainable use of native or indigenous plants and microorganisms present the least human and environmental risk. Examples include monitored natural attenuation or applications based on gradual succession of native or indigenous ecosystems. Sustainable uses by definition include

TABLE 3-1 Strengths and Weaknesses of Different Kinds of Plants used in Phytoremediation Applications

Plant type	Applications[a]	Strong points[a]	Drawbacks	Final disposition
Thlaspi, *Alyssum*, and *Brassica* spp.	Phytoextraction and phytomining	Metal hyperaccumulation and metal tolerance (Brooks 1998)	Low adaptability and low biomass production; toxic metals can leach from fallen leaves, dead roots, and other plant parts. It may take more than 10 years to remediate a site (Salt *et al.* 1995)	Subjected to volume reduction by incineration or composting; disposed in hazardous waste landfills and economical recovery of metals by smelting
Grasses [*i.e.*, alfalfa (*Medicago sativa*)] and food crops	Stabilization and rhizodegradation	Rapid growth, processing of contaminants results in a modest plant contamination but the fibrous rooting, trace elements may serve as animal feed supplements (Anderson *et al.* 1993, Schnoor 1997)	Risk for accumulation of contaminants or by-products into food chain	Composting, and animal, feeding when contamination does not lead to toxicity
Atriplex and *Salicornia*	Salt extraction and volume reduction of oil-field brines	Salt tolerant plants may serve as animal feed and as a source of salt	Possibly invasive	Edible (animal feed and essential oils) or any disposal seems acceptable except on salt sensitive soils

(*continues*)

TABLE 3-1 (*continued*)

Plant type	Applications[a]	Strong points[a]	Drawbacks	Final disposition
Poplar (*Populus* spp.), willow (*Salix* spp.), and cottonwood (*Populus deltoides*)	Contaminant biotransformation, barrier for hydraulic containment, and vegetative cover for landfills and riparian buffers	Phreatophytic (takes water from aquifers), with rapid growth, deep rooting, and high transpiration rate	Long time required for biotransformation, with potential plant contamination and risks of food chain accumulation (Sandermann 1994)	Production of energy by short rotation coppice
Duckweed (*Lemma minor*) and pennywort (*Hydrocotyle* spp.)	Rhizofiltration and phytotransformation	Adapted to wetlands with rapid growth	Invasive in some settings	Collection by skimming and some feeding to animals is possible depending on contaminants
Reeds (*Phragmites* spp.), bamboo (*Bambusa* spp.), and cattails (*Typha* spp.)	Constructed wetlands	Adapted to wetlands with rapid growth	Invasive in some settings	Biomass collection for fuel, fiber, and paper production

[a]See terminology defined in McCutcheon and Schnoor (this book).

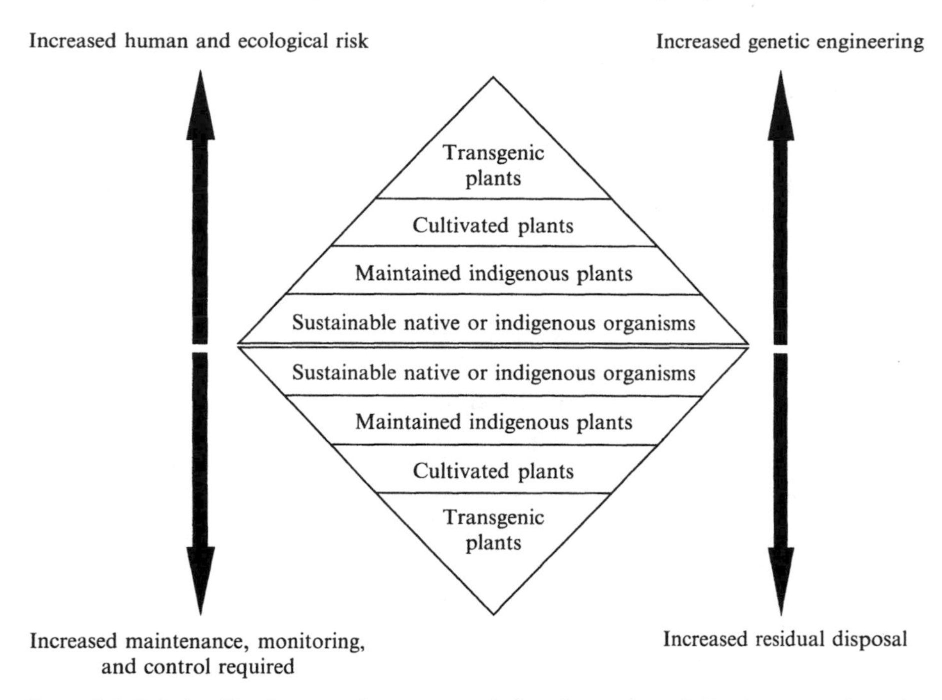

Figure 3-1 Relationships between plant types and the advantages and disadvantages of use for phytoremediation.

ecosystems that degrade or adequately disperse pollutants without intensive human intervention.

Limited maintenance of indigenous plants by selective cutting or harvest, fertilization, irrigation, pest management, or other ecological restoration and limited forestry practices tends to decrease diversity, limit habitat, and decrease ecological functions. Supplementing monitored natural attenuation by planting or some applications of phytostabilization (see McCutcheon and Schnoor this book for definitions) are good examples of these types of applications. Some contaminants such as metals may accumulate in leaves and require raking and collection, or organic contaminants temporarily may accumulate, requiring fencing and other limited controls for a finite period.

Intensive cultivation of domesticated plants involving soil preparation, fertilization, irrigation, harvest, disposal of residuals, and periodic replanting and rotations sustains typical monocultures that provide very little ecological diversity and limited habitat. Applications that typically must resort to intensive agronomic practices involve difficult-to-handle residuals, such as high metals content and accumulated organic compounds and by-products that require more control to prevent food chain poisoning. Typical forestry practices and intensive cultivation can also easily introduce invasive or disease-prone plants. In the U.S., many states restrict the importation and use of some plants. In Florida, for example, *Eucalyptus* originally introduced for dewatering and land reclamation from wetlands has invaded many more areas than originally intended and is now banned. Programs in the U.S. limit the wetland plants that may be introduced on federal lands.

Largely because of unknown competitive advantages and allergenic concerns, transgenic plants are generally thought to represent the largest human and ecosystem risks (Käppeli and Auberson 1998, European Environment Agency 1999). Concerns about accumulation of toxic or allergenic materials have only begun to be explored. More information is becoming available on the ecosystem effects of transgenic plants, such as that from 81 European Union–sponsored research projects (http://europa.eu.int/comm/research/quality-of-life/gmo/index.html).

Related to this issue is the progressively increased degree of maintenance, monitoring, and control required for maintained indigenous, cultivated, or transgenic plants. Generally, greater control is necessary to handle residuals that are more dangerous. Limited maintenance or forestry practices, intensive cultivation, and genetic engineering are being resorted to because existing native and indigenous ecosystems have been unable to assimilate all existing contamination. As a result, forestry, ecological restoration, cultivation, and genetic engineering practices are being explored to stabilize, accumulate, and transform contaminants in concert with progressive levels of cultivar development and control. Progressive control is necessary to prevent undesired exposures temporarily until contaminant intensity can be dispersed or reduced to the point that natural, sustainable degradation and geological recycling can take over.

Genetic control is expected to vary from simple herbicide development and use to the control of evolution of nonproductive weeds, to classical genetic breeding to produce highly manageable cultivars, to development of transgenic plants that treat or control contaminants that no existing organism can handle fully to achieve sustainable treatments. The range of genetic control derives from cultivation of natural plants (*e.g.*, metal hyperaccumulator plants illustrated in Table 3-1), selection, breeding, *in vitro* propagation of suitable cultivars, and genetic engineering (Cunningham and Ow 1996, Raskin 1996, Rugh *et al.* 1998, Doty *et al.* 2000). Examples of breeding include hybrid poplars (*Populus* spp.) and willows (*Salix* spp.). Examples of genetic engineering include transplantation of the *merA* and *merB* genes from a bacterium into *Arabidopsis thaliana* and yellow poplar (*Liriodendron tulipifera*) to provide mercury tolerance, demethylation, and phytovolatilization.

Many advantages offset the greater human and ecosystem risks and required control, maintenance, and monitoring of the use of managed indigenous, cultivated, and transgenic plants. Several cultivated plants grow faster, are hardier, and are more easily maintained than indigenous plants. Other manageable plants also transpire great amounts of water and thus could be tailored for specific applications such as phytohydraulic containment (defined in McCutcheon and Schnoor this book). In addition, many of the same plants, especially trees, offset remediation costs when harvested to provide short-rotation biomass (fast-growing trees harvested every three to 5 years) for fuel (Cunningham and Ow 1996), paper, fiber, wood products, resins, and other raw materials for industrial processing (Glass 1997), drug manufacturing, and animal feeding. Facultative phreatophytes, such as eastern cottonwood (*Populus deltoides*), aspen (*Populus tremula*), hybrid poplar (*Populus* spp.), willow (*Salix* spp.), salt cedar (*Tamarix* spp.), and others have a great capacity for extracting contaminated water either from the soil or from the water table (Schnoor *et al.* 1995, DiTomaso 1998). Commodity crops such as Indian mustard (*Brassica juncea*), tobacco (*Nicotiana tabacum*), flax (*Linum usitatissimum*), hemp (*Cannabis sativa*), and bamboo (*Bambusa* spp.) are used in several ways. These plants are useful in removing heavy metals from soil and water (COST Action 837 2000), and in treating animal wastes, more so than native or indigenous plants, while producing recoverable energy and biomass.

PRACTICAL ASPECTS

The application of plants for environmental remediation requires the evaluation of a number of practical issues that have been divided into what are referred to in this chapter as pre-harvest and post-harvest plans or strategies. Pre-harvest activities include the selection, design, implementation, and maintenance of phytoremediation applications. Post-harvest activities involve the disposal of plant and contaminant residues, which must also be taken into account fully during the design phase. The primary reason for distinguishing these two sets of activities is that residual disposal practices and benefical reuse have been neglected in the past and are not as advanced as the design and implementation strategies.

Pre-harvest planning, implementation, and maintenance issues include the following:

1. Site characterization
2. Plant selection
3. Pilot investigation of the fate of the contaminants in the plant system (Sandermann 1994, Salt *et al.* 1998)
4. Determining irrigation and amendment requirements
5. Estimating the capability of the roots
6. Establishing the rate of contaminant uptake and time required for remediation

Characterization normally consists of measurement or estimation of levels of contaminants and soil properties, as well as vertical and horizontal distributions throughout saturated and unsaturated soils (Lepp 1981). Plant selection is based on evaluation of growth rate, hardiness, evapotranspiration capacity, contaminant transformation and degradation capabilities, accumulation capacities, tolerance to contaminants, soil fertility requirements, soil moisture requirements, salinity tolerance, and adaptability to other biotic and abiotic stresses (Cunningham and Ow 1996). Irrigation and amendment requirements include fertilization, pH adjustment, adding organic matter, adding chelates, use of herbicides and pesticides, and making any other agronomic adjustment (Schnoor 1997, ITRC 1999, Thomas and Buck 1999). Roots intercept and transpire contaminated ground and soil water as well as depress the water table during evapotranspiration (Schnoor *et al.* 1995, Hong *et al.* 2001, Landmeyer 2001, Quinn *et al.* 2001, Eberts *et al.* this book, Hirsh *et al.* this book, Weaver *et al.* this book). The rate of uptake and time involved can be estimated from contaminant transfer rates (Salt *et al.* 1995, Boyd 1996) observed at similar sites, from pilot investigations, or from treatability studies.

Post-harvest strategies, on the other hand, are normally concerned with the following:

1. Mechanical or manual collection of plant materials and other residues
2. Risks that may exist for equipment operators and other harvest personnel
3. Fate of any remaining plant material
4. Fate of roots and underground residues
5. Disposal and transformation of harvested material, which is rarely taken into consideration

Risks for harvest personnel arise from breathing contaminated dust, accidental injuries, contact with contaminated material, and other dangers. Remaining plant material include shed leaves, branches, and flowers, which are seasonally available and where release is sometimes unpredictable due to external events (fallen contaminated parts can be dispersed off-site by wind and animals). Roots and underground residues are rarely accessible and are often difficult or expensive to eliminate from the soil but the consequences can be very deleterious to the next crop.

The disposal of more or less toxic plant residues in landfills without any recovery is the very unfortunate but necessary (especially in the U.S.) current practice for disposal of accumulating plants (Chaney *et al.* 2000). Composting for volume reduction before landfill disposal of the metal-laden residue or the more sustainable complete destruction of many organic contaminants has been proposed but not well investigated. Incineration or burning with energy recovery is sometimes possible for woody plants if the excessive release of metals is prevented. Economical recovery of material such as metals from plant tissues (Brooks *et al.* 1998) has been found feasible for biomining of nickel (Nicks and Chambers 1995). Sweden uses willow (*Salix* spp.) for wastewater assimilation and then fuel (Jordahl *et al.* this book). China uses reeds and trees for wastewater treatment and paper production in a number of ecological engineering projects. The U.S. has investigated the energy content of a number of fast-growing trees under biomass for fuels programs but these same trees used for phytoremediation have not been in use long enough to see if the final crops will be harvested for energy or fiber. Recovery of energy, wood products, fiber, food, feed, various industrial products (*e.g.*, resins), and other secondary or artifactual products have not been explored for the conceivable range of phytoremediation residues.

Among the greatest advantages of phytoremediation are the following:

1. Presumed lower costs at some sites (*e.g.*, see the section "Economics, Investment, and Intellectual Property Constraints," but note that monitoring requirements have not been fully defined for U.S. applications)
2. *In situ* nature of the treatment
3. Applicability to extensive areas of limited contamination
4. Capability to address multiple objectives, including erosion control, ecological restoration, site maintenance, and secondary production of biomass, energy, and industrial raw materials
5. Attractiveness as a green technology

The greatest limitations are as follows:

1. Phytotoxicity to many of the contaminants exists
2. *In situ* applications are relatively slow for many applications
3. Depth of treatment is limited for contaminated soils and waters
4. Contaminant mass transport in soils may limit availability and require longer treatment times (*i.e.*, several growing seasons up to decades)
5. Some regulatory policies are not consistent with the use of phytoremediation (Miller 1996, Schnoor 1997, Glass 2000a)

CONSTRAINTS

Phytoremediation at contaminated sites is constrained in a number of ways. The most important constraints are public opinion and current regulations, competition with standard cleanup methods, lack of sufficient investments in innovative treatments, and proprietary rights.

Public Opinion and Regulatory Constraints

According to the Interstate Technology and Regulatory Council (a confederation of several U.S. states that tracks innovative waste cleanup developments, ITRC 1999), stakeholders are concerned about a number of factors. Stakeholders include the public, nongovernmental organizations, and regulatory agencies, among others. These factors of concern include the following:

1. Toxicity and bioavailability of transformation products and the mobilization into groundwater or into the food chain
2. Lack of research on the fate of contaminants in plant metabolic cycles
3. Release of contaminants from animal droppings or from wood
4. Disposal of harvested material
5. Depth of the treatment zone
6. Dependence of the phytoremediation process on season and climate
7. Possible contamination of wildlife and pets

Public acceptance of phytoremediation commonly is listed as an advantage of the method, based on the belief that green technologies are more widely accepted in a community than alternatives involving unsightly excavations and heavy equipment. Nevertheless, this acceptance normally requires assessment of risks to human health and the environment (ITRC 1999), dissemination of reliable information, and education (Egri and Pinfield 1996, Collins-Jarvis 1997). One notable educational initiative is the "Citizens' Guides" describing innovative treatments, which describe advantages and problems (U.S. EPA 1998).

The introduction of transgenic plants requires attention to both public opinion and regulatory constraints. In this context, the use of transgenic plants is better than that of genetically modified microorganisms to prevent widespread dispersion in the environment (Garbisu and Alkorta 1999). Public acceptance of "environmental" biotechnology is usually higher in contrast to an unfavorable public consensus on transgenic technology in general (Beringer 1991, Glass 1997, Miller 1997, Gunter *et al.* 1999).

Given the failures of current economic systems to protect the environment of this planet, regulations are necessary. The political implementation of environmental restrictions both creates and limits the remediation market (Wrubel *et al.* 1997). Tougher rules on landfills restrict disposal to only certain types of wastes, and air and water discharges have become more controlled. These changes are causing waste generators to bear more of the environmental costs. In response, waste generators are exploring lower-cost phytoremediation and other innovations for *in situ* treatment. As a result, demonstrated remediation efficacy and cost-effectiveness are vital economic factors (Glass 1999, ITRC 1999). Cost-effectiveness, treatment efficacy, and time required for remediation are now leading factors in technology selection by site owners. Natural attenuation, which is the ultimate low-cost approach, has gained acceptance, particularly for chlorinated solvents and the more readily biodegradable contaminants such as petroleum hydrocarbons.

Effectiveness and costs are also important to regulatory agencies but only after ensuring public and ecosystem health. For example, the U.S. Environmental Protection Agency (EPA) requires, in order of preference, that the following nine criteria be used to evaluate alternatives for remediation (Flechas 2002, http://www.epa.gov/hudson/ninecriteria.htm):

1. Overall protection of human health and the environment
2. Compliance with applicable or relevant and appropriate requirements (ARARs)
3. Long-term effectiveness and permanence
4. Reduction of contaminant toxicity, mobility, or volume through treatment
5. Short-term effectiveness
6. Implementability
7. Costs
8. State acceptance
9. Community acceptance

The best regulatory systems incorporate community involvement in the actual decision-making process as in land use decisions in New Zealand, or at least as in public comment periods used in the U.S. Regulators play at least two roles in the acceptance of an innovative technology. Regulatory scientists and engineers determine the assessment methods and the requirements in the ultimate disposal of plant residues, and the staffs of regulatory agencies monitor or direct the consideration of innovative approaches at each site cleaned up.

Barriers and biases in environmental laws and regulations favor traditional technologies over innovations such as phytoremediation (Lester 1989, Timian and Connolly 1996). Conservative by practice, regulators, risk assessors, and design engineers naturally develop more confidence in standard practices. Developers of innovative technologies, working closely with regulatory offices and public interest groups to expedite evaluations of innovative treatments, normally overcome this natural bias. In countries where waste management legislation is prescriptive, legislative or judicial relief may be necessary to introduce cost-saving, protective innovations. For example, vegetative capping does not seem to meet the prescriptive infiltration requirements of the U.S. Resources Conservation and Recovery Act (Schnoor 2002).

Site remediation and control are based on at least the following three types of regulatory requirements: (1) site-specific risk assessment, (2) achievement of cleanup levels based on air, water, and soil quality standards that may or may not be based on risk, and (3) best demonstrated available technology (BDAT). Ideally, a site-specific risk assessment should govern each clean up, but these investigations can be expensive. Instead, many countries like the U.S. prescribe conservative air and water quality standards based on generic risks determined by using typical exposures to a 70 kilogram individual with standard daily consumptions of water and certain foods. Simple dose-response toxicology studies have been used as a rudimentary risk assessment in the past. Depending on the extent and quality of the data available for risk assessment and potential toxicity, safety factors of 10 to 10^6 are applied. When these conservative cleanup standards are unfavorable to innovative biotreatments, an option is to request a site-specific risk assessment to govern remedy selection and cleanup levels.

When the cleanup levels or maximum allowable concentrations of contaminants at treated sites (*i.e.*, end points) are based on BDAT, these regulations are normally biased unfavorably against innovative biotreatments, including phytoremediation. The reason is that BDAT is a traditional or standard technology (Glass 1999, COST Action 837 2000). In comparison with standard mechanical methods, many phytoremediation technologies cannot remove 100 percent of the contaminants as quickly. The rates of contaminant uptake by plants and other biotreatments tend to decrease as contaminant concentrations decrease.

Therefore, adoption of more realistic end points based on human and ecological risk assessments and future land use will undoubtedly favour better consideration of phytoremediation and other *in situ* biotreatments compared to more traditional technologies (Rock and Sayre 2000). Risk assessment (Commission of the European Communities 2000) has the following five components:

1. Identification of the agents that may have adverse effects and, therefore, present risks (Morgan 1993)
2. Quantitative determination of the severity of those possible adverse effects (Fischhoff 1997)
3. Estimation of the probability of adverse exposure to the environment and to the population (Slovic *et al.* 1979)
4. Characterization of inherent uncertainties with the precautionary approach of opting for a worst-case scenario
5. Communication of the risk to stakeholders and other citizens (Krimsky and Plough 1988, Patterson and Allen 1997, Williams *et al.* 1999)

For phytoremediation, the European Union has chosen to apply the precautionary principle, *i.e.*, “Where there are threats of serious or irreversible damage, lack of full scientific certainty shall not be used as a reason for postponing cost-effective measures to prevent environmental degradation." This precautionary principle affects phytoremediation in positive ways. For example, phytoremediation is one measure that does no further damage, is inexpensive, and at least ecologically restores a site. Risk evaluators, however, are cautious in the face of uncertainty. For certain contaminants, therefore, the cleanup level could be "as low as reasonably achievable" (ALARA), which is the precautionary approach when there are no scientific data about adverse effects to human health and the environment (Commission of the European Communities 2000). Compared to other technologies, ALARA levels may be more difficult to achieve using phytoremediation. Therefore, even risk-driven remediation has constraints for innovative biotreatments.

Competition with Other Technologies and Investments in Innovative Technologies

In the marketplace of pollution cleanup, phytoremediation must compete with existing technologies, many of which are not based on the same constraints as biotreatments. These competing standard treatments do not require pilot investigations or treatability studies as frequently as phytoremediation and other innovative biotreatments. Thus, phytoremediation suffers a disadvantage during remedy selection because of a lack of investment in generic or general pilot studies to adequately establish the limits of application and the costs associated. However, there is guidance from Glass (2000a) that provides an extensive comparison of the costs for phytoremediation of metals-contaminated soils and for the competing standard cleanup technologies. On the basis of this overview, phytoremediation may be economically useful under specific conditions, depending on the soil and climate, the level of contamination, and the chemical and physical properties of the metals involved. Similar guidance is needed for phytoremediation of a range of organic contaminants in soil, water, and air.

An important niche for phytoremediation can be defined by getting site owners, cleanup contractors, and regulators to realize that multiple approaches are necessary for most sites. If the selection process properly takes advantage of phytoremediation, the approach can avoid significant costs in cleaning up large areas of moderate contaminant levels, whereas energy-intensive methods can be used to clean up phytotoxic hot spots quickly. Low-cost phytoremediation can also be a useful restoration technique to maintain and further clean up a site for which decontamination standards were achieved but not all residues were removed. In addition, phytoremediation can be proactively applied in the green design of industrial facilities such as the new Ford Rouge River automobile factory in Detroit, Michigan (U.S.) to prevent the buildup of contaminants over long periods thereby preventing the creation of future urban brownfields.

Economics, Investment, and Intellectual Property Constraints

The total world remediation market in the year 1998 was 15×10^9 to 18×10^9 U.S. dollars. In the European Union that market in 1998 was 2×10^9 to 4×10^9 U.S. dollars; and in Canada 1×10^9 to 2×10^9 U.S. dollars. The 1999 markets for phytoremediation were estimated at 1×10^6 to 2×10^6 U.S. dollars in Canada, 2×10^6 to 5×10^6 U.S. dollars in Europe, and 1×10^6 to 2×10^6 U.S. dollars in the rest of the world. The estimates for phytoremediation shares of the total bioremediation market in the U.S. were 30×10^6 to 49×10^6 U.S. dollars in 1999, and 235×10^6 to 400×10^6 U.S. dollars in 2005. In 1999, the largest part of the market share involved the remediation of organic contaminants in groundwater (7×10^6 to 12×10^6 U.S. dollars), the control of landfill leachates (5×10^6 to 8×10^6 U.S. dollars), the remediation of organic chemicals in soil (5×10^6 to 7×10^6U.S. dollars), and the remediation of metals-contaminated soils (5×10^6 to 6×10^6 U.S. dollars). These amounts doubled in comparison to the previous year. This was primarily the result of an increase in the number of remediation companies (Glass 1999, Glass 2000a). For the estimated 400 000 contaminated sites in Western Europe, nations within the European Union may spend up to 400×10^9 Euros (352×10^9 U.S. dollars) for site remediation over the next 20 to 25 years (Glass 1999). The 30-year cleanup costs in the U.S. have been estimated as high as 1.7×10^{12} U.S. dollars (Timian and Connolly 1996).

Several preliminary cost analyses establish that phytoremediation should capture a significant portion of the market for waste management. Strand *et al.* (1995) showed that tree transpiration of trichloroethene-contaminated groundwater was not only feasible at shallow depths of about 6 meters (20 feet), but only cost 20 percent of the costs of standard pump-and-treat operations. In 1999, Phytotech (Monmouth Junction, New Jersey, U.S.) estimated the cost for phytoremediation of metals-contaminated soil and compared that to an estimated cost for conventional excavation and disposal of that same soil in a hazardous waste landfill. The estimated cost for phytoremediation was 150 000 to 250 000 U.S. dollars per hectare (60 000 to 100 000 U.S. dollars per acre) compared to an estimated cost of 1×10^6 U.S. dollars per hectare (400 000 U.S. dollars per acre) or more for excavation and disposal in a landfill.

The market projections by Glass (1999) must be viewed as approximate given the small, unstable phytoremediation market. Similar to bioremediation (Pool 2001), a number of different phytoremediation applications are in advanced stages of development, but commercialization or proof of principle in the field has not been achieved.

At least 19 companies in the U.S. (see Table 3-2 for the 15 dedicated companies), at least 1 in Canada, at least 8 in Europe (see Table 3-3 for some of the companies), and others elsewhere are involved in the development of phytoremediation (Glass 1999). These companies are

TABLE 3-2 Capabilities and Intellectual Properties of Dedicated Phytoremediation Companies in the U.S. as of 2002

Company and date of creation (URL)	Plants and pollutants	Phytotechnology	Protection of intellectual property rights	Research connections
Applied Natural Sciences, Inc., 1993 (www.treemediation.com)	Hybrid poplars and willow (Salicaceae) to treat chlorinated solvents, pesticides, other organic contaminants, nutrients, and metals	Deep planting, hydraulic control, transformation, and rhizodegradation	Utility patents for deep planting (TreeWell® and TreeMediation®) (U.S. #5 829 191 and #5 829 192)	U.S. Department of Energy, Argonne National Laboratory
Applied PhytoGenetics, Inc., 1999 (www.appliedphytogenetics.com)	Naturally occurring plants for organic contaminants, transgenic plants for elemental contaminants	Transformation, extraction, and hyperaccumulation	Patent-protected mercury phytoremediation (U.S.#5 668 294 and U.S.#5 965 796) and trade secrets	University of Georgia
Ecolotree, Inc., 1990 (www.ecolotree.com)	Hybrid poplar trees (*Populus* spp.), legumes, and grasses to treat organic contaminants, landfill leachates, agrochemical spills, contaminated soil and groundwater, brownfields, municipal and industrial wastewater, and animal feed lot drainage, and to stabilize stream riparian areas	Hydraulic control, vegetative covers, phytoirrigation, and riparian buffers	Utility patents for vegetative capping, ECap® (U.S. #5 947 041) and EBuffer® (U.S. #6 250 237) plus trade secrets	University of Iowa and Oregon State University
Ecoscience, Inc., 1978 (www.ecoscnc.com)	Ecological restoration and wetlands to treat wastewater	Stabilization, rhizodegradation, and transformation	Development transgenic plants	
Edenspace Systems Corporation, acquired Phytotech 1999, which began 1993 (www.edenspace.com)	Hyperaccumulators [Indian mustard (*Brassica juncea*) and sunflower (*Helianthus annuus*)] to extract metals, arsenic, and radionuclides	Phytoextraction, hyperaccumulation, phytostabilization, and rhizofiltration	11 patents for phytoextraction (U.S. #05 364 451), hyperaccumulation, and rhizofiltration; various applications pending; and seeds, soil amendments, hyperaccumulation inducing agents	Rutgers University

(*continues*)

TABLE 3-2 (*continued*)

Company and date of creation (URL)	Plants and pollutants	Phytotechnology	Protection of intellectual property rights	Research connections
Lemna Technologies, Inc., 1983 (www.lemnatechnologies.com)	Duckweed (*Lemna* spp.) for wastwater treatment	Plant lagoons that remove nutrients, organic wastes, and solids	Patents dating to 1985 for LemTec™	
Living Technologies, Inc., ca. 1992 (www.living machines.com)	Aquatic plants for wastewater treatment in reactors	Stimulated bioremediation, transformation, and rhizofiltration	Unique bioreactor design, Living Machine™	
Phytokinetics, Inc., 1994 (www.phytokinetics.com)	Trees and grasses to treat organic contaminants, and nutrients	Rhizodegradation, hydraulic control, and transformation	Trade secrets	Utah State and other universities
Planteco, 2000 (www.planteco.com)	Trees, grasses, and aquatic plants to treat soils and groundwater contaminated with chlorinated solvents, perchlorate, and hydrocarbons	Hydraulic control, transformation, and rhizodegradation	Trade secrets and unique microbial mat bioreactors	University of Gengia
Sustainable Strategies, 1973 (www.ecological-engineering.com)	Aquatic plants in greenhouses or water gardens for treatment of domestic, industrial, or animal wastewaters	Transformation, rhizolfdegradion, and rhizofiltration	Patented Solar Aquatics System® and wastewater garden, washwater garden, and Bamboo Forest Manure Management System	Tufts University and University of Toronto
Thomas Consultants, Inc., 1989 (www.thomasconsultants.com)	Hybrid poplars (*Populus* spp.) to treat metals, organic contaminants, and nutrients	Nutrient uptake and stabilization	Trade secrets	
TreeTec Environmental Corp., 1995 (www.treetec.com)	Salicaceae for soil, water, and air decontamination	Phytoirrigation and carbon dioxide removal from air	Selection of varieties, proprietary Sagitta trees, and extensive inventory of hybrids	Washington State University

Verdant Technologies, Inc., 1996 (www.verdanttech.com)	Trees to treat chlorinated solvents and organic contaminants	Transformation	Patents pending and trade secrets	University of Washington, University of South Carolina, and U.S. Department of Energy Savannah River Ecology Laboratory
Viridian Environmental, LLC, 1998	Hyperaccumulators to remove metals	Phytomining and stabilization	Utility patent by assignment	U.S. Department of Agriculture, University of Maryland, and University of Sheffield
Wolverton Environmental Services, Inc., ca. 1992 (www.wolvertonenvironmental.com)	Reeds and aquatic plants for wastewater treatment and indoor plants for air purification	Transformation, rhizodegradation, and rhizofiltration	Patented Micro-Agro® process and other patents	U.S. National Aeronautics and Space Administration

Note: Data obtained from Glass (1999, 2000b), and company web sites.

TABLE 3-3 Dedicated Phytoremediation Companies in Canada and Europe

Company and date of startup (URL)	Plants	Pollutants and waste	Phytotechnology
Canada			
Abydoz Environmental Inc., 1997 (www.abydoz.com)	Reeds (*Phragmites* spp.) and other wetland plants	Domestic and industial wastewater	Patented wetlands: Kickuth BioReactor™ and PhytoKlare™
Europe			
BioPlanta GmbH, Germany, 1991 (www.bioplanta-leipzig.de)	Reeds (*Phragmites* spp.) and other plants	Wastewater, sludge, phenolic coal wastes, animal waste, and mine wastes	Wetland treatment, vegetative capping, revegetation, and rhizodegradation
Consulagri S.r.l., Italy (consugri@tin.it)	Not specified	Metals	Extraction
Eco-Pest S.L., Spain (www.ecopest-sl.com)	Aquatic plants	Domestic and industrial wastewaters	Extraction and transformation in bioreactors with water-immersed roots (Living Machine™)
Körte-Organica Ecotechnologies, Hungary, 1989 and 1998 (www.korteorganica.hu)	Aquatic plants	Industrial wastewater and groundwater	Extraction and transformation in bioreactors with water-immersed roots (Living Machine™)
Oceans ESU, U.K., 1991 (www.oceans-esu.com)	Reed (*Phragmites* spp.) beds	Industrial, agricultural, and domestic wastewater and runoff	Wetland treatment

Note: Data obtained from Glass (1999, 2001b), and company web sites.

1. Dedicated solely to phytoremediation development
2. Diversified specialty companies with experience in other biotechnology areas, including mostly plant breeding for micropropagation (*i.e.*, plant nurseries) and for developing agricultural biotechnology
3. Diversified consulting companies, mainly engineering companies
4. Industrial companies, such as oil or chemical companies primarily developing treatments for oil spills
5. Companies involved in more generic activities, such as producers and suppliers of nutrients, amendments, instrumentation, and other equipment

In addition, a few research institutions are involved in the development of phytoremediation. For phytoremediation, however, a clear trend does not seem to have evolved of universities, research institutions, and government agencies learning to collaborate with large companies to conduct phytoremediation research and develop some innovations (NRC 1997). One exception is the development of rhizodegradation of petroleum hydrocarbons through the Remedial Technology Demonstration Forum (Rock this book).

If the phytoremediation marketplace develops as the market has for other innovative remediation industries, small, innovative companies dedicated to providing only phytoremediation services will be dominant initially, but large, diversified companies will eventually take over the market (OECD 1996, Watanabe 1997). Bioremediation is following this track. The last two decades of research and development has resulted in a bioremediation market now dominated by diversified engineering consulting firms, and not by dedicated firms.

Venture capital is typically the source of funding for entrepreneurial, start-up companies. However, private capitalization requires that a technology start-up have protected intellectual property rights in the form of patents and trade secrets. With private venture capital and the right intellectual property, high-tech development has rapidly occurred globally in areas such as biotechnology, telecommunication, and semiconductors in the last three to four decades. The investment risk for an innovation or product, however, is generally inversely proportional to the stage of development. Newer, less proven ventures must offer a very high potential return. In comparison with other possible investments, the importance and magnitude of the problem addressed, the proposed solution, and the team intellectual resources available to manage and achieve the development must be evaluated. Normally the goal is to recover the invested capital through a stock exchange or other public offerings. For biotechnology in general, very large agro-businesses and other types of companies have been buying small start-up companies since 1990, primarily for the intellectual property rights (NRC 1997).

Lacking a sustainable, proprietary basis or marketing alliances, some dedicated U.S. phytoremediation firms have already lost market share in favor of engineering consulting firms that can design and implement projects with access to the right plants through nurseries and appropriate agricultural biotechnology institutions.

The strategies used by small, dedicated companies in the U.S. to develop and sustain a market share involving phytoremediation seem to include combinations of the following: (1) lease from university and U.S. government biotechnology laboratories or develop utility patents (those based on the U.S. 1790 Patent Law as expanded in the 1980s by the courts and the U.S. Patent and Trademark Office to plants and plant processes) on processes and improvements involving plants, and (2) rely on trade secrets and offerings of experience with certain approaches, including soil amendments. Plant patents (established in the U.S. in 1930) and plant variety certifications (from U.S. Department of Agriculture as of 1970 and as amended in 1994) do not seem to be used directly in phytoremediation to develop market shares. Current European legislation does not permit the patenting of plant varieties or the biological and breeding methods used to produce new plants (Directive 98/44/EC, European Commission 1998). The inability to obtain patents involving new plants could be a hurdle for commercialization in Europe if methods patents for using plants in remediation are not sufficient.

Global commercialization opportunities should also be kept in mind despite the differences in patent opportunities. Canada is considering certifications of biotreatments that would allow products to indicate that any residues can be treated with phytoremediation and other treatments under certain circumstances (McIntyre this book). Adherence to the standards of the International Organization for Standardization (ISO; http://www.iso.ch) will also be of importance in globalization. For example, the ISO 14001 standard is the "environmental policy" which must be defined by the top managers of a company. The environmental policy must include commitment to compliance with environmental laws and company policies as well as commitment to continuous improvement in pollution prevention and control. This commitment involves planning, implementation and operations, checking and corrective action, and management review.

Some dedicated U.S. phytoremediation firms do maintain vital research connections with the university and government institutes where critical intellectual property was discovered (*e.g.*, Ecolotree, Viridian Environmental, and Phytokinetics). A few patents such as deep planting in borehole casings by Applied Natural Sciences (TreeWell®) have arisen from ongoing market development. A few U.S. firms have obtained private capital, U.S. Small Business Administration research grants, or support to work with U.S. national laboratories (*e.g.*, Phytotech, Phytoworks, Phytokinetics, and Applied Natural Sciences). In most cases known to these authors, the limited research support that these entrepreneurs were able to secure was shared with the originating research institutes or important new research alliances were formed. Unfortunately, one U.S. company known to have sought venture capital has gone out of business.

Given the general trends in biotechnology, developments in the allied bioremediation field, and difficulty in attracting private capital and government research support (see McCutcheon and Rock 2001), the long-term sustainability of dedicated U.S. firms is in doubt. Experience has shown that the small profit margins of the environmental cleanup market have not supported advanced research. This is consistent with the current state of dedicated U.S. phytoremediation companies now that Phytotech has reorganized. With some exceptions, dedicated U.S. companies seem to be developing phytoremediation approaches on a site-by-site basis, as applications seem profitable to site owners and cleanup managers. This notoriously short-sighted approach bodes ill for a rational, rapid development of phytoremediation and may be the reason that bioremediation required over two decades to achieve consistent, practical applications (*e.g.*, natural attenuation).

The necessary intellectual property rights seem to arise from two activities. Some patentable approaches or trade secrets derive from experience and practice in the field, whereas most patents seem to arise from research in land grant universities with fertile biotechnology institutes that license discoveries to dedicated companies. A few patents have been issued to U.S. government laboratories, at least one of which has been licensed (Wolfe *et al.* 1998, Chaney *et al.* 1999). These licensed or assigned patents also come with constraints. The original research team may be involved but usually lacks adequate business experience for commercialization. Additionally, research organizations often overvalue the original discoveries and undervalue development costs in a way that inhibits raising private development capital. The interest of the U.S. government (also a major client in cleanups) in many patents supported by federal research funding remains unclear to many, despite the U.S. Bayh–Dole Patent Policy Act of 1980 and the Stevenson–Wydler Technology Innovation Act of 1980 as amended (NRC 1997). All of these factors contribute to lags in development.

A major hurdle to the commercialization of phytoremediation and many biotechnologies is the fundamental tension between open research and protected development. The conflict is that between the development of a consensus on engineering practice to facilitate selection and application of phytoremediation *versus* the need to have patents and trade secrets for dedicated companies to survive. In particular, aggressive marketing by companies engaged in phytoremediation development may be contrary to having well-defined technology limitations that can be used by site owners and their consulting engineers to select phytoremediation as a cleanup option to be evaluated with other candidate remedies. Attempted applications outside the poorly defined realm of established feasibility should be properly communicated to site owners and regulators. This will avoid perceived failures to achieve a field proof of principle. In addition, a heuristic or experience-related niche based on trade secrets that some dedicated and consulting companies have adopted can be counterproductive to the development of ecological engineering design guidance. The fact that dedicated U.S. companies have failed indicates that some mistakes have already been made in assessing the early commercialization potential of the proprietary and heuristic information that has flowed from science discoveries to date. A related factor is that universities and government laboratories are not skilled generally at developing and licensing proprietary inventions in phytoremediation that companies can develop and commercialize easily. Technology certification programs, in misjudging the state-of-the-science, also have made mistakes (U.S. AEC 1999a, 1999b) that could be affecting commercialization.

At least one U.S. company received private capitalization. This company was sold. Further, those companies that seem to rely on patents to establish a market share do not seem to be doing business on the basis of fees to apply the technologies; rather, most dedicated companies seem to be making money from payments for the time taken to design and implement site-specific applications. Yet, the market may not be saturated with expert applicators, as Glass (1999, 2000a) notes that the phytoremediation market doubled between 1998 and 1999 due to the entry of new firms. Billing for time and using a few patents to control competition for some applications implies that the commercial potential of phytoremediation is far from being realized, and may not be, if the inadequate capital returns of the environmental cleanup market are to be relied upon. Therefore, it seems that both fundamental research support to national laboratories and universities is vital, along with the mastery and funding of new technology transfer efforts. Without these investments, the pioneering firms and those to follow may not be successful in commercialization of sustainable low-cost phytoremediation applications.

The retrenchment of well-conceived, well-operated companies such as Phytotech (see Table 3-2) not only implies that assigned patents and private capitalization are not fully adequate, but that the state-of-the-science may not be sufficient. This company had a very strong connection to Rutgers University in New Jersey (U.S.) where the original patents were issued, and up to 1.3×10^6 U.S. dollars were directed by the company. Despite the apparent strong science connection and more research and development than most dedicated firms, the science basis of phytoextraction is still not established well enough to achieve proof of principle in the field for lead extraction (Rock this book). (Lead is the most prevalent metal contaminant in the U.S.)

IMPROVING PHYTOREMEDIATION APPLICATIONS

Expanding the future application of phytoremediation is dependent on research (Ow 1996, Raskin 1996, Salt *et al.* 1998, Pletsch *et al.* 1999) and development. Both research and development must be focused on both pre- and post-harvest improvements.

Forecast of Needed Pre-Harvest Improvements

Developments since the early 1990s have increased knowledge of the pre-harvest process but, as noted in Figure 3-2, much remains to be done to expand

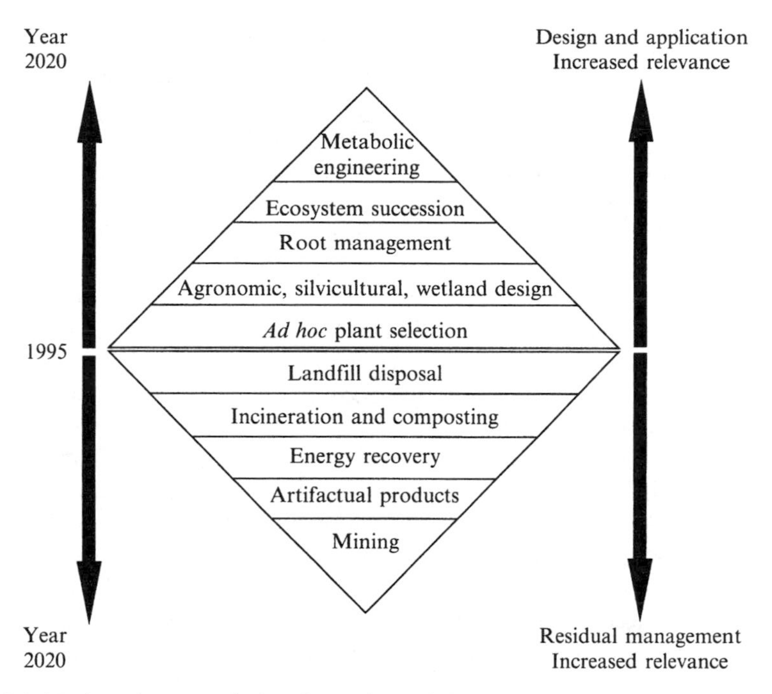

Figure 3-2 Moving phytoremediation forward: needed improvements in pre-harvest and post-harvest strategies.

the applicability of phytoremediation. Some of the potential steps include (Cunningham and Ow 1996, Raskin 1996, InterCOST 2000) the following:

1. Select and optimize candidate plants by genetic breeding and other approaches
2. Optimize agronomic and silvicultural uses of fertilizers, irrigation, pest control, and other practices
3. Optimize the rhizosphere environment using root selection and training with amendments, and inoculation with selected bacteria and fungi
4. Master ecosystem succession and the self-engineering of plant communities to change local biogeochemical conditions to promote growth and survival
5. Determine where existing genetic diversity is inadequate to assimilate wastes and use metabolic or genetic engineering to create cost-effective safe transgenic plants

In addition, detailed mass balances and pathway analyses are necessary for each element and compound subject to phytoremediation to quantify risk reduction during the process. Interim toxicity and bioaccumulation testing and monitoring (ITRC 1999) must be performed during applications until the pathways are fully mapped and dose–response relationships defined for each significant product for the range of ambient conditions.

Since the mid 1990s, the practice has improved upon the *ad hoc* selection of natural plants (those not domesticated for agriculture or silviculture) for applications. Reeves (1995) developed field kits to identify natural metal hyperaccumulator plants that might be useful for selecting indigenous hyperaccumulator plants for applications. Best *et al.* (1997) reported enzyme-linked immuno-specific assays (ELISA) for nitroreductase activity of native and indigenous plants conducted in the field at the Iowa Army Ammunition Plant. Medina *et al.* (2000) and McCutcheon *et al.* (this book) established simpler plant testing and selection protocols for wetland plants to treat explosives. The most important advance is that a certification protocol has been established for Canada, along with the introduction of two comprehensive databases for the selection of plants for metals accumulation and rhizodegradation of petroleum hydrocarbons (McIntyre this book).

Nevertheless, much remains to be done to advance the art of plant selection. First, additional collecting (Baker *et al.* 2000) and cataloging of useful plants is very necessary for all applications of phytoremediation. The full cataloging and archiving of seed of native, hyperaccumulating plants near mineral outcrops and from other areas is especially crucial. Reeves (1995) and Baker (1995) highlight the danger of extinction to these vital genetic resources due to surface mining (especially in Africa near copper deposits) and urban development. In addition, definition of a plant ideotype (the phenotype or genotype that a breeder signifies as "ideal," see Medina *et al.* this book), mass selection (recurring cycles of selection and cross-pollination), and *in vivo* and *in vitro* clonal selection for specific phytoremediation applications are needed.

In addition to some progress in plant selection, the last several years have seen steady advances in applications of agronomic practices and amendments (chelators) to manage crops of Indian mustard (*Brassica juncea*) and sunflower (*Helianthus annuus*) to extract lead from soil (Blaylock 2000). Banks *et al.* (1998) and Hutchinson *et al.* (this book) illustrate how agronomic practices for growing grass have been adopted for the rhizodegradation of oil-contaminated soils. Jordahl *et al.* (this book) shows that good silviculture practices have been adopted for growing trees to control groundwater contamination and landfill leachates. McCutcheon *et al.* (this book) and Thompson *et al.* (this book) demonstrate that some wetland design approaches have proven useful for treatment of explosives-contaminated waters. See Horne (2000) and COST Action 837 (2001a) for a review of wetland treatment techniques for other wastes.

The next important steps in applying agronomic and silvicultural techniques involve proteomic and genomic explorations of phytoremediation plants, recently approached by COST Action 837 (2001b). Understanding the enzymology of plants from the proteins produced and genes expressing those enzymes will allow a better understanding of basic plant functions and optimization of phytoremediation crops and tree plantations. A number of techniques are available, and new techniques are being created at a rapid pace in biotechnology and health research. Characterization of the functions of proteins (*e.g.*, cytochrome P-450, peroxidases, and laccases) and some smaller molecules (*e.g.*, lignin and cellulose) involved in xenobiotic chemical transformation, transfer, and conjugation (*i.e.*, tolerance and detoxification) already can be accomplished by classical isolation methods or reverse genetics, and from genetic transformation and phenotypic effects of mutations in protective functions. Genes that express the various enzymes can be identified by classical methods, complementation studies, map-based cloning, molecular-marker-assisted selection, differential display reverse transcriptase polymerase chain reaction, microarrays, chromosome-based cloning, serial analysis of gene expression, and use of genetically modified organisms. The genomes of Brassicaceae (*Arabidopsis thaliana*), rice (*Dryza sativa*), legumes (*Pisum* spp.), and loblolly pine (*Pinus taeda*) (The *Arabidopsis* Genome Initiative 2000) and synteny studies (the search in different species for conserved chromosome or subchromosome fragments carrying similar genes in the same order) will be vital to relating plant responses during phytoremediation to gene regulation and expression (Sandermann 1994) that produce the enzymes involved. Cell and tissue cultures, somaclonal variation, and recombinant dioxyribonucleic acid (DNA) methods also will contribute in identifying *in vitro* strategies for improving crops and other plants for phytoremediation (Raskin 1996).

If the full promise of phytoremediation is to be realized, then at least three more design strategies must be explored with intensive fundamental and applied research. The first strategy involves root management. Early on, J.L. Schnoor realized that augering 2 meters (8 feet) into the soil and planting poles or 3-year-old hybrid poplar (*Populus* spp.) trees accelerated root growth over depths up to 4 meters (12 feet). See Hong *et al.* (2001) for the application of this technique. Landmeyer (2001) and Ferro *et al.* (2001) have further explored the natural rooting of hybrid poplar (*Populus* spp.) in humid and arid climates, respectively. Hutchinson *et al.* (this book) reviews the root ecology of grasses and other plants. The latest innovation is the deep planting of trees down to 12 to 15 meters (40 to 50 feet) deep, with and without borehole casings (Quinn *et al.* 2001; Negri, Gatliff *et al.* this book). Nevertheless, many more options, including root selection and training, must be explored to access and treat all contaminated areas in typical soils. Root selection may involve optimization of root and microorganisms interactions; increases in root surface, such as for hairy roots involving *Agrobacterium rhizogenes*; and root inoculations for various purposes, including production of natural surfactants and defense against toxicity. Root growth into all areas of contaminated soils can be enhanced by placement of subsurface irrigation and nutrient release systems, subsurface bedding of compost and other nutrient rich mixtures, chemical training, and use of physical barriers such as deep planting in a borehole casing or polyethylene sleeve (Quinn *et al.* 2001).

In the second strategy, the greater self-engineering capabilities of plant communities and ecosystems must be brought to bear in contaminated soil treatment so that a succession of pioneer species to climax plant species will change soil properties over time. For example, better ecological management and restoration techniques are already known to provide favorable reducing conditions to spur microbial dechlorination (Eberts *et al.* this book). Another example is root inoculation with ecto- or endo-mycorrhizae to increase both uptake and tolerance of toxic contaminants (Leyval *et al.* 1997). Changing plant species dominance will alter oxygen pumping to the soil (0 to 30 centimeters deep), transpiration, exudation, root turnover, organic matter and soil genesis, nutrients, redox conditions, and the microbial community over time to facilitate anaerobic–aerobic microbial cycles to degrade some persistent xenobiotic contaminants and stabilize others. Plant succession also provides diverse enzymatic transformation pathways not available to monocultures. Communities provide more genetic and phenotypic diversity for tolerance and transformation than a single hybrid clone (Medina *et al.* this book).

The need to involve the methods of paleoecology to certify the usefulness of slower phytoremediation processes is of particular importance (McCutcheon and Rock 2001, Fletcher 2002, Olson *et al.* this book). This includes the testing of rhizo and phytodegradation of petroleum hydrocarbons and polychlorinated biphenyls by looking back at treatments that have occurred spontaneously with revegetation of oil-laden sludge pits and other decommissioned disposal areas at refineries and chemical plants.

The third new strategy recognizes that plants generally transform organic pollutants, and that the genetic diversity of natural and genetically bred plants is finite. This limited genetic diversity is not expected to facilitate fully sustainable phytoremediation of the universe of conceivable xenobiotic chemicals. Some of these compounds [*e.g.*, DDT or 1,1,1-trichloro-2,2-*bis*(*p*-chlorophenyl) ethane and polychlorinated biphenyls] seem to lack natural analogs and thus tend to persist. Because microbial metabolism generally results in mineralization, transgenic engineering will be necessary to combine the autotrophic energy independence of plants with the heterotrophic enzymatic processes needed for complete mineralization. Thus, the newest developments from plant metabolic and genetic engineering (Cunningham and Ow 1996, Raskin 1996, Rugh *et al.* 1998) are expected to be very necessary to achieve not only more sustainable phytodegradation but also more sustainable treatments of organic chemicals, metals, and contaminant mixtures. Morikawa *et al.* (1998) and Morikawa *et al.* (this book) provides an excellent justification for developing transgenic plants to remove nitrous oxides from the atmosphere of polluted cities. *Eucalyptus*, *Tillandsia*, *Solidago*, and *Magnolia* species are being explored for this purpose.

Developments in molecular biology, classical plant breeding, and plant genetic engineering will therefore be necessary to increase the applicability of phytoremediation (Ow 1996, Pletsch *et al.* 1999) and to achieve sustainable treatments. Treatments must be extended to the use of halophytes for arid and coastal remediation where high salt levels occur in water, sediment, and soil (*e.g.*, Negri, Hinchman *et al.* this book). Both protein and gene selection, as described earlier for characterization and for extension and optimization of agronomic practices, will be necessary.

Forecast of Needed Post-Harvest Improvements

Unfortunately, post-harvest strategies have not developed consistently with pre-harvest approaches (Miller 1996, Schnoor 1997, ITRC 1999, COST Action 837 2000). Much work is necessary to forecast and certify when plants have detoxified contaminants removed from soil, water, or air such that acceptable residual risks to humans and the environment are attained. Until sustainable phytoremediation is developed for all contaminants of concern, residual management will remain an important constraint and design issue. Available management options in order of relevance from Figure 3-2 are as follows:

1. Landfill disposal of contaminated plant residues
2. Incineration to destroy organic transformation products or reduce metals-laden biomass and landfill volume if the metals are not released to the air in significant quantities
3. Composting to degrade xenobiotic residues and to reduce the volume before landfill disposal
4. Energy recovery from fast-growing woody plants
5. Production of fiber, paper, wood products, lumber, animal feed, industrial raw materials (*e.g.*, resins), materials for drug manufacture, and food (where the contaminants are excess plant nutrients) and other artifactual products
6. Smelting for economical recovery of rare metals

COMPANY STRATEGIES FOR THE FUTURE

Phytoremediation is a niche treatment with limited applicability and limited resource requirements. Companies involved in phytoremediation should benefit from strategies, approaches, and experience, which include the following:

1. Placing more focus on niche markets and only accepting projects specifically suited for phytoremediation (*e.g.*, large areas or fields with relatively low levels of contamination, such as brownfields, contaminated groundwater or surface water, and wastewater)
2. Pursuing increased partnerships with engineering consulting firms

3. Increasing diversification into consistent standard technologies (*e.g.*, chemical treatment, soil washing, thermal treatment, landfilling, incineration, and electrokinetics) to provide treatment packages
4. Actively pursuing mergers and acquisitions

In this stage of development public financial support for research and development is needed, especially for conducting expensive, long-term research that cannot be supported by operating profits alone.

SUMMARY AND CONCLUSIONS

When phytoremediation becomes a proven remediation technology, significant cost savings, better community relations, and esthetic ecologically based restorations are expected for large areas of moderate contamination. The ideal niche will probably be use of phytoremediation in concert with other more expensive treatments of contaminated hot spots. As such, phytoremediation should be ideal for restoration of the thousands of brownfields (abandoned industrial and commercial facilities usually mildly contaminated with mixtures of chemicals) found in Europe and the U.S. However, the applications of phytoremediation are anticipated to require substantially more time for treatment than do conventional cleanup technologies. To establish phytoremediation as a reliable cleanup method, documented accurate definition of the range of applicability and potential profit margins for various applications are necessary. Furthermore, until plants are genetically modified (conventionally or through recombinant DNA) to completely degrade organic contaminants and post-harvest methods developed to recycle elemental contaminants such as metals and some transformation products, phytoremediation applications will not be as sustainable as possible. Interim applications will continue to require laboratory and field pilot testing to define the management intervention necessary for a required level of performance. Monitoring and testing for toxicity and bioaccumulation of transformation products and the pollutants involved will continue to be necessary until all significant pathways are defined.

The high potential for technology improvement from genetics and agronomy is an important factor in moving phytoremediation into the forefront of remediation technology. Recent achievements in plant genomic and proteomic research significantly enrich the potential of phytoremediation. This intellectual enrichment is comparable to that which occurred in solid-state technology several decades ago. Unfortunately, enrichment by advances in biotechnology is not adequate to ensure that the cost savings of phytoremediation can be realized. Fundamental plant research in the U.S. and Europe is not being supported to the degree necessary. Many questions remain regarding technology transfer and investments in development.

The present annual market in phytoremediation is on the order of 10^7 U.S. dollars compared to 10^9 U.S. dollars for other biological technologies, (*e.g.*, pharmaceuticals), or to the order of 10^{12} U.S. dollars for the great successes of solid-state electronics (*i.e.*, mobile phones and microprocessors). Further, the environmental remediation market has a notoriously low return on investment. There are indications that fundamental phytoremediation research is woefully underfunded, especially in the U.S. Experience with phytoremediation and other biotreatments already demonstrates that additional government investment in development is necessary, given the extensive cleanup requirements at government facilities worldwide. In addition, the future of the pioneering dedicated phytoremediation companies is not clear. Certainly, these companies need continual research findings from universities and government institutes for the long-term, and better development support in the short-term. Better intellectual property transfer and protection, which does not hamper the open research climate that sustains phytoremediation, is also necessary (NRC 1997). As phytoremediation matures, all companies and organizations involved must be ready for market restructuring and other changes that seem inevitable if biotechnology developments are a useful guide. With these critical investments, phytoremediation promises to be an integral and sustainable waste management option for the coming century.

Acknowledgments

The authors are indebted to all the participants of COST Action 837 and the Chair, Dr. Jean-Paul Schwitzguébel, Swiss Federal Institute of Technology, Laboratory for Environmental Biotechnology, Lausanne, Switzerland, for a ready source of scientific information and inspiration. For his invaluable discussions, Nelson Marmiroli is also indebted to Dr. Stanislaw W. Gawronski, Warsaw Agricultural University, Department of Basic Natural Sciences in Horticulture, Warsaw, Poland. David Glass, who also provided additional information on dedicated companies, Paul Flathman, Edward Mead, and Robert Swank reviewed this chapter. Jerald L. Schnoor evaluated the reviews and accepted the chapter for this book. Steve McCutcheon is pleased to acknowledge financial support from the U.S. Strategic Environmental Research and Development Program, and the U.S. Environmental Protection Agency Office of Research and Development Contaminated Sites Program. This paper has been reviewed in accordance with the U.S. Environmental Protection Agency peer and administrative review policies and approved for publication. Mention of trade names or commercial products does not constitute endorsement or recommendation for use.

REFERENCES

Anderson, T.A., E.A. Guthrie, and B.T. Walton (1993) Bioremediation in the rhizosphere. *Environ. Sci. Technol.* **27**: 2630–2636.

Baker, A.J.M. (1995) Metal hyperaccumulation by plants: our present knowledge of the ecophysiological phenomena. In: *Proceedings/Abstracts of the Fourteenth*

Annual Symposium, Current Topics in Plant Biochemistry, Physiology and Molecular Biology – Will Plants Have a Role in Bioremediation? Interdisciplinary Plant Group, University of Missouri, Columbia, Missouri, April 19–22, pp. 7–8.

Baker, A.J.M., S.P. McGrath, R.D. Reeves, and J.A.C. Smith (2000) Metal hyperaccumulator plants: a review of the ecology and physiology of a biological resource for phytoremediation of metal-polluted soils. In: *Phytoremediation of Contaminated Soil and Water*. N. Terry and G. Bañuelos, eds. Lewis Publishers, Boca Raton, Florida, pp. 85–107.

Banks, M.K., A.P. Schwab, R.S. Govindaraju, P. Kulakow, S.L. Lewis, and K.I. Rathbone (1998) Craney Island phytoremediation project. Final Report. Rice University, Advanced Applied Technology Demonstration Facility for Environmental Technology, Houston, Texas.

Beringer, J.E. (1991) The release of genetically modified organisms. In: *Innovation and Environmental Risk*. L. Roberts and A. Weale, eds. Belhaven Press, New York, pp. 57–63.

Best, E.P.H., M.E. Zappi, H.L. Fredrickson, S.L. Sprecher, and J.L. Miller (1997) Screening of aquatic and wetland plant species for phytoremediation of explosives-contaminated groundwater from the Iowa Army Ammunition Plant. Technical Report EL-97-2, U.S. Army Engineer Waterways Experiment Station, Vicksburg, Mississippi.

Blaylock, M.J. (2000) Field demonstration of phytoremediation of lead-contaminated soils. In: *Phytoremediation of Contaminated Soil and Water*. N. Terry and G. Bañuelos, eds. Lewis Publishers, Boca Raton, Florida, pp. 1–12.

Boyd, V. (1996) Pint-sized plants pack a punch in fight against heavy metals. *Environ. Protect*. **7**(5): 38–39.

Brooks, R.R., ed. (1998) *Plants that Hyperaccumulate Heavy Metals*. CAB International, Wallingford, England.

Brooks, R.R., M.F. Chambers, L.J. Nicks, and B.H. Robinson (1998) Phytomining. *Trends Plant Sci*. **3**: 359–362.

Chaney, R.L., J.S. Angle, A.J.M. Baker, and Y.-M. Li (1999) Method for phytomining of nickel, cobalt and other metals from soil. Patent 5 944 872. U.S. Patent and Trademark Office, Washington, D.C. (http://www.uspto.gov).

Chaney, R.L., Y.-M. Li, S.L. Brown, F.A. Homer, M. Malik, J.S. Angle, A.J.M. Baker, R.D. Reeves, and M. Chin (2000) Improving metal hyperaccumulator wild plants to develop commercial phytoextraction systems: approaches and progress. In: *Phytoremediation of Contaminated Soil and Water*. N. Terry and G. Bañuelos, eds. Lewis Publishers, Boca Raton, Florida, pp. 129–158.

Collins-Jarvis, L. (1997) Participation and consensus in collective action organizations: the influence of interpersonal *versus* mass-mediated channels. *J. Appl. Commun. Res*. **25**: 1–16.

Commission of the European Communities (2000) Communication from the Commission on the precautionary principle. COM(2000) 1 (http://europa.eu.int/comm/dgs/health_consumer/library/pub/pub07_en.pdf).

COST Action 837 (2000) Phytoremediation. state of the art in Europe (an intercontinental comparison). Proceedings of the First Scientific Workshop, held in Hersonissos, Crete, Greece, April 6–8 (http://lbewww.epfl.ch/COST837/).

COST Action 837 (2001a) Wastewater treatment and plants as a "green liver:" the European approach, experience and trends held in Larnaca, Cyprus, May 18–19 (http://lbewww.epfl.ch/COST837/).

COST Action 837 (2001b) Genomics, proteomics and other -omics approaches within COST Action 837. held in Montpellier, France, Nov. 30–Dec. 1 (http://lbewww.epfl.ch/COST837/).

Cunningham, S.D. and W. R Berti (1993) Remediation of contaminated soil with green plants: an overview. *In Vitro Cell. Dev. Biol.* **29P**: 207–212.

Cunningham, S.D. and D.W. Ow (1996) Promises and prospects of phytoremediation. *Plant Physiol.* **110**: 715–719.

DiTomaso, J.M. (1998) Impact, biology, and ecology of saltcedar (*Tamarix* spp.) in the southwestern United States. *Weed Technol.* **12**: 326–336.

Doty, S.L., T.Q. Shang, A.M. Wilson, J. Tangen, A.D. Westergreen, L.A. Newman, S.E. Strand, and M.P. Gordon (2000) Enhanced metabolism of halogenated hydrocarbons in transgenic plants containing mammalian cytochrome P-450 2E1. *Proc. Nat. Acad. Sci. USA* **97**(12): 6287–6291.

Egri, C.P. and L. Pinfield (1996) Organizations and the biosphere: ecology and environments. In: *Handbook of Organization Studies*. S.R. Clegg, C. Hardy, and W.R. Nord, eds. Sage Publications, Thousand Oaks, California, pp. 459–483.

European Commission (1998) Directive 98/44/EC of the European Parliament and of the Council of 6 July 1998 on the legal protection of biotechnological inventions. *Official J.* **L 213**: 13–21.

European Environment Agency (1999) Genetically modified organisms, Chapter 3.9. Environment in the European Union at the turn of the century. Environmental Assessment Report No.2, GH-18-98-784-EN-C, European Environment Agency, Copenhagen, Denmark (http://www.eea.eu.int/92-9157-202-0/en).

Ferro, A., J. Chard, R. Kjelgren, B. Chard, D. Turner, and T. Montague (2001) Groundwater capture using hybrid poplar trees: evaluation of a system in Ogden, Utah. *Int. J. Phytorem.* **3**(1): 87–104.

Fischhoff, B. (1997) Ranking risks. In: *Environment, Ethics, and Behavior: the Psychology of Environmental Valuation and Degradation*. M.H. Bazerman, D.M. Messick, A.E. Tenbrunsel, and K.A. Wade-Benzoni, eds. New Lexington Press, San Francisco, California, pp. 342–371.

Flathman, P.E. and G.R. Lanza (1998) Phytoremediation: current views on an emerging green technology. *J. Soil Contam.* **7**: 415–432.

Flechas, F.W. (2002) Case study: Union Pacific Railroad-regulatory considerations for implementation of phytoremediation. In: *Proceedings of the EPA Phytoremediation State of the Science Conference*, U.S. Environmental Protection Agency Report EPA 625R-01/011a and b, Cincinnati, Ohio, held in Boston, Massachusetts, May 1–2, 2000 (http://www.epa.gov/ORD/NRMRL/Pubs/625R01011b/625R01011bchap31.pdf).

Fletcher, J.S. (2002) Rhizosphere remediation of recalcitrant soil contaminants: an important component of long-term sustained biosystem treatment. In: *Phytoremediation State of the Science Conference*. U.S. Environmental Protection Agency Report EPA 625R-01/011a and b, Cincinnati, Ohio, held in Boston, Massachusetts, May 1–2, 2000 (http://www.epa.gov/ORD/NRMRL/Pubs/625R01011b/625R01011 bchap6.pdf).

Garbisu, C. and I. Alkorta (1999) Utilization of genetically engineered microorganisms (GEMs) for bioremediation. *J. Chem. Technol. Biotechnol.* **74**: 599–606.

Glass, D.J. (1997) Prospects for use and regulation of transgenic plants in phytoremediation. In: *In Situ and On-site Bioremediation*. B.C. Alleman and A. Leeson, eds. Vol. 4, Battelle Press, Columbus, Ohio, pp. 51–56.

Glass, D.J. (1999) U.S. and international markets for phytoremediation, 1999–2000. D. Glass Associates, Inc., Needham, Massachusetts (www.channel1.com/dglassassoc/).

Glass, D.J. (2000a) Economic potential of phytoremediation. In: *Phytoremediation of Toxic Metals: Using Plants to Clean Up the Environment*. I. Raskin and B.D. Ensley eds. John Wiley, New York, pp. 15–31.

Glass, D.J. (2000b) The 2000 phytoremediation industry. D. Glass Associates, Inc., Needham, Massachusetts (www.channel1.com/dglassassoc/INFO/directory.htm).

Gunter, B., J. Kinderlerer, and D. Beyleveld (1999) The media and public understanding of biotechnology – a survey of scientists and journalists. *Sci. Commun.* **20**: 373–394.

Horne, A.J. (2000) Phytoremediation by constructed wetlands. In: *Phytoremediation of Contaminated Soil and Water*. N. Terry and G. Bañuelos, eds. Lewis Publishers, Boca Raton, Florida, pp. 13–39.

Inter-COST (2000) Workshop on bioremediation held in Sorrento, Italy, Nov. 15–18 (http://lbewww.epfl.ch/COST837/intercost_report.htm).

Interstate Technology and Regulatory Council (ITRC) (1999) Phytoremediation decision tree. Washington, D.C. (http://www.itrcweb.org).

Käppeli, O. and L. Auberson (1998) How safe is safe enough in plant genetic engineering? *Trends Plant Sci.* **3**: 276–281.

Krimsky, S. and A. Plough (1988) *Environmental Hazards. Communicating Risks as a Social Process*. Auburn House Publishers, Dover, Massachusetts.

Landmeyer, J.E. (2001) Monitoring the effect of poplar trees on petroleum-hydrocarbon and chlorinated-solvent contaminated ground water. *Int. J. Phytorem.* **3**(1): 61–85.

Larcher, W. (1995) *Physiological Plant Ecology*. Springer-Verlag, New York.

Lepp, N.W. (1981) *Effect of Heavy Metal Pollution on Plants. Vol. 2. Metals in the Environment*. Elsevier Applied Science Publishers, New York.

Lester, J.P., ed. (1989) *Environmental Politics and Policy: Theories and Evidence*. Duke University Press, Durham, North Carolina.

Leyval, C., K. Turnau, and K. Haselwandter (1997) Effect of heavy metal pollution on mycorrhizal colonization and function: physiological, ecological and applied aspects. *Mycorrhiza* **7**: 139–153.

McCutcheon, S.C. and W.J. Mitsch (1994) Ecological and environmental engineering: potential for progress. *J. Environ. Eng.* **120**(3):479–480, and *Ecol. Eng.* **3**:107–109.

McCutcheon, S.C. and S.A. Rock (2001) Phytoremediation: State of the Science Conference and other developments, editorial introduction and special commentary. *Int. J. Phytorem.* **3**(1): 1–11.

Medina, V.F., S.L. Larson, A.E. Bergstedt, and S.C. McCutcheon (2000) Phytoremoval of trinitrotoluene from water using batch kinetics studies. *Water Res.* **34**(10): 2713–2722.

Miller, H. (1997) The EPA's war on bioremediation. *Nat. Biotechnol.* **15**: 486.

Miller, R.R. (1996) *Phytoremediation.* Ground-Water Remediation Technologies Analysis Center Technology Overview Report TO-96–03, National Environmental Technology Applications Center and the University of Pittsburgh, Pennsylvania (http://www.gwrtac.org/pdf/phyto-0.pdf).

Mohr, H. and P. Schopfer. (1995) *Plant Physiology.* Springer-Verlag, New York.

Morgan, M.G. (1993) Risk analysis and management. *Sci. Am.* **269**: 32–41.

Morikawa, H., A. Higaki, M. Nohno, M. Takahashi, M. Kamada, M. Nakata, G. Toyohara, Y. Okamura, K. Matsui, S. Kitani, K. Fujita, K. Irifune, and N. Goshima (1998) More than a 600-fold variation in nitrogen dioxide assimilation among 217 plant taxa. *Plant Cell Environ.* **21**: 180–190.

National Research Council (NRC) (1997) *Intellectual Property Rights and Plant Biotechnology.* National Academy Press, Washington, D.C.

Nicks, L.J. and M.F. Chambers (1995) Farming for metals? *Mining Environ. Manag.* **3**(3): 15–18.

Organization for Economic Cooperation and Development (OECD) (1996) The global environmental goods and services industry. Paris.

Ow, D.W. (1996) Heavy metal tolerance genes: prospective tools for bioremediation. *Res. Conserv. Recycling* **18**: 135–149.

Patterson, J.D. and M.W. Allen (1997) Accounting for your actions: how stakeholders respond to the strategic communication of environmental activist organizations. *J. Appl. Commun. Res.* **25**: 293–316.

Pletsch, M., B. Santos de Araujo, and B.V. Charlwood (1999) Novel biotechnological approaches in environmental remediation research. *Biotechnology Advances* **17**: 679–687.

Pool, R. (2001) Environmental contamination, biotechnology, and the law: the impact of emerging genomic information. U.S. National Research Council, National Academy Press, Washington, D.C.

Quinn, J.J., M.C. Negri, R.R. Hinchman, L.M. Moos, J.B. Wozniak, and E.G. Gatliff (2001) Predicting the effect of deep-rooted hybrid poplars on the groundwater flow system at a phytoremediation site. *Int. J. Phytorem.* **3**(1): 41–60.

Raskin, I. (1996) Plant genetic engineering may help with environmental cleanup. *Proc. Nat. Acad. Sci. USA* **93**: 3164–3166.

Reeves, R.D. (1995) Chemical aspects of metal hyperaccumulation. In: *Proceedings/Abstracts of the Fourteenth Annual Symposium, Current Topics in Plant Biochemistry, Physiology, and Molecular Biology—Will Plants Have a Role in Bioremediation*? Interdisciplinary Plant Group, University of Missouri, Columbia, Missouri, April 19–22, pp. 9–10.

Rock, S. and P.G. Sayre (2000) Regulatory considerations for phytoremediation. In: *Phytoremediation of Toxic Metals: Using Plants to Clean Up the Environment.* I. Raskin and B.D. Ensley, eds. John Wiley, New York, pp. 43–49.

Rugh, C.L., J.F. Senecoff, R.B. Meagher, and S.A. Merkle (1998) Development of transgenic yellow poplar for mercury phytoremediation. *Nat. Biotechnol.* **16**: 925–928.

Salt, D.E., M. Blaylock, N.P.B.A. Kumar, V. Dushenkov, B.D. Ensley, I. Chet, and I. Raskin (1995) Phytoremediation: a novel strategy for the removal of toxic metals from the environment using plants. *Biotechnology* **13**: 468–474.

Salt, D.E., R.D. Smith, and I. Raskin (1998) Phytoremediation. In: *Annual Review of Plant Physiology and Plant Molucular Biology*, Annual Reviews, Palo Alto California. **49**: 643–668.

Sandermann, H., Jr. (1994) Higher plant metabolism of xenobiotics: the "green liver" concept. *Pharmacogenetics* **4**: 225–241.

Schnoor, J.L. (1997) Phytoremediation. Ground-Water Remediation Technologies Analysis Center Technology Evaluation Report TE-98-01, National Environmental Technology Applications Center and the University of Pittsburgh, Pennsylvania (http://www.gwrtac.org/pdf/phyto-e.pdf).

Schnoor, J.L. (2002) Phytoremediation of soil and groundwater. Ground-Water Remediation Technologies Analysis Center Technology Evaluation Report TE-02-01, National Environmental Technology Applications Center and the University of Pittsburgh, Pennsylvania (http://www.gwrtac.org/pdf/phyto-e-2002.pdf).

Schnoor, J.L., L.A. Licht, S.C. McCutcheon, N.L. Wolfe, and L.H. Carreira (1995) Phytoremediation of organic and nutrient contaminants. *Environ. Sci. Technol.* **29**: 318A-323A.

Slovic, P., B. Fischhoff, and S. Lichtenstein (1979) Rating the risks. *Environment* **21**(4): 14–20, 36–39. Reprinted in (1979) Chemtech **9**: 738–744; (1979) *Electrical Perpectives*; Y. Haimes, ed. (1981) *Risk/benefit Analysis in Water Resources Planning and Management* Plenum, New York; C. Hohenemser and J.X. Kasperson, eds. (1982) *Risk in the Technological Society.* Westview, Boulder, Colorado; T.S. Glickman and M. Gough, eds. (1990) *Readings in Risk*. Resources for the Future, Washington, D.C.; and P. Slovic, ed. (2000) *The Perception of Risk*. Earthscan, London.

Strand, S.E., L. Newman, M. Ruszaj, J. Wilmoth, B. Shurtleff, M. Brandt, N. Choe, G. Ekuan, J. Duffy, J.W. Massman, P.E. Heilman, and M.P. Gordon (1995) Removal of trichloroethylene from aquifers using trees. In: *Innovative Technologies for Site Remediation and Hazardous Waste Management, Proceedings of the National Conference*, R.D. Vidic and F.G. Pohland, eds, Environmental Engineering Division, American Society of Civil Engineers, Pittsburgh, Pennsylvania, July 26, pp. 597–604.

The *Arabidopsis* Genome Initiative (2000) Analysis of the genome sequence of the flowering plant *Arabidopsis thaliana. Nature* **408**: 796–815.

Thomas, P.R. and J.K. Buck (1999) Agronomic management for phytoremediation. In: *Phytoremediation and Innovative Strategies for Specialized Remedial Applications*. A. Leeson and B.C. Alleman, eds. Battelle Press, Columbus, Ohio, pp. 115–120.

Timian, S.J. and D.M. Connolly (1996) The regulation and development of bioremediation. *Risk* **7**: 279 (http://www.fplc.edu/risk/vol7/summer/timian.htm).

U.S. Army Environmental Center (U.S. AEC) (1999a) The use of constructed wetlands to phytoremediate explosives-contaminated groundwater at the Milan Army Ammunition Plant, Milan, Tennessee. Report to the Department of Defense, Environmental Security Technology Certification Program, Arlington, Virginia.

U.S. Army Environmental Center (U.S. AEC) (1999b) Draft first year report—phytoremediation of lead contaminated soils, demonstration at the Twin Cities Army Ammunition Plant. Report to the Department of Defense, Environmental Security Technology Certification Program Arlington, Virginia.

U.S. Department of Energy (U.S. DOE) (1994) Summary report of a workshop on phytoremediation research needs. Held at Santa Rosa, California, June 24–26, DOE/EM-0224. Oak Ridge, Tennessee.

U.S. Environmental Protection Agency (U.S. EPA) (1998) A citizen's guide to phytoremediation. Technology Fact Sheet, EPA/542-F-98-011. Technology Innovation Office, Washington, D.C.

Watanabe, M.E. (1997) Phytoremediation on the brink of commercialization. *Environ. Sci. Technol.* **31**: 182A–186A.

Williams, B.L., S. Brown, and M. Greenberg (1999) Determinants of trust perceptions among residents surrounding the Savannah River Nuclear Weapons Site. *Environ. Behav.* **31**: 354–371.

Wolfe, N.L., L.H. Carreira, and M.C. Delgado (1998) Method for remediating environmental contaminants. Patent 5 711 020. U.S. Patent and Trademark Office, Washington, D.C. (http://www.uspto.gov).

Wrubel, R.P., S. Krimsky, and M.D. Anderson (1997) Regulatory oversight of genetically engineered microorganisms: has regulation inhibited innovation? *Environ. Manag.* **21**: 571–586.

SECTION II

FUNDAMENTALS OF PHYTOTRANSFORMATION AND CONTROL OF CONTAMINANTS

4

SOME FUNDAMENTAL ADVANCES FOR XENOBIOTIC CHEMICALS

J.-P. Schwitzguébel and T. Vanek

SUMMARY

Higher plants have a remarkably versatile system that protects against the phytotoxicity of natural compounds and man-made synthetic chemicals (xeno biotic compounds). The detoxification process often includes a primary step or phase I. Commonly an oxidation or hydrolysis of the pollutant serves to provide a functional group suitable for subsequent conjugation to endogenous small molecules, such as glutathione or sugars (phase II). During phase III, metabolites can either end up as bound residues in the outer cell wall or be stored in the vacuole. Using a combination of classical biochemistry, chemistry, and modern molecular biology techniques, different European laboratories participating in COST Action 837 have recently defined a number of detoxification processes. Among these advances highlighted in the present chapter is the use of cell cultures as a tool for screening the potentialities of different plant species to accumulate and detoxify xenobiotic chemicals. Special emphasis is placed on nitroaromatics, nitroesters, and sulfonated aromatic compounds. Current knowledge, the areas of uncertainty, and needs for future research and development are presented to improve the efficiency and applicability of phytoremediation.

INTRODUCTION

Plants contain a complex array of enzymes that are able to detoxify herbicides and other organic pollutants (Komossa *et al.* 1995, Kreuz *et al.* 1996, Coleman *et al.* 1997, Cole and Edwards 2000). The exact number of plant enzymes potentially involved in the metabolism of xenobiotic compounds

Phytoremediation: Transformation and Control of Contaminants,
Edited by Steven C. McCutcheon and Jerald L. Schnoor.
ISBN 0-471-39435-1 (cloth)

remains unknown. Furthermore, enzymes reported in the literature may correspond to newly discovered specific enzymes, or are already known non-specific enzymes with newly discovered functions. This enzymatic complexity has been revealed by a combination of classical biochemistry and, more recently, through screening for tentatively identified genome-coding sequences, and by searching databases describing various genomes. While a very powerful tool in identifying the sequences of these enzymes, molecular biology cannot yet unravel the precise catalytic properties or functions in the metabolism of xenobiotic chemicals. However, the identification of genes possibly encoding detoxifying enzymes can "fast-track" the characterization of complex families of enzymes that may be very difficult to study *in planta* due to copurification, compounded by low abundance and instability.

In this context, the present chapter aims to (1) highlight recent progress made in different European laboratories participating in COST Action 837 on "Plant biotechnology for the removal of organic pollutants and toxic metals from wastewaters and contaminated sites," supported by the European Commission (http://lbewww.epfl.ch/COST837) and (2) stress the usefulness of plant-cell cultures as a tool for screening the potentialities of different species to accumulate and detoxify xenobiotic chemicals, with a special emphasis on nitroaromatics, nitroesters, and sulfonated aromatic compounds. However, relating the xenobiochemical functions of enzymes to the roles in endogenous metabolism remains an important challenge (Rea *et al.* 1998, Gordeziani *et al.* 1999, Palmer 1999, Edwards *et al.* 2000, Werck-Reichhart *et al.* 2000, Jones and Vogt 2001). The successful implementation of phytoremediation also relies on a better understanding of the physical, chemical, and biological interactions of plants with the environment.

GENES AND ENZYMES INVOLVED IN THE PHYTOREMEDIATION OF ORGANIC POLLUTANTS

Oxygenases and Peroxidases

The addition of oxygen to hydrophobic compounds increases both solubility and chemical reactivity, and can be achieved by dioxygenases, monooxygenases, and peroxidases. The cytochrome P-450 represents a class of enzymes, present in most plants and mammals, that are involved in different types of oxygenation reactions for many natural and xenobiotic compounds. The P-450 enzymes are membrane-bound proteins, usually acting as monooxygenases. These enzymes use electrons from reduced nicotinamide adenine dinucleotide phosphate (NADPH) to activate molecular oxygen, incorporate one of the atoms into an organic compound, and reduce the second atom into water. Electrons from NADPH are transferred to P-450 enzymes *via* flavoproteins called cytochrome P-450 reductases. Both plant P-450 enzymes and the associated reductases are usually bound to the cytoplasmic surface of the endoplasmic reticulum. The P-450 proteins have a molecular mass ranging from 45 to 62 kilodaltons, and may have as little as 16 percent amino acid identity. However, the overall tridimensional structure is conserved, as are a few residues on both sides of the iron-containing heme structure. Depending on the specific P-450 protein and the contaminant molecule, the result of catalysis in most cases is hydroxylation, but heteroatom dealkylation, epoxidation, deamination, isomerization, C–C or C=N cleavage, dimerization, ring formation or extension, dehydration, and dehydrogenation or reduction have also been reported (Halkier 1996, Schuler 1996, Werck-Reichhart *et al.* 2000).

The primary function of P-450 enzymes in higher plants seems to be the biosynthesis of sterols, oxylipins, hormones, and specific metabolites, such as phenylpropanoids, isoprenoids, alkaloids, or amino-acid derivatives. Probably, a secondary function of P-450 enzymes catalyzing some of these reactions is the metabolism of exogenous molecules like herbicides, insecticides, or other pollutants, usually resulting in detoxification (Schuler 1996, Khatisashvili *et al.* 1997, Schalk *et al.* 1997, Stiborova *et al.* 2000). Some of these enzymes are present in most plant tissues, while the activity of others is restricted to specific tissues and organs. Some P-450 enzymes are always active (or constitutive), while the activity of others becomes important only after chemical treatment with metals, herbicides, or herbicide safeners (compounds that reduce phytotoxicity, which are also inducible). The existence of multiple forms of the enzyme controlled by an outstanding number of inducers is a key factor in the metabolism of xenobiotic compounds. The interplay of specific P-450 forms and inducers determines which xenobiotic chemicals will be metabolized in a given tissue and at what rate (Durst *et al.* 1997, Robineau *et al.* 1998).

In recent years, an increasing number of P-450 genes have been identified, and several have been successfully expressed in recombinant systems such as yeasts. Systematic sequencing has recently revealed that the CYP gene superfamily for the cytochrome P-450 enzyme is composed of at least 51 families and 104 subfamilies. Over 400 plant CYP genes have now been reported, which belong to either CYP51 (also present in mammals), CYP71 to CYP99, or CYP701 to CYP721 (Chapple 1998). For example, the CYP76B1 gene from Jerusalem artichoke (*Helianthus tuberosus*) has been isolated using the purification of a highly inducible protein, microsequencing, and isolation of the gene most strongly inducible by aminopyrine. Production of CYP76B1 was optimized in yeast by coexpression with different P-450 reductases, alone, or as fusion proteins. Yeast-expressed CYP76B1 very efficiently metabolized alkoxycoumarins, alkoxyphenoxazones, and herbicides of the class of phenylurea. The latter were di-dealkylated to nonphytotoxic metabolites. Metabolism of phenylurea and the other xenobiotic compounds by CYP76B1 proceeded with a high turnover rate, comparable to that observed with natural substances (Batard *et al.* 1998). The gene was also overexpressed in tobacco (*Nicotiana tabacum*) and in *Arabidopsis thaliana*. Preliminary results indicated that CYP76B1 confers to transgenic plants an increased tolerance to various phenylurea herbicides (Hehn *et al.* 1999).

The next steps should be to (1) prepare fusion proteins between the plant cytochrome P-450 and the associated electron donor (NADPH cytochrome P-450 reductase), (2) express these in other species, and (3) determine the detoxification capacity in a high throughput system using a range of organic pollutants. Genes of potentially interesting enzymes could be used in genetic engineering of target plants suitable for phytoremediation (Werck-Reichhart *et al.* 2000).

Both fungal and plant peroxidases reduce hydrogen peroxide to water at the expense of aromatic reductants, for which these enzymes appear to be relatively nonspecific. Intracellular and extracellular peroxidases have been found in both membrane-bound and soluble fractions. Several forms of these enzymes are known, both constitutive as well as inducible. Plant peroxidases are noted for coupling reactions that lead to lignin synthesis and formation of humic residues, whereas fungal peroxidases catalyze the oxidative depolymerization and aromatic ring opening in the delignification process.

Interesting projects aim to develop a simple, inexpensive, and highly effective *in situ* remediation technique for organically contaminated soils, based on the exploitation and optimization of a naturally occurring process known as oxidative coupling that is involved in the synthesis of humic substances in soil. The structural resemblance of many xenobiotic compounds to humic acid building blocks means that bioavailability may be reduced using peroxidase *via* a two-step mechanism, phytotransformation and phytostabilization due to bound residue formation. Of particular interest is the peroxidase liberated from horseradish (*Armoracia rusticana*) roots (Setti *et al.* 1998, Coyle *et al.* 1999).

In vitro studies with crude horseradish peroxidase extracts have investigated a selection of phenolic compounds that may be targeted on the basis of a structure–reactivity relationship. Phenolics and other aromatic compounds have been the focus of several studies because these are some of the most common soil pollutants worldwide as a result of the low biodegradability and widespread application in the form of pesticides. Soil pollutants are normally composed of mixtures of both inorganic and organic contaminants, entailing possible synergistic or antagonistic effects. Inhibition of the peroxidase reaction by cadmium, zinc, nickel, and lead at concentrations of 100 milligrams per liter is of concern because these metals are often present in contaminated soils. A number of soil parameters, such as pH and temperature, could also affect peroxidase activity under field conditions (Coyle *et al.* 1999).

Glutathione Transferases

Detoxification of xenobiotic compounds is often achieved by conjugation to small natural molecules like glutathione, sugars, and malonate. Several enzyme classes are known to catalyze these reactions, and glutathione

S-transferases (GSTs) and glycosyl-transferases are abundant enzymes found in all organs of plants (Pflugmacher and Sandermann 1998, Schrenk *et al.* 1998, Pflugmacher *et al.* 1999, Pflugmacher *et al.* 2000).

Glutathione *S*-transferases were first identified in plants nearly 30 years ago from the detoxification of herbicides by conjugation with the tripeptide glutathione (Lamoureux *et al.* 1991, Andrews *et al.* 1997). Many of these GSTs appear to be central to the defense mechanisms of higher plants against xenobiotic compounds and natural toxic organic molecules (Cole *et al.* 1997, Edwards *et al.* 2000). These proteins are a ubiquitous family of isoenzymes encompassing a wide range of functions, among which are (1) the detoxification of xenobiotic chemicals as well as natural endogenous and potentially toxic compounds, (2) a protective action against oxidative stress, and (3) a role in the normal secondary metabolism. More recently, however, advances in the understanding of the structure, classification, and function of these enzymes have clarified the roles in the metabolism of xenobiotic compounds. Current insights now offer the potential to engineer GSTs with enhanced detoxifying activities (Cummins *et al.* 1997, Dixon *et al.* 1998).

Active GSTs are composed of two subunits with molecular masses in the range of 25 to 27 kilodaltons. Subunits can be identical (homodimers) or different (heterodimers), each containing an active site composed of a highly conserved binding domain for glutathione (G-site) and a more variable binding domain for a hydrophobic compound (H-site). Plant GSTs are surprisingly diverse, but can be grouped into one of four distinct types on the basis of similarities in amino acid sequences (Rossini *et al.* 1996, Frova *et al.* 1997, Edwards *et al.* 2000).

A unified plant GST nomenclature based on assignment to these four classes has been recently established and include the following: phi (former Type I), zeta (former Type II), tau (former Type III) and theta (former Type IV). Two of these classes, theta and zeta, are also found in animals and appear to have very specific functions. Zeta class transferases act as isomerases in the catabolism of aromatic amino acids, while theta GSTs possess high glutathione peroxidase activity and are probably involved in detoxifying products of oxidative stress. The other two classes, phi and tau, occur only in plants. Enzymes in these two classes are the most abundant plant GSTs, and are responsible for the large majority of conjugation reactions involving xenobiotic compounds (Edwards *et al.* 2000).

To determine the detoxification and selectivity of herbicides in crops and weeds, the complement deoxyribonucleic acid (DNA) of GSTs have been cloned and analyzed for several major crops and competing weeds. Weeds contain a similar range of GSTs as crops, but produce key herbicide-detoxifying enzymes at 10- to 20-fold lower levels (Dixon *et al.* 1998). Treatment of maize (*Zea mays*) and wheat (*Triticum aestivum*) with herbicide safeners, compounds that increase herbicide tolerance in cereals, radically altered the GST complement of these crops (Cole and Edwards 2000). Both herbicide-safener-inducible phi and tau classes of GSTs can also act as glutathione peroxidases, and are thus able to reduce organic hydroperoxides arising from oxidative stress.

The phi and tau classes of GSTs are good candidates for use in engineering organisms for increased herbicide detoxification and phytoremediation activity. The GSTs can be tailored to perform specific detoxification roles either by searching for an existing but as yet uncharacterized GST with the appropriate activity, or by altering the activity of a known GST. To alter GST activities the following methods have been used (Dixon *et al.* 1999a, 1999b):

1. Formation of heterodimers (union of different subunits). A modified bacterial expression vector was constructed to allow GST coexpression; combinations of maize (*Zea mays*) GSTs were thus expressed and analyzed for heterodimer formation. The heterodimers were then assayed for activities
2. Synthesis of chimeric enzymes (union of subunits from different plant species). Chimeric GSTs were constructed by engineering a restriction site into the linker region between different domains of the enzymes, and by using this site to swap domains between enzymes. The chimeras were expressed and analyzed for changes in GST activity
3. Random mutagenesis and recombination. Glutathione *S*-transferase sequences were randomly mutated and recombined using DNA shuffling to generate a large pool of altered enzymes. These mutant enzymes were expressed in bacteria and screened for increased herbicide-detoxifying activity. Screening of 3000 mutant enzymes gave four mutants showing up to 20-fold higher rates of herbicide detoxification than the parent enzymes

Another interest is the development of plants capable of coping with stresses from the presence of xenobiotic chemicals. For such a purpose, two approaches are possible. The first approach is based on the development of new strategies of selection: gametophytic selection (selection at pollen level) and genetic dissection of plant tolerance by linkage analysis with molecular markers so as to identify and localize on chromosomes the genetic factors conferring tolerance. This information can be used for marker-assisted selection, a procedure much more powerful than traditional breeding methods. This work resulted in the production of improved maize (*Zea mays*) genotypes and the detection of quantitative trait loci (see definition at http://www.ndsu.nodak.edu/instruct/mcclean/plsc731/quant/quant1.htm) in the same species. This work also explains the high proportion of plant variability for tolerance of the herbicide alachlor (Sari-Gorla *et al.* 1997, Rossini *et al.* 1998).

The second approach to the development of genotypes is based on the detection of tolerance mechanisms—in particular for the detoxification of xenobiotic chemicals and the identification of the controlling genetic system. The GSTs involved in the response to different stresses have a precise pattern of expression. To determine the pattern, the genetic and molecular characterization of the GST gene family in maize (*Zea mays*) was completed. Three of the genes have been localized on maize (*Zea mays*) chromosomes. The functional characterization of the different isoforms defined the possible role in plant defense against herbicides. Furthermore, expression analysis at transcript- and gene-product level indicated that the enzyme isoforms are developmentally regulated and some are induced by specific herbicides (Jepson *et al.* 1997, Sari-Gorla *et al.* 1997).

Rice (*Oryza sativa*) is a model species for cereals, because of the global agronomic importance, amenability for biotechnology, and simple, mapped genome. The rice (*Oryza sativa*) GST gene family is presently under investigation. Results obtained from maize (*Zea mays*) and rice (*Oryza sativa*) should be useful to forecast the resistance to xenobiotic compounds in species not genetically well known. Thus, the characterization at genetic and molecular levels of the biological system of resistance in these species may provide important tools for phytoremediation (Soranzo *et al.* 2000).

Glycosyltransferases

The addition of sugars to endogenous and exogenous organic molecules is widespread among plants and animals. The conjugation to carbohydrate, usually glucose, increases water solubility, improves chemical stability, reduces chemical reactivity, and alters the biological activity of organic molecules. Using nucleotide-activated sugars, *O*-, *N*-, *S*- or *C*-glycosyltransferases (GTs) are able to attach sugars at an oxygen (O), nitrogen (N), sulfur (S), and carbon (C) atom, respectively, on natural plant products or xenobiotic compounds. Foreign compounds, originating from other organisms and man-made chemicals, are also glycosylated by plants, and O-glycosylation is the most common conjugation of agrochemicals in plants (Cole and Edwards 2000, Jones and Vogt 2001).

Plant glycosyltransferases comprise a large family of enzymes involved in glycosylation of both endogenous metabolites and xenobiotic substances. However, the number of expressed GTs involved in plant metabolism and the specificities of these enzymes remain largely unknown. A survey of more than 50 plant species showed a wide distribution of *O*-, *N*-, and *S*-GT activities in crude enzyme preparations. Interestingly, certain species showed particularly high activity with several specific xenobiotic chemicals (Pflugmacher and Sandermann 1998). However, there is little information at the molecular level concerning either the expression of individual enzymes and contaminant specificities, or the roles in different plant defense and detoxification processes. The genome sequencing of the model plant *Arabidopsis thaliana* now provides all genes of a single plant species. According to the available information, the number of glycosyltransferases seems to exceed a hundred different proteins (Messner *et al.* 2000). Preliminary results obtained with a few recombinant expressed enzymes indicate that these proteins glycosylate quite a number of endogenous and xenobiotic substances. Interestingly, xenobiotic chemicals such as trichlorophenols tend to be good candidates for glycosylation, even when crude enzyme preparations containing an unknown mixture of individual proteins are investigated (Messner *et al.* 2000).

The use of combined approaches should provide additional insight into the interactions between endogenous metabolic pathways and detoxification of xenobiotic substances, and the developmental and spatial expression patterns of glycosyltransferases. In addition, the inducibility by xenobiotic substances should also be revealed.

Transport Systems Across the Vacuolar Membrane

Phytotransformation of herbicides and other xenobiotic compounds and conjugation to glycosides or glutathione are usually considered a detoxification process, but the metabolites and conjugates produced may eventually exert further biological activity. Therefore, efficient detoxification requires the accumulation of the modified or conjugated pollutant into the vacuole, or secretion into the apoplast (Blake-Kalff *et al.* 1997, Schroeder 1997).

According to the prevailing chemiosmotic model for energy-dependent solute transport, the primary energizers for solute transport into the plant vacuole are two proton pumps, which generate an electrochemical gradient across the membrane, using either adenosine triphosphate (ATP) or pyrophosphate. A first exception was observed in the case of herbicides conjugated to glutathione that were transported into the vacuole directly energized by ATP. It was shown that the uptake of glutathione conjugates of metolachlor, dinitrobenzene, and oxidized glutathione (the latter being a special case of a glutathione conjugate) is strongly ATP-dependent. Treatment of barley (*Hordeum vulgare*) with so-called safeners, or *Arabidopsis thaliana* with chlorodinitrobenzene, conjugated to glutathione within the cell, increases the activity of the glutathione conjugate pump (Rea *et al.* 1998).

Direct ATP-dependence has been shown for a large range of substances, such as chlorophyll catabolites (Lu *et al.* 1998, Tommasini *et al.* 1998), herbicides conjugated to glucose (Klein *et al.* 1996), and sulfonated compounds (Klein *et al.* 1997). Recent studies have demonstrated the existence of a group of organic solute transporters, belonging to the ATP-binding cassette (ABC) superfamily directly energized by magnesium-ATP rather than by a transmembrane proton–electrochemical potential difference (Rea *et al.* 1998). Originally identified in microbial and animal cells, the ABC superfamily is one of the largest and most widespread protein families known. Capable of transporting a broad range of substances, including sugars, peptides, alkaloids, inorganic anions, and lipids, all ABC transporters consist of one or two copies each of an integral membrane sector and cytosolically oriented ATP-binding domain. To date, two major subclasses, the multidrug resistance-associated proteins (MRPs) and multidrug resistance proteins (MDRs), have been identified in plants. However, only the MRPs have been defined functionally. These proteins participate in the transport of exogenous and endogenous anions and glutathione-conjugated compounds from the cytosol into the vacuole. The functions of MRPs include herbicide detoxification, cell pigmentation, alleviation of oxidative damage, storage of antimicrobial compounds, and transport of heavy metal chelates (Rea *et al.* 1998).

Many plant secondary products and xenobiotic compounds are glycosylated and then accumulated within the large, central vacuole. It has been shown that plants contain at least two different transport systems for glycosylated substances. Some plant specific glycosylated flavonoids such as isovitexin are transported by a proton chemiosmotic mechanism. Using another transport system hydroxyprimisulfuronglucoside, a herbicide modified within the plant by the action of P-450 monooxygenases and glycosyltransferases, is taken up by a process directly energized by ATP, which can be inhibited by vanadate (Klein *et al.* 1996).

Testing mutant yeast strains lacking individual ABC transporters for the capacity to transport glutathione conjugates revealed that a cadmium-sensitive yeast strain with a deletion in the yeast cadmium factor 1 (YCF1) gene has a strongly reduced glutathione conjugate transport activity, indicating that the protein expressed by YCF1 is a cadmium–glutathione conjugate transporter. The YCF1 protein is highly homologous to another member of the ABC transporter family, namely the human MRP 1. Heterologous expression of MRP 1 in the YCF1 mutant restores the glutathione transport activity and tolerance to cadmium. A database search for plant proteins homologous to MRP 1 and YCF1 revealed that several proteins of *Arabidopsis thaliana* are very similar to MRP 1 and YCF1 and, therefore, must be considered as possible glutathione-conjugate transporters. Two members of this family were cloned and expressed in the yeast YCF1 mutant. In the case of AtMRP3, cadmium tolerance and glutathione conjugate transport activity are partially restored. The plant protein produced by the gene is also able to transport chlorophyll catabolites. The transport into the vacuole is kinetically similar to that of the ABC-transporter, but does not interfere with glutathione conjugates. The level of the gene transcript is strongly increased in response to treatment with herbicides (Tommasini *et al.* 1997, Lu *et al.* 1998).

To assign a function to the different MRP-like enzymes, expression studies used plants treated with xenobiotic chemicals detoxified through conjugation to glutathione or glucose. However, these studies did not reveal a correlation between increased transcript levels and the respective detoxification (Tommasini *et al.* 1997). Currently, three MRP-like genes of *Arabidopsis thaliana* have been characterized in detail. All of the three transporters expressed from the genes exhibit glutathione conjugate transport activity, but only the AtMRP3 gene confers cadmium resistance (Lu *et al.* 1998, Rea *et al.* 1998, Tommasini *et al.* 1998). The next step should be to elucidate which other xenobiotic or plant-derived compounds are recognized by transporters directly energized by ATP.

USE OF PLANT-CELL CULTURES IN PHYTOREMEDIATION

Plant cell cultures (see Ellis 1988 for a review) grown as a callus on solid media or cell suspensions in liquid media could be a very important source of many secondary metabolites such as the anticancer drug taxol (Wang *et al.* 2001a), adaptogenes from *Panax ginseng* (Wu and Zhong 1999), and the potential antimalarial drug artemisin (Wang *et al.* 2001b). For phytoremediation, cell cultures are a marvelous screening tool and a model system for phytotransformation (Sandermann *et al.* 1984, Harms 1992). Plant cell cultures offer the following advantages (Fowler 1987, Harms 1992):

1. Independence from environmental factors, including climate, pests, and geographical and seasonal constraints
2. Defined cultivation system with greater process control and available when required
3. More consistent product quality and yield
4. Faster and lesser cost than growing whole plants
5. Rapid means to investigate toxicological and metabolic fate of chemicals in plants

Nevertheless, cell culture utilization has limitations. Although totipotency (capability of a single cell to develop into a complete plant) theoretically guarantees that cell cultures have the same intracellular metabolism as the whole plant, the effect of cell morphological and biochemical specialization or differentiation in the whole plant cannot be fully replicated in cell cultures. Therefore, more effective screening for remediation and for synthesis of metabolites may require (1) addition of specific precursors to better mimic whole plant metabolism and (2) use of more than one cell culture to better mimic whole plant detoxification.

Phytotransformation

For phytotransformation (transformation of natural or xenobiotic chemicals using whole plants or plant cells), only a limited number of enzymes is necessary to be active in the metabolic pathway, in comparison with *de novo* production of natural compounds in which the whole metabolic pathway must be available. Usually, phytotransformation by plant cell suspension has been studied with the aim of achieving the formation of natural secondary products that can be potentially used in industrial production or, for environmental purposes, to degrade undesired xenobiotic chemicals.

Cell culture research contains a wealth of information about how numerous natural and synthetic organic chemicals are transformed by a number of plants (see for example Stepan-Sarkissian 1991). The most important transformation reactions discovered to date include glycosylation, hydroxylation, reduction, esterification, epoxidation, isomerization, oxidation, and transfer of a methoxy group. Depending upon the conditions chosen, an individual compound may be converted to different products. For example, Stohs (1980) achieved the conversion of progesterone from nine different chemical reactions using fourteen different plant cell cultures. On the other hand, one specific cell line is able to express the enzymes necessary to transform different added substances. In the case of transformation of cyclic alcohols and monoterpenes using tobacco (*Nicotiana tabacum*) cells, the following reactions occur: regioselective hydroxylation of a C=C double bond, enantioselective hydroxylation, stereospecific reduction of a keto-group, and reciprocal conversion of cyclic alcohols and ketones (Suga *et al.* 1982). Other examples of natural and synthetic chemicals transformed include glucosidation of digitoxigenin by a Grecian foxglove (*Digitalis lanata*) cell suspension (Vanek *et al.* 1986), oxidation of verbenol by free and immobilized *Solanum aviculare* cells (Vanek *et al.* 1989a), reduction of 4-(methoxybenzyl)-1-cyclohexanone by *Solanum aviculare* free-cell suspension (Vanek *et al.* 1989b), and transformation of limonene to carvone *via* carveol (hydroxylation and oxidation) by *Solanum aviculare* and *Dioscorea deltoidea* cells (Vanek *et al.* 1999).

Phytotransformation of Xenobiotic Compounds

Cell suspension cultures of plants have been repeatedly demonstrated as a very suitable model system for the metabolism of xenobiotic compounds (Sandermann *et al.* 1984, Harms 1992, Wilken *et al.* 1995, Duc *et al.* 1999b). Such cells rapidly take up foreign compounds, possess high rates of metabolite formation, and grow free from microorganisms. Cell cultures are also very suitable for investigating polar hydrophilic molecules, which can be taken up from liquid media without problems. Generally, the spectrum of resulting metabolites in plants and plant cell cultures is, in principle, identical, though quantitative differences may occur (Harms 1992).

Phytotransformation of selected pesticides and herbicides have been extensively studied to identify the structure and biological activity of the resulting metabolites. For example, position-specific aryl hydroxylation reactions (6-hydroxy- *versus* 8-hydroxy-compound) of the herbicide bentazon in various cell cultures reflected the differences found between resistant and susceptible plant species (Sterling and Balke 1989). The structural activity data has been utilized for the selection of pesticides that are metabolized quickly to nontoxic products so that ecological and health risks can be assessed, and new, active, and safe analogues of pesticides and herbicides can be developed. The information can also be used to design phytoremediation approaches to clean up pesticide spills (Gianfreda and Nannipieri 2001).

In the case of other xenobiotic chemicals, the same biochemical investigative protocols using axenic cultures are being used to design other phytoremediation techniques. For example, plant tissue cultures have been used to define the degradation of 2-(4-methoxybenzyl)-1-cyclohexanone (Vanek *et al.* 1989b), trinitrotoluene (TNT) (Lucero *et al.* 1999), 4-nitrophenol (Malcherek *et al.* 1998), hydroxytoluene (Flores *et al.* 1994), fluoranthene (Kolb and Harms 2000), 4-n-nonylphenol (Bokern *et al.* 1998), 1,4-dichlorobenzene (Wang *et al.* 1996), nitroglycerine (Goel *et al.* 1997), formaldehyde (Giese *et al.* 1994), parathion (Vanderkrol *et al.* 1995), polychlorinated biphenyls (Wilken *et al.* 1995), phenoxyacetic acid (Laurent and Scalla 1999), pentachlorophenol (Schafer and Sandermann 1988), and anthraquinonesulfonic acids (Duc *et al.* 1999b). These cultures were mainly derived from crops but axenic aquatic plants, hybrid poplars (*Populus* spp.), and other plants are being used more and more.

EXPLOSIVES

Nitroaromatic Compounds

Only a few nitroaromatics have been isolated from natural sources. In contrast, synthetic nitroaromatics are produced extensively as explosives, dyes, polymer additives, pesticides, and pharmaceuticals (Higson 1992). These chemicals also serve as solvents or feedstocks in the preparation of aminoaromatic derivatives. Some of these compounds such as nitrobenzene, nitrotoluene, trinitrophenol, 2,4,6-trinitro- or 2,4-dinitro and 2,6-dinitrotoluene were or are produced on a massive scale (Figure 4-1). Inadequate management of the wastes from this massive production pollutes the air, water, and soils in many countries (see Medina *et al.* this book), with extensive soil and water pollution by explosives (mainly trinitrotoluene) being the worst (Spain 1995). Many studies have shown that nitroaromatics and the resulting degradation products are toxic, mutagenic, and suspected as being carcinogenic (see Spanggord *et al.* 1982, Medina *et al.* this book). As a consequence, nitroaromatics are listed as priority pollutants by several environmental protection agencies.

Bioremediation for nitroaromatics-contaminated sites (Bennett 1994), especially involving TNT, has been the subject of many studies in recent years (Higson 1992, Spain 1995) with only limited field-scale success. This includes limited field-testing of phytoremediation in the U.S. (see McCutcheon *et al.* and Thompson *et al.* this book). Only composting of explosives is currently accepted and yet many questions remain about the sequestered metabolites of

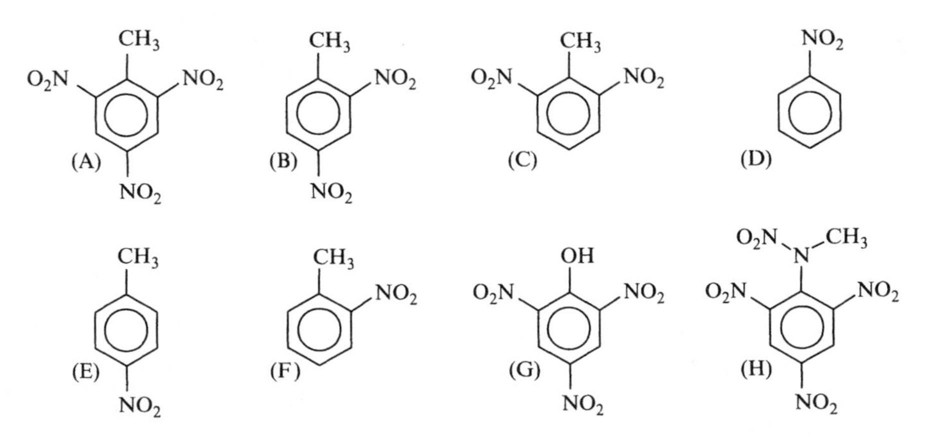

Figure 4-1 The structure of nitroaromatic compounds: (A) 2,4,6-trinitrotoluene, (B) 2,4-dinitrotoluene, (C) 2,6-dinitrotoluene, (D) nitrobenzene, (E) 4-nitrotoluene, (F) 2-nitrotoluene, (G) 2,4,6-trinitrophenol, and (H) 2,4,6-trinitrophenylmethylnitramine (tetryl).

nitroaromatics (Griest *et al.* 1998). In contrast, the European experience with TNT phytoremediation seems more promising (Koehler *et al.* 2001, Thomas *et al.* 2001).

To develop the appropriate phytoremediation process, it is necessary to study (1) the uptake of TNT, (2) the degradation pathway, and (3) the toxicity of resulting products. Scheidemann *et al.* (1998) have observed no significant differences in quality or quantity of uptake of nitroaromatic compounds by dicotyledonous and monocotyledonous plants. Gong *et al.* (1999) found that the monocotyledonous oat (*Avena sativa*) and wheat are less sensitive to TNT-contaminated soil than turnip (*Brassica campestris*) and cress (*Nasturtium officinale*). However, low concentrations of TNT (5 to 25 milligrams per kilogram of soil for cress and 25 to 50 milligrams per kilogram of soil for turnip) stimulated seedling growth. The phytotoxicity of TNT, dinitrotoluenes (DNTs), and metabolites was also tested using algae (*Selenastrum capricornutum*). The most toxic were the following hydroxyaminonitrotoluenes: 2-hydroxyaminodinitrotoluene (2-HADNT), 2,4-dihydroxyaminonitrotoluene (2,4-DHANT), 2,6-diaminonitrotoluene (2,6-DANT), hydroxyaminonitrotoluenes (HANTs), aminonitrotoluenes (ANTs), and diaminotoluene (DAT) (Tadros *et al.* 2000).

Sun *et al.* (2000) recently studied a distribution of [^{14}C] activity in axenic plants and reported that bromegrass (*Bromus inermis* Leyss) exposed to [^{14}C]-TNT accumulated [^{14}C] in roots (21.3 percent) and in shoots (3.8 percent), and the rest of the activity remained in the medium. Earlier studies with radiolabeled TNT in hybrid poplars (*Populus* spp.) have shown that radiolabeled [^{14}C]-TNT was translocated from the hydroponic solution to the roots (78 percent), stem (13 percent), and leaves (9 percent) (Thompson *et al.* 1998). The partial mineralization of TNT to CO_2 (0.003 percent of [^{14}C]) described and discussed by Sun *et al.* (2000) is too low to be physiologically significant. Many conjugates were identified. This aspect could explain why the stoichiometry of TNT degradation is elucidated only partially. The plant metabolic pathway of TNT degradation is similar to the microbial pathway, especially the first steps leading to the formation of hydroxyamino- and aminodinitrotoluenes (Figure 4-2).

In some studies (Drzyzga *et al.* 1998, Hawari *et al.* 1998), 2,4,6-triaminotoluene (TAT) was found as a product of TNT transformation, especially under anaerobic conditions by the bacterium *Clostridium bifermentans* (Lewis *et al.* 1996). However, the degradation of TNT to TAT in plants was only proposed (Rivera *et al.* 1998) (Figure 4-3). Rieger and Knackmuss (1995) stated that the main reason the formation of TAT in plants is less probable, is due to the strong reducing potential of TAT ($E_h = -200$ millivolts), that prevents the accumulation of TAT in cells by autoxidation and polymerization.

The scheme of TNT degradation has evolved (Rieger and Knackmuss 1995, Thompson *et al.* 1998) with the identification of other transformation products. The existence of a wide spectra of as yet unidentified conjugates in different degradation stages might explain the problems in achieving an exact stoichiometric relation between doses of TNT and the subsequent breakdown products (Pavlostathis *et al.* 1998, Burken *et al.* 2000).

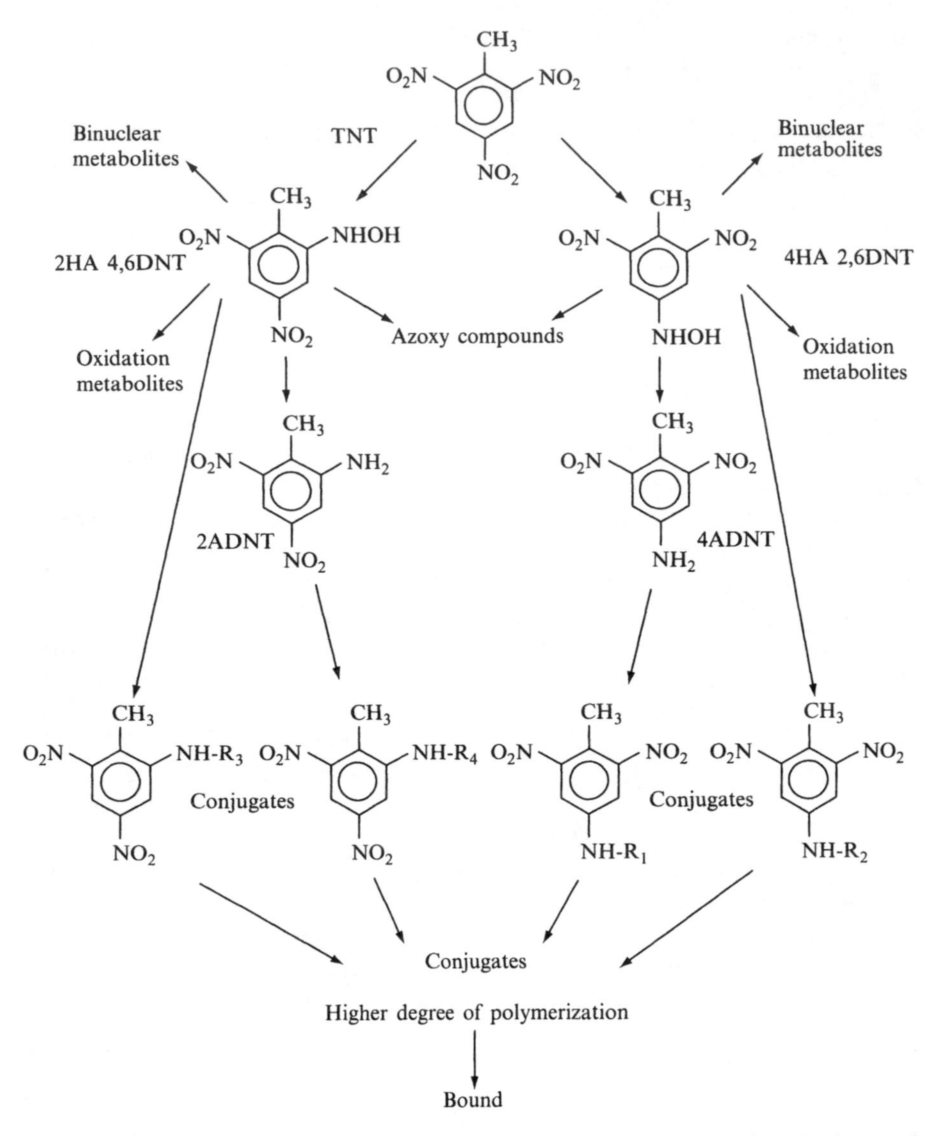

Figure 4-2 Trinitrotoluense reduction pathways and formation of conjugates from Burken *et al.* (2000, Copyright CRC Press). R_1, R_2, R3, and R_4 represent sugars with six carbon atoms.

The oxidation of TNT on the methyl group represents a second alternative process of TNT degradation in plant tissues (Figure 4-3). The oxidation of the methyl group seems more probable than direct cleavage. This hypothesis is based on the identification and characterization of unique metabolites from the reduction products of TNT in aquatic phytoremediation systems of parrot feather (*Myriophyllum aquaticum*). These compounds include 2-amino-4, 6-dinitrobenzoic acid, 2,4-dinitro-6-hydroxybenzyl alcohol,

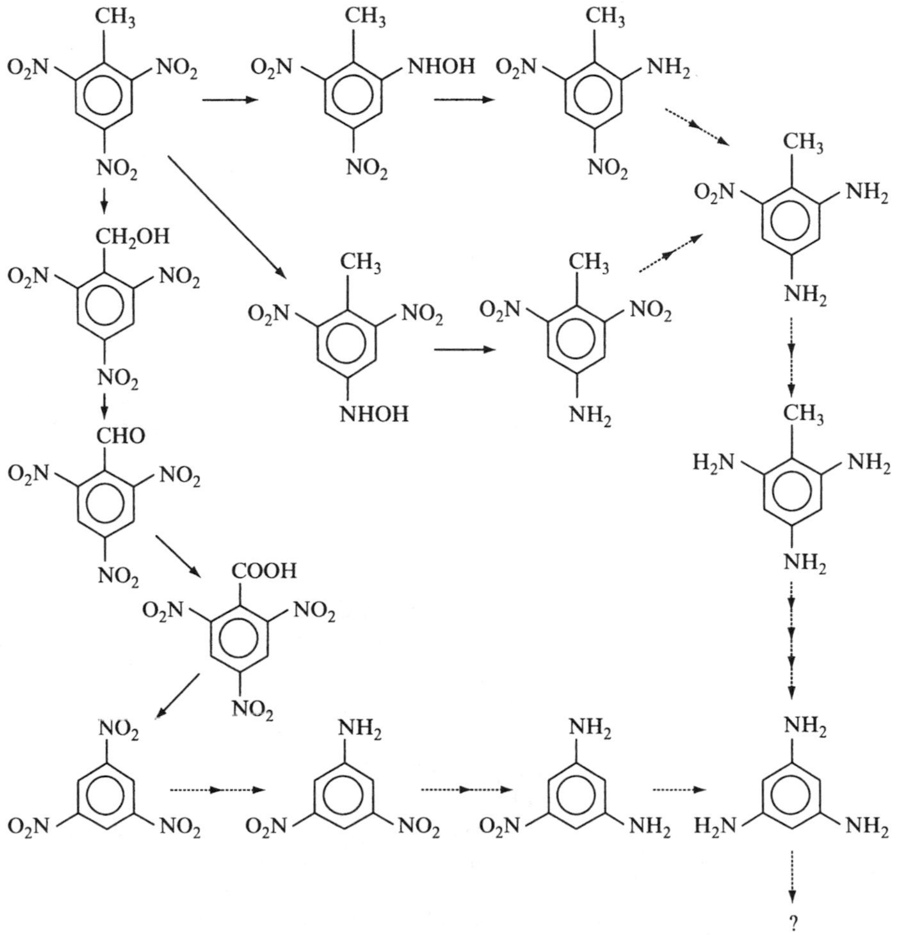

Figure 4-3 Proposed reduction and oxidation pathways for trinitrotoluene. Adapted from Rivera *et al.* (1998) and Vanek *et al.* (2003).

2-*N*-acetoxyamino-4, 6-dinitrobenzaldehyde, and 2,4-dinitro-6-hydroxytoluene (Bhadra *et al.* 1999a). This alternative degradation pathway (Figure 4-3) was supported by Vanek *et al.* (2000b) in the same laboratory where trinitrobenzene was identified using liquid chromatography coupled to mass spectrometry during TNT-degradation studies in plant cell culture systems. Trinitrobenzene was detected in cell-free extract prepared from a suspension culture of poplar (*Populus simonii*) and incubated for 3 hours with TNT at 28 °C. The oxidation of the methyl group was not caused by light, because incubations were done in the dark (Nepovím *et al.* 1999). A suspension culture of poplar (*Populus simonii*) maintained in the dark (Vanek *et al.* 2000b) also transformed TNT into trinitrobenzene.

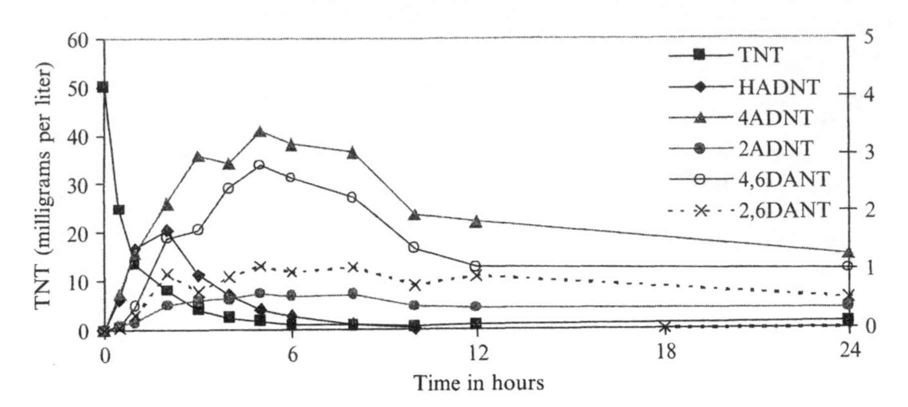

Figure 4-4 Total trinitrotoluene degradation by *Solanum aviculare* cells. Abbreviations—TNT: trinitrotoluene; HADNT: hydroxyaminodinitrotoluene; 4ADNT: 4-aminodinitrotoluene; 2ADNT: 2-aminodinitrotoluene; 4,6DANT: 4,6-diaminonitrotoluene; and 2,6DANT: 2,6-diaminonitrotoluene.

In contrast, during the degradation of TNT in plant suspension cultures of *Solanum aviculare* and rhubarb (*Rheum palmatum*), only the products of nitro-group reduction could be extracted from cells or detected after excretion back into the medium (Figure 4-4): aminodinitrotoluenes (ADNTs), diaminonitrotoluenes (DANTs), and hydroxyaminonitrotoluene (HADNT) (Hubálek and Vanek 1998). During a similar study of TNT degradation, Bhadra *et al.* (1999b) have observed only the presence of aminodinitrotoluenes in the medium and in plant tissue of a periwinkle hairy root culture of periwinkle (*Catharanthus roseus*). In light of the above mentioned results, it is clear that the alternative pathway (*via* trinitrobenzene) needs more study before being incorporated into the plant TNT transformation pathway.

Nitroesters

Over the last 100 years a variety of technological and medical applications have been found for nitrate esters, notably as high explosives and vasodilators. The most important nitrate esters are nitroglycerin (glycerol trinitrate, GTN) and pentaerythritol tetranitrate (PETN). Although useful therapeutically in low doses, nitrate esters and the resulting metabolites are generally toxic at higher levels. Thus, a significant environmental impact occurs. Wendt *et al.* (1978) found acute mammalian toxicity levels of 30 to 1300 milligrams per kilogram for GTN, whereas Urbanski (1984) determined the lethal dose affecting 50 percent of a population (LD50) of fish to be 1 milligram per liter. These toxicity levels are of concern in wastewaters and soil contaminated in the area of industrial plants manufacturing GTN and PETN, and waste containing GTN and nitrocellulose released during handling or storage.

All reports of degradation of nitrate esters by prokaryotic and eukaryotic systems involve sequential denitration steps, resulting in multiple partially

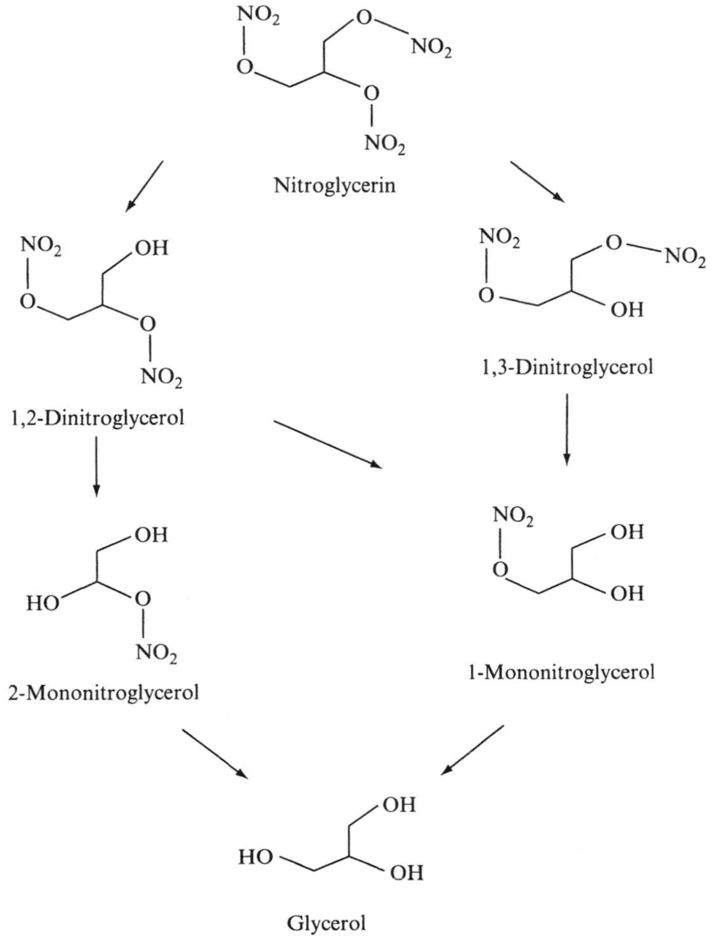

Nitroglycerin

1,2-Dinitroglycerol

1,3-Dinitroglycerol

2-Mononitroglycerol

1-Mononitroglycerol

Glycerol

Figure 4-5 Degradation pathways of nitroglycerin.

denitrated products shown in Figures 4-5 and 4-6. All bacteria where nitrate ester degradation has been characterized have very similar enzymes. The enzymes catalyze the nicotinamide cofactor-dependent reductive cleavage of nitrate esters that produces alcohol and nitrate. Purification of the PETN reductase from *Enterobacter cloacae* yielded a monomeric protein of around 40 kilodaltons, which required NADPH as a cofactor for activity. Similar enzymes were responsible for the nitrate ester degrading activity in *Agrobacterium radiobacter* (Snape *et al.* 1997)—"nitrate ester reductase"—and in the strains of *Pseudomonas fluorescens* and *Pseudomonas putida* (Blehert *et al.* 1999)—"xenobiotic reductases." All utilize a noncovalently bound flavine mononucleotide as a redox cofactor.

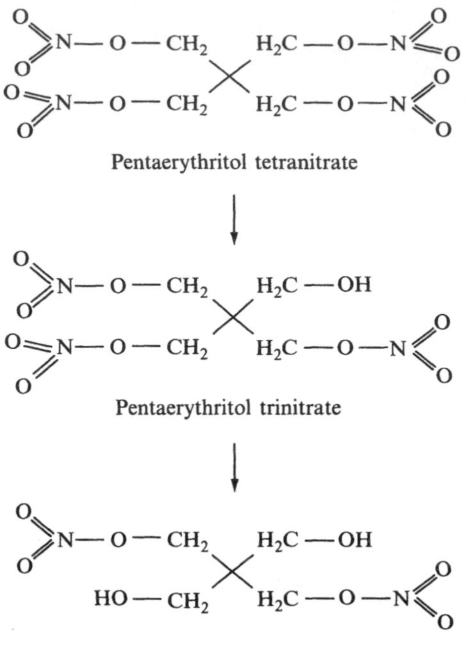

Pentaerythritol tetranitrate

↓

Pentaerythritol trinitrate

↓

Pentaerythritol dinitrate

Figure 4-6 Scheme of pentaerythritol tetranitrate degradation. Not shown in the pathway, the degradation continued *via* pentaerythritol mononitrate to pentaerythritol, which might be utilized as a carbon source.

By contrast, there are few reports concerning nitroester degradation in plants. Goel *et al.* (1997) demonstrate that beetroot (*Beta vulgaris*) denitrates GTN, and the denitration is enhanced by expression of PETN reductase in transgenic seedlings, which also enhances the denitration of glyceroldinitrate (GDN) to glycerolmononitrate (GMN). French *et al.* (1999) followed the same approach and developed transgenic tobacco (*Nicotiana tabacum*) plants that express PETN reductase to degrade nitrate ester explosives and TNT. Seeds from this transgenic plant were able to germinate and grow in media containing GTN and TNT levels toxic to the wild type tobacco.

Rheum palmatum tissue culture proceeded along the pathway shown in Figures 4-5 and 4-6 (Vanek *et al.* 2000a). The concentration of GTN was decreased to 39 percent of the initial concentration of 50 milligrams per liter within 10 days, while the same concentration of PETN was totally transformed. From the analysis of products and stoichiometry of both processes, glycerol and pentaerythritol were formed as end products that can eventually be used by plant cells as a carbon source (Maestri *et al.* 1991).

To check the possibility of large-scale application of these results, hydroponic poplar (*Populus simonii*) and aspen (*Populus tremula*) plants were tested with wastewaters from an ammunition factory containing 600 milligrams per liter of nitroesters (GTN and GDN). Results obtained show that both plants were able to grow under these conditions and to eliminate and transform all of the nitroesters within 15 days.

Sulfonated Aromatic Compounds

Around a million tons of more than 10 000 different synthetic dyes and pigments are produced annually worldwide. Approximately 10 percent of this amount is released into the environment, mainly *via* industrial effluents. Synthetic sulfonated anthraquinones are very important starting material to produce a large palette of dyes. Naphthalene sulfonic acids, benzene sulfonate, and *p*-toluene sulfonate are intermediates in the manufacture of optical brighteners, pickling agents, dyestuffs, tanning agents, insecticides, surfactants, antioxidants, wetting agents, and many other products. Containing at least one sulfonic group and often varying substitutions such as nitro groups, these xenobiotic compounds are not uniformly susceptible to decolorization and biodegradation in conventional wastewater treatment plants. The organosulfonate group plays an important role not only in altering the solubility and dispersion properties of the xenobiotic molecule, but also in increasing recalcitrance to microbial breakdown because of the thermodynamically stable carbon–sulfur bond (Kertesz *et al.* 1994). Consequently, effluents from detergent, dye, and textile industries are often contaminated with sulfonated aromatic compounds. These loads are major sources of pollutants to the environment, especially water systems (Greim *et al.* 1994, Young and Yu 1997). It has also been recently reported that benzene- and naphthalene-sulfonates can now be found in leachates from landfills (Riediker *et al.* 2000).

Because dyes usually contain a wide variety of substituted sulfonated aromatic compounds, the spectrum of bacteria that could degrade these chemicals has thus been intensively investigated for possible application to treat this type of industrial wastewater. Microbial degradation of such dyes and by-products often requires unusual catabolic activities rarely found in a single species. Thus, bacterial isolates only degrade a few dyes and the accumulation of dead end products often occurs (Contzen *et al.* 1996, Dangmann *et al.* 1996, Cook *et al.* 1999). The decolorization of several synthetic dyes including azo- and anthraquinone-derivatives has also been examined in white-rot fungal cultures, which are known to produce powerful manganese peroxidase and lignin peroxidase. However, an inhibition occurs at rather low concentrations, depending on the individual dye structure (Muralikrishna and Renganathan 1993, Young and Yu 1997).

Because of the poor capability of microorganisms to degrade sulfonoaromatic compounds, conventional wastewater treatment plants and biofilter processes are ineffective in managing this significant class of contaminants (Robinson *et al.* 2001). In this context, green-plant-based treatments are being explored. The options include (1) constructed wetlands, (2) hydroponic reactors, and (3) processes involving plant materials such as ground horseradish

(*Armoracia* spp.) or agricultural residues. All options offer a potentially low cost, low maintenance biological method for wastewater treatment. Most of the plant based systems currently in use have been designed to treat domestic wastewater, but have a great potential to treat industrial effluents containing recalcitrant organics such as priority pollutants and dyes (Furukawa and Fujita 1993, Davies and Cottingham 1994, Fernandez *et al.* 1996, Bhunia *et al.* 2001).

Sulfonated Anthraquinones

Anthraquinones occur naturally in several plant genera like *Rheum*, *Rumex*, and *Rubia* (Trease and Evans 1983, van den Berg and Labadie 1989, Matsuda *et al.* 2001). Furthermore, the pathway leading to the biosynthesis of natural anthraquinones has been recently unraveled and several enzymes involved in the process have been characterized (Hagendoorn *et al.* 1994, Leduc *et al.* 1997, Ramos-Valdivia *et al.* 1997). It was therefore assumed that the hardy rhubarb (*Rheum palmatum*) might possess enzymes capable of transforming sulfonated anthraquinones that could be harnessed to treat wastewater from the dye and detergent industries. As a first step, cells were isolated from rhubarb (*Rheum palmatum*) and grown in bioreactors in the presence of 700 to 800 milligrams per liter (2 micromoles per liter) anthraquinones with sulfonate groups in different positions (Figure 4-7). At different times, cells were separated from the growth medium by filtration and the filtrate was analyzed by high performance liquid chromatography (HPLC), capillary electrophoresis (CE), and spectrophotometry for the parent compounds and potential metabolites (Schwitzguébel *et al.* 1995, 1996, Schwitzguébel 1996, 1998). The most significant results are summarized in Figures 4-8 and 4-9, and Table 4-1.

Because the expression of enzymes could vary as a function of the cell culture growth phase, anthraquinone-1-sulfonate (AQ-1-S) was added at time zero, day 12 (exponential phase of growth), and day 22 (stationary phase). The highest uptake rate was observed when the sulfonated compound was present from the beginning of the growth (0.13 micromoles per liter of culture and per day), leading to an overall decrease of 95 percent within 15 days of incubation (Figure 4-8). If added after 12 or 22 days of cultivation, the concentration of AQ-1-S decreased at a rate of 0.07 micromoles per liter of culture per day. Neither derivatives nor metabolites of AQ-1-S were observed by HPLC. In contrast, analyses done with CE showed the appearance of two unidentified compounds in the medium. The first metabolite appeared after 5 days of incubation, reached a maximum after 12 days, and then totally disappeared. The second metabolite appeared after 8 days of incubation and increased as a function of time until the end of cultivation at 21 days. The identification of these metabolites is in progress. Transformation of AQ-1-S by rhubarb (*Rheum palmatum*) cells was also suggested by spectrophotometric measurements of the filtrate. The ratio of absorbance at 256 nanometers and 326 nanometers markedly decreased as a function of time. However, no desulfonation of AQ-1-S occurred, because no release of sulfate was observed (Figure 4-9).

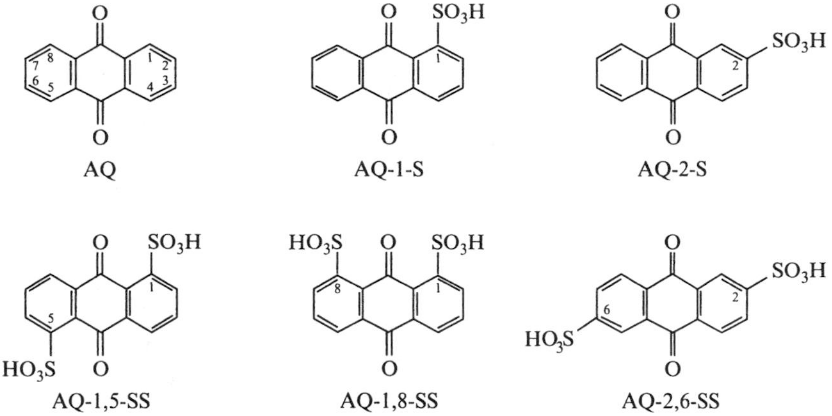

AQ AQ-1-S AQ-2-S

AQ-1,5-SS AQ-1,8-SS AQ-2,6-SS

Figure 4-7 Structure of sulfonated anthraquinones.

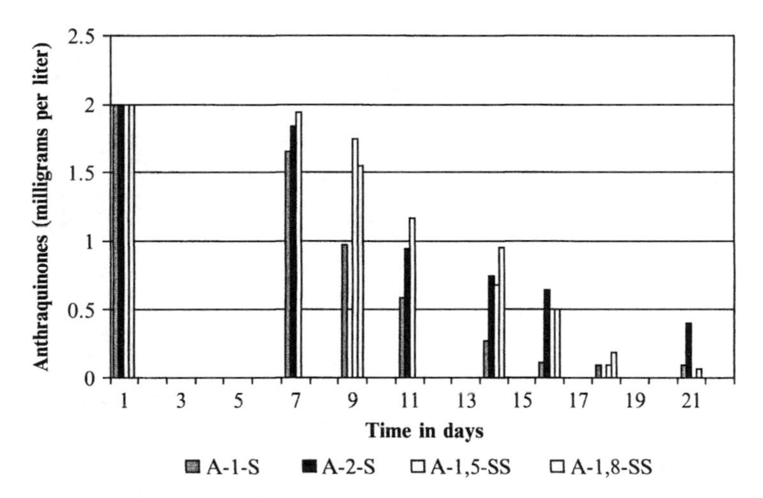

Figure 4-8 Removal of sulfonated anthraquinones by rhubarb (*Rheum palmatum*) cells cultivated *in vitro*. Results are expressed as the concentration of sulfonated anthraquinones remaining in the medium of cultivation. See definitions of AQ-1-S, AQ-2-S, AQ-1,5-SS, and AQ-1,8-SS in Table 4-1 and Figure 4-7.

Starting with 2 micromoles per liter from the beginning of growth, the concentration of anthraquinone-2-sulfonate (AQ-2-S) decreased to 0.64 micromoles per liter within 15 days, for an uptake rate of 0.08 to 0.09 micromoles per liter of culture per day (Figure 4-8). Even if the uptake rate of AQ-2-S was slower than the value observed for AQ-1-S, a phytotransformation also occurred. Unlike AQ-1-S, an HPLC peak of an unknown compound appeared in the medium after 10 days of incubation, increasing linearly with time. The occurrence of a phytotransformation process was also confirmed by results obtained with CE and spectrophotometry. A significant

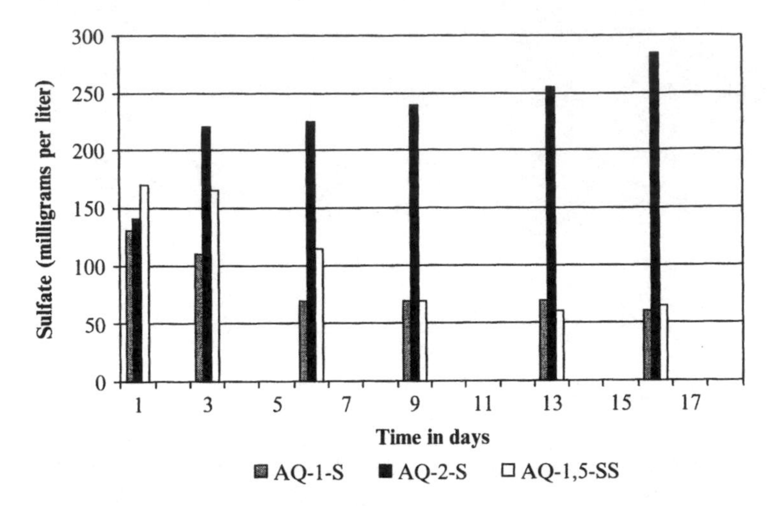

Figure 4-9 Transformation of sulfonated anthraquinones by rhubarb (*Rheum palmatum*) cells cultivated *in vitro*. Results are expressed as sulfate concentration in the medium of cultivation. The initial concentration of sulfate was 170 milligrams per liter. See definitions of AQ-1-S, AQ-2-S, and AQ-1,5-SS in Table 4-1 and Figure 4-7.

TABLE 4-1 Present Knowledge of the Fate of Different Aromatic Sulfonated Compounds by Microorganism Free Rhubarb (*Rheum palmatum*) Cells

Compound	Uptake	Transformation	Desulfonation	Mineralization
AQ-1-S	++++	++	?	?
AQ-2-S	++	+++	++	?
AQ-1,5-SS	++++	?	?	?
AQ-1,8-SS	++++	?	?	?
AQ-2,6-SS	++	?	?	?
I	++	+	?	?
II	++++	++	?	?
III	+++	+++	++	?
IV	++	?	?	?
V	++	++	?	?

The following abbreviations are used—AQ-1-S: anthraquinone-1-sulfonate; AQ-2-S: anthraquinone-2-sulfonate; AQ-1,5-SS: anthraquinone-1,5-disulfonate; AQ-1,8-SS: anthraquinone-1,8-disulfonate; AQ-2,6-SS: anthraquinone-2,6-disulfonate;

Note: I: 2-chloro-5-nitro-benzene sulfonate; II: 2-hydroxy-4-sulfo-1-naphthalenediazonium; III: 2-hydroxy-4-sulfo-6-nitro-1-naphthalenediazonium; IV: 7-nitro-1,3-naphthalene disulfonate; and V: 7-amino-1,3-naphthalene disulfonate. Anthraquinone sulfonates were added at 2 micromoles per liters, compounds I to V at 1 micromoles per liters. Each "+" means 25 percent of the added compound taken up, transformed, or desulfonated and "?" means unknown.

desulfonation occurred during the phytotransformation of AQ-2-S, as indicated by the increasing release of sulfate (Figure 4-9).

Also starting with a concentration of 2 micromoles per liter from the beginning of growth, the concentration of anthraquinone-1,5-disulfonate

(AQ-1,5-SS) decreased at a rate of 0.08 to 0.11 micromoles per liter of culture per day. Neither HPLC, CE, nor spectrophotometry could detect intermediates or metabolites of A-1,5-SS in the medium. This compound appeared to be simply accumulated by rhubarb (*Rheum palmatum*) cells, without any phytotransformation (Figures 4-8 and 4-9 and Table 4-1). Starting at a concentration of 2 micromoles per liter supplied from the beginning of the growth, the concentration of anthraquinone-1,8-disulfonate (AQ-1,8-SS) decreased to 0.06 micromoles per liter after 20 days, showing an uptake rate of about 0.10 micromoles per liter of culture per day. Starting at a concentration of 2 micromoles per liter from the beginning of the growth, the concentration of anthraquinone-2,6-disulfonate (AQ-2,6-SS) decreased only to 1.52 micromoles per liter after 17 days, for an uptake rate of 0.028 micromoles per liter of culture per day (Table 4-1).

Other Sulfonated Aromatic Compounds

Based on promising results obtained with sulfonated anthraquinones, the ability of rhubarb (*Rheum palmatum*) cells to accumulate and transform other sulfonated aromatic compounds was investigated. Results obtained showed that cultured rhubarb (*Rheum palmatum*) cells were also able to efficiently accumulate 2-chloro-5-nitro-benzene sulfonate, 2-hydroxy-4-sulpho- and 2-hydroxy-4-sulfo-6-nitro-naphthalene-diazonium, as well as 1,3-naphthalene disulfonates, containing either an amino or a nitro group in position 7 (Table 4-1). Furthermore, rhubarb (*Rheum palmatum*) cells were able to transform four of these compounds, without releasing the metabolites into the medium. When present at an initial concentration of 1 micromole per liter (250 to 350 milligrams per liter), the four naphthalene derivatives almost totally disappeared from the medium within a week and were not detrimental to rhubarb (*Rheum palmatum*) cells cultivated *in vitro*. At higher concentrations, or with 2-chloro-5-nitro-benzene sulfonate, the removal achieved was less complete or took more time (Schwitzguébel *et al.* 1998, Duc *et al.* 1999a, 1999b). The position of the sulfonated group on the aromatic ring structure, rather than the number of such groups, appears to be the main factor controlling the fate of these pollutants in rhubarb (*Rheum palmatum*) cells. Differences observed between the fate of several similar sulfonated aromatic compounds in plant cells cultivated *in vitro* look extremely interesting and could be of great significance for possible environmental applications (treatment of industrial effluents and phytoremediation of polluted soils or groundwater).

DEVELOPMENT OF A BIOLOGICAL PROCESS TO TREAT WASTEWATER CONTAINING SULFONATED AROMATIC POLLUTANTS

On the basis of the promising results already obtained, two different but complementary issues are presently under investigation at the Swiss Federal Institute of Technology, Lausanne, including the following:

1. It is a prerequisite that living plants used in most phytoremediation or rhizofiltration processes are resistant or tolerant to the level of pollutants present. For optimal applications, the capability of whole plants to germinate, grow, and develop in the presence of pollutant amounts comparable to those found in industrial effluents must be investigated. Therefore, the project started with the screening of plant material against several sulfonated aromatic pollutants of current concern. Not only rhubarb (*Rheum palmatum*), but also other plants producing anthraquinones will be tested, such as water dock (*Rumex hydrolapatum*). As a comparison, plants not producing anthraquinones, such as maize (*Zea mays*), rape (*Brassica napus*), and celery (*Apium graveolens*), will be grown in the presence of different sulfonoaromatics. Plants selected for tolerance will be tested for growth under hydroponic conditions and the capacity to remove sulfonoaromatics from contaminated water will be evaluated. If the pollutant under investigation disappears from the liquid medium, then the possible adsorption, accumulation, transformation, and degradation by the plant will be quantified. The rates and yields of pollutant removal calculated per unit biomass will be used as one of the criteria for the choice of plant species for future applications.

2. In higher plants producing natural anthraquinones, these compounds often exist as glycosylated derivatives (van der Plas *et al.* 1998, Matsuda *et al.* 2001). Therefore, the possible involvement of glycosyltransferases in the metabolism of sulfonated anthraquinones will be determined, along with the capacity to degrade these xenobiotic compounds. In addition, the capability of the enzymes that naturally biosynthesize anthraquinones to metabolize synthetic sulfonated aromatic compounds will be evaluated. As a comparison, enzymes from plants not producing anthraquinones, but known to transform or conjugate other organic pollutants, will be tested. The possible use of sulfonated organic pollutants as a sulfur source, after desulfonation, will be determined for different plant species (Hell 1997). Finally, sulfonated aromatic pollutants could also be directly accumulated into the vacuole, without being transformed or conjugated (Klein *et al.* 1997). Vacuoles will thus be isolated from selected plant species and the capability to accumulate sulfonated aromatic pollutants will be evaluated. A better understanding of biochemical mechanisms leading to the accumulation into the vacuole, conjugation, transformation, and degradation of sulfonated aromatic compounds is expected for different plant species—those that produce natural anthraquinones and those that do not.

CURRENT KNOWLEDGE, AREAS OF UNCERTAINTY, AND NEEDS FOR FUTURE RESEARCH AND DEVELOPMENT

The objective of phytoremediation is to use plants to remove organic pollutants and toxic metals from the environment and convert these compounds to harmless products. To achieve this, plants must be able to

1. Take up the pollutants from the environment
2. Transfer the pollutants to the site of metabolism or storage

3. Detoxify the xenobiotic compounds by transformation, conjugation, or storage

An extensive knowledge is presently available on genes and enzymes involved in the detoxification of xenobiotic compounds, which occur in many plants. One of the most important challenges is how to use this basic scientific information to improve the efficiency of phytoremediation in the field. This could imply the use of genetically modified plants, specially designed and constructed for remediation purposes.

Another critical factor for the successful implementation of phytoremediation is to better understand and control the uptake and transfer of organic pollutants. Numerous organic pollutants are hydrophobic [logarithm of the octanol–water partitioning coefficient (log K_{ow}) greater than 4] and thermodynamically stable in the environment. Bioavailability is thus often a major limiting factor for phytoremediation of persistent organic pollutants. Of particular interest are the roles of root exudates and arbuscular mycorrhizal fungi in the modification of the capability of plants to take up pollutants from contaminated soils and wastewaters. Therefore, more research is needed to explore and exploit the subtle and complex interactions between pollutants, soil material, plant roots, dead fine roots, and microorganisms in the rhizosphere (Bromilow and Chamberlain 1995, Trapp 2000). The transfer of pollutants from the root to the upper parts of the plant is also not well known, and studies on xylem and phloem mobility of xenobiotic compounds in relation to physicochemical properties are scarce (Bromilow and Chamberlain 1995, Grimm *et al.* 1995, Wright *et al.* 1996, Trapp 2000).

Physical, chemical, and biological stress seems to result in rapid activation of a variety of oxido-reduction enzymes, including those associated with the respiratory system, to minimize oxidative damage. Growing plants in the presence of high concentrations of pollutants that interact with these enzymes may also cause significant stress to plants. Therefore, the mutual interactions between xenobiochemical metabolism and the whole plant metabolic network *in vivo* must also be addressed (Gordeziani *et al.* 1999, Palmer 1999). It is especially necessary to (1) characterize the effects of xenobiotic compounds on the metabolism of plants, and (2) detect and localize pollutants and the associated degradation products at the subcellular and tissue levels. For such a purpose, the application of noninvasive nuclear magnetic resonance methods should be very useful (Ratcliffe and Roscher 1998). In conclusion, it appears that the combined use of adequate approaches at the laboratory scale is very helpful to increase tolerance and the long-term survival of the plant, and thus to improve the efficiency of phytoremediation in the field.

Acknowledgments

The work at the Institute of Organic Chemistry and Biochemistry of the Czech Academy of Science was supported by COST 837.10, GACR grant No. 206/99/1252, 206/99/P034, and Z4 055 905 research project. J.-P. Schwitzguébel acknowledges the Swiss

Federal Office for Education and Science for financial support (Grants COST C98.0047, C99.0016, and C01.0079).

REFERENCES

Andrews, C.J., I. Jepson, M. Skipsey, J.K. Townson, and R. Edwards (1997) Glutathione transferase activities toward herbicides used selectively in soybean. *Pestic. Sci.* **51**: 213–222.

Batard, Y., M. Le Ret, M. Schalk, T. Robineau, F. Durst, and D. Werck-Reichhart (1998) Molecular cloning and functional expression in yeast of CYP76B1, a xenobiotic-inducible 7-ethoxycoumarin *O*-deethylase from *Helianthus tuberosus*. *Plant J.* **14**: 111–120.

Bennett, J.W. (1994) Prospects for fungal bioremediation of TNT munition waste. *Int. Biodeter. Biodegr.* **34**: 21–34.

Bhadra, R., R.J. Spanggord, D.G. Wayment, J.B. Hughes, and J.V. Shanks (1999a) Characterization of oxidation products of TNT metabolism in aquatic phytoremediation systems of *Myriophyllum aquaticum*. *Environ. Sci. Technol.* **33**: 3354–3361.

Bhadra, R., D.G. Wayment, J.B. Hughes, and J.V. Shanks (1999b) Confirmation of conjugation processes during TNT metabolism axenic plant roots. *Environ. Sci. Technol.* **33**: 446–452.

Bhunia, A., S. Durani, and P.P. Wangikar (2001) Horseradish peroxidase catalyzed degradation of industrially important dyes. *Biotechnol. Bioeng.* **72**: 562–567.

Blake-Kalff, M.M.A., R.A. Randall, and J.O.D. Coleman (1997) Compartmentation of detoxified xenobiotics in plant cells. In: *Regulation of Enzymatic Systems Detoxifying Xenobiotics in Plants*. K.K. Hatzios, ed. Kluwer Academic Publishers, Dordrecht, The Netherlands, pp. 245–259.

Blehert, D.S., B.G. Fox, and G.H. Chambliss (1999) Cloning and sequence analysis of two *Pseudomonas* flavoprotein xenobiotic reductases. *J. Bacteriol.* **181**: 6254–6263.

Bokern, M., P. Raid, and H. Harms (1998) Toxicity, uptake and metabolism of 4-n-nonylphenol in root cultures and intact plants under septic and aseptic conditions. *Environ. Sci. Pollut. Res.* **5**: 21–27.

Bromilow R.H. and K. Chamberlain (1995) Principles governing uptake and transport of chemicals. In: *Plant Contamination—Modeling and Simulation of Organic Chemical Processes*. S. Trapp, J.C. McFarlane, eds. Lewis Publishers, Boca Raton, Florida, pp. 37–68.

Burken, J.G., J.V. Shanks, and P.L. Thompson (2000) Phytoremediation and plant metabolism of explosives and nitroaromatic compounds. In: *Biodegradation of Nitroaromatic Compounds and Explosives*. J.C. Spain, J.B. Hughes, and H.J. Knackmuss, eds. Lewis Publishers, Boca Raton, Florida, pp. 240–275.

Chapple, C. (1998) Molecular-genetic analysis of plant cytochrome P450-dependent monooxygenases. In: *Annual Review of Plant Physiology and Plant Molecular Biology*. Annual Reviews, Palo Alto, California. **49**: 311–343.

Cole, D.J. and R. Edwards (2000) Secondary metabolism of agrochemicals in plants. In: *Metabolism of Agrochemicals in Plants*. T. Roberts, ed. John Wiley, Chichester, United Kingdom, pp. 107–154.

Cole, D.J., I. Cummins, P.J. Hatton, D. Dixon, and R. Edwards (1997) Glutathione transferases in crops and major weeds. In: *Regulation of Enzymatic Systems Detoxifying Xenobiotics in Plants*. K.K. Hatzios, ed. Kluwer Academic Publishers, Dordrecht, The Netherlands, pp. 139–154.

Coleman, J.O.D., M.M.A. Blake-Kalff, and T.G.E. Davies (1997) Detoxification of xenobiotics by plants: chemical modification and vacuolar compartmentation. *Trends Plant Sci.* **2**: 144–151.

Contzen, M., R.M. Wittich, H.J. Knackmuss, and A. Stolz (1996) Degradation of benzene 1,3-disulfonate by a mixed bacterial culture. *FEMS Microbiol. Lett.* **136**: 45–50.

Cook, A.M., H. Laue and F. Junker (1999) Microbial desulfonation. *FEMS Microbiol. Rev.* **22**: 399–419.

Coyle, C., P. Duggan, M. Godinho, and D. McGrath (1999) The development of a phytoremediation technique for the detoxification of soils contaminated with phenolic compounds using horseradish peroxidase (*Armoracia rusticana*): preliminary results. *Int. J. Phytorem.* **1**: 189–202.

Cummins, I., D.J. Cole, and R. Edwards (1997) Purification of multiple glutathione transferases involved in herbicide detoxification from wheat (*Triticum aestivum* L.) treated with the safener fenchloarazole-ethyl. *Pestic. Biochem. Physiol.* **59**: 35–49.

Dangmann, E., A. Stolz, A.E. Kuhm, A. Hammer, B. Feigel, N. Noisommit-Rizzi, M. Rizzi, M. Reuss, and H.J. Knackmuss (1996) Degradation of 4-aminobenzenesulfonate by a two-species bacterial coculture—Physiological interactions between *Hydrogenophaga palleronii* S1 and *Agrobacterium radiobacter* S2. *Biodegradation* **7**: 223–229.

Davies, T.H. and P.D. Cottingham (1994) The use of constructed wetlands for treating industrial effluents (textile dyes). *Water Sci. Technol.* **29**: 227–232.

Dixon, D.P., I. Cummins, D.J. Cole, and R. Edwards (1998) Glutathione-mediated detoxification systems in plants. *Curr. Opin. Plant Biol.* **1**: 258–266.

Dixon, D.P., D.J. Cole, and R. Edwards (1999a) Dimerisation of maize glutathione transferases in recombinant bacteria. *Plant Mol. Biol.* **40**: 997–1008.

Dixon, D.P., D.J. Cole, and R. Edwards (1999b) The role of plant glutathione transferases in xenobiotics metabolism. In: *Plant Enzymes Involved in the Metabolism of Organic Pollutants*. COST Action 837, held in Geneva, Switzerland, Dec. 9–11, p. 8 (http://lbewww.epfl.ch/COST837/).

Drzyzga, O., D. Bruns-Nagel, T. Gorontzy, K.H. Blotevogel, D. Gemsa, and E. von Low (1998) Mass balance studies with C-14-labeled 2,4,6-trinitrotoluene (TNT) mediated by an anaerobic *Desulfovibrio* species and an aerobic *Serratia* species. *Curr. Microbiol.* **37**: 380–386.

Duc, R., T. Vanek, P. Soudek, and J.P. Schwitzguébel (1999a) Experimentation with rhubarb in Europe. *Soil and Groundwater Cleanup* February/March: 27–30.

Duc, R., T. Vanek, P. Soudek, and J.P. Schwitzguébel (1999b) Accumulation and transformation of sulfonated aromatic compounds by rhubarb cells (*Rheum palmatum*) *Int. J. Phytorem.* **1**: 255–271.

Durst, F., I. Benveniste, A. Lesot, J.P. Salaün, and D. Werck-Reichhart (1997) Induction of plant cytochrome P450. In: *Regulation of Enzymatic Systems Detoxifying Xenobiotics in Plants*, K.K. Hatzios, ed. Kluwer Academic Publishers, Dordrecht, The Netherlands, pp. 19–34.

Edwards, R., D.P. Dixon, and V. Walbot (2000) Plant glutathione *S*-transferases: enzymes with multiple functions in sickness and in health. *Trends Plant Sci.* **5**: 193–198.

Ellis, B.E. (1988) Natural products from plant-tissue culture. *Nat. Prod. Rep.* **5**: 581–612.

Fernandez, N., E. Chacin, C. Garcia, N. Alastre, F. Leal, and C.F. Forster (1996) The use of seed pods from *Albizia lebbek* for the removal of alkyl benzene sulphonates from aqueous solution. *Process Biochem.* **31**: 383–387.

Flores, H.E., Y.R. Dai, A.J. Freyer, and P.J. Michaels (1994) Biotransformation of butylated hydroxytoluene in hairy root cultures. *Plant Physiol. Biochem.* **32**: 511–519.

Fowler, M.W. (1987) Plant cell biotechnology. In: *Basic Biotechnology*. J. BuLock and B. Kristiansen, eds. Academic Press, San Diego, pp. 525–544.

French, C.E., S.J. Rosser, G.J. Davies, S. Nicklin, and N.C. Bruce (1999) Biodegradation of explosives by transgenic plants expressing pentaerythritol tetranitrate reductase. *Nat. Biotechnol.* **17**: 491–494.

Frova, C., M. Sari-Gorla, M.E. Pe, A. Greenland, I. Jepson, and L. Rossini (1997) Role of the different GST isozymes of maize in herbicide tolerance: genetic and biochemical analysis. In: *Regulation of Enzymatic Systems Detoxifying Xenobiotics in Plants*. K.K. Hatzios ed. Kluwer Academic Publishers, Dordrecht, The Netherlands, pp. 171–181.

Furukawa, K. and M. Fujita (1993) Advanced treatment and food production by hydroponic type wastewater treatment plant. *Water Sci. Technol.* **28**: 219–228.

Gianfreda, L. and P. Nannipieri (2001) Basic principles, agents and feasibility of bioremediation of soil polluted by organic compounds. *Minerva Biotecnol.* **13**: 5–12.

Giese M., U. Bauerdoranth, C. Langebartels, and H. Sandermann, Jr. (1994) Detoxification of formaldehyde by the spider plant (*Chlorophytum comosum* L.) and by soybean (*Glycine max* L.) cell-suspension cultures. *Plant Physiol.* **104**: 1301–1309.

Goel, A., G. Kumar, G.F. Payne, and S.K. Dube (1997) Plant cell biodegradation of a xenobiotic nitrate ester, nitroglycerin. *Nat. Biotechnol.* **15**: 174–177.

Gong, P., B.M. Wilke, and S. Fleischmann (1999) Soil-based phytotoxicity of 2,4,6-trinitrotoluene (TNT) to terrestrial higher plants. *Arch. Environ. Contam. Toxicol.* **36**: 152–157.

Gordeziani, M., G. Khatisashvili, T. Ananiashvili, T. Varazashvili, M. Kurashvili, G. Kvesitadze, and P. Tkhelidze (1999) Energetic significance of plant monooxygenase individual components participating in xenobiotics degradation. *Int. Biodeter. Biodegr.* **44**: 49–54.

Greim, H., J. Ahlers, R. Bias, B. Broecker, H. Hollander, H.P. Gelbke, H.J. Klimisch, I. Mangelsdorf, A. Paetz, N. Schön, G. Stropp, R. Vogel, C. Weber, K. Ziegler-Skylakakis, and E. Bayer (1994) Toxicity and ecotoxicity of sulfonic acids: structure-activity relationship. *Chemosphere* **28**: 2203–2236.

Griest, W.H., A.J. Stewart, A.A. Vass, and C.H. Ho (1998) Chemical and toxicological characterization of slurry reactor biotreatment of explosives-contaminated soils. Final Report SFIM-AEC-ET-CR-96186. U.S. Army Environmental Center, Aberdeen Proving Ground, Maryland.

Grimm, E., A. Grube, S. Jahnke, and S. Neumann (1995) Retention of xenobiotics along the phloem path. *Planta* **197**: 11–18.

Hagendoorn, M.J.M., L.H.W. van der Plaas, and G.J. Segers (1994) Accumulation of anthraquinones in *Morinda citrifolia* cell suspensions. a model system for the study of the interaction between secondary and primary metabolism. *Plant Cell, Tissue Org. Culture* **38**: 227–234.

Halkier, B.A. (1996) Catalytic reactivities and structure/function relationships of cytochrome P450 enzymes. *Phytochemistry* **43**(1): 1–21.

Harms, H.H. (1992) *In vitro* systems for studying phytotoxicity and metabolic fate of pesticides and xenobiotics in plants. *Pestic. Sci.* **35**: 277–281.

Hawari, J., A. Halasz, L. Paquet, E. Zhou, B. Spencer, G. Ampleman, and S. Thiboutot (1998) Characterization of metabolites in the biotransformation of 2,4,6-trinitrotoluene with anaerobic sludge: role of triaminotoluene. *Appl. Environ. Microbiol.* **64**: 2200–2206.

Hehn, A., L. Didierjean, Y. Batard, T. Robineau, and D. Werck-Reichhart (1999) Chemically-inducible P450 involved in phenylurea metabolism in higher plants. In: Plant Enzymes Involved in the Metabolism of Organic Pollutants. COST Action 837, held in Geneva, Switzerland, Dec 9–11, p. 10 (http://lbewww.epfl.ch/COST837/).

Hell, R. (1997) Molecular physiology of plant sulfur metabolism. *Planta* **202**: 138–148.

Higson, F.K. (1992) Microbial degradation of nitroaromatic compounds. *Adv. Appl. Microbiol.* **37**: 1–19.

Hubálek, M. and T. Vanek (1998) Degradation of 2,4,6-trinitrotoluene by cell suspension of *Solanum aviculare. Int. Biodeter. Biodegr.* **42**: pp 251.

Jepson, I., D.C. Holt, V. Roussel, S.Y. Wright, and A.J. Greenland (1997) Transgenic plant analysis as a tool for the study of maize glutathione *S*-transferases. In: *Regulation of Enzymatic Systems Detoxifying Xenobiotics in Plants.* K.K. Hatzios, ed. Kluwer Academic Publishers, Dordrecht, The Netherlands, pp. 313–323.

Jones, P. and T. Vogt (2001) Glycosyltransferases in secondary plant metabolism: tranquilizers and stimulant controllers. *Planta* **213**: 164–174.

Kertesz, M.A., A.M. Cook, and T. Leisinger (1994) Microbial metabolism of sulfur- and phosphorus-containing xenobiotics. *FEMS Microbiol. Rev.* **15**: 195–215.

Khatisashvili, G., M. Gordeziani, G. Kvesitadze, and F. Korte (1997) Plant monooxygenases: participation in xenobiotic oxidation. *Ecotoxicol. Environ. Safe.* **36**: 118–122.

Klein, M., G. Weissenboeck, A. Dufaud, C. Gaillard, K. Kreuz, and E. Martinoia (1996) Different energization mechanisms drive the vacuolar uptake of a flavonoid glucoside and an herbicide glucoside. *J. Biol. Chem.* **271**: 29666–29671.

Klein, M., E. Martinoia, and G. Weissenboeck (1997) Transport of lucifer yellow CH into plant vacuoles—Evidence for direct energization of a sulphonated substance and implications for the design of new molecular probes. *FEBS Lett.* **420**: 86–92.

Koehler, H., J. Warrelmann, P. Behrend, T. Frische and T. Lorenzen (2001) *In situ* phytoremediation of TNT-contaminated soil. Abstracts of ISEB Phytoremediation Meeting. held in Leipzig, Germany, May 15–17, p. 35.

Kolb, M. and H. Harms (2000) Metabolism of fluoranthene in different plant cell cultures and intact plants. *Environ. Toxicol. Chem.* **19**: 1304–1310.

Komossa, D., C. Langebartels, and H. Sandermann, Jr. (1995) Metabolic processes for organic chemicals in plants. In: *Plant Contamination—Modeling and Simulation of Organic Chemical Processes*. S. Trapp and J.C. McFarlane, eds. Lewis Publishers, Boca Raton, Florida, pp. 69–103.

Kreuz, K., R. Tommasini, and E. Martinoia (1996) Old enzymes for a new job—herbicide detoxification in plants. *Plant Physiol.* **111**: 349–353.

Laurent, F.M.G. and R. Scalla (1999) Metabolism and cell wall incorporation of phenoxyacetic acid in soybean cell suspension culture. *Pestic. Sci.* **55**: 3–10.

Lamoureux, G.L., D.G. Rusness, P. Schroeder, and H. Rennenberg (1991) Diphenyl ether herbicide metabolism in a spruce cell suspension culture: the identification of two novel metabolites derived from a glutathione conjugate. *Pestic. Biochem. Physiol.* **39**: 291–301.

Leduc, C., I. Birgel, R. Muller, and E. Leistner (1997) Isochorismate hydroxymutase from a cell-suspension culture of *Galium mollugo. Planta* **202**: 206–210.

Lewis, T.A., S. Goszczynski, R.L. Crawford, P.A. Korus, and W. Admassu (1996) Products of anaerobic 2,4,6-trinitrotoluene (TNT) transformation by *Clostridium bifermentans. Appl. Environ. Microbiol.* **62**: 4669–4674.

Lu, Y.P., Z.S. Li, Y.M. Drozdowicz, S. Hoertensteiner, E. Martinoia, and P.A. Rea (1998) AtMRP2, an *Arabidopsis* ATP binding cassette transporter able to transport glutathione *S*-conjugates and chlorophyll catabolites: functional comparisons with AtMRP1. *Plant Cell.* **10**: 267–282.

Lucero, M.E., W. Mueller, J. Hubstenberger, G.C. Phillips, and M.A. O'Connell (1999) Tolerance to nitrogenous explosives and metabolism of TNT by cell suspensions of *Datura innoxia. In Vitro Cell. Dev. Biol.: Plant* **35**: 480–486.

Maestri, E., F.M. Restivo, M. Gulli, and F. Tassi (1991) Glutamate-dehydrogenase regulation in callus-cultures of *Nicotiana plumbaginifolia*—effect of glucose feeding and carbon source starvation on the isoenzymatic pattern. *Plant Cell Environ.* **14**: 613–618.

Malcherek, K., J. Breuer, I. Schuphan I, and B. Schmidt (1998) Metabolism of 4-nitrophenol in aseptically cultivated plants of the species wheat (*Triticum aestivum* L.), soybean (*Glycine max* L.), wild oat (*Avena fatua* L.) and corn cockle (*Agrostemma githago* L.). *J. Plant Physiol.* **153**: 192–199.

Matsuda, H., T. Morikawa, I. Toguchida, J.Y. Park, S. Harima, and M. Yoshikawa (2001) Antioxidant constituents from rhubarb: structural requirements of stilbenes for the activity and structures of two new anthraquinone glucosides. *Bioorg. Med. Chem.* **9**: 41–50.

Messner, B., H. Sandermann, A. Schaeffner, O. Thulke, and S. Wegener (2000) Functional genomics of *Arabidopsis* glycosyl transferases. In: Phytoremediation 2000—State of the Art in Europe—An Intercontinental Comparison. COST Action 837 Workshop, held in Hersonissos, Crete, Greece, April 6–8, pp. 85–86 (http://lbeww.epfl.ch/COST837/).

Muralikrishna, C. and V. Renganathan (1993) Peroxidase-catalyzed desulfonation of 3,5-dimethyl-4-hydroxy and 3,5-dimethyl-4-aminobenzenesulfonic acids. *Biochem. Biophys. Res. Comm.* **197**: 798–804.

Nepovím, A., M. Hubálek, and T. Vanek (1999) Enzymatic degradation of 2,4,6-TNT by cell suspension culture of *Rheum palmatum*. In: Plant Enzymes Involved in the

Metabolism of Organic Pollutants. COST Action 837, held in Geneva, Switzerland, Dec. 9–11, p. 13 (http://lbeww.epfl.ch/COST837/).

Palmer, J.M. (1999) The nature and regulation of terminal oxidases in plants—their role in integrated metabolism and response to stress. In: Plant Enzymes Involved in the Metabolism of Organic Pollutants. COST Action 837, held in Geneva, Switzerland, Dec. 9–11, pp. 14–15 (http://lbeww.epfl.ch/COST837/).

Pavlostathis, S.G., K.K. Comstock, M.E. Jacobson, and F.M. Saunders (1998) Transformation of 2,4,6-trinitrotoluene by the aquatic plant *Myriophyllum spicatum. Environ. Toxicol. Chem.* **17**: 2266–2273.

Pflugmacher, S. and H. Sandermann, Jr. (1998) Taxonomic distribution of plant glucosyltransferases acting on xenobiotics. *Phytochemistry* **49**: 507–511.

Pflugmacher, S., K. Geissler, and C. Steinberg (1999) Activity of phase I and phase II detoxification enzymes in different cormus parts of *Phragmites australis. Ecotoxicol. Environ. Safe.* **42**: 62–66.

Pflugmacher, S., P. Schroeder, and H. Sandermann, Jr. (2000) Taxonomic distribution of plant glutathione *S*-transferases acting on xenobiotics. *Phytochemistry* **54**: 267–273.

Ramos-Valdivia, A., R. van der Heijden, and R. Verpoorte (1997) Elicitor-mediated induction of anthraquinone biosynthesis and regulation of isopentenyl diphosphate isomerase and farnesyl diphosphate synthase activities in cell suspension cultures of *Cinchona robusta. Planta* **203**: 155–161.

Ratcliffe, R.G. and A. Roscher (1998) Prospects for *in vivo* NMR methods in xenobiotic research in plants. *Biodegradation* **9**: 411–422.

Rea, P.A., Z.S. Li, Y.P. Lu, Y.M. Drozdowicz, and E. Martinoia (1998) From vacuolar GS-X pumps to multispecific ABC transporters. In: *Annual Review of Plant Physiology and Plant Molecular Biology*. Annual Reviews, Palo Alto, California. **49**: 727–760.

Riediker, S., M.J.F. Suter, and W. Giger (2000) Benzene and naphthalenesulfonates in leachates and plumes of landfills. *Water Res.* **34**: 2069–2079.

Rieger, P.G. and H.J. Knackmuss (1995) Basic knowledge and perspectives on biodegradation of 2,4,6-trinitrotoluene and related nitroaromatic compounds in contaminated soil. In: *Biodegradation of Nitroaromatic Compounds*. J.C. Spain, ed. Plenum Press, New York, pp. 1–18.

Rivera, R., V.F. Medina, S.L. Larson, and S.C. McCutcheon (1998) Phytotreatment of TNT-contaminated groundwater. *J. Soil Contam.* **7**: 511–529.

Robineau, T., Y. Batard, S. Nedelkina, F. Cabello-Hurtado, M. Le Ret, O. Sorokine, L. Didierjean, and D. Werck-Reichhart (1998) The chemically inducible plant cytochrome P450 CYP76B1 actively metabolizes phenylureas and other xenobiotics. *Plant Physiol.* **118**: 1049–1056.

Robinson, T., G. McMullan, R. Marchant, and P. Nigam (2001) Remediation of dyes in textile effluent: a critical review on current treatment technologies with a proposed alternative. *Bioresource Technol.* **77**: 247–255.

Rossini, L., I. Jepson, A.J. Greenland, and M. Sari-Gorla (1996) Characterization of glutathione *S*-transferase isoforms in three maize inbred lines exhibiting differential sensitivity to alachlor. *Plant Physiol.* **112**: 1595–1600.

Rossini, L., C. Frova, M.E. Pe, L. Mizzi, and M. Sari-Gorla (1998) Alachlor regulation of maize glutathione *S*-transferase genes. *Pestic. Biochem. Physiol.* **60**: 205–211.

Sandermann, H. Jr., D. Scheel, and T. Von der Trenck (1984) Use of plant-cell cultures to study the metabolism of environmental chemicals. *Ecotoxicol. Environ. Safe.* **8**: 167–182.

Sari-Gorla, M., P. Krajewski, G. Binelli, C. Frova, G. Taramino, and M. Villa (1997) Genetic dissection of herbicide tolerance in maize by molecular markers. *Mol. Breeding* **3**: 481–493.

Schafer, W. and H. Sandermann, Jr. (1988) Metabolism of pentachlorophenol in cell-suspension cultures of wheat (*Triticum aestivum* L.)—tetrachlorocatechol as a primary metabolite. *J. Agric. Food Chem.* **36**: 370–377.

Schalk, M., M.A. Pierrel, A. Zimmerlin, Y. Batard, F. Durst, and D. Werck-Reichhart (1997) Xenobiotics: substrates and inhibitors of the plant P450s. *Environ. Sci. Pollut. Res.* **4**: 229–234.

Scheidemann, P., A. Klunk, C. Sens, and D. Werner (1998) Species dependent uptake and tolerance of nitroaromatic compounds by higher plants. *J. Plant Physiol.* **152**: 242–247.

Schrenk, C., S. Pflugmacher, R. Brüggemann, H. Sandermann, Jr., C.E.W. Steinberg, and A. Kettrup (1998) Glutathione *S*-transferase activity in aquatic macrophytes with emphasis on habitat dependence. *Ecotoxicol. Environ. Safe.* **40**: 226–233.

Schroeder, P. (1997) Fate of glutathione *S*-conjugates in plants. In: *Regulation of Enzymatic Systems Detoxifying Xenobiotics in Plants.* K.K. Hatzios, ed. Kluwer Academic Publishers, Dordrecht, The Netherlands, pp. 233–244.

Schuler, M.A. (1996) Plant cytochrome P450 monooxygenases. *Crit. Rev. Plant Sci.* **15**: 235–284.

Schwitzguébel, J.P. (1996) Biodegradation of xenobiotics by plants. In: *Kurzfassungen der DECHEMA-Jahrestagungen '96, Band I.* DECHEMA, Frankfurt am Main, Germany, pp. 442–443.

Schwitzguébel, J.P. (1998) From plant cells to phytoremediation. In: *Biotec'98—Book of abstracts.* M. Mota and E.C. Ferreira, eds. Barbosa and Xavier, Braga, Portugal, ISBN 972-97810-0-1, p. 8.

Schwitzguébel, J.P., T. Vanek, N. Mathieu, J.L. Thirot, and M. Novotny (1995) Biotransformation of sulfonated xenobiotic compounds by plant cells. In: *Biosorption and Bioremediation.* T. Macek, K. Demnerova and M. Mackova, eds. Czech Society for Biochemistry and Molecular Biology, ISBN 80-902013-0-X, Prague, pp. L 2–10.

Schwitzguébel, J.P., J.L. Thirot, N. Mathieu, M. Novotny, and T. Vanek (1996) Biotransformation of sulfonated aromatic compounds by plant cells. *Experientia* **52**: A22.

Schwitzguébel, J.P., R. Duc, and T. Vanek (1998) Development of a biological process to treat industrial effluents and by-products using plant cells. In: *Abstracts of Third Annual International Conference on Phytoremediation.* held in Houston, Texas, June 22–25, p. 10.

Setti, L., S. Scali, I. Degli Angeli, and P.G. Pifferi (1998) Horseradish peroxidase-catalyzed oxidative coupling of 3-methyl 2-benzothiazolinone hydrazone and methoxyphenols. *Enzyme Microb. Technol.* **22**: 656–661.

Snape, J.R., N.A. Walkley, A.P. Morby, S. Nicklin, and G.F. White (1997) Purification properties and sequence of glycerol trinitrate reductase from *Agrobacterium radiobacter*. *J. Bacteriol.* **179**: 7796–7802.

Soranzo, N., C. Frova, E. Pe, R. Rizzardi, and M. Sari-Gorla (2000) Characterisation of GST gene family in rice. In: *Phytoremediation 2000—State of the Art in Europe—An Intercontinental Comparison*. COST Action 837 Workshop. p. 151 (http://lbeww.epfl.ch/COST837/).

Spain, J.C. (1995) Biodegradation of nitroaromatic compounds. In: *Annual Review of Microbiology*. Palo Alto, California. **49**: 523–555.

Spanggord, R.J., K.E. Mortelmans, A.F. Griffing, and V.F. Simmon (1982) Mutagenicity in *Salmonella typhimurium* and structure-activity relationship of wastewater components emanating from the manufacture of trinitrotoluene. *Environ. Mutagen.* **4**: 163–179.

Stepan-Sarkissian, G. (1991) Biotransformation by plant cell cultures. In: *Plant Cell and Tissue Culture*. A. Stafford and G. Warren, eds. Open University Press, Buckingham, United Kingdom, pp. 163–204.

Sterling, T.M. and N.E. Balke (1989) Differential bentazon metabolism and retention of bentazon metabolites by plant-cell cultures. *Pestic. Biochem. Physiol.* **34**: 39–48.

Stiborova, M., H.H. Schmeiser, and E. Frei (2000) Oxidation of xenobiotics by plant microsomes, a reconstituted cytochrome P450 system and peroxidase: a comparative study. *Phytochemistry* **54**(4): 353–362.

Stohs, S.J. (1980) Biotransformations in plant cell cultures. In: *Plant Cell Cultures*. A. Fiechter, ed. Akademie-Verlag, Berlin, p. 85.

Suga, T., T. Hirata, Y. Aoki, S. Lee, H. Hamada, and M. Futatsugi (1982) Biotransformation of foreign substrates with callus tissues—transformation of monoterpenes with tobacco suspension cells. In: *Proc. 5th International Congress on Plant Tissue and Cell Culture*. A. Fujivara, ed. Maruzen, Tokyo, p. 381.

Sun, W.H., G.L. Horst, R.A. Drijber, and T.E. Elthon (2000) Fate of 2,4,6-trinitrotoluene in axenic sand culture systems containing smooth bromegrass. *Environ. Toxicol. Chem.* **19**: 2038–2046.

Tadros, M.G., A. Crawford, A. Mateo-Sulivan, C. Zhang, and J.B. Hughes (2000) Toxic effects of hydroxyamino intermediates from microbial transformation of trinitrotoluene and dinitrotoluenes on algae *Selenastrum capricornutum*. *Bull. Environ. Contam. Toxicol.* **64**: 579–585.

Thomas, H., A. Gerth, B. Eulering, and A. Böhler (2001) Neue Erkenntnisse zur biologischen *in situ* Sanierung TNT-kontaminierter Böden. *TerraTech.* **2**: 52–54.

Thompson, P.L., L.A. Ramer, and J.L. Schnoor (1998) Uptake and transformation of TNT by hybrid poplar trees. *Environ. Sci. Technol.* **32**: 975–980.

Tommasini, R., E. Vogt, J. Schmid, M. Fromenteau, N. Amrhein, and E. Martinoia (1997) Differential expression of genes coding for ABC-transporters after treatment of *Arabidopsis thaliana* with xenobiotics. *FEBS Lett.* **411**: 206–210.

Tommasini, R., E. Vogt, M. Fromenteau, S. Hoertensteiner, P. Matile, N. Amrhein, and E. Martinoia (1998) An ABC-transporter of *Arabidopsis thaliana* has both glutathione-conjugate and chlorophyll catabolite transport activity. *Plant J.* **13**: 773–780.

Trapp, S. (2000) Modeling uptake into roots and subsequent translocation of neutral and ionisable organic compounds. *Pest Manag. Sci.* **56**: 767–778.

Trease, G.E. and W.C. Evans (1983) Phenols and phenolic glycosides. In: *Pharmacognosy*, 12th ed. Baillière Tindall, London, pp. 368–414.

Urbanski, T. (1984) *Chemistry and Technology of Explosives*. Pergamon Press Ltd., Oxford, United Kingdom.

van den Berg, A.J.J. and R.P. Labadie (1989) Quinones. In: *Methods in Plant Biochemistry, Vol. 1, Plant Phenolics*. J.B. Harborne, ed. Academic Press, London, pp. 451–491.

van der Plas, L.H.W., M.J.M. Hagendoorn, and D.C.L. Jamar (1998) Anthraquinone glycosylation and hydrolysis in *Morinda citrifolia* cell suspensions: regulation and function. *J. Plant Physiol.* **152**: 235–241.

Vanderkrol, D., I. Schuphan, B. Thiede, and B. Schmidt (1995) Metabolism of [ring-2,6-C-14]parathion in plant-cell suspension-cultures of carrot (*Daucus carota*), purple foxglove (*Digitalis purpurea*), soybean, thorn apple (*Datura stramonium*) and wheat (*Triticum aestivum*). *Pestic. Sci.* **45**: 143–152.

Vanek, T., T. Macek, and J. Harmatha (1986) Glucosidation of digitoxigenin by tissue culture of *Digitalis lanata*. *Biotechnol. Lett.* **8**: 859–862.

Vanek, T., T. Macek, K. Stranský, and K. Ubik (1989a) Plant cells immobilized in pectate gel: biotransformation of verbenol isomers by *Solanum aviculare* free and immobilized cells. *Biotechnol. Tech.* **6**: 411–414.

Vanek, T., Z. Wimmer, T. Macek, D. Šaman, A. Svatoš, and M. Romanuk (1989b) Stereochemistry of the enzymatic reduction of 2-(4-methoxybenzyl)-1-cyclohexanone by *Solanum aviculare* cells *in vitro*. *Biocatalysis* **2**: 265–272.

Vanek, T., I. Valterová, and T. Vaisar (1999) Biotransformation of (*S*)-(−)- and (R)-(+)-limonene using *Solanum aviculare* and *Dioscorea deltoidea* plant cells. *Phytochemistry* **50**(8): 1347–1351.

Vanek, T., A. Nepovim, and S. Zeman (2000a) Phytoremediation of selected explosives in a model system of plant tissue cultures. *Kor. J. Plant Tissue Culture* **27**: 379–385.

Vanek, T., A. Nepovim, M. Hubálek, Z. Jalova, and S. Zeman (2000b) Phytoremediation of explosives from laboratory experiments to field applications. In: Phytoremediation 2000—State of the Art in Europe—An Intercontinental Comparison. COST Action 837 Workshop, held in Hersonissos, Crete, Greece, April 6–8, pp. 26–28 (http://lbewww.epfl.ch/COST837/).

Vanek, T., A. Nepovím, R. Podlipná, S. Zeman, and M. Vágner (2003) Phytoremediation of selected explosives. *Water, Air, Soil Poll.* (in press).

Wang, C.G., J.Y. Wu, and X.G. Mei (2001a) Enhancement of taxol production and excretion in *Taxus chinensis* cell culture by fungal elicitation and medium renewal, *Appl. Microbiol. Biotechnol.* **55**: 404–410.

Wang, J.W., Z. Zhang, and R.X. Tan (2001b) Stimulation of artemisinin production in *Artemisia annua* hairy roots by the elicitor from the endophytic *Colletotrichum* sp. *Biotechnol. Lett.* **23**: 857–860.

Wang, M.J., M. Bokern, C. Boehme, K.C. Jones, and H. Harms (1996) Phytotoxicity, uptake and metabolism of 1,4-dichlorobenzene by plant cells. *Environ. Toxicol. Chem.* **15**: 1109–1114.

Wendt, T.M., J.H. Cornell, and A.M. Kaplan (1978) Microbial degradation of glycerol nitrates. *Appl. Environ. Microbiol.* **36**: 693–699.

Werck-Reichhart, D., A. Hehn, and L. Didierjean (2000) Cytochromes P450 for engineering herbicide tolerance. *Trends Plant Sci.* **5**: 116–123.

Wilken, A., C. Bock, M. Bokern, and H. Harms (1995) Metabolism of different PCB congeners in plant-cell cultures. *Environ. Toxicol. Chem.* **14**: 2017–2022.

Wright, K.M., R.W. Horobin, and K.J. Oparka (1996) Phloem mobility of fluorescent xenobiotics in *Arabidopsis* in relation to their physicochemical properties. *J. Exp. Bot.* **47**: 1779–1787.

Wu, J.Y and J.J. Zhong (1999) Production of ginseng and its bioactive components in plant cell culture, current technological and applied aspects. *J. Biotechnol.* **68**: 89–99.

Young, L. and J. Yu (1997) Ligninase-catalysed decolorization of synthetic dyes. *Water Res.* **31**: 1187–1193.

5

ENZYMES USED BY PLANTS AND MICROORGANISMS TO DETOXIFY ORGANIC COMPOUNDS

N. L. Wolfe and C. F. Hoehamer

SUMMARY OF PRACTICAL IMPLICATIONS

A variety of plant and microbial enzymes are involved in phytoremediation, including nitroreductases, glycosyl and glutathione transferases, oxidases, phosphatases, nitrilases, and dehalogenases. These enzymes are involved in the transformation of toxic xenobiotic compounds such as explosives, pesticides, nerve gases, and halogenated organic compounds. Whereas nitroreductases and glycosyltransferases play an important role in the transformation and conjugation of explosives, dehalogenases and many diverse organophosphatases detoxify other contaminants by reducing either halogen groups or organically bound phosphate, respectively. In addition, plants and many microorganisms contain an abundance of oxidases such as laccases and peroxidases. These enzymes are involved in many plant processes including lignification and defense, and may play a role in the metabolism of explosives. Finally, nitrilases catalyze the biodegradation of endogenous plant hormones and commercial herbicides. Knowledge of these metabolic enzymes that catalyze detoxification provide important fundamental insight into metabolic pathways for many organic contaminants and ultimately determine the effectiveness of phytoremediation.

GLOSSARY

Acetylation: A reaction that involves the addition of an acetyl group (CH_3COO) to a substrate or xenobiotic molecule.

Phytoremediation: Transformation and Control of Contaminants,
Edited by Steven C. McCutcheon and Jerald L. Schnoor.
ISBN 0-471-39435-1 (cloth)

Dimer: An enzyme or protein containing two subunits. Each subunit can be similar or different, a homo- or hetero-dimer, respectively.
Epicotyl: The upper portion of the axis of an embryo or seedling, above the cotyledons (seed leaf) and below the next leaf.
Esterification: A reaction that involves the addition of an ester group (RCOOR) to a substrate.
Expression: The production of an enzyme from the associated gene.
Flavin: An intracellular cofactor in many macromolecules commonly used with many enzymes as a source of electrons. The most common flavins are flavin mononucleotide and flavin adenine dinucleotide.
Glutathione: The common tripeptide present in living cells often involved in conjugation processes during metabolism. The peptide units are expressed as Glu-Cys-Gly. Homoglutathione consists of peptides designated Glu-Cys-Ala in which Glu is glutamic acid, Cys is cysteine, Gly is glycine, and Ala is alanine.
Glycosylation: A reaction that involves the addition of a sugar or other carbohydrate to a substrate usually through an oxygen (O)- or nitrogen (N)- linked complex.
Heterocyclic triazinyl ring: A six-member carbon cyclic ring containing three nitrogen heteroatoms.
Heme: An iron-porphyrin complex found in hemoglobin and cytochromes.
Hydrolysis: A reaction that involves the breaking of a bond by the splitting of water.
Hydroxylation: A hydrolytic reaction that involves the addition of a hydroxyl group (OH^-) to a substrate molecule.
Isoform: Different enzymes with similar functionalities and structure being expressed by one gene.
Lignification: The process by which a very hydrophobic aromatic polymer is synthesized in the middle lamellae of the plant cell wall, to povide structural support and which aids in a variety of plant functions.
Metalloenzyme: Enzymes that contain metals within the molecular structure.
Overexpressed: The increased expression of enzymes by genetic manipulation.
Trimer: An enzyme or protein with three subunits.

Note: For information on the enzymes used in this chapter, see (http://www.chem qmw.ac.uk/iubmb/enzyme). For example, see EC 2.5.1.18 for glutathione transferase.

INTRODUCTION

Over the course of evolution, plants have acquired metabolic pathways for a variety of structurally diverse xenobiotic compounds. Plant degradation pathways are compound-specific, frequently induced, and generally occur in three stages including transformation, conjugation, and compartmentalization

(Sandermann 1992). In the first phase of this metabolic process, an organic contaminant is taken up into the plant tissue, generally through root tissue, and subjected to a variety of enzymes capable of catalyzing transformations such as oxidation, hydroxylation, dehalogenation, reduction, esterification, and hydrolysis. The most common transformation is oxidation due to the abundance and diversity of multifunctional oxidases (Sandermann 1992). The second phase of xenobiotic degradation in plants, subsequent to transformation, is frequently conjugation. This metabolic step involves endogenous enzymes that react with the newly transformed organic products, forming metabolites derivatized with carbohydrates, amino acids, and other molecules (Hatzios and Penner 1982, Bhadra *et al.* 1999). Conjugation is important in the degradation of certain intracellular plant hormones and commercial herbicides in several plants (Hatzios and Penner 1982). Finally, if the modified contaminant is not completely metabolized or released from the plant, the last stage of metabolism observed for many organic compounds is compartmentalization (Sandermann 1992). During this phase of metabolism, the transformed or conjugated organic product is stored or sequestered. The physiological localization of organic metabolites within the plant primarily depends on the contaminant.

An important feature of plant enzyme–mediated transformation is whether the enzyme is constitutive or induced. While most constitutive enzymes regulate cellular "housekeeping" functions and are generally expressed in low concentrations at consistent rates, induced enzymes are synthesized at varying rates in response to the growth conditions of the cell. Induced expression of some enzymes can be enhanced or activated due to the presence or absence of organic compounds during cellular growth and differentiation. For example, an intracellular nitroreductase in the bacteria *Enterobacter cloacae* can be enhanced five- to tenfold by the addition of 2,4,6-trinitrotoluene (TNT) to the culture medium (Bryant and DeLuca 1990). Additionally, aquatic macrophytes and microorganisms grown under phosphate deprivation produce increased levels of phosphatases (EC 3.1.3). Furthermore, the production of metabolic enzymes *via* induction not only allows these organisms to have enhanced metabolic capability but also the expression enables researchers to study important enzymes which are randomly expressed or are present in very low amounts.

The primary transformation reactions within plants are generally catalyzed by endogenous enzymes. These proteins are usually induced and attack common functional substituents, including nitro (NO_2), amino (NH_2), hydroxyl (OH), carboxyl (COO), or halogen (normally Cl, Br, and I) groups. Substituents such as these are present on many aromatic and aliphatic xenobiotic compounds (Sandermann 1992). For trinitro-substituted explosives such as TNT; 1,3,5-trinitrobenzene; 2,4,6-trinitro-*m*-xylene; and 2,4,6-trinitrophenol (picric acid), oxidation of the aromatic ring has not been observed in any biological degradation pathway. This resistance to chemical oxidation can be attributed to the strong electron-withdrawing character of the nitro groups, which causes the aromatic ring to be electron-deficient. Therefore, any electrophilic (electron-seeking) attack on the π-electron system is generally unfavorable. Instead, aquatic and terrestrial plants (Palazzo and Leggett 1986, Hughes *et al.* 1997, Thompson *et al.* 1998), microorganisms (McCormick *et al.* 1976, Naumova *et al.* 1988, Boopathy and Kulpa 1994, Gorontzy *et al.* 1994, Regan and Crawford 1994, Fiorella and Spain 1997), and aquatic sediments (van Beelen and Burris 1995) initially transform TNT and other nitroaromatics through nitroreduction.

NITROREDUCTASES

Reduction of a nitro group is the first step in the physiological or abiotic transformation of many different nitroaromatic compounds. Nitroreduction initiates through a series of one- or two-electron transfers, and for TNT reduction starts with a nitro group positioned at C_2 or C_4. Complete reduction of a nitro group to an amino group requires the transfer of six electrons, usually observed through a series of two-electron transfers. The first pair of electrons produces a nitroso-intermediate (R-NO), while the next electron pair forms a hydroxylamino-derivative (R-NHOH). Finally, the transfer of the last pair of electrons yields a fully reduced amino group (R-NH_2). This nitroreduction scheme for TNT is represented in Figure 5-1.

Enzymes that catalyze the reduction of aromatic nitro groups, the so-called nitroreductases, are classified by activity in the presence of molecular oxygen into two classes, Types I and II, each with distinctive properties (Peterson *et al.* 1979). Type I nitroreductases are defined as oxygen-insensitive and catalyze the reduction of a nitro (NO_2) group *via* a series of two-electron transfers, producing nitroso (NO), hydroxylamino (NHOH), and amino (NH_2) derivatives. Examples of these enzymes are few but include nitroreductases from enteric bacteria, NADPH–quinone oxidoreductase (EC 1.6.5.5) from a variety of mammals, and possibly xanthine dehydrogenase (EC 1.1.1.204) from rat liver (Bryant and McElroy 1991). Type II nitroreductases are oxygen-sensitive enzymes that catalyze the reduction of a nitro group *via* a series of single-electron transfers, initially forming a radical anion. If produced under aerobic conditions, this radical is immediately re-oxidized by molecular oxygen to yield

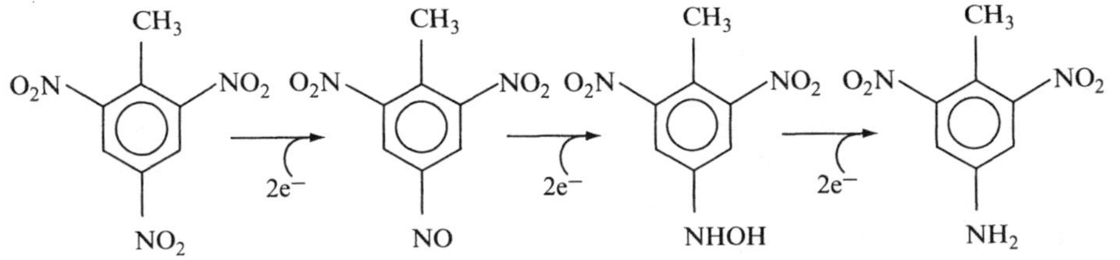

Figure 5-1 Reductive transformation of 2,4,6-trinitrotoluene (TNT) to 4-amino-2,6-dinitrotoluene (4-ADNT).

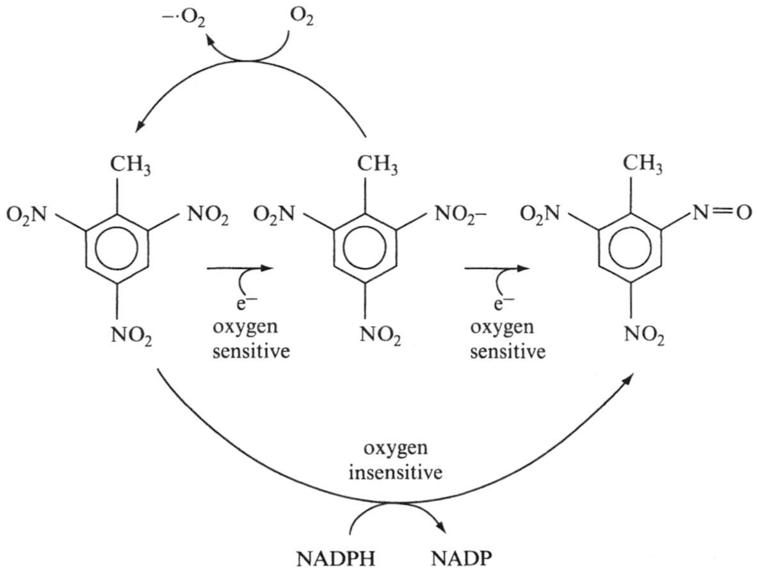

Figure 5-2 Oxygen-sensitive and insensitive electron reduction of 2,4,6-trinitrotoluene (TNT) by nitroreducates (adapted from Bryant and McElroy 1991). Note that NADPH and NADP are reduced and oxidized forms, respectively, of nicotinamide-adenine dinucleotide poosphate.

a superoxide anion radical and the original nitroaromatic compound (Peterson *et al.* 1979, Bryant and McElroy 1991). This incomplete reaction is called a "futile cycle" because the reducing equivalents, usually reduced forms of nicotinamide-adenine dinucleotide (NADH) or nicotinamide-adenine dinucleotide phosphate (NADPH), are oxidized, resulting in no reduction of the aromatic nitro group (Müller 1983). Therefore, the oxygen-sensitive enzymes can only reduce nitroaromatic compounds under anaerobic conditions (Peterson *et al.* 1979). The distinct enzymatic mechanisms by which TNT and other nitroaromatic explosives are reduced by nitroreductases, are represented in Figure 5-2. The enzymes already known to have Type II nitroreductase activity include the cytochrome P-450 reductase (EC 1.6.2.4), cytochrome *c* reductase (EC 1.6.2.4), aldehyde oxidase (EC 1.6.2.4), glutathione reductase (EC 1.6.2.4), xanthine dehydrogenase (EC 1.1.1.204), succinic dehydrogenase (EC 1.3.99.1), and quinone reductase (EC 1.6.99.2) (Bryant and McElroy 1991).

In spite of the apparent ubiquitous presence of nitroreductases in many species and sediments, surprisingly little attention has been focused on these enzymes in higher organisms. However, much is known about the mammalian oxygen-insensitive NAD(P)H–quinone oxidoreductase (EC 1.6.99.2), which is capable of reducing many different quinones and nitroaromatic compounds. (Note that NAD(P)H is NADPH or NADH.) This enzyme has been purified and characterized from liver tissues of cow, rabbit, mouse, and rat (Bryant and McElroy 1991), and even documented in fungi (Buswell *et al.* 1979). Although crude, cell-free extracts with nitroreductase activity were obtained from spinach (*Spinacia oleracea*) (Wilson 1998), white-rot fungus (*Panerochaete chrysosporium*) (Rieble *et al.* 1994), and aquatic sediments (van Beelen and Burris 1995), the well-characterized nitroreductases have been purified from bacteria (Tatsumi *et al.* 1981, Kinouchi and Ohnishi 1983, Watambe *et al.* 1989, Bryant and DeLuca 1990, Somerville *et al.* 1995). Initial studies of both Type I and Type II nitroreductases used cell-free extracts of *Escherichia coli* (Tatsumi *et al.* 1981). Expanding on that work, other researchers have been successful in purifying and characterizing both types of these enzymes from a variety of other bacteria, including *Enterobacter cloacae* (Bryant and DeLuca 1990), *Bacteroides fragilis* (Kinouchi and Ohnishi 1983), *Pseudomonas pseudocalagenes* (Somerville *et al.* 1995), and *Salmonella typhimurium* (Watambe *et al.* 1989). Thus, the study of microbial nitroreductases appears to have received much attention, while information regarding the physiological function, structure, and localization of plant nitroreductases remains largely unreported.

Purification and characterization of these bacterial nitroreductases reveal that many of these enzymes share similarities in function. First, most of these enzymes are inducible and are capable of reducing a variety of nitroaromatics to nitroso (NO), hydroxylamino (NHOH), and amino (NH_2) metabolites utilizing either NADH or NADPH as a source of reducing equivalents. While most nitro-reducing enzymes are capable of using both dinucleotide electron donors, certain others specifically prefer either NADPH or NADH for enzyme activity (Kinouchi and Ohnishi 1983, Bryant and DeLuca 1990). In addition, several known nitroreductases require flavin mononucleotide or flavin adenine dinucleotide for nitroreduction. This observation led to important recombinant studies using the *Salmonella typhimurium* nitroreductase. That work revealed that these flavin cofactors are probably involved in the transfer of electrons from the active site of the enzyme to the nitroaromatic substrate (Watambe *et al.* 1990). Although bacterial nitroreductases are generally induced in liquid cultures by the addition of TNT or nitrobenzene (Bryant and DeLuca 1990, Somerville *et al.* 1995), these enzymes are capable of reducing a variety of diverse nitroaromatics. Thus, nitroreductases appear to have a broad range of substrate specificity for nitroaromatic compounds but not for nitro-substituted aliphatics. Furthermore, the transformation of TNT by nitroreductases forms aminated metabolites which are less toxic than TNT in certain test organisms (Won *et al.* 1976, Honeycutt *et al.* 1996), but more important, these enzymes generate metabolites that are required substrates for subsequent conjugation reactions (Hatzios and Penner 1982).

Conjugation, the second phase of xenobiotic metabolism in plants, is the addition of a sugar moiety or tripeptide to the contaminant or new transformation product (Sandermann 1992). Conjugation can be catalyzed by endogenous enzymes called transferases, forming glucosyl, malonyl, glutathione, or homoglutathione derivatives. While certain compounds such as (2,4-dichlorophenoxy)acetic acid (2,4-D), and a variety of herbicides undergo *O*-glycosylation during metabolism, reduced nitroaromatic metabolites involve

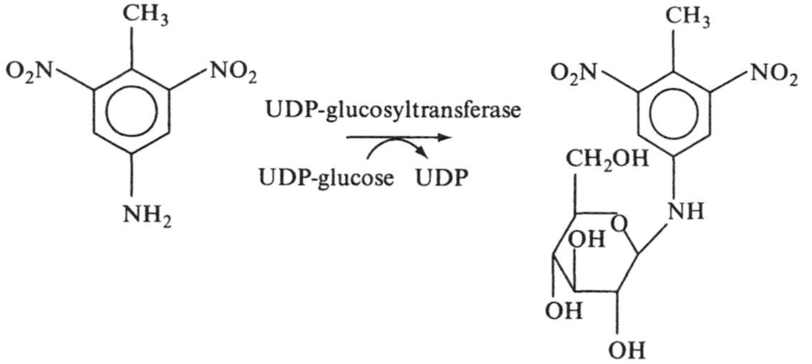

Figure 5-3 Conjugation of glucose to 4-amimo-2,6-dinitrotoluene (4-ADNT) by UDP-glucosyltransferase (EC 2.4.1.71). Note: UDP is uridine dinueleotide phosphate.

the addition of a molecule of glucose or malonate to the newly formed amino groups *via* *N*-glycosylation (Hatzios and Penner 1982). This reaction was demonstrated when Bhadra *et al.* (1999) showed that in periwinkle (*Catharanthus roseus*) hairy root cultures, TNT is transformed into 4-amino-2,6-dinitrotoluene (4-ADNT), following the *N*-linked addition of glucose to the newly formed amino group of 4-ADNT as shown in Figure 5-3. Furthermore, Bhadra *et al.* (1999) also noted that other distinct conjugated TNT products were formed.

An enzyme that catalyzes the transfer of glucose to the aminated compound, UDP-glucosyltransferase, has been purified from soybean (*Glycine max*) and several other species of plants (Frear 1968). The source of glucose for this reaction is uridine dinucleotide phosphate (UDP). The biological function of the enzyme and the sugar molecule on the metabolite has attracted much speculation but unfortunately remains undetermined. However, glycosylated metabolites are considered dead end products in the detoxification of many organic compounds in certain plants (Hatzios and Penner 1982). When produced, conjugated products are typically less toxic, more stable, soluble and polar, and higher in molecular weight than the original compound. This alteration in structure and identity causes many conjugated metabolites to have different chemical and physical properties compared to the parent molecule, making many of these products difficult to detect and study. A number of commercial herbicides that undergo glucose conjugation reactions in metabolism include 2,4-D, 3-amino-2,5-dichlorobenzoic acid (chloramben), and 3-isopropyl-1,*H*-2,1,3-benzothiadiazin-4(3*H*)-one 2,2-dioxide (bentazon) (Hatzios and Penner 1982).

A different type of conjugation reaction involves the transfer of amino acids to the contaminant molecule. For example, several studies report the conjugation of aspartic acid to certain plant hormones including indoleacetic acid in pea (*Pisum sativum*) epicotyls (Andreae and Good 1957) and 2,4-D acid in wheat (*Triticum aesitivum* L.) (Klambt 1961). The intracellular products formed within these plants are indoleacetyl-aspartic acid and

2,4-dichlorophenoxyacetyl-aspartic acid. The enzyme that catalyzes the transfer of this amino acid to the substrate is aspartyltransferase (EC 2.3.2.7). Although a number of different amino acids have been observed in conjugation, the most common are aspartic acid, glutamic acid (Glu), and serine. Additionally, it is very possible that each amino acid transferred in a conjugation event requires a specific transferase enzyme. The herbicide 3-amino-*s*-triazole) (amitrole) is known to involve the addition of serine during metabolism, forming β-(3-amino-1,2,4-triazol-1-yl)-α-alanine (Carter 1975). The conjugated metabolite of amitrole was not only determined to be the terminal product formed but also considered nonphytotoxic. In general, conjugation reactions involving amino acid transferases occur less frequently compared to carbohydrate and glutathione conjugations (Hatzios and Penner 1982).

Tripeptides like glutathione (Glu-Cys-Gly) and homoglutathione (Glu-Cys-Ala) are molecules, which are also transferred to metabolites in another type of conjugation. These tripeptide compounds are crucial for cellular maintenance and are present in most living organisms. Glutathione transferase (EC 2.5.1.18) catalyzes the transfer of these tripeptides to certain organic contaminants (Chasseaud 1974, Clarke and Smith 1975). Although purified from corn (*Zea mays*), this enzyme has been studied in a variety of plant species, including sugarcane *Saccharin officinarum* Johnson grass (*Sorghum halepense*) (Frear and Swanson 1970, Guddwear and Dauterman 1979, Hatzios and Penner 1982). In addition to being in plants, glutathione transferase has also been identified and studied in mammalian tissue (Guddwear and Dauterman 1979). Data from those studies reported that the plant glutathione transferase was a trimer, and the enzyme from mouse liver extracts was a dimer, showing that these enzymes differ in plants and mammals. Whereas glutathione transferase is inducible and has unique substrate specificity for 2-chloro-substituted *s*-triazines, the best substrates for these enzymes are compounds with structures containing a heterocyclic triazinyl ring. Some herbicides that have already been shown to be conjugated with glutathione include (Hatzios and Penner 1982):

2-chloro-2′,6′-diethyl-*N*-(methoxymethyl)acetanilide (atrazine)

2-chloro-4,6-*bis*(ethylamino)-*s*-triazine (simazine)

2-{[4-chloro-6-(ethylamino)-2-triazine-2-yl]amino}-2-methyl propionitrile (cyanazine)

If the organic by-products are not volatilized or released from the plant in the form of an exudate, the third phase of xenobiotic metabolism may involve compartmentalization or storage of the conjugated product into the plant-storage organelles. While some soluble organic conjugates are often stored in vacuoles, other conjugated metabolites form covalent complexes with lignin or other biopolymers in the plant cell wall (Sandermann 1992). Indeed, studies investigating explosives degradation in aquatic plants have revealed that TNT metabolites become increasingly incorporated into the plant tissue over time, forming irreversible complexes with plant tissue (Vanderford *et al.* 1997). Studies of TNT degradation using periwinkle (*Catharanthus roseus*) revealed that approximately 22 percent of the initial TNT could be extracted from the hairy root cultures and identified as conjugated TNT metabolites, with the remaining nitroaromatic metabolites irreversibly associated with tissue (Bhadra *et al.* 1999). The same study concluded that compartmentalization or bioaccumulation of nitroaromatic products was the final step in the metabolism of TNT by plants. A plausible scheme illustrating the metabolic steps and the enzymes involved in the detoxification of TNT by periwinkle (*Catharanthus roseus*) is represented in Figure 5-4. However, Bhadra *et al.* (1999) did not demonstrate whether endogenous plant enzymes were involved in the compartmentalization of TNT metabolites.

Instead of being shuttled to the cell wall for covalent deposition, some conjugated xenobiotic compounds have been shown to be routed to vacuoles for storage. Coleman *et al.* (1997) reported that certain xenobiotic compounds compounds conjugated with glutathione or carbohydrates both require adenine triphosphate (ATP)–dependent enzymes to gain entry into the vacuolar organelles. Enzymes that catalyze the entry of glutathione and glycosyl-derivatives are called glutathione- and glucoside-conjugate transporters, respectively. These enzymes are similar to other transporters (Davies and Coleman 2000) which are transmembrane proteins that may display "flippase" or "pump" mechanisms when transferring the conjugated pollutant molecule through the membrane (Coleman *et al.* 1997) at the expense of ATP. In spite of the vast knowledge that exists on transmembrane proteins, this section will not discuss

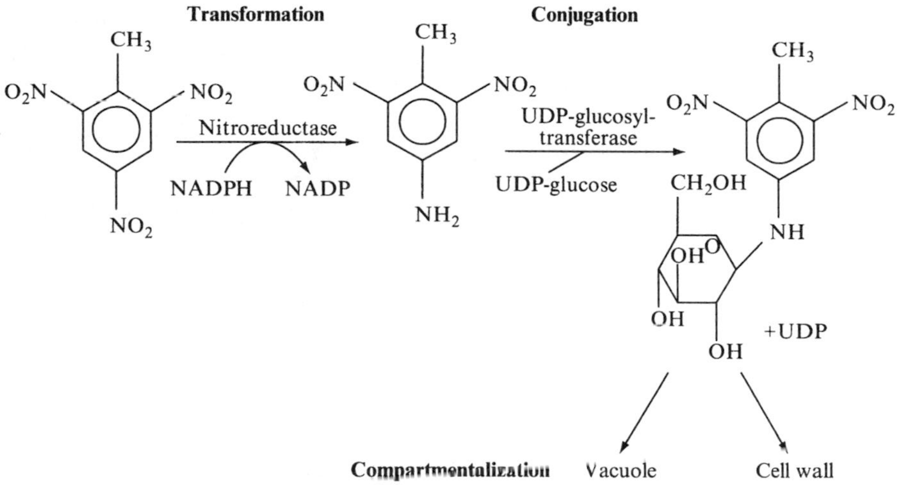

Figure 5-4 Representation of transformation, conjugation, and compartmentalization of 2,4,6-trinitrotoluene (TNT) by periwinkle (*Catharanthus roseus*). Note that NADPH and NADP are reduced and oxidized forms respectively, of nicotinamide-adenine dinucleotide phosphate and UDP is uridine dinucleotide phosphate.

these complex enzymes in any detail. However, because TNT and other xenobiotic metabolites become conjugated with glucose and possibly glutathione, a possibility exists that these nitroaromatic conjugates could also be shuttled into vacuoles by transporters for storage and eventually bioaccumulation. The low recoveries of reduced nitroaromatic products that were extracted in previous plant TNT transformation studies could have been due to conjugated nitroaromatics stored in vacuoles. Furthermore, conjugated contaminants may be stored in vacuoles before reacting with other enzymes to covalently associate with the plant cell wall.

OXIDASES

Oxidative enzymes in both plants and microorganisms have potential roles in TNT detoxification. Briefly, laccases (EC 1.10.3.2) and peroxidases (EC 1.11.17) are oxidative enzymes expressed in plants (Sterjiades *et al.* 1993) in various isoforms and have a broad specificity for aromatic substrates containing hydroxyl- or amino-constituents. Laccases (*p*-diphenol: O_2 oxidoreductases), are blue copper oxidases that are involved in lignin degradation. Laccases have been purified from several sources, including the Japanese lacquer tree (*Rhus vernicifera*) (Reinhammar 1970), and the fungi *Polyporus versicolor* (Mosbach 1963), and *Neurospora crassa* (Lerch *et al.* 1978). Although the original substrate for this enzyme is syringaldezine, laccases are capable of oxidizing many aromatic compounds using dioxygen as the terminal electron acceptor, forming an oxidized aromatic product and two molecules of water. In addition, studies have shown that this enzyme is highly glycosylated and can contain several different copper cations for enzyme activity (Mosbach 1963, Reinhammar 1970). There is no direct evidence of this enzyme being involved in xenobiotic degradation, but some reduced nitroaromatics like monoamino- and diamino-derivatives of TNT would be suitable substrates for this enzyme.

Peroxidases (EC 1.11.17) are heme-containing enzymes that are widely distributed among most plants and microorganisms. In plants, peroxidases have been considered as multifunctional enzymes ranging in cellular functions from lignification (Lagrimini 1991) and the crosslinking of polysaccharides (carbohydrate biopolymers) in the cell wall (Fry 1986), to pathogen resistance (Salzer and Hager 1993) and plant hormone metabolism (Gunse and Elstner 1992). Peroxidases are inducible enzymes that are expressed as soluble, extracellular, and membrane-bound proteins throughout certain plant tissues where oxidation of aromatic substrates occurs using hydrogen peroxide as a cosubstrate. Because of the broad specificity for aromatic compounds containing hydroxyl or amino groups, peroxidases have been considered as possible catalysts involved in TNT metabolism.

In shake flask cultures of two white-rot fungi, *Phanerochaete chrysosporium* (Fernando *et al.* 1990, Stahl and Aust 1993) and *Phlebia radiata* (van Aken *et al.* 1997), enzymatic oxidation of reduced TNT metabolites has been observed in the culture medium under lignolytic conditions. Lignolytic culture conditions, defined by limited nitrogen in the growth media, trigger the induction of two nonspecific peroxidases, manganese peroxidase and lignin peroxidase. In both studies, mineralization of [^{14}C]-TNT by *Phanerochaete chrysosporium* and *Phlebia radiata* was only observed in the presence of these peroxidases, and no mineralization of TNT was observed when these enzymes were absent from the growth media. Furthermore, in a similar study, Valli *et al.* (1992) demonstrated that in *Phanerochaete chrysosporium*, under lignolytic conditions, 2,4-dinitrotoluene could be oxidized by manganese peroxidase to form a nitrobenzoquinone. Thus, the oxidation of reduced nitroaromatic metabolites is due to the presence of these two well-known peroxidases.

Other oxidative enzymes that attack reduced nitroaromatics have been investigated in bacteria. In particular, transformation studies using *Pseudomonas* strain MOA1 demonstrated that certain oxidative enzymes are capable of transforming radiolabeled 2-amino-4,6-dinitrotoluene (2-ADNT) and 4-ADNT into more polar metabolites *via* an oxygen-dependent reaction (Alverez *et al.* 1995). The study revealed that the enzymes catalyzing these extracellular oxidative reactions could possibly be a number of oxidases such as peroxidases or mono- and oxygenases dioxygenases. Although the enzymes and the structures of these polar metabolites were undetermined, this route of metabolism is another example of how oxidative and reductive enzymes are coupled during the detoxification of nitroaromatics by microorganisms.

In a fashion similar to conjugation in plants, some bacteria are capable of transforming TNT, followed by a subsequent modification of the amino-derivatives. When grown in shake flask cultures, the bacterium *Mycobacterium vaccae* excretes a certain enzyme into the culture medium that oxidizes 4-A-DNT into 4-amino-2,6-nitrobenzoic acid (Vanderberg *et al.* 1995). In addition, this bacterium is also known for expressing a different enzyme that catalyzes the transfer of a methoxy group (CH_3O) to 4-nitro-2,4-diamino-toluene to form 2,4-diamino-6-nitrobenzyl methylether. Similarly, the aerobic bacterium, *Pseudomonas fluorescens* strain B-3468, was shown to transform TNT into 4-nitro-2,4-diaminotoluene, followed by an acetylation reaction (Naumova *et al.* 1988). The physiological role of this acetylation is unknown, but the reaction yielded a dead-end TNT metabolite, 4-*N*-acetylamino-2-amino-6 nitrotoluene that accumulated in the growth medium.

Contrary to most documented TNT degradation pathways, the aromatic ring of TNT has been reduced by a strain of *Mycobacteria*. Vorbeck *et al.* (1994) observed that the addition of a hydride (H^-) to TNT forms a Meisenheimer complex. This route of metabolism occurs much less frequently than nitroreduction, only being observed in TNT degradation studies using a few mixed bacteria. The enzyme that catalyzes this unique transformation is pentaerythritol tetranitrate reductase (French *et al.* 1999). Once the Meisenheimer complex forms, subsequent breakdown likely involves enzymes that sequentially catalyze the release of nitrite groups (NO_2^-) from the aromatic ring, generating nitrogen-free metabolites. In addition, 2,4-dihydroxylphenol was detected in the cultures. Because the pentaerythritol tetranitrate reductase initiated a unique metabolic pathway for effective TNT mineralization, researchers purified, characterized, cloned, and expressed this microbial enzyme in tobacco (*Nicotiana rustica*) plants (French *et al.* 1999). The presence of this microbial enzyme in the tobacco (*Nicotiana rustica*) genome enables these plants to metabolize TNT extensively and uniquely, and possibly other nitroaromatic explosives. However, it is likely that complete mineralization of TNT to carbon dioxide (CO_2) and other small metabolites by pentaerythritol tetranitrate reductase would require subsequent enzyme-mediated metabolic steps, possibly multiple steps. If the plant did not harbor such gene products, incomplete degradation would occur. Thus, identifying the additional enzymes and obtaining the DNA sequences for gene insertion is also necessary.

PHOSPHATASES

Some of the most abundant and ubiquitous enzymes in the environment are phosphatases (EC 3.1.3). The general name "phosphatase" was adopted to describe a broad group of enzymes that transform organophosphate compounds by hydrolyzing the ester (POR) and anhydride (POOPO) linkages to produce the corresponding phosphoric acid. Two types of phosphatases are classified according to the pH-activity optima of enzymatic activity and are termed either alkaline phosphatase (EC 3.1.3.1) or acid phosphatase (EC 3.1.3.2). Depending on the structure of the organophosphate substrate, the enzymes have been further classified into three major groups, including phosphoric monoester hydrolases (EC 3.1.3), phosphoric diester hydrolases (EC 3.1.4), and triphosphoric monoester hydrolases (EC 3.1.5).

Alkaline phosphatases are widespread in the environment and are often referred to as nonspecific phosphomonoesterases that hydrolyze phosphate from substrates containing phosphomonoesters (COP bonds). These enzymes have been reported in a variety of organisms, including bacteria, fungi, algae, and many marine macrophytes. An alkaline phosphatase from the bacterium *Escherichia coli* has been purified to homogeneity and extensively characterized (Applebury and Coleman 1969). Data obtained from that study indicated that this enzyme was capable of catalyzing the hydrolysis of a variety of organophosphorous substrates and required Zn^{2+} or Co^{2+} for enzyme activity. This alkaline phosphatase was determined to be a homo- or hetero-dimer (dependent on isoenzyme) with a molecular mass of approximately 100 kilodaltons and functioned at mildly acidic pH ranges. Other work investigating alkaline phosphatase includes the purification and characterization of this enzyme from the halotolerant bacterium *Halomonas elongata* (Bylund *et al.* 1990) and *Cladosporium cucumerinum* (Pedregosa *et al.* 1990). Similar regulatory and enzymatic characteristics were observed when studying the properties of these enzymes. Both phosphatases were induced by phosphate deprivation, were characterized by alkaline pH optima activity, and were localized to the cell surface.

In addition to bacteria, numerous distinct alkaline phosphatase activities have been detected in many rhizosphere fungi (Ho 1988) and in benthic algae (Atkinson 1987). While alkaline phosphatase activities were detected in 44 species of marine macrophytes (Hernandez *et al.* 1994), only an excreted alkaline phosphatase activity was determined in lake water (Berman 1970). In addition, Morita *et al.* (1996) isolated a 50 kilodaltons glycosylinositol-phospholipid-anchored alkaline phosphatase from the aquatic macrophyte giant duckweed (*Spirodela oligorrhiza*). They observed induction of two distinct isoforms of alkaline phosphatase by phosphate deprivation and subsequently purified one of these enzymes. The enzyme displayed similar functionality to bacterial alkaline phosphatases and was determined to be lipid bound through a ceramide group. A unique characteristic of the giant duckweed (*Spirodela oligorrhiza*) alkaline phosphatase was resistance to the enzymatic phosphotidyl-inositol specific lipase C cleavage, indicating a strong linkage to the plant cell wall. Studies with various alkaline phosphatases reveal supportive evidence indicating that these enzymes are not only similar within plants and microorganisms but are also released and stabilized in lakes and other water bodies (Berman 1970). Alkaline phosphatases have been more frequently observed than acid phosphatases.

Acid phosphatases (EC 3.1.3.2), classified as monophosphoric-monoester-phosphohydrolases, catalyze the hydrolysis of C–O–P bonds similar to alkaline phosphatases. The physiological role of these enzymes is to generate available phosphate for cellular growth of many organisms. These enzymes optimally function at lower pH, and over the last decade, acid phosphatases have received considerable attention. One acid phosphatase was purified from potato tuber (*Solanum tuberosom*) and characterized extensively (Gellatly *et al.* 1994). Multiple isoforms of this enzyme were induced by Mg^{2+} and these isoforms hydrolyzed a variety of organophosphorous compounds at a mildly acidic pH. Gel filtration chromatography and SDS PAGE (sodium dodecyl sulfate polyacrylamide gel electophoresis) demonstrated that this acid phosphatase was 100 kilodaltons containing two subunits; thus this enzyme was likely to be a homodimer. The enzyme truned out to be intracellular, but low amounts of this phosphatase were excreted into the culture medium. Zinc chloride and molybdate inhibited the potato tuber (*Solanum tuberosom*) acid phosphatase. Interestingly, this enzyme was determined to have phosphotyrosine activity, which is unique among acid phosphatases (Gellatly *et al.* 1994). In addition to plants, acid phosphatases have been purified and characterized from microorganisms including *Halomonas elongata* (Bylund *et al.* 1990), *Cladosporium cucumerinum* (Pedregosa *et al.* 1990), and a variety of rhizosphere fungi (Ho 1988).

Of more interest to phytoremediation are plants and microorganisms containing phosphatases capable of hydrolyzing the di- and triphosphate esters. A large majority of the widely used organophosphorus pesticides contains triester linkages, and many of these compounds contain sulfur in the place of oxygen as shown in Figure 5-5.

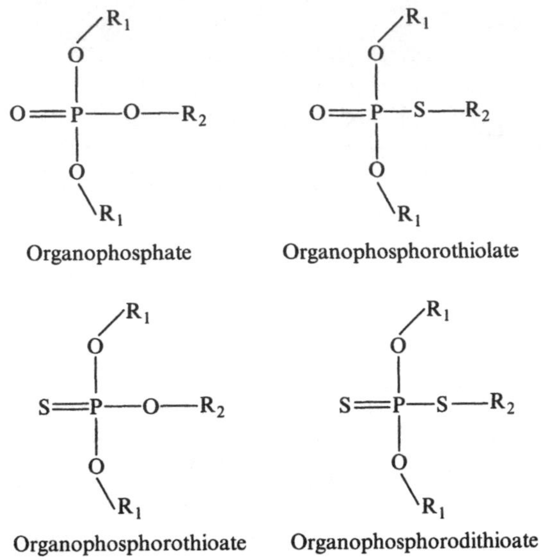

Organophosphate Organophosphorothiolate

Organophosphorothioate Organophosphorodithioate

Figure 5-5 Major groups of organophosphorus pesticides, where R_1 is generally an alkyl moiety such as methyl (CH_3) and ethyl (C_2H_5) groups and R_2 is usually an alkyl or aryl moiety.

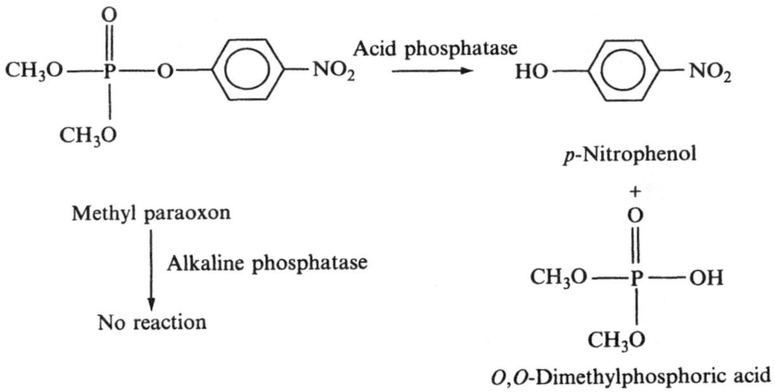

Acid phosphatase

p-Nitrophenol

Methyl paraoxon

Alkaline phosphatase

No reaction

O,O-Dimethylphosphoric acid

Figure 5-6 Production of *p*-nitrophenol from methyl paraoxon by acid phosphatase (EC 3.1.3.2).

Using purified samples of alkaline phosphatase (EC 3.1.3.1) provided by Morita *et al.* (1996), C. Mazur (U.S. Environmental Protection Agency, Athens, Georgia, written communication, 1999) reported that this alkaline phosphatase had no specificity for *O,O*-dimethyl-*O-p*-nitrophenylphosphate (methyl paraoxon) or *O,O*-dimethyl-*O-p*-phenylphosphorothioate (methyl parathion) (Figure 5-6). However, in the same study, a purified acid phosphatase hydrolyzed these organophosphate pesticides. Thus, it appears evident that acid and alkaline phosphatases have different organophosphate substrate requirements.

Among the most dangerous organophosphate compounds are the nerve agents and certain xenobiotic pesticides that act as acetylcholinesterase inhibitors. Such organophosphorous toxins tightly bind the active site of acetylcholinesterase (EC 3.1.1.7) to form an irreversible complex with the enzyme, thus blocking the required production of an essential neurotransmitter, acetylcholine. Methylphosphonoflurodic acid 1-methyl ethyl ester (sarin), methylophosphonofluordic acid 1,2,2-trimethyl propyl ester (soman), and the structurally-related analogues phosphorofluoridic acid *bis*-(1-methyl ethyl) ester (DFP) and *O,O*-diisopropyl-*S*-(2-diiospropylaminoethyl) phosphorothiolate (tetriso) are examples of these highly toxic organophosphorous compounds which generally have a chemical backbone resembling $R_1R_2P(=O)$-F, where R_1 and R_2 are varied to enhance acetylcholinesterase inhibition (Hoskin and Walker 1997). Interests in these deadly agents have led researchers to study the potential for biodegradation. Certain bacteria have been reported to express unique induced enzymes capable of degrading DFP. Such enzymes have been referred to as DFPases (diisopropyl-flurophosphatase, EC 3.1.8.2) and are present in both prokaryotes and eukaryotes. Briefly, DFPases are divided into two classes, Mn^{2+} induced and Mn^{2+} inhibiting. Based on limited information, both of these enzymes are effective in detoxifying organophosphorous acetylcholinesterase inhibitors. The DFPases have been purified from bacteria, including *Flavobacterium* ATCC 27551 (PAR^+) and *Pseudomonas diminuta* (Attaway *et al.* 1987). Coincidentally, an unknown enzyme with similar function in *Pseudomonas diminuta* also hydrolyzed another nerve agent analogue, tetriso (Hoskin and Walker 1997).

Based on the work of Hoskin *et al.* (1999), hydrolysis studies of DFP and soman were investigated using duckweed (*Lemna minor*), giant duckweed (*Spirodela oligorrhiza*), germinated mung bean (*Vigna radiata*), and the slime mold (*Dictyostelium discoideum*) (Figure 5-7). The study revealed that soman

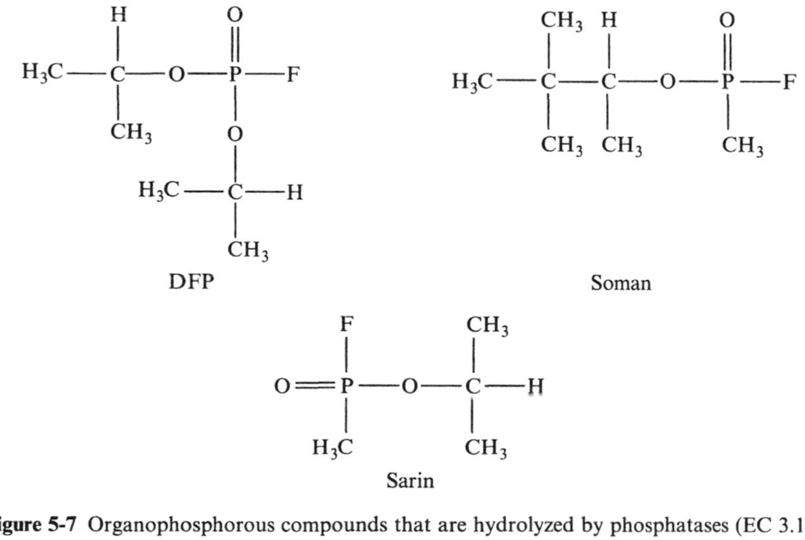

DFP

Soman

Sarin

Figure 5-7 Organophosphorous compounds that are hydrolyzed by phosphatases (EC 3.1.3).

was hydrolyzed by giant duckweed (*Spirodela oligorrhiza*), mung bean (*Vigna radiata*), and slime mold (*Dictyostelium discoideum*), but not duckweed (*Lemna minor*). The only organism capable of hydrolyzing DFP was slime mold (*Dictyostlieum discoideum*). Interestingly, in both mung bean (*Vigna radiata*) and slime mold (*Dictyostelium discoideum*), soman was more rapidly degraded by the addition of Mn^{2+} but inhibited by EDTA (ethylenediamine-tetraacetate). Because EDTA inhibited this enzyme, it is believed that divalent cations are important active site components for these enzymes. Because many other purified phosphatases did not hydrolyze DFP *in vitro*, Hoskin *et al.* (1999) concluded that the transformation of DFP was likely to be catalyzed by a novel plant phosphatase.

Using the aquatic plants parrot feather (*Myriophyllum aquaticum*), giant duckweed (*Spirodela oligorrhiza*), and Canadian waterweed (*Elodea canadensis*), Gao *et al.* (2000) investigated the capability of these plants to transform three organophosphate compounds including malathion [(dimethoxyphosphinothioyl)thio]butanedioic acid diethyl ester (malathion), *O*,*O*-methyl *S*-[2-(ethylthio) ethyl]phosphorothioate (demeton-*S*-methyl), and methylphosphramidic acid 2-chloro-4-(1,1-dimethyl-ethyl)phenyl methyl ester (crufomate) (Figure 5-8). For each plant, rapid transformation of all organophosphate compounds was observed, and the controls indicated that this transformation step was enzyme-mediated. In the same study, when comparing the degradation rates of the organophosphates by intact giant duckweed (*Spirodela oligorrhiza*) to a cell-free enzyme extract of the same plant, the enzyme extract displayed a higher rate of transformation (Table 5-1).

Another group of nonspecific, multifunctional phosphatases includes the organophosphorous hydrolases (EC 3.1.8.1). These hydrolases catalyze the transformation of toxic organophosphate compounds containing P–O, P–S, P–CN, and P–F bonds. The lack of specificity in hydrolyzing a variety of

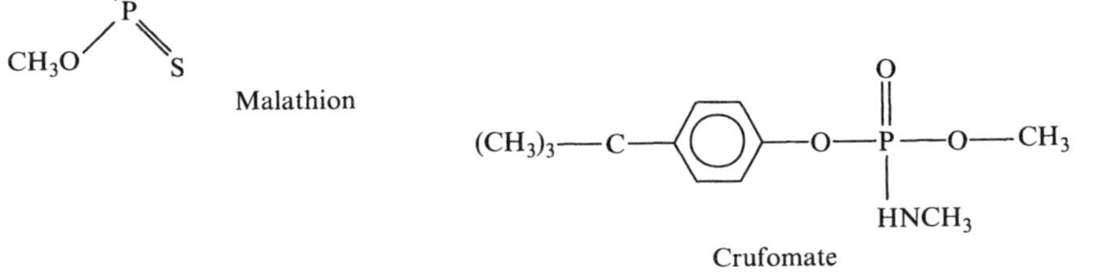

Malathion

Demeton-*S*-methyl

Crufomate

Figure 5-8 Organophosphate pesticides known to undergo acid phosphatase (EC 3.1.3.2) hydrolysis.

TABLE 5-1 Disappearance Rate Constants (*k*) and Half-Lives ($t_{1/2}$) of Organophosphorous Compounds Incubated with Parrot Feuther (*Myriophyllum aquaticum*), Giant Duckweed (*Spirodela oligorrhiza,*), Canadian Waterweed (*Elodea canadensis*), and a Cell-Free Extract of Giant Duckweed (*Spirodela oligorrhiza*)

Plants	Malathion®				Demeton-*S*-methyl				Crufomate			
	C_0 (milligrams per liter)	k (per hour)	r^2	$t_{1/2}$ (hour)	C_0 (milligrams per liter)	k (per hour)	r^2	$t_{1/2}$ (hour)	C_0 (milligrams per liter)	k (per hour)	r^2	$t_{1/2}$ (hour)
Parrot feather	0.5	0.012	0.91	58	5.0	0.011	0.93	69	5.0	0.005	0.86	139
Giant duckweed	0.5	0.014	0.92	49	5.0	0.012	0.89	57	5.0	0.007	0.90	99
Canadian waterweed	0.5	0.006	0.88	116	5.0	0.056	0.87	12	5.0	0.003	0.91	2302
Enzyme extract of giant duckweed	3.03[a]	0.019	0.94	36	38.71[a]	0.013	0.92	53	34.28[a]	0.009	0.90	78

[a]Concentration in micromoles per liter instead of milligrams per liter.

Note: The kinetics of organophosphorous compound disappearance was described by a first-order rate expression: $\ell n\ (C_t/C_0) = -kt$ (C_0 = the zero time substrate concentration, C_t = the substrate concentration at time t, and k = first-order disappearance rate constant). This equation was used to determine the value of k based on organophosphorous concentration data over the incubation period using liner regression. r^2 is the correlation coefficient. The half-life of disappearance was calculated from $t_{1/2} = (\ell n\ 2)/k$.

phosphate linkages allows these enzymes to react with a broad spectrum of organophosphorous compounds. Indeed, a purified organophosphorous hydrolase was capable of hydrolyzing a variety of organophosphate compounds containing P–O and P–S bonds (Sakai and Tadano 1993, Hoskin *et al.* 1995). Purified organophosphorous hydrolase was infused into the cell wall of *Escherichia coli* cells, and as a result this bacterium was capable of degrading parathion (*O,O*-diethyl-*O*-*p*-phenylphosphorothioate) and paraoxon (*O,O*-diethyl-*O*-*p*-nitrophenylphosphate) (Richins *et al.* 1997). That study was expanded, and a later report demonstrated that *Escherichia coli*, with the same organophosphorous hydrolase infusion protein, was capable of rapidly detoxifying methyl parathion, (*O,O*-diethyl-*O*-2-isopropyl-6-methyl-pyrimidin-4-yl phosphorothiote (diazinon), and *O*-3-chloro-4methyl-2-oxo-2*H*-chromen-7-yl *O,O*-dirthyl phosphorothioate (coumaphos) (Mulchandani *et al.* 1999). Another study reported the purification and characterization of organophosphorous hydrolase from *Pseudomonas diminuta* (Hoskin *et al.* 1995). The work revealed that this organophosphorous hydrolase mainly cleaved P–S linkages and hydrolyzed tetriso, soman, and another acetylcholinesterase (EC 3.1.1.7) inhibitor *O*-ethyl-*S*-(2-diisopropylaminoethyl) methyl phosphonothiolate. This enzyme required Mg^{2+} and EDTA significantly inhibited the activity. Organophosphorous hydrolase activity has also been shown to be present in several different cell-free extracts of aquatic plants (Hoskin *et al.* 1995, Morita *et al.* 1996).

In addition to organophosphorous hydrolase (EC 3.1.8.1), other enzymes may also mediate the degradation of organophosphate compounds. These enzymes may involve the mixed-function oxidases (Bond and Bradley 1997), flavin-containing monooxygenases (Levi and Hodgson 1992) that catalyze P=S bond oxidation, glutathione transferase (EC 2.5.1.18) that catalyzes a C–S bond cleavage (Edwards and Onen 1998), and carboxylesterase (EC 3.1.1.1) that is known to perform C–O bond cleavage and *O*-dealkylation (Lan *et al.* 1983).

Organophosphate acid anhydrases (EC 3.1.8.2) are another group of non-specific enzymes that catalyze the degradation of many organophosphorous substrates, releasing phosphate for cellular growth. These enzymes have demonstrated the capability to effectively hydrolyze both DFP and *N,N'*-diisopropylphosphorodiamidofluoridate (mipafox) (Yu and Sakurai 1995). Organophosphate acid anhydrases degraded acetylcholinesterase inhibitors (DFP and mipafox) at slightly acidic pH. Multiple organophosphate acid anhydrase isoforms have been purified from seedlings of mung bean (*Vigna radiata*) and these enzymes displayed molecular masses from 22 to 500 kilodaltons. Three major isoforms were induced by EDTA, whereas zinc chloride inhibited enzyme activity. One of the enzymes was localized throughout all tissues of the plant. Cheng *et al.* (1996) investigated cloning and expression of a 35-kilodalton organophosphate acid anhydrases from the bacterium *Pseudomonas diminuta*. This enzyme required Zn^{2+} or Co^{2+} for organophosphorous hydrolysis, clearly indicating a major difference from plant organophosphate acid anhydrases. Some of the toxic organophosphorous compounds that were degraded include sarin, soman, and pesticides. Researchers have concluded that these enzymes are very similar to the organophosphorous hydrolases (EC 3.1.8.1).

Another group of phosphatases that are present in many organisms are the phosphotriesterases (EC 3.1.8.1). These enzymes are capable of cleaving organophosphorous compounds containing phosphotriester bonds. Phosphotriesterases, like many other phosphatases, are capable of hydrolyzing many organophosphorous substrates that subsequently release phosphate. One such phosphotriesterase was purified from the bacterium *Pseudomonas diminuta*, characterized extensively, and crystallized to a resolution of 2 angstroms (Benning *et al.* 1995). Enzymological studies demonstrated that this phosphatase could not hydrolyze any mono- or diphosphate containing organophosphorous compounds (Donarski *et al.* 1989). Thus, unlike many other phosphatases with unselective broad substrate reactivities, this enzyme requires a specific triphosphoester structure in the substrate. However, phosphotriesterases, like many other phosphatases, contain a Zn^{2+} center for enzyme activity and EDTA inhibits these enzymes. The active center of this enzyme prefers Zn^{2+} but still has enzyme activity using Cd^{2+}, Ni^{2+}, and Mn^{2+}. Similar enzymatic characteristics can be observed when comparing this enzyme to the phosphotriesterase purified from *Escherichia coli* (Benning *et al.* 1995). Future studies with these enzymes will focus on the capability to transform toxic nerve agents.

It is well known that organophosphate compounds are transformed by a variety of different processes in the environment. In contrast to photochemical (Kamiya and Kameyama 1998) or abiotic hydrolysis (Lacorte *et al.* 1995) reactions, enzymatic transformation on plant or soil surfaces or in natural waters will depend on the type of phosphatase and the contaminant substrate. Because a majority of the studies investigating organophosphate remediation has been limited to bacteria and plant-associated microorganisms (Levanon 1993, Burken and Schnoor 1996, Zayed *et al.* 1998), the capability of plants to transform organophosphorous compounds without the participation of associated microorganisms remains arguable (Schnoor *et al.* 1995, van der Krol *et al.* 1995). However, Gao *et al.* (2000) demonstrated that axenically grown *Myriophyllum aquaticum* (parrot feather), in the absence of culturable microorganisms, rapidly transformed several organophosphate compounds.

NITRILASES

Most of the literature regarding nitrilases and nitrile hydratases resulted from studies investigating the regulation of cellular levels of the hormone indole acetic acid. Plants synthesize and metabolize this hormone *via* multiple enzymatic pathways. The final step in the metabolic pathway of indole acetic acid is the hydrolysis of the nitrile (CN) functional group to a carboxylic acid.

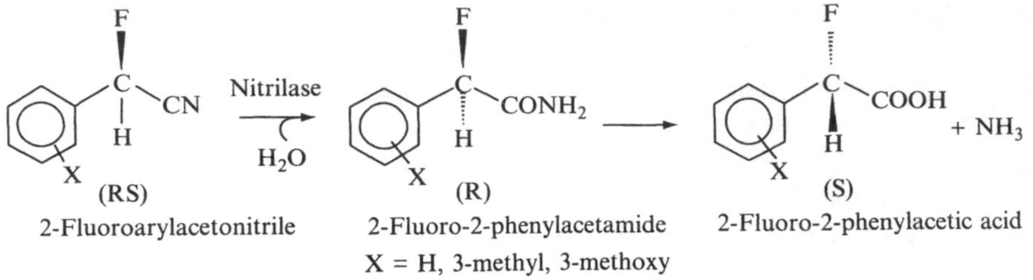

Nitrilase $\xrightarrow{H_2O}$; + NH_3

(RS) 2-Fluoroarylacetonitrile — (R) 2-Fluoro-2-phenylacetamide — (S) 2-Fluoro-2-phenylacetic acid

X = H, 3-methyl, 3-methoxy

Figure 5-9 Hydrolysis of 2-fluoroarylacetonitriles by a nitrilase isolated from *Arabidopsis thaliana*. R and S represent the sterochemical configuration at the asymetric carbon atoms in the molecule.

There are two distinct nitrile-hydrolyzing enzymes (Cowan *et al.* 1998). The first type is the nitrile aminohydrolase (EC 3.5.5.1), which is a sulfur-containing enzyme that hydrolyzes a nitrile group to the corresponding carboxylic acid. The second type of nitrilase is a nitrile hydratase (EC 4.2.1.84), which hydrolyzes the nitrile group to the corresponding amide (NH_2). Both of these metalloenzymes contain iron (Fe) or cobalt (Co) cations in a thiolate ligand network. In addition, a related cyanate hydrolase (EC 3.5.5.3) was shown to hydrolyze cyanic acid to form carbon dioxide and ammonia.

Four nitrilases have been identified in *Arabidopsis thaliana*. The physiological role of these enzymes is thought to be involved in the biosynthetic pathway of indole acetic acid from tryptophan. The importance of this enzyme is evident, but it is unclear whether nitrilases detoxify xenobiotic compounds. To examine the potential of nitrilases in phytoremediation, these enzymes were tested against six substituted 2-fluoroarylacetonitriles as substrates (Effenberger and Osswald, 2001). The 2-fluoronitriles were transformed to give the corresponding (R)-2-fluoroarylacetiamides as major products (see Figure 5-9).

DEHALOGENASES

Halogenated compounds are ubiquitous contaminants in the environment. One of the most common halogenated compounds that contaminate ecosystems is trichloroethene. This pollution has led to numerous *in vitro* laboratory studies showing that plants (Newman *et al.* 1997) and microorganisms (Slater *et al.* 1995) are capable of degrading trichloroethene as well as many other diverse halogenated compounds. Trichloroethene degradation studies using both the aquatic plant and a cell-free extract from Canadian waterweed (*Elodea canadensis*) demonstrated rapid dehalogenation of the chlorinated aliphatic substrates shown in Table 5-2 (Wolfe and Datta, unpublished data). The dehalogenation of trichloroethene to dichloroethylene is illustrated in Figure 5-10. As shown, a chloride anion is removed from trichloroethene to form dichloroethylene and hydrochloric acid. This dehalogenation was not observed when using halogenated aromatics as substrates.

TABLE 5-2 Degradation of Various Halogenated Compounds by Cell-Free Extracts of Canadian Waterweed (*Elodea Canadensis*) after 24 Hours

Substrate	Percent degradation
Hexachloroethane	100
Tetrachloroethylene	59
Trichloroethylene	49
Pentachlorophenol	0
1,3,5-Trichlorobenzene	30
1,4-Dichlorobenzene	16
Carbon tetrachloride	100
Toxaphene	0
1,1,1-Trichloro-2,2-*bis* (4-chlorophenyl)ethane (DDT)	97

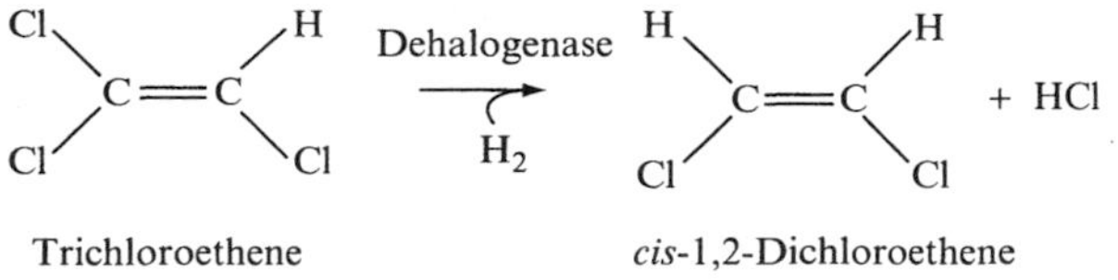

Trichloroethene —Dehalogenase, H_2→ *cis*-1,2-Dichloroethene + HCl

Figure 5-10 Dehalogenation of trichloroethene to *cis*-1,2-dichloroethene by dehalogenase.

The primary step in the biodegradation of halogenated compounds is catalyzed by dehalogenases. These enzymes catalyze the hydrolytic cleavage of a carbon-halogen bond by removing a halogen group such as bromine, iodine, or chlorine from either an aromatic or an aliphatic substrate. The addition of halogenated compounds to the growth media of certain microorganisms has shown that various isoforms of dehalogenases can be induced (Janssen and Witholt 1992). While no purified and characterized plant dehalogenase has been reported, most well-known dehalogenases have been purified and characterized from five different species of *Pseudomonas* (Schneider *et al.* 1991, Jones *et al.* 1992, Murdiyatmo *et al.* 1992, Kawasaki *et al.* 1994) and from *Xanthobacter antotrophicus* GJ10 (van der Ploeg *et al.* 1991). Although hydride (H^-) can be an electron donor for dehalogenase reactions, other reported dehalogenases are capable of utilizing different cosubstrates such as glutathione, NADH, and oxygen (Fetzner and Lingens 1994). Whether or not dehalogenases require a metal species or prosthetic cofactor for enzyme activity is unknown. A comparison of known gene sequences indicates that there is significant homology among different microbial dehalogenases. See Janssen *et al.* (1994), Fetzner and Lingens (1994), and Hardman (1991) for more detailed information on the structure, genetics, and biological application of microbial dehalogenases.

CONCLUSIONS

Because phytoremediation of organic compounds occurs in the presence of plants and a variety of associated microorganisms, a considerable amount of work on xenobiotic degradation of environmental contaminants has been accomplished for a variety of organisms. Results from these studies have collectively demonstrated that the detoxification steps for many organic compounds are enzyme-mediated. Therefore, a variety of important plant and microbial enzymes that detoxify xenobiotic compounds are known to be important.

Whether or not phytoremediation will be an alternate strategy for the detoxification of soil and water contaminated with organic pollutants is still largely unknown. A major concern regarding phytoremediation is the lack of information available on the terminal products formed by plants. Before adding plants into a xenobiotic-contaminated ecosystem, it must be shown that the products are less toxic than the original pollutant compounds. Some researchers have concluded that conjugated metabolites are terminal products formed in plants (Carter 1975). In periwinkle (*Catharanthus roseus*) cultures, glucose-conjugated TNT metabolites were extracted, isolated, and identified, but the relative toxicity as compared to TNT and other TNT metabolites was not reported (Bhadra *et al.* 1999). It is essential that the toxicities of these conjugated metabolites are known and compared to TNT. Perhaps, an easy way to generate the same TNT metabolites for such toxicity studies would be enzymatic production *in vitro* after the addition of 4-ADNT, UDP-glucose, and purified UDP-glucosyltransferase (EC 2.4.1.71). By forming known products from a purified enzyme, a time-consuming and laborious purification procedure of conjugated metabolites from intact plants or cell cultures could be eliminated. Studying and applying the enzymes that catalyze the detoxification reactions would thus be key to answering much that is unknown about phytoremediation.

Although plants contain many enzymes for transformation and conjugation of xenobiotic chemicals, microorganisms contain a wider variety of enzymes that mineralize many recalcitrant organic compounds. Many of these enzymes have been critically studied and have been applied using gene technology. Due to recent advances in plant genetics, proteomics, and completion of the *Arabidopsis thaliana* genome, researchers are now able to identify specific genes that encode various detoxification enzymes in *Arabidopsis thaliana*. Gene identification in *Arabidopsis thaliana* is possible by obtaining a known sequence of an enzyme or a gene and by searching for homology on a database containing known *Arabidopsis thaliana* sequences.

Recently, several bacterial genes have been shown to be successful when cloned and expressed in plants. This type of work seeks a genetic means of phytoremediation improvement. First, the *Escherichia coli* genes, *merA* and *merB*, which produce enzymes that reduce oxidized mercury to elemental mercury, were overexpressed in *Liriodendron tulipifera* (yellow poplar) and

Arabidopsis thaliana for enhanced mercury reduction and phytovolatilization (Rugh *et al.* 1996, Bizily *et al.* 2000). Additionally, Naested *et al.* (1999) reported the cloning and expression of a bacterial haloalkane dehalogenase (EC 3.8.1.5), *dhlA*, in *Arabidopsis thaliana*. By inserting this gene into the plant genome, *Arabidopsis thaliana* would theoretically be capable of effectively degrading halogenated and other related haloalkane pollutants. In another study, the microbial pentaerythritol tetranitrate reductase from *Mycobacterium vaccae* was cloned into tobacco (*Nicotiana rustica*) plants. With this microbial gene, tobacco (*Nicotiana rustica*) plants would be capable of highly effective TNT transformation which could produce unique microbial metabolites, possibly leading to mineralization. Because mineralization of TNT has only been observed by microorganisms under strict anaerobic or lignolytic conditions, transgenic tobacco (*Nicotiana rustica*) would possess a novel route of TNT degradation, different from what has ever been observed by plants. Based on these and other upcoming molecular studies in xenobiotic phytoremediation, it is evident that investigating both enzymology and gene technology can offer many advantages. In future projects, other microbial genes expressing detoxification enzymes can be cloned into plants and expressed, allowing the host plant to have a wider range of phytoremediation capabilities. When acting in combination or synergy, enzymes from both plants and rhizosphere microorganisms could potentially make phytoremediation a remarkable, cost-effective alternate means of xenobiotic detoxification.

Disclaimer

This paper has been reviewed in acordance with the U.S. Environmental Protection Agency peer and administrative review policies and approved for publication. Mention of trade names or commercial products does not constitute endorsement or recommendation for use.

REFERENCES

Alverez, M.A., L.K. Christopher, J.L. Botsford, and P.J. Unkefer (1995) *Pseudomonas aeruginosa* strain MAO1 aerobically metabolizes the aminodinitrotoluene produced by 2,4,6-trinitrotoluene group reduction. *Can. J. Microbiol.* **41**: 984–991.

Andreae, W.A. and N.E. Good (1957) The formation of indoleacetylaspartic acid in pea seedlings. *Plant Physiol.* **30**: 380–382.

Applebury, M.L. and J.E. Coleman (1969) *Escherichia coli* alkaline phosphatase. *J. Biol. Chem.* **244**: 308–318.

Atkinson, M.J. (1987) Alkaline phosphatase activity of coral reef benthos. *Coral Reefs*. **6**: 59–62.

Attaway, H., J.O. Nelson, A.M. Baya, M.J. Voll, W.E. White, D.J. Grimes, and R.R. Colwell (1987) Bacterial detoxification of diisopropyl fluorophosphate. *Appl. Environ. Microbiol.* **53**: 1685–1689.

Benning, M.M., J.M. Kuo, F.M. Raushel, and H.M. Holden (1995) Three-dimensional structure of the binuclear metal center of phosphotriesterase. *Biochem.* **34**: 7973–7978.

Berman, T. (1970) Alkaline phosphatases and phosphorus availability in Lake Kinneret. *Limnol. Oceanogr.* **15**: 663–674.

Bhadra, R., D.G. Wayment, J.B. Hughes, and J.V. Shanks (1999) Confirmation of conjugation processes during TNT metabolism by axenic plant roots. *Environ. Sci. Technol.* **33**: 446–452.

Bizily, S.P., C.L. Rugh, and R.B. Meagher (2000) Phytoremediation of hazardous organomercurials by genetically engineered plants. *Nat. Biotechnol.* **18**: 213–217.

Bond, J.A. and B.P. Bradley (1997) Resistance to malathion in heat-shocked *Daphnia Magna. Environ. Toxicol. Chem.* **16**: 705–712.

Boopathy, R. and C.F. Kulpa (1994) Biotransformation of a *Methanococcus* sp. (strain B) isolated from a lake sediment. *Can. J. Microbiol.* **40**: 273–278.

Bryant, C. and M. DeLuca (1990) Purification and characterization of an oxygen-insensitive NAD(P)H nitroreductase from *Enterobacter cloacae. J. Biol. Chem.* **266**: 4119–4125.

Bryant, C. and W.D. McElroy (1991) Nitroreductases. In: *Chemistry and Biochemistry of Flavoenzymes.* F. Muller, ed. CRC Press, Boca Raton, Florida, pp. 291–304.

Burken, J.G. and J.L. Schnoor (1996) Phytoremediation: plant uptake of atrazine and role of root exudates. *J. Environ. Eng.* **122**: 958–963.

Buswell, J.A., S. Hamp, and K.E. Eriksson (1979) Intracellular quinone reduction in *Sporotrichum pulverulentum* by a NADPH quinone reductase. *FEBS. Lett.* **108**: 229–232.

Bylund, J.E., J.K. Dyer, D.E. Feely, and E. L. Martin (1990) Alkaline and acid phosphatases from the extensively halotolerant bacterium *Halomonas elongata. Curr. Microbiol.* **20**: 125–131.

Carter, M.C. (1975) Amitrole. In: *Herbicides: Chemistry, Degradation, and Mode of Action.* P.C. Kearney and D.D. Kaufman, eds. Marcel Dekker. New York, pp. 337–398.

Chasseaud, L.F. (1974) Glutathione *S*-transferases. In: *Glutathione*, Proc. 16th Conf. German Soc. Chem. L. Flohe, H.C. Benohr, H. Sies, H.D. Waller, and A. Wendel, eds. Thieme, Stuttgart, Germany, pp. 90–108.

Cheng, T.C., S.P. Harvey, and G.L. Chen (1996) Cloning and expression of a gene encoding a bacterial enzyme for decontamination of organophosphorous nerve agents and nucleotide sequence of the enzyme. *Appl. Environ. Microbiol.* **62**: 1636–1641.

Clarke, A.G. and J.N. Smith (1975) Glutathione *S*-aryltransferase as a model for glutathione *S*-transferases. *Environ. Qual. Safety, Suppl.* **3**: 346–350.

Coleman, J.O.D., M.A. Mechteld, B. Kalff, and T.G.E. Davies (1997) Detoxification of xenobiotics by plants: chemical modification and vacuolar compartmentation. *Trends Plant Sci.* **2**: 1144–1151.

Cowan, D., R. Cramp, R. Pereira, D. Graham, and Q. Almatawah (1998) Biochemistry and biotechnology of mesophilic and thermophilic nitrile metabolizing enzymes. *Extrophil. Life Extreme Cond.* **2**: 207–216.

Davies, T.G.E and J.O.D. Coleman (2000) The Arabidopsis thaliana ATP-biding cassette proteins: an emerging superfamily. *Plant Cell Environ.* **23**: 431–443.

Donarski, W.J., D.P. Dumas, D.P. Heitmeyer, V.E. Lewis, and F.M. Raushel (1989) Structure-activity relationships in the hydrolysis of substrates by the phosphotriesterase from *Pseudomonas diminuta. Biochemistry* **28**: 4650–4655.

Edwards, R. and W.J. Onen (1998) Regulation of glutathione *S*-transferases of *Zea mays* in plants and cultures. *Planta* **175**: 99–106.

Effenberger, F. and S. Osswald (2001) Enantioselective hydrolysis of (RS)-2-fluoroarylacetonitriles using nitrilase from *Arabidopsis thaliana. Tetra. Asymm.* **12**: 279–285.

Fernando T., J.A. Bumpus, and S.D. Aust (1990) Biodegradation of TNT (2,4,6-trinitrotoluene) by *Phanerochaete chrysosporium. Appl. Environ. Microbiol.* **56**: 1666–1671.

Fetzner, S. and F. Lingens (1994) Bacterial dehalogenases: biochemistry, genetics and biotechnological applications. *Microbiol. Rev.* **58**: 641–685.

Fiorella, P.D. and J.C. Spain (1997) Transformation of 2,4,6-trinitrotoluene by *Pseudomonas pseudocaligenes* JS52. *Appl. Environ. Microbiol.* **63**: 2007–2015.

Frear, D.S (1968) Herbicide metabolism in plants: purification and properties of the UDP-glucose: arylamine *N*-glucosyl-transferase from soybean. *Phytochemistry* **7**: 381–390.

Frear, D.S. and H.R. Swanson (1970) Biosynthesis of *S*-(4-ethylamino-6-isopropyl-2-*s*-triazino) glutathione: partial purification and properties of a glutathione *S*-transferase from corn. *Phytochemistry* **9**: 2123–2132.

French, C.E., S.J. Rosser, G.J. Davies, S. Nicklin, and N.C. Bruce (1999) Biodegradation of explosives by transgenic plants expressing pentaerythritol tetranitrate reductase. *Nat Biotechnol.* **17**: 491–494.

Fry, S. (1986) Cross-linking of matrix polymers in the growing cell walls of angiosperms. *Annu. Rev. Plant Physiol.* **37**: 165–186.

Gao, J.A., W.A. Garrison, C. Mazur, C. Hoehamer, and N.L. Wolfe (2000) Uptake and phytotransformation of organophosphorus pesticides by axenically cultivated aquatic plants. *J. Agric. Food Chem.* **48**: 6121–6127.

Gellatly, K.S., G.B.G. Moorhead, S.M.G. Duff, D.D. Lefebvre, and W.C. Plaxton (1994) Purification and characterization of a potato tuber acid phosphatase having significant phosphotyrosine phosphatase activity. *Plant Physiol.* **106**: 223–232.

Gorontzy, T.O., M.W. Drzyzga, D. Kahl, J. Bruns-Nagel, E.V. Breitung, R.K. Loew, and K.H. Blotevogel (1994) Microbial degradation of explosives and related compounds. *Crit. Rev. Microbiol.* **20**: 265–284.

Guddwear, M.B. and W.C. Dauterman (1979) Purification and properties of a glutathione *S*-transferase from corn which conjugates *s*-triazine herbicides. *Pestic. Biochem. Physiol.* **12**: 1–9.

Gunse, B. and E.F. Elstner (1992) Formation of activated states of indole-acetic acid and cytokinins—an experimental approach to a hypothesis concerning signal transduction. *J. Plant Physiol.* **140**: 536–540.

Hardman, D.J. (1991) Biotransformation of halogenated compounds. *Crit. Rev. Biotechnol.* **11**: 1–40.

Hatzios, K. and D. Penner (1982) *Metabolism of Herbicides in Higher Plants.* Burgess Publishing Company, Minneapolis, Minnesota, pp. 43–57.

Hernandez, I., F.X. Niell, and J.A. Fernandez (1994) Alkaline phosphatase activity in marine macrophytes: histochemical localization in some widespread species in southern Spain. *Mar Biol.* **120**: 501–509.

Ho, I. (1988) Acid phosphatase, alkaline phosphatase, and nitrate reductase activity of selected ectomycorrhizal fungi. *Can. J. Bot.* **67**: 750–753.

Honeycutt, M.E., A.S. Jarvis, and V. McFarland (1996) Cytotoxicity of trinitrotoluene and its metabolites. *Ecotoxicol. Environ. Safe.* **36**: 282–287.

Hoskin, F.C.G. and J.E. Walker (1997) Malathion as a model for the enzymatic hydrolysis of the neurotoxic agent, VX. *Bull. Environ. Contam. Toxicol.* **59**(1): 9–13.

Hoskin, F.C.G, J.E. Walker, W.D. Dettbarn, and J.R. Wild (1995) Hydrolysis of tetriso by an enzyme derived from *Pseudomonas diminuta* as a model for the detoxication of *O*-ethyl *S*-(2-diisopropylaminoethyl) methylphosphonothiolate (VX). *Biochem. Pharmacol.* **49**(5): 711–715.

Hoskin, F.C.G., J.E. Walker, and C.M. Mello (1999) Organophosphorous acid anhydrolase in slime mold, duckweed and mung bean: a continuing search for a physiological role and a natural substrate. *Chem.-Biol. Interact.* **119/120**: 399–404.

Hughes, J.B., J. Shanks, M. Vanderford, J. Lauritzen, and R. Bhadra (1997) Transformation of TNT by aquatic plants and plant tissue cultures. *Environ. Sci. Technol.* **31**: 266–271.

Janssen, D.B. and B. Witholt (1992) Aerobic and anaerobic degradation of halogenated aliphatics. In: *Metal Ions in Biological Systems.* H. Sigel, ed. Marcel Decker, New York, pp. 229–327.

Janssen, D.B., F. Pries, and J.R. van der Ploeg (1994) Genetics and biochemistry of dehalogenating enzymes. In: *Annual Review of Microbiology.* Annual Reviews, Palo Alto, California. **48**: 163–191.

Jones, D.H.A., P.T. Barth, D. Byrom, and C.M. Thomas (1992) Nucleotide sequence of a structural gene encoding a 2-haloacid dehalogenase of *Pseudomonas putida* strain AJ1 and purification of an encoded protein. *J. Gen. Microbiol.* **138**: 675–683.

Kamiya, M. and K. Kameyama (1998) Photochemical effects of humic substances on the degradation of organophosphorus pesticides. *Chemosphere* **36**: 2337–2344.

Kawasaki, H., T. Toyama, T. Maeda, H. Nishino, and K. Tonomura (1994) Cloning and sequence analysis of a plasmid encoded 2-haloacid dehalogenase gene from *Pseudomonas putida* no. 109. *Biosci. Biotech. Biochem.* **50**(1): 160–163.

Kinouchi, T. and Y. Ohnishi (1983) Purification and characterization of 1-nitropyrene nitroreductases from *Bacteroides fragilis. Appl. Environ. Microbiol.* **46**: 596–604.

Klambt, H.D. (1961) Wachstumsinduktion und wuchsstoffmetabolismus in weizenkoleoptilzylinder. *Planta* **57**: 339–353.

Lacorte, S., S.B. Lartiges, P. Garrigues, and D. Barceló (1995) Degradation of organophosphorous pesticides and their transformation products in estuarine waters. *Environ. Sci. Technol.* **29**: 431–438.

Lagrimini, L.M. (1991) Wood induced deposition of polyphenols in transgenic plants overexpressing peroxidase. *Plant Physiol.* **96**: 577–583.

Lan, P.T., R. Main, N. Motoyama, and W.C. Dauterman (1983) Hydrolysis of malathion by rabbit liver oligomeric and monomeric carboxylesterases. *Pestic. Biochem. Physiol.* **20**: 232–237.

Lerch, K., J. Deinum, and B. Reinhammar (1978) The state of copper in *Neurospora* laccase. *Biochim. Biophys. Acta.* **54**: 7–14.

Levanon, D. (1993) Roles of fungi and bacteria in the mineralization of the pesticides atrazine, alachlor, malathion, and carbofuran in soil. *Soil Biol. Biochem.* **25**: 1097–1105.

Levi, P.E. and E. Hodgson (1992) Metabolism of organophosphorous compounds by the flavin-containing monooxygenase. In: *Organophosphorous: Chemistry, Fate, and Effects.* J.E. Chambers and P.E. Levi, eds. Academic Press, San Diego, California, pp.141–145.

McCormick, N.G., F.E. Feeherry, and H.S. Levinson (1976) Microbial transformation of 2,4,6-trinitrotoluene and other nitroaromatic compounds. *Appl. Environ. Microbiol.* **31**: 949–958.

Morita, N., H. Nakazato, H. Okuyama, Y. Kim, and G.A. Thompson (1996) Evidence for a glycosylinositolphospholipid-anchored alkaline phosphatase in the aquatic plant *Spirodela oligorrhiza. Biochim. Biophys. Acta.* **1290**: 53–52.

Mosbach, R. (1963) Purification and some properties of laccase from *Polyporus versicular. Biochim. Biophys. Acta.* **73**: 204–211.

Mulchandani, A., I. Kaneva, and W. Chen (1999) Detoxification of organophosphate nerve agents by immobilized *Escherichia coli* with surface-expressed organophosphorus hydrolase. *Biotechnol. Bioeng.* **63**: 216–223.

Müller, M. (1983) Mode of action of metronidazole on anaerobic bacteria and protozoa. *Surgery* **93**: 165–171.

Murdiyatmo, U., W. Asmara, J.S.H. Tsang, A.J. Baines, A.T. Bull, and D.J. Hartman (1992) Molecular biology of 2-haloacid halidohydratase IVa from *Pseudomonas cepacia* MBA4. *Biochem. J.* **284**(1): 87–93.

Naested, H., M. Fennema, L. Hao, M. Andersen, D.B. Janssen, and J. Mundy (1999) A bacterial haloalkane dehalogenase gene as a negative selectable marker in *Arabidopsis. Plant J.* **18**: 571–576.

Naumova, R.P., S.Y. Selivanovskaya, and F.A. Mingatina (1988) Possibility of deep bacterial destruction of 2,4,6,-trinitrotoluene. *Mikrobiologiya.* **57**(2): 218–222. (translated from Russian to English by Plenum Publishing, UDC 579.841.11-222.2, pp. 169–173, ISBN 0026-2617/88/5702-0169).

Newman, L.A., S.E. Strand, N. Choe, J. Duffy, B.B. Shurtleff, J. Wilmoth, and M.P. Gordon (1997) Uptake and biotransformation of trichloroethylene by hybrid poplars. *Environ. Sci. Technol.* **31**: 1062–1067.

Palazzo, A.J. and D.C. Leggett (1986) Effect and disposition of TNT in a terrestrial plant. *J. Environ. Qual.* **15**: 49–52.

Pedregosa, A.M., F. Pinto, I.F. Monistrol, and F. Laborda (1990) Regulation of acid phosphatases of *Cladosporium cucumerinum* by inorganic phosphate. *Mycol. Res.* **95**: 720–724.

Peterson, F.J., R.P. Mason, J. Hovsepian, and J.L. Holtzman (1979) Oxygen-sensitive and insensitive nitroreduction by *E. coli* and rat hepatic microsomes. *J. Biol. Chem.* **254**: 4009–4014.

Regan, K.M. and R.L. Crawford (1994) Characterization of *Clostridium bifermentens* and its biotransformation of 2,4,6-trinitrotoluene (TNT) and 1,3,5-triaza-1,3,5 trinitrocyclohexane. *Biotechnol. Lett.* **16**: 1081–1086.

Reinhammar, B. (1970) Purification and properties of laccase and stellacyanin from *Rhus vernicifera. Biochim. Biophys. Acta.* **205**: 35–41.

Richins, R.D., I. Kaneva, A. Mulchandani, and W. Chen (1997) Biodegradation of organophosphorus pesticides by surface-expressed organophosphorus hydrolase. *Nat. Biotechnol.* **15**: 984–987.

Rieble, S., D.K. Joshi, and M. Gold (1994) Aromatic nitroreductase from the basidiomycete *Phanerochaete chrysosporium. Biochem. Biophys. Res. Comm.* **205**: 298–304.

Rugh, C.L., H.D. Wilde, N.M. Stack, D.M. Thompson, A.O. Summers, and R.B. Meagher (1996) Mercuric ion reduction and resistance in transgenic *Arabidopsis thaliana* plants expressing a modified bacterial *merA* gene. *Proc. Natl. Acad. Sci. USA* **93**: 3182–3187.

Sakai, H. and T. Tadano (1993) Characteristics of response of acid phosphatase secreted by the roots of several crops to various conditions in the growth media. *Soil Sci. Plant Nutr.* **39**: 437–444.

Salzer, P. and A. Hager (1993) Effect of auxins and ectomycorrhizal elicitors on wall bound proteins and enzymes of spruce [*Picea abies* (L) Karst] cells. *Trees-Struct. Funct.* **8**: 49–55.

Sandermann, H., Jr. (1992) Plant metabolism of xenobiotics. *Trends Biol. Sci.* **17**: 82–84.

Schneider, B., R. Miller, R. Frank, and F. Lingens (1991) Complete nucleotide sequences and comparison of the structural genes of two haloalkanoic acid dehydrogenases from *Pseudomonas* spp. strain CBS3. *J. Bacteriol.* **173**: 1530–1535.

Schnoor, J.L., L.A. Licht, S.C. McCutcheon, N.L. Wolfe, and L.H. Carreira (1995) Phytoremediation of organic and nutrient contaminants. *Environ. Sci. Technol.* **29**: 318A–323A.

Slater, J.H., A.T. Bull, and D.J. Hardmann (1995) Microbial dehalogenation. *Biodegradation* **6**: 181–189.

Somerville, C.C., S.F. Nishino, and J.C. Spain (1995) Purification and characterization of a nitrobenzene nitroreductase from *Pseudomonas pseudocaligenes* JS45. *Appl. Environ. Microbiol.* **177**: 3837–3842.

Stahl, J. and S.D. Aust (1993) Metabolism and detoxification of TNT by *Phanerochaete chrysosporium. Biochem. Biophys. Res. Comm.* **192**: 477–482.

Sterjiades, R., J.F. Dean, G. Gamble, D.S. Himmelsbach, and K.E.L. Eriksson (1993) Extracellular laccases and peroxidases from sycamore maple (*Acer pseudoplatanus*) cell-suspension cultures. *Planta* **190**: 75–87.

Tatsumi, K., T. Doi, N. Koga, H. Yoshimura, H. Koga, and T. Horiuchi (1981) Studies on the nitrofuran nitroreductase from *E. coli. J. Biochem.* **89**: 855–869.

Thompson, P.L., L.A. Ramer, and J.L. Schnoor (1998) Uptake and transformation of TNT by hybrid poplar trees. *Environ. Sci. Technol.* **32**: 975–980.

Valli, K., B.J. Brock, D.K. Joshi, and M.H. Gold (1992) Degradation of 2,4-dinitrotoluene by the lignin degrading fungus *Phanerochaete chrysosporium. Appl. Environ. Biol.* **58**: 221–228.

van Aken, B.V., K. Skubisz, H. Naveau, and S.N. Agathos (1997) Biodegradation of 2,4,6-trinitrotoluene (TNT) by the white-rot basidiomycete *Phlebia radiata*. *Biotechnol. Lett.* **19**: 813–817.

van Beelen, P. and D.R. Burris (1995) Reduction of the explosive 2,4,6-trinitrotoluene by enzymes from aquatic sediments. *Environ. Toxicol. Chem.* **14**: 2115–2123.

Vanderberg, L.A., J.J. Perry, and P.J. Unkefer (1995) Catabolism of 2,4,6-trinitrotoluene by *Mycobacterium vaccae*. *Appl. Microbiol. Biotechnol.* **43**: 937–945.

Vanderford, M., J.V. Shanks, and J.B. Hughes (1997) Phytotransformation of trinitrotoluene (TNT) and distribution of metabolic products in *Myriophyllum aquaticum*. *Biotechnol. Lett.* **19**: 277–280.

van der Krol, D., I. Schuphan, B. Thiede, and B. Schmidt (1995) Metabolism of [ring-2,6-^{14}C]parathion in plant cell suspension cultures of carrot (*Daucus carota*), purple foxglove (*Digitalis purpurea*), soybean, thorn apple (*Datura stramonium*) and wheat (*Triticum aestivum*). *Pestic. Sci.* **45**(2): 143–152.

van der Ploeg, J., G. Van Hall, and D.B. Jansson (1991) Characterization of the haloacid dehydrogenase from *Xanthobacter autotrophicus* GJ10 and sequencing of the *dblh* gene. *J. Bacteriol.* **173**: 7925–7933.

Vorbeck, C., H. Lenke, P. Fischer, and H.J. Knackmuss (1994) Identification of a hydride-Meisenheimer complex as a metabolite of 2,4,6-trinitrotoluene by a *Mycobacterium* strain. *J. Bacteriol.* **176**: 932–934.

Watambe, M., T. Nishino, K. Takio, T. Sofuni, and T. Nohmi (1989) Purification and characterization of wild type and mutant classical nitroreductases of *Salmonella typhimurium*. *J. Biol. Chem.* **273**(3): 23922–23928.

Watambe, M., M. Ishidate, and T. Nohmi (1990) Nucleotide sequence of the *Salmonella typhimurium* gene. *Nucleic Acids Res.* **18**: 1059.

Wilson, E. (1998) Spinach enzymes breakdown explosives. *Chem. Eng. News.* August: 40.

Won, W.D., L.H. Disalvo, and J. Ng (1976) Toxicity and mutagenicity of 2,4,6-trinitrotoluene and its microbial metabolites. *Appl. Environ. Microbiol.* **31**: 576–580.

Yu, M.H. and S. Sakurai (1995) Diisopropylfluorylphosphate (DFP)-hydrolyzing enzymes in mung bean (*Vigna radiata*) seedlings. *Environ. Scie.* (Toyoko) **3**(2): 103–111.

Zayed, A., S. Gowthaman, and N. Terry (1998) Phytoaccumulation of trace elements by wetland plants. *J. Environ. Qual.* **27**: 715–721.

6

PLANT TOLERANCES TO CONTAMINANTS

V. F. Medina, E. Maestri, M. Marmiroli, A. C. Dietz, and S. C. McCutcheon

SUMMARY OF PRACTICAL IMPLICATIONS

Current phytoremediation techniques require that plants live in the zone of contamination. Consequently, plant viability is a critical issue in the successful application of phytoremediation. Transformation is often essential to detoxification and phytoremediation of organic contaminants. For example, parrot feather (*Myriophyllum aquaticum*) is capable of quickly transforming trinitrotoluene at concentrations that could otherwise result in the death of this aquatic plant. Hybrid poplar (*Populus deltoides* × *Populus nigra*) continues to transform chlorinated solvents and explosives even as transpiration rates decline. Contaminant stress usually leads to metabolic changes in the plants and the expression of unique biochemicals. Understanding plant responses to contaminant stress defines whether a plant has the potential to transform and tolerate a wide range of organic chemicals, which would enhance phytoremediation. The fundamental mechanisms of plant tolerance and resistance to xenobiotic chemicals are related to the phytotoxicity of two major classes of contaminants—chlorinated aliphatic compounds and explosives. The relation to phytotoxicity provides a practical understanding of how tolerance and resistance may affect phytoremediation applications.

GLOSSARY

Apoplast: The solution phase outside living cells in an intact plant, including cell walls and xylem conducting cells.

Carotenoid: Highly unsaturated orange or red pigments synthesized in all green tissues, also required by animals, and characterized chemically by a long aliphatic polyene chain composed of eight isoprene units.

Phytoremediation: Transformation and Control of Contaminants,
Edited by Steven C. McCutcheon and Jerald L. Schnoor.
ISBN 0-471-39435-1 (cloth)

Chloroplast: An organelle containing various types of chlorophylls where photosynthesis and biopolymer synthesis occur.
Chlorosis: The yellowing of leaves and other tissues due to loss of chlorophyll.
Cytoskeleton: The highly structured and oriented framework of complex protein fibrils that is responsible for the mechanical properties of cytoplasm.
Cytosol: The soluble portion of the cytoplasm around the nucleus, organelles, and membranes.
Electrophilic: Being attracted to electrons.
Genotype: A specified genetic makeup of a class or group of individuals.
Heme: Compounds derived from protoporphyrin and ferrous or ferric iron that constitute the nonprotein groups of hemoproteins.
Mitochondria: The organelles that produce ATP (adenosine triphosphate), which produces energy for the cell through respiration.
Nucleophilic: Having an affinity for atomic nuclei or donating electrons.
Pectins: Complex colloidal carbohydrate derivatives.
Phenotype: The physical or outward appearance generated from a genotype.
Photosystems I and II: Two photochemical reactions that occur during photosynthesis in chloroplasts; I proceeds best in long-wavelength light and II proceeds best in short-wavelength light (less than 690 nanometers).
Phytochelatin: Heavy-metal complexing peptides of higher plants.
Plastoquinone: A paraquinone with two methyl substituents and a side chain of one or more isoprene units, related to vitamin K, occurring mostly in plant chloroplasts that play a role in photosynthetic phosphorylation [formation of ATP from ADP (adenosine diphosphate) during photosynthesis].
Synthase: An enzyme that catalyzes synthesis without breaking a high-energy phosphate bond.
Synthetase: An enzyme that catalyzes the biosynthesis of a compound requiring the concurrent breaking of a diphosphate bond in a triphosphate such as adenosine triphosphate.
Thylakoid: The chloroplast membrane where photosynthesis occurs.
Tubulin: A structural globular protein that polymerizes to form microtubules.

Note: For information on the enzymes used in this chapter, see http://www.chem.qmul.ac.uk/iubmb/enzyme/. See http://www.helrss.demon.co.uk/class-pesticides.html for the formal chemical names of the common pesticides names used in this chapter.

TOLERANCE: RESISTANCE, SENSITIVITY, AND OTHER RESPONSES TO TOXICITY

The understanding of tolerance and plant stress is important in applying green plants to control or treat hazardous wastes. Phytoremediation often involves high contaminant concentrations that may cause stress or toxicity beyond natural or background levels. All these stresses play a role in assessing whether application of phytoremediation is successful or not. Plant stress has been defined as "environmental constraints, shortages and excesses in the supply of solar energy, water, and mineral nutrients" including also "sub- or supraoptimal temperatures and growth inhibiting toxins" (Grime 1979). Two factors in the ambient environment capable of creating stress are pollutants, especially environmental toxins and excessive micronutrients. As primary stressors, toxins and nutrients can interact with other stressors including biotic stress from competition, predation, parasitism, and related stresses, altering the susceptibility of the affected organisms. For chemicals in the environment, as for several other factors, there is a range that allows growth of all individuals within a species without injurious effects, and a range that does not allow growth of any individuals. Between healthy conditions and death is a range of effects and injuries over which not all individuals are affected to the same degree (Fitter and Hay 1987). Defining this range of tolerance is critical to most applications.

A plant is 'sensitive' if the stress results in injury or death, and 'resistant' if the plant is able to survive and reproduce, contributing genetically to the next generation (Baker 1987). Resistance occurs when an organism undergoes metabolic changes; starting from changes in hormone production and mobilization, rate-controlling enzymes are affected, thereby modifying substrate concentrations and enzyme activity. Gene transcription and protein synthesis may also ultimately be affected (Hoffman and Parsons 1991). Because of these changes, a resistant plant can sequester chemicals or exclude and control toxins in order to respond to the detrimental stress.

There are important distinctions between resistance at the individual level and at the population level (Hoffman and Parsons 1991). In general, both individuals and populations are capable of self-engineering ambient conditions (McCutcheon and Mitsch 1994). Self-engineering occurs when individual organisms, populations, communities, or ecosystems modify or 'engineer' the surrounding environment to improve survivability or alter conditions to reduce or eliminate toxic effects. For example, plants (both individuals and communities) may respond to environmental stresses by lowering the pH or by excretion of substances to bind with or spur degradation of toxic agents (Fitter and Hay 1987).

Individual organism responses that contribute to resistance and tolerance include the following:

1. Exclusion: the plant or other organism recognizes the toxic compound and prevents (Fitter and Hay 1987) or reduces uptake (*e.g.*, by adsorption to the cuticle or skin)
2. Amelioration: the plant or other organism absorbs the toxic compound but minimizes the effects on critical organs by chelation, covalent bonding, dilution, degradation, metabolism, compartmentalization, volatilization, diffusion, and excretion

3. Avoidance: the organism produces physico–chemical barriers in order to insulate critical organs from the effects of the stress (Baker 1987)
4. Evasion: exposure to the stress ends or is reduced by responses that remove the organism from the stressful environment

Population and community responses include the following:

1. Succession: groups of organisms change species mix, numbers, and diversity to achieve communities that are more stress resistant
2. Natural Selection: selection from generation to generation of genetic variants differing in stress response will eventually occur from stressed populations

Exclusion, amelioration, and avoidance can be difficult to distinguish for some plant responses. Examples and definitions have been provided by several authors, *e.g.*, Levitt (1980), Baker (1987), Fitter and Hay (1987), Hoffmann and Parsons (1991) and Salisbury and Ross (1992). Plant exclusion is a more active process of controlling ion pumps and other active uptake processes. Plant amelioration generally involves a series of active internal metabolic and detoxification processes that segregate or transform toxins to protect critical organs, tissues, and metabolic functions. Avoidance is typically conceived to be a passive response involving insulation of critical internal tissues, organs, and metabolic functions.

When considering resistance, some authors distinguish between 'tolerance' and 'avoidance.' A tolerant organism is capable of decreasing or repairing the deleterious effects, whereas with avoidance the stressful agents are prevented from reaching critical tissues (Levitt 1980) or organs. Others think of resistance as resulting from (1) metabolic processes that remove nitrogen, phosphorus, and other critical growth elements from complex molecules, and (2) detoxification.

Not all of the highly evolved metabolic processes that plants use for self-engineering are available to prokaryotes (archaebacteria, bacteria, and cyanobacteria). Thought to have evolved to counter harmful allelopathic compounds from other plants, natural insecticides, decay products, and other natural biogeochemical insults (Coleman *et al.* 1997), plant resistance or tolerance should not be expected for every class of xenobiotic compound that has been created. In fact, an entire agribusiness frequently and profitably designs new xenobiotic herbicides to overcome evolving plant resistances.

Because of lack of motility, plants have limited capabilities to evade stress, especially the root-bound species. Evasion responses include the switching to dormant or quiescent stages until the insult has been removed, and modification of root growth in order to reach less contaminated soil (Hoffman and Parsons 1991), or by directional growth above ground and re-rooting into cleaner soil. For instance, the emergent aquatic plant parrot feather (*Myriophyllum aquaticum*) will not place roots into 2,4,6-trinitrotoluene (TNT) contaminated sediments, leaving the plant free to float to cleaner areas (Young 1995).

Community and population responses to stress occur over much longer times (Bradshaw and McNeilly 1991, Hoffmann and Parsons 1991). Succession occurs more quickly than selection and makes use of a greater diversity of metabolic processes, including growth enzymes and senescence enzymes in different locations at different times. Succession and selection are not necessarily complimentary. Succession may truncate selection before a population achieves the best resistance to an insult and might be the reason, for example, that plants have not adapted to all heavy metal outcrops.

A particular asset of plant populations in dealing with stress is the common occurrence of a property known as 'phenotypic plasticity,' which is defined as the extent of variation in phenotypic expression of traits by a single genotype in different environments (Hoffman and Parsons 1991). By reversible changes in physiological processes and metabolic reactions, plant populations of the same genotype can express different phenotypes in response to different environmental insults. Examples of phenotypic plasticity involve several aspects of plant life, such as: (1) closure of stomata; (2) development of roots, shoots, and leaves; (3) flowering; and (4) changes in metabolic rates. Excessive alkali or metal content in soils, and nutrient depletion can induce similar modifications (Bradshaw 1965). Because phenotypic plasticity is due to genetic variation in species and populations, natural selection can cause changes in plasticity levels. Generally, high levels of plasticity are associated with low levels of stress resistance in individuals, and with low levels of genetic variation in populations (Hoffman and Parsons 1991). Plastic responses are not very common if the stress is persistent, as is the case with most contaminants, unless a single individual encounters micro spatial variation in soil concentrations. Therefore, the occurrence of anthropogenic contamination is more likely to select for resistant genetic variants (see the section "Genetic Basis of Tolerance").

Another phenomenon linked to stress is hormesis, defined as the stimulatory effect on organism performance occurring in response to low levels of exposure to agents that are harmful at high levels of exposure (Forbes 2000). This phenomenon has been demonstrated for several contaminants and several species, especially TNT and parrot feather (*Myriophyllum aquaticum*). However, the stimulation usually occurs on some aspects of performance at the individual level and does not affect the population. It has been suggested that organisms become adapted to many deleterious xenobiotic compounds because these chemicals are similar to natural molecules (Siciliano and Germida 1998), and that homeostatic mechanisms can respond by changing the rate of uptake, exclusion, transformation, segregation, and other metabolic process that have evolved. Therefore, hormesis could be the by-product of normal responses in counteracting the effects of inhibitors (Stebbing 1998). Effects on general fitness have not been described, and therefore hormesis may have no consequences on population dynamics and evolution (Forbes 2000).

BASIS OF PLANT TOLERANCE

Effects of Contaminants

Phytotoxicity of a contaminant depends on the uptake potential, biochemical reactivity, and exposure dose. A given exposure dose may correspond to different degrees of 'internal dose' in different species or individuals, according to rates of entry, distribution within the plants, environmental conditions, and many other factors (McLaughlin and Norby 1991). The 'effective internal dose' will be affected by metabolism and detoxification, which transform, conjugate, compartmentalize, or even excrete the contaminant. The rates of chemical transformation and the type of metabolites produced will depend on the plant species.

Metabolism of xenobiotic contaminants has two similar phases in animals and plants. Phase I is the functionalization of the molecules, and Phase II is the 'conjugation' with other molecules utilizing the functional groups created in Phase I (Heflich 1991, Sandermann 1994). The products of Phase I reactions are usually less toxic and more hydrophilic, with reactive sites such as hydroxyl or carboxyl groups (see Coleman *et al.* 1997 and references therein, Burken this book). Occasionally, the products are reactive species with the capability of interacting with macromolecules and therefore are more toxic than the parent compound. The products of Phase II are water-soluble conjugates with covalent links to endogenous molecules such as glucose, malonate, or glutathione, and the conjugation reactions are catalyzed by transferases (Coleman *et al.* 1997). In Phase III, the resulting water-soluble derivatives can be compartmentalized by export from the cytosol into the vacuole or the apoplast through membrane-localized transport proteins (Coleman *et al.* 1997). Phase III in animals involves excretion rather than storage.

Plants can mineralize only a few xenobiotic chemicals. One example is trichloroethylene (TCE), which can be degraded by poplar (*Populus* spp.) trees and aquatic plants to water, carbon dioxide, and chloride due to the presence of oxidative enzymes (Gordon *et al.* 1998, Newman *et al.* 1999, Nzengung and Jeffers 2001, Nzengung *et al.* this book, Shang *et al.* this book). In numerous cases, metabolism of organic xenobiotic molecules is similar to that in mammalian and microbial systems. Degradation of TNT has been described, leading to metabolites such as 2-amino-4,6-dinitrotoluene or 4-amino-2,6-dinitrotoluene (4-ADNT) (Medina and McCutcheon 1996, Coleman *et al.* 1997, Hughes *et al.* 1997, Meagher 2000, Burken this book, Jacobson *et al.* this book, McCutcheon *et al.* this book, Schwitzguebel and Vanek this book, Subramanian and Shanks this book, Wolfe and Hoehamer this book). Transformation of polychlorinated biphenyls by plant cell cultures has led to oxygenated and dechlorinated products (Butler *et al.* 1992, Lee and Fletcher 1992, Wilken *et al.* 1995, Bock and Harms 1996, Mackova *et al.* 1997, Macek *et al.* 2000, Harms *et al.* this book), but the enzymatic activities involved have not yet been fully characterized (Meagher 2000). In any case, the chlorine content affects the degradability of the congeners (Puri *et al.* 1997, Nzengung and Jeffers 2001, Harms *et al.* this book). Limited metabolism of fluoranthene to different forms of hydroxyfluoranthene has been observed in cell cultures of several plant species, leading to formation of conjugates with glucose and glucuronic acid (Kolb and Harms 2000). A different polyaromatic hydrocarbon, phenanthrene, can be rapidly metabolized to phenols by the roots of plants such as sunflower (*Helianthus annuus*) and pea (*Pisum sativum*) (Liste and Alexander 1999). Even the highly recalcitrant pentachlorophenol has been metabolized by wheat (*Triticum aestivum* L.) and soybean (*Glycine max* L.) cell cultures to glucosides and nonextractable residues (Langebartels and Harms 1984). In summary, plants can transform a wide range of xenobiotic compounds into less or nontoxic metabolites.

Growth and Development

Seed germination is a very sensitive phase of plant growth, and many classical toxicity tests are based upon seed germination or seedling emergence in the presence of contaminants (for an extensive review see Kapustka 1997 and references therein). After germination, roots are usually the tissues that are most sensitive to damage.

Root and shoot lengths are often used as measures of growth in plants. For example, exposure to TNT-contaminated soils affected root and shoot length, and dry weight in tall fescue (*Festuca arundinacea*) and smooth bromegrass (*Bromus inermis*). Tissue differentiation and respiration in roots are also affected by TNT concentrations (Krishnan *et al.* 2000). Polychlorinated biphenyls are other xenobiotic compounds that inhibit growth in higher plants (Puri *et al.* 1997). Other toxicity symptoms typically recorded are chlorosis, necrosis, and wilting of leaves, all of which are indicative of interference with photosynthesis and water transport (see the following section "Physiological Processes").

Physiological Processes

Herbicides are paradigmatic organic pollutants with several phytotoxic effects (see Table 6-1). The effects include inhibition of photosynthesis, respiration, lipid biosynthesis, and amino acid synthesis, plus cyanide accumulation from ethylene biosynthesis, and nodulation (Böger and Sandmann 1998).

Several mechanisms inhibit photosynthetic activity. The D1 protein of photosystem II is a target for the toxic action of several compounds (Schwenger-Erger and Barz 2000). Encoded by the chloroplast genome, the D1 protein binds directly to the photosystem II center (P-680) and to plastoquinones. Some herbicide molecules attack this system by binding to the D1 protein, replacing normal ligands and interrupting the electron transport (Salisbury and Ross 1992).

Protoporphyrinogen oxidase (Protox) catalyzes the common step in the chlorophyll and heme biosynthetic pathways in plastids and mitochondria.

TABLE 6-1 Physiological Effects of Selected Classes of Herbicides

Mode of action	HRAC group (WSSA group) [a]	Class: subclass [b]	Examples	Reference
Photosynthesis inhibitors				
Photosystem II D1 protein inhibition	C1 (5)	Triazine: chlorotriazine	Atrazine and simazine	Alfonso *et al.* (1996)
	C1 (5)	Triazinone	Metribuzin	Schwenger-Erger and Barz (2000)
	C2 (7)	Urea: phenylurea	Diuron and linuron	Mengistu *et al.* (2000)
Protoporphyrinogen IX oxidase inhibition	E (14)	Diphenyl ether: nitrophenyl ether	Acifluorfen and oxyfluorfen	Lermontova and Grimm (2000)
		Triazolone	Carfentrazone	Dayan *et al.* (1998)
		Unclassified [c]	Cinidon-ethyl	Grossmann and Schiffer (1999)
Proton transport inhibition	D (22)	Quaternary ammonium	Paraquat	Po and Ho (1997)
Pigment synthesis inhibition (bleaching)				
1. *p*-hydroxyphenyl-pyruvate dioxygenase	F2 (28)	Cyclopropylisoxazole	Isoxaflutole	Pallett *et al.* (1998)
2. Phytoene desaturase	F1 (12)	Pyridazinone	Norflurazon	La Rocca *et al.* (2000)
Respiration inhibitors				
Uncoupling or membrane disruption in mitochondria	M (24)	Dinitrophenol	Dinitro-*o*-cresol	Vicente *et al.* (1998)
Lipid biosynthesis inhibitors				
Acetyl-CoA carboxylase inhibition	A (1)	Cyclohexene oxime	Sethoxydim	Kuk *et al.* (1999)
		Phenoxy: aryloxyphenoxy-propionic	Fenoxaprop-P and diclofop	Incledon and Hall (1997)
				Böger *et al.* (2000)
Fatty acid elongation inhibition	K3 (15)	Amide: chloroacetanilide		Böger *et al.* (2000)
	K3 (15)	Triazole	Cafenstrole	Kern *et al.* (1997)
	N (8)	Thiocarbamates	Tri-allate	
	N (26)	Halogenated aliphatic: trichloroacetic acid	Dalapon	

Amino acid synthesis inhibitors				
Asparagine synthetase		Unclassified	Cinmethylin	Romagni *et al.* (2000)
Glutamine synthetase	H (10)	Organophosphorous	Glufosinate	Coruzzi and Last (2000)
Acetohydroxyacid synthase— branched chain amino acids	B (2)	Imidazolinone	Imazapyr and imazethapyr	Foes *et al.* (1999)
		Triazolopyrimidine	Flumetsulam	Wright et al. (1998)
		Urea: sulfonylurea	Chlorosulfuron and nicosulfuron	Foes et al. (1999)
		Unclassified	Pyrithiobac	Wright and Penner (1998)
5-Enolpyruvyl-3-phospho-shikimic acid synthase—aromatic amino acids	G (9)	Organophosphorous	Glyphosate	Saroha et al. (1998)
Ethylene biosynthesis stimulators			Dicamba	
Induction of 1-aminocyclopropane-1-carboxylic acid synthase	O (4)	Aromatic acid: benzoic acid	Quinclorac and quinmerac	Grossmann and Kwiatkowski (2000)
		Aromatic acid: quinolinecarboxylic acid		Grossmann and Kwiatkowski (2000)
		Phenoxy: phenoxyacetic	2,4-D	Wei et al. (2000)
Cell division inhibitors				
Microtubule polymerization	K1 (3)	Dinitroaniline	Oryzalin and trifluralin	Anthony and Hussey (1999)
Cellulose biosynthesis	L (21)	Amide	Flupoxam and isoxaben	Sabba and Vaughn (1999)
	L	Aromatic acid: quinolinecarboxylic acid	Quinclorac	Sabba and Vaughn (1999)
	L (20)	Nitrile	Dichlobenil	Sabba and Vaughn (1999)
Unknown		Halogenated aliphatic	Methyl bromide	

[a]HRAC (Herbicide Resistance Action Committee) group designation. In parentheses is the Weed Science Society of America (WSSA) group designation. See http://www.plantprotection.org/hrac/moa2001.htm

[b]Classification according to "Compendium of Pesticide Common Names," http://www.hclrss.demon.co.uk/class_herbicides.html

[c]HRAC Chemical Family is *N*-phenylphthalimides.

Inhibition of this enzyme by peroxidizing herbicides leads to accumulation of the substrate protoporphyrinogen IX, which is responsible for light-dependent damages such as lipid peroxidation and cellular leakage (Grossman and Schiffer 1999, Lermontova and Grimm 2000). Herbicides and other compounds that generate superoxide radicals can produce other oxidative stresses, and the reactive species can cause lipid peroxidation in thylakoid membranes, affecting light-induced proton transport (Po and Ho 1997).

Carotenogenesis, the synthesis of carotenoid pigments, can be blocked by bleaching herbicides through inhibition of 4-hydroxyphenylpyruvate dehydrogenase and phytoene desaturase, two enzymes involved in carotenoid synthesis (Pallett *et al.* 1998, La Rocca *et al.* 2000). The former catalyzes the biosynthesis of homogentisate, a precursor of plastoquinone, which is an essential cofactor for the latter. In both cases, the result is a decrease in carotenoids and photosynthesis.

Several xenobiotic compounds affect photosynthetic processes. Anthracene becomes especially toxic after photooxidation, which yields compounds such as anthraquinone and hydroxyanthraquinones (Mallakin *et al.* 2000). The primary site of action is electron transport in photosystem I followed by inhibition of photosystem II, and thereby inhibiting net photosynthesis. The anthracene phototransformation products also affect chlorophyll fluorescence and induce chlorophyll degradation with prolonged exposure (Huang *et al.* 1997). Xenobiotic compounds such as pentachlorophenol also affect photosynthesis, as evidenced by inhibition of oxygen evolution by isolated chloroplasts (Langebartels and Harms 1984).

Organic compounds such as 2,4-dinitrophenol and the herbicide dinitro-*o*-cresol also inhibit respiration (Vicente *et al.* 1998). These chemicals affect the mitochondrial membrane, making it more permeable to protons, thereby causing the collapse of the transmembrane electrical potential and the uncoupling of electron transport from phosphorylation (the transfer of phosphate to form organic phosphate molecules).

Another toxic effect is the inhibition of the activity of the plastid acetyl-CoA carboxylase. This, in turn, leads to inhibition of lipid (fatty acid) biosynthesis and to the formation of reactive oxygen species. The latter induces oxidative stress and subsequent ethylene production, causing irreversible damage to meristematic tissues of sensitive plants (Incledon and Hall 1997). Lipid biosynthesis can also be inhibited during fatty acid elongation, and particularly by the action on the condensing enzyme system elongase (Hickey and McNeilly 1975, Kern *et al.* 1997).

Amino acid synthesis is another critical step in plant metabolism. As shown in Table 6-1, herbicides can inhibit asparagine synthetase and glutamine synthetase (Romagni *et al.* 2000). These enzymes are very important in plant nitrogen metabolism. Acetohydroxyacid synthase, also called acetolactate synthase, catalyzes the first step in the biosynthesis of branched chain amino acids and is the target for several herbicides (Wright and Penner 1998, Foes *et al.* 1999). The shikimate pathway leads to the synthesis of chorismate, which is the precursor for the synthesis of aromatic compounds in plants, including essential amino acids and secondary metabolites. Chorismate is also the precursor of lignin biosynthesis. The herbicide glyphosate specifically inhibits the penultimate enzyme of the shikimate pathway, 5-enolpyruvyl-3-phosphoshikimic acid synthase, leading to death in susceptible plants (Saroha *et al.* 1998, Herrmann and Weaver 1999).

Auxinic herbicides (those that promote selected tissue growth) stimulate the activity of the 1-aminocyclopropane-1-carboxylic acid synthase, which in turn increases the synthesis of ethylene and the accumulation of cyanide as a co-product, resulting in the phytotoxic effects of chlorosis and necrosis in shoots and roots. Accumulation of abscisic acid can also be a consequence, leading to reduction of growth and ultimately senescence (Grossman and Kwiatkowski 2000).

One physiological process which seems to be very sensitive to xenobiotic compounds is the nodulation by Rhizobia in leguminous plants. Inhibiting concentrations are always lower for nodulation than for other manifestations of toxicity (Neumann and Werner 2000).

Cells and Organelles

Several aromatic hydrocarbons, including benzene, toluene, and acridine, can interact with membranes to disrupt function and increase permeability to ions. Damage to membranes seems to derive from direct interaction of hydrocarbons with membrane lipids, displacing the molecules in the membrane, and also from production of reactive oxygen species after photolysis of the xenobiotic compounds (Sikkema *et al.* 1994). For instance, creosote increases dose-dependent ion leakage in Eurasian milfoil (*Myriophyllum spicatum*) (McCann and Solomon 2000).

Alterations in chloroplasts often accompany toxic effects on photosynthetic processes (Table 6-1). These alterations include disappearance of starch grains, swelling and disarrangement of thylakoid grana (lamellar stacks of chlorophyll-containing material in chloroplasts), and lowering of chlorophyll and carotenoid content (La Rocca *et al.* 2000).

The organization of cell walls can be compromised by the inhibition of cellulose biosynthesis caused by different herbicide classes, including amides, quinolinecarboxylic acid, and nitriles (see Table 6-1). These classes of herbicides also inhibit cell division at the stage of cell plate formation. Cell walls then buildup higher proportions of extensins (proteins) or the polysaccharides pectin and callose (glucose polymers) (Sabba and Vaughn 1999). Cell division is also affected by herbicides that inhibit polymerization of tubulins to form cytoskeleton microtubules (Anthony and Hussey 1999).

Macromolecules

Many organic chemicals released into the environment are mutagens and carcinogens, including chlorinated aliphatic hydrocarbons, aromatic amines, and polycyclic aromatic hydrocarbons (Heflich 1991). Chemical carcinogens generally form covalent bonds with DNA (deoxyribonucleic acid), RNA (ribonucleic acid), and proteins (forming macromolecular adducts) due to the electrophilic properties of the chemical or of the metabolic derivatives (Casciano 1991). For example, peroxidases in plants can produce electrophilic intermediates from several xenobiotic compounds (Sandermann 1994) that covalently bind to DNA, RNA, and proteins. Planar molecules, such as acridines, are capable of interactions involving intercalation into double-stranded DNA without covalent bonding (Ripley 1991). Indirect effects of organic and inorganic pollutants upon macromolecules and cell structures are also exerted through the formation of reactive oxygen species, hydroxyl radicals, and singlet oxygen, starting from superoxide and hydrogen peroxide (Clemens *et al.* 1999).

Nucleophilic molecules directly attack macromolecules such as lignins (insoluble, highly cross-linked aromatic polymers that cement cellulose to form cell walls), pectins, hemicelluloses (insoluble polysaccharides found with cellulose and lignin in plant cell walls), and proteins (Sandermann 1994). Nonpolar metabolites of TNT in wheat (*Triticum aestivum* L. cv. Haven) and a species of bean given as *Phaseolus vulgaris* isolated by cell fractionation were associated with lignin, pectin, and hemicellulose, whereas polar metabolites were associated with the soluble fractions (Sens *et al.* 1999).

Genetic Basis of Tolerance

Anthropogenic changes in the environment lead to variable phenotypic responses between plant species and within species. One of the underlying causes of phenotypic variability in tolerance to xenobiotic chemical exposure results from genotypic variation occurring in cultivated and natural populations (Taylor *et al.* 1991). The ultimate cause of genetic variation is mutation, ranging from point mutations to chromosomal aberrations. A genetic basis for tolerance exists assumedly because of resistance arising from physiological mechanisms that confer to the individual the ability to develop and function normally in stressful conditions (Baker 1987).

Inheritance Mechanisms: Quantitative Versus a Single Gene

The first stage in determining the inherited quantity of plant tolerance to pollutants is to ascertain the relative contributions of genotype and environment to the observed variation. This can done in controlled experiments where environmental influence is usually kept at a minimum by performing the experiments over a limited range of conditions. From such controlled studies, several inherited quantitative traits have been observed to influence tolerance (Roose 1991). However, it can be hypothesized that specific enzyme variants can confer different characteristics to tolerant and sensitive species (see later section "Molecular Basis of Tolerance"). In this case, tolerance will depend on a single gene and will cosegregate with the gene variant in following generations. Alternatively, the genes responsible for tolerance could encode for modifier or regulatory proteins, affecting the activity and expression of several enzymes involved in xenobiotic metabolism. In both cases, Mendelian inheritance of resistance would be expected. As a result, both quantitative and single gene control over tolerance to pollutants is likely, especially from the several different genes contributing to the complex processes of metabolism, detoxification, and inactivation occurring within plants.

Resistance to toxins such as herbicides can involve a simple genetic basis, including structural modifications at the target site of the toxic action (see Table 6-1), increases in activity of detoxification pathways, or exclusion of the toxin from roots or leaves (Hoffman and Parsons 1991). A different mechanism of resistance can be the detoxification of a toxic product, such as the action of beta-cyanoalanine synthase on HCN, that confers to rice resistance to the herbicide quinclorac (Grossman and Kwiatkowski 2000). However, if the contaminant enters the cells and tissues and is translocated to various organs, several different mechanisms and pathways could contribute to the existence and evolution of resistance (see the section "Molecular Basis of Tolerance").

Population Genetics in Natural Conditions

Individual organisms react to anthropogenic stress at a point that varies in time according to the individual sensitivity, but population exposure varies spatially and dynamically. The concentrations of contaminants in soils may vary over short distances but remain relatively constant with time, whereas contaminants in water and air are distributed more uniformly in space but can vary rapidly in time. The resulting stresses will reduce critical plant functions in relation to varying severity and sensitivity, exerting selection on the existing genotypic variation within the population. According to the Darwinian view, less tolerant individuals will be gradually lost from the population or will bring a limited contribution to the gene pool of subsequent generations, and therefore the population structure will change over time (Bradshaw and McNeilly 1991). This phenomenon can be of limited spatial scale (Antonovics *et al.* 1971) or can extend over larger areas depending on the mode of pollination, seed dispersal, and other reproductive processes.

A paradigmatic example of genetic mechanisms involves resistance to herbicides in agricultural weeds. Herbicide resistance has become a problem in proportion to herbicide use (Bradshaw and McNeilly 1991). The basis of the different sensitivities to herbicides of species and biotypes has been traced to alterations in the target sites. For example, sequence analysis of genes isolated from six diuron tolerant biotypes of annual bluegrass (*Poa annua*) has revealed a substitution of amino acids from valine to isoleucine at position 219 in the D1 protein (Mengistu *et al.* 2000). In addition, dinitroaniline resistance in invasive barnyard grass (*Eleusine indica*) biotypes has been traced to a point mutation in the tubulins constituting the microtubules (Anthony and Hussey

1999). Finally, a summer cypress (*Kochia scoparia*) biotype, simultaneously resistant to triazine, imidazolinone, and sulfonylurea herbicides, was found to be the carrier of two mutations altering the target sites consisting of the D1 protein and acetohydroxyacid synthase (Foes *et al.* 1999).

Trade-Offs between Tolerance and Other Fitness-Related Traits

At the individual level, resistance to a particular pollutant could have different effects on resistance to other pollutants. Instances have been described in which the same mechanism confers resistance to several pollutants. For example, classes of herbicides with the same site of action could be detoxified by the same enzyme or countered by the same modification in the target site. Resistance to more than one heavy metal at a time can be conferred by heat-shock treatment (Neumann *et al.* 1994) or by phytochelatin induction (turning on plant synthesis of proteins that bind metals, Hoffman and Parsons 1991) even if usually plants show different degrees of tolerance towards different metals. Genes involved in avoidance mechanisms by excluding contaminants from the root will probably affect resistance to several chemical compounds at the same time (Hoffman and Parsons 1991).

However, deleterious effects have been associated with genes providing resistance to herbicides, decreasing fitness in the absence of stress (Hoffman and Parsons 1991). Whenever resistance is acquired through alterations in the target sites, it is likely that normal functioning of the protein is affected. For instance, Lettuce (*Lactuca sativa*) resistant to sulfonylurea herbicides carries a mutation in the gene coding for acetohydroxyacid synthase. The resistant allele (trait of a genotype) has detrimental effects on the enzyme function because activity in the resistant lines of plants is reduced as compared with sensitive lines, thereby altering the regulation of the branched chain amino acid biosynthesis (Eberlein *et al.* 1999). Similarly, some plants resistant to heavy metals showed reduced growth rates in optimal conditions on uncontaminated soils (Hoffman and Parsons 1991). In the case of resistance by exclusion, for instance, the mechanisms preventing uptake of toxic ions could also prevent uptake of necessary ions (Fitter and Hay 1987). These deleterious effects of 'tolerance genes' could depend on secondary effects of the same genes conferring tolerance (pleiotropy) or on other genes which are linked on the chromosomes to these genes (linkage disequilibrium). For example, metal resistance could be associated with increased metal requirement, leading to growth impairment on normal soils. However, constitutively tolerant species such as *Arabidopsis halleri* show that tolerance can also occur without disadvantages in uncontaminated conditions (Bert *et al.* 2000).

At the population level, changes in the genetic structure due to anthropogenic stress may coincide with physiological changes, which in turn can affect the species distribution and interaction with the environment. Therefore, pollution resistance can be associated with fitness costs (Taylor *et al.* 1991). A very basic trade-off effect of resistance to pollutants can be seen in the higher energy expenditures usually required to activate resistance mechanisms and to maintain normal function in stressful environments (Baker 1987, Hoffman and Parsons 1991). Moreover, strong selection for pollution resistance can eventually eliminate a large part of the original genetic variability, with consequences for plasticity and response capacity towards other stresses.

Genetic Variation for Tolerance in Plant Species

Species differ in capability to evolve resistance to anthropogenic stress. Not fully understood, these evolved tolerances are species- and stress-specific (Taylor *et al.* 1991). Evolution depends on the simultaneous occurrence of selection and heritable variation, and it is therefore possible that species failing to adapt to a particular anthropogenic stress were limited by the absence of the appropriate genetic variation (genostasis) (Baker 1987). From several reviews of plant resistance to metal toxicity, for example, only a limited number of species can be found on contaminated sites after years of selection (Baker *et al.* 1988, Bradshaw and McNeilly 1991). The knowledge concerning resistance to organic pollutants is much more limited, but examples are cited in other chapters of this book. Another example involves plant tolerance to TNT after a screening of different grass species on contaminated soil. The various grass species differed in the capability to grow in the presence of the contaminant and were differently susceptible to the presence of a *Pseudomonas* strain that metabolized TNT to amino metabolites. Some grass species showing tolerance to TNT were killed in the presence of *Pseudomonas*, whereas other species grew better when inoculated. In other studies, tolerance to TNT in terms of shoot growth was not increased by previous exposures. The genetic basis of this difference in tolerance has not yet been elucidated (Siciliano and Greer 2000).

Relying upon existing variation, selection for resistance can exhaust the genetic resources of a population. This explains why even resistant populations have limits on the level of stress that can be tolerated. The rapidity with which this capacity is saturated depends on the number of genes involved (Bradshaw and McNeilly 1991). Populations which have been under directional selection for stress responses reach a limit because low levels of genetic variation remain in the population, in contrast to unselected populations (Hoffman and Parsons 1991). An additional factor affecting selection for resistance is the frequency of resistant genotypes in the different generations. Due to genetic drift, the contribution to the following generation by resistant genotypes present at very low frequency in a small population could be very low or null, leading to the disappearance of the required genetic variation for the evolution of any further degree of tolerance (Bradshaw and McNeilly 1991).

Resistance can be controlled by few or by many genes (see the earlier section "Inheritance Mechanisms: Quantitative *versus* a Single Gene"). In the latter case, the maximum level of resistance is reached after several cycles of selection. However, as shown for a resistant *Agrostis* population on a copper refinery site, even 70 years of selection did not lead to an increase in the frequency of individuals resistant to very high metal concentrations, presumably because of other traits with a negative impact on fitness (Wu *et al.* 1975). Selection could therefore be visualized as occurring in three stages (Bradshaw and McNeilly 1991): (1) elimination of sensitive genotypes, (2) elimination of all but the most resistant genotypes, and (3) interbreeding of survivors to yield even more resistant genotypes and further selection. The first two stages may be limited by the extent of genetic variation, leading to extinction of the population; the third may be limited by reproductive success relative to other species. As outlined in the introduction ("Tolerance: Resistance, Sensitivity, and Other Responses to Toxicity"), there is also the possibility that 'apparently resistant' species are in fact successfully avoiding or escaping, and competition from these unaffected individuals can lead to additional selection burden on the species evolving real tolerance to the contaminant. For this reason, weed scientists have proposed leaving untreated strips of weeds within pollination range of fields to slow or prevent the evolution of resistance to herbicides (Herbicide Resistance Action Committee 1999, http://plantprotection.org/HRAC/howtomi1.html).

Molecular Basis of Tolerance

Resistance to a pollutant can involve mechanisms that are highly specific for that particular pollutant, such as induction of particular detoxifying enzyme, or mechanisms that confer resistance to several different compounds or stresses because of a common physiological or biochemical basis. For example, diversion of energy from normal metabolic processes to processes protecting cells or repairing damages could be considered as a general resistance mechanism (Hoffman and Parsons 1991). Energy conservation in dormant stages could in fact be considered as an extreme form of resistance.

Isolation and Characterization of Genes and Proteins Involved in Tolerance

When plant exposure to a contaminant occurs, a background set of enzymatic conditions exists. If very specific conditions exist, these enzymes could interact with the contaminant immediately. But this is uncommon. In general, the most important response relies upon the induction of new protein synthesis and modification of gene expression to resist toxicity.

Constitutive Expression

Several contaminants give rise to free radicals and hydrogen peroxide within plant cells. Scavenging systems to manage these free radicals are present in several subcellular compartments and involve enzymes such as superoxide dismutase, catalase, peroxidase, ascorbate peroxidase, monodehydroascorbate reductase, dehydroascorbate reductase, glutathione reductase, and glutathione peroxidase (Foyer *et al.* 1994). These enzymes are constitutively expressed, but pollutants can also induce expression. For example, peroxidase content was correlated to polychlorinated biphenyl disappearance in cell cultures, and polychlorinated biphenyls can affect peroxidase activity and isoenzyme (enzymes of similar function encoded by different genes) pattern (Macek *et al.* 2000).

The cytochrome P-450 enzymes are oxidative heme proteins that are bound to the cytoplasmic surface of the endoplasmic reticulum, and are involved in Phase I functionalization reactions (see section "Effects of contaminants"). These enzymes catalyze mono-oxygenation reactions using oxygen activated by electron transfer through cytochrome P-450 reductases (Werck-Reichhart *et al.* 2000). This xenobiotic inactivation (sometimes leading to mutagenic or carcinogenic metabolites) has been extensively studied in mammals (Smith *et al.* 1995). Recently, several genes encoding for members of this large family have been isolated and characterized in plants. More than 280 genes related to the cytochrome P-450 have been identified in *Arabidopsis thaliana*, and almost 500 genes have been identified in all plants investigated so far (Coleman *et al.* 1997, Werck-Reichhart *et al.* 2000). According to the green liver hypothesis (Sandermann 1994, Burken this book), different variants of these enzymes with different efficiencies in the metabolism of specific xenobiotic compounds can lead to different susceptibilities of individual plants in coping with chemical stress. For example, increases in metabolic efficiency have been correlated with resistance to herbicides in weed populations (Werck-Reichhart *et al.* 2000). Difficulties in isolating specific cytochrome P-450 isoforms (set of proteins with a similar function encoded by different genes) have delayed the demonstration that these enzymes are involved in the metabolism of xenobiotic compounds such as herbicides; apparently, herbicides and biological substrates can be metabolized by the same isoform or isoenzyme (Werck-Reichhart *et al.* 2000). Specific isoforms metabolizing phenylurea herbicides have been isolated from Jerusalem artichoke (*Helianthus tuberosus*) and soybean (*Glycine max*), and shown to confer tolerance when expressed in transgenic yeast (*Saccharomyces* spp.) and tobacco (*Nicotiana tabacum*) (Werck-Reichhart *et al.* 2000).

Glutathione *S*-transferases (GSTs) are responsible for metabolic degradation of xenobiotics by Phase II conjugation reactions (see section "Effects of Contaminants"). These enzymes are abundant cytoplasmic proteins, and the corresponding genes belong to a superfamily comprising 25 or more members (Edwards *et al.* 2000). Several electrophilic herbicides are detoxified by GST-catalyzed glutathione conjugation, and mutants overproducing this enzyme are resistant to herbicides because the detoxification is accelerated (Edwards *et al.* 2000 and references therein). Moreover, herbicide safeners (molecules that protect crops from herbicides) stimulate the induction of GSTs (Sandermann 1994, Coleman *et al.* 1997). Individual GST isoenzymes can selectively detoxify specific xenobiotic chemicals, with species-specific differences (Edwards *et al.* 2000). Expression of GST genes is regulated during the plant life cycle and also by exposure to pathogens, environmental stresses, and chemical treatments, although the role in metabolism of endogenous substances has not been elucidated. The GSTs can also function as a glutathione peroxidase and reduce organic hydroperoxides. Increased expression of the gene *AmGST2* confers multiple resistances to peroxidation by herbicides such as paraquat in the grass *Alopecurus myosuroides* (Cummins *et al.* 1999). Further, GSTs can function in isomerization reactions without the intervention of glutathione. In at least one example, isomerization is required for the activation of inactive herbicides (Edwards *et al.* 2000). Other Phase II enzymes also form xenobiotic conjugates, including *O*-glucosyltransferases, *O*-malonyltransferases, *N*-glucosyltransferases, and *N*-malonyltransferases. The respective malonyl and glucosyl conjugates are soluble and can be accumulated in vacuoles, avoiding toxicity (Sandermann 1994, Macek *et al.* 2000).

Transport proteins in the tonoplast (vacuole membrane) or in the plasmalemma (outer cell membrane) are essential during the Phase III compartmentalization of conjugated xenobiotics. The tonoplast contains ATP-dependent transporters for the compartmentation of glutathione conjugates in vacuoles. Expression of these proteins is also inducible by herbicide safeners, suggesting a role in detoxification (Coleman *et al.* 1997). Other tonoplast transporters are members of the ATP-binding cassette transporter superfamily, which has been described in all types of organisms (Coleman *et al.* 1997, Rea *et al.* 1998). This superfamily can transport a wide range of substances, both organic and inorganic. Similar proteins confer resistance to cytotoxic drugs in animal cells and to cadmium in yeast. The class named multi-drug resistance associated proteins has been characterized in plants. These proteins are energized by Mg-ATP directly and are involved in herbicide detoxification, alleviation of oxidative damage, and transport of heavy metal chelates (Rea *et al.* 1998). Carboxypeptidases and dipeptidases in the vacuole metabolize the glutathione in xenobiotic conjugates, producing cysteine conjugates, which are then transported back to the cytoplasm where the metabolites are exported to the apoplast for binding to lignin and cellulose (Coleman *et al.* 1997).

The NADPH:thioredoxin reductase is part of a redox system that plays roles in light-mediated enzyme regulation in photosynthesis. The isolated complex from *Arabidopsis* catalyzes the oxidation of NADPH [reduced form of nicotinamide–adenine dinucleotide (phosphate), the common agent for hydride reduction in cells] by reducing TNT and other nitroaromatic compounds and initiating redox cycling (Miskiniene *et al.* 1998). The complex therefore has nitroreductase activity. Redox active nitroaromatics, such as TNT and the herbicide 3,5-dinitro-*o*-cresol, can confer prooxidant properties to the antioxidant enzyme, explaining some of the toxicity of these compounds. The enzyme also can modify some compounds to less cytotoxic metabolites.

Induction by One or More Contaminants

Response to stresses, including chemical stresses, involves a shift in the pattern of transcription and translation, leading to changes in the profile of synthesized proteins. Proteins specifically expressed during a stress are commonly considered to be involved in tolerance to and survival of the stress, and therefore of adaptive value, even though in many cases this assumption rests on correlative evidence. Protein expression also appears to be an important process in plant tolerance to metals even if the process seems different from that of animals and the involvement of metallothioneins is not yet clear (Robinson *et al.* 1993).

Induction by other Environmental Factors

Cells and organisms respond to increases in temperature above optimal values by synthesizing a group of proteins collectively known as 'heat-shock proteins' (Buchanan *et al.* 2000). This response, involving evolutionarily conserved genes and proteins, seems to have a general protective function against the damages exerted by high temperatures on cellular structures and functions. Individual heat-shock proteins can also be induced by contaminants, including heavy metals, organic compounds, and oxidative agents (Hoffman and Parsons 1991). Recent evidence suggests that the several classes of heat-shock proteins can protect proteins from denaturation, can renature damaged proteins, and can shield sensitive structures from damage due to denatured proteins. With these general protective functions, heat-shock proteins can confer tolerance to all those stresses that can affect protein conformation either directly or indirectly. In cell cultures of wild tomato (*Lycopersicon peruvianum*) a short heat stress preceding the exposure to metal stress induced metal tolerance and prevented metal-induced damages to membranes observed in unstressed cells that were not prestressed by heat (Neumann *et al.* 1994).

Other Molecules Involved in Tolerance

Glutathione (gamma-glutamyl-cysteinyl-glycine) is an example of a homeostatic system which prevents oxidation of lipids, membranes, proteins, and macromolecules. Therefore, high endogenous or internal levels of glutathione can be beneficial as a first response towards some of the effects exerted by pollutants. Additional glutathione molecules can be produced following the induction of specific enzymatic activities (Edwards *et al.* 2000). Other scavenging molecules with antioxidant functions are ascorbate and carotenoids. Phytochelatins are important in tolerance of metals (Grill *et al.* 1985, Hoffman and Parsons 1991, Zenk 1996).

Genomics and Proteomics Approaches

Determining the physical location of genes involved in a particular process is a prerequisite for fully understanding the biochemistry of resistance and tolerance. Moreover, an extensive knowledge of resistance and tolerance requires a catalogue of all genes involved. New approaches are allowing exploration of the whole array of genes which change in expression level in specific conditions.

One approach starts from the isolation of gene sequences, which then have to be associated with specific biological functions by means of literature information, predictive algorithms, or experiments with transgenic plants. An alternative approach starts from proteins and the associated sequences, leading to isolation of the corresponding genes by designing oligonucleotides (chain formed from nucleic acids) to be used in cloning strategies. Still another approach involves isolating and characterizing proteins that show degradative activity towards various chemical substrates (Schnoor *et al.* 1995). The latter approach is similar to searches of tropical plants and insects to discover medicinal properties.

PHYTOTOXICITY OF CHLORINATED ALIPHATIC COMPOUNDS

Basic knowledge of the toxicity of chlorinated solvents to plants used in phytoremediation is vital for the successful application of this technology. Yet, most of the research to date has focused on the capability of plants to take up and metabolize chlorinated aliphatic compounds rather than the cause of toxicity. Nevertheless, a few studies have dealt with toxic effects of trichloroethylene (TCE) and other chlorinated aliphatic compounds (also see Langebartels and Harms 1986 and Harms *et al.* this book for some phytotoxicity information on pentachlorophenol, dichloroaniline, and chloroaniline).

Schubert *et al.* (1995) examined the germination of tobacco (*Nicotiana tabacum* L. var. xanthi nc) pollen exposed to vapors of six chlorinated solvents (1,2-dichloroethane, 1,1,1-trichloroethane, 1,1,2,2-tetrachloroethane, *trans*-dichloroethylene, trichloroethylene, and tetrachloroethylene) for 2 hours. The results indicated that the toxicity of a solvent increased with increasing number of chlorine atoms within a series of ethanes or ethenes, and that ethanes were more toxic than similarly chlorinated ethenes. Biernacki *et al.* (1995) examined the exposure of the aquatic macrophyte American wild celery (*Vallisneria americana*) to hydroponic solutions of 66, 132, or 396 milligrams of TCE per liter. Leaf area and leaf production decreased as TCE exposure concentration increased. Exposure of hybrid poplar (*Populus trichocarpa* × *Populus deltoides* H11-11) to 50 milligrams of TCE per liter for 8 months caused reduced growth, lower root mass, and fewer fine roots when compared to controls (Newman *et al.* 1997). Poplar (*Populus* spp.) cell cultures continued to grow when exposed to 260 milligrams of TCE per liter for 3 or 5 days. Ryu *et al.* (1996) investigated the toxicity of gaseous TCE to several plants [soybean (*Glycine max* L. Merr), tobacco (*Nicotiana tabacum*), tomato (*Lycopersicon esculentum*), pumpkin (*Cucurbita maxima*), sweet potato (*Dioscoria batata*), and alfalfa (*Medicago sativa*)]. Toxicity occurred at a vapor concentration of 0.2 micromoles per liter, which corresponded to 500 micromoles per liter/TCE (66 milligrams per liter) in the aqueous phase. Electrolyte leakage was evidence that the TCE impacted cell membranes.

Toxicity Investigation

A series of nine chlorinated aliphatic compounds (Table 6-2) were tested for phytotoxicity to hybrid poplar (*Populus deltoides* × *Populus nigra* 'DN34') (Dietz 2000, Dietz and Schnoor 2001). Pre-rooted 20-centimeter (8-inch) cuttings of hybrid poplar (*Populus deltoides* × *Populus nigra* 'DN34') were grown hydroponically with the lower root portion in sealed reactors to minimize volatilization (Thompson *et al.* 1998). Chemical solutions were replaced every 2 days to maintain a constant exposure concentration.

Phytotoxicity tests were conducted with triplicate reactors dosed for a period of 2 weeks. At the higher solvent concentrations (above the zero-growth levels provided in Table 6-2), wilting of shoots and damage to roots were observed. At concentrations between zero-growth and half zero-growth levels, fine root formation was arrested, similar to other studies (Newman *et al.* 1997). Reductions in total biomass and transpiration were monitored as indicators of acute toxicity, and both showed similar patterns. Highly chlorinated aliphatic compounds were more toxic to poplar (*Populus* spp.) cuttings than compounds with fewer chlorine atoms within the set of five ethenes or four ethanes tested (see Table 6-2). The ethenes were more toxic than the corresponding ethanes, contrasting with results by Schubert *et al.* (1995). Because the octanol–water partition coefficient (K_{ow}) is strongly correlated

TABLE 6-2 Toxic Effects on Hybrid Poplar (*Populus deltoides* × *Populus nigra* DN34) From Chlorinated Aliphatic Compounds (Dietz and Schnoor 2001)

Chemical	log K_{ow}[a]	Zero-growth concentration log K_{ow} (milligrams per liter)	50 percent transpiration concentration log K_{ow} (milligrams per liter)
Tetrachloroethylene	3.4	45 ± 3	38 ± 6
Trichloroethylene	2.42	118 ± 12	131 ± 22
trans-Dichloroethylene	2.06	465 ± 50	349 ± 74
cis-Dichloroethylene	1.86	582 ± 57	494 ± 83
1,1-Dichloroethylene	2.13	543 ± 54	281 ± 56
1,1,2,2-Tetrachloroethane	2.39	151 ± 17	151 ± 34
1,1,2-Trichloroethane	2.07	307 ± 20	253 ± 36
1,1,1-Trichloroethane	2.49	267 ± 29	160 ± 33
1,1-Dichloroethane	1.79	1059 ± 109	802 ± 165

[a]*Source*: Howard, P.H., ed. (1990) *Handbook of Environmental Fate and Exposure Data for Organic Chemicals*. Lewis Publishers, Chelsea, Michigan, U.S. Note that log K_{ow} is the logarithm of the octanol–water partioning coefficient.

with the number of chlorine atoms, solvents with a high log K_{ow} are more toxic than those with lower log K_{ow} (Table 6-2).

Basis of Tolerance to Chlorinated Aliphatic Compounds

Several possible mechanisms cause the acute toxicity of chlorinated aliphatic compounds to hybrid poplar (*Populus deltoides* × *Populus nigra* 'DN34'). Because root damage was one of the major adverse effects of these solvents, the mechanism may simply be physical damage to root membranes due to solvent molecule destabilization of the lipid bilayer, leading to reduced ability to transpire water. This type of damage has been reported for phytotoxic oils (van Overbeek and Blondeau 1953).

Metabolites (trichloroethanol, trichloroacetic acid, and dichloroacetic acid) form in poplar tissues after exposure to TCE. Cytochrome P-450 may be involved in this oxidative transformation. Trichloroacetic acid has been used as a commercial herbicide so, one hypothesis is that these metabolites are toxic. But if trichloroacetic acid and dichloroacetic acid are toxic, the point of action may be outside the leaves. Nzengung and Jeffers (2001) found accumulations of up to 2900 milligrams of trichloroacetic acid per kilogram of wet weight, and 1700 milligrams of dichloroacetic acid per kilogram of weight in the leaves of live willow (*Salix nigra*). Nevertheless, laboratory studies with transgenic tobacco (*Nicotiana tabacum*) plants containing human cytochrome P-450 2E1 demonstrated increased uptake of TCE with no buildup of these metabolites (Doty *et al.* 2000). A different study showed no visible toxic effects in pine (*Pinus sylvestris* L.) seedlings exposed to trichloroacetic acid *via* root uptake or mist treatments, although levels of peroxidase and glutathione *S*-transferase were higher in the exposed seedlings (Schroder *et al.* 1997). Peroxidase and glutathione *S*-transferase may also play a role in detoxification of chlorinated aliphatic compounds and may increase plant tolerance. While enzyme activity was not monitored, a study with American wild celery (*Vallisneria americana*) found that plants which survived TCE treatment seemed to be naturally resistant and were just as likely to flower as untreated plants (Biernacki *et al.* 1995). This suggests that a constitutive enzyme system may be present in certain plants, thus increasing tolerance. Further research is needed to determine which enzyme systems are most important for plant tolerance to chlorinated aliphatic compounds. Such knowledge could be used to genetically engineer species capable of growth at sites with high concentrations of these solvents.

In addition to the general mechanism of chlorinated aliphatic phytotoxicity, the trends in relative toxicity are not yet understood. Studies with hybrid poplar (*Populus deltoides* × *Populus nigra* 'DN34') and tobacco (*Nicotiana tabacum*) pollen both show that toxicity increases with increasing chlorination within a set of ethenes or ethanes, perhaps because the more highly chlorinated aliphatic compounds require more enzymatic steps for metabolism (Nzengung and Jeffers 2001). In hybrid poplar (*Populus deltoides* × *Populus nigra* 'DN34') cuttings, chlorinated ethenes were more toxic than the corresponding ethanes. This could be due to epoxide intermediates that form during metabolism of chlorinated ethenes. Chlorinated ethane metabolism would more likely form a carbanion intermediate, which may be more easily metabolized by plants. However, studies with tobacco (*Nicotiana tabacum*) pollen germination showed the opposite trend; chlorinated ethanes were more toxic than the corresponding ethene. Therefore, chlorinated aliphatic compounds may affect germination differently than whole plant growth, or the two species may have different enzyme capabilities.

TOXICOLOGY AND PHYTOTOXICITY OF EXPLOSIVES

Explosives Contamination

About 2000 U.S. Department of Defense facilities are contaminated with explosives in both soil and groundwater (McCutcheon *et al.* 1995, Moore 1995, GIT 1995). The most well-known and widespread contaminant and therefore the most widely studied is TNT. However, many sites are also contaminated with RDX (hexahydro-1,3,5-trinitro-1,3,5-triazine), HMX (octahydro-1,3,5.7-tetranitro-1,3,5,7-tetrazocine), dinitrotoluene (which consists of two isomers, 2,4-dinitrotoluene and 2,6-dinitrotoluene), and trinitrobenzene. Any country historically involved in the production of explosives is likely to have significant contamination. Contamination also has been reported in Canada, Germany, the Czech Republic, Poland, Albania, China, Australia, and Mexico. Therefore, the problem is truly international in scope.

Toxicology of Explosives

Most toxicological studies of explosives have involved TNT. The toxic effects of TNT have been studied involving microorganisms (Klausmeier *et al.* 1973, Won *et al.* 1976, Simini *et al.* 1995, Bethe-Conti *et al.* 1998), algae (Smock *et al.* 1976, Won *et al.* 1976, Hitchcock *et al.* 2003), tide pool copepods (*Tigriopus californicus*), oyster larvae (*Crassotrea gigas*; Won *et al.* 1976), a macroinvertebrate (*Daphnia magna*; Liu *et al.* 1976, Bailey *et al.* 1984), insects (Griest *et al.* 1995), fathead minnows (*Pimephales promelas*; Liu *et al.* 1976, Smock *et al.* 1976, 1983), earthworms (*Lumbricus terrestris*; Simini *et al.* 1995), plants (Andrews 1996, Peterson *et al.* 1996, Thompson *et al.* 1998), mammalian cell cultures (Bethe-Conti *et al.* 1998), and rats (Kaplan and Kaplan 1982, Gordon and Hartely 1992). In each case, toxic or reproductive effects were demonstrated. Shown to be mutagenic and a U.S. Environmental Protection Agency Group C suspected carcinogen (Gordon and Hartley 1992), TNT causes liver damage and aplastic anemia in humans (Kaplan and Kaplan 1982). As a result, the lifetime health advisory for TNT is 2 micrograms per liter. Studies involving minnows indicate that nonpolar nitroaromatics (TNT and dinitrotoluene) have higher toxicity than the associated polar transformation products (amino-derivatives) (Liu *et al.* 1983). One product of TNT breakdown, 4-hydroxlamino-2,6-dinitrotoluene, inhibits lignin peroxidases from white rot fungi (*Phanerochaete chrysosporium*) and thereby stops subsequent transformation of amino-derivatives (Bumphus and Tatarko 1994, Michels and Gottschalk 1995). In the case of nitro-polyaromatic hydrocarbons such as 1-nitropyrene, reductive reactions increase mutagenicity (Rafii *et al.* 1994).

Toxicity studies for RDX and HMX are few in number, but both are listed as U.S. Environmental Protection Agency Class C carcinogens (possible or suspected carcinogens) (Hawari 2000). Although not mutagenic, RDX causes nausea, vomiting, unconsciousness, and seizures in humans, and is a tumorigen and poison for rats (Kaplan *et al.* 1965, Levine *et al.* 1981, Etnier 1989, Levine *et al.* 1990, Lewis 1992). Concentrations of RDX in excess of 105 micrograms per liter are unacceptable (Etnier 1989). The toxicity of HMX is not well defined (Sealock 2002).

Aqueous-Phase Phytotoxicity of Explosives

The degree of phytotoxicity depends on plant species, stage of growth, amount of biomass or plant density (Andrews 1996), contaminant bioavailability, and other environmental factors (Burken *et al.* 2000). Some of the causes of phytotoxicity have been explored in various investigations of phytoremediation for TNT and other explosives.

Lauritzen (1998) investigated the relationship between growth stage and TNT phytotoxicity using axenic hairy root cultures (*Catharanthus roseus* clone LBE 6-1). In an active growth phase, growth rates decreased almost immediately after exposure to 20, 30, and 40 milligrams of TNT per liter and approximately to the same extent in each case. The TNT transformation rates corrected for the amount of biomass were about the same during active growth as during a stationary phase (culture was metabolically active but not growing, Burken *et al.* 2000).

Andrews (1996) investigated different densities of parrot feather (*Myriophyllum aquaticum*) in batch and continuous flow systems. The contaminated groundwater contained approximately 1 to 4 milligrams of TNT per liter, undetectable to 22 milligrams of dinitrotoluene per liter, 0.2 to 0.5 milligrams of ADNT per liter (did not distinguish between 2- and 4-ADNT), 0.1 to 0.4 milligrams of trinitrobenzene per liter, and undetectable to 46 milligrams of nitrotoluene per liter. The U.S. Volunteer Army Ammunition Plant in Chattanooga, Tennessee, was the source of contaminated groundwater for these treatability investigations.

Parrot feather (*Myriophyllum aquaticum*) density was determined to be an important factor in tolerance up to approximately 50 grams of wet parrot feather (*Myriophyllum aquaticum*) per liter. Lower plant densities of 20 grams of wet parrot feather (*Myriophyllum aquaticum*) per liter were found by Andrews (1996) to be less healthy than those at 100 grams per liter. At 20 grams per liter, parrot feather (*Myriophyllum aquaticum*) initially wilted and died in 3 days exposed to 45 milligrams of nitrotoluene per liter, 20 milligrams of dinitrotoluene per liter, 2.7 milligrams of TNT per liter, 0.4 milligrams of ADNT per liter, and 0.4 milligrams of trinitrobenzene per liter. Parrot feather (*Myriophyllum aquaticum*) at 100 grams per liter survived these concentrations with just browning and yellowing of leaves. In Volunteer groundwater free of detectable nitrotoluene and having less than 0.9 milligrams of dinitrotoluene per liter, 2 milligrams of TNT per liter, 0.5 milligrams of ADNT per liter, and 0.2 milligrams of trinitrobenzene per liter, there was no visible stress except that the parrot feather (*Myriophyllum aquaticum*) did not grow any new roots or leaves as compared to the controls.

Medina and Tano (unpublished) put these *ad hoc* observations for parrot feather (*Myriophyllum aquaticum*) into context in Figure 6-1 using 7-day exposures at different plant densities. Chlorophyll analysis was used to monitor plant health. Plant stress and local tissue necrosis were observed to occur from the tip of the emergent portion of the plant, downwards in both studies. The tips dried out and became brittle, resembling a plant that had been killed by water stress. The brittleness progressively continued down the stem until the plant was virtually dried out.

Batch studies (Medina *et al.* 2000) investigating removal of 1, 15, 30, 64, and 128 millligrams of TNT per liter found that parrot feather (*Myriophyllum aquaticum*) exposed to concentrations of 128 milligrams of TNT per liter died during the 2-day experiment. The plants exposed to 64 milligrams of TNT per liter were stressed, showing such symptoms as wilting. However, TNT was removed from solution even by the dead and dying plants exposed to high concentrations (also observed by Young 1995, Wolfe and McCutcheon unpublished, Jacobson *et al.* this book), whereas, plants killed by exposure to high temperature did not remove TNT from solution (Medina *et al.* 2000). Vanderford (1996) and Vanderford *et al.* (1997) found that parrot feather (*Myriophyllum aquaticum*) could tolerate initial concentrations less than or equal to 100 milligrams of TNT per liter. Over an exposure period of weeks,

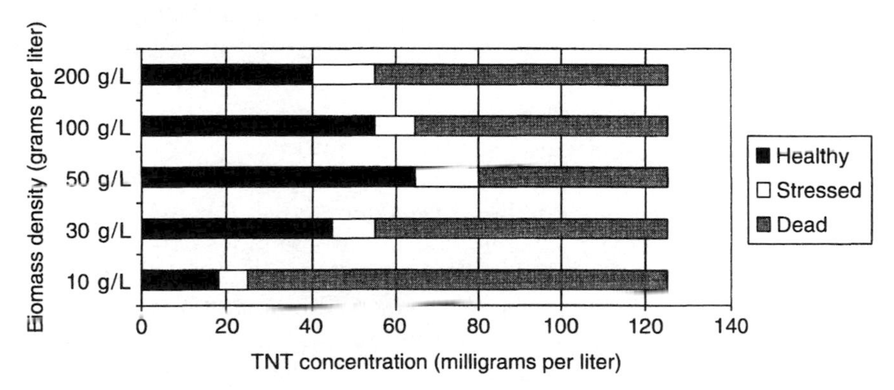

Figure 6-1 Effects of aquatic plant density on parrot feather (*Myriophyllum aquaticum*) toxicity in trinitrotoluene (TNT) contaminated water.

parrot feather (*Myriophyllum aquaticum*) exposed to TNT in other reactors actually seemed healthier than in the control reactors (Andrews 1996). Whether the transformation of TNT might have provided some nitrogen or hormesis of parrot feather (*Myriophyllum aquaticum*) to TNT occurred, similar observations have also been noted (Wolfe and McCutcheon, unpublished observations). In addition, parrot feather (*Myriophyllum aquaticum*) exposed continuously to TNT for periods of several weeks continued to grow after removal to tanks receiving no TNT. Parrot feather (*Myriophyllum aquaticum*) also continued to reproduce vegetatively (regrowth from broken segments of the plant) after one-time exposures of up to 32 milligrams of TNT per liter. However, exact comparisons to unexposed plants were not made (Medina and Tano, unpublished). In addition, parrot feather (*Myriophyllum aquaticum*) exposed to residual concentrations for up to a year have continued to produce new roots and leaves (Anderson 1996).

Medina *et al.* (2002) determined that parrot feather (*Myriophyllum aquaticum*) in continuous flow-through reactors receiving a loading of 2.5 grams per cubic meter per day) survived and grew throughout a 70-day test. At a loading of 0.97 grams of TNT per cubic meter per day) and at day 36, concentrations ranged from 3.5 milligrams of TNT per liter to undetectable levels from the influent to effluent, and ADNT from undetectable in the influent to approximately 1.2 milligrams per liter through most of the reactor.

Schott and Worthley (1974) found reduced growth of duckweed (*Lemna perpusilla*) at relatively low concentrations (about 1 milligram of TNT per liter) over an exposure of 30 days. Pavlostathis *et al.* (1998) studied the effect of aqueous phase TNT exposure on Eurasian milfoil (*Myriophyllum spicatum*), a close relative of parrot feather (*Myriophyllum aquaticum*). At concentrations of greater than 60 milligrams of TNT per liter, the milfoil (*Myriophyllum spicatum*), showed toxic effects such as chlorosis and leaf loss within 7 to 27 hours of incubation. At lower concentrations (from 11 to 29 milligrams of TNT per lilter), the milfoil (*Myriophyllum spicatum*) was healthier for a longer period of *time*, but showed chlorosis and leaf loss within 123 hours of exposure. At still lower concentrations (up to 4.5 milligrams of TNT per liter), the plants were healthier still. Some replicates eventually suffered toxic effects, while others continued to grow. None of the controls (which had no TNT exposure) had any sign of ill health. However, even as the plants exhibited signs of ill health, transformation of TNT continued.

Best *et al.* (1998) measured growth rates of the aquatic plants coontail (*Ceratophyllum demersum*), pondweed (*Potamogeton pectinatus* L.), American pondweed (*Potamogeton pectinatus* Poir.), and common arrowhead (*Sagittaria latifolia* Willd.), exposed to TNT and RDX concentrations ranging from 0 to 2 milligrams per liter. Consistent with Andrews (1996), but not Schott and Worthley (1974), concentrations of up to 2 milligrams of TNT per liter actually appeared to enhance the growth rate for each plant, with the greatest stimulation occurring at a concentration of 0.5 milligram of TNT per liter.

Palazzo and Leggett (1986) found deleterious effects of TNT on yellow nutsedge (*Cyperus esculentus* L.) at concentrations as low as 5 milligrams of TNT per liter over a 42-day exposure. Root growth was the parameter most affected, followed by leaf growth and rhizome weight. Increased effects were found with increasing concentrations up to the maximum concentration of 20 milligrams of TNT per liter. Görge *et al.* (1994) also found that concentrations of 5 milligrams of TNT per liter reduced root growth for chive (*Allium schoenoprasum*). Cell suspensions of Jimsonweed (*Datura innoxia*) exposed to TNT had results that were more variable (Lucero *et al.* 1999). Cells exposed to concentrations less than or equal to 20 milligrams of TNT per liter had no toxic effect or notable effect on viability. Concentrations of 30 and 50 milligrams of TNT per liter had lethal effects. At 50 milligrams of TNT per liter, up to 60 percent of the cells died. However, these lethal effects were completely different if the cell suspensions were more than 18 months old.

Thompson *et al.* (1998) found reduced transpiration accompanied by leaf chlorosis and abscission in hydroponically grown poplar (*Populus* spp.) trees at levels as low as 5 milligrams of TNT per liter. The transpiration rates steadily decreased throughout the range tested (0 to 15 milligrams of TNT per liter). The results at both bench and pilot scale agreed well.

Peterson *et al.* (1996) found no effect on germination and early seedling development of tall fescue (*Festuca arundinacea*) hydroponically exposed to concentrations up to 30 milligrams of TNT per liter. Higher concentrations up to 60 milligrams of TNT per liter, the limit of the study, delayed seedling development, reduced secondary root and shoot growth, and caused abnormal radicle (lower portion of the axis of a plant embryo or seedling) tissue development. No effect was found at breakdown product 4-ADNT concentrations less than or equal to 7.5 milligrams of ADNT per liter; however, higher 4-ADNT concentrations resulted in measurable toxic effects. The effect threshold of 4-ADNT being lower than that of TNT suggests that breakdown products could play an important role in TNT phytotoxicity to fescue (*Festuca arundinacea*), depending on transformation rates.

Only Tadros *et al.* (2000) seem to have investigated the toxic effects of hydroxylamino- and amino-transformation products in studying the effects on the alga *Selenastrum carpricornutum*. The hydroxylamine by-products used in their study came from the bacterial transformation of TNT, but similar transformational products have been found in plant systems as well (Pavlostathis *et al.* 1998).

Sun *et al.* (2000) found that concentrations of 36 milligrams of TNT per liter inhibited shoot length growth of smooth bromegrass (*Bromus inermis*) by 40 percent as compared to unexposed controls. However, no significant difference was found in comparison of photosynthetic rates or respiration rates. No positive or negative effect on growth or TNT degradation was found by pre-exposure of bromegrass (*Bromus inermis*) to TNT.

Peterson *et al.* (1998) investigated the effect of aqueous phase concentrations of up to 60 milligrams of TNT per liter on germination and seeding development of two closely related grasses: switch grass (*Panicum virgatum*) and smooth bromegrass (*Bromus inermis*). Switch grass (*Panicum virgatum*) germination was unaffected by TNT concentrations tested, but bromegrass (*Bromus inermis*) germination rates steadily decreased with increasing TNT concentrations. Furthermore, bromegrass (*Bromus inermis*) was more sensitive than switch grass (*Panicum virgatum*) to TNT concentrations in terms of root growth and shoot growth.

The effects of TNT on ecological communities have not been fully investigated. However, Young (1995) and Rivera *et al.* (1998) report that snails, insects, and algae (species unknown) grow in reactors with active parrot feather (*Myriophyllum aquaticum*) exposed to TNT. Jacobson *et al.* (this book) tested TNT exposure to wetland plant communities and did not note any obvious effects on the plants involved. In the only known study of TNT effects on communities, Hitchcock *et al.* (2003) investigated simple parrot feather (*Myriophyllum aquaticum*), rotifer (*Brachionus calyciflorus*), and algae (*Nannochloropsis* spp.) systems. Algal growth was unaffected over 96 hours with initial concentrations of up to 21.2 milligrams of TNT per liter. Growth rates declined during exposure to 31.8 milligrams of TNT per liter and over the range of 42.4 to 52.9 milligrams of TNT per liter. Growth rates did not significantly change with exposures up to 17 milligrams of ADNT per liter. Similarly, Smock *et al.* (1976) found that the growth of the green algae *Microstis aeruginosa* and *Selenastrum capricornutum* suffered a 50 percent inhibition at exposures of 15 and 5 milligrams of TNT per liter, respectively. Rotifer (*Brachionus calyciflorus*) survivorship, reproduction, growth rates, and life expectancy were consistently better in reactors with parrot feather (*Myriophyllum aquaticum*) than without (96 hours exposure). The no observable adverse effect level on rotifers (*Brachionus calyciflorus*) alone was 2.3 milligrams of TNT per liter but parrot feather (*Myriophyllum aquaticum*) consistently prevented toxicity at higher initial exposures. Only at initial concentrations of 25 milligrams of TNT per liter did rotifers (*Brachionus calyciflorus*) show increased generation time and decreased life expectancy in the presence of parrot feather (*Myriophyllum aquaticum*). Parrot feather (*Myriophyllum aquaticum*) initially prevented measurable effects at 8 milligrams of TNT per liter. Compared to the controls, rotifers (*Brachionus calyciflorus*) actually increased in numbers more when parrot feather (*Myriophyllum aquaticum*) was present, indicating that the plant actually spurs growth of rotifers even in the presence of TNT.

The phytotoxicity of explosives other than TNT is even less understood, but RDX seems less toxic than TNT (Burken *et al.* 2000). Best *et al.* (1998) measured growth rates of the aquatic plants coontail (*Ceratophyllum demersum* L.), American pondweed (*Potamogeton pectinatus* Poir.), and common arrowhead (*Sagittaria latifolia* Willd.) exposed to concentrations ranging from undetected to 2 milligrams of RDX per liter. Similar to the effects of TNT, RDX also enhanced growth. The only inhibition of growth was found for arrowhead at a concentration of 2 milligrams of RDX per liter. Larson *et al.* (1999) found transformation of RDX to hydroxyl amino derivatives in various aquatic species. However, Thompson *et al.* (1999) reported that RDX did not transform in poplar (*Populus deltoides* × *Populus nigra* DN34) trees, hydroponically exposed to the compound but rather accumulated. No toxic effects occurred for the 7-day exposure time. Chen (1993) as described in Burken *et al.* (2000) did measure the phytotoxicity for RDX in hydroponic systems. Estimated toxicity occurred at 21 milligrams of RDX per liter (the highest investigated) for maize (*Zea mays*) and wheat (*Triticum aestivum*) grown hydroponically. The no observable effect level was 13 milligrams of RDX per liter. Broom corn (*Sorghum vulgare*) and soybean (*Glycine max*) showed no toxic effects at 21 milligrams of RDX per liter after 30 days exposure. Growth and transpiration rates of hybrid poplar (*Populus deltoides* × *Populus nigra* 'DN34') cuttings were also not affected at 21 milligrams of RDX per liter after 14 days.

Cell suspensions of Jimsonweed (*Datura innoxia*) rapidly degraded TNT, RDX, and HMX (Lucero *et al.* 1999). During these studies, the cell suspensions showed no effect on viability from HMX and RDX exposures in excess of the reported solubility limits (5 milligrams of HMX per liter and up to 60 milligrams of RDX per liter). These results imply that phytotoxicity to RDX may result primarily from the transformation products.

Fletcher *et al.* (1990) studied the effect of nitrobenzene exposure on soybean (*Glycine max*) plants at concentrations ranging from 0.02 to 2 milligrams of nitrobenzene per liter and one concentration at 100 milligrams of nitrobenzene per liter. No measurable effects were found at concentrations of 2 milligrams of nitrobenzene per liter or less. Reduced root growth was found for the plants exposed to 100 milligrams of nitrobenzene per liter; however, transpiration and photosynthetic rates were not measurably affected. McFarlane *et al.* (1990) studied the effects of 72 hours of 8 milligrams of nitrobenzene per liter exposure on eight species of terrestrial plants: (1) green ash (*Fraxinus pennsylvanica* Marsh.), (2) hybrid poplar (*Populus* × *robusta* C. K. Schneid.), (3) autumn olive (*Elaeagnus umbellata* Thund.), (4) Russian olive (*Elaeagnus augustifolia* L.), (5) honeysuckle (*Lonicera tatarica* L.), (6) barley (*Hordeum vulgare* cv. Klages), (7) soybean (*Glycine max* [L.] Merr. cv. Fiskby v), and (8) lettuce (*Lactuca sativa* L. cv. Grand Rapids). Green ash (*Fraxinus pennsylvanica* Marsh.) and lettuce (*Lactuca sativa* L. cv. Grand Rapids) showed no visible sign of toxicity, but transpiration and photosynthetic rates were inhibited. Green ash (*Fraxinus pennsylvanica* Marsh.) recovered within 10 hours, while lettuce (*Lactuca sativa* L. cv. Grand Rapids) needed 60 hours to recover after termination of exposure. Autumn olive (*Elaeagnus umbellata* Thund.) and Russian olive (*Elaeagnus augustifolia* L.) had more severe effects, including visible impacts on health. The autumn olive (*Elaeagnus umbellata* Thund.) was killed by the exposure but Russian olive (*Elaeagnus augustifolia* L.) was able to recover within about 10 hours after the exposure ended. Soybeans (*Glycine max* [L.] Merr. cv. Fiskby v), barley (*Hordeum vulgare* cv. Klages), honeysuckle (*Lonicera tatarica* L.), and hybrid poplar (*Populus* ×*robusta* C. K. Schneid.) showed virtually no measurable toxic effect. All of the plants maintained the capability to transform as well as volatilize the nitrobenzene.

French *et al.* (1999) compared the toxic effects of munitions on regular tobacco (*Nicotiana tabacum*) and a transgenic tobacco capable of expressing of the pentaerythritol tetranitrate reductase, an enzyme capable of reducing nitro groups commonly found in munitions. Concentrations of 0.05 millimoles per liter (11 milligrams of TNT per liter) and 1 millimole per liter glycerol trinitrate inhibited the germination and growth of the normal tobacco (*Nicotiana tabacum*) plants, but did not affect the transgenic plants, presumably because the enzymes could transform the compound to less toxic forms.

For different aquatic species, the no observable adverse effect levels range from 1 to 65 milligrams of TNT per liter. For different terrestrial species, no-effect levels were observed to range from 5 to 36 milligrams of TNT per liter. However, these no-effect levels also vary with length of exposure. Nitrobenzene seems to have similar no-effect levels as TNT but fewer observations are available, while RDX seems less phytotoxic than TNT.

Phytotoxicity and Bioavailability of Explosives in Soils

A few *ad hoc* field observations and some laboratory investigations provide an incomplete picture of phytotoxicity of explosives in soil. Young (1995) found essentially sterile conditions on TNT-contaminated bald spots at the Alabama Army Ammunition Plant near Talladega, but vegetation was noted to have encroached onto the outer edges of these spots over the previous three decades. In the bald areas where no plants or microorganisms grew, the concentrations were 4000 milligrams of TNT per kilogram of dry soil or greater. Near the margins where pioneering weeds and shrubs (species unknown) were growing on the former bald areas, concentrations were on the order of hundreds of milligrams of TNT per kilogram of dry soil. Beyond the transition area on the former bald areas, pine (*Pinus* spp.) forests were growing in concentrations that varied from tens of milligrams of TNT per kilogram of dry soil to detection limits. Similar conditions were observed at the Crane Naval Weapons Station in Indiana. Only yucca (*Yucca aloifolia*) plants were growing in bare areas with concentrations of total explosives (TNT, RDX, and HMX) of approximately 520 to 760 milligrams of TNT per kilogram of dry soil. Despite these pioneering plants accumulating the highest levels of explosives known (292 milligrams of TNT per kilogram of dry weight, 34 milligrams of HMX per kilogram, and 300 milligrams of RDX per kilogram of dry weight), seeds from the yuccas sprouted to form viable offspring.

In the laboratory, Mueller *et al.* (1995) found that Jimsonweed (*Datura innoxia*) and wild tomato (*Lycopersicon peruvianum*) tolerated soil concentrations up to 750 milligrams of TNT per kilogram and reduced these TNT concentrations by 90 percent in 2 weeks. Shojaee (1994) mixed cell suspensions of Jimsonweed (*Datura innoxia*) with TNT contaminated soils and found that the *Datura innoxia* cells can survive soil concentrations up to 900 milligrams of

TNT per kilogram of dry soil. Peterson *et al.* (1998) found that switch grass (*Panicum virgatum*) and smooth bromegrass (*Bromus inermis*) appeared to survive only soil concentrations of less than 50 milligrams of extractable TNT per kilogram of dry soil. Gong *et al.* (1999) studied the toxicity of TNT-contaminated soils on four other common plants. Two of the plants were dicotyledons [cress (*Lepidium sativum*) and turnip (*Brassica rapa*)], and two were monocotyledons [oat (*Acena sativa*) and wheat (*Triticum aestivum*)].

Wilson (1995) investigated treatment of TNT-contaminated soil from the Alabama Army Ammunition Plant in Childersburg (725 milligrams of TNT per kilogram of dry soil) using a batch wetland system with parrot feather (*Myriophyllum aquaticum*) as the plant species. The initial planting of parrot feather (*Myriophyllum aquaticum*) immediately died, presumably from high concentrations of TNT and transformation products in the aqueous phase. However, the dead plants reduced the TNT concentration in solution, and a subsequent replanting of the reactor resulted in survival and growth of the parrot feather (*Myriophyllum aquaticum*) as well as about a 90 percent reduction in the TNT concentration in the soil. Dead plants left to settle to the top of the contaminated sediments served as a buffer to reduce TNT leaching from the sediment. This same technique was applied at the Iowa Army Ammunition Plant by using local wetland sediments to cover the interface between contaminated soils and the wetland (Thompson *et al.* this book).

Lacking extensive soil toxicity investigations, extrapolation is possible using hydroponic toxicity and soil partitioning characteristics of explosives. The degree of partitioning is also an important determinant and surrogate for bioavailability.

Several studies have investigated the sorption of TNT and the associated metabolites to soils and humic materials (Pennington and Patrick 1990, Comfort e*t al.* 1995, Li *et al.* 1997, Sheremata *et al.* 1999). Most strongly sorbed to humic and other organic materials, soil type also affected the sorption of TNT; sandy soils yielded higher levels of extractable TNT than illite or topsoil (Sheremata *et al.* 1999). Reductive transformation products of TNT, 4-ADNT, and diaminonitrotoluene (DANT) have higher sorptive capacities than TNT: DANT > ADNT > TNT (Sheremata *et al.* 1999). A portion of TNT (about 30 percent) is irreversibly bound to soils (Comfort *et al.* 1995, Sheremata *et al.* 1999). It is likely that the same degree of soil sorptive capacities will hold for trinitrobenzene and dinitrotoluene, as these compounds are similar in structure to TNT.

Folsom e*t al.* (1988) conducted studies on plant uptake of TNT under various soil conditions. Plant concentrations of TNT and the associated transformation products were statistically the same for soil pH ranging from 5 to 8. Initial soil concentrations ranging from 100 to 400 milligrams of TNT per kilogram of dry soil did not affect the TNT and transformation product concentrations in the plants investigated. Although the overall effectiveness can vary, it is clear that plants can significantly reduce the concentration of

TNT in the soils studied (Folsom *et al.* 1988, Cataldo *et al.* 1989, Wilson 1995, Thompson *et al.* 1998).

Singh e*t al.* (1998) concluded that, even after an aging period of 168 days, most RDX is available for transport. Given this bioavailability, they concluded that RDX contamination in soils should be treated as rapidly as possible.

UNCERTAINTIES IN DETERMINING TOLERANCE AND IMPLICATIONS FOR PHYTOREMEDIATION

Toxicity of Transformation Products

The treatment of organic contaminants often results in the transformation of the contaminants to other compounds of unknown toxicity. In many cases, the transformation products are difficult to identify. Therefore, it is not surprising that the toxicities of transformation products are relatively unknown for many contaminants. Studies identifying these products and determining the phytotoxicity would enhance treatment system design. Until the transformation pathways are defined for all contaminants in all the significant plants used in phytoremediation, toxicity testing (Griest *et al.* 1995) of the plant, soil, and water residues may be necessary.

Soil Toxicity and Bioavailability

The paucity of soil phytotoxicity observations for hydrophobic contaminants limits toxicity and bioavailability projections, requiring pilot testing for most applications. However, hydroponically determined dose–response studies that dominate current understanding for parrot feather (*Myriophyllum aquaticum*) and a few other plants are useful. The dose–response relationships and limited knowledge of sorption allow some estimates of bioavailability and toxicity, and will be useful in extrapolating results from future tests of soil toxicity.

Reproductive Studies

Recent work with low levels of herbicide exposure suggests that reproductive toxicity may occur in certain plant species without noticeable morphological symptoms or effects on biomass. Exposure of peas (*Pisum sativum*) and soybean (*Glycine max*) to low levels of chlorsulfuron resulted in flower abortion without other symptomology (Fletcher *et al.* 1995, Fletcher *et al.* 1996). Thus, plants may be seemingly tolerant to a toxicant because no extrinsic signs of toxicity are apparent, but the life cycle may be disrupted by failure to form seeds.

In the phytoextraction of metals, a reproductive effect may be insignificant, as plants should be harvested before reproduction occurs. In many applications with organic compounds, degradation and transformation of dispersed material over a long time frame are the ultimate goals. In such a long-term application, reproductive effects might be important. Therefore, studies in this area would have value for improving phytoremediation designs to clean up organic contaminants in sustainable approaches.

Food Chain Effects

Tolerant plants exposed to sublethal levels may accumulate contaminants such as explosives without effects on vegetative growth and reproductive performance (Fletcher *et al.* 1990, Thompson *et al.* 1998). Although not observed to date, biomagnification in the fatty tissues of herbivores (Fletcher *et al.* 1990) and carnivores, and temporally limited exposures, are possible.

In the phytoremediation of organic compounds, the compounds are often transformed into more complex final products such as the polymeric compounds found in the transformation of explosives or chloroacetic acids in the case of TCE (Hughes *et al.* 1997, Nzengung and Jeffers 2001). Ambiguous metabolic products, which are often called bound residues, are also found in the degradation of pesticides, explosives, and other contaminants (Klein and Scheunert 1982, Khan and Dupont 1987). These unknown compounds further complicate the assessment of food chain effects. Further, it is unlikely that a particular phytoremediation system could be fenced and managed to completely eliminate herbivorous activity. Therefore, studies investigating the food chain effects of selected organic contaminants and associated breakdown products are needed.

Acknowledgments

Phil Sayre, Larry Davis, Hans Harms, and Chris Hoehamer provided insightful and through peer reviews used in the evaluation and acceptance of this chapter by Jerald L. Schnoor. Bob Swank also reviewed the chapter. Steven L. Larson, Daniel Tano, and Sumona Das Gupta supported or contributed to the work reported by Victor Medina in this chapter. Marta Marmiroli and Elena Maestri acknowledge support by the Italian Ministry of Education, University, and Research and by the National Research Council Target Project "Biotechnology." Annette Dietz acknowledges support from the Center for Biocatalysis and Bioprocessing and the U.S. National Institute of Health training grant program at the University of Iowa, and from the U.S. Environmental Protection Agency Hazardous Substance Research Center: Great Plains/Rocky Mountains centered at Kansas State University. Support for the work of Steve McCutcheon that went into this chapter came from the U.S. Environmental Protection Agency Office of Research and Development Contaminated Sites Program, the Strategic Environmental Research and Development Program (including Project 720 managed by the Army Waterways Experiment station) of the Departments of Defense and Energy and the U.S. Environmental Protection Agency, an internal grant from the National Exposure Research Laboratory, and cooperative interagency research agreements with the U.S. Air Force Restoration Division at Wright Patterson Air Force Base and Human Systems Centre at Books Air Force Base, U.S. Navy Southern Command, and U.S. Army Waterways Experiment Station. This paper has been reviewed in accordance with the U.S. Environmental Protection

Agency peer and administrative review policies and approved for publication. Mention of trade names or commercial products does not constitute endorsement or recommendation for use.

REFERENCES

Alfonso, M., J.J. Pueyo, K. Gaddour, A.L. Etienne, D. Kirilovsky, and R. Picorel (1996) Induced new mutation of D1 serine-268 in soybean photosynthetic cell cultures produced atrazine resistance, increased stability of S(2)Q(B)(-) and S(3)Q(B)(-) states, and increased sensitivity to light stress. *Plant Physiol.* **112**: 1499–1508.

Andrews, S.L. (1996) The effects of temperature on the phytoremediation of munitions (TNT) contaminated water. Unpublished M.S. thesis. Department of Biological and Agricultural Engineering, University of Georgia, Athens.

Anthony, R.G. and P.J. Hussey (1999) Dinitroaniline herbicide resistance and the microtubule cytoskeleton. *Trends Plant Sci.* **4**: 112–116.

Antonovics, J., A.D. Bradshaw, and R.G. Turner (1971) Heavy metal tolerance in plants. *Adv. Ecol. Res.* **7**: 1–85.

Bailey, H.C., R.J. Spanggord, H.S. Javitz, and D.H.W. Liu (1984) Toxicity of TNT (trinitrotoluene) wastewaters to aquatic organisms. Volume 4. Chronic Toxicity of 2,4-Dinitrotoluene and Condensate Water. SRI International. Report Number SRI-LSU- 4262 ADA 1535368. Menlo Park, California, 91 pp.

Baker, A.J.M. (1987) Metal tolerance. *New Phytol.* **106**: 93–111.

Baker, A., R. Brooks, and R. Reeves (1988) Growing for gold... and copper... and zinc. *New Sci.* March 10: 44–48.

Bert, V., M.R. Macnair, P. De Laguerie, P. Saumitou-Laprade, and D. Petit (2000) Zinc tolerance and accumulation in metallicolous and nonmetallicolous populations of *Arabidopsis halleri* (Brassicaceae). *New Phytol.* **146**: 225–233.

Best, E.P.H., J.L. Miller, H.L. Fredrickson, S.L. Larson, M.E. Zappi, and T.H. Streckfuss (1998) Explosives removal from groundwater of the Iowa Army Ammunition Plant in continuous-flow laboratory systems planted with aquatic and wetland plants. U.S. Army Corps of Engineers, Waterways Experiment Station. Technical Report EL-98-13. Vicksburg, Mississippi.

Bethe-Conti, L., H. Jacobi, S. Kleihauer, and I. Witte (1998) Cytotoxicity and mutagenicity of a 2,4,6-trinitrotoluene (TNT) and hexagen contaminated soil in *S. typhimurium* and mammalian cells. *Chemosphere* **37**(2): 209–218.

Biernacki, M., J.L. Doust, and L.L. Doust (1995) Effects of trichloroethylene, plant sex and site of origin on modular demography in *Vallisneria americana*. *J. Appl. Ecol.* **32**: 761–777.

Bock, C. and H. Harms (1996) Metabolism of selected polychlorobiphenyl congeners in plant cell cultures. *Organohalogen Compounds.* **28**: 53–57.

Böger, P. and G. Sandmann (1998) Action of modern herbicides. In: *Photosynthesis: A Comprehensive Treatise*. A.S. Raghavendra, ed. Cambridge University Press, Cambridge, England, pp. 337–351.

Böger, P., B. Matthes, and J. Schmalfuss (2000) Towards the primary target of chloroacetamides—new findings pave the way. *Pest Manag. Sci.* **56**: 497–508.

Bradshaw, A.D. (1965) Evolutionary significance of phenotypic plasticity in plants. *Adv. Genet.* **13**: 115–153.

Bradshaw, A.D. and T. McNeilly (1991) Evolution in relation to environmental stress. In: *Ecological Genetics and Air Pollution.* G.E. Taylor, Jr., L.F. Pitelka, and M.T. Clegg, eds. Springer-Verlag, New York, pp. 11–31.

Buchanan, B.B., W. Grussem, and R.L. Jones (2000) *Biochemistry and Molecular Biology of Plants.* American Society of Plant Physiology, Rockville, Maryland, pp. 1198–1201.

Burken, J., J.V. Shanks, and P. Thompson (2000) Phytoremediation and plant metabolism of explosives and nitroaromatic compounds. In: *Biodegradation of Nitroaromatic Compounds and Explosives.* J.C. Spain, J.B. Hughes, and H.-J. Knackmuss, eds. Lewis Publishers, Boca Raton, Florida, pp. 239–275.

Bumphus, J.A. and M. Tatarko (1994) Biodegradation of 2,4,6-trinitrotoluene by *Phanerochaete chrysosporium*: identification of initial degradation products and the discovery of a TNT metabolite that inhibits lignin peroxidases. *Curr. Microbiol.* **28**: 185–190.

Butler, J.M., A.W. Groeger, and J.S. Fletcher (1992) Characterization of monochlorinated biphenyl products formed by Paul's Scarlet rose cells. *Bull. Environ. Contam. Toxicol.* **49**: 821–826.

Casciano, D.A. (1991) Introduction: historical perspectives of genetic toxicology. In: *Genetic Toxicology.* A.P. Li and R.H. Heflich, eds. CRC Press, Boca Raton, Florida, pp. 1–12.

Cataldo, D.A., S.D. Harvey, R.J. Fellows, R.M. Bean, and B.D. McVeety (1989) An evaluation of the environmental fate and behavior of munitions materiel (TNT, RDX) in soil and plant systems. Final Report. Project Number 88P8853. Pacific Northwest Laboratory, Richland, Washington.

Chen , D. (1993) Plant uptake and soil adsorption of RDX. M.S. thesis. University of Illinois, Urbana-Champaign, Illinois.

Clemens, S., E.J. Kim, D. Neumann, and J.I. Schroeder (1999) Tolerance to toxic metals by a gene family of phytochelatin synthases from plants and yeast. *EMBO J.* **18**: 3325–3333.

Coleman, J.O.D., M.M.A. Blake-Kalff, and T.G.E. Davies (1997) Detoxification of xenobiotics by plants: chemical modification and vacuolar compartmentation. *Trends Plant Sci.* **2**: 144–151.

Comfort, S.D., P.J. Shea, L.S. Hundal, Z. Li, B.L. Woodbury, J.L. Martin and W.L. Powers (1995) TNT transport and fate in contaminated soil. *J. Environ. Qual.* **24**: 1174–1182.

Coruzzi, G. and R. Last (2000) Amino acids. In: *Biochemistry & Molecular Biology of Plants.* B.B. Buchanan, W. Gruissem, and R.L. Jones, eds. American Society of Plant Physiologists, Rockville, Maryland, pp. 358–410.

Cummins, I., D.J. Cole, and R. Edwards (1999) A role for glutathione transferases functioning as glutathione peroxidases in resistance to multiple herbicides in black-grass. *Plant J.* **18**: 285–292.

Dayan, F.E., B.M. Armstrong, and J.D. Weete (1998) Inhibitory activity of sulfentrazone and its metabolic derivatives on soybean (*Glycine max*) protoporphyrinogen oxidase. *J. Agric. Food Chem.* **46**: 2024–2029.

Dietz, A.C. (2000) Phytotoxicity of chlorinated aliphatics to hybrid poplar cuttings and axenic tissue cultures, Ph.D. Dissertation. Department of Civil and Environmental Engineering, University of Iowa, Iowa City.

Dietz, A.C. and J.L. Schnoor (2001) Phytotoxicity of chlorinated aliphatics to hybrid poplar (*Populus deltoides* × *nigra*, DN-34). *Environ. Toxicol. Chem.* **20**(2): 389–393.

Doty, S.L., T.Q. Shang, A.M. Wilson, J. Tangen, A.D. Westergreen, L.A. Newman, S.E. Strand, and M.P. Gordon (2000) Enhanced metabolism of halogenated hydrocarbons in transgenic plants containing mammalian cytochrome P450 2E1. *Proc. Nat. Acad. Sci. USA*. **97**: 6287–6291.

Eberlein, C.V., M.J. Guttieri, P.H. Berger, J.K. Fellman, C.A. Mallory Smith, D.C. Thill, R.J. Baerg, and W.R. Belknap (1999) Physiological consequences of mutation for ALS-inhibitor resistance. *Weed Sci.* **47**: 383–392.

Edwards, R., D.P. Dixon, and V. Walbot (2000) Plant glutathione *S*-transferases: enzymes with multiple functions in sickness and health. *Trends Plant Sci.* **5**: 193–198.

Etnier, E.L. (1989) Water quality criteria for hexahydro-1,3,5-trinitro-1,3,5-triazine (RDX). *Regul, Toxicol, Pharmacol.* **9**(2): 147–157.

Fitter, A.H. and R.K.M. Hay (1987) *Environmental Physiology of Plants*. Academic Press, London.

Fletcher, J.S., J.C. McFarlane, T. Pfleeger, and C. Wickliff (1990) Influence of root exposure on the fate of nitrobenzene in soybean. *Chemosphere* **20**(5): 513–523.

Fletcher, J.S., T.G. Pfleeger, and H.C. Ratsch (1995) Chlorsulfuron influence on garden pea reproduction. *Physiol. Plantarum.* **94**: 261–267.

Fletcher, J.S., T.G. Pfleeger, H.C. Ratsch, and R. Hayes (1996) Potential impact of low levels of chlorsulfuron and other herbicides on growth and yield of nontarget plants. *Environ. Toxicol. Chem.* **15**(7): 1189–1196.

Foes, M.J., L.X. Liu, G. Vigue, E.W. Stoller, L.M. Wax, and P.J. Tranel (1999) A kochia (*Kochia scoparia*) biotype resistant to triazine and ALS-inhibiting herbicides. *Weed Sci.* **47**: 20–27.

Folsom, B.L., J.C. Pennington, C.L. Teeter, M.R. Barton, and J.A. Bright (1988) Effects of soil pH and treatment level on persistence and plant uptake of 2,4,6-trinitrotoluene. U.S. Army Corps of Engineers, Waterways Experiment Station. Technical Report EL-88-22. Vicksburg, Mississippi.

Forbes, V.E. (2000) Is hormesis an evolutionary expectation? *Funct. Ecol.* **14**: 12–24.

Foyer, C.H., P. Descourvieres, and K.J. Kunert (1994) Protection against oxygen radicals: an important defense mechanism studied in transgenic plants. *Plant Cell Environ.* **17**: 507–523.

French, C.E., S.J. Rosser, G.J. Davies, S. Nicklin, and N.C. Bruce (1999) Biodegradation of explosives by transgenic plants expressing pentaerythritol tetranitrate reductase. *Nat. Biotechnol.* **17**: 491–494.

Georgia Institute of Technology (GIT) (1995) Outcome of the bioremediation of explosives-contaminated sites working meeting. Atlanta, Georgia.

Gong, P., B.-M. Wilke, and S. Fleischmann (1999) Soil-based phytotoxicity of 2,4,6-trinitrotoluene (TNT) to terrestrial higher plants. *Arch. Environ. Contam. Toxicol.* **36**: 152–157.

Gordon, L. and W.R. Hartley (1992) 2,4,6-Trinitrotoluene. In: *Drinking Water Health Advisor: Munitions*. Lewis Publishers, Boca Raton, Florida, pp. 327–398.

Gordon, M., N. Choe, J. Duffy, G. Ekuan, P. Heilman, I. Muiznieks, M. Ruszaj, B.B. Shurtleff, S. Strand, J. Wilmoth, and L. A. Newman (1998) Phytoremediation of trichloroethylene with hybrid poplars. *Environ. Health Persp.* **106**(Supplement 4): 1001–1004.

Görge, E., S. Brant, and D. Werner (1994) Uptake and metabolism of 2,4,6-trinitrotoluene in higher plants. *Environ. Sci. Pollut. Res. Int.* **21**: 229–233.

Griest, W.H., R.L. Tyndall, A.J. Stewart, J.E. Caton, A.A. Vass, C.-H. Ho, and W.M. Caldwell (1995) Chemical characterization and toxicological testing of windrow composts from explosives-contaminated sediments. *Environ. Toxicol. Chem.* **14**(1): 51–59.

Grill, E., E.-L. Winnacker, and M.H. Zenk (1985) Phytochelatins: the principal heavy-metal complexing peptides of higher plants. *Science* **230**: 674–676.

Grime, J.P. (1979) *Plant Strategies and Vegetation Processes*. John Wiley, Chichester, United Kingdom.

Grossman, K. and J. Kwiatkowski (2000) The mechanism of quinclorac selectivity in grasses. *Pestic. Biochem. Physiol.* **66**: 83–91.

Grossmann, K. and H. Schiffer (1999) Protoporphyrinogen oxidase-inhibiting activity of the new, wheat-selective isoindoldione herbicide, cinidon-ethyl. *Pestic. Sci.* **55**: 687–695.

Hawari, J. (2000) Biodegradation of RDX and HMX: from basic research to field application. In: *Biodegradation of Nitroaromatic Compounds and Explosives*. J.C. Spain, J.B. Hughes, and H.-J. Knackmuss, eds. Lewis Publishers, Boca Raton, Florida, pp. 277–310.

Heflich, R.H. (1991) Chemical mutagens. In: *Genetic Toxicology*. A.P. Li and R.H. Heflich, eds. CRC Press, Boca Raton, Florida, pp. 143–202.

Herrmann, K.M. and L.M. Weaver (1999) The shikimate pathway. In: *Annual Review of Plant Physiology and Plant Molecular Biology*, Annual Reviews, Palo Alto, California.**50**: 463–503.

Hitchcock, D.R., M.C. Smith, and S.C. McCutcheon (2003) Using rotifer population demographic parameters to assess impacts of the degradation products from trinitrotoluene phytoremediation by *Myriophyllum aquaticum*. *Ecotoxicol. Environ. Safe.* **54** (in press).

Hickey, D.A. and T. McNeilly (1975) Competition between metal tolerant and normal plant populations: a field experiment on normal soil. *Evolution* **29**: 458–464.

Hoffmann, A.A. and P.A. Parsons (1991) *Evolutionary Genetics and Environmental Stress*. Oxford University Press, Oxford, United Kingdom.

Huang, X.D., B.J. McConkey, T.S. Babu, and B.M. Greenberg (1997) Mechanisms of photoinduced toxicity of photomodified anthracene to plants: inhibition of photosynthesis in the aquatic higher plant *Lemna gibba* (duckweed). *Environ. Toxicol. Chem.* **16**: 1707–1715.

Hughes J.S., J. Shanks, M. Vanderford, J. Lauritzen, and R. Bhadra (1997) Transformation of TNT by aquatic plants and plant tissue cultures. *Environ. Sci. Technol.* **31**: 266–271.

Incledon, B.J. and J.C. Hall (1997) Acetyl-coenzyme A carboxylase: quaternary structure and inhibition by graminicidal herbicides. *Pestic. Biochem. Physiol.* **57**: 255–271.

Kaplan, D.L. and A.M. Kaplan (1982) 2,4,6-Trinitrotoluene-surfactant complexes: decomposition, mutagenicity and soil leaching studies. *Environ. Sci. Technol.* **16**: 566–571.

Kaplan, A.S., C.F. Berghout and L.A. Peczenik (1965) Human intoxication from RDX. *Arch. Environ. Health.* **10**: 877–883.

Kapustka, L.A. (1997) Selection of phytotoxicity tests for use in ecological risk assessment. In: *Plants for Environmental Studies.* W. Wang, J.W. Gorsuch, and J.S. Hughes, eds. Lewis Publishers, Boca Raton, Florida, pp. 516–540.

Kern, A.J., L.L. Jackson, and W.E. Dyer (1997) Fatty acid and wax biosynthesis in susceptible and triallate-resistant *Avena fatua* L. *Pestic. Sci.* **51**: 21–26.

Khan, S.U. and S. Dupont (1987) Bound pesticide residues and their bioavailability. In: *Pesticide Science and Biotechnology: Proceedings of the Sixth International Conference.* R.M. Greenhalgh and T.R. Roberts, eds. Blackwell Science, Boston, Massachusetts, pp. 417–420.

Klausmeier, R.E., J.L. Osmon, and D.R. Walls (1973) The effect of trinitrotoluene on microorganisms. *Dev. Ind. Microbiol.* **15**: 309–317.

Klein, W. and I. Scheunert (1982) Bound pesticide residues in soil, plants, and food with particular emphasis on the application of nuclear techniques. In: *Agrochemicals: Fate in Food and Environment.* International Atomic Energy Agency. IAEA-SM-263/38. Vienna, Austria, pp. 177–205.

Kolb, M. and H. Harms (2000) Metabolism of fluoranthene in different plant cell cultures and intact plants. *Environ. Toxicol. Chem.* **19**: 1304–1310.

Krishnan, G., G.L. Horst, S. Darnell, and W.L. Powers (2000) Growth and development of smooth bromegrass and tall fescue in TNT-contaminated soil. *Environ. Pollut.* **107**: 109–116.

Kuk, Y.I., J.R. Wu, J.F. Derr, and K.K. Hatzios (1999) Mechanism of fenoxaprop resistance in an accession of smooth crabgrass (*Digitaria ischaemum*). *Pestic. Biochem. Physiol.* **64**: 112–123.

La Rocca, V., F. Dalla Vecchia, R. Barbato, A. Bonora, E. Bergantino, and N. Rascio (2000) Plastid photodamage and *Cab* gene expression in barley leaves. *Physiol. Plantarum.* **109**: 51–57.

Langebartels, C. and H. Harms (1984) Metabolism of pentachlorophenol in cell suspension cultures of soybean and wheat: pentachlorophenol glucoside formation. *J. Plant Physiol.* **113**(3): 201–211.

Langebartels, C. and H. Harms (1986) Plant cell suspension cultures as test systems for anecotoxicologic evaluation of chemicals: growth inhibition effects and comparison with the metabolic fate in intact plants. *Angew. Bot.* **60**: 113–123.

Larson, S.L., R.P. Jones, L. Escalon, and D. Parker (1999) Classification of explosives transformation products in plant tissue. *Environ. Toxicol. Chem.* **18**: 1270–1276.

Lauritzen, J.R., III (1998) *Catharanthus roseus* as a model for the study of the phytoremediation of TNT. M.S. Thesis. Chemical Engineering, Rice University, Houston, Texas.

Lee, I. and J.S. Fletcher (1992) Involvement of mixed function oxidase systems in polychlorinated biphenyl metabolism by plant cells. *Plant Cell Rep.* **11**: 97–100.

Lermontova, I. and B. Grimm (2000) Overexpression of plastidic protoporphyrinogen IX oxidase leads to resistance to the diphenyl-ether herbicide acifluorfen. *Plant Physiol.* **122**: 75–83.

Levine, B.S., E.M. Furedi, D.E. Gordon, J.M. Burns, and P.M. Lish (1981) Thirteen week toxicity study of hexahydro-1,3,5-trinitro-1,3,5-triazine (RDX) in Fischer 344 rats. *Toxicol. Lett.* **8**(4–5): 241–245.

Levine, B.S., E.M. Furedi, D.E. Gordon, J.J. Barkley, and P.M. Lish (1990). Toxic interactions of the munitions compounds TNT and RDX in F344 rats. *Fundam. Appl. Toxicol.* **15**(2): 373–380.

Levitt, J. (1980) *Responses of Plants to Environmental Stress.* Academic Press, New York.

Lewis, R.J. (1992) *Sax's Dangerous Properties of Industrial Materials.* Van Nostrand Reinhold, New York.

Li, A.Z., K.A. Marx, J. Walker, and D.L. Kaplan (1997) Trinitrotoluene and metabolites binding to humic acid. *Environ. Sci. Technol.* **31**(2): 584–589.

Liste, H.H. and M. Alexander (1999) Rapid screening of plants promoting phenanthrene degradation. *J. Environ. Qual.* **28**: 1376–1377.

Liu, D.H.W., R.J. Spanggord, H.C. Bailey, and D.C.L. Jones (1976) Toxicity of TNT wastewater (pink water) to aquatic organisms. annual report. 15 Apr–31 Dec 1975. Stanford Research Institute. Report number AD-A-031067. Menlo Park, California.

Liu, D.H.W., R.J. Spanggord, H.C. Bailey, and D.C.L. Jones (1983) Toxicity of TNT wastewaters to aquatic organisms, volume 2. acute toxicity of condensate and 2,4-dinitrotoluene. SRI International. Report LSU-4262-VOL-2. Menlo Park, California.

Lucero, M.E., W. Mueller, J. Hubstenberger, G.C. Phillips, and M.A. O'Connell (1999) Tolerance to nitrogenous explosives and metabolism of TNT by cell suspensions of *Datura innoxia. In Vitro Cell. Dev. Biol.*: *Plant* **35**: 480–486.

Macek, T., M. Mackova, and J. Kas (2000) Exploitation of plants for the removal of organics in environmental remediation. *Biotechnol. Adv.* **18**: 23–34.

Mackova, M., T. Macek, J. Ocenaskova, J. Burkhard, K. Demnerova, and J. Pazlarova (1997) Biodegradation of polychlorinated biphenyls by plant cells. *Int. Biodeter. Biodegr.* **39**: 317–325.

Mallakin, A., D.G. Dixon, and B.M. Greenberg (2000) Pathway of anthracene modification under simulated solar radiation. *Chemosphere* **40**: 1435–1441.

McCann, J.H. and K.R. Solomon (2000) The effect of creosote on membrane ion leakage in *Myriophyllum spicatum* L. *Aquat. Toxicol.* **50**: 275–284.

McCutcheon, S.C. and W.J. Mitsch (1994) Ecological and environmental engineering: potential for progress. *J. Environ. Eng.* **120**(3): 479–480, and *Ecol. Eng.* **3**: 107–109.

McCutcheon, S.C., N.L. Wolfe, L.H. Carreira, and T.Y. Ou (1995) Phytoremediation of hazardous wastes. In: *Innovative Technologies for Site Remediation and Hazardous Waste Management,*. Proceedings of the National Conference R.D. Vidic and F.G. Pohland, eds. American Society of Civil Engineers, New York. Held in Pittsburgh, Pennsylvania, July 26, pp. 597–604.

McFarlane, C., T. Pfleeger, and J. Fletcher (1990) Effect of uptake and disposition of nitrobenzene in several terrestrial plants. *Environ. Toxicol. Chem.* **9**: 513–520.

McLaughlin, S.B. and R.J. Norby (1991) Atmospheric pollution and terrestrial vegetation: evidence of changes, linkages, and significance to selection processes. In: *Ecological Genetics and Air Pollution.* G.E. Taylor, Jr., L.F. Pitelka, and M.T. Cleggs, eds. Springer-Verlag, New York, pp. 61–101.

Meagher, R.B. (2000) Phytoremediation of toxic elemental and organic pollutants. *Curr. Opin. Plant Biol.* **3**: 153–162.

Medina, V.F. and S.C. McCutcheon (1996) Phytoremediation: modeling removal of TNT and its breakdown products. *Remediation* **6**(4): 31–45.

Medina, V.F., S.L. Larson, A.E. Bergstedt, and S.C. McCutcheon (2000) Phyto-removal of trinitrotoluene from water with batch kinetic studies. *Water Res.* **34**(10): 2713–2722.

Medina, V.F., S.L. Larson, and S.C. McCutcheon (2002) Evaluation of continuous flow-through phytoreactors for the treatment of TNT-contaminated water. *Environ. Prog.* **21**(1): 29–36.

Mengistu, L.W., G.W. Müller-Warrant, A. Liston, and R.E. Barker (2000) psbA mutation (valine(219) to isoleucine) in *Poa annua* resistant to metribuzin and diuron. *Pestic. Manag. Sci.* **56**: 209–217.

Michels, J. and G. Gottschalk (1995) Pathway of 2,4,6-trinitotoluene (TNT) degradation by *Phanerochaete chrysosporium.* In: *Biodegradation of Nitroaromatic Compounds.* J.C. Spain, ed. Plenum Press, New York, pp. 135–149.

Miskiniene, V., J. Sarlauskas, J.-P. Jaquot, and N. Cenas (1998) Nitrobenzene reactions of *Arabidopsis thaliana* thioredoxin reductase. *Biochim. Biophys. Acta.* **1366**: 275–283.

Moore, T. (1995) Rising to the challenge in military site cleanups. *Environ. Eng. World.* May–June Issue. pp. 28–33.

Mueller, W., G.W. Bedell, S. Shojaee, and P.J. Jackson (1995) Bioremediation of TNT wastes by higher plants. *Proceedings of the 10th Annual Conference on Hazardous Waste Research.* L.E. Erickson, D.L. Tillison, S.C. Grant, and J.P. McDonald, eds. U.S. Environmental Protection Agency Great Plains-Rocky Mountain Hazardous Substance Research Center, Kansas State University, Manhattan, Kansas, pp. 222–231.

Neumann, H. and D. Werner (2000) Gene expression of *Medicago sativa* inoculated with *Sinorhizobium meliloti* as modulated by the xenobiotics cadmium and fluoranthene. *Z. Naturforsch, C.J. Biosci.* **55**: 222–232.

Neumann, D., O. Lichtenberger, D. Guenther, K. Tschiersch, and L. Nover (1994) Heat shock proteins induce heavy-metal tolerance in higher plants. *Planta* **194**: 360–367.

Newman, L.A., S.E. Strand, N. Choe, J. Duffy, G. Ekuan, M. Ruszaj, B.B. Shurtleff, J. Wilmoth, P. Heilman, and M.P. Gordon (1997) Uptake and biotransformation of trichloroethylene by hybrid poplars. *Environ. Sci. Technol.* **31**(4): 1062–1067.

Newman, L.A., X.P. Wang, I.A. Muiznieks, G. Ekuan, M. Ruszaj, R. Cortellucci, D. Domroes, G. Karscig, T. Newman, R.S. Crampton, R.A. Hasmonay, M.G. Yost, P.E. Heilman, J. Duffy, M.P. Gordon, and S.E. Strand (1999) Remediation of trichloroethylene in an artificial aquifer with trees: a controlled field study. *Environ. Sci. Technol.* **33**: 2257–2265.

Nzengung, V.A. and P. Jeffers (2001) Sequestration, phytoreduction, and phytooxidation of halogenated organic chemicals by aquatic and terrestrial plants. *Int. J. Phytorem.* **3**: 13–40.

Palazzo, A.J. and D.C. Leggett (1986) Effect and disposition of TNT in a terrestrial plant. *J. Environ. Qual.* **15**: 49–52.

Pallett, K.E., J.P. Little, M. Sheekey, and P. Veerasekaran (1998) The mode of action of isoxaflutole. I. physiological effects, metabolism, and selectivity. *Pestic. Biochem. Physiol.* **62**: 113–124.

Pavlostathis, S.G., K.K. Comstock, M.E. Jacobson, and F.M. Saunders (1998) Transformation of 2,4,6-trinitrotoluene by the aquatic plant *Myriophyllum spicatum*. *Environ. Toxicol. Chem.* **17**: 2266–2273.

Pennington, J.C. and W.H. Patrick, Jr. (1990) Adsorption and desorption of 2,4,6-trinitrotoluene. *J. Environ. Qual.* **19**: 559–567.

Peterson, M.M., G.L. Horst, P.J. Shea, S.D. Comfort, and R.K.D. Peterson (1996) TNT and 4-amino-2,6-dinitrotoluene influence on germination and early seedling development of tall fescue. *Environ. Pollut.* **93(1)**: 57–62.

Peterson, M.M., G.L. Horst, P.J. Shea, and S.D. Comfort (1998) Germination and seedling development of switchgrass and smooth bromegrass exposed to 2,4,6-trinitrotoluene. *Environ. Pollut.* **99**: 53–59.

Po, E.S.M. and J.W. Ho (1997) Paraquat affects light-induced proton transport through chloroplast membranes in spinach. *Comp. Biochem. Physiol., C* **118**: 65–69.

Puri, R.K., Y. Qiuping, S. Kapila, W.R. Lower, and V. Puri (1997) Plant uptake and metabolism of polychlorinated biphenyls (PCBs). In: *Plants for Environmental Studies*. W. Wang, Gorsuch, J.W., and J.S. Hughes, eds. CRC Lewis Publishers, Boca Raton, Florida, pp. 481–513.

Rafii, F., A.L. Selby, R.K. Newton, and C.E. Cerniglia (1994) Reduction and mutagenic activation of nitroaromatic compounds by a *Mycobacterium* sp. *Appl. Environ. Microbiol.* **60(12)**: 4263–4267.

Rea, P.A., Z.S. Li, Y.P. Lu, Y.M. Drozdowicz, and E. Martinoia (1998) From vacuolar GS-X pumps to multispecific ABC transporters. In: *Annual Review of Plant Physiology Plant and Molecular Biology*, Annual Reviews, Palo Alto, California. **49**: 727–760.

Ripley, L.S. (1991) Mechanisms of gene mutations. In: *Genetic Toxicology*. A.P. Li and R.H. Heflich, eds. CRC Press, Boca Raton, Florida, pp. 13–40.

Rivera, R., V.F. Medina, S.L. Larson, and S.C. McCutcheon (1998) Phytoremediation of TNT contaminated groundwater. *J. Soil Contam.* **7**: 511–529.

Robinson, N.J., A.M. Tommey, C. Kuske, and P.J. Jackson (1993) Plant metallothioneins. *Biochem. J.* **295**: 1–10.

Romagni, J.G., S.O. Duke, and F.E. Dayan (2000) Inhibition of plant asparagine synthetase by monoterpene cineoles. *Plant Physiol.* **123**: 725–732.

Roose, M.L. (1991) Genetics of response to atmospheric pollutants. In: *Ecological Genetics and Air Pollution*. G.E. Taylor, Jr., L.F. Pitelka, and M.T. Clegg, eds. Springer-Verlag, New York, pp. 111–126.

Ryu, S.B., L.C. Davis, J. Dana, K. Selk, and L.E. Erickson (1996) Evaluation of Toxicity of Trichloroethylene for Plants. Proceedings of the HSRC/WERC Joint

Conference on the Environment. Great Plains/Rocky Mountain Hazardous Waste Research Center, Kansas State University, Manhattan, Kansas.

Sabba, R.P. and K.C. Vaughn (1999) Herbicides that inhibit cellulose biosynthesis. *Weed Sci.* **47**: 757–763.

Salisbury, F.B. and C.W. Ross (1992) *Plant Physiology*. Wadsworth Publishing Company, Belmont, California.

Sandermann, H., Jr. (1994) Higher plant metabolism of xenobiotics: the 'green liver' concept. *Pharmacogenetics* **4**: 225–241.

Saroha, M.K., P. Sridhar, and V.S. Malik (1998) Glyphosate-tolerant crops: genes and enzymes. *J. Plant Biochem. Biotechnol.* **7**: 65–72.

Schnoor, J.L., L.A. Licht, S.C. McCutcheon, N.L. Wolfe, and L.H. Carriera (1995) Phytoremediation of organic and nutrient contaminants. *Environ. Sci. Technol.* **29**: 318A–323A.

Schott, C.D. and E.G. Worthley (1974) The toxicity of TNT and related wastes to an aquatic flowering plant: *Lemna perpusilla* Toor. Edgewood Arsenal Technical Report EB-TR-74016. Aberdeen Proving Ground, Maryland.

Schroder, P., S. Juuti, S. Roy, H. Sandermann, and S. Sutinen (1997) Exposure to chlorinated acetic acids: responses of peroxidase and glutathione *S*-transferase activity in pine needles. *Environ. Sci. Pollut. Res.* **4**: 163–171.

Schubert, U., L. Wisanowsky, and U. Kull (1995) Determination of phytotoxicity of several volatile organic compounds by investigating the germination pattern of tobacco pollen. *J. Plant Physiol.* **145**: 514–518.

Schwenger-Erger, C. and W. Barz (2000) Decreased rate of degradation of the D1 protein in metribuzin-resistant photoautotrophic *Chenopodium rubrum* cell cultures. *J. Plant Physiol.* **156**: 458–461.

Sealock, G.A., Jr. (2002) Phytoremediation of explosives using *Populus deltoides*. M.S. thesis. Daniel B. Warnell School of Forest Resources, University of Georgia, Athens.

Sens, C., P. Scheidemann, and D. Werner (1999) The distribution of ^{14}C-TNT in different biochemical compartments of the monocotyledonous *Triticum aestivum*. *Environ. Pollut.* **104**: 113–119.

Sheremata, T.W., S. Thiboutot, G. Ampleman, L. Paquet, A. Halasz, and J. Hawari (1999) Fate of 2,4,6-trinitrotoluene and its metabolites in natural and model soil systems. *Environ. Sci. Technol.* **33**: 4002–4008.

Shojaee, S. (1994) Bioremediation of TNT-contaminated soils by higher plants. Ph.D. dissertation. New Mexico State University, Las Cruxes, New Mexico.

Siciliano, S.D. and J.J. Germida (1998) Mechanisms of phytoremediation; biochemical and ecological interactions between plants and bacteria. *Environ. Rev.* **6**: 65–79.

Siciliano, S.D. and C.W. Greer (2000) Plant-bacterial combination to phytoremediate soil contaminated with high concentrations of 2,4,6-trinitrotoluene. *J. Environ. Qual.* **29**: 311–316.

Sikkema, J., J.A.M. de Bont, and B. Poolman (1994) Interactions of cyclic hydrocarbons with biological membranes. *J. Biol. Chem.* **269**: 8022–8028.

Simini, M., R.S. Wentsel, R.T. Checkai, C.T. Phillips, N.A. Chester, M.A. Major, and J.C. Amos (1995) Evaluation of soil toxicity at Joliet Army Ammunition Plant. *Environ. Toxicol. Chem.* **14**: 623–630.

Singh, J., S.D. Comfort, L.S. Hundal and P.J. Shea (1998) Long-term RDX sorption and fate in soil. *J. Environ. Qual.* **27**: 572–577.

Smith, G., C.A.D. Smith, and C.R. Wolf (1995) Pharmacogenetic polymorphisms. In: *Environmental Mutagenesis*. D.H. Phillips and S. Venitt, eds. Bios Scientific Publishers, Oxford, England, pp. 86–106.

Smock, L.A., D.L. Stoneburner, and J.R. Clark (1976) The toxic effects of trinitrotoluene (TNT) and its primary degradation products on two species of algae and the fathead minnow. *Water Res.* **10**: 537–543.

Stebbing, A.R.D (1998) A theory for growth hormesis. *Mutat. Res.* **403**: 249–258.

Sun, W.-H., G.L. Horst, R.A. Drijber, and T.E. Elthon (2000) Fate of 2,4,6-trinitrotoluene in axenic sand culture systems containing smooth bromegrass. *Environ. Toxicol. Chem.* **19**: 2038–2046.

Tadros, M.G., A. Crawford, C. Mateo-Sullivan, and J.B. Hughes (2000) Toxic effects of hydroxylamino intermediates from microbial transformation of trinitrotoluene and dinitrotoluenes on algae *Selenastrum carpricornutum. Bull. Environ. Contam. Toxicol.* **64**: 579–585.

Taylor, G.E., Jr., L.F. Pitelka, and M.T. Clegg (1991) Introduction. In: *Ecological Genetics and Air Pollution*. G.E. Taylor, Jr., L.F. Pitelka, and M.T. Clegg, eds. Springer-Verlag, New York, pp. 1–9.

Thompson, P.L., L.A. Ramer, A.P. Guffey, and J.L. Schnoor (1998) Decreased transpiration in poplar trees exposed to 2,4,6-trinitrotoluene. *Environ. Toxicol. Chem.* **17**(5): 902–906.

Thompson, P.L., L.A. Ramer, and J.L. Schnoor (1999) Hexahydro-1,3,5-trinitro-1,3,5-triazine translocation in poplar trees. *Environ. Toxicol. Chem.* **18**(2): 279–284.

Vanderford, M. (1996) Aquatic phytotransformation of trinitrotoluene in contaminated media. M.S. thesis. Rice University, Houston, Texas.

Vanderford, M., J.V. Shanks, and J.B. Hughes (1997) Phytotransformation of trinitrotoluene (TNT) and distribution of metabolite products in *Myriophyllum aquaticum. Biotechnol. Lett.* **54**: 936–944.

van Overbeek, J. and R. Blondeau (1953) Mode of action of phytotoxic oils. *Weeds* **3**: 55–63.

Vicente, J.A.F., M.S. Santos, A.E. Vercesi, and V.M.C. Madeira (1998) Comparative effects of the herbicide dinitro-*o*-cresol on mitochondrial bioenergetics. *Pestic. Sci.* **54**: 43–51.

Wei, Y.D., H.G. Zheng, and J.C. Hall (2000) Role of auxinic herbicide-induced ethylene on hypocotyls elongation and root/hypocotyls radial expansion. *Pest Manag. Sci.* **56**: 377–387.

Werck-Reichhart, D., A. Hehn, and L. Didierjean (2000) Cytochromes P-450 for engineering herbicide tolerance. *Trends Plant Sci.* **5**(3): 116–123.

Wilken, A., C. Bock, M. Bokern, and H. Harms (1995) Metabolism of different PCB congers in plant cell cultures. *Environ. Toxicol. Chem.* **14**(12): 2017–2022.

Wilson, E.C. (1995) Field analysis of biochemical remediation in TNT contaminated soils. M.S. thesis. Department of Civil Engineering, Auburn University, Auburn, Alabama.

Won, W.D., L.H. DiSalvo, and J. Ng (1976) Toxicity and mutagenicity of 2,4,6-trinitrotoluene and its microbial metabolites. *Appl. Environ. Microbiol.* **31**(4): 576–589.

Wright, T.R. and D. Penner (1998) Corn (*Zea mays*) acetolactate synthase sensitivity to four classes of ALS-inhibiting herbicides. *Weed Sci.* **46**(1): 8–12.

Wright, T.R., N.F. Bascomb, S.F. Sturner, and D. Penner (1998) Biochemical mechanism and molecular basis for ALS-inhibiting resistance in sugarbeet (*Beta vulgaris*) somatic cell selections. *Weed Sci.* **46**(1): 13–23.

Wu, L., A.D. Bradshaw, and D.A. Thurman (1975) The potential for evolution of heavy metal tolerance in plants. 3. the rapid evolution of copper tolerance in *Agrostis stolonifera. Heredity* **34**: 165–187.

Young, D.G. (1995) The biochemical remediation of a TNT contaminated soil. Ph.D. dissertation. Civil Engineering, Auburn University, Alabama.

Zenk, M.H. (1996) Heavy metal detoxification in higher plants—a review. *Gene* **179**: 221–230.

7

ROOT DEVELOPMENT AND ROOTING AT DEPTHS

M. C. Negri, E. G. Gatliff, J. J. Quinn, and R. R. Hinchman

SUMMARY OF PRACTICAL IMPLICATIONS

Phytoremediation can often be extended to contamination at depths below the "normal" rooting zone of 1 meter (3 feet). Root growth at the desired depth is dependent on the spatial availability of water, nutrients, and oxygen and is limited by the mechanical impedance of the soil overburden. Natural rooting at depths as deep as 50 meters (160 feet) occurs when plants that are genetically predisposed to root deeply (phreatophytes) do not find sufficient water and nutrient resources in the surface soil to satisfy growth needs. Because of this, subarid climates with porous soil may be the best for phytoremediation of deeper contamination with minimal manipulation or engineering of root growth. In temperate regions, however, where water is easily accessible in surface soil horizons, rooting at depth may never occur spontaneously and increasingly complex rooting manipulation or engineering techniques need to be utilized to achieve root development in the target zone. Case studies establish that engineered deep rooting is feasible to at least 12 meters (40 feet).

INTRODUCTION

Phytoremediation is a broad term that encompasses the utilization of higher plants to remediate a contaminated medium. Most of the research has focused on contamination of surficial [up to 2 to 3 meters (6 to 9 feet)] soil or groundwater, while research on remediating contaminants in the subsurface (deeper vadose zone and groundwater) has received far less attention. Field applications and research alike have clearly focused on the visible aboveground portion of the phytoremediation system, typically assuming that roots would grow into a certain subsoil, with little information available on the likelihood of this

---

Phytoremediation: Transformation and Control of Contaminants,
Edited by Steven C. McCutcheon and Jerald L. Schnoor.
ISBN 0-471-39435-1 (cloth)

happening or on the factors that would predict or favor root growth into the target area of the subsoil. Although satisfactory knowledge is emerging on uptake and fate once contaminants contact the root, very little is known about (1) how this crucial contact between the contaminated medium and the root occurs and develops as a function of soil conditions and groundwater depth, (2) what implications are entailed, and (3) how the rooting capabilities of plants can be best utilized. In the meantime, the number of field applications increases every day, flagging not only the interest in the technology, but also the risk of failure entailed in the use of an approach not appropriately understood. This chapter will explore the implications and potential for the remediation of contamination occurring below the commonly accepted range of a typical rooting zone.

Phytoremediation, like bioremediation and natural attenuation, has often drawn criticism and is not widely applied because of the unpredictability of biological or natural systems compared to *ex situ* traditional chemical or physical approaches. The deployment of any *in situ* technology depends largely on site-specific conditions. Usually, increasingly complex designs are necessary with increasing depths of contamination. A lack of a comprehensive understanding of root distribution and interactions with groundwater and soil stratigraphy results in a poor prediction of phytoremediation effectiveness and potential improvements. This lack of understanding of the dynamics of root growth causes design guesses with an increased risk of failure, or a large safety factor (in spite of higher costs) to offset the unknowns.

Success in phytoremediation below 1 meter (3 feet) relies on effective rooting at those depths. Although a significant amount of information exists, especially on the relationship between root growth and soil conditions, this research has focused mostly on crops, plants of economic interest, and natural environments. Not much information is available on rooting at depths. Sufficient data, however, is available to hypothesize the behavior of a root system in a specific subsoil and to determine what level of design complexity is necessary to optimize the likelihood of success in reaching a contaminated zone.

Several planting approaches are available to phytoremediate contamination that is located below what is commonly accepted as the "normal" rooting depth of plants. Deep rooting reaches depths beyond 1 meter (3 feet) and usually up to 12 meters (40 feet) below ground surface. When deep rooting is required, phreatophytic trees that obtain water from an aquifer such as poplars (*Populus* spp.) and willows (*Salix* spp.) are the usual choice because of the natural capability to develop roots to these and greater depths. However, even trees with the capability to root deeply will only do so if site conditions allow. Willows (*Salix* spp.) and poplars (*Populus* spp.) may not develop roots to the depths required for remediation if the site conditions are not amenable to deep rooting, most often because there is no strategic need or benefit for the plant to spend the extra energy required to do so (this is commonly referred to as the principle of optimization). This is especially true for climates that are more humid and where the water budget for these trees can be adequately supplied by rainwater stored in the near-surface soil.

To ensure rooting below 1 meter (3 feet), trees are usually planted as deep as possible in trenches or boreholes that are often equipped with "root engineering props" to direct and accelerate root growth. These props are typically proper backfill materials to ensure favorable conditions for root growth, aeration tubes to supply the growing roots with additional oxygen at depth, and casings to prevent the growing root from utilizing water from the surrounding soil, thereby "directing" root growth to the targeted zone and enhancing the remedial effect of the tree. Proper casing can also serve an important purpose of eliminating the potential for surface water flowing preferentially ("short circuiting") through the borehole or trench, causing contamination of the groundwater by surface water. This short circuiting can be problematic in certain settings, especially where the backfill material is more porous than the surrounding soil conditions. In addition, regulations in some states may require casing or comparable measures to prevent short circuiting.

Often debated as to necessity, the discussion that follows provides sufficient evidence of the importance of these root-engineering methodologies. However, as in most *in situ* remediation approaches, a determination of the ratio of costs *versus* benefits is vital and is largely dependent on site-specific conditions such as stratigraphy, depth of contamination, climate, time requirements to achieve or demonstrate cleanup, and regulatory issues. Several patents have been awarded to phytoremediation companies, some specifically related to the delivery and installation methods for establishing deep-rooting trees.

This chapter illustrates the main variables that influence root growth and that are taken into account when designing a phytoremediation installation targeting deeper contamination. The complexity of such a design, associated with the highest likelihood of success, is discussed using several existing phytoremediation sites.

ROOT FUNCTIONS AND TERMINOLOGY

For most land plants, the entire root system is visible only if the plant is carefully removed from the soil; consequently, the root system is often called the "hidden half" (Waisel *et al.* 1996). Root development extremes range from the very deep-penetrating roots of xerophytic (adapted to grow in very dry or desert conditions) woody perennials to the shallow root systems of short-season annuals.

Plant roots have multiple functions. The two most evident are to anchor and support the plant in the soil and to ensure the uptake and transport of soil solution (soil water plus all of its dissolved constituents) throughout the plant body. Other functions include the metabolism and storage of carbohydrates and other metabolites, the synthesis of plant hormones, the secretion of exudates that provide a growth medium for free-living rhizosphere organisms, and the formation of intercellular air spaces that permit diffusion of air (particularly oxygen) to the respiring root tissues.

As this chapter mostly focuses on the development of roots to access deep contamination, it will mainly examine water, nutrient, and oxygen acquisition. Root morphology and physiology are discussed elsewhere in this book (Hutchinson *et al.*).

The most commonly used root terminology adopted throughout this chapter follows the definitions cited by Lynch (1995). Root morphology refers to the surface features of a single root axis as an organ, including characteristics of the epidermis such as root hairs, root diameter, and root cap, and typically excluding features related to cell and tissue organization. Root topology refers to how individual root axes are connected to each other through branching (stable to deformation or rotation of axes, root topology is easier to study than root architecture). "Herringbone" and "dichotomous" are the two main topologies and consist, respectively, in a main axis with primary laterals and two main axes with greater branching. Root distribution is the presence of roots, rather than the orientation in a positional gradient or grid. Distribution is usually defined as root biomass or length as a function of depth or distance from the stem. Root architecture represents the spatial configuration of complex assemblages of subunits, with the implication that the overall configuration has some functional significance. Root architecture usually is not concerned with morphology but entails knowledge of topology and distribution. Root architecture plays a significant role in plant growth and productivity as it is correlated with the capability of a plant to access resources in a spatially heterogeneous environment.

Often, roots are also described as primary, secondary, and tertiary. While losing favor in the scientific community, these adjectives are often still found in the literature. Primary roots form at the time of seed germination. Lateral roots, also called secondary and tertiary roots, emerge as branch roots from primary roots. Adventitious roots are additional laterals that develop between those of the regular acropetal (developing sequence from the base toward the apex) series (Charlton 1996). Topologically, monocotyledons usually have simpler root systems, whereas dicotyledon root systems (which include those of most deciduous trees, including Salicaceae) are typically more diverse due to secondary root development.

FACTORS CONTROLLING ROOT GROWTH

The root–soil interface is a major site of interaction between the plant and the ambient environment. Plant roots have complex and often poorly understood capabilities to adapt to changing conditions during growth. Many factors cause changes during root growth. Of these, the most relevant for root development are the presence and location of essential growth factors (water, nutrients, and oxygen), and the physical properties of the soil that make root development more or less affordable to the plant in relation to the resources available. These factors, alone and in conjunction, have in nature a primary role in determining the depth of rooting and the architecture of the root system. These "tools" also are the main phytoremediation design parameters used to direct root growth to the area of interest.

Genetically induced differences in root development and architecture have been clearly recognized, albeit not studied in depth. Genetic traits responsible for root development are poly- or multigenic (Zobel 1996). These traits are mostly responsible for (1) modulating the response of the plant to water stress (depth of rooting), (2) root tolerance of mechanical impedance, (3) response to nutrient deficiency and low temperatures, (4) the capability to enter into symbiotic relationships, and (5) the capability to develop aerenchyma for survival in waterlogged environments.

Water

Large terrestrial plants with extensive root systems are able to move large amounts of soil solution into the plant body through the roots, transport it through the stem to the aboveground parts, and transpire this water out of the leaves as water vapor. Plants transpire water to move nutrients from the soil solution to leaves and stems where photosynthesis occurs, and to cool the plant. Contaminants present in the soil solution can also be taken up and sequestered, metabolized, or volatilized along with the transpired water. Water available to plants comes from precipitation and groundwater. Low water use is a trait considered desirable in most economically important plants; however, the ideal phytoremediation plants for groundwater capture may be poor at water conservation, but capable of taking up and "processing" large, unrestricted volumes of soil solution or groundwater. This trait is usually a consequence of adaptation to growing in conditions of relatively unlimited access to water [*e.g.*, poplar (*Populus* spp.), willow (*Salix* spp.), Bulrush (*Scirpus* spp.), and marsh grasses], which in turn may be a result of the capability of these plants to tap into more permanent sources of water like aquifers.

Water moves through a plant in response to a potential gradient, which is started by the loss of water in the leaf by transpiration dependent on radiant energy. This water potential is transferred to the xylem and roots, and results in the roots taking up water from the soil. Plants differ in the capability to "extract" water from the soil against forces that hold the water to the soil. Plants adapted to drought or saline conditions typically have the capability to extract water from soils under the most negative (lowest) water potentials.

Research has suggested that marked differences in rooting patterns among species are adaptations to maximize access to water and to minimize competition for water during drought periods (Dawson and Pate 1996). Stable hydrogen isotope analyses of groundwater, rainwater, and leaf tissue water allow for the distinction of different water sources utilized by plants. By using differences in hydrogen isotope ratios found in xylem water and source water, Ehleringer *et al.* (1991) were able to demonstrate that different species cohabiting a desert habitat used water from different sources. Dawson (1996) noted similar findings for large and small sugar maples in a temperate forest. Busch *et al.* (1992) found that isotopic ratios of water extracted from *Populus* and *Salix* species growing in the southwestern U.S. were not significantly different from groundwater or saturated soil, suggesting that phreatophytes share a similar pattern of water uptake.

Phreatophytes are generally described as upland plants that have the capability of extending root systems to deeper saturated soil horizons in search of water. In 1923, Meinzer defined a phreatophyte as a plant that habitually obtains water from the zone of saturation, either directly or through the capillary fringe (cited in Erdman and Christenson 2000). The capability to grow deep roots is a characteristic of plants commonly responding to a limited supply of water, which explains why many phreatophytes are plants growing in arid or subarid climates.

Roots of various phreatophytes extend from 4 meters to over 15 meters (13 to 49 feet) up to 53 meters (174 feet) into the soil profile (Nobel 1996). Hybrid poplars (*Populus* spp.) and willows (*Salix* spp.) are the most relevant phreatophytes that are being utilized in engineered plant systems to achieve hydraulic control (contaminated groundwater capture and removal) (Nyer and Gatliff 1996, Quinn *et al.* 2001) and, concomitantly, to transform and transpire organic contaminants (see Doucette *et al.* this book, Eberts *et al.* this book, Hirsh *et al.* this book, Shang *et al.* this book). In addition, a number of other phreatophytic species [(such as eucalyptus (*Eucalyptus* spp.) and saltcedar (*Tamarix* spp.)] reach significant rooting depths but have not yet been hydraulically evaluated for use in phytoremediation systems. Several general reviews are available on rooting depths of plants (Canadell *et al.* 1996, Jackson *et al.* 1996).

Some phreatophytes are obligate users of groundwater while others such as salt cedar (*Tamarix* spp.) can also rely on soil moisture from precipitation during wet periods (Smith *et al.* 1998). Typically, these phreatophytes growing under "natural" conditions have a dimorphic root system. A dimorphic system involves both surficial roots (anchoring the plant and usually taking up the bulk of the nutrients and water when available at the soil surface) and a deep root (sinker and tap root). The function of the deep root is to ensure a constant water supply when surface water is scarce by tapping into the more moist soil horizons and the water table. Dawson and Pate (1996) found that dimorphic root systems of native Australian phreatophytes were obtaining water from two distinct sources over the course of a wet winter–dry summer annual cycle. Shallow lateral roots, restricted to the top 40 centimeters (16 inches) of soil, contributed largely to the total water uptake during the wet season, but the deep-rooting taproot [sinker, found to reach the water table at 2 to 5 meters (6 to 15 feet)], was responsible for the highest uptake of water as the season turned to dry. Moreover, these authors demonstrated that the shallow lateral roots remained recharged by a continuous supply of groundwater provided by the sinker root throughout the very dry summer.

Although it is well established that phreatophytic roots are capable of exploiting deep-water sources during drought periods (Canadell *et al.* 1996), the dynamics of water uptake and movement throughout an entire wet–dry cycle are still largely unexplored (Dawson and Pate 1996), as are the main mechanisms that promote root growth deep into dry soil. Also largely unknown is the theoretical and practical maximum depth of rooting. Knowing this could resolve substantial uncertainties in the phytoremediation of groundwater. Related to rooting depth is whether phreatophytes have particularly sophisticated systems to lift water from these depths, as significant decreases in water potential must occur. Having a lower axial resistance compared to more surficially rooting plant species may be a mechanism adopted by phreatophytes to facilitate such transport (Taylor and Klepper 1978). Axial resistance slows water flow by friction between the moving water column and the xylem walls.

A number of studies indicate that deep root initiation occurs in response to low water availability or high transpirational demands; however, there is considerable debate on this issue. While some authors found a decline in root growth after canopy senescence, others found substantial deep root growth in white oak (*Quercus alba*) and black walnut (*Juglans nigra*) after the leaves had fallen (Kuhns *et al.* 1995 in Hendrick and Pregitzer 1996). Others report deep root initiation by black walnut (*Juglans nigra*) occurred only after the leaves had fully expanded and moisture in the surface soil had begun to decline. These authors also showed that root mortality and turnover was lower at depths than at the surface of the soil. Roots below the depth of 75 centimeters or 30 inches were about 11 percent of the annual production and 4 percent of the annual mortality. Also, growth often continued well after leaf senescence, indicating that deep rooting is a long-term investment by the plant, often initiated to overcome drought during the mid-summer months.

Relatively little information is available on rooting characteristics of phreatophytic Salicaceae such as the poplars and the willows used in phytoremediation systems. The root system of the genus *Populus* is generally dichotomous with strongly developed horizontal roots growing radially away from the taproot. About 80 percent of the roots of hybrid poplar *Populus trichocarpa* × *Populus deltoides* exhibit angles relative to the horizontal plane between 0° and 30°. Horizontal roots could be found at depths of 5 to 20 centimeters (2 to 8 inches), several tree lengths away from the base of the stem (Pregitzer and Friend, 1996). In addition to the horizontal roots, vertical or sinker roots branching from the horizontal roots were found to explore the soil below 1 meter (3 feet) depths. Roots of cottonwoods (*Populus* spp.) have been found in natural ecosystems where groundwater, the only source of water, was 9 meters (27 feet) deep below surface (Robinson 1958 in Erdman and Christenson 2000). Deep sinker roots seem to be common within the genus *Populus*. Fine roots of the genus *Populus* have been found to be long, thin, and relatively unbranched with very high specific root length (about 50 meters per gram), and often associated with both ectomychorrizal and, preferentially, vesicular–arbuscular mychorrhizal fungi (Pregitzer and Friend 1996). Root growth of the genus *Populus* is very rapid (over 10 millimeters or 0.4 inches per day), under high moisture and nutrient conditions. Allocation of carbon to root growth in the genus *Populus* seems to be genetically determined; however, it appears that inherent root-growth plasticity and, therefore, adaptability to the environment may be an important attribute of fast-growing genotypes (Pregitzer and Friend 1996).

Increased rooting depths, a decreased shoot-to-root ratio, and a decrease in the specific leaf areas (to conserve water through smaller transpiration surfaces) were hypothetically caused by drought on Salicaceae. Increased root depth of black poplar (*Populus nigra*) in deeper soil layers after drought was considered a consequence of shortage of moisture or possibly a result of the removal of the root-inhibiting soil saturation (Van Splunder *et al.* 1996).

Phreatophytic shrubs grow in areas where groundwater is moderately deep [up to 11 meters (33 feet) and perhaps 18 meters (54 feet) for greasewood (*Sarcobatus vermiculatus*) with a requirement of at least 1 meter (3 feet) of unsaturated soil, most of the time]. Sagebrush (*Artemisia* spp.) has taproots as deep as 4 meters (12 feet) and usually penetrates the capillary fringe above the saturation zone (Nichols 1994).

While the flexibility of the dimorphic root system is highly beneficial in natural conditions, insofar as it allows the plant to maintain a viable root system and access to water year-round, it may not be ideal in phytoremediation applications where the main goal is to draw water only from a specific water source (typically the groundwater, or a particular water table in case of multiple aquifers or distinct horizons). This may be especially true for humid regions where abundant rainfall maintains moisture in the surface soil to provide the plant with water (thus, in fact, limiting deep root growth). To promote deep root growth in humid regions, intervention may be necessary to ensure that roots grow towards the target water source and circumvent easier-to-access, uncontaminated water sources. Discussions of these steps are in the rest of this section.

Water redistribution within the plant, the root system, and adjacent soil has also been the subject of a number of studies. Early research by Breazeale and Cryder (1934), cited by Caldwell *et al.* (1998), provided the initial proof of water redistribution by demonstrating that seedling roots of catclaw acacia (*Acacia greggii*) were able to grow from moist soil through an air gap isolated with a double layer of paraffin and beeswax into a very dry soil, which eventually became more moist concomitantly with root development. The main mechanism of water redistribution is called "hydraulic lift" and may have significant implications for phytoremediation. A schematic of the hydraulic lift process is depicted in Figure 7-1.

Hydraulic lift is the process of water movement from relatively moist, deeper soil to shallower, dry soil layers using plant root systems as a conduit. Water released into these shallow soil layers from roots during periods when

Figure 7-1 Schematic of the process of hydraulic lift. The roots in drier surface soil release water absorbed at depth.

transpiration ceases is then absorbed the next day and transpired (Richards and Caldwell 1987). Described in detail by Caldwell *et al.* (1998), hydraulic lifting occurs in many plant species (including the genus *Populus*), some of which use this mechanism to displace significant amounts of water daily. These authors hypothesize that the vertical, upwards transfer of water is the most evident phenomenon, but that the same mechanism could be used by plants to move water downwards or laterally if the principal driver of this displacement were different water potentials. Burgess *et al.* (1998) confirmed that roots of the phreatophytes silky oak (*Grevillea robusta*) and river red gum (*Eucalyptus camaldulensis*) were able to transport water from wetter surface layers to the dry subsoil in a reverse hydraulic lift that better fits the definition of "hydraulic redistribution." Hydraulic redistribution could explain, at least in part, how plant roots succeed in perforating often thick dry soil strata (Fabiao *et al.* 1985, Glenn and Welker 1993), searching for more reliable moisture sources in distant water bearing units. It is important to note that the deep-reaching root system need not be extensively ramified or developed to ensure a reliable water supply; often only a thin, deep root is necessary to provide the plant with an adequate water supply. Not fully understood, hydraulic lift is mainly attributed to reverse flow or water leaking from root tissue at elevated xylem water potentials (such as those at night when transpiration declines) into soil at lower water potential. Apoplastic gaps and aquaporins (water channels in the symplast) may contribute to the flow of water from the root to the drier, surrounding soil (Caldwell *et al.* 1998).

Hydraulic redistribution alleviated the effect of water deficits nearer the root base, as deeper roots continued to absorb and transport water at night. During the day, uptake of this water by lateral shallow roots supplemented water uptake by deep roots (Caldwell and Richards 1989). When hydraulic lift was prevented, these authors found from 25 to 52 percent reduction of the transpiration expected in the presence of hydraulic lift.

Hydraulic lift does not necessarily occur only in arid or semiarid environments: the importance in mesic environments subjected to periodic soil water deficits is discussed by Dawson (1993). Postulated (but not sufficiently investigated) effects and evolutionary reasons for hydraulic lift are improved rhizosphere processes, especially activity and life span of fine roots and associated microorganisms. One particular effect is the increased dissolution and mobilization of nutrients (and possibly contaminants) from dry soil for plant and microbial utilization when the water is released (Caldwell *et al.* 1998).

While not completely understood, hydraulic lift has potential implications for phytoremediation. In case of soil cleanup in dry climates where all of the soil depth is the target, hydraulic lift may prove to be beneficial for contaminant and nutrient mobilization and transport, and for spurring rhizosphere degradation. For the remediation of deeper, moist soil or groundwater, the hydraulic redistribution or hydraulic lift of water may be inhibited altogether when surficial roots are prevented from growing. If surficial roots are present, hydraulic lift may be neutral or beneficial to the desired outcome, depending on the planting method. Water redistributed into surficial dry zones from deep subsoil may be "wasted" for the purpose of mobilizing nutrients when a medium is used near the surface that does not support roots, but may still contribute to improved total transpiration rates as water lifted to the surface is further transpired by surface vegetation or surficial roots.

Hydraulic redistribution may also be a mechanism by which selected plants may transfer "survival" water at depths to promote additional root development in the dry subsoil to reach the water-bearing units. However, a clear, quantitative understanding of the effectiveness of this mechanism in ensuring that deep rooting occurs compatible with the desired remedial action is currently not available.

Nutrients

The basic mechanisms that cause nutrients and roots to come into contact include nutrient transport within the soil, and root development and soil exploration. As a first consideration, uptake by a plant can only occur if the mineral nutrient is dissolved in the soil solution or groundwater. Dissolution rates and speciation depend on soil conditions.

Nutrient transport within the soil occurs by mass flow and diffusion. The advective transport of the nutrients dissolved in the soil solution is highly determined by evapotranspiration rates. As plant transpiration and evaporation from the soil surface displace water, a localized flow is created from the soil towards the root or upwards towards the soil surface due to capillary rise. Capillary rise occurs when surface tension causes a negative pressure of suction of water into dry soil above the water table. The total flux of nutrient will be dependent on the rate of water movement and nutrient concentration. Diffusion, the spontaneous thermodynamic movement of molecules along concentration gradients, is caused by the natural heterogeneity in the soil (presence of localized patches of decaying organic matter and localized fertilization) and by the relative depletion or enrichment of nutrients caused by selective plant uptake. A summary of the processes involved in nutrient transfer from soil to plants is modified from Jungk (1996) in Table 7-1.

TABLE 7-1 Processes Involved in Nutrient Transfer from Soil to Plant

Process	Factors affecting process
Nutrient uptake	Concentrations at root surfaces Kinetics of uptake
Transport from soil to root → Advection, diffusion	Concentration of soil solution Concentration gradients Diffusion coefficients
Mobilization by roots → Desorption, dissolution → Hydrolysis of organic compounds	Depletion of soil solutions Root die back Root exudates (H^+, HCO_3, reducing agents, chelating agents, and organic anions) Soil chemical composition pH of soil solutions Enzymes
Mobilization by associated organisms	Mycorrhizal infection Bacteria

Note: Modified from Jungk 1996.

In general, bioavailable nutrients are most abundant in the surface soil horizons and are typically scarcer below the average rooting zone. In unfertilized soils, bioavailable nutrients are normally present in patches that coincide with decaying organic matter; these patches contribute significantly to soil heterogeneity and may be as large as bodies of dead animals, layers of leaf litter, or deposits of animal feces, or as small as dead bacteria or fungi (Tibbett 2000). Plant roots need to access these patches in competing with soil microorganisms for the acquisition of nutrients, especially nitrogen.

From the many schools of thought in this field, a guiding principle is that root branching and growth follow the principle of optimization (or the avoidance of unnecessary resource allocations) for nutrient acquisition. Acquisition is based on genetic predisposition and plastic responses (adaptation) to nutrient levels and other soil conditions. Soil and plant reserves of energy, water, and nutrients are used to acquire additional resources. The responses are different for perennial and annual plants, and for enriched and poor soils. In addition, the potential for developing mycorrhizal or rhizobial associations is also integral to plant resource acquisition and therefore plays a role in root development. Root systems consisting of only a main axis and one order of laterals (herringbone topology) was considered by Fitter (1985) as the topology that is most effective at resource acquisition, and may be preferentially found in perennial plants and in conditions of low nutrient availability, in spite of requiring more resources. Annual plants and plants growing in nutrient-rich soils do not need an efficient system and may utilize less resource-intensive dichotomous topologies. Fitter (1985) also predicted, and Taub and Goldberg (1996) confirmed, that root systems of a given species should become more herringbone-like as soil resources become less available (plastic response).

Nutrient levels in the soil can also directly influence root morphology and physiology. Longer, thinner roots (having a higher specific root length, or length of roots for a specific mass in meters per milligram) occur in young root systems and in low-nutrient soils (Fitter 1985). In conditions of phosphorus deficiency, rice (*Oryza sativa*) plants growing in reduced soils were able to increase phosphorus availability indirectly by secreting oxygen (which oxidizes iron and acidifies the soil solution), and directly by releasing hydrogen ions from the root (Kirk and Le Van Du 1997). While the phosphorus deficiency reduced plant dry mass by 50 percent, root mass was increased approximately 2-fold. During phosphorus deficiency, the fine root portion of root surface area increased 10 percent, while root porosity increased 25 to 40 percent, supposedly to improve oxygen transport.

Roots colonize well-fertilized soil micro sites relatively quickly. Sword (1998) showed that fertilization stimulated the initiation of branched lateral roots in loblolly pine (*Pinus taeda*). However, various studies found that roots are poor competitors for nitrogen in the presence of soil microbial populations, and usually respond quickly only to large zones enriched with mineralized nutrients. Experiments conducted by Liu and Dickmann (1992) on

Populus hybrids suggest that under uniform enrichment of soil nitrogen, more fine roots grow to fully exploit available soil resources. Roots, however, may not have the morphological plasticity to respond to small-scale nutrient patches, or to nutrients still bound into complex organic matter. These nutrients are exploited quickly and competitively by root-associated mychorrizal fungi (Tibbett 2000) through the production of appropriate exo-enzymes. Mychorrizal associations are of special importance for the acquisition of nutrients of limited mobility, such as phosphate.

Little research has been conducted on nutrient acquisition of deep-rooted poplars (*Populus* spp.) or willows (*Salix* spp.), and what is available clearly shows that nutrient acquisition by naturally growing trees is mostly a function of surface roots and the mychorrhizal associations. Mychorrhizal fungi associated with roots are also found at depths; however, under typical circumstances trees are unlikely to derive a significant nutritional advantage from direct or mychorrhizal nutrient acquisition at depth because deeper soil generally has lower nutrient concentrations than near-surface soils. This has fundamental implications for phytoremediation installations (compared to natural ecosystems). Low nutrient levels will slow down the overall growth rates, including root and shoot growth. Unless proper levels of fertilizer are supplied to the subsoil where rooting is sought, the poorer or unbalanced nutritional status of the deeper horizons could limit deeper root proliferation. Depending on the modality, fertilization of the deeper subsoil may favor direct or mychorrizal-mediated nutrient uptake.

Despite the limited knowledge, fertilization is a vital approach in phytoremedition. Localized fertilization is a flexible tool to achieve growth rates that are acceptable and to obtain rooting directions oriented strategically where needed to stimulate root growth and to avoid rooting in uncontaminated areas.

Air

Oxygen is an essential element for roots, as a significant portion of the carbohydrates produced in photosynthesis is respired in the roots. Rates of root respiration are species-specific (Lambers *et al.* 1996). Respiration rates are higher in root areas where significant biosynthesis occurs, consistent with increasing demands for adenosine triphosphate (ATP). Active ion uptake is among the many biological functions of roots that are highly demanding of respiratory energy.

Root respiration rates are affected by soil pH, temperature, and most substantially by oxygen levels in the root and surrounding soil solution, which in turn is a function of soil porosity and aeration. Deeper soil horizons are typically more deprived of oxygen because of soil pedogenic removal of oxygen and organic matter, leaching of reduced products, soil compaction and overburden, and especially more limited and difficult gas exchange with the atmosphere. Soil texture and structure, and flooding events (including a fluctuating water table) are the main regulators of the oxygen levels in the soil solution. Roots and soil microorganisms competitively use oxygen. Depending on the initial levels, temperature, and root and microbial biomass, dissolved oxygen in the soil solution can be depleted very quickly. However, a negative impact to root functions and growth may occur well before the point of anaerobiosis or anoxia.

Plants vary in adaptability to reduced root oxygen levels. The most adapted plants are those typical of wetlands that have developed morphological and biochemical adaptation mechanisms to cope with periods of flooding. Morphological adaptations may be constitutional in angiosperm (seed plant) roots growing in constantly wet environments (constitutive aerenchyma, or air-conducting tissue), or adaptive, when flooding is periodical. Adaptive aerenchyma may develop from plant cell death and dissolution (lysigenous aerenchyma) or by separation of cells during development (schyzogenous aèrenchyma). Other morphological changes occur in wetland plants, such as increases in suberization, cuticularization, and lignification of the exodermis walls (cork, wax, or callus formation) to prevent radial losses of oxygen from the roots to the soil (Clark and Harris 1981). Biochemical adaptations also occur in plants as an adaptive mechanism to oxygen deficiency. Altered biochemical pathways in many wetland species provide sufficient energy by using alternate terminal electron acceptors.

A number of phreatophytes appears to be able to grow under partly anaerobic conditions (because of the depth of rooting and fluctuating water table levels), which may mean that these plants can provide the associated roots with the oxygen necessary for survival by developing aerenchyma. Flooded roots of willow (*Salix* spp.) (contrary to unsaturated controls) showed lysigenous aerenchyma formation by death of a major portion of the root cortical cells (Kawase and Whitmoyer 1980).

Very limited information is available on whether roots can actively develop into the saturated zone. Optimization principles would suggest that, although occasional or even prolonged submersion by water could be accepted and tolerated by plant roots, actual growth into saturated zones might be too resource-intensive for the plant while providing little advantage in terms of nutrient acquisition and respiration. Water is usually freely available at the capillary fringe with no need for the root to actively develop in the reduced saturated zone. Van Splunder *et al.* (1996) found willows (*Salix* spp.) growing deep in saturated soil, whereas black poplar (*Populus nigra*) roots remained above the water table. Segelquist 1993 (cited in Van Splunder *et al.* 1996) also reported a diminished root growth and enhanced root die-off of black poplar (*Populus nigra*) under saturated conditions.

Root growth of terrestrial plants is undoubtedly faster and more abundant in soil areas of higher oxygen. Because of shorter pathways for air exchange, even in potted plants, root growth on pot sides is usually noticeable and more extensive than in the center of the pot. To aid in the development of deep rooted phytoremediation trees, aeration tubing supplied to growing roots have the purpose of favoring preferential diffusion of air from the ground surface to feed the deep root system. While trees can survive without these tubes, it is usually accepted that aeration tubes are an influential factor in increasing rooting rates.

Soil Physical Conditions

Soil physical conditions are a significant component in determining root growth, direction, and morphology. These effects are direct (soil mechanical impedance) and indirect as soil physical conditions such as bulk density dramatically influence soil capacity to contain air and water. The mechanical strength of a soil is the resistance force that allows the roots to provide the necessary support to the plant. Soil mechanical strength also significantly influences root growth. Roots tend to grow in areas of least resistance. When the bulk density of the soil is low, such as when the soil is uncompacted and highly structured, roots will develop around particles with limited effort, and the architecture of the root system will be primarily a function of genetic predisposition and of soil chemical conditions (*e.g.*, nutrients, water, air, and minor factors). As bulk density increases, roots will seek the existing openings (*i.e.*, pores, fractures, and biopores created by soil fauna and dead roots) as preferential root growth directions.

There is little information available on the minimum pore size to allow for root penetration, but 0.2 millimeters was found to be the minimum width necessary for the roots of grasses of temperate climates (Wiersum 1957). Russell (1973) emphasized that, depending on soil water potential, pore spaces large enough to allow for quick water drainage may not be large enough for root growth to occur without significant pressures being applied by the tip. The same author reported that roots could rarely grow in sandy, light soils with bulk density higher than 1.8 grams per cubic meter, and in heavier clayey or silty soils having a bulk density higher than 1.6 grams per cubic meter. Texture plays a significant role in determining mechanical impedance as different particle size components of the soil exhibit different interparticle binding forces and friction. At a specific bulk density, mechanical impedance is also inversely correlated with soil water content.

In the presence of a highly compacted substrate, elongating roots exert high pressure on the soil particles in contact with the root tip. It was clear since 1893 when Pfeiffer conducted his experiments (cited in Waisel *et al.* 1996) that root elongation is only possible when the mechanical impedance of the soil acting against the cross-section of the root is less than the mechanical pressure exerted by the root itself (Gill and Bolt 1955, cited in Waisel *et al.* 1996).

Maximum root growth pressures vary, with axial root pressure (caused by the advancing root cap) being usually higher than radial ones (caused by the radially expanding root behind the root cap). Highest mean values reviewed by Misra *et al.* (1986) were around 1.4 megapascals with maximum values reaching 2.65 megapascals for garden pea (*Pisum sativum*) roots (axial). Experimental determinations of root pressure are dependent on the method used. Values that correlate with a significant decrease in elongation rates have been reported between 0.05 megapascals and 0.5 megapascals (Russell and Goss 1974 cited in

Richards and Greacen 1986, Bengough and Mullins 1990, Bennie 1996). The model described by Richards and Greacen (1986) predicted lower mechanical impedances for finer roots, and suggested that radial expansion of the root behind the tip can effectively relieve soil pressure ahead of the elongating tip.

Russell and Goss (1974) [cited in Bennie (1996)] also noted that roots growing under high external pressures in a pressurized artificial medium exhibited clear morphological differences compared to roots that were not subjected to the same pressure. While the total root volume remained similar in these two cases, the pressure-exposed roots were considerably larger in diameter and had a higher number of lateral roots per unit of length of the primary root.

Mechanical impedance to root growth is predicted in practice by using a penetrometer, which measures the pressure that is necessary to advance a metal probe into a soil sample. Many different types of penetrometers and methods are available; however, not all are suitable to predict resistance to root growth. Constant rate penetrometers, which have been found to have the best correlation with root pressures, are static and have a blunt probe (30 ° angle) that exerts both axial and radial pressures to the soil (Whiteley *et al.* 1981). A selection of acceptable instruments includes the pocket penetrometer and the cone penetrometer. Detailed descriptions of instruments and acceptable methods are given in Bradford (1986).

Other factors that influence mechanical impedance are aggregate size and surface roughness, and soil macrostructure. As the soil is composed of many aggregate sizes and each one has a characteristic influence on the resistance to penetration, Grant *et al.* (1985) coupled micropenetrometer readings with spectral analysis (the analysis of the variance of penetrometer tip resistance spectra) in evaluating the influence of aggregate size in root resistance in more structured soils. These authors found that larger aggregates provided greater resistance to penetration and that by comparing variance spectra among several aggregate sizes, information on the structural makeup of the soil aggregates could be revealed. This approach was found useful in evaluating the path of least resistance between zones of high- and low-penetrometer resistance that roots may use for growth.

At or near the ground surface, the bulk density of a soil is dependent on the degree of compression, which can be affected by external factors mostly represented by vehicular or animal traffic over the soil (Grimes *et al.* 1978). In cultivated fields, the greatest cause for compaction and reduced root penetration is typically the repeated tramping of the soil with farm implements, especially if conducted when the soil is too wet. At training fields, repeated traffic with military equipment is the predominant cause of ecological disruption and prevention of root establishment.

Deeper in the subsoil, mechanical impedance of roots and water can be naturally high as a consequence of pedogenic features such as hardpans or horizons of sharply contrasting texture and structure. At those depths, measurements of mechanical impedance by penetrometer are further complicated by overburden pressure, as a greater mass of overburden inhibits particle displacement at the tip of the probe. The overburden also restricts root penetration. Bradford *et al.* (1971) determined that penetrometer readings for deeper soil horizons are not comparable with others taken from surficial, unconfined soil horizons.

Stratigraphic Influence on Root Systems

Stratigraphy, or the natural distribution of soil strata, incorporates all soil and water distribution factors that influence root growth. The lateral and vertical extent of roots is restricted, for example, by fine-grained overconsolidated units and bedrock because of mechanical impedance. Soil moisture conditions and the depth to groundwater will also govern phreatophyte root distribution. While a wide range of subsurface conditions may be present in the immediate vicinity of a particular phreatophytic plant, several fundamental situations may be used to illustrate a variety of possible root growth scenarios. A phreatophyte growing above a surficial, unconfined aquifer may extend roots to the capillary fringe (Figure 7-2A). Silty or clayey lenses within the unsaturated and saturated zones may impede the downward extent of the roots. In a residual soil above fractured, weathered bedrock, roots may extend downward through fractures to seek out the saturated zone (Figure 7-2B). In glacial or lacustrine deposits of thick surficial clay unit, roots may be restricted to the weathered surficial soil under natural conditions (Figure 7-2C) because the plants are not able to penetrate a significant thickness of low-moisture, fine-grained sediments. In this situation, phytoremediation would not appear to be a likely remedial strategy for the underlying aquifer; however, several solutions (Figure 7-2D) may be implemented to force phreatophytic trees to rely on a deep contaminated aquifer as a source of water. The case studies that follow describe these approaches in more detail.

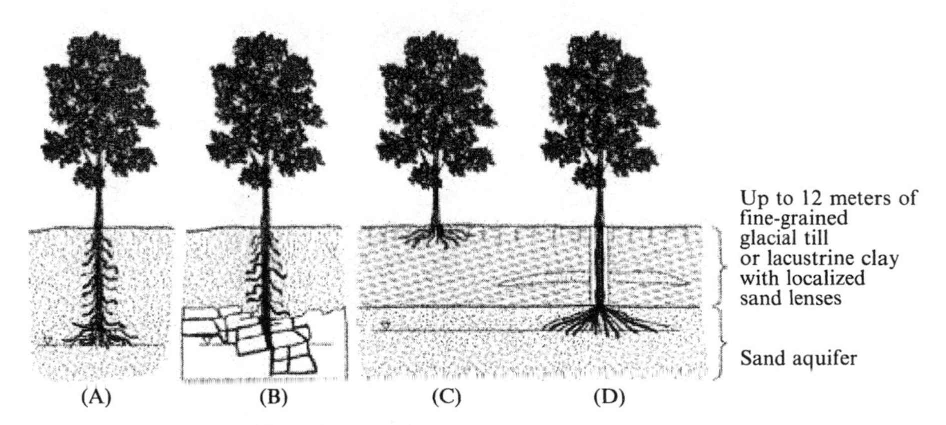

Figure 7-2 Examples of phreatophyte root development in different stratigraphic settings: (A) A sandy, unconfined surficial aquifer. (B) Residual soil developed over fractured bedrock. (C) Soil developed on top of a thick, fine-grained deposit. (D) Engineered deep planting designed to enable trees to reach underlying contaminated aquifer.

PHYTOREMEDIATION DESIGN AS A FUNCTION OF SITE CONDITIONS: CASE STUDIES

As discussed in the previous sections, root growth into a contamination zone will be dependent principally on the following:

1. Depth to contamination
2. Depth to water
3. Climate and precipitation patterns
4. Distribution of water bearing horizons through the soil profile and stratigraphy
5. Soil mechanical strength in different horizons
6. Presence of nutrients (usually limited to surface)
7. Availability of oxygen in groundwater or vadose zone

For several conditions, especially in drier climates, plant roots may, with time, reach the required depth without intervention. However, a number of approaches exist to accelerate or ensure vertical root development and optimal removal in relatively short times. Figures 7-3 and 7-4 provide an example of root growth rates at depth in the same amount of time for plants confined

This H.P. 510 poplar (*Popular spp.*) was planted in 1992 and harvested in 1994. The 1.5 meter (5-foot) casing forced roots to quickly develop down toward the capillary fringe of the aquifer.

An enhanced aeration study was attempted by forcing air down a PVC pipe with vent holes in the lower section. The "forced air tube" was installed to 3 meters (10 feet) below the ground surface.

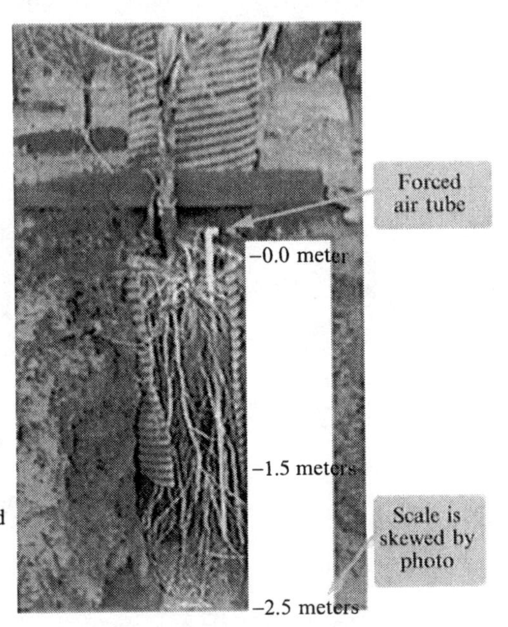

Figure 7-3 Two-year development of a cased root. (photograph courtesy of Applied Natural Sciences, Inc.). Note the aeration tube arising from the galvanized steel casing just to the right of the tree trunk.

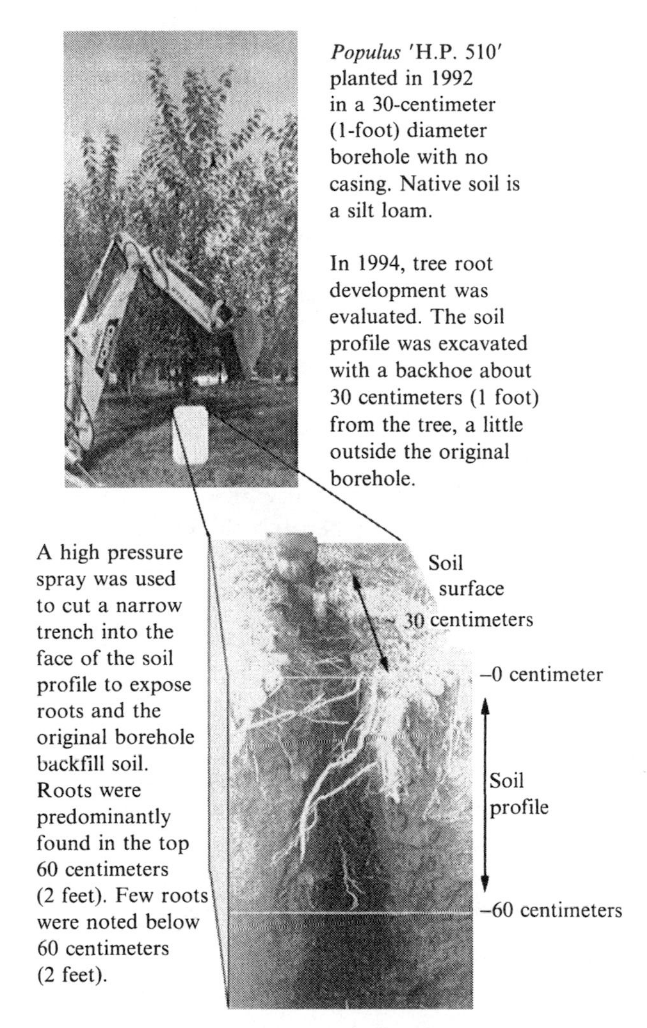

Figure 7-4 Two-year development of a root free of casing. Photograph courtesy of Applied Natural Sciences.

in a casing (Figure 7-3) and not cased (Figure 7-4). From an engineering standpoint, better certainty on where the roots are deriving water provides better quantification of performance.

In other more complex cases, the probability of roots naturally reaching the target depth decreases sharply because of conditions that either favor surficial rooting or prevent deep rooting. A few examples of these cases, described next, include one installation in the U.S. Midwest (Argonne National Laboratory, Illinois). The Argonne site has highly heterogeneous glacial terrain and temperate moist climate that favor a dimorphic root system. A clean, perched aquifer above the contaminated aquifer further prevents roots from growing at depths. Two other case study sites are located along the eastern U.S. coast (Aberdeen Proving Grounds, Maryland, and Staten Island, New York).

Argonne Site

Volatile organic compounds in soil and groundwater and low levels of tritium in the downgradient groundwater contaminate a former waste treatment area at Argonne National Laboratory. Phytoremediation was deployed to prevent groundwater migration and remove and degrade the volatile organic compounds and tritium within and downgradient of the source areas. The stratigraphy at this site consists of approximately 18 meters (54 feet) of glacial tills interlaced with sands, gravels, and silts of varying character, thickness, and lateral extent, over Silurian dolomite bedrock. The glacial sequence is composed of Lemont drift overlain by the Wadsworth Formation. Both units are dominated by fine-grained, low-permeability till.

Permeable zones of varying character and thickness are, however, present in each. These materials range from silty sands to sandy, clayey gravels to gravelly sands. In some locations, pure silt was encountered. If deep enough, this silt was saturated and thus assumed to play an important role in the flow of groundwater in the area. The permeable zones have a wide range of shapes; thicknesses range from less than 0.3 to 4.5 meters (approximately 1 to 15 feet) and have limited lateral extent. Based on this information and a preliminary agronomic assessment that determined soil physical and chemical conditions over the depth, golden weeping willow (*Salix alba* × *Salix babylonica*), hybrid prairie cascade willow [*Salix pentandra* × *Salix* × *blanda* (*Salix babylonica* × *Salix Fragilis*)], laurel-leaved willow (*Salix pentandra*), and hybrid poplar (*Populus trichocarpa* × *Populus deltoides*) trees were selected for planting. In the summer of 1999, a total of approximately 800 trees were planted in the following areas:

1. In the source area, to treat soil contamination
2. In the "hydraulic control area" downgradient from the source, to contain and treat groundwater that lies approximately 10 meters (30 feet) below ground surface

From a rooting design perspective, the source area is relatively uncomplicated. Homogenized by previous soil mixing, the structureless, hydrated soil had minimal mechanical strength. At this site, the target zone of rooting lies throughout the first 10 meters (30 feet) from the surface, with the main limitations being the saturated condition and the presence of volatile organic contaminants in varying concentrations. To overcome these unfavorable conditions, hybrid willow (*Salix* spp.) trees were planted in boreholes, free of any casing, for rooting throughout the soil profile, with an adequate supply of nutrients in the backfill material (the soil lacked chemical and biological fertility), and an aeration tube to provide supplemental aeration of the saturated substrate. Roots are expected to develop over a 10-meter (30-foot) depth with an herbaceous ground cover and gradually deplete water from the more surficial horizons. This dewatering has been observed at other sites where vegetation was used to remediate amorphous-hydrated soils. At this time, presence of the volatile organic compounds contaminants and the associated degradation product trichloroacetic acid is detectable in plant tissue, indicating that the roots are indeed growing in contact with the contaminated soil. Further study is ongoing by the authors of this chapter in determining root architecture and distribution in the soil, with particular focus on directional root growth as a function of local soil conditions and contaminant concentration. Although contaminant levels are, in general, well below what is generically considered a phytotoxic value, in conditions of soil heterogeneity roots may preferentially grow in less contaminated areas rather than in those of greater contamination.

The downgradient hydraulic control area, on the other hand, is a good example of conditions in which deep rooting may never occur spontaneously. The presence of a perched aquifer shallower than the target aquifer would offer a sufficient supply of water to plant roots, thereby eliminating the need to develop into the subsoil to the target depth. In this area, poplar (*Populus* spp.) trees were planted in boreholes drilled down to the contaminated aquifer using a patented technology. This technology was selected, in consideration of the hydrogeological setting of the site, to target root growth in the contaminated glacial drift permeable unit approximately 10 meter (30 foot) deep. The poplars (*Populus* spp.) were planted in 0.6-meter (2-foot) diameter caisson boreholes lined with plastic sleeves to act as a barrier to the shallow aquifer and to direct the roots exclusively to the deeper contaminated aquifer. These boreholes were filled with a mixture of topsoil, sand, peat, and manure to promote root growth and tree development. Aeration tubes were also provided to accelerate the rates of root growth at depth (Figure 7-5). Ongoing monitoring of volatile organic compounds and tritium in plant tissue and transpirate will indicate when the roots have reached the contaminated aquifer.

Aberdeen Proving Grounds Site

A system was installed to treat trichloroethene in groundwater that is 1.5 to 4.5 meters (5 to 15 feet) below ground surface and to prevent off site migration of contaminant plume into coastal wetlands. A site assessment was performed by collecting and evaluating soil and groundwater data and reviewing available site environmental data. These data show that a coarse sand horizon is located above the lowest seasonal water table level. The sand horizon drastically limits the capillary fringe. Boreholes were drilled through this horizon and backfilled to maintain consistent soil medium that would allow a gradual lowering of the capillary fringe as the water table subsided, thereby preventing the trees from experiencing a sharp break in water availability from the groundwater. Hybrid poplars (*Populus deltoides* × *Populus trichocarpa*) were planted in the cased boreholes to create a zone of influence that would limit

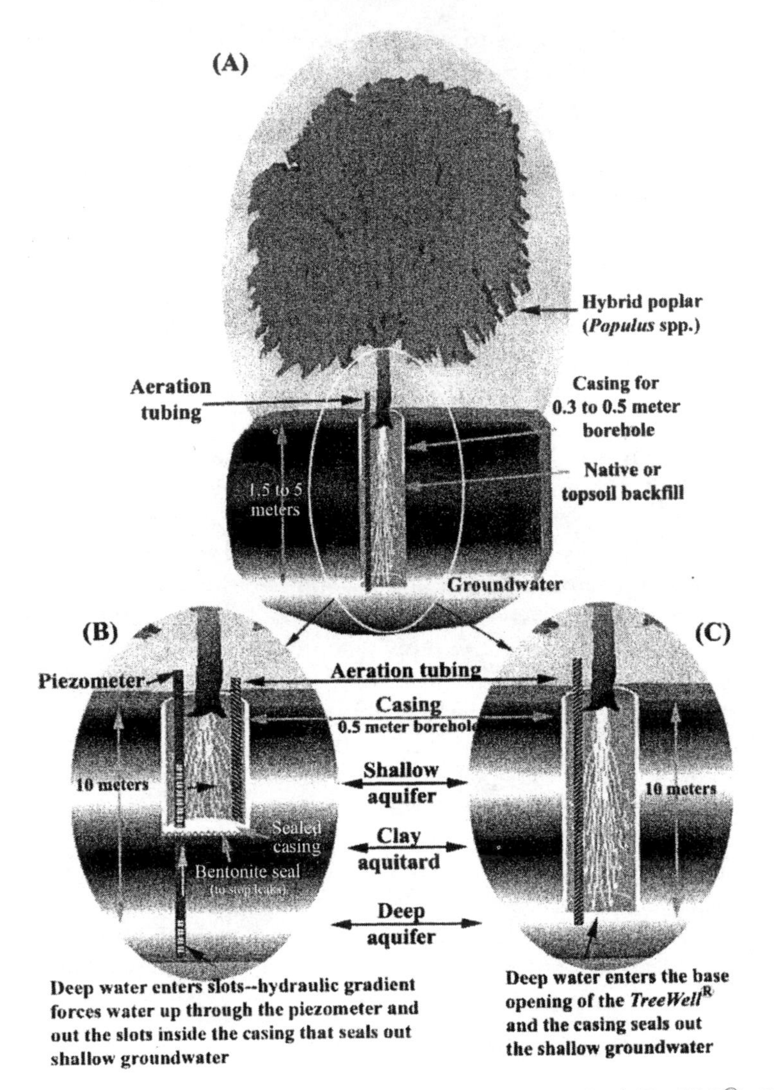

Figure 7-5 Installation diagrams for three case studies: (A) Staten Island (TreeWell®), (B) Aberdeen Proving Ground (TreeMediation®), and (C) Argonne National Laboratory hydraulic control area (TreeWell®). TreeMediation® and TreeWell are patented and trademarked by Applied Natural Sciences, Inc., 4129 Tonya Trail, Hamilton, Ohio.

the off site migration of the contaminated groundwater (Figure 7-5B). Due to constraints to site access, backfill soil consisted of the original drilling cuttings which have a somewhat plastic consistency when wet. Preferred sand was not added above the root crown to provide ballast and stability to the borehole planted trees. These limitations apparently contributed to an uncommon occurrence of tree lodging during a storm with hurricane force winds, a situation which was corrected after the fact.

Sampling of leaf and stem tissues has been performed along with evaluation of upgradient and downgradient groundwater. This project is described in detail by Hirsch *et al.* (this book).

Staten Island, New York Site

In the downgradient perimeter of a landfill, ammonium ions and heavy metals were found in two aquifers where the shallow groundwater is 2.4 to 4.5 meters (8 to 15 feet) below the ground surface and the deep groundwater is 10 to 11 meters (30 to 33 feet) below the ground surface. The objective of the phytoremediation installation was the prevention of off-site migration of contaminants into coastal wetlands.

A site assessment was performed by collecting and evaluating soil and groundwater data and reviewing available site environmental data. These data indicated that the two hydraulically separate aquifers have similar hydrostatic head and flow, but are separated by approximately 5 meters (15 feet) of dense clay subsoil. Over 500 trees were installed to hydraulically intercept and contain the groundwater flow of both aquifers (Figure 7-5A). To remediate the shallow aquifer, boreholes were drilled through porous fill material making up the vadose zone above the shallow aquifer. These 30-centimeter (12-inch) diameter boreholes were cased in the upper 1.5 meters (5 feet) and backfilled with a better quality growth medium. In this situation, casing hastened the root growth to the groundwater but was not designed to totally stop shallower root development over time due to the porous nature of the vadose soil.

To remediate the deeper artesian aquifer, 60-centimeter (24-inch) diameter boreholes were drilled into the upper layer of the dense clay horizon about 5 meters (15 feet) below ground surface. A bottom-sealed bag (to stop infiltration of the shallow aquifer water) was placed in each hole and an aeration tube was placed in the bag, which was then backfilled with high quality topsoil. A hybrid poplar (*Populus* 'H.P. 308 × H.P. 510') tree was planted in each bag, sand was added to provide long-term ballast, and the bag was then closed at the top to prevent rainwater infiltration. A tube of PVC was installed inside the bag by making a 3.7-centimeter (1.5-inch) diameter borehole through the bag, the dense clay below, and into the deep aquifer. The piezometer tube was configured to allow water to enter the bottom of the borehole from the deep aquifer and flow up into the sealed bag for use by the tree. The planting configuration was designed to intercept and control the flow of each aquifer.

To date, water level measurements have shown a distinct diurnal fluctuation of the groundwater in response to the water use by the trees. Since planting in mid-summer (June and July) 1998, both sets of trees have grown 2 to 2.5 meters (6 to 8 feet) per year, which indicates aggressive growth and water use.

METHODS FOR MONITORING ROOT DEVELOPMENT

The study of root distribution is approximately 250 years old. Historical improvements occurred in the 19th century, but formalization of root excavation techniques can be dated to the early 1900s, and tracer techniques were developed in the 1950s (Canadell *et al.* 1996). Recent advances in digitalization, mapping, and image processing have dramatically improved techniques necessary to study root development.

Methods for examining root systems vary in complexity, precision, and accuracy, depending on the size of the root system being examined and whether the study is being conducted *in situ* or *ex situ*. The biggest challenges obviously occur when studying roots nondestructively in natural conditions. A very exhaustive but somewhat dated review of methods for studying root systems is given in Bohm (1979). A more recent and comprehensive review of root methods was edited by Smit *et al.* (2000).

Excavation, augering, and the preparation of a profile wall or a monolith are commonly adopted methods for studying roots under field conditions. Because these are destructive and drastic methods, the temporal progressions of root growth cannot be determined. Furthermore, these methods can only be used to outline root architecture unless precautions are taken to ensure that the spatial distribution of roots is not altered by the removal of the surrounding soil. These methods, however, have the advantage of allowing for the retrieval of root samples.

Common methods for the separation of roots from soil involve rinsing procedures where water is either sprayed on the roots, or roots are immersed in a water-filled container and floated or sieved. Smucker *et al.* (1982) developed a root washer-elutriator in which roots are freed from soil particles by the kinetic energy of pressurized spray jets and air flotation. While this equipment is considered an improvement to previous methods (Box 1996), several problems connected with root washing are still unresolved. The problems are mainly related to the accurate recovery of roots, loss of cell components, and presence of soil still adhering to the root surface. These problems may or may not constitute a limitation, depending on the type of study to be conducted.

Nondestructive, alternative methods include the study of roots through long used rhizotrons and minirhizotrons. Rhizotrons are glass-walled structures through which roots can be observed in soil. These containers are used mostly for experimental purposes and usually not considered for the study of roots growing under field conditions. Minirhizotrons consist of clear, transparent tubes, usually approximately 5 centimeters (2 inches) in diameter that are inserted into the ground, usually at an angle. A camera or video recorder is then inserted into the tube to observe root development at different times. Minirhizotrons, relatively inexpensive to install, can be utilized more easily under field conditions if properly installed. Careful installation and use includes the following:

1. The proper contact between the tube and the soil so that the clear exterior surface is not smeared or scratched during insertion (and thus made opaque)

2. Proper sealing so that light and air do not diffuse to depths that alter rooting patterns
3. The image recording device used is sensitive enough to capture the root size of interest

The use of minirhizotrons and most other root analysis techniques have been revitalized with electronic imaging techniques and powerful software to analyze the digital information. Older, time-consuming techniques relying on manual measurement and estimation of root mass, length, diameter, and other parameters, have been replaced during the last ten years by video image analyzers, scanners, and digital video cameras, and by software specifically designed for this purpose. Root count per unit area of the minirhizotrons is considered by Box (1996) as the quickest and easiest video-image data to acquire.

A number of software products have been developed in recent years (*e.g.*, Rhizoscan, WinRhizo, AMAP mod, and others). Rhizoscan is an example of these products that was used by Berzin *et al.* (2000) to measure total root length (millimeters), number of tips, root volume (cubic millimeters), fresh and dry biomass, and pigment concentration (by analyzing hue-saturation values) in hairy root cultures of beet (*Beta vulgaris*). The root cultures were scanned and bitmap images were produced for analysis with the Rhizoscan software. While this technique was not used in the field, the Rhizoscan software may be amenable to use with video-recorded images collected from specific "stations" or locations along the minirhizotron walls. Box (1996) highlights the need for careful, *ex situ* calibration of image analysis techniques adopted in conjunction with minirhizotrons. Calibration means understanding how a system is able to discriminate between a root, organic residues, and other soil components, and to document the accuracy and precision of measurements of root length, diameter, and branching. Some of these calibration protocols are available from Smucker (1991).

Danjon *et al.* (1999) were able to measure above and below ground architecture and biomass by using a three-dimensional, low magnetic field digitizing device. Five-year old pine (*Pinus pinaster*) trees were previously uprooted with a lumbering crane. Output data were analyzed using software (AMAP mod) specifically designed by Godin *et al.* (1997).

GROUNDWATER MODELING APPLIED TO PHYTOREMEDIATION DESIGN

Phytoremedial designs for groundwater may have goals of destroying contaminants and providing hydraulic containment. In either case, an important aspect of the planning is the evaluation of the relationship among the expected water use of the phreatophytes, the planting density, the varying seasonal hydrologic conditions (*e.g.*, recharge, potentiometric levels, and discharge), and the stratigraphic complexity of the subsurface. Numerical groundwater modeling is well suited for addressing these interactions and providing a basis for designing the plantation because these models can account for transient groundwater flux and transpiration, three-dimensional subsurface features, and various hydraulic boundary conditions.

One numerical code useful in groundwater phytoremediation evaluation and design is the finite difference program MODFLOW (McDonald and Harbaugh 1988) from the U.S. Geological Survey. The code has been used to evaluate hydraulic containment by eastern cottonwoods (*Populus deltoides*) and hybrid poplar (*Populus deltoides* × *Populus trichocarpa*) planted surficially (Eberts *et al.* 2000, Eberts *et al.* this book, Hirsh *et al.* this book, Weaver *et al.* this book) and hybrid poplars (*Populus trichocarpa* × *Populus deltoides*) in engineered deep plantings (Quinn *et al.* 2001).

An important aspect of evaluating a phytoremedial plantation is a conceptual understanding of the depth of root penetration and the lateral extent of the root system. As previously described, this information is generally not known with precision. However, continued work in the observation of root distribution should provide a better insight on the most accurate approach to incorporating phreatophytic water use in numerical models. Currently, a key factor in the use of MODFLOW for evaluating the transpiration by phreatophytes is the depth that roots are assumed to penetrate within each model cell containing a phreatophyte. The evapotranspiration package of the code allows the user-specified maximum transpiration rate to take place in a model cell when the hydraulic head is at or above the top of the aquifer. Transpiration decreases linearly with decreasing head to an assigned cutoff depth within the cell. By improving the understanding of how roots interact with the aquifer media and the fluctuating water table, incorporating water use calculations within existing or future codes will be more accurate.

Acknowledgments

The submitted manuscript has been created by the University of Chicago as Operator of Argonne National Laboratory ("Argonne") under Contract No. W-31-109-Eng-38 with the U.S. Department of Energy. The U.S. Government retains for itself, and others acting on behalf, a paid-up, nonexclusive, irrevocable worldwide license and said article to reproduce, prepare derivative works, distribute copies to the public, and perform publicly and display publicly, by or on behalf of the Government.

REFERENCES

Bengough, A.G. and C.E. Mullins (1990) The resistance experienced by roots growing in a pressurized cell, a reappraisal. *Plant Soil* **123**: 73–82.

Bennie, A.T.P. (1996) Growth and mechanical impedance. In: *Plant Roots—The Hidden Half*. Y. Waisel, A. Eschel, and U. Kafkafi, eds. Marcel Dekker, New York, pp. 453–470.

Berzin, I., B. Cohen, D. Mills, I. Dinstein, and J.C. Merchuk (2000) RhizoScan: a semiautomatic image processing system for characterization of the morphology and secondary metabolite concentration in hairy root cultures. *Biotechnol. Bioeng.* **70**: 17–24.

Bohm, W. (1979) *Methods for Studying Root Systems.* Springer-Verlag, New York.

Box, J.E., Jr. (1996) Modern methods for root investigation. In: *Plant Roots—The Hidden Half.* Y. Waisel, A. Eschel, and U. Kafkafi, eds. Marcel Dekker, New York, pp. 193–239.

Bradford, J.M. (1986) Penetrability. In: *Methods of Soil Analysis, Part 1, Physical and Mineralogical Methods.* A. Klute, ed., 2nd ed. Soil Science Society of America Book Series 5, Madison, Wisconsin.

Bradford, J.M., D.A. Farrell, and W.E. Larson (1971) Effect of soil overburden pressure on penetration of fine metal probes. *Soil Sci. Soc. Am. Proc.* **35**: 12–15.

Burgess, S.S.O., M.A. Adams, N.C. Turner, and C.K. Ong (1998) The redistribution of soil water by tree root systems. *Oecologia* **115**(3): 306–311.

Busch, D.E., N.L. Ingraham, and S.D. Smith (1992) Water uptake in woody riparian phreatophytes of the southwestern United States: a stable isotope study. *Ecol. Appl.* **2**(4): 450–459.

Caldwell, M.M. and J.H. Richards (1989) Hydraulic lift: water efflux from upper roots improves effectiveness of water uptake by deep roots. *Oecologia* **79**: 1–5.

Caldwell, M.M., T.E. Dawson, and J.H. Richards (1998) Hydraulic lift: consequences of water efflux from the roots of plants. *Oecologia* **113**: 151–161.

Canadell, J., R.B. Jackson, J.R. Ehleringer, H.A. Mooney, O.E. Sala, and E.D. Schulze (1996) Maximum rooting depth of vegetation types at the global scale. *Oecologia* **108**: 583–595.

Charlton, W.A. (1996) Lateral root initiation. In: *Plant Roots—The Hidden Half.* Y. Waisel, A. Eschel, and U. Kafkafi, eds. Marcel Dekker, New York.

Clark, L.H. and W.M. Harris (1981) Observation on the root anatomy of rice. *Am. J. Bot.* **68**: 154–161.

Danjon, F., D. Bert, C. Godin, and P. Trichet (1999) Structural root architecture of 5-year-old *Pinus pinaster* measured by 3D digitizing and analyzed with AMAPmod. *Plant Soil* **217**: 49–63.

Dawson, T.E. (1993) Hydraulic lift and water use by plants: implications for water balance, performance, and plant-plant interactions. *Oecologia* **95**: 565–574.

Dawson, T.E. (1996) Determining water use by trees and forests from isotopic, energy balance, and transpiration analyses: the roles of tree age and hydraulic lift. *Tree Physiol.* **16**: 263–272.

Dawson, T.E. and J.S. Pate (1996) Seasonal water uptake and movement in root systems of Australian phraeatophytic plants of dimorphic root morphology: a stable isotope investigation. *Oecologia* **107**: 13–20.

Eberts, S.M., C.W. Schalk, J. Vose, and G.J. Harvey (2000) Hydrologic effects of cottonwood trees on a shallow aquifer containing trichloroethene. *Hydrol. Sci. Technol.* **15**(1–4): 115–121.

Ehleringer, J.R., S.L. Phillips, W.F.S. Schuster, and D.R. Sandquist (1991) Differential utilization of summer rains by desert plants. *Oecologia* **88**: 430–434.

Erdman, J.A. and S. Christenson (2000) Elements in cottonwood trees as an indicator of groundwater contaminated by landfill leachate. *Ground Water Manag. Res.* Winter 2000: 120–126.

Fabiao, A., H.A. Persson, and E. Steen (1985) Growth dynamics of superficial roots in Portuguese plantations of *Eucalyptus globulus* labil. studied with a mesh bag technique. *Plant Soil* **83**: 233–242.

Fitter, A.H. (1985) Functional significance of root morphology and root system architecture. In: *Ecological Interactions in Soil: Plant, Microbes, and Animals.* A.H. Fitter, D. Atkinson, D.J. Read, and M.B. Usher, eds. Blackwell Scientific Publications, Oxford, United Kingdom.

Glenn, D.M. and W.V. Welker (1993) Water transfer diminishes root competition between peach and tall fescue. *J. Am. Soc. Hort. Sci.* **118**: 570–574.

Godin, C., E. Costes, and Y. Caraglio (1997) Exploring plant topological structure with the AMAP Mod software: an outline. *Silva Fenn.* **31**: 355–366.

Grant, C.D., B.D. Kay, P.H. Groenevelt, G.E. Kidd, and G.W. Thurtell (1985) Spectral analysis of micropenetrometer data to characterize soil structure. *Can. J. Soil Sci.* **65**: 789–804.

Grimes, D.W., W.R. Sheesley, and P.L. Wiley (1978) Alfalfa root development and shoot regrowth in compact soil of wheel traffic patterns. *Agron. J.* **70**: 955–958.

Hendrick, R.L. and K.S. Pregitzer (1996) Temporal and depth-related patterns of fine root dynamics in northern hardwood forests. *J. Ecol.* **84**: 167–176.

Jackson, R.B., J. Canadell, J.R. Ehleringer, I.F. Sala, and E.D. Schulze (1996) A global analysis of root distribution for terrestrial biomes. *Oecologia* **108**: 389–411.

Jungk, A.O. (1996) Dynamics of nutrient movement at the soil-root interface. In: *Plant Roots—The Hidden Half.* Y. Waisel, A. Eschel, and U. Kafkafi, eds. Marcel Dekker, New York.

Kawase, M. and R.E. Whitmoyer (1980) Aerenchyma development in waterlogged plants. *Am. J. Bot.* **67**: 18–22.

Kirk, G. and G.J.D. Le Van Du (1997) Changes in rice root architecture, porosity, and oxygen and proton release under phosphorus deficiency. *New Phytol.* **135**: 191–200.

Lambers, H., O.K. Atkin, and I. Scheurwater (1996) Respiratory patterns in roots in relation to their functioning. In: *Plant Roots—The Hidden Half.* Y. Waisel, A. Eschel, and U. Kafkafi, eds., Marcel Dekker, New York, pp. 323–362.

Liu, Z. and D.I. Dickmann (1992) Responses of two hybrid *Populus* clones to flooding, drought, and nitrogen availability. I. morphology and growth. *Can. J. Bot.* **70**: 2265–2270.

Lynch, J. (1995) Root architecture and plant productivity. *Plant Physiol.* **109**: 7–13.

McDonald, M.G. and A.W. Harbaugh (1988) A modular three-dimensional finite-difference groundwater flow model. *Techniques of Water Resources Investigations*, Book 6, Chapter A1. Reston, Virginia, U.S. Geological Survey, 528 pp.

Misra, R.K., A.R. Dexter, and A.M. Alston (1986) Maximum axial and radial growth pressures of plant roots. *Plant Soil* **95**: 315–326.

Nichols, W.D. (1994) Groundwater discharge by phreatophyte shrubs in the Great Basin as related to depth to groundwater. *Water Resources Res.* **30**: 3265–3274.

Nobel, P.S. (1996) Ecophysiology of roots of desert plants, with special emphasis on agoves and cacti. In: *Plant Roots—The Hidden Half.* Y. Waisel, A. Eschel, and U. Kafkafi, eds. Marcel Dekker, New York, pp. 823–844.

Nyer, E.K. and E.G. Gatliff (1996) Phytoremediation. *Ground Water Monit. Rem.* **16**(1): 58–62.

Pregitzer, K.S. and A.L. Friend (1996) The structure and function of *Populus* root systems. In: *Biology of Populus and its Implications for Management and Conservation*, Part II, Chapter 14. NRC Research Press, National Research Council of Canada, Ottawa, Ontario, Canada, pp. 331–354.

Quinn, J.J., M.C. Negri, R.R. Hinchman, L.M. Moos, J.B. Wozniak, and E.G. Gatliff (2001) Predicting the effect of deep-rooted hybrid poplars on the groundwater flow system at a phytoremediation site. *Int. J. Phytorem.* **3**(1): 41–60.

Richards, B.G. and E.L. Greacen (1986) Mechanical stresses on an expanding cylindrical root analogue in granular media. *Aust. J. Soil Res.* **24**: 393–404.

Richards, J.H. and M.M. Caldwell (1987) Hydraulic lift: substantial nocturnal water transport between soil layers by *Artemisia tridentata* roots. *Oecologia* **73**: 486–489.

Russell, E.W. (1973) *Soil Conditions and Plant Growth.* Longman Group, London.

Smit, A.L., A.G. Bengough, C. Engles, M. Van Noordwijk, S. Pellerin, and S.C. Van De Gieijn, eds. (2000) *Root Methods: A Handbook.* Springer-Verlag, Berlin-Heidelberg.

Smith, S.D., D.A. Devitt, A. Sala, J.R. Cleverly, and D.E. Busch (1998) Water relations of riparian plants from warm desert regions. *Wetlands* **18**(4): 687–696.

Smucker, A.J.M. (1991) Contemporary analytical methods for quantifying plant root dynamics. In: *Root Ecology and Its Practical Applications.* L. Kuchera, E. Hubbl, E. Lichtenegger, H. Persson, and M. Sobotik, eds. Verein Für Wurzelforschung, Klagenfurt, Austria, pp. 721–726.

Smucker, A.J.M., S.L. McBurney, and A.K. Srivastava (1982) Quantitative separation of roots from compacted soil profiles by hydropneumatic elutriation system. *Agron. J.* **74**: 500–503.

Sword, M.A. (1998) Seasonal development of loblolly pine lateral roots in response to stand density and fertilization. *Plant Soil* **200**: 21–25.

Taub, D.R. and D. Goldberg (1996) Root system topology of plants from habitats differing in soil resource availability. *Funct. Ecol.* **10**: 258–264.

Taylor, H.M. and B. Klepper (1978) The role of rooting characteristics in the supply of water to plants. In: *Advances in Agron.* N.C. Brady, ed. Academic Press, New York.

Tibbett, M. (2000) Roots, foraging, and the exploitation of soil nutrient patches. the role of mycorrhizal symbiosis. *Funct. Ecol. 2000* **14**: 397–399.

Van Splunder, I., L.A.C.J. Voesenek, H. Coops, X.J.A. De Vries, and C.W.P.M. Bloom (1996) Morphological responses of seedlings of four species of Salicaceae to drought. *Can. J. Bot.* **74**: 1988–1995.

Waisel, Y., A. Eschel, and U. Kafkafi, eds. (1996) *Plant Roots—The Hidden Half.* Marcel Dekker, New York.

Whiteley, G.M., J.S. Hewitt, and A.R. Dexter (1981) The buckling of plant roots. *Physiol. Plant* **54**: 333–342.

Wiersum, L.K. (1957) The relationship of the size and structural rigidity of pores to their penetration by roots. *Plant Soil* **9**: 75–85.

Zobel, R.W. (1996) Genetic control of root systems. In: *Plant Roots—The Hidden Half*. Y. Waisel, A. Eschel, and U. Kafkafi, eds. Marcel Dekker, New York, pp. 21–30.

8

MEASURING AND MODELING TREE AND STAND LEVEL TRANSPIRATION

J. M. Vose, G. J. Harvey, K. J. Elliott, and B. D. Clinton

SUMMARY OF PRACTICAL IMPLICATIONS

Transpiration is a key process in the application of phytoremediation to soil or groundwater pollutants. To be successful, vegetation must transpire enough water from the soil or groundwater to control or take up the contaminant. Transpiration is driven by a combination of abiotic (climate, soil water availability, and groundwater depth) and biotic (leaf area, stomatal functions, root amount and distribution, and hydraulic characteristics) that need to be evaluated when considering appropriate site and species combinations. The protocols are not trivial, but transpiration can be measured at a variety of scales using techniques such as direct measurements of sap flow on individual trees, eddy flux gradient analyses, or gauged watersheds. Alternatively, models can be used to estimate transpiration, but these usually require on-site calibration or parameterization to produce accurate predictions. Case study analyses across a range of site conditions and species indicate a maximum transpiration capacity of approximately 7.5×10^6 liters of water per hectare per year (8×10^5 gallons of water per acre per year), with a range of 1.5×10^6 to 7.5×10^6 liters per hectare per year (1.6×10^5 to 8×10^5 gallons per acre per year). Variation among sites is related to species, tree size, and stocking (*i.e.*, vegetation density) differences. Application of a physiologically based and site-specific parameterized model suggests reasonable agreement between measured and predicted transpiration estimates for the Air Force Plant 4 site in central Texas.

Phytoremediation: Transformation and Control of Contaminants,
Edited by Steven C. McCutcheon and Jerald L. Schnoor.
ISBN 0-471-39435-1 (cloth)

IMPORTANCE OF ACCURATE MEASUREMENTS OF TRANSPIRATION

Transpiration—the amount of water used by a tree or stand of trees—is one of the key processes in the application of phytoremediation of soil water or groundwater pollutants. To be successful, native or planted vegetation must transpire enough water from the soil or groundwater layer containing the pollutant to control the transport or decrease the mass of contaminant. Hence, quantifying current and future transpiration and determining the principal location of water uptake by native and planted vegetation on the site must be the evaluation criteria for applying phytoremediation. Quantifying transpiration requires a thorough and accurate assessment of water use patterns such as, transpiration rates, depth of soil water uptake, interactions with climate, and soil water availability. Measuring current transpiration or predicting future transpiration is not trivial. Because transpiration is an integrated response of the atmosphere–plant–soil continuum, measurements and predictions of transpiration capacity must account for (1) variation in climatic driving variables (*i.e.*, solar radiation, water vapor saturation deficit, precipitation, wind speed, and temperature), (2) structural and physiological (leaf stomatal function) characteristics of the vegetation (leaf surface area, and root area and extent), and (3) soil water dynamics (water-holding capacity, and permeability).

Evapotranspiration and transpiration are often used interchangeably, but these processes are different. Evapotranspiration includes the amount of water transpired by the vegetation, and losses due to evaporation of intercepted precipitation and soil surface evaporation. In forests, interception evaporation is a function of rainfall intensity and leaf and branch surface area, ranging from about 10 to 50 percent (Helvey 1971, Myers and Talsma 1992, Vose and Swank 1992). In closed canopied forests, soil evaporation is a minor component of the overall water budget (Vose and Swank 1992), but may become increasingly important in open stands. From a phytoremediation perspective, transpiration is the key factor to consider because interception evaporation does not involve soil water or groundwater.

The process of transpiration involves water movement through the soil, roots, stems, and leaves into the atmosphere in response to water potential gradients—always moving in the direction of smaller potential or negative gradients. Water potential is near zero when water is freely available and decreases to negative values when water becomes more limiting. The movement of water from the leaf interior to the atmosphere occurs through small openings in the leaf called stomata, which open and close in response to external (*e.g.*, climatic factors) and internal (*e.g.*, water potentials of leaves) driving variables. Species vary considerably in stomatal responses to these driving variables and provide opportunities for selecting species to optimize transpiration in different climatic environments.

Five methods are used to quantify transpiration: (1) precipitation minus runoff on gaged watersheds, (2) energy balance (*e.g.*, Penman-Monteith equation), (3) eddy covariance, (4) hydrologic models, and (5) direct sap-flow measurements. The first three methods are integrated estimates for the entire vegetation–soil complex and provide estimates of evapotranspiration not transpiration. Hence, those methods do not directly partition water losses based on transpiration *versus* evaporation and provide no information on the source of water (*i.e.*, shallow *versus* deep soil layers) for transpiration. Estimating transpiration with methods 1, 2, and 3 requires an independent analysis of the contribution of interception and soil surface evaporation. Hydrologic models vary considerably in complexity, ranging from very simple models [*e.g.*, Thornthwaite (1948) indices of potential evapotranspiration] to detailed physiologically based models that link vegetation, soils, and the atmosphere (Vose and Swank 1992). In contrast, sap-flow measurements provide a direct measure of transpiration (after correcting for time lags) under field conditions at the individual tree level (Hinckley *et al.* 1994, Martin *et al.* 1997, Vose *et al.* 2000). However, modeling or other scaling approaches are required to extrapolate tree-level measurements to the stand.

In summary, there are numerous approaches to quantifying transpiration in native or plantation-derived vegetative ecosystems. However, these methods vary considerably in accuracy, in data and measurement requirements, and in the capability to predict future transpiration rates as stands develop. In this chapter, we review approaches to quantifying forest transpiration from the leaf level to the stand and discuss the pros and cons of different approaches. We then provide applications of a subset of these approaches from phytoremediation case studies in Texas, Colorado, and Florida.

OVERVIEW OF CONTROLS ON TRANSPIRATION

Transpiration rates vary considerably among species and geographic regions (Figure 8-1). Which factors contribute to this variation? At large scales (*i.e.*, regions), climate is an overriding control. The strong relationship between evapotranspiration and precipitation (Figure 8-1) suggests that transpiration is principally limited by soil water supply. However, other climatic factors such as temperature, atmospheric vapor pressure deficit, and solar radiation also play important roles and interact with soil water availability and physiological status of the plants (Figure 8-2). For example, one of the key effects of temperature is through the influence on the length of growing season, in which longer periods with temperatures above freezing promote longer leaf area duration and hence, surface area available for transpiration. Frozen or cold soils also restrict transpiration (Fahey 1979) by limiting the permeability of cell membranes (Kaufmann 1977, Kozlowski *et al.* 1991). Solar radiation provides the energy for transpiration and regulates stomatal opening. As a

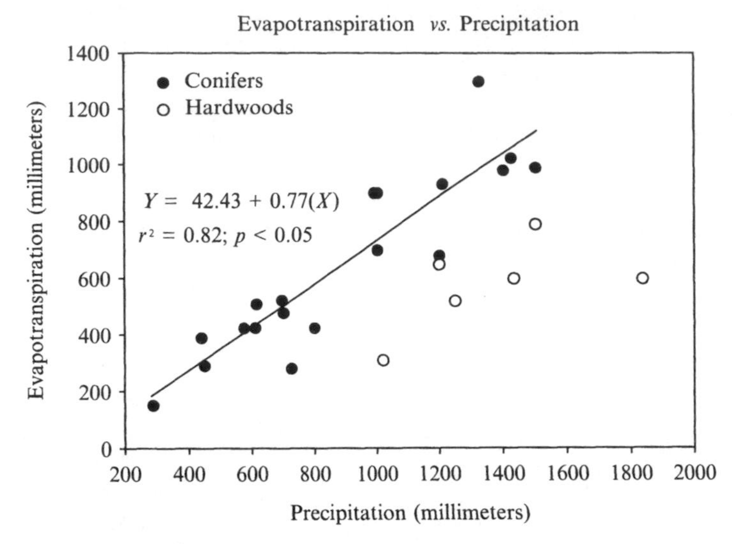

Figure 8-1 Relationship between annual evapotranspiration (Y) and precipitation (X) for hardwood and confier species (redrawn from Vose and Swank 1992, Knight *et al.* 1994 and data from this chapter). The regression line represents the data for the conifer species only. Note that r^2 is the correlation coefficient and p is probability.

result, a strong relationship generally occurs between solar radiation and transpiration, estimated as sap flow in Figure 8-2. Atmospheric vapor pressure deficit provides the gradient to which leaf-water vapor responds through the leaf stomata (Figure 8-2), and wind speed has a direct influence on the leaf boundary layer (Gates 1980). Optimal climatic conditions for transpiration include high soil water availability, high solar radiation, high vapor pressure deficits, warm temperatures for extended periods, and high wind speed. In most cases, these conditions do not occur simultaneously because increased soil water availability is usually a result of high rainfall that decreases solar radiation (due to increased cloud cover) and vapor pressure deficit (due to higher humidity). Species that have the ability to utilize deeper sources of soil or groundwater [*i.e.*, phreatophytic vegetation such as poplar (*Populus* spp.) and willow (*Salix* spp.)] are an especially attractive option in hot, dry, and windy environments in the southwestern U.S., because transpired water can be derived from groundwater (Dawson and Ehleringer 1991, Busch *et al.* 1992). Several studies have evaluated the influence of phreatophytes on surface and groundwater (*e.g.*, Robinson 1970, Van Hylckama 1980, Allen *et al.* 1999) from the perspective of negative impacts on streamflow and groundwater recharge. From a phytoremediation standpoint however, the high water consumption of phreatophytes has a positive effect to decrease aquifer recharge and influence the movement of contaminated shallow groundwater.

The structure, morphology, and physiological characteristics of the vegetation are also important regulators of transpiration. For example, at equal

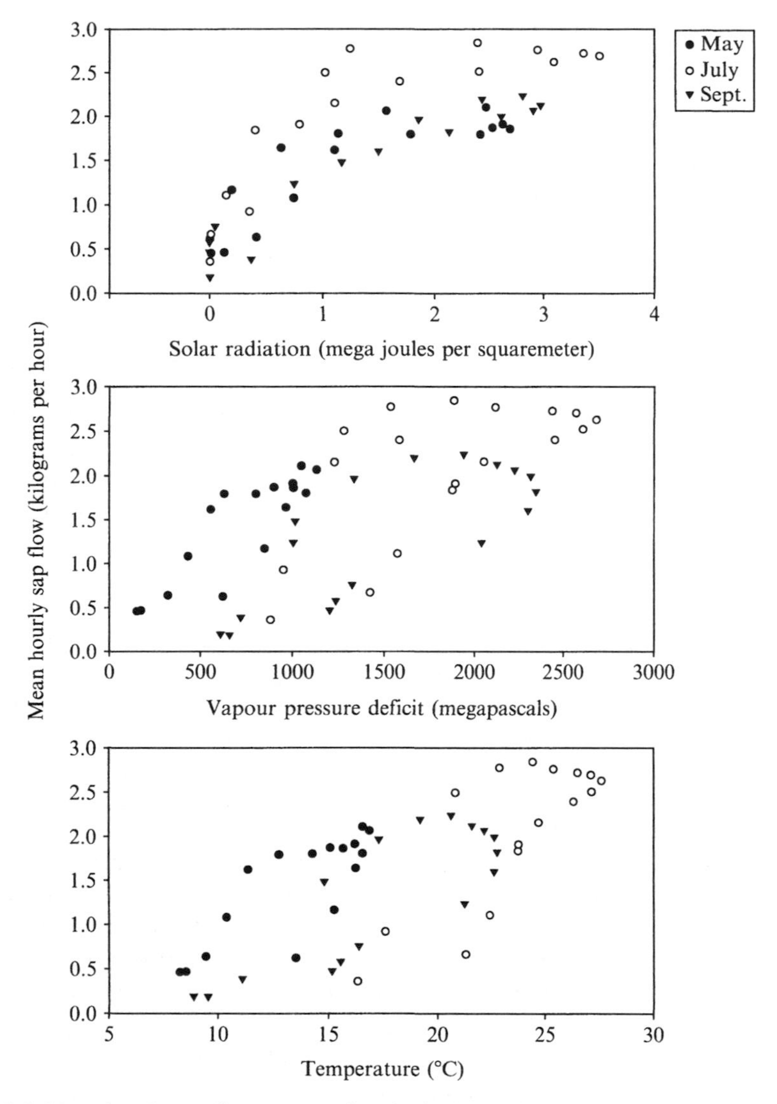

Figure 8-2 Mean hourly sap flows *versus* climatic driving variables for three seasonal measurement (May, July, and September) periods in central Colorado.

precipitation inputs, there are large differences in transpiration between conifer and hardwood species (Figure 8-1), with hardwoods generally lower than conifers. Causes for these coarse scale differences are generally well known. The single greatest controlling factor is the quantity of leaves, expressed as leaf area index (in square meters per square meter). Site water availability and leaf area are related in that, sites with the greatest water availability typically have the highest leaf area index (Gholz 1982, Long and Smith 1990), although nutrient availability (Vose and Allen 1988, Colbert *et al.* 1990) and temperature also play a role (Gholz 1986, Cropper and Gholz 1994) in determining the maximum leaf area index. Watershed studies have documented strong relationships between leaf area and streamflow, with streamflow increasing exponentially as leaf area decreases (Douglass and Swank 1975, Swank *et al.* 2000). Because precipitation minus streamflow is an estimate of evapotranspiration at watershed scales, the implication is a direct control of stand-level transpiration by leaf area. Other structural and physiological factors regulating transpiration include the amount and permeability of sapwood and stomatal characteristics such as conductance and responsiveness to climatic variation and overall plant water status (*i.e.*, water potential of leaves). Differences among species in leaf area, the rate of attainment of maximum leaf area, and physiological characteristics regulating the rate of water movement through the plant (sapwood amount and permeability, and stomatal conductance) provide opportunities for manipulating vegetation composition and structure to optimize transpiration. Optimal structural and physiological conditions for high transpiration amounts include rapid development of high leaf area, high stomatal conductance and sapwood permeability, and physiological characteristics that facilitate rapid responses to climatic conditions promoting transpiration.

Because transpiration is a function of root uptake from the soil and groundwater, soil characteristics are an important factor determining transpiration. Root growth and volume of soil occupied by roots are also important because water movement is slow when soils are drier than field capacity. Several factors determine soil water availability. First, the amount of precipitation entering the soil is a function of infiltration rate. Soils with low infiltration rates due to factors such as compaction or fine texture will have lower soil water availability because some precipitation may move across the soil surface in overland flow. Once in the soil, soil water availability is a function of water holding capacity and unsaturated hydraulic conductivity, both of which are determined by soil texture. Texture impacts water availability in different ways. Heavy clay soils (*e.g.*, pore size less than 0.2 micrometer) have limited soil water availability because of very low rates of movement in the soil (*i.e.*, conductivity) due to the fine pore space. In contrast, coarse textured sandy soils (*e.g.*, pore size greater than 50 micrometers) have low water availability because of rapid drainage. Rooting volume and the presence or absence of restrictive layers are also important soil factors determining transpiration. For example, compacted soils provide a physical barrier to root growth, limiting root extension (Heilman 1981).

QUANTIFYING TRANSPIRATION

Leaf Level

Because water exits the plant primarily through leaf stomata (a small amount of cuticular transpiration may also occur in stems of some species), leaf–water relations are a key factor determining whole-plant transpiration (Schulze 1991). The concentration gradient of water vapor between the interior of the leaf and the atmosphere at the leaf boundary layer defines the maximum transpiration rate. Vapor exchange is also determined by the opening size of the stomata. When stomata are wide open, transpiration occurs at about 20 percent to 40 percent of the rate of evaporation of open water (Waring and Schlesinger 1985), whereas closed stomata limit transpiration to less than 1 percent of open water. Stomatal opening is controlled by guard cell turgor, which responds to light, temperature, vapor pressure, and water potential of the leaves. The rate of movement of water through the stomata is the stomatal conductance. The rate of stomatal response to climatic conditions varies by species, but generally reflects responses to current conditions, whereas stomatal responses to water potential in leaves may reflect previous climatic and environmental conditions.

Because of the tight linkage between transpiration, leaf stomatal conductance (hereafter referred to as leaf conductance), and water potential of leaves, knowledge of all three parameters is useful for evaluating transpiration capacity. For example, species that exhibit high leaf-level transpiration and conductance, and maintain high water potential in the leaves have the capacity to transpire large quantities of water. Similarly, the relationship between water potential in the leaves and conductance is often threshold dependent; *i.e.*, species that maintain high leaf conductance at low water potential have the capacity to transpire more water under dry conditions (Zhang *et al.* 1997). Because of the importance of factors such as leaf area index and distribution, sapwood amount and permeability, and the difficulty in extrapolating spatially and temporally from the leaf to stand level, there may be no direct correspondence between leaf-level transpiration and overall stand transpiration. Typically, leaf conductance and tree and stand-level transpiration are most highly related in young stands with simple canopy architecture (Vose *et al.* 2000), such as closely spaced, even-aged monocultures. However, as stands develop, the linkage between leaf conductance and tree or stand-level sap flow declines due to shifts in the importance of stomatal *versus* boundary layer conductance to total vapor phase conductance (Heilman *et al.* 1996, Martin *et al.* 1999). Hence, leaf-level measurements should only be used as an indicator of transpiration capacity.

Tree Level

Transpiration at the whole-tree level represents the integrated movement of water vapor from all the leaves in the crown of the tree. As mentioned in the previous section, spatial and temporal variation severely limits extrapolation of individual leaf measurements to the tree, so more direct measurements at the tree level are required. Two approaches have typically been used. In a few instances, entire trees have been enclosed in a cuvette and the flux of water vapor calculated based on the rate of increase in humidity within the enclosure. This approach is severely limited by methodological constraints such as the size of trees, heat buildup within the cuvette, and alterations in the boundary layer and vapor pressure gradients.

Sap-flow rate and volume have also been used as an estimate of transpiration (Steinberg *et al.* 1989). Because of lags between water movement in the stem and leaf-level transpiration, sap flow is not a direct measure of transpiration, but can be corrected after accounting for lags (Schulze *et al.* 1985, Philips *et al.* 1997). Typically, a 1- to 2-hour lag correction is applied to real time sap-flow data to account for this temporal difference (Philips *et al.* 1997, Vose *et al.* 2000).

Two sap-flow techniques have been utilized; heat balance and heat pulse. For the heat-balance approach, collars consisting of a heating element and thermocouples above and below the heating element are placed around the stem and the entire stem section is heated. Sap flow is calculated using the heat-balance principle based on the difference in temperature between thermocouples above and below the heated stem section, after subtracting for heat loss due to conduction by stemwood (Baker and van Bavel 1987). An advantage of this approach is that it integrates sap flow along the entire stem and does not require an independent estimate of sapwood area. For larger trees, paired probes are inserted vertically into the sapwood (Granier 1987). The upper probe is heated and both contain thermocouples. The probes measure heat dissipation, which increases with sap flow and the resultant cooling of the heat source, as the apparent thermal conductance of sapwood increases with sap velocity. To convert sap velocity to sap flow rate, the cross-sectional area of sapwood must also be determined. Typically, trees are cored and sapwood to heartwood ratios quantified. Because sap-flow probes measure sap flow velocity at only one location, multiple probes are required to adjust for the variation in sapwood thickness and permeability in the stem section. Despite this, unaccounted for variation in horizontal and vertical variation in sapwood thickness and permeability introduces some error into sap-flow estimates obtained with probes. The magnitude of error can be determined experimentally and corrected for in small trees by comparing sap flow with actual transpiration using procedures such as weighing lysimeters. In large trees, corrections are much more difficult and hence, predictions have more uncertainty. In contrast to the heat-balance method, the heat pulse method estimates sap flow based on the time lag between pulses of heat and the distance between the sensors (Swanson 1962).

Stand Level

While it is informative to understand transpiration at the leaf and tree level to help evaluate species and environments suitable for phytoremediation, stand-level transpiration ultimately determines how much soil water and groundwater are removed. However, unlike leaf and tree measurements, no methods directly measure stand transpiration. Instead, three indirect measurement approaches have been utilized. These approaches involve gaged watersheds, extrapolation of individual tree measurements, and eddy flux estimates. Gaged watersheds require a combination of well-defined watershed boundaries, tight bedrock, and well-constructed weirs or gages to provide accurate transpiration estimates. If these criteria are met, then evapotranspiration (*ET*) is estimated by the equation

$$ET = P - RO \pm \text{soil water storage} \quad (8\text{-}1)$$

where P = precipitation and RO = runoff, determined from weirs or gages. Because P is a component of the equation, the accuracy of precipitation measurements will also influence evapotranspiration estimates. Changes in soil water storage are usually assumed to be negligible at annual time steps, although this is clearly not the case over shorter intervals. Hence, using this approach at time steps less than a year requires determining changes in soil water storage. Because evapotranspiration is estimated, interception evaporation must be determined and subtracted to estimate transpiration.

Extrapolating individual tree measurements to the stand can be done in a number of ways. For example, instruments that measure sap flow can be installed on trees representing the averaged sized tree and mean sap flow multiplied times the number of trees in the stand (*i.e.*, a "mean-tree" approach). Considerable uncertainty in stand-level estimates can accompany this approach where sites are variable. Alternatively, relationships between tree diameter, sapwood area, or basal area and sap flow at the individual tree level can be applied to all trees. In both approaches, repeated sampling is required to account for seasonal variability.

The eddy flux method uses water vapor gradients at fixed intervals above and below the canopy to calculate evapotranspiration. The technique is based on the assumption that water vapor flux is proportional to the vertical gradient of water vapor between two measurement points (averaged over several minutes). Typically, measurements are conducted from towers extending through the canopy. To be useful for estimating transpiration of a particular stand, the stand must be large enough to encompass most of the footprint measured by the sensors. In many phytoremediation applications conceived as of 2003, the stands are too small for an eddy flux approach to be appropriate.

Modeling

The use of modeling provides a potentially powerful tool for predicting current transpiration of native or planted vegetation and for projecting future transpiration capacity as a function of stand development. At the coarsest level of forecasting, gross measures of plant water demand and use can be derived from empirical estimates of potential evapotranspiration (Thornthwaite 1948, Mongaran 1973). These approaches usually consider climate and soils to some extent, but do not consider vegetation effects such as leaf area index, rooting depth, or leaf-level physiological characteristics. Hence, empirical approaches are useful for gross estimates of transpiration, but have limited utility for evaluating actual effects on the groundwater. At the other extreme, detailed physiological models that link the soil–plant–atmosphere continuum provide much more accurate estimates of transpiration. Depending upon the structure, models may also provide estimates of specific uptake locations within the soil profile (Huff and Swank 1985, Vose and Swank 1992, Vose and Swank 1994). Using detailed physiologically based models results in significantly greater data requirements. The most accurate application of these models requires site-specific estimates of soils, climate, and physiological characteristics of the major species on the site. However, large-scale application of detailed models with generalized parameters may provide estimates sufficiently accurate to be used in evaluating phytoremediation applications.

MEASURING AND MODELING TRANSPIRATION: CASE STUDY APPLICATIONS

Study Site Descriptions

Sap flow was measured at sites in Texas, Florida, and Colorado as components of larger studies evaluating the efficacy of using phytoremediation technology to clean up shallow groundwater contaminants. The north–central Texas study site was located, about 15 kilometers west of Fort Worth. The climate of this area is characterized as subhumid, with mild winters and hot, humid summers. The average annual precipitation is 80 centimeters per year with most rainfall occurring between May and October. Average annual temperature is 18.6 °C. Study plots were located on the U.S. Naval Air Station, which adjoins U.S. Air Force Plant 4. A plume containing trichloroethylene was detected in the terrace alluvial aquifer in 1985. To demonstrate phytoremediation potential, eastern cottonwood (*Populus deltoides* Marsh.) trees were planted in two plantations over the TCE plume. One plantation was planted with vegetative cuttings (whips) and the other with 1-year-old nursery grown seedlings. Each plantation was approximately 80 by 20 meters and located perpendicular to groundwater flow in the alluvial aquifer. Sap-flow measurements were conducted using the heat-balance method (collars) in the first and second year after plantation establishment.

The eastern Florida site was located in the city of Orlando. The climate of the area is humid, with mild winters and hot, humid summers. The average annual temperature is 22.6 °C and the average annual rainfall is 123 centimeters. Native vegetation of interest was located on the U.S. Naval Training Center. Trichloroethylene and tetrachloroethylene, which originated from a dry-cleaning facility that is no longer in operation, contaminate shallow groundwater. The plume extends under a 2-hectare forest and seepage wetland before reaching Lake Druid that borders the forest. A dense and diverse mix of overstory and understory species occur in the forest (density of 107 trees per hectare), with red bay [*Persea borbonia* (L.) Spreng.], camphor [*Cinnamomum camphora* (L.) Nees & Eberm.], slash pine and longleaf pine (*Pinus* spp.), sweet bay (*Magnolia virginiana* L.), and live oak and laurel oak

(*Quercus* spp.) most abundant in the overstory. The most abundant understory species are skunk vine (*Paederia foetida* L.), saw palmetto [*Serenoa repens* (Bartr.) Small], cinnamon fern (*Osmunda cinnamomea* L.), and Christmas fern [*Polystichum acrostichoides* (Michx.) Schott.].

The central Colorado site is located approximately 20 kilometers southwest of Denver. The climate of the area is dry, with warm summers and cold winters. Annual precipitation averages approximately 44 centimeters, with 30 percent of this amount received in April and May. The average annual temperature is 12 °C. Study plots were located on the U.S. Air Force Plant PJKS. Trichloroethylene and dichloroethylene from a variety of sources contaminate the site. Measurements were conducted in two existing stands of natural vegetation: cottonwood–willow (*Populus* spp.–*Salix* spp.) and Gambel oak (*Quercus gambelii* Nutt.) The cottonwood–willow (*Populus* spp.–*Salix* spp.) stand is restricted to riparian areas (approximately 1 percent of the total land area of the site), while the Gambel oak (*Quercus gambelii*) stand is on more midslope locations (approximately 30 percent of the total land area of the site).

Methods

The sampling approach and methods varied among the three studies based on study objectives, species composition, and tree sizes. For the Texas study, sap flow from saplings in the plantation was estimated using sap-flow gauges (Dynamax Inc., Houston, TX) on 14 to 16 trees (divided equally among whips and 1-year-old trees) in May, June, July, August, and October over a 2-year period. During each measurement period, sap-flow measurements were taken every minute for 2 to 3 consecutive days. Data presented in this chapter represent averages of both plantations. In addition, sap flow was measured on nine larger native trees growing near the plantations using thermal dissipation probes (Dynamax, Inc., Houston, TX). Species sampled were: eastern cottonwood (*Populus deltoides* Marsh.), American elm (*Ulmus americana* L.), black willow (*Salix nigra* Marsh.), sugarberry [or large hackberry, (*Celtis laevigata* Willd.)], Eastern red cedar (*Juniperus virginiana* L.), and mesquite (*Prosopis pubescens* Benth.). At the end of sampling, increment cores were taken from the nine large trees for determining sapwood area.

For the Orlando study, sap flow was estimated using thermal dissipation probes installed on nine trees representative of major canopy species. Species sampled were: slash pine (*Pinus elliottii* Engelm.), longleaf pine (*Pinus palustris* Mill.), live oak (*Quercus virginiana* Mill.), laurel oak (*Quercus hemisphaerica* Bartram ex. Willd.), sweet bay (*Magnolia virginiana* L.), and camphor [*Cinnamomum camphora* (L.) Nees & Eberm.]. Two probe sets were installed into the sapwood on the north and south sides of sample trees, and sampling was conducted in November, March, and July for 2 to 3 consecutive days over a 1-year period. At the end of sampling, increment cores were taken and sapwood area determined.

For the Colorado study, sap flow was estimated using thermal dissipation probes on eight trees representing three species: eastern cottonwood (*Populus deltoides* Marsh.), narrow-leaf cottonwood (*Populus angustifolia* James.), and Gambel oak (*Quercus gambelii* Nutt.). Two probe sets were installed into the sapwood on the north and south side of sample trees and sampling was conducted in May, July, and September over a 1-year-period. At the end of sampling, increment cores were collected and sapwood area determined.

For all three studies, data were summarized to provide average hourly sap flow rates (kilograms per hour) or daily totals (kilograms per day). In addition, climate was measured at all three studies with climate stations located on-site. Measurements included: hourly rainfall (centimeters), wind speed (meters per second), solar radiation (Watts per square meter), temperature (°C), and relative humidity (percentage). Relative humidity and air temperature were used to calculate vapor pressure deficit (megapascals).

For the Texas plantation site, we parameterized and applied a mechanistic model of sap flow (PROSPER) and compared the results to sap flow measurements. Evapotranspiration at the Texas site was simulated because data were available to parameterize the model (Vose *et al.* 2000). The PROSPER model has been described in detail elsewhere (Goldstein *et al.* 1974, Huff and Swank 1985), so only a general description is provided here. The PROSPER code is a phenomenological, one-dimensional model that links the atmosphere, vegetation, and soils. Plant and soil characteristics are combined into a single evapotranspiration surface that is characterized by a resistance to water vapor loss. This resistance is analogous to the relationship between stomatal resistance and water potential of the leaves and is a function of the water potential of the evapotranspiration surface. Evapotranspiration is predicted by a combined energy balance–aerodynamic method (Penman-Monteith equation modified as described in Swift *et al.*, 1975) that is a function of the surface resistance to vapor loss described previously. The PROSPER model uses electrical network equations (Goldstein *et al.* 1974) to balance water allocation among vegetation and soil horizons. The flow of water within and between soil and plant is a function of soil hydraulic conductivity, soil water potential, root characteristics in each soil layer, and surface water potential. The PROSPER model predicts evapotranspiration, transpiration, and soil water distribution between soil layers daily, but monthly data are most accurate. The PROSPER model requires the following climatic data: solar radiation, precipitation, wind speed, air temperature, and vapor pressure. Initial model parameters include surface resistance to vapor loss, leaf area index, root distribution and surface area, soil moisture release, and several other parameters listed in Goldstein *et al.* (1974).

Transpiration Estimates

Maximum transpiration rates for the study sites indicate large variation in transpiration potentials among sites (Table 8-1). On a per tree basis, rates ranged from 8 to 120 kilograms per tree per day. Much of this variation was related to differences in tree size that reflects differences in leaf area and sapwood area.

TABLE 8-1 Midsummer Peak Sap-flow Rates Averaged across Species and Measurement Days

Site	Sapwood area (square centimeters per tree)	Sap-flow rates			
		Kilograms per day per square meter of sapwood	Kilograms per tree per day	Liters per hectare per year	Gallons per acre per year
Texas					
Plantation	30	2600	8	3 620 000	387 200
Native trees	820	1463	120	7 551 000	807 600
Colorado	234	1043	24	1 510 000	161 500
Florida	710	1535	109	6 859 000	733 600

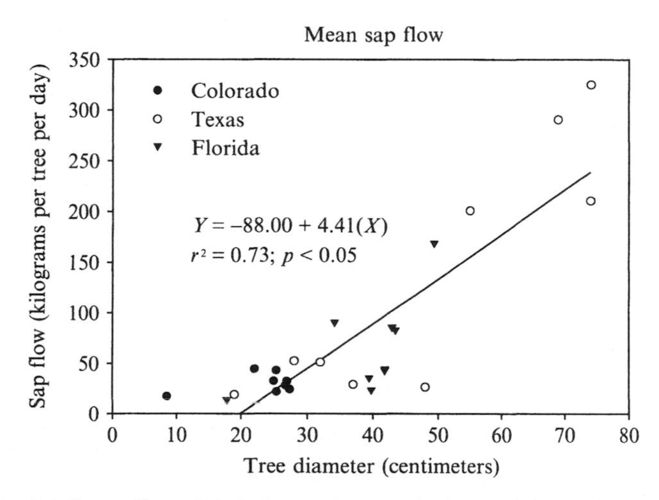

Figure 8-3 Total daily sap flows (Y) during peak transpiration periods (midsummer) *versus* tree diameter (X) across a range of species and site. Note that r^2 is the correlation coefficient and p is probability.

For example, when pooling the data across the sites, a significant proportion of the variation in transpiration rates among and within sites can be explained by tree diameter (Figure 8-3). Larger trees typically have greater sapwood volume resulting in more water transporting vessels (angiosperms) and tracheids (conifers) for sap-flow movement in the stems. Because leaf area is also related to sapwood area, larger trees will typically have greater leaf area index; and hence, greater surface area for transpiration.

When sap-flow rates are adjusted based on sapwood area (*i.e.*, kilograms per day per square meter of sapwood), the variation in transpiration reflects species related differences in physiology (leaf, stem, and root), leaf area to sapwood area ratios, and site-dependent factors such as soil water availability and climate driving variables. Because species composition varies among sites and physiological and physical factors influence transpiration simultaneously, these studies cannot separate physiological and climatological effects; to do so requires an evaluation of transpiration rates of the same species and genotype in differing climatic and soil water availability conditions. For example, species sampled at the three sites represent a mixture of conifers, and ring porous and diffuse porous hardwood species, resulting in large differences in sapwood permeability and specific conductivity among sites and among species within sites (Figure 8-4). In general, sap-flow velocity is lower in conifers and diffuse-porous species because sap flow moves through a number of annual rings, whereas water moves through only one or two annual rings in ring-porous species (Kramer and Kozlowski 1979, Kozlowski *et al.* 1991). Despite the limitations of the current approach, some notable patterns emerge when evaluating transpiration after adjusting for differences in sapwood area. For example, the cottonwood (*Populus deltoides*) plantation in Texas had the highest transpiration rate per unit of sapwood area, followed by the Florida stand, large trees in Texas, and the Colorado stand (Table 8-1). The high transpiration rate for cottonwood (*Populus deltoides*) in the plantation is a function of species characteristics that promote high transpiration, high leaf area per unit sapwood in the developing canopy, and access to shallow groundwater. In contrast, transpiration rates per unit of sapwood area were lowest in Colorado, even though the site contained cottonwood (*Populus deltoides*) and several of the measured trees occurred in the riparian zone. The combination

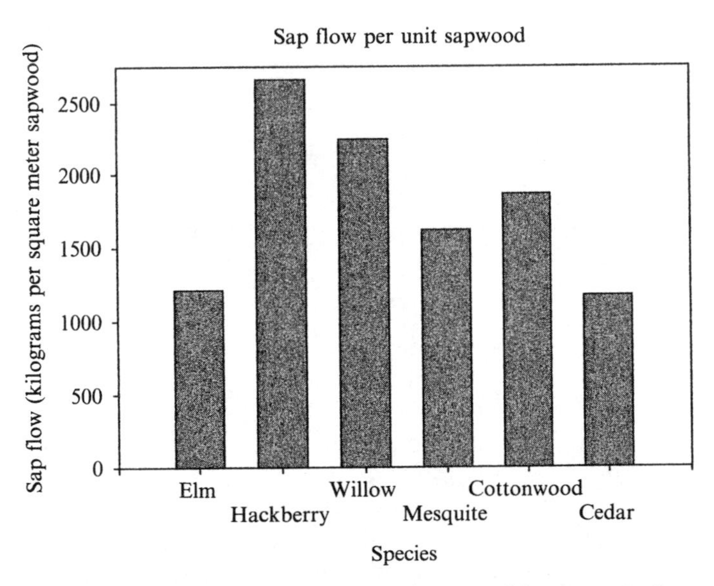

Figure 8-4 Mean growing season sap-flow rate per unit sapwood for six species in north–central Texas. Black willow (*Salix nigra* March.), eastern cottonwood (*Populus deltoides* Marsh.), and eastern red cedar (*Juniperus virginiana*, L.) are diffuse porous species, while American elm (*Ulmus americana* L.), hackberry (*Celtis laevigata* Willd.), and mesquite (*Prosopis pubescens* Benth.) are ring porous.

of species composition and climate characteristics were not as conducive to high sap-flow rates per unit sapwood area relative to the other sites.

Using sap-flow techniques to predict actual stand transpiration requires frequent sampling to account for seasonal variation. Ideally, sap flow should be measured continuously for the entire growing season on a large number of trees. Because this approach is often impractical, an alternative is to measure sap flow at shorter frequencies and calculate bounds or maximum values as a tool to evaluate phytoremediation potentials. Because the sampling frequency varied among the case studies described here, we focused only on measurements during the highest transpiration period (midsummer). These estimates can be used as a "best-case scenario" approach—that is, if these rates occurred on the site, would transpiration be sufficient to control the plume? To estimate maximum potential transpiration at the stand level (*i.e.*, kilograms per hectare or gallons per acre), we extrapolated the tree transpiration data (kilogram per tree per day) assuming a 180 days transpiration period and a stem density of 350 stems per hectare, except for the plantation where actual tree density was used (Table 8-1). We emphasize that these data provide estimates of maximum transpiration capacity under *in situ* climate conditions because the peak sap-flow rates were used in the extrapolation and previous studies have shown considerable seasonal variation in sap flow (Vose *et al.* 2000). The 350 stems per hectare is representative of a fully stocked stand under most forest conditions and is consistent with full canopy closure and maximum leaf area index.

When comparing results from the sites with mature trees, the variation in maximum transpiration capacity is considerable. The Texas site has a maximum transpiration capacity of approximately 7.5×10^6 liters of water per hectare per year (8×10^5 gallons of water per acre per year) if the site was fully stocked with the sampled species. By contrast, the Colorado site has a maximum transpiration capacity of approximately 1.5×10^6 liters per hectare per year (1.6×10^5 gallons per acre per year). The plantation site in Texas currently has a maximum transpiration capacity of approximately 3.7×10^6 liters per hectare per year (4×10^5 gallons per acre per year). However, we anticipate that transpiration will equal or exceed the estimate from mature trees on the site (*versus* 7.5×10^6 liters per hectare per year or 8.0×10^5 gallons per acre per year) once the canopy develops and achieves the maximum leaf area.

Comparison of Measured *versus* Modeled Transpiration

A critical need for phytoremediation is the development and application of a tool to provide species and site-based estimates of transpiration. While a powerful tool for measuring transpiration from vegetation already on-site or quantifying transpiration of planted vegetation, sap-flow measurements at every phytoremediation site may not be practical. One potential tool for application across sites is the development or application of models. In most cases, however, models need to be calibrated or parameterized for specific site and species conditions. To evaluate the use of such a tool, we parameterized

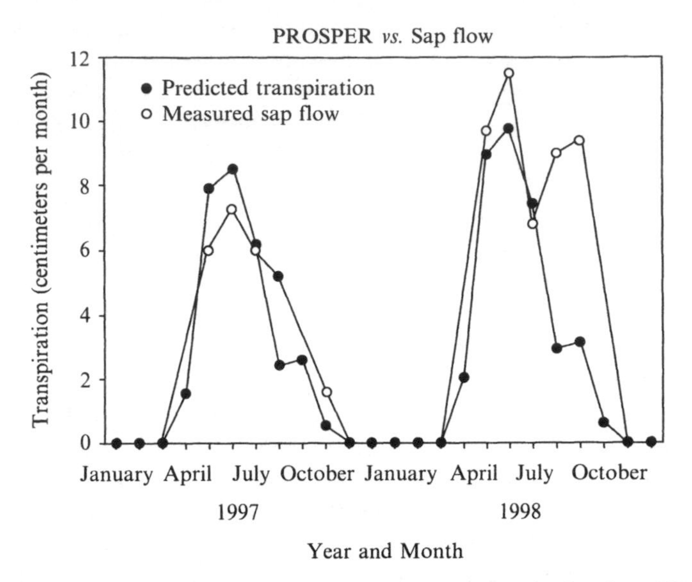

Figure 8-5 Comparison of measured sap flow and predicted transpiration from PROSPER for cottonwood (*Populus deltoides* Marsh.) plantations in north–central Texas.

PROSPER for the Texas plantation using intensive site (soils, climate, and root distribution) and leaf-level measurements (stomatal characteristics and leaf area index) (see Vose *et al.* 2000). We compared monthly transpiration estimates obtained with PROSPER to transpiration estimated from sap-flow measurements over a 2-year period (Figure 8-5). Comparisons indicated generally good agreement between predicted and measured values, except during the late summer that was coincident with some of driest and hottest periods (August and September 1998) in our study. During this period, PROSPER predicted a considerable decline in transpiration, while measured values showed an increase. We attribute this discrepancy to an inability of PROSPER to adequately simulate root uptake from shallow groundwater during drought conditions, since the original formulation of PROSPER was designed to only simulate surface and soil water dynamics (Goldstein *et al.* 1974, Huff and Swank 1985). The results of this comparison are consistent with other studies that have shown that PROSPER provides reasonable estimates of either evapotranspiration or transpiration (Vose and Swank 1992, Vose and Swank 1994). However, refinements in the subsurface water and groundwater hydrology and subsequent availability to tree roots might improve the predictive capability and usefulness as a phytoremediation evaluation tool.

CONCLUSIONS AND RECOMMENDATIONS

The importance of transpiration to the success of phytoremediation applications suggests that accurate estimates of current and potential transpiration should be a high priority when considering this approach for site management. Both measuring and modeling transpiration are important. Assessments require a substantial sampling commitment for direct measurements or parameterizing physiologically based models. This requires detailed knowledge of local site conditions and physiological parameters for the major species. For screening assessments, we recommend that published estimates be used to set the bounds for maximum transpiration capacity based on general climate and vegetation characteristics of the location. If these general transpiration rates are great enough to influence groundwater hydrology, then evaluations of current, enhanced (*e.g.*, manipulating the structure and species composition of current vegetation), or new vegetation transpiration capacity should proceed. Technology and models exist to provide reasonable estimates and predictions of transpiration. However, the accuracy of the estimates depends on the investment in accounting for the spatial and temporal variation or in providing site and species-specific estimates for physiologically based transpiration models.

Acknowledgments

This work was supported in part by the Department of Defense Environmental Security Technology Certification Program Project E95BRAB004, Gregory J. Harvey, Program Manager and in part by the Department of Defense, U.S. Navy (SOUTHNAVFACENGCOM), Charleston, South Carolina. We thank Chris Sobek for her assistance in collecting sap-flow data, summarizing data, and generating graphs. Several other members of the Coweeta staff assisted with field sampling at various times, including: Bob McCollum, Nathan Stover, Patsy Clinton, Laura Phillips, and Josh Ward.

REFERENCES

Allen, S.J., R.L. Hall, and P.T. Rossier (1999) Transpiration by two poplar varieties grown as coppice for biomass production. *Tree Physiol.* **19**: 493–501.

Baker, J.M. and C.H.M. van Bavel (1987) Measurement of mass flow of water in stems of herbaceous plants. *Plant Cell Environ.* **10**: 777–782.

Busch, D.E., N.L. Ingraham, and S.D. Smith (1992) Water uptake in woody riparian phreatophytes of the southwestern United States: a stable isotope study. *Ecol. Appl.* **2**(4): 450–459.

Colbert, S.R., E.J. Jokela, and D.G. Neary (1990) Effects of annual fertilization and sustained weed control on dry matter partitioning, leaf area, and growth efficiency of juvenile loblolly and slash pine. *Forest Sci.* **36**: 995–1014.

Cropper, W.P., Jr. and H.L. Gholz (1994) Evaluating potential response mechanisms of a forest stand to fertilization and night temperature: a case study using *Pinus elliottii*. *Ecol. Bull.* (Copenhagen) **43**: 165–160.

Dawson, T.E. and J.R. Ehleringer (1991) Streamside trees do not use stream water. *Nature* **350**(6316): 335–337.

Douglass, J.E. and W.T. Swank (1975) Effects of management practices on water quality and quantity: Coweeta Hydrologic Laboratory, North Carolina. USDA Forest Service Gen. Tech. Rep. SE-13, Asheville, North Carolina.

Fahey, T.J. (1979) The effect of night frost on the transpiration of *Pinus contorta* sp. *latifolia. Oecologia* **14**: 483–490.

Gates, D.M. (1980) *Biophysical Ecology*. Springer-Verlag, New York, p. 611.

Gholz, H.L. (1982) Environmental limits on aboveground net primary production, leaf area and biomass in vegetation zones of the Pacific Northwest. *Ecology* **53**: 469–481.

Gholz, H.L. (1986) Canopy development and dynamics in relation to primary production. In: *Crown and Canopy Structure in Relation to Primary Production*. T. Fujimori and D. Whitehead, eds. Forest and Forest Products Research Institute, Ibaraki, Japan. pp. 224–242.

Goldstein, R.A., J.B. Mankin, and Luxmoore, R.J. (1974) Documentation of PROSPER: a model of atmosphere-plant-soil water flow. EDFB/IBP-73/9, U.S. Department of Energy Oak Ridge National Laboratory, Oak Ridge, Tennessee.

Granier, A. (1987) Evaluation of transpiration in a Douglas fir stand by means of sap flow measurements. *Tree Physiol.* **3**: 309–320.

Heilman, P. (1981) Root penetration of Douglas-fir seedlings into compacted soil. *Forest Sci.* **27**: 660–666.

Heilman, P.E., T.M. Hinckley, D.A. Roberts, and R. Ceulemans (1996) Production physiology. In: *Biology of Populus and Its Implications for Management and Conservation*. R.F. Stettler, H.D. Bradshaw, Jr., P.E. Hielman, and T.M. Hinckley, eds. NRC Research Press, Ottawa, Canada, pp. 459–489.

Helvey, J.D. (1971) A summary of rainfall interception by certain conifers of North America. In: *Biological Effects of the Hydrologic Cycle*. E.J. Monke, ed. Proc. Third Int. Seminar for Hydrology Professors, Dept. of Agricultural Engineering, Purdue University, West Lafayette, Indiana, pp. 103–113.

Hinckley, T.M., J.R. Brooks, J. Cermak, R. Ceulemans, J. Kucera, F.C. Meinzer, and D.A. Roberts (1994) Water flux in a hybrid poplar stand. *Tree Physiol.* **14**: 1005–1018.

Huff, D.W. and W.T. Swank (1985) Modeling changes in forest evapotranspiration. In: *Hydrological Forecasting*. M.G. Anderson and T.P. Burt, eds. John Wiley, New York, pp. 121–151.

Kaufmann, M.R. (1977) Soil temperature and the drying effects on water relations of *Pinus radiata. Can. J. Bot.* **55**: 2413–2418.

Knight, D.H., J.M. Vose, V.C. Baldwin, K.C. Ewel, and Grodzinska, K. (1994) Contrasting patterns in pine forest ecosystems. *Ecol. Bull.* (Copenhagen) **43**: 9–19.

Kozlowski, T.T., P.J. Kramer, and S.G. Pallardy (1991) *The Physiological Ecology of Woody Plants*. Academic Press, New York, p. 657.

Kramer, P.J. and T.T. Kozlowski (1979) *Physiology of Woody Plants*. Academic Press, New York, p. 811.

Long, J.N. and F.W. Smith (1990) Determinants of stemwood production in *Pinus contorta* var. *latifolia* forests: the influence of site quality and stand structure. *J. Appl. Ecol.* **27**: 847–856.

Martin, T.A., K.J. Brown, J. Cermak, R. Ceulemans, J. Kucera, F.C. Meinzer, J.S. Rombold, D.G. Sprugel, and T.M. Hinckley (1997) Crown conductance and tree and stand transpiration in second-growth *Abies amabilis* forest. *Can. J. Forest Res.* **27**: 797–808.

Martin, T.A., T.M. Hinckley, F.C. Meinzer, and D.G. Sprugel (1999) Boundary layer conductance, leaf temperature and transpiration in *Abies amabilis* branches. *Tree Physiol.* **19**: 435–443.

Meyers, B.J. and T. Talsma (1992) Site water balance and tree water status in irrigated and fertilized stands of *Pinus radiata*. *Forest Ecol. Manag.* **50**: 17–42.

Mongaran, C., (1973) Economic feasibility of irrigating southern pines. *Water Resources Res.* **9**: 1485–1495.

Phillips, N., A. Nagchaudhuri, R. Oren, and G. Katul (1997) Time constant for water transport in loblolly pine trees estimated from time series of evaporative demand and stem sapflow. *Trees* **11**: 412–419.

Robinson, T.W. (1970) Evapotranspiration by woody phreatophytes in the Humbolt River Valley near Winnemucca, Nevada. U.S. Geological Survey Professional Paper 491–D, U.S. Government Printing Office, Washington, D.C.

Schulze, E.-D. (1991) Water and nutrient interactions with plant water status. In: *Responses of Plants to Multiple Stresses*. H.A. Mooney, W.E. Winner, and E.J. Pell, eds. Academic Press, New York, pp. 89–101.

Schulze, E.-D., J. Cermak, R. Matyessek, M. Penka, R. Zimmerman, F. Vasicek, W. Gries, and J. Kucera (1985) Canopy transpiration and water fluxes in the xylem of the trunk of *Larix* and *Picea*—a comparison of xylem flow, prorometer, and cuvette measurements. *Oecologia* **66**: 475–483.

Steinberg, S.L., C.H.M. van Bavel, and M.J. McFarlan (1989) A gauge to measure mass of flow of sap in stems and trunks of woody plants. *J. Am. Soc. Hort. Sci.* **114**: 466–472.

Swank, W.T., J.M. Vose, and K.J. Elliott (2000) Long-term hydrologic and water quality responses following commercial clearcutting of mixed hardwoods on a southern Appalachian catchment. *Forest Ecol. Manag.* **5267**: 1–16.

Swanson, R.H. (1962) An instrument for detecting sap movement in woody plants. U.S. Department of Agriculture Forest Service Rocky Mountain Forest and Range Experiment. Station Paper. 68.

Swift, L.W., Jr., W.T. Swank, J.B. Mankin, R.J. Luxmoore, and R.A. Golstein (1975) Simulation of evapotranspiration and drainage from mature and clear-cut forest and young pine plantations. *Water Resources Res.* **11**(5): 667–673.

Thornthwaite, C.W. (1948) An approach towards a rationale classification of climate. *Geogr. Rev.* **38**: 55–94.

Van Hylckama, T.E.A. (1980) Weather and evapotranspiration studies in a saltcedar thicket in Arizona. U.S. Geological Survey Professional Paper 491–494, U.S. Government Printing Office, Washington, D.C.

Vose, J.M. and H.L. Allen (1988) Leaf area, stemwood growth, and nutrition relationships in loblolly pine. *Forest Sci.* **34**: 546–563.

Vose, J.M. and W.T. Swank (1992) Water balances. In: *Atmospheric Deposition and Forest Nutrient Cycling: A Synthesis of the Integrated Forest Study*. D.W. Johnson and S.E. Lindberg, eds. Ecological Studies 91, Springer-Verlag, New York, pp. 27–49.

Vose, J.M. and W.T. Swank (1994) Effects of long-term drought on the hydrology and growth of a white pine plantation in the southern Appalachians. *Forest Ecol. Manag*. **64**: 25–39.

Vose, J.M., W.T. Swank, G.J. Harvey, B.D. Clinton, and C. Sobek (2000) Leaf water relations and sapflow in eastern cottonwood (*Populus deltoides* Bartr.) trees planted for phytoremediation of a groundwater pollutant. *Int. J. Phytorem*. **2**(1): 53–73.

Waring, R.H. and W.H. Schlesinger (1985) *Forest Ecosystems: Concepts and Management*. Academic Press, New York, 340 pp.

Zhang, H., L.P. Simmonds, J.I.L. Morison, and D. Payne (1997) Estimation of transpiration by single trees: comparison of sap flow measurements with a combination equation. *Agric. For. Meteorol*. **87**: 155–169.

SECTION III

SCIENCE AND PRACTICE FOR AROMATIC, PHENOLIC, AND HYDROCARBON CONTAMINANTS

9

TRANSFORMATION OF ORGANIC CONTAMINANTS BY DIFFERENT PLANT SYSTEMS

H. Harms, M. Bokern, M. Kolb, and C. Bock

SUMMARY OF PRACTICAL IMPLICATIONS

A significant body of research, using plant cell cultures and hairy root cultures as well as whole plants grown either under aseptic or septic conditions, establishes that plants metabolize by common metabolic pathways a number of persistent nonpolar xenobiotic pollutants. These contaminants include polycyclic aromatic hydrocarbons, polychlorinated biphenyls, pentachlorophenol, 4-chloroaniline, and the surfactant 4-*n*-nonylphenol. Uptake depends on the plant species and on the physicochemical properties of the chemical. The main metabolites are polar conjugates with carbohydrates and amino acids. Some plant species incorporated large amounts of the chemicals and the associated metabolites into bound residues, which may require harvest and disposal. The four-step protocol pioneered in this work is also very useful for screening plants and nonpolar compounds to determine phytotoxicity, metabolism, and the formation of bound residues during feasibility testing at a site. The selection protocol will be particularly useful for rhizodegradation to ensure that any plant uptake and residues can be managed.

INTRODUCTION

Phytoremediation has evolved to become a potential technology for cleanup of contaminated sites (Schnoor *et al.* 1995, Macek *et al.* 2000). Successful phytoremediation of contaminants requires biological availability for absorption or uptake and metabolism by plants or associated microbial systems

Phytoremediation: Transformation and Control of Contaminants,
Edited by Steven C. McCutcheon and Jerald L. Schnoor.
ISBN 0-471-39435-1 (cloth)

(Cunningham *et al.* 1995). Many studies of organic pollutants reveal that uptake and metabolism are controlled by molecular size and configuration (Harms 1983) but perhaps even more by the capability of the tested plant species (Bokern and Harms 1993, Harms 1996). Therefore, in order to use this environmental remediation technique most efficiently, knowledge of the metabolic potential of suitable plant species for transforming and degrading chemicals is essential. To provide this knowledge, a four-step standardized protocol has been developed. Besides tests with intact plants under septic conditions for soil–plant uptake studies, plants under aseptic conditions in hydroponic cultures are used to study uptake and the formation of bound residues. Additional *in vitro* techniques such as plant cell suspension cultures or hairy root cultures have been used for investigating the phytotoxicity, metabolic fate, and persistence of xenobiotic chemicals by [^{14}C] tracer experiments (Harms and Langebartels 1986). Especially, plant cell, tissue and, recently, root cultures have become a major tool in the study of an increasing number of fundamental and applied problems in plant sciences. Thus, these *in vitro* systems have contributed to a better understanding of the action of agrochemicals and the fate of organic pollutants. In addition, these *in vitro* tests are invaluable to screen and select plants for an optimal capability for the remediation of pollutants.

TEST SYSTEMS

Cell Suspension and Hairy Root Cultures

Both cell and root cultures offer many advantages for metabolism tests during the last 1 to 3 days of the late logarithmic growth phase (Langebartels and Harms 1984, Harms 1973). Cell suspension cultures of about 20 different crop plants have been established (Bokern and Harms 1997, Kolb 2000). In addition, because of the organized structure and the development of secondary cell walls, root cultures are a very good intermediate metabolic system between cell cultures and whole plants. The cell cultures represent an aseptic *in vitro* system growing in a defined media under standardized conditions of temperature, illumination, and gaseous exchange. The uniform cell material shows a rapid growth and an easy uptake of exogenously applied chemicals. The turnover and metabolism rates are normally high, which simplifies the isolation and identification of chemical compounds and metabolites.

Root cultures grow under the same standardized conditions but without phytohormones. In contrast to cell cultures, differentiated tissues such as hairy roots are less sensitive to microbial contamination. It is, therefore, possible to perform tests under aseptic as well as septic conditions and thus draw a parallel between axenic experiments and those using unsterilized plants. Root cultures are an *in vitro* model for the rhizosphere, integrating plants, and microorganisms. Furthermore, previous investigations (Harms and Kottutz 1991, Bokern *et al.* 1998) demonstrated that organ cultures, particularly root cultures, are very suitable for studying bound residue formation.

Aseptically Grown Plants

Despite many advantages over whole plants, questions remain about the full validity of extrapolating xenobiotic metabolism *in vitro* to metabolism *in vivo*. Therefore, plants were grown under aseptic conditions (Langebartels and Harms 1986). Surface sterilized seeds germinated on a nutrient agar for microorganisms. Aseptic seedlings were transferred to glass tubes, mounted on glass forks or fixed with cotton wool into a polyurethane ring. The roots of the seedlings extend into Knop's nutrient solution (Ruge 1951) with radiolabeled test compound. Thirty-six plants were placed on a tray that was put into a glass container covered with an airtight Plexiglas® lid. The plants grew under a 15-hour light regime (23 000 lux from HQJ-T 1000 W/D lamps, Osram Berlin) at 20 ± 1 °C. During the 5-day incubation period, the container was aerated with sterile air, which passed through 1 mole per liter sodium hydroxide solution to trap released $^{14}CO_2$.

Uptake of Xenobiotic Chemicals by Plants from Soil

To investigate the uptake of xenobiotic compounds by crop plants from soils, the plants were grown in a rectangular Plexiglas® bioreactor (Figure 9-1) containing 1.2 kilograms of soil polluted with 5 or 10 milligrams of the selected radiolabeled contaminants per kilogram (Bokern *et al.* 1998). To follow the

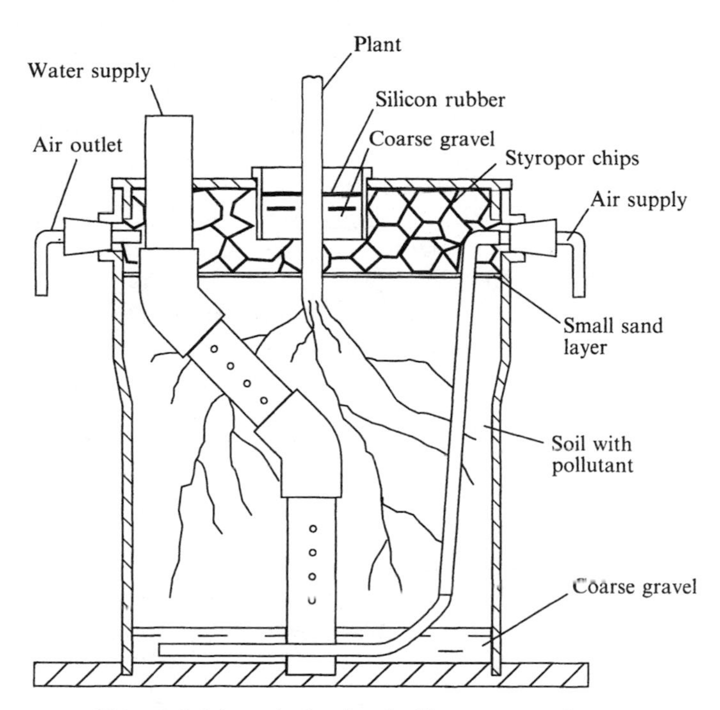

Figure 9-1 Schematic showing the bioreactor experiments.

evolution of $^{14}CO_2$ or other labeled volatiles from degradation processes in the soil and to distinguish between uptake of the original compound and a possible photosynthetic incorporation of released labeled carbon dioxide, an airtight lid separated the rhizo and aerial spheres. The shoots passed through appropriate openings in the lid and the stems were sealed into the openings with silicon rubber. A circulating air stream allowed continuous sampling of the gases from the root zone for subsequent analysis. This air stream was sucked through an identical bioreactor that in contrast contained uncontaminated soil and a plant in order to determine the uptake of that part of the radiolabeled compound (*e.g.*, compound fixation of carbon dioxide in the dark) released by the aforementioned experimental bioreactor. As the uptake by roots was very slow, the incubation period lasted 3 to 4 weeks.

UPTAKE AND METABOLISM BY PLANTS

Polychlorinated Biphenyl

Polychlorinated biphenyls (PCBs) are used as capacitor and transformer oils because of the good dielectric properties. Toxic to animals and humans (*e.g.*, causing Yusho disease), the production of PCBs was stopped in most of the industrial countries. Due to the persistence and lipophilicity, these compounds still belong to the group of most important pollutants in the environment. Sites contaminated with PCBs are a severe problem, which possibly can be solved by phytoremediation. Realization of the possibility depends on the discovery and selection of plants that metabolize PCBs. Cell suspension and hairy root cultures are the vital tools to screen numerous plants of various plant families within an appropriate time.

Plants take up PCBs from contaminated soils by the roots or by assimilation into leaves close to the ground when significant amounts of volatile PCBs are released from the soil. Animal ingestion of PCB-contaminated soil and fodder plants can contaminate humans by ingestion of fish, milk products, and animal fat, along with ingestion of contaminated vegetables. Humans accumulate PCBs in the fat tissue and in the fetus, causing headaches, acne, blood changes, liver and kidney damage, labored breathing, and weight loss. Higher chlorinated PCBs accumulate in fat tissue easier than lower chlorinated PCBs (Müller and Korte 1973, Ballschmiter *et al.* 1978, Brunn *et al.* 1989, Brunn *et al.* 1990, Safe 1990).

Wilken *et al.* (1995) describe the metabolism of 10 different PCB congeners in cell cultures. Two mixtures of five unlabeled congeners were applied to cell cultures of 12 different plant species (Table 9-1). The difference (measured by means of gas-chromatography-electron capture detector analysis) in the recovery of the parent congener between intact and inactivated cells is expressed as plant metabolism magnitude.

The decrease in PCBs, obviously attributable to metabolism, is dependent on the plant species tested and on the tested PCB congener. As for plants,

TABLE 9-1 Metabolism (in Percent) of Different Polychlorinated Biphenyl Congeners (5 Micromoles per liter) by 13 Different Plant Cell Cultures During 4-Day Exposures

PCB congener[a]	PSR (1)	GM (2)	TR (3)	HV (4)	TA (5)	PA (6)	DC (7)	LE (8)	AH (9)	Others (10 to 13)
2,2′	100	23	–	10	–	–	–	20	14	–
2,4′	90	56	70	48	–	64	10	–	–	–
4,4′	–	–	–	–	–	–	–	–	–	–
2,2′,5	90	45	30	17	83	–	–	–	–	–
2,4,4′	37	24	–	–	–	12	–	–	–	–
2,4′,5	90	27	18	–	–	20	–	–	–	–
2,2′,4,4′	–	–	10	–	–	–	–	–	–	–
2,2′,5,5′	35	12	–	–	20	–	–	–	–	–
2,2′,4,4′,5	–	–	–	–	–	–	–	–	–	–
2,2′,4,4′,5,5′	–	–	–	–	–	–	–	–	–	–

[a] Chlorine substitution location provided in the following diagram. Note the locations of the *ortho* (*o*), *meta* (*m*), and *para* (*p*) positions used elsewhere in the text:

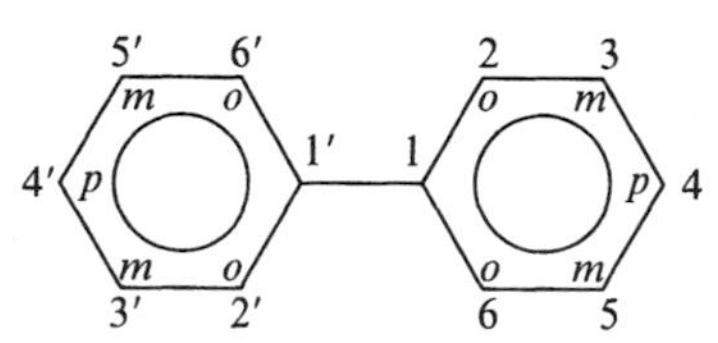

Note: Rosaceae — (1) PSR: *Rosa* spp. (Paul's Scarlet rose). Fabaceae — (2) GM: *Glycine max* L. Merr. 'Mandarin' (soybean) and (3) TR: *Trifolium repens* L. (clover). Poaceae — (4) HV: *Hordeum vulgare* L. (barley), (5) TA: *Triticum aestivum* L. 'Heines Koga II' (wheat), and (6) PA: *Pennisetum americanum* Rich. (pennisetum). Apiaceae — (7) DC: *Daucus carota* L. 'Nantaise' (carrot). Solanaceae — (8) LE: *Lycopersicon esculentum* Mill. (tomato). Chenopodiaceae — (9) AH: *Atriplex hortensis* L. (garden orach), (10) BVa: *Beta vulgaris* var. *altissima* L. 'Monogold' (sugar beet), (11) BVe: *Beta vulgaris* var. *esculenta* L. 'Bush Mono Leaves' (fodder beet), (12) CR: *Chenopodium rubrum* L. (red goosefoot), and (13) CQ: *Chenopodium quinoa* L. (oryza orach). Cultures (10) to (13) did not metabolize of the PCB congeners listed. Data for *Rosa* spp. (Paul's Scarlet rose) from Lee and Fletcher (1992) and the other cultures from Wilken *et al.* (1995).

two cultures of leguminous species (Fabaceae)—soybean (*Glycine max*) and clover (*Trifolium repens*)—were able to metabolize the largest numbers of congeners with six and four, respectively. The three grass species (Poaceae)—pennisetum (*Pennisetum americanum*), wheat (*Triticum aestivum*), and barley (*Hordeum vulgare*)—metabolized a smaller number of different congeners, sometimes to a very high degree, *e.g.*, 83 percent of PCB 18 is metabolized by wheat (*Triticum aestivum*) cultures. Carrot (*Daucus carota*) and tomato (*Lycopersicon esculentum*) cultures showed limited metabolism of PCBs, each only metabolizing one of the tested congeners to a maximum of 20 percent. Except for garden orach (*Atriplex hortensis*), none of the tested orach species (Chenopodiaceae)—sugar beet (*Beta vulgaris* var. *altissima*), fodder beet (*Beta vulgaris* var. *esculenta*), red goosefoot (*Chenopodium rubrum*) and oryza orach (*Chenopodium quinoa*)—were able to metabolize any of the tested PCBs.

Garden orach (*Atriplex hortensis*) showed a decrease in PCB 4 of 14 percent, due to plant metabolism. Clearly, the metabolic capability is strongly dependent on the plant species tested, although the limited number of tested species does not allow a generalization to whole plant families.

There seems to be a relationship between metabolism and the structural properties of PCBs. With lower chlorinated compounds, a higher plant metabolism was observed, especially for those that did not possess chlorine substituents in the *ortho*- or *meta*-position of the molecule. Plants metabolize best those PCBs that possess a free *ortho–meta* and *meta–para* position, while those with no free *ortho–meta* or *meta–para* positions are metabolized by only some of the cultures to a lesser extent. If chlorine atoms are substituted into these positions, no metabolism is detectable. None of the tested cell cultures metabolized the penta- and the hexachlorinated PCBs 101 and 153. Also, higher water solubility was related to better metabolism. An exception to these trends, PCB 15 (4,4′) was not metabolized despite a low degree of chlorination and relatively high water solubility. The coplanar configuration sterically hinders an attack of enzymes. The data obtained with cell cultures in the present study are consistent with those found for Paul's Scarlet rose (*Rosa* spp.) (Lee and Fletcher 1992).

A special extraction scheme (Figure 9-2) was used to investigate the metabolism and to separate applied [^{14}C]-PCB and the associated metabolites. The plant material was incubated with the [^{14}C]-labeled PCB. After incubation for 96 hours, the medium was separated from the cells by filtration. The cells were ground in liquid nitrogen and extracted by the addition of dichloromethane and methanol (Bligh–Dyer extraction, Bligh and Dyer 1959). After filtration and separation, an inextricable residue and a monophase extract

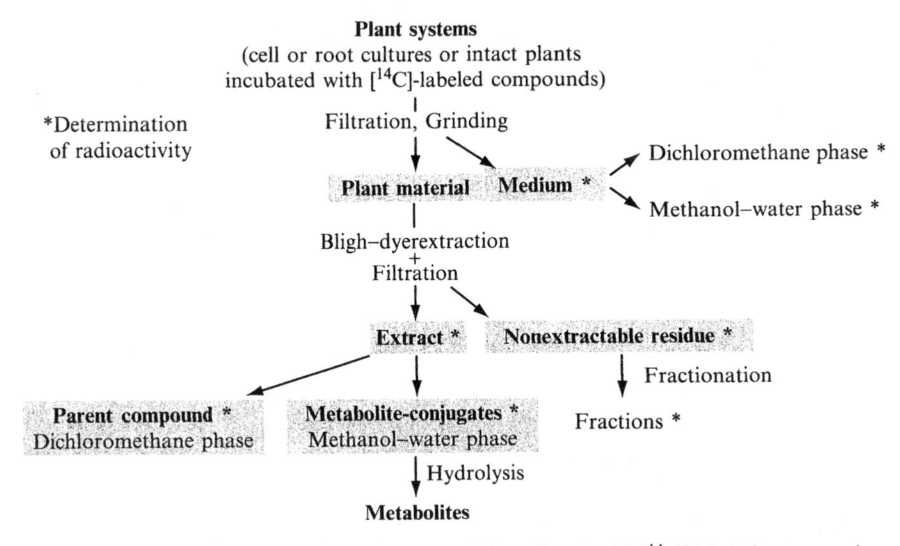

Figure 9-2 Extraction protocol for plant material incubated with [^{14}C]-labeled compounds.

TABLE 9-2 Distribution of Radioactivity (in Percent of [^{14}C]) in Different Fractions of Cell Cultures after 96 Hours Exposure to Three Different Polychlorinated Biphenyls

Fraction	2-Chloro-biphenyl (Soybean)	2,2′,5,5′-Tetra-chlorobiphenyl (Wheat)	3,3′,4,4′-Tetrachlorobiphenyl		
			(Tomato)	(Lettuce)	(Rose)
Medium	7	7	13	<1	6
Cell extract	58	69	81	95	81
Methanol–water phase	**36**	**57**	**2**	**8**	**18**
Dichloromethane phase	**22**	**12**	**79**	**87**	**63**
Inextricable residue	10	8	1	1	2
Recovery	75	84	95	<97	89

Note: *Glycine max* (soybean), *Triticum aestivum* (wheat), *Lycopersicon esculentum* (tomato), *Lactuca sativa* L. (lettuce), and *Rosa* spp. (Paul's Scarlet rose). With permission from Bock and Harms (1996) *Organohalogen Compounds* p. **28:** 55.

resulted. In a second step, a solution of dichloromethane and water was added to the monophase extract resulting in a two-phase system. The system involved a more polar methanol and water phase, which contain the conjugated metabolites and a less polar dichloromethane phase with the extracted PCB. The metabolic activity was estimated by the amount of radioactivity in the methanol and water phase. The same separation was performed for the medium.

For investigating the metabolism of PCBs (Bock and Harms 1996) [^{14}C]-labeled 2-chlorobiphenyl (PCB 1) and 2,2',5,5'-tetrachlorobiphenyl (PCB 52) as well as 3,3′,4,4′-tetrachlorobiphenyl (PCB 77) were applied to five plant cell cultures. Table 9-2 shows the distribution of radioactivity.

In all cases, the major amount of radioactivity has been detected in cell extracts. Whereas for PCB 1 and PCB 52, higher amounts were present in the methanol and water fraction, PCB 77 occurred in higher amounts in the dichloromethane phase. The latter was due to the parent compound, whereas radioactivity in the methanol and water phase was associated with polar metabolites. Paul's Scarlet rose (*Rosa* spp.) was the only plant culture, which showed noteworthy metabolism. Analysis of the methanol and water phase revealed several polar metabolites that yielded a number of products after hydrolysis with hydrochloric acid. On the basis of cochromatography with standards, these products were identified to be hydroxylated PCBs. The hydroxylation reactions seem to be similar to that described for the metabolism of PCBs in animals (Heeschen *et al.* 1993). Cytochrome P-450 dependent monooxygenases oxidize PCBs *via* epoxide intermediates to phenols. Depending on the actual structure of a PCB, the enzymes are induced differently. Different cytochrome P-450 isoenzymes catalyze the hydroxylation reactions forming different metabolites. Induced by planar PCBs, P-450 dependent monooxygenases [3-methylcholanthrene (MC)-type] catalyze a hydroxylation in the *ortho* position. The metabolism of PCB 77 that has been studied in detail (Bock 1999) is an example of this kind of hydroxylation reaction. This tetrachlorinated PCB 77, which has low water solubility (1.88×10^{-9} moles per liter), was metabolized by only three of the tested cultures. These were tomato (LE) (*Lycopersicum esculentum*), lettuce (LS) (*Lactuca sativa*), and to the greatest extent, Paul's Scarlet rose (PSR) (*Rosa* spp.). No other tested cultures metabolized PCB 77. A chromatographic separation shows that several polar compounds formed. The gas chromatography–mass spectrometry analysis reveals that these metabolites are monohydroxy and dihydroxy conjugates of PCB 77 (Figure 9-3). The main metabolites identified are 2-hydroxy-3,3′,4,4′-tetrachlorobiphenyl and 6-hydroxy-3,3′,4,4′-tetrachlorobiphenyl. Furthermore, a 5-hydroxy-3,3′,4,4′-tetrachlorobiphenyl has been identified in lettuce (*Lactuca sativa*) cultures.

Those P-450-dependent monooxygenases [phenobarbital (PB)-type] that are induced by nonplanar PCBs catalyze a hydroxylation in the *para* position in the lower chlorinated ring of the molecule. Examples for this are the metabolism of PCB 1 and PCB 52, which have been described elsewhere (Bock and Harms 1996, Bock 1999). The main metabolite of PCB 1 transformation has been identified as 2-chloro-5-biphenylol. Another transformation product is 3-chloro-2-biphenylol, while the most polar product was 2′-chloro-3,

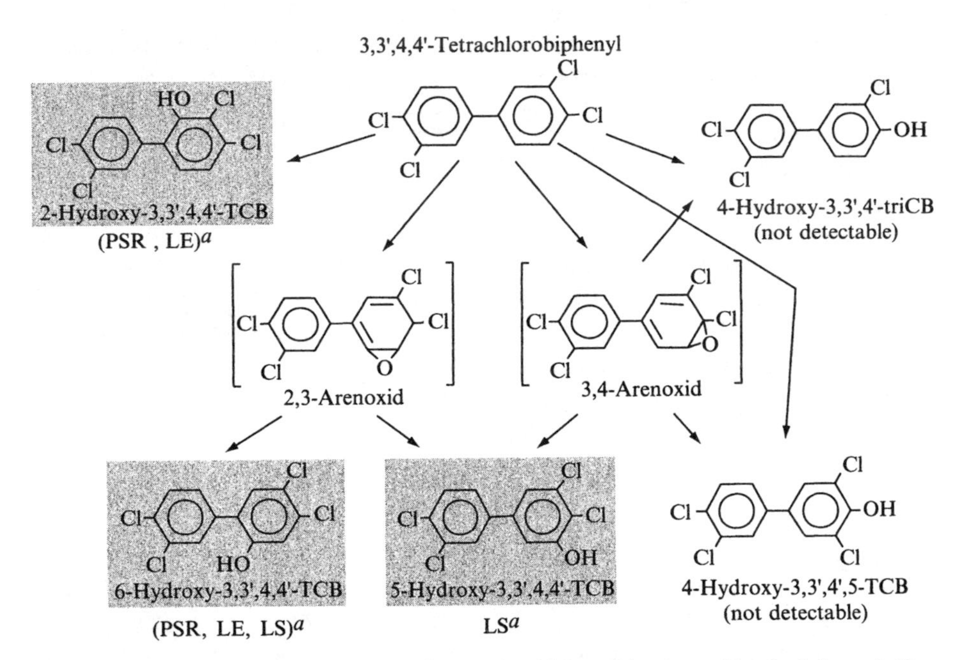

Figure 9-3 Metabolism of PCB 77 (3,3′,4,4′-tetrachlorobiphenyl) in plants. The shaded metabolites have been detected in cell cultures of various plant species, the others are known from animal metabolism but are not detectable in the investigated cell cultures. *Note*: [a]Abbreviations for cell cultures are PSR: *Rosa* spp. (Pauls's Scarlet rose); LE: tomato (*Lycopersicon esculentum*); and LS: lettuce (*Lactuca sativa* L.).

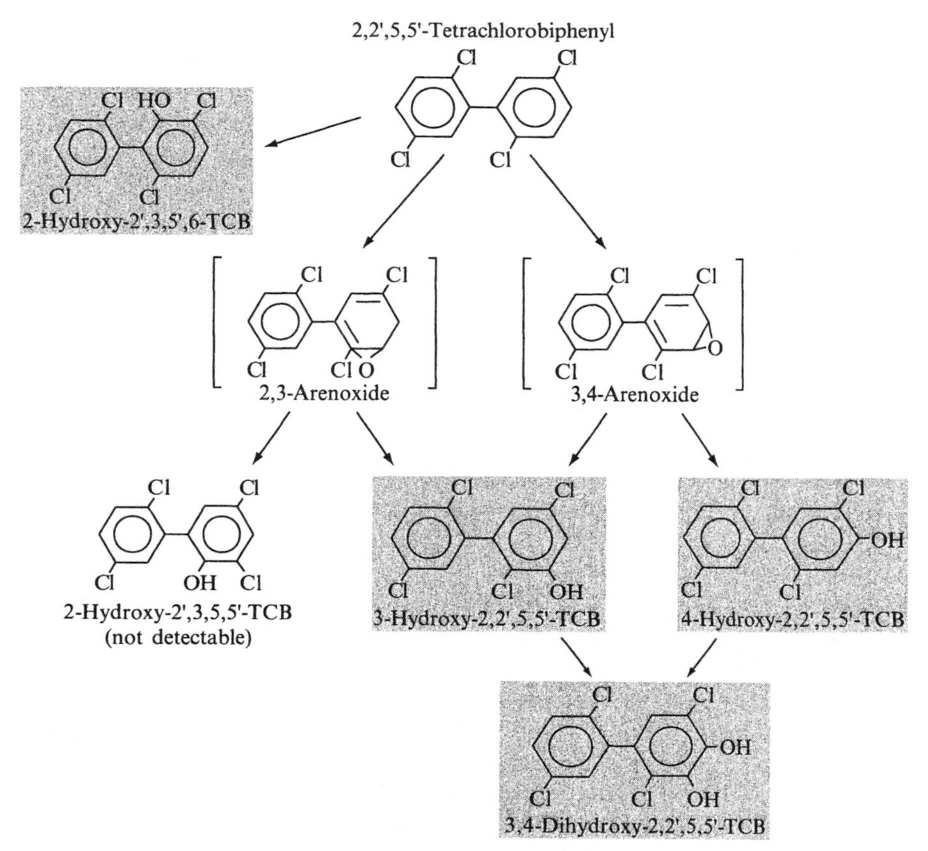

Figure 9-4 Metabolism of PCB 52 (2,2′,5,5′-tetrachlorobiphenyl) in plants. The shaded metabolites have been detected in cell cultures of various plant species, the others are known from animal metabolism but are not detectable in the investigated cell cultures.

4-biphenyldiol. Figure 9-4 illustrates the transformation pathway for PCB 52. For PCB 52, 2,2′,5,5′-tetrachloro-4-biphenylol was the main metabolite, but also monohydroxylated and dihydroxylated biphenyls could be detected.

These aforementioned investigations demonstrate that plant cells metabolize even nonpolar PCB congeners. The metabolism seems to be very species-specific. Chenopodiaceae cell cultures are not able to metabolize any of the tested PCBs (see Table 9-1), while Paul's Scarlet rose (*Rosa* spp.) cultures metabolize most of the tested congeners. Therefore, Paul's Scarlet rose (*Rose* spp.) should be investigated further for application of phytoremediation to clean up PCB contaminated sites.

Polycyclic Aromatic Hydrocarbons

Polycyclic aromatic hydrocarbons (PAHs) are undoubtedly the most widespread carcinogens in the environment. Environmental contamination by

PAHs arises mainly from industrial releases, household disposal, and motor vehicle exhausts, in addition to related activities of modern civilization. The pyrolysis or incomplete charring of organic substances at temperatures between 500 and 700 °C release an array of more than 160 PAHs. The distribution of PAHs occurs mainly by waste releases and transport in the water and air. Rainfall removes PAHs from the atmosphere, leaving the contaminant in surface waters and soils and on plant surfaces.

Benzo(a)pyrene

Benzo(*a*)pyrene is an indicator for the risk due to total PAHs because this compound is known to be the most carcinogenic. The uptake and metabolism of this compound has been a subject of numerous investigations (Harms 1983, Harms and Kottutz 1991, 1992). Experiments with cell cultures derived from vertebrates, alveolar macrophages, or bacteria showed that benzo(*a*) pyrene is metabolized to oxygenated derivatives. Plants too, are able to metabolize this compound to quinones and other oxygenated derivatives. Kinetic studies (Harms *et al.* 1977) with [^{14}C]-labeled benzo(*a*)pyrene and cell suspension cultures of red goosefoot (*Chenopodium rubrum*) have shown that benzo(*a*)pyrene-3,6-quinone and benzo(*a*)pyrene-1,6-quinone, respectively, are the first detectable metabolites formed (Figure 9-5). These metabolites point to the involvement of cytochrome P-450 type enzymes. With increasing time of incubation, the amount of quinones decreased and radioactivity could be detected in the diol and polar compound fractions instead. With extended incubation, the amount of radioactivity in the extracted residue also increased. There seems to be a relationship between the amount of oxygenated metabolites formed and incorporation into insoluble fractions.

Fluoranthene

Fluoranthene is one of the most abundant PAHs detected in air, water, soils, sediments, and biota, including humans. This particular PAH has been reported to be mutagenic (Kaden *et al.* 1979, Thilly *et al.* 1980, Barfknecht *et al.* 1982, LaVoie *et al.* 1982), cocarcinogenic with benzo(*a*)pyrene (Van Duuren and Goldschmidt 1976), and potentially carcinogenic (Busby *et al.* 1984, LaVoie *et al.* 1994). In contrast to the widely accepted view that PAHs are not taken up by plants, but absorb only onto surfaces, Fritz (1983) and Wetzel *et al.* (1994) reported uptake of fluoranthene and other PAHs into root cores and shoots. Especially in cases of heavy-metal-stressed soils, PAH uptake was stimulated (Haas *et al.* 1990). To judge possibilities of phytoremediation for nonpolar compounds such as fluoranthene, detailed investigations on the uptake and metabolism were performed. To assess food quality, the metabolic fate of fluoranthene in plants must also be known. Metabolic pathways of fluoranthene have been studied in investigations of microbial degradation and of the toxic effects on mammals. However, information about the metabolic fate of fluoranthene in plant systems is lacking.

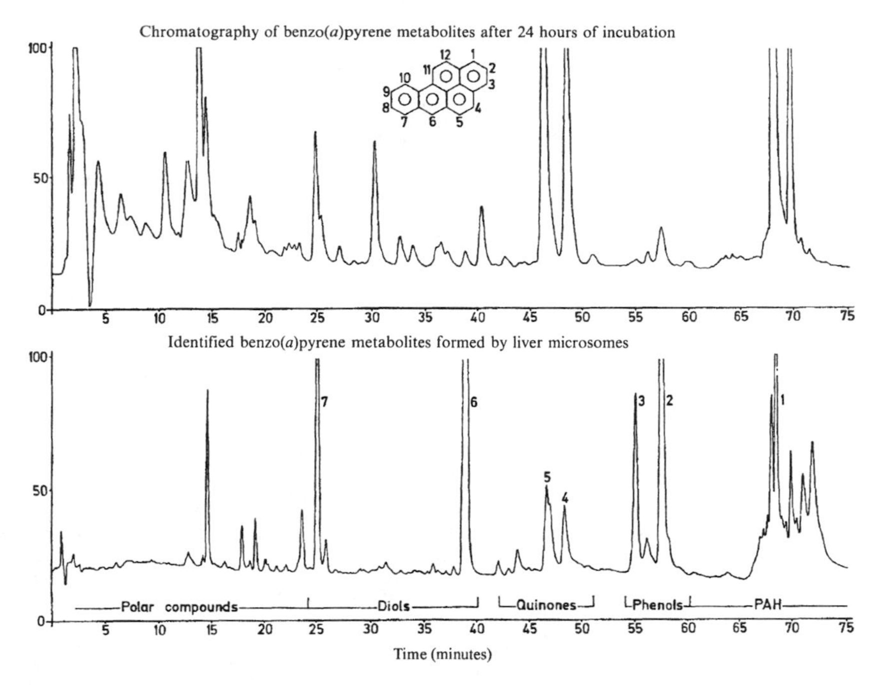

Figure 9-5 High performance liquid chromatography traces of benzo(*a*)pyrene metabolites. The numbered peaks are: (1) benzo(*a*)pyrene, (2) 3-hydroxybenzo(*a*)pyrene, (3) 9-hydroxybenzo(*a*)pyrene, (4) benzo(*a*)pyrene-3,6-quinone, (5) benzo(*a*)pyrene-1,6-quinone, (6) 7,8-dihydro-7,8-dihydroxybenzo(*a*)pyrene, and (7) 9,10-dihydro-9,10-dihydroxybenzo(*a*)pyrene with permission from Harms *et al.* (1997) *Zeitschrift für Naturforschung* **32c**: p. 325 with permission.

In analyzing the metabolism of this hazardous compound, cell suspension cultures of various plant species have been incubated with 1 milligram per liter of fluoranthene. The solution for incubation was a mixture of [^{14}C]-3-fluoranthene and unlabeled fluoranthene in methanol (Kolb and Harms 2000). Table 9-3 shows the distribution of the radioactivity after extraction and subsequent liquid–liquid partitioning (see Figure 9-2) between nonpolar (dichloromethane) and polar solvents (methanol–water) and inextricable (bound) residues.

Thin layer chromatographic analysis of the nonpolar dichloromethane phase demonstrates that most of the applied activity is still present as parent compound, whereas the radioactivity recovered in the methanol and water phase was associated with metabolic products of fluoranthene. Only the cultures of lettuce (*Lactuca sativa*), tomato (*Lycopersicon esculentum*), wheat (*Triticum aestivum*), and rose (*Rosa* spp.) revealed remarkable turnover rates. Autoclaved plant cells were the control. These autoclaved cells did not produce any metabolites. This demonstrates that no abiotic reactions occurred.

Further chromatographic analysis was performed on the cell extracts of those cultures showing higher turnover rates. The pattern of metabolites seems to be similar for all cultures tested. While tomato (*Lycopersicon*

TABLE 9-3 Distribution of Radioactivity (in Percent) in Different Fractions after Application of [^{14}C]-3-Fluoranthene to Various Cell Cultures for 2 Days

	Cell Extractions and Residue				
Cell culture[a]	Dichloromethane phase	Methanol–water phase	Bound residue	Medium	Recovery
Garden orach	90.3	4.4	0.3	4.8	99.8
Sugar beet	70.8	1.2	2.2	4.3	78.5
Fodder beet	91.4	2.7	0.8	3.9	98.8
Rape	67.0	1.5	1.7	8.7	78.9
Carrot	85.1	1.9	1.1	4.1	92.2
Soybean	74.6	4.2	4.5	8.7	92.0
Barley	82.5	5.3	0.4	4.2	92.4
Lettuce	66.5	6.3	4.9	7.9	85.6
Tomato	72.4	15.0	0.7	5.4	92.6
Paul's Scarlet rose	16.1	49.9	1.6	27.2	95.8
Wheat	73.0	9.1	1.7	4.7	88.4

[a]Rape is *Brassica nupus* L. Other scientific names provided in Table 9.1. and 9.2.

Notes: All data are given in percent of applied [^{14}C]-3-fluoranthene and are mean values of three replicate tests. Reprinted with permission from M. Kolb and H. Harms (2000) Metabolism of fluoranthene in different plant cell cultures and intact plants. *Environ. Toxicol. Chem.* **19**: p. 1306. Copyright SETAC, Pensacola, Florida, USA.

esculentum) and wheat (*Triticum aestivum*) produce two main metabolite peaks and two other small ones, the rose culture only formed two peaks corresponding with the main metabolites.

After hydrochloric acid hydrolysis to release the metabolites from conjugation with cell components (*e.g.*, glucose and glucuronic acid), a clean up with gel permeation chromatography was performed. The samples were separated by semipreparative high performance liquid chromatography into two [Paul's Scarlet rose (*Rosa* spp.)] or four [tomato (*Lycopersicon esculentum*)], fractions. Because of the intensive fluorescence, it was possible to detect the metabolites very accurately without any radioactive label by comparing the peak pattern with the pattern of corresponding radioactive samples.

For identification of the metabolites, the concentrated fractions were analyzed by electron-impact ionization and high performance liquid chromatography diode array detection. Upon gas chromatography–mass spectrometry analysis of chromatographically separated fractions 3 and 4 of the Paul's Scarlet rose (*Rosa* spp.) and tomato (*Lycopersicon esculentum*), mass spectra with a base peak of the mass to charge ratio of 218 were detected (Table 9-4). These base peaks could be attributed to the molecular ions [M^+] of monohydroxylated fluoranthene isomers. In addition, the mass spectra contained fragment ions with a mass to charge ratio of 189 [M^+-CHO]. These data agree with those from the gas chromatography–mass spectra specification of Rice *et al.*

TABLE 9-4 Gas Chromatography–Mass Spectra (GC–MS) Data for the Metabolite Fractions of Paul's Scarlet rose (*Rosa* spp.) and Tomato (*Lycopersicon esculentum*) Isolated by Semipreparative High Performance Liquid Chromatography

Cell culture	High performance liquid chromatography fraction (retention time window)	GC–MS retention index[a]	Mass to charge ratio (relative intensity)
Paul's Scarlet rose	Fraction 3 (24 to 26 minutes)	2340	218[b] (100), 189 (50)
	Fraction 4 (27 to 29 minutes)	2320	218 (100), 189 (62)
Tomato	Fraction 3 (24 to 26 minutes)	2342	218 (100), 189 (50)
	Fraction 4 (27 to 29 minutes)	2330	218 (100), 189 (53)
		2358	218 (100), 189 (58)

[a]Retention time in relation to the retention of *n*-alkanes analyzed under the same conditions.
[b]Base peak of the mass spectrum.

Note: Reprinted with permission from M. Kolb and H. Harms (2000) Metabolism of fluoranthene in different plant cell cultures and intact plants. *Environm. Toxicol. Chem.* **19**: p. 1308. Copyright SETAC, Pensacola, Florida, USA.

(1983) for fluoranthenols. Using spectra libraries of the ultraviolet-visible light spectrum (Rice *et al.* 1983, Babson *et al.* 1986), high performance liquid chromatography diode array detection analyses were compared to confirm the formation of fluoranthenols and to identify the position of the OH-group. Figure 9-6 shows the detected spectra.

The comparisons of the ultraviolet–visible light spectra with those of standards confirm that the metabolites formed in tomato (*Lycopersicon esculentum*) cultures correspond to 8-hydroxyfluoranthene, 1-hydroxyfluoranthene, and 3-hydroxyfluoranthene. In cultures of Paul's Scarlet rose (*Rosa* spp.), only 8-hydroxyfluoranthene and 1-hydroxyfluoranthene are detectable.

In a second approach, the metabolism of fluoranthene was studied with intact plants. For that purpose, aseptically grown tomato (*Lycopersicon esculentum*) plants were incubated with [^{14}C]-fluoranthene in hydroponic solution. Four weeks after sterilization of the seeds, aseptic seedlings were transferred into fresh nutrient solution containing [^{14}C]-fluoranthene mixed with unlabeled fluoranthene (1 milligram per liter nutrient solution). The distribution of the applied radioactivity after an incubation time of 14 days was analyzed. The highest proportion of the radioactivity (62 percent) was recovered from the roots. Of this activity, 92 percent (59 percent of the total) was extracted with dichloromethane. Some proportion of the 62 percent might be adsorbed at the root surface, especially as the concentration was higher than the fluoranthene water solubility. However, the 8 percent radioactivity (3 per cent of the total) recovered in the polar methanol and water phase proves that the roots took up and metabolized some fluoranthene. The shoots contained small proportions of the radioactivity. More than two thirds of these small proportions were polar metabolites. Nevertheless, comparing the absolute values, larger quantities of metabolites were formed in the roots than in the

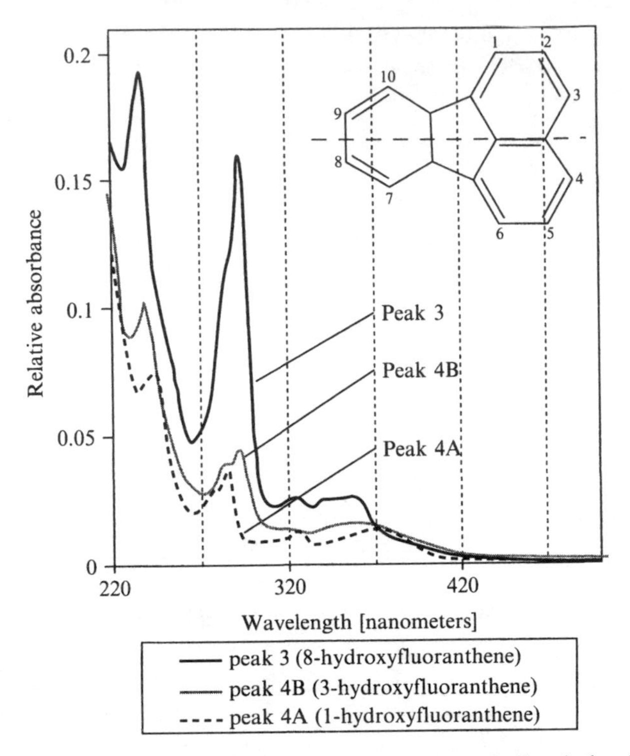

Figure 9-6 Ultraviolet–visible light spectra of fluoranthene metabolites isolated from tomato (*Lycopersicon esculentum*) incubated with fluoranthene for 96 hours. Peak 3: ultraviolet visible light spectrum of the peak detected in chromatographically separated fraction 3 (see Table 9-4). Peak 4A/4B: ultraviolet–visible light spectra of the peaks detected in chromatographically separated fraction 4 (see Table 9-4). Reprinted with permission from M. Kolb and H. Harms (2000) Metabolism of fluoranthene in different plant cell cultures and intact plants. *Environ. Toxicol. Chem.* **19**: p. 1308. Copyright SETAC, Pensacola, Florida, USA.

shoots. No mineralization to $^{14}CO_2$ took place in 14 days. As the system was sterile, this confirms that plants do not mineralize fluoranthene.

The hydrolyzed metabolite extracts of shoots and roots were analyzed with high performance liquid chromatography and radiochemical detection. Figure 9-7 compares the chromatograms for shoot and roots with the chromatogram of the hydrolyzed metabolites formed by tomato (*Lycopersicon esculentum*) cell cultures.

The qualitative peak patterns are identical. Consequently, the fluoranthenols identified in cell cultures are also formed in intact plants. Only the quantitative levels differed. Related to the monohydroxylated fluoranthenes, higher fractions of the more polar metabolites (retention time window 15 to 20 minutes) were found in the shoots than in the roots. In the roots, the monohydroxylated fluoranthene isomers dominated the metabolites as in the cell culture. The same metabolites that we identified (*i.e.*, 1-hydroxyfluoranthene, 3-hydroxyfluoranthene, and 8-hydroxyfluoranthene) were also identified

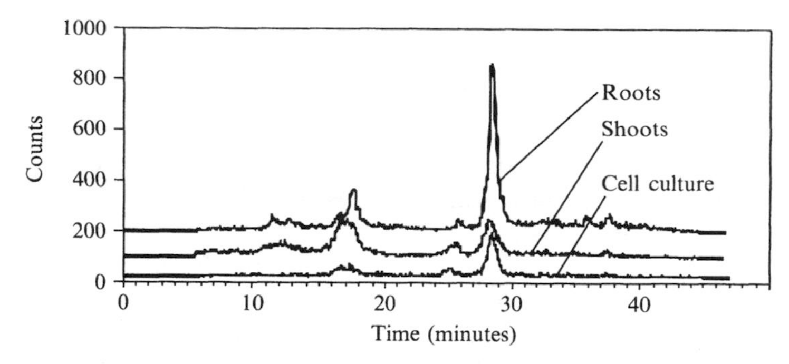

Figure 9-7 High performance liquid chromatography traces of [^{14}C]-fluoranthene metabolites after hydrolysis in tomato cell cultures and roots and shoots of intact tomato (*Lycopersicon esculentum*). Reprinted with permission from M. Kolb and H. Harms (2000) Metabolism of fluoranthene in different plant cell cultures and intact plants. *Environ. Toxicol. Chem.* **19**: p. 1309. Copyright SETAC, Pensacola, Florida, USA.

in metabolism studies with rat hepatic microsomal enzymes (Babson *et al.* 1986). This finding agrees with the green-liver concept of Sandermann (1994) regarding similarity in the metabolic and enzymatic fate of xenobiotic chemicals in plants and animals. Pothuluri *et al.* (1992) found 1-hydroxyfluoranthene and 8-hydroxyfluoranthene to be formed by rat liver microsomes and Rice *et al.* (1983) determined 7-hydroxyfluoranthene as a major metabolite of hepatic enzymes. Unlike our plant studies, *trans*-2,3-dihydroxy-2,3-dihydrofluoranthene was observed to be the major hepatic metabolite in the animal studies. This metabolite and the corresponding 2,3-quinone were involved in the major metabolic activation pathway for the mutagenicity of fluoranthene (Babson *et al.* 1986). The monohydroxylated compounds were not claimed to be biologically active. Furthermore, Palitti *et al.* (1986) reported the presence of 8-hydroxyfluoranthene, 7-hydroxyfluoranthene, and 3-hydroxyfluoranthene metabolites in very low quantities in serum and blood of mice treated with fluoranthene. Some toxic effects on the proliferation of bone marrow cells were noticed after this treatment.

In conclusion, the results of these experiments confirm that fluoranthene is taken up slightly in roots and shoots of plants as reported by Fritz (1983) and Wetzel *et al.* (1994). Further, hydroxylation is the most important metabolic pathway for fluoranthene. Even though uptake and metabolism are negligible, plants should be selected for rhizodegradation to avoid temporary accumulation. Contamination of the plant material with parent compounds and potentially toxic metabolites, especially those that are not covered by normal residue analyses, must be considered.

As uptake rates are dependent on the plant species, the tests with the cell suspension cultures were a very useful tool for the selection of the most suitable species for further studies with intact plants. Additionally, using cell cultures better enabled the identification of the metabolites.

Phenolic Compounds and Aromatic Amines

Pentachlorophenol and 4-Chloroaniline

Pentachlorophenol and the associated salts are wood preservatives because of the antimicrobial, herbicidal, and insecticidal properties. The compound has also been used as an herbicide in rice. A known degradation product of a variety of substituted phenylurea compounds which are used mainly as herbicides, You *et al.* (1982) thoroughly studied the fate of 4-chloroaniline in the soil and in soil microorganisms. A common feature of the metabolism of these compounds in soil and plants is a large proportion of inextricable residues bound to high molecular weight compounds such as humic substances in soil or lignin in plants.

The validity of extrapolating data obtained with cell culture techniques to those of intact plants is still a matter of debate. In order to compare these two systems, cell suspension cultures and wheat (*Triticum aestivum*) seedlings of the same cultivar were incubated with pentachlorophenol and 4-chloroaniline (Langebartels and Harms 1984, 1985). Figure 9-8 shows the metabolic fate of these compounds in the two differentiated plant systems.

Both cell cultures and whole plants take up and metabolized pentachlorophenol and 4-chloroaniline. Intact wheat (*Triticum aestivum*) plants translocate the [^{14}C]-label of both compounds from the roots to the shoots. Cell cultures adsorbed pentachlorophenol very rapidly and formed significant amounts of polar metabolites, which were mainly associated with the cells. Of the radiolabeled pentachlorophenol, 41 percent was conjugated into the inextricable residue. The [^{14}C]-label was bound mainly to lignin and to a high molecular weight hemicellulose fraction. Polar conjugates were also extracted from roots and shoots, and pentachlorophenol glycosides were predominant as found in the cultured cells. More than 16 percent of the total pentachlorophenol radioactivity from shoots and roots were bound residues. These were fractionated into several cell wall components similar to that in cell cultures.

In cell cultures, 72 percent of 4-chloroaniline was bound residue. Further studies showed that this high proportion of [^{14}C]-label was mainly associated with the pectin and lignin fractions of the cell wall. In addition, the roots of wheat (*Triticum aestivum*) plants rapidly absorbed and metabolized 4-chloroaniline to polar conjugates. The radioactivity of the bound residue fraction reached more than 61 percent which is comparable to the percentage found in cell cultures.

Nonylphenol

One of the persistent products in the degradation of alkylphenol-polyethoxylates (nonionic surfactants), 4-nonylphenol is frequently found in sewage sludge and effluents of sludge treatment. The compound has been of concern to the public since aquatic toxicity was proven in the 1980s (Giger *et al.* 1981, Giger *et al.* 1984, BUA 1988, DiCorcia *et al.* 1994, Sweetman 1994, Jobst 1995). More recent discussions on bans or plans to introduce environmental quality

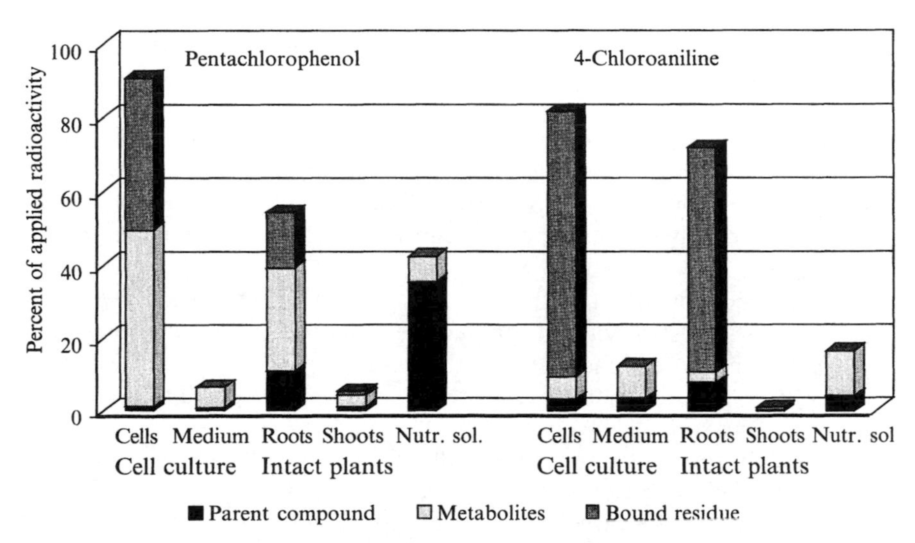

Figure 9-8 Metabolism of pentachlorophenol and 4-chloroaniline in wheat (*Triticum aestivum*) cell suspension cultures and intact plants. Note that nutr. sol. is nutrient solution. The data plotted represent averages from tables in Harms and Langebartels (1986) for cell cultures exposed for 48 hours and in Langebartels and Harms (1986) for intact (whole) plants exposed for 72 hours to 1 milligram per liter of each compound. Harms (1992, 1996) provide similar but slightly different plots based on repeated observations.

standards have been triggered by findings that nonylphenols are weakly estrogenic (Soto *et al.* 1991, Purdom *et al.* 1994, White *et al.* 1994, Jobling *et al.* 1996, Servos 1999). The example of nonylphenol accumulating during the degradation of nonylphenol-polyethoxylates shows that the disappearance of the parent compound does not necessarily prove complete degradation, and metabolites that are even more toxic may accumulate. Information on plant metabolism of nonylphenol is scarce, although application of sludge to agricultural land and the use of the associated wastewater in irrigation may have an impact on plant growth and use for food or feedstuff. Data concerning toxicity, uptake, and the metabolic pathways will contribute to the evaluation of certain plants for phytoremediation and animal and human consumer risks.

Technical grades of 4-nonylphenol consist of numerous differently branched isomers. In order to reduce the complexity of the experiments we used 4-*n*-nonylphenol (Figure 9-9). Investigation of plant uptake and metabolism focused on intact plants under aseptic conditions and in soil plant systems using radioactively labeled [^{14}C]-compound.

In metabolism studies, the distribution of radioactivity following incubation (with cell suspension cultures usually for 48 hours) was investigated in different fractions (see Figure 9-2). The distribution of radioactivity was a function of the concentration in cell suspension cultures of carrot (*Daucus carota*). With increasing concentrations of 4-*n*-nonylphenol, increasing amounts of 4-*n*-nonylphenol-equivalents (calculated on the basis of radioactivity) occurred in plant cells (Figure 9-9). With higher concentrations, higher proportions of 4-*n*-nonylphenol equivalents occurred as inextricable residues, indicating an important role of segregation in detoxification (Bokern and Harms 1997). This finding may be important for phytoremediation, as uptake and metabolism appear to be strongly dependent on the concentration of the chemical available to the plant cell.

The influence of the tested plant species on the metabolism of 4-*n*-nonylphenol was tested with 14 different cell cultures from seven plant families. The cultures were incubated with 4-*n*-nonyl[^{14}C]phenol at a nontoxic concentration of 1 milligram per liter (4.5 micromoles per liter). Table 9-5 documents the results.

All cell cultures took up and metabolized 4-*n*-nonylphenol. No $^{14}CO_2$ occurred in the gaseous phase indicating that plant cells do not mineralize 4-*n*-nonylphenol to carbon dioxide. Except for soybean (*Glycine max*), the major proportion of radioactivity was recovered from the polar fraction of the soluble cell extract (Bokern and Harms 1997). This indicates that the metabolites formed are more polar than the parent 4-*n*-nonylphenol, which was extracted into the nonpolar phase. Figure 9-10 shows a high performance liquid chromatographic separation of soluble metabolites formed by wheat (*Triticum aestivum*) cell cultures (A) and the same extracts after acid hydrolysis (B).

The metabolites were partly purified and the structures elucidated by means of mass spectrometric techniques. The two major high performance liquid chromatography peaks appearing in the hydrolyzed extract were identified to be monohydroxy-4-*n*-nonylphenols (peak II) and dihydroxy-4-*n*-nonylphenols (peak I). Except for C-1 and C-9, the hydroxylations occurred at all carbon atoms of the alkyl side chain (Bokern *et al.* 1996). Table 9-6 shows the composition of the 4-*n*-nonylphenol metabolites that have not undergone hydrolysis.

The analysis of the compounds that had not been hydrolyzed gave for fraction [a] (Figure 9-10) several dihydroxylated 4-*n*-nonylphenols (positions of the OH groups in the aliphatic side chain of 4-*n*-nonylphenol at the carbon atoms 5,7; 3,7; and 4,7), which were conjugated with glucose and glucuronic acid. Fraction [b] contained a compound isomeric to the dihydroxylated compounds in fraction [a], with OH groups in positions 4 and 8 of the aliphatic side chain. Fraction [c] was heterogeneous, containing monodihydroxylated derivatives [position of the OH groups at carbon atom(s) 7-H_2O, 4,7; 5,7; 4,8]

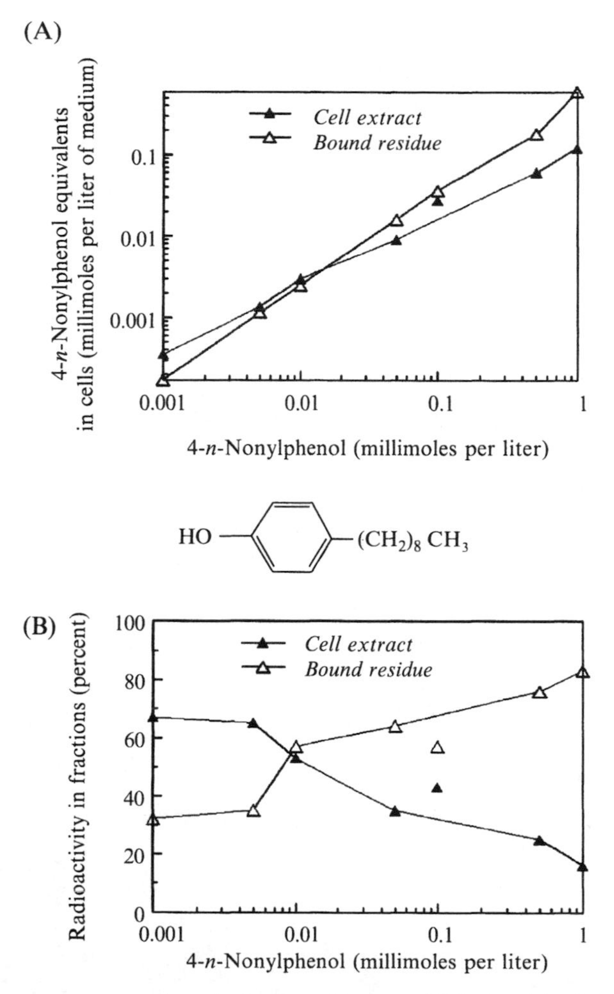

Figure 9-9 Concentration-dependent distribution of 4-*n*-nonyl[^{14}C]phenol in different fractions of carrot (*Daucus carota*) cell suspension cultures exposed to an initial concentration of 1 milligrams per liter for 2 days. (A) Absolute amounts of 4-*n*-nonylphenol equivalents detected in the soluble extract and in the inextricable residue calculated on the basis of radioactivity. (B) Relative distribution of radioactivity (percent) in cell extracts and bound residue. Adapted from Bokern and Harms (1997).

conjugated with glucose and different organic acids, *e.g.*, malonic acid. Fraction [d] yielded a dominant mass spectroscopic signal corresponding to monohydroxylated 4-*n*-nonylphenols (position of the OH group at carbon atoms 4, 5, 6, 7, or 8 of the aliphatic side chain), conjugated with a glucuronosyl–glucose moiety. In all cases, the phenolic–hydroxy group was conjugated (Bokern *et al.* 1996).

TABLE 9-5 Distribution of Radioactivity (in Percent) 2 Days after Application of 1 Milligram per Liter 4-*n*-Nonylphenol to Cell Suspension Cultures of Various Plant Species

	Cells[b]			Medium		
Cell culture[a]	Dichloromethane phase	Methanol–water phase	Bound residue	Dichloromethane phase	Methanol–water phase	Recovery (percent)
Lupinus polyphyllus	5.4	62.5	25.7	5.1	1.8	100.5
Lupinus hartwegii	8.8	77.2	11.5	0.8	1.7	100.0
Glycine max	6.2	25.0	31.7	(39.1)[d]		102.0
Daucus carota	12.7	33.4	22.4	14.0	10.9	93.4
Lactuca sativa	3.7	18.8	38.4	0.6	15.8	77.3
Lycopersicon esculentum	0.0	87.0	6.6	0.0	5.7	99.3
Atriplex hortensis	42.8	45.1	4.8	4.6	1.6	98.9
Chenopodium quinoa	70.8	7.9	2.5	7.5	0.5	89.2
Chenopodium rubrum	13.9	71.3	3.8	2.4	3.5	94.9
Beta vulgaris var. *esculenta*	23.9	56.8	11.8	0.0	5.6	98.1
Beta vulgaris var. *altissima*	0.0	68.5	2.0	0.0	5.5	76.0
Hordeum vulgare	20.4	52.6	6.5	(3.5)[c]		83.0
Pennisetum americanum	0.0	64.0	10.1	2.8	6.2	83.1
Triticum aestivum	2.1	60.6	15.8	(5.0)[c]		83.5

[a]See Tables 9-1 and 9-2 for common plant names of cell cultures; note the following: *Lupinus polyphyllus* (lupin, perennial) and *Lupinus hartwegii* (lupin, annual). The seven families involved include Leguminosae, Fabaceae, Apiaceae, Compositae, Solanaceae, Chenopodiaceae, and Poaceae.
[b]All data are given in percent of applied [^{14}C]-4-nonylphenol and are average values of five combined parallel experiments.
[c]Medium not separated into two phases.

Note: Adapted from Bokern and Harms (1997).

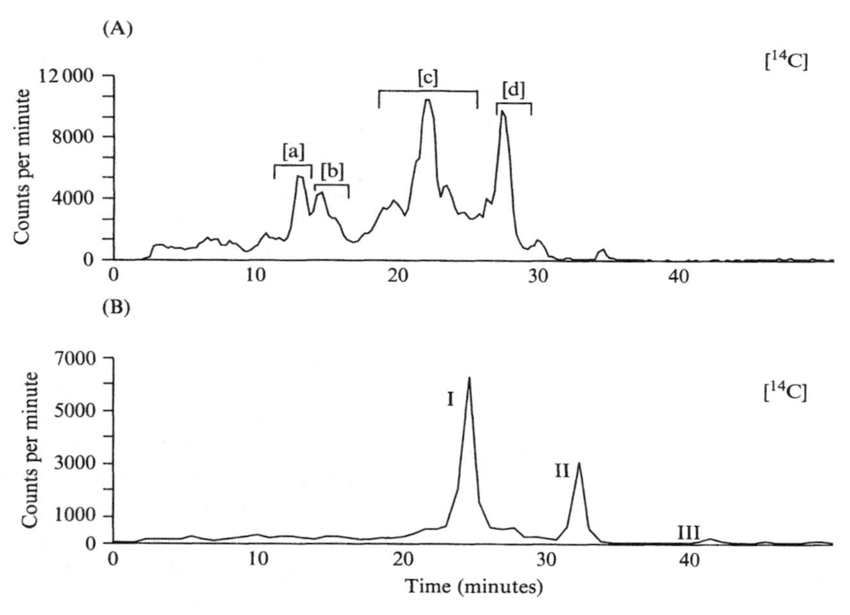

Figure 9-10 High performance liquid chromatographic traces for a crude extract of a wheat (*Triticum aestivum*) cell suspension culture after 2 days exposure to 1 milligram per liter of 4-*n*-nonylphenol (A) and after hydrolysis (B). Reprinted with permission from Bokern and Harms (1997). Copyright (1997) American Chemical Society.

TABLE 9-6 Metabolites Formed by Wheat (*Triticum aestivum*) Cells 2 Days after Exposure to 1 milligram per liter of 4-*n*-Nonylphenol

Fraction[a]	Glucose	Glucuronic acid	Position of OH-groups on 4-*n*-nonylphenol
[a]	++[b]	+[c]	5,7 (70 percent)[d]; 3,7 (20 percent); 4,7 (10 percent)
[b]	++	+	4,8 (80 percent)
[c]	++	–[e]	7-H_2O (5 percent); 4,7 (40 percent); 5,7 (25 percent); 4,8 (30 percent)
[d]	++	+	4 (50 percent); 5 (10 percent); 6 (15 percent); 7 (20 percent); 8 (3 percent)

[a]Fractions defined chromatographically in Figure 9-10 (A).
[b]++: Major constituent of the metabolite.
[c]+: Minor constituents of the metabolite.
[d] Relative amount (percent) of a compound in the actual fraction.
[e] –: Not detected.

Note: Reprinted with permission from Bokern and Harms (1997). Copyright American Chemical Society.

Metabolism in cell cultures of all tested plant species seems to follow the same pathway as described for wheat (*Triticum aestivum*). The majority of plant metabolites show the identical characteristics with respect to enzymatic and chemical hydrolysis, and chromatographic behavior. However, there are differences in the capability of certain plants to oxidize 4-*n*-nonylphenol, which appear to be related to plant taxonomic classifications, *e.g.*, all tested Chenopodiaceae seemed to lack enzyme systems leading to dihydroxylated compounds (Bokern and Harms 1997). So far, there are no data concerning the toxicity of the metabolites formed, which obviously is quite important, as even more toxic compounds may accumulate.

The aforementioned investigations demonstrate that plant cell cultures are a very useful tool in the search for suitable plants or groups of plants for phytoremediation. At the same time, cell cultures provide information about the metabolic pathways involved and the metabolic potential of a plant species. However, cell cultures cannot provide data about uptake rates and the distribution of a chemical and the associated metabolites in intact plants, which obviously are much more complex. To obtain a more "realistic" estimation of these processes, the uptake and metabolism of 4-*n*-nonyl[^{14}C]phenol were tested with aseptically grown plants (Table 9-7).

TABLE 9-7 Distribution of Radioactivity (Percent) in Aseptically Grown Plants 4 Days After Application of 1 Milligram per Liter of 4-*n*-Nonylphenol

	Lycopersicon esculentum ($n=9$)	*Triticum aestivum* ($n=18$)	*Atriplex hortensis* ($n=70$)
Medium	5.0 ± 1.1	51.9 ± 4.7	5.6 ± 2.0
Dichloromethane phase	0.7	26.2	3.6
Methanol–water phase	4.3	25.7	2.0
Shoot	4.2 ± 0.9	1.1 ± 0.5	3.2 ± 0.4
Dichloromethane phase	0.1	n.e	0.1
Methanol–water phase	1.9	n.e	0.5
Inextricable residue	2.2	0.4	2.6
Root	59.4 ± 5.1	27.1 ± 3.1	78.3 ± 5.3
Dichloromethane phase	13.1	4.3	6.2
Methanol–water phase	11.1	6.5	7.7
Inextricable residue	35.2	16.3	64.4

Abbreviation used—n.e.: not examined.

Note: Plants were analyzed in three replicates with three plants for tomato (*Lycopersicon esculentum*), in three replicates with six plants for wheat (*Triticum aestivum*), and in seven replicates with ten plants for garden orach (*Atriplex hortensis*). After the first fractionation in medium, shoot and root extracts of the respective plants were combined for phase separation and bound residue determination. Therefore, standard deviations are not given for these values. Reproduced with permission of *Environ. Sci. Pollut. Res.* **5**(1): p. 26 (1998).

Plants grown aseptically in nutrient media containing 4-*n*-nonyl[^{14}C]phenol incorporated and metabolized the contaminant. The amount taken up differed enormously, depending on the plant species. None of the tested plants mineralized 4-*n*-nonylphenol to $^{14}CO_2$. Radioactivity occurred in shoots indicating that 4-*n*-nonylphenol or the associated metabolites were transported within the plants. Sufficiently high uptake rates into garden orach (*Atriplex hortensis*) plants allowed a further extraction and analysis according to the standardized method in Figure 9-2 (Harms and Langebartels 1986). Chromatographic comparison of the extracts with those obtained from the respective cell suspension culture indicated, qualitatively, the same metabolism. Quantitatively, the bound residue constitutes the major fraction showing radioactivity in aseptically grown plants, whereas the polar cell extract contained the major proportion of radioactivity in cell cultures (Bokern *et al.* 1998).

To take into account more complex interactions between soil and plants, soil–plant transfer studies with 4-*n*-nonyl[^{14}C]phenol were undertaken. Mineralization of 4-*n*-nonylphenol occurred (up to 30 percent), which can be attributed to microbial activity in the soil. In contrast, medium grown plants do not show any carbon dioxide evolution (Table 9-8).

TABLE 9-8 Distribution of Radioactivity (Percent) in Plant and Soil Systems Incubated with 4-*n*-Nonyl[^{14}C]phenol (4-*n*-NP) for 21 Days ($n = 3$)

Plant species	Experiment	Shoot	Root	Soil	CO_2	Recovery
Tomato (*Lycopersicon esculentum*)	5 milligrams of 4-*n*-nonylphenol per kilogram of soil	0.15	0.10	64.8	30.8	95.8
	10 milligrams of 4-*n*-nonylphenol per kilogram of soil	0.25	0.12	57.5	43.9	101.8
Control without plant:	5 milligrams of 4-*n*-nonylphenol per kilogram of soil	–	–	62.0	29.8	91.8
Garden orach (*Atriplex hortensis*)	5 milligrams of 4-*n*-nonylphenol per kilogram of soil	0.49	0.33	48.9	10.6	60.3
	10 milligrams of 4-*n*-nonylphenol per kilogram of soil	0.69	0.55	62.5	17.3	81.0
Control without plant:	5 milligrams of 4-*n*-nonylphenol per kilogram of soil	–	–	49.0	11.6	60.6

Note: Due to the low amounts of radioactivity in plants, the respective materials of three experiments were combined and mixed to obtain a reliable measurement. Relative standard deviations for the measurements were less than 12 percent. The unit mg 4-*n*-NP/Kg is the milligrams of 4-*n*-nonylphenol per kilogram of soil. Reproduced with permission of *Environ. Sci. Pollut. Res.* **5** (1): p. 26 (1998).

The two plant species tested [tomato, (*Lycopersicon esculentun*) and tomato, garden orach (*Atriplex hortensis*)] took up 4-*n*-nonylphenol at both 5 and 10 milligrams per kilogram. However, over a period of 21 days, the amount did not exceed 1.5 percent of the applied radioactivity. Despite only limited uptake in both types of plants, higher concentrations lead to higher uptake (which has also been shown for cell suspension cultures). This is both in absolute and in relative amounts (percent). As has been shown for aseptically grown plants, radioactivity is translocated into the shoot. Due to the experimental technique used, photosynthetic fixation of released $^{14}CO_2$ can be excluded (Bokern *et al.* 1998). Uptake rates depended on the plant species tested, which also has been shown for aseptically grown plants and cell cultures with 4-*n*-nonylphenol as well as with other chemicals (Wilken *et al.* 1995, Wang *et al.* 1996, Bokern and Harms 1997, Bokern *et al.* 1998). About 60 percent of the applied compound remained in the soil. The soil used in experiments with garden orach (*Atriplex hortensis*) was extracted according to the procedure for the plant material. Depending on the test concentration, 10 to 25 percent of radioactivity in the soil was soluble. The major part of the radioactivity in soil is associated with the bound residue. The extractable amount of 4-*n*-nonylphenol in soil is within the range determined by Kirchmann *et al.* (1991). For predictions of the bioavailability of 4-*n*-nonylphenol from soils, studies that are more detailed are required to understand and predict the very complex interactions between soil microorganisms, soil physics, soil chemistry, and plants (Novak *et al.* 1995).

INEXTRICABLE (BOUND) RESIDUES

Following the application of radiolabeled compounds to cell cultures or intact plants, large amounts of the radioactivity were often associated with insoluble plant components. These residues are referred to as "bound" or "inextricable residues," which typically cannot be released from the plant matrix by extraction with organic solvents. The pattern of binding depends on the plant species, as well as the physicochemical properties of the compound. Bound residues were assigned to defined cell wall fractions using a sequential fractionation procedure (Langebartels and Harms 1986). Table 9-9 shows the distribution of the radioactivity in various cell wall components of wheat (*Triticum aestivum*) cultures after treatment with different [^{14}C]-labeled chemicals.

With all chemicals tested, the lignin fraction contained a major proportion of radioactivity in the bound residue. With 4-chloroaniline, nearly 50 percent of the radioactivity in the bound residue was located in the pectin fraction and 27 percent in the lignin fraction, whereas this distribution was reversed for 3,4-dichloroaniline. With pentachlorophenol, most of the radioactivity was not only in the hemicellulose fraction, but also in lignin and pectin fractions. Pentachloronitrobenzene, however, was predominantly associated with lignin and only a smaller portion was in pectin and protein. In addition

TABLE 9-9 Radioactivity Released from Cell Wall Fractions of Wheat (*Triticum aestivum*) Cell Suspension Cultures Treated with Various [^{14}C]-Labeled Xenobiotic Chemicals

Reagent	Material liberated	Radioactivity (percent) released from cell-wall fractions of plants incubated with:				
		4-CA	3,4-DCA	PCP	PCNB	4-NP
α-Amylase	Starch	2.6	3.4	2.8	4.3	11.5
Pronase E	Proteins	5.6	3.4	23.1	19.8	5.2
EGTA	Pectin	48.3	20.9	6.9	13.8	14.4
Dioxane-HCl (2 moles per liter)	Lignin	27.4	52.1	21.5	42.5	31.2
KOH (24 percent)	Hemicellulose	4.8	11.5	35.5	6.0	28.3
H_2SO_4 (72 percent)	Cellulose	1.6	0.2	1.9	1.8	9.2
	Residual	2.8	0.3	2.8	2.8	0.2

The following abbreviations are used—4-CA: 4-chloroaniline; 3,4-DCA: 3,4-dichloroaniline; PCP: pentachlorophenol; PCNB: pentachloronitrobenzene; 4-NP: 4-*n*-nonylphenol; EGTA: ethylenedioxy*bis*(ethylenenitrilo)tetra(acetic acid).

Note: Residual is the amount of measured radioactivity remaining after the starch, protein, pectin, lignin, hemicellulose, and cellulose were removed from cell wall material. Adapted from Harms (1992).

to being bound to lignin, 4-*n*-nonylphenol was bound primarily to hemicellulose and pectin.

Phytotoxicity of a compound may restrict the suitability of certain plants for phytoremediation. Screening of cell suspension cultures is an excellent way of estimating the relative sensitivity of a plant species in a relatively short time. Phytotoxicity and formation of bound residues as related to plant species was tested using 14 different cell cultures with 4-*n*-nonylphenol. Table 9-10 documents the results.

From the data presented in Table 9-10, it is obvious that the formation of bound residues and the distribution of radioactivity in cell wall fractions are species-specific. The data for phytotoxicity and inextricable residues of cell cultures suggest that the capacity to form such residues may be associated with higher tolerance to 4-*n*-nonylphenol. This confirms the hypothesis that the bound residue fraction (*e.g.*, plant cell wall) is one of the important detoxification sites in plant cells (Sandermann 1994). From the data listed in Table 9-10, it is obvious that the species with the highest tolerance segregate high proportions of radioactivity into bound residues. The most sensitive species segregate only low proportions of radioactivity to the bound residue fraction. A similar phenomenon occurred when [^{14}C]-metribuzin was applied to different cultivars of soybean (*Glycine max*). The tolerant cultivar formed greater amounts of bound residues than the susceptible one (Dupont and Khan 1992). However, for species exhibiting EC_{50} (effect concentration toxic to 50 percent of a population) values between 50 and 500 micromoles per liter 4-*n*-nonylphenol such a relationship is less obvious. Thus, there are obviously

TABLE 9-10 Distribution of Radioactivity in Fractions of Bound Residues from Cell Cultures of Different Plant Species after 2-Day Incubations with 1 Milligram per Liter of 4-*n*-Nonyl[^{14}C]phenol

Cell culture	Concentration range for EC_{50} (micromoles per liter)	Total radioactivity in bound residue (percent)	Relative distribution in fractions (percent)					
			Starch	Protein	Pectin	Lignin	Hemicellulose	Cellulose
Solanaceae: *Lycopersicon esculentum*	50 to 100	6.6	28.4	43.3	11.0	10.8	4.6	1.9
Compositae: *Latuca sativa*	500 to 1000	38.4	8.1	12.6	26.4	43.8	7.8	1.4
Apiaceae: *Daucus carota*	>1000	22.4	4.6	9.4	18.3	45.6	15.7	6.1
Fabaceae: *Glycine max*	>1000	31.7	16.4	25.3	12.7	34.8	5.3	5.5
Leguminosae:								
Lupinus polyphyllus	>1000	25.7	0.0	39.5	14.3	25.6	19.2	1.4
Lupinus hartwegii	50 to 100	11.5	17.0	4.3	4.9	31.4	27.1	15.2
Chenopodiacea:								
Atriplex hortensis	100 to 500	4.8	2.4	6.7	6.7	12.3	44.2	27.6
Chenopodium rubrum	50 to 100	3.8	0.0	27.0	17.8	17.2	14.5	23.5
Beta vulgaris var. *esculenta*	100 to 500	11.8	2.2	10.5	18.1	47.1	1.3	2.8
Poaceae:								
Hordeum vulgare	100 to 500	6.5	0.0	42.5	8.8	39.6	7.3	1.7
Pennisetum americanum	50 to 100	10.1	2.0	22.5	12.9	50.9	2.8	9.0
Triticum aestivum	100 to 500	15.8	11.5	5.2	14.4	31.2	28.3	9.2

Note: See Tables 9-1, 9-2, and 9-5 for common names of plant cells. Relative distributions are based on five parallel experiments.

additional processes influencing sensitivity or tolerance. Phytotoxicity and the bound residue formation again seem to be characteristic of plant families. Among the plant species tested, lettuce (*Latuca sativa*), carrot (*Daucus carota*), soybean (*Glycine max*), and perennial lupin (*Lupinus polyphyllus*) exhibited high capacities for residue formation and were at the same time quite tolerant to 4-*n*-nonylphenol, whereas the tested Chenopodiaceae species formed only limited amounts of bound residues and were relatively sensitive.

The use of plants for animal or human food raises the question of bioavailability, which normally means availability of the soluble fraction. For inextricable residues, bioavailability is still a matter of discussion, which has not been investigated well enough (*e.g.*, Sandermann *et al.* 1990). The few investigations (Sandermann *et al.* 1990, 1992, Goll *et al.* 1996) concerning animal bioavailability of bound residues reveal that the binding mechanism and binding site of a compound to residue fractions is of importance for the bioavailability of the compound. Inextricable residues of 12 different cell suspension cultures were characterized with respect to the main binding sites of radioactivity derived from 4-*n*-nonylphenol. All cell cultures exhibited a specific distribution of radioactivity in different fractions of bound residues (Table 9-10). In most of the cultures (7 out of 12), lignin was the fraction to which the major part of radioactivity was associated, but in some of the cultures, the protein fraction (*Hordeum vulgare*) [barley, red goosefro (*Chenopodium rubrum*)], and tomato (*Lycopersicon esculentum*) and the hemicellulose fraction [garden orach (*Atriplex hortensis*)] were the main fractions containing radioactivity. The distribution of radioactivity in cell wall fractions seems to be species-specific. All tolerant species incorporate considerable amounts of radioactivity into the lignin fraction. However, at present there is no clear evidence of a relation between binding site of radioactivity and tolerance.

The different patterns of binding of a chemical and the metabolites associated with various cell wall components suggest that bound residues may differ in bioavailability. Thus, the identification of chemicals bound to certain cell wall components should enable an estimate of the long-term risks of these chemicals as plants die and are consumed.

CONCLUSIONS

Effective and economical phytoremediation is highly dependent on finding suitable plants for cleaning contaminated sites. The results of our four-step protocol demonstrate that intact plants and different *in vitro* systems assimilate all the tested compounds, even nonpolar ones. Uptake depends on the plant species and on the physicochemical properties of the chemical. In this respect, care should be taken in predicting the behavior of a plant species on the basis of other species. From the results of soil–plant translocation studies, the uptake of 4 to 5 ring (and higher) PAHs, higher chlorinated PCBs, and 4-*n*-nonylphenol appear to be low. The soil plant transfer studies with [^{14}C]-labeled 4-*n*-nonylphenol reveal that microorganisms mineralize an appreciable amount of that compound. The amount of released $^{14}CO_2$ is much higher in soil reactors with plants as compared to unplanted soil reactors. This indicates that root necrosis and exudation foster the degradation of xenobiotic compounds. Bearing this in mind, phytoremediation can be an appropriate technology for sites contaminated even with nonpolar PAHs, nonpolar PCBs, and other nonpolar compounds. This is especially true, when the roots of plants can economically reach contamination to a certain depth of the soil.

In vitro systems such as cell suspension and hairy root cultures efficiently provide information on the phytotoxicity, metabolism, and the formation of bound residues of a xenobiotic compound with one test in a reasonable time. Those cultures are, therefore, very suitable for screening and selecting plants with an optimal capacity for the remediation of certain xenobiotic compounds. The results of our studies reveal that the metabolic pathways for degrading a certain chemical are similar in *in vitro* cultures and intact plants. There are certain differences in the quantity of metabolites formed but the pathways are the same. In most cases, plants do not mineralize the xenobiotic contaminants. The only case we experienced so far involved studies (Berlin *et al.* 1971) with [^{14}C]-labeled *ortho*-dihydroxy phenols for which ring cleavage and the release of $^{14}CO_2$ occurred. The main metabolites are typically polar conjugates with carbohydrates, amino acids, and other plant constituents. When the storage capacity of the vacuole is exhausted, the parent compound and the metabolites are bound to polymer structures of the cell wall. The bound residue fraction is thus one of the important detoxification sites in plants. Species with higher tolerance toward a certain chemical incorporate greater proportions of that compound into bound residues. For phytoremediation, this indicates that this plant material may have to be harvested and disposed separately in landfills, or treated by composting or incineration as required for heavy metal accumulating plants. On the basis of this understanding of the metabolism of PCBs, PAHs, aromatic compounds, and surfactants, phytoremediation pilot studies are necessary to determine if the botanical processes can be managed to clean up field sites in a reasonable time to reach appropriate risk reduction requirements.

REFERENCES

Babson, J.R., S.E. Russo-Rodriguez, R.V. Wattley, P.L. Bergstein, W.H. Rastetter, H.L. Liber, B.M. Liber, B.M. Andon, W.G. Thilly, and G.N. Wogan (1986) Microsomal activation of fluoranthene to mutagenic metabolites. *Toxicol. Appl. Pharm.* **85**: 355–366.

Ballschmiter, K., M. Zell, and J. New (1978) Persistence of PCBs in the ecosphere: will some PCB components never degrade? *Chemosphere* **1**: 173–176.

Barfknecht, T.R., R.A. Hites, E.L. Cavaliers, and W.G. Thilly (1982) Human cell mutagenicity of polycyclic aromatic hydrocarbon components of diesel emissions. *Dev. Toxicol. Environ. Sci.* **10**: 2777–2294.

Berlin, J., W. Barz, H. Harms, and K. Haider (1971) Degradation of phenolic compounds in plant cell cultures. *FEBS Lett.* **16**: 141–146.

Bligh, E.G., and W.J. Dyer (1959) A rapid method for total lipid extraction and purification. *Can. J. Biochem. Physiol.* **37**: 911–917.

Bock, C. (1999) Untersuchungen des Metabolismus von 3,3',4,4'-Tetrachlorbiphenyl (PCB 77) und 2,2',5-Trichlorbiphenyl (PCB 18) in ausgewählten pflanzlichen In-Vitro-Systemen. *Landbauforschung Voelkenrode.* Sonderheft **207**: 1–106.

Bock, C., and H. Harms (1996) Metabolism of selected polychlorobiphenyl congeners in plant cell cultures. *Organohalogen Compounds* **28**: 53–57.

Bokern, M. and H. Harms (1993) Toxicity and metabolism of 4-nonylphenol in different plant systems. *Med. Fac. Landbouww. Univ. Gent.* **58**(H. 2a): 217–224.

Bokern, M., and H. Harms (1997) Toxicity and metabolism of 4-*n*-nonylphenol in cell suspension cultures of different plant species. *Environ. Sci. Technol.* **31**: 1849–1854.

Bokern, M., M. Nimtz, and H. Harms (1996) Metabolites of 4-*n*-nonylphenol in wheat cell suspension cultures. *J. Agric. Food Chem.* **44**: 1123–1127.

Bokern, M., P. Raid, and H. Harms (1998) Toxicity, uptake and metabolism of 4-*n*-nonylphenol in root cultures and intact plants under septic and aseptic conditions. *Environ. Sci. Pollut. Res.* **5**: 21–27.

Brunn, H., S. Georgii, V. Stojanovic, R. Flemming, and R. Thalacker (1989) Foreign substances in foods—determination of the daily dietary intake. part 1. polychlorinated biphenyls in selected foods (in German). *Dtsch. Lebensm.-Rundsch.* **85**(8): 239–246.

Brunn, H., S. Georgii, and J. Prucha (1990) Polychlorinated biphenyls (PCB) in human adipose tissue (in German). *Z. Lebensm. Unters. Forsch.* **190**(2): 108–111.

BUA-Stoffbericht 13 (1988) 4-Nonylphenol. *Beratergremium für umweltrelevante Altstoffe (BUA) der Gesellschaft Deutscher Chemiker*, eds. VCH, Weinheim, Germany.

Busby, W.F., M.E. Goldman, P.M. Newberne, and G.N. Wogan (1984) Tumorigenicity of fluoranthene in newborn mouse lung adenoma bioassay. *Carcinogenesis* **5**: 1311–1316

Cunningham, S.D., W.R. Berti, and J.W. Huang (1995) Phytoremediation of contaminated soils. *Trends Biotechnol.* (*TIBTECH*) **13**: 393–397.

Di Corcia, A., R. Samperi, and A. Marcomini (1994) Monitoring aromatic surfactants and their biodegradation intermediates in raw and treated sewages by solid-phase extraction and liquid chromatography. *Environ. Sci. Technol.* **28**: 850–858.

Dupont, S., and S.U. Khan (1992) Bound (non-extractable) ^{14}C-residues in soybean treated with [^{14}C]metribuzin. *J. Agric. Food Chem.* **40**: 890–893.

Fritz, W. (1983) Model tests on the passage of benzo(a)pyrene from the soil into crops (in German). *Zeitschr. Gesundh. Hyg.* **29**: 370–373.

Giger, W., E. Stephanou, and C. Schaffner (1981) Persistent organic chemicals in sewage sludge effluents: 1. identification of nonylphenols and nonylphenolethoxylates by glass capillary gas chromatography/mass spectrometry. *Chemosphere* **19**: 1253–1263.

Giger, W., P.H. Brunner, and C. Schaffner (1984) 4-Nonylphenol in sewage sludge: accumulation of toxic metabolites from nonionic surfactants. *Science* **225**: 623–625.

Goll, M., M. Bokern, H. Valenta, and H. Harms (1996) Bioverfuegbarkeit von ochratoxin A aus verschiedenen gebundenen Rückstandsfraktionen. In: *Proceedings 18. Mykotoxin Workshop*, M. Gareis and R. Scheurer, eds. Publ. Bundesforschungsanstalt fuer Fleischforschung, Kulmbach, Germany. ISBN 3-00-000933-7, pp.112–117.

Haas, R., I. Schreiber, and G. Koss (1990) Uptake of PAHs and heavy metals in cereal – mutual influence (in German). *UWSF – Z. Umweltchem. Ökotox.* **2**: 66–70.

Harms, H. (1975) Metabolisierung von benz(*a*)pyren in pflanzlichen zellsuspensionskulturen und Weizenkeimpflanzen. *Landbauforschung Voelkenrode.* **25**: 83–90.

Harms, H. (1983) Uptake and conversion of three different 5-ring polycyclic aromatic hydrocarbon (PAHs) in cell suspension cultures of various Chenopodiaceae-species. *Z. Naturforsch.* **38c**: 382–386.

Harms, H. (1992) In-vitro systems for studying phytotoxicity and metabolic fate of pesticides and xenobiotics in plants. *Pestic. Sci.* **35**: 277–281.

Harms, H. (1996) Bioaccumulation and metabolic fate of sewage sludge derived organic xenobiotics in plants. *Sci. Total Environ.* **185**: 83–92.

Harms, H. and E. Kottutz (1991) Tissue culture tests for studying phytotoxicity and metabolic fate of pesticides and xenobiotics in plants. In: Plant tier testing: a workshop to evaluate nontarget plant testing in subdivision pesticide guidelines, U.S. Environmental Protection Agency Report EPA/600/9–91/041, Corvallis, Oregon, pp. 70–79.

Harms, H. and E. Kottutz (1992) Uptake, fate and persistence of organic contaminants in different plant systems – cell suspension cultures, root cultures and intact plants. In: *Effect of Organic Contaminants in Sewage Sludge on Soil Fertility, Plants and Animals.* J.E. Hall, D. Sauerbeck, and P. L'Hermite, eds. EC-Cost Proc., Brussels, pp. 125–133.

Harms, H. and C. Langebartels (1986) Standardized plant cell suspension test systems for an ecotoxicologic evaluation of the metabolic fate of xenobiotics. *Plant Sci.* **45**: 157–165.

Harms, H., W. Dehnen, and W. Moench (1977) Benzo(a)pyrene metabolites formed by plant cells. *Z. Naturforsch.* **32c**: 321–326.

Heeschen, W., A. Blüthgen, and J. Steimer (1993) Allgemeines über Aufnahme, Verteilung Metabolismus, Retention und ausscheidung polychlorierter biphenyle beim Säuger. In: *Carry Over von polychlorierten Biphenylen. Schriftenreihe des Bundesministeriums für Ernährung, Landwirtschaft und Forsten, Reihe A: Angewandte Wissenschaft. Landwirtschaftsverlag GmbH, Münster, pp. Heft* **418**: 33–43.

Jobling, S., D. Sheahan, J.A. Osborne, P. Matthiessen, and J.P. Sumpter (1996) Inhibition of testicular growth on rainbow trout (*Oncorhynchus mykiss*) exposed to estrogenic alkylphenol chemicals. *Environ. Toxicol. Chem.* **15**: 194–202

Jobst, H. (1995) Chlorophenols and nonylphenols in sewage sludges. part 1. occurrence in sewage sludges of Western German treatment plants from 1987 to 1989. *Acta Hydrochim. Hydrobiol.* **23**: 20–25.

Kaden, D.A., R.A. Hites, and W.G. Thilly (1979) Mutagenicity of soot and associated polycyclic aromatic hydrocarbons to *Salmonella typhimurium. Cancer Res.* **39**: 4159–4159.

Kirchmann, H., H. Aastroem, and G. Joensaell (1991) Organic pollutants in sewage sludge. 1. effect of toluene, naphthalene, 2-methylnaphthalene, 4-*n*-nonylphenol and di-2-ethylhexyl phthalate on soil biological processes and their decomposition in soil. *Swed. J. Agric. Res.* **21**(3): 107–113.

Kolb, M. and H. Harms (2000) Metabolism of fluoranthene in different plant cell cultures and intact plants. *Environ. Toxicol. Chem.* **19**: 1304–1310.

Langebartels, C. and H. Harms (1984) Metabolism of pentachlorophenol in cell suspension cultures of soybean and wheat: pentachlorophenol glucoside formation. *Z. Pflanzenphysiol.* **113**: 201–211.

Langebartels, C. and H. Harms (1985) Analysis for non-extractable (bound) residues of pentachlorophenol in plant cells using a cell wall fractionation procedure. *Ecotoxicol. Environ. Safe.* **10**: 268– 279.

Langebartels, C. and H. Harms (1986) Plant cell suspension cultures as test systems for an ecotoxicologic evaluation of chemicals: growth inhibition effects and comparison with the metabolic fate in intact plants. *Angew. Bot.* **60**: 113–123.

LaVoie, E.J., S.S. Hecht, V. Bedenko, and D. Hoffmann (1982) Identification of the mutagenic metabolites of fluoranthene, 2-methylfluoranthene and 3-methylfluoranthene. *Carcinogenesis* **3**(8): 841–846.

LaVoie, E.J., Z.-W. Cai, C.L. Metscher, and E.H. Weyand (1994) Tumorigenic activity of fluoranthene, 2-methylfluoranthene and 3-methylfluoranthene in newborn CD-1 mice. *Carcinogenesis* **15**(10): 2131–2135.

Lee, I. and J.S. Fletcher (1992) Involvement of mixed function oxidase systems in PCB metabolism by plant cells. *Plant Cell Rep.* **11**: 97–100.

Macek, T., M. Mackova, and J. Kás (2000) Exploitation of plants for the removal of organics in environmental remediation. *Biotechnol. Adv.* **18**: 23–34.

Müller, W. and F. Korte (1973) Polychlorierte Biphenyle – Nachfolger des DDT? *Chem. Unserer Zeit* **7**: 112–119.

Novak, J.M., K. Jayachandran, T.B. Moorman, and J.B. Weber (1995) Sorption and binding of organic compounds in soils and their relation to bioavailability. In: *Bioremediation: Science and Applications.* Soil Science Society of America, American Society of Agron., Crop Science Society, eds. Madison, Wisconsin, pp. 13–31.

Palitti, F., R. Cozzi, M. Fiori, F. Palombo, C. Polcaro, G. Perez, and E. Possagno (1986) An *in vitro* and *in vivo* study on mutagenic activity of fluoranthene: comparison between cytogenetic studies and HPLC analysis. *Mutat. Res.* **174**(2): 125–130.

Pothuluri, J.V., R.H. Heflich, and P.P. Fu (1992) Fungal metabolism and detoxification of fluoranthene. *Appl. Environ. Microbiol.* **58**: 937–941.

Purdom, C.E., P.A. Hardiman, V.J. Bye, N.C. Eno, C.R. Tyler, and J.P. Sumpter (1994) Estrogenic effects of effluents from sewage sludge treatments works. *Chem. Ecol.* **8**: 275–285.

Rice, J.E., E.J. LaVoie, and D. Hoffmann (1983) Synthesis of isomeric phenols and the *trans*-2,3-dihydrodiol of fluoranthene. *J. Org. Chem.* **48**: 2360–2363.

Ruge, U. (1951) *Uebungen zur Wachstums-und Entwicklungsphysiologie der Pflanze.* Springer-Verlag, Heidelberg, Germany.

Safe, S. (1990) Polychlorinated biphenyls (PCBs), dibenzo-p-dioxins (PCDDs), dibenzofurans (PCDFs), and related compounds: environmental and mechanistic considerations which support the development of toxic equivalency factors (TEFs). *CRC Crit. Rev. Toxicol.* **21**: 51–88.

Sandermann, H., Jr. (1994) Higher plant metabolism of xenobiotics: the "green liver" concept. *Pharmacogenetics* **4**: 225–241.

Sandermann, H., Jr., M. Arjmand, L. Gennity, R. Winkler, C.B. Struble, and P.W. Aschbacher (1990) Animal bioavailability of defined xenobiotic lignin metabolites. *J. Agric. Food Chem.* **38**: 1877–1870.

Sandermann, H., Jr., T.J. Musick, and P.W. Aschbacher (1992) Animal bioavailability of a 3,4-dichloroaniline-lignin metabolite fraction from wheat. *J. Agric. Food Chem.* **4**: 2001–2007.

Schnoor, J.L., L.A. Licht, S.C. McCutcheon, N.L. Wolfe, and L.H. Carreira (1995) Phytoremediation of organic and nutrient contaminants. *Environ. Sci. Technol.* **29**: 318A–323A.

Servos, M.R. (1999) The aquatic toxicity, estrogenic responses and bioaccumulation of alkylphenols and alkylphenol polyethoxylates. *Water Qual. Res. J. Can.* **34**: 123–177.

Soto, A.M., H. Justicia, J.W. Wray, and C. Sonnenschein (1991) *p*-Nonylphenol: an estrogenic xenobiotic released from "modified" polystyrene. *Environ. Health Persp.* **92**: 167–173.

Sweetman, J. (1994) Development and application of a multi-residue analytical method for the determination of n-alkanes, linear alkylbenzenes, polynuclear aromatic hydrocarbons and 4-nonylphenol in digested sewage sludges. *Water Res.* **28**: 343–353.

Thilly, W.G., J.G. DeLuca, E.E. Furth, H. Hoppe, D.A. Kaden, J.J. Krolewske, H.L. Liber, T.R. Slopek, S.A. Sleppikoff, R.J. Tizard, and B.W. Perman (1980) Gene-locus mutation assays in diploid human lymphoblas lines. In: *Chemical Mutagenicity*. vol. 6. F.J. de Serres and A. Hollaender, eds. Plenum Publishing, New York, pp. 331–364.

Van Duuren, B.L. and B.M. Goldschmidt (1976) Carcinogenic and tumor-promoting agents in tobacco carcinogenesis. *J. Natl. Cancer Inst.* **56**: 1237–1242.

Wang, M.-J., M. Bokern, C. Böhme, K.C. Jones, and H. Harms (1996) Phytotoxicity, uptake and metabolism of 1,4-dichlorobenzene by plant cells. *Environ. Toxicol. Chem.* **15**: 109–104.

Wetzel, A., T. Alexander, S. Brandt, R. Haas, and D. Werner (1994) Reduction by fluoranthene of copper and lead accumulation in *Triticum aestivum* L. *Bull. Environ. Contam. Toxicol.* **53**: 856–862.

White, R., S. Jobling, S.A. Hoare, J.P. Sumpter, and M.G. Parker (1994) Environmentally persistent alkylphenolic compounds are estrogenic. *Endocrinology* **135**: 175–182.

Wilken, A., C. Bock, M. Bokern, and H. Harms (1995) Metabolism of different PCB congeners by plant cell cultures. *Environ. Toxicol. Chem.* **14**: 2017–2022.

You, I.S., R.A. Jones, and R. Bartha (1982) Evaluation of a chemically defined model for the attachment of 3,4-dichloroaniline to humus. *Bull. Environ. Contam. Toxicol.* **29**: 476–482

10

ECOLOGY OF RHIZOSPHERE BIOREMEDIATION

P. E. Olson, K. F. Reardon, and E. A. H. Pilon-Smits

SUMMARY

Rhizosphere bioremediation or rhizodegradation is the enhanced biodegradation of recalcitrant organic pollutants by root-associated bacteria and fungi under the influence of select plant species. Vegetation can increase the total numbers of beneficial fungi and bacteria in contaminated soil from a general rhizosphere effect. In addition, some root-released compounds have the capability to induce genes for enzymes responsible for microbial metabolism of recalcitrant organic pollutants. From the successive use of select vegetation and sound plant management practices, the total proportion of pollutant degraders will increase in numbers and activity in the rhizosphere, leading to enhanced rhizodegradation of recalcitrant organic pollutants and faster site recovery. Thus, the application of fundamental ecological principles to rhizosphere bioremediation designs is critical for success.

GLOSSARY

Biodegradation: The biologically mediated transformation of a molecule, eventually resulting in small molecules with little or no toxicity (*e.g.*, water and carbon dioxide). The term can be used to refer to either a single transformation step ("primary biodegradation") or complete conversion (often termed "mineralization").

Bioremediation: The application of biodegradation to reduce the concentration (and toxicity) of pollutants in soil, water, or air.

Dissipate: The decrease in contaminant concentrations from abiotic (*e.g.*, irreversible sorption) and biotic (*e.g.*, biodegradation) influences.

Metabolism: The complete set of enzymatic reactions that enable an organism to convert nutrients into energy and biomass. Metabolism can be divided

Phytoremediation: Transformation and Control of Contaminants,
Edited by Steven C. McCutcheon and Jerald L. Schnoor.
ISBN 0-471-39435-1 (cloth)

into two parts: catabolism, the reactions that yield energy, and anabolism, the reactions that consume energy to synthesize new cellular material.

Rhizodegradation: The enhanced biodegradation of recalcitrant organic pollutants by root-associated bacteria and fungi under the influence of select plant species.

Rhizodeposition: The release of carbon-containing (photosynthetically derived) compounds from plant roots into the rhizosphere.

Rhizosphere: The zone of soil under the direct influence of a plant root.

Rhizosphere effect: The observation of higher microbial biomass and activity in the rhizosphere.

INTRODUCTION

The manufacture, use, and disposal of a variety of toxic compounds have led, directly and indirectly, to human and environmental health risks. In response to the growing threat of such toxic compounds, the U.S. Congress enacted the Resource Conservation and Recovery Act (RCRA) in 1976 to regulate the safe production, use, transport, and disposal of hazardous materials. In 1980, the Comprehensive Environmental Response, Compensation, and Liability Act (CERCLA), also known as "Superfund," was enacted to promote the safe cleanup and remediation of inactive sites contaminated with hazardous materials that pose a serious threat to human health and the environment. Under the direction of the U.S. Environmental Protection Agency (EPA), sites across the U.S. are inspected to determine which locations pose an immediate threat to persons living or working near the contaminated site. The EPA inventoried and prioritized these sites that were in greatest need of cleanup, on the National Priorities List (NPL). In 1999, the CERCLA List of Priority Hazardous Substances at NPL sites, complied by the EPA and Agency for Toxic Substances and Disease Registry, identified 275 toxic substances posing the greatest threat to human and environmental health. The toxic substances on the list are ranked and prioritized based on frequency, toxicity, and potential for human exposure at NPL sites. Nearly 70 percent of the 275 toxic substances on the CERCLA priority list are organic compounds. Most of these organic compounds are not amenable to rapid biodegradation and therefore are long-lived or recalcitrant in the environment. Two important classes of recalcitrant organic compounds on the CERCLA priority list are polychlorinated biphenyls (PCBs) and polycyclic aromatic hydrocarbons (PAHs), ranked sixth and ninth in priority, respectively. In addition, numerous individual PCBs and PAHs, and mixtures of both, are ranked on the priority list (Table 10-1).

In response to the growing concerns about environmental pollution and costs associated with cleaning up polluted sites, alternative remedial strategies are being explored and employed. Bioremediation is the biologically mediated cleanup of contaminated sites encompassing the use of bacteria, fungi, plants,

TABLE 10-1 Representative Individual and Mixtures of Polychlorinated Biphenyls and Polycyclic Aromatic Hydrocarbons on the 1999 CERCLA List of Priority Hazardous Substances[a]

Polychlorinated biphenyls	1999 Rank[b]	Polycyclic aromatic hydrocarbons	1999 Rank[b]
Polychlorinated biphenyls	6	Benzo(*a*)pyrene	8
Aroclor 1260	13	Polycyclic aromatic hydrocarbons	9
Aroclor 1254	14	Benzo(*b*)fluoranthene	10
Aroclor 1248	25	Dibenz(*a,h*)anthracene	17
Aroclor 1242	28	Creosote	21
Aroclor 1221	41	Benz(*a*)anthracene	35
Aroclor 1016	45	Benzo(*k*)fluoranthene	57
Aroclor	58	Benzofluoranthene	68
Aroclor 1232	63	Naphthalene	75
Aroclor 1240	78	Fluoranthene	101
Aroclor 1268	189	Chrysene	115

[a]Agency for Toxic Substances and Disease Registry http://www.atsdr.cdc.gov/99list.html
[b]1999 Priority List Ranking: prioritization of substances based on a combination of frequency, toxicity, and potential for human exposure at National Priority List sites.

and other biological agents. Phytoremediation is a broad technology that capitalizes on the innate qualities of vegetation for the enhanced sequestration, removal, or elimination of a given pollutant. Although relatively inexpensive and less intrusive compared to other remedial processes, phytoremediation is perhaps the most complex remedial technology to fully implement and utilize today. Microbial bioremediation and phytoremediation can be combined in a complementary manner ("rhizosphere bioremediation") by using select vegetation and root-associated microorganisms in the enhanced cleanup of polluted environments. This chapter focuses on the plant–microbe interactions that are critical to rhizosphere bioremediation, with particular attention to rhizosphere biodegradation of PCBs and PAHs.

Polychlorinated Biphenyls

Polychlorinated biphenyls (PCBs) are a group of manufactured chemicals comprised of 209 individual congeners (Mackay *et al.* 1992a). Each congener is composed of two bridged phenyl rings that are substituted with up to ten chlorine atoms (Figure 10-1). Due to the low flammability and good insulating properties, PCBs were widely used as coolants and lubricants in transformers, capacitors, and other electrical equipment. As a consequence of frequent use, PCBs have been released into the environment from leaking equipment, poor handling and storage, spills, and improper disposal. Polychlorinated biphenyls may cause irritation of the human nose, throat, and skin. In addition, some PCBs have been shown to cause cancer in laboratory

Polychlorinated Biphenyls (PCBs)

Cl Cl Cl Cl

Cl

2,2′-Dichlorobiphenyl

2,5,2′-Trichlorobiphenyl

Cl Cl Cl Cl

Cl Cl Cl

Cl Cl

3,4,3′,4′-Tetrachlorobiphenyl

2,4,5,2′,5′-Pentachlorobiphenyl

Polycyclic Aromatic Hydrocarbons (PAHs)

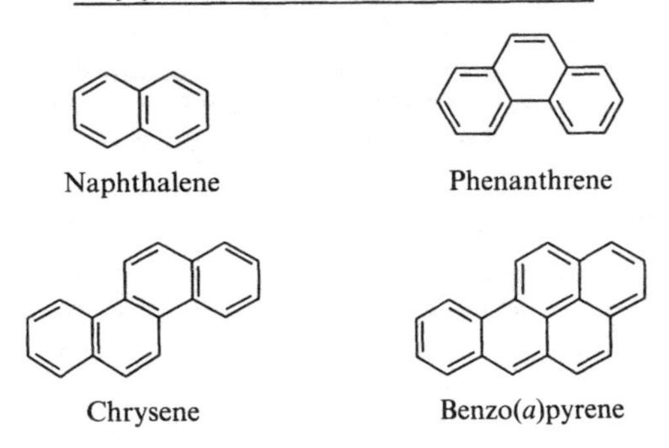

Figure 10-1 Representative polychlorinated biphenyl and polycyclic aromatic hydrocarbon structures. Adapted from Mackay *et al.* (1992a, 1992b).

animals. As a result of the toxicity and persistence in the environment, PCB manufacture was banned in the U.S. in 1977. Polychlorinated biphenyls are found at over 25 percent of the 1430 NPL sites in the U.S. as of 1999.

In the U.S., mixtures of PCB congeners were manufactured under the industrial trade name Aroclor, with each mixture named according to the average number of chlorine substitutions. The higher the average number of chlorine substitutions in the PCB mixture, the greater the industrial properties as coolants and lubricants. However, more highly chlorinated PCB congeners are not only resistant to thermal decomposition, but are also highly resistant to biodegradation. Due to the physicochemical characteristics (Table 10-2), PCBs tend to sorb to organic material in the environment and to have low bioavailability. Biotransformations of the more chlorinated PCBs are restricted to anaerobic reductive dechlorination or cometabolic removal of chlorine atoms under aerobic conditions. After a sufficient number of chlorine

TABLE 10-2 Physical and Chemical Properties of Representative Polychlorinated Biphenyls and Polycyclic Aromatic Hydrocarbons[a]

Polychlorinated biphenyl congener	Chlorine number[b]	Solubility (milligrams per liter)[c]	log K_{ow}^{d}	log K_{oc}^{e}
2,2′	2	5.41×10^{-6}	5.02	4.93
2,5,2′	3	1.72×10^{-6}	5.64	5.54
3,4,3′,4′	4	5.99×10^{-7}	6.52	6.41
2,4,5,2′,5′	5	3.65×10^{-8}	6.85	6.73
2,4,5,2′,4′,5′	6	8.32×10^{-9}	7.44	7.31
Polycyclic aromatic hydrocarbons	**Number of rings[f]**	**Solubility (milligrams per liter)**	**log K_{ow}**	**log K_{oc}**
Naphthalene	2	31.0	3.37	3.31
Phenanthrene	3	1.29	4.46	4.38
Chrysene	4	2.0×10^{-3}	5.61	5.51
Benzo(*a*)pyrene	5	3.8×10^{-3}	6.04	5.94
Benzo(*g,h,i*)perylene	6	2.6×10^{-4}	7.23	7.11

[a]Adapted from Mackay *et al.* (1992a, 1992b).
[b]Number of chlorine residues substituted on biphenyl structure.
[c]Aqueous water solubility (25 °C).
[d]Octanol–water partition coefficient.
[e]Soil organic carbon–water partition coefficient (soil sorption coefficient) in the units of cubic centimeters of solution volume per gram of dry sorbent.
[f]Number of fused benzene rings.

atoms have been removed, lower chlorinated PCBs are relatively biodegradable aerobically *via* oxygenase pathways.

Polycyclic Aromatic Hydrocarbons

Polycyclic aromatic hydrocarbons (PAHs) are a group of more than 100 different organic compounds (Mackay *et al.* 1992b) comprised of fused aromatic rings (Figure 10-1) that are resistant to biodegradation in the environment. Polycyclic aromatic hydrocarbons occur in coal tar, wood preservatives, and wastes from manufacturing and petrochemical plants. In addition, PAHs enter the environment *via* incomplete combustion, including from fires and vehicles. As a result, PAHs are widespread soil contaminants usually found as mixtures of low molecular weight (two to three fused aromatic rings) and high molecular weight (four or more rings) compounds. All of the PAHs are toxic, including the high molecular weight PAHs, which are potent carcinogens. Polycyclic aromtic hydrocarbons occur at nearly 50 percent of the 1430 NPL sites in the U.S. as of 1999.

In general, low molecular weight PAHs are moderately biodegradable under aerobic conditions (Pothuluri and Cerniglia 1994, Shuttleworth and Cerniglia 1995, Sutherland *et al.* 1995, Cerniglia 1997, Kanaly and Harayama

2000). In contrast, high molecular weight PAHs undergo very slow aerobic biodegradation. Furthermore, high molecular weight PAH biodegradation is often limited to cometabolism, as the presence of a low molecular weight PAH such as naphthalene is required to induce the oxygenases responsible for the initial attack on the high molecular weight PAH molecules. The low molecular weight PAHs also provide carbon and energy sources to the degrading cells because the low solubilities of high molecular weight PAHs cannot support active microbial populations. Because low molecular weight PAHs are more readily and more rapidly degraded, high molecular weight PAHs are often quite recalcitrant in the environment. This is a concern because the larger PAHs are more carcinogenic.

Persistence of Polychlorinated Biphenyls and Polycyclic Aromatic Hydrocarbons in the Environment

Polychlorinated biphenyls and PAHs are recalcitrant organic pollutants in the terrestrial environment (Mackay *et al.* 1992a, 1992b). Both compounds share similar physicochemical characteristics resulting in poor biodegradation and bioavailability (Table 10-2). Polychlorinated biphenyls are more water insoluble, and generally have higher log K_{ow} and log K_{oc} than PAHs. However, both PCBs and PAHs sorb strongly to organic material in soil and water environments or become tightly bound inside soil particles. Furthermore, the low aqueous solubility and lipophilic nature of these compounds limit movement in the environment, including the transport into the cytoplasm of microorganisms and plant cells capable of biodegradation.

BIODEGRADATION OF POLYCHLORINATED BIPHENYLS AND POLYCYCLIC AROMATIC HYDROCARBONS

The biodegradation of recalcitrant organic compounds (both naturally occurring and xenobiotic) in aquatic and terrestrial environments is, by definition, limited in rate and typically in extent (Leahy and Colwell 1990, Huesemann 1997, Kelsey and Alexander 1997, Webster *et al.* 1998, Jones and de Voogt 1999, Reid *et al.* 2000). Limits to biodegradation can include the physicochemical characteristics of the compound, low bioavailability, insufficient microbial biomass, limited genetic induction, or the abundance of alternative carbon sources for preferential metabolism. However, given favorable conditions and the necessary metabolic potential, even recalcitrant organics are amenable to eventual biodegradation in the environment. A general understanding of the metabolism of recalcitrant organic compounds in the environment, including the ecological processes involved in the natural induction of metabolic pathways, can lead to strategies for the enhanced biodegradation of recalcitrant pollutants.

Bacterial Metabolism of Polychlorinated Biphenyls and Polycyclic Aromatic Hydrocarbons

The biodegradation of PCBs and PAHs by bacteria has been extensively studied and several reviews on the subject are available (Brown *et al.* 1987, Abramowicz 1990, Pothuluri and Cerniglia 1994, Abramowicz 1995, Shuttleworth and Cerniglia 1995, Sutherland *et al.* 1995, Seeger *et al.* 1997, Juhasz and Naidu 2000, Kanaly and Harayama 2000). Although these studies have focused on anthropogenic pollutants, one must consider that many bacteria indigenous to the soil environment have evolved the capability of biodegrading naturally occurring organic compounds that are structurally similar to these pollutants. Bacteria capable of metabolizing aromatic compounds are ubiquitous in most environments. The aerobic metabolic pathways (typically featuring oxygenases) involved in aromatic compound biodegradation generally show broad substrate specificity.

The general mode of PCB biodegradation by bacteria involves two main respiratory processes (Figure 10-2) (Brown *et al.* 1987, Abramowicz 1990, Abramowicz 1995, Seeger *et al.* 1997). Highly chlorinated (more than four chlorine substitutions) PCBs are typically restricted to anaerobic reductive dechlorination, an energy yielding process. Chlorine atoms are preferentially removed from the *meta*- and *para*-positions on the biphenyl structure, leaving

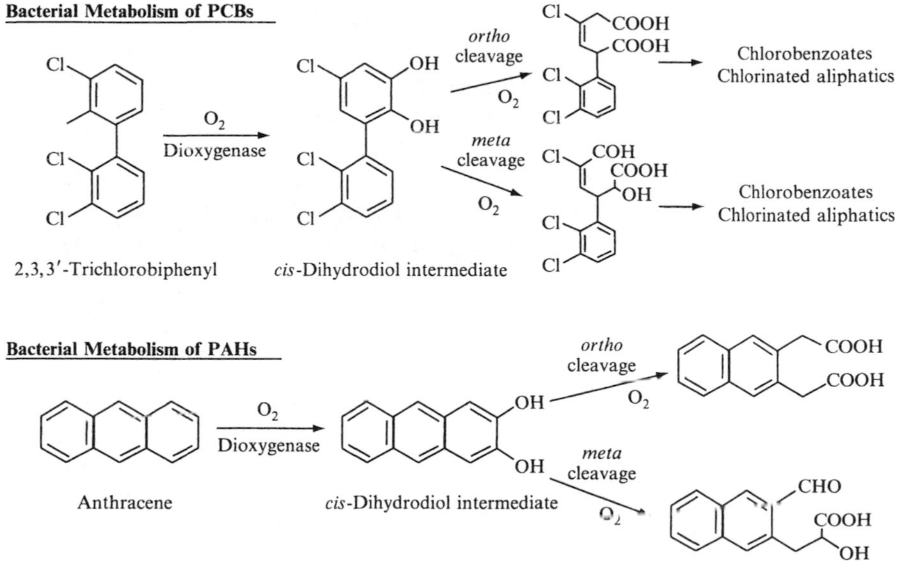

Bacterial Metabolism of PCBs

2,3,3′-Trichlorobiphenyl — O_2 Dioxygenase → *cis*-Dihydrodiol intermediate — *ortho* cleavage O_2 → Chlorobenzoates, Chlorinated aliphatics; *meta* cleavage O_2 → Chlorobenzoates, Chlorinated aliphatics

Bacterial Metabolism of PAHs

Anthracene — O_2 Dioxygenase → *cis*-Dihydrodiol intermediate — *ortho* cleavage O_2; *meta* cleavage O_2

Figure 10-2 Aerobic bacterial metabolism of polychlorinated biphenyls and polycyclic aromatic hydrocarbons. Adapted from Abramowicz (1990, 1995), Seeger *et al.* (1997), Juhasz and Naidu (2000), Kanaly and Harayama (2000).

less chlorinated *ortho*-substituted congeners. The dechlorination of PCBs reduces the potential toxicity, carcinogenicity, and bioaccumulation of the original congener. Under favorable redox conditions, the dechlorination of PCBs leads to the second mode of bacterial biodegradation: aerobic metabolism of the less chlorinated PCBs, often a cometabolic process fueled by structural analogs such as biphenyl. Aerobic oxidative pathways utilize dioxygenases to convert less chlorinated PCBs to chlorobenzoic acids and chlorinated aliphatics. The general mode of metabolism involves the incorporation of molecular oxygen into the phenyl ring. Following subsequent dehydrogenation (removal of hydrogen), catechol ($C_{15}H_{14}O_6$) is formed, followed by ring cleavage or fission to form chlorobenzoic acid and chlorinated aliphatics. Chlorobenzoic acids and chlorinated aliphatics can be further metabolized to carbon dioxide by a number of aerobic microorganisms.

The biodegradation of PAHs under aerobic conditions has been well studied (Pothuluri and Cerniglia 1994, Shuttleworth and Cerniglia 1995, Sutherland *et al.* 1995, Juhasz and Naidu 2000, Kanaly and Harayama 2000). Most studies have focused on lower molecular weight PAHs. Typically, catabolism starts with a monooxygenase or dioxygenase reaction on one ring that adds hydroxy (OH) groups. The hydroxylated ring then undergoes ring fission, producing a substituted PAH with one ring less than the parent has (Figure 10-2). Subsequent oxygenase reactions are used to eventually mineralize the PAH. Far less is known about the biodegradation of high molecular weight PAHs by aerobic bacteria, although it is assumed to follow the same mechanism. The biodegradation of pyrene, chrysene, benzo(*a*)pyrene, and others has been demonstrated in few microbial species (Heitkamp *et al.* 1988, Mahaffey *et al.* 1988, Chen and Aitken 1999). It is not known whether high molecular weight PAH degradation is a common capability of soil microorganisms, nor is it known whether these compounds are able to support microbial growth. However, given the range of naturally occurring organic compounds in the environment, it is likely that high molecular weight PAHs can be biodegraded by indigenous microorganisms. Conclusions regarding microbial growth on high molecular weight PAHs are complicated by the low bioavailability of these large molecules.

The anaerobic biodegradation of PAHs is less well understood. Naphthalene loss has been reported under nitrate-reducing conditions (Mihelcic and Luthy 1988, Langenhoff *et al.* 1996) and under sulfate-reducing conditions (Thierrin *et al.* 1992, Coates *et al.* 1997). Similarly, Rockne and Strand (1999) reported biodegradation of phenanthrene with nitrate and sulfate as electron acceptors. In addition, Zhang and Young (1997) enriched anaerobic consortia that mineralized phenanthrene under sulfate-reducing conditions. Finally, results from an anaerobic digester study imply that PAHs may be degradable under methanogenic conditions (Parker 1995).

At sites contaminated with PAH mixtures, the lower molecular weight compounds are usually preferentially degraded, leaving the high molecular weight PAHs (Pothuluri and Cerniglia 1994, Shuttleworth and Cerniglia

1995, Sutherland *et al.* 1995, Juhasz and Naidu 2000, Kanaly and Harayama 2000). The degradation of these remaining PAHs is often extremely slow. Although these low rates are often attributed to the very low bioavailability of the high molecular weight PAHs, Cornelissen *et al.* (1998) reported that the biodegradation of a range of PAHs in sediment was limited by microbial factors, not bioavailability. Furthermore, many aerobic PAH-degrading bacteria have been shown to have a relatively broad substrate range, and when grown on one PAH, those organisms are typically able to degrade others, including high molecular weight PAHs (Sutherland *et al.* 1995, Chen and Aitken 1999). In some cases, the degradation of high molecular weight PAH has been attributed to cometabolism by cells growing on low molecular weight PAHs (Kanaly *et al.* 1997, Chen and Aitken 1999). These findings suggest the importance of naturally occurring aromatic compounds in stimulating and selecting for PAH degraders in the environment.

Fungal Metabolism of Polycyclic Biphenyls and Polycyclic Aromatic Hydrocarbons

Fungi are critically important in the decomposition of naturally occurring organic material and the cycling of carbon in the environment (Muncnerova and Augustin 1994, Cerniglia 1997, Gunther *et al.* 1998, Braun-Lullemann *et al.* 1999, Meharg and Cairney 2000). Fungi typically exist in the terrestrial environment either as free-living saprophytic forms or in symbiotic associations with vegetation and the plant roots (mycorrhizae). Although fungi are regarded as important decomposers in the environment, limited information exists for application in the biodegradation of recalcitrant organic pollutants. In general, fungi have a broad substrate range for metabolizing organic compounds, and typically employ monooxygenases to metabolize aromatic structures. The monooxygenases utilized to degrade naturally occurring compounds have also been shown to metabolize recalcitrant organic pollutants such as PCBs and PAHs (Muncnerova and Augustin 1994, Cerniglia 1997, Braun-Lullemann *et al.* 1999, Meharg and Cairney 2000). Due to the immense size and very high surface interface with soil, it has been suggested that mycorrhizae may enhance the biodegradation of recalcitrant organic pollutants (Donnelly and Fletcher 1994, Braun-Lullemann *et al.* 1999, Meharg and Cairney 2000).

Fungi, both saprophytic and plant-associated, are active in the degradation of lignin and other complex aromatic hydrocarbons (Figure 10-3) (Muncnerova and Augustin 1994, Cerniglia 1997, Braun-Lullemann *et al.* 1999, Meharg and Cairney 2000). In these processes, many fungi use cytochrome P 450 monooxygenase to oxidize both PCBs and PAHs to arene oxides, which can then be transformed by an epoxide hydrolase to *trans*-dihydrodiols. These reactions have been reported for a range of PCBs (Abramowicz 1990) and PAHs, including benz(*a*)anthracene (four rings) and benzo(*a*)pyrene (five rings) (Sutherland *et al.* 1995). The white-rot fungi (*e.g., Phanerochaete*

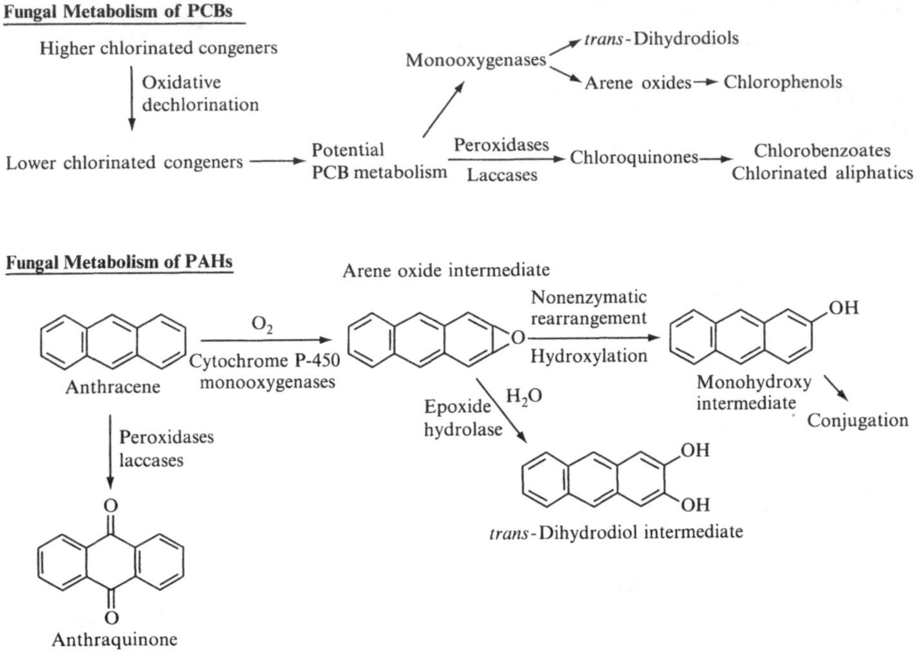

Fungal Metabolism of PCBs

Higher chlorinated congeners
Oxidative dechlorination
Lower chlorinated congeners
Potential PCB metabolism
Monooxygenases
trans-Dihydrodiols
Arene oxides
Chlorophenols
Peroxidases
Laccases
Chloroquinones
Chlorobenzoates
Chlorinated aliphatics

Fungal Metabolism of PAHs

Anthracene
O_2
Cytochrome P-450 monooxygenases
Arene oxide intermediate
O
Nonenzymatic rearrangement
Hydroxylation
OH
Monohydroxy intermediate
Conjugation
Epoxide hydrolase
H_2O
OH
OH
trans-Dihydrodiol intermediate
Peroxidases laccases
O
O
Anthraquinone

Figure 10-3 Fungal metabolism of polychlorinated biphenyls and polycyclic aromatic hydrocarbons. Adapted from Abramowicz (1990, 1995), Cerniglia (1997).

chrysosporium) also metabolize PCBs and PAHs, but generally use lignin peroxidases to form quinones (Muncnerova and Augustin 1994, Cerniglia 1997, Gunther *et al.* 1998, Cameron *et al.* 2000, Novotny *et al.* 2000).

Plant Metabolism of Polychlorinated Biphenyls and Polycyclic Aromatic Hydrocarbons

Organic compounds found in the soil around vegetation may have several fates (Sandermann 1992, Simonich and Hites 1995, Macek *et al.* 2000, Zaalishvili *et al.* 2000). If the compound has physicochemical properties such as low water solubility and high log K_{ow}, it is unlikely that the compound will be transported into and within the plant. However, if the organic compound has favorable transport characteristics (log K_{ow} between 0.5 and 3), uptake and transport of the compound into the plant is likely (Burken and Schnoor 1996, Burken and Schnoor 1997, Burken and Schnoor 1998). Typically, the uptake of organic compounds is restricted to low molecular weight compounds or to higher molecular weight compounds that have been transformed outside of the plant in ways that favor uptake (*e.g.*, increased solubility following hydroxylation). Once the organic compound has been transported from the soil into the plant, several alternative fates exist (Sandermann 1992, Simonich and Hites 1995, Macek *et al.* 2000, Zaalishvili *et al.* 2000). In general, the two most likely fates include the oxidation or biodegradation of the organic compound *in vivo*, and conjugation and sequestration within the plant, either in the vacuole or by irreversibly binding to cell wall material. The first step in both processes involves hydroxylation of the organic compound *via* monooxygenases (cytochrome P-450) or peroxidases within the plant (Gunther *et al.* 1998). This step will make nonpolar organics more mobile and easily transportable within the plant tissue. Depending on the organic compound, these same monooxygenases or peroxidases can ultimately lead to the oxidation of the compound *via* various pathways to carbon dioxide and water. The second scenario involves the transformation (*e.g.*, hydroxylation), conjugation to another molecule, and transport and storage of the organic compound into the plant vacuole or cell wall material. The conjugation step can utilize sugars (*e.g.*, glycosylation), amino acids, and small peptides (*e.g.*, glutathione) (Field and Thurman 1996).

The metabolism of PCBs in plant cell cultures has been well studied. In early studies, axenic *Rosa* spp. 'Paul's Scarlet' cells metabolized between 20 to 80 percent of a variety of PCB congeners, with the most water-soluble PCBs being metabolized to the greatest extent (Lee and Fletcher 1992). In addition, chlorine substitutions at the *ortho*-position increased solubility and metabolism of the congeners. In a later study, twelve different plant species, representing five plant families (Fabaceae, Poaceae, Apiaceae, Solanaceae, Chenopodiaceae), were evaluated in axenic tissue cultures for the metabolism of ten different congeners (Wilken *et al.* 1995). In general, the metabolism of defined PCB congeners was found to be highly dependent on the plant species. Two species of Fabaceae, soybean (*Glycine max*), and creeping white clover (*Trifolium repens*), exhibited the highest capacity to metabolize a wide range of PCB congeners, with hydroxylation and conjugation steps being particularly important. Hairy root cultures of black nightshade (*Solanum nigrum*) have been shown to transform a complex mixture of 59 PCB congeners, with residual PCBs decreasing in solution over time (Mackova *et al.* 1997). In addition, hairy root cultures of black nightshade (*Solanum nigrum*) were recently demonstrated to metabolize a wide range of 22 individual congeners (Kucerova *et al.* 2000). Nearly 72 percent of the original PCB congeners were metabolized, with hydroxylation being common. Despite these intriguing results with plant cells, little is known about PCB metabolism in intact plants.

The metabolism of PAHs in vegetation has been well characterized (Harms *et al.* this book). In addition to removing up to 20 percent of PCBs from solution, *in vitro* cultured cells of black nightshade (*Solanum nigrum*), wheat (*Triticum aestivum*), barley (*Hordeum vulgare*), soybean (*Glycine max*), tomato (*Lycopersicon esculentum*), mulberry (*Morus rubra*), and birch (*Betula pendula*) significantly removed PAHs after 14 days of incubation (Kucerova *et al.* 2001). Axenic tissue cultures from eleven different plant species were used to study the metabolism of fluoranthene (Kolb and Harms 2000). In general, less than 5 percent of the parent fluoranthene was metabolized by the plant species. However, the results suggest that hydroxylations were significant transformations for PAHs. Trees and other plants have been suggested as important instruments for sorbing PAHs from polluted air (Wagrowski and Hites 1997, Bakker *et al.* 1999, Kipopoulou *et al.* 1999, Howsam *et al.* 2000), as the waxy leaf surface of vegetation acts as a competitive sink for the lipophilic PAHs partitioned in the atmosphere. In addition, hydroxypyrene conjugates have also been found in woody vegetation, suggesting that plants may readily metabolize PAHs from the atmosphere (Nakajima *et al.* 1996).

Combined Plant–Microbial Metabolism of Recalcitrant Organics

The biodegradation of PAHs, PCBs, and other recalcitrant organic pollutants has been demonstrated with fungi, bacteria, and even plants. Despite this potential, these pollutants are highly persistent in the environment. During the last 10 years, phytoremediation has been suggested as an ecologically and economically sound cleanup tool (Anderson *et al.* 1993, Lee and Banks 1993, Shimp *et al.* 1993, Erickson *et al.* 1994, Schnoor *et al.* 1995, Salt *et al.* 1998, Siciliano and Germida 1998). Schnoor *et al.* (1995) defined phytoremediation as "the use of vegetation for the *in situ* treatment of contaminated soils and sediments." These authors identified several potential mechanisms for the phytoremediation of organic pollutants, including release of exudates that stimulate microbial growth. Other mechanisms that reduce organic contaminant levels in soil may also occur, but the goal of supporting rhizosphere bioremediation by providing a stimulatory environment for enhanced biodegradation of pollutants by root-associated microorganisms should be considered further.

OBSERVATIONS OF RHIZODEGRADATION OF RECALCITRANT ORGANICS

The soil environment is a dynamic ecosystem with continual changes in physical, chemical, and biological structure (Brady 1990). Within this ecosystem, the growth of plant roots in the soil has profound effects. The rhizoplane is regarded as the region at which the plant root and soil interface, while the rhizosphere is the enlarged zone of soil under the direct influence of a plant root (Curl and Truelove 1986, Lynch 1990, Bowen and Rovira 1991, Bowen and Rovira 1999, Klein 2000). The abiotic and biotic characteristics in the rhizosphere differ greatly from the rootless, bulk soil. From the growth of vegetation, plant roots improve soil quality and deposit large amounts of organic material into the surrounding soil that is utilized by bacteria and fungi. As a result, the rhizosphere provides a unique terrestrial environment that harbors a remarkable diversity of soil microorganisms.

Rhizosphere microorganisms are actively involved in the decomposition and cycling of organic carbon in the environment (Lynch and Whipps 1990, Grayston *et al.* 1996, Jones 1998, Kuzyakov and Domanski 2000, Toal *et al.* 2000). Within the rhizosphere, consortia of organisms mineralize simple and complex naturally occurring organic compounds. As a result, researchers have focused on the rhizosphere as a zone of enhanced biodegradation (Sims and Overcash 1983, Anderson *et al.* 1993, Erickson *et al.* 1994), and on the plant–microbe interactions that occur there (Walton and Anderson 1992, Siciliano and Germida 1998). Rhizosphere bioremediation can capitalize on the innate qualities of vegetation and the potential for the mineralization of organic pollutants in the rhizosphere. Understanding the dynamic nature and ecological processes within the rhizosphere is essential to employing rhizodegradation as a remedial option.

Laboratory and Greenhouse Rhizodegradation Studies

The capability of vegetation to stimulate the biodegradation of PCBs by root-associated fungi has been demonstrated (Donnelly and Fletcher 1994). Ectomycorrhizal fungi, which can exist in the soil as free-living organisms or in plant-specific associations, have been shown to metabolize a variety of PCB congeners in the laboratory (Donnelly and Fletcher 1995). Out of 21 ectomycorrhizal fungi species screened for PCB metabolism, 14 were able to degrade a mixture of 11 PCB congeners. Results from this work confirm previous studies that show less chlorinated congeners are metabolized aerobically more readily than the more chlorinated PCBs.

The observation that naturally occurring plant compounds can stimulate PCB-degrading bacteria provides compelling evidence for the importance of specific rhizosphere interactions in the rhizodegradation of recalcitrant PCBs (Fletcher *et al.* 1995). Aerobic metabolism of PCBs by bacteria generally takes place only in the presence of a cometabolite that resembles the biphenyl structure of the PCB backbone. A variety of secondary plant compounds, including catechin, coumarin, and flavonoids, resemble the biphenyl structure. Research into the phytoremediation of PCBs has shown that these plant compounds support the growth and degradative activity of known PCB-degrading bacteria (Donnelly *et al.* 1994). To further confirm the role of vegetation in stimulating PCB degradation in the soil, 17 different perennial plant species were grown in sand cultures and screened for the production of phenolic compounds that could stimulate PCB-degrading bacteria (Fletcher and Hegde 1995). Results from this screening suggested that only a few plant species have the characteristics (*e.g.*, release of sufficient quantities of stimulatory phenols) most desirable for enhanced rhizodegradation of PCBs. Two of these plants, mulberry (*Morus rubra*) and Osage orange (*Maclura pomifera*), are native trees in the U.S. from the Moraceae family. Mulberries have the capacity to release sufficient quantities of phenolic compounds into soil throughout the year, as well as a massive release of these stimulatory compounds corresponding to leaf senescence at the end of the growing season (Hegde and Fletcher 1996, Leigh 1997, Leigh *et al.* 2002).

Specific chemicals from other plant species have been shown to enhance metabolism of PCBs by soil bacteria (Gilbert and Crowley 1997, Hernandez *et al.* 1997). For example, a terpenoid, carvone, from spearmint (*Mentha spicata*), has been shown to induce PCB-degrading *Arthrobacter* to cometabolize PCBs (Gilbert and Crowley 1997). Furthermore, higher levels of biphenyl-utilizing bacteria were found in soils amended with orange (*Citrus sinensis*) peels, *Eucalyptus* leaves, pine needles (*Pinus* spp.), and ivy leaves (*Hedera* spp.), suggesting that plant compounds may enhance the biodegradation of PCBs (Hernandez *et al.* 1997). The biodegradation of PCBs from the cometabolism of carvone can be further enhanced by the addition of surfactants (Singer *et al.* 2000). The combined influence of root-released compounds and soil additives in enhancing the biodegradation of recalcitrant organic pollutants holds great promise for the cleanup of highly contaminated sites.

The role of vegetation in the enhanced rhizodegradation of PAHs in the rhizosphere has been actively studied over the last 10 years. In a seminal paper on the subject, a mixture of eight common prairie grasses was evaluated in the greenhouse for enhanced disappearance of four- and five-ringed PAHs in the rhizosphere (Aprill and Sims 1990). The presence of the prairie grasses, each having a deep, fibrous root system, significantly lowered the concentration of high molecular weight PAHs in the rhizosphere after 219 days when compared to the unplanted controls. In another landmark paper, pyrene was significantly dissipated in the rhizosphere of fescue (*Festuca arundinacea*), Sudan grass (*Sorghum vulgare*), switch grass (*Panicum virgatum*), and alfalfa (*Medicago sativa*) after 24 weeks compared to the unplanted control (Reilley *et al.*, 1996). Pyrene removal was further stimulated in the rhizosphere by the addition of simple organic acids (*e.g.*, formic and succinic acid), suggesting the importance of root-released compounds and rhizosphere microorganisms in the biodegradation of PAHs. The concentrations of several recalcitrant hydrocarbons were significantly reduced in the rhizosphere of ryegrass (*Lolium perenne*) by a consortium of microorganisms (Gunther *et al.* 1996). The total number of phenanthrene degraders in the rhizosphere of slender oat (*Avena barbata*) was approximately three times higher than unplanted, bulk soil after only 32 days following planting (Miya and Firestone 2000). The higher number of phenanthrene degraders corresponded to a significant decrease in pollutant concentrations, suggesting the stimulatory effect of vegetation on pollutant degradation.

Rhizodegradation of more recalcitrant PAHs has also been observed. For example, the enhanced degradation of [^{14}C]-benzo(*a*)pyrene was demonstrated in the rhizosphere of fescue (*Festuca arundinacea*) even though the pollutant was highly sorbed to the organic material present in the soil (Banks *et al.* 1999). Further evidence supporting rhizodegradation of recalcitrant PAHs was reported after significant disappearance of three- to six-ringed PAHs occurred in the rhizosphere of perennial ryegrass (*Lolium perenne*) in soils aged for over 6 months after spiking the soil with eight PAHs (Binet *et al.* 2000a). Nine plant species were evaluated for enhanced removal of pyrene (Liste and Alexander 2000). The rhizosphere of each plant significantly dissipated pyrene after 56 days when compared to unplanted controls.

More importantly, the results from this plant-screening experiment revealed that pepper (*Capsicum annuum*) and radish (*Raphanus sativus*) promoted a faster decrease in pyrene concentrations, suggesting the importance of plant selection. A most-probable-number analysis that quantified bacteria growing on a mixture of phenanthrene, chrysene, pyrene, and benzo(*a*)pyrene (Wrenn and Venosa 1996) showed that the number of PAH degraders increased over a 1-year period in PAH-contaminated soil planted with perennial ryegrass (*Lolium perenne*), fescue (*Festuca arundinacea*), kleingrass (*Panicum coloratum*), and (annual) Indian mustard (*Brassica juncea*), but not in soils planted with mulberry (*Morus rubra*), *Geum* spp., or *Potentilla* spp. Furthermore, root-released compounds from those plants were shown to support the growth of PAH-degrading bacteria to different degrees (*i.e.*, plants supported growth of some bacteria, but not others) (Reardon *et al.* 2001). Although plant selection appears to be critical for rhizodegradation of recalcitrant organic pollutants, intraspecific variation among agronomic genotypes of alfalfa (*Medicago sativa*) suggests that specific cultivars of plants may be favored over others within the same species (Wiltse *et al.* 1998). The previous experiments illustrate the role of plants in the rhizodegradation of PAHs, and suggest an important role for species-specific root-released compounds.

Fungi inhabiting the rhizosphere of select plant species may play an important role in the rhizodegradation of PAHs. The establishment and maintenance of plant species growing on PAH-contaminated soils may be greatly improved by symbiotic relationships with arbuscular mycorrhizal fungus (Leyval and Binet 1998). Significant removal of PAHs was demonstrated in the presence of ryegrass (*Lolium perenne*) inoculated with the arbuscular mycorrhizal fungi *Glomus mosseae* (Binet *et al.* 2000b). In addition to enhancing PAH removal, the amount of PAHs absorbed by the root surfaces were significantly lower for ryegrass (*Lolium perenne*) treated with arbuscular mycorrhizal fungi, suggesting a role for the fungus in plant survivability in PAH-contaminated soils. Similarly, mycorrhizae may improve the survival of trees and enhance the rhizoremediation of soils contaminated with crude oil (Nicolotti and Egli 1998). In none of these studies was the true role of the mycorrhizae delineated, *i.e.*, whether the fungi degraded PAHs, enhanced plant growth, and thereby stimulated bacterial PAH biodegradation, or both. In addition to root-specific mycorrhizae, free-living forms of fungi [*e.g.*, white-rot fungi, (*Phanerochaete chysosporium*)] can be associated with the rhizosphere. White-rot fungi (*Phanerochaete chysosporium*) have been shown to partially oxidize PAHs to intermediates having higher water solubility, leading to increased bioavailability and biodegradation of the metabolized PAH (Meulenberg *et al.* 1997). The combined influence of fungi and bacteria may greatly enhance the degradation of PAHs (Canet *et al.* 2001), and such microbial interactions may be stimulated in the rhizosphere of select plant species.

In summary, the overall mechanisms responsible for enhanced removal of recalcitrant organic pollutants in the rhizosphere are complex and not well understood. Plants are capable of direct uptake, accumulation, and transformation of organics (Burken and Schnoor 1996, Burken and Schnoor 1997, Burken and Schnoor 1998), but given the low aqueous water solubility and significant amount of sorption, plant uptake of PCBs and PAHs is unlikely to be a significant factor in biodegradation. However, because root-associated microorganisms play a major role in the decomposition of recalcitrant and naturally occurring organic compounds, it is reasonable to assume that enhanced rhizodegradation of organic pollutants should proceed at appreciable rates within the rhizosphere of select vegetation.

Field Studies of Polycyclic Aromatic Hydrocarbon-Contaminated Sites

A 3-year field-based study to evaluate the rhizodegradation of PAHs was initiated in 1992 by the Union Carbide Corporation (Qiu *et al.* 1997). The growth of prairie buffalo grass (*Buchloe dactyloides*) in aged, contaminated soils resulted in significant reductions in PAH levels in the rhizosphere compared to unplanted, control soil. In a parallel experiment, twelve warm season grasses were evaluated for growth and rhizodegradation potential. On the basis of this 3-year plant screening experiment, kleingrass (*Panicum coloratum*) was determined to have the best potential for growth in contaminated soils and PAH removal in the rhizosphere compared to the other plant species. Analysis of both root and shoot material from all test plots demonstrated no evidence of PAH uptake and bioaccumulation into the plant tissue. Thus, microbial metabolism was presumed responsible predominately for PAH removal in these field studies.

A field demonstration on the rhizodegradation of diesel and PAHs was conducted at the U.S. Navy Craney Island Fuel Terminal in Portsmouth, Virginia (Hutchinson *et al.* this book). Three vegetated treatments and an unplanted control were evaluated on soils contaminated with aged diesel fuel. The vegetated treatments included plots planted with (1) a cool-season grass [tall fescue, (*Festuca arunclinacea*)], (2) a mixture of a cool-season grass [(annual ryegrass (*Lolium perenne*)] and a warm-season grass [(Bermuda grass, (*Cynodon dactylon*)], and (3) a shallow-rooted legume [(white clover, (*Trifolium repens*)]. After 2 years, the vegetated plots had higher microbial numbers and pollutant degraders in the soil, and significantly higher removal of total petroleum hydrocarbons and PAHs compared to controls free of vegetation. In addition, the authors obtained similar results from greenhouse studies and provided field designs and cost projections for utilizing phytoremediation at other field locations.

In a different field study, the disappearance of total petroleum hydrocarbons (TPHs) was monitored over three growing seasons (Nedunuri *et al.* 2000). Treatments included seeding the field plots with sorghum (*Sorghum bicolor*), ryegrass (*Lolium perenne*), or St. Augustine grass (*Stenotaphrum secundatum*). Unseeded control plots had volunteer vegetation removed by spraying with herbicides. Although the field sites were inherently heterogeneous in terms of soil properties and contaminant distributions, vegetated plots yielded significantly enhanced removal of TPHs and recalcitrant PAHs over the three growing seasons. Furthermore, the ryegrass (*Lolium perenne*) and St. Augustine grass (*Stenotaphrum secundatum*) treatments had higher reduction in TPHs compared to the other treatments. In addition to presenting these results, Nedunuri *et al.* (2000) discussed the importance of spatial heterogeneity in evaluating phytoremediation at field sites.

Currently, the TPH subgroup of the Remedial Technologies Development Forum Phytoremediation Action Team (http://www.rtdf.org/public/phyto/) has initiated eleven, 3-year field trials throughout the U.S. to evaluate plant growth and phytoremediation performance at sites contaminated with weathered petroleum hydrocarbons (Rock this book). A standard experimental protocol used at each of the sites highlights effects from differences in climates, soil conditions, and hydrocarbon concentrations, including PAHs. The standard treatments include a cool season grass-legume mixture, a locally optimized treatment with native plants, and an unplanted, unfertilized control. The goal of these cooperative field trials is to evaluate the efficacy of rhizodegradation of weathered petroleum hydrocarbons, including PAHs, as a remedial treatment option. Results from such long-term phytoremediation investigations around the globe will provide reasonable information for regulatory agencies on the effectiveness of rhizodegradation of TPHs and PAHs under field conditions.

Forensic Phytoremediation Investigations of Polychlorinated Biphenyl and Polycyclic Aromatic Hydrocarbon Contaminated Sites

Compelling evidence exists regarding the beneficial use of phytoremediation for the treatment of recalcitrant organic pollutants, but the underlying mechanisms are not readily understood. In addition, the choice of plants suitable for the rhizodegradation of organic pollutants has received little attention, as easy-to-grow grasses and agronomic plants are usually employed for greenhouse-based experiments and field demonstration sites. From an ecological context, plant selection should be directed toward vegetation that can not only tolerate the contaminated environment at a given locale (including climatic and soil conditions), but also has the potential to form rhizosphere-specific associations between plant roots and pollutant-degrading microorganisms. Furthermore, no phytoremediation studies have been completed under field conditions in a time frame consistent with the normal degradation of recalcitrant organic compounds found naturally in the environment. As a result, little attention has been paid to the ecological and environmental factors that are likely to be most important for the successful development of a rhizosphere-based bioremediation technology of recalcitrant organic pollutants.

Examinations of naturally revegetated terrestrial sites contaminated with recalcitrant organic pollutants provide compelling evidence to support the long-term efficacy of rhizodegradation (Fletcher 2001, Pivetz 2001). These forensic investigations of natural rhizodegradation describe the temporal, spatial, and ecological events critical for the reestablishment of vegetation at highly disturbed, contaminated sites. Similar to evaluating long-term (greater than 3 years) phytoremediation demonstrations, investigations of natural phytoremediation can reveal the ecological factors important for (1) tolerant vegetation reestablishment, root growth, and rhizosphere interactions; (2) contaminant fate; and (3) forecasts for future restoration of the site. Applying sound ecological principles and appropriate planting and management practices at contaminated sites can accelerate, maximize, and sustain important natural processes that will lead to the rhizodegradation of recalcitrant organic pollutants and remediation of polluted locales.

The role of vegetation in the removal of PAHs from contaminated soil has been demonstrated from forensic phytoremediation investigations (Olson and Fletcher 1999, Olson and Fletcher 2000, Olson *et al.* 2001). Through natural plant invasion and succession, tolerant vegetation was instrumental in the enhanced removal of PAHs at a former industrial sludge basin. Although a diverse community of plants was found growing at the contaminated site (51 species from 22 families), only a few species [*e.g.*, common sunflower (*Helianthus annuus*), Bermuda grass (*Cynodon dactylon*), and mulberry (*Morus* spp.)] were dominant in terms of relative cover and abundance (Olson and Fletcher 2000). Root growth and penetration of tolerant vegetation into the contaminated sludge depended on age and species of plant. For example, trees on an average penetrated the sludge at greater depths (90 centimeters) compared to forbs (60 centimeters) and grasses (50 centimeters). The levels of recalcitrant PAHs in the rhizosphere of an older mulberry (*Morus* spp.) tree were significantly reduced to depths of greater than 1 meter in the rhizosphere, suggesting the importance of long-term biological associations formed in the rhizosphere (Olson and Fletcher 1999). However, regardless of the vegetation, root growth into the sludge correlated with a significant reduction in all PAHs compared to the unrooted sludge (Olson *et al.* 2001). These findings demonstrate that investigations of revegetated contaminated sites can provide useful field-based information that validates the potential for the rhizodegradation of organic pollutants.

Limited information exists regarding the rhizodegradation of PCBs in field-based sites. A forensic phytoremediation investigation was initiated to evaluate vegetation and root-associated microorganisms occurring naturally at two geographically distinct sites contaminated with PCBs in the Czech Republic (Leigh *et al.* 2001). Results to date demonstrate that plants can invade, colonize, and tolerate soils contaminated with PCBs. A diverse community of trees, grasses and forbs, representing 51 plant species from 23 families, is currently growing at the two sites. These plant species appear to interact with rhizosphere microorganisms to enhance rhizodegradation of PCBs. Plants currently growing at these PCB-contaminated sites represent ideal candidates for PCB rhizodegradation evaluations.

In an effort to identify plants suitable for phytoremediation and rhizodegradation of oil-contaminated soils, a floristic investigation was conducted in Canada (Frick *et al.* 1999, Godwin and Thorpe 2000). Plants growing on soils contaminated with petroleum hydrocarbons were collected and identified throughout various geographic regions in Canada. In addition, plants growing in adjacent, unpolluted environments were identified, as plants growing in unpolluted sites were probably the seed donors for vegetation that has invaded and colonized the oil-contaminated sites. Differences in vegetation patterns between the contaminated and uncontaminated sites reveal which plants are tolerant of polluted environments. These tolerant plants have been naturally selected to colonize oil-contaminated sites under differing ecological and climatic conditions, and should provide a valuable resource for future phytoremediation and rhizodegradation demonstrations.

ECOLOGICAL CONTEXT OF RHIZODEGRADATION

Phytoremediation capitalizes on the innate abilities of vegetation to remediate and restore sites contaminated with a variety of pollutants (Anderson *et al.* 1993, Lee and Banks 1993, Shimp *et al.* 1993, Erickson *et al.* 1994, Schnoor *et al.* 1995, Salt *et al.* 1998, Siciliano and Germida 1998). Regardless of the contaminant(s) found at a given polluted site, vegetation and root-associated microorganisms will act together to influence the ultimate fate of the pollutant. Organic and inorganic contaminants are under the biotic influence of both the plant growing at the site and the microorganisms inhabiting the root zone. It is important to stress that optimizing the dynamic interactions between plants and microbes will largely determine the success of phytoremediation strategies such as volatilization, extraction, stabilization, transformation, and degradation of the contaminant in the plant (phytodegradation) or rhizosphere (rhizodegradation). In most remedial strategies for organic pollutants, the complete destruction (*e.g.*, biodegradation) of the pollutant molecule is highly desirable. Due to the persistence, toxicity, and widespread abundance of recalcitrant organic pollutants in the terrestrial environment, efficient removal of the contaminant will probably be a result of the

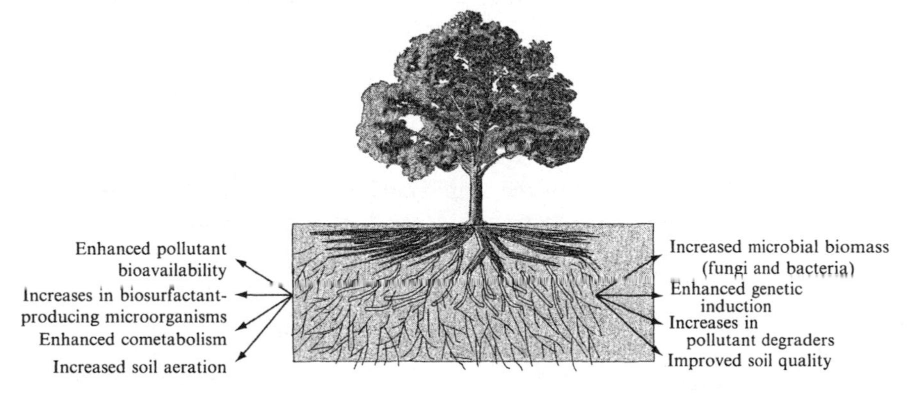

Figure 10-4 Plant–microorganisms interactions and mechanisms of enhanced rhizosphere bioremediation.

combined influence of both plants and root-associated microbes through rhizodegradation (Figure 10-4).

Rhizodegradation of recalcitrant organic compounds is not a new concept. For many decades, ecologists, soil scientists, and agriculturalists have been actively engaged in research concerning the role of vegetation and root-associated microorganisms in the cycling of carbon in the environment. As a result, a tremendous amount of information exists on a subject pertinent to rhizosphere bioremediation. Furthermore, ecologists have recognized the importance of vegetation in restoring disturbed environments. Through succession, vegetation has the remarkable capacity to restore disturbed sites to conditions resembling those prior to the disturbance. Thus, phytoremediation need not be considered a "new technology," but a technology that can build on a long history of observations and research.

Root-Released Compounds and Rhizodeposition

The rhizosphere is a complex and dynamic environment inhabited by a wide diversity of microorganisms (Curl and Truelove 1986, Lynch 1990, Bowen and Rovira 1991, Bowen and Rovira 1999). The driving force in this ecological zone is the growth of plant roots and the release of photosynthetically derived compounds into the surrounding soil. The sum total of carbon released from plant roots into the rhizosphere is rhizodeposition (Whipps and Lynch 1985, Lynch and Whipps 1990). The release of organic compounds into the rhizosphere supports the growth and metabolic activities of rhizosphere microorganisms (Bowen and Rovira 1976, Barber and Lynch 1977, Richards 1987).

Plant roots have a secondary function as storage organs for photosynthetically derived carbon from primary and secondary plant metabolism (Salisbury and Ross 1992, Taiz and Zeiger 1998, Buchanan *et al.* 2000) (Figure 10-5). However, plants differ in the types and amounts of organic compounds being stored in the root (Whipps and Lynch 1985, Lynch and Whipps 1990). Annual plants transfer approximately 30 to 60 percent of the net fixed carbon to the roots, and 40 to 90 percent of this stored carbon is lost *via* rhizodeposition. In contrast, perennial vegetation maintains roots over many growing seasons and rhizodeposition tends to be cyclic, depending on the time of year. Approximately 30 to 70 percent of net fixed carbon is transferred to perennial roots, with roughly 25 to 80 percent of the stored carbon lost annually through rhizodeposition. In field experiments, annual rhizodeposition accounts for less than 1.5 megagrams of carbon per hectare for annual plants to greater than 7.5 megagrams of carbon per hectare for perennial vegetation, including trees. Diverse rhizosphere communities of microorganisms continually metabolize the tremendous annual flow of carbon from vegetation into the soil.

A wide assortment of plant-derived organic compounds is released into the terrestrial environment through rhizodeposition (Lynch and Whipps 1990, Grayston *et al.* 1996, Jones 1998, Kuzyakov and Domanski 2000, Toal *et al.*

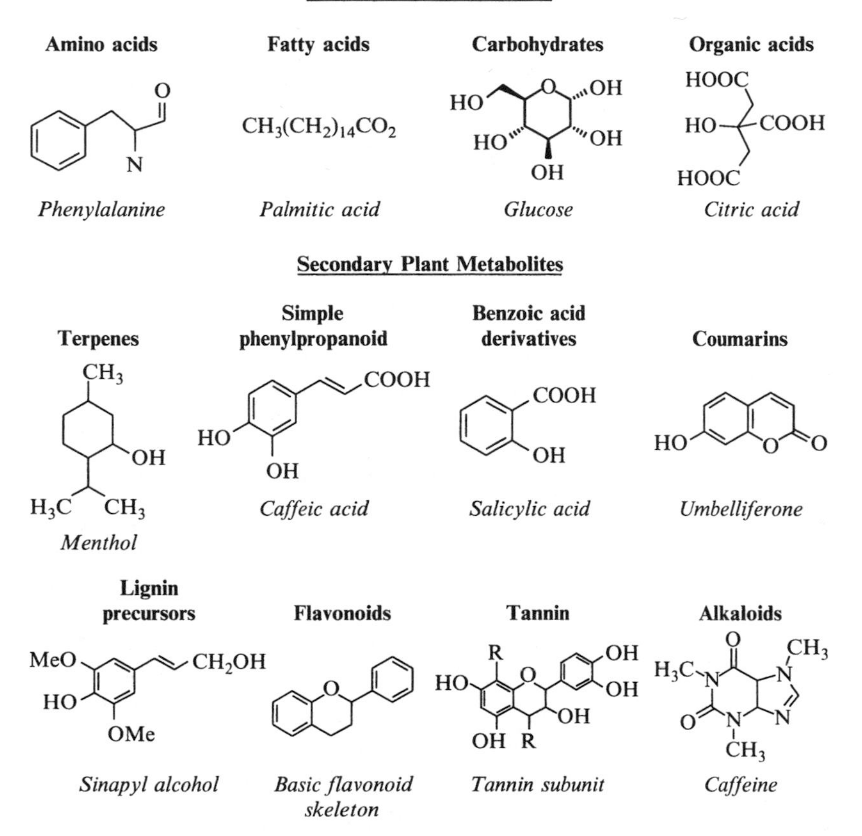

Figure 10-5 Representative primary and secondary plant metabolites involved in rhizodegradation. Adapted from Salisbury and Ross (1992), Taiz and Zeiger (1998), Buchanan *et al.* (2000).

2000). Five general classes of organic compounds have been identified as rhizodeposits (Table 10-3) (Rovira *et al.* 1979). Exudates are low molecular weight inorganic and organic compounds that passively leak from root cells. Root exudates include simple carbohydrates, amino acids, and small organic acids. Secretions are both low and high molecular weight organic compounds that are actively released from the root surface *via* metabolic processes. Examples of root secretions include enzymes and polymeric carbohydrates. Plant mucilages are gelatinous materials secreted by root cap cells, epidermal tissue, and root hairs. In addition, plant mucilages include sloughed root cap cells and dead epidermal cells. Mucigel includes gelatinous material from both plant and bacterial origin. Finally, root lysates are compounds deposited into the soil by the lysis of root cells. Root lysates comprise the contents of vacuoles (storage organelles for secondary plant metabolites), as well as cytoplasmic content, cell membranes, and cell wall material. All five general

TABLE 10-3 Five Classes of Rhizodeposits and the Potential Roles in Enhanced Rhizodegradation[a]

Rhizodeposit class	Metabolic origin	Organic compound examples	Mode of rhizodeposition	Potential influence on rhizodegradation
Root exudates	Primary plant metabolism	Low molecular weight (*e.g.*, carbohydrates, amino acids, and organic acids)	Passive leakage from root cells	Increased microbial biomass from "rhizosphere effect"
Root secretions	Primary and secondary plant metabolism	Low and high molecular weight (*e.g.*, enzymes, and polymeric carbohydrates)	Active secretion from root cells	Increased bioavailability, increased microbial biomass
Plant mucilages	Primary plant metabolism	Low and high molecular weight (*e.g.*, carbohydrates, and cellular debris)	Active secretion and passive loss from root	Increased bioavailability, increased microbial biomass
Mucigel	Primary plant and bacterial metabolism	Low and high molecular weight (*e.g.*, carbohydrates, and cellular debris)	Active secretion and passive loss from root and bacteria	Increased bioavailability, increased microbial biomass
Root lysates	Primary and secondary plant metabolism	Low and high molecular weight (*e.g.*, phenolic compounds, and cellular debris)	Passive loss from the autolysis of root cells	Increased microbial biomass, gene induction, cometabolism

[a] Adapted from Rovira (1979).

classes of rhizodeposits are involved in supporting the growth of microorganisms in the root zone.

Several factors, acting separately or in combination, affect the rhizodeposition of plant compounds into the soil (Richards 1987, Lynch and Whipps 1990, Grayston *et al.* 1996, Jones 1998, Kuzyakov and Domanski 2000, Toal *et al.* 2000). The quality and quantity of organic compounds released into the soil are plant species dependent. All plants release primary plant metabolites (*e.g.*, simple carbohydrates, amino acids, and organic acids) into the rhizosphere. These compounds are rapidly metabolized by soil microorganisms and are cited as the cause of the higher microbial biomass in the root zone—the rhizosphere effect. Secondary plant metabolites have a more restricted distribution in the plant kingdom because these molecules generally occur in individual plant species or taxonomically related groups of species. Due to the structural similarity of secondary plant metabolites and anthropogenic pollutants, naturally occurring compounds may stimulate microorganisms involved in the rhizodegradation of recalcitrant pollutants. Because the quality and quantity of rhizodeposits are also strongly influenced by the lifecycle and growth stage of the individual plant, as well as the conditions in which the plant is growing, these factors may impact the effectiveness of the rhizodeposition process. Finally, microorganisms in the rhizosphere directly influence rhizodeposition. Thus, the ecological dynamics in the rhizosphere and response of individual plant species in rhizodeposition will have a great bearing on the success of rhizodegradation of recalcitrant organic pollutants.

MECHANISMS OF ENHANCED RHIZOSPHERE BIOREMEDIATION

General Rhizosphere Effect

A major factor limiting the biodegradation of recalcitrant organic compounds is an insufficient microbial population capable of metabolizing the compound. Although it has been shown that indigenous microorganisms in virtually all terrestrial environments are capable of degrading xenobiotic pollutants (Fulthorpe *et al.* 1996), the numbers of such microorganisms are generally too low to reduce contamination at any given site within a reasonable time. As a result, a primary goal for microbiologically based biodegradation of recalcitrant organic pollutants should be to increase the total numbers of those microorganisms capable of metabolizing the pollutant at appreciable rates.

Organic compounds released into the environment from vegetation, particularly the various root released compounds, increase the total numbers of both fungi and bacteria associated with the root zone (Bowen and Rovira 1976, Barber and Lynch 1977, Curl and Truelove 1986, Richards 1987, Lynch 1990, Bowen and Rovira 1999). One of the putative functions of root-released compounds is to support the growth and metabolic activities of a diverse microbial community in the rhizosphere (Shimp *et al.* 1993, Erickson *et al.* 1994, Schnoor *et al.* 1995, Salt *et al.* 1998, Siciliano and Germida 1998). As a result, the microbial density in rhizosphere soil may be 10 to 10 000 fold higher than in bulk soil (Salt *et al.* 1998, Siciliano and Germida 1998). Fostering this microbial community is advantageous to the plant, because rhizosphere microbes have been shown to promote plant growth, enhance mineral and water uptake, produce antibiotics to inhibit soil pathogens, and produce plant growth regulators (Kapulnik 1996). For several decades, researchers have described this increase in soil microorganism numbers in the root zone as a "general rhizosphere effect."

A general rhizosphere effect can be the result of many combined influences (Curl and Truelove 1986, Lynch 1990, Bowen and Rovira 1991, Bowen and Rovira 1999). The growth of plant roots in the soil improves soil conditions and provides an ideal environment for microorganisms to inhabit. The repeated demonstration that total microbial biomass is increased in the rhizosphere of a variety of plant species is probably the result of simple, low molecular weight, organic and inorganic compounds from primary plant metabolism being released into the soil environment *via* root exudation. The composition of basic exudate is similar among most plant taxa (Lynch and Whipps 1990, Grayston *et al.* 1996, Jones 1998, Kuzyakov and Domanski 2000, Toal *et al.* 2000). Therefore, a general rhizosphere effect results in increased microbial biomass, but not necessarily an increase in the activity of microbes biodegrading the recalcitrant organic pollutant.

Specific Population Increases and Induction of Degradative Pathway Genes

Through the release of low molecular weight organic and inorganic compounds into the rhizosphere, a general rhizosphere effect can increase the numbers of all microbes in the soil (Curl and Truelove 1986, Klein *et al.* 1988, Lynch 1990, Bowen and Rovira 1991, Shimp *et al.* 1993, Schnoor *et al.* 1995, Bowen and Rovira 1999, Klein 2000). Although total microbial biomass may be higher in the rhizosphere soil at contaminated sites (Jordahl *et al.* 1997, Nichols *et al.* 1997), those microbes with the capability to biodegrade the recalcitrant organic pollutant must be stimulated, *i.e.*, the genes encoding the enzymatic pathways for the pollutant degradation must be induced. The release of higher molecular weight compounds (from lysates and other types of root-released compounds) from plants may induce pollutant-degrading genetic pathways in microorganisms. For example, phenolic compounds in rhizodeposits promote the growth of certain soil bacteria that degrade organic compounds, such as PCBs and PAHs (Donnelly *et al.* 1994, Fletcher and Hegde 1995, Fletcher *et al.* 1995, Hegde and Fletcher 1996, Gilbert and Crowley 1997, Siciliano and Germida 1998). Limited information exists on the role of naturally occurring compounds in rhizodegradation of contaminants, but due to the structural similarity of naturally occurring plant compounds and anthropogenic pollutants, it is reasonable to assume that the former can induce the appropriate microbial degradative pathways (Leigh *et al.* 2002).

The biodegradation of PCBs is stimulated by naturally occurring plant compounds (Donnelly *et al.* 1994, Fletcher and Hegde 1995, Hegde and Fletcher 1996, Gilbert and Crowley 1997, Singer *et al.* 2000). Phenolic compounds and terpenoids collected from a variety of plant species or utilized in pure forms supported the growth of PCB-degrading bacteria and stimulated the biodegradation of PCBs. Plant phenolics and terpenoids have the capacity to serve as growth substrates for the bacteria, as cometabolites for aerobic PCB degradation, and as inducers for biphenyl metabolism.

An important aspect of PAH degradation by aerobic bacteria is the induction of the genes for the catabolic enzymes. All genes involved in the early steps of PAH degradation are inducible. Salicylate, an intermediate in the naphthalene dioxygenase pathway, induces the *nah* genes (for naphthalene degradation) (Yen and Serdar 1988). Salicylate also induces degradation of benz(*a*)anthracene by *Sphingomonas yanoikuyae* (Mahaffey *et al.* 1988) and the degradation of several high molecular weight PAHs by *Pseudomonas saccharophila* P15 (Chen and Aitken 1999). Salicylate is an essential signaling compound in plants and may play an important role in the stimulation of pollutant degraders and the rhizodegradation of PAHs. Evidence for this was recently obtained using the bioluminescent reporter strain *Pseudomonas fluorescens* HK44, which contains the *lux* gene fused to the *nahG* gene (encoding salicylate hydroxylase). When incubated with ryegrass (*Lolium perenne*) root-released compounds, strain HK44 produced light, indicating that one or more plant-released chemicals (possibly salicylate) stimulated the production of PAH-degrading enzymes (Reardon *et al.* 2001). Therefore, the careful selection of appropriate plant species could preferentially select for and maintain the growth of microorganisms capable of degrading recalcitrant organic pollutants in the rhizosphere.

Enhanced Bioavailability of Recalcitrant Organic Pollutants

Select vegetation may have the innate capability to enhance the bioavailability of recalcitrant organic compounds in the rhizosphere (Siciliano and Germida 1998). Root-released compounds, including cationic chelators and organic acids, alter nutrient bioavailability. These organic compounds may also increase the bioavailability of contaminants by increasing the cation exchange capacity of the rhizosphere and competing for binding sites within the soil. The resulting decrease in pollutant complexation with soil matrices and organic matter would lead to increased bioavailability of the recalcitrant contaminant. Furthermore, plant roots release compounds that may act as surfactants. Various glycerolipids and glycoproteins released from roots can reduce interfacial surface tension of pollutants and increase aqueous water solubility. In addition, other lipophilic compounds from roots, such as membranes, lipids, and sterols, may enhance the solubility of a pollutant and increase partitioning to the aqueous phase in the soil solution.

In contrast to the direct influence of root-released compounds in enhancing pollutant bioavailability, select vegetation may stimulate and select for biosurfactant-producing populations of bacteria in the rhizosphere. Utilizing a simple method for enumerating biosurfactant-producing bacteria on blood agar plates (Carrillo *et al.* 1996), Reardon et al. (2001) suggested that more biosurfactant-producing bacteria were present in rhizosphere soil from contaminated and uncontaminated soil compared to controls free of vegetation, respectively. A higher proportion of biosurfactant-producing organisms in the rhizosphere would probably increase the solubility of the recalcitrant organic pollutants and thus enhance rhizodegradation.

Plant Compounds and Cometabolites

Even when a sufficient level of microbial biomass is present and bioavailability is not an issue, organic pollutants still may not degrade. In many cases, this occurs because either the genes for degradative pathways have not been induced, or there is insufficient energy for cell growth. Issues of gene induction and the role that root-released compounds can play were discussed previously. Energy production by cells is also critical, and the biodegradation of some contaminants may not supply enough energy for the cells, either because the pollutant is present at levels too low to support a specific microbial population or because the microorganisms can acquire little or no energy from metabolizing the compound. Plant chemicals have the potential to ameliorate these limitations.

An important concept in the biodegradation of recalcitrant organic pollutants is cometabolism. Cometabolism is loosely defined as the degradation of a compound that occurs only in the presence of other organic compounds that serve as inducers of the necessary genes and the primary energy source (McCarty 1987). Bioremediation studies have illustrated the importance of cometabolism for recalcitrant organic pollutants, such as PCBs and PAHs.

Polychlorinated biphenyls are amenable to biodegradation under aerobic conditions, but this is generally limited to PCBs with one to three chlorine substitutions. The aerobic biodegradation of the more chlorinated PCBs is slow or nonexistent. However, the addition of an artificial cometabolite (*e.g.*, biphenyl) allows for the metabolism of higher chlorinated PCBs, and even enhances the biodegradation rate of less chlorinated PCBs. Naturally occurring plant compounds have also been demonstrated to stimulate PCB metabolism and act as cometabolites, similar to biphenyl (Donnelly *et al.* 1994, Fletcher and Hegde 1995, Hegde and Fletcher 1996, Gilbert and Crowley 1997, Singer *et al.* 2000).

The aerobic biodegradation of higher molecular weight PAHs is under similar constraints as that of PCBs. In other words, the high molecular weight PAHs are typically recalcitrant in the environment because of difficulties in obtaining sufficient energy from metabolism. This is apparent after an oil or diesel fuel spill, as easily degradable compounds are readily metabolized, leaving behind high molecular weight PAHs. However, the addition of a cometabolite results in the biodegradation of the high molecular weight PAHs (Bouchez *et al.* 1995). Although firm experimental evidence is lacking, it is reasonable to assume that structurally similar plant compounds can provide the energy source (and gene induction) for cometabolism of high molecular weight PAHs.

Recalcitrant organic contaminants are often present at concentrations low enough that the rate of energy production through biodegradation would not be sufficient to meet the maintenance energy demands of a microbial population (Bouwer and McCarty 1983). However, research has shown that even trace organics can be consumed as "secondary substrates" when readily biodegradable carbon sources are added (Stratton *et al.* 1983). Root-released compounds may provide the necessary characteristics to drive the metabolism of PAHs, PCBs, and other recalcitrant organics as secondary substrates.

Plant Management Practices for Enhanced Rhizodegradation

Plant ecologists have recognized the role of vegetation and ecological events responsible for improving site conditions and ecosystem recovery after disturbances (*e.g.*, Gleason 1926, Tansley 1935, Clements 1936, Whittaker 1962, Barbour *et al.* 1987). Just as many natural events (*e.g.*, volcanic activity, fires, and weather) lead to distressed lands, environments contaminated with anthropogenic pollutants can be considered from an ecological perspective as disturbed sites. Ecological events, such as plant invasion, competition, succession, and community structure change (Pickett 1982), have been observed to be instrumental in restoring polluted environments (Olson and Fletcher 2000). Sound plant management strategies, including watering schedules, fertilization, and herbicide usage, can be implemented to (1) optimize the rhizosphere for improved plant growth (Bowen and Rovira 1999); (2) sustain the health, quality, and biotic component of the soil (Doran and Zeiss 2000); (3) prime the contaminated soil for the enhanced turnover of organic carbon (Kuzyakov *et al.* 2000); and (4) ultimately stimulate the biodegradation of anthropogenic pollutants by rhizosphere microorganisms.

Concepts of plant succession should be applied in phytoremedial designs for the rhizodegradation of recalcitrant organic pollutants (Olson and Fletcher 2000). Succession is the continuous, sequential change in species composition of natural communities (Barbour *et al.* 1987). Autogenic succession is the process in which continual changes in the biotic component (*i.e.* plants and rhizosphere microorganisms) at a given site modifies the environment and improves site conditions (Tansley 1935). From this perspective, a single plant species cannot foster the rhizosphere interactions for optimal rhizodegradation, nor can a single species be equally useful under all soil and climatic conditions. Unfortunately, phytoremediation to date has generally employed single species or, occasionally, a few related plant species grown together. These management practices are at odds with the natural ecological recovery of disturbed sites, which involves multiple, diverse plant species in an evolving plant community (*e.g.*, Gleason 1926, Tansley 1935, Clements 1936, Whittaker 1962, Barbour *et al.* 1987).

Plant selection for enhanced rhizodegradation should be patterned after native vegetation of the given climatic area and plants that have been demonstrated to grow on previously contaminated soils (Frick *et al.* 1999, Godwin and Thorpe 2000, Olson and Fletcher 2000, Leigh *et al.* 2001). Furthermore, plants should be selected based on inherent abilities to produce secondary metabolites (*e.g.*, root phenolics and flavonoids, see Rao 1990, Glasby 1991) that may foster rhizosphere microbes capable of biodegrading recalcitrant organic pollutants. The contaminated site should first be stabilized with a variety of pioneer plants capable of withstanding severe environments. As site conditions improve, additional native plants would be planted at the site and ultimately the site would be revegetated with a variety of annual and perennial grasses and forbs, as well as shrubs and trees. Appropriate plant management strategies, including watering, fertilization, and plant clipping could be devised to assure the success of this engineered form of natural succession. Furthermore, the time frame should be consistent with natural successional events and normal rhizodegradation of naturally occurring compounds in uncontaminated environments (*i.e.*, on the order of years). The successive utilization of a variety of plants is more likely to foster the activity of those root-associated microorganisms in contaminated soil most instrumental in biodegrading PCBs and PAHs.

Acknowledgments

Preparation of this chapter was supported in part by Research Grant SP-99 from the U.S. Environmental Protection Agency and the Great Plains/Rocky Mountain Hazardous Substance Research Center. The authors are also grateful for input from our colleagues John Fletcher (University of Oklahoma, Department of Botany and Microbiology), Albert Venosa (U.S. EPA), and Donald Klein and Nancy DuTeau (Colorado State University, Department of Microbiology).

REFERENCES

Abramowicz, D.A. (1990) Aerobic and anaerobic biodegradation of PCBs: a review. *Crit. Rev. Biotechnol.* **10**(3): 241–251.

Abramowicz, D.A. (1995) Aerobic and anaerobic PCB biodegradation in the environment. *Environ. Health Persp.* **103**(Supplement 5): 97–99.

Anderson, T.A., E.A. Guthrie, and B.T. Walton (1993) Bioremediation in the Rhizosphere. *Environ. Sci. Technol.* **27**(13): 2630–2636.

Aprill, W. and R.C. Sims (1990) Evaluation of the use of prairie grasses for stimulating polycyclic aromatic hydrocarbon treatment in soil. *Chemosphere* **20**: 253–265.

Bakker, M.I., M. Vorenhout, D.T.H.M. Sijm, and C. Kolloffel (1999) Dry deposition of atmospheric polycyclic aromatic hydrocarbons in three *Plantago* species. *Environ. Toxicol. Chem.* **18**(10): 2289–2294.

Banks, M.K., E. Lee, and A.P. Schwab (1999) Evaluation of dissipation mechanisms for benzo(*a*)pyrene in the rhizosphere of tall fescue. *J. Environ. Qual.* **28**: 294–298.

Barber, D.A. and J.M. Lynch (1977) Microbial growth in the rhizosphere. *Soil Biol. Biochem.* **9**: 305–308.

Barbour, M.G., J.H. Burk, and W.D. Pitts (1987) *Terrestrial Plant Ecology*. 2nd ed. Benjamin/Cummings Publishing Company, Menlo Park, California.

Binet, P., J.M. Portal, and C. Leyval (2000a) Dissipation of 3–6 ring polycyclic aromatic hydrocarbons in the rhizosphere of ryegrass. *Soil Biol. Biochem.* **32**: 2011–2017.

Binet, P., J.M. Portal, and C. Leyval (2000b) Fate of polycyclic aromatic hydrocarbons (PAH) in the rhizosphere and mycorrhizosphere of ryegrass. *Plant Soil* **227**: 207–213.

Bouchez, M., D. Blanchet, and J.P. Vandecasteele (1995) Degradation of polycyclic aromatic hydrocarbons by pure strains and defined strain associations: inhibition phenomena and cometabolism. *Appl. Microbiol. Biotechnol.* **43**: 156–164.

Bouwer, E.J. and P.L. McCarty (1983) Transformations of 1- and 2-carbon halogenated aliphatic organic compounds under methanogenic conditions. *Appl. Environ. Microbiol.* **45**: 1286–1294.

Bowen, G.D. and A.D. Rovira (1976) Microbial colonization of plant roots. In: *Annu. Review of Phytopathol*ogy. Annual Reviews, Palo Alto, California. **14**: 121–144.

Bowen, G.D. and A.D. Rovira (1991) The rhizosphere—the hidden half of the hidden half. In: Plant *Roots—The Hidden Half*. Y. Waisel, A. Eshel, and U. Kaffkafi, eds. Marcel Dekker, New York, pp. 641–669.

Bowen, G.D. and A.D. Rovira (1999) The rhizosphere and its management to improve plant growth. In: *Advances in Agronomy*, D.L. Sparks, ed. Academic Press, San Diego, California. **66**: 1–102.

Brady, N.C. (1990) *The Nature and Properties of Soils*. 10th ed. Macmillan Publishing Company, New York.

Braun-Lullemann, A., A. Huttermann, and A. Majcherczyk (1999) Screening of ectomycorrhizal fungi for degradation of polycyclic aromatic hydrocarbons. *Appl. Microbiol. Biotechnol.* **53**: 127–132.

Brown, J.F., R.E. Wagner, H. Feng, D.L. Bedard, M.J. Brennan, J.C. Carahan, and R.J. May (1987) Environmental dechlorination of PCBs. *Environ. Toxicol. Chem.* **6**: 579–593.

Buchanan, B.B., W. Gruissem, and R.L. Jones (2000) *Biochemistry and Molecular Biology of Plants*. American Society of Plant Physiologists, Rockville, Maryland.

Burken, J.G. and J.L. Schnoor (1996) Phytoremediation: plant uptake of atrazine and role of plant exudates. *J. Environ. Eng.* **122**(11): 958–963.

Burken, J.G. and J.L. Schnoor (1997) Uptake and metabolism of atrazine by poplar trees. *Environ. Sci. Technol.* **31**(5): 1399–1406.

Burken, J.G. and J.L. Schnoor (1998) Predictive relationships for uptake of organic contaminants by hybrid poplar trees. *Environ. Sci. Technol.* **32**(21): 3379–3385.

Cameron, M.D., S. Timofeevski, and S.D. Aust (2000) Enzymology of *Phanerochaete chrysosporium* with respect to the degradation of recalcitrant compounds and xenobiotics. *Appl. Microbiol. Biotechnol.* **54**: 751–758.

Canet, R., J.G. Birnstingl, D.G. Malcolm, J.M. Lopez-Real, and A.J. Beck (2001) Biodegradation of polycyclic aromatic hydrocarbons (PAHs) by native microflora and combinations of white-rot fungi in a coal-tar contaminated soil. *Bioresource Technol.* **76**(2): 113–117.

Carrillo, P.G., C. Mardaraz, S.I. Pitta-Alvarez, and A.M. Giulietti (1996) Isolation and selection of biosurfactant producing bacteria. *World J. Microbiol. Biotechnol.* **12**: 82–84.

Cerniglia, C.E. (1997) Fungal metabolism of polycyclic aromatic hydrocarbons: past, present and future applications in bioremediation. *J. Ind. Microbiol. Biotechnol.* **19**: 324–333.

Chen, S.H. and M.D. Aitken (1999) Salicylate stimulates the degradation of high-molecular weight polycyclic aromatic hydrocarbons by *Pseudomonas saccharophila* P15. *Environ. Sci. Technol.* **33**: 435–439.

Clements, F.E. (1936) Nature and structure of the climax. *J. Ecol.* **24**: 252–284.

Coates, J.D., J. Woodward, J. Allen, P. Philp, and D.R. Lovley (1997) Anaerobic degradation of polycyclic aromatic hydrocarbons and alkanes in petroleum-derived marine harbor sediments. *Appl. Environ. Microbiol.* **62**: 3589–3593.

Cornelissen, G., H. Rigterink, M.M.A. Ferdinandy, and P.C.M. van Noort (1998) Rapidly desorbing fractions of PAHs in contaminated sediments as a predictor of the extent of bioremediation. *Environ. Sci. Technol.* **32**: 966–970.

Curl, F.A. and B. Truelove (1986) *The Rhizosphere*. Springer-Verlag, New York.

Dittmer, H.A. (1937) A quantitative study of roots and root hairs of a winter rye-plant (*Secale cereale*). *Am. J. Bot.* **24**: 417–420.

Donnelly, P.K. and J.S. Fletcher (1994) Potential use of mycorrhizal fungi as bioremediation agents. In: *Bioremediation Through Rhizosphere Technology*. J.R. Coats, ed. American Chemical Society, Washington, D.C., pp. 93–99.

Donnelly, P.K. and J.S. Fletcher (1995) PCB metabolism by ectomycorrhizal fungi. *Bull. Environ. Contam. Toxicol.* **54**: 507–513.

Donnelly, P.K., R.S. Hegde, and J.S. Fletcher (1994) Growth of PCB-degrading bacteria on compounds from photosynthetic plants. *Chemosphere* **28**(5): 981–988.

Doran, J.W. and M.R. Zeiss (2000) Soil health and sustainability: managing the biotic component of soil quality. *Appl. Soil Ecol.* **15**: 3–11.

Erickson, L.E., M.K. Banks, L.C. Davis, A.P. Schwab, N. Muralidharan, K. Reilley, and J.C. Tracy (1994) Using vegetation to enhance *in-situ* bioremediation. *Environ. Prog.* **13**(4): 226–231.

Field, J.A. and E.M. Thurman (1996) Glutathione conjugation and contaminant transformation. *Environ. Sci. Technol.* **30**(5): 1413–1418.

Fletcher, J.S. (2001) Rhizosphere remediation of recalcitrant soil contaminants: an important component of long-term sustained biosystem treatment. In: State of the

Science Conference. U.S. Environmental Protection Agency Report EPA/625/R-01/011b, held May 1–2, 2000, Boston, Massachusetts, pp. 139–147 (http://www.epa.gov/ORD/NRMRL/Pubs/625R01011b/625R01011bchap6.pdf).

Fletcher, J.S. and R.S. Hegde (1995) Release of phenols by perennial plant roots and their potential importance in bioremediation. *Chemosphere* **31**(4): 3009–3016.

Fletcher, J.S., P.K. Donnelly, and R.S. Hegde (1995) Biostimulation of PCB-degrading bacteria by compounds released from plant roots. In: *Bioremediation of Recalcitrant Organics*. R.E. Hinchee, D.B. Anderson and R.E. Hoeppel, eds. Battelle, Columbus, Ohio, pp. 131–136.

Frick, C.M., R.E. Farrell, and J.J. Germida (1999) Assessment of phytoremediation as an *in-situ* technique for cleaning oil-contaminated sites. Report to the Petroleum Technology Alliance of Canada (PTAC), Calgary, Alberta.

Fulthorpe, R.R., A.N. Rhodes, and J.M. Tiedje (1996) Pristine soils mineralize 3-chlorobenzoate and 2,4-dichlorophenoxyacetate via different microbial populations. *Appl. Environ. Microbiol.* **62**(4): 1159–1166.

Gilbert, E.S. and D.E. Crowley (1997) Plant compounds that induce polychlorinated biphenyl biodegradation by *Arthrobacter* sp. Strain B1B. *Appl. Environ. Microbiol.* **63**(5): 1933–1938.

Glasby, J.J. (1991) *Dictionary of Plants Containing Secondary Metabolites*. Taylor and Francis, London.

Gleason, H.A. (1926) The individualistic concept of the plant association. *Bull. Torrey Bot. Club*. **53**: 7–26.

Godwin, B. and J. Thorpe (2000) Assessment of phytoremediation as an *in-situ* technique for cleaning oil-contaminated sites—phase II final report. R.E. Farrell, ed. Petroleum Technology Alliance of Canada (PTAC), Calgary, Alberta.

Grayston, S.J., D. Vaughan, and D. Jones (1996) Rhizosphere carbon flow in trees, in comparison with annual plants: the importance of root exudation and its impact on microbial activity and nutrient availability. *Appl. Soil Ecol.* **5**: 29–56.

Gunther, T., U. Dornberger, and W. Fritsche (1996) Effects of ryegrass on biodegradation of hydrocarbons in soil. *Chemosphere* **33**: 203–215.

Gunther, T., U. Sack, M. Hofrichter, and M. Latz (1998) Oxidation of PAH and PAH-derivatives by fungal and plant oxidoreductases. *J. Basic Microbiol.* **38**(2): 113–122.

Hegde, R.S. and J.S. Fletcher (1996) Influence of plant growth stage and season on the release of root phenolics by mulberry as related to development of phytoremediation technology. *Chemosphere* **32**(12): 2471–2479.

Heitkamp, M.A., W. Franklin, and C.E. Cerniglia (1988) Microbial metabolism of polycyclic aromatic hydrocarbons: isolation and characterization of a pyrene-degrading bacterium. *Appl. Environ. Microbiol.* **54**: 2549–2555.

Hernandez, B.C., S.C. Koh, M. Chial, and D.D. Focht (1997) Terpene-utilizing isolates and their relevance to enhanced biotransformation of polychlorinated biphenyls in soil. *Biodegradation* **8**: 153–158.

Howsam, M., K.C. Jones, and P. Ineson (2000) PAHs associated with the leaves of three deciduous tree species. I—concentrations and profiles. *Environ. Pollut.* **108**: 413–424.

Huesemann, M.H. (1997) Incomplete hydrocarbon biodegradation in contaminated soils: limitations in bioavailability or inherent recalcitrance? *Biorem. J.* **1**(1): 27–40.

Jones, D.L. (1998) Organic acids in the rhizosphere—a critical review. *Plant Soil* **205**: 25–44.

Jones, K.C. and P. de Voogt (1999) Persistent organic pollutants (POPs): state of the science. *Environ. Pollut.* **100**: 209–221.

Jordahl, J.L., L. Foster, J.L. Schnoor, and P.J.J. Alvarez (1997) Effect of hybrid poplar trees on microbial populations important to hazardous waste bioremediation. *Environ. Toxicol. Chem.* **16**: 1318–1321.

Juhasz, A.L. and R. Naidu (2000) Bioremediation of high molecular weight polycyclic aromatic hydrocarbons: a review of the microbial degradation of benzo[*a*]pyrene. *Int. Biodeter. Biodegr.* **45**: 57–88.

Kanaly, R.A. and S. Harayama (2000) Biodegradation of high-molecular-weight polycyclic aromatic hydrocarbons by bacteria. *J. Bacteriol.* **182**(8): 2059–2067.

Kanaly, R., R. Bartha, S. Fogel, and M. Findlay (1997) Biodegradation of [^{14}C]benzo[*a*]pyrene added in crude oil to uncontaminated soil. *Appl. Environ. Microbiol.* **63**: 4511–4515.

Kapulnik, Y. (1996) Plant growth promotion by rhizosphere bacteria. In: *Plant Roots: the Hidden Half.* Y. Waisel, A. Eshel, and U Kaflazi, eds. Marcel Dekker, New York, pp. 769–781.

Kelsey, J.W. and M. Alexander (1997) Declining bioavailability and inappropriate estimation of risk of persistent compounds. *Environ. Toxicol. Chem.* **16**(3): 582–585.

Kipopoulou, A.M., E. Manoli, and C. Samara (1999) Bioconcentration of polycyclic aromatic hydrocarbons in vegetables grown in an industrial area. *Environ. Pollut.* **106**: 369–380.

Klein, D.A. (2000) The rhizosphere. In: *Encyclopedia of Microbiology*, 2nd ed., Vol. 4 (Q–Z). J. Lederberg, ed. Academic Press, New York, pp. 117–126.

Klein, D.A., B.A. Frederick, M. Biondini, and M.J. Trlica (1988) Rhizosphere microorganism effects on soluble amino acids, sugars, and organic acids in the root zone of *Agropyron cristatum, Agropyron smithii* and *Bouteloua gracilis. Plant Soil* **110**: 19–25.

Kolb, M. and H. Harms (2000) Metabolism of fluoranthene in different plant cell cultures and intact plants. *Environ. Toxicol. Chem.* **19**(5): 1304–1310.

Kucerova, P., M. Mackova, L. Chroma, J. Burkhard, J. Triska, K. Demnerova, and T. Macek (2000) Metabolism of polychlorinated biphenyls by *Solanum nigrum* hairy root clone SNC-9O and analysis of transformation products. *Plant Soil* **225**: 109–115.

Kucerova, P., C. in der Wiesche, M. Wolter, T. Macek, F. Zadrazil, and M. Mackova (2001) The ability of different plant species to remove polycyclic aromatic hydrocarbons and polychlorinated biphenyls form incubation media. *Biotechnol.Lett.* **23**: 1355–1359.

Kuzyakov, Y. and G. Domanski (2000) Carbon input by plants into the soil. *J. Plant Nutr. Soil Sci.* **163**: 421–431.

Kuzyakov, Y., J.K. Friedel, and K. Stahr (2000) Review of mechanisms and quantification of priming effects. *Soil Bio. Biochem.* **32**: 1485–1498.

Langenhoff, A.A.M., A.J.B. Zehnder, and G. Schraa (1996) Behavior of toluene, benzene, and naphthalene under anaerobic conditions in sediment columns. *Biodegradation* **7**: 267–274.

Leahy, J.G. and R.R. Colwell (1990) Microbial degradation of hydrocarbons in the environment. *Microbiol. Rev.* **54**(3): 305–315.

Lee, E. and M.K. Banks (1993) Bioremediation of petroleum contaminated soil using vegetation: a microbial study. *J. Environ. Sci. Health* A **28**: 2187–2198.

Lee, I. and J.S. Fletcher (1992) Metabolism of polychlorinated biphenyls (PCBs) by plant tissue cultures. In: *Progress in Plant Cellular and Molecular Biology*. H.J.J. Nijkamp, L.H.W. van der Plas, and J. van Aartrijk, eds. Kluwer Academic Publishers, Dordrecht, The Netherlands, pp. 656–660.

Leigh, M.B. (1997) The release of phenolic compounds into the soil by root turnover of mulberry (*Morus rubra* L.). Masters thesis, University of Oklahoma, Norman, Oklahoma.

Leigh, M.B., J.S. Fletcher, D.P. Nagle, M. Mackova, and T. Macek (2001) Vegetation and fungi at Czech PCB-contaminated sites as bioremediation candidates. In: *Phytoremediation, Wetlands and Sediments*. A. Leeson, E.A. Foote, M.K. Banks and V. Magar, eds. Sixth International Symposium on In Situ and On-Site Bioremediation. June 4–7, San Diego, California. Battelle Press, Columbus, Ohio, pp. 61–68.

Leigh, M.B., J.S. Fletcher, X. Fu, and F.J. Schmitz (2002) Root turnover: an important source of microbial substances in rhizosphere remediation of recalcitrant contaminants. *Environ. Sci. Technol.* **36**: 1579–1583.

Leyval, C. and P. Binet (1998) Effect of polyaromatic hydrocarbons in soil on arbuscular mycorrhizal plants. *J. Environ. Qual.* **27**: 402–407.

Liste, H.H. and M. Alexander (2000) Plant-promoted pyrene degradation in soil. *Chemosphere* **40**: 7–10.

Lynch, J.M. (1990) *The Rhizosphere*. Wiley, New York.

Lynch, J.M. and J.M. Whipps (1990) Substrate flow in the rhizosphere. *Plant Soil* **129**: 1–10.

Macek, T., M. Mackova, and J. Kas (2000) Exploitation of plants for the removal of organics in environmental remediation. *Biotechnol. Adv.* **18**: 23–34.

Mackay, D., W.Y. Shui, and K.C. Ma (1992a) *Illustrated Handbook of Physical-Chemical Properties and Environmental Fate for Organic Chemicals. Volume I. Monoaromatic Hydrocarbons, Chlorbenzenes, and PCBs*. Lewis Publishers, Chelsea, Michigan.

Mackay, D., W.Y. Shui, and K.C. Ma (1992b) *Illustrated Handbook of Physical-Chemical Properties and Environmental Fate for Organic Chemicals. Volume II. Polynuclear Aromatic Hydrocarbons, Polychlorinated Dioxins, and Dibenzofurans*. Lewis Publishers, Chelsea, Michigan.

Mackova, M., T. Macek, P. Kucerova, J. Burkhard, J. Pazlarova, and K. Demnerova (1997) Degradation of polychlorinated biphenyls by hairy root culture of *Solanum nigrum*. *Biotechnol. Lett.* **19**(8): 787–790.

Mahaffey, W.R., D.T. Gibson, and C.E. Cerniglia (1988) Bacterial oxidation of chemical carcinogens: formation of polycyclic aromatic acids from benz[*a*]anthracene. *Appl. Environ. Microbiol.* **54**(10): 2415–2423.

McCarty, P.L. (1987) Bioengineering issues related to *in-situ* remediation of contaminated soils and groundwater. In: *Environmental Biotechnology*. G.S. Omenn, ed. Plenum Press, New York, pp. 143–162.

Meharg, A.A. and J.W.G. Cairney (2000) Ectomycorrhizas—extending the capabilities of rhizosphere remediation. *Soil Biol. Biochem.* **32**: 1475–1484.

Meulenberg, R., H.H.M. Rijnaarts, H.J Doddema, and J.A. Field (1997) Partially oxidized polycyclic aromatic hydrocarbons show an increased bioavailability and biodegradability. *FEMS Microbiol. Lett.* **152**(1):45–49.

Mihelcic, J.R. and R.G. Luthy (1988) Degradation of polycyclic aromatic hydrocarbon compounds under various redox conditions in soil-water systems. *Appl. Environ. Microbiol.* **54**: 1182–1187.

Miya, R.K. and M.K. Firestone (2000) Phenanthrene-degrader community dynamics in rhizosphere soil from a common annual grass. *J. Environ. Qual.* **29**: 584–592.

Muncnerova, D. and J. Augustin (1994) Fungal metabolism and detoxification of polycyclic aromatic-hydrocarbons—a review. *Bioresource Technol.* **48**: 97–106.

Nakajima, D., E. Kojima, S. Iwaya, J. Suzuki, and S. Suzuki (1996) Presence of 1-hydroxypyrene conjugates in woody plant leaves and seasonal changes in their concentrations. *Environ. Sci. Technol.* **30**(5): 1675–1679.

Nedunuri, K.V., R.S. Govindaraju, M.K. Banks, A.P. Schwab, and Z. Chen (2000) Evaluation of phytoremediation for field-scale degradation of total petroleum hydrocarbons. *J. Environ. Eng.* **126**(6): 483–490.

Nichols, T.D., D.C. Wolf, H.B. Rogers, C.A. Beyrouty, and C.M. Reynolds (1997) Rhizosphere microbial populations in contaminated soils. *Water Air Soil Pollut.* **95**: 165–178.

Nicolotti, G. and S. Egli (1998) Soil contamination by crude oil: impact on the mycorrhizosphere and on the revegetation potential of forest trees. *Environ. Pollut.* **99**(1): 37–43.

Novotny, C., P. Erbanova, T. Cajthaml, N. Rothschild, C. Dosoretz, and V. Sasek (2000) *Irpex lacteus*, a white-rot fungus applicable to water and soil bioremediation. *Appl. Microbiol. Biotechnol.* **54**: 850–853.

Olson, P.E. and J.S. Fletcher (1999) Field evaluation of mulberry root structure with regard to phytoremediation. *Biorem. J.* **3**(1): 27–33.

Olson, P.E. and J.S. Fletcher (2000) Ecological recovery of vegetation at a former industrial sludge basin and its implication to phytoremediation. *Environ. Sci. Pollut. Res.* **7**(4): 195–204.

Olson, P.E., J.S. Fletcher, and P.R. Philp (2001) Natural attenuation/phytoremediation in the vadose zone of a former industrial sludge basin. *Environ. Sci. Pollut. Res.* **8**(4): 243–249.

Parker, W.J. (1995) Fate of polynuclear aromatic hydrocarbons during anaerobic digestion of municipal wastewater sludges. *Water Environ. Res.* **67**(7): 1052–1059.

Pickett, S.T.A. (1982) Population patterns through twenty years of old field succession. *Vegetatio* **49**: 45–59.

Pivetz, B.E. (2001) Phytoremediation of contaminated soil and ground water at hazardous waste sites. *Ground Water Issue*, EPA/540/S-01/500. Robert S. Kerr Environmental Research Laboratory, U.S. Environmental Protection Agency, Ada, Oklahoma.

Pothuluri, J.V. and C.E. Cerniglia (1994) Microbial metabolism of polycyclic aromatic hydrocarbons. In: *Biological Degradation and Bioremediation of Toxic Chemicals*. G.R. Chaudhry, ed. Dioscorides Press, Portland, Oregon, pp. 92–124.

Qiu, X., T.W. Leland, S.I. Shah, D.L. Sorensen, and E.W. Kendall (1997) Field study: grass remediation for clay soil contaminated with polycyclic aromatic hydrocarbons. In: *Phytoremediation of Soil and Water Contaminants*. E.L. Kruger, T.A. Anderson, and J.R. Coats, eds. American Chemical Society, Washington, D.C., pp. 186–199.

Rao, A.S. (1990) Root flavonoids. *Bot. Rev.* **56**(1): 1–84.

Reardon, K.F., P.E. Olson, J. Lowell, E. Pilon-Smits, K. Hale, B. Page, N.M. DuTeau, and D.A. Klein (2001) Stimulation of microbial polycyclic aromatic hydrocarbon biodegradation by plant chemicals. Abstr. Pap. S 221:49-BIOT Part 1. American Chemical Society, Washington, D.C., April 3.

Reid, B.J., K.C. Jones, and K.T. Semple (2000) Bioavailability of persistent organic pollutants in soils and sediments-a perspective on mechanisms, consequences and assessment. *Environ. Pollut.* **108**: 103–112.

Reilley, K.A., M.K. Banks, and A.P. Schwab (1996) Dissipation of polycyclic aromatic hydrocarbons in the rhizosphere. *J. Environ. Qual.* **25**: 212–219.

Richards, B.N. (1987) Microbiology of the rhizosphere. In: *The Microbiology of Terrestrial Ecosystems*. B.N Richards, ed. Longman Scientific Technical/John Wiley, New York, pp. 222–254.

Rockne, K.J. and S.E. Strand (1999) Biodegradation of bicyclic and polycyclic aromatic hydrocarbons in anaerobic enrichments. *Environ. Sci. Technol.* **32**(24): 3962–3967.

Rovira, A.D., R.C. Foster, and J.K. Martin (1979) Origin, nature and nomenclature of the organic materials in the rhizosphere. In: *The Root–Soil Interface*. J.L. Harley and R.S. Russell, eds. Academic Press, London, pp. 1–4.

Salisbury, F.B. and C.W. Ross (1992) *Plant Physiology*. 4th ed. Wadsworth Publishing Company, Belmont, California.

Salt, D.E., R.D. Smith, and I. Raskin (1998) Phytoremediation. In: *Annual Review of Plant Physiology and Plant Molecular Biology*. Annual Reviews, Palo Alto, California. **49**: 643–668.

Sandermann, H., Jr. (1992) Plant metabolism of xenobiotics. *Trends Biochem. Sci.* **17**: 82–84.

Schnoor, J.L., L.A. Licht, S.C. McCutcheon, N.L. Wolfe, and L.H. Carreira (1995) Phytoremediation of organic and nutrient contaminants. *Environ. Sci. Technol.* **29**: 318A–323A.

Seeger, M., K.N. Timmis, and B. Hofer (1997) Bacterial pathways for the degradation of polychlorinated biphenyls. *Mar. Chem.* **58**: 327–333.

Shimp, J.F., J.C. Tracy, L.C. Davis, E. Lee, and W. Huang (1993) Beneficial effects of plants in the remediation of soil and groundwater contaminated with organic materials. *Environ. Sci. Technol.* **23**: 41–77.

Shuttleworth, K.L. and C.E. Cerniglia (1995) Environmental aspects of PAH biodegradation. *Appl. Biochem. Biotechnol.* **54**: 291–302.

Siciliano, S.D. and J.J. Germida (1998) Mechanisms of phytoremediation: biochemical and ecological interactions between plants and bacteria. *Environ. Rev.* **6**: 65–79.

Simonich, S.L. and R.A. Hites (1995) Organic pollutant accumulation in vegetation. *Environ. Sci. Technol.* **29**(12): 2905–2914.

Sims, R.C. and M.R. Overcash (1983) Fate of polynuclear aromatic compounds (PNAs) in soil-plant systems. *Residue Rev.* **88**: 1–68.

Singer, A.C., E.S. Gilbert, E. Luepromchai, and D.E. Crowley (2000) Bioremediation of polychlorinated biphenyl-contaminated soil using carvone and surfactant-grown bacteria. *Appl. Microbiol. Biotechnol.* **54**: 838–843.

Stratton, R., E. Namkung, and B.E. Rittmann (1983) Biodegradation of trace-organic compounds by biofilms on porous medium. *J. Am. Water Works Assoc.* **75**: 463–469.

Sutherland, J.B., F. Rafii, A.A. Khan, and C.E. Cerniglia (1995) Mechanisms of polycyclic aromatic hydrocarbon degradation. In: *Microbial Transformation and Degradation of Toxic Organic Chemicals.* L.Y. Young and C.E. Cerniglia, eds. Wiley-Liss, New York, pp. 269–306.

Taiz, L. and E. Zeiger (1998) *Plant Physiology*. 2nd ed. Sinauer Associates Publishers, Sunderland, Massachusetts.

Tansley, A.G. (1935) The use and abuse of vegetational concepts and terms. *Ecology* **16**: 284–307.

Thierrin, J., G.B. Davis, C. Barber, B.M. Patterson, F. Pribac, T.R Power, and M. Lambert (1993) Natural degradation rates of BTEX compounds and naphthalene in a sulfate reducing groundwater environment. *Hydrol. Sci. J.* **38**(4): 309–322.

Toal, M.E., C. Yeomans, K. Killham, and A.A. Meharg (2000) A review of rhizosphere carbon flow modeling. *Plant Soil* **222**: 263–281.

Wagrowski, D.M. and R.A. Hites (1997) Polycyclic aromatic hydrocarbon accumulation in urban, suburban, and rural vegetation. *Environ. Sci. Technol.* **31**(1): 279–282.

Walton, B.T. and T.A. Anderson (1992) Plant-microbe treatment systems for toxic waste. *Curr. Opin. Biotechnol.* **3**: 267–270.

Webster, E., D. Mackay, and F. Wania (1998) Evaluating environmental persistence. *Environ. Toxicol. Chem.* **17**(11): 2148–2158.

Whipps, J.M. and J.M. Lynch (1985) Energy losses by the plant in rhizodeposition. In: *Plant Products and the New Technology*. K.W. Fuller and J.R. Gallon, eds. Annu. Proc. Phytochem. Soc. Eur. Oxford Univ. Press, Oxford, United Kingdom. **26**: 59–71.

Whittaker, R.H. (1962) Classification of natural communities. *Bot. Rev.* **28**: 1–239.

Wilken, A., C. Bock, M. Bokern, and H. Harms (1995) Metabolism of different PCB congeners in plant cell cultures. *Environ. Toxicol. Chem.* **14**(12): 2017–2022.

Wiltse, C.C., W.L. Rooney, Z. Chen, A.P. Schwab, and M.K. Banks (1998) Greenhouse evaluation of agronomic and crude oil-phytoremediation potential among alfalfa genotypes. *J. Environ. Qual.* **27**: 169–173.

Wrenn, B.A. and A.D. Venosa (1996) Selective enumeration of aromatic and aliphatic hydrocarbon degrading bacteria by a most-probable-number procedure. *Can. J. Microbiol.* **42**: 252–258.

Yen, K.M. and C.M. Serdar (1988) Genetics of naphthalene catabolism in pseudomonads. *CRC Crit. Rev. Microbiol.* **15**: 247–268.

Zaalishvili, G.V., G.A. Khatisashvili, D.S. Ugrkhelidze, M.S. Gordeziani, and G.I. Kvesitadze (2000) Plant potential for detoxification. *Appl. Biochem. Microbiol.* **36**(5): 443–451.

Zhang, X. and L.Y. Young (1997) Carboxylation as an initial reaction in the anaerobic metabolism of naphthalene and phenanthrene by sulfidogenic consortia. *Appl. Environ. Microbiol.* **63**: 4759–4764.

11

BIODEGRADATION OF PETROLEUM HYDROCARBONS IN THE RHIZOSPHERE

S. L. Hutchinson, A. P. Schwab, and M. K. Banks

SUMMARY OF PRACTICAL IMPLICATIONS

With proper site design and management, the presence of vegetation can accelerate the bioremediation of petroleum hydrocarbons in soil by enhancing the soil microbial population and facilitating the movement of gases, required nutrients, and water through the soil profile. Considerations for site design include the contaminant (type and amount), climate, soil properties, and vegetation selection. Important issues for site management and process optimization are water and nutrient availability, which can be controlled through fertilization and irrigation.

GLOSSARY

Aging: The process of contaminants changing composition with time in part due to volatility, solubility, abiotic and biotic transformation (weathering), and diffusing into and reacting with dynamic soil aggregates to change sorption–desorption characteristics.

Least significant difference (LSD): A procedure for making pairwise comparisons among a set of population means (Fisher 1949).

Mycorrhizae: A mutualistic symbiosis between plant and fungus localized in a root or root-like structure in which energy moves primarily from plant to fungus and inorganic resources move from fungus to plant (Allen 1991).

Root axis: The centerline of the root along which primary root growth results from meristematic or root-tip growth, producing new organs and root elongation; secondary root growth results from lateral roots breaking

Phytoremediation: Transformation and Control of Contaminants,
Edited by Steven C. McCutcheon and Jerald L. Schnoor.
ISBN 0-471-39435-1 (cloth)

through the endodermis and cortex during cell division and growing away from the primary root axis.

Soil structure: The arrangement and organization of particles in the soil

INTRODUCTION

Processes for the removal of organic contaminants from soil and water involve direct and indirect plant effects. Direct plant effects include uptake into the plant system where the contaminant can be transformed in the roots or translocated into the plant shoots. Once in the aboveground biomass, the compound may be transformed, transpired from the plant, or accumulated in the biomass for harvest. Indirect plant effects include increased microbial activity and degradation due to rhizosphere effects as well as polymerization of the compound into the organic soil fraction (Cunningham *et al.* 1997).

This chapter will focus on the use of plants to enhance the bioremediation of petroleum hydrocarbons in soil. Topics to be covered include how plants affect the soil profile and remediation process, current research in this area, several case studies, and system design and optimization issues.

PLANT-INDUCED CHANGES IN SOIL

Growing plants have profound impacts on the physical and chemical properties of soil. Roots explore the upper horizons of the soil and remove water, assimilate cations and anions, exude soluble organic and inorganic substances, and slough dead cells. Most plants have an extensive rooting system that extends into the soil, penetrates micropores, disrupts some soil aggregates, creates channels for air and water penetration, enhances the development of soil structure, and exposes more soil surface area for microbial interaction. Plants excrete amino acids, carboxylic acids, carbohydrates, nucleic acid derivatives, growth factors, enzymes, and other related products (Alexander 1977, Olson *et al.* this book) that spur the rhizosphere microbial community, allowing for increased degradation of organic substances. These compounds can promote flocculation of clays in soil and solubilize metals through chelation. In response to the nutrient status of the soil, roots may decrease pH, increase pH, or alter the oxidation–reduction potential of the rhizosphere. All these root-related factors can impact the degradation of petroleum hydrocarbons and other recalcitrant organic pollutants.

Rooting Structures, Growth, and Physiology

Stimulated activities of rhizosphere microorganisms and exploration of the soil are thought to be the most important root-induced factors in accelerating the degradation of soil contaminants. Thus, understanding the physiology, structure, and growth habit of roots is critical to optimizing phytoremediation systems. Under field conditions, roots are frequently more variable in appearance and development than shoots. Root proliferation in favorable portions of the soil can fully compensate for poor growth in other areas of the soil.

Although the appearance and overall physiology of mature roots may vary due to varietal differences and environmental influences, all roots have the same beginnings from a primary root. Root development and growth result from cell division and elongation of the apical meristem (Figure 11-1). Older roots of broad-leafed plants (dicotyledons) continue to increase in diameter because of cell division perpendicular to the root axis. This allows the plant to support heavy shoot systems. In contrast, the older roots in grasses (monocotyledons) are relatively uniform in thickness because the root attains maximum diameter when the cells are fully expanded.

The general structure of the apex of roots is similar for all flowering plants (Figure 11-1). In the elongation zone, the division of the cells proceeds away from the base of the root, *i.e.*, into the soil, resulting in increased root length. Cells generated into the soil form the root cap and are surrounded by a gelatinous material called mucigel. Just above the elongation zone is the zone of maturation. Cells in this zone mature and differentiate into the following plant tissues:

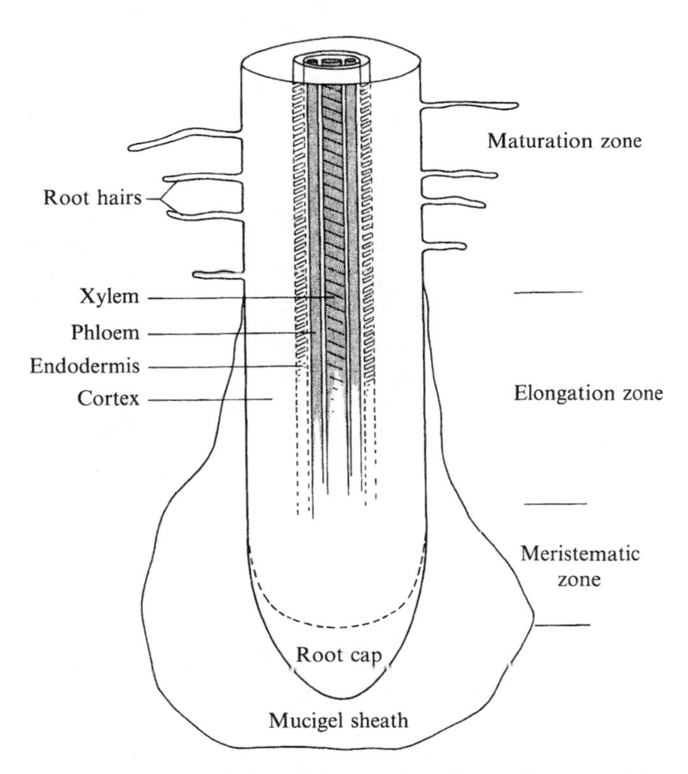

Figure 11-1 Longitudinal section of the apical zone of a primary plant root. Adapted from R.S. Russell (1977) *Plant Root Systems*. Copyright McGraw-Hill.

1. The epidermis (*i.e.*, the outer layer of cells that surround the root and from which root hairs develop)
2. The xylem (*i.e.*, the conducting tissue that transports water and solutes to the shoot)
3. The phloem, which passes plant metabolites downward to the roots
4. The endodermis, which surrounds the xylem and phloem and
5. The *cortex*, which is found between the endodermis and the epidermis

Although highly variable, roots extend into the soil at a rate of approximately 1 centimeter per day (0.4 inches per day). Specific rates of growth are the following (Lungley 1973): root axes, 2 centimeters per day (0.78 inches per day); primary laterals, 0.5 centimeters per day (0.2 inches per day); and secondary laterals, 0.1 centimeters per day (0.04 inches per day). Greater rates of root elongation (greater than 6 centimeters per day, 2.4 inches per day) have been reported for maize (*Zea mays*) (Taylor *et al.* 1970).

Important Environmental Factors in Root Growth

Roots provide physical support for plants, assimilate water and nutrients, store molecules that are high in energy, extend laterally and downward in the soil, and chemically interact with all soil phases. The chemical and physical activities will dictate the root zone of influence in the soil, and this is determined by many environmental factors as well as the genetics of the plant. Some of the most important root properties that may contribute to phytoremediation include root length, surface area, mass, depth of penetration, quantity and composition of dead roots and exudates, proliferation of root hairs, and bacterial and fungal associations. Factors that affect these root properties include air temperature, soil temperature, water availability, nutrient concentrations, plant species and variety, and the physical properties of the soil.

For example, consider the influence of moisture on root growth. Roots tend to grow rapidly in the presence of adequate moisture, and root growth slows down when moisture content becomes limiting (Figure 11-2). Similarly, excessive moisture can negatively impact root growth. When soils are saturated with water, oxygen becomes depleted because of limited diffusion. Without adequate oxygen, root metabolic processes and root growth cease. Managing water to maintain optimal water conditions is an obvious necessity, but other important implications emerge. For example, there is often interest in cleaning up soils with shallow groundwater (or the groundwater itself), but root growth is limited in saturated soils unless the plant roots have the ability to supply oxygen with the aerenchyma, *i.e.*, plant tissues capable of transporting oxygen from the leaves to the roots.

The growth and development of roots, including the response to environmental influences, is highly dependent on the genetic makeup of the plant species or variety. Some plants, such as carrots (*Daucus carota* L.) and sugar beets (*Beta vulgaris* L.), have a main taproot with relatively limited branching (Figure 11-3),

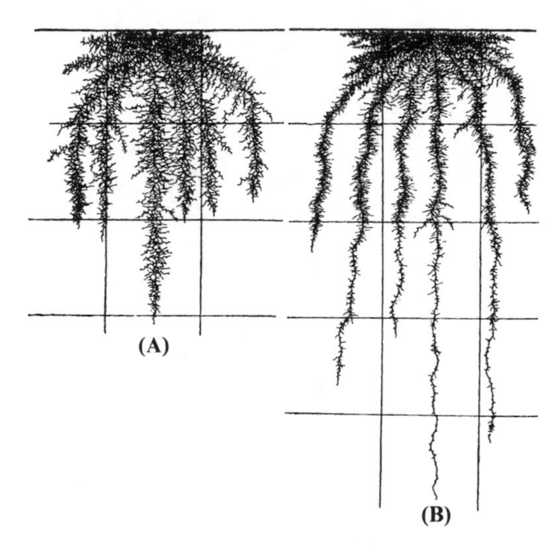

Figure 11-2 Roots of mature wheat (*Triticum aestivum* L.) grown (A) on dry land and (B) on irrigated land. Grid lines are 30 centimeters (1 foot) apart. Adopted from R.S. Russell (1977) *Plant Root Systems.* Copyright. McGraw-Hill.

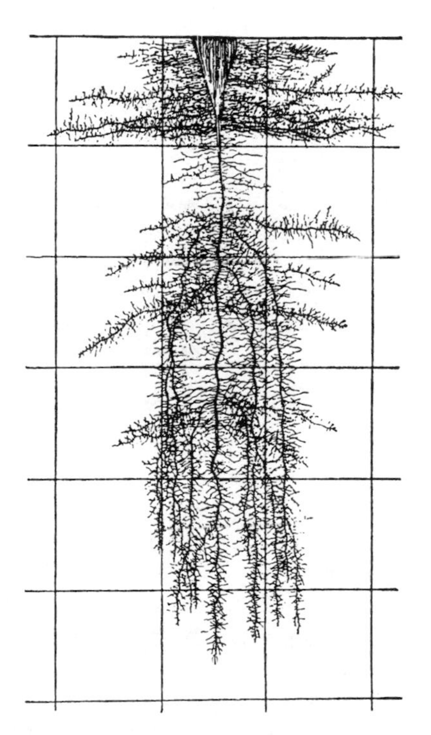

Figure 11-3 Mature root system of sugar beet (*Beta vulgaris* L.) grown for approximately 5 months without moisture limitation. Grid lines are 30 centimeters (1 foot) apart. Adopted from R.S. Russell (1977) *Plant Root Systems.* Copyright. McGraw-Hill.

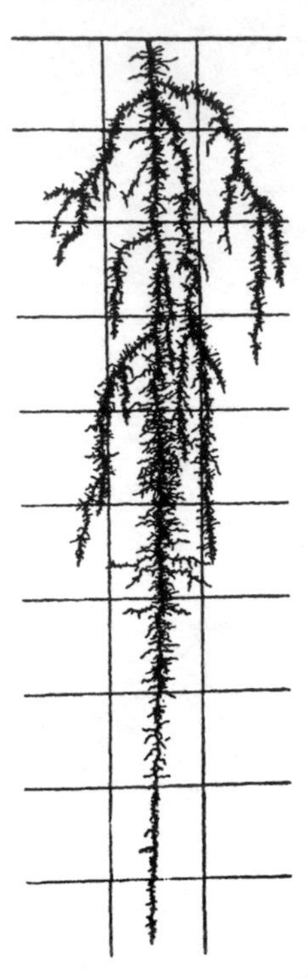

Figure 11-4 Root system of alfalfa (*Medicago sativa* L.) in second year of growth. Grid lines are 30 centimeters (1 foot) apart. Adopted from R.S. Russell (1977) *Plant Root Systems*. Copyright. McGraw-Hill.

but still significant fine roots. Other plants, such as clover (*Trifolium repens* L.) and alfalfa (*Medicago sativa* L.), for example, have a less dominant taproot, but have root systems with considerable ability for dee exploration of the soil (Figure 11-4). Grasses are known to have highly branched, fine root systems that explore a large volume of soil, particularly the micropores.

Interaction Between Roots and Polycyclic Aromatic Hydrocarbons

For hydrophobic large molecular contaminants, such as polycyclic aromatic hydrocarbons (PAHs) and other petroleum hydrocarbons, the likely mechanism for removal from soil is rhizosphere degradation (also known as plant-assisted bioremediation, see McCutchean and Schnoor this book). Uptake of these large molecules into the plant is improbable. The uptake of an organic compound by plants is dependent on the water solubility and octanol–water partition coefficient (K_{ow}). Schwarzenbach *et al.* (1993) present these and other properties of selected PAHs.

Phytoremediation of PAHs occurs primarily in the rhizosphere, *i.e.*, the small volume of soil immediately surrounding and most influenced by the root. The rhizosphere cannot be defined unambiguously, but compared to the bulk soil, this root zone generally has larger concentrations of plant exudates, elevated carbon dioxide (CO_2) gas pressures, greater microbial activities, and is strongly subject to the depletion of cations and anions due to plant uptake. As a result, strong concentration gradients are established, creating a chemically dynamic system.

Bacteria degrade PAHs in the rhizosphere in the presence of oxygen and nutrients. In aged soils contaminated with PAHs, the contaminants may be strongly sorbed to soil surfaces and are often found in very small interstices (less than or equal to 1 micrometer in diameter) of the soil. Roots can enhance microbial degradation by exploring otherwise inaccessible microzones of the soil.

The most important property controlling the uptake of organics by plant roots is lipophilicity, *i.e.*, the distribution of a chemical between the soil solution and the lipids in the plant cells (Trapp and McFarlane 1995). Lipophilicity is directly related to the octanol–water partition coefficient, K_{ow}. The values of $\log K_{OW}$ for organic solutes range from -1 to 10, and from 3.4 to 6.5 for PAHs. Concentration of these chemicals in plant roots can be described by the equation

$$\log(RCF - 0.82) = 0.77 \log K_{ow} - 1.52 \tag{11-1}$$

where RCF is the root concentration factor (Briggs *et al.* 1982), *i.e.*, the ratio of the concentration of the chemical in the roots to the concentration in the external solution. Translocation from roots to shoots, which is a critical step for direct plant degradation of contaminants, is described by the equation

$$TSCF = 0.784 \exp\left[\frac{-(\log K_{ow} - 1.78)^2}{2.44}\right] \tag{11-2}$$

where $TSCF$ is the transpiration stream concentration factor. This equation has a maximum value at $\log K_{ow} = 1.8$ (Figure 11-5; Briggs *et al.* 1982) for the translocation of pesticides in barley (*Hordeum vulgare*).

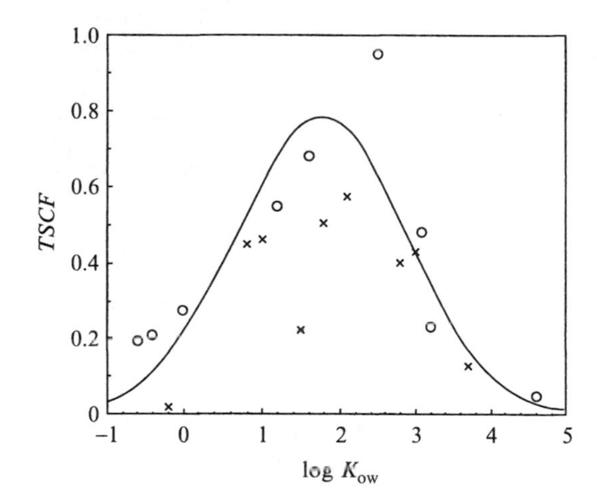

Figure 11-5 Relationship between the logarithm octanol–water partition coefficient ($\log K_{ow}$) and translocation of chemicals from barley (*Hordeum vulgare*) roots to shoots. Note that o = *O*-methylcarbamoyloximes and x = substituted phenylureas. Adapted from Briggs *et al.* (1982) *Pesticide Science*. Copyright Society of Chemical Industry. Reproduced with permission. Permission is granted by John Wiley and Sons Ltd. on behalf of the SCI.

For PAHs, uptake is quite limited because PAHs have low water solubility and high lipophilicity. The low concentrations of PAHs in the aqueous phase will rapidly and strongly partition onto the root cell membranes. As a result, plant uptake of PAHs is extremely low.

Effect of Microorganisms

The primary factor influencing the phytoremediation of petroleum hydrocarbon contaminants in soil is microbial activity. Alexander (1977) discusses in detail the factors influencing microbial activity. Microbial populations and activity are strongly associated with the water and nutrient contents of the soil, the plant species, and the type of contaminant. Microbial activity is affected by soil pH (Figure 11-6) and by temperature (Figure 11-7). In addition, the extensive root surface area of monocot grasses creates an optimal environment for the contaminant-degrading microorganisms (Tate 1995).

Total microbial activity varies with the available water in soil and varies from almost no activity at low water potential to maximum activity at an optimum water-filled pore space of 60 percent (Tate, 1995). Soil bacteria have been found to function in soil with water potentials as low as −3 megapascals (Wilson and Griffin 1975, Wildung *et al.* 1975). As the water-filled pore space increases toward saturation and available oxygen becomes limited, microbial activity declines (Figure 11-8).

Soil Physical Properties and Contaminant Degradation

The presence of vegetation affects many physical properties of soil, including structure, porosity, hydraulic conductivity, and infiltration rate. These properties, in turn, influence microbial activity by regulating the transport of required water and nutrients through the soil profile and by controlling soil

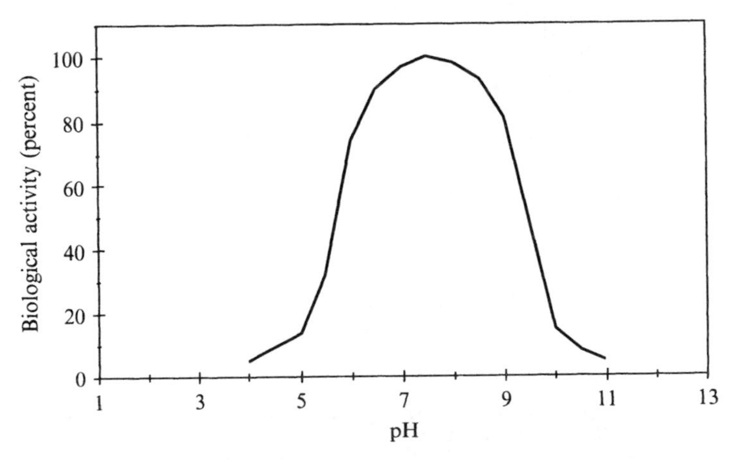

Figure 11-6 Variation of hypothetical biological activity with changes in soil pH. Adapted from R.L. Tate (1995) *Soil Microbiology*. Copyright. Reprinted with permission of John Wiley.

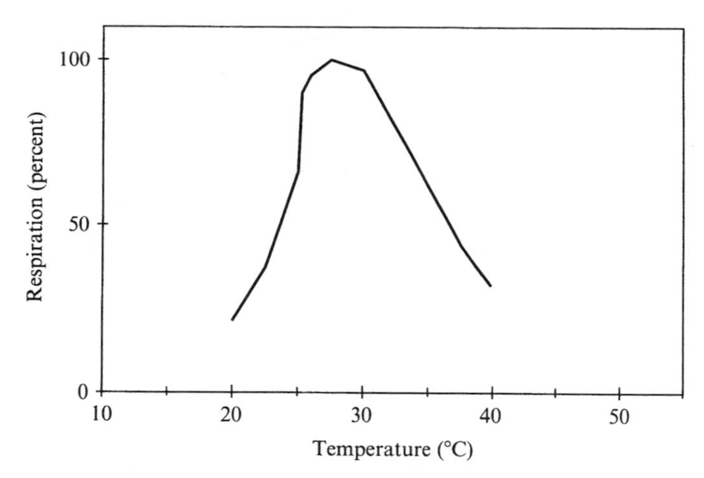

Figure 11-7 Typical response of a mesophilic bacteria population to temperature variation. An optimum temperature of 25 to 30 °C is depicted. Adapted from R.L. Tate (1995) *Soil Microbiology*. Copyright. Reprinted with permission of John Wiley.

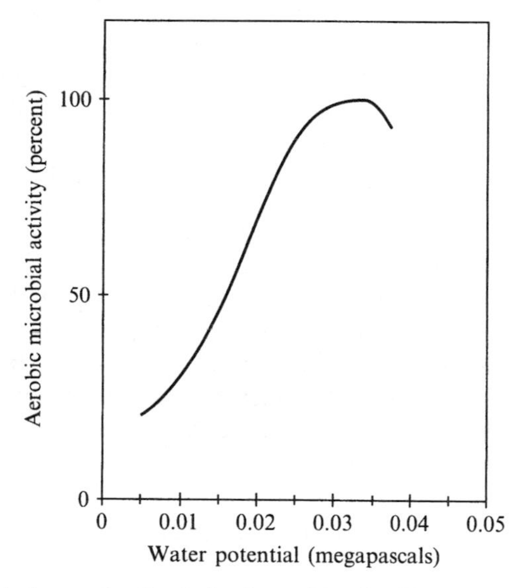

Figure 11-8 Hypothetical example of variation in aerobic microbial activity in soil related to available moisture. Adapted from R.L. Tate (1995) *Soil Microbiology*. Copyright. Reprinted with permission of John Wiley.

aeration. Understanding how the presence of vegetation impacts soil structure is an important step toward identifying the mechanisms of phytoremediation.

Bioremediation and rhizodegradation of petroleum hydrocarbons are driven by microorganisms, which in turn depend heavily on adequate moisture and oxygen. The physical properties of soil influence both oxygen and water transport. Soils with high clay content tend to have lower hydraulic conductivities and lower diffusion coefficients than soils with low clay content. Soils with high clay content also tend to be dominated by very small interstices in which contaminants may become trapped and are inaccessible to microorganisms. The growth of plants can increase infiltration by promoting development of soil structure, and when roots decay, open channels remain. Fine roots, root hairs, and mycorrhizae have the capability of penetrating extremely small pores in soil and accessing contaminants contained within.

DEGRADATION OF POLYCYCLIC AROMATIC HYDROCARBONS IN THE RHIZOSPHERE

Previous Research

Bioremediation of petroleum hydrocarbons in soil by indigenous microorganisms has been established as feasible treatment option. The biodegradation rate of the more recalcitrant and potentially toxic petroleum hydrocarbon contaminants such as PAHs, however, is initially high but quickly declines as the PAH-contaminated waste ages. Biodegradation of PAHs is limited by the strong adsorption potential and low water solubility. Vegetation plays an important role in the biodegradation of these toxic organic chemicals in soil because the presence of rhizosphere microorganisms accelerates biodegradation.

Plants can influence the biodegradation of contaminants in several ways. A large fraction of petroleum hydrocarbon contaminants can be strongly adsorbed on organic matter. Plants exude appreciable quantities of carbon through the root system. By increasing the organic matter content in contaminated soils, plants may affect contaminant bioavailability through sorption. Many petroleum contaminants are not readily desorbed and, consequently, are not available for bioremediation (Hatzinger and Alexander 1995). Also, the effect of humification in the rhizosphere may be considerable. Free radicals in humus may react with PAHs forming electron acceptor–donor complexes which may initiate covalent bonding of the PAHs or the associated metabolites to humic material (Mahro *et al.* 1994). Increased mineralization (*i.e.*, conversion into carbon dioxide and water) and formation of inextricable, bound residues have been observed after compost addition to soil. Metabolic by-products that are formed from PAH biodegradation (*e.g.*, hydroxy carbonic acids and phenolic compounds) can chemically interact with soil organics and can be incorporated into humic material, thereby forming these bound residues.

Physical restrictions may contribute to reduced bioavailability of contaminants in soil. Contaminant entrapment may be the primary reason that bioavailability of contaminants is reduced after aging. Nonionic contaminants are physically trapped within soil micropores and diffusion is slowed by contaminant partitioning between pore liquid and the pore wall, and by the tortuous flow path between the micropore and the outer soil particle. The diameters of these micropores are less than 1 micrometer and may be as small as 0.02 micrometer; consequently, entrapped material would not be accessible to microorganisms (Pignatello 1989, Farrell and Reinhard 1994).

Roots of all plants have at least some capability to penetrate soil aggregates and small pores. The fine roots of aggressive plants can disrupt soil aggregates, increase exposed surface area, and enhance biodegradation of entrapped hydrophobic contaminants. Root biomass may be as high as 20 grams per kilogram of soil with total root length as high as 100 meters per kilogram of soil (Detling 1979).

Plants may indirectly contribute to the microbial degradation of highly sorbed contaminants such as PAHs. Rhizosphere soil adjacent to plant roots has been observed to contain greater microbial densities than those observed outside the soil rhizosphere (Paul and Clark 1989). In fact, Rovira and Davey (1974) determined that the number of bacteria quantified in rhizosphere soil was as much as 20 times greater than that normally quantified outside the soil rhizosphere. Short, gram-negative rods (specifically *Pseudomonas* spp., *Flavobacterium* spp., and *Alcaligenes* spp.) are the most common microorganisms found in the rhizosphere (Barber 1984). The presence of plant exudates and seasonal root dieback are primarily responsible for the increased microbial population densities generally observed in rhizosphere soil. These materials serve as sources of energy, carbon, nitrogen, and growth factors for these populations. The activity of microorganisms in the root zone stimulates root exudation that further stimulates microbial activity (Barber and Martin 1976).

The effect of plants and the associated rhizosphere on the fate of petroleum hydrocarbon contaminants has been evaluated in several studies (Aprill and Sims 1990, Ferro *et al.* 1994, Schwab *et al.* 1995, Reilley *et al.* 1996). In general, plants enhanced the removal of contaminants from soil. Also, in studies using [^{14}C]-labeled contaminants in closed plant chambers, mineralization was observed to be greater in rhizosphere soils than in soils without vegetation (Lee 1996).

Aprill and Sims (1990) investigated the effects of using deep-rooted prairie grasses for remediation of PAH-contaminated soil. They suggested that due to the fibrous nature, the roots of these perennial grasses might be more effective in stimulating microorganisms in the rhizosphere. Big bluestem (*Andropogon gerardii*), Indian grass (*Sorghastrum nutans*), switchgrass (*Panicum virgatum*), Canada wild rye (*Elymus canadensis*), little bluestem (*Schizachyrium scoparius*), side oats grama (*Bouteloua curtipendula*), western wheatgrass (*Agropyron smithii*), and blue grama (*Bouteloua gracilis*) were evaluated. After 219 days of growth, PAH removal was greater in the rhizosphere soil than outside the soil rhizosphere. The order of removal among the four PAHs that were investigated correlated with the water solubility of the compound; the most water-soluble PAH exhibited the greatest extent of degradation.

Ferro *et al.* (1994) evaluated the mineralization of pentachlorophenol and phenanthrene in rhizosphere soil using [^{14}C]. Crested wheatgrass (*Agropyron desertorum*) was grown in flow-through plant chambers. In vegetated pentachlorophenol-contaminated soil, 22 percent of the initial pentachlorophenol was converted to $^{14}CO_2$ after 155 days. In the soil without vegetation, only 5 percent of the pentachlorophenol was mineralized after 155 days.

Schwab and Banks (1994) investigated the degradation of PAHs in the rhizosphere of a variety of plants grown in petroleum hydrocarbon–contaminated soil. Alfalfa (*Medicago sativa*), tall fescue (*Festuca arundinacea*), big bluestem (*Andropogon gerardii*), and Sudan grass (*Sorghum vulgare* var *sudanense*) were used. Pyrene was one of the target PAH compounds assessed in this study. After 4 weeks of plant growth, pyrene concentrations had decreased from an initial level of 100 milligrams per kilogram of soil to less than 12.6 milligrams per kilogram of soil. During week 24, concentrations of pyrene were less than 2.4 milligrams per kilogram of soil. Pyrene degradation was appreciably greater in vegetated soil than in barren soil.

Lee (1996) investigated the fate of benzo(*a*)pyrene in soil planted with Sudan grass (*Festuca arundinacea*). In a 190-day laboratory study using [^{14}C]-labeled benzo(*a*)pyrene in sealed plant chambers, greater mineralization was observed in vegetated soil than in a barren soil control. In the vegetated soil chamber, 1.3 percent was mineralized ($^{14}CO_2$), 0.121 percent was associated with plant tissue (0.115 percent with the root fraction and 0.006 percent with shoots), and 92 percent of the radiolabel remained in the soil. In the soil control chamber without plants, 0.7 percent was mineralized and 96 percent of the [^{14}C] remained in the soil. Further analysis of the soil revealed that 50 percent of the original benzo(*a*)pyrene remained in the barren soil control whereas only 44 percent of the parent compound was found in vegetated rhizosphere soil.

Qui *et al.* (1994) assessed the removal of naphthalene in the presence of prairie buffalo grass (*Buchloe dactyloides* var. *prairie*) in a field test. The rate of naphthalene removal was greater in the presence of plants. Additional data generated during the investigation suggested that the presence of Verde kleingrass (*Panicum coloratum* var. *verde*), common buffalo grass (*Buchloe dactyloides*), and Meyer zoysiagrass (*Zoysia japonica* var. *meyer*) might also be effective in reducing concentrations of PAHs in soil.

Gunther *et al.* (1996) reported that ryegrass (*Lolium multiflorum*) effectively enhanced the degradation of petroleum hydrocarbons in the rhizosphere. After 22 weeks of plant growth, the extractable hydrocarbon concentration in the phytoremediated soil had been reduced by 97 percent, while only an 82 percent reduction was noted in the soil control without vegetation. Microbial analyses established that microorganisms in the rhizosphere were responsible for the degradation.

Pradhan *et al.* (1998) determined that alfalfa (*Medicago sativa* L.), switchgrass (*Panicum virgatum*), and little bluestem (*Schizachyrium scoparius*) were effective in enhancing the degradation of PAHs in soil from a manufactured gas plant. Their results showed that alfalfa (*Medicago sativa* L.) removed 56 percent, switchgrass (*Panicum virgatum*) removed 57 percent, and little bluestem (*Schizachyrium scoparius*) removed 47 percent of the total PAHs in soil. The PAHs were reduced by only 26 percent in the control without vegetation.

The microbial densities of petroleum hydrocarbon degraders have been shown to be consistently greater in the rhizosphere soil than in soil without vegetation (Epuri and Sorensen 1997, Nichols *et al.* 1997). In addition, the microbial densities of petroleum hydrocarbon degraders have historically been shown to be greater in rhizosphere soil exposed to petroleum hydrocarbon contaminants than in soil with no or very little prior exposure (Radwan *et al.* 1998).

Enhancement of petroleum hydrocarbon degradation in the rhizosphere is possible using a variety of plants. In addition to plant selection, site-specific issues such as soil characteristics and proper management of fertilization and irrigation are important in maintaining an efficient phytoremediation process.

Case Studies

Our group has performed several field pilot-scale studies focusing on the phytoremediation of petroleum hydrocarbon contaminated soil. Three of the pilot-scale studies are presented in the following sections. The first pilot-scale or case study involved the remediation of crude oil contaminated soil immediately following the release from an underground pipeline near the Gulf of Mexico. A second field study was performed in a small portion of a 6.0-hectare (15-acre) land treatment facility in Virginia where diesel fuel contaminated soil was being biologically treated. The third and final field study was performed in California at a site where soil was contaminated with aged diesel fuel and heavy motor oil.

Gulf Coast Crude Oil Contaminated Site

A rupture in a pipeline under a rural field contaminated several acres (hectares) of agricultural land with crude oil. The pipeline company responsible for remediating the contaminated soil evaluated phytoremediation on a small portion of this site. A 100 by 70 meters (328 by 230 feet) rectangular area, which was subdivided into four 50 by 35 meters (164 by 115 feet) plots, was used for the pilot tests. Selected chemical and physical properties of this soil are given in Table 11-1. One of the four plots remained barren during the entire test as a control. A second plot was planted with sorghum–sudan grass (*Sorghum bicolor* L. and *Sorghum vulgare* var. *sudanense*) each spring, and each fall the plants were disked into the soil. A third plot was initially seeded with cowpeas [*Vigna unguiculata (L.) Walp. subsp. unguiculata*], but this legume did not survive the saturated soil conditions encountered during the rainy months. St. Augustine grass [*Stenotaphrum secundatum* (Walter) Kuntze] was substituted for cowpea [*Vigna unguiculata (L.) Walp. subsp. unguiculata*]. The vegetation was rotated in the last plot. This plot was seeded with soybean [*Glycine max* (L.) Merrill] each spring and summer, and with annual rye [*Lolium multiflorum* Lam. (syn. *Lolium italicum* A. Braun.)] each fall. Each plot was fertilized with 23 kilograms or 50 pounds of nitrogen (130 kilograms of nitrogen per hectare, 116 pounds of nitrogen per acre) and 11 kilograms or 25 pounds of phosphorus (65 kilograms of phosphorus per hectare, 58 pounds of phosphorus per acre every 2 months. The duration of the field test was 21 months.

Twenty samples were collected from each of the four plots at the beginning of the test and were analyzed for total petroleum hydrocarbons (TPHs). At

TABLE 11-1 Soil Properties from Field Phytoremediation Sites

Soil property	Gulf Coast crude oil	Virginia diesel	California heavy motor oil
pH	6.5	7.4	7.2
Bray P (milligrams per kilogram of soil)[a]	71	21	6.0
Organic C (milligrams per kilogram of soil)	25	22	1.4
Inorganic N (milligrams per kilogram of soil)	55	2.7	1.5
Electrolytic conductivity (decisiemens per meters)	7.1	3.5	1.2
Texture:			
Sand (percent)	35	60	59
Silt (percent)	30	21	26
Clay (percent)	35	19	15
Initial TPH (milligrams per kilogram of soil)	8200 to 16 000	1000 to 3600	3100 to 1400

[a]Bray P is measured using a soil extraction procedure that involves a solution consisting of ammonium fluoride and hydrochloric acid. The phosphorus extracted using this solution on acid soils is correlated to plant-available phosphorus.

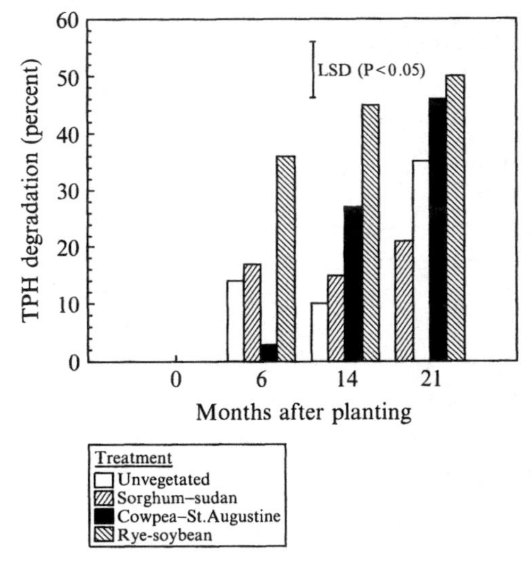

Figure 11-9 Degradation of total petroleum hydrocarbons (TPH) in a crude-oil contaminated soil near the Gulf of Mexico (Nedunuri *et al.* 2000). Least significant difference (LSD) shown at probability of error less than 5 percent. Planting occurred in August 1994.

months 6, 14, and 21, samples were also similarly collected and analyzed (Figure 11-9). The soil samples were extracted prior to analysis according to U.S. EPA Method 3540C, Soxhlet Extraction (U.S. EPA 1986), and analyzed for TPHs according to U.S. EPA Method 418.1, Total Recoverable Petroleum Hydrocarbons, Spectrophotometric, Infrared (U.S. EPA 1983).

Concentrations of TPH in soil as a function of time in each of the four test plots are presented in Figure 11-9. By month 6, degradation was significant (probability of error less than 5 percent) in each of the plots except for the cowpea–St. Augustine grass plot. By month 14, a pattern in the treatment of TPH had emerged that remained unchanged until the end of the field test at month 21. Degradation TPH in all of the vegetated test plots was greater than in the barren control. The extent of TPH degradation in the barren control did not change appreciably after month 6. From months 14 to 21, the extent of TPH degradation in all test plots had increased. Based on the data presented in Figure 11-9, TPH degradation was significantly (probability of error less than 5 percent) more extensive in the annual rye (*Lolium multiflorum* Lam.)–soybean [*Glycine max* (L.) Merrill] and cowpea [*Vigna unguiculata* (L.) Walp. sulosp. *unguiculata*]–St. Augustine [*Stenotaphrum secundatum* (Walter) Kuntze] grass test plots than in the control and sorghum (*Sorghum bicolor* L.)–Sudan (*Sorghum vulgare* var. *sundanense*) grass test plot. The least significant difference was estimated using paired observations at each time; this project did not have a statistical design and so a retrospective approach was necessary to interpret the results. Using geostatistical concepts to characterize the variability in the degradation of TPH, Nedunuri *et al.* (2000) similarly concluded that the extent of TPH degradation in the rye (*Lolium multiflorum* Lam.)–soybean [*Glycine max* (L.) Merrill] and cowpea [*Vigna unguiculata* (L.) Walp. sulosp. *unguiculata*]–St. Augustine [*Stenotaphrum secundatum* (Walter) Kuntze] grass test plots was significantly greater than in the control and sorghum (*Sorghum bicolor* L.)–Sudan (*Sorghum vulgare* var. *sudanense*) grass test plot.

Virginia Diesel Fuel Contaminated Site

In cooperation with the U.S. Navy, a field pilot-scale project was set up to determine the potential of phytoremediation to remediate diesel fuel contaminated soil. The soil used for this investigation had been excavated from the vicinity of offsite storage tanks and transported to a fully contained bioremediation cell which was owned by the U.S. Navy and used for land treatment of petroleum hydrocarbon contaminated soils. The area of the entire treatment cell was approximately 6.0 hectares (15 acres). The project was performed on a field plot that was approximately 0.2 hectare (0.5 acre) in size. Selected properties of the soil are given in Table 11-1. The field testing area was divided into six blocks of equal size and the following four treatments were prepared within each block: unplanted, barren control, tall fescue (*Festuca arundinacea*), white clover (*Trifolium repens* L.), and Bermuda grass (*Cynodon dactylon* L.) appiled as sod. To ensure proper emergence of the seedlings, a sprinkler system was used to initially irrigate the site. Once the plants were established, irrigation was discontinued.

The site was fertilized based on a carbon:nitrogen:phosphorus ratio of 100:10:4 (Chang *et al.* 1996) reported to be successful for bioremediation.

Based on the reported recommendation, 50 kilograms of nitrogen per hectare (44 pounds of nitrogen per acre) and 20 kilograms of nitrogen per hectare (18 pounds of nitrogen per acre) were added every 2 months for 2 years. Six times per year, four soil samples were appropriately collected from each of the 24 field-treatment areas and analyzed for TPH (Schwab *et al.* 1999).

As there were a large number of samples to be analyzed for TPH at each sampling event (*i.e.*, 96 samples), an effort was made to reduce the time required for sample analysis. A more rapid shake-flask method of extraction was developed, which was compared to the Soxhlet extraction of TPH. This method was shown to be both precise and accurate (Schwab *et al.* 1999). Diesel range organic compounds (*i.e.*, C_{10} to C_{28}) in the extracts were then quantified by gas chromatography.

Changes in TPH concentration in the field plots as a function of time are presented in Figure 11-10. As previously noted at the Gulf Coast crude oil contaminated site, a pattern of treatment of TPH emerged after 6 months, which did not change by the end of the 24-month test. At the end of 24 months, differences among the treatments were appreciable. Degradation of TPH in all vegetated plots was significantly (probability of error less than 5 percent) greater than treatment of TPH in the barren control, and treatment of TPH was most effective in those treatment areas seeded with clover (*Trifolium repens* L.).

One of the more noticeable observations was the rapid increase in TPH degradation in the clover (*Trifolium repens* L.) plots during the last months of the trial. Between months 21 and 24, TPH degradation in the clover (*Trifolium repens* L.) plots increased from 30 to 50 percent. This increased TPH

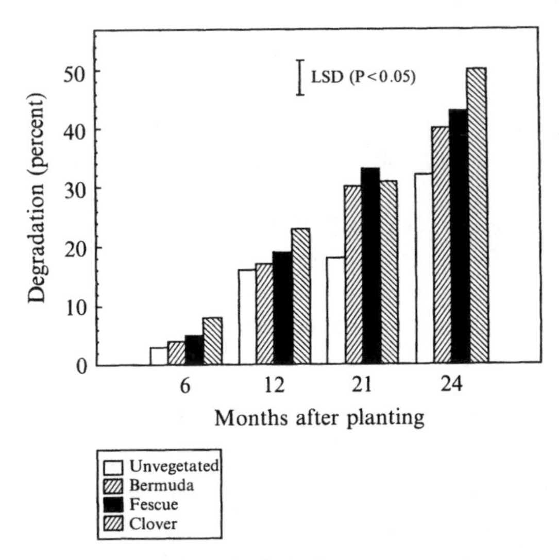

Figure 11-10 Changes in total petroleum hydrocarbon concentration over time in soil for the Virginia diesel fuel contaminated site. Each of the four treatments was replicated six times in a randomized complete block experimental design. Least significant difference (LSD) shown at probability of error less than 5 percent. Planting occurred in October 1995.

degradation occurred in spite of the fact that the clover (*Trifolium repens* L.) died during the winter and few plants survived in those plots.

A possible explanation for the observed increased TPH degradation between months 21 and 24 in the clover (*Trifolium repens* L.) plots was the possibility that the clover (*Trifolium repens* L.) roots of the dead plants provided a source of readily available carbon for enhanced microbial activities and subsequent degradation of TPH. Whereas the root length densities (length of root per unit volume of soil) of all plant species was reduced between months 21 and 24, as presented in Figure 11-11, those densities for clover (*Trifolium repens* L.) had decreased nearly tenfold. Fletcher and Hegde (1995) had observed that the roots of certain plants contain phenolic compounds that are similar in structure to PAHs and selectively foster the growth of some microorganisms while inhibiting the growth of others.

California Heavy Oil Contaminated Site

Soil from a U.S. West Coast naval base, which was contaminated with highly weathered diesel fuel and heavy motor oil, was treated in a 20 by 35 by 0.9

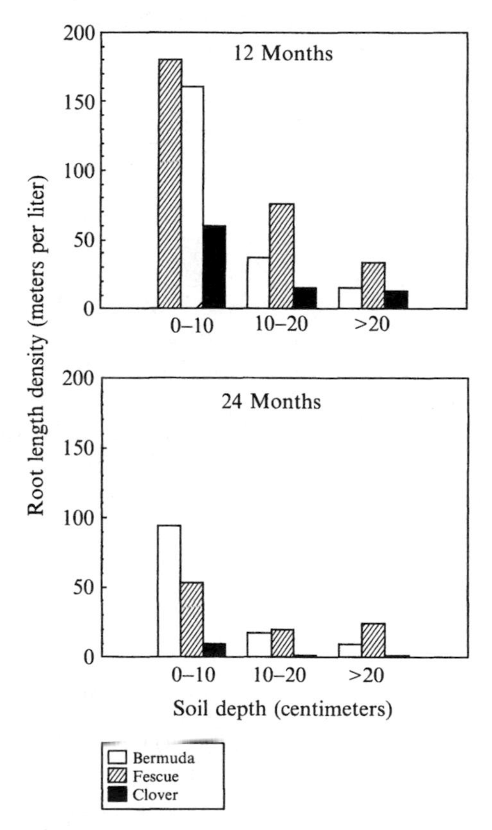

Figure 11-11 Root length density (root length per unit volume soil) as a function of depth and time for the three plant species.

meters deep (66 by 115 by 3 feet deep) treatment cell. The surface area of the cell was 0.07 hectare (0.17 acre) and was capable of treating 630 cubic meters (820 cubic yards) of contaminated soil. A geotextile was used to line the cell, and a leachate collection system was constructed within the cell.

A preliminary analysis (U.S. EPA 1986) of the soil for petroleum hydrocarbons showed that the contaminants were distributed in a stratified pattern within the soil. Based on this pattern of petroleum hydrocarbon distribution, four parallel strips were established as blocks with three treatments in each for a blocked statistical experimental design. These treatments were prepared as an unplanted control, a mixture of native California grasses, and a mixture of grasses and broadleaf species that are typically used for the revegetation of roadsides.

The petroleum hydrocarbons were highly weathered, as shown by the fact that less than 10 percent of that extractable in hydrophobic solvents were quantifiable by gas chromatography. The remaining 90 percent were polar compounds. These polar compounds retained much of the original chemical structure of petroleum hydrocarbons but have hydroxyl and carboxyl groups substituted within the structure, which yield the polar properties observed.

Fertilization for the 2-year field test was based on the total amount of extractable material (mass of material extractable in dichloromethane) and on a target carbon:nitrogen:phosphorus ratio of 100:8:4 (Hutchinson *et al.* 2001). Because of the highly weathered and recalcitrant nature of this material, the rates of degradation were expected to be less than those observed in either the Gulf Coast or the Virginia sites.

An irrigation system was constructed and was used whenever soil moisture became too low for good plant growth. Unlike the Virginia site, the California site required frequent irrigation throughout the trial. Soils were sampled quarterly and analyzed for petroleum hydrocarbons. Plant growth parameters (biomass production, height, percent ground cover) were analyzed at the same time. Contaminants were extracted by a shake-flask method and quantified by gas chromatography according to modifications of U.S. Environmental Protection Agency (EPA) SW-846 Methods 8310, "Polynuclear Aromatic Hydrocarbons, and 8015B, Nonhalogenated Organics Using GC/FID" (U.S. EPA 1986, Schwab *et al.* 1999).

As presented in Figure 11-12, the patterns of hydrocarbon degradation at the California site were very similar to those observed at both the Gulf Coast and the Virginia sites. During the first 6 months of the feasibility investigations, significant (probability of error less than 5 percent) petroleum hydrocarbon degradation was observed for all treatments. At the end of the 24-month field study, the two vegetated treatments had significantly (probability of error less than 5 percent) greater rates of degradation than the unplanted control. As anticipated, the extent of petroleum hydrocarbon degradation at the California site was less than that observed at the previously described Gulf Coast and Virginia sites. Petroleum hydrocarbons at both the Gulf Coast and Virginia sites were less weathered, and degradation proceeded both at a greater rate and to a greater extent than at the California site.

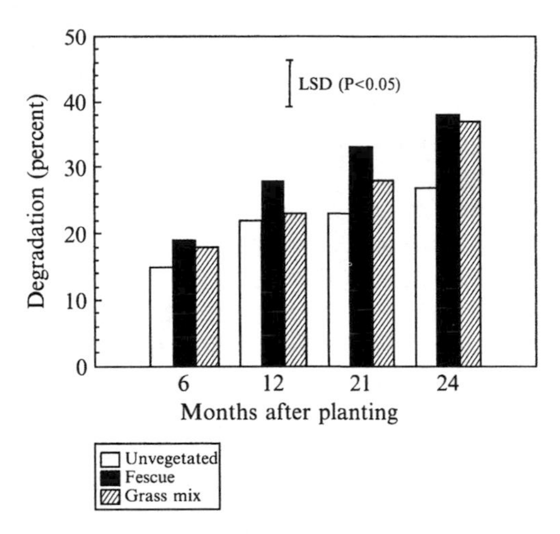

Figure 11-12 Degradation of total petroleum hydrocarbons in the California soil. Least significant difference (LSD) shown at probability of error less than 5 percent. Planting occurred in September.

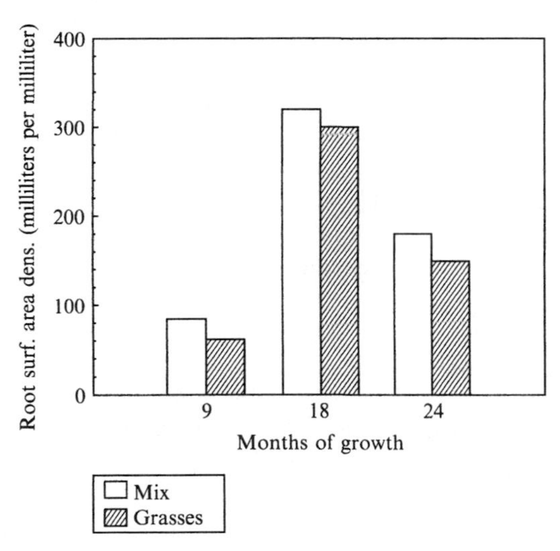

Figure 11-13 Changes in root surface area density as a function of time for the California motor oil contaminated site. Data are presented for the upper 30 centimeters (1 foot) of the soil profile. Time 0 for this field test was September.

The diesel fuel and motor oil were weathered or aged, *i.e.*, partial degradation occurred, and only 10 percent of the carbon compounds, which were originally associated with the spilled material, remained as unaltered hydrocarbon material. The remaining 90 percent of this weathered material was primarily comprised of polar compounds, which were not quantified by gas chromatography. Polar compounds are, in general, considered to be recalcitrant, but as presented in Figure 11-12, were degraded more by phytoremediation.

The observed pattern of root growth in the upper 30 centimeters (12 inches) of soil at this site (Figure 11-13) was typical for perennial species. Seasonal cycling followed a consistent increase in root surface area. Physical examination of the soil clearly showed that the roots were exploring the entire volume of the soil even though the densest rooting was found in surface soil [*i.e.*, upper 30 centimeters (12 inches)].

An encouraging result from the investigation at the California site was the continued degradation of remaining hydrocarbons in the soil. These results establish the field proof of concept for the phytoremediation of fresh and aged oil in soil. Given time the soil may return to a clean condition, but that field proof of principle has not been established. (See Rock this book for the program expected to establish the proof of principle for the phytoremediation of oil-contaminated soil.)

DESIGN AND OPTIMIZATION

In many ways, the phytoremediation of petroleum hydrocarbon–contaminated soil is quite simple. Plants are established, fertilizer is added, the site is maintained, and the contaminants continuously disappear. However, there are challenges to the application of this technology that are not apparent until the process is actually initiated at a given site. Attention must be given to plant selection as well as to optimization of both fertilization and irrigation. An improperly designed and executed phytoremediation system may not support plants, have poor growth, be subject to pests and diseases, and may not meet the desired remediation goal.

Plant Selection

The selection and mix of plant species have a substantial impact on the success of phytoremediation. A growing body of published literature reports that not all plant species have the same potential for enhancing remediation. The rhizosphere of hybrid poplar (*Populus deltoides* × *Populas nigra* 'DN34' 'Imperial Carolina')trees, for example, was reported not to be as enriched in PAH-degrading microorganisms as soil outside the rhizosphere (Jordahl *et al.* 1997). The roots of certain trees, however, are reported to be capable of sustaining bacteria that have the capability to degrade hydrocarbon contaminants (Fletcher and Hedge 1995). Although the reason for these differences among trees and other plant species is not clear, these examples illustrate the point that species selection can impact the success of the phytoremediation process. In certain circumstances, the capability of the plants to survive in a given environment may override considerations of the potential for phytoremediation. After vegetation is established, the site is generally more suitable for plant growth, and plant species with more phytoremediation potential can be planted.

Table 11-2 presents a listing of a number of plant species that have been used to enhance the degradation of petroleum hydrocarbons at several locations. A more complete compilation of this literature is available through the U.S. EPA at the following website: (http://www.rtdf.org//public/phyto/phytobib/biba-b.html). McIntyre (this book) presents a database that lists all known plants that enhance the treatment of TPH-contaminated soils.

Selection Criteria

There are a number of factors that must be considered prior to final selection of phytoremediation as a treatment alternative at a site. These factors, which include, for example, climate, availability of water, salinity and other phytotoxins, consultation with local experts, and screening procedures with candidate plant species, are discussed in this and later in the section "Selection Process."

For phytoremediation to have any chance for success at a site, the selected plants must be able to germinate, establish, and flourish under the environmental constraints of the site. Understanding the impact of site properties on both plant growth and contaminant degradation is critical.

Climate

The length of the growing season, as well as rainfall and temperature patterns, will have a strong influence on the plant species considered for use at a given site. Plant species that thrive in cool regions of high rainfall may not survive a single growing season in a hot, dry climate. In arid and semiarid areas, plant species that cannot survive periods of drought and high temperatures should be eliminated from consideration. Similarly, plant species that cannot tolerate periods of water saturation may be unsuitable for areas subject to extended rainfall.

TABLE 11-2 Plant Species Previously Used Successfully in Phytoremediation Studies for Aged Petroleum Contaminants

Species	Location	Reference
Verde kleingrass (*Panicum coloratum* var. *verde*), prairie buffalo grass (*Buchloe dactyloides* var. *prairie*)	Texas Gulf Coast	Qiu et al. (1994)
Ryegrass (*Lolium multiflorum*), St. Augustine grass [*Stenotaphrum secundatum* (Walter) Kuntze]	U.S. Gulf Coast	Schwab and Banks (1994)
Bermuda grass (*Cynodon dactylon* L.), tall fescue (*Festuca arundinacea*), white clover (*Trifolium repens* L.)	Virginia	Banks *et al*, (1998), Gunther *et al.* (1996), Ferro *et al.* (1994)
Prairie grasses (mix)	Greenhouse	Aprill and Sims (1990)

Soil Effects on Water Availability and Control
Although the water balance is related to climate, many other aspects of importance are related to soil and topography. Bulk density, soil texture, and hydraulic conductivity affect the infiltration of water into the soil and the subsequent availability. For example, dense, clay soils will be resistant to infiltration, and runoff will be greater. Although sandy soils often exhibit greater infiltration, the large hydraulic conductivities and low water retention do not generally allow for appreciable storage of water. These properties of soil could influence which plant species are considered for phytoremediation.

High water tables are not unusual at industrial sites. If the water table rises to the surface, plants must be selected that can tolerate flooded conditions. Many of these plants, *e.g.*, willow (*Salix* spp.), trees may also be able to effectively lower the water table through transpiration. Degradation of petroleum hydrocarbons can proceed effectively under aerobic conditions, but generally proceeds only very slowly under flooded conditions.

Salinity and Other Phytotoxins
Petroleum hydrocarbon contaminated soil is frequently associated with high concentrations of soluble salts. The salinity, or concentration of soluble salts in a soil sample, is quantified by (1) obtaining the extract from the saturated paste of the soil (Rhoades, 1982), (2) measuring the electrical resistance of the extract, and (3) converting the measured resistance to electrical conductivity. Soils with an electrical conductivity greater than 2 decisiemens per meter are classified as salt affected. Most plant species are not significantly impacted until the electrical conductivity is greater than 4 decisiemens per meter (Bohn *et al.* 1985), which is unfortunate because the conductivity for most petroleum hydrocarbon contaminated soils is greater than 4 decisiemens per meter. Compilations of salt-resistant plant species (*e.g.*, Bohn *et al.* 1985) will aid in the selection of the most appropriate plant species for remediation of saline, hydrocarbon contaminated soils. With proper site preparation and management, the impact of salts on plant growth will be reduced.

The cause of phytotoxicity in soil at contaminated sites is not limited to salts. Metals, anions, and some organic compounds can be toxic at some level. Certain components in petroleum, particularly volatile organic compounds, are also toxic to plants. Fortunately, most of these compounds volatilize into the air or degrade as the petroleum ages or weathers. Information is available in the published literature on the resistance of some plant species to volatile organic compounds but not in the context of phytoremediation. As with salinity, most metal toxicities can be overcome by a combination of species selection and site management.

Selection Process

As part of the selection process in considering phytoremediation as a treatment alternative at a site, plant species will be selected based on a review of relevant information, including the published literature, the advice of experts, and the results of screening studies.

Consulting Local Experts
Local expertise is generally available at horticultural suppliers (*i.e.*, plant nurseries, commercial greenhouses, and turf farms), commercial seed suppliers, agricultural cooperatives, extension services, land grant universities, and government agencies. These experts are invaluable in suggesting plant species that grow well locally and the selection of which may be very appropriate in meeting the goals of phytoremediation at a particular site.

Plant species that are generally considered for phytoremediation exhibit the following desirable characteristics: perennial growth, extensive root system, long growing season, and hardiness. By selecting a perennial plant, management is reduced considerably. Frequent reseeding is inconvenient, particularly for remote sites, and may even be an unacceptable option (*e.g.*, due to gaps in erosion control).

Although only limited data are available, a significant correlation exists between root mass and rate of phytoremediation. This relationship is very apparent for plant varieties or genotypes (group of plants defined by specified, internally coded, inheritable information) within a species but not as apparent between plant species. With a long growing season, acceptable root growth continues and rhizosphere microorganisms remain active.

Screening Procedures
The survivability of a plant species in the soil to be cleaned up can be evaluated in a greenhouse or growth chamber using a simple screening procedure (Kulakow *et al.* 2000). The species of interest are planted in small pots with as little as 100 grams of contaminated soil. Following germination, the plants are then allowed to develop for 6 weeks. The plant species exhibiting the best growth (*i.e.*, greatest biomass) should then be kept as possible candidates for use in the phytoremediation of that site. Notable however, is that certain plant species take a long time to develop, as much as two growing seasons, and may be mistakenly eliminated by the short-term screening procedure. In addition, this procedure is inappropriate for evaluating the phytoremediation potential of a plant species. Plant vigor is not necessarily related to the degradation of contaminants in the rhizosphere (Kyle 1998, Wiltse *et al.* 1998).

Fertilization

The inorganic mineral nutrients that are most often reported to limit the bioremediation of petroleum hydrocarbon contaminants in soil are nitrogen (N) and phosphorus (P). While not all bioremediation systems respond to the addition of mineral nutrients, the addition of nitrogen fertilizer has been observed to enhance the bioremediation of hydrocarbon contamination in both aquatic and soil environments (Glaser 1991, Rasiah *et al.* 1992, Churchill *et al.* 1995,

Lin and Mendelssohn 1998). Mineralization rates of hydrocarbon waste in soil have been observed to increase in response to the addition of fertilizer, thus indicating the importance of nutrient availability in the bioremediation process (Rasiah *et al.* 1992). Graham *et al.* (1995) conducted a laboratory study to determine the optimum application rates of various nutrients in enhancing the biodegradation of xylenes, anthracene, phenanthrene, and *n*-hexadecane in soil. Biodegradation was enhanced by nutrient addition, and each of the four contaminants responded differently to the various application rates of the fertilizers tested.

A site must be optimally fertilized to support plant growth and to maximize an active microbial population capable of degrading the contaminants. As reported, nutrients are generally added in a carbon:nitrogen:phosphorus ratio of 100:10:1 milligrams of nutrient per kilogram of soil (Cookson 1995; Hutchinson *et al.* 2001), where carbon is the available total organic carbon concentration in the soil. At most contaminated sites, the organic carbon of the contaminants dominate measured total organic carbon. Mineral nutrient addition based on optimal plant growth underestimates the total nutrient requirement because the additional support of a contaminant-degrading microbial population must also be considered. However, because not all organic carbon is readily available, the potential to contaminate both surface and groundwater with excessive nitrogen and phosphorus mineral nutrient additions must also be taken into account.

For the purpose of illustration, consider a site at which the soil is contaminated to a depth of 30 centimeters (12 inches) with petroleum hydrocarbons at a concentration of 3000 milligrams of TPH per kilogram of soil. The 100:10:1 ratio of carbon:nitrogen:phosphorus translates into 300 milligrams of nitrogen per kilogram of soil and 30 milligrams of phosphorus per kilogram of soil for supporting microbial and plant growth on the petroleum hydrocarbons. For a bulk density of 1500 kilograms per cubic meter of soil, this would translate to fertilizer applications of 1350 kilograms per hectare (1200 pounds per acre) for nitrogen and 134 kilograms per hectare (120 pounds per acre) for phosphorus. However, if these very high rates of fertilization were applied at the time of planting, an appreciable amount of the applied nitrogen and phosphorus could be lost prior to incorporation into microbial and plant biomass. These water-soluble mineral nutrients will be lost by leaching from the soil and by transport off the soil surface with runoff. These very high rates of fertilization would also leave an excess of the applied nitrogen and phosphorus, which may not be used immediately by the microorganisms for hydrocarbon degradation. Microbial communities in the soil will use nitrogen and phosphorus for rapid production of biomass from readily available soil carbon with minimal use of the contaminant carbon.

When the available nitrogen and phosphorus in the applied fertilizer decreases to the point of limiting microbial growth on available organic contaminants such as petroleum hydrocarbons, the elevated indigenous microbial population will start to decline. The nitrogen and phosphorus assimilated by indigenous microorganisms will be in an organic form and is not immediately available. To become available, the microorganisms must lyse and decay must take place through biogeochemical cycles such as the soil–nitrogen cycle. Referring to the nitrogen cycle, when cells lyse, the released proteins and nucleic acids must be biologically converted by the indigenous microorganisms into inorganic ammonium and nitrate. It is the ammonium and nitrate that are assimilated by indigenous microorganisms for growth on available organics. Ammonium nitrate (NH_4NO_3), for example, is a mineral nutrient fertilizer that is used in supporting both plant and microbial growth.

A better approach for nutrient addition is to split the required amount of fertilizer and apply that fertilizer on a bimonthly or a quarterly basis. Consider the previously described illustration of soil contaminated with petroleum hydrocarbons at a concentration of 3000 milligrams of TPH per kilogram of soil. For a 3 year project, a better approach is to apply nitrogen at the rate of 112 kilograms of nitrogen per hectare (100 pounds per acre) and phosphorus at the rate of 11 kilograms of phosphorus per hectare (10 pounds per acre) every 3 months in excess of that required by the plants.

Water Management

In areas with low seasonal or annual rainfall, installation of an irrigation system at a site may dramatically increase contaminant degradation rates by maintaining optimal soil moisture content. All biochemical processes require liquid water. For microorganisms, water provides the essential medium for colony formation and is a primary ingredient in a variety of cell processes such as hydrolysis and hydroxylation (Tate 1995). Available water also indirectly affects microbial activity by controlling diffusion of gases such as oxygen and carbon dioxide, transport of nutrients, soil pH, and soil temperature (Atlas and Bartha 1993). Aerobic microbial activity increases from essentially zero at very low water contents to an optimum at 60 percent water-filled pore space (Figure 11-14).

The water content in soils is continually changing. Fluctuations in soil moisture content not only affect microbial populations but can also alter nutrient availability and the physical structure of the soil. Disruption of soil structure and the death of microorganisms and fine roots due to the cycling water content of soil can liberate occluded organic matter and result in increased microbial activity (Tate 1995). Van Gestel *et al.* (1993) demonstrated that drying and rewetting of soils enhanced mineralization of carbon and nitrogen where the sources were inactive and dead biomass, and other organic residues. The magnitude of release of carbon and nitrogen appeared to be influenced by soil properties such as carbon content and texture. West *et al.* (1992) observed similar results using four different soil types. Rewetting of dried soils resulted in increased respiration for all soil types. The similarity of microbial response in contrasting soils suggests that microbial communities

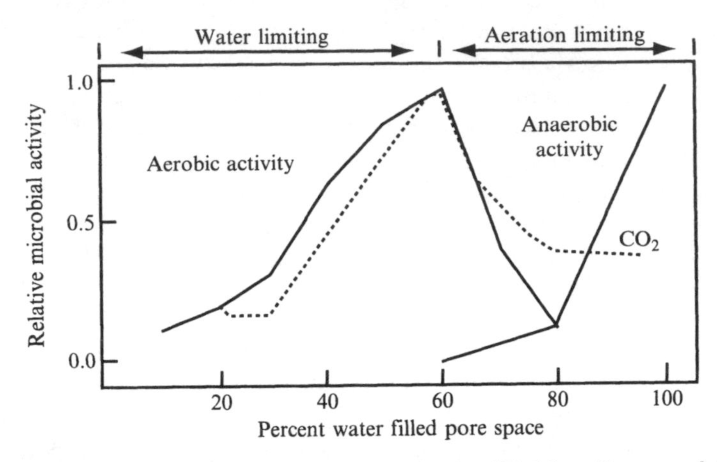

Figure 11-14 Effect of soil water on microbial activity. Modified from Doran *et al.* (1988).

have similar survival strategies to resist desiccation. Jager and Bruins (1975) concluded that the loss of carbon due to microbial respiration in soils with cycling water contents was proportional to the temperature at which the soils were dried, suggesting that more severe treatments resulted in increased availability of organic matter for microbial growth.

Changes in soil water content occur naturally through precipitation, flooding, and evaporation. On a macro scale, soil water can be controlled with proper irrigation and drainage. The five basic methods of applying irrigation water are flooding, furrow irrigation, sprinkling, trickle irrigation, and subirrigation (Linsley and Franzini 1979). The most common irrigation method used for turf grass is sprinkler irrigation. However, new subsurface irrigation methods are gaining in popularity in some areas (Jensen *et al.* 1983).

MEASURES OF SUCCESS: ACCEPTABLE END POINTS

Bioavailability and Residual Toxicity

Over the past few years, regulatory concerns have turned towards an understanding of the level of soil contamination that may be left on a site without posing a significant threat to the environment. This effort has led to attempts to correlate contaminant concentrations in soils with toxicity and bioavailability. Extraction procedures for organic contaminants from soils and sediments have primarily centered on maximizing removal of all the contaminants through exhaustive extraction procedures. Although the total contaminant concentration in soil is required for site design and regulating compliance, the relationship between total contaminant concentration and bioavailability or toxicity is rarely obvious. Information is available to estimate the effect of various solvents on contaminant solubility and release. What has not been sufficiently investigated, however, is how soil type, contaminant matrix, and aging will impact the efficacy of a given solvent and, in turn, how this can be used in developing chemical indices for predicting bioavailability (Alexander 1999).

Hatzinger and Alexander (1995) examined the extractability of phenanthrene by Soxhlet extraction from muck, loam soil, and aquifer sand as a function of contaminant aging. Butanol was used as the extraction solvent. The amount of butanol-extractable material declined in both muck and loam soil with aging and was greater at all times in loam soil than in muck. Total extractable phenanthrene (butanol with Soxhlet extraction) remained relatively constant. Data from the laboratories of the coauthors of this chapter (Wetzel *et al.* 1997, Schwab *et al.* 1998) have established that the concentration of PAHs extracted is strongly dependent on soil type, solvent properties, and aging.

Kelsey *et al.* (1997) investigated the correlation of the method of solvent extraction of atrazine and phenanthrene from a loam soil to microbial mineralization and to the uptake of the same compounds by worms. Key observations were (1) mineralization was approximately 2.5 times greater than worm uptake, (2) aging reduced both availability and extraction efficiency, (3) the solute mass extracted varied with solvent type, and (4) the solute mass extracted with a given solute mixture was predictable.

A risk-based approach to quantify environmentally acceptable end points includes measurements of chemical toxicity. A complete toxicity assessment of a contaminated soil involves assays using plants, soil microorganisms, and soil invertebrates.

Plant toxicity assessments are particularly relevant when phytotoxic contaminants are present in soil. In germination studies, seeds are exposed to the contaminated soil and seedlings are counted after an extended incubation period. Results are compared to germination studies in uncontaminated control soil. Seed germination has been observed to increase appreciably in contaminated soil after extended aging (Wang and Bartha 1990, Baud-Grasset *et al.* 1994), thus indicating a reduction in toxicity due to contaminant sorption and chemical complexation. Bioaccumulation of contaminants in aboveground stem and leaves and in belowground root biomass of plants may be assessed in longer greenhouse studies.

Soil invertebrate toxicity assessments are important because earthworms have been directly linked to soil health and are an essential part of the terrestrial food web. Earthworms are exposed to contaminants when the contaminants are dissolved in water and sorbed to soil. Toxicity may be quantified by measuring changes in worm biomass and population density over time. Bioaccumulation can also be easily assessed using earthworms. The U.S. EPA (1994) recommends toxicity testing for contaminated soil using both earthworm (*Eisenia foetida*) and lettuce (*Lactuca sativa*) germination assays. In field studies, earthworm population density has been observed to increase after long-term aging of contaminants (Loehr *et al.* 1985).

The U.S. EPA (1994) also suggested including microbial toxicity tests such as the Microtox® assay when developing a protocol to assess the toxicity of soil. Microtox® is a commonly used toxicity test, which uses the phosphorescent bacterium *Vibrio fischeri* as the test organism. Bacterial bioluminescence is often inhibited by exposure to toxic compounds. Microtox is used primarily to evaluate toxicity in aqueous matrices; however, exposing the bacteria to contaminated soil in a water suspension is also used as a method. Microtox toxicity has been observed to decrease with aging of petroleum hydrocarbon–contaminated soil (Wang and Bartha 1990).

Soil Health

Karlin *et al.* (1997) defined soil quality as "the fitness of a specific kind of soil, to function within its capacity and within natural or managed ecosystem boundaries, to sustain plant and animal productivity, maintain or enhance air and water quality, and support human health and habitation." The goal of remediating contaminated soil is to return the soil to a healthy state within the parameters of this definition.

Key attributes of soil quality include the interrelated biological, chemical, and physical characteristics of the soil. For example, the chemical state of the soil determines nutrient availability, which in turn affects the biological community; the physical structure of the soil affects the water holding capacity, which in turn controls biological and chemical reactions.

Several physical properties of soil can be measured to determine soil quality. Some of these properties, such as texture, are relatively static and do not readily change with different management practices. Other properties, including aggregate stability, bulk density, penetration resistance, porosity, and hydrophobicity are subject to measurable changes and can be monitored as an assessment of the physical quality of the soil.

The presence of vegetation affects several physical properties of the soil, including soil structure, porosity, hydraulic conductivity, and infiltration rate. These properties, in turn, affect microbial activity by regulating the transport of required water, nutrients, and soil gases through the soil profile. Understanding how the presence of vegetation impacts soil quality is an important part of understanding the mechanisms of phytoremediation.

REFERENCES

Alexander, M. (1977) *Introduction to Soil Microbiology*. 2nd ed. John Wiley, New York, pp. 423–437.

Alexander, M. (1999) *Biodegradation and Bioremediation*. 2nd ed., Academic Press, San Diego, California.

Allen., M.F. (1991) *The Ecology of Mycorrhizae*. Cambridge University Press, New York, p. 184.

Aprill, W. and R.C. Sims (1990) Evaluation of the use of prairie grasses for stimulating polycyclic aromatic hydrocarbon treatment in soils. *Chemosphere* **20**: 253–265.

Atlas, R.M. and R. Bartha (1993) *Microbial Ecology: Fundamentals and Applications.* Benjamin/Cummings, Redwood City, California, pp. 383–412.

Banks, M.K., A.P. Schwab, R.S. Govindaraju, P. Kulakow, S.L. Lewis, and K.I. Rathbone (1998) Craney Island phytoremediation project final report. Rice University, Advanced Applied Technology Demonstration Facility, Houston, Texas.

Barber, D.A. and J.K. Martin (1976) The release of organic substances by cereal roots into soil. *New Phytol.* **76**: 68.

Barber, S.A. (1984) *Soil Nutrient Bioavailability.* Wiley-Interscience, New York.

Baud-Grasset, F., S.I. Safferman, S. Baud-Garasset, and R.T. Lamar (1994) Demonstration of soil bioremediation and toxicity reduction by fungal treatment. In: *Bioremediation of Chlorinated and Polycyclic Aromatic Hydrocarbons* R. Hinchee, A. Leeson, L. Semprini, and S.K. Ong, eds. Lewis Publishers, Boca Raton, Florida, pp. 496–500.

Bohn H.L., B.L. McNeal, and G.A. O'Connor (1985) *Soil Chemistry.* 1st ed. Wiley-Interscience, New York.

Briggs, G.G., R.H. Bromilow, and A.A. Evans (1982) Relationships between lipophilicity and root uptake and translocation of non-ionized chemicals by barley. *Pestic. Sci.* **13**: 405–504.

Chang, Z.Z., R.W. Weaver, and R.L. Rhykerd (1996) Oil bioremediation in a high and a low phosphorus soil. *J. Soil Contam.* **5**(3): 215–224.

Churchill, S.A., R.A. Griffin, L.P. Jones, and P.F. Churchill (1995) Biodegradation rate enhancement of hydrocarbons by an oleophilic fertilizer and a rhamnolipid biosurfactant. *J. Environ. Qual.* **24**: 19–28.

Cookson, J.T. (1995) *Bioremediation Engineering: Design and Application.* McGraw-Hill, New York.

Cunningham, S.D., J.R. Shann, D.E. Crowley, and T.A. Anderson (1997) Phytoremediation of contaminated water and soil. In: *Phytoremediation of Soil and Water.* ACS Symposium Series No. 664. E.L. Kruger, T.A. Anderson, and J.R. Coats, eds. American Chemical Society, Washington, D.C., pp. 2–17.

Detling, J.K. (1979) Processes controlling blue grama production on the shortgrass prairie. In: *Perspectives in grassland ecology.* N. French, ed. Springer-Verlag, New York, p. 25.

Doran, J.W., L.N. Mielke, and S. Stamatiadis (1988) Microbial activity and N cycling as regulated by soil water-filled pore space. Paper No. 132. Proc. Int. Soil Tillage Research Organization, July 11–15.

Epuri V. and D.L. Sorensen (1997) Benzo(*a*)pyrene and hexachlorobiphenyl contaminated soil: phytoremediation potential. In: *Phytoremediation of Soil and Water.* ACS Symposium Series No. 664. E.L. Kruger, T.A. Anderson, and J.R. Coats, eds. American Chemical Society, Washington, D.C., pp. 200–222.

Farrell, J. and M. Reinhard (1994) Desorption of halogenated organics from model solids, sediments, and soil under unsaturated conditions. *Environ. Sci. Technol.* **28**: 53–62.

Ferro, A.M., R.C. Sims, and B. Bugbee (1994) Hycrest crested wheatgrass accelerates the degradation of pentachlorophenol in soil. *J. Environ. Qual.* **23**: 272–281.

Fisher, R.A. (1949) *The Design of Experiments.* Oliver and Boyd, Edinburgh.

Fletcher, J.S. and R.S. Hegde (1995) Release of phenols by perennial plant-roots and their potential importance in bioremediation. *Chemosphere* **31**(4): 3009–3016.

Glaser, J.A. (1991) Nutrient-enhanced bioremediation of oil-contaminated shoreline: the Valdez experience. In: *On-site Remediation Processes for Xenobiotic and Hydrocarbon Treatment*. R.E. Hinchee and R.F. Offenbuttel, eds. Butterworth-Heinemann, Stoneham, Massachusetts, pp. 366–384.

Graham, D.W., V.H. Smith, and K.P. Law (1995) Application of variable nutrient supplies to optimize hydrocarbon biodegradation. In: *Bioremediation of Recalcitrant Organics*. R.E. Hinchee, R.E. Hoeppel, and D.B. Anderson eds. Battelle Press, Columbus, Ohio, pp. 331–340.

Gunther T., U. Dornberger, and W. Fritsche (1996) Effects of ryegrass on biodegradation of hydrocarbons in soil. *Chemosphere* **33**: 203–215.

Hatzinger, P.B. and M. Alexander (1995) Effect of aging of chemicals in soil on their biodegradability and extractability. *Environ. Sci. Technol.* **29**(12): 537–545.

Hegde R.S. and J.S. Fletcher (1996) Influence of plant growth stage and season on the release of root phenolics by mulberry as related to development of phytoremediation technology. *Chemosphere* **32**: 2471–2479.

Hutchinson, S.L., M.K. Banks, and A.P. Schwab (2001) Phytoremediation of petroleum sludge: impact of inorganic fertilizer. *J. Environ. Qual.* **30**: 395–403.

Jager, G. and E.H. Bruins (1975) Effect of repeated drying at different temperatures on soil organic matter decomposition and characteristics, and on the soil microflora. *Soil Biol. Biochem.* **7**(2): 153–159.

Jensen, M.E., D.S. Harrison, H.C. Korven, and F.E. Robinson (1983) The role of irrigation in food and fiber production. In: *Design and Operation of Farm Irrigation Systems*. American Society of Agricultural Engineers Monograph Number 3. St. Joseph, Michigan, pp. 15–41.

Jordahl, J.L., L. Foster, J.L. Schnoor, and P.J.J. Alvarez (1997) Effect of hybrid poplar trees on microbial populations important to hazardous waste bioremediation. *Environ. Toxicol. Chem.* **16**: 1318–1321.

Karlin, D.L., M.J. Mausbach, J.W. Doran, R.G. Clive, R.F. Harris, and G.E. Schuman (1997) Soil quality: a concept, definition, and framework for evaluation. *Soil Sci. Soc. Am. J.* **61**: 4–10.

Kelsey, J.W., B.D. Kottler, and M. Alexander (1997) Selective chemical extractants to predict bioavailability of soil-aged organic chemicals. *Environ. Toxicol. Chem.* **16**: 582–585.

Kulakow, P.A., A.P. Schwab, and M.K. Banks (2000) Screening plant species for growth on weathered sediments contaminated with petroleum hydrocarbons. *Int. J. Phytoremed.* **2**(4): 297–317.

Kyle, W.A. (1998) Heritability of the alfalfa cultivar Riley for the enhanced bioremediation of crude-oil contaminated soil. M.S. thesis, Dept. of Agron., Kansas State University, Manhattan.

Lee, E. (1996) The fate of polycyclic aromatic hydrocarbons in the rhizosphere of *Festuca arundinacea*. Ph.D. dissertation, Dept. of Civil Engineering, Kansas State University, Manhattan.

Lin, Q.X. and I.A. Mendelssohn (1998) The combined effects of phytoremediation and biostimulation in enhancing habitat restoration and oil degradation of petroleum contaminated wetlands. *Ecol. Eng.* **10**: 263–274.

Linsley, R.K. and J.B. Franzini (1979) *Water Resources Engineering*. 3rd ed. McGraw-Hill, New York, pp. 27–34.

Loehr, R.C., J.H. Martin, E.F. Neuhauser, R.A. Norton, and M.R. Malecki (1985) Land treatment of an oily waste–degradation, immobilization, and bioaccumulation. Report EPA/600/2–85/009. U.S. Environmental Protection Agency, Washington, D.C.

Lungley, D.R. (1973) The growth of root systems–a numerical computer simulation model. *Plant Soil* **38**: 145–159.

Mahro, B., G. Schaefer, and M. Kastner (1994) Pathways of microbial degradation of polycyclic aromatic hydrocarbons in soil. In: *Bioremediation of Chlorinated and Aromatic Hydrocarbon Compounds*. R.E. Hinchee, A. Leeson, L. Semprini, and S.K. Ong, eds. Lewis Publishers, Boca Raton, Florida, pp. 203–217.

Nedunuri, K., R.S. Govindaraju, M.K. Banks, A.P. Schwab, and Z. Chen (2000) Evaluation of phytoremediation for field-scale degradation of total petroleum hydrocarbons. *J. Environ. Eng.* **126**: 483–490.

Nichols, T.D., D.C. Wolf, H.B. Rogers, C.A. Beyrouty, and C.M. Reynolds (1997) Rhizosphere microbial populations in contaminated soils. *Water, Air Soil Pollut.* **95**: 165–178.

Paul, E.A. and F.E. Clark (1989) *Soil Microbiology and Biochemistry*. Academic Press, San Diego, California, pp. 81–84.

Pignatello, J.J. (1989) Reactions and movement of organic chemicals. In: *Soils*. B.L. Sawhney and K. Brown, eds. Soil Science Society of America, Madison, Wisconsin, pp. 45–80.

Pradhan, S.P., J.R. Conrad, J.R. Paterek, and V.J. Srivastava (1998) Potential of phytoremediation for treatment of PAHs in soil at MGP sites. *J. Soil Contam.* **7**: 467–480.

Qui, X., S.I. Shah, E.W. Kendall, D.L. Sorenson, R.C. Sims, and M.C. Engelke (1994) Grass-enhanced bioremediation for clay soils contaminated with polynuclear aromatic hydrocarbons. In: *Bioremediation through Rhizosphere Technology*. ACS Symposium Series No. 563. T.A. Anderson and J.R. Coats, eds. American Chemical Society, Washington, D.C., pp. 142–157.

Radwan S.S., H. Al-Awadhi, N.A. Sorkhoh, and I.M. El-Nemr (1998) Rhizospheric hydrocarbon-utilizing microorganisms as potential contributors to phytoremediation for the oily Kuwaiti desert. *Microbiol. Res.* **153**: 247–251.

Rasiah, V., R.P. Voroney, and R.G. Kachanoski (1992) Biodegradation of an oily waste as influenced by nitrogen forms and sources. *Water, Air Soil Pollut.* **65**: 143–151.

Reilley, K., M.K. Banks, and A.P. Schwab (1996) Dissipation of polynuclear aromatic hydrocarbons in the rhizosphere. *J. Environ. Qual.* **25**: 212–219.

Rhoades, J.D. (1982) Soluble salts. In: *Methods of Soil Analysis Part 2–Chemical and Microbiological Properties*. 2nd ed. A.L. Page, ed., Soil Science Society of America, Madison, Wisconsin, pp. 167–179.

Rovira, A.D. and C.B. Davey (1974) Biology of the rhizosphere. In: *The Plant Root and its Environment*. E.W. Carson, ed., University Press of Virginia, Charlottesville, pp. 153–204.

Russell, R.S. (1977) *Plant Root Systems*. McGraw-Hill, New York. 298 pp.

Schwab, A.P. and M.K. Banks (1994) Biologically mediated dissipation of polyaromatic hydrocarbons in the root zone. In: *Bioremediation through Rhizosphere Technology*. ACS Symposium Series No. 563. T.A. Anderson and J.R. Coats, eds. American Chemical Society, Washington, D.C., pp. 132–141.

Schwab, A.P., M.K. Banks, and M. Arunuchalam (1995) Biodegradation of polycyclic aromatic hydrocarbons in rhizosphere soil. In: *Bioremediation of Recalcitrant Organics*. R.E. Hinchee, R.H. Hoeppel, and D.B. Anderson, eds. Battelle Press, Columbus, Ohio, pp. 23–29.

Schwab, A.P., A.A. Al-Assi, and M.K. Banks (1998) Adsorption of naphthalene to plant roots. *J. Environ. Qual.* **27**: 169–174.

Schwab, A.P., J. Su, S. Wetzel, S. Pekarek, and M.K. Banks (1999) Extraction of petroleum hydrocarbons from soil by mechanical shaking. *Environ. Sci. Tech.* **33**: 1940–1945.

Schwarzenbach, R.P., P.M. Gschwend, and D.M. Imboden (1993) *Environmental Organic Chemistry*. John Wiley, New York, pp. 682.

Tate, R.L. (1995) *Soil Microbiology*. John Wiley, New York. p. 107.

Taylor, H.M., M.G. Huck, B. Klepper, and Z.F. Lund (1970) Measurement of soil grown roots in a rhizotron. *Agron. J.* **62**: 807–809.

Trapp, S. and C. McFarlane (1995) *Plant Contamination: Modeling and Simulation of Organic Processes*. Lewis Publishers, Boca Raton, Florida, 254 pp.

U.S. Environmental Protection Agency (1983) Methods for chemical analysis of water and wastes. Environmental Monitoring and Support Laboratory, Office of Research and Development, Cincinnati, Ohio.

U.S. Environmental Protection Agency (1986) Test methods for evaluating solid waste, physical/chemical methods. 3rd ed. Office of Solid Waste and Emergency Response, Washington, D.C.

U.S. Environmental Protection Agency (1994) ECO update, U.S. EPA 540-F-94-012, Washington, D.C.

Van Gestel, M., R. Merckx, and K. Vlassak (1993) Microbial biomass responses to soil drying and rewetting: the fate of fast- and slow-growing microorganisms in the soils from different climates. *Soil Biol. Biochem.* **25**(1): 109–123.

Wang, X. and R. Bartha (1990) Effects of bioremediation on residues, activity and toxicity in soil contaminated by fuel spills. *Soil Biol. Biochem.* **22**: 501–505.

West, A.W., G.P. Sparling, C.W. Feltham, and J. Reynolds (1992) Microbial activity and survival in soils dried at different rates. *Aust. J. Soil Res.* **30**: 209–222.

Wetzel, S.C., M.K. Banks, and A.P. Schwab (1997) Rhizosphere effects on the degradation of pyrene and anthracene in soil. In: *Phytoremediation of Soil and Water*. ACS Symposium Series No. 664. E.L. Kruger, T.A. Anderson, and J.R. Coats, eds. American Chemical Society, Washington, D.C., pp. 254–262.

Wildung, R.E., T.R. Garland, and R.L. Buschbom (1975) The interdependent effects of soil temperature and water content on soil respiration rate and plant root decomposition in arid grassland soils. *Soil Biol. Biochem.* **7**: 373–378.

Wilson, J.M. and D.M. Griffin (1975) Water potential and the respiration of microorganisms in the soil. *Soil Biol. Biochem.* **7**: 199–204.

Wiltse, C.C., W.L. Rooney, Z. Chen, A.P. Schwab, and M.K. Banks (1998) Detection of variability for agronomic and phytoremediation potential among alfalfa clones grown in crude oil contaminated soil. *J. Environ. Qual.* **24**: 169–173.

SECTION IV

TRANSFORMATION AND CONTROL OF EXPLOSIVES

12

ROLE OF PLANTS IN THE TRANSFORMATION OF EXPLOSIVES

M. Subramanian and J. V. Shanks

SUMMARY

The proven uptake and complete removal of low levels of explosives in aqueous plant systems suggest that plants show promise in the field for remediation of trinitrotoluene (TNT). Elucidation of TNT transformation pathways helps in developing quantitative models for removal of explosives and aids in understanding the kinetics of uptake and removal. Further work on toxicity of the final bound residues, and the effect on the ecosystem is important. The discovery of a significant amount of oxidative metabolites during TNT transformation in an aquatic system suggests alternative pathways for TNT degradation. Research to explore the transformation of oxidative metabolites is warranted.

INTRODUCTION

The manufacturing, handling, and storage of explosives have resulted in soil, sediment, and groundwater contamination at several documented sites in the U.S., Germany, U.K., Canada, and Australia, with anecdotal evidence of a significant number of sites worldwide (Spain 2000). The explosives 2,4,6-trinitrotoluene (TNT), hexahydro-1,3,5-trinitro-1,3,5-triazine (RDX) and octahydro-1,3,5,7-tetranitro-1,3,5,7-tetrazine (HMX) are marginally biodegradable by soil bacteria under natural conditions and hence persist in the environment (Spain 2000). No microorganism has been shown to use TNT, RDX, or HMX as a sole carbon source, although biotransformation of TNT to aminodinitrotoluenes is common. These transformation metabolites persist in the environment as well. Trinitrotoluene and the associated metabolites have documented toxicity (Yinon 1990) to humans (Nagel *et al.* 1999a),

Phytoremediation: Tansformation and Control of Contaminants,
Edited by Steven C. McCutcheon and Jerald L. Schnoor.
ISBN 0-471-39435-1 (cloth)

plants (Palazzo and Leggett 1986), and microorganisms (Won *et al.* 1976). The potential for seepage of explosives and the associated metabolites into the groundwater and the re-vegetation of polluted sites have given rise to many concerns about the entry of explosives like TNT into the food chain (Palazzo and Leggett 1986, Harvey *et al.* 1990, Nagel *et al.* 1999b). These concerns have resulted in intense and worldwide efforts to remediate contaminated land and water.

Excavation and incineration have been used to remediate explosive-contaminated soils in the U.S. (Spain 2000) and Western Europe (Fritsche *et al.* 2000). However, because incineration is both an expensive and destructive process, alternatives are being sought. Composting recently has been implemented in place of incineration at several sites in the U.S. (Spain 2000). Composting is environmentally less invasive and less expensive as compared to incineration. Currently, explosive-laden groundwater is also being cleaned up in the U.S. using pump-and-treat methods with carbon filters (Spain 2000) and ultraviolet oxidation systems. High material and labor costs for both composting and pump-and-treat methods underscore the need for additional remediation options.

Phytoremediation shows considerable promise for explosives remediation, especially for treatment of large volumes of lightly contaminated soil and groundwater (Burken *et al.* 2000). Phytoremediation has several advantages, mainly of being a low-cost, ecologically superior process and having strong public acceptance. In the mid-1990s, reports of rapid TNT disappearance in aquatic plant systems (Wolfe *et al.* 1994, Schnoor *et al.* 1995) stimulated research and field evaluations of phytoremediation of explosives using aquatic and terrestrial plants (for reviews see Burken *et al.* 2000, Jacobson *et al.* this book, McCutcheon *et al.* this book, Thompson *et al.* this book). Commercial success of phytoremediation for explosives contamination hinges on many variables (Burken *et al.* 2000), including insights into the actual biochemistry of the process (the enzymes involved and definitive reaction network of pathways) and plant tolerance to contaminants.

Most research to date on phytoremediation of explosives has concentrated on elucidating the capability of plants to transform TNT. Several small-scale studies show that plants possess an inherent capacity to remove TNT and other nitroaromatics from soil and water (Palazzo and Leggett 1986, Gorge *et al.* 1994, Best *et al.* 1997, Larson 1997, Vanderford *et al.* 1997, Burken and Schnoor 1998, Pavlostathis *et al.* 1998, Rivera *et al.* 1998, Salt *et al.* 1998, Scheidemann *et al.* 1998, Thompson *et al.* 1998, Best *et al.* 1999, Bhadra *et al.* 1999a, Bhadra *et al.* 1999b, Larson *et al.* 1999a, Larson *et al.* 1999b, Sens *et al.* 1999, Wayment *et al.* 1999). These small-scale studies need to be supplemented with large-scale field studies and mid-scale semi-basic transformation studies to determine if scale-up has an impact on TNT removal and metabolism. Some work on a larger-scale is discussed elsewhere in this book.

One of the reasons for the prevalence of remediation methodologies utilizing microbial remediation is the deep understanding of the biochemistry behind the action of microbes (Rieger and Knackmuss 1995) and hence the consequential incorporation of that knowledge into engineering design criteria. For phytoremediation to reach a similar level of acceptance, much work is required in gleaning information about the enzymes involved, metabolites produced, and final fates of compounds in the transformation pathway. Progress in understanding pollutant transformation pathways, determining carbon fates *via* mass balances, and identifying associated metabolites formed will go a long way in helping phytoremediation evolve into an industrially accepted procedure.

This chapter summarizes our current understanding of plant metabolism of explosives, mainly TNT, and the implications for further research. The chapter, while emphasizing TNT metabolism by plants, also delves into the removal of associated explosives RDX and HMX by plant tissue cultures. Because most of the literature describes experiments that were performed with TNT or some of the associated transformation metabolites, understanding metabolism by plants has reached significant levels. Based on the results of these experiments, a hypothetical pathway scheme for TNT transformation has been proposed. This pathway has been elucidated in some detail as to the transformation of the TNT molecule to other products, but not precisely for specific reaction steps catalyzed by enzymes. The various mechanism that govern the pathway are consistent with the "green-liver" model. Carbon-14 mass balances of TNT removal in various systems have been included, which help quantify the TNT flux and the various end-points of TNT metabolism. This chapter emphasizes much of the basic work on the TNT transformation pathways using the parrot feather (*Myriophyllum aquaticum*) plant system and periwinkle (*Catharanthus roseus*) hairy root cultures.

BASIS OF PLANT TRANSFORMATION STUDIES

Motivation

Understanding the plant transformation pathways at three different levels—metabolites, enzymes, and genes—is important for several reasons. The final fates of the pollutant and the transformation metabolites are crucial for regulatory reasons and as a measure of the effectiveness of the process. Some studies indicate that plant growth may be severely inhibited by explosives even at low concentrations (Palazzo and Leggett 1986), while many other studies report an absence of conspicuous deleterious effect of explosives on plants (Salt *et al.* 1998, Bhadra *et al.* 1999a). Plant health after prolonged exposure to pollutants, and the effectiveness of plants in pollution control over continued periods of time, are further factors that can be better understood after carrying out basic studies of transformation. The maximum concentration of pollutants a given plant species can effectively remove is another important parameter that imposes an upper limit on the applicability of phytoremediation. Plant biomass and bioavailability of the contaminants are other important variables that will influence the success of the process.

Understanding the types of enzymes and corresponding genes involved is important in suggesting not only a mechanism for transformation, but perhaps also a way of improving efficiency by generating genetically modified plants with the sole purpose of transforming pollutants (French *et al.* 1999). Transgenic plants would have several inherent advantages, one being more efficient pollutant removal since the key enzyme(s) involved in flux-limitation(s) in the pathway can be over-expressed. Ideally, this would result in a faster rate of pollutant removal, lower biomass requirements, and well-characterized fate of the pollutants. Another potential advantage would be phytotoxicity management. Some of the TNT transformation products in the natural environment, both from soil bacteria and from plants, are phytotoxic (detrimental to plant growth) (Tadros *et al.* 2000). Metabolic engineering of the pathways to decrease levels of these compounds may be possible, resulting in plants that may tolerate higher initial levels of offending compounds. In addition, new routes of degradation through the application of foreign genes may be emphasized, resulting in less-toxic metabolites than those produced by the existing pathway.

Choice of Model Plant Systems: *Periwinkle* and *Parrot Feather*

In the studies being described here, three different plant systems have been used for fate and mass balance studies of TNT transformation pathways—axenic tissue cultures of the terrestrial plant periwinkle (*Catharanthus roseus*), and axenic and native plants of the aquatic species parrot feather (*Myriophyllum aquaticum*). These systems offer sets of advantages and drawbacks. Hairy root cultures of periwinkle (*Catharanthus roseus*), in particular, present a number of significant advantages over using whole plants in TNT-transformation pathway studies. These axenic plant cultures (free of all other living organisms) are genetically stable and the results are highly reproducible (Shanks and Morgan 1999). Experimentally versatile, large amounts of biomass can be generated in a controlled setting, as hairy root cultures can be grown in shake-flasks and bioreactors (Shanks and Morgan 1999). However, in the context of phytoremediation, the key advantage in using axenic cultures would be to obtain a definite verification of the capacity of plants to transform explosives because the cultures are fully free of microbial contamination. Therefore, all results obtained using periwinkle (*Catharanthus roseus*) are indicative of the potential of plant roots to metabolize TNT and do not include any significant microbial or symbiotic relationships. Another significant advantage in using root cultures to carry out pathway studies would be the simplified plant physiology involved, relative to using whole plants. Phenomena such as the transport of substrates into the vascular bundle and leaves, and photosynthesis are conveniently excluded, while greater emphasis would be placed on the biological mechanisms roots adopt to counter nitroaromatics. This would help in better elucidating pathway information from the observations.

Axenic cultures of the aquatic plant species parrot feather (*Myriophyllum aquaticum*) have been grown (Hughes *et al.* 1997), but this system is problematic as the axenic plants are not as robust as native counterparts. The important advantage of native parrot feather (*Myriophyllum aquaticum*) is the documented TNT removal capabilities, and the use in various field studies for wetland systems (Best *et al.* 1997, Larson *et al.* 1999a). The whole plant may give different results than a specific organ culture, namely difference between a tissue in isolation *versus* an intact plant. Any similarities observed between results among parrot feather (*Myriophyllum aquaticum*) and periwinkle (*Catharanthus roseus*) have interesting implications for plant selection criteria, as these are two widely different plant species. Similarities in TNT transformation point to the possibility that several different plant species may utilize similar metabolic schemes.

The basic studies of periwinkle (*Catharanthus roseus*) and parrot feather (*Myriophyllum aquaticum*) focused on the identification of the transformation products, the chemical nature of the products, and the sequence of transformation. These studies did not determine the enzymes involved, pathway regulation at the enzymatic level or genetic level, or the kinetic properties of the enzymes. A molecular biology approach is useful for these latter studies, and plants for which the genome has been sequenced (genomes for periwinkle (*Catharanthus roseus*) and parrot feather (*Myriophyllum aquaticum*) have not been sequenced) would be a better choice. The transformation pathway (*i.e.* network structure) here does, however, open up the possibility of a genetic and biochemical approach to the study of TNT transformation pathways using a plant species that is better characterized genetically.

TRINITROTOLUENE TRANSFORMATION STUDIES

General Methods

A number of TNT exposure experiments have been performed with both *Catharanthus roseus* hairy roots and parrot feather (*Myriophyllum aquaticum*) to gauge the capacity of these plants to remove TNT from solution and to study the various mechanisms that govern removal (Hughes *et al.* 1997, Bhadra *et al.* 1999a, Bhadra *et al.* 1999b, Wayment *et al.* 1999). The periwinkle (*Catharanthus roseus*) roots were in the stationary phase of growth when [ring-U-^{14}C] TNT was added. Generally, reversed-phase high pressure liquid chromatography with a C8 column was used to separate the TNT metabolites, while several techniques such as photodiode array detection, nuclear magnetic resonance spectroscopy, mass spectrometry, and enzymatic assays were used to identify the metabolites. A variety of analytical and extraction procedures were used to isolate the transformation products. Part of the intracellular portion was comprised of compounds that could be extracted using sonication with a solvent and the rest were fractions that were irreversibly attached to the biomass, henceforth referred to as bound residues (Hughes *et al.* 1997). Oxidation of the biomass was necessary to complete the [^{14}C] mass balance. All studies showed nearly complete TNT uptake and transformation by plants, thereby demonstrating the inherent capacity of both these plant systems to remove TNT from wastewater.

Working Model of the Trinitrotoluene Transformation Pathway

A working model of the TNT transformation pathway is proposed and illustrated in Figure 12-1. This working model is based on a composite of experiments with periwinkle (*Catharanthus roseus*) and parrot feather (*Myriophyllum aquaticum*). The abbreviations for the various metabolites present in the pathway have been listed in the figure caption, along with the International Union of Pure and Applied Chemistry (IUPAC) names. All the metabolites mentioned in the pathway have been identified in plant systems amended with TNT. Basically, studies of periwinkle (*Catharanthus roseus*) hairy roots produced reduced monoamino (NH_2, NHR, or NR_2) products and conjugates, but no oxidized products (Hughes *et al.* 1997, Vanderford *et al.* 1997, Bhadra *et al.* 1999b, Wayment *et al.* 1999). Studies of parrot feather (*Myriophyllum aquaticum*) resulted in the production of a wide array of oxidized, reduced, conjugated, and binuclear products (Bhadra *et al.* 1999a). The TNT transformation pathway is comprised of these distinct reduced, oxidized, conjugated, and azoxy products. The oxidative metabolites are outlined on the periphery of the pathway in Figure 12-1, while the reductive products are inside. This arrangement has been used for clarity and is not indicative of flux directions.

Most of the metabolites, with the exception of the conjugates 2A-1 and 4A-1, and the monoamines, were identified in parrot feather (*Myriophyllum aquaticum*). Significantly, conjugation (TNT-1 and TNT-2) was observed in these two different plant species (Vanderford *et al.* 1997). Reductive metabolites are observed in several aquatic and terrestrial species (Burken *et al.* 2000). Plants may use different or additional pathways to achieve TNT transformation; more work is needed in this area. Oxidative products have only been reported for parrot feather (*Myriophyllum aquaticum*) and not in any other plant-based study, although these products have been identified in microbial systems (Nagel *et al.* 1999b). With a significant percentage of transformation metabolites remaining unidentified in most studies, various other mechanisms conceivably are involved in TNT transformation (Hughes *et al.* 1997, Thompson *et al.* 1998, Sens *et al.* 1999).

The importance of this proposed pathway, however, results from a fairly complete picture of TNT transformation, and hence the first good understanding of the biochemistry behind the transformation process. Moreover, with a nearly complete mass balance in parrot feather (*Myriophyllum aquaticum*), a substantially complete TNT-fate balance can be modeled from the information presented in the pathway. Determination of the key branches in the pathway and modification of pathways to increase the rate of removal of TNT are some of the plausible offshoots from the information.

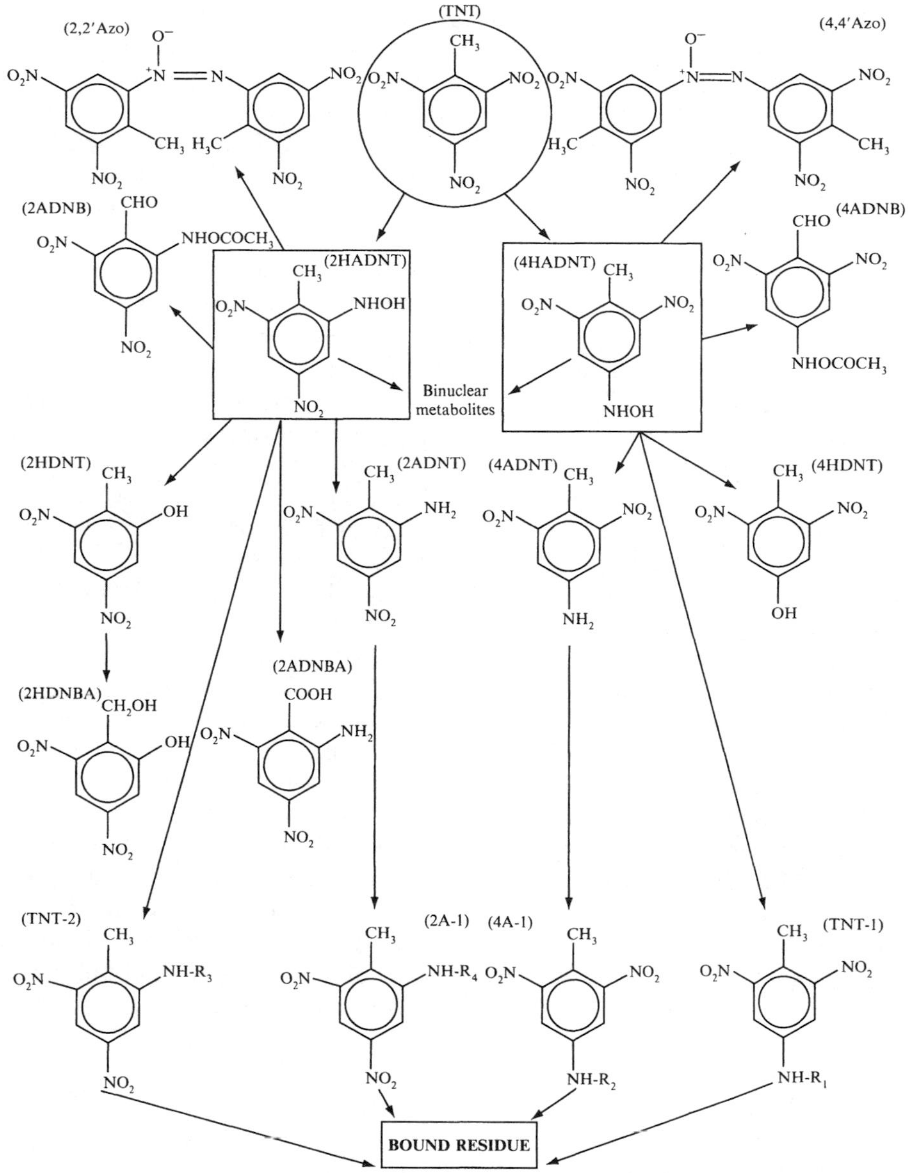

Figure 12-1 Hypothetical TNT transformation pathway in plants, encompassing all known metabolites produced. Arrows indicate the general directions of the pathway and do not necessarily represent a single step reaction. Abbreviations—TNT: 2,4,6-trinitrotoluene; 2ADNT: 2-amino-4,6-dinitrotoluene; 4ADNT: 4-amino-2,6-dinitrotoluene; 2HADNT: 2-hydroxylamino-4,6-dinitrotoluene; 4HADNT: 4-hydroxylamino-2,6-dinitrotoluene; 4,4′Azo: 2,2′, 6,6′-tetranitro-4,4′-azoxytoluene; 2,2′Azo: 4,4′,6,6′-tetranitro-2,2′-azoxytoluene; 2HDNT: 2-hydroxy-4,6-dinitrotoluene; 4HDNT: 4-hydroxy-2,6-dinitrotoluene; 2ADNB: 2-*N*-acetamido-4,6-dinitrobenzaldehyde; 2ADNB: 4-*N*-acetamido-2,6-dinitrobenzalydehyde; 2HDNBA: 2-hydroxy- 4,6-dinitrobenzyl alcohol; and 2ADNBA: 2-amino-4,6-dinitrobenzoic acid. Note that TNT-1, TNT-2, 2A-1 and 4A-1 represent various conjugates observed. R_1, R_2, R_3, and R_4, are six carbon sugars.

Two of the metabolites shown in the pathway are 2-hydroxylamino-4,6-dinitrotoluene and the corresponding 4-hydroxylamino isomer. The reduction of a nitro-group (NO_2) to an amine-group involves two intermediates, the nitroso-group (NO) and the hydroxylamino group (NHOH) (Ahmad and Hughes 2000). Anaerobic microbial studies of TNT transformation have shown that these compounds are fairly central in the bacterial transformation of TNT (Ahmad and Hughes 2000). Generally regarded as short-lived intermediates, both intermediates were shown to persist in anaerobic microbial systems (Ahmad and Hughes 2000). In aerobic media, these metabolites are highly unstable and degrade quickly (Wang and Hughes 1998). The 4-hydroxylamino isomer has been reported in plant systems (Pavlostathis *et al.* 1998), but reliable estimates of concentration have not yet been obtained, perhaps due to the instability of the isomer. Quantification of these metabolites is one of the current challenges in TNT phytoremediation studies. Derivatization of the hydroxylamino isomers in a measurement assay has facilitated detection of these compounds (Wang and Hughes 1998). Recent research with periwinkle (*Catharanthus roseus*) exposed to high TNT concentrations (greater than 50 milligrams per liter) and the use of a modified derivatization procedure have conclusively proved the presence of both hydroxylamines in the plant transformation pathway (Subramanian and Shanks this book). Addition of high levels of TNT to periwinkle (*Catharanthus roseus*) may have served in raising intermediate levels and additionally slowing plant metabolism due to the toxicity of high TNT concentrations (Subramanian and Shanks this book).

The potential importance of these hydroxylamines can be gauged by looking at the position in the pathway. Based upon an analogy to the bacterial studies, the hydroxylamines appear before the reduction products and azoxy compounds in the pathway. Condensation of the hydroxylamino (NHOH) bond with an nitroso (NO) bond gives rise to an azoxy compound, and reduction of the hydroxylamino (NHOH) group gives rise to the monoamines (Ahmad and Hughes 2000). Justification for the positioning of hydroxylamines early in the pathway, before oxidation products and some of the conjugation products, was obtained from precursor exposure studies with the monoamines. Monoamine precursor exposure studies, in which 2-amino-4,6-dinitro-toluene and the 4-amino isomers were added separately to the plant tissues (Bhadra *et al.* 1999a, Bhadra *et al.* 1999b), did not form any oxidative metabolites and formed only a couple of the conjugated metabolites. These results indicate that the split in oxidative, reductive, and conjugative pathways occurred before formation of monoamines, probably with the formation of hydroxylamines. In addition, the absence of some of the conjugates that were previously identified in TNT studies indicated that conjugation did not have to succeed the reductive pathway. Both these findings seem to indicate that hydroxylamines, formed before the monoamines, are the basic precursors for oxidative and some conjugative compounds. These experiments, therefore, were of crucial importance in obtaining useful information about the TNT transformation pathway.

Green-Liver Model

The presence of reduced, oxidized, and conjugated products and the final formation of polymerized lignin-like inextricable bound residues is in keeping with the proposed "green-liver model" for detoxification of xenobiotics by plants (Sandermann 1994). According to this model, plants cannot use xenobiotic compounds as substrates for growth, unlike microorganisms. Hence, plants have to metabolize xenobiotic compounds using detoxification pathways akin to the way the human liver metabolizes the same contaminants. The pathway in Figure 12-1 is consistent with the green-liver concept (Burken *et al.* 2000, Burken this book) in demonstrating the existence of various schemes that help inactivate the reactive functionalities and yet do not degrade the explosive.

Reduction

The reduction products, 2-amino-4,6-dinitrotoluene and the corresponding 4-amino isomer, are two of the primary transformation products and have been identified in most TNT-removal studies, although in very low concentrations (Palazzo and Leggett 1986, Harvey *et al.* 1990, Best *et al.* 1997, Hughes *et al.* 1997, Larson 1997, Vanderford *et al.* 1997, Burken and Schnoor 1998, Pavlostathis *et al.* 1998, Rivera *et al.* 1998, Thompson *et al.* 1998, Bhadra *et al.* 1999a, Bhadra *et al.* 1999b, Nagel *et al.* 1999a, Sens *et al.* 1999, Wayment *et al.* 1999). In some cases, 2,4-diamino-6-nitrotoluene and the corresponding 4,6-diamino isomer have been reported, although these are rarer products that require quite strong reducing conditions (Thompson *et al.* 1998, Sens *et al.* 1999). Azoxy compounds, like 2,2′-azoxy-tetranitrotoluenes have been observed in traces, indicating formation by the condensation of hydroxylamines (Pavlostathis *et al.* 1998). Two unidentified binuclear metabolites have also been observed in studies with parrot feather (*Myriophyllum aquaticum*) in fairly significant quantities—5.6 molar percent of initial TNT (Bhadra *et al.* 1999a). Reduction, although probably not the primary phenomenon, appears to be a significant and ubiquitous phenomenon in transformation of nitroaromatics. Nitroreductase, an NADPH-(nicotinamide adenine dinucleotide phophate) dependent enzyme requiring anaerobic conditions, may be one of the enzymes involved in the reduction of the nitro group (Hatzios and Penner 1982). Because aerobic conditions generally prevail in these systems, activity of the nitroreductases may have been low and hence low concentrations of monoamines result. Other studies with ferredoxin:NADP oxidoreductase (EC 1.18.1.2, www.chem.qmul.ac.uk/iubmb/enzyme) from spinach (*Spinacia oleracea*) leaves seem to indicate the capability to carry out nitro–hydroxylamino reductions efficiently (Shah and Campbell 1997), although this seems to require anaerobic conditions too. Wolfe and Hoehamer (this book) describe an oxygen-insensitive nitroreductase.

Oxidation

Six oxidized metabolites, consisting of nearly 30 percent of the transformation, were identified in experiments with parrot feather (*Myriophyllum aquaticum*), the only plant-based study so far to report oxidized metabolism products (Bhadra *et al.* 1999a). Oxidation of the methyl groups (CH_3) and hydroxylation (introduction of OH) of the aromatic ring after removal of a nitro group are some of the reactions that may be involved (Bhadra *et al.* 1999a). In addition, carboxylation (introduction of COOH), hydroxylation, addition of aldehydes (CHO), and addition of an acetamido group ($NHOCOCH_3$) were observed, which seem to suggest the interplay of many different enzymes in TNT transformation (Bhadra *et al.* 1999a). Cytochrome P-450 monooxygenases (EC 1.14.14.1, www.chem.qmul.ac.uk/iubmb/enzyme) are a diverse group of enzymes that are perhaps involved in detoxification of many herbicides by oxidizing lipophilic substrates, increasing polarity and allowing further conjugation and storage in vacuoles or cell walls (Schuler 1996, Schalk *et al.* 1997). These enzymes that require NADPH or NADH (nicotinamide adenine dinucleotide), split dioxygen into two monooxygens, one oxygen atom being incorporated into the compound and the other becoming a part of a water molecule (Hatzios and Penner 1982). Cytochrome-P-450 monooxygenases have been implicated in the detoxification of many xenobiotics (Schuler 1996). Cytochrome-P450 monooxygenases catalyzing the addition of hydroxyl groups and oxidizing methyl groups have been isolated and characterized in various plants (Schuler 1996).

Conjugation and Sequestration

Conjugation occurs when a plant molecule is added to the original or modified xenobiotic contaminant. The plant molecule could be a sugar ($C_nH_{2n}O_n$) fatty acid ($C_nH_{2n+1}COOH$), amino acid [$RCH(NH_2)COOH$], thiol (RSH), or even glutathione ($C_{10}H_{17}N_3O_6S$) (Hatzios and Penner 1982). The molecule is added onto one of the functional groups of the contaminant, usually the more reactive centers. In studies with periwinkle (*Catharanthus roseus*) and parrot feather (*Myriophyllum aquaticum*), four unique conjugates were observed that had sugar molecules being bonded to the amino group on the ring (Bhadra *et al.* 1999a, Bhadra *et al.* 1999b, Wayment *et al.* 1999). The structures of these conjugates have not been determined, but two bear similarity to the 2-amino-DNT and the remaining two are similar to the 4-amino-DNT, which is reason to believe that the monoamines are the substrates for conjugative enzymes. However, the monoamine exposure studies seem to indicate the possibility of direct addition of plant sugars to the hydroxylamine group. Hence, both monoamines and hydroxylamines seem subject to conjugation, which suggests a wide array of plant enzymes catalyzing the conjugative reactions. These conjugates are further polymerized to form inextricable bound residues, which are irreversibly incorporated into the plant biomass (Hatzios and Penner 1982, Hatzios 1991, Bhadra *et al.* 1999a, Bhadra *et al.* 1999b, Sens *et al.* 1999, Burken *et al.* 2000). Studies have found that nearly 20 percent of the total pollutant added passes through the conjugative pathway and greater than 30 percent ends up as bound residues (Khan 1988, Hughes *et al.* 1997, Sens *et al.* 1999). One study found that a significant portion of the bound residues was found in the lignin fraction of the biomass (27 percent) and lesser amounts were found in the pectin, hemicellulose, and protein fractions of the biomass (Sens *et al.* 1999). Therefore, the conjugates appear to polymerize and become bound to the lignin, or else be isolated in the vacuole to form the other end products of TNT metabolism. Further characterization of these bound residues, long-term fate, reactivity, and distribution in the various plant organs are areas that need to be actively explored.

MASS BALANCES

Mass balances were performed by adding [ring-U-^{14}C]-2,4,6-trinitrotoulene, and quantifying the [^{14}C] radioactivity from the extracellular media and intracellular extracts (Bhadra *et al.* 1999a, Bhadra *et al.* 1999b, Wayment *et al.* 1999). Radioactivity from the intracellular inextricable bound residues was analyzed by breaking down the tissue with a bio-oxidizer. Nearly complete recovery of the [^{14}C] occurred, although significant portions of the extracted metabolites were not separated and identified. Figure 12-2 gives transient concentrations of all the metabolites, conjugates, bounds, and unknowns in periwinkle (*Catharanthus roseus*) amended with TNT (Bhadra *et al.* 1999b, Wayment *et al.* 1999). This figure clearly indicates high concentration of the unknown metabolites, decreasing concentration of the nitroaromatics, and increasing concentration of bound residues. In periwinkle (*Catharanthus roseus*), the concentration of TNT decreases to below detectable limits within 75 hours (Figure 12-2) and is accompanied by a corresponding, although small increase in concentration of the monoamines. The levels of the monoamines and conjugates increase initially as TNT is metabolized and then tend to undergo a subsequent decrease as bound residues are formed. This inextricable mass of several end products cannot, by definition, be further separated and distinguished at this time (Bhadra *et al.* 1999b, Wayment *et al.* 1999).

Figures 12-3 and 12-4 give estimates of the molar distributions of carbon in periwinkle (*Catharanthus roseus*) and parrot feather (*Myriophyllum aquaticum*), although these balances should not be considered a final distribution. As exposure time to TNT increases, the percentage of inextricable bound residues increases (Vanderford *et al.* 1997) as the transformation process continues. The effects of parameters such as weight of biomass, time of harvesting of biomass, and initial concentrations of TNT need to be investigated over longer periods of time in both systems before meaningful comparisons can be made. The identification of the oxidative metabolites in parrot feather (*Myriophyllum aquaticum*) (6 percent unknowns) helped complete the mass balance to a greater extent than periwinkle (*Catharanthus roseus*) (45 percent unknowns).

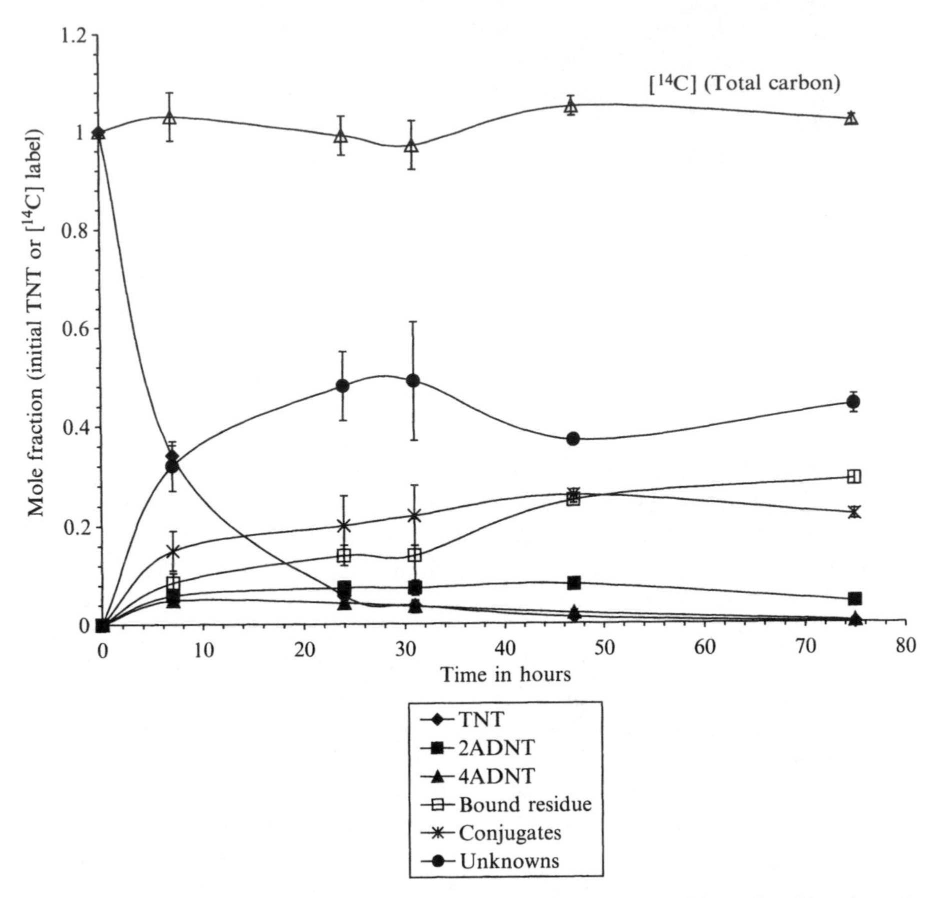

Figure 12-2 Transient mole fractions of TNT, ADNTs, conjugates, and bound residue, in periwinkle (*Catharanthus roseus*), upon addition of TNT, and total [^{14}C] fraction. Data from Bhadra *et al.* (1999b). See definitions of abbreviations in Figure 12-1.

COMPARISONS OF PLANT METABOLISM

From a comparative analysis of TNT transformation by periwinkle (*Catharanthus roseus*) and parrot feather (*Myriophyllum aquaticum*), significant differences exist in metabolites formed and the final fate of carbon. Additionally there could be large variations in the kinetics of the two systems, the polymerization and binding mechanisms, and other such parameters, but these have not been the emphases of the studies. Both plant systems did not show significant deterioration in health due to the addition of TNT at these levels, showing that both plant systems are capable of attenuating the toxic effects of TNT (Hughes *et al.* 1997, Vanderford *et al.* 1997, Bhadra *et al.* 1999a, Bhadra *et al.* 1999b, Wayment *et al.* 1999). The main difference lies in the type of metabolites produced by both the systems (see Figures 12-3 and 12-4). Aquatic parrot feather (*Myriophyllum aquaticum*) produces significant amounts of oxidative metabolites and trace amounts of azoxy compounds in addition to the reductive and conjugated metabolites, whereas periwinkle (*Catharanthus roseus*) produces only the latter two classes of metabolites. The differences in the metabolic response of the two systems to TNT is suggestive of the vast nature of defenses that the plant may possess in dealing with foreign compounds, which may be indicative of the capability to detoxify a wide spectrum of pollutants (Hatzios and Penner 1982, Hatzios 1991, Durst *et al.* 1997, Salt *et al.* 1998). Another significant difference between the species is the rate of TNT transformation being much faster in parrot feather (*Myriophyllum aquaticum*) than in periwinkle (*Catharanthus roseus*). A kinetic analysis of the two species will quantify these differences.

Metabolite Analysis

Analysis of the unknown concentration distribution, in periwinkle (*Catharanthus roseus*) reveals a higher percentage of unknowns in the intracellular portion as compared to the extracellular (Figure 12-3 provides the total—intra

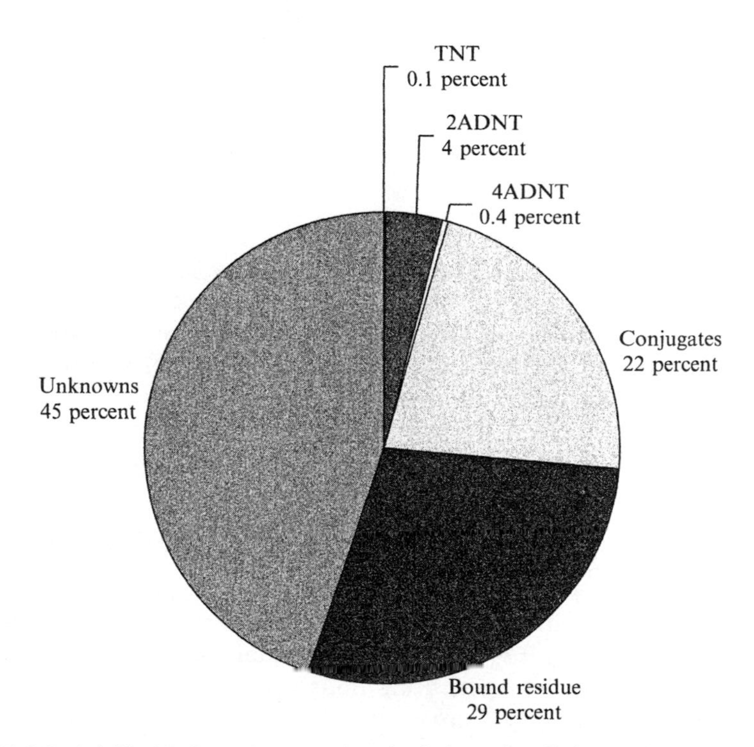

Figure 12-3 Periwinkle (*Catharanthus roseus*) molar balance for all (intracellular and extracellular) metabolites 75 hours after amendment with TNT. Of the total [^{14}C] present, 16 percent was extracellular. Adapted from Bhadra *et al.* (1999b). See definitions of abbreviations in Figure 12-1.

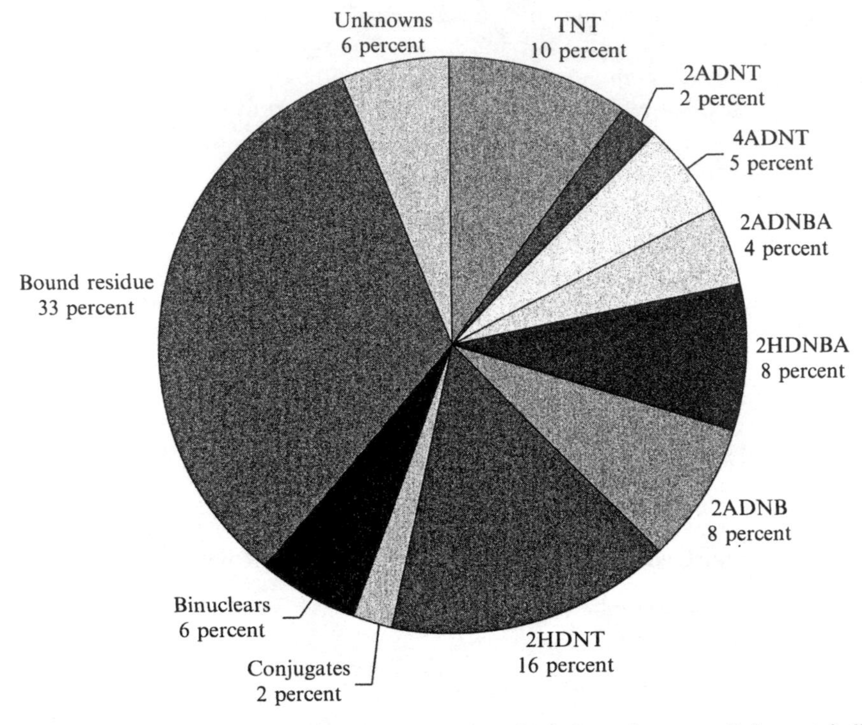

Figure 12-4 Parrot feather (*Myriophyllum aquaticum*) molar balance for extracellular metabolites 12 days after amendment with TNT. Adapted from Bhadra *et. al.* 1999a). See definitions of abbreviations in Figure 12-1.

and extracellular—distribution of [^{14}C]) (Hughes *et al.* 1997, Bhadra *et al.* 1999b). Initially, the unknown extracellular concentration is around 15 percent, but this falls to less that 10 percent at the end of the experiment. In contrast, the unaccounted intracellular fraction increases from around 17 percent to 36 percent (Bhadra *et al.* 1999b). This seems to indicate that the unidentified product in the extracellular phase is continuously transforming into products in the intracellular phase. The unaccounted portion of the carbon could be unidentified conjugates, other reduced or oxidized metabolites, or conjugates formed by bonding of the methyl group to plant organic molecules. The methyl group could be modified, *e.g.*, oxidized, before conjugation. A very dynamic state also exists in the roots and the surrounding media, with the concentration of all the metabolites changing constantly. One of the final transformation residues formed could be a inextricable bound residue, with all other products being transformed. This is again in keeping with the green-liver model, which predicts a sequestration of products as the final detoxification result.

The absence of identified oxidative products in periwinkle (*Catharanthus roseus*) could be due to one of several reasons. Studies with hairy roots have shown that growing in clumps could significantly reduce the concentration of oxygen available (Yu *et al.* 1997). Experiments have shown oxygen deprivation of the roots in shake-flask cultures is a natural phenomena, and this manifests as deficient growth of the roots (Yu *et al.* 1997). Cytochrome P-450 monooxygenases need dioxygen, both as a substrate and probably as an inducer (Schuler 1996). Hence, the observed low oxygen levels could prevent activation of these enzymes. Other reasons for the absence of oxidative products could include the fact that oxidative products could have a high reactivity and hence a short lifetime. Alternatively, the conditions for reduction may have been far more favorable than for the oxidative reductions and hence precluded any oxidative reactions. There may also be significant differences in the enzymes implicated while using whole plants of parrot feather (*Myriophyllum aquaticum*) as opposed to using roots of periwinkle (*Catharanthus roseus*). All these criteria can be manipulated under experimental conditions and hence the hypotheses tested.

In the concentration of inextricable residues and nitroaromatics, parrot feather (*Myriophyllum aquaticum*) also shows much the same trend as periwinkle (*Catharanthus roseus*). However, significantly high concentrations of oxidative products exist (36 percent) (Figure 12-4), which indicates the importance in the transformation process. Demonstration of the involvement in the oxidative reactions will prove useful in further understanding metabolism since cytochrome P-450 has been extensively studied and characterized in herbicide detoxification. Long-term fate and toxicity of the oxidative metabolites is necessary to understand the usefulness of oxidation in the long-term removal of TNT.

REMEDIATION OF OTHER EXPLOSIVES

While TNT remediation has received the most attention and research, other widely used explosives such as RDX and HMX are indeed significant causes for concern. Thus the efficacy of plants in cleaning up these explosives is the topic of this section. Both RDX and HMX are much more recalcitrant to any biologically induced modification and hence eliminating these compounds is much more of a challenge. Various aquatic, axenic, and terrestrial plants have been screened for RDX and HMX removal including parrot feather (*Myriophyllum aquaticum*), Canadian water weed (*Elodea canadensis*), sago pondweed (*Potamogeton pectinatus*), wool grass (*Scirpus cyperinus*), and axenic periwinkle (*Catharanthus roseus*) roots (Best *et al.* 1999, Larson *et al.* 1999a, Larson *et al.* 1999b, Bhadra *et al.* 2001) with highly contrasting results.

Results with HMX have indicated significant resistance to biological transformation. No significant uptake or transformation of HMX occurred after 16 days of exposure to parrot feather (*Myriophyllum aquaticum*), while periwinkle (*Catharanthus roseus*) exhibited minimal uptake after 60 days of exposure. Removal by dead plants and photolytic degradation of HMX was also below 10 percent, further indicative of high chemical stability (Bhadra *et al.* 2001). The significant difference observed in HMX removal in live roots *versus* heat-killed controls after such an extended length of time may mean that plant uptake of HMX is an induced effect, not a constitutive property (Bhadra *et al.* 2001). No metabolites of HMX transformation were identified and no mineralization was observed.

On the other hand, RDX can be removed from aqueous media by various plant species. Aquatic parrot feather (*Myriophyllum aquaticum*) removed more than 65 percent of initial RDX from the medium, while as Figure 12-5 shows, periwinkle (*Catharanthus roseus*) roots removed only, on an average, 50 percent of the initial RDX (Bhadra *et al.* 2001). In other submergent and emergent aquatic species, an average of 20 to 30 percent of RDX was removed from the groundwater, although in the case of sago pondweed (*Potamogeton pectinatus*) and wool grass (*Scirpus cyperinus*), nearly complete RDX uptake was seen (Best *et al.* 1999). This makes these two species promising candidates for further remediation studies, and may also be an indication of different vascular and uptake systems or difference in enzyme activities. Further investigation of the factors governing RDX uptake is warranted.

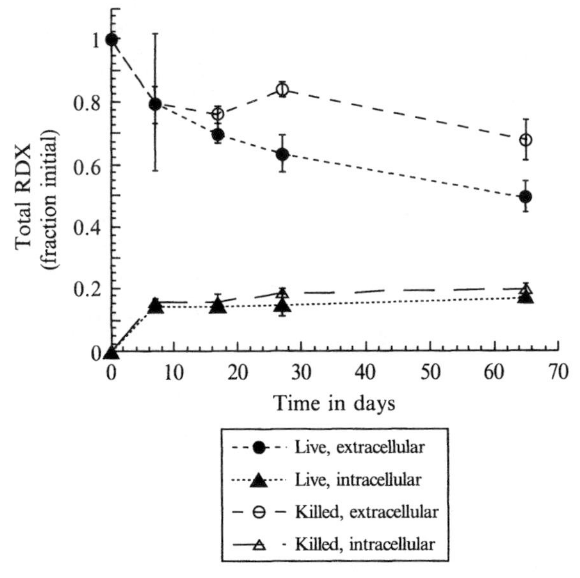

Figure 12-5 Levels of extracellular and intracellular RDX in axenic root cultures of the terrestrial plant periwinkle (*Catharanthus roseus*) and in heat-killed controls. RDX recovery from both aqueous and heat-killed controls was close to initial amendment over the duration of the experiment. The initial RDX level was 7.28 milligrams per liter. Data are the average of triplicate root cultures or controls. Reprinted with permission from Bhadra *et al.* (2001) *Chemosphere* 44 (2001) Studies on plant mediated fate of the explosives RDX and HMX. **44**(5): 1259–1264. Copyright. Elsevier Science.

In laboratory studies, RDX accumulates in plant tissue (Larson *et al.* 1999a, Burken *et al.* 2000, Bhadra *et al.* 2001). In periwinkle (*Catharanthus roseus*) hairy roots, [^{14}C] mass balances indicate that bound residues from RDX also accumulate within root tissue (Bhadra *et al.* 2001). Sampling of plants that had been exposed to RDX at the Iowa Army Ammunition Plant, however, revealed that RDX was below detection limits (Burken *et al.* 2000). Additional mechanisms may have removed RDX from aqueous media before uptake, or perhaps additional factors in the field that are not present in the lab, *i.e.*, sunlight, may result in degradation in plant tissue (Burken *et al.* 2000).

These preliminary investigations of RDX and HMX remediation indicate the potential of plants to remediate RDX contaminated waters, although to limited extents. Further identification of metabolites formed, schemes to improve RDX uptake by plants, and complete plant mass balances are necessary to fully understand the phenomena, especially in plants exposed to sunlight. However, HMX remediation presents a greater challenge. Further studies on these explosives and other related explosives might answer why some explosives are readily transformed, and others are more recalcitrant.

CONCLUSION

Pathway studies of TNT transformation have helped shed light on the mechanisms, bottlenecks, and future strategies for using phytoremediation in the field. Chief amongst these findings are that plants are indeed an efficient system for detoxifying land and water resources. Nevertheless, the potential for commercial phytoremediation will require additional basic research. Plants do not mineralize the explosive TNT and various multi-natured metabolites are formed with varying toxicities and final fates. Enzymatic transformation of TNT does indeed take place, and elucidating these enzymes and developing mutants are of immediate interest. Removal of RDX occurs to a lesser extent, and HMX displays much greater recalcitrance. Toxicity of the metabolites formed and the products are of concern and need further investigation. Various plant systems deal with explosives differently, although some fundamental steps appear the same. Plants that are selected or engineered for enhanced rates of transformation and decreased levels of phytotoxicity are likely to be valuable in a commercial phytoremediation process.

Acknowledgments

J. Shanks extends thanks to S. McCutcheon for providing the opportunity to work in this field, to J. Hughes for a fruitful and fun collaboration, and to R. Bhadra, D. Wayment, M. Vanderford, J. Lauritzen, M. Subramanian, R. Williams, M. Barman, and M. Stone for their dedicated work. The authors extend thanks to Cargill, for fellowship support for M. Subramanian.

REFERENCES

Ahmad, F. and J.B. Hughes (2000) Anaerobic transformation of TNT by *Clostridium*. In: *Biodegradation of Nitroaromatic Compounds and Explosives*. J.C. Spain, J.B. Hughes, and H.J. Knackmuss, eds. Lewis Publishers, Boca Raton, Florida, pp. 185–212.

Best, E.P.H., M.E. Zappi, H.L. Fredrickson, S.L. Sprecher, S.L. Larson, and M. Ochman (1997) Screening of aquatic and wetland plant species for phytoremediation of explosives-contaminated groundwater from the Iowa Army Ammunition Plant. In: *Bioremediation of Surface and Subsurface Contamination*. R.K. Bajpai and M.E. Zappi eds. *Ann. New York Acad. Sci.*, New York, **829**: 179–194.

Best, E.P.H., S.L. Sprecher, S.L. Larson, H.L. Fredrickson, and D.F. Bader (1999) Environmental behavior of explosives in groundwater from the Milan army ammunition plant in aquatic and wetland plant treatments. Removal, mass balances and fate in groundwater of TNT and RDX. *Chemosphere* **38**(14): 3383–3396.

Bhadra, R., R.J. Spanggord, D.G. Wayment, J.B. Hughes, and J.V. Shanks (1999a) Characterization of Oxidation products of TNT metabolism in aquatic phytoremediation systems of *Myriophyllum aquaticum*. *Environ. Sci. Technol.* **33**: 3354–3361.

Bhadra, R., D.G. Wayment, J.B. Hughes, and J.V. Shanks (1999b) Confirmation of conjugation processes during TNT metabolism by axenic roots. *Environ. Sci. Technol.* **33**: 446–452.

Bhadra, R., D.G. Wayment, R.K. Williams, S.N. Barman, M.B. Stone, J.B. Hughes, and J.V. Shanks (2001) Studies on plant-mediated fate of the explosives RDX and HMX. *Chemosphere* **44**(5): 1259–1264.

Burken, J.G. and J.L. Schnoor (1998) Predictive relationships for uptake of organic contaminants by hybrid poplar trees. *Environ. Sci. Technol.* **32**: 3379–3385.

Burken, J.G., J.V. Shanks, and P.L. Thompson (2000) Phytoremediation and plant metabolism of explosives and nitroaromatic compounds. In: *Biodegradation of Nitroaromatic Compounds and Explosives*. J.C. Spain, J.B. Hughes, and H.J. Knackmuss, eds. Lewis Publishers, Boca Raton, Florida, pp. 239–275.

Durst, F., I. Benveniste, A. Lesot, J.P. Salaun, and D.W. Reichhart (1997) Induction of plant cytochrome P-450. In: *Regulation of Enzymatic Systems Detoxifying Xenobiotics in Plants*. K.K. Hatzios, ed. Kluwer Academic Publishers, Dordrecht, the Netherlands, pp.19–34.

French, C.E., S.J. Rosser, G.J. Davies, S. Nicklin, and N.C. Bruce (1999) Biodegradation of explosives by transgenic plants expressing pentaerythritol tetranitrate reductase. *Nat. Biotechnol.* **17**: 491–494.

Fritsche, W., K. Scheibner, A. Herre, and M. Hofrichter (2000) Fungal degradation of explosives: TNT and related nitroaromatic compounds. In: *Biodegradation of Nitroaromatic Compounds and Explosives*. J.C. Spain, J.B. Hughes, and H.J. Knackmuss, eds. Lewis Publishers, Boca Raton, Florida, pp. 213–237.

Gorge, E., S. Brandt, and D. Werner (1994) Uptake and metabolism of 2,4,6 TNT in higher plants. *Environ. Sci. Pollut. Res.* **1**: 229–233.

Harvey, S., R.J. Fellows, D.A. Cataldo, and R.M. Bean (1990) Analysis of 2,4,6-trinitrotoluene and its transformation products in soils and plant tissues by high performance liquid chromatography. *J. Chromatogr.* **518**: 361–374.

Hatzios, K.K. (1991) Biotransformation of herbicides in higher plants. In: *Environmental Chemistry of Herbicides*. A.J. Cessna, ed. CRC Press, Boca Raton, Florida, **2**: 141–185.

Hatzios, K.K. and D. Penner (1982) *Metabolism of Herbicides in Higher Plants*. Bugess Publishing Company, Minneapolis, Minnesota.

Hughes, J.B., J.V. Shanks, M. Vanderford, J. Lauritzen, and R. Bhadra (1997) Transformation of TNT by aquatic plants and plant tissue cultures. *Environ. Sci. Technol.* **31**(1): 266–271.

Khan, S.U. (1988) Bound Residues. In: *Environmental Chemistry of Herbicides*. A.J. Cessna, ed. CRC Press, Boca Raton, Florida, **2**: 265–280.

Larson, S.L. (1997) Fate of explosive contaminants in plants. In: *Bioremediation of Surface and Subsurface Contamination*. R.K. Bajpai and M.E. Zappi, eds. *Ann. New York Acad. Sci.*, New York, **829**: 195–201.

Larson, S.L., R.P. Jones, L. Escalon, and D. Parker (1999a) Classification of explosives transformation products in plant tissue. *Environ. Toxicol. Chem.* **18**(6): 1270–1276.

Larson, S.L., C.A. Weiss, L.B. Escalon, and D.B. Parker (1999b) Increased extraction efficiency of acetonitrile/water mixtures for explosive determination in plant tissues. *Chemosphere* **38**(9): 2153–2162.

Nagel, D.B., S. Scheffer, B. Casper, H. Garn, O. Drzyzga, E.V. Low, and D. Gemsa (1999a) Effect of 2,4,6-trinitrotoluene and its metabolites on human monocytes. *Environ. Sci. Technol.* **33**: 2566–2570.

Nagel, D.B., T.C. Schmidt, O. Drzyzga, E.V. Low, and K. Steinbach (1999b) Identification of oxidized TNT metabolites in soil samples of a former ammunition plant. *Environ. Sci. Pollut. Res.* **6**(1): 7–10.

Palazzo, A.J. and D.C. Leggett (1986) Effect and disposition of TNT in terrestrial plants. *J. Environ. Qual.* **15**(1): 49–52.

Pavlostathis, S.G., K.K. Comstock, M.E. Jacobson, and M.F. Saunders (1998) Transformation of 2,4,6-trinitrotoluene by the aquatic plant *Myriophyllum spicaticum*. *Environ. Toxicol. Chem.* **17**(11): 2266–2273.

Rieger, P.-G. and H.-J. Knackmuss (1995) Basic knowledge and perspectives on biodegradation of 2,4,6-trinitrotoluene and related nitroaromatic compounds in contaminated soil. In: *Biodegradation of Nitroaromatic Compounds*. J.C. Spain, ed. Plenum Press, New York, pp. 1–18.

Rivera, R., V.F. Medina, S.L. Larson, and S.C. McCutcheon (1998) Phytotreatment of TNT-contaminated groundwater. *J. Soil Contam.* **7**(4): 511–529.

Salt, D.E., R.D. Smith, and I. Raskin (1998) Phytoremediation. In: *Annual Review of Plant Physiology and Plant Molecular Biology*. Annual Reviews, Palo Alto, California: **49**: 643–648.

Sandermann, J.H. (1994) Higher plant metabolism of xenobiotics: the 'green liver' concept. *Pharmacogenetics* **4**: 225–241.

Schalk, M., M.A. Pierrel, A. Zimmerlin, Y. Batard, F. Durst, and D.W. Reichhart (1997) Xenobiotics: substrates and inhibitors of the plant cytochrome P-450. *Environ. Sci. Pollut. Res.* **4**(4): 229–234.

Scheidemann, P., A. Klunk, C. Sens, and D. Werner (1998) Species dependent uptake and tolerance of nitroaromatic compounds by higher plants. *J. Plant Physiol.* **152**(2–3): 242–247.

Schnoor, J.L., L.A. Licht, S.C. McCutcheon, N.L. Wolfe, and L. H. Carreira (1995) Phytoremediation of organic and nutrient contaminants. *Environ. Sci. Technol.* **29**: 318A–323A.

Schuler, M.A. (1996) Plant cytochrome P-450 monooxygenases. *Crit. Rev. Plant Sci.* **15**(3): 235–284.

Sens, C., P. Scheidemann, and D. Werner (1999) The distribution of ^{14}C-TNT in different biochemical compartments of the monocotyledonous *Triticum aestivum*. *Environ. Pollut.* **104**: 113–119.

Shah, M.M. and J.A. Campbell (1997) Transformation of nitrobenzene by ferredoxin NADP oxidoreductase from spinach leaves. *Biochem. Biophys. Res. Comm.* **241**: 694–796.

Shanks, J.V. and J. Morgan (1999) Plant 'hairy root' culture. *Curr. Opin. Biotechnol.* **10**(2): 151–155.

Spain, J.C. (2000) Introduction. In: *Biodegradation of Nitroaromatic Compounds and Explosives*. J.C. Spain, J.B. Hughes, and H.J. Knackmuss, eds. Lewis Publishers, Boca Raton, Florida, pp. 1–5.

Subramanian, M. and J.V. Shanks. Derivatization of 2 and 4-hydroxylamine-dinitrotoluenes for detection in plant tissue cultures for TNT-transformation studies. *Manuscript in preparation.*

Tadros, M.G., A. Crawford, A. Mateo-Sullivan, C. Zhang, and J.B. Hughes (2000) Toxic effects of hydroxylamino intermediates from microbial transformation of trinitrotoluene and dinitrotoluenes on algae *Selenastrum capricornutum*. *Bull. Environ. Contam. Toxicol.* **64**: 579–585.

Thompson, P.L., L. Ramer, and J.L. Schnoor (1998) Uptake and transformation of TNT by hybrid poplar trees. *Environ. Sci. Technol.* **32**: 975–980.

Vanderford, M., J.V. Shanks, and J.B. Hughes (1997) Phytotransformation of trinitrotoluene (TNT) and distribution of metabolic products in *Myriophyllum aquaticum*. *Biotechnol. Lett.* **19**(3): 277–280.

Wang, C. and J.B. Hughes (1998) Derivatization and separation of 2,4,6-trinitrotoluene metabolic products. *Biotechnol. Tech.* **12**(11): 839–842.

Wayment, D.G., R. Bhadra, J. Lauritzen, J.B. Hughes, and J.V. Shanks (1999) A transient study of formation of conjugates during TNT metabolism by plant tissues. *Int. J. Phytoremed.* **1**(3): 227–239.

Wolfe, N.L., T.Y. Ou, L. Carriera, and D. Gunnison (1994) Alternative methods for biological destruction of TNT: a preliminary feasibility assessment of enzymatic degradation. U.S. Army Corps of Engineers Waterways Technical Report IRRP-94-3, Vicksburg, Mississippi.

Won, W.D., L.H. DiSalvo, and J. Ng (1976) Toxicity and mutagenicity of 2,4,6-trinitrotoluene and its microbial metabolites. *Appl. Environ. Microbiol.* **31**: 575–580.

Yinon, J. (1990) *Toxicity and Metabolism of Explosives*. CRC Press, Boca Raton, Florida.

Yu, S., M.G.P. Mahagamasekera, G.R.C. Williams, K. Kanokwaree, and P.M. Doran (1997) Oxygen effects in hairy root culture. In: *Hairy Roots-Culture and Applications*. P.M. Doran, ed. Harwood Academic Publishers, Amsterdam, the Netherlands, pp. 139–150.

13

TRANSFORMATION KINETICS OF TRINITROTOLUENE CONVERSION IN AQUATIC PLANTS

M. E. Jacobson, S. Y. Chiang, L. Gueriguian, L. R. Westholm, J. Pierson, G. Zhu, and F. M. Saunders

SUMMARY OF PRACTICAL IMPLICATIONS

Engineered aquatic plant consortia are considered good candidates for treating groundwater contamination and waste streams. Laboratory and outdoor test were used in developing bench and field-scale pilot-testing protocols for the design of aquatic plant lagoons to treat water contaminated with 2,4,6-trinitrotoluene (TNT). Pilot testing evaluated different species of plants as well as different timescales of exposure. The common aquatic plants that were tested, removed TNT from contaminated water rapidly but with the accumulation of aminodinitrotoluene (ADNT). The accumulation is consistent with other observations and is probably the result of reactor volumes, flow rates, as well as the initial influent concentrations (10 milligrams of TNT per liter) used in these tests. The results from pilot scale testing indicated that created treatment wetlands are ready to treat groundwater with monitoring of by-products and optimization of some design parameters. Designs should involve optimization of plant density, residence time, and flow rate and require at least 2 years for by-product monitoring.

INTRODUCTION

Phytoremediation is an alternative method for decontamination of groundwater and waste streams generated during the production of explosives

Phytoremediation: Transformation and Control of Contaminants,
Edited by Steven C. McCutcheon and Jerald L. Schnoor.
ISBN 0-471-39435-1 (cloth)

(Cataldo *et al.* 1989, Harvey *et al.* 1991, Harvey *et al.* 1993, Schnoor *et al.* 1995, Hughes *et al.* 1997, Hughes *et al.* 1998, Pavlostathis *et al.* 1998). Found in waste streams at munition sites, TNT rapidly disappears from contaminated water in the presence of selected plants (Palazzo and Leggett 1986, Harvey *et al.* 1991, Wolfe *et al.* 1994, Schnoor *et al.* 1995, Hughes *et al.* 1997). Products of TNT metabolism by plants (phytotransformation) temporarily accumulate in tissue and some are released into the aqueous phase over time (Palazzo and Leggett 1986, Medina *et al.* 1996, Hughes *et al.* 1997, Price *et al.* 1997, Vanderford *et al.* 1997, Bhadra *et al.* 1999a). These plant-associated transformation products are dominated by aminonitrotoluenes (Bhadra *et al.* 1999a). As plant exposure times increase, inextricable and conjugated components derived from the initial TNT introduction become associated with plant material (Bhadra *et al.* 1999a).

Aquatic plant communities are good candidates for treating groundwater contamination in engineered systems because of the large surface area, ease of culture, and high organic uptake potential (Brix 1989, McIntyre and Lewis 1997, Schnoor 1997). Plants are photosynthetic autotrophs; consequently, energy requirements are easily met in natural and engineered environments. Some aquatic plants offer the additional benefit of an opportunistic lifestyle of rapid growth and metabolism. In addition, communities of submerged aquatic macrophytes exhibit similarity in metabolic function but diversity in light, temperature, and nutrient requirements, enabling development of remediation schemes to be applied in a natural setting of species succession on a seasonal and longer-term basis. This chapter evaluates the capabilities of selected monocultures and communities of plants exposed to TNT in both short (days) and long-term (weeks) laboratory kinetic studies, as well as in an extended term (more than a month) pilot-scale field study. These approaches were taken to compare the effect of TNT on different species of plants over varying periods of treatment.

For short-term studies, aqueous phase information is presented from laboratory experiments on five aquatic plant species: waterweed (*Elodea canadensis* Rich.), floating moss (*Salvinia rotundifolia*), yellow iris (*Iris pseudacorus* L.), pickerelweed (*Pontederia cordata* L.), and Chinese sedge (*Eleocharis tuberosa*). Information from longer-term laboratory exposures on the transformation of TNT and retention of transformation products by plant tissues is presented for *Myriophyllum spicatum* (Eurasian water milfoil). Temporary plant tissue accumulation is based on both aqueous phase removal of TNT and analysis of extracted plant materials. Results from pilot-scale field studies conducted in engineered freshwater lagoons are also presented. Plants in this study include two monocultures of waterweed (*Elodea canadensis* Rich.) and parrot feather (*Myriophyllum aquaticum*), mixed cultures of Nymphaeaceae (water lilies) including chinese or East Italian lotus (*Nelumbo nucifera*), American or white fragrant water lily (*Nymphea odorata*), and *Lotus* spp., and a mixed marsh community.

MATERIALS AND METHODS

Short-Term Laboratory Tests

Short-term testing involved waterweed (*Elodea canadensis* Rich.), floating moss (*Salvinia rotundifolia*), yellow iris (*Iris pseudacorus* L.), pickerelweed (*Pontederia cordata* L.), and Chinese water chestnut (*Eleocharis tuberosa*). All plants were obtained from Rocky Shoals Nursery in Fairburn, Georgia, and maintained in an outdoor nursery on the Georgia Institute of Technology campus prior to testing. Plants were chosen for ease of culture, to be characteristic of a number of ecological habitats, and to represent a variety of root to shoot ratios.

Batch experiments were conducted at temperatures between 23° and 25°C. A cycle of 16 hours of light and 8 hours of dark was maintained throughout experiments using 75-watt Gro and Show lamps. Solutions were prepared at 10 milligrams of TNT per liter (ChemService, West Chester, Pennsylvania, 99 percent purity) and buffered to a pH of approximately 7.5 with an excess of sodium bicarbonate. This excess of bicarbonate (approximated by calculations of carbon fixation rates and biomass accumulation) acted as a carbon source for some submerged species. All experiments were run in triplicate in either 1- or 2-liter polypropylene beakers depending on the plant size. Control tests, with no plant, were initiated in parallel with plant experiments to measure other processes related to the transformation of TNT (*e.g.*, adsorption, photodegradation, and bacterial transformation). Plant mass was measured at the beginning of the experiment, and the growth of plants was negligible through the course of the batch experiments. All experimental reactors were gently bubbled with a small aeration stone to simulate mixing found in natural, slow-flow environments. The pH of the aqueous phase was measured daily.

Removal of TNT and release of transformation products was measured by analysis of 2 -milliliter aqueous samples taken at selected times from each reactor throughout the study for treatment times of up to 200 hours. Samples were centrifuged (Eppendorf® model 5415C) at 14 000 revolutions per minute (relative centrifugal field of 27 000 times standard gravity) and the resulting supernatant transferred to high performance liquid chromatography vials. All samples were immediately refrigerated in the dark until analyzed. Analysis was performed within a week of sampling.

Concentrations of TNT and transformation products were measured using a Hewlett Packard™ Series 1050 high performance liquid chromatography system. The system consisted of a gradient controller and mobile phase pump, an automatic injector, a photodiode array detector, and a personal computer loaded with Hewlett Packard® ChemStation™ software to perform the analysis of samples. A 250 millimeters by 4.6 millimeters Supelco™ LC-18-DB reverse-phase column (Supelco, Inc.), interfaced with an All-Guard Adsorbosphere C-18-5U, 150 millimeters guard column (Altech Associates, Inc.), was used for

TNT and transformation product identification. The method operated with a run time of 10 minutes and a mobile phase flow of 1.0 millimeter per minute. The mobile phase consisted of 70 percent acetonitrile and 30 percent 0.005 moles of potassium chloride per liter, the latter used to minimize tailing and stabilize retention times. The detection of TNT was based on absorbance measurements between 200 to 450 nanometers.

A TNT calibration standard and a blank were run at the beginning and after every 20 sample injections as a quality assurance check for retention times and instrument precision. The detection limit for TNT was 0.01 milligram per liter. In addition to TNT, the following transformation products were also measured by high performance liquid chromatography analysis: 2-amino-4,6-dinitrotoluene, 4-amino-2,6-dinitrotoluene, 4-hydroxylamino-2,6-dinitrotoluene, 2,4-diamino-6-nitrotoluene, 2,2′-azoxy-4,4′,6,6′-tetranitrotoluene, 2,4′-azoxy-2′,4,6,6′-tetranitrotoluene, and 4,4′-azoxy-2,2′,6,6′-tetranitrotoluene. The isomers 2-amino-4,6-dinitrotoluene and 4-amino-2,6-dinitrotoluene were observed to coelute and, therefore, will be referred to collectively in this chapter as ADNT. All standards for these analyses were purchased from SRI International™, Menlo Park, California.

Long-Term Studies and Biomass Analysis

A batch exposure investigation of 21 days duration was initiated using Eurasian water milfoil (*Myriophyllum spicatum*). The initial exposure concentration was 10 milligrams of TNT per liter or approximately 40 micromoles per liter. The tests were conducted in triplicate batch reactors. Reactors were sacrificed at 0, 0.5, 3, 7, 24, 48, 73, 170, 335, and 506 hours. At each time, accumulation of TNT and associated transformation products found in the plant tissue and concentrations in the aqueous phase were measured. For this test, adsorbed and transformed plant-associated materials were treated as one pool. In order to determine plant-associated TNT analytes, extraction of plant material was done. This consisted of freezing plants in liquid nitrogen, grinding, and adding acetonitrile (5 milliliters per gram of plant wet weight) to a weighed aliquot, shaking overnight, centrifuging and removing the aqueous extract, followed by fractionating of the subsample with a C-18 separation cartridge. Because excess chlorophyll interfered with the performance by binding to the chromatography column irreversibly, pretreatment of samples using a separation cartridge was necessary. The separation cartridge was preconditioned with solvent at the same ratio of solvent to water as the sample, and fractions were collected at time intervals determined by running standards of TNT and by-products through the cartridge. Extraction efficiency ranging from 69 percent to 75 percent was determined by adding a known concentration of standard to the sample. After separation, the aliquots were analyzed by high performance liquid chromatography as outlined previously.

Pilot-Scale Field Study Reactor Design and Sampling

The pilot-scale field system was composed of a main structure, 6.1 by 3.7 by 1.2 meters (20 by 12 by 4 feet), containing several internal cells (Figure 13-1). Five internal cells manufactured from high-density polyethylene and having an approximate working volume for each reactor of up to 662.4 liters (175 galons), contained the aquatic vegetation used to treat TNT. Cell volumes were adjusted based on plant growth requirements. Control cells, containing plants grown with no TNT, included four separate 60.8 cubic centimeters (1000 cubic inches) low-density polyethylene basins and three separate 94.6-liter (25-galon) high-density polyethylene cylindrical basins. These seven cells contained monocultures as well as mixed marsh communities, and functioned as controls for any environmental effects on the plants not related to TNT exposure. The plants tested included (1) monocultures of waterweed (*Elodea canadensis*), (2) monocultures of Hydrilla (*Hydrilla verticillata*), (3) monocultures of parrot feather (*Myriophyllum aquaticum*), (4) a mixed culture of Nymphaeaceae (water lilies) including Chinese or East Indian lotus (*Nelumbo nucifera*), American white or fragrant water lily (*Nymphaea odorata*), and lotus (*Lotus* spp.), and (5) a mixed marsh community consisting of blue broad leaf pickerelweed (*Pontederia cordata*), Chinese water chestnut (*Eleocharis tuberosa*), Nardoo (*Marsilea drummondii*), and iris (*Iris* spp.) Influent and effluent flows to and from the individual cells were controlled and subsurface water flow percolated through fieldstones, the

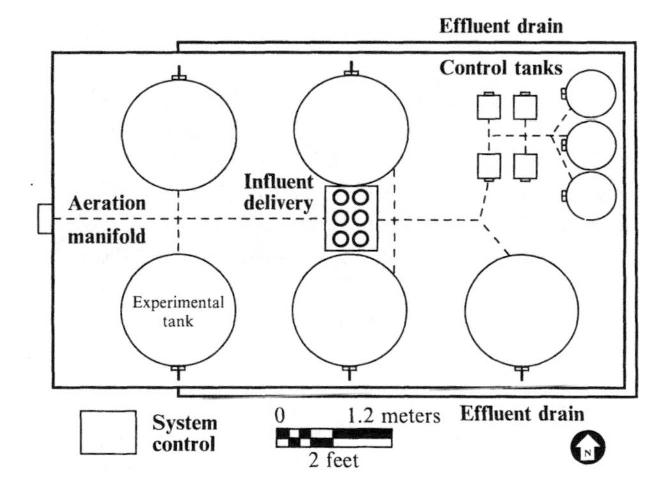

Figure 13-1 Plan view of outdoor field reactors. A temperature-controlled, rock-filled containment vessel surrounded all reactors. The five large tanks contained plants as described in the section "Materials and Methods," "Pilot-Scale Field Study Reactor Design and Sampling," that were continuously fed a constant flow of 10 milligrams of TNT per liter for 1200 hours. Other tanks in the containment area held plants that were not subjected to TNT exposure but were exposed to natural climate. Sampling was done through an effluent drain at the bottom of each cell and from above. All reactors were gently mixed by wind and aeration.

latter used to dampen the effects of temperature. Aeration provided cell mixing so as to distribute more uniformly the influent wastewater and to control vectors such as mosquitoes.

The continuous flow introduced to each tank (including the small control tanks) a soluble TNT concentration of 1 milligram of TNT per liter during August 21 to September 28, 1995 and 10 milligrams per liter during October 1 to November 28, 1995. The hydraulic retention time was 10 days in all tanks. The influent feed system stored and distributed TNT to the plant lagoons simultaneously while maintaining matching hydraulic residence times. Data acquisition and control systems were utilized to provide TNT loading and monitor temperature, pH, rainfall, and sunlight intensity. A LI-COR® LI-190SA terrestrial sensor, placed approximately 2.4-meters (7.3-feet) above the surface of fieldstone in the aquatic plant lagoon, measured ambient sunlight and served as a reference reading. A LI-COR® LI-192SA underwater sensor measured sunlight intensity experienced at various depths within the different tanks. Both the data from radiation sensors and the rain gauge data were recorded using a LI-COR® LI-1000 datalogger. A more detailed description of the aquatic plant lagoon design and data is available from J. Pierson (john.pierson@gtri.gatech.edu).

BENCH AND PILOT RESULTS

Laboratory Studies

Control Reactors

Plant-free control reactors containing TNT were monitored for up to 200 hours. These control reactors did not show a significant decrease in TNT concentration with time. Decrease of TNT when observed ranged from a low of 1.8 percent to a high of 6.2 percent and may be a result of photodegradation, adsorption to the container, and microbial activity. No change in pH was observed in the control or test reactors throughout the experiment.

Short-Term Reactors: Removal from the Aqueous Phase

The disappearance of TNT and subsequent appearance of TNT transformation products in the aqueous phase were observed in all plant species (Table 13-1 and Figure 13-2). The most abundant and consistent transformation product observed was ADNT. The data for TNT disappearance from the aqueous phase were fitted to a single, two-parameter exponential decay function. The underlying basis of this model was that no significant biomass change occurred during the tests. The ADNT appearance was fit to a single, two-parameter exponential function for the rise to a maximum. This was a simple best fit and was not linked to the removal of TNT from solution. The equation for TNT disappearance is

$$[TNT] = [TNT]_0 \exp(-kt) \tag{13-1}$$

and for ADNT appearance

$$[ADNT] = [ADNT]_{max} [1 - \exp(k_2 t)] \tag{13-2}$$

where $[TNT]$ or $[ADNT]$ are the concentrations of TNT or ADNT (micromoles per liter or milligrams per liter) at time t (hours), $[TNT]_0$ is the TNT concentration at time zero (micromoles per liter or milligrams per liter), k is the pseudo first order rate constant (per hour), $[ADNT]_{max}$ is the maximum observed ADNT concentration (micromoles per liter or milligrams per liter), and k_2 is a pseudo first order rate constant (per hour) for ADNT appearance. Furthermore, Gueriguian (1996) and Pavlostathis *et al.* (1998) noted that plant density affects the rate of TNT removal from reactors as well as transformation product formation. Therefore, the pseudo first order rate constant k must be equal to k^*B for TNT disappearance and k_2 is equal to $k_2^* B$ for ADNT appearance, in which B is the plant density (grams per liter). As a result, k^* is equal to k/B that is the second order or biomass-normalized TNT disappearance rate constant (liters per gram per hour) and k_2^* is the biomass-normalized ADNT appearance rate constant (liters per gram per hour) (Table 13-1).

These expressions of the best fit do not necessarily represent physiological mechanisms involved in transformation or TNT removal. Plant density affected the rate of TNT removal from reactors as well as transformation product formation (Gueriguian 1996, Pavlostathis *et al.* 1998). All kinetic data were normalized for plant mass by dividing the fitted pseudo first order kinetic rate constants by plant density (Table 13-1). Once the kinetic rate constants were normalized

TABLE 13-1 Pseudo First Order Removal Kinetics and Kinetics Normalized to Total Biomass for Plants Exposed to Trinitrotoluene

Plant species	Trinitrotoluene added, $[TNT]_0$ (milligrams per liter)	First order rate constant, k (per hour)[a]	Normalized rate constant for total plant biomass, k^*(liters per gram per hour)
Chinese water chestnut (*Eleocharis tuberosa*)	11.0	0.027 (0.97)	0.0008
Floating moss (*Salvinia rotundifolia*)	9.3	0.059 (0.94)	0.0015
Yellow iris (*Iris pseudacorus* L.)	11.0	0.034 (0.97)	0.0010
Blue broad leaf pickerelweed (*Pontederia cordata* L.)	11.0	0.020 (0.96)	0.0006
Waterweed (*Elodea canadensis* Rich.)	9.3	0.10 (0.99)	0.0031

[a]Values in parentheses represent the correlation coefficient r^2.

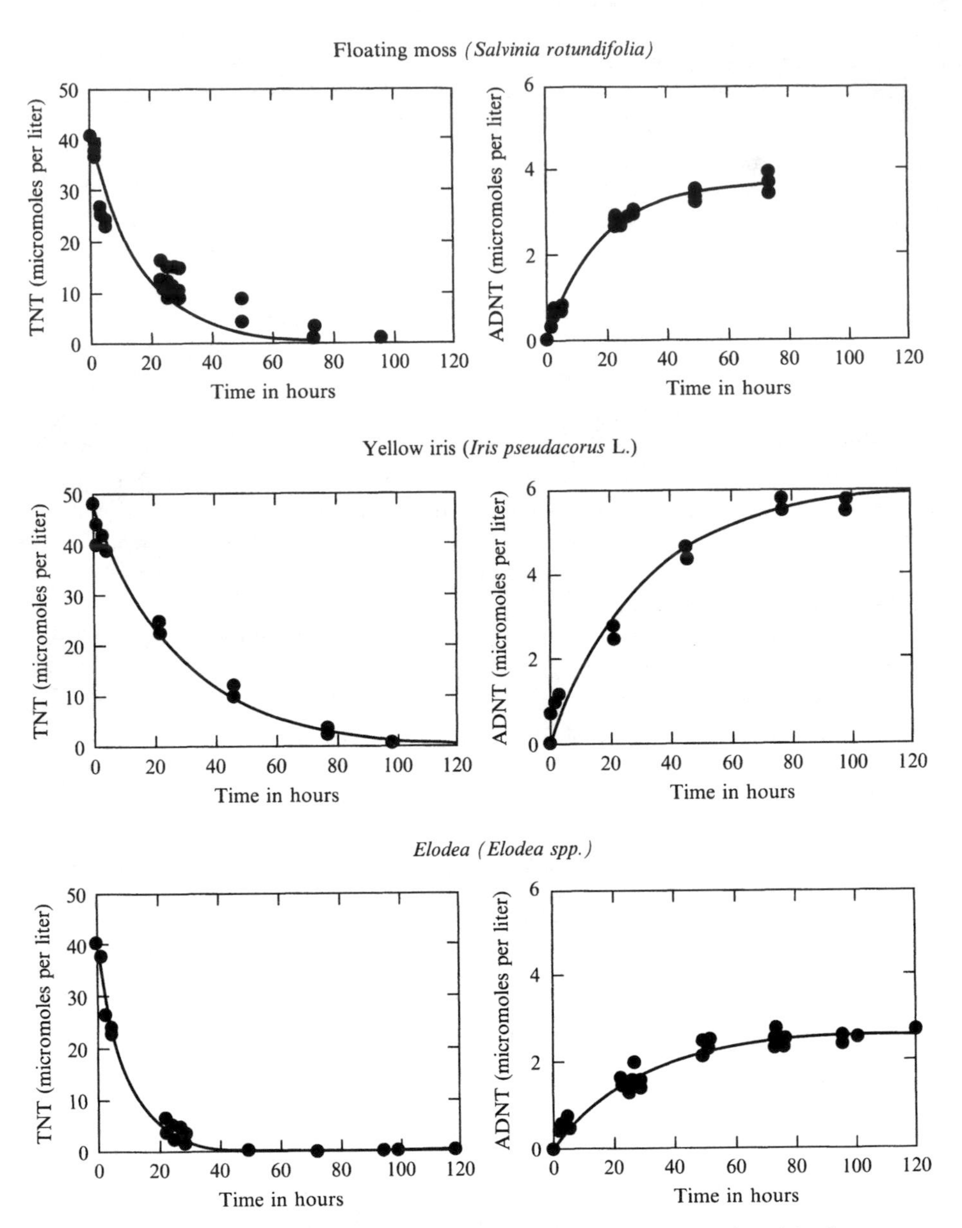

Figure 13-2 Trinitrotoluene removal and aminodinitrotoluene production found in the aqueous phase of three representative species exposed to 10 milligrams of TNT per liter for 120 hours. Three replicate reactors were used for each plant species. Lines of fit are for TNT, generated using the Equation (13-1) that is based on starting concentrations of TNT in the reactor, while the fit for ADNT data is a two-parameter fit of Equation (13-2) based on the maximum observed concentrations over the duration of the tests. Actual rate constants and correlation coefficients (r^2) are found in Tables 13-1 and 13-2 for the lines shown. Note that 1 micromole of TNT = 0.22713 milligram of TNT per liter and 1 micromole of ADNT = 0.18215 milligram of ADNT per liter.

for plant mass, the following pattern emerged among the species from highest to lowest biomass-normalized rate constant k^*: *Elodea canadensis* > *Salvinia rotundifolia* > *Iris pseudacorus* > *Pontederia cordata* > *Eleocharis tuberosa.*

The removal of TNT normalized to belowground plant biomass better expresses the effect of biomass that has direct contact with the contaminant in the aqueous phase but these measurements were not available for plants used in these tests. This effect naturally varies with shoot-to-root biomass ratios, which is different for different species. Some variability in a given species may also occur due to different levels of ambient carbon dioxide, nutrients, sediment type, season, and other factors (Roman and Daiber 1984, Thorne-Miller and Harlin 1984, Hsieh 1988, Siddique *et al.* 1990, Hunt 1991, Xu and Kumar 1992, Ericsson 1995, Rogers *et al.* 1996, Franzaring *et al.* 2000). To illustrate the effect of root-to-shoot ratios, salt marsh plants have two-thirds of the total production submerged (Schubauer and Hopkinson 1984). Therefore, waterweed (*Elodea canadensis*), a totally submerged plant, is expected to be more metabolically active on a biomass basis compared to emergent plants.

Short-Term Reactors: Appearance of By-Products in the Aqueous Phase

Aminodinitrotoluene (ADNT) appeared in the aqueous phase within the first 2 hours (Figure 13-2). This indicates a rapid transformation response initiated by the plant or plant-associated epiphytes (Hoehamer 2000). The buildup of ADNT in the aqueous phase is the net sum of plant reduction plus any

TABLE 13-2 Kinetics of Aminodinitrotoluene Appearance for Plants exposed to Trinitrotoluene

Plant species	Trinitrotoluene added, $[TNT]_0$ (milligrams per liter)	Maximum aminodinitrotoluene concentration, $[ADNT]_{max}$ (milligrams per liter)		First order rate constant, k_2 (per hour)[a]
		Observed	Estimated	
Chinese water chestnut (*Eleocharis tuberosa*)	11.0	ND	ND	ND
Floating moss (*Salvinia rotundifolia*)	9.3	0.72	0.67	0.058 (0.99)
Yellow iris (*Iris pseudacorus* L.)	11.0	1.25	1.11	0.33 (0.97)
Blue broad leaf pickerekweed (*Pontederia cordata* L.)	11.0	1.39	ND	ND
Waterweed (*Elodea canadensis* Rich.)	9.3	0.51	0.49	0.37 (0.97)

[a]Correlation coefficient in parentheses.

Abbreviation used—: ND: no data available.

microbial reduction minus uptake into the plant. Waterweed (*Elodea canadensis*), yellow iris (*Iris pseudacorus*), and floating moss (*Salvinia rotundifolia*) exhibited high net releases of ADNT into the aqueous phase (Table13-2). The net release of ADNT into the aqueous phase over time indicates that the uptake and transformation of TNT *via* reductive processes occur in the plant environment. Diaminonitrotoluenes and 2,2′-azoxy-4,4′,6,6′-tetranitrotoluene were also measured in the aqueous phase (data not shown). The diaminonitrotoluenes were a transient detected after 21 hours of incubation in all plants, and then tissue-accumulated diaminonitrotoluenes were detected transiently after 21 hours in all plants. Diaminonitrotoluenes dropped below detection limits for all other measurement times. Neither of these products represents a significant amount (less than 1 percent of the TNT added). In fact, these products did not close the mass balance derived from the initial TNT concentrations, implying that further transformation products are yet to be found in the plant (*e.g.*, see Subramanian and Shanks this book).

The ADNT represented 10 percent or less of the TNT taken up by the plants. Most of the transformation products were not found in the aqueous phase, nor readily extractible from plant materials without strong oxidation. This observation suggests that the material may be conjugated or otherwise tightly bound to plant tissues (Hughes *et al.* 1997, Vanderford *et al.* 1997, Bhadra *et al.* 1999a).

Long-Term Reactors: Accumulation of Transformation Products in Plant Biomass

After long-term exposure, the two detectable nitroaromatics present in the aqueous phase were ADNT and TNT (Figure 13-3). After 48 hours, TNT was no longer detectable in the aqueous phase. With increasing time of exposure, ADNT decreased but was still measurable in the aqueous phase after 21 days.

Trinitrotoluene, ADNT, and azoxy compounds were found in plant extracts in a total amount that was low in comparison to the amount of TNT loaded to the system (Figure 13-3). The mass of extractable TNT and products accumulated for approximately the first 73 hours, and then decreased with time until the end of the test at 21 days. The plants appeared healthy throughout the course of the experiment.

Pilot-Scale Field Reactor Studies

Figure 13-4 presents the removal of TNT from the aqueous phase and appearance of transformation products with time in the field reactor system. The 10-day hydraulic residence time was the same for the five tanks (Figure 13-1), but the estimated amount of biomass differed considerably from tank to tank. However, the amount of biomass present was not the sole controlling factor for the rate of TNT disappearance and ADNT appearance in the aqueous phase.

At the end of the test, the amount of aboveground biomass (amount above the field stone) was estimated by dividing the reactors into measurable quadrants,removing the aboveground biomass in selected quadrants, and weighing

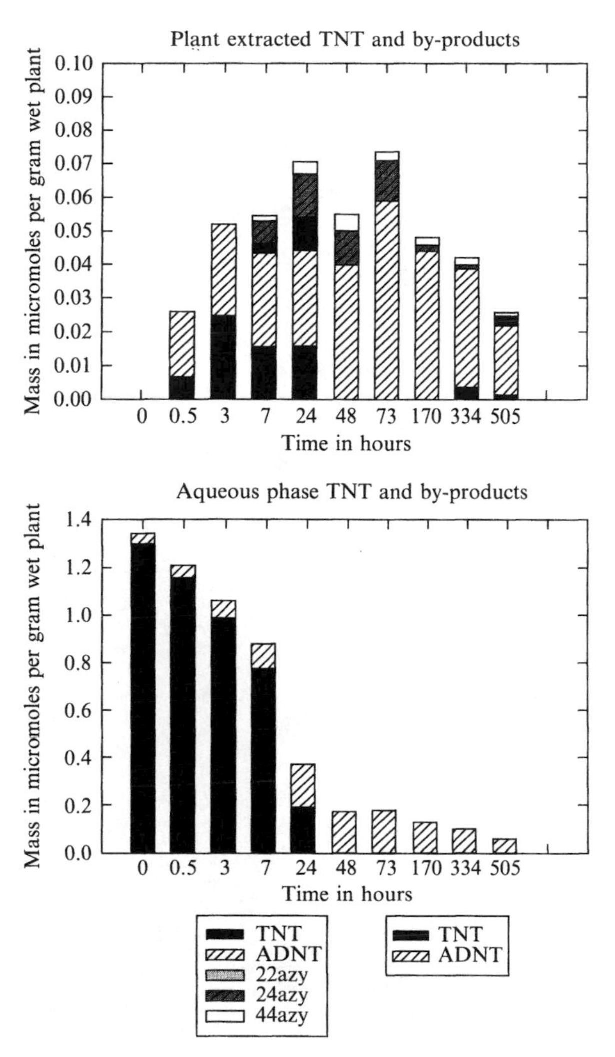

Figure 13-3 Tissue and aqueous phase measurements from Eurasian water milfoil (*Myriophyllum spicatum*) exposures of initially 10 micromoles per liter (2.3 milligrams per liter) of TNT over 505 hours. Compounds found include: TNT (2,4,6-trinitrotoluene), ADNT (aminodinitrotoluene), 22azy (2,2′-azoxy-4,4′,6,6′-tetranitrotoluene), 24azy (2,4′-azoxy-2′,4,6,6′-tetranitrotolue), and 44azy (4,4′-azoxy-2,2′,6,6′-tetranitrotoluene).

the shoots. The plant density per quadrant was extrapolated to estimate the plant density of each container. Table 13-3 presents the measurements of aboveground biomass for future comparison to natural or larger reactor systems.

During the constant flow of 10 milligrams of TNT per liter into the tanks of plants, the maximum consistent residual concentration measured was 1.6 milligrams of TNT per liter (7 micromoles per liter), found in the mixed marsh system. Declining TNT concentration from 500 to 1200 hours (Figure 13-4)

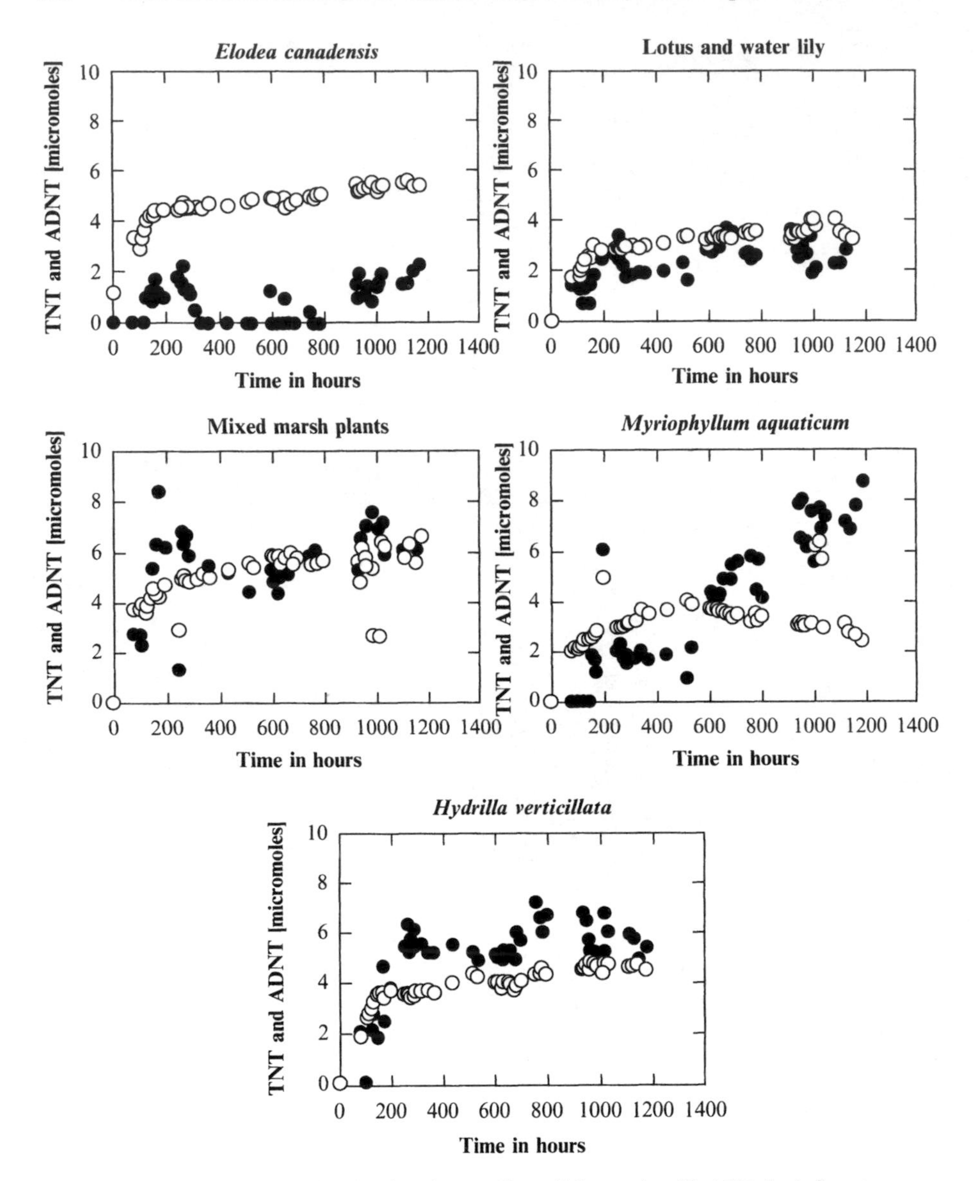

Figure 13-4 Field reactor results. During August 21 until September 28, 1995 the influent concentration was 1 milligram of TNT per liter and during October 9 to November 28, 1995, 10 milligrams of TNT per liter. Open circles are TNT (2,4,6–trinitrotoluene) accumulation with time, and the closed circles are ADNT (aminodinitrotoluene) production in the system. No other significant transformation products were detectable. Note that 1 (micromoles per liter) of TNT = 0.22713 milligrams of TNT per liter and 1 (micromoles per liter) ADNT = 0.18215 milligram of ADNT per liter.

may be partially attributable to changes in biomass of parrot feather (*Myriophyllum aquaticum*) in the treatment tank. The marsh system also had the greatest biomass (almost double the other reactor systems—see Table 13-3).

TABLE 13-3 Plant Biomass in Pilot-Scale Reactors

Species	Total biomass in reactor (kilograms)	Biomass density (grams per liter)
Elodea canadensis	36	45
Lotus and water lily	5.9	13
Mixed marsh community	25	87
Myriophyllum aquaticum	27	41
Hydrilla verticillata	32	40

The reactor containing lotus and water lily showed the lowest residual level of TNT in the water column and had the lowest total biomass. The waterweed (*Elodea canadensis*) reactor had the lowest ADNT accumulation, followed by the lotus and water lily reactor > mixed marsh reactor ≈ *Hydrilla verticillata* reactor > *Myriophyllum aquaticum* reactor (Figure 13-4).

SYNTHESIS AND CONCLUSIONS

Plants rapidly reduced the initial concentrations of 10 milligrams of TNT per liter in the water column during the 30-hour to 4-day tests presented in this chapter. Species differences in both uptake rates and net transformation rates found in this study may be related to a combination of differences in plant morphology, metabolism, internal and external microbial associations, and phytotoxicity and partitioning of TNT. Waterweed (*Elodea canadensis*) proved to be the fastest at TNT removal. This submerged species provided a high surface area to the aqueous media per unit biomass. Cell surfaces are more accessible in waterweed (*Elodea canadensis*), compared to the other species in this study. This increase in surface area of exposure, plus less differentiated tissue specialization, may result in the high uptake rates in waterweed (*Elodea canadensis*). Therefore, the relative surface area and partitioning most likely control TNT uptake in the plant species tested.

The octanol–water partitioning coefficients (K_{ow}) generally govern passive, nonselective removal of nonionic contaminants from water. The base 10 logarithm of K_{ow} for TNT has been estimated as between 1.6 and 2 (Urbanski 1964, Ainsworth *et al.* 1983), while both ADNT isomers have a base 10 logarithm of 1.3 to 1.6 (Sens *et al.* 1999). Materials with a base 10 logarithm of octanol–water partition coefficients in the range of 0.5 to 3 are most likely to be taken-up across the plant membranes (Briggs *et al.* 1983, Ryan, *et al.* 1988). Other physical parameters, such as pH, pKa, and the lipid and water content also play a role in uptake and transmembrane transport of contaminants (Ryan, *et al.* 1988, Anderson *et al.* 1993, Bockers *et al.* 1994, Salt *et al.* 1998).

Because this study did not directly measure relative plant metabolic activity, transformation rate differences are not easily explained. Assumedly, those plants with higher metabolism and greater enzymatic capacity will take up and transform contaminants more rapidly and potentially have a higher tolerance to the contaminants. This assumption remains to be tested across species by use of a number of parallel methods such as the photosynthetic-respiratory quotient, selective enzyme activity, and carbon turnover rate. As exposure concentration or exposure time increases, uptake and transformation rates of TNT decrease (Gueriguian 1996, Benton 1997) in a predictable fashion. In waterweed (*Elodea canadensis*), this effect is associated with inhibition of photosystem II (second photochemical system that plants use to convert longer wave length sunlight into chemical energy), specifically, but not necessarily exclusively, at the Qb binding site (similar to the effect of the pesticide atrazine on plants). The TNT also acts as an electron donor to the oxygen-evolving complex (similar to the effects of hydroxylamine) and quenches chlorophyll fluorescence (like many quinone analogs). The kinetics of carbon fixation in the presence of TNT seems to follow the form of an inhibition function. Although TNT negatively affects carbon fixation, plants keep growing and transforming TNT (Benton 1997).

Both biomass and initial exposure concentration affect rates of TNT uptake (Gueriguian 1996, Benton 1997, Pavlostathis *et al.* 1998) as seen in comparable studies on plant uptake of TNT (Medina and McCutcheon 1996, Rivera *et al.* 1998, Medina *et al.* 2000, McCutcheon *et al.* this book). These studies vary in the initial TNT concentrations, plant types, initial biomass of the exposed plant, and physical conditions. General removal rates reported by other studies are slower than the rates presented in this chapter, but are comparable in order of magnitude. Pilot studies at the Alabama Army Ammunition Plant in Childersburg (Young 1995) demonstrate toxic impact (illustrated by plant death) at levels (5 to 7 milligrams of TNT per liter) lower than in this study. Cataldo *et al.* (1989) subjected bush beans (*Phaseolus vulgaris*) to 10 milligrams of TNT per liter (the same concentration used in the current study) and found that bush beans (*Phaseolus vulgaris*) were able to take up 0.35 milligram of TNT per kilogram wet weight of plant in shoots and leaves and 1.5 milligrams of TNT per kilogram wet weight of plant in roots.

This study involved extraction of TNT and the associated transformation products at different times. The dominant, extractable by-product found in this study (*i.e.*, ADNT) appeared, reached a maximum, and then began to decrease with time. Other studies that extracted [^{14}C]-labeled TNT and transformation products from aquatic plants as a function of time of exposure (Hughes *et al.* 1997, Price *et al.* 1997, Vanderford *et al.* 1997, Hughes *et al.* 1998) were not able to provide details on the nature of all the transformation products. The work on periwinkle axenic hairy root cultures (*Catharanthus roseus*) as well as the work on axenic parrot feather (*Myriophyllum aquaticum*) illustrated that approximately one-third to one-half of the radiolabeled carbon remained in solution, whereas the greater proportion is typically associated with plant tissue. In killed controls, 60 to 85 percent of the [^{14}C]-label remained in the culture media (Hughes *et al.* 1997). Although TNT rapidly disappeared from the aqueous media containing active plants, reduction products were observed only at trace concentrations in the media. Over time, fewer plant-associated products were extractable.

One other study (Bhadra *et al.* 1999a) examined TNT transformation processes in plants, focusing on the formation and fate of products beyond the aminated reduction products, 2-amino-4,6-dinitrotoluene, and 4-amino-2,6-dinitrotoluene. Bhadra *et al.* (1999a) provided the first direct evidence of the involvement of the primary reduction products of TNT, 2-amino-4,6-dinitrotoluene, and 4-amino-2,6-dinitrotoluene, in conjugation processes during plant detoxification of TNT. A similar process perhaps occurred during the tests presented in this chapter. Reduction followed by conjugation would result in a decrease in aqueous phase products and an increase in transformation products in the extractable plant phase, followed by a slow disappearance in the extractable fraction as binding occurs to cell walls.

In defining the role of plants in the remediation of organic contaminants, some studies reported that most compounds and the associated by-products bind irreversibly within the plant (Komossa *et al.* 1995, Trapp and McFarlane 1995, Field and Thurman 1996), while others found that compounds are stored in vacuoles. Bound residues are acid-hydrolyzable conjugates (Coleman *et al.* 1997). Residues remaining in the plant after extraction could be toxic to other species of plants or animals when either left to decompose or be ingested. Bound residues of TNT derivatives have been found in parrot feather (*Myriophyllum aquaticum*) and axenic root cultures of periwinkle (*Catharanthus roseus*) (Hughes *et al.* 1997, Vanderford *et al.* 1997). A more detailed follow up mass balance of the same periwinkle (*Catharanthus roseus*) root cultures (Bhadra *et al.* 1999a) indicates that by the 75th hour after TNT amendment, extractable conjugates comprised 22 percent of the total [^{14}C] originally added, bound residues comprised another 29 percent, ADNT was 5 percent, and the rest remained unidentified. This study provides direct evidence for conjugation processes in plant detoxification of TNT. No information is currently available on the long-term stability of these residues and complete mineralization of TNT by plants has not been demonstrated.

In this study, pilot-scale reactors were exposed to a constant supply of TNT at an initial concentration of 10 milligrams per liter. The highest residual levels of TNT in the aqueous phase ranged from 3 to 6 milligrams per liter, with most reactors containing TNT at the lower end of the range. An increase in hydraulic residence time, decrease in TNT dosage, and increase in plant biomass, should decrease the residual levels of TNT even further in these type reactors during the future optimization that is necessary. Accumulation of ADNT occurred in all field reactors, and by the end of the exposure trial was observed to be at equal or higher levels than the TNT present. The TNT residual levels as well as ADNT accumulation may not necessarily correlate only with plant biomass (Table 13-3). Root metabolism, sorption, and storage capacity may also play important roles in the ultimate transformation rates measured, as may toxicological effects. The potential of plants to remove contaminants from water has been explored for metals and metalloids

(Banyos *et al.* 1993, Rai *et al.* 1995, Salt *et al.* 1998), and for sewage treatment, and less so to date for industrial organic contaminants. The type of pilot testing introduced in this chapter is an important step in determining both plant tolerances for contaminants and potential removal efficiencies.

Both submerged and emergent aquatic macrophytes were selected to assess the potential for remediation of explosive compounds. These results and those of Thompson *et al.* (this book) suggest that phytoremediation is useful in an aquatic remediation scheme when groundwater is pumped to the surface and is fed to either hydroponic or sediment-associated aquatic systems. The groundwater and plant material from these systems should be monitored to determine rates of contaminant removal and to confirm that significant accumulation does not occur.

This emerging technology seems to be cost effective; however, the efficiency of different species of plants as well as the toxicity of the contaminant and transformation products to plants and animals must be considered. Life habit, leaf and stem morphology, and biochemical differences among aquatic plant species may result in a variety of uptake and transformation, efficiencies. Plants can be used to remediate organic contaminants by uptake, transformation and subsequent accumulation of metabolites (Trapp and McFarlane 1995). If the contaminant is not fully mineralized by the plants, harvesting after a selected time, leaving plants in holding tanks to permit further metabolism, or when necessary, further treatment (*e.g.* fungal, bacterial, or chemical treatments) may be required.

Acknowledgments

The authors acknowledge the Strategic Environmental Research and Development Program of the U.S. Departments of Defense and Energy, and the U.S. Environmental Protection Agency that supported a National Research Council fellowship for the first author (Project 720) and the U.S. Environmental Protection Agency Hazardous Substance Research Center/South and Southwest, through which the work was funded. Steve McCutcheon was the project officer and N.L. Wolfe arranged the fellowship and provided inspiration. Steve McCutcheon provided valuable discussions and editorial support, Jim Price gave technical support with plants, and Donna Brown shared technical and inspirational support.

REFERENCES

Ainsworth, C.C, S.D. Harvey, J.E. Szecsody, M.A. Simmons, V.I. Cullinan, T.C. Resch, and G.M. Mong (1983) Relationship between the leachability characteristics of unique energetic compounds and soil properties. Technical Report 93–17415. U.S. Army Biomedical Research and Development Laboratory, Frederick, Maryland.

Anderson, T.A., E.A. Guthrie, and B.T. Walton (1993) Bioremediation. *Environ. Sci. Technol.* **27**: 2632–2636.

Banyos, G.S., G. Cardon, B. Mackey, J. Ben-Asher, L. Wu, P. Beuselinck, S. Akohoue, and S. Zambruski (1993) Plant and environment interactions, boron and selenium removal in boron-laden soils by four sprinkler irrigated plant species. *J. Environ. Qual.* **22**: 786–792.

Barko, J.W. and R.M. Smart (1983) Effects of organic matter additions to sediment on the growth of aquatic plants. *J. Ecol.* **71**: 161–175.

Benton, M. (1997) Effects of 2,4,6, trinitrotoluene (TNT) on carbon fixation rates in *Elodea nutalli.* M.S. Thesis, School of Civil and Environmental Engineering, Georgia Institute of Technology, Atlanta, Georgia.

Bhadra, R., D.G. Wayment, J.B. Hughes, and J.V. Shanks (1999a) Confirmation of conjugation processes during TNT metabolism by axenic plant roots. *Environ. Sci. Technol.* **33**: 446–452.

Bhadra, R., R.J. Spanggord, D.G. Wayment, J.B. Hughes, and J.V. Shanks (1999b) Characterization of oxidation products of TNT metabolism in aquatic phytoremediation systems of *Myriophyllum aquaticum. Environ. Sci. Technol.* **33**(19): 3354–3361.

Bockers, M., C. Rivero, T. Thiede, T. Jankowski, and B. Schmidt (1994) Uptake, translocation and metabolism of 3,4–dichloroaniline in soybean and wheat plants. *Biosciences* **49**: 719–726.

Briggs, G.S., R.H. Bromilow, A.A. Evans, and M. Williams (1983) Relationships between lipophilic distribution of non-ionised chemicals in barley shoots following uptake by the roots. *Pestic. Sci.* **4**: 492–500.

Brix, H. (1998) The use of aquatic macrophytes in water-pollution control. *Ambio* **18**(2): 100–107.

Cataldo, D.A., S. Harvey, R. Fellows, R.M. Bean, and G.D. McVeety (1989) An evaluation of the environmental fate and behavior of munitions materials (TNT, RDX) in soil and plant systems. U.S. Department of Energy, Pacific Northwest Laboratory Final report. PNL–7370 UC-402, Richland, Washington.

Coleman, J.O.D., M.M.A. Blake-Kalff, and T.G.E. Davies (1997) Detoxification of xenobiotics by plants: chemical modification and vacuolar compartmentation. *Trends Plant Sci.* **2**: 144–151.

Ericsson, T. (1995) Growth and shoot: root ratio of seedlings in relation to nutrient availability. *Plant Soil* **168/169**: 205.

Field, J.A. and E.M. Thurman (1996) Glutathione conjugation and contaminant transformation. *Environ. Sci. Technol.* **30**(5): 1415–1418.

Franzaring, J., A.E.G. Tonneijck, A.W.N. Kooijman, and T.A. Dueck (2000) Growth response to ozone in plant species from wetlands. *Environ. Exp. Bot.* **44**(1): 39–48.

Gueriguian, L.F. (1996) Kinetics of 2,4,6-trinitrotoluene (TNT) phytotransformation using *Elodea densa.* M.S. Thesis, Georgia Institute of Technology, School of Civil and Environmental Engineering, Atlanta, Georgia, 162 pp.

Hartman, W.J. (1975) An evaluation of land treatment of municipal wastewater and physical siting of facility installations. U.S. Dept. Army, Washington, D.C.

Harvey, S.D., R.J. Fellows, A. Cataldo, and R.M. Bean (1991) Fate of the explosive hexahydro-1,3,5-trinitro-1,2,5-triazine (RDX) in soil and bioaccumulation in bush bean hydroponic plants. *Environ. Toxicol. Chem.* **10**: 845–855.

Harvey, S.D., R.J. Fellows, A. Cataldo, and R.M. Bean (1993) Analysis of the explosive 2,4,6-trinitrophenylmethyllnitramine (tetryl) in bush bean plants. *J. Chromatogr.* **630**: 167–177.

Hoehamer, C.F. (2000) Biotransformation of 2,4,6 trinitrotoluene (TNT) by wild and axenically grown isolates of *Myriophyllum aquaticum* and a plant-associated fungus *Fusarium oxysporum*. Ph.D. dissertation, Biochemistry, University of Georgia, Athens.

Hsieh, Y.P. (1988) Dynamics of carbon, nitrogen, and phosphorus cycling in a sawgrass tidal marsh with special reference to the aboveground primary production. *J. Environ. Qual.* **17**(4): 676–781.

Hughes, J.B., J.V. Shanks, M. Vanderford, J. Lauritzen, and R. Bhadra (1997) Transformation of TNT by aquatic plants and plant tissue cultures. *Environ. Sci. Technol.* **31**: 266–271.

Hughes, J.B., C.Y. Wang, R. Bhadra, and A. Richardson (1998) Bamberger rearrangement during TNT metabolism by *Clostridium acetobutylicum*. *Environ. Toxicol. Chem.* **17**: 343.

Hunt, L.A. (1991) Shoot and root dry weight and soil water in wheat, triticale and rye. *Can. J. Plant Sci. Rev. Can.* **71**(1): 4–12.

Komossa, D.C., C. Langebartels, and H. Sandermann, Jr. (1995) Metabolic processes for organic chemicals in plants. In: *Plant Contamination*. S. Trapp and J.C. McFarlane, eds., CRC Press, Boca Raton, Florida.

McIntyre, T. and G.M. Lewis, (1997) The advancement of phytoremediation as an innovative environmental technology for stabilization, remediation or restoration of contaminated sites in Canada: a discussion paper. *J. Soil Contam.* **6**(3): 27–232.

Medina, V.F. and S.C. McCutcheon (1996) Phytoremediation: modeling removal of TNT and its breakdown products. *Remediation* Winter: 31–45.

Medina, V.F., R. Rivera, S. Larson, and S.C. McCutcheon (1996) Phytoreactors show promise in treating munitions contaminants. *Soil Groundwater Cleanup* Feb/Mar: 19–24.

Medina, V.F., S.L. Larson, A.E. Bergstedt, and S.C. McCutcheon (2000) Phytoremoval of trinitrotoluene from water with batch kinetic studies. *Wat. Res.* **34**(10): 2713–2722.

Palazzo, A.J. and D.C. Leggett (1986) Effect and disposition of TNT in terrestrial plants. *J. Environ. Qual.* **15**: 49–52.

Pavlostathis, S.G., K.K. Comstock, M.E. Jacobson, and F.M. Saunders (1998) Transformation of 2,4,6-trinitrotoluene (TNT) by the aquatic plant *Myriophyllum spicatum*. *Environ. Toxicol. Chem.* **17**(11): 2266–2273.

Price, R.A., J.C. Pennington, S.L. Larson, D. Naumann, and C.A. Hayes (1997) Plant uptake of explosives from contaminated soil and irrigation water at the former Nebraska Ordnance Plant, Mead, Nebraska. Technical Report EL-97-11. U.S. Army Engineer Waterways Experimental Station, Vicksburg, Mississippi.

Rai, U.N., S. Sinha, R.D. Tripathi, and P. Chandra (1995) Wastewater treatability potential of some aquatic macrophytes: removal of heavy metals. *Ecol. Eng.* **5**: 5–12.

Rivera, R., V.F. Medina, S.L. Larson, and S.C. McCutcheon (1998) Phytotreatment of TNT-contaminated groundwater. *J. Soil Contam.* **7**(4): 511–520.

Rogers, H.H., S.A. Prior, and R.J. Mitchell (1996) Root shoot ratio of crops as influenced by CO_2. *Plant Soil* **187**(2): 229–239.

Roman, C.T. and F.C. Daiber (1984) Aboveground and belowground primary production dynamics of two Delaware Bay tidal marshes. *Bull. Torrey Bot. Club* **111**(1): 34–41.

Ryan, J.A., R.M. Bell, J.M. Davidson, and G.A. O'Connor (1988) Plant uptake of non-ionic chemicals from soils. *Chemosphere* **17**: 2299–2323.

Salt, D.E., R.D. Smith, and I. Raskin (1998) Phytoremediation. In: *Annual Review* of *Plant Physiology and Plant Molecular Biology*. Annual Reviews, Palo Alto, California. **49**: 643–668.

Schnoor, J.L. (1997) Phytoremediation. Technology Evaluation Report TE-98-01, Ground-Water Remediation Technology Analysis Center, Concurrent Technologies Corp. Pittsburgh, Pennsylvania, p. 37.

Schnoor, J.L., L.A. Licht, S.C. McCutcheon, N.L. Wolfe, and L.H. Carreira (1995) Phytoremediation of organic and nutrient contaminants. *Environ. Sci. Technol.* **29**: 318A–323A.

Schubauer, J.P. and C.S. Hopkinson (1984) Above and belowground emergent macrophyte production and turnover in a coastal marsh ecosystem, Georgia. *Limnol. Oceanor.* **29**(5): 1052–1065.

Sens, C., P. Scheidemann, and D. Werner (1999) The distribution of ^{14}C-TNT in different biochemical compartments of the monocotyledonous *Triticum aestivum*. *Environ. Pollut.* **104**: 113–119.

Siddique, K.H., R.K. Belford, and D. Tennant (1990) Root: shoot ratios of old and modern, tall and semi-dwarf wheats in a mediterranean environment. *Plant Soil* **121**(1): 89–95.

Thorne-Miller, B. and M.M. Harlin (1984) The production of *Zostera marina* L. and other submerged macrophytes in a coastal lagoon in Rhode Island. *Bot. Mar.* **27**: 539–346.

Trapp, S. and J.C. McFarlane (1995) *Plant Contamination: Modeling and Simulation of Organic Chemical Processes.* Lewis Publishers, Boca Raton, Florida, 254 pp.

Urbanski, T. (1964) *Chemistry and Technology of Explosives.* MacMillan, New York.

Vanderford, M., J.V. Shanks, and J.B. Hughes (1997) Phytotransformation of trinitrotoluene (TNT) and distribution of metabolic products in *Myriophyllum aquaticum*. *Biotechnol. Lett.* **19**: 277–280.

Wolfe, N.L., T.Y. Ou, L. Carriera, and D. Gunnison (1994) Alternative methods for biological destruction of TNT: a preliminary feasibility assessment of enzymatic degradation. Technical Report IRRP-93-3, U.S. Army Corps of Engineers Waterways Experimental Station. Vicksburg, Mississippi.

Xu, J.G., and N.G. Kumar (1992) Aboveground and belowground net primary production of four barley (*Hordeum vulgare* L.) cultivars in western Canada. *Can. J. Plant Sci.* **72**(4): 1131–1140.

14

PROOF OF PHYTOREMEDIATION FOR EXPLOSIVES IN WATER AND SOIL

S. C. McCutcheon, V. F. Medina, and S. L. Larson

SUMMARY OF PRACTICAL IMPLICATIONS

Field pilots prove the concept of using created treatment wetlands to restore groundwaters and flooded soils contaminated with explosives. Wetland treatment of concentrated wastewater is ready for field testing. Batch- and continuous-flow treatability protocols for contaminated groundwater are ready to be formalized to relate the rapid removal to the effects of temperature, indigenous plant density, and phytotoxicity of 2,4,6-trinitrotoluene (TNT) to wetland design. Dissolved oxygen, pH, and enzymatic cofactors do not seem to be important in healthy treatment wetlands with organic-rich sediments. The kinetics of hexahydro-1,3,5-trinitro-1,3,5-triazine (RDX) or TNT by-products are design limiting in many cases. The second order kinetics of TNT uptake and transformation must be confirmed with treatability testing at each site for the plants and density selected. The kinetics of RDX and other explosives have not been fully described, especially beyond the use of aquatic monocultures. Although pilot optimization is pending, enough experience is available to design wetlands that take up and transform TNT and similar nitroaromatic contaminants. Designs involving RDX and other explosives require bench treatability, sediment and plant community optimization, and field testing. Transformation product sequestration in plants indicates significant risk reduction. Nevertheless, the unknown fate of some sequestered products requires tissue monitoring for at least 2 years in field applications.

Phytoremediation: Transformation and Control of Contaminants,
Edited by Steven C. McCutcheon and Jerald L. Schnoor.
ISBN 0-471-39435-1 (cloth)

PHYTOREMEDIATION OF EXPLOSIVES

Potential for Using Green Plant Transformation of Explosives

The practice of using plants to transform explosives wastes in water and soil is one of the success stories in the development of phytoremediation in the 1990s. The observations by Palazzo and Leggett (1986), Cataldo *et al.* (1989), and Harvey *et al.* (1990, 1991, 1992) established that plants not only take up TNT from water and soil but also transform the contaminant. These studies did not initially trigger interest in phytoremediation because the focus was on the food-chain effects of bioaccumulation for explosives in the plants (Pennington 1988) as opposed to development of a treatment technology. Independently, Wolfe *et al.* (1993) traced the fast reactions of TNT and other nitroaromatic compounds in reduced sediments to stable enzymes that originated with plants (see Table 1-4, McCutcheon and Schnoor this book). Very quickly, these reactions were recognized as the conceptual basis of a new approach to waste management and were part of the basis for expanding the term phytoremediation to include the cleanup of organic compounds (Schnoor *et al.* 1995, McCutcheon *et al.* 1995). These early studies at the U.S. Environmental Protection Agency (EPA) laboratory in Athens, Georgia, led to a number of seminal investigations of the pathways and kinetics of plant and sediment-derived transformation of TNT (van Beelen and Burris 1995, Best *et al.* 1997a, Best *et al.* 1997b, Best *et al.* 1997c, Best *et al.* 1997d, Best *et al.* 1999a, Best *et al.* 1999b, Bhadra *et al.* 1999a, Bhadra *et al.* 1999b, Burken *et al.* 2000), Hughes *et al.* 1997, Thompson 1997, Thompson *et al.* 1998, Pavlostathis *et al.* 1998, Vanderford *et al.* 1997.

Cataldo *et al.* (1989), Harvey *et al.* (1990, 1991), Harvey *et al.* (1992), and Wolfe *et al.* (1993) also noted that plants take up and transform RDX, octahydro-1,3,5,7-tetranitro-1,3,5,7-tetraazocine (HMX), tetryl, and other explosives. However, the state of the art has not progressed as rapidly for these other explosives. Just as the fate and transport of RDX in soil and groundwater is markedly different from TNT (Pennington *et al.* 1999a, Pennington *et al.* 1999b, Pennington *et al.* 1999c), so is the fate in plants (*e.g.*, compare Harvey *et al.* 1991 and Thompson *et al.* 1998, this book).

The Department of Defense (1995) notes that explosives, primarily TNT, contaminate about 2000 U.S. Federal sites. The problem is worldwide (Polish State Inspectorate for Environmental Protection 1995, Spain 2000, Medina *et al.* this book). Given the expense of cleanup of these national defense facilities (Jerger and Woodhull 2000), phytoremediation has been pursued to save money and to begin to ecologically restore sites at the same time.

Conceptual Approach to Integrating Phytoremediation Research and Engineering

The development of phytoremediation at the U.S. EPA laboratory in Athens followed four principles. First, trial and error was used to discover fast reactions of classes of contaminants in anoxic sediments. The reaction kinetics were characterized at the same time (Jafvert and Wolfe 1987, Weber and Wolfe 1987, Tratnyek and Wolfe 1990, 1993, Metwally and Wolfe 1990, Peijnenburg *et al.* 1992a, 1992b, Masunaga *et al.* 1995, Berger and Wolfe 1996). Figure 14-1 is an example of the reactions observed at different TNT concentrations.

Second, these reactions were related presumptively to stable plant enzymatic activities in the sediments, including nitroreductase, nitrilase, and dehalogenase activities (Schnoor *et al.* 1995, van Beelen and Burris 1995). Structure–activity relationships indicated that the reactions were related to the organic fraction of the sediments (Wolfe and Delgado 1991). Various columns segregated components of the sediment organic matter until a narrow range of molecular weights could be associated with the observed activity (see van Beelen and Burris 1995 for a similar extraction). These crude extracts were introduced to solutions of the pure contaminant in distilled water to confirm the reactions (Figure 14-2) and to characterize the reactions to a limited degree. Double monoclonal antibodies were developed from the extracts (Wolfe *et al.* 1993) by the University of Georgia Monoclonal Antibodies Facility and used to create an ELISA (enzyme linked immunosorbent assay) to begin tracing the origin of these stable activities (Best *et al.* 1997b). In each of the three isolations attempted to date, the ELISA indicated that the source of the activities was probably plant enzymes that settled into the sediments after the plant died. It was surprising that three out of three

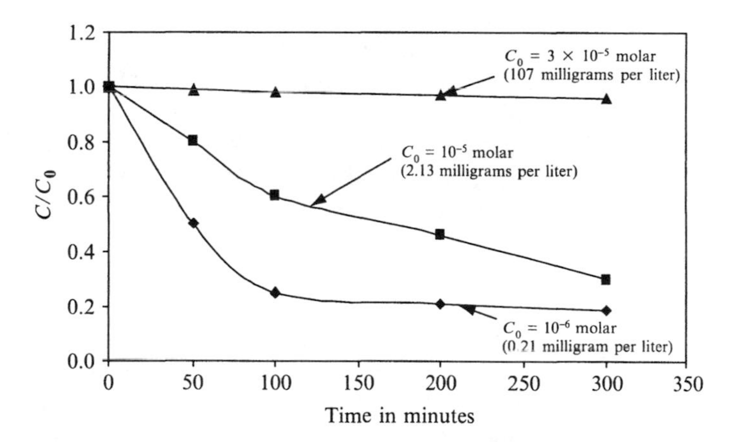

Figure 14-1 Example of fast aqueous reactions observed for trinitrotoluene at different concentrations when mixed with local sediments from a pond in Cherokee Park, Clark County, Georgia. The contributing watershed seemed free of sources of nitroaromatic compounds. Note that C is the concentration of trinitrotoluene as a function of time and C_0 is the initial concentration in solution. Analysis by high performance liquid chromatography with a detection limit of 0.01 milligram per liter and relative standard deviation of 2 percent (Jenkins 1989, U.S. EPA 1992, Wolfe *et al.* 1993). Unpublished data from N.L. Wolfe and T.Y. Ou, U.S. EPA laboratory, Athens, Georgia.

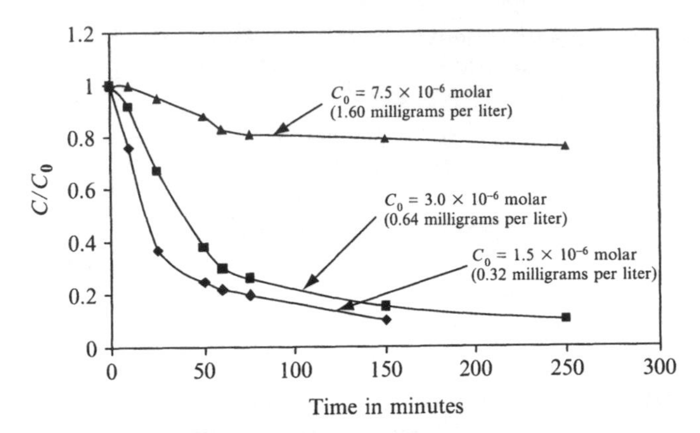

Figure 14-2 Batch reactivity of a 1 microgram sediment organic matter crude extract exposed to different trinitrotoluene concentrations. Technology Applications, Inc. derived this reactive component of the sediment using extraction, precipitation, hydrophobic interaction, ion exchange, sizing, and zinc affinity columns under contract to the U.S. EPA laboratory, Athens, Georgia. Note that C is the concentration of trinitrotoluene as a function of time and C_0 is the initial concentration. Analysis by high performance liquid chromatography with a detection limit of 0.01 milligram per liter and relative standard deviation of 2 percent (Jenkins 1989, U.S. EPA 1992, Wolfe *et al.* 1993).

isolations led to enzymatic activity of plant origin, given the dominant role of microorganisms in biogeochemical cycling of organic materials. Perhaps the segregation techniques favored plant-derived activity, or stable plant enzymes are more widely distributed than originally suspected by remediation engineers. Nevertheless, these isolations are now suspected to have involved enzymes conjugated with large humic molecules, on which the monoclonal antibodies fortuitously attached to the conjugated activity. The small amounts of activity directly isolated from plants indicate that much smaller molecules are involved and imply that the original findings previewed in Schnoor *et al.* (1995) probably involved enzymes conjugated with humic molecules. Despite the conjugation, the proteins isolated from sediments did have many characteristics of plant enzymes, including being glycosylated and iron-centered.

van Beelen and Burris (1995) reported the isolation of ubiquitous nitroreductase activity from sediments, aquatic plants [arrowhead (*Sagittaria* spp.)], and macrophytic algae (*Chara* spp.). Furthermore, nitroreductases capable of reducing various explosives were isolated from spinach (*Spinacia oleracea*) (Shah and Spain 1996, Shah and Campbell 1997, Wilson 1998). Wolfe and Hoehamer (this book) note the two types of plant nitroreductases. The sediment extractions in Athens and by van Beelen and Burris (1995) seem to involve type I oxygen-insensitive nitroreductases, whereas Shah and Spain (1996) isolated type II oxygen-sensitive nitroreductases.

In the process of confirming the role of plants in these fast reactions, Vanderford *et al.* (1997) and Hughes *et al.* (1997) established rapid transformation due to whole plants in aseptic (or anexic) tissue cultures. Lauritzen (1998) provided more evidence of plant-dominated transformation using axenic hairy root cultures of periwinkle (*Catharanthus roseus*, 'LBE 6-1'). Studies comparing anexic or aseptic and septic whole plant kinetics conducted at the U.S. EPA laboratory in Athens indicated that the reaction rates of TNT disappearance were very similar (Figure 14-3). Studies using bleach to remove surface microorganisms from parrot feather (*Myriophyllum aquaticum*) also indicated that removal was virtually identical to untreated plants (Medina *et al.* 2000a).

The third component to this work was based on early explorations of reaction kinetics and pathways using mass balance investigations. The initial hypothetical pathway of TNT reduction by plants (Medina and McCutcheon 1996) became a framework that has been continually improved as better analytic, anexic, and radiolabeled techniques were applied (Bhadra *et al.* 1999a, Bhadra *et al.* 1999b, Burken *et al.* 2000, Subramanian and Shanks this book).

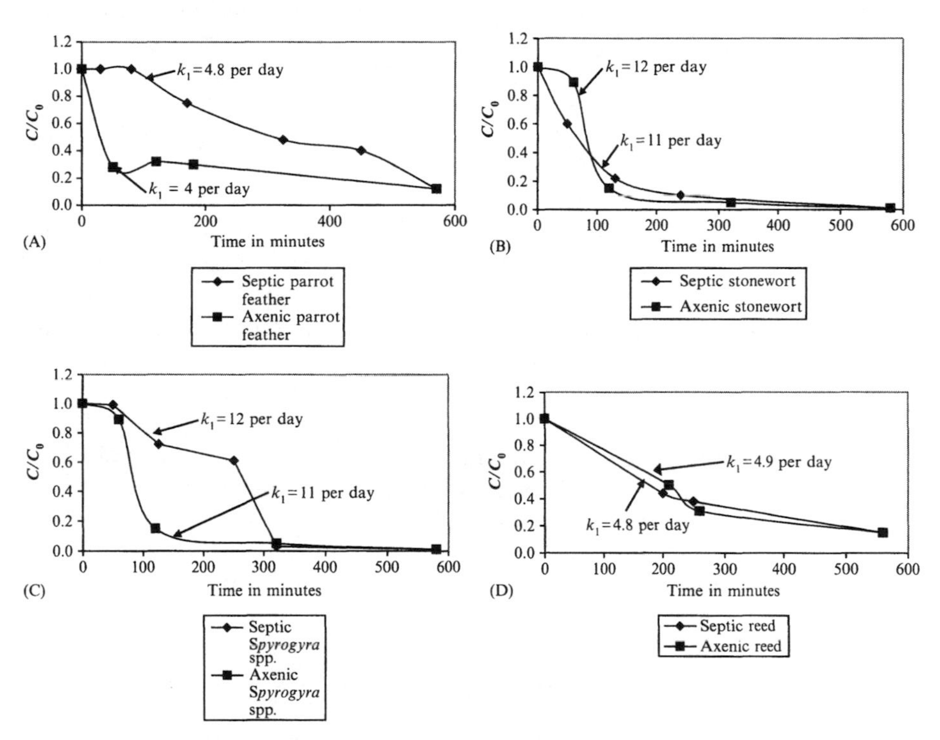

Figure 14-3 Similarity in removal of trinitrotoluene by axenic (or aseptic) and septic whole plants: (A) parrot feather (*Myriophyllum aquaticum*), (B) stonewort (*Nitella* spp.), (C) *Spirogyra* spp., and (D) reed (*Eleocharis* spp.). Note that C is the concentration of trinitrotoluene as a function of time, C_0 is the initial concentration, and k_1 is the first order disappearance rate constant. Analysis by high performance liquid chromatography with a detection limit of 0.01 milligram per liter and relative standard deviation of 2 percent (U.S. EPA 1992, Jenkins 1989, Wolfe *et al.* 1993). Experiments conducted at the U.S. EPA laboratory, Athens, Georgia by Wolfe *et al.*

The fourth and final part of this work has been to start the ecological and environmental engineering investigations to take advantage of the rapid transformation reactions for waste management at a very early stage. Typically, kinetics and pathways investigations were used to confirm the fast reactions over wide ranges of pH, temperature, and other conditions, rather than just duplicate the original experiments that revealed the reactions (McCutcheon *et al.* 1995, van Beelen and Burris 1995).

Created Wetland Concept

The initial discoveries of the rapid transformation, combined with the lack of TNT contamination observed in natural wetlands draining highly contaminated areas of the Alabama Army Ammunition Plant near Childersburg, led to the development of the created wetlands concept to treat contaminated water, sediments, and soils (McCutcheon, seminar presented to U.S. EPA laboratory in Cincinnati, Ohio, 1991). The created wetland concept also encompasses aquatic plant lagoons, a term later adopted by the Georgia Institute of Technology in Atlanta, Georgia (HSRC 1997, Jacobson *et al.* this book) in follow-up development of field pilot procedures. More recently these types of wetlands are being referred to as constructed or treatment wetlands to avoid confusion with mitigation wetlands. Mitigation wetlands are not used for treatment but are those that replace natural wetlands lost due to land development. Treatment, constructed, or created wetlands require National Pollution Discharge Elimination System Permits. The created wetland concept involves introducing plants after flooding contaminated soils using dikes, embankments, small dams, and other structures if the soil infiltration rate, topography, and availability of water permits. In some cases, contaminated soil may be excavated and placed in created wetlands. Treatment wetlands may also be established where soil is excavated. The concept also includes the treatment of wastewaters, groundwater that is pumped to the land surface, or groundwater flows that intersect the land surface.

Experience in the laboratory and field demonstrated that wetlands with standing water adequately buffer floating plants from highly concentrated explosives that diffuse from sediments or flooded soils, at least on a short-term basis of weeks or months. Introduction of adequate plant biomass and clean water, before explosives build up to saturation levels, is normally protective of viable plants. In a wetland, the initial shock can be managed by

1. The controlled introduction of explosives-contaminated water from groundwaters or industrial processes by equalization
2. Filling treatment wetlands having contaminated sediments with both plants and clean water before explosives become saturated in the water column
3. Planting or releasing densely spaced plants over contaminated sediments

4. Covering the contaminated soil with a thin veneer of natural wetland sediment

The purpose of this chapter is to present the state of the art in using constructed treatment wetlands to apply phytoremediation to contaminated groundwater, wastewater, and soil.

In order to establish the proof of concept (a term defined in McCutcheon and Schnoor this book) for using constructed treatment wetlands, a comprehensive research and development program was implemented. The program consisted of work done at a variety of facilities, both government laboratories and universities. The sections that follow present these studies with emphasis on the work at the U.S. EPA laboratory in Athens, Georgia. Some of the U.S. EPA work is appearing in publication for the first time in this chapter. The following chapter (Thompson *et al.* this book) presents the field proof of concept achieved at the Iowa Army Ammunition Plant (Middleton).

BATCH INVESTIGATIONS: GROUNDWATER AND WASTEWATER

To establish the wetland conceptual design, investigations of TNT, RDX, HMX, and other explosives in groundwater and wastewater started with bench-top batch reactor tests. In this engineering development phase, investigations using 200-milliliter (0.05-gallon) reactors defined performance over ranges of key design parameters (Medina *et al.* 2000b). Key operating parameters were identified by modeling the sequential transformation of TNT in plug flow or batch settings (Medina and McCutcheon 1996). All of these bench-top feasibility and optimization studies generally consisted of growing the plants in solutions of explosives in deionized water. Performance was evaluated by monitoring the disappearance of TNT and the buildup and disappearance of the transformation products [including aminodinitrotoluene (ADNT) and diaminonitrotoluene (DANT)] in the aqueous phase by the methods in Medina *et al.* (2000b). At the end of some of the batch tests, plant tissue samples were analyzed for TNT and the associated transformation products at the U.S. Army Waterways Experiment Station (Vicksburg, Mississippi) using the methods described in Larson *et al.* (1999).

Pilot testing for specific sites focused on TNT and RDX contaminated waters to establish the proof of concept for both wastewater and groundwater treatment. Pilot testing using explosives-contaminated wastewaters from the McAlester (Texas) and the Milan (Tennessee) Army ammunition plants led to recommendations for pilot field testing of phytoremediation (Concurrent Technology Corp. 1997). At the Iowa Army Ammunition Plant, two full-scale pilots successfully restored groundwater discharging into two created wetlands (Thompson *et al.* this book).

Laboratory Proof of Principle: Treatment of Trinitrotoluene-Contaminated Water

Summary of the Proof of Concept and Some Design Implications

Transformation Products

The original work on phytoremediation of TNT at the U.S. EPA laboratory in Athens was performed with the attached macrophytic algae stonewort (*Nitella* spp.) that demonstrated fast removal of TNT with a half-life of 71 minutes (Figure 14-4). By comparison, microbial uptake and transformation of TNT is approximately two orders of magnitude slower for typical biomass densities (4320 minutes; see Table 1-4, McCutcheon and Schnoor this book). Pathway analyses established that transformation products would be important in design. Medina and McCutcheon (1996) demonstrated that analytical equations based on sequential first order reactions adequately describe the uptake and transformation of TNT and the subsequent appearance and disappearance of ADNT and DANT in batch reactors with a plant density of 15 grams wet weight per liter of stonewort (*Nitella* spp.) and parrot feather (*Myriophyllum aquaticum*) in two separate tests. The results indicated that TNT reached the maximum contaminant level of 0.002 milligrams per liter (or the drinking water standard, which was accepted as the cleanup level for injection back into the groundwater) in 13 days. By contrast, ADNT plus DANT did not reach 0.050 milligrams per liter (total explosives criteria used by the U.S. Army in the absence of health-based standards for each by-product)

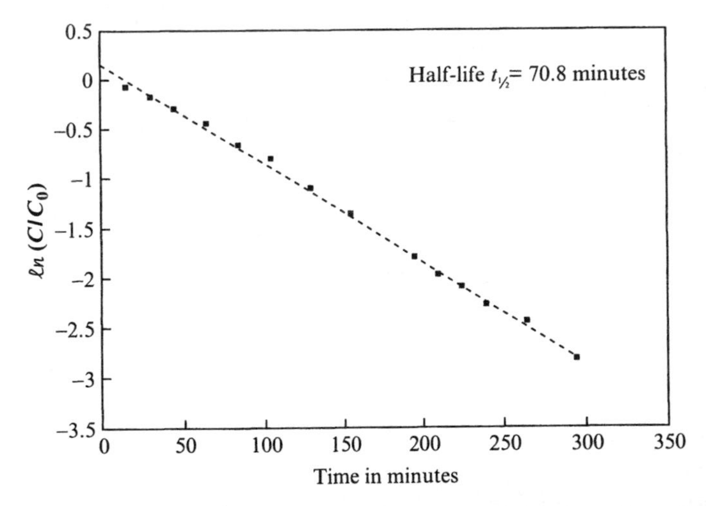

Figure 14-4 First order disappearance of trinitrotoluene from solution in an 800-milliliter beaker with 20 grams wet weight of stonewort (*Nitella* spp.) per liter. Note that ℓn is the natural logarithm, C is the concentration of trinitrotoluene as a function of time, and C_0 is the initial concentration. Analysis by high performance liquid chromatography. Unpublished data from N.L. Wolfe and T.Y. Ou, U.S. EPA laboratory, Athens, Georgia.

until 26 days after plants were introduced. No other transformation products besides ADNT and DANT were identified in these early batch investigations. However, hydroxylamines and conjugation products were detected in other batch reactor tests conducted under somewhat different conditions (Pavlostathis *et al.* 1998, Bhadra *et al.* 1999a, Bhadra *et al.* 1999b, Burken *et al.* 2000, Subramanian and Shanks this book).

Additional studies were conducted at the U.S. EPA laboratory in Athens involving the phytotransformation of radiolabeled TNT by stonewort (*Nitella* spp.). The tests were conducted under highly reducing conditions and resulted in the development of radiolabeled chromatograms, which identified ADNT, DANT, and triaminotoluene as transformation products (Figure 14-5). The triaminotoluene was distributed mostly in the stonewort (*Nitella* spp.), but some occurred in solution. Follow-up work established that triaminotoluene was rapidly broken down into short chain fragments by autoxidation, plant laccase enzymes, or stonewort (*Nitella* spp.) (Figure 14-6). Subsequently, the fragments slowly disappeared from solution and from plant tissue, probably due to conjugation and sequestration.

Nevertheless, subsequent studies conducted by a variety of research groups indicate that triaminotoluene formation under natural conditions is unlikely. For example, Hughes *et al.* (1997), Thompson (1997), Vanderford *et al.* (1997), Bhadra *et al.* (1999a, Bhadra *et al.* 1999b) conducted detailed mass balances on plant material after exposure to TNT. These investigators did not detect triaminotoluene, but instead found large, polar fractions which could not be identified. Thus, triaminotoluene formation only occurs under highly reducing conditions that are not expected in most phytotreatment scenarios.

Consequently, further tests were conducted to characterize transformation products associated with phytoremediation. Larson *et al.* (1999) used a combination of analytical techniques, including liquid scintillation counting, high-performance liquid chromatography, and gel permeation chromatography, to characterize explosives and metabolites of explosives in plant tissues. Analyzing tissues of yellow nutsedge (*Cyperus esculentus*), corn (*Zea mays*), lettuce (*Lactuca sativa*), tomato (*Lycopersicon esculentum*), radish (*Raphanus sativus*), and parrot feather (*Myriophyllum aquaticum*) exposed to TNT and RDX showed that (1) extensive transformation of the explosives occurred, (2) the products had significantly higher polarity and water solubility than the parent compounds, and (3) the molecular weights of the transformation products were significantly greater than those of the parent compounds. These findings suggest that, even though levels of the parent compound in exposed plants are low (less than 10 milligrams per kilogram of plant wet weight), the concentration of extractable higher molecular weight residues of explosives in these plant tissues is high (approximately 3000 milligrams per kilogram of plant wet weight).

These high molecular weight fractions of TNT transformation have not been identified. Nor are there any published references on the toxicity of TNT by-product contaminated plant material. However, the most likely fate of the high molecular weight fraction is conjugation and storage in the cell wall or cell vacuoles. The potential for release of these by-products after cell death and the associated toxicity needs to be determined.

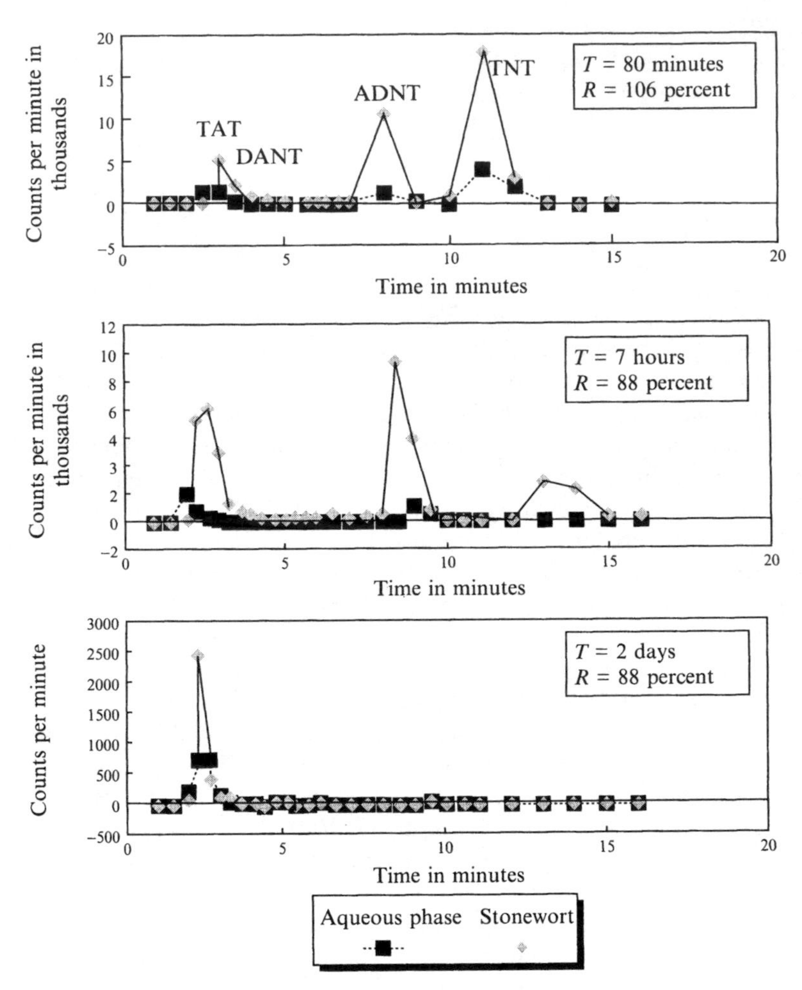

Figure 14-5 [^{14}C] radioactivity in solution and within stonewort (*Nitella* spp.) over time during trinitrotoluene (TNT) transformation to aminodinitrotoluene (ADNT), diaminonitrotoluene (DNT), and triaminotoluene (TAT). The testing was conducted under highly reducing conditions to produce triaminotoluene, which is not observed in photoautotrophic systems otherwise. Labeled trinitrotoluene activity of 47 900 counts per minute per 50 microliters was added to a solution containing 20 grams wet weight of stonewort (*Nitella* spp.) per liter and sealed in three flasks that were sacrificed after 80 minutes, 7 hours, and 48 hours of exposure. The activities of the labeled trinitrotoluene, aminodinitrotoluene, diaminonitrotoluene, and triaminotoluene in solution and in stonewort (*Nitella* spp.) were first separated chromatographically over 16 minutes as shown and then counted. Recoveries, *R*, ranged from 88 to 106 percent. Note the change in scale on each y-axis. Unpublished data from N.L. Wolfe and T.Y. Ou, U.S. EPA laboratory, Athens, Georgia.

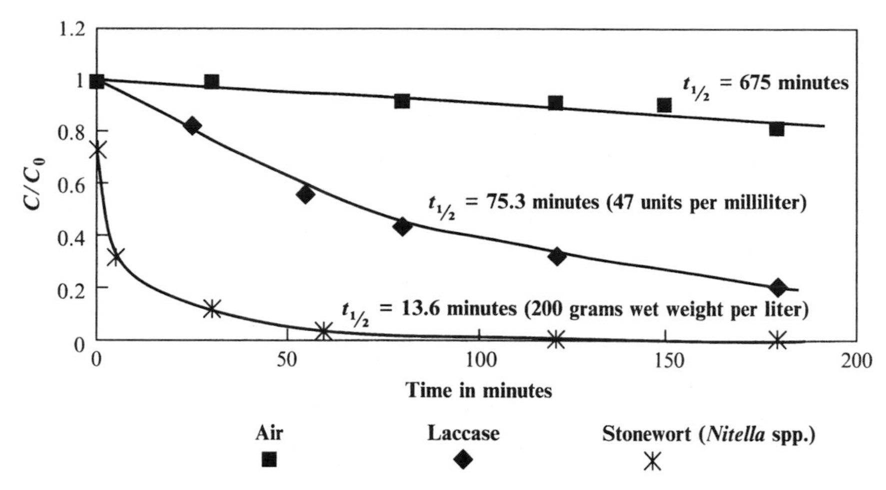

Figure 14-6 Half-lives of triaminotoluene removal from solution in three reactors including in an aerobic solution, in solution with commercially available plant laccase (47 units per milliliter), and in solution with 20 grams wet weight of stonewort (*Nitella* spp.) per liter. Note that C is the concentration of triaminotoluene as a function of time, C_0 is the initial concentration, and $t_{1/2}$ is the half-life of triaminotoluene in solution. Analysis by high performance liquid chromatography with a detection limit of 0.01 milligram per liter and relative standard deviation of 2 percent (Jenkins 1989, U.S. EPA 1992, Wolfe *et al.* 1993). Unpublished data from T.Y. Ou and N.L. Wolfe, U.S. EPA laboratory, Athens, Georgia.

Characterization of Enzymatic Activity

Limited characterization of the enzymatic activity extracted from sediment and identified in plants (*i.e.*, the type I nitroreductase) established that pH (confirmed by Comstock 1996) or dissolved oxygen in the range of most wetlands did not radically affect the reactivity with nitroaromatic compounds (van Beelen and Burris 1995). However, temperature did influence the kinetics leading to investigations to define an Arrhenius–van't Hoff relationship (Andrews 1996, Medina *et al.* 2000b). In addition, proteases or bacteria quickly deactivated the crude extracts of nitroreductase activity (Wolfe *et al.* 1993, van Beelen and Burris 1995). However, the deactivation seems to be an artifact of the crude extract isolation. Indications of nitroreductase activity in long-buried aquifer material imply long-term stability in sediment (also see van Beelen and Burris 1995 on the stability of activity in sediment). Plants also conserve nitroreductase activity (Comstock 1996) even after death (Young 1995). Furthermore, some metals negatively affect the activity, but whole plants that self-engineer pH to neutral also seem to handle mildly elevated zinc and lead in explosives wastes (Wolfe *et al.* 1993). The cofactors, NADPH (ß-nicotinamide-adenine dinucleotide phosphate, reduced) and ferredoxin, are important to increase the reactivity (Wolfe *et al.* 1993), but these factors were readily available in a properly maintained shallow wetland rich in organic material.

Engineering Batch Tests: Ranges of Operating Parameters for Monocultures

After discovery that whole plants and extracted sediment activity removed TNT from solution, Medina and McCutcheon (1996) formulated and used a model of TNT transformation to determine some design parameters. These and other important parameters determined by Andrews (1996), Riveva *et al.* (1998), Medina *et al.* (2000), and others include (1) phytotoxicity, (2) formation and bioavailability of transformation products, (3) concentration effects on uptake and transformation kinetics, (4) monoculture density, (5) prior exposure and acclimation, (6) seasonal temperature effects on monoculture uptake and transformation kinetics, (7) photoperiod, (8) plant nutrition requirements, and (9) effects of enzymatic cofactors. The sequential transformation kinetics determine the retention time of a created wetland or plant lagoon necessary to achieve water discharge standards. Typical cleanup objectives are 0.002 milligram of TNT per liter, 0.01 milligram of RDX per liter, and 0.05 milligram of the total of all explosives present per liter, although these can vary from site to site due to different U.S. state standards and uses. Investigations to determine the effects of sunlight intensity, nutrient concentrations, plant community synergy, and other factors were funded and deferred to the U.S. EPA Hazardous Substance Research Center/South and Southwest to develop a protocol for pilot field testing (Qaisi *et al.* 1996a, 1996b, HSRC 1996, HSRC 1997, Valsaraj *et al.* 1998, Jacobson *et al.* this book). The mixture effects of different explosives, initial degree of organic rich sediment loading in creating the treatment wetland, and many other operational challenges were deferred to pilot field testing for the Volunteer (Best *et al.* 2000a, 2000b), the Milan (ESTCP 1999), and the Iowa Army ammunition plants (Thompson *et al.* this book).

Medina *et al.* (2000b) reported on the initial pilot investigations (bench scale) of the effects of TNT concentration, plant density, and temperature. Andrews (1996) and Best *et al.* (1997b), Best *et al.* (1998a), Best *et al.* (1998b), Best *et al.* (1999a, 1999b) applied and elaborated upon these pilot methods in treatability investigations using contaminated groundwater from the Volunteer, the Milan, and the Iowa Army ammunition plants. In addition, these studies established alternative protocols for plant selection. Best *et al.* (1997b, Appendix A) established that ELISA testing for nitroreductase activity is feasible. However, results in Medina *et al.* (2000b) indicated that 18-hour batch monitoring of single or representative numbers of smaller plants in 200-milliliter (0.05-gallon) beakers for TNT removal from solution was simpler and more adaptable to current design procedures (U.S. EPA 1988). Similar approaches are described by Best *et al.* (1997a), Best *et al.* (1997b), Best *et al.* (1999a, 1999b).

Phytotoxicity to Wetland Plants

Burken *et al.* (2000) and Medina *et al.* (this book) establish that the degree of phytotoxicity depends on the plant species, stage of growth, amount of biomass (Andrews 1996) in contact with the TNT (Comstock 1996, Jacobson *et al.* this book), contaminant bioavailability, and other environmental factors. Because of these factors, the no-observed-adverse-effect-levels range from 1 to 65 milligrams of TNT per liter (Medina *et al.* this book). Nevertheless, this range may be over estimated for treatment wetlands with continuous flows. Batch reactor tests with continuously declining concentrations are the basis of most TNT phytotoxicity data. Lower concentration effects are anticipated in continuous flow reactors due to sustained exposure over long periods.

There is a tradeoff between sensitivity to TNT and rate of uptake and transformation. Parrot feather (*Myriophyllum aquaticum*) is more tolerant of TNT, but the related Eurasian water milfoil (*Myriophyllum spicatum*) typically takes up and transforms TNT faster. Medina *et al.* (this book) describe the phytotoxicity of TNT (undetectable to saturation at 128 milligrams of TNT per liter) for different parrot feather (*Myriophyllum aquaticum*) densities exposed to rapidly declining concentrations for 7 days. Observations by Medina *et al.* (2000b) are consistent in that 200 grams wet weight of parrot feather (*Myriophyllum aquaticum*) per liter did not show obvious phytotoxicity to TNT until concentrations reached 64 milligrams of TNT per liter when severe wilting occurred after 17 hours. The uptake rate constant was 0.96 per day. At 128 milligrams of TNT per liter, parrot feather (*Myriophyllum aquaticum*) died in 14 hours, but also continued to transform TNT (1.2 per day) after death occurred. Eurasian water milfoil (*Myriophyllum spicatum*) densities of 33 grams of wet weight per liter (Comstock 1996, Pavlostathis *et al.* 1998) had a lower threshold to TNT stress (as little as approximately 5 milligrams of TNT per liter), similar to observations for other plants like yellow nutsedge (*Cyprus esculentus*) (Palazzo and Leggett 1986). Within 19 days after the 5-day exposure to 5 milligrams of TNT per liter or more, all milfoil (*Myriophyllum spicatum*) triplicates with 33 grams of wet weight per liter died. During both 5-day and 2-day exposures (long-term plant viability was not investigated), milfoil (*Myriophyllum spicatum*) transformed TNT faster than parrot feather (*Myriophyllum aquaticum*) (Comstock 1996) except at an initial TNT concentration of 1 milligram per liter. Given the limited effects information, treatability studies are necessary for each application until enough information is available to optimize plant species selection, sensitivity, exposure, and biomass density.

Transient Aminodinitrotoluene and Other By-Product Formation

Results from the intensive investigation of TNT phytotransformation pathways (Subramanian and Shanks this book) was informative but may not yet be complete enough for all design and operational decisions. Batch studies established that the transformations of ADNT and DANT were typically slower than the depletion of TNT from solution. These studies involved water milfoil (*Myriophyllum spicatum*) (Pavlostathis *et al.* 1998), parrot feather (*Myriophyllum aquaticum*) (Medina and McCutcheon 1996, Hughes *et al.* 1997, Vanderford *et al.* 1997, Medina *et al.* 2000b), and other aquatic plants (Carter 1996, Gueriguian 1996, Rivera *et al.* 1998, Medina *et al.* 2000b). The ADNT and DANT usually occur in low concentration in both solution and tissue, and represent a small fraction of the total TNT mass removed (Pavlostathis *et al.* 1998, Bhadra *et al.* 1999a, Bhadra *et al.* 1999b, Burken *et al.* 2000, Medina *et al.* 2000b). Furthermore, the known by-products were no more and usually less toxic than TNT. Therefore, TNT transformation at least significantly reduced the risk associated with aqueous exposure. For optimal treatment, however, by-product transformation limits wetland design.

To better quantify the limitations of by-product transformation, batch studies also were conducted to investigate the removal of ADNT isomers (2-ADNT and 4-ADNT) from solution by parrot feather (*Myriophyllum aquaticum*) at a plant density of 20 grams of wet weight per liter (Rivera *et al.* 1998). The removal rate constants of both of these isomers were nearly equal: 0.17 per day for 4-ADNT and 0.20 per day for 2-ADNT. Thus, routine monitoring may not need to distinguish isomers of ADNT.

Later studies (Medina, Bergstedt, and McCutcheon, unpublished) compared ADNT removal by different plants at densities of 50 grams of weight wet per liter. The pseudo first order removal rate constant for ADNT removal by waterweed (*Elodea canadensis*) was 0.76 per day, compared to 0.18 per day for parrot feather (*Myriophyllum aquaticum*). This is particularly interesting because the TNT removal rate constant (at plant densities of 50 grams of wet weight per liter) for waterweed (*Elodea canadensis*) (0.72 per day) was approximately half that for parrot feather (*Myriophyllum aquaticum*) (1.8 per day) (Rivera *et al.* 1998, Medina *et al.* 2000b). This suggests that plant communities should ultimately be the most effective phytotreatment technology for compounds that undergo sequential transformation, replacing monocultures that are more at risk to disease, pests, and weather. Alternatively, segmented treatment wetlands could be designed with smaller compartments of plants expected to rapidly remove TNT followed by treatment compartments with different plants to remove by-products faster. More sustainable, elongated plug-flow wetlands could also be planted with communities and allowed to self-select the most durable succession of plants along the concentration gradients of TNT and by-products.

Until the exact fate and toxicity of ADNT, DANT, hydroxylaminodinitrotoluenes, and other transformation products are fully known, some care will be necessary to monitor by-products in the field. The fate of ADNT in plant tissue (Rivera *et al.* 1998) must be monitored to be certain that this toxic transformation product and DANT become further reduced or irreversibly bound into plant tissue. Monitoring the treatment wetlands installed at the Iowa Army Ammunition Plant (Thompson *et al.* this book) establish that 2 years of quarterly plant tissue monitoring may be sufficient for some aquatic plants.

Concentration Effects on Kinetic Rate Constants

The removal kinetics for TNT are typically first order for a constant amount of plant biomass (Figure 14-4; Young 1995, Comstock 1996, Pavlostathis *et al.*

1998, Medina *et al.* 2000b). However, the rate constant is actually second order relative to plant density (surrogate for enzymatic activity) and decreases with increasing TNT concentrations (Figure 14-7). Different plants follow different relationships, but most test results available to date have the same general trend of going from high values for the rate constant at TNT concentrations less than 2 to 3 milligrams per liter to small rate constants at higher concentrations. Phytotoxic or inhibitory effects (Comstock 1996, Gueriguian 1996, Pavlostathis *et al.* 1998, Jacobson *et al.* this book) and saturation of the enzymatic activity of the plants (Medina *et al.* 2000b) were suggested causes of the relationship between rate constants and TNT concentration, but multiple enzymes also may be involved. Until the kinetics are better understood, treatability and pilot investigations are necessary for almost all applications. The design implications for a constructed wetland treating TNT contaminated water are

1. Plug flow in a single constructed wetland or a series of small wetlands should require smaller detention times compared to continuously mixed reactors and thus require a lagoon with smaller area and volume
2. Hardier plants may be needed or higher maintenance plant replacement may need to take place at the head (influent) of the lagoon or wetland system where the highest concentrations occur

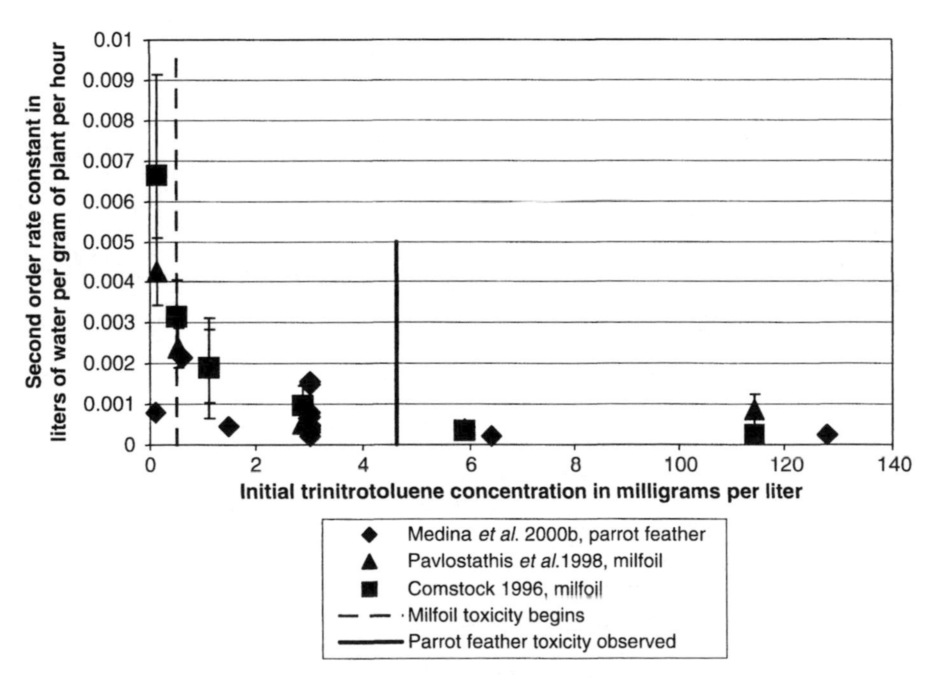

Figure 14-7 Effect of trinitrotoluene concentration on batch second order rate constants for trinitrotoluene uptake and transformation. All tests were conducted at 22 °C. Plants: Parrot feather (*Myriophyllum aquaticum*) and Eurasian water milfoil (*Myriophyllum spicatum*).

Plant Density Effects

Young (1995), Andrews (1996), Pavlostathis *et al.* (1998) and Medina *et al.* (2000b) confirmed early observations that increased plant density increases TNT removal rate constants in batch reactors (Figure 14-8). Different relationships exist for each combination of initial concentration and species of plant. Generally, the highest possible sustainable plant density should be the design and maintenance objective. The appearance of a red–pink color in the reactor due to the photolysis of TNT and the associated transformation products is an indication that plant density is too sparse (Young 1995, Comstock 1996).

Effect of Previous Exposure

No lag in first order TNT uptake and transformation has been observed (McCutcheon *et al.* 1995). Single generation plant cultures simply cannot undergo the genetic selection that several generations of bacteria may undergo to acclimate to some toxic conditions (see Medina *et al.* this book on plant tolerance). Furthermore, no protein expression seems necessary for TNT uptake and transformation. However, Comstock (1996), Pavlostathis *et al.* (1998), and Medina *et al.* (2000b) have investigated the effects of prior exposure to TNT. Comstock (1996) noted that as multiple TNT doses were added to milfoil (*Myriophyllum spicatum*) and allowed to decrease to near detection levels (0.1 milligram per liter), the transformation rate constant decreased

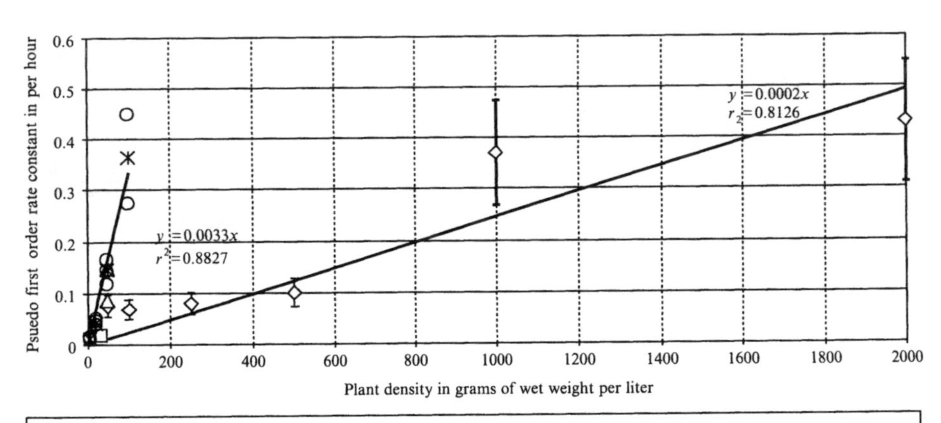

Figure 14-8 Relationships between plant density (x) and pseudo first order rate constants (y) for trinitrotoluene. All tests were conducted at 22 °C. Plants and initial concentrations: parrot feather (*Myriophyllum aquaticum*) exposed to 30 milligrams of trinitrotolune per liter and Eurasian water milfoil (*Myriophyllum spicatum*) exposed to 29 milligrams of trinitrotoluene per liter. The correlation coefficient, r^2, describes how well a linear fit relates plant density and the rate constant.

from one dose to the next. Similarly, for 200 grams wet weight of parrot feather (*Myriophyllum aquaticum*) per liter exposure to 30 milligrams of TNT per liter for 48 hours, prior exposure reduced the transformation rate. Medina *et al.* (2000b) found faster TNT transformation (3.8 per day or a half-life of 4.4 hours) than the same 48-hour exposure plus exposure to TNT-contaminated water for 3 months prior (2.3 per day or a half-life of 7.3 hours). Brief 3.5-hour exposures of milfoil (*Myriophyllum spicatum*) to 1.2 milligrams of TNT per liter did not seem to induce any more enzymatic activity or reduce the second order transformation rate constant normalized by plant biomass (Comstock 1996). However, high variability from plant to plant (17 to 48 percent in pseudo first order rate constants) may have obscured any effect of brief, prior exposure.

Young (1995) found that care is necessary in introducing high concentrations of TNT to parrot feather (*Myriophyllum aquaticum*) and other aquatic plants. Parrot feather (*Myriophyllum aquaticum*) died when introduced to concentrations in excess of 100 milligrams of TNT per liter. However, when introduced before high concentrations buildup from the same source, the plants were able to reduce concentrations to those that were not phytotoxic.

Temperature Effects on Kinetics

Andrews (1996) and Medina *et al.* (2000b) determined that TNT transformation rates of parrot feather (*Myriophyllum aquaticum*) follow the expected Arrhenius–van't Hoff relationship between 2 and 34 °C. Between 34 and 42 °C, the rate coefficient was approximately constant with temperature, but parrot feather (*Myriophyllum aquaticum*) was visibly stressed at the higher temperature. At 54 °C, the rate of transformation was zero and the plant died due to denaturing of the enzymatic processes involved (Medina *et al.* 2000b). van Beelen and Burris (1995) found a similar response of the nitroreductase activity isolated from sediments.

Photoperiod Effects

Bench tests by Comstock (1996) suggest that reductive transformation of TNT by aquatic plants does not seem to be directly related to photosynthesis over short periods (that is of 5 days), except for perhaps the initial diel (24-hour light and dark) period. Nevertheless, Jacobson *et al.* (this book) noted a longer-term toxic effect of TNT on photosystem II of plants (chlorophyll *a*-protein complex with maximum light absorption at 680 nanometers) that may explain the death of Eurasian water milfoil (*Myriophyllum spicatum*) within 19 days of exposure.

Plant Nutrition

Nutrient requirements are an important consideration in plant selection. Pavlostathis *et al.* (1998) noted that Eurasian water milfoil (*Myriophyllum spicatum*) required an inorganic carbon source that caused changes in pH

TABLE 14-1 Nitroaliphatics, Nitroaromatics, Triazines, Tetraazocines, and Some Associated Transformation Products Associated with Explosives Wastes that Undergo Phytotransformation

Structure or formula	Functional group or compound name	Pseudo first order rate constant[a] (per day)	Source
CH_3NO_2	Nitromethane	8.4	Wolfe *et al.* unpublished
$CH_3CH_2NO_2$	Nitroethane	7.6	Wolfe *et al.* unpublished
 X; NO_2	Function groups X: NO_2, $CH_3C{=}O$, CN, Cl, Br, H, CH_3O, CH_3, ethyl – CH_3CH_2, butyl – $CH_3CH_2CH_2CH_2$, and octyl – $CH_3CH_2CH_2CH_2CH_2CH_2CH_2CH_2$	1.6 to 16	Wolfe *et al.* unpublished
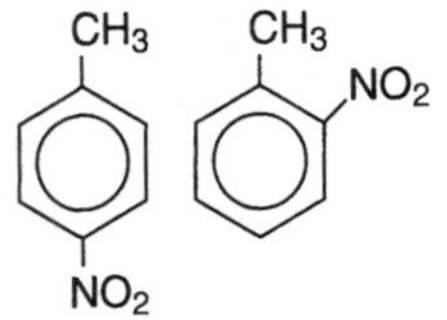 CH_3; NO_2; CH_3; NO_2	Nitrotoluene	0.56	Andrews (1996)
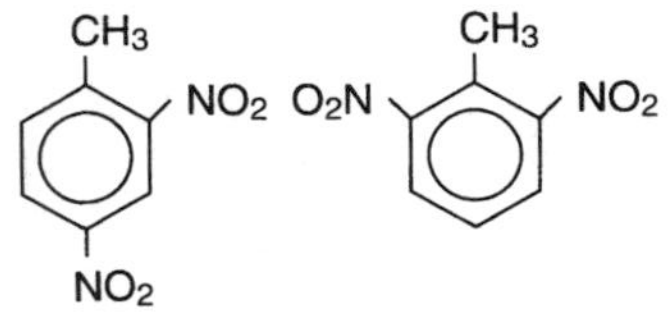 CH_3; NO_2; NO_2; CH_3; O_2N; NO_2	Dinitrotoluene	0.5	Andrews (1996), Todd (1996)
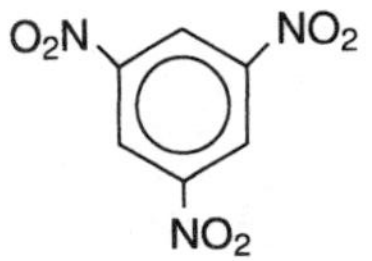 O_2N; NO_2; NO_2	1,3,5-Trinitrobenzene	3.2 to 4.8	Andrews (1996), Wolfe *et al.* unpublished

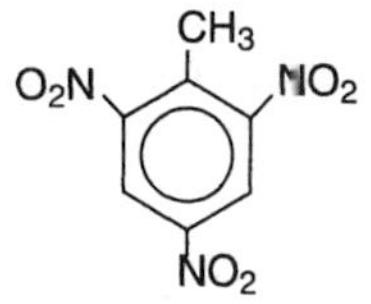 CH_3, O_2N, NO_2, NO_2	2,4,6-Trinitrotoluene	0.12^b to 50	Andrews (1996), Best *et al.* (1997a), Medina *et al.* (2000b), Wolfe *et al.* unpublished
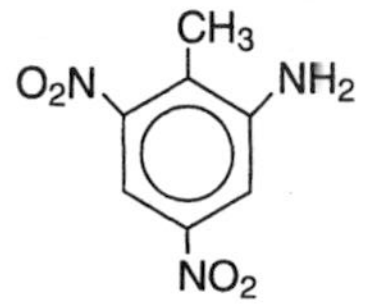 CH_3, O_2N, NH_2, NO_2	2-Amino-4,6-dinitrotoluene	0.17 to 0.46	Andrews (1996), Rivera *et al.* (1998)
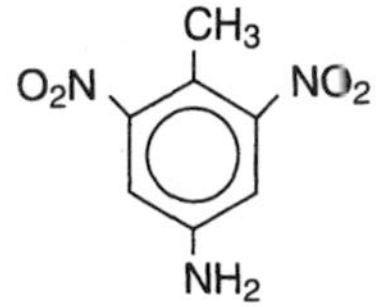 CH_3, O_2N, NO_2, NH_2	4-Amino-2,6-dinitrotoluene	0.17 to 0.76	Andrews (1996), Rivera *et al.* (1998), Medina, Bergstedt, and McCutcheon, unpublished
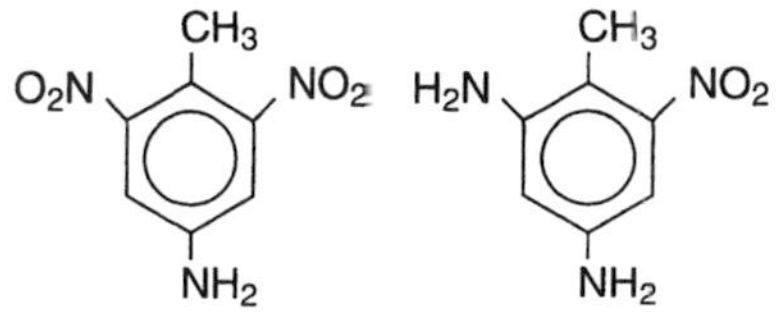 CH_3, O_2N, NO_2, NH_2; CH_3, H_2N, NO_2, NH_2	Diaminonitrotoluene	0.0015	Rivera *et al.* (1998)

(continues)

TABLE 14-1 ***(continued)***

Structure or formula	Functional group or compound name	Pseudo first order rate constant[a] (per day)	Source
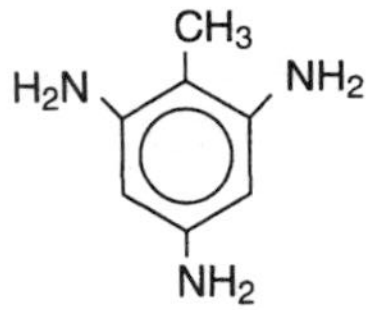	Triaminotoluene	73	Wolfe *et al.* unpublished
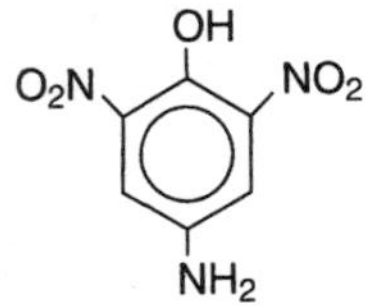	2,4,6-Trinitrophenol (picric acid)	0.50	Wolfe *et al.* unpublished
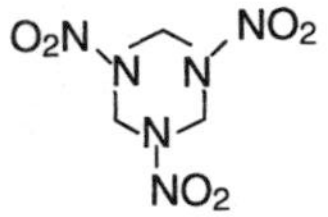	Hexahydro-1,3,5-trinitro-1,3,5-triazine (RDX)	0.001 to 12[c]	Best *et al.* (1997a), Concurrent Technologies Corp. (1997), Rivera *et al.* (1998), Wolfe *et al.* unpublished
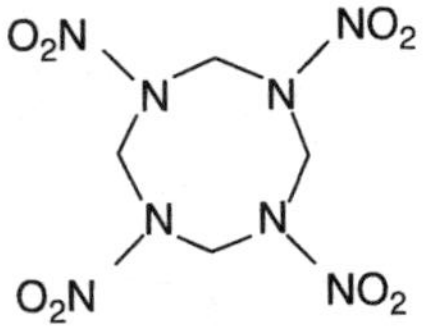	Tetranitro-1,3,5,7-tetraazocine (HMX)	0.24 to 0.52	Rivera *et al.* (1998), Wolfe *et al.* unpublished

[a]The reported rate constants are quite sensitive to several factors, and the ranges presented primarily derive from screening with a range of different plant densities and other factors.
[b]Lower values due to very low plant biomass, high toxicity, and low temperature. Typical value = 14 per day.
[c]Higher rate constants from pilot optimization of plant and sediment systems. Typical value = 0.64 per day.

as bicarbonate was taken up by the growing plants (Comstock 1996). By contrast, parrot feather (*Myriophyllum aquaticum*), which was slower in transforming TNT, seemed much hardier; no nutrients or bicarbonate were necessary before or during batch testing (Medina *et al.* 2000b), some of which occurred over several months. Furthermore, even though stonewort (*Nitella* spp.) transformed TNT faster, parrot feather (*Myriophyllum aquaticum*) was chosen for further studies because stonewort (*Nitella* spp.) was more difficult to obtain and had sensitive nutritional requirements (Medina *et al.* 2000b).

Enzymatic Cofactors

Comstock (1996) and Pavlostathis *et al.* (1998) found that sodium azide treatments of 2000 milligrams per liter shut down all TNT transformation by water milfoil (*Myriophyllum spicatum*). Because sodium azide blocks electron flow and adenosine triphosphate (ATP) molecular synthesis, this is an independent confirmation of the importance of NADPH or other electron sources for the reduction of TNT in plant systems.

Laboratory Studies of Water Contaminated with Other Explosives

Table 14-1 presents the structure of various explosives and some plant metabolites of these compounds, including RDX, HMX, trinitrobenzene, and picric acid, along with experimentally obtained pseudo first order rate constants for the disappearance of each compound from water. Not included are tetryl (Harvey *et al.* 1992), nitrocellulose, nitroglycerine, and pentaerythritol tetranitrate (Schwitzguébel and Vanek this book). Most notable from Table 14-1 is that structurally different HMX (a tetraazocine) and RDX (a triazine) are approximately an order of magnitude slower in uptake and transformation by plants than nitroaromatic compounds. In fact, at low plant densities and for short reaction periods, RDX can appear nonreactive (Carter 1996). Therefore, RDX kinetics are expected to be design limiting for phytoremediation of wastewaters containing mixed explosives of comparable concentrations and may require coupled phyto and microbial remediation with significant redox control. Because of lesser mobility, HMX does not occur in concentrations comparable to those for TNT and RDX. This implies that HMX kinetics may not be limiting despite the slower kinetics compared to TNT.

Less is know about the transformation pathways of other explosives besides TNT. Nevertheless, Larson *et al.* (1999) finds that all transformation products known to date (2003) seem to be sequestered. Neither Larson *et al.* (1999) nor Thompson *et al.* (this book) observed RDX accumulation in aquatic plants.

Pilot Bench Testing: Milan and McAlester Army Ammunition Plant Wastewaters

The Georgia Institute of Technology conducted pilot testing of phytoremediation to manage explosives-contaminated wastewaters for the National Defense Center for Environmental Excellence (NDCEE) (Concurrent Technologies Corp. 1997, Qazi and Kanaras 1997). The NDCEE screened 34 treatment technologies to replace the activated carbon sorption currently used at active U.S. Army ammunition plants. Of the five technologies selected by NDCEE for pilot testing, only phytoremediation, a thermophilic microbial transformation on granular activated carbon, and a fluidized bed bioreactor were recommended for pilot field testing. The recommendations were based upon criteria selected by NDCEE prior to screening and bench testing. The criteria were weighted as follows: (1) total life cycle costs—35 percent, (2) effluent quality—25 percent, (3) process operability—15 percent, (4) flexibility—10 percent, (5) health and safety—10 percent, and (6) commercial availability—5 percent. The phytoremediation system tested by the Georgia Institute of Technology in batch reactors reduced the total amounts of TNT, RDX, HMX, dinitrotoluene, and trinitrobenzene in the wastewaters of the McAlester (Texas) and the Milan (Tennessee) Army ammunition plants by 98 to 99 percent, respectively. The system consisted of parrot feather (*Myriophyllum aquaticum*), Canadian waterweed (*Elodea canadensis*), bulrush (*Scirpus* spp.), and sediment. The NDCEE expected to meet a criterion for surface water discharge of 0.05 milligram per liter for all explosives present (typically called nitrobodies) after scale-up to an on-site pilot test for phytoremediation.

Full-Scale Pilot Testing: Contaminated Groundwater at Iowa Army Ammunition Plant

Thompson *et al.* (this book) details the successful operation of the two created treatment wetlands at the Iowa Army Ammunition Plant, which are fed by groundwater contaminated with TNT and RDX. Since the wetlands became operational in 1998, all of the TNT and RDX have been completely contained and removed by the wetland processes. The stored waters in the treatment units increased in explosives concentration during the winter recharge period and then decreased during the spring plant growth phase until undetectable levels of the explosives were achieved by the plant–microbial ecosystem during the late summer and fall. Except for screening pilot investigations by Best *et al.* (1997a), Best *et al.* (1997b), Best *et al.* (1997d), Best *et al.* (1998b) and Thompson (1997), the U.S. Army deferred bench and intermediate pilot testing at the Iowa Plant in favor of full-scale pilot testing in the field. Until the optimization of the wetlands to treat more groundwater and surface water contaminated with explosives, this field testing is classified as phytorestoration. However, the application to date does establish the field proof of concept for groundwater treatment and has the potential to establish the field proof of principle if optimization goes forward. Nevertheless, phytorestoration or cleanup of widespread light to moderate levels of contamination has been the primary objective of the development of this technology from the beginning.

BATCH AQUEOUS TREATMENT OF TRINITROTOLUENE DIFFUSING FROM CONTAMINATED SOIL

Screening investigations of the phytotransformation of explosives in six soils were conducted in 125-milliliter and 800-milliliter laboratory batch reactors (Wolfe *et al.* 1993). The results were used to design a pilot field test with 200-liter (50-gallon) batch reactors in the field at the Alabama Army Ammunition Plant (Young 1995). The laboratory and field batch reactors involved different soil depths and TNT concentrations.

Bench Tests: Six Army and Navy Facilities

Soils with TNT concentrations ranging from 0.4 to 1.3 milligrams of TNT per kilogram of dry soil were used to demonstrate the feasibility of using wetland systems to treat soils contaminated with explosives. The soils came from the Radford Army Ammunition Plant (Virginia), the Lone Star Army Ammunition Plant (Texas), the Iowa Army Ammunition Plant, and the Submarine Base Bangor (Washington State). The testing occurred in 125-milliliter batch reactors with 10 grams (0.02 pound) of soil and 50 milliliters of distilled water. The reactors also had 10 grams (0.02 pound fresh weight) of stonewort (*Nitella* spp.) added, while control reactors were identically prepared and maintained except that no plants were added. In addition, similar tests were set up using soils from the McAlester Army Ammunition Plant in Texas and the Nebraska Ordinance Plant at Hastings. The McAlester soil contained 610 milligrams of TNT per kilogram of dry soil, 61 milligrams of ADNT per kilogram of dry soil, and 44 milligrams of DANT per kilogram of dry soil. The Hastings soil contained 1200 milligrams of TNT per kilogram of dry soil, 19 milligrams of ADNT per kilogram of dry soil, and 2.4 milligrams of DANT per kilogram of dry soil. As these latter two soils had higher TNT concentrations, 100 milliliters of distilled water and 30 grams (0.07 pound of wet weight) of stonewort (*Nitella* spp.) were added to 250-milliliter reactors. The water and sediment in the unmixed reactors were sampled at days 0, 3, 7, 11, 29, and 43 (Wolfe *et al.* 1993).

The batch reactors achieved a typical risk based cleanup level of 30 milligrams of TNT per kilogram of dry soil by day 3 for the McAlester soil, and by day 43 for the Hastings soil (Figure 14-9). The four lightly contaminated soils started with TNT levels below 30 milligram per kilogram of dry soil and reached undetectable levels (0.01 milligram of TNT per kilogram) by day 43,

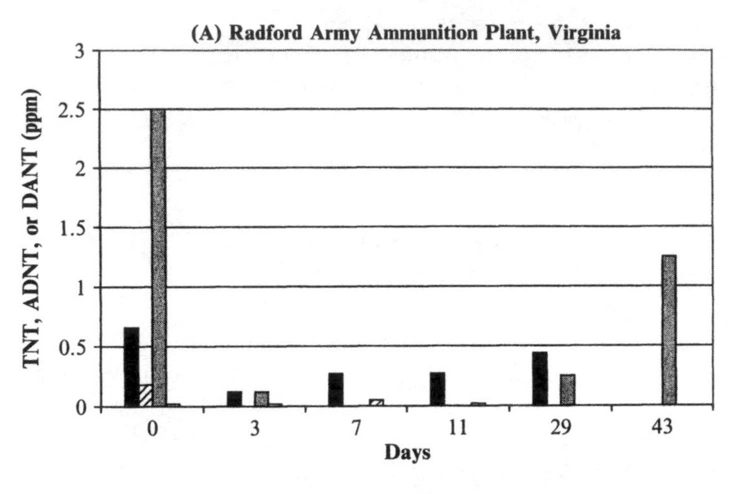

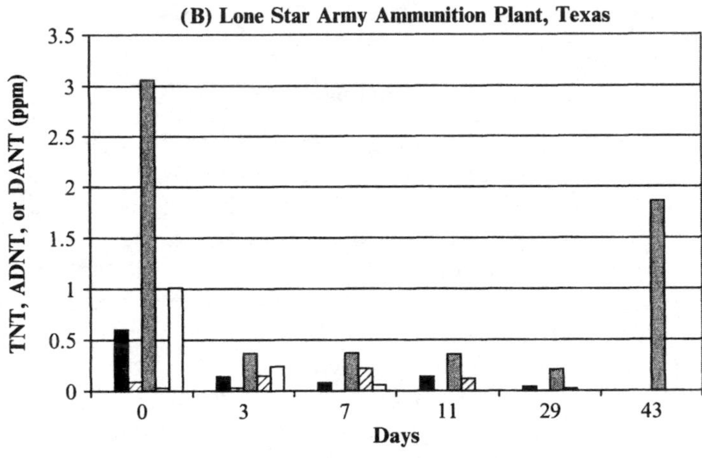

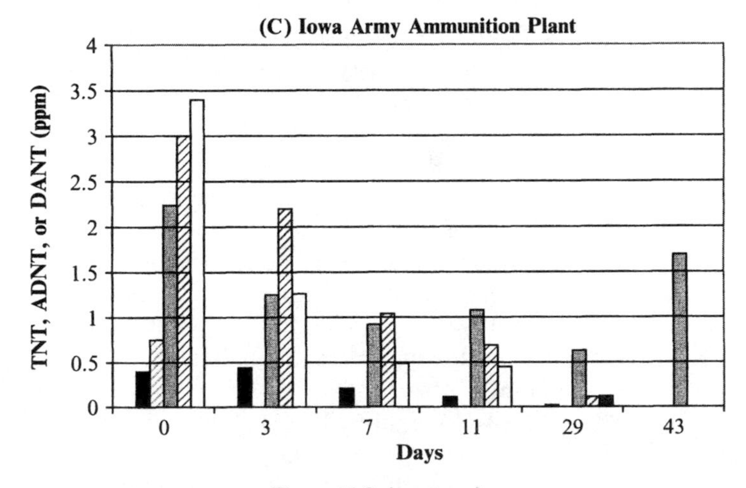

Figure 14-9 (*continues*)

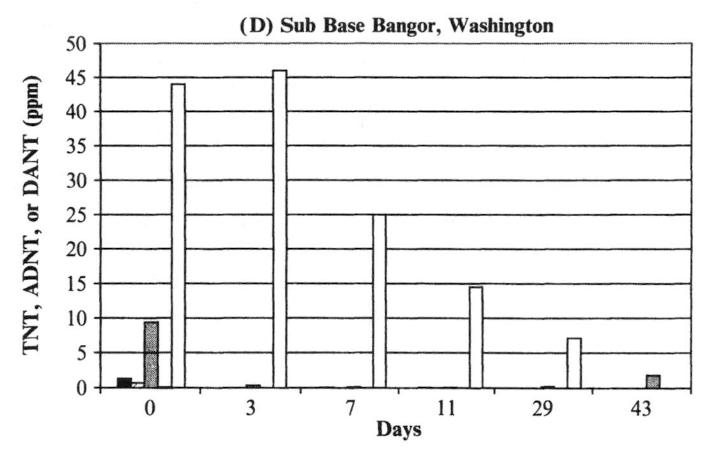

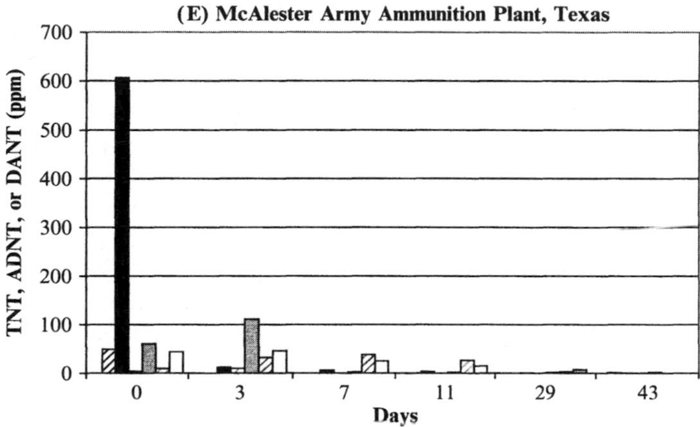

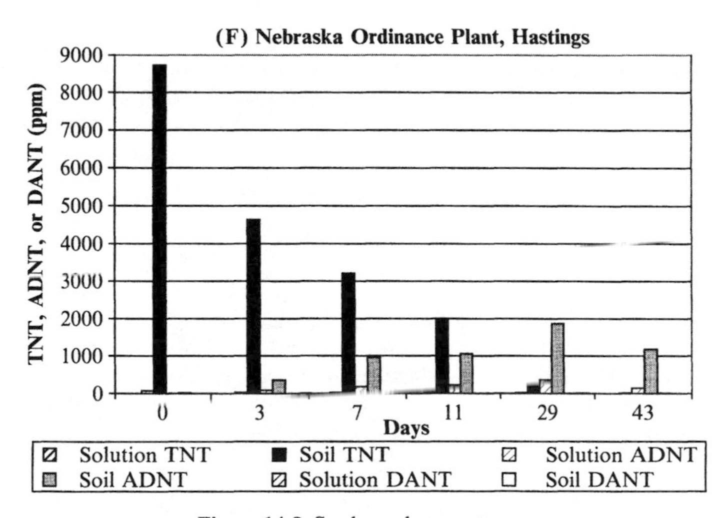

Figure 14-9 See legend on next page

as did the McAlester soil. For the highly contaminated soils, longer treatment times (beyond 43 days) may be necessary to achieve typical cleanup levels for water, even after attaining target soil-cleanup levels. Longer operational times were also necessary to achieve transformation of the dinitro- and diamino-isomers produced by TNT breakdown. The products in the lightly contaminated soils do not build up to levels of concern because the initial TNT concentrations are so low. In the McAlester soil, the products build up and begin decreasing sequentially, reflecting the pathway for transformation. For the highest levels of soil TNT contamination found at the Nebraska Ordinance Plant, the products continually increased. Clearly, the sequential product formation and rates of transformation are the process-limiting steps, requiring additional pilot investigation before phytoremediation can be applied to the flooded Nebraska Ordinance Plant soils.

Bench Proof of Concept: Alabama Army Ammunition Plant

Four 1-liter (0.26-gallon) laboratory bench tests were conducted using homogenized soil from the Alabama Army Ammunition Plant in Childersburg with an initial soil TNT concentration of 6000 milligrams per kilogram of dry soil (Young 1995). The mass of soil added to each reactor was 47 grams (0.104 pound, approximately 0.5 millimeters or 0.2 inch deep) in two reactors, 84 grams (0.19 pound, approximately 10 millimeters or 0.4 inch deep), and 165 grams (0.36 pound, approximately 20 millimeters or 0.8 inch deep) with 800 milliliters of distilled water. The reactor soil diffused TNT under quiescent conditions until a saturation level of approximately 128 milligrams of TNT per liter occurred after 210 hours in all four reactors (Figure 14-10). After 210 hours, 50 grams (0.11 pound) of stonewort (*Nitella* spp.) was added to each reactor. Within 40 hours, the aqueous concentrations of TNT dropped to near detection limits. Within 95 hours, all aqueous concentrations of TNT were below detection limits.

The initial 210-hour equilibration of TNT in solution over the three depths of contaminated soil (5, 10, and 20 millimeters or 0.2, 0.4, and 0.8 inch) indicated that the release of TNT was most likely controlled by diffusion at the soil–water interface. A numerical simulation of the concentrations in solution and in the soil (Criscenti 1995) confirmed that molecular diffusion was the dominant transport mechanism but that some soil mixing and diffusion of plant activity into the soil was also possible. Over the fourfold range of

Figure 14-9 Kinetics of trinitrotoluene and by-product removal from soil and water for six contaminated U.S. Army and U.S. Navy facilities (A to F). Analysis was done by high performance liquid chromatography with a detection limit of 0.01 milligram per liter for days 0, 3, 7, 11, and 29. Day 43 samples were analyzed by micellar electrokinetic capillary electrophoresis with a detection limit of 0.5 milligram per liter. The relative standard deviation for analysis by chromatography and electrophoresis was 2 percent (Wolfe *et al.* 1993). Note the different scale for each bar chart. Abbreviations—ppm: parts per million or milligrams per liter for concentrations in solution, and milligrams per kilogram for contaminant concentrations in soil; TNT: trinitrotoluene; ADNT: aminodinitrotoluene; and DANT: diaminonitrotoluene.

sediment depths in the reactors, the diffusive flux was not sensitive to sediment depth, substantiating the conclusion that molecular diffusion was the dominant transport mechanism.

In the initial investigations, the health of the plants and algae were not closely monitored. Nutrients were not added and the light levels from normal fluorescent lighting were not supplemented, nor was an attempt made to mimic normal photoperiods. Organic carbon was approximately 0.8 percent and soil pH was 4.85 (Criscenti 1995). For a sample collected in the same area, Young (1995) found that the TNT-contaminated soil contained 0.2 percent nitrogen and 0.9 percent carbon. Potassium was 59 milligrams per kilogram and phosphorus 6 milligrams per kilogram of soil on a dry weight basis. Because of these factors, the shock of being added to water at near saturation levels of TNT, and the cumulative effects of TNT and the metabolites, stonewort (*Nitella* spp.) died in each reactor after approximately 500 hours (3 weeks). Both live and dead stonewort (*Nitella* spp.) continued to transform TNT until after 700 hours, when the transformation rate decreased to less than the rate of diffusion from the contaminated soil (Figure 14-10).

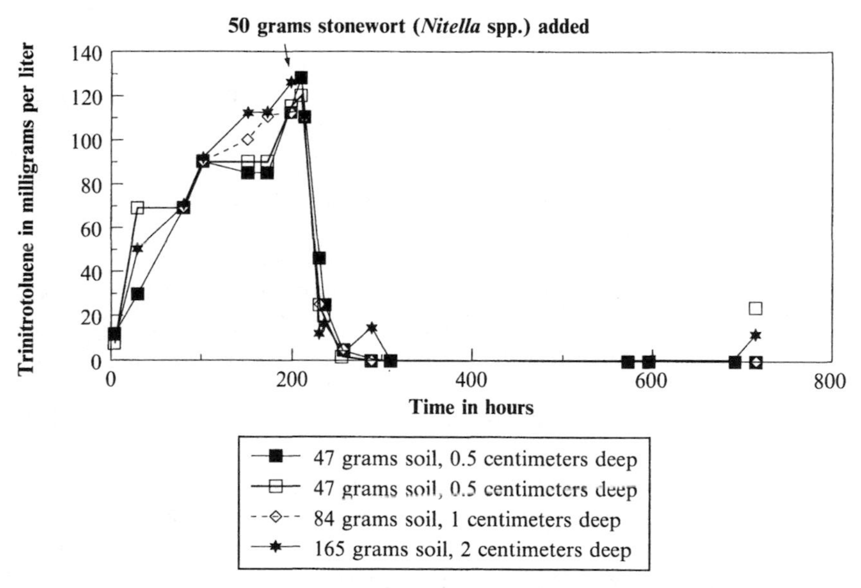

Figure 14-10 Bench proof of concept testing of created wetland concept using Alabama Army Ammunition Plant (Childersburg) contaminated soil (6000 milligrams of 2,4,6-trinitrotoluene per kilogram of dry soil) at three depths of soil. Stonewort (*Nitella* spp.) density was 63 grams of wet weight per liter. Analysis by high performance liquid chromatography with a detection limit of 0.01 milligram of 2,4,6-trinitrotoluene per liter and relative standard deviation of 2 percent (Jenkins 1989 U.S. EPA 1992, Wolfe *et al.* 1993. Unpublished data from N.L. Wolfe *et al.*, U.S. EPA laboratory, Athens, Georgia.

These investigations and others (Young 1995) established that nitroreductase activity continues after the death of macrophytic algae and plants. The activity was initially sufficient to outstrip the mass-transport limited diffusion from the sediments, but eventually bacterial proteases and other factors reduced the activity (van Beelen and Burris 1995) until the TNT buildup in the water column started anew.

Pilot Field Investigation: Alabama Army Ammunition Plant

Young (1995) adopted the concept for treatment of TNT-contaminated soil using created wetlands and scaled up to 200-liter (50-gallon) batch reactors. Field Test 1 consisted of two plant-free controls (pond water and contaminated soil) and six reactors containing contaminated soil, pond water, and parrot feather (*Myriophyllum aquaticum*) in a temporary greenhouse at the Alabama Army Ammunition Plant. The greenhouse was open-ended to maintain natural temperature fluctuations while keeping rainfall out of the reactors.

The reactors were filled with 64 kilograms (140 pounds) of contaminated soil to a depth of 40 to 50 millimeters (1.5 to 2 inches) and 190 to 230 liters (50 to 60 gallons) of unfiltered pond water. After 1 hour for settling, 4.5 kilograms (10 pounds) of parrot feather (*Myriophyllum aquaticum*) collected from a nearby pond was added to six of the eight reactors (randomly selected) to achieve a plant density of 20 grams of wet weight per liter.

The initial TNT concentration in the soil placed in the reactors was approximately 4000 milligrams per kilogram of dry soil, although inhomogeneity occurred, primarily due to small nuggets of crystalline TNT. The contaminated loam was prepared for testing by thoroughly mixing and screening through a 6.4-millimeters (0.25-inch) mesh to remove twigs, leaves, stones, large TNT nuggets, and soil clumps.

Water samples were collected from the reactors approximately weekly for 13 weeks from November 26, 1993 to February 28, 1994. During each sampling effort, one sediment sample also was randomly collected from each quadrant of the reactor.

The results were mixed. Statistically, the soil concentrations of TNT did not change during testing in the plant reactors and were not distinguishable from the soil concentration in the plant-free reactors. This was due to the short testing times and because of the mass-transport limitation of TNT diffusing across the soil–water interface. However, the plant reactors did maintain lower aqueous TNT concentrations compared to the plant-free reactors (Young 1995). By week 6, the plants began to darken in color and showed no evidence of plant growth thereafter. By week 8, all plants in all reactors were dead, requiring transplantation of new plants. Problems with plant survival could have been related to the cumulative toxicity of TNT and the products of transformation, light levels, perhaps nutrients, and the handling of the plants during collection and transplantation.

Winter temperatures, which ranged from – 6 to 27 °C, influenced the results as reactor temperature affects parameters such as TNT solubility, TNT diffusion, and plant activity. However, because plant reactor TNT concentrations were normally less than the detection limits, the effect of the lower temperature was not significant for this particular experimental design.

Field Test 2 was an expanded test with five planted reactors and five unplanted controls (from May 19 to September 6, 1994). The same plant density of 20 grams of plant (wet weight) per liter was used. Figure 14-11 shows the water concentrations of TNT over time in the five reactors with plants and in the five plant-free reactors. From week 1 until week 4, small amounts of TNT were measured in solution in the parrot feather (*Myriophyllum aquaticum*) reactors, and after that the aqueous TNT levels were undetectable (less than 0.01 milligram per liter). In the plant-free reactors, TNT in solution increased during the 15-week investigation, once again implying that TNT transformation is mass-transport limited at the sediment–water interface.

The traces of TNT observed during the first weeks in the plant reactors may have occurred due to unintended mixing. After the rate of parrot feather (*Myriophyllum aquaticum*) removal exceeded the rate of TNT diffusion from the sediments, the water column concentrations quickly dropped to undetectable levels (0.01 milligram per liter). Observed aqueous TNT concentrations in the parrot feather (*Myriophyllum aquaticum*) reactors in Test 2 were less than those of Test 1 (5 to 9 milligrams of TNT per liter; Young 1995). This was probably due to the effect of summer temperature on the biological rates of transformation. The formation of ADNT and DANT, combined with low numbers of microorganisms (on the order of 80 to 200 per milliliter; Young 1995) established that plant transformation dominated TNT remediation in these field reactors.

In the field controls, Young (1995) reported that some residual enzymatic activity remained in the pond water used in the field tests, probably associated with colloidal particles. This activity prolonged the achievement of saturated levels of TNT by diffusion from the contaminated soil in the field plant-free reactors. Approximately 14 percent of saturation (128 milligrams of TNT per liter at 20 °C) was achieved in 15 weeks (Figure 14-11).

All plant-free reactors in both series of pilot field tests were pink in color, characteristic of photolytically induced complexes of TNT occurring when TNT-contaminated water is in sunlight. The parrot feather (*Myriophyllum aquaticum*) reactors were colorless and slightly turbid. Another striking difference was that tadpoles, snails, and various insect larvae (spp. unknown) grew in the parrot feather (*Myriophyllum aquaticum*) reactors but not in the plant-free reactors.

Laboratory toxicity tests established that the tadpoles, most likely of the southern toad (*Bufo* spp.), survived very well in parrot feather (*Myriophyllum aquaticum*) reactors but died within hours in plant-free reactors (Young 1995). In both field Tests 1 and 2, a green mat of unidentified algae and

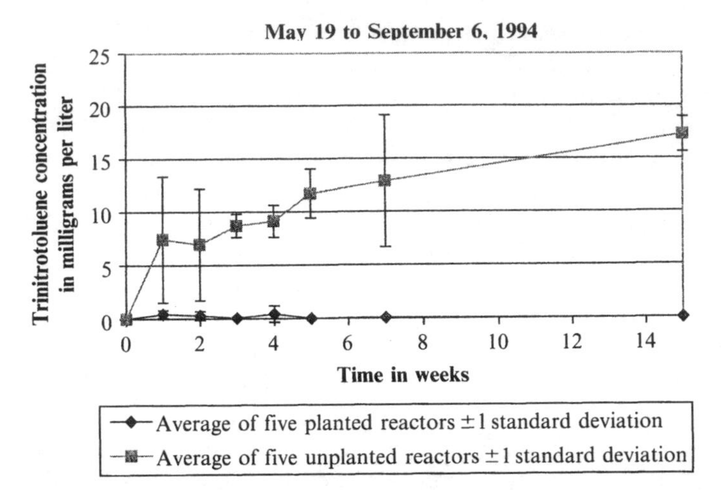

Figure 14-11 Field batch test of created treatment wetland approach to clean up trinitrotoluene-contaminated soil from the Alabama Army Ammunition Plant (Childersburg). The high performance liquid chromatography detection limit was 0.01 milligram trinitrotoluene per liter (Jenkins 1989, U.S. EPA 1992, Wolfe *et al.* 1993). Data from Young (1995).

microorganisms grew in the plant reactors but not in the plant-free reactors, most likely due to TNT toxicity.

The parrot feather (*Myriophyllum aquaticum*) in Field Test 2 reactors also died within 5 to 10 weeks, and the reactor was replanted. A contaminant-free reactor outside the greenhouse was a nursery for plant replacement. The transplants thrived for 5 to 10 weeks but were not subject to the same handling, drying, and weighing process as the original plantings. As a result, longer-term phytotoxicity to TNT is indicated but cannot be established from this observation alone.

CONTINUOUS FLOW INVESTIGATIONS

In practice, created wetlands treat continuous flows of contaminated water. Therefore, continuous-flow reactors were used to scale up from batch reactors to establish created wetlands as viable alternatives for treating industrial process waste streams and contaminated groundwater. These investigations started with one 16-liter (4.2-gallon) reactor to study the effects of temperature, pH, and other operational factors. Although the 16-liter (4.2-gallon) reactor provided promising results, this pilot setup lacked sufficient control to isolate all the parameters involved in the transformation of explosives. Therefore, a series of bench studies using 7-liter (1.8-gallon) and 8-liter (2.1-gallon) engineering reactors were conducted using both synthetic wastewaters and contaminated groundwater from the Volunteer and the Milan Army ammunition plants, both in Tennessee.

Bench Continuous Flow Reactor Tests

The original 16-liter (4.2-gallon) reactor tests established the proof of concept for treating continuous waste streams of TNT, RDX, and both TNT and RDX together. Sediment (6 liters or 2 gallons) and 100 grams wet weight of parrot feather (*Myriophyllum aquaticum*) per liter were used to maintain a healthy monoculture. The explosives TNT and RDX were quantified at strategic times during testing using high performance liquid chromatography. Influent concentrations of TNT were 0.04, 0.4, 1, 10, 30, 45, and 60 milligrams per liter for seven tests conducted with a steady flow rate of 2 milliliters per minute (0.8 gallon per day) to achieve an average residence time of 50 hours. Three other tests were run with a 30 milligrams of TNT per liter influent concentration with residence times of 33, 17, and 5 hours (flow rates of 3, 6, and 21 milliliters per minute or 1.1, 2.3, and 8.0 gallons per day). During the 50-hour residence time, all seven tests reached a steady aqueous concentration. The higher influent concentrations typical of TNT-contaminated wastewaters were reduced 98 percent to more than 99 percent, but the effluent concentrations were still two to three orders of magnitude greater than the typical drinking water standard of 0.002 milligram of TNT per liter. As the flow rate was increased and retention time decreased, the reactor efficiency decreased as expected to as little as 50 percent removal at 21 milliliters per minute (8.0 gallons per day). Figure 14-12 establishes that at constant plant biomass the reactor kinetics are pseudo first order at 2 milliliters per minute (0.8 gallon per day). Figure 14-13 shows that for constant biomass of parrot feather (*Myriophyllum aquaticum*) maintained over the long

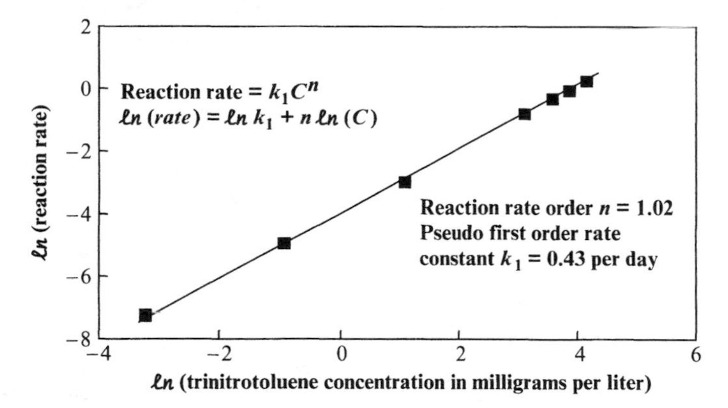

Figure 14-12 Reaction order and rate constant from plug flow in a 16-liter or 4.2 gallon (6 liters or 2 gallons of clean sediment) continuous flow through reactor with a parrot feather (*Myriophyllum aquaticum*) density of 10 grams of wet weight per liter treating initial trinitrotoluene concentrations of 0.04, 0.4, 1, 10, 30, 45, and 60 milligrams per liter at a constant flow rate of 2 milliliters per minute (0.8 gallon per day). Unpublished data from T.Y. Ou and N.L. Wolfe, U.S. EPA laboratory, Athens, Georgia.

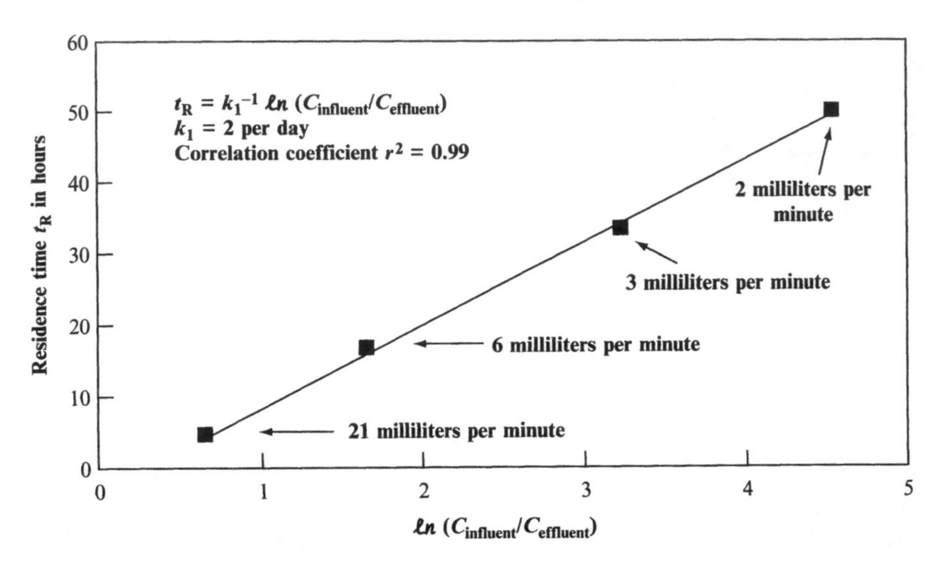

Figure 14-13 Plug-flow reaction rate plot for a 16-liter or 4.2-gallon (6 liters or 2 gallons of clean sediment) continuous flow through reactor treating an inlet trinitrotoluene concentration of 30 milligrams per liter and with a parrot feather (*Myriophyllum aquaticum*) density of 10 grams of wet weight per liter. Flow rate was incrementally increased from 2 to 21 milliliters per minute (0.8 to 8.0 gallons per day). Unpublished data from T.Y. Ou and N.L. Wolfe, U.S. EPA laboratory, Athens, Georgia.

term, treatment efficiency for this bench reactor could be described with a plug flow kinetic model. The pseudo first order reaction rate constant of 2.1 per day is very similar to batch rate constants observed by Medina *et al.* (2000b) for the 30 milligrams per liter initial concentration of TNT and 100 grams wet weight of parrot feather (*Myriophyllum aquaticum*) per liter density.

Initial Bench Testing for Milan Army Ammunition Plant Groundwater Cleanup

The feasibility of using created wetlands to treat groundwater from the Milan Army Ammunition Plant was tested using the 16-liter (4.2-gallon) reactor containing parrot feather (*Myriophyllum aquaticum*). The groundwater obtained from the site contained 2.25 milligrams of TNT per liter and 2.35 milligrams of RDX per liter. The plant density was approximately 100 grams wet weight of parrot feather (*Myriophyllum aquaticum*) per liter. Throughout the 57 hours of operation with a 2 milliliters per minute (0.8 gallon per day) flow rate, TNT in the reactor effluent was undetectable (less than 0.1 milligram per liter). After one residence time in the reactor, RDX reached a stable concentration of 0.86 milligram per liter (about 64 percent reduction). Therefore, consistent with the batch bench-scale testing by Carter (1996), RDX did not interfere with the treatment of TNT, but was slower than TNT uptake and transformation.

Having established the feasibility of continuously treating contaminated groundwater from the Milan Army Ammunition Plant, some bench optimization followed. Batch testing with 2 milligrams of RDX per liter and 200 grams wet weight of parrot feather (*Myriophyllum aquaticum*) per liter, plus reducing sediments with inorganic nutrients and zero-valent iron, showed that at least a sixfold increase in the second order removal rates constant was possible. This increase placed RDX transformation more in line with nominal rates for the aquatic phytoremediation of TNT (Table 14-1). The optimized technique was successful in the bench testing for wastewaters from both the McAlester and the Milan Army ammunition plants (Concurrent Technologies Corp. 1997).

Bench Proof of Concept for Trinitrotoluene Removal

Continuous-flow reactor investigations also were conducted to define design parameters (Rivera *et al.* 1998). These studies used 7-liter (1.8-gallon) reactors operating in an upflow mode. These studies monitored the transformation of TNT both in solution and in parrot feather (*Myriophyllum aquaticum*) tissue.

Rivera *et al.* (1998) indicated that continuous-flow reactors could effectively remove TNT from wastes (greater than 95 percent) with concentrations of 1, 5, and 10 milligrams of TNT per liter at hydraulic residence times ranging from 12 to 76 days. However, the formation of ADNT hampered the performance of these reactors. During the 12-day and 30-day residence times, ADNT concentrations were higher in the effluent than in the influent. Although longer residence times decreased the effluent concentration, ADNT was still at 0.062 milligram per liter at the 76-day residence time. Although small, this concentration exceeds the drinking water standard of 0.002 milligram per liter derived for TNT and 0.050 milligram per liter for the sum of all explosives (nitrobodies). Of course, treated effluents need not always meet drinking water standards, particularly if the treated waste is discharged into a sanitary sewer system.

Bench Treatability Investigations: Volunteer Army Ammunition Plant

Bench treatability studies using contaminated groundwater from the Volunteer Army Ammunition Plant in Chattanooga, Tennessee, demonstrated the capability of parrot feather (*Myriophyllum aquaticum*) to deplete TNT to below detection limits of 0.0005 milligrams per liter, even in the presence of other nitroaromatic contaminants (Andrews 1996). These other contaminants—dinitrotoluene, ADNT, nitrotoluene, and trinitrobenzene—were also transformed within the reactor.

At room temperature with a plant density of 20 grams of wet weight per liter, reaction constants were calculated for standard batch solutions of dinitrotoluene, ADNT, nitrobenzene, and trinitrobenzene. All these compounds were taken up and transformed, but at much slower rates than TNT

(3.5 per day or a half-life of 4.8 hours). Removed relatively fast, trinitrobenzene had a pseudo first order reaction constant of 3.2 per day (half-life of 5.3 hours). Similar rate constants of 0.461 per day (half-life of 36.2 hours) and 0.463 per day (half-life of 35.8 hours) characterized the transformation of 2,4-dinitrotoluene and 2-ADNT, respectively. A slow reaction constant of 0.18 per day (half-life of 90 hours) characterized the transformation of 4-ADNT. While 2,6-dinitrotoluene was transformed, the trend was erratic.

Temperature studies also were conducted on the contaminated water from the Volunteer Army Ammunition Plant to model the efficiency during seasonal fluctuations and provide design guidance. Transformation occurred over a temperature range from just above freezing, 2.5 to 40 °C, with maximum reaction rate constants occurring at approximately 35 °C. For a constant plant–water mass ratio of 1 to 50 (20 grams of wet weight per liter) and initial TNT concentration of approximately 2.02 milligrams per liter, pseudo first order reaction constants varied from a low of 0.62 per day (half-life of 27 hours) at 2.5 °C to a high of 6.3 per day (half-life of 2.6 hours) at 35 °C. At temperatures greater than 35 °C, the uptake and transformation slowed yielding a rate constant of 4.5 per day (half-life of 3.7 hours) at 40 °C.

Pilot Field Investigation: Volunteer Army Ammunition Plant

A 115-day pilot investigation (Miller *et al.* 1997, Best *et al.* 2000a, 2000b) at the Volunteer Army Ammunition Plant evaluated the uptake and transformation of groundwater contaminated with TNT, 2,4-dinitrotoluene, 2,6-dinitrotoluene, and 4-nitrotoluene by narrowleaf cattail (*Typha angustifolia* L.) and waterweed (*Elodea canadensis* Rich. in Michx.). These plants were evaluated in nine 600-liter (160-gallon) rectangular polyethylene continuous-flow reactors (1.9-meter long by 1.0-meter wide by 0.8-meter deep or 6.2 by 3.2 by 2.6 feet) both in full sunlight and under ultraviolet screens. The retention time was 7 days. Canadian waterweed (*Elodea canadensis* Rich. in Michx.) did not survive total explosives concentrations ranging from 2 to 30 milligrams per liter (the latter being mainly nitrotoluene). The reactors were replanted with American pondweed (*Potamogeton nodosus* Poir.) and coontail (*Ceratophyllum demersum* L.). These species also died. Only the emergent cattail (*Typha angustifolia* L.) was able to establish a robust biomass. Nitroaromatic compounds did not accumulate in sediments and only transient traces of TNT products occurred in plants. Unfortunately, the Volunteer Army Ammunition Plant bench treatability tests (Andrews 1996) were not used in setting up this pilot field test.

Scale-Up Attempt: Milan Army Ammunition Plant

A two-cell wetland of 11 by 32 meters and 11 by 11 meters (36 by 106 feet and 36 by 36 feet), approximately 0.8 meter (2.5 feet) deep, and with a water surface area of 0.036 hectare (0.088 acre) was constructed at the Milan

Army Ammunition Plant (ESTCP 1999). Parrot feather (*Myriophyllum aquaticum*) was established in both wetland cells. A continuous flow of 27 cubic meters per day (5 gallons per minute) with 1.25 to 4.44 milligrams of TNT per liter and 3.25 to 9.2 milligrams (of total exposives) per liter maintained an approximate 10-day residence time in the wetland. Unfortunately, an adequate population of parrot feather (*Myriophyllum aquaticum*) was not maintained. Tadpoles grew in the basins and defoliated the plants. Photolytic products formed in the wetland lagoons due to inadequate plant mass and may have reduced the sunlight reaching the plants. A hailstorm decimated the parrot feather (*Myriophyllum aquaticum*). Before the problems with tadpoles and hail, the treatment wetland system met the 0.002 milligram of TNT per liter drinking water standard, but not the criteria for total explosives, 50 milligrams per liter. Part of the failure was due to the fact that system was not designed originally to treat RDX and other explosives, and that the treatment of TNT transformation products was not taken into account in the design.

DESIGN AND PROCESS KINETICS FOR SCALE-UP TO PHYTOREMEDIATION WETLANDS

Design Overview

The important wetland design parameters include flow rate, type and placement of vegetation, and surface area and depth. The waste management objectives define the flow rate based on one of the following:

1. Pumping rate for deeper extraction of explosives-contaminated groundwater
2. Groundwater seepage rate into surface waters
3. Wastewater discharge rate from munitions production or demilitarization

For emergent species in large treatment wetlands, hydraulic loading should be increased by the amount of evapotranspiration. Species and plant community selections are based on the following:

1. Availability of indigenous or native aquatic wetland plants to rapidly take up, transform, and conjugate or assimilate all explosive wastes present and for the full range of expected concentrations
2. State and federal regulations for invasive species, sound weed and plant control practices, climatic constraints, and disease susceptibility at the site

Required wetland surface area, depth, and configuration are all derived from the flow rate, habitat requirements of the plants selected, and the kinetics of explosives uptake, transformation, conjugation, and sequestration. The current state-of-the-practice is to design elongated wetlands to approximate continuous plug flow until the effects of concentration and plant density are more fully investigated. Restoration wetlands that trap and transform explosives like those at the Iowa Army Ammunition Plant are fitted to the local topography with some limited earth movement in order to provide a range of depths to sustain the plant communities selected for treatment.

Depths of excavated treatment wetlands vary throughout to match habitat requirements of the selected plant community and to exclude ineffective or undesirable invasive species. Depth, sediment placement (source of seed and nutrients), and nutrient supplements are primary controls on gradation of communities from hardy pioneers to withstand the initial toxicity of the waste, to fast-reacting species to minimize the surface area requirements such that construction costs are reduced. Hardy pioneering rooted vegetation may be useful in all recharge zones for initial treatment of intercepted contaminated groundwater.

These same design variables are also selected to maintain as much plant activity as possible in warm and cool seasons. For design simplicity, segmented, excavated, or diked wetland cells may be used to grow different plants or communities for high, moderate, and low contaminant concentrations expected at different times of the year. Design safety factors may be in the form of extra wetland treatment cells. Provisions for maintenance may also be addressed with extra cells that are rotated out of the treatment train for replanting, sediment replenishment or supplementation, winterization with cold weather species, plant maintenance, and pest and disease control.

Phytokinetics

The kinetics of phytotransformation (Medina *et al.* 2000b) have been observed to be second order

$$-\frac{dC}{dt} = k_{\mathrm{E}}CE \approx k_{\mathrm{P}}CP = k_1 C \tag{14-1}$$

where C is the concentration of the contaminant, t the time, k_{E} the second order rate constant with respect to enzyme concentration E, k_{p} the second order rate constant with respect to plant density P and $k_1 = k_{\mathrm{p}}P$ the pseudo first order rate constant. Equation (14-1) is based on the observation that enzymatic activity seems to be proportional to plant density P (the mass of plant material per unit volume of waste solution). Medina and McCutcheon (1996) and Pavlostathis *et al.* (1998) assumed this form in earlier analysis of batch tests for TNT transformation. When plant matter is in excess, or plant growth (or loss) is negligible over a discrete reaction time as usually found in batch tests, plant density P is a constant, allowing in either case, the use of a pseudo first order equation. The use of the second order rate constant k_p provides the advantage of normalizing the results of tests performed at different plant densities. However, the pseudo first order kinetic rate constant k_1 and the associated half-life allows for easy comparison with a range of treatment technologies for which first order kinetic data are available (*e.g.*, see Table 1-4, McCutcheon and Schnoor this book for comparison of phyto and bioremediation treatment rates).

Henri–Michaelis–Menten kinetics also are useful to characterize enzymatic kinetics

$$v = \frac{V_m C}{K_m + C} \tag{14-2}$$

where v is the reaction velocity at C, the contaminant (substrate) concentration, V_m is the maximum reaction velocity or rate, and K_m is the half-saturation constant. Equation (14-2) can be linearized to estimate V_m and K_m. Lineweaver–Burk and Eadie–Hofstee linearizations are typically used (Gueriguian 1996), but other approaches may be useful under different circumstances. Medina *et al.* (2000b) used an average of the results from the Lineweaver–Burk and Eadie–Hofstee linearizations to obtain values of the maximum reaction velocity V_m of 5.8 milligrams of TNT per liter per hour) and half-saturation constant K_m of 94 milligrams of TNT per liter for parrot feather (*Myriophyllum aquaticum*) over the range of TNT concentrations of 1 to 128 milligrams per liter. As can be seen in Medina *et al.* (2000b), this fit was very good over the range of TNT concentrations from 1 to 30 milligrams per liter. However, at higher concentrations some deviation occurred, perhaps due to transformation inhibition (Gueriguian 1996) or involvement of multiple enzymes.

The Henri–Michaelis–Menten equation is a more general kinetic description that encompasses both zero and first order kinetics for one enzyme reacting with one substrate or compound. The Henri–Michaelis–Menten equation reduces to zero order kinetics when the contaminant concentration is much greater than the half-saturation constant. For this condition, the reaction velocity is approximately equal to the maximum velocity. Because the half-saturation constant K_m of 94 milligrams of TNT per liter cannot be significantly smaller than C, the initial concentration of TNT, zero order kinetics are not applicable for parrot feather (*Myriophyllum aquaticum*) over the TNT concentration range of 0.002 to 128 milligrams per liter (solubility at room temperature) and higher (i.e., supersaturated as found in some waste streams). Therefore, zero order kinetics would not be appropriate for sizing treatment wetlands dominated by parrot feather (*Myriophyllum aquaticum*).

At low substrate (contaminant) concentrations in which the concentration C is much less than the half-saturation constant K_m then the small concentration plus the half-saturation constant is approximately equal to the half-saturation constant ($C + K_m \approx K_m$). Substituting this into the Henri–Michaelis–Menten Equation (14-2) and expressing the reaction velocity v as dC/dt gives

$$v = \frac{V_m}{K_m} C = -\frac{dC}{dt} \tag{14-3}$$

For lower contaminant concentrations, the Henri–Michaelis–Menten equation can be used to estimate the pseudo first order rate constant. Over the full range of 1 to 128 milligrams of TNT per liter, Medina *et al.* (2000b) found $V_m/K_m = k_1 = 1.46$ per day. As a comparison, Medina *et al.* (2000b) experimentally defined a range of k_1 from 0.96 to 3.8 per day. Given the range of the rate constants, neither the Henri–Michaelis–Menten equation nor the pseudo first order equation seems to be general enough for industrial wastewaters. Because of the high temperature and steam washout procedures, wastewaters can be supersaturated with all explosives (*e.g.*, up to 200 milligrams of TNT per liter). Until the kinetics are generalized with the effects of inhibition on specific processes, the accepted procedure should be to design plug-flow reactor configurations using pseudo first order kinetics for the highest expected influent concentration and lowest sustained plant density during the coldest period of the year. The pseudo first order rate constant for the highest expected concentrations of TNT should be derived from bench treatability studies (Andrews 1996, Best *et al.* 1997a, Best *et al.* 1997b, Best *et al.* 1998b, Rivera *et al.* 1998, Medina *et al.* 2000b, Medina *et al.* 2002). If the resulting conservative design is not cost-effective, then a quasi-dynamic analysis using piecewise constant empirical relationships between the pseudo first order rate constant and TNT concentration and plant density will be necessary. The quasi-dynamic analysis is best confirmed with a pilot field test (Thompson *et al.* this book).

Typical groundwater concentrations up to 10 milligrams of TNT per liter may be accurately described by pseudo first order kinetics. If x is the acceptable variance to the pseudo first order rate constant expressed as a fraction (i.e., 5 percent is 0.05), then the highest concentration (in milligrams of TNT per liter) for which a first order simplification is valid (C_s) is

$$C_s = \frac{94.75 \text{ milligrams of TNT per liter}}{1 - x} - 94.2 \text{ milligrams of TNT per liter} \tag{14-4}$$

Medina *et al.* (2000b) measured the coefficient of variance of the pseudo first order rate constant as 28 percent. This value is similar to the variability measured by Carter (1996), Comstock (1996), Gueriguian (1996), and Pavlostathis *et al.* (1998). At this variance, C_s is equal to 37 milligrams of TNT per liter. Consistent with this estimate, Medina *et al.* (2000b) observed some nonlinearity for 30 milligrams of TNT per liter in 200 grams wet weight of parrot feather (*Myriophyllum aquaticum*) per liter acclimation investigations, and this range matches the 1 to 30 milligrams of TNT per liter range of the best fit of the Henri–Michaelis–Menten equation. Comstock (1996) observed significant deviations from pseudo first order batch kinetics for initial concentrations of 115 milligrams of TNT per liter transformed by 21 to 26 grams wet weight of Eurasian water milfoil (*Myriophyllum spicatum*) per liter. Nevertheless, 28 percent variance is quite large, dictating bench treatability investigations for each design until the kinetics can be more fully defined for the typical range of groundwater TNT concentrations.

For phytotreatments in which a contaminant like RDX may be limiting, the required hydraulic residence time of the wetland can be estimated as

$$t = \frac{\ell n\left(\frac{C_{\text{effluent}}}{C_{\text{influent}}}\right)}{k_1} \tag{14-5}$$

where C_{effluent} is based on the surface water discharge standards, *e.g.*, 0.05 milligram of total explosives per liter, and C_{influent} is the maximum expected concentration from the wastewater stream or contaminated groundwater. The pseudo first order transformation rate constant should be determined from bench treatability or pilot investigations, and selected as the lowest expected during the coldest period of operation. If plant density is expected to vary, then the minimum second order rate constant should be used instead with appropriate modification of Equation (14-1). The effects of concentration, plant density, temperature, and other design factors on the kinetics of uptake and transformation for other explosives besides TNT has not been determined. Thus, more comprehensive treatability (Andrews 1996, Best *et al.* 1997a, Best *et al.* 1997b, 1998b, Rivera *et al.* 1998, Medina *et al.* 2000b, Medina *et al.* 2002) and pilot investigations (Thompson *et al.* this book) are necessary.

Estimates of the wetland residence time for TNT transformation to ADNT and then DANT and other by-products (Subramanian and Shanks this book) must be based on the kinetics of the observed sequential transformations. The calculation requires an iterative numerical solution illustrated by Medina and McCutcheon (1996). However, only a few studies have defined the rate constants for ADNT and DANT transformation and it is not known if the effects of concentration, plant density, and water temperature are similar to the effects on TNT kinetics.

Required treatment wetland volume is then the design flow multiplied by the residence time, compensating for plant volume and using a typical design safety factor of 25 percent for wetlands (Medina and McCutcheon 1996), or

$$V = 1.25Qt\left(1 + \frac{P}{1000}\right) \tag{14-6}$$

The water surface area of the treatment wetland is then volume V divided by average depth. The required wetland footprint includes the water surface area plus the area occupied by the surrounding dike and other buffers, including fencing to exclude wildlife and humans (if necessary). Equalization may be useful to optimize surface area based on seasonally higher transformation rate constants. Any equalization basin should be seeded with organic-rich sediment and planted with hardy floating plants that can survive variable depths.

STATE-OF-THE-ART: TREATMENT OF EXPLOSIVES IN WATER AND SOIL

From the pilot investigations introduced and reviewed in this chapter, it is evident that the field proof of concept has been obtained for some explosives waste management options that involve green plants. Table 14-2 is a summary of the state of the practical arts for TNT cleanup, for which the most work has been undertaken. Table 14-3 summarizes progress in achieving the proof of concept for phytoremediation of RDX.

Monitoring of two created treatment wetlands since 1998 at the Iowa Army Ammunition Plant establishes the field proof of principle for constructed wetlands to restore typical RDX and TNT contaminated groundwater (Thompson *et al.* this book). Since construction in 1998, no contaminated water has been released from either wetland. Limited monitoring should continue until the groundwater and wetland is free of RDX to be sure the proof of principle remains valid. In addition, because the created wetlands at the Iowa Army Ammunition Plant are full-scale pilots, optimization is possible that may decrease the cost per volume (normally expressed as U.S. dollars per 1000 gallons) for groundwater treatment, and may provide the proof of principle for treating contaminated surface waters. The wetlands zero discharge *de facto* meets the drinking water standards of 0.002 milligram of RDX per liter and 0.002 milligram of TNT per liter. Release of water when the levels are below these standards would drain more groundwater and could eventually accommodate some diversion of surface water into the Line 1 wetland for treatment (Thompson *et al.* this book). The operation of the Iowa Army Ammunition Plant wetlands should be optimized to first treat as much contaminated groundwater as possible and then as much contaminated surface water as possible. Such an optimization can achieve the first proof of principle for wetland treatment of groundwaters and surface waters.

Because the Iowa Army Ammunition Plant evaluation is a full-scale pilot, new applications will still require screening and treatability investigations before final application decisions. Draft guidance and a design manual is possible based on the extensive phytotoxicity and phytotransformation screening and wetland design work at several Army ammunition plants. Nevertheless, the guidance has not been written.

Attenuation has been observed in natural wetlands. Therefore, documentation of any ongoing attenuation at other sites should be possible based on the design and monitoring that was necessary to gain acceptance of the Iowa

TABLE 14-2 Basis of the State of the Practice in Phytoremediation of Trinitrotoluene-Contaminated Wastes

			Soil treatments		
	Ground and surface water interception or pumping into created wetland	Wastewater (pink water) treatment lagoons	Terrestrial plantation	Wetland flooding	Intrinsic attenuation
Laboratory proof of concept	√–Batch and continuous flow tests reviewed in this chapter	√–Batch and continuous flow observations reviewed in this chapter	√–Hybrid poplar (*Populus deltoides* × *Populus nigra* 'DN34') (Thompson 1997); need to compile studies of phytotoxicity, efficiency, and extent	√–Created wetland based on batch observations reviewed in this chapter. *X*–Flat-bed sheet flow reactor (Qaisi *et al.* 1996a, Qaisi *et al.* 1996b, Valsaraj *et al.* 1998) proposed, but not yet proven conceptually	Not applicable
Pilot: bench screening or optimization	√–Bench testing for Iowa Army Ammunition Plant (Thompson *et al.* this book) and Milan Army Ammunition Plant (this chapter)	√–McAlester and Milan Army ammunition plants (Concurrent Technologies Corp. 1997)		√–Bench scale, soil slurry reactors with plants (Wolfe *et al.* 1993)	Not applicable

TABLE 14-2 Basis of the State of the Practice in Phytoremediation of Trinitrotoluene-Contaminated Wastes

	Ground and surface water interception or pumping into created wetland	Wastewater (pink water) treatment lagoons	Soil treatments: Terrestrial plantation	Soil treatments: Wetland flooding	Intrinsic attenuation
Pilot: field investigations	√–Bench and field pilots at Volunteer Army Ammunition Plant (see this chapter) and full scale pilots at Iowa Army Ammunition Plant (Thompson *et al.* this book)	Proposed for McAlester and Milan Army ammunition plants (Concurrent Technologies Corp. 1997) and should be implemented as soon as possible		√–Batch investigation at Alabama Army Ammunition Plant (Young 1995), contaminated soil flooded and treated with parrot feather (*Myriophyllum aquaticum*)	Observations at Alabama (natural wetlands, Young 1995) and Joliet Army ammunition plants (soil and groundwater, Wolfe *et al.* 1999), and Crane Naval Weapons Station (soil and groundwater) not sufficient to establish concept
Field validation or demonstration	√–Iowa Army Ammunition Plant groundwater restoration successful since 1998 (Thompson *et al.* this book); optimization required to achieve field proof of principle for surface and groundwater cleanup *X*–Milan Army Ammunition Plant groundwater remediation not successful (this chapter)			Should be undertaken as soon as possible and use experience at the Iowa Army Ammunition Plant	Investigation and validation urgently needed
Design manual or guidance	Not yet planned but should be drafted from experience at Iowa Army Ammunition Plant (Thompson *et al.* this book) and other studies in this chapter				

Note: √–proof achieved; *X*–testing failed or not yet sufficiently proven.

TABLE 14-3 Basis of the State of the Practice in Phytoremediation of Wastes Contaminated with Hexahydro-1,3,5-trinitro-1,3,5-triazine (RDX)

	Ground and surface water interception or pumping into created wetland	Wastewater (pink water) treatment lagoons	Soil treatments		Intrinsic attenuation
			Terrestrial plantation	Wetland flooding	
Laboratory proof of concept	√–Batch and continuous flow testing in this chapter	√–Batch and continuous flow testing in this chapter	*X*–Concern about accumulation in leaves – need studies of phytotoxicity, efficiency, and extent	No known studies	Not applicable
Pilot: bench screening or optimization	√–Iowa Army Ammunition Plant (Thompson *et al.* this book) and Milan Army Ammunition Plant (this chapter)	√–McAlester and Milan Army ammunition plants (Concurrent Technologies Corp. 1997)	*X*–Laboratory low intensity lighting may lead to unrepresentative accumulation in leaves (Thompson 1997)		Not applicable
Pilot: field investigations	√–Full scale pilots at Iowa Army Ammunition Plant (Thompson *et al.* this book)	Proposed for McAlester and Milan Army ammunition plants (Concurrent Technologies Corp. 1997)			

TABLE 14-3 Basis of the State of the Practice in Phytoremediation of Wastes Contaminated with Hexahydro-1,3,5-trinitro-1,3,5-triazine (RDX)

	Ground and surface water interception or pumping into created wetland	Wastewater (pink water) treatment lagoons	Soil treatments		Intrinsic attenuation
			Terrestrial plantation	Wetland flooding	
Field validation or demonstration	√–Iowa Army Ammunition Plant groundwater restoration successful since 1998 (Thompson *et al.* this book) *X*–Milan Army Ammunition Plant groundwater remediation not successful (this chapter)				No known observations
Design manual or guidance	Not yet planned but should be drafted from experience at Iowa Army Ammunition Plant (Thompson *et al.* this book) and this chapter				

Note: √–proof achieved; *X*–testing failed.

Army Ammunition Plant wetland pilot treatment tests. Moreover, a natural attenuation protocol should be considered for vegetative attenuation of explosives in water and soil based on observations at the Alabama and the Joliet (Illinois) Army ammunition plants.

Future laboratory and field investigation may focus on the use of algae-dominated wetlands or treatment lagoons. Macrophytic algae such as stonewort (*Nitella* spp.) was effective at treating TNT, as described in the section "Bench Proof of Principle: Alabama Army Ammunition Plant" and "Bench Tests: Six Army and Navy Facilities." Furthermore, "control" reactors colonized by planktonic algae (spp. unknown) transformed TNT, and the degree of treatment was similar to that of planted reactors (Rivera *et al.* 1998, Medina *et al.* 2002). Studies of cyanobacteria (*Anabaena spiroides*); (commonly known as blue-green algae) also had promising results (Jackson 1996). Algae might have some kinetic advantages in that the surface area to volume ratio is higher than that for plants, and uptake may be simpler. Given the more elaborate structure of plants, the enzymatic activity of unicellular algae may be brought to bear quicker and in greater quantity of activity per unit of biomass. Nevertheless, algae and bacteria both cannot concentrate as much biomass and specialized activity as plants. Despite this, algae grow faster than plants, and could replace some biomass and enzymatic activity if pests or diseases strike the plants involved. Like plants, algae use light as an energy source and carbon dioxide as a carbon source, but are more sensitive to water depth and shading.

Other areas of need include fully defining the kinetics of TNT, RDX, other explosives, and the principal transformation products, particularly ADNT. The final gaps in knowledge about the sequestration and ultimate fate of TNT, RDX, and other explosives transformation products in plants must be addressed to finalize design procedures and reduce long-term monitoring requirements.

Most treatment wetland tests to date involve maintaining a single species of plants. However, plants that remove TNT the fastest may not remove ADNT the fastest (Medina, Bergstedt, and McCutcheon unpublished). Some system optimization of sediment and plants has enhanced limiting transformation rates of RDX to be more in line with faster TNT and other nitroaromatic transformations. Eventually, communities living in a self-optimized sediment–water system must be developed to lessen the risks of disease, pests, and weather. To date, most testing of wetland treatments for explosives compounds has involved tests of a single generation of plants. Toxicity effects on successive generations and the potential for improved kinetics do not seem to have been investigated for explosives.

Another opportunity for research and development includes bench testing of methods to remediate explosives-contaminated soil using green plants. A series of bench tests and pilot field tests for phytotoxicity, product formation, cleanup efficiency, and optimization of agroeconomic parameters in soil–plant systems are urgently needed.

Acknowledgments

The authors appreciate the reviews of Robert Mueller, Tomas Vanek, Philip Thompson, Edward Mead, Lee Wolfe, and Robert Swank. Robert Mueller noted an important distinction with mitigation wetlands, and Tomas Vanek noted that the basis of the Henri–Michaelis–Menten equation should explain some kinetic discrepancies. Philip Thompson noted the importance of genetic selection in microbial acclimation. Lee Wolfe was especially helpful in noting semantic errors involving rate constants. Jerald L. Schnoor evaluated the reviews and accepted this chapter for the book. T.Y. Ou and Laura Carriera collected much of the early data reported, under contact to the U.S. EPA laboratory in Athens, Georgia, based on the experimental designs and scientific guidance of Lee Wolfe. Lee Wolfe discovered the series of fast reactions in reducing sediments with several post doctoral fellows and collaborators. Laura Carreira pioneered the extraction of activity from sediments, thus introducing a sound biochemistry basis for explosives phytoremediation.

Louis Thibodeaux, then director of the U.S. EPA Hazardous Substance Research Center/South and Southwest at Louisiana State University, and Dale Manty, U.S. EPA Research Centers project officer expanded the wetland concept to include confirmation of the science basis of explosives phytoremediation. Mindy Vanderford, while an M.S. candidate at Rice University, pioneered the use of anexic plants to establish the dominance of plant metabolism. Joe Hughes and Jackie Shanks, then on the faculty at Rice, supervised a series of rigorous anexic and mass balance analyses that confirmed that plants were responsible for the fast reactions and mapped the complex transformation pathways with advanced chemical and biochemical analysis. Mark Zappi, the originator of Project 720, solidified the long-term Strategic Environmental Research and Development Program (SERDP) support for the development of phytoremediation and provided seminal advice, along with his successor, Rakesh Bajpai, on leave from the University of Missouri. Elly Best of the Waterways Experiment Station pioneered some indigenous plant screening and pilot field testing techniques. Ted Streckfuss of the U.S. Army Corps of Engineers Omaha District first sketched out some of the elements of design for the treatment wetlands constructed at the Iowa Army Ammunition Plant.

Lt. Colonel David Young of the U.S. Air Force advanced the created wetland concept while working on a dissertation at Auburn University in Alabama, along with two masters students who followed—Eva Wilson and Scott Todd. Rod Jenkins and Clifford Lange supervised development of these theses and the resulting contributions to the art.

Support for the work of Steven McCutcheon that went into this chapter and the comprehensive development program came from the U.S. EPA Office of Research and Development Contaminated Sites Program, SERDP (including Project 720 managed by the Army Waterways Experiment Station), and an internal grant from the U.S. EPA National Exposure Research Laboratory. Project 720 supported a National Research Council fellowship for Victor Medina. This paper has been reviewed in accordance with the U.S. EPA peer and administrative review policies and approved for publication. Mention of trade names or commercial products does not constitute endorsement or recommendation for use.

REFERENCES

Andrews, S.L. (1996) The effects of temperature on the phytoremediation of munitions (TNT) contaminated water. Unpublished M.S. thesis. Biological and Agricultural Engineering. University of Georgia, Athens, Georgia.

Berger, B. and N.L. Wolfe (1996) Hydrolysis of sulfonylurea herbicides in water and sediment. *Environ. Toxicol. Chem.* **15**: 1500–1507.

Best, E.P.H., M.E. Zappi, H.L. Fredrickson, S.L. Sprecher, S.L. Larson, and M. Ochman (1997a) Screening of aquatic and wetland plant species for phytoremediation of explosives-contaminated groundwater from the Iowa Army Ammunition Plant. In: *Bioremediation of Surface and Subsurface Contamination*. R.K. Bajpai and M.E. Zappi, eds. *Ann. New York Acad. Sci.*, New York. **829**: 179.

Best, E.P.H., J.L. Miller, M.E. Zappi, H.L. Fredrickson, S.L. Sprecher, and S.L. Larson (1997b) Screening of aquatic and wetland plant species for phytoremediation of explosives-contaminated groundwater from the Iowa Army Ammunition Plant. U.S. Army Engineer Waterways Experiment Station Technical Report EL-97-2. Vicksburg, Mississippi.

Best, E.P.H., S.L. Sprecher, S.L. Larson, H.L. Fredrickson, and D. Bader (1997c) Fate and mass balances of [^{14}C]-TNT and [^{14}C]-RDX in aquatic and wetland plants in groundwater from the Milan Army Ammunition Plant. In: Abstracts Book, 12th Annual Conference on Hazardous Waste Research. Hazardous Substance Research Center/Great Plains, Kansas State University, Manhattan, Kansas. Held in Kansas City, Missouri, May 19–22, pp. 10–11.

Best, E.P.H., J.L. Miller, M.E. Zappi, H.L. Fredrickson, S.L. Sprecher, S.L. Larson, and T. Streckfuss (1997d) Degradation of TNT and RDX in groundwater from the Iowa Army Ammunition Plant in flow-through systems planted with aquatic and wetland plants. In: Abstracts Book, 12th Annual Conference on Hazardous Waste Research. Hazardous Substance Research Center/Great Plains, Kansas State University, Manhattan, Kansas. Held in Kansas City, Missouri, May 19–22, pp. 8–9.

Best, E.P.H., S.L. Sprecher, S.L. Larson, and H.L. Fredrickson (1998a) Environmental behavior and fate of explosives in groundwater from the Milan Army Ammunition Plant in aquatic and wetland plants: fate of TNT and RDX. U.S. Army Environmental Center Report No. SFIM-AEC-ET-CR-97060, by U.S. Army Engineer Waterways Experiment Station. Vicksburg, Mississippi.

Best, E.P.H., J.L. Miller, M.E. Zappi, and T.H. Streckfuss (1998b) Explosives removal from groundwater of the Iowa Army Ammunition Plant in continuous-flow laboratory systems planted with aquatic and wetland plants. U.S. Army Corps of Engineers Waterways Experiment Station Technical Report EL-98-13. Vicksburg, Mississippi.

Best, E.P.H., S.L. Sprecher, S.L. Larson, H.L. Fredrickson, and D.F. Bader (1999a) Environmental behavior of explosives in groundwater from the Milan Army Ammunition Plant in aquatic and wetland plant treatments: removal, mass balances and fate of TNT and RDX in groundwater. *Chemosphere* **38**: 3383–3396.

Best, E.P.H., S.L. Sprecher, S.L. Larson, H.L. Fredrickson, and D.F. Bader (1999b) Environmental behavior of explosives in groundwater from the Milan Army Ammunition Plant in aquatic and wetland plant treatments. uptake and fate of TNT and RDX in plants. *Chemosphere* **39**: 2057–2072.

Best, E.P.H., J.L. Miller, and S.L. Larson (2000a) Removal of explosives in constructed wetlands. In: *7th International Conference on Wetland Systems for Water Pollution Control*. Lake Buena Vista, Florida, Nov. 11–16. **3**: 1373–1382.

Best, E.P.H., J.L. Miller, and S.L. Larson (2000b) Explosives removal from groundwater at the Volunteer Army Ammunition Plant, Tennessee, in small-scale wetland modules. In: *Wetlands and Remediation: An International Conference*. J.L. Means and R.E. Hinchee, eds. Battelle Press, Columbus, Ohio. Held in Salt Lake City, Utah, Nov. 16–17, 1999. pp. 365–373.

Bhadra, R., D.G. Wayment, J.B. Hughes, and J.V. Shanks (1999a) Confirmation of conjugation processes during TNT metabolism by axenic plants. *Environ. Sci. Tech.* **33**(3): 446–452.

Bhadra, R., R.J. Spanggord, D.G. Wayment, J.B. Hughes, and J.V. Shanks (1999b) Characterization of oxidation products of TNT metabolism in aquatic phytoremediation systems of *Myriophyllum aquaticum*. *Environ. Sci. Tech.* **33**(19): 3354–3361.

Burken, J.G, J.V. Shanks, and P.L. Thompson (2000) Phytoremediation and plant metabolism of explosives and nitroaromatic compounds. Chapter 10. In: *Biodegradation of Nitroaromatic Compounds and Explosives*. J. C. Spain, J.B. Hughes, and H.J. Knackmuss, eds. Lewis Publishers, Boca Raton, Florida, pp. 239–276.

Carter, W.H. (1996) Phytoremediation of explosives-contaminated ground water. M.S. thesis, Georgia Institute of Technology, Atlanta, Georgia.

Cataldo, D.A., S. Harvey, R.J. Fellows, R.M. Bean, and B.D. McVeety (1989) An evaluation of the environmental fate and behavior of munitions materiel (TNT, RDX) in soil and plant systems: environmental fate and behavior of TNT. Pacific Northwest Laboratory, Report number PNL-7370, UC-402, Richland, Washington. Final Report for U.S. Medical Research and Development Command, Fort Detrick, Frederick, Maryland. Project Order Number 88PP8853.

Comstock, K.K. (1996) Transformation of 2,4,6-trinitrotoluene (TNT) by the aquatic plant *Myriophyllum spicatum*. M.S. thesis, Georgia Institute of Technology. Atlanta, Georgia.

Concurrent Technologies Corporation (1997) Pink water treatment: technology options. Technical Report for Task No. N.0015, National Defense Center for Environmental Excellence, Johnstown, Pennsylvania.

Criscenti, L.J. (1995) Preliminary coupled diffusion/reaction model for the bioremediation of TNT through the use of plant enzymes. Battelle Pacific Northwest Laboratory, Richland, Washington. Letter Report to the U.S. Environmental Protection Agency Environmental Research Laboratory, Athens, Georgia.

Department of Defense (1995) Defense environmental restoration program—annual report to Congress for fiscal year 1994. U.S. Department of Defense, Washington, D.C.

Environmental Security and Technology Certification Program (ESTCP) (1999) The use of constructed wetlands to phytoremediate explosives-contaminated groundwater at the Milan Army Ammunition Plant, Milan, Tennessee. U.S. Department of Defense, Washington, D.C.

Gueriguian, L.F. (1996) Kinetics of 2,4,6-trinitrotoluene (TNT) phytotransformation using *Elodea densa*. M.S. thesis, Georgia Institute of Technology, Atlanta, Georgia.

Harvey, S.D., R.J. Fellows, D.A. Cataldo, and R.M. Bean (1990) Analysis of 2,4,6-trinitrotoluene and its transformation products in soils and plant tissues by high performance liquid chromatography. *J Chromatogr.* **518**: 361–374.

Harvey, S.D., R.J. Fellows, D.A. Cataldo, and R.M. Bean (1991) Fate of the explosive hexahydro-1,3,5-trinitro-1,3,5-triazine (RDX) in soil and bioaccumulation in bush bean hydroponic plants. *Environ. Toxicol. Chem.* **10**: 845–855.

Harvey, S.D., R.J. Fellows, J.A. Campbell, and D.A. Cataldo (1992) Determination of the explosive 2,4,6-trinitrophenylmethylnitramine (tetryl) and its transformation products in soils. *J Chromatogr.* **605**: 227–240.

Hazardous Substance Research Center South and Southwest (HSRC) (1996) Using vegetation to clean up TNT-contaminated sites. *Centerpoint*. Louisiana State University, Georgia Institute of Technology, and Rice University. **3**(1): 4–6.

Hazardous Substance Research Center South and Southwest (HSRC) (1997) Aquatic plant lagoon systems for explosives remediation. Research Brief #14, Louisiana State University, Georgia Institute of Technology, and Rice University.

Hughes, J.B., J. Shanks, M. Vanderford, J. Lauritzen, and R. Bhadra (1997) Transformation of TNT by aquatic plants and plant tissue cultures. *Environ. Sci. Tech.* **31**: 266–271.

Jackson, G.H. (1996) Biotransformation of 2,4,6-trinitrotoluene (TNT) by the cyanobacterium *Anabaena spiroides*. M.S. thesis, Georgia Institute of Technology, Atlanta, Georgia.

Jafvert, C.T. and N.L. Wolfe (1987) Degradation of selected halogenated ethanes in anoxic sediment-water systems. *Environ. Toxicol. Chem.* **6**(11): 827–837.

Jenkins, T.F. (1989) Development of an analytical method for the determination of extractable nitroaromatics and nitramines in soil. Ph.D. dissertation, University of New Hampshire, Durham.

Jerger, D.E. and P.M. Woodhull (2000) Applications and cost for biological treatment of explosives-contaminated soils in the U.S. Chapter 14. In: *Biodegradation of Nitroaromatic Compounds and Explosives*. J.C. Spain, J.B. Hughes, and H.J. Knackmuss, eds. Lewis Publishers, Boca Raton, Florida, pp. 395–424.

Larson, S.L., R.P. Jones, L. Escalon, and D. Parker (1999) Classification of explosives transformation products in plant tissue. *Environ. Toxicol. Chem.* **18**: 1270–1276.

Lauritzen, J.R., III (1998) *Catharanthus roseus* as a model system for the study of the phytoremediation of TNT. M.S. thesis, Chemical Engineering, Rice University, Houston, Texas.

Masunaga, S., N.L. Wolfe, and L.H. Carriera (1995) Transformation of benzonitriles in anaerobic sediment and in sediment extract. *Environ. Toxicol. Chem.* **14**: 1827–1838.

McCutcheon, S.C., N.L. Wolfe, L.H. Carriera, and T.Y. Ou (1995) Phytoremediation of hazardous waste. In: *Innovative Technologies for Site Remediation and Hazardous Waste Management Proceedings of the National Conference*. R. D. Vidic and F.G. Pohland, eds. Environmental Engineering Division, American Society of Civil Engineers, New York. Held in Pittsburgh, Pennsylvania, July 26, pp. 597–604.

Medina, V.F. and S.C. McCutcheon (1996) Phytoremediation: modeling removal of TNT and its breakdown products. *Remediation* **6**: 31–45.

Medina, V.F., P. Jeffers, S.L. Larson, and W. Perez (2000a) Sterilization of plants for phytoremediation studies by bleach treatment. *Int. J. Phytoremed.* **2**: 287–295.

Medina, V.F., S.L. Larson, A.E. Bergstedt, and S.C. McCutcheon (2000b) Phytoremoval of trinitrotoluene from water using batch kinetics studies. *Water Res.* **34**(10): 2713–2722.

Medina, V.F., S.L. Larson, and S.C. McCutcheon (2002) Evaluation of continuous flow-through phytoreactors for the treatment of TNT-contaminated water. *Environ. Prog.* **21**(1): 1–8.

Metwally, M. El-S. and N.L. Wolfe (1990) Hydrolysis of chlorostilbene oxide II: modeling of hydrolysis in aquifer samples and in sediment-water systems. *Environ. Toxicol. Chem.* **9**: 963–973.

Miller, J.L., E.P.H. Best, and S.L. Larson (1997) Degradation of explosives in ground water at the Volunteer Army Ammunition Plant in flow-through systems planted with aquatic and wetland plants. In: *Abstracts Book: 12th Annual Conference on Hazardous Waste Research.* Hazardous Substances Research Center/Great Plains, Kansas State University, Manhattan, Kansas. Held in Kansas City, Missouri, May 19–22, pp. 9–10.

Palazzo, A.J. and D.C. Leggett. (1986) Effect and disposition of TNT in a terrestrial plant. *J. Environ. Qual.* **15**: 49–52.

Pavlostathis, S.G., K.K. Comstock, M. Jacobson, and F.M. Saunders (1998) Transformation of 2,4,6-trinitrotoluene by the aquatic plant *Myriophyllum spicatum. Environ. Toxicol. Chem.* **17**(11): 2266–2273.

Peijnenburg, W. J.G.M., M.J.T. Hart, H.A. den Hollander, D. van de Meent, H.H. Verboom, and N.L. Wolfe (1992a) QSARs for predicting reductive transformation rate constants of halogenated aromatic hydrocarbons in anoxic sediment systems. *Environ. Toxicol. Chem.* **11**(3): 301–314.

Peijnenburg, W. J.G.M., M.J.T. Hart, H.A. den Hollander, D. van de Meent, H.H. Verboom and N.L. Wolfe (1992b) Reductive transformations of halogenated aromatic hydrocarbons in anaerobic water-sediment systems: kinetics, mechanisms and products. *Environ. Toxicol. Chem.* **11**(3): 289–300.

Pennington, J.C. (1988) Plant uptake of 2,4,6-trinitrotoluene, 4-amino-2,6-dinitrotoluene, and 2-amino-4,6-dinitrotoluene using ^{14}C-labeled and unlabeled compounds. U.S. Army Engineer Waterways Experiment Station. Technical Report EL-88–90. Vicksburg, Mississippi.

Pennington, J.C., R. Bowen, J.M. Brannon, M. Zakikhani, D.W. Harrelson, D. Gunnison, J. Mahannah, J. Clarke, T.F. Jenkins, and S. Gnewuch (1999a) Draft protocol for evaluating, selecting, and implementing monitored natural attenuation at explosives-contaminated sites. U.S. Army Corps of Engineers Engineering Research and Development Center Technical Report EL-99-10. Vicksburg, Mississippi.

Pennington, J.C., D. Gunnison, M. Zakikhani, H.L. Fredrickson, and D.B. Ringelberg (1999b) Natural attenuation of explosives in soil and water systems at Department of Defense sites: final report. U.S. Army Corps of Engineers Engineering

Research and Development Center Technical Report SERDP-99-1. Vicksburg, Mississippi.

Pennington, J.C., D. Gunnison, D.W. Harrelson, J.M. Brannon, M. Zakikhani, T. F. Jenkins, J.U. Clarke, C.A. Hayes, T. Myers, E. Perkins, D. Ringelberg, D.M. Townsend, H. Fredrickson, and J.H. May (1999c) Natural attenuation of explosives in soil and water systems at Department of Defense sites: interim report. U.S. Army Corps of Engineers Engineering Research and Development Center Technical Report EL-99-8. Vicksburg, Mississippi.

Polish State Inspectorate for Environmental Protection (1995) Guide-book of ecological damages and remediation works at the areas abandoned by stationing in Poland Russian Federation troops. Warsaw.

Qaisi, K.M., L.J. Thibodeaux, K.S. Ro, K.T. Valsaraj, and D.D. Adrian (1996a) A proposal for a field-scale pilot demonstration unit for bioremediation of TNT contaminated soil, *J. Environ. Sci. Health* **A31**(9): 2287–2294.

Qaisi, K.M., K.S. Ro, D. Reible, L.J. Thibodeaux, K.T. Valsaraj, and W.D. Constant (1996b) A transport process of TNT from flooded highly contaminated surface soil bed. *J. Environ. Sci. Health* **A3**(10): 2515–2532.

Qazi, M. and L. Kanaras (1997) Pink water treatment technology evaluations. In: *Proceedings Second Tri-Service Environmental Technology Workshop*. U.S. Army Environmental Center, June 10–12, St. Louis, Missouri, pp. 422–431.

Rivera, R., V.F. Medina, S.L. Larson, and S.C. McCutcheon (1998) Phytotreatment of TNT contaminated groundwater. *J. Soil Contam.* **7**(4): 511–529.

Schnoor, J.L., L.A. Licht, S.C. McCutcheon, N.L. Wolfe, and L.H. Carreira (1995) Phytoremediation: an emerging technology for contaminated soils. *Environ. Sci. Tech.* **29**(7): 318A–323A.

Shah, M.M. and J.C. Spain (1996) Elimination of nitrite from the explosive 2,4,6-trinitrophenylmethylnitramina (tetryl) catalyzed by ferredoxin NADP oxidoreductase from spinach. *Biochem. Biophys. Res. Comm.* **220**: 563–568.

Shah, M.M. and J.A. Campbell (1997) Transformation of nitrobenzene by ferredoxin NADP oxidoreductase from spinach leaves. *Biochem. Biophys. Res. Comm.* **241**: 794–796.

Spain, J.C. (2000) Introduction. Chapter 1. In: *Biodegradation of Nitroaromatic Compounds and Explosives*. J.C. Spain, J.B. Hughes, and H.J. Knackmuss, eds. Lewis Publishers, Boca Raton, Florida, pp. 1–5.

Thompson, P.L. (1997) Phytoremediation of munitions (RDX, TNT) waste at the Iowa Army Ammunition Plant with hybrid poplar trees. Ph.D. dissertation, Civil and Environmental Engineering, University of Iowa, Iowa City.

Thompson, P.L., L.A. Ramer, and J.L. Schnoor (1998) Uptake and transformation of TNT by a poplar hybrid. *Environ. Sci. Tech.* **32**(7): 975–980.

Todd, S.R. (1996) Phytoremediation of dinitrotoluene contaminated soils using parrot feather (*Myriophyllum aquaticum*). M.S. thesis, Civil Engineering, Auburn University, Alabama.

Tratnyek, P.G. and N.L. Wolfe (1990) Characterization of the reducing properties of anaerobic sediment slurries using redox indicators. *Environ. Toxicol. Chem.* **9**: 289–295.

Tratnyek, P.G. and N.L. Wolfe. (1993) Oxidation and acidification of anaerobic sediment-water systems by autoclaving. *J. Environ. Qual.* **22**: 375–378.

U.S. Environmental Protection Agency (EPA) (1988) Design manual: constructed wetlands and aquatic plant systems for municipal wastewater treatment. Report EPA/625/1-88/022. Cincinnati, Ohio.

U.S. Environmental Protection Agency (EPA) (1992) Test Methods for Evaluating Solid Wastes, Physical-Chemical Methods SW-846. 3rd ed. Proposed Update II. Office of Solid Waste and Emergency Response, Washington, D.C.

Valsaraj, K.T., K.M. Qaisi, W.D. Constant, L.J. Thibodeaux, and K.S. Ro (1998) Diffusive transport of 2,4,6-trinitrotoluene (TNT) from contaminated soil to overlying water. *J. Haz. Mat.* **59**: 1–12.

van Beelen, P. and D.R. Burris (1995) Reduction of the explosive 2,4,6-trinitrotoluene by enzymes from aquatic sediments. *Environ. Toxicol. Chem.* **14**(12): 2115–2123.

Vanderford, M., J.V. Shanks, and J.B. Hughes (1997) Phytotransformation of trinitrotoluene (TNT) and distribution of metabolic products in *Myriophyllum aquaticum*. *Biotechnol. Lett.* **19**: 277–280.

Weber, E.J. and N.L. Wolfe (1987) Kinetic studies of aromatic azo compounds in anaerobic sediment/water systems. *Environ. Toxicol. Chem.* **6**(12): 911–919.

Wilson, E. (1998) Spinach enzymes break down explosives. *Chem. Eng. News* August Issue: 40.

Wolfe, N.L. and M.C. Delgado (1991) Isolation and characterization of bioorganic reducing agents in anaerobic sediments. 21st Annual Symposium on Environmental-Analytical Chemistry, Jekyll Island, Georgia.

Wolfe, N.L., T.Y. Ou, and L. Carreira (1993) Biochemical remediation of TNT-contaminated soils. Report to the U.S. Army Engineer Waterways Experiment Station. Vicksburg, Mississippi.

Wolfe, N.L., S. McCutcheon, T.Y. Ou, M. Jacobson, P.L. Thompson, and J. Weaver (1999) Feasibility of using natural attenuation coupled with phytoremediation as a remedial alternative for explosives contamination at Site L1, Joliet Army Ammunition Plant, Joliet, Illinois. Report submitted to U.S. Army Industrial Operations Command. Rock Island, Illinois. U.S. Environmental Protection Agency, Athens, Georgia.

Young, D.G. (1995) The biochemical remediation of a TNT contaminated soil. Ph. D. dissertation, Civil Engineering, Auburn University, Auburn, Alabama.

15

PHYTORESTORATION AT THE IOWA ARMY AMMUNITION PLANT

P. L. Thompson, D. D. Moses, and K. M. Howe

SUMMARY OF PRACTICAL IMPLICATIONS

The U.S. Army Corps of Engineers constructed the first, full-scale treatment wetlands with the purpose of containing and treating residual TNT (2,4,6-trinitrotoluene) and RDX (hexahydro-1,3,5-trinitro-1,3,5-triazine) contamination at the Iowa Army Ammunition Plant (IAAP). Prior to construction, approximately 38 000 kilograms of explosives were excavated to achieve regulatory cleanup goals at the site. The wetlands were then constructed to treat the residual contamination. Pilot-screening studies were conducted to establish vegetative uptake, fate, and toxicity of TNT and RDX. Results from the first 2 years of field pilot testing indicated that TNT concentrations were undetectable in surface water or plant tissues. Concentrations of RDX in wetland surface waters were below the human health advisory level of 2 micrograms per liter and were undetectable in plant tissues during the summer months. Wetland performance in the winter months often exceeded 0.002 milligrams of RDX per liter, but at no time were these waters discharged into adjacent surface waters. Furthermore, after the initial start-up period, the concentrations of RDX in the treatment wetlands never exceeded the ecological risk concentration of 0.19 milligram of RDX per liter. The cost of constructing, planting, and monitoring the wetlands was comparable to that of covering the residual contamination with native topsoil.

CLEANUP OF THE IOWA ARMY AMMUNITION PLANT

From 1943 to 1975, the IAAP (Middletown, Iowa, U.S.) produced large volumes of wastewater containing TNT and RDX at the Line 1 and Line 800 production areas. In 1948, the U.S. Army constructed an earthen embankment and

Phytoremediation: Transformation and Control of Contaminants,
Edited by Steven C. McCutcheon and Jerald L. Schnoor.
ISBN 0-471-39435-1 (cloth)

concrete spillway across Brush Creek forming a 1.4-hectare (3.6-acre) impoundment near the Line 1 production facility. Explosives-contaminated wastewater generated from the Line 1 production facility was discharged into the Line 1 Impoundment. Similarly, in 1943, the Army constructed a leaching field and evaporation furrows for the discharge of explosives-contaminated wastewater from the Line 800 production facility. The leaching field and evaporation furrows were not adequate to handle the volumes of generated wastewater, and in the mid-1960s, earthen berms were constructed to form a lagoon. The lagoon served as a settling pond to reduce particulates prior to discharge into a Brush Creek tributary. Contaminated sludges were also trucked in from other operations around IAAP and disposed at the Line 800 Pinkwater Lagoon.

The Line 1 Impoundment and Line 800 Pinkwater Lagoon were considered to be the greatest sources of explosives contamination at the IAAP. Both the Line 1 Impoundment and the Line 800 Pinkwater Lagoon drain into Brush Creek, which reportedly flowed red in color as it discharged past the boundary of the IAAP.

These disposal practices ceased in the mid-1970s, but the soils in both basins were highly contaminated with TNT and RDX. It should be noted that TNT contamination has historically been confined to soils at source locations at the IAAP. Found not only in soils at source locations, RDX is also commonly found in groundwater and in down-gradient surface waters. For example, the surface water in Brush Creek below the basins is contaminated with RDX in the 10 milligrams per liter range, and groundwater below the Line 800 Pinkwater Lagoon is contaminated with RDX in the 10 milligrams per liter range. Groundwater located beyond the boundaries of the IAAP in the floodplain of the Mississippi and Skunk Rivers is also contaminated with RDX on the order of 0.001 milligram per liter. Hence, the IAAP is on the U.S. Comprehensive Environmental Response and Compensation Liability Act (CERCLA) National Priority List.

The Line 1 Impoundment and the Line 800 Pinkwater Lagoon sites were undisturbed for more than 20 years during which time native grasses and wetland species colonized the areas. In 1995, the U.S. Army Corps of Engineers established a soil cleanup plan, which involved excavating approximately 76 000 cubic meters (100 000 cubic yards) of soil from Line 1 (18 000 cubic meters or 24 000 cubic yards containing over 1769 kilograms of explosives) and Line 800 (58 000 cubic meters or 76 000 cubic yards containing over 36 780 kilograms of explosives) at an estimated cost of 5 000 000 U.S. dollars. In 1997, about 10 percent of the excavated soil (believed to represent 80 percent of the contamination mass) was disposed of in a lined temporary stockpile called a Corrective Action Management Unit. Composting treatment of the material in the Corrective Action Management Unit began in 2001. Less-contaminated soils were relocated into a lined landfill or located below the low permeable layer of the landfill cover. Concentrations of 0.002 milligram per liter for TNT and RDX in groundwater were estimated to represent 10^{-6} levels of groundwater exposure risk. Based on these groundwater concentrations, the remediation goals for soil excavation of 1.3 milligrams of RDX per kilogram of dry soil and 47.6 milligrams of TNT per kilogram of dry soil were established.

The U.S. Army Corps of Engineers realized that exposing soils contaminated with residual amounts of explosives near the cleanup levels would result in contaminated surface water. The U.S. Environmental Protection Agency (EPA) approved the use of phytoremediation as a means to treat explosives-contaminated surface water in both basins. The construction of engineered wetlands at the IAAP was the first full-scale implementation of a phytoremediation system designed to treat explosives. Wetlands were chosen over terrestrial phytoremediation, because the vast majority of residual contamination was in areas that would be inundated with shallow water. The wetlands would provide the ancillary benefits of establishing a biologically diverse habitat with aesthetic value; however, the treatment wetlands were not considered for the regulatory purpose of mitigation banking. Low operation, energy, and maintenance costs were also a driving force behind the construction of the treatment wetlands. The U.S. EPA approval for the restoration plan hinged on the necessity of monitoring ecological risks, which included both plant tissue and surface water sampling.

This chapter begins with a brief summary of the literature, which includes laboratory work performed expressly for the IAAP project. Then, details relating to design, implementation, and operation of the treatment wetlands will be presented and discussed. Finally, a cost comparison between the backfilling and phytoremediation alternatives is provided.

REVIEW OF EXPLOSIVES UPTAKE AND PHYTOTOXICITY

Uptake of 2,4,6-Trinitrotoluene

The first successful characterization of the plant uptake of TNT was for yellow nutsedge (*Cyperus esculentus*) (Palazzo and Leggett 1986). The associated products 4-amino-2,6-dinitrotoluene (4-ADNT), 2-amino-4,6-dinitrotoluene (2-ADNT), and 2,4,6-trinitrotoluene were qualitatively determined to be in all plant tissues (leaves, roots, rhizomes, and tubers) with 4-ADNT appearing most frequently. It was proposed that 4-ADNT and 2-ADNT were formed in the plant because care was taken to ensure no metabolites were detected in the hydroponic nutrient solution. These products are most likely formed *via* the interaction of TNT with plant enzymes and the reduced forms of nicotinamide-adenine dinucleotide (NADH) or nicotinamide-adenine dinucleotide phosphate (NADPH) (Schnoor *et al.* 1995, van Beelen and Burris 1995). Unfortunately, biological conversion of TNT to these transformation products may not be an acceptable end point because these compounds are also explosive and toxic.

The U.S. Pacific Northwest Laboratory (Richland, Washington) studied the fate of both TNT and RDX in bush bean (*Phaseolus vulgaris*), wheat (*Triticum aestivum*), and blando brome (*Bromus hordeaceus*) (Cataldo *et al.* 1989, Cataldo *et al.* 1990, Harvey *et al.* 1990, 1991). These investigators performed mass balances using [U-^{14}C]-TNT both hydroponically and in soils and concluded that all three species had a high capacity to absorb TNT, with the bush bean (*Phaseolus vulgaris*) showing the highest rate of uptake. All three plants showed similar results. After 60 days of exposure, as much as 77 percent of the TNT-related label had accumulated in bush bean (*Phaseolus vulgaris*) root tissues with less than 20 percent of that fraction being recovered as TNT. The remainder consisted of 4-ADNT, 2-ADNT, a highly polar unknown compound, and bound residues. About 15 percent of the radiolabel was recovered in the stem and was again identified as 4-ADNT, 2-ADNT, and a number of highly polar components which, upon acid hydrolysis, yielded 4-ADNT and 2-ADNT. This indicated that some of the TNT transformation products could be transported as conjugates. The accumulation of the applied label ranged from 6 to 13 percent in leaf tissues. About 1 percent of the radiolabel was recovered in the pod and seed of the bean (*Phaseolus vulgaris*) plant. Overall, more than 80 percent of the TNT-related radiolabel was determined to be conjugated or transformed in the tissues of each plant species. No volatilization of $^{14}CO_2$ or volatile organic transformation products was observed.

Other studies have drawn similar conclusions with a variety of plants. Exposure of chives (*Allium schoenoprasum*) and alfalfa (*Medicago sativa*) to TNT that did not contain radioactive [^{14}C] as a tracer illustrated that the majority of detectable nitrotoluenes were present in root tissues with 10 percent being identified as TNT and 80 percent as 4-ADNT and 2-ADNT (Göerge *et al.* 1994). Scheidemann *et al.* (1998) assessed TNT uptake by eleven plant species that were cultivated in 10, 100, and 500 milligrams of TNT per kilogram of dry soil, respectively. Alfalfa (*Medicago sativa*) showed the greatest amount of uptake with the majority of the explosives-related compounds (TNT, 4-ADNT, and 2-ADNT) remaining in the roots. Four of six cultivars of wheat (*Triticum aestivum*) were able to significantly reduce the TNT concentration in the rhizosphere soil. This was attributed to both plant uptake as well as plant-enhanced microbial activity.

Hughes *et al.* (1997) used mass balances to assess the fate of [U-^{14}C]-TNT in aquatic plants such as parrot feather (*Myriophyllum aquaticum*) and Eurasian watermilfoil (*Myriophyllum spicaticum*), and in plant-tissue cultures periwinkle (*Catharanthus roseus*). They too found TNT to be quickly absorbed and transformed by all plant species. Their average [^{14}C] recoveries ranged from 93 ± 4.8 to 99 ± 9.5 percent with no significant mineralization to $^{14}CO_2$ over 7-day periods. These researchers were able to extract an average of 27 percent of the radiolabel from parrot feather (*Myriophyllum aquaticum*) tissues and were only able to identify about 6 percent of the radiolabel taken up by the plants as TNT, 4-ADNT, or 2-ADNT. Hence, about 98 percent of the [^{14}C] taken up was unidentified with about 48 percent of that being inextricable. Because TNT was detected in plant tissues on only one occasion, they hypothesized that TNT may be transformed during transport into root tissues. The results were similar to those of Vanderford *et al.* (1997).

In other research, the capability of Eurasian watermilfoil (*Myriophyllum spicaticum*) to transform TNT was tested in batch studies (Pavlostathis *et al.* 1998). Rapid disappearance of TNT was observed and reduced degradation products were detected. However, the majority of the added TNT could not be recovered from the plant biomass. More recently, previously unknown TNT transformation products have been determined to be conjugates of 2-ADNT and 4-ADNT (Bhadra *et al.* 1999a), and oxidative metabolites of TNT (Bhadra *et al.* 1999b). The conjugates are thought to be produced by the plant as a means of detoxification and are possibly precursors to bound residues (Bhadra *et al.* 1999a).

Phytotoxicity of 2,4,6-Trinitrotoluene

The subacute toxicity (*i.e.*, experienced in less than 30 days) of TNT to duckweed (*Lemna perpusilla* Torr.) was found at concentrations of and greater than 1 milligram per liter (Schott and Worthley 1974). A 42-day hydroponic study of TNT toxicity to yellow nutsedge (*Cyperus esculentus*) resulted in deleterious effects at concentrations of TNT equal to or greater than 5 milligrams of TNT per liter (Palazzo and Leggett 1986). Plant yields were 54 to 74 percent higher in the controls when compared to the tests with TNT concentrations of 5, 10, and 20 milligrams of TNT per liter. The greatest reduction in biomass occurred in the roots where the dry weights were 95 to 97 percent less than the controls. Leaf growth differed from 51 to 74 percent whereas rhizome (the underground part of the stem) weight, which was least affected, showed a 29 percent reduction.

Whole-plant studies of Jimsonweed (*Datura innoxia*) and wild tomato (*Lycopersicon peruvianum*) showed that these plants were able to grow well in soils contaminated with 750 milligrams of TNT per kilogram of dry soil (Mueller *et al.* 1995). Both plants were negatively affected by concentrations in the 1000 milligrams of TNT per kilogram of dry soil range. Peterson *et al.* (1996) studied the germination and early development of tall fescue (*Festuca arundinacea* Schreb.) seeds exposed to TNT and 4-ADNT. Germination and growth were affected by concentrations greater than 30 milligrams of TNT per liter. Germination was not significantly affected by the maximum concentration of 4-ADNT tested (15 milligrams per liter), whereas growth was reduced. Switch grass (*Panicum virgatum*) and smooth bromegrass (*Bromus inermis*) germination and growth were also evaluated for tolerance to TNT (Peterson *et al.* 1998). The authors concluded that both species were capable of withstanding soil-solution concentrations of less than 15 milligrams of TNT per liter. Toxicity testing has also revealed that the common brush bean (*Phaseolus vulgaris*) was the most tolerant species of the 11 tested, growing in soils with TNT levels up to 500 milligrams of TNT per kilogram of dry soil (Scheidemann *et al.* 1998).

Uptake of Hexahydro-1,3,5-Trinitro-1,3,5-Triazine

Much more recalcitrant than TNT, RDX degrades biologically under anaerobic conditions but rarely under aerobic conditions (McCormick *et al.* 1981, Spanggord *et al.* 1983). Hence, transformation products of RDX are not commonly found. Harvey *et al.* (1991) studied the uptake of RDX by plants and found results with respect to translocation and transformation that were quite different from those for TNT. Mass balance results averaging 82 ± 1.6 percent for bush bean (*Phaseolus vulgaris*) demonstrated that 69 percent of the absorbed label accumulated in leaf tissues after 7 days. Roughly 18 and 13 percent of the absorbed label were found in stem and root tissues, respectively. Radiochromatographic analysis of plant extracts revealed that the majority (up to 86 percent) of the radiolabel remained as RDX. There was also an increasing trend of polar metabolite production with time. After 60 days, plant extracts showed that only 21 to 51 percent of the radiolabel remained as RDX, 8 percent to 30 percent existed as unknown polar compounds, and 20 to 50 percent was not extractable (Cataldo *et al.* 1990). Similar to TNT, no volatilization of $^{14}CO_2$ or volatile organic transformation products was observed.

Chen (1993) studied the uptake of RDX by four different plant species: maize [*Zea mays* (L.) 'LH74 × FR600'], soybean [*Glycine max* (L.) Merr. 'Harper 87'], sorghum or Sudan grass (*Sorghum sudanese* L. 'SSG'), and wheat [*Triticum aestivum* (L.) 'Becker']. Hydroponic concentrations of 0, 6, 13, and 21 milligrams of RDX per liter were tested. The concentration of RDX in leaf tissue for the 21 milligrams of RDX per liter dosage was 168, 200, 408, and 436 milligrams of RDX per kilogram of dry leaf tissue for soybeans [*Glycine max* (L.) Merr. 'Harper 87'], maize [*Zea mays* (L.) 'LH74 × FR600'], wheat [*Triticum aestivum* (L.) 'Becker'], and sorghum (*Sorghum sudanese* L.), respectively. Root tissue concentrations for the same RDX solution were 80, 85, 120, and 156 milligrams of RDX per kilograms for soybeans [*Glycine max* (L.) Merr. 'Harper 87'], wheat [*Triticum aestivum* (L.) 'Becker'], sorghum (*Sorghum sudanese* L.), and maize [*Zea mays* (L.) 'LH74 × FR600'], respectively. Similar to Cataldo *et al.* (1990), Chen determined that RDX concentrations in plant tissues increased linearly with the increase in RDX levels in hydroponic solution. A similar trend was found for soils spiked with 0, 12.5, 25, 50, 100, 220, 494, and 903 milligrams of RDX per kilogram of soil. Labeled [U-^{14}C]-RDX was not used for these experiments although mass balances were conducted for the soil studies. These mass balance recoveries ranged from 73 to 97 percent.

Phytotoxicity of Hexahydro-1,3,5-Trinitro-1,3,5-Triazine

Chen (1993) estimated the hydroponic toxicity of RDX to maize [*Zea mays* (L.) 'LH74 × FR600'] and wheat [*Triticum aestivum* (L.) 'Becker'] to be approximately 21 milligrams of RDX per liter (the highest concentration tested) with the no-observed-effect-level being 13 milligrams of RDX per liter (probability less than 0.05). No toxicity was observed for sorghum (*Sorghum sudanese* L.) or soybean [*Glycine max* (L.) Merr. 'Harper 87'] plants for RDX concentrations up to 21 milligrams of RDX per liter for these 30-day exposures.

IOWA ARMY AMMUNITION PLANT SCREENING STUDIES

In 1995, the U.S. Army Corps of Engineers selected the University of Iowa (Iowa City, Iowa) and the U.S. Army Waterways Experiment Station (Vicksburg, Mississippi) to conduct pilot-screening studies of plant species for implementation at the site. An enzyme-linked immunosorbent assay (ELISA) was developed by the U.S. EPA Laboratory (Athens, Georgia) and used to determine which native plant species had significant nitroreductase activity. Wetland plants from Stump Lake were screened and selected for further testing by the Waterways Experiment Station. The Army was also interested in the use of poplar (*Populus* spp.) trees in upland areas of residual contamination because poplar (*Populus* spp.) trees had been used for previous phytoremediation efforts (Schnoor *et al.* 1995). These pilot-screening investigations were important because, at the time, only a few (mostly agricultural) plant species had ever been tested for TNT (Palazzo and Leggett 1986, Cataldo *et al.* 1989, Cataldo *et al.* 1990) or RDX (Cataldo *et al.* 1989, Cataldo *et al.* 1990, Chen 1993) uptake.

Best *et al.* (1997a) and Best *et al.* (1997b) studied explosives uptake kinetics by using native wetland plants and contaminated IAAP groundwater samples. These investigators evaluated two submersed plant species: American pondweed (*Potamogeton nodosus* Poir.) and coontail (*Ceratophyllum demersum* L.). In addition, they tested the emergent species: water plantain (*Alisma subcordatum* Raf.), common arrowhead (*Sagittaria latifolia* Willd.), fox sedge (*Carex vulpinoidea* Michx.), wool grass [*Scirpus cyperinus* (L.) Kunth], blunt spike rush [(*Eleocharis obtusa* (Willd.)], reed canary grass (*Phalaris arundinacea* L.), narrowleaf cattail (*Typha angustifolia* L.), and parrot feather [*Myriophyllum aquaticum* (Vell.) Verdc.]. At 25 °C, 10-day plant incubations demonstrated 94 to 100 percent TNT disappearance, whereas groundwater sediment, and autoclaved sediment controls ranged from 62 to 85 percent. Uptake resulted in concentrations that were near the detection level of 0.0001 milligram of TNT per liter of extract, indicating that TNT did not bioaccumulate in the plant tissues. They concluded that TNT removal followed first order kinetics with rate constants ranging from 0.038 per hour for reed canary grass (*Phalaris arundinacea* L.) to 0.012 per hour for parrot feather [*Myriophyllum aquaticum* (vell.) verdc.]. These same investigators found similar results with wetland species using groundwater from the U.S. Milan Army Ammunition Plant (Milan, Tennessee) (Best *et al.* 1999a, 1999b).

Thompson *et al.* (1998a) found uptake of TNT by hybrid poplar (*Populous deltoides* × *Populus nigra*, 'DN34') trees from hydroponic solutions but sorption retarded uptake from contaminated soil. Once in the tree, up to 75 percent of the absorbed explosive remained in root tissues, with up to 10 percent eventually being translocated to the leaves. Hybrid poplars

(*Populus deltoides* × *Populus nigra*, 'DN34') were able to transform TNT to 4-ADNT, 2-ADNT, and possibly 2,4-diamino-6-nitrotoluene with only polar, unknown products being found in the leaves. The tree may have also transformed TNT to at least four additional unidentified compounds. These unknowns were more polar than TNT as indicated by retention times on the high-performance liquid chromatograph and at least one of the unknowns was the only TNT-related product found to translocate to leaf tissues.

Thompson *et al.* (1998b) also determined that TNT was toxic to hybrid poplar (*Populus deltoides* × *Populus nigra* 'DN34') at concentrations greater than 5 milligrams of TNT per liter in lab studies. Larger hybrid poplars (*Populus deltoides* × *Populus nigra* 'DN34') grown under greenhouse conditions showed an increased tolerance to TNT with no significant (probability less than 0.05) effects upon exposure to 5 milligrams of TNT per liter for a 4-week period (Thompson *et al.* 2000). This increased tolerance was possibly a result of having light intensities that were three to four times those used under laboratory conditions. Hence, plant metabolism and toxin sequestration due to growth may have been augmented under more natural growing conditions.

Best *et al.* (1997a), and Best *et al.* (1997b) also evaluated RDX disappearance with the aforementioned aquatic species tested for TNT uptake. At 25 °C, plant incubations achieved an average removal rate of less than 15 percent for RDX. The plant removal rate was comparable to the groundwater (11 percent), sediment (21 percent), and autoclaved sediment (25 percent) controls. Concentrations in the plants averaged 200 milligrams per kilogram dry weight. The short 10-day trial did not distinguish whether RDX was permanently bioaccumulated or if the RDX accumulations were transient. The removal of RDX was modeled by zero order kinetics, and removal by reed canary grass (*Phalaris arundinacea* L.) proceeded with a rate constant of 13 micrograms of RDX per liter per hour. For the Milan Army Ammunition Plant study, Best *et al.* (1999a, 1999b) reported limited biotransformation of RDX to unidentified polar products during 13-day incubations.

Thompson *et al.* (1999) found RDX to be taken up relatively slowly by hybrid poplar (*Populus deltoides* × *Populus nigra* 'DN34') as compared to TNT. Mass balances were comparable to those of other studies in that up to 60 percent of the RDX removed (about 25 percent of that applied) was translocated to the leaves. The RDX taken up was not shown to be transformed to any appreciable extent under laboratory conditions for 7-day periods. Moreover, Thompson (1997) found no significant difference in growth or transpiration rates between controls and plants exposed to 7, 12, and 21 millligrams of RDX per liter over acute exposure periods of 14 days.

WETLAND DESIGN AT THE IOWA ARMY AMMUNITION PLANT

The results from the previously discussed screening studies of Best *et al.* (1997a), and Best *et al.* (1997b) demonstrated that phytoremediation of TNT-contaminated water was feasible, whereas the findings for RDX showed limited removal. However, treatment wetlands were constructed with the understanding that full-scale field implementation would likely be more efficient compared to laboratory screenings. This increased removal efficiency was expected for a number of reasons. First, an increase in plant-assisted degradation was anticipated because the plant density targeted for the field would be denser than that tested in the laboratory. Second, synergistic effects between wetland plants, sediment, and sediment-based microorganisms were expected to increase removal. Third, the Line 1 Impoundment and the Line 800 Lagoon were designed and constructed to regulate hydraulic retention times and to prevent any discharges of contaminated water. Therefore, increasing the treatment period from 10-day screenings to a full 6-month growing season would also increase the removal efficiency for explosives. Finally, it was anticipated that direct photolysis of the contaminants would play a significant role under natural growing conditions. (Spanggord *et al.* 1983, Sikka *et al.* 1980).

Site Preparation

Prior to excavating contaminated soils from the Line 1 area, the small Brush Creek that was running through the contaminated zone was permanently diverted around the Line 1 area. Contaminated soils were excavated and as an alternative to backfilling, an engineered wetland was created to first contain and then treat residual contamination. Because 100 percent of the contaminated soil was not removed by excavation, the remaining residual was expected to leach into the wetlands and be removed by plants, active sediment, microorganism, and photolysis.

The created wetland was seeded with a 15-centimeter (6-inch) layer of nutrient-rich topsoil (from the upper reaches of Brush Creek) and silt (from nearby Stump Lake). Because the pilot screenings (Best *et al.* 1997a), Best *et al.* (1997b) had shown that no particular plant species was superior at removing explosives from water, species selection was primarily based upon choosing a community of plants that could survive the climatic conditions of Iowa. For this reason, the topsoil and Stump Lake silt were also relied upon as a source of seeds for submergent and emergent vegetation.

Two reinforced-concrete control structures were constructed to regulate water into and out of the Line 1 treatment wetland. The upstream hydraulic control structure was designed to divert water into the impoundment from Brush Creek. The downstream hydraulic control structure was designed to be used for water discharge (if necessary) from the system *via* a slide gate and stop logs. A gated crossing was also located on Brush Creek to provide access to the hydraulic control structures and to divert Brush Creek flows into the Line 1 Impoundment wetland. Use of both hydraulic controls was expected to be very minimal. The system was also designed to divert water from Brush Creek to fill the wetland during drought conditions. The downstream hydraulic structure was used to regulate the water-surface elevation of the wetland and to discharge water from the impoundment in future optimization studies. The initial operating hydraulic residence time of the wetland was designed to approach infinity. By design, water can be released from the system after levels drop below 0.002 millligram of RDX per liter.

The Line 800 Pinkwater Lagoon was also excavated and lined with seed-laden silt from Stump Lake. Seeds for plants such as reed canary grass (*Phalaris arundinacea* L.), pondweed (*Potamogeton nodosus* Poir.), and arrowhead (*Sagittaria latifolia* Willd.), which were identified during the aforementioned nitroreductase-screening were contained in the Stump Lake silt. Unlike Brush Creek, there is little drainage into the Line 800 Lagoon. Water in the lagoon varies from 0 to 1.8 meters (0 to 6 feet) in depth. A single hydraulic control structure was constructed at the Line 800 Lagoon for discharge, if necessary, during high water levels.

Phytoremediation Monitoring

The 0.002 milligram of RDX per liter U.S. EPA Health Advisory Lifetime level was used as a discharge standard. The water quality level allowable in the impoundment was based upon the ecological risk value of 0.19 milligram of RDX per liter (Talmage *et al.* 1999).

The treatment wetlands were monitored to demonstrate the effectiveness of control to see if degradation occurred and to ensure that the process did not pose an ecological risk. Surface water, sediments, and plant tissues were monitored for explosives. Data was collected during the first two growing seasons, 1998 and 1999. A vigorous growth of diverse upland, fringe, and wetland plants naturally established in both the Line 1 and line 800 ponds. Approximately 50 species were observed at each wetland. Reed canary (*Phalaris arundinacea* L.) and barnyard (*Echinochloa crusgalli* Michx.) grasses dominated along the banks of both wetlands. Pondweed (*Potamogeton nodosus* Poir.), *Spirogyra* spp., coontail (*Ceratophyllum demersum* L.), and arrowhead (*Sagittaria latifolia Willd.*) were also established, especially along the wetland perimeters.

Replicate quality assurance testing by a third party revealed major commercial laboratory errors for all of the plant-tissue data collected for the 1998 growing season. Hence, that data was deemed to be unusable. This experience pointed to the complexity of the analytical methods and the need to ensure that commercial laboratories are certified for plant-tissue analyses. During the 1999 growing season, plant-tissue samples were obtained in June, (at the beginning of the growing season), August (during the mid-point of the growing season), and in late September (toward the end of the growing season). The plant tissues obtained were whole plant samples with stems and leaves, but the roots did not remain intact. The plants were washed to remove sediments and preserved at 4 °C until prepared for analysis. Split samples of plant tissue samples were collected and sent to the U.S. Army Waterways Experiment Station and the University of Iowa for explosives analysis by U.S. EPA high-performance liquid chromatography Method 8330. Sample extracts from both the Waterways Experiment Station and the University of Iowa were sent to the U.S. Army Corps of Engineers Quality Assurance Laboratory (Omaha, Nebraska).

Pondweed (*Potamageton nodosus* Poir.), freshwater algae (*Spirogyra* spp)., coontail (*Ceratophyllum demersum* L.), reed canary grass (*Phalaris arundinacea* L.), duckweed (*Lemna perpusilla* Torr.), and arrowhead (*Sagittaria latifolia* Willd.) were sampled at the Line 800 wetland. Pondweed (*Potamageton nodosus* Poir.), freshwater algae (*Spirogyra* spp)., coontail (*Ceratophyllum demersum* L.), arrowhead (*Sagittaria latifolia* Willd.), barnyard grass (*Echinochloa crusgalli* Michx.), and water plantain (*Alisma subcordatum* Raf.) were sampled at the Line 1 wetland. These species were chosen for analysis because the plants represented the most abundant biomass in the treatment wetlands.

None of the three sampling rounds in both Line 1 and Line 800 treatment wetlands detected explosives in the plant tissues. This result is quite important, because it illustrates that natural growing conditions may have allowed the plants to metabolize RDX more than under laboratory conditions. As a positive confirmation of the analytical methods, plant tissue samples were obtained from the plants growing in the highly contaminated soils that were located in the Corrective Action Management Unit (temporary stockpile) at the IAAP. These plants were rooted in soils containing 1.6 milligrams of RDX per kilogram dry soil and 2040 milligrams of TNT per kilogram of dry soil. The Corrective Action Management Unit barnyard grass (*Echinochloa crusgalli* Michx.) sample yielded an RDX level of 0.52 milligram of RDX per kilogram of dry plant, a 4-ADNT level of 0.47 milligram of 4-ADNT per kilogram of dry plant, a 2-ADNT level of 0.56 milligram of 2-ADNT per kilogram of plant, and a dinitrobenzene level of 0.37 milligram per kilogram of plant; TNT was undetectable.

Surface Water Quality

Composite surface water samples were obtained monthly at both treatment wetlands. The primary contaminant found in the surface water at both wetlands was RDX. The first surface water quality samples were obtained in January of 1998 when the wetlands had first filled with water and had little vegetation established. The initial RDX levels were the highest recorded with 0.26 milligram of RDX per liter at the Line 800 wetland and 0.778 milligram of RDX per liter at the Line 1 wetland, respectively.

The initial elevated RDX levels were attributed to the fresh wetting of the recently disturbed soils containing residual contamination of 1.3 milligrams of RDX per kilograms of dry soil and 47.6 milligrams of TNT per kilograms of dry soil after excavation. The residual contamination was in a clayey soil with partition coefficients of 1.0 and 5.3 liters of water per kilogram of dry soil for RDX and TNT, respectively (Thompson 1997). Not strongly bound to the soil, as indicated by a low partition coefficient, RDX readily partitioned into the groundwater.

Figures 15-1 and 15-2 illustrate the results of the monthly sampling effort for both Line 1 and Line 800 treatment wetlands for RDX; 2,4,6-trinitrotoluene

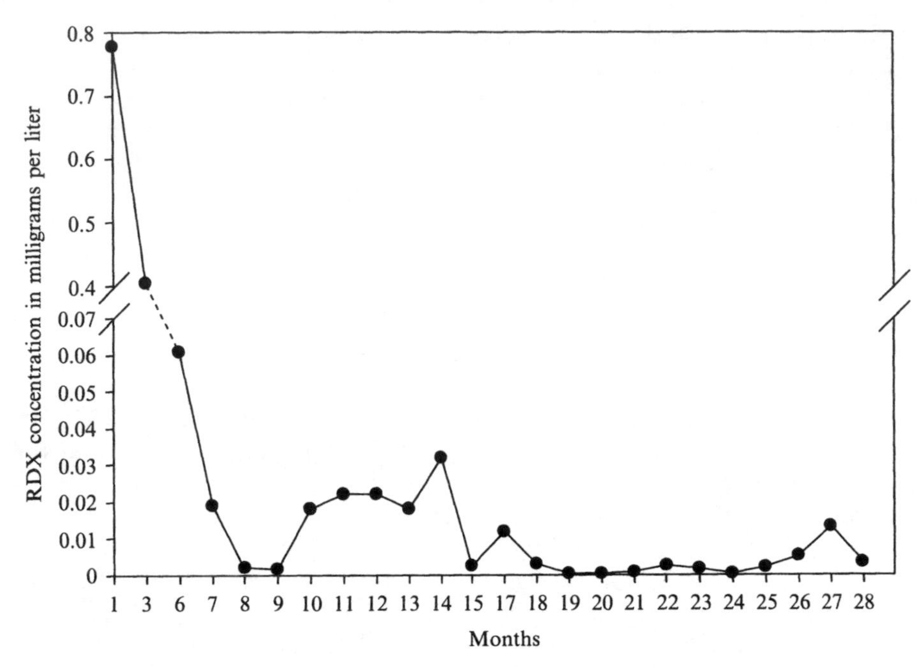

Figure 15-1 Line 1 surface water RDX concentrations from January 1997 to April 2000.

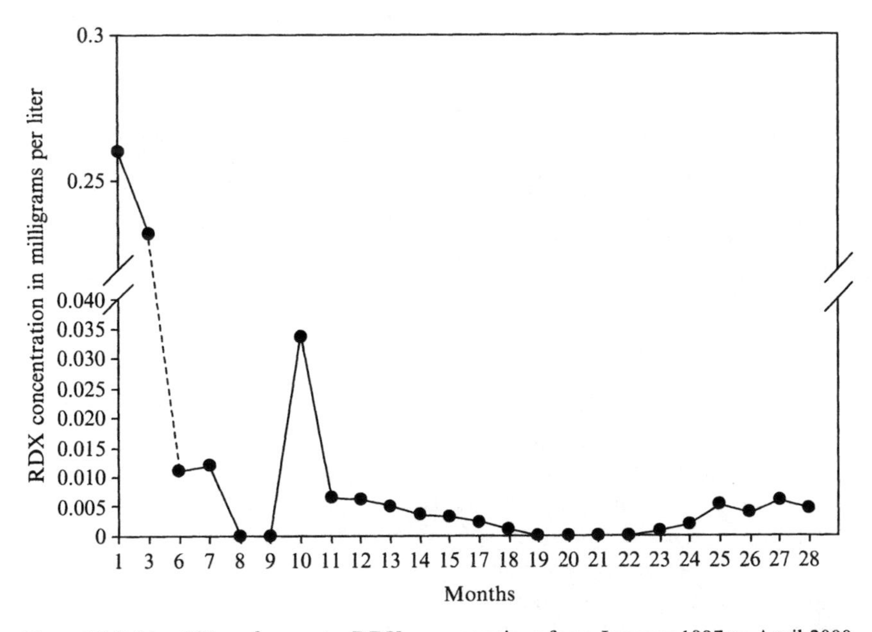

Figure 15-2 Line 800 surface water RDX concentrations from January 1997 to April 2000.

was not observed in the water at either wetland. During the first growing season (1998), the RDX in the Line 1 Impoundment decreased from the initial value of 0.778 milligram of RDX per liter in January to less than 0.002 milligram of RDX per liter in August. During the winter, the RDX level increased to 0.032 milligram of RDX per liter in February of 1999. During the second growing season (1999), RDX in the Line 1 wetland decreased from the winter high of 0.032 milligram of RDX per liter to near undetectable levels in July of 1999. In the second winter (1999–2000), the RDX level increased to as high as 0.0133 milligram of RDX per liter in March of 2000.

The pattern for RDX concentrations in the Line 800 wetland was similar. During the first growing season (1998), the RDX in the Line 800 wetland decreased from the initial value of 0.26 milligram of RDX per liter in January to undetectable levels in August, but in the fall, the RDX level increased to 0.034 milligram of RDX per liter in October of 1998. During the second growing season (1999), the RDX in the Line 800 wetland decreased from the fall high of 0.034 milligram of RDX per liter to undetectable levels in July of 1999. During the second winter (1999 to 2000), the RDX level increased to 0.0060 milligram of RDX per liter in March of 2000.

Verification samples were taken to confirm the bottom of the excavations were in compliance with the soil remediation goals of 1.3 milligrams of RDX per kilogram of dry soil and 47.6 milligrams of TNT per kilogram of dry soil. The bottom of the 1.2-hectare (3-acre) Line 1 Impoundment excavation met these criteria. The bottom of the 2.2-hectare (5.5-acre) Line 800 wetland excavation generally met these criteria with the exception of an isolated area that had approximately 20 milligrams of RDX per kilogram of dry soil. Two sediment-sampling rounds in 1998 and three rounds in 1999 were performed to verify sampling results. Sediment samples were obtained from surficial sediments and from a depth of approximately 0.3 meter (1 foot) below ground surface. These samples established that the treatment wetland was consistently attenuating discharge to surface waters. All sampling rounds indicated that the explosives levels were at or below the remediation goals except for an area in the Line 800 treatment wetland where a sample had 10 milligrams of RDX per kilogram of dry soil and 67 milligrams of TNT per kilogram of dry soil. The residual soil contamination measured in 1999 for both wetlands averaged approximately 1 milligram of RDX per kilogram of dry soil and 5 milligrams of TNT per kilogram of dry soil.

Cost Analysis

The cost to backfill approximately 76 000 cubic meters (100 000 cubic yards) into the excavated areas at both the Line 1 Impoundment and the Line 800 Lagoon was estimated to be 700 000 U.S. dollars. The initial cost of reclaiming both excavations as treatment wetlands was approximately 560 000 U.S. dollars. Specifically, diverting Brush Creek around the Line 1 Impoundment cost approximately 200 000 U.S. dollars, building the upper and lower water control structures at the Line 1 Impoundment was approximately 300 000 U.S. dollars, and blanketing the Line 800 Lagoon with approximately 2300 cubic meters (3000 cubic yards) of sediment from Stump Lake was approximately 60 000 U.S. dollars.

During the first two growing seasons, approximately 120 000 U.S. dollars per year was spent monitoring the plant uptake, sediments, and water quality. Monitoring plant uptake and sediments have been terminated. Monthly water quality data will be monitored (15 000 U.S. dollars per year) in the future until winter rebound levels drop below the allowable discharge level of 0.002 milligram of RDX per liter. Consequently, the monitoring and site preparation costs for phytorestoration were about the same to slightly more than the estimated cost for backfilling.

CONCLUSIONS

During the first 2 years of operation, the phytoremediation wetlands at both Line 1 and Line 800 have been successful in treating RDX and TNT contamination while protecting the local ecosystem. Aqueous phase TNT was never detected in either treatment wetland, and RDX levels in the wetlands were consistently below the ecological water quality criteria of 0.19 milligram of RDX per liter. Because the treatment wetlands were operated on a zero discharge basis, the 0.002 milligram of RDX per liter discharge requirement was also satisfied. Results further illustrated that if these treatment units were used for groundwater treatment, one attractive management strategy would be to release treated water during the summer months when RDX concentrations were consistently below 0.002 milligram of RDX per liter while storing and treating during other times of the year.

Other natural processes aided the reduction of explosives in the surface water besides phytoremediation. The explosives removal mechanisms, although not completely established, are most likely a synergistic combination of photolysis, plant metabolic activity, and waterborne and soil microorganism degradation. Even though monitoring costs were slightly more than originally planned, phytoremediation proved to be of comparable cost to backfilling.

REFERENCES

Best, E.P.H., M.E. Zappi, H.L. Fredrickson, S.L. Sprecher, and J.L. Miller (1997a) Screening of aquatic and wetland plant species for phytoremediation of explosives-contaminated groundwater from the Iowa Army Ammunition Plant. Technical Report EL-97-2. U.S. Army Engineer Waterways Experiment Station, Vicksburg, Mississippi.

Best, E.P.H., M.E. Zappi, H.L. Fredrickson, S.L. Sprecher, S.L. Larson, and M. Ochman (1997b) Screening of aquatic and wetland plant species for phytoremediation of explosives-contaminated groundwater from the Iowa Army Ammunition Plant.

In: *Bioremediation of Surface and Subsurface Contamination*. R.K. Bajpai and M.E. Zappi, eds. *Ann. New York Acad. Sci*, New york. **829**: 179–194.

Best, E.P.H., S.L. Sprecher, S.L. Larson, H.L. Fredrickson, and D.F. Bader (1999a) Environmental behavior of explosives in groundwater from the Milan Army Ammunition Plant in aquatic and wetland plant treatments: removal, mass balances and fate of TNT and RDX in groundwater. *Chemosphere* **38**: 3383–3396.

Best, E.P.H., S.L. Sprecher, S.L. Larson, H.L. Fredrickson, and D.F. Bader (1999b) Environmental behavior of explosives in groundwater from the Milan Army Ammunition Plant in aquatic and wetland plant treatments: uptake and fate of TNT and RDX in plants. *Chemosphere* **39**: 2057–2072.

Bhadra, R., D.G. Wayment, J.B. Hughes, and J.V. Shanks (1999a) Confirmation of conjugation processes during TNT metabolism by axenic plants. *Environ. Sci. Technol.* **33**(3): 446–452.

Bhadra, R., R. Spanggord, D.G. Wayment, J.B. Hughes, and J.V. Shanks (1999b) Characterization of oxidation products of TNT metabolism in aquatic systems of *Myriophyllum aquaticum*. *Environ. Sci. Technol.* **33**(19): 3354–3361.

Cataldo, D.A., S.D. Harvey, R.J. Fellows, R.M. Bean, and B.D. McVetty (1989) An evaluation of the environmental fate and behavior of munitions materiel (TNT, RDX) in soil and plant systems: TNT. U.S. Department of Energy Pacific Northwest Laboratory Report PNL-7370, NTIS# AD-A223 546. U.S. Army Medical Research and Development Command, Fort Detrick, Maryland.

Cataldo, D.A., S.D. Harvey, and R.J. Fellows (1990) An evaluation of the environmental fate and behavior of munitions materiel (TNT, RDX) in soil and plant systems: RDX. U.S. Department of Energy Pacific Northwest Laboratory Report PNL-7529. U.S. Army Medical Research and Development Command, Fort Detrick, Maryland.

Chen, D. (1993) Plant uptake and soil adsorption of RDX. M.S. thesis. Agronomy, University of Illinois at Urbana-Champaign.

Göerge, E., S. Brandt, and D.J.G. Werner (1994) Uptake and metabolism of 2,4,6-trinitrotoluene in higher plants. *Environ. Sci. Pollut. Res.* **1**(4): 229–233.

Harvey, S.D., R.J. Fellows, D.A. Cataldo, and R.M. Bean (1990) Analysis of 2,4,6-trinitrotoluene and its transformation products in soils and plant tissues by high performance liquid chromatography. *J. Chromatogr.* **518**: 361–374.

Harvey, S.D., R.J. Fellows, D.A. Cataldo, and R.M. Bean (1991) Fate of the explosive hexahydro-1,3,5–trinitro-1,3,5-triazine (RDX) in soil and bioaccumulation in bush bean hydroponic plants. *Environ. Toxicol. Chem.* **10**: 845–855.

Hughes, J.B., J. Shanks, M. Vanderford, J. Lauritzen, and R. Bhadra (1997) Transformation of TNT by aquatic plants and plant tissue cultures. *Environ. Sci. Technol.* **31**: 266–271.

McCormick, N.G., J.H., Cornell, and A.M. Kaplan (1981) Biodegradation of hexahydro-1,3,5-trinitro-1,3,5-triazine. *Appl. Environ. Microbiol.* **42**: 817–823.

Mueller, W., G.W. Bedell, S. Shojaee, and P.J. Jackson (1995) Bioremediation of TNT wastes by higher plants. In: *Proceedings of the 10th Annual Conference on Hazardous Waste Research*, L.E. Erickson, D.L. Tilllison, S.C. Grant, and J.P. McDonald, eds. U.S. Environmental Protection Agency Great Plains-Rocky Mountain Hazardous Substance Research Center, Kansas State University, Manhattan, Kansas, pp. 222–230.

Palazzo, A.J. and D.C. Leggett (1986) Effect and disposition of TNT in a terrestrial plant. *J. Environ. Qual.* **15**(1): 49–52.

Pavlostathis, S.G., K.K. Comstock, M.E. Jacobson, and M.F. Saunders (1998) Transformation of 2,4,6-trinitrotoluene by the aquatic plant *Myriophyllum spicatum*. *Environ. Toxicol. Chem.* **17**(11): 2266–2273.

Peterson, M.M., G.L. Horst, P.J. Shea, S.D. Comfort, and R.K.D. Peterson (1996) TNT and 4-amino-2,6-dinitrotoluene influence on germination and early seedling development of tall fescue. *Environ. Pollut.* **93**: 57–62.

Peterson, M.M., G.L. Horst, P.J. Shea, and S.D. Comfort (1998) Germination and seedling development of switchgrass and smooth bromegrass exposed to 2,4,6-trinitrotoluene. *Environ. Pollut.* **99**(1): 53–59.

Scheidemann, P., A. Klunk, C. Sens, and D. Werner (1998) Species dependent uptake and tolerance of nitroaromatic compounds by higher plants. *J. Plant Physiol.* **152**(2/3): 242–247.

Schnoor, J.L., L. Licht, S.C. McCutcheon, N.L. Wolfe, and L.H. Carriera (1995) Phytoremediation: an emerging technology for contaminated soils. *Environ. Sci. Technol.* **29**(7): 318A–323A.

Schott, C.D. and E.G. Worthley (1974) The toxicity of TNT and related wastes to an aquatic flowering plant, *Lemna perpusilla* Torr. Report AD-778 Edgewood Arsenal. U.S. Army Aberdeen Proving Ground, Maryland.

Sikka, H.C., S. Banerjee, E.J. Pack, and H.T. Appleton (1980) Environmental fate of RDX and TNT. Contract no. DAMD17-77-C-7026. U.S. Army Medical Research and Development Command, Fort Detrick, Maryland, Syracuse Research Corporation, Syracuse, New York.

Spanggord, R.J., W.R. Mabey, T. Mill, T.W. Chou, J.H. Smith, and S. Lee (1983) Environmental fate studies on certain munition wastewater constituents: final report, phase II—laboratory studies. SRI International, Menlo Park. Report AD A099256. U.S. Army Medical Research and Development Command, Fort Detrick, Maryland.

Talmage S.S., D.M. Opresko, C.J. Maxwell, C.J.E. Welsh, F.M. Cretella, P.H. Reno, and F.B. Daniel (1999) Nitroaromatic munition compounds: environmental effects and screening values. *Rev. Environ. Contam. Toxicol.* **161**: 1–156.

Thompson, P.L. (1997) Phytoremediation of munitions (RDX, TNT) waste at the Iowa Army Ammunition Plant, Ph.D. Dissertation, University of Iowa, Iowa City.

Thompson, P.L., L.A. Ramer, and J.L. Schnoor (1998a) Uptake and transformation of TNT by a poplar hybrid. *Environ. Sci. Technol.* **32**(7): 975–980.

Thompson, P.L., L.A. Ramer, A.P. Guffey, and J.L. Schnoor (1998b) Decreased transpiration in poplar trees exposed to 2,4,6-trinitrotoluene. *Environ. Toxicol. Chem.* **17**(5): 902–906.

Thompson, P.L., L.A. Ramer, and J.L. Schnoor (1999) 1,3,5-trinitro-1,3,5-triazaine (RDX) translocation in hybrid poplar trees. *Environ. Toxicol. Chem.* **18**(2): 279–284.

Thompson, P.L., L.A. Ramer, P. Yong, and J.L. Schnoor (2000) Pilot-scale treatment of TNT-spiked groundwater by hybrid poplars. In: *Emerging Technologies in Hazardous Waste Management VIII*. D.W. Tedder and F.G. Pohland, eds. Kluwer Academic/Plenum, New York, pp. 109–118.

van Beelen, P. and D.R. Burris (1995) Reduction of the explosive 2,4,6-trinitrotoluene by enzymes from aquatic sediments. *Environ. Toxicol. Chem.* **14**(12): 2115–2123.

Vanderford, M., J.V. Shanks, and J.B. Hughes (1997) Phytotransformation of trinitrotoluene (TNT) and distribution of metabolic products in *Myriophyllum aquaticum*. *Biotechnol. Lett.* **19**: 277–280.

SECTION V

FATE AND CONTROL OF CHLORINATED SOLVENTS AND OTHER HALOGENATED COMPOUNDS

16

SEQUESTRATION AND TRANSFORMATION OF WATER SOLUBLE HALOGENATED ORGANIC COMPOUNDS USING AQUATIC PLANTS, ALGAE, AND MICROBIAL MATS

V. A. Nzengung, W. L. O'Niell, S. C. McCutcheon, and N. L. Wolfe

SUMMARY OF PRACTICAL IMPLICATIONS

Bench and greenhouse testing establish that aquatic plants, macrophytic algae, and microbial mats all separately transform and mineralize common halogenated contaminants such as trichloroethylene, tetrachloroethylene, hexachloroethane, and carbon tetrachloride. Even dead plants and algae continue to transform halogenated organic compounds—a potential link between plants and dehalogenase activity in sediments. Plants, algae, and mats also are good biosorbents that may damp uneven loading. The phytotransformation pathways combine phytoreduction and phytooxidation informing carbon dioxide, chloride, and bound residues. The more highly halogenated contaminants (*e.g.*, tetrachloroethylene) are phytoreduced, whereas phytooxidation dominates the transformation of trichloroethylene and less chlorinated organic contaminants. Microbial mats applied in bioreactors and treatment wetlands can take advantage of both reductive and oxidative reactions in easily engineered aerobic and anoxic zones. In addition, monitored natural attenuation in contaminated wetlands is feasible. Therefore,

---

Phytoremediation: Transformation and Control of Contaminants,
Edited by Steven C. McCutcheon and Jerald L. Schnoor.
ISBN 0-471-39435-1 (cloth)

evaluation of on-site remediation must consider sequestration and phyto-transformation along with microbial degradation of halogenated organic contaminants during treatment of contaminated wastewaters, soils, and groundwaters. Based on these results, treatment wetlands and bioreactors are a viable option ready for field-testing. The next developments should include bench scale continuous flow testing and applications of microbial mats in created treatment wetlands.

INTRODUCTION

The U.S. Air Force reports that chlorinated solvents are the most common contaminants of concern at U.S. hazardous waste sites. Approximately, a quarter of the over 5000 sites seems to involve surficial groundwater plumes. At several hazardous waste sites, groundwater plumes with chlorinated organic contaminants upwell into natural wetlands and subsequently flow into nearby streams, ponds, lakes, and estuaries. As these contaminants encounter vegetation, a significant fraction is sequestered and biotransformed by live plants, by dead and decaying plant matter, and by the diverse consortia of microorganisms associated with plants within these ecosystems. At sites where vegetative biochemical processes significantly contribute to the attenuation of the contaminants of concern, then a final remedial action plan may simply involve the enhancement and monitoring of the natural plant-mediated processes. Also, providing a better understanding of the mechanisms by which dissolved organic contaminants are removed from water by macrophytic algae, microbial mats, and aquatic plants aids in the design of efficient treatment wetlands or bioreactors for treating wastewaters. The potential applications include treating groundwater that flows into wetlands or can be pumped into bioreactors and wetlands. However, most phytoremediation studies have focused instead on the decontamination of groundwater and soil by terrestrial grasses and woody plants with a root zone (rhizosphere) naturally enriched with a diversity of microbial consortia (Anderson and Walton 1995, Strand *et al.* 1995, Schnabel *et al.* 1997, Burken and Schnoor 1997, 1998, Newman *et al.* 1997, Doucette *et al.* 1998, Newman *et al.* 1999, Orchard *et al.* 2000).

Several investigations document the capability of selected aquatic plants, macrophytic algae, and microbial mats to function as *in situ* and on-site biosinks and biofilters of halogenated organic compounds (Bender *et al.* 1995, Adey *et al.* 1996, Jeffers *et al.* 1998, O'Niell 1999, O'Niell *et al.* 1999, Nzengung *et al.* 2000, Garrison *et al.* 2000, Nzengung and Jeffers 2001). The contribution of photoautotrophs in the removal of organic contaminants from water must be significant given that plants make up the bulk of biomass present in many aquatic environments (Uhlmann 1979) and that plants present a large surface area of lipid-rich (lipophilic) cuticle. Thus, even if the biotransformations of anthropogenic organic compounds by photoautotrophs were negligible, the role as biosorbents still would be significant.

Although microbial mats primarily consist of cyanobacteria (formerly termed blue-green algae that are neither a true algae or a plant), O'Niell *et al.* (1999) pointed out the potential for applications that is similar to other phytoremediation techniques. Nzengung *et al.* (1999), Nzengung *et al.* (2000), and Nzengung and Jeffers (2001) investigated the phytotransformation kinetics, pathways, and mass balances for halogenated compounds using aquatic plants and macrophytic algae. Gao *et al.* (2000) investigated the capability of three aquatic plants to accumulate and transform 1,1,1-trichloro-2-(*o*-chlorophenyl)-2-(*p*-chlorophenyl)ethane (*o,p*'-DDT) and 1,1,1-trichloro-2,2-*bis*(*p*-chlorophenyl)ethane (*p,p*'-DDT).

The mechanisms by which aquatic, terrestrial, and woody plants metabolize the same halogenated organic compounds may vary due to differences in the physiology and the local biogeochemistry. A growing understanding of the mechanisms (phytoprocesses) used by green plants and other photoautotrophs to metabolize halogenated organic compounds is emerging. First, transformation in the root zone (rhizodegradation) or associated sediments result in the mineralization of a fraction of the parent organic compound or formation of intermediate products. Next, the plant takes up and metabolizes, assimilates, and volatilizes some of the organic pollutants dissolved in water. Vroblesky *et al.* (1999) showed that the parent compound and the intermediate products are taken up into trees through the transpiration stream. Burken and Schnoor (1997), Nietch *et al.* (1999), Burken (this book), and Davis *et al.* (this book) discussed the physiological and biochemical mechanisms by which trees and other plants take up chlorinated aliphatics and other organic contaminants. Following uptake into the plant, the fate of the parent compounds and the intermediates will be determined by various phytoprocesses including phytotransformation, phytoaccumulation, phytoassimilation, and phytovolatilization (see Table 1-2 of McCutcheon and Schnoor this book for definitions of phytoprocesses). Phytotransformation refers to two specific types of plant reactions. First, phytoreduction is the reductive transformation of oxidized organic compounds by plant reducing enzymes, and second phytooxidation is oxidative transformations catalyzed by plant oxidizing enzymes (Nzengung and Jeffers 2001). Plants also mineralize some chlorinated contaminants by phytoöxidation and phytoreduction.

This chapter focuses on the network of phytoprocesses in four aquatic systems where photoautotrophic and bacterial transformations of common chlorinated and brominated organic contaminants occur. These aquatic systems included: (1) microbial mats, (2) live aquatic plants and macrophytic algae, (3) microorganism-free (aseptic) live aquatic plants, and (4) dead aquatic plants and macrophytic algae, and dried microbial mats. The goals of these studies were to: (1) determine if both physiologically stressed and unstressed aquatic plants transformed chlorinated aliphatic pollutants in the presence and absence of microorganisms; (2) determine the extent of sorption and transformation of chlorinated aliphatic compounds by both living and dead plants and macrophytic algae, and wet and dry microbial mats; and (3) identify the phytoremediation mechanisms used by these aquatic systems. Particular emphasis was on determining reaction rates, identifying transformation products, and estimating mass balance distributions of contaminants and metabolites. This information will be useful in the evaluation and potential application of constructed treatment wetlands or stand-alone bioreactors for chlorinated solvent plumes in wetland areas and for industrial wastewaters, respectively. In addition, the incorporation of plant transformation into contaminant fate models, especially for wetland ecosystems, should greatly improve the accuracy and reliability of predictive exposure models under development by the U.S. Environmental Protection Agency and other agencies.

OVERVIEW OF EXPERIMENTAL METHODS

Selection of Photoautotrophs and Probe or Test Compounds

Microbial mats, self-sustaining assemblages of mixed aquatic microbial species, were tested for the capability to remove the common contaminants, tetrachloroethylene (PCE) and trichloroethylene (TCE) from water. The mats were primarily composed of photoautotrophic cyanobacteria (*e.g.*, *Oscillatoria* spp., *Nostoc* spp., and *Anabaena* spp.) with related heterotrophic bacteria. Microbial mats have great utility in waste management because of the easily achievable growth requirements, the capability to support a mixed community of microorganisms, aerobic and anaerobic zones in close proximity, and survival capability in a wide range of freshwater and saltwater environments (O'Niell *et al.* 1999). The heterotrophic bacteria include *Pseudomonas* spp., *Flavobacterium* spp., sulfur-reducing bacteria, and purple autotrophic bacteria (can be photoheterotrophs, photoautotrophs or chemoheterotrophs, depending on the environment). These bacteria can biodegrade chlorinated organics through reductive dehalogenation, whereas the near-surface layers of the mat are photosynthetic and offer the potential to degrade halogenated compounds through oxidative pathways. The mixed-species mats used in these tests were grown in-house using a patented technique developed at Clark-Atlanta University by J. Bender and P. Phillips (U.S. patent numbers 5 522 985, 5 614 097, 6 008 028, and 6 033 559 at www.uspto.gov and Bender *et al.* (1995).

The macrophytic algae and plants used in other tests include freshwater algae [*Spirogyra* spp. and stonewort (*Nitella* spp.)], the submergent wetland plant Canadian waterweed (*Elodea canadensis*), the emergent wetland plant parrot feather (*Myriophyllum aquaticum*), and the terrestrial plant spinach (*Spinacia oleracea*). The filamentous algae and free floating aquatic plants are widespread, aquatic macrophytes, which are a source of food for waterfowl and a shelter for small aquatic invertebrates. Waterweed (*Elodea canadensis*) was selected as a representative aquatic plant for sorption isotherm tests. Spinach (*Spinacia oleracea*) is typically a good source of many plant enzymes.

Two chlorinated alkanes served as probe or representative compounds in the investigation of aquatic plants and macrophytic algae—hexachloroethane (HCA) and carbon tetrachloride. Hexachloroethane is a fully halogenated ethane that undergoes reductive dehalogenation to form PCE nearly quantitatively, and is easy to monitor. Carbon tetrachloride, which is present in wastewaters and in many aquifers, is a carcinogenic greenhouse gas that depletes atmospheric ozone.

Batch Sorption Tests

Nzengung *et al.* (1999), O'Niell (1999), and O'Niell *et al.* (1999) described the methods used in the batch sorption investigations. Predetermined masses of wet or dry microbial mat (5 grams wet weight and 0.5, 1, 1.5, and 2 grams dry weight), or entire aquatic plants [variety of wet weights of waterweed (*Elodea canadensis*)] were immersed in aqueous solutions of contaminants contained in 20 and 60 milliliter crimp-capped vials or screw-capped test tubes with silicon-faced septa. Headspace in the vials and test tubes was minimized to reduce the effects of volatilization on mass balance calculations. Additionally, tests relied upon controls to determine volatilization and other losses. At the end of the equilibration period [conservatively selected as 6 hours for PCE and 12 hours for TCE sorbing to mats, and 24 hours for HCA and PCE sorbing to waterweed (*Elodea canadensis*)], the liquid phase was analyzed for the parent compound by gas chromatography with an electron capture detector. The sorbed mass of solute was estimated by the following two methods:

1. As the difference between the initial and final concentration in solution after equilibrium occurred (difference method)
2. The direct measurement of the sorbed concentration in the solvent extract of the plant or mat (direct method)

In addition, radiolabeled compounds were used to trace the distribution of contaminants and metabolites after reaching sorption equilibrium by direct measurement in the aqueous phase and by combusting a subsample of the biomass, as described in the section below "Mass Balance Tests" (also see Nzengung *et al.* 1999).

Microbial Mat Transformation Tests in Sealed Bioreactors

Transformation testing was performed in three sealed glass bioreactors that were equipped with pH, Eh (pHoenix Electrode™, Co., Houston, Texas), and dissolved oxygen probes (Cole-Palmer Instrument Co., Chicago, Illinois). A nonjacketed fermentation vessel (Applikon™ Analyzers, Inc., Malverne, New York) served as the bioreactor (Figure 16-1). Each bioreactor was filled with 2.5 liters of Allen–Arnon growth media and 200 grams wet weight of mat, leaving a headspace volume of 0.4 liters (O'Niell 1999). The reactors were

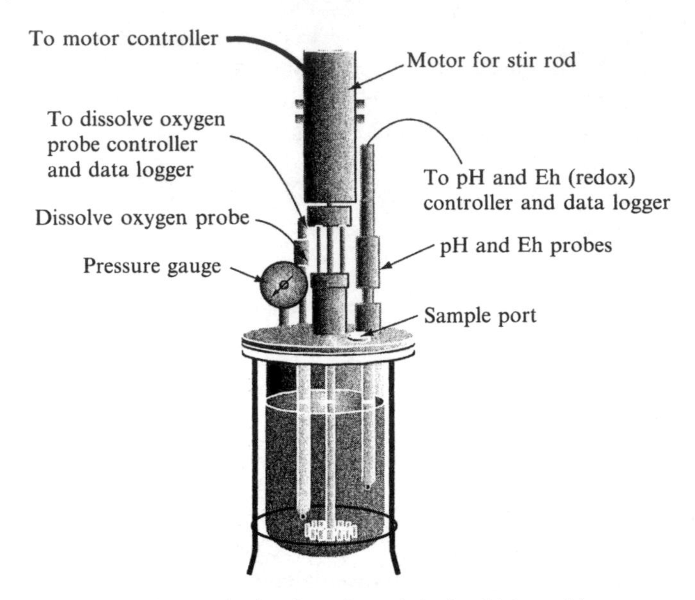

Figure 16-1 Schematic drawing of a sealed microbial mat bioreactor.

maintained under static conditions (no mechanical mixing) in a greenhouse at $25 \pm 2\,^{\circ}C$ under 50 percent shaded ambient light (approximately 500 micromoles of photons or microeinsteins per second per square meter maximum). Before collecting a liquid phase sample, the reactor contents were slowly mixed for 1 minute using a built-in impeller. Each liquid sample was taken using a gas-tight syringe with a 15-centimeter (6-inch) Luer™ tip needle inserted through a septum in the reactor head plate.

After adding the Allen–Arnon media and mat, the bioreactors were maintained in the greenhouse ($25 \pm 2\,^{\circ}C$) while the mat acclimated to these conditions. Dissolved oxygen, Eh (estimated using a probe to measure oxidation–reduction potential), and pH data were collected throughout the tests. After 72 hours acclimation, PCE or TCE was added to the bioreactors to achieve an initial solution of 10 milligrams per liter, a concentration that is consistent with some industrial waste stream concentrations but higher than many groundwater concentrations. However, 10 milligrams per liter assures measurable product formation for pathway analysis. One-milliliter aqueous phase samples were then collected at 24- and 48-hour intervals, extracted with solvents, and analyzed for parent compounds and metabolites using gas chromatography with an electron capture detector.

Tests with Septic, Axenic, and Irradiated Aquatic Plants

The experimental details of these studies have already appeared in Nzengung *et al.* (1999). This section includes brief comments about the general nature of the batch vial and air-flow-through tests. Entire aquatic plants were immersed in aqueous solutions of HCA and PCE in crimp-capped vials and in screw-capped test tubes. Tests were also conducted in 4.5-liter plant growth chambers with continuous airflow. The glass chambers were maintained under a plant-grow light at an intensity of 100 micromoles of photons per second per square meter during a 12-hour light and 12-hour dark cycle. The airflow rate was set at 25 milliliters per minute (minimum for this system), which resulted in the daily exchange of at least 60 volumes of headspace. A ball and socket joint connected the growth chamber outlet to a series of three screw-top traps to enable the quantification of losses due to volatilization (Figure 16-2).

The role of plants in metabolism of halogenated organic compounds in the absence of bacteria, fungi, and other plant-associated microorganisms is important in understanding the performance of phytoremediation systems. For different testing, plants were sterilized by: (1) dilute hypochlorite solution to surface-sterilize Canadian waterweed (*Elodea canadensis*) and parrot feather (*Myriophyllum aquaticum*), (2) intense gamma irradiation doses sufficient to ensure external and internal sterilization of waterweed (*Elodea canadensis*) shoots, and (3) use of axenically grown parrot feather (*Myriophyllum aquaticum*). Anexic plants were matched with septic plants of the same species to determine microorganism effects. Plate counts confirmed that gentle bleaching and thorough rinsing virtually eliminated bacteria, especially from leaf surfaces. Gamma radiation of 3000 grays (300 000 rads) eliminated all plate counts of bacteria.

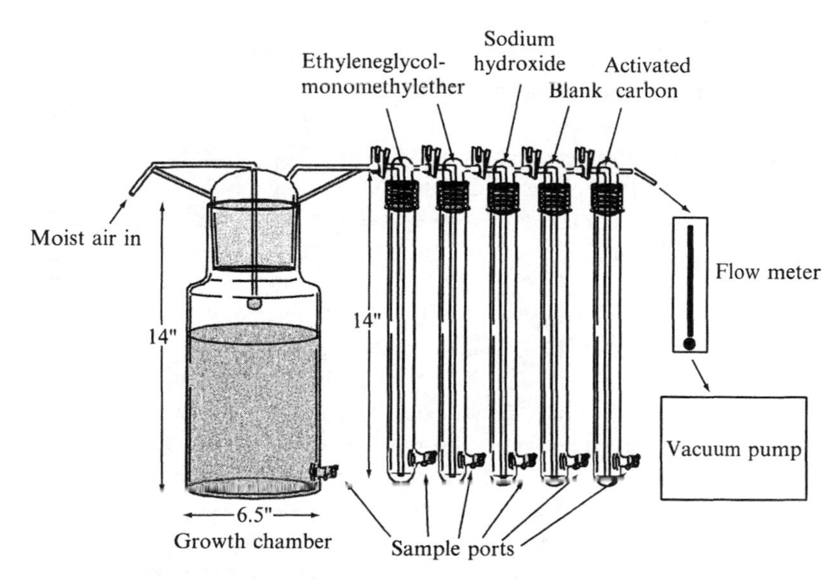

Figure 16-2 Air-flow-through plant-growth chamber used in some tests to stimulate environmental conditions (Nzengung *et al.* 1999 Copyright, CRC Press). Note that 1 inch = 2.54 centimeters.

Mass Balance Tests

Mass balance tests were conducted by dosing microbial mat, freshwater algae (*Spirogyra* spp.), or Canadian waterweed (*Elodea canadensis*) in a single sealed bioreactor with both unlabeled and [^{14}C]-labeled TCE, carbon tetrachloride, and tetrabromoethylene (4.4×10^8 becquerel or 0.12 millicurie to obtain 1200 counts per minute in 1 milliliter of sample), respectively. At the end of each test, the aqueous and biomass phases were removed from the bioreactor and [^{14}C] was measured in the various media. The total aqueous phase radioactivity was measured directly by adding liquid samples to a scintillation cocktail and analyzing with a liquid scintillation counter. Radioactivity in the biomass phase was extracted with solvents that were then added to a cocktail and assayed by liquid scintillation. Vapor phase activity or concentration was estimated using the Henry's Law coefficient. An acid-base method was used to measure the radioactivity of purgeable volatile compounds and carbon dioxide *versus* nonvolatile components in the aqueous phase (Vogel and McCarty 1985). Biomass radioactivity was determined directly in some tests; duplicate subsamples were combusted in a biological oxidizer and the evolved $^{14}CO_2$ was trapped and assayed by liquid scintillation.

SEQUESTRATION AND TRANSFORMATION RESULTS

Microbial Mat Results

Sequestration

The sorption of PCE and TCE by mixed species microbial mats followed a linear isotherm over the range of 1 to 10 milligrams per liter, as demonstrated by the correlation coefficients r^2 ranging from 0.95 to 0.99 (Figure 16-3 and Table 16-1). Thus, the sequestration of chlorinated ethylenes by microbial mats increases with the solute concentration.

For the living microbial mat, average sorption coefficients of 6.63 ± 0.30 ($\pm$standard error) milliliters of solution per gram of plant wet weight was determined for PCE and 2.34 ± 0.60 ($\pm$standard error) milliliters of solution per gram of plant wet weight for TCE. The difference between the sorption coefficients of PCE and TCE was due to the greater aqueous solubility of TCE. The average sorption coefficient of PCE to dried mat was 58.2 ± 3.57 ($\pm$standard error) milliliters of solution per gram of mat dry weight and 16.2 ± 1.89 ($\pm$ standard error) milliliters of solution per gram of mat dry weight for TCE (Table 16-1). Because the mat was composed of approximately 90 percent water, 5 grams of living mat (wet weight) should sorb an amount equivalent to that sorbed by 0.5 gram of dry mat. Nevertheless, the measured sorption coefficients for dried mat were approximately an order of magnitude greater than those for living mat (Table 16-1), which was consistent with the trends observed for sorption of organic compounds to wet

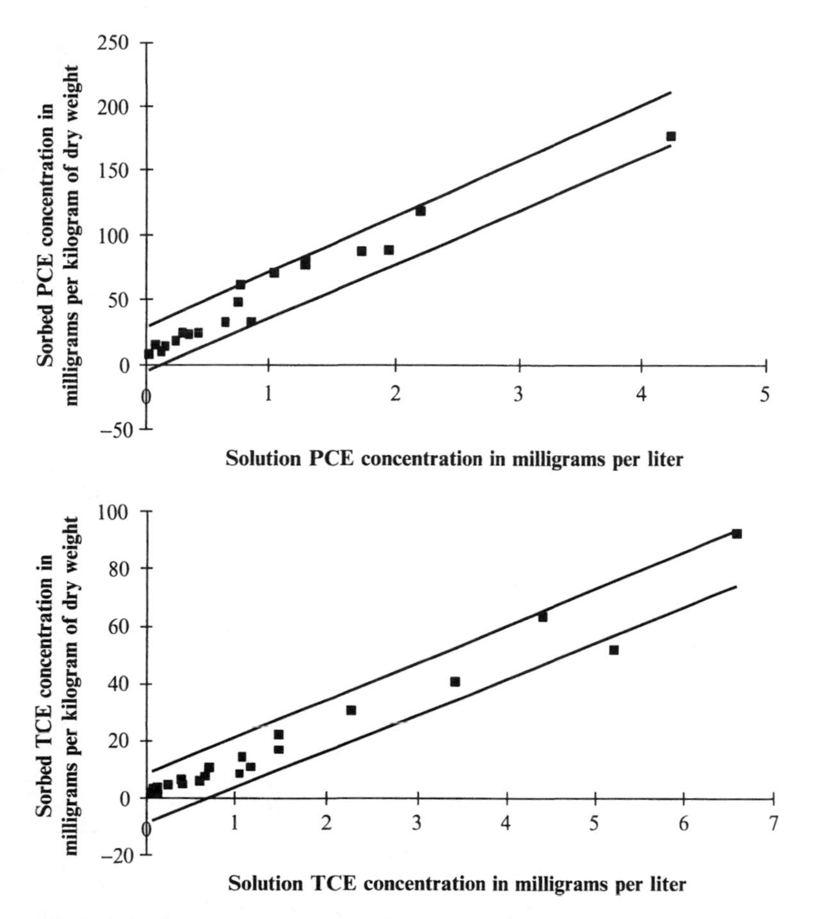

Figure 16-3 Pooled dried-mat sorption data for tetrachloroethylene (PCE) (top) and trichloroethylene (TCE) (bottom) with upper and lower 95 percent prediction bands. A prediction band (or prediction interval) is the certainty of the scatter about a certain regression line. A 95 percent prediction band indicates that, in general, 95 percent of the points will be contained within the bands. Reprinted with permission from O'Niell *et al.* (1999).

and dry soils (Chiou 1990). At least two explanations are obvious. First, the dry mat was ground into smaller particles before each test. Thus, the ground, dried mat provided a greater surface area for adsorption than the wet, living mat. Second, the dried mat seemed to be only partially hydrated and biologically active during the 6 to 12 hours used to achieve sorption equilibrium. The partially hydrated mat seemed to present a more hydrophobic surface for sorption of PCE, TCE, and the associated hydrophobic metabolites.

The results of an analysis of variance established that the population means of the PCE or population means of the TCE sorption coefficients remained equal as the mass of the mat varied (O'Niell *et al.* 1999). (This statistical method is one way to test the linearity of sorption over the range of

TABLE 16-1 Batch Sorption to Microbial Mat Dosed with Tetrachloroethylene (PCE) or Trichloroethylene (TCE) in 20-Milliliter Vials

		Sorption coefficient ± standard error (milliliters of solution per gram of wet or dry weight of mat)				
Material	Mass (grams)	Difference method	r^2	Direct method	r^2	Average
Solute: tetrachloroethylene (PCE)						
Living Mat (wet weight)	5.0	6.33 ± 0.30	0.97	6.93 ± 0.30	0.99	6.63 ± 0.30
Mat (dried weight)	0.5	73.1 ± 4.61	0.97	67.4 ± 3.88	0.99	58.2 ± 3.57
	1.0	49.6 ± 0.85	0.99	58.6 ± 1.47	0.99	
	1.5	46.1 ± 4.22	0.95	50.5 ± 8.26	0.99	
	2.0	62.8 ± 5.31	0.96	57.6 ± 0.01	0.97	
Dried mat [^{14}C]-PCE	0.5	52.6 ± 5.91	0.99	—	—	57.30 ± 4.66
	2.0	62.0 ± 3.40	0.98	—	—	
Silage [^{14}C]-PCE	0.25	41.3	0.99	—	—	41.3
Solute: trichloroethylene (TCE)						
Living Mat (wet weight)	5.0	2.93 ± 0.60	0.99	1.75 ± 0.60	0.98	2.34 ± 0.60
Mat (dried weight)	0.5	15.6 ± 1.40	0.99	23.9 ± 3.33	0.99	16.2 ± 1.89
	1.0	11.4 ± 0.53	0.99	18.6 ± 1.36	0.99	
	1.5	18.2 ± 3.68	0.99	10.9 ± 3.68	0.97	
	2.0	14.0 ± 0.22	0.99	17.3 ± 0.92	0.99	
Dried mat [^{14}C]-TCE	0.5	12.8 ± 1.40	0.99	—	—	13.8 ± 1.21
	2.0	14.8 ± 1.01	0.99	—	—	
Silage [^{14}C]-TCE	0.25	12.9	0.99	—	—	12.9

Note:

$$\text{Sorption coefficient}\left(\frac{\text{milliliters of solution}}{\text{gram of wet or dry mat}}\right) = \frac{\left[\text{Sorbed concentration}\left(\frac{\text{milligrams of contaminant}}{\text{gram of wet or dry mat}}\right)\right]\left(\frac{1000\ \text{milliliters}}{\text{liter}}\right)}{\text{Equilibrium solution concentration}\left(\frac{\text{milligrams of contaminant}}{\text{liter of solution}}\right)}$$

The sorbed concentration was the difference between the initial concentration minus the solution concentration attained after equilibrium (difference method), or was by direct measurements of the sorbed mass removed by solvents. The direct method was not applied to all sampples as denoted by—. The averages were for all of the sorption coefficients measured by both the difference method and direct method. The correlation coefficient r^2 is a measure of the goodness of fit of the linear isotherm. Sorption coefficients were expressed on a wet weight basis for living mat tests, and on a dry weight basis for dried mat, but the dried mat was of course at least partially hydrated during the aqueous sorption experiments. Silage, a significant starter component of microbial mats, was tested to determine if this material dominated mat sorption. Reprinted with permission from O'Niell *et al.* (1999).

concentrations examined.) The probabilities or levels of significance calculated for the PCE data (0.15 for the direct method and 0.44 for the difference method) were greater than 0.05, indicating that the mean values of the partitioning coefficients calculated for the four masses of mat were not statistically different. Similar results were obtained for TCE, with probabilities of 0.14 for the direct method and 0.67 for the difference method.

The [^{14}C]-PCE and [^{14}C]-TCE sorption tests yielded similar results. Average sorption coefficients of 57.3 ± 4.66 ($\pm$ standard error) milliliters of solution per gram of dry weight and 13.8 ± 0.99 ($\pm$ standard error) milliliters of solution per gram of dry weight were calculated for sorption of PCE and TCE to dried mat, respectively (Table 16-1). Although desorption of PCE and TCE from the mats was not studied extensively, two treatments using hexane and sonication for 30 to 60 minutes were required to recover the extractable sorbed PCE and TCE from the mats. In addition, the sustained transformation of PCE and TCE, suggests that desorption of these compounds may not be significant in a sealed bioreactor treatment system.

Because published values of sorption data for PCE do not seem to exist, only the TCE sorption data calculated for the mats were compared to published sorption data for other media (Table 16-2). Both dried and live microbial mat had a lower organic carbon normalized partitioning coefficient (based on an estimated faction organic carbon for the mat of 0.4) than the other materials listed in Table 16-2, but this comparison alone ignores several advantages of the mat.

Microbial Mat Bioreactor Treatment Systems

Figure 16-4 shows the changes in pH, Eh (measure of oxidation–reduction potential), and dissolved oxygen in a microbial mat bioreactor dosed with PCE during a representative 2-week period of the 84-day test. Anaerobic conditions developed in the mat each night, persisted on cloudy days, and aerobic conditions developed during some sunny days. Mat health did not appear to suffer during these tests. No visual changes, signs of dying, or putrefaction were observed. The dissolved oxygen data indicated that the mat produced similar amounts of oxygen (12 milligrams per liter or greater at the maximum) at the end of the test on day 84 as at the beginning. At night, the mat consumed all dissolved oxygen in the bioreactor for respiration. Maximum dissolved oxygen concentrations of more than 12 milligrams per liter were detected on sunny days, but on cloudy days the dissolved oxygen concentrations were less than 1 milligram per liter, and after several cloudy days, the dissolved oxygen was sometimes undetectable. On sunny days, Eh values varied from approximately 250 millivolts during peak sunlight to less than −400 millivolts at night. The pH remained between 11 and 13 for the duration of the test with maximum pH values measured when dissolved oxygen levels were highest each day. After 84 days, the wet weight of the mat in one representative reactor increased from 200 grams at the start of the experiment to 248 grams at the end. The pH,

TABLE 16-2 Comparison of Trichloroethylene (TCE) Sorption Coefficients for Different Media (Sorbents)

Medium	Sorption coefficient ± standard error (milliliters of solution per gram of dry weight)[a]	Fraction organic carbon (dimensionless)	Organic carbon normalized sorption coefficient ± standard error (milliliters of solution per gram of dry weight)[a]	Concentration range (milligrams per liter)
Wet microbial mat	2.34 ± 6.0	0.4	$5.85 \pm 15.$	1.0 to 10
Dry microbial mat	16.2 ± 1.89	0.4	40.6 ± 4.73	1.0 to 10
Humin[b]	1.28	0.0088	145	≈ 2
Oxidized humin[b]	0.4	0.0014	286	≈ 2
Fats, waxes, and resins[b]	348	0.833	418	≈ 2
Algae[c]	1400 to 5370[d]	0.4	3500 to 13 400[d]	0.04 to 4.4
Humic coated alumina[e]	0.308	0.0045	68.4	—[e]
Sediment (< 1 millimeter in diameter)[e]	0.12 (0.11 to 0.13)	0.0019	63.2 (57 to 68)	—[e]
Sandy loam soil[f]	0.5	0.001	500	175 to 875
Organic topsoil[f]	13.5	0.117	115	175 to 875
Peat moss[f]	93.4	0.494	189	175 to 875
Granular activated carbon[f]	81 100	1	81 100	175 to 875

[a]The organic carbon normalized sorption coefficient is the sorption coefficient divided by the fraction organic carbon. The dry microbial mat was only partially hydrated during the time required to achieve sorption equilibrium (O'Niell *et al.* 1999). The sorption coefficient for the live mat on a wet weight basis from Table 16-1 was converted to a dry weight basis based on the average mat water content of 90 percent (10 percent biological solids).

[b]Garbarini and Lion (1986).

[c]*Chlorella vulgaris*, *Scenedesmus quadricauda*, and *Selenastrum capricornutum* from Smets and Rittmann (1990).

[d]Range of data reported.

[e]Allen-King *et al.* (1996). Concentration range was not reported.

[f]Zytner (1992).

Note: Reprinted with permission from O'Niell *et al.* (1999)

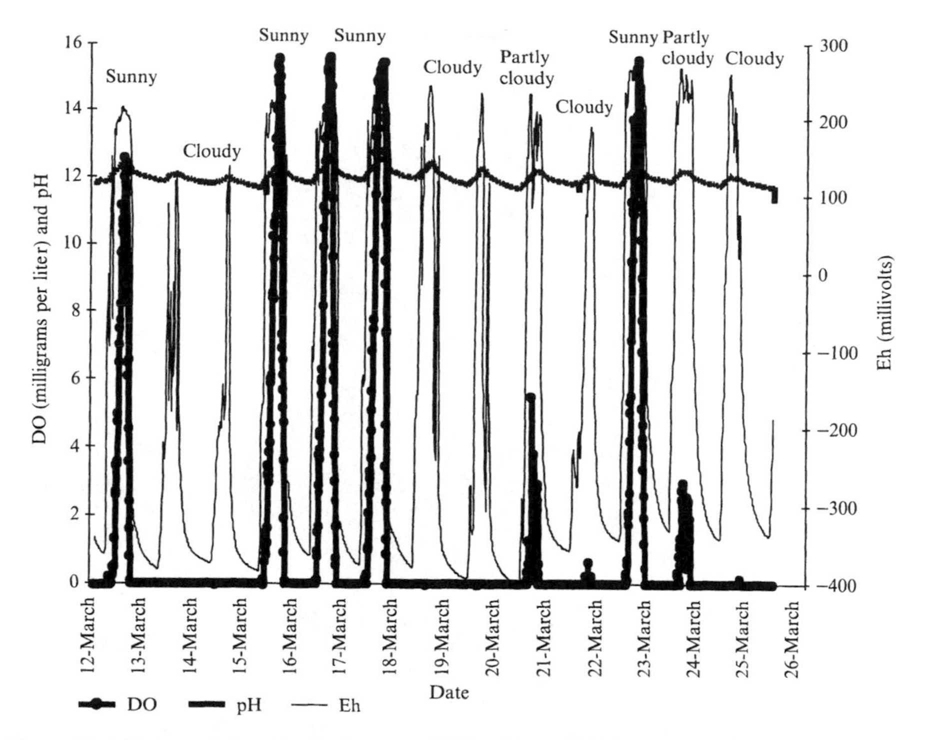

Figure 16-4 Representative dissolved oxygen (DO), pH, and Eh (a measure of redox) in a microbial mat bioreactor (see Figure 16-1) dosed with tetrachloroethylene (PCE). These 14 days of the 84-day test (March 5 to May 28, 1998) show diel changes due to sunlight changes in a greenhouse (25 ± 2 °C) located in Athens, Georgia. Adapted from O'Niell (1999).

oxidation–reduction potential (Eh), and dissolved oxygen data suggested that suitable conditions for oxidative and reductive transformation of PCE occurred in the bioreactors on most days.

The initial aqueous phase concentration of PCE measured in the bioreactor was 12.5 milligrams per liter (probably due to incomplete mixing *versus* 10 milligrams per liter expected from the amount of PCE added to solution). This concentration decreased to 5.1 milligrams per liter in 24 hours, a decrease attributed mainly to the initially very rapid sorption to the mat. The high initial concentration of PCE was chosen to determine if relatively high concentrations of this compound would affect mat health and to ensure that the metabolites would be formed at detectable concentrations. The aqueous phase PCE concentration subsequently decreased to less than 1 milligram per liter within 37 days. On the final day 84 of the test, PCE was not detected (Figure 16-5). Zero order and pseudo first order models best described the PCE initial sorption, and uptake and transformation, respectively. The PCE sorption rate constant for the bioreactor tests was 4.9 milligrams per liter per day. The PCE transformation coefficient was 0.0797 ± 0.0043 (± standard

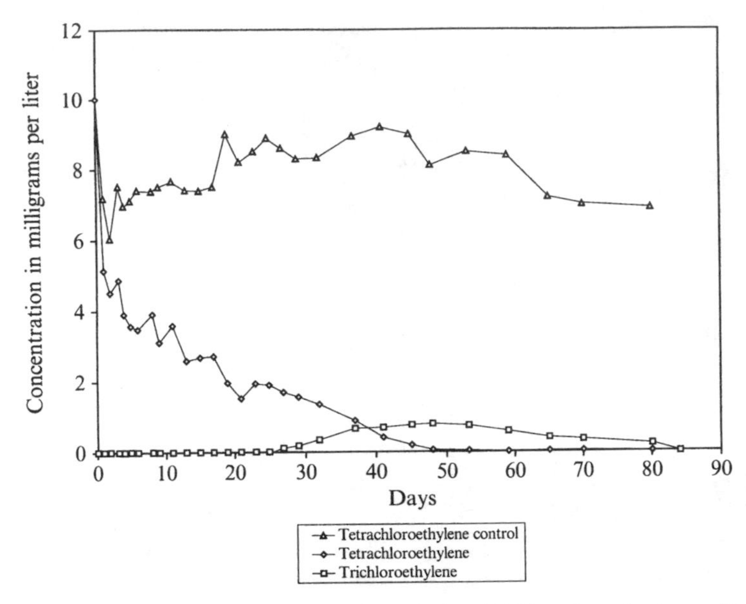

Figure 16-5 Transformation of aqueous tetrachloroethylene (PCE), and production and transformation of aqueous trichloroethylene (TCE) in a sealed microbial mat bioreactor and a control bioreactor. Note that with the mat, the initial concentration of tetrachloroethylene was measured as 12.5 milligrams per liter *versus* 10 milligrams per liter targeted. The tetrachloroethylene in the control reactor (no mat) rapidly sorbed to the reactor and volatilized into the headspace but did not form trichloroethylene. The detection limits were 0.001 milligram per liter for both PCE and TCE.

error) per day. The half-life of aqueous phase PCE uptake and transformation in the bioreactors was 8.7 days. The transformation product TCE appeared after 25 days, but also transformed to undetectable levels before the test was terminated at 84 days.

Reductive and Oxidative Metabolites

Microbial mats transformed PCE into reductive and oxidative products. Trichloroethylene, which generally appeared in samples after several days or weeks of incubation, was the only reductive transformation product detected in the mat bioreactors dosed with PCE. The reductive products *cis*-dichloroethylene and vinyl chloride either did not accumulate from reductive transformation or daily photosynthetic production (as much as 12 milligrams of dissolved oxygen per liter formed in solution most days) oxidized most of the TCE. The lack of *cis*-dichloroethylene and vinyl chloride (a known carcinogen) formation is a significant advantage of mats because these products are more hazardous than the original PCE or TCE.

Monochloroacetic, dichloroacetic, and trichloroacetic acids and chloroethanols were measured with a derivatization technique using diazomethane to convert the acids to esters, which were then analyzed using gas chromatography with an electron capture detector. All three chloroacetic acids were detected in batch vials dosed with PCE or TCE and exposed to light intensities greater than 120 micromoles of protons per second per square meter. Dichloroacetic acid was detected after 18 hours in batch vials filled with 8 grams wet weight of mat and dosed with 8 milligrams per liter of TCE. As the tests proceeded, the TCE concentration decreased, trichloroacetic acid was detected, and the concentration of dichloroacetic acid began to increase. After 60 days, the dichloroacetic acid concentration decreased to less than 0.02 milligram per liter. See O'Niell (1999 Figure 5-7, Chapter 5). The mats may also form other unidentified metabolites. See Shang *et al.* (this book) on conjugated metabolites and later in "Radiolabeled Mass Balances" of the section "Microbial Mat" for information on bound residues. In bioreactor tests for PCE removal, dichloroacetic acid and TCE were the only metabolites identified in the aqueous phase.

Monochloroacetic acid, dichloroacetic acid, and trichloroacetic acid were not detected in mat samples that were not dosed with PCE or TCE. Therefore, the oxidation or chlorination of natural compounds in the mats did not form these acids.

Aquatic Plant Results

Sequestration

Preliminary kinetic tests indicated that plants very rapidly removed HCA and PCE from the liquid phase (within 15 minutes of dosing). The initial rapid removal of the chlorinated compounds from solution was consistent with previous observations. These observations have shown that organic compounds sorb strongly to algal and microbial mat biomass in a few minutes and partition to bacteria in a few hours (Johnson and Kennedy 1973, Grimes and Morrison 1975, Harding and Phillips 1978, Matter-Müller *et al.* 1980, Baughman and Paris 1981, McRae 1985, Smets and Rittmann 1990, O'Niell *et al.* 1999). Nzengung *et al.* (1999) showed that HCA and PCE sorb strongly to heat-killed Canadian waterweed (*Elodea canadensis*) (plant phase). Sorption coefficients of 8 ± 0.3 ($\pm$ standard deviation) milliliters of solution per gram of wet weight and 23 ± 0.7 ($\pm$ standard deviation) milliliters of solution per gram of wet weight were calculated for PCE and HCA, respectively. As observed with microbial mats, the sorption by plant biomass is an important fate process in the removal of chlorinated aliphatic compounds from aquatic environments. The higher sorption coefficient of HCA than PCE by an approximate order of magnitude corresponds to the greater aqueous solubility of PCE.

Aquatic Plant Uptake and Transformation

Nzengung *et al.* (1999) observed that both septic and stressed aquatic plants (stressed from being sealed in batch vials) metabolize halogenated aliphatic compounds. The results of these tests provided insight into the role of aquatic and wetland plants in the metabolism of halogenated organic compounds. The removal of halogenated aliphatics from water by partially submerged [*e.g.*, emergent parrot feather (*Myriophyllum aquaticum*)] or wholly submerged [*e.g.*, Canadian waterweed (*Elodea canadensis*)] aquatic plants occurred because of several phytoprocesses. The very rapid initial uptake of HCA and PCE, observed immediately after dosing the sealed serum bottles (usually within the first 15 to 30 minutes), was due to sequestration (Nzengung *et al.* 1999, Nzengung and Jeffers 2001). Sequestration of halogenated organic compounds, as applied here, includes multiple physical (absorption, adsorption, and partitioning) and chemical processes (complexation and reaction with cuticular and membrane components) by which plants remove anthropogenic organic chemicals from air and liquid media, either temporarily or permanently. Sequestration may be synonymous with biosorption used more frequently in the published literature. Due to the very fast kinetics of sequestration, uptake of the parent compound by transpiration was considered negligible during this process. This rapid initial step was followed by the relatively slow phytotransformation step. Thus, two pseudo first order equations were used by Nzengung *et al.* (1999) to describe the initial, rapid sequestration and the slow transformation (indicated by metabolite formation) (Table 16-3). Because these sorption tests were conducted in sealed vials with no headspace, the effects of the air phase was not considered; however, volatilization and transpiration would have to be taken into account in natural systems. The magnitude of volatilization and transpiration will depend on the water surface area covered by biomass and the relative amount of emergent biomass.

To simulate phytotransformation of halogenated aliphatic pollutants by aquatic plants in natural environments, Nzengung *et al.* (1999) and Rennels (1998) investigated the removal of HCA and PCE from aqueous solution by Canadian waterweed (*Elodea canadensis*) and parrot feather (*Myriophyllum aquaticum*) grown in continuous air-flow-through bioreactors (Figure 16-2). Hexachloroethane exhibited pseudo first order kinetics until completely depleted from solution (Figure 16-6). For a plant density of less than or equal to 63 grams of wet weight per liter of water used in some of the air-flow-through tests, the high initial rapid sorption did not occur. For the larger plant density of 220 grams of wet weight per liter of water, the initial rapid removal by sequestration occurred. The removal of HCA and PCE from solution in the growth chamber tests exhibited faster kinetics than in the batch studies. The sealed batch vials represented a more stressful environment for the plants and the plant density was different than in the chambers. The faster kinetics observed in growth chamber studies, complicated by the different plant density, suggested that the metabolic activity of healthy growing aquatic plants was higher than that of the stressed plants in sealed vials. Given the limited data obtained in the latter air-flow-through tests, the complicating differences in plant density, and the fact that a fraction of the parent compounds and the associated metabolites volatilized and collected in the ethylene glycol traps, additional confirmation of these findings is necessary. The sodium hydroxide trap captured any $^{14}CO_2$ that flowed from the growth

TABLE 16-3 Different Aquatic Plant and Macrophytic Algae Mediated Sorption and Transformation Rate Constants for Hexachloroethane (HCA) and Carbon Tetrachloride (CCl_4)

Contaminant	Type of algae or plant (in batch vials unless otherwise noted)	Wet weight of algae or plant (grams)	Initial concentration (milligrams per liter)	Normalized zero order sorption rate coefficient[a] (per day)	Pseudo first order transformation constant ± standard error for pooled data (per day)[b]	r^2
HCA	Live Canadian waterweed (*Elodea canadensis*)	3.5	1	9.8	0.96 ± 0.29	0.85
HCA	Live Canadian waterweed (*Elodea canadensis*)	7	1	7.0	0.31 ± 0.05	0.93
HCA	Live Canadian waterweed (*Elodea canadensis*)	7	5	6.5	0.048 ± 0.02	0.71
HCA	Live Canadian waterweed (*Elodea canadensis*)	7	10	2.6	0.048 ± 0.01	0.98
HCA	Live Canadian waterweed (*Elodea canadensis*)	11	1	6.5	0.29 ± 0.02	0.99
HCA	Live Canadian waterweed (*Elodea canadensis*) in growth chamber	260	10	—[c]	0.002 ± 0.00[d]	0.93
HCA	Live Canadian waterweed (*Elodea canadensis*) in growth chamber	345	10	—	0.0010 ± 0.00[d]	0.88
HCA	Dead Canadian waterweed (*Elodea canadensis*)	0.5	5	—	7.63 ± 1.3	0.92
HCA	Dead Canadian waterweed (*Elodea canadensis*)	7	1	6.5	0.072 ± 0.01	0.99
HCA	Live parrot feather (*Myriophyllum aquaticum*)	5.3	1	0.24	0.048 ± 0.00[d]	0.99
HCA	Live parrot feather (*Myriophyllum aquaticum*) in growth chamber	528	6	0.024	0.0012 ± 0.00[d]	0.98
HCA	Live stonewort (*Nitella* spp.)	5.3	2.6	—	0.34 ± 0.07	0.92
HCA	Live stonewort (*Nitella* spp.)	3.2	2.6	—	0.22 ± 0.07	0.88
HCA	Live stonewort (*Nitella* spp.)	2.8	2.6	—	0.34 ± 0.02	0.97
HCA	Live stonewort (*Nitella* spp.)	3	1	—	0.096 ± 0.01	0.93
HCA	Live freshwater algae (*Spirogyra* spp.)	7	1	4.1	0.14 ± 0.02	0.91
HCA	Live freshwater algae (*Spirogyra* spp.)	13	2.6	0.96	0.024 ± 0.01	0.95
HCA	Dead freshwater algae (*Spirogyra* spp.)	10	1.6	4.1	0.38 ± 0.14	0.76
CCl_4	Live Canadian waterweed (*Elodea canadensis*)	5	5	1.7	0.072 ± 0.01	0.96
CCl_4	Live freshwater algae (*Spirogyra* spp.)	8.5	5	—	0.017[e] ± 0.00[d]	0.91
CCl_4	Dead freshwater algae (*Spirogyra* spp.)	5	7.3	—	0.072 ± 0.01	0.92

[a]Zero order sorption rate coefficients were normalized by dividing by the mass of plant or algae used in the respective test.

[b]Rate constant for which the correlation coefficient r^2 is shown.

[c]—means that the normalized sorption coefficient was not measurable.

[d]Standard error is less than 0.005 per day.

[e]This transformation was zero order with the units of milligrams per liter per day.

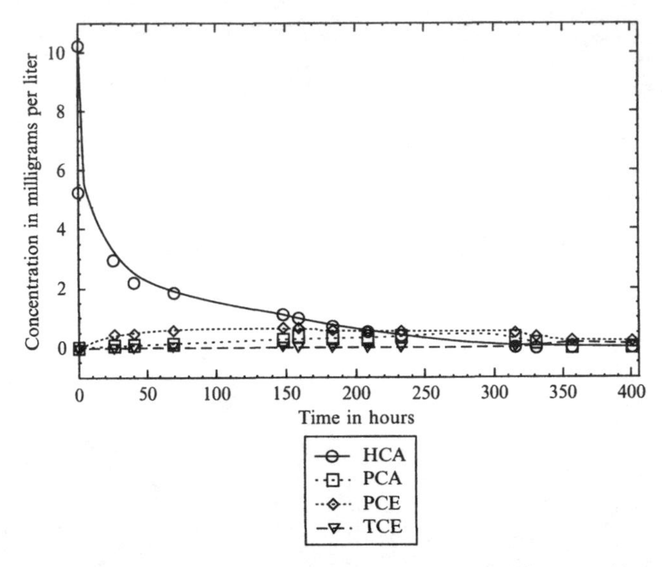

Figure 16-6 Uptake and transformation of an initial dose of 10 milligrams of hexachloroethane per liter by Canadian waterweed (*Elodea canadensis*) in an air-flow-through-plant-growth chamber maintained under a 12-hour light and 12-hour dark cycle. Hexachloroethane (HCA) and the transformation products tetrachloroethylene (PCE), pentachloroethane (PCA), and trichloroethylene (TCE) were measured by gas chromatography with an electron capture detector and gas chromatography with a mass spectrometer (Nzengung *et al.* 1999 Copyright, CRC Press).

chamber. Liquid scintillation counting was used to quantify the amount of captured $^{14}CO_2$.

Microorganism-Free Plant Transformations

Studies conducted with axenic parrot feather (*Myriophyllum aquaticum*) verified that plants could metabolize fully chlorinated aliphatic pollutants (methanes and ethanes) in the absence of microorganisms and fungi. After 5 days of incubating axenic plantlets with carbon tetrachloride and HCA separately, all samples were sacrificed and analyzed to confirm the presence or absence of the parent compound and transformation products. More than 70 percent removal of carbon tetrachloride and HCA occurred. Both reductive and oxidative transformation products formed in solution and in the plant tissue extracts. However, no transformation occurred in killed controls (autoclaved plants) during the same incubation period of 5 days. Losses of 20 percent attributed to sorption to plant biomass occurred in control samples.

Additionally, the complete elimination of bacteria in the batch vials by gamma irradiation with 3000 grays (300 000 rads), shown in Table 16-4, did not completely inhibit the live-plant-mediated transformation of HCA. Instead, only a decrease in the metabolic activity of live Canadian waterweed (*Elodea canadensis*) occurred, which was attributed to the sensitivity of this

TABLE 16-4 Bacteria Counts and Corresponding Rate Constants for Hexachloroethane (HCA) Uptake and Transformation by 7 Grams of Wet Weight of Surface Sterilized (Gamma Irradiated) Canadian Waterweed (*Elodea canadensis*) Exposed to a 52 Milliliter Solution with an Initial Concentration of 1 Milligram of Hexachloroethane per Liter in 60 Milliliter Vials

Live or dead Canadian waterweed (*Elodea canadensis*)[a]	Gamma radiation (grays)[b]	Incubation period for microbial culturing (hours)	Number of bacteria per milliliter		Normalized zero order sorption rate constant ± standard error (per day)[c]	Pseudo first order rate transformation constant ± standard error (per day)
			Aerobes	Anaerobes		
Live	0	24	$> 100\,000$	$> 71\,000$	33.8	2.1 ± 0.43
		48	$> 100\,000$	$> 71\,000$		
Live	1000	48	43 000	45 000	35.0	1.6 ± 0.26
Live	3000	24	0	0	29.8	0.96 ± 0.14
		48	0	0		
Live, duplicate	3000	24	4000	6000	ND	ND
		48	4000	6000		
Dead	0	48	$> 10^6$	ND	45.4	0.48 ± 0.05
Dead	1000	48	60 000	ND	34.8 ± 1.44	1.9 ± 0.48
Dead	3000	48	0	0	25.7	0.48 ± 0.05

[a]The number of bacteria were determined at the beginning of plant exposure to 1 milligram of hexachloroethane per liter in 60 milliliter vials. After hexachloroethane was added to live or dead plants and water in the vials and selected vials were gamma irradiated, samples were taken from the vials, plated, incubated for 24 or 48 hours, and the number of colonies counted. The dead Canadian waterweed (*Elodea canadensis*) tests involved sealing plants and water in the vials for 2 weeks until the plants died as evidenced by the brown color instead of the normal green and the rotting smell.

[b]1 rad = 0.01 grays. Note that radiation was applied at the beginning of the tests.

[c]Zero order sorption rate coefficients in milligrams per liter per day were normalized by dividing by the density of plant or algae in milligrams per liter used in the respective test.

Note: Abbreviation used — ND: not determined. Replicate total bacterial colony counts performed after standard (24 or 48 hour) incubations of cultures (Rennels 1998) on basic blood agar (aerobes and anaerobes), cooked meat broth (anaerobes), and thioglycolate broth with indicator (aerobes). Two unplanted controls were monitored in parallel that contained 60 milliliters of 1 milligram of hexachloroethane per liter irradiated with 1000 and 3000 grays of gamma radiation. No bacteria were observed in these control solutions and the hexachloroethane did not decrease in concentration during the testing. See Nzengung *et al.* (1999) for additional details.

plant to high doses of gamma irradiation (Forsythe and Evangelou 1993, Holst and Nagel 1997). The tests involved the same plant density of 135 grams of wet weight per liter, initial solute concentration of 1 milligram of HCA per liter, and plant harvest and preparation during the months of March and April.

Dead and decaying biomass make up a significant fraction of organic matter in aquatic environments. Therefore, plant-mediated transformation of HCA and carbon tetrachloride by natural and sterilized dead Canadian waterweed (*Elodea canadensis*) plants was verified. The only observed trend in tests conducted with the dead plants was a decrease in the rate and amount of sorption with increased gamma-irradiation dose. The dead plants were first killed by the stress of being sealed, when freshly harvested, in serum bottles for at least 14 days before gamma irradiation. Meanwhile, autoclaved controls of aquatic plants and macrophytic algae sequestered but did not metabolize HCA and carbon tetrachloride. The autoclave sterilization is a harsh treatment that destroys or changes the enzymatic activity used by plants in metabolic reactions. For the same initial HCA concentration and wet weight of Canadian waterweed (*Elodea Canadensis*) harvested in the same month, the sorption rate coefficients for the live and dead plants were the same and the transformation constants for live plants was similar to that of dead waterweed (*Elodea canadensis*) (Table 16-4). This result suggested that the plant metabolic activity of waterweed (*Elodea canadensis*) did not increase with the expected increase of bacteria and fungi growth in the sealed vials. The latter comparisons were applied only to plants harvested in the same growing season because the measured transformation for the same plant changed seasonally.

Because many microorganisms are associated with aquatic plant surfaces, the metabolic rates of surface-sterilized samples in sealed vials were anticipated to decrease by several orders of magnitude. This hypothesis proved not to be the case. Also anticipated was that the metabolic rate of irradiated samples would increase with increased reaction time if the number of bacteria increased in the serum bottles, which again was not the case. The fact that both axenic parrot feather (*Myriophyllum aquaticum*) and surface sterilized Canadian waterweed (*Elodea canadensis*) metabolized both HCA and carbon tetrachloride, established that aquatic plants play an active and direct role in the transformation of halogenated aliphatics in aquatic environments. Jeffers *et al.* (1998) reported similar observations in experiments in which they used the foliage of 9 herbaceous, 18 deciduous, and 12 coniferous plants to remove methyl bromide from air.

Dhankher *et al.* (1999) showed that crude and partially purified plant extracts also reductively transformed chlorinated alkanes and alkenes. Aqueous suspensions of minced spinach (*Spinacia oleracea*) reductively transformed HCA to pentachloroethane, PCE, and TCE (unpublished work by Nzengung). These findings provided additional direct evidence that plants have the capability to directly metabolize halogenated organic compounds in the absence of plant-associated or colonizing microorganisms.

Phytotransformation Products and Pathways

Phytooxidation in plant tissues of chlorinated ethylenes to chloroethanols and chloroacetic acids in plant tissues has been widely reported in previous phytoremediation studies (Nzengung and Jeffers 2001). However, few studies have focused on phytoreduction catalyzed by plants in general, and aquatic plants, specifically. Nzengung *et al.* (1999) showed that freshwater algae (*Spirogyra* spp)., Canadian waterweed (*Elodea canadensis*), and axenic parrot feather (*Myriophyllum aquaticum*) metabolized carbon tetrachloride to chloroform. Tests of 560 grams wet weight of freshwater algae (*Spirogyra* spp). per liter exposed to 18 milligrams of $^{14}CCl_4$ per liter showed that the chloroform formed during the test did not exceed 12 percent of the initial $^{14}CCl_4$ added to the vials (Figure 16-7). Thus, chloroform persisted in the solution, and presumably reacted slowly with reducing enzymes to form a carbene, or reacted with oxidative enzymes to form phosgene. Both intermediates are very unstable, and can possibly react with water to form carbon dioxide and hydrochloric acid or with plant and macrophytic algae tissue to form bound residue. Liquid scintillation assays of [^{14}C] in combusted biomass quantified the difference between bound carbon tetrachloride and the associated metabolites extractable by organic solvents *versus* the assimilated metabolites (irreversibly bound). The bound fraction, extracted from plant tissue with organic

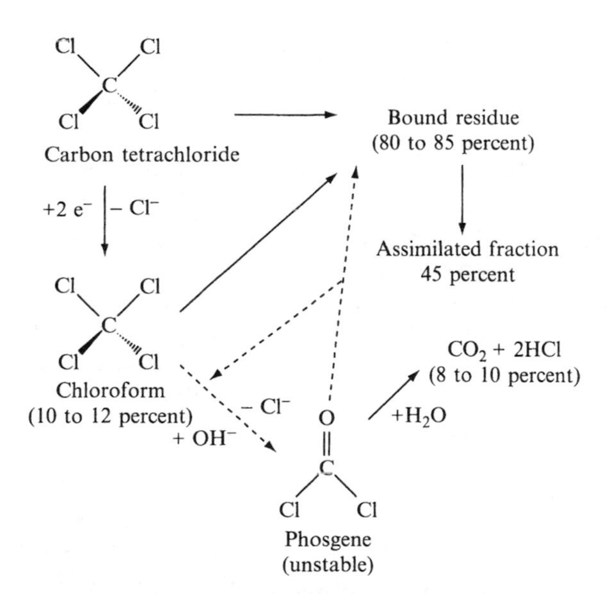

Figure 16-7 Pathway for macrophytic algae and aquatic plant mediated transformation of carbon tetrachloride (modified from Nzengung *et al.* 1999). At the conclusion of this test, the averaged mass recovery was 95.3 ± 7.6 ($\pm$ standard error) percent. Note that the proton and hydrogen balance was not completed. The dashed arrows indicated that phosgene formation and sequestration into plant tissue was not observed but presumed. Hypothetical carbene formation from chloroform and mineralization to carbon dioxide and hydrochloric acid was not included.

solvents, exceeded 35 percent of the original [^{14}C] during most of the test. During the test, the plants transformed and assimilated approximately 45 percent of the initial $^{14}CCl_4$ into tissue. The concentration of tissue-extractable chloroform and carbon tetrachloride increased and then decreased with increased incubation time. These results established that carbon tetrachloride was not simply sequestered, but transformed to other unidentified metabolites or was sequestered, transformed, and assimilated. Because of the heterogeneous composition of the plant extracts, hydrophilic metabolites such as formic acid could not be identified. In a similar manner, Kim *et al.* (1997) showed that organic chemicals could be incorporated within humic substances by oxidative enzymes and rendered nontoxic.

Aquatic plants and macrophytic algae reductively transformed HCA to pentachloroethane and PCE as the main dehalogenation products, with the subsequent formation of relatively small amounts (maximum of less than 0.5 milligram per liter) of TCE and trace amounts (less than 0.1 milligram per liter) of 1,1,2,2-tetrachloroethane and 1,1,2-trichloroethane. The photoautotrophs used in these test included 130 to 280 grams wet weight of freshwater algae (*Spirogyra* spp). per liter, 130 grams wet weight of stonewort (*Nitella* spp.) per liter, 62 to 220 grams wet weight of waterweed (*Elodea canadensis*) per liter, and 200 grams wet weight of parrot feather (*Myriophyllum aquaticum*) per liter. Upon dosing with 1 to 10 milligrams of HCA per liter, the concentrations of pentachloroethane and PCE in aquatic plants and macrophytic algae increased to a maximum within the first 200 hours. The concentrations then slowly decreased to below or approaching the gas chromatography detection limits of 0.005 to 0.02 milligram per liter after 145 to 590 hours depending on plant densities and initial concentrations. (Detection limits increased over time as organic acids accumulated.) The measured PCE concentration in solution was generally greater than that of pentachloroethane, but did not exceed 30 percent of the initial HCA concentration of 1 to 10 milligrams per liter in either sealed batch or growth chamber tests. Figure 16-6 illustrates that PCE typically did not exceed 6 to 10 percent of the original HCA concentration. The amount of extractable phytoreduction products for HCA transformation decreased with length of time into the test, suggesting either that further transformation into polar metabolites or that covalently binding to the macrophytic algae and plant tissues occurred. The metabolism of HCA within axenic parrot feather (*Myriophyllum aquaticum*) plantlets produced detectable (*i.e.*, on the order of 0.001 milligram per liter) quantities of trichloroacetic acid and dichloroacetic acid after 5 days. This result indicates that the metabolism of HCA by photoautotrophs involves both reductive and oxidative transformation pathways. The microbial and plant reductive transformation products, *trans*-dichloroethylene, *cis*-dichloroethylene, and vinyl chloride, were not identified in these tests despite exhaustive analysis.

Figure 16-8 provides the phytotransformation pathways for HCA, PCE, and TCE that account for the formation of observed reductive and oxidative transformation products. According to the proposed pathway, the first step

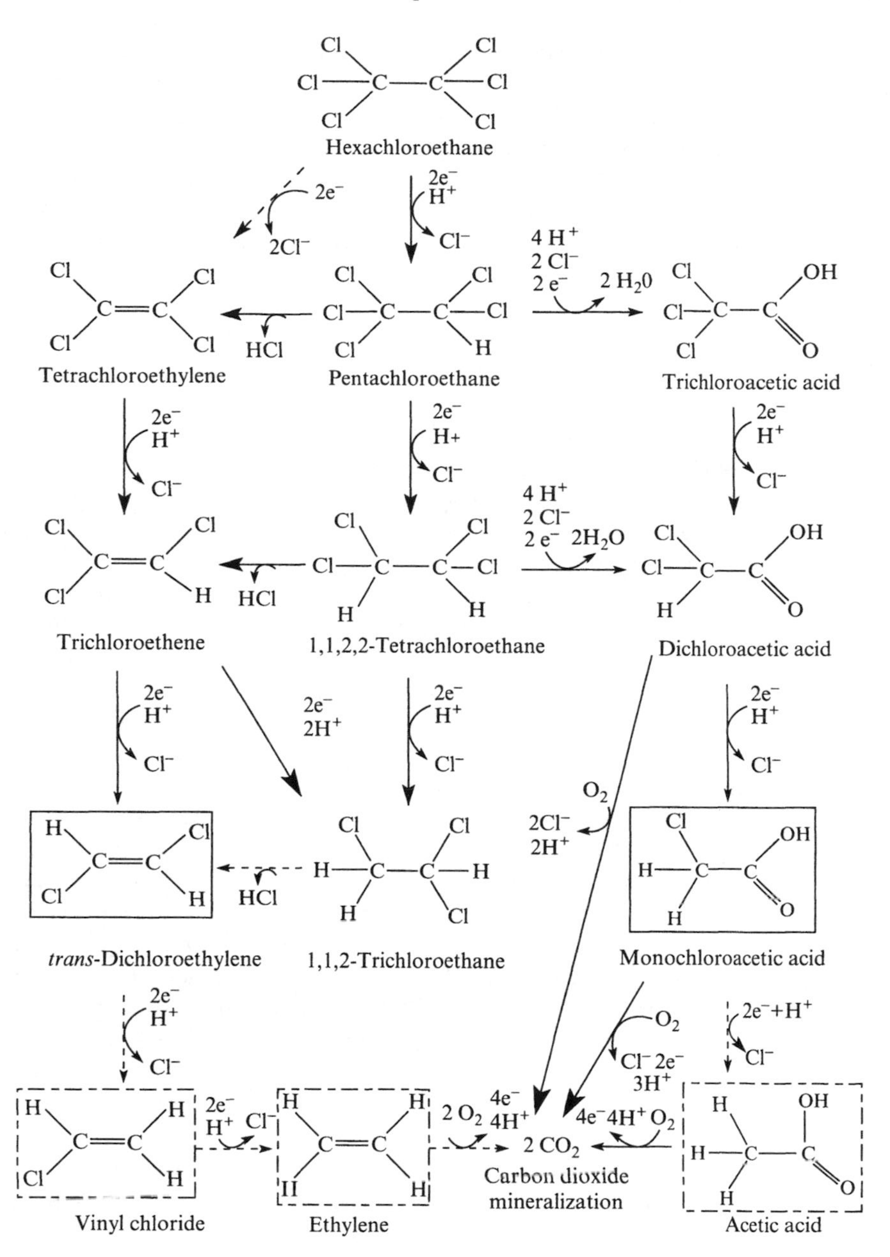

Figure 16-8 Pathway for photoautotrophic (freshwater plants, macrophytic algae, and microbial mats) mediated transformation of hexachloroethane (HCA), tetrachloroethylene (PCE), and trichloroethylene (TCE) based on identified products in solution and in tissue (modified from Nzengung *et al.* 1999). Monochloroacetic acid and *trans*-dichloroethylene have only been observed with microbial mats during cycling between highly oxidizing and highly reducing conditions. This is denoted by the enclosure in boxes. Vinyl chloride, ethylene, and acetic acid are presumed to occur but none of these metabolites have been detected as denoted by enclosure in

(*continues*)

for a plant-mediated transformation of HCA by aquatic plants may involve a two-electron reduction, with the loss of chloride to form a pentachloroethyl carbanion as an intermediate or vicinal chloride elimination to produce PCE. The carbanion can either pick up a proton from water to form pentachloroethane or lose a second chloride to form a carbene. Carbenes are very reactive intermediates (that have not been measured under these circumstances), which can react with water to give trichloroacetic acid as a stable product or react with plant matter to form the observed bound residue. The reaction of the carbanion with a proton produces pentachloroethane. Similarly, pentachloroethane can undergo a two-electron reduction to form a tetrachloroethyl carbanion that can react with a proton to form 1,1,2,2-tetrachloroethane, which can further react to form 1,1,2-trichloroethane, or undergo hydrochloric acid (HCl) elimination to form TCE as observed (Figure 16-8). As the number of chlorine substituents of chlorinated aliphatics decreased, the capability to undergo reductive dehalogenation decreased (Bradley and Chapelle 1996). Pentachloroethane and tetrachloroethane undergo hydrolysis to trichloroacetic and dichloroacetic acids, respectively. Because only trace amounts of 1,1,2,2-tetrachloroethane and 1,1,2-trichloroethane were detected in solution the most favored phytotransformation pathway for HCA are the reactions of the pentachloroethyl carbanion by vicinal elimination and with a proton to give PCE and pentachloroethane, respectively.

The similarity of phytoreduction products of HCA in tests conducted with septic, stressed, and dead aquatic plants and macrophytic algae, plus microbial mats suggested that the same dehalogenase activity catalyzed the reactions. Additionally, the metabolism of a given halogenated contaminant to similar products by axenic parrot feather (*Myriophyllum aquaticum*) and surface-sterilized Canadian waterweed (*Elodea canadensis*) sustains the conclusion that the same dehalogenase activity mediated the reductive dehalogenations, and that oxidative enzymes (possibly cytochrome P-450 with monooxygenase, glutathione, or laccase activity) mediated binding through nucleophilic addition. Other studies have provided evidence of phytooxidation in similar or other plant systems (Lamoreaux *et al.* 1970, Roper *et al.* 1996, Kim *et al.* 1997, Newman *et al.* 1997, Schnabel *et al.* 1997 Newman *et al.* 1999). Metabolites of phytoreduction reactions remained mainly in solution, whereas products of phytooxidation except for chloride that can easily diffuse or be excreted from cells (Newman *et al.* 1999) occurred only in the extracted plant tissues. The latter observations suggested that plant activities that catalyze phytoreduction might be associated with the plant membrane allowing diffusion of the volatile reductive products back into solution. The other plausible explanation is that greater toxicity of some reduced metabolites induces exudation before the oc-

---

Figure 16-8 (***continued***) dashed boxes. The dashed arrows therefore denote hypothetical reactions. Bound residues in aquatic plants and microbial mats have been measured using radiolabeled compounds but the pathways have not been defined. In terrestrial plants, Shang *et al.* (2001) observed a trichloroethanol glucoside formed from TCE that is probably incorporated into plant tissue.

currence of phytooxidation, but excretion by plants instead of vacuole or cell wall storage is not common. Overt toxicity and the associated exudation did not occur in any tests other than highly stressed sorption tests in vials.

RADIOLABELED MASS BALANCES

Microbial Mat

After 30 days, an average of 51.7 percent of the [^{14}C]-HCA and the resulting metabolites was detected in the liquid phase, 15.6 percent was extracted from the solid phase (wet mat) using solvents, 1.7 percent was calculated to be in the vapor phase (based on Henry's law), 15 percent was sorbed to the chamber, and 2.8 percent was assimilated or irreversibly bound to the wet mat. The resulting mass recovery was 86.8 ± 7.7 ($\pm$ standard error) percent. Analysis of the liquid phase using an acid-base extraction procedure indicated that 44 percent of the total liquid phase radioactivity consisted of volatile compounds other than carbon dioxide (presumably TCE), 30 percent of this total was due to nonvolatile metabolites, and 26 percent of this total was attributed to carbon dioxide (mineralized TCE). Analysis of the solvent extract of the wet mat phase using gas chromatography with an electron capture detector indicated that 63 percent of the mat phase radioactivity was due to TCE and 37 percent was due to metabolites of TCE. The unquantified, sorbed metabolites in the mat may be composed partially or wholly of chloroacetic acids or other metabolites. Accurate measurement of $^{14}CO_2$ evolution due to mineralization in the mat was not possible because of the ongoing assimilation and turnover of carbon dioxide during photosynthesis and respiration. Thus, the assimilated fraction actually includes some mineralized [^{14}C].

Macrophytic Algae and Aquatic Plants

Mass balances based on gas chromatography with an electron capture detector that quantified phytotransformation products of nonradioactive halogenated contaminants in solution, routinely showed progressively poorer recoveries with increasing reaction time. The poorer recovery resulted from the formation of metabolites not identified by gas chromatography and from the assimilation of the chlorinated organic contaminants into biomass. In addition, Nzengung *et al.* (2000) observed that mass balance recoveries based on chloride in solution from reductive dehalogenation was also poor due to high background concentrations of chloride. As an alternative, the quantitative dehalogenation of tetrabromoethylene by Canadian waterweed (*Elodea canadensis*) to bromide was measured using a Dionex™ Ion Chromatograph. After 21 days of exposure, 65 percent of the initial 10 milligrams of tetrabromoethylene per liter was transformed to tribromoethylene and 30 percent of total bromine in the initial tetrabromoethylene was measured as dissolved bromide anions in solution, giving a mass balance recovery for bromine of 95 percent.

The assimilated metabolites were quantified by combusting the plant residue in a biological oxidizer and trapping the $^{14}CO_2$ from the radiolabeled contaminants in a carbosorb scintillation cocktail. For freshwater algae (*Spirogyra* spp)., all detectable solution phase $^{14}CCl_4$ was taken up and transformed in 10 days with 10 to 12 percent of the initial solute identified as chloroform, and 8 percent converted to carbon dioxide. Extractable bound metabolites in freshwater algae (*Spirogyra* spp). accounted for 40 percent of the total radioactivity; 45 percent was inextricable and considered to be assimilated or irreversibly bound.

CONCLUSIONS

Like terrestrial plants, freshwater macrophytic algae and aquatic plants sequester and metabolize chlorinated alkanes and alkenes. Sequestration of the various halocarbons is separable from transformation by the vastly different time scales for the processes. For the selected plants and contaminants investigated, sequestration appeared to reach equilibrium within a few hours if this process occurred, whereas biochemical transformation was characterized by pseudo first order kinetics for periods ranging from hours up to weeks. Both septic and stressed (as a result of axenic conditions, gamma irradiation, and being sealed in serum bottles) aquatic plants and macrophytic algae all metabolized HCA or carbon tetrachlorine to the same products of HCA or carbon tetrachloride, respectively. The enzymatic activity observed in dead macrophytic algae and dead plants suggests that these photoautotrophs could be one of many sources of natural metabolic activity in sediments (McCutcheon *et al.* this book). The responsible factor(s) or enzyme(s) seems to become associated with the sediment as the algae and plants decay.

The suite of identified metabolites further suggested that phytotransformation of chlorinated contaminants involved more than one pathway, possibly involving more than one enzyme or plant component. These pathways are very similar to those observed for hybrid poplar clone H11-11 (*Populus trichocarpa* × *Populus deltoides*; Newman *et al.* 1997, Newman *et al.* 1999), tobacco (*Nicotiana tabacum*; Shang *et al.* 2001), cottonwood (*Populus deltoides*), and black willow (*Salix nigra*) (Nzengung and Jeffers 2001). The following are the important processes for aquatic plant, algal, and microbial mat transformations of halogenated contaminants:

1. Sequestration
2. Phytoreduction of highly oxidized halogenated alkenes and alkanes with release of some of the metabolites back into solution
3. Phytooxidation of the lower oxidized phytoreduction products
4. Assimilation

In the case of microbial mats, the terms bioreduction and biooxidation may be more appropriate. Overall, this chapter indicated that freshwater macrophytic algae, aquatic plants, and microbial mats play an important role in the attenuation and treatment of chlorinated organic contaminants and should be used in the remediation of industrial waste streams, contaminated groundwaters, contaminated wetlands, and other aquatic environments. Bench and field pilots are necessary, but these can also serve as treatability studies until the principle of phytoremediation of chlorinated organic contaminants is fully proven in a field setting.

Sequestration and transformation of the chlorinated ethylenes PCE and TCE were observed in the microbial mat tests. The microbial mat sequestration data indicated that such mats have a lower value of the organic carbon normalized sorption coefficient than those values for other carbonaceous materials listed in Table 16-2, but such a comparison ignores several advantages of mats. Firstly, mats can be dried for easy transport and quickly revived upon rehydration. Secondly, mats are a self-replenishing resource that is easy to produce with naturally available nutrients and solar energy. Thirdly, mats can withstand very harsh fresh water and saline environmental conditions. Fourthly, mats are a cost-effective biosorbent that can contain contaminants if spikes in concentration overwhelm the rate of transformation or toxic conditions kill the organisms involved. Finally, the primary advantage of the microbial mats is the capability to transform halogenated contaminants such as PCE and TCE through multiple pathways, resulting in the mineralization rather than a simple phase transfer from one media to another. Therefore, new and retrofit microbial mat bioreactors or created treatment wetlands of photoautotrophs should be a more sustainable, effective, and economical treatment than volatile contaminant stripping and activated carbon sorption treatments used at many groundwater cleanup sites. In addition, the capability to sorb a wide range of contaminants means that microbial mats and other photoautotrophs can easily treat mixtures of wastes, including organic contaminants and metals if the biological material is harvested and disposed after the sorption capacity is reached.

Acknowledgments

Water O'Niell and Valentine Nzengung thank Judy Bender and Peter Phillips for providing the inocula and the sealed bioreactors used in the microbial mats investigations. The authors appreciate the insightful and detailed reviews of Sergei Ostroumov, Milton Gordon, Bill Doucette, and Bob Swank. Jerald L. Schnoor served as editor to accept this chapter for the book. Vera Madison provided exceptional editorial support. Financial support for this work came from the U.S. Department of Energy through a Florida A&M University subcontract to the University of Georgia (Contract #C-9594/10-21-RR176-234), the Strategic Environmental Research and Development Program (including Project 720 managed by the Army Waterways Experiment Station) of the U.S. Departments of Defense and Energy and the U.S. Environmental Protection Agency, and cooperative interagency research agreements with the U.S. Air Force Restoration Division at Wright Patterson Air Force Base and the U.S. Navy Southern Command. This chapter has been reviewed in accordance with U.S. Environmental Protection Agency peer and administrative review policies and approved for publication. Mention of trade names or commercial products does not constitute endorsement or recommendation for use.

REFERENCES

Adey, W.H., C. Luckett, and M. Smith, (1996) Purification of industrially contaminated ground waters using controlled ecosystems. *Ecol. Eng.* **7**: 191–212.

Allen-King, R.M., D. McKay, and M.R. Trudell (1996) Organic carbon-dominated trichloroethylene sorption in a clay-rich glacial deposit. *Ground Water* **35**(1): 124–130.

Anderson T. and B. Walton (1995) Comparative fate of [^{14}C] trichloroethylene in the root zone of plants from a former solvent disposal site. *Environ. Tox. Chem.* **14**: 2041–2047.

Baughman, G.L. and D.F. Paris (1981) Microbial bioconcentration of organic pollutants from aquatic systems—a critical review. *CRC Crit. Rev. Microbiol.* **8**: 205–227.

Bender, J., P. Phillips, R. Lee, S. Rodrigues-Eaton, G. Saha, B. Longanathan, and L. Sonnenberg (1995) Degradation of chlorinated organic compounds by microbial mats. In: *Biological Unit Processes for Hazardous Waste Treatment*. R.E. Hinchee, G.D. Sayles, and R.S. Skeen, eds. Third International *In Situ* and On-Site Bioreclamation Symposium. Battelle Press, Columbus, Ohio. Held in San Diego, California, April 24–27. **3**(9): 299–310.

Bradley, P.M. and F.H. Chapelle (1996) Anaerobic mineralization of vinyl chloride in Fe (III)-reducing aquifer sediments. *Environ. Sci. Technol.* **30**(6): 2084–2086.

Burken J.G. and J.L. Schnoor (1997) Uptake and metabolism of atrazine by poplar trees. *Environ. Sci. Technol.* **31**(5): 1399–1406.

Burken, J.G. and J.L. Schnoor (1998) Predictive relationships for uptake of organic contaminants by hybrid poplar trees. *Environ. Sci. Technol.* **32**(21): 3379–3385.

Chiou, C.T. (1990) Sorption of nonionic compounds in soil. In: *Humic Substances in Soil and Crop Sciences: Selected Readings*. P. MacCarthy, C.E. Clapp, R.L. Malcom, and P.R. Bloom, eds. American Society of Agron. and Soil Science Society of America, Madison, Wisconsin, pp. 111–160.

Dhankher, O.P., J. Tucker, V.A. Nzengung, and N.L. Wolfe (1999) Isolation, purification and partial characterization of plant dehalogenase-like activity from waterweed (*Elodea canadensis*). In: *Phytoremediation and Innovative Strategies for Specialized Remedial Applications*. A. Leeson and B.C. Alleman, eds. 5th Int. Symp. *In Situ* and On-Site Bioremediation: Phytoremediation. Battelle Press, Columbus, Ohio, Held in San Diego, California, April 19–22. **5**(6): 145–150.

Doucette, W.J., B. Bugbee, S. Hayhurst, W.A. Plaehn, D.C. Downey, S.A. Taffinder, and R. Edwards (1998) Phytoremediation of dissolved-phase trichloroethylene using mature vegetation. In: *Bioremediation and Phytoremediation—Chlorinated and Recalcitrant Compounds*. G.B. Wickramanayake and R.E. Hinchee, eds. Battelle Press, Columbus, Ohio. Held in Monterey, California, May 18–21. C1(4): 251–256.

Forsythe, K.W., Jr. and P. Evangelou (1993) Cost and benefits of irradiation and other selected quarantine treatments for fruit and vegetable imports to the United

States of America. In: *Cost-Benefit Aspects of Food Irradiation Processing*. International Atomic Energy Agency, Vienna, Austria.

Garbarini, D.R. and L.W. Lion (1986) Influence of the nature of soil organics on toluene and trichloroethylene, *Environ. Sci. & Technol.* **20**: 1263–1269.

Garrison, A.W., V.A. Nzengung, J.K. Avants, J. Ellington, and N.L. Wolfe (2000) The phytodegradation of *p,p′*-DDT and enantiomers of *o,p′*-DDT. *Environ. Sci. Technol.* **34**: 1663–1670.

Grimes, D.J. and S.M. Morrison (1975) Bacterial bioconcentration of chlorinated hydrocarbon insecticides from aqueous systems. *Microbiol. Ecol.* **2**: 43–59.

Harding, L.W., and J.M. Phillips (1978) Polychlorinated biphenyl uptake by marine phytoplankton. *Mar. Biol.* **49**: 103–111.

Holst, R.W. and D.J. Nagel (1997) Radiation effects on plants. Chapter 2. In: *Plants for Environmental Studies*. W. Wang, J.W. Gorsuch, and J.S. Hughes, eds., CRC-Lewis Publishers, Boca Raton, Florida.

Jeffers, P.M., N.L. Wolfe, and V.A. Nzengung (1998) Green plants: a terrestrial sink for atmospheric CH_3Br. *Geophys. Res. Lett.* **25**(1): 43–46.

Johnson, B.T. and J.O. Kennedy (1973) Biomagnification of *p,p′*-DDT and methoxychlor by bacteria. *Appl. Microbiol.* **26**(1): 66–71.

Kim, J-E., E. Fernandez, and J.M. Bollag (1997) Enzymatic coupling of the herbicide bentazon with humus monomers and characterization of reaction products. *Environ. Sci. Technol.* **31**: 2392–2398.

Lamoreaux, G.L., R.H. Shimabukuro, H.R. Swanson, and D.S. Frear (1970) Metabolism of 2-chloro-ethylamino-6-isopropylamino-*s*-triazine in excised sorghum sections. *J. Agric. Food Chem.* **18**(1): 81–87.

Matter-Müller, C., W. Gujer, W. Giger, and W. Strumm (1980) Nonbiological elimination mechanisms in a biological sewage treatment plant. *Prog. Wat. Technol.* **12**: 299–314.

McRae, I.C. (1985) Removal of pesticides in water by microbial cells adsorbed to magnetite. *Wat. Res.* **19**: 825–830.

Newman, L.A., S.E. Strand, N. Choe, J. Duffy, G. Ekuan, M. Ruszai, B.B. Shurtleff, J. Wilmoth, and M.P. Gordon (1997) Uptake and biotransformation of trichloroethylene by hybrid poplars. *Environ. Sci. Technol.* **31**: 1062–1067.

Newman, L.A., X. Wang, I.A. Muiznieks, G. Ekuan, M. Ruszaj, R. Cortellucci, D. Domroes, G. Karscig, T. Newman, R.S. Crampton, R.A. Hashmonay, M.G. Yost, P.E. Heilman, J. Duffy, M.P. Gordon, and S.E. Strand (1999) Remediation of TCE in an artificial aquifer with trees: a controlled field study. *Environ. Sci. Technol.* **33**(13): 2257–2265.

Nietch, C.T., J.T. Morris, and D.A. Vroblesky (1999) Biophysical mechanisms of trichloroethylene uptake and loss in baldcypress growing in shallow contaminated groundwater. *Environ. Sci. Technol.* **33**: 2899–2904.

Nzengung, V.A., N.L. Wolfe, D. Rennels, and S.C. McCutcheon (1999) Use of aquatic plants and algae for decontamination of waters polluted with chlorinated alkanes. *Int. J. Phytoremed.* **1**(3): 203–226.

Nzengung, V.A. and P.M. Jeffers, (2001) Sequestration, phytoreduction, and phytooxidation of halogenated organic chemicals by aquatic and terrestrial plants. *Int. J. Phytoremed.* **3**(1): 13–40.

Nzengung, V.A., C. Wang, and S. Box (2000) Phytotransformation pathways and mass balances for chlorinated alkanes and alkenes. In: *Proceedings of EPA Phytoremediation State of the Science Conference*. U.S. Environmental Protection Agency Report EPA/625/R-01/011a and b, Cincinnati, Ohio. Held in Boston, Massachusetts, May 1–2 (http://www.epa.gov/ORD/NRMRL/Pubs/625R01011b/625R01011bchap13.pdf).

O'Niell, W.L. (1999) Biosorption and transformation of tetrachloroethylene and trichloroethylene using mixed species microbial mats. Ph.D. dissertation, University of Georgia, Athens.

O'Niell, W., V.A. Nzengung, J. Noakes, J. Bender, and P. Phillips (1999) Biosorption accompanied by biodegradation of PCE and TCE using mixed-species microbial mats. *J. Haz. Waste Res.* **2**(2): 2–1 to 2–16 (http://www.engg.ksu.edu/HSRC/JHSR/vol2no2.pdf).

Orchard, B.J., W.J. Doucette, J.K. Chard, and B. Bugbee (2000) Uptake of TCE by hybrid poplar trees grown hydroponically in flow through plant growth chambers. *Environ. Toxicol. Chem.* **19**(4): 895–903.

Rennels, D.E. (1998) Aquatic plants—mediated transformation of halogenated aliphatic organic compounds. Masters thesis, University of Georgia, Athens.

Roper, J.C., J. Dec, and J.M. Bollag (1996) Using minced horseradish roots for the treatment of polluted waters. *J. Environ. Qual.* **25**: 1242–1247.

Schnabel, W.E., A.C. Dietz, J.G. Burken, J.L. Schnoor, and P.J. Alvarez (1997) Uptake and transformation of trichloroethylene by edible garden plants. *Water Res.* **31**(4): 816–824.

Shang, T.Q., S.L. Doty, A.M. Wilson, W.N. Howald, and M.P. Gordon, (2001) Trichloroethylene oxidative metabolism in plants: the trichloroethanol pathway. *Phytochemistry* **58**(7): 1055–1065.

Smets, B.F. and B.E. Rittmann (1990) Sorption equilibria for trichloroethane on algae. *Wat. Res.* **24**: 335–360.

Strand, S.E., L. Newman, M. Ruszaj, J. Wilmoth, B. Shurtleft, M. Grandt, N. Choe, G. Ekuan, J. Duffy, J.W. Massman, P.E. Heilman, and M.P. Gordon (1995) Removal of trichloroethylene from aquifers using trees. In: *Innovative Technologies for Site Remediation and Hazardous Waste Management, Proceedings of the National Conference*, Vidic, R.D. and Pohland, F.G., eds. Environmental Engineering Division, American Society of Civil Engineers, New york. Held in Pittsburgh, Pennsylvania, July 26. pp. 605–612.

Uhlmann, D. (1979) *Hydrobiology*. John Wiley, New York.

Vogel, T.M. and P.L. McCarty (1985) Biotransformation of tetrachloroethylene to trichloroethylene, dichloroethylene, vinyl chloride, and carbon dioxide under methanogenic conditions. *Appl. Environ. Microb.* **49**(5): 1080–1083.

Vroblesky, D.A, C.T. Nietch, and J.T. Morris (1999) Chlorinated ethenes from groundwater in tree trunks. *Environ. Sci. Technol.* **33**: 510–515.

Zimdahl, R.H. and S.M. Gwynn (1977) Soil degradation of three dinitroanilines. *Weed Sci.* **25**: 247–251.

Zytner, R.G. (1992) Adsorption–desorption of trichloroethylene in granular media. *Water, Air, Soil Pollut.* **65**: 245–255.

17

FATE OF TRICHLOROETHYLENE IN TERRESTRIAL PLANTS

T. Q. Shang, L. A. Newman, and M. P. Gordon

SUMMARY OF PRACTICAL IMPLICATIONS

Tissue cultures, sealed chambers with whole plants, and greenhouse and large-scale field containers with hybrid poplar (*Populus* spp.) establish that trees oxidatively transform trichloroethylene (TCE) and conjugate some products. These studies prove that the concept of using trees to clean up and control shallow contaminated groundwater is valid, but other studies are underway to optimize the process. Trees significantly mineralize TCE to chloride, which the roots excrete back into the groundwater and transpire very little TCE into the atmosphere. Plant uptake and metabolism can remove more TCE than microbial reduction. Normally, oxidative products do not accumulate in trees, but each prospective application must be pilot-tested until this is clearly established. Metal cleaning, dry cleaning, and many other industrial and commercial activities contaminate thousands of sites worldwide with TCE. Given the hepatotoxicity and suspected carcinogenicity, a low-cost, effective treatment is necessary to manage risks from TCE and other solvents at the many sites involved, and phytoremediation using trees meets that need.

INTRODUCTION

Trichloethylene Metabolism in Mammals and Bacteria

In mammals, TCE is metabolized by hepatic cytochrome P-450 mixed-function oxygenases [enzymes that catalyze the addition of one or two oxygen atoms to substrates using molecular oxygen (O_2) as the oxidizer], but

Phytoremediation: Transformation and Control of Contaminants,
Edited by Steven C. McCutcheon and Jerald L. Schnoor.
ISBN 0-471-39435-1 (cloth)

primarily by one enzyme in the P-450 family designated as 2E1. Trichloroethylene is oxidized to chloral, which undergoes further transformation into either trichloroethanol or trichloroacetic acid. Trichloroethanol may conjugate to glucuronic acid. Trichloroacetic acid is unstable in aqueous conditions and can decompose to carbon dioxide and chloroform (Merck Index). Other oxidative metabolites of TCE in mammals include carbon monoxide, formate, glyoxylate, and dichloroacetic acid. Trichloroethylene may also form glutathione conjugates in mammals; the metabolite, dichlorovinyl-L-cysteine, has been found in urine of patients exposed to TCE. The oxidative pathway is, however, predominant in mammals. Figure 17-1 from Lash *et al.* (2000) summarizes the metabolism of TCE in mammals.

Both a hepatotoxin and suspected carcinogen, the effects of TCE are associated with metabolic activation for the oxidative and conjugative pathways. The metabolites are both toxic and carcinogenic. However, activation also occurs in a less defined way, in which either a TCE epoxide intermediate or other acyl chloride intermediates covalently bind to proteins and nucleic acids (Cai and Guengerich 2001).

To our knowledge, bacteria are incapable of using TCE as the sole source of energy (Wackett 1994). However, many species of bacteria are capable of degrading TCE by either oxidative or reductive cometabolism (use of more than one substrate for energy). The bacterial enzymes that use TCE as a substrate are the soluble methane monoxygenase, toluene 2-monoxygenase, and the toluene dioxygenase. The metabolites resulting from these enzymes are similar to the oxidative metabolites in mammals. Reductive metabolism of TCE by anaerobic bacteria leads to the intermediates *cis*-dichloroethylene and vinyl chloride. The reductive pathway can eventually lead to complete dechlorination to form ethene. In some cases, however, reductive dechlorination by bacteria leads to the accumulation of the intermediates, especially vinyl chloride, which is more toxic than TCE and is a known carcinogen. Figure 17-2 summarizes the bacterial metabolism of TCE. Detailed reviews of TCE metabolism in mammalian and bacterial systems can be found in Wackett (1994) and Ensley (1991).

Analytical Challenges

Like any xenobiotic substance, the metabolism of chlorinated solvents in plants can vary greatly with plant age and size, season, growth conditions, stress, and nutrient and soil condition. The studies reviewed in this section discuss the effects of these variables, and the influence on phytotransformation (Table 17-1).

Volatile organic compounds present a unique set of analytical challenges. First, the compounds easily partition into the atmosphere. The dimensionless Henry's law constant for TCE is 0.437 (Schwarzenbach *et al.* 1993). In a stirred aqueous solution one-half of the equilibration concentration in the headspace is reached in only 20 minutes (Davis *et al.* 1998a). Carbon tetrachloride is even

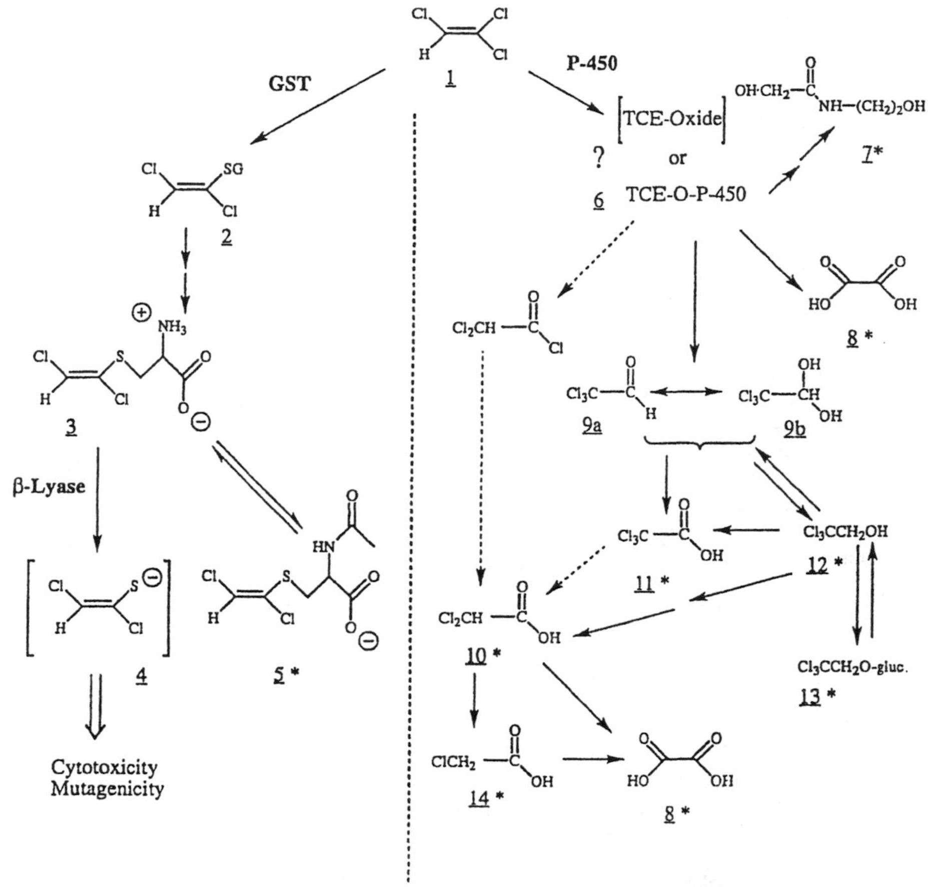

Figure 17-1 Mammalian metabolism of trichloroethylene. Reproduced with permission from Lash *et al.* (2001). Metabolites marked with an asterisk are known urinary metabolites. Metabolites: 1 = trichloroethylene, 2 = dichlorovinyl glutathione, 3 = dichlorovinyl cysteine, 4 = 1,2-dichlorovinylthiol, 5 = *N*-acetyl-dichlorovinyl-cysteine, 6 = trichloroethylene-P-450 or trichloroethylene-oxide intermediate, 7 = *N*-(hydroxyacetyl)-aminoethanol, 8 = oxalic acid, 9a = chloral, 9b = chloral hydrate, 10 = dichloroacetic acid, 11 = trichloroacetic acid, 12 = trichloroethanol, 13 = trichloroethanol glucuronide, and 14 = monochloroacetic acid. Note that GST is glutathione *S*-transferase.

more volatile, with a dimensionless Henry's constant of 1.25 (U.S. EPA, http://www.epa.gov/oerrpage/superfund/resources/soil/attachc.pdf).

Second, TCE adsorbs readily on many types of material, including soil, plastic, silicone grease, and sealants. Therefore, whether in the construction of exposure chambers, or in the storage and handling of stock solutions and samples, precautions must be taken to use proper materials. Whenever possible, glass and Teflon™ should be used; stainless steel is also acceptable except where rust is a potential problem. Sample vials should be fitted with screw caps with Teflon™-lined septa. When repeated sampling is necessary,

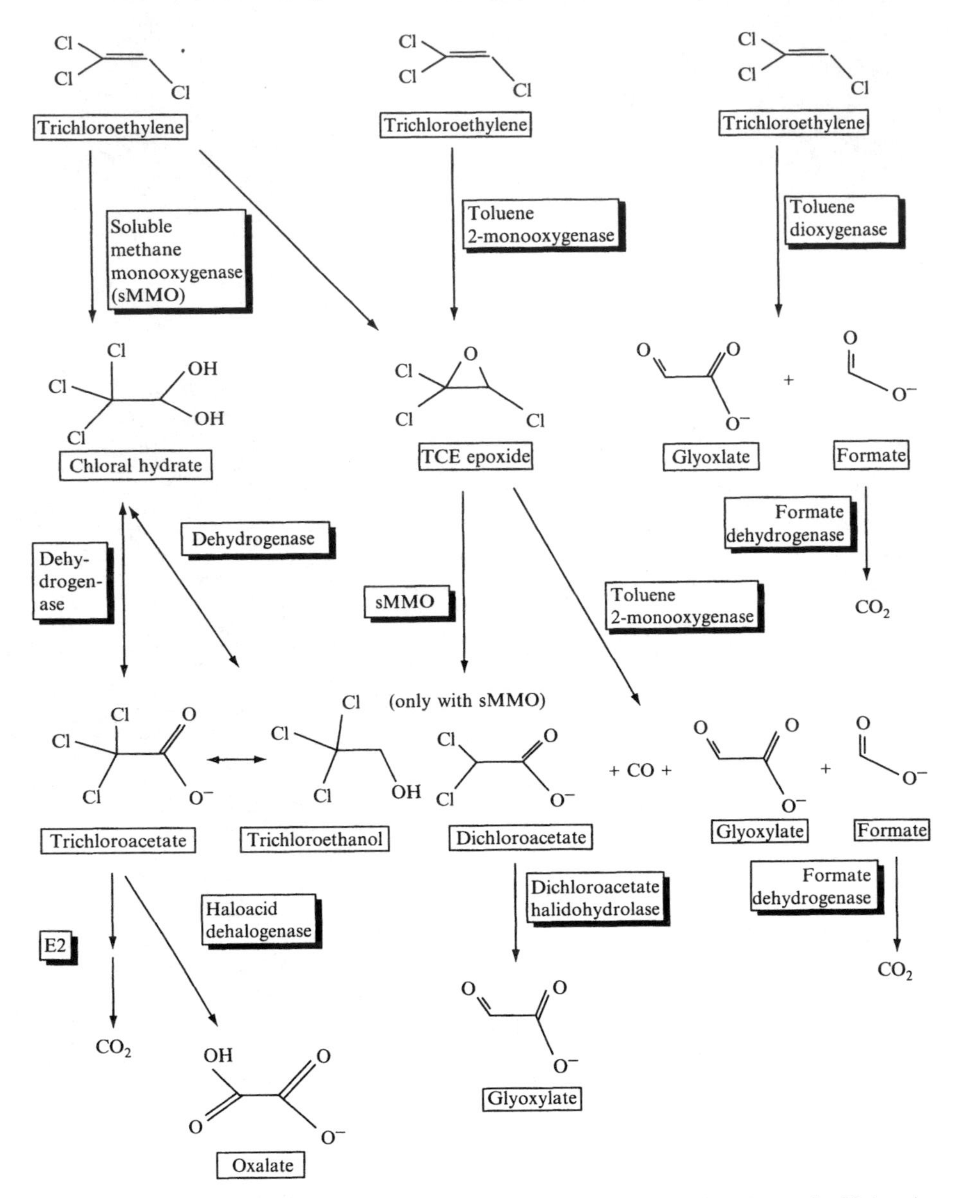

Figure 17-2 Bacterial oxidative cometabolism pathways of trichloroethylene. From the University of Minnesota Biocatalysis/Biodegradation Database (Ellis *et al.* 2001) (http://umbbd.ahc.umn.edu/tce/tce_image_map.html).

mini-nert valves should be used in place of septa. Long-term samples should be kept at −80 °C to minimize volatilization.

A physical seal is necessary for mass balance studies, at the plant–glass interface and glass–glass interface. The choice of sealant should be tested for

TABLE 17-1 Metabolism Studies using [^{14}C]-Trichloroethylene

Investigator	Plants	Container separating root and foliage	Dose and duration	Controls and replicates	Airflow, humidity, and carbon dioxide (CO_2)	Analysis for [^{14}C] fate	Mass recovery and results[a]
Schroll *et al.* (1994)	Carrot (*Daucus carota* var. *sativa*) and radish (*Raphanus sativus*) in 450 to 900 grams of loess	No separation for plants with root exposure "Special closed cover" for plants exposed to air only	Labeled and unlabeled TCE at 0.175 milligram per kilogram of dry soil, allowed to equilibrate in the headspace of chamber Duration: 7 days	Plants in uncontaminated soil but exposed to contaminated air	Continuous airflow, rate not reported	Exit air from reactor: TCE and CO_2 traps Soil: solvent extraction. Plant tissue: oxidizer	Recovery not mentioned [^{14}C] found in plants with root exposure or foliar exposure; [^{14}C] higher than that in soil TCE not found in plant tissue
Anderson and Walton (1995)	1. *Lespedeza cuneata* 2. Bahia grass (*Paspalum notatum*) 3. Goldenrod (*Solidago* spp.) 4. Loblolly pine (*Pinus taeda*) 5. Soybean (*Glycine max*) Plants 1 to 4 were previously exposed to TCE and plant 5 was unexposed	Root zone in 125-milliliter flasks sealed by glass stopper and silicone-rubber seal Whole plants in sealed glass chambers	Soil already contaminated with nonradioactive TCE; [^{14}C]-TCE dissolved in water and injected into soil; and concentration not reported Duration: 9 to 32 days	Contaminated soil without vegetation Autoclaved soil Triplicates for each plant	Foliage chamber: continuous airflow, rate not reported; humidity 74 percent CO_2 at ambient concentration monitored by infrared detector Root chamber flushed with air every 24 hours	Growth chamber exit air: activated charcoal trap Root zone exit air: Aquasol TCE trap, then CO_2 trap Soil: solvent extraction Plant tissue: oxidation	[^{14}C] recovery greater than 70 percent Increased $^{14}CO_2$ (up to 30 percent) in vegetated soil *vs.* soil without vegetation (10 to 18 percent) or sterile soil (approximately 10 percent) *in Lespedeza cuneata*, soybean (*Glycine max*), and pine (*Pinus taeda*) but no significant $^{14}CO_2$ increase in goldenrod (*Solidago* spp.) and Bahia grass (*Paspalum notatum*) Plateau of $^{14}CO_2$ evolution in approximately 5 days

(*continues*)

TABLE 17-1 (*continued*)

Investigator	Plants	Container separating root and foliage	Dose and duration	Controls and replicates	Airflow, humidity, and carbon dioxide (CO_2)	Analysis for [^{14}C] fate	Mass recovery and results[a]
							Tissue incorporation 1 to 21 percent: highest in soybean (Glycine max), pine (Pinus taeda), and goldenrod (*Solidago* spp.) with higher incorporation in roots than in shoots Significant [^{14}C] in foliage chamber exit air with pine, soybean (Glycine max), and goldenrod (*Solidago* spp.) $^{14}CO_2$ from foliage air not tested or reported
Schnabel *et al.* (1997)	Carrots (*Daucus carota* var. *sativa*), spinach (*Spinacia oleracea*), and tomatoes (*Lycopersicon* spp. 'Epoch') grown in garden soil with no detectable TCE	None—plants in ceramic pots (3 to 4 kilograms of soil) with drainage dishes, placed in sealed glass aquaria (416 liters for tomato (*Lycopersicon* spp. 'Epoch'), and 114 liters for carrot (*Daucus carota* var. *sativa*), spinach (*Spinacia oleracea*), and soil control)	Plants irrigated with 0.14 and 0.56 miligram of [^{14}C] and unlabeled TCE per liter Duration: 3 to 7 days	Three sterile soils and three septic soils only dosed at 0.56 milligram per liter. Four replicates per treatment	Continuous airflow, 3 to 6 volumes exchanged per day Condenser coil to remove excess water vapor in bioreactor	Reactor exit air: activated carbon trap for TCE, then NaOH trap for CO_2 Soil: oxidized Plant tissue: oxidized Condensate water and reactor rinsate: liquid scintillation counter	Recovery: 50 to 70 percent, higher in high dose experiments [^{14}C] recovered: 74 to 95 percent in TCE traps; 5 to 25 percent sorbed to soil; 1 to 2 percent in plant tissue (including root, stalk, and edible parts) [^{14}C] higher in plants than in soil No TCE and volatile metabolites found in plant tissue and [C^{14}] in tissue could not be extracted with carbon disulfide or 10 normal sulfuric acid

Newman *et al.* (1997)	Tumor cell culture of hybrid-poplar (*Populus deltoides* × *Populus trichocarpa* 'H-11-11')	Cells and media in suspension in sealed 1-liter glass flasks	Total concentration: 10 milligrams per liter in solution Duration: 3 to 5 days	Growth media alone Killed cells Replicates not reported	50 percent volume headspace	Headspace air: TCE and CO_2 traps Cells: extracted with solvents Media: liquid scintillation counter	Trichloroethanol, dichloroacetic acid, and trichloroacetic acid found in cells 82 to 92 percent of recovered [^{14}C] was unchanged TCE 1.5 percent of recovered [^{14}C] was $^{14}CO_2$ 3.6 percent [^{14}C] incorporated in cells was extractable by acid and methanol
Burken and Schnoor (1998)	Hybrid poplar (*Populus deltoides* × *Populus nigra*), hydroponic, rooted for 2 weeks in 200 milliliters of medium	Root zone in 270-milliliter culture flask, sealed with septum and acrylic sealant, and foliage enclosed in 1-liter inverted flask in 1-liter inverted flask	Unlabeled: as saturated aqueous solution Labeled: as 50 milligrams per liter methanol solution Duration: 4 to 8 days	Plant-free controls and decapitated controls (lower stem and root only). Four replicates closed with TCE	Continuous airflow: 0.8 to 1.1 liter per minute	Foliage chamber air: activated charcoal trap for TCE and CO_2 trap Root zone solution: liquid scintillation counter Plant tissue: oxidized	Recovery: 83 ± 8 percent ($\pm$ one standard deviation) Approximately 50 percent recovered [^{14}C] in root solution and headspace, approximately 20 percent in TCE trap of foliage chamber, 11 percent in sealant, and less than 3 percent in tissue

(*continues*)

TABLE 17-1 (*continued*)

Investigator	Plants	Container separating root and foliage	Dose and duration	Controls and replicates	Airflow, humidity, and carbon dioxide (CO_2)	Analysis for [^{14}C] fate	Mass recovery and results[a]
Gordon *et al.* (1998)	Hybrid poplar (*Populus deltoides* × *Populus trichocarpa* 'H-11-11') rooted cuttings in peat and vermiculite	Root zone and foliage individually enclosed, separated by a third chamber to prevent leakage	Water solution of 5 milligrams per liter injected into soil Duration: 7 days	Glass rod controls with soil to test leakage Replicates not reported	Airflow: 1 to 1.5 liters per minute	Exit air from foliage, root, and middle chambers: TCE and CO_2 traps Soil and plant tissue: oxidized	Recovery: approximately 75 percent Leakage to middle chamber: up to 40 percent Approximately 1 percent [^{14}C] in foliage chamber air
Orchard *et al.* (2000b)	Hybrid poplar (*Populus deltoides* × *Populus nigra*, 'DN34') cuttings rooted 4 to 5 weeks in hydroponic solution	Dual chambers separated by physical seal and pressure differential	Concentration maintained constant by continuous application of labeled and unlabeled TCE to the root chamber: 1. 0.6 milligram per liter in root solution Duration: 13 days	One zero dose control and three replicates.	Airflow: foliage chambers: 4 to 7 liters per minute Root chambers: 0.02 to 0.05 liters per minute	Exit air in foliage and root chamber: various volatile organic compound traps and CO_2 traps Root zone solution: liquid scintillation counter Plant tissue: oxidizer Chamber rinsates: liquid scintillation counter	Recovery: 92 to 94 percent Greater than 90 percent in root zone volatile organic compound traps (volatilization) Approximately 1 percent in root zone CO_2 traps but CO_2 source not confirmed Less than 0.1 percent in foliar TCE traps; none in foliar CO_2 traps 0.2 to 0.5 percent in plant roots, less in leaves and stems Approximately 0.3 percent in root zone solution Transpiration stream concentration factor: 0.08 to 0.18

	Same as above	Same as above	2. 0.9 milligram per liter Duration: 11 days	Two steel rod controls, two replicates	Same as above except one plant and one control had N_2 circulating through root zone instead of air	Same as above	Recovery: 93 to 97 percent Greater than 90 percent in root zone volatile organic compound traps No [^{14}C] in foliar TCE or CO_2 traps Approximately 0.1 percent in roots, less in leaves and stems No significant difference between air *vs.* N_2 root zone conditions Transpiration stream concentration factor: 0.09
	Same as above, but 26-day-exposed plants were smaller in size	Same as above	3. 10 milligrams per liter and 70 milligrams per liter Duration: 12 or 26 days	None: One plant exposed to each of four combinations of concentration and duration	Same as 1.	Same as above	[^{14}C] incorporation similar to above 70-milligrams per liter plants showed depressed transpiration, and less [^{14}C] transpired or transformed Dichloroacetic acid found in roots and shoots of

(*continues*)

TABLE 17-1 (*continued*)

Investigator	Plants	Container separating root and foliage	Dose and duration	Controls and replicates	Airflow, humidity, and carbon dioxide (CO_2)	Analysis for [^{14}C] fate	Mass recovery and results[a]
							10-milligrams per liter treatments as the major chlorinated metabolite and detected metabolites were a small fraction of [^{14}C] in tissue Transpiration stream concentration factor: 0.2, except 70 milligrams per liter, 26-day treatment that showed toxicity
Nzengung and Jeffers (2001)	Willow (*Salix nigra*)	Root zone in 2.2-liter flask with 10 milliliters of headspace and foliage in inverted 5-liter Erlenmeyer flask; sealed with silicone grease	Duration: 20 to 30 days	One zero dose control and duplicates	Airflow rate: 0.025 liter per minute at constant humidity	Growth solution (rhizosphere): liquid scintillation counter Headspace of the rhizosphere solution: gas chromatography–flame ion detector and CO_2 traps Foliage chamber air: TCE and CO_2 traps	[^{14}C] recovery: 67 to 86 percent $^{14}CO_2$ found in rhizosphere solution (0.5 to 1 percent of added [^{14}C]), but not in the foliage headspace 47 to 48 percent added [^{14}C] recovered in rhizosphere solution 10 to 28 percent added [^{14}C] recovered in foliage chamber air, presumably as TCE 10 percent added [^{14}C] recovered in plant tissue, mostly in roots

[a]Values for [^{14}C] are not equivalent to concentrations of TCE but represents the parent compound plus metabolites.

Abbreviation used—: TCE : trichloroethylene.

adsorption and leakage, but Doty *et al.* (2000) found the best sealants to be fluorolube (glass–glass) and plumber's putty (plant–glass). Orchard *et al.* (2000) used rope caulk for a flexible, gas-tight, nontoxic seal with greater than 90 percent recovery of [^{14}C]-TCE. They reported, however, that silicone caulking could be toxic to plant growth. Others reported that silicone rubber sealants are permeable to TCE (Davis *et al.* 1998a). Wax (Davis *et al.* 1998a) and acrylic sealants (Burken and Schnoor 1998) have also been used to seal around plant stems.

Labeled TCE is often used to trace the fate of the carbon atoms. We found that [^{14}C]-TCE often contains 1 to 3 percent nonvolatile impurities, sometimes higher, even when the supplier claims greater than 99 percent purity by gas chromatography and high pressure liquid chromatography. The amount of impurities tends to increase over time, but can be minimized by distillation. This observation was confirmed by the technical support department of Sigma-Aldrich (Brian Dulle, February 2001). In cases where a low percentage of transformation is investigated, the presence of such impurities is an important consideration.

OVERVIEW OF EXPERIMENTAL DESIGNS

Biological Systems

Cell and Tissue Cultures

Newman *et al.* (1997) reported studies with suspensions of tumor cells from a hybrid poplar (*Populus trichocarpa* × *Populus deltoides* 'H-11-11'). Initially, no other tissue culture techniques such as the use of hairy roots or protoplasts were published in connection with terrestrial plant metabolism of chlorinated solvents. The failure to use these cultures was perhaps justified for a number of reasons. Cell cultures, especially undifferentiated cells, can have a metabolism dramatically different from that of actual plants, depending on the growth phase and the contaminants involved. Long-term investigations are difficult because of the need to contain the cultures in relatively small vessels and to dilute the solutions with fresh media regularly. In addition, oxygen consumption is rapid with callus cells in sealed containers.

Despite the disadvantages, cell cultures have been shown to share many properties of the whole plant. For example, the oxidative metabolites that were identified in field-grown trees (Newman *et al.* 1997) and hydroponic trees (Orchard *et al.* 2000b) were the same as those found in cell cultures. The extent of [^{14}C]-TCE mineralization in cell cultures was similar to that in hydroponically maintained plants (Gordon *et al.* 1998, Orchard *et al.* 2000b). Therefore, the cell culture system is very useful in confirming the results of whole plant studies, and the aseptic nature ensures that any metabolism observed is not from rhizosphere degradation. Moreover, cell and tissue cultures provide the unique advantage of being relatively homogenous, easy to handle, and are readily contained. As such, far more controls and replicates are possible with cell and tissue cultures than with whole plants, and a better mass balance is achieved (close to 100 percent).

Hydroponic Plants

Generally, cuttings of trees or other plants are surface sterilized, rooted in Hoagland's or other nutrient solutions with rooting hormones, allowed to grow to certain sizes in sterile media, checked for sterility, and then transferred to experimental apparatus containing TCE. Like cell cultures, hydroponic plants can grow axenically from sterilized seeds, thus excluding the effects of microorganisms. The root zone solution provides an easy way to distribute TCE or other contaminants. Thus, the exposure is homogenous and easily monitored. Symptoms of root stress, such as discoloration, can be observed quickly.

The disadvantage of hydroponic testing is the limited size of the sealed test vessels. The plants used in these exposure experiments were perhaps no more than 4 to 5 weeks old and about 20 centimeters high (Orchard *et al.* 2000a, 2000b). In addition, root biology in hydroponics may be different from soil-grown plants. In solution, roots easily become oxygen depleted. Roots are also directly exposed to the contaminated water; in the field, plant roots are likely to be exposed to concentrations far lower than that in the groundwater (Davis *et al.* 1998a).

Several approaches can offset the disadvantages. Orchard *et al.* (2000a) added silicate to the root solution, a component ubiquitous in soil, which may regulate micronutrient uptake. They also introduced airflow through the sealed root chamber to minimize oxygen depletion. Finally, a TCE solution was constantly injected to maintain a constant level more akin to groundwater exposure.

Interestingly, although in many cases hydroponic plants are surface sterilized to eliminate many bacterial effects, Orchard *et al.* (2000a) argue that hydroponic roots do not necessarily have decreased microbial activity compared to roots growing in soil. Therefore, hydroponic plants may be used in two ways: surface sterilized and unsterilized to represent different root environments. Because of the possibility of endophytes, it is important to check for sterility.

Soil-Grown Plants

Plants grown in soil imitate field conditions best because the soil can come directly from a field site of interest to accurately represent the soil type, initial microorganism population, and even contaminant and nutrient concentrations likely to be encountered. Exposing plants to pollutants in soil can be tricky because contaminants are difficult to distribute evenly. Several methods have been used for dosing: (1) by using already contaminated soil from a field site at a known contaminant level (Anderson and Walton1995), (2) soil can be dosed directly by injecting a water solution into an enclosed, relatively small volume of soil (Gordon *et al.* 1998), and (3) groundwater contamination can be simulated by injecting the contaminant solution into a sand layer underneath the soil, through a watering tube or well (Narayanan *et al.* 1995, Gordon *et al.* 1998, Newman *et al.* 1999,).

One of the complicating factors when studying TCE metabolism in a soil-plant system is soil adsorption. Organic compounds, including TCE, can adsorb strongly to soil and become unavailable to plants. However, the extent of adsorption is difficult to define, and may depend upon many factors such as soil type, moisture, and organic content. Therefore, the actual levels of TCE to which the plants are exposed are difficult to determine. In addition, careful controls are needed to ensure that disappearance of TCE is not simply the result of soil adsorption.

Studying plants in the presence of the rhizosphere is a realistic but challenging method to ascertain the relative roles of plants and microorganisms. This is because plant roots, microorganisms, and the soil are so closely linked that one may not be isolated without altering the others. For example, sterilizing soil by various methods can change the properties of the soil and the resulting adsorption of TCE. Irradiation seems to change the soil to the least extent compared to autoclaving or chemical sterilization; nevertheless, change still occurs (Sheremata *et al.* 1997). Removing plants from the soil, on the other hand, reduces the microorganism community because plant roots exude nutrients that supplement surrounding soil nutrition (Jordahl *et al.* 1997).

Most investigators studying TCE fate used both hydroponic and soil-grown plants to separate plant effects and rhizosphere effects. Additional controls included imitating the soil environment as closely as possible without the presence of plants. For example, Anderson and Walton (1995) used soil-grown plants as well as soil removed from the root zone of plants to investigate the effect of plants upon microorganisms. Similarly, Narayanan *et al.* (1995) studied TCE removal by soil-grown plants followed by the removal of aboveground portions of the plants.

Transgenic Plants

Doty *et al.* (2000) generated tobacco, (*Nicotiana tabacum* var. *xanthi*, n. c.) plants, that expressed human cytochrome P-450 2E1 in the leaf, stem, and root. These plants displayed heightened ability to oxidatively metabolize TCE, as well as another P-450 2E1 substrate, ethylene dibromide (Doty *et al.* 2000). The level of the TCE oxidation product, trichloroethanol, correlates with the level of 2E1 expression. Although genetically altered (transgenic) tobacco (*Nicotiana tabacum* var. *xanthi*, n. c.) plants are not intended for use as phytoremediation plants for chlorinated solvents, this work as proof of concept shows the potential of genetic engineering to dramatically improve the ability of plants to oxidatively degrade contaminants.

In an artificial aquifer study, poplar (*Populus trichocarpa* × *Populus deltoides* 'H-11-11') trees removed more than 90 percent of TCE (Newman *et al.* 1999). However, the removal was accompanied by consumption of most of the water that was injected, and the concentration of TCE in the simulated groundwater did not decrease as dramatically as did the mass of TCE. Increased ability of genetically engineered trees to mineralize TCE may achieve decreased contaminant concentrations without the depletion of groundwater. Transgenic tobacco (*Nicotiana tabacum* var. *xanthi*, n. c.) plants have shown the capability to reduce the concentration of ethylene dibromide in root solutions (Doty *et al.* 2000). In that study, the genetically engineered tobacco (*Nicotiana tabacum* var. xanthi, n. c.), expressing P-450 2E1 showed a two-fold increase over control tobacco plants in capability to remove ethylene dibromide from hydroponic solutions, while consuming the same amount of water. This translates to a twofold decrease in ethylene dibromide concentration in the growth solution. In addition, transgenic plants with higher contaminant degradation rates may also minimize the area and time required for remediation. Another potentially desirable use of genetic engineering is that low-expression plant enzymes capable of degrading a contaminant may be identified and over expressed (produced in excess) in the same species. Thus, plants that are already adapted to the particular site can be improved for remediation. Further work along these lines holds promise of trees that are custom tailored to detoxify specific classes of pollutants.

Potential disadvantages to the transgenic approach include difficulty, cost, and the time involved. Even when the gene is successfully incorporated in the genome, the protein may not be expressed, or only produced by the plant at very low levels. For example, our lab transformed poplar (*Populus* spp.) trees by incorporating dehalogenases and cytochrome P-450 2E1 successfully into the genome, but the transgene products were not detectable or barely detectable (Sharon Doty, University of Washington, personal communication 2002). The transferred gene may be unstable. Other concerns about the general use of transgenic plants relate to public acceptance issues reviewed by Wolfenbarger and Phifer (2000).

Analytical Systems

Mass-Balance Chambers

The volatile nature of chlorinated solvents makes mass-balance approaches particularly critical and also challenging. Most groups that study TCE in plants have designed some type of mass-balance chamber to enclose the plants and, sometimes, to separate the roots from the foliage. Table 17-1 is a compilation of mass-balance studies using [C^{14}] as a tracer. Table 17-2 compiles the mass-balance studies that only use unlabeled TCE. Technical aspects of the various experimental designs are discussed next in this section.

TABLE 17-2 Trichloroethylene Uptake and Transpiration Studies in Sealed Chambers without Radioactive Tracers

Investigator	Plants	Container separating root and foliage	Dose and duration	Controls and replicates	Airflow, humidity, and carbon dioxide (CO_2)	Sampling and analysis	Mass recovery and results
Wolverton *et al.* (1989)	Nine plant species grown in potting soil in pots: 1. Bamboo palm (*Chamaedorea seifrizii*) 2. English ivy (*Hedera helix*) 3. *Ficus benjamina* 4. Gerbera daisy (*Gerbera jamesoni*) 5. Janet Craig (*Dracaena deremensis*) 6. Madagascar dragon tree (*Dracaena marginata*) 7. Mass cane (*Dracaena massangeana*) 8. Mother-in-law's tongue (*Sansevieria laurentii*) 9. Peace Lily	No separation: entire plant placed in Plexiglas® chamber (0.4 to 0.8 cubic meters)	Neat TCE injected into chamber and allowed to evaporate and equilibrate with air to a concentration in air of 15 to 20 parts per million or 0.1 to 0.3 parts per million Duration: 24 hours	Leak test controls with empty chamber Soil controls with potting soil in pots	Air circulated within chamber Refrigerated condenser coil	Chamber air: sampled through Gastec® or Tenax tube and analyzed by desorption and gas chromatography–flame ionization detection	Certain plants removed more TCE from air than soil controls at both high and low concentrations

(*continues*)

TABLE 17-2 (*continued*)

Investigator	Plants	Container separating root and foliage	Dose and duration	Controls and replicates	Airflow, humidity, and carbon dioxide (CO_2)	Sampling and analysis	Mass recovery and results
	(*Spathiphyllum* spp. 'Mauna Loa' i)						
Narayanan *et al.* (1995)	Alfalfa (*Medicago sativa*) grown in sandy silt previously exposed to phenol	No separation of root and foliage: alfalfa (*Medicago sativa*) grown in U-shaped channels, cut to 5 centimeters height regularly and channels contained in closed chamber	TCE and trichloroacetic acid at 0.05 to 0.1 milligram per liter continuously introduced to saturated zone of the treatment chamber (simulated groundwater) for over 3 months, then washed with water for 1 month	Undosed alfalfa (*Medicago sativa*) Dosed reactors with aboveground portion removed One channel each with multiple alfalfa (*Medicago sativa*) plants	Not mentioned	1. Groundwater sampled at various points from inlet to outlet 2. Headspace of chamber 3. Plant tissue 4. Soil chloride analyzed at the end of the experiment	1. Mass recovery not known 2. 80 percent TCE mass removed from groundwater daily, 27 percent decrease in concentration 3. 36 percent of added TCE was unchanged 4. 16 percent TCE in the headspace 5. Methane detected in groundwater; estimated 23 percent added TCE degraded anaerobically 6. Chloride accumulation in the top layer of the soil and in groundwater 7. Removal of aboveground plants had similar results

							8. No detectable TCE (less than 1 percent added TCE) in plant tissue
Davis *et al*. (1998)	Hybrid poplar (*Populus deltoides* × *Populus nigra*), salt cedar (*Tamarix parviflora*), and sunflower (*Helianthus annuus*) grown in hydroponic solution	Root system in 600-milliliter side arm flasks sealed with cork and wax and foliage in 78.5-liter chamber	132 milligrams per liter TCE, alone or with trichloroethane or methanol, in 250-milliliter root zone solution by adding enough TCE to headspace to equilibrate with solution Duration: up to 30 hours	1. Cut plants 2. Deuterated water (D_2O) used to trace plant transpiration 3. One or two plants for each treatment	Static root zone with greater than 50 percent headspace for oxygen Foliar chamber: airflow at 1 liter per minute and CO_2 level at half ambient in most cases Refrigerated condenser coils and plants heated to 22 to 25 °C	Headspace air in foliage chamber: Fourier transform infrared detector Root zone headspace: gas chromatography Solution concentration: inferred from root zone headspace concentration	1. TCE moves through plants readily: TCE concentrations in chamber headspace reached steady state at about 600 minutes for whole plants, 500 minutes for cut plants 2. Compared to deuterated water, there is a time lag for appearance of TCE in the foliar headspace 3. Transpiration stream concentration factor: 0.1 to 0.58 for whole plants. 4. Transpiration stream concentration factor in cut sunflowers (*Helianthus annuus*) close to 1; cut hybrid poplar (*Populus nigra*), 0.2 to 0.35

(*continues*)

TABLE 17-2 (*continued*)

Investigator	Plants	Container separating root and foliage	Dose and duration	Controls and replicates	Airflow, humidity, and carbon dioxide (CO_2)	Sampling and analysis	Mass recovery and results
Cornejo *et al.* (1999)	Spider plant (*Chlorophytum comosum*) grown in soil	No separation: entire plants placed in 40-liter glass aquarium	Neat TCE injected into chamber to evaporate and equilibrate with air Duration: 8 hours	TCE, benzene, or pentane injected into chambers without plants	Air circulated at 0.15 liters per minute	Chamber air analyzed by gas chromatography–mass spectroscopy	TCE removed from air
Nzengung *et al.* (2001)	Willow (*Salix nigra*) and cottonwood (*Populus deltoides*) grown hydroponically	Root zone in 2.2-liter flask with 0.01 liter headspace and foliage exposed to air	Doses varied Duration: 40 to 70 days	Duplicates and one unexposed control	Airflow rate: 0.025 liter per minute Bioreactors located in greenhouse with constant humidity	Growth solution: hexane extraction and gas chromatography–electron capture detection Headspace of the growth solution (rhizosphere): gas chromatography–flame ionization detection and gas chromatography–mass spectroscopy Plant tissues: extraction with hexane and methyl *tertiary*-butyl ether and analyzed with gas chromatography–electron capture detector	Rate of TCE removal increased with water uptake of the tree Reductive metabolites found in rhizosphere headspace: *trans*-dichloroethylene, ethene, ethane, and methane Oxidative metabolites found in tissue, mostly trichloroacetic acid

Abbreviation used—: TCE: trichloroethylene.

Dosing Method and Duration

Trichloroethylene was added neat or by injecting a stock solution dissolved in water, methanol, or acetone. For [^{14}C] tracer studies, most of the exposure durations were 1 to 2 weeks. For healthy and actively growing plants, a prerequisite for studying metabolism, the use of a mass-balance chamber limits exposure duration. Therefore, accommodations are necessary when long-term experiments are undertaken. For example, Orchard *et al.* (2000b) used smaller plants for 26-day experiments, and Naranyanan *et al.* (1995) harvested plants regularly during the 3 months of exposure.

Plant Type and Adaptation

We are not aware of any evidence that prior exposure to TCE changes the response of plants. In the work by Anderson and Walton (1995), both previously unexposed and previously exposed plants accelerated rhizosphere mineralization of TCE. Most of the other known studies use previously unexposed plants. Hybrid poplar (*Populus* spp.) is one of the more commonly tested plants, although a variety of clones have not been tested. Plant selection has not been a focus of laboratory-scale experiments, but may be crucial in field applications. The importance of selecting the right clone of hybrid poplar (*Populus* spp.) for any particular purpose is discussed in Dickmann and Isebrands (1999).

Controls

Because of the potential for leakage and adsorption, many types of controls have been employed to properly assess the metabolism of TCE. The minimal control for a mass-balance study is leakage tests with the empty chamber. When soil-grown plants are used, many studies use unplanted, autoclaved soil as controls to separate the effect of soil adsorption from the actions of plants and microorganisms. When the root zone is separated from foliage, the plant is usually sealed around the stem; both stainless steel and glass rods have been used to simulate a stem in associated unplanted controls. Burken and Schnoor (1998) also used decapitated plants to represent the absence of foliage. Anderson and Walton (1995) used autoclaved soil to assess the involvement of microorganisms. Lack of appropriate controls can make experimented results difficult to interpret. For example, Cornejo *et al.* (1999) showed a decrease of TCE concentration in chamber air with potted plants. However, lacking soil or dead plant controls, this reduction could be due to plant action, soil adsorption, or even leakage.

Separation of Root and Foliar Sections

The root system can be sealed apart from the crown in order to investigate root uptake. The proper seal is very important because TCE can easily leak into the foliar section and be mistaken as root uptake. In our laboratory, we found that, without proper precautions, up to 40 percent of added TCE radiolabel can leak into the foliar chamber (Newman, unpublished). Therefore, we used a ring section between the root zone and the foliar chamber to trap any leakage. Orchard *et al.* (2000a) used a high flow rate in the foliar zone and a pressure differential to prevent such leakage.

Plant Growth Conditions

Orchard *et al.* (2000a) recently reviewed the factors important in the construction of a chamber that could not only ensure measurable mass recovery but also minimize the effect of enclosure on normal plant growth and metabolism. Some of these factors are (1) keeping the humidity of the foliage chamber below 80 percent, such that transpiration is not inhibited, (2) ensuring an adequate supply of carbon dioxide for normal photosynthesis; (3) replenishing oxygen in the root zone, and (4) maintaining a photosynthetic photon flux of at least 15 percent of full sunlight. Most of the mass-balance studies have addressed these factors to varying degrees.

Sample Analysis

A mass balance requires the analysis of every component of the exposed system. Therefore, the chamber air, root solution and soil, and plant tissue must be routinely analyzed for TCE and metabolites. In the chamber headspace, TCE is usually measured by pumping the air through a sorbent tube, and subsequently extracting or desorbing the sorbent for measurement using gas chromatography. Activated carbon is the most frequently used sorbent. However, a high airflow rate or a high water content can decrease sorption efficiency. To measure the mineralization of radiolabeled TCE, an alkaline trap follows the TCE trap to collect the carbon dioxide produced. The chemical form of radioactivity in the carbon dioxide trap must be confirmed by barium precipitation to rule out the possibility of a breakthrough of TCE. Whenever radiolabeled TCE is used, plant tissue is most often oxidized to carbon dioxide for measurement because radioactivity incorporated in plants is often not extractable by other means. Schroll *et al.* (1994) and Anderson and Walton (1995) measured the [^{14}C] in soil by solvent extraction, while other groups oxidized the soils.

Laboratory Scale Open Systems

When mass balance is not an objective, plants exposed to a contaminant in open systems may allow growth and metabolism that is more realistic. Open systems are also more feasible when large numbers of plants need to be treated and compared in parallel. Toxicity screens and other preliminary tests are routinely performed with plants grown in 1 to 2 liter beakers of soil. A glass tube is typically inserted to the bottom of the beaker to allow injection of contaminant solutions of varying concentrations.

One problem with open systems is the evaporation of TCE from the soil and root solutions. In Newman *et al.* (1997), TCE was detected in leaf bags placed on both dosed and control plants, making it difficult to ascertain the actual leaf transpiration of TCE. Evaporation of TCE from the contaminated soil has been detected by absorption on carbon tubes and should be monitored. Such air traces of contamination have been observed in other experiments at the University of Washington and can lead to ambiguous results. This problem has also been encountered in the field with natural vegetation (Doucette *et al.* 1998).

Field Studies

Newman *et al.* (1999) conducted full-scale studies of trees growing under field conditions to determine the capability for removal of TCE from an aquifer. Double-walled compartments were constructed of high-density polyethylene liners, measuring approximately 1.5 meters deep by 3 meters wide by 5.7 meters long. Each compartment had an artificial aquifer composed of coarse sand occupying the bottom 0.3 meters. The aquifer was overlaid with 1.1 meters of Sultan silty clay loam. Contaminant solutions were injected into the sand layer through wells, and the gravity distribution of water facilitated by the bottom slightly sloping towards the exit pipe. A variety of trees was planted in these compartments, and the artificial aquifers contaminated with a range of chlorinated compounds, including TCE. The most advanced studies were done with hybrid poplars (*Populus trichocarpa* × *Populus deltoides* 'H-11-11') and TCE.

Other groups have planted trees on contaminated sites around the U.S. Discussions of some of these sites can be found in Eberts *et al.* (this book), Hirsh *et al.* (this book), and Jordahl *et al.* (this book).

FATE OF TRICHLOROETHYLENE

Uptake and Transpiration

Several studies have conflicting conclusions regarding the capability of plants to take up TCE from soil or groundwater and transpire this contaminant through foliage into the atmosphere. These studies are summarized in Tables 17-1 and 17-2, and reviewed in Orchard *et al.* (2000b).

Transpiration into Air

The transpiration of TCE by plants can only be proven in those studies that separated roots and foliage, so that evaporation from root solutions or soil could not be mistaken as transpiration. Such studies include Anderson and Walton (1995), Burken and Schnoor (1998), Davis *et al.* (1998a) Gordon *et al.* (1998), Orchard *et al.* (2000b), and Nzengung and Jeffers (2001). All of these studies determined that plants are capable of taking up and transpiring TCE *via* foliage into the air. However, the extent of TCE transpiration reported was rather different. Burken and Schnoor (1998) reported that hybrid poplar

(*Populus deltoides* × *Populus nigra*) foliage transpired 20 percent of the TCE added to the root zone hydroponic solution. In contrast, Gordon *et al.* (1998) and Orchard *et al.* (2000b) found that less than 1 percent of TCE added to the sealed root zone was transpired, using soil-grown and hydroponic poplar (*Populus* spp.) plants, respectively. Orchard *et al.* (2000b) varied dosage, root zone oxygen level, and exposure time, but none of these changes increased the percentage of TCE transpiration by hydroponic hybrid poplars (*Populus deltoides* × *Populus nigra* 'DN34'). In a controlled field study using poplar (*Populus trichocarpa* × *Populus deltoides* 'H-11-11') trees (Newman *et al.* 1999), TCE was found in leaf bags during the first two growing seasons, with the transpiration of TCE accounting for 9 percent of the total TCE lost from the treatment cells. However, during the third season, no TCE was detected in leaf bags. Anderson and Walton (1995) found 1 to 20 percent additional [^{14}C]-TCE in the exit air of the foliage chambers, depending on the plant species. The highest transpiration percentage was observed for loblolly pine (*Pinus taeda*).

Several factors may contribute to the transpiration of TCE by plants. First, TCE may be partially excluded by the plant roots so that the concentration entering the transpiration stream is limited. Generally, however, exclusion is only expected with hydrophilic compounds (logarithm of the octanol–water partitioning coefficient less than 1.8), and therefore TCE (logarithm of the octanol–water partitioning coefficient equal to 2.3) is expected to readily pass through the lipid membranes of the root epidermal layer.

Second, sorption by plant tissue may delay the appearance of TCE in the transpiration stream. Davis *et al.* (1998b) did find that aspen (*Populus* spp.) wood splints, crushed and dried alfalfa (*Medicago sativa*), and topsoil displayed significant sorption of TCE, whereas sucrose, glass beads, and Whatman™ and brown kraft paper did not. They also showed that in hydroponic plants, increasing exposure time from 1 week to 3 weeks did not increase the percentage of TCE transpired. Therefore, sorption and desorption equilibrium is quickly established and little TCE is accumulated in the plant before transpiration occurs.

Third, metabolism within the plant (see the section "Identification and Characterization of Metabolites") is another factor that may cause concentrations of TCE in the transpiration stream to be lower than expected. This factor is only noticeable if the equilibration of TCE into the plant transpiration stream is much slower than the rate of metabolism.

Trichloroethylene Uptake into Plant Tissue

Regardless of whether the reported TCE transpiration is high or low, there seems to be agreement that the steady state concentrations of TCE in plant tissues (especially in aboveground tissue) are usually low. For example, Schnabel *et al.* (1997) did not find any TCE in carrots (*Daucus carota* var. *sativa*), spinach (*Spinacia oleracea*), and tomatoes (*Lycopersicon* spp. 'Epoch') treated with TCE at 0.14 to 0.56 milligrams per liter in soil, and the radioactivity that was taken up was apparently transformed. Narayanan *et al.* (1995) also found no TCE in alfalfa (*Medicago sativa*) plants exposed to 0.05 to 0.1 milligrams of TCE per liter of groundwater. Nzengung and Jeffers (2001) measured 0.07 and 0.17 milligram of TCE per kilogram wet weight in the leaves of mature cottonwood (*Populus deltoides*) trees and willow (*Salix nigra*) trees, respectively, which grew above TCE-polluted groundwater.

Newman *et al.* (1997) and Newman *et al.* (1999) found the highest TCE levels in stems, followed by roots, and low or often no detectable TCE concentrations in leaves. Davis *et al.* (1999) studied the diffusive loss of TCE from hybrid poplar (*Populus* spp.) and corkscrew willow (*Salix* spp.) tissue. They found that TCE was lost rapidly through the stem because of radial diffusion. One would also expect that the large leaf surface area and efficient gas exchange during normal growth could deplete TCE from the leaves rapidly. Metabolism is another mechanism that could result in low TCE concentrations in plant tissue (see the section "Identification and Characterization of Metabolites").

There are some possible exceptions and variations between different plant species investigated. Anderson and Walton (1995) reported up to 20 percent of the [^{14}C] incorporated in plant tissue, with the highest percentage in roots. Whether the radiolabel was TCE or metabolites was not determined. Loblolly pine (*Pinus taeda*) was found to show the highest rate of uptake and mineralization of TCE. This variation in TCE uptake was mirrored in a recent study. Vroblesky *et al.* (1999) looked for TCE in tree trunks on a contaminated site, where no TCE was detected in the ambient air. They found that loblolly pine (*Pinus taeda*) and two other coniferous tree species had comparable levels of TCE in the trunk, whereas nearby oak (*Quercus* spp.) and sweet gum (*Liquidambar styraciflua*) trees contained TCE levels below or just above the detection limit. These authors proposed that coniferous trees might have had higher capacity for TCE uptake than other trees, through different rhizosphere interactions and different water use physiology.

From the perspective of field applications, transpiration may pose air contamination problems, and uptake in tissues presents possible contamination of the food chain. However, Davis *et al.* (1998a) and Davis *et al.* this book) pointed out that buildup of transpired TCE in air is unlikely because of rapid mixing and dilution. In the controlled field study by Newman *et al.* (1999), TCE could not be detected in the atmosphere around the canopy, and after the second year, not in the air directly surrounding the leaves either. As implied previously in this section, TCE does not appear to accumulate significantly in most plant tissues. Nevertheless, because differences between plant species have been observed, candidate species intended for phytoremediation must be tested for possible transpiration and tissue accumulation of TCE.

Transformation of Radiolabeled Trichloroethylene

Radiolabeled TCE has been used as a tracer to determine the fate of TCE in sealed environments (Table 17-1). The resolution of these experiments to differentiate between various forms of the radiolabel is limited. Generally, volatile chemicals are purged from the system and trapped or sorbed. The carbon dioxide that passes through these traps is subsequently trapped with an alkaline solution. The [^{14}C] in the volatile components is generally assumed to be unchanged TCE. This has been in some cases confirmed by extracting the trapped volatile compounds for gas chromatographic analysis. To our knowledge, no volatile metabolites of TCE have been found in the headspace of plant foliage in bench studies. Nzengung and Jeffers (2001), however, did find *trans*-dichloroethylene and ethene in the headspace of a rhizosphere solution. Nonvolatile forms of [^{14}C] in the tissue are assumed to be metabolites, but may also consist of TCE tightly bound to plant tissue.

Various mass balance studies using [^{14}C]-TCE are mostly in agreement on the fate of TCE. More than 75 percent of TCE is unaltered over the experimental duration (7 to 32 days). All three studies using hydroponic hybrid poplar (*Populus deltoides* × *Populus nigra* 'DN34') or hybrid poplar (*Populus trichocarpa* × *Populus deltoides* 'H-11-11') tumor cells (Newman *et al.* 1997, Gordon *et al.* 1998, Orchard *et al.* 2000b) found 1 to 2 percent [^{14}C] transformation to carbon dioxide, showing that mineralization of TCE occurs in plants. Most studies found less than 3 percent of the [^{14}C] incorporated in plant tissue (Newman *et al.* 1997, Schnabel *et al.* 1997, Burken and Schnoor 1998, Orchard *et al.* 2000b), and that most of the radiolabel cannot be extracted by solvents (Schroll *et al.* 1994, Schnabel *et al.* 1997, Orchard *et al.* 2000b).

Studies by Anderson and Walton (1995) found a much higher transformation of [^{14}C]-TCE than the other studies. Tissue incorporation was as high as 20 percent in loblolly pine (*Pinus taeda*) and soybean (*Glycine max*) plants. However, the other plants had levels of tissue incorporation similar to studies using poplar (*Populus* spp.) trees and garden plants. Although most results from radiolabel studies seem to suggest low levels of transformation, caution is warranted because in these studies much of the TCE can be expected to volatilize into the headspace or be stripped from the root zone by airflow.

Identification and Characterization of Metabolites

Various oxidative metabolites of TCE, similar to the products of mammalian and bacterial enzymes, have been found in plants. Table 17-3 summarizes the results from various studies where metabolites were detected. Newman *et al.* (1997) Newman *et al.* (1999) and Orchard *et al.* (2000b) reported trichloroethanol, trichloroacetic acid, and dichloroacetic acid in hydroponically grown poplar (*Populus* spp.) as well as in soil-grown poplar (*Populus* spp.) exposed to TCE. The levels reported are in the same range. Newman *et al.* (1997) also found the same metabolites in poplar (*Populus trichocarpa* × *Populus deltoides* 'H-11-11') tumor cell cultures, as did Doucette *et al.* (1998) in plant tissue collected at a TCE-contaminated site. In tobacco (*Nicotiana tabacum* var. *xanthi* n. c.) plants, the major metabolite of TCE is trichloroethanol, which exists both in the free form and as a sugar conjugate

TABLE 17-3 Known Metabolites of Trichloroethylene in Various Plants

Metabolites	Plants containing metabolites and the growing conditions	Highest concentration reported (milligrams per kilogram)[a]	Investigator	Notes[b]
Trichloroethanol	Live oak (*Quercus virginiana*), saw palmetto (*Serenoa repens*), and castor bean (*Ricinus communis*), indigenous vegetation	0.3 (dry weight)	Doucette *et al.* (1998)	GC–ECD
	Willow (*Salix nigra*), indigenous	1.57	Nzengung and Jeffers (2001)	GC–ECD and GC–MS
	Hybrid poplar (*Populus deltoides* × *Populus trichocarpa* 'H-11-11'), field test	0.56	Newman *et al.* (1999)	GC–ECD and GC–MS
	Tobacco (*Nicotiana tabacum* var. *xanthi*, n. c.), hydroponic	20	Doty *et al.* (2000)	GC–ECD and GC–MS
	Hybrid poplar (*Populus deltoides* × *Populus trichocarpa* 'H-11-11') callus cells, suspension	0.2	Newman *et al.* (1997)	GC–ECD and GC–MS
Trichloroacetic acid	Live oak (*Quercus virginiana*), saw palmetto (*Serenoa repens*), and castor bean (*Ricinus communis*), indigenous vegetation	1 (dry weight)	Doucette *et al.* (1998)	GC–ECD
	Willow (*Salix nigra*), indigenous	31.34	Nzengung and Jeffers (2001)	GC–ECD and GC–MS
	Hybrid poplar (*Populus deltoides* × *Populus trichocarpa* H-11-11) trees, field test	40	Newman *et al.* (1999)	GC–ECD and GC–MS
	Hybrid poplar (*Populus deltoides* × *Populus trichocarpa* 'H-11-11') callus cells, suspension	7.2	Newman *et al.* (1997)	GC–ECD and GC–MS
Dichloroacetic acid	Live oak (*Quercus virginiana*), indigenous vegetation	0.723 (dry weight)	Doucette *et al.* (1998)	GC–ECD
	Cottonwood (*Populus deltoides*) tree, indigenous	2.32	Nzengung and Jeffers (2001)	GC–ECD and GC–MS
	Hybrid poplar (*Populus deltoides* × *Populus trichocarpa* 'H-11-11') trees, field test	19	Newman *et al.* (1999)	GC–ECD

(*continues*)

TABLE 17-3 (*continued*)

Metabolite	Plants containing metabolites and the growing conditions	Highest concentration reported (milligrams per kilogram)[a]	Investigator	Notes[b]
	Hybrid poplar (*Populus deltoides* × *Populus trichocarpa* 'H-11-11') trees, green house	0.3	Newman *et al.* (1997)	GC–ECD and GC–MS
	Hybrid poplar (*Populus deltoides* × *Populus nigra*, 'DN34') cuttings, hydroponic	10 (dry weight)	Orchard *et al.* (2000)	GC–MS
	Hybrid poplar (*Populus deltoides* × *Populus trichocarpa* 'H-11-11') callus cells, suspension	39	Newman *et al.* (1997)	GC–ECD and GC–MS
Trichloroethanol glycoside	Tobacco (*Nicotiana tabacum* var. *xanthi*, n. c.), hydroponic; and poplar (*Populus deltoides* × *Populus trichocarpa* H-11-11) callus cells, suspension	30	Shang *et al.* (2001)	GC–MS
Carbon dioxide[c]	Hybrid Poplar (*Populus deltoides* × *Populus nigra*, 'DN34') cuttings, hydroponic	Less than 1 percent of total TCE added	Orchard *et al.* (2000)	LS, not confirmed
	Hybrid Poplar (*Populus deltoides* × *Populus trichocarpa* 'H-11-11') callus cells, suspension	1.5 percent of total TCE added	Newman *et al.* (1997)	LS
Chloride ion	Hybrid Poplar (*Populus deltoides* × *Populus trichocarpa* 'H-11-11') trees, field test	65 percent of total TCE removed	Newman *et al.* (1999)	Found in soil
	Alfalfa (*Medicago sativa*), laboratory channels	17 percent of total TCE removed	Narayanan *et al.* (1995)	Found in soil water

[a]On a wet weight basis unless otherwise noted.
[b]Abbreviation used—GC: gas chromatography; ECD: electron capture detector; MS: mass spectroscopy; and LS: liquid scintillation.
[c]Anderson and Walton (1995) detected carbon dioxide from [^{14}C]-TCE in flasks containing soils with various plants. In that study, the soil–root compartment was sealed from the rest of the plant, and the only carbon dioxide measured was metabolism in the soil–root compartment. They concluded that the carbon dioxide formation was because of the rhizosphere effect and not attributed to plants due to metabolism. It cannot be ruled out, however, that plant roots contributed to the degradation and mineralization of TCE.

(Shang *et al.* 2001). In hybrid poplar (*Populus trichocarpa* × *Populus deltoides* 'H-11-11') trees and cell cultures as well as in a tropical legume, (*Leuceana leucocephala*), trichloroethanol was also found to be highly glycosylated. The identities of trichloroethanol, trichloroacetic acid, dichloroacetic acid, and trichloroethanol glycoside have been confirmed by gas chromatography–mass spectroscopy. The fact that these oxidative metabolites have been found in comparable levels in axenic cell cultures, hydroponically grown plants, and soil-grown plants confirm that these are the products of plant metabolism, not of microbial activity.

Both in the controlled field study (Newman *et al.* 1999) and greenhouse experiments (Newman *et al.* 1997) with soil-grown plants, the major metabolite was trichloroacetic acid, with highest levels present in leaves. However, in the cell culture study (Newman *et al.* 1997) and in hydroponically grown hybrid poplars (*Populus deltoides* × *Populus nigra*, 'DN34') by Orchard *et al.* (2000b) as well as in mature cottonwood (*Populus deltoides*) trees growing at a TCE-contaminated field site (Nzengung and Jeffers 2001), the primary metabolite was dichloroacetic acid. It is not clear whether this difference is due to the physiology of soil-grown trees *versus* axenic plants and cultures, methods and duration of exposure, or some other factors. Despite the difference in metabolite production, the overall amounts of these chlorinated oxidative metabolites were low in plant tissue (see Table 17-3). The highest level of a chlorinated oxidative metabolite was about 40 milligrams of trichloroacetic acid per kilogram wet weight of plant tissue. In the field study by Newman *et al.* (1999), the total amount of oxidative metabolites in plant tissue constituted less than 0.04 percent of the TCE removed from the planted containers. Orchard *et al.* (2000b) confirmed that oxidative metabolites represented only a small fraction of the [^{14}C] incorporated into plant tissue in hydroponic plants exposed to TCE.

In contrast to the small amounts of chlorinated metabolites located in plant tissue, Newman *et al.* (1999) found significantly elevated levels of chloride ions in soils that were planted and exposed to TCE, compared to the soils that were unplanted and exposed to the same amounts of TCE. The increased chloride in the soil accounted for 65 percent of total chlorine from the TCE removed by the trees. The chloride concentration was highest 0.4 to 0.6 meters below the ground surface, near the effluent withdrawal well. Narayanan *et al.* (1995) also reported elevated chloride in soils planted with alfalfa (*Medicago sativa*) and exposed to TCE through simulated groundwater, although the concentration was only significantly elevated in the top layer. The increased chloride accounted for an estimated 17 percent of the TCE removed.

Many hypotheses exist to explain the removal and degradation of TCE in planted systems. Narayanan *et al.* (1995) proposed a 30 percent TCE removal by reductive degradation of TCE in the groundwater under anaerobic and methanogenic conditions, supported by the detection of methane in the water drained from the test channels. They also proposed that the remaining 70 percent of removed TCE is drawn from groundwater through plant transpiration into the vadose zone, where soil microorganisms, possibly methanotrophs, aerobically degraded the contaminant. No methane was detected in the test chamber headspace. By contrast, Newman *et al.* (1999) proposed that some TCE was taken up by plant roots, degraded in plant tissue, and the excess chloride excreted to the soil. The chlorinated metabolites may be transient intermediates of the degradation process undergoing constant turnover. Newman *et al.* (1999) argued that active plant uptake and metabolism, rather than enhanced bacterial degradation, was the principal mechanism of TCE removal, because the rhizosphere soil did not show enhanced mineralization of TCE in lab studies. However, the lab studies in closed chambers may not fully represent critical field conditions, as is evident by the low rate of mineralization by plants in such chambers compared to field observations. Therefore, it is still unclear whether plant metabolism or enhanced microbial activity may be the principal mechanism for TCE removal in closed chambers. Whatever the mechanism, the nearly complete removal of TCE from simulated groundwater was only observed in planted compartments, compared to only about 30 percent removal in unplanted ones. Therefore, whether or not most of the TCE transformed, the presence of trees was essential for the enhanced removal.

Nzengung and Jeffers (2001) found anaerobic, reductive metabolites of TCE (*trans*-dichloroethylene and ethane) in the headspace of the rhizosphere solution of hydroponic plants. Newman *et al.* (1999) found *cis*-dichloroethylene in the simulated groundwater in the controlled field study, and Vroblesky *et al.* (1999) found *cis*-dichloroethylene and vinyl chloride in the root zone of the trees at a field site. The reductive metabolites were not found in plant tissue, and could be the products of anaerobic, microbial activity in the rhizosphere in these studies. However, see Nzengung *et al.* (this book).

Some of the oxidative and reductive metabolites of TCE are toxic. Temporary accumulation of TCE metabolites in plant tissue could potentially introduce food chain contamination problems. In poplar (*Populus* spp.) trees, the levels of metabolite found were low, in the milligrams per liter range in most cases. Furthermore, no toxicity was found when TCE-treated poplar (*Populus* spp). tissue was fed to soil insects (Sorbet 1998). In tobacco (*Nicotiana tabacum* var. *xanthi*, n.c.) once the source of TCE was removed, trichloroethanol disappeared from plant tissue in about 2 weeks (Doty *et al.* 2000). When significant accumulation of metabolites is observed in a plant intended for phytoremediation (Nzengung and Jeffers 2001), pilot studies are necessary to establish the stability and persistence of such metabolites, and to investigate bioaccumulation and ecological risk.

Trichloroethylene Metabolism in the Context of the "Green-Liver" Concept

The "green-liver" concept refers to the similarity between plant and animal transformation of xenobiotic contaminants. Similar to xenobiotic metabolism processes in the liver, organic chemicals introduced into plants are modified

TABLE 17-4 Parallels between Mammalian and Plant Metabolism of Trichloroethylene

	Mammalian	Plant
Phase I: oxidation	Trichloroethanol, trichloroacetic acid, and dichloroacetic acid	Trichloroethanol, trichloroacetic acid, and dichloroacetic acid
Phase II: conjugation	Trichloroethanol glucuronic acid conjugate	Trichloroethanol glycoside
Phase III: secretion	Macromolecular adducts and urine secretion (no counterpart in plants)	Bound residue

in three stages. The stages include: phase I activation, which involves the introduction of a functional group, often through oxidation; phase II conjugation of the parent compound or the phase I-activated compound to various native compounds such as glutathione or glucose; and phase III, which involves compartmentation in the vacuole or further processing into bound residue (Coleman *et al.* 1997).

Although TCE metabolic pathways in all plants have not been completely delineated, all three phases of transformation are evident. Trichloroethanol and trichloroacetic acid are oxidative products of TCE, trichloroethanol is conjugated to glucose, and bound residues originating from TCE have been reported (Nzengung and Jeffers 2001, Shang *et al.* 2002). As shown in Table 17-4, all three stages have equivalents in the mammalian hepatic pathways, making the "green-liver" description particularly apt for TCE.

FIELD PROOF OF CONCEPT FOR THE PHYTOREMEDIATION OF TRICHLOROETHYLENE

The ultimate proof of phytoremediation as a tool for the cleanup of sites contaminated with chlorinated solvents is a successful field application. Unfortunately, the application of almost any technology to the cleanup of contaminated groundwater can take years, if not decades, to accomplish, and phytoremediation takes longer than most to confirm the proof of principle. An additional analytical complication is that there are relatively few sites that are contaminated only with TCE; most sites are contaminated with mixtures of compounds. Even on those sites that were originally contaminated over time, the TCE has degraded into biotic and abiotic by-products.

Field studies that are currently under way to achieve proof of principle are given in Eberts *et al.* (this book) and Hirsh *et al.* (this book). Multi-year monitoring is needed to determine the success at each unique site. Although these studies may require several years before providing conclusive information, results to date are promising.

The artificial aquifer system (see the section "Field Studies") provided a controlled environment that mimics field conditions. This study represents the important proof of concept for phytoremediation of TCE in a field setting. Over 98 percent of the TCE injected into the simulated groundwater layer was removed from the aquifer system, with less than 8 percent being transpired over the first two growing seasons. After the second year, no TCE was measured in the transpired vapor. Tissue analysis showed the same oxidative metabolites and similar concentrations as seen in the laboratory studies (Newman *et al.* 1997, Newman *et al.* 1999). These investigations are a vital link between the laboratory (Newman *et al.* 1997, Gordon *et al.* 1998, Orchard *et al.* 2000b) and the field (Newman *et al.* 1999). In the field proof of concept tests, the trees removed almost all of the water from the aquifer system when the water-application rates were less than the plant water uptake rates. The trees also removed nearly 100 percent of the contaminants. When water application rates exceeded plant-uptake rates, the removal of contaminants was significantly lower. This establishes that the hydrology of the site and the physiology of the plants must be fully understood before a successful application can take place.

REFERENCES

Anderson, T.A. and B.T. Walton (1995) Comparative fate of [^{14}C] trichloroethylene in the root zone of plants from a former solvent disposal site. *Environ. Toxicol. Chem.* **14**(12): 2041–2047.

Burken, J.G. and J.L. Schnoor (1998) Predictive relationships for uptake of organic contaminants by hybrid poplar trees. *Environ. Sci. Technol.* **32**: 3379–3385.

Cai, H. and F.P. Guengerich (2001) Reaction of trichloroethylene oxide with proteins and DNA: instability of adducts and modulation of functions. *Chem. Res. Toxicol.* **14**: 54–61.

Coleman, J.O.D., M.M.A. Blake-Kalff, and T.G.E. Davies (1997) Detoxification of xenobiotics by plants: chemical modification and vacuolar compartmentation. *Trends Plant Sci.* **2**(4): 144–151.

Cornejo, J.J., F.G. Munoz, C.Y. Ma, and A.J. Stewart (1999) Studies on the decontamination of air by plants. *Ecotoxicology* **8**: 311–320.

Davis, L.C., S. Vanderhoof, J. Dana, K. Selk, K. Smith, G. Goplen, and L.E. Erickson (1998a) Movement of chlorinated solvents and other volatile organics through plants monitored by Fourier transform infrared (FT-IR) spectrometry. *J. Hazard. Subs. Res.* **1**(4): 1–26.

Davis, L.C., S. Vanderhoof, and D. Lupher (1998b) Sorption of trichloroethylene (TCE) and other halogenated aliphatics on plant materials. In: *Proc. 1998 Conf. Hazard. Waste Res.* L.E. Erickson and M.M. Rankin, eds. U.S. Environmental Protection Agency Great Plains/Rocky Mountains Hazardous Substance Research Center, Kansas State University, Manhattan, Kansas. Held at Snowbird, Utah, May 18–21. pp. 319–325.

Davis, L.C., D. Lupher, J. Hu, and L.E. Erickson (1999) Transport of trichloroethylene through living plant tissues. In: *Proc. 1999 Conf. Hazard. Waste Res.* L.E. Erickson and M.M. Rankin, eds. U.S. Environmental Protection Agency Great Plains/Rocky Mountains Hazardous Substance Research Center, Kansas State University, Manhattan, Kansas. Held at St. Louis, Missouri, May 24–27. pp. 203–209.

Dickmann, D.I. and J.G. Isebrands (1999) *Caveat emptor. Am. Nurseryman* **189**(5): 61–65.

Doty, S.L., T.Q. Shang, A.M. Wilson, J. Tangen, A.D. Westergreen, L.A. Newman, S.E. Strand, and M.P. Gordon (2000) Enhanced metabolism of halogenated hydrocarbons in transgenic plants containing mammalian cytochrome P-450 2E1. *Proc. Nat. Acad. Sci. USA* **97**(**12**): 6287–6291.

Doucette, W.J., B. Bugbee, S. Hayhurst, W.A. Plaehn, D.C. Downey, S.A. Taffinder, and R. Edwards (1998) Phytoremediation of dissolved-phase trichloroethylene using mature vegetation. In: *Bioremediation and Phytoremediation: Chlorinated and Recalcitrant Compounds.* G.B. Wickramanayake and R.E. Hinchee, eds. Battelle Press, Columbus, Ohio. Held in Monterey, California, May 18–21. pp. 251–256.

Ellis, L.B.M., C.D. Hershberger, E.M. Bryan, and L.P. Wackett (2001) The University of Minnesota Biocatalysis/Biodegradation Database: emphasizing enzymes. *Nucleic Acids Res.* **29**: 340–343 (http://umbbd.ahc.umn.edu/).

Ensley, B.D. (1991) Biochemical diversity of trichloroethylene metabolism. In: *Annual Review of Microbiology* **45**: 283–299.

Gordon, M., N. Choe, J. Duffy, G. Ekuan, P. Heilman, I. Muiznieks, M. Ruszaj, B.B. Shurtleff, S. Strand, J. Wilmoth, and L.A. Newman (1998) Phytoremediation of trichloroethylene with hybrid poplars. *Environ. Health Persp.* **106**(supplement 4): 1001–1004.

Jordahl, J.L., L. Foster, J.L. Schnoor, and P.J.J. Alvarez (1997) Effect of hybrid poplar trees on microbial populations important to hazardous waste bioremediation. *Environ. Toxicol. Chem.* **16**(6): 1318–1321.

Lash, L.H., J.W. Fisher, J.C. Lipscomb, and J.C. Parker (2000) Metabolism of trichloroethylene. *Environ. Health Persp.* **108**(supplement 2): 177–200.

Narayanan, M., L.C. Davis, and L.E. Erickson (1995) Fate of volatile chlorinated organic compounds in a laboratory chamber with alfalfa plants. *Environ. Sci. Technol.* **29**(9): 2437–2444.

Newman, L.A., S.E. Strand, N. Choe, J. Duffy, G. Ekuan, M. Ruszaj, B.B. Shurtleff, J. Wilmoth, P. Heilman, and M.P. Gordon (1997) Uptake and biotransformation of trichloroethylene by hybrid poplars. *Environ. Sci. Technol.* **31**(4): 1062–1067.

Newman L.A., X. Wang, I.A. Muiznieks, G. Ekuan, M. Ruszaj, R. Cortellucci, D. Domroes, G. Karscig, T. Newman, R.S. Crampton, R.A. Hashmonay, M.G. Yost, P.E. Heilman, J. Duffy, M.P. Gordon, and S.E. Strand (1999) Remediation of TCE in an artificial aquifer with trees: a controlled field study. *Environ. Sci. Technol.* **33**(13): 2257–2265.

Nzengung, V.A. and P. Jeffers (2001) Sequestration, phytoreduction, and phytooxidation of halogenated organic chemicals by aquatic and terrestrial plants. *Int. J. Phytoremed.* **3**(1): 13–40.

Nzengung, V.A., C. Wang, and S. Box (2001) Phytotransformation pathways and mass balances for chlorinated alkanes and alkenes. In: *Proceedings: EPA*

Phytoremediation State of the Science Conference. U.S. Environmental Protection Agency Report EPA/625/R-01/011a and b, Cincinnati, Ohio. Held in Boston Massachusetts, May 1–2, 2000. (http://www.epa.gov/ORD/NRMRL/Pubs/625R01011b/625R01011bchap13.pdf).

Orchard, B.J., W.J. Doucette, J.K. Chard, and B. Bugbee (2000a) A novel laboratory system for determining fate of volatile organic compounds in planted systems. *Environ. Toxicol. Chem.* **19**(4): 888–894.

Orchard, B.J., W.J. Doucette, J.K. Chard, and B. Bugbee (2000b) Uptake of TCE by hybrid poplar trees grown hydroponically in flow through plant growth chambers. *Environ. Toxicol. Chem.* **19**(4): 895–903.

Schnabel, W.E., A.C. Dietz, J.G. Burken, J.L. Schnoor, and P.J. Alvarez (1997) Uptake and transformation of trichloroethylene by edible garden plants. *Water Res.* **31**(4): 816–824.

Schroll, R., B. Bierling, G. Cao, U. Dorfler, M. Lahaniati, T. Langenbach, I. Scheunert, and R. Winkler (1994) Uptake pathways of organic chemicals from soil by agricultural plants. *Chemosphere* **28**(2): 297–303.

Schwarzenbach, R.P., P.M. Gschwend, and D.M. Imboden (1993) *Environ. Org. Chem.* John Wiley, New York, p. 681.

Shang, T.Q., S.L. Doty, A.M. Wilson, W.N. Howald, and M.P. Gordon (2001) Trichloroethylene oxidative metabolism in plants: the trichloroethanol pathway. *Phytochemistry* **58**(7): 1055–1065.

Sheremata, T.W., R.N. Yong, and S.R. Guiot (1997) Simulation and sterilization of a surrogate soil organic matter for the study of the fate of trichloroethylene in soil. *Comm. Soil Sci. Plant Anal.* **28**(13 and 14): 1177–1190.

Sorbet, M. (1998) Impact of contaminated litter from a TCE phytoremediation project and of major metabolites of TCE on terrestrial isopods and aquatic amphipods. Masters thesis. Clemson University, Clemson, South Carolina.

Vroblesky, D.A., C.T. Nietch, and J.T. Morris (1999) Chlorinated ethenes from groundwater in tree trunks. *Environ. Sci. Technol.* **33**: 510–515.

Wackett, L.P. (1994) Bacterial cometabolism of halogenated organic compounds. In: *Microbial Transformation and Degradation of Toxic Organic Chemicals*. L.Y. Young and C. Cerniglia, eds. John Wiley, New York.

Wolfenbarger L.L. and P.R. Phifer (2000) The ecological risks and benefits of genetically engineered plants. *Science* **290**: 2088–2093.

Wolverton, B.C., A. Johnson, and K. Bounds (1989) Interior landscape plants for indoor air pollution abatement. Final Report, National Aeronautics and Space Administration, John C. Stennis Space Center, Mississippi, September 15.

18

UPTAKE, METABOLISM, AND PHYTOVOLATILIZATION OF TRICHLOROETHYLENE BY INDIGENOUS VEGETATION: IMPACT OF PRECIPITATION

W. J. Doucette, B. G. Bugbee, S. C. Smith, C. J. Pajak, and J. S. Ginn

SUMMARY OF PRACTICAL IMPLICATIONS

Field studies in Florida and Utah showed that indigenous vegetation contributed to the natural attenuation of trichloroethylene (TCE) in shallow groundwater plumes. However, despite similar exposure to TCE, plant uptake appeared greater at the Utah site where summer precipitation is minimal. Trichloroethylene was detected in all exposed plants at both sites, but plant concentrations were 10 to 100 times higher at the Utah site. Plant metabolites of TCE were also detected. Using a novel sampling technique, TCE was identified in transpiration samples collected at the Utah site, but not the Florida site. The lack of phytovolatilization and significantly lower TCE plant concentrations at the Florida site are most likely due to the smaller fraction of contaminated groundwater used by plants for transpiration because of the frequent precipitation. However, additional studies are necessary before definite conclusions can be drawn regarding the influence of climate on plant uptake. The impact of plants relative to other attenuation mechanisms was not directly evaluated at these sites and should be addressed in future studies.

Phytoremediation: Trnasformation and Control of Contaminants,
Edited by Steven C. McCutcheon and Jerald L. Schnoor.
ISBN 0-471-39435-1 (cloth)

INTRODUCTION

Impact of Vegetation on Trichloroethylene in Soil and Shallow Groundwater

Plants have profound effects on physical, chemical, and biological processes in soils and can significantly impact the fate of organic chemicals in soil. Understanding the fate of organic contaminants in plants and soils is critical in evaluating the effectiveness of phytoremediation, performing risk assessments, and quantifying the relative impact of plants on natural attenuation.

Increased microbial and chemical activity occurs on and around root surfaces and plants can take up, metabolize, and transpire organic chemicals. These activities are impacted by the extent of water that flows to the root surface *via* transpiration from plant leaves. If the water flow to the root is large enough, hydraulic control of contaminant plumes occurs. Plants can also come into direct contact with contaminants because of extensive root systems. In addition, root exudates may enhance cometabolic degradation in the rhizosphere. Laboratory studies using TCE have identified enhanced rhizosphere degradation, uptake and transpiration (phytovolatilization), and uptake and metabolism as potential plant mediated processes (Walton and Anderson 1990, Schroll *et al.* 1994, Anderson and Walton 1995, Narayanan *et al.* 1995, Gordon *et al.* 1997, Newman *et al.* 1997, Schnabel *et al.* 1997, Burken and Schnoor 1998, Orchard *et al.* 2000a, 2000b).

While the literature is generally consistent regarding the identification of phytoremediation processes, a consensus has not been reached about the magnitude of these processes and the effect on the overall fate of TCE. The lack of consensus is especially noticeable in describing the extent of plant uptake and phytovolatilization. For example, minimal root uptake and phytovolatilization of TCE have been reported in some studies (Schroll *et al.* 1994, Schnabel *et al.* 1997, Orchard *et al.* 2000a, 2000b) while extensive uptake and phytovolatilization have been described by others (Burken and Schnoor 1998).

Pilot data (Newman *et al.* 1997, Shang *et al.* this book) and field-scale data (Compton *et al.* 1998, Jones *et al.* 1998, Eberts *et al.* this book, Hirsh *et al.* this book) generally support the mechanistic observations reported in laboratory studies. However, most of these studies examined young trees that had been planted recently. Fewer studies have focused on the potential impact of mature vegetation growing within contaminated groundwater plumes. Yet this information is necessary to predict the long-term effectiveness of phytoremediation, determine the potential impact of vegetation on the overall natural attenuation process, and evaluate potential food-chain contamination.

The objective of this chapter is to compare the uptake, metabolism, and phytovolatilization of TCE by indigenous vegetation growing over contaminated groundwater at two sites that vary significantly in precipitation received. The two sites, located at Cape Canaveral Air Station (CCAS), Florida, and Hill Air Force Base (Hill AFB), Utah, were similar in groundwater TCE concentration, depth from the ground surface to the water table, and in the relative duration of exposure to contaminants. These sites are, however, very different in climate, especially in the amount and frequency of precipitation. This difference in precipitation was used to examine the hypothesis that trees growing in semi-arid regions are more likely to obtain a greater fraction of water requirements from contaminated groundwater resulting in a higher uptake of TCE. Unique procedures for the collection and analysis of phytovolatilization and plant tissue samples were developed during these studies and will also be described.

Organic Contaminant Uptake: Transpiration Stream Concentration Factor

Because of commercial interests, much of the plant uptake data and related theory for organic xenobiotic chemicals have been generated for pesticides. Much less information is available for other industrial organic chemicals. Plasma-membrane transporters mediating the uptake of sugar, amino acid, peptide, glutathione, and other biotic chemicals have been identified (*e.g.*, Delrot *et al.* 2001). However, the uptake of xenobiotic organic compounds by plants is believed to be a passive process (McFarlane 1995) related, at least in part, to the lipophilicity of the contaminant (Briggs *et al.* 1982) as described by the octanol–water partition coefficient (K_{ow}).

The transpiration stream concentration factor (*TSCF*) has been widely used to describe organic contaminant uptake by plants. The *TSCF* is a dimensionless ratio of the concentration in the xylem sap to bulk concentration in the root-zone solution (Russell and Shorrocks 1959)

$$TSCF = \frac{\text{Concentration in xylem sap}}{\text{Root-zone solution concentration}} \tag{18-1}$$

Because xylem sap concentrations are difficult to measure directly for intact plants, *TSCF* is often determined from measured shoot concentrations; the shoot-tissue concentration is normalized to the amount of water transpired during exposure to the chemical (*i.e.*, concentration in xylem sap is equal to milligrams of compound in shoots per liter of water transpired) assuming no phytovolatilization of the chemical. Almost all experimental *TSCF* values have been obtained in the laboratory where the root zone concentrations can be more easily measured and controlled. Accurate *TSCF* values are extremely difficult to determine in the field because of the inherent variability associated with root zone exposure (*i.e.*, soil and groundwater heterogeneity, variability in contaminant concentration, and groundwater *versus* surface water use).

A compound may be actively or passively taken up by plants or may be excluded. Active uptake (*TSCF* greater than 1) generally occurs with nutrient cations (NH_4^+, PO_4^+, and K^+) and requires the expenditure of metabolic energy. With the possible exception of some hormone-like chemicals (2,4-dichlorophenoxyacetic acid or 2,4-D), there is no evidence of active uptake of anthropogenic chemicals (McFarlane 1995). Passive uptake (*TSCF* equal to 1) occurs when a chemical is taken up directly with water because of the gradient of water potential resulting from evapotranspiration (McFarlane 1995). A chemical is said to be excluded (*TSCF* less than 1) when uptake is not directly proportional (1:1) to water uptake, although the mechanism of uptake is still thought to be a passive process. It has been suggested that the passive uptake of organic contaminants may be driven by the concentration gradient formed as the solute accumulates on the outside of the root membrane (Marschner 1995). However, factors such as membrane permeability and xylem sap solubility of the contaminant may limit the extent or kinetics of passive uptake (Hsu *et al.* 1990). Sorption and rapid metabolism of contaminants within the tree would also act to reduce xylem concentrations and keep *TSCF* values from reaching 1.

One of the most widely cited studies involving the measurement and prediction of *TSCF* was performed by Briggs *et al.* (1982) who evaluated the uptake and translocation of two series of nonionic organic chemicals (*O*-methylcarbamoyloximes and substituted phenylureas) by 10-day-old intact barley (*Hordeum vulgare*) plants. All the chemicals having a value of the logarithm (log) of K_{ow} ranging from −0.57 to 4.6 were found to have a *TSCF* less than 1 and it was concluded that the chemicals moved passively into the shoot with the transpiration water (Briggs *et al.* 1982). The bell-shaped data relating *TSCF* to the log K_{ow} were fit using a Gaussian curve

$$TSCF = 0.784 \exp - \frac{(\log K_{ow} - 1.78)^2}{2.44} \qquad (18\text{-}2)$$

suggested an optimal lipophilicity for maximum uptake and translocation at log $K_{ow} = 1.8$ ($TSCF = 0.784$). The Gaussian fit inferred that compounds that were either highly polar (log K_{ow} less than 0.5) or highly lipophilic (log K_{ow} greater than 4.5) would not be significantly taken up by plants. The highly polar organic compounds were thought to be selectively rejected at the root lipid membranes, while the highly lipophilic molecules were bound to the roots and not translocated. Based on the curve, only organic chemicals of intermediate lipophilicity (log K_{ow} between 0.5 and 4.5) are expected to pass to the xylem unhindered upon establishment of equilibrium with the root tissues. More recently, similar relationships between *TSCF* and log K_{ow} have been reported by Hsu *et al.* (1990), Sicbaldi *et al.* (1997), and Burken and Schnoor (1998).

While the empirical relationships for uptake and translocation presented by Briggs *et al.* (1982) and others are useful for initial estimates, numerous examples involve significant over or under prediction of plant uptake. For example, McFarlane *et al.* (1987) found *TSCF* values lower than predicted for three chemicals having similar log K_{ow} values, while significant uptake and translocation of a water miscible compound (1,4-dioxane) has been reported (Aitchison *et al.* 2000). In addition, for a given compound and $\log K_{ow}$, several different *TSCF* values can often be found in the literature. Indeed, Briggs *et al.* (1982) suggest that it is a "gross oversimplification" to conclude that plant uptake and translocation are influenced only by reversible partitioning of the compound into the nonaqueous phases of the root.

Estimating the Uptake of Trichloroethylene by Plants on a Field Scale

The significance of plant uptake in the removal of TCE from contaminated groundwater is still unclear, especially considering the wide range of *TSCF* values (0.02 to 0.75) that have been reported in the literature for TCE (*e.g.*, Burken and Schnoor 1998, Orchard *et al.* 2000b). The following simplified illustration highlights the critical variables involved in estimating TCE uptake by plants on a field scale.

Annual TCE uptake from a shallow aquifer per unit area per year can be estimated as follows

$$\text{Mass of TCE removed by plant uptake} = (TSCF)(C_{TCE})(T)(f) \qquad (18\text{-}3)$$

where *TSCF* is assumed to be constant, C_{TCE} is the average groundwater concentration of TCE (milligrams per liter), T is the cumulative volume of water transpired per unit area per year (liters per square meter per year), and f is the fraction of the plant water needs met by contaminated groundwater. This expression assumes that C_{TCE} is constant. A more realistic calculation would incorporate the change in C_{TCE} over time as a function of the physical, chemical, or biological processes that may be occurring and possible changes in *TSCF* with exposure concentration.

Transpiration rates in the field vary widely depending on the soil-water availability and evaporative demand. Potential transpiration rates, calculated from pan evaporation rates, are widely used to schedule irrigation of crops. The potential transpiration rate can be 10 liters per square meter per day on hot days in well-watered soils in dry climates, but the cumulative annual transpiration rate is more useful in long-term phytoremediation calculations. The potential annual transpiration rate can be as high as 1800 liters per square meter per year in hot desert climates such as Arizona, and as low as 200 liters per square meter per year in cool, moist environments like Alaska (Camp *et al.* 1996, Allen *et al.* 1998). However, even well-watered crops can fail to attain the potential transpiration rate in the summer because of partial stomatal closure during periods of high evaporative demand. The difference between the actual and potential transpiration rates is larger in hot, desert environments. During winter months, deciduous trees drop leaves and evergreen trees have low transpiration rates as the result of shorter days, lower light levels, and colder temperatures. When forced to use groundwater, phreatophytic plants typically do not achieve the high transpiration rates that occur with vegetation that uses surface water (Camp *et al.* 1996, Allen *et al.* 1998). Thus, the actual annual transpiration rate is usually below the potential rate. Depending on the climate, 200 to 1400 liters per square meter per year probably represents a reasonable range of values for annual transpiration (also see McCutcheon and Schnoor this book). In a recent review of 52 water-use studies since 1970, Wullschleger *et al.* (1998) found that 90 percet of the observations for maximum rates of daily water use were between 10 and 200 liters per day for individual trees that averaged 21 meters (70 feet) in height.

The fraction of groundwater used by plant roots is difficult to measure and is poorly characterized. As expected, groundwater use tends to decrease as the availability of surface water increases (Nilsen and Orcutt 1996). Additional studies using stable isotope techniques (Nilsen and Orcutt 1996) are necessary to determine a reasonable range of values for this parameter. Until such data are available, a range of groundwater use fractions from 0.1 to 0.5 is probably realistic for climates with more than 40 centimeters (16 inches) of precipitation per year.

Using a groundwater concentration of 1 milligram of TCE per liter, an average *TSCF* value of 0.12 (Orchard *et al.* 2000b), and the high and low estimates for transpiration rate and fraction of groundwater used, yearly plant uptake values ranging from 2.4 to 84 milligrams of TCE per square meter per year can be calculated using Equation 18-3. As additional information regarding transpiration rates and the fraction of groundwater used by plants becomes available, the estimated range can be narrowed. Using the *TSCF* value of 0.75 reported by Burken and Schnoor (1998) would increase the upper range value to 525 milligrams of TCE per square meter per year.

Calculations of this type should be used to determine if plant uptake might be a significant removal mechanism at a particular site. These calculations also show that contaminant uptake is likely more important in climates where rainfall is low during summer months, requiring plants to obtain a higher percentage of moisture from the water table.

Scope and Objectives of Project

Field studies were conducted at CCAS and Hill AFB to determine if indigenous vegetation was involved in the uptake, volatilization, and metabolism of TCE from shallow contaminated groundwater plumes. Historical depth to groundwater (1.5 to 2.3 meters or 5 to 8 feet below ground surface) and groundwater TCE concentrations (1 to 10 milligrams of TCE per liter) were similar at the two sites but the amounts (CCAS: 127 centimeters per year or 50 inches per year and Hill AFB: 48 centimeters per year or 19 inches per year) and frequency of precipitation were very different. This difference between the two sites enabled us to qualitatively test the hypothesis that trees growing in semi-arid regions are more likely to obtain a greater fraction of water from contaminated aquifers resulting in higher tissue concentrations of TCE. A secondary objective was the development of a novel field sampling technique for the determination of phytovolatilization and development of analytical methods for determining plant-tissue concentrations of TCE and TCE metabolites [2,2,2-trichloroethanol (TCEt) 2,2,2-trichloroacetic acid (TCAA), and 2,2-dichloroacetic acid (DCAA)].

To better understand the potential exposure to and the fate and transport of TCE in the plants and soil, groundwater and soil samples were collected and analyzed for TCE and the potential metabolites. Surface flux measurements from the soil also were collected to quantify the amount of TCE volatilizing directly from the groundwater and soil surface within the system.

SITE DESCRIPTIONS

Cape Canaveral Air Station, Florida

Bounded by the Atlantic Ocean on the east and the Banana River on the west, CCAS is located on a barrier island in Brevard County, Florida. The main complex occupies about 6500 hectares (25 square miles) of assembly and launch facilities for missiles and space vehicles. The southern boundary is a manmade shipping channel, and the John F. Kennedy Space Center adjoins CCAS to the north.

The climate at CCAS is semi-tropical, with rainfall (127 centimeters or 50 inches) evenly spread throughout the year (Mallander 1990). This is equal to 1270 liters per square meters per year. Evapotranspiration has been previously estimated at approximately 75 percent of the annual rainfall or 950 liters per square meters per year (Mallander 1990).

The site of interest is the Ordnance Support Facility denoted CCAS Site 1381. From 1968 to 1977, the facility housed acid and solvent dip tanks for cleaning metal components. The chlorinated solvent plume at CCAS Site 1381 is thought to be a result of accidental spills and poor handling practices of the drummed solvents.

The CCAS Site 1381 has a shallow surficial aquifer with measured depths to groundwater between 1.0 and 1.5 meters (3.5 and 4.8 feet) above mean sea level (U.S. National Geodetic Vertical Datum), and approximately 1.5 meters (5 feet) below ground surface. Historical sampling events identified a region of TCE-contaminated groundwater (1 to 10 milligrams per liter) near the southwestern corner of the facility. Low concentrations of vinyl chloride and *cis*-1,2-dichloroethene have also been reported.

Hill Air Force Base, Utah

Hill AFB is located in northern Utah about 40 kilometers (25 miles) north of Salt Lake City and about 8 kilometers (5 miles) south of Ogden. The Great Salt Lake is approximately 19 kilometers (12 miles) to the west of Hill AFB and the Wasatch Mountain Range is about 6.5 kilometers (4 miles) to the east. The surface elevation ranges from approximately 1400 to 1500 meters (4600 to 5000 feet) above National Geodetic Vertical Datum of 1929. The climate in the vicinity of Hill AFB is temperate and semi-arid. The frost-free growing season is from May through September. The average annual precipitation recorded during the period 1978 to 1990 at a gage located 1 kilometers (0.5 mile) northeast of the base was 50.3 centimeters (19.8 inches). The majority of precipitation falls from October through May. May is usually the wettest month, and June and July are the driest months. Potential average annual evaporation is approximately 114 centimeters (45 inches) of water per year.

The specific site of interest at Hill AFB is designated as Operable Unit 2 (OU2). The disposal of chlorinated degreasing solvents, mainly TCE, in trenches between of 1967 and 1975 has resulted in the contamination of a shallow aquifer (not used as a drinking water source) at this site. Historically, groundwater concentrations of TCE within the plume have ranged from 0.4 to 10 milligrams per liter. Several other chlorinated solvents (1,1,1-trichloroethane, perchloroethylene, and *cis*-1,2-dichloroethylene) have also been found in groundwater samples collected at the site, at much lower concentrations than TCE.

SAMPLE COLLECTION AND ANALYSIS

Sampling Design

Cape Canaveral Air Station

The initial field sampling at CCAS was performed during the second week of June 1997. Follow-up plant tissue sampling occurred near the end of July and beginning of August 1997. Re-sampling of the transpiration gas and surface flux occurred at this site on September 24 and 25, 1997. Tissue and transpiration gas samples were collected from three plant species, identified within the current extent of dissolved-phase TCE contamination. These species included a live oak (*Quercus virginiana*), a castor bean (*Ricinus communis*), and a saw palmetto (*Serenoa repens*). The live oak (*Quercus virginiana*) was the largest and was expected to have the highest transpiration rate and deepest rooting depth of the three species sampled. The saw palmetto (*Serenoa repens*) is a shrub that is common to the region. The castor bean (*Ricinus communis*), is considered a "nuisance" plant growing at the site.

A second live oak (*Quercus virginiana*) was sampled east of the known extent of dissolved-phase TCE and was to originally serve as a "control." However, results from the initial sampling indicated that this tree may have been previously exposed to TCE and, therefore, not be an appropriate control. Therefore, leaf tissue was collected and analyzed from an off-base live oak (*Quercus virginiana*) (Orlando, Florida, 60 kilometers or 40 miles from the site) as a true control for these analyses.

In addition, after completion of the field sampling in June 1997 and analysis of transpiration gas and surface flux samples, results indicated that the carrier gas (compressed air of breathing quality) used during the initial sampling contained measurable levels of TCE (approximately 4.6 micrograms per cubic meter). Subsequently, sampling used carrier gas with lower levels of TCE (approximately 0.33 microgram per cubic meter).

Groundwater, soil, surface flux, and soil gas samples were also collected at several locations at the site. Soil cores used to determine root density were also collected near each of the sampled plants. The general sampling locations are illustrated in Figure 18-1.

Hill Air Force Base

Sampling took place during a 2-week period in late August and early September 1999 just prior to the trees at the seep location being removed as part of an effort to improve the efficiency of an existing seep water collection system. The removal of the trees enabled us to collect stem cores at 1-meter intervals along the entire length of the trees to document the vertical distribution of TCE within the trees at the seep location. In addition, the concentration of TCE in the trees removed from the seep area was monitored periodically over several months to examine the stability of TCE in the wood. Because of the short time (about 2 weeks) between the project initiation and the tree removal, the sampling effort focused on collection and analysis of plant tissue

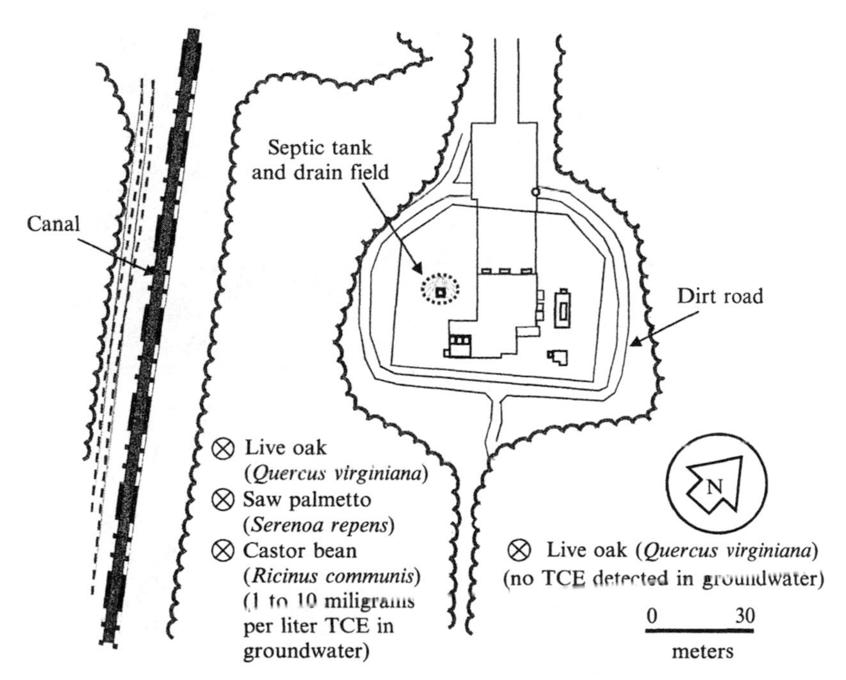

Figure 18-1 Plant sampling locations at Cape Canaveral Air Station Ordnance Support Facility 1381.

and transpiration gas samples from three species of trees located in three different places at OU2.

Trees from three areas (two within the historically defined groundwater TCE plume and one outside) located down gradient (east) of the source area were sampled. The seep location was situated on a steep hillside; the seasonal seeps are active mainly during the spring and after significant precipitation events. The canal location was at the bottom of the hillside just east of the Weber canal that supplies irrigation water to nearby agricultural interests.

Plant tissue and transpiration samples were collected from three species of trees, poplar (*Populus* spp.), Russian olive (*Elaeagnus angustifolia*), and willow (*Salix* spp.), at three different locations (nine total trees), two within the TCE-contaminated groundwater plume (seep and canal locations) and one outside (control location). The sample locations are illustrated in Figure 18-2. The age of the trees sampled in the seep location ranged from about 15 [willow

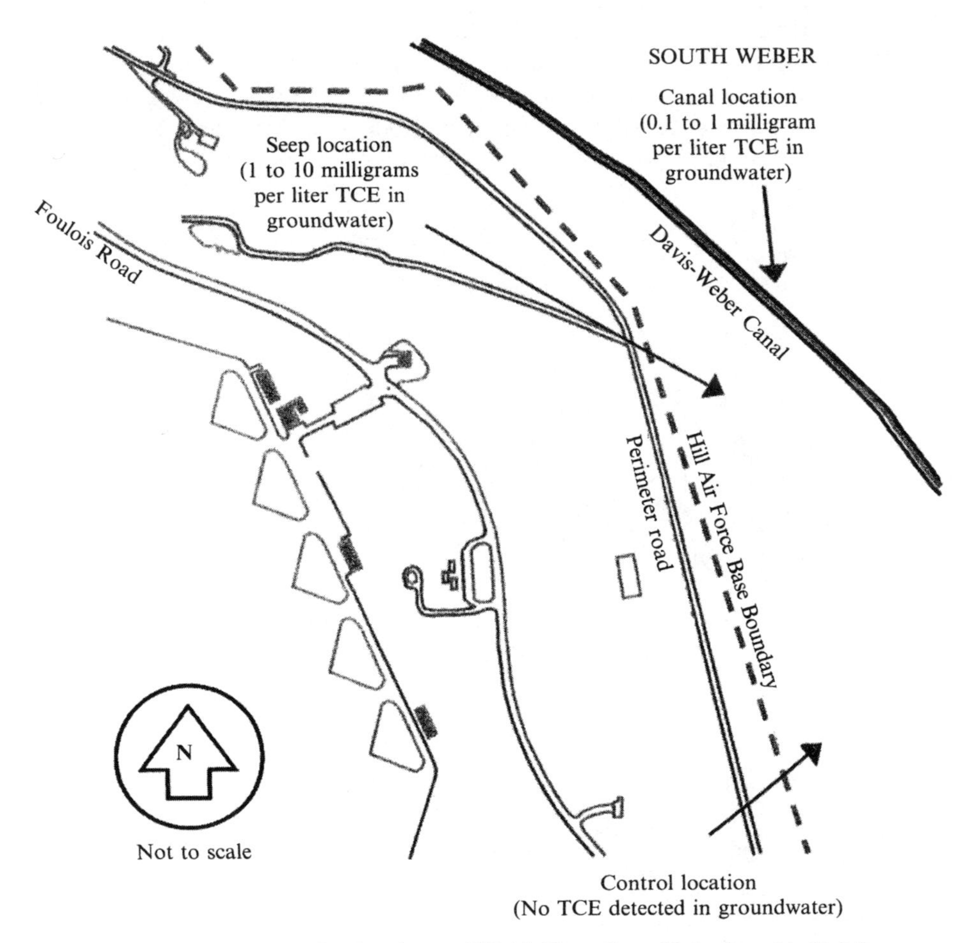

Figure 18-2 Plant sampling locations at Hill Air Force Base, Utah, Operable Unit 2.

(*Salix* spp.)] to 30 [poplars (*Populus* spp.)] years based on ring counts. The heights of these trees ranged from 13.4 meters [(44 feet, poplar (*Populus* spp.) [to 3 meters [(10 feet, willow (*Salix* spp.)]. The two locations within the plume were chosen mainly because of the depth to groundwater, groundwater TCE concentrations, and availability of mature trees. In the first location within the plume, the trees were growing in an area of seasonal groundwater seeps and at the time of sampling, the depth to groundwater was about 2.3 meters (7.5 feet, seep location). Sampling in the previous years had indicated that TCE concentrations in groundwater at this site ranged from 1 to 10 milligrams per liter. The second location within the plume was about 30 meters (100 feet) downgradient from the first location, just below a concrete-lined irrigation canal (canal location). Historically, groundwater TCE concentrations below the canal location have been about a factor of 10 lower than at the seep location. Some leakage of irrigation water from the canal into the shallow aquifer has been documented and contributes to the seasonal variability in groundwater TCE concentrations at the canal location. The control location was about 100 meters (330 feet) south of the seep location in an area where no TCE has been detected in the groundwater.

Plant Sample Collection

Because of the differences in physical–chemical properties, separate collection and extraction procedures were used for the volatile TCE and the three semi-volatile chlorinated metabolites (TCEt, TCAA, and DCAA). In addition, slightly different sample collection and analysis procedures for TCE were used at the CCAS and Hill AFB sites. At CCAS, methanol extracts of the plant tissue were analyzed for TCE using purge and trap gas chromatography with an electron-capture detector. However, because of problems associated with excessive foaming during purging (mainly leaves and roots), a headspace gas chromatography method was subsequently used to determine TCE in plant tissue collected from Hill AFB.

Trichloroethylene Samples at Cape Canaveral Air Station

Leaf samples (2 to 5 grams fresh mass) were removed from the branches by gloved hand and placed directly in pre-weighed 40 milliliter glass vials equipped with Teflon®-lined rubber septa. After the plant tissue samples were added to the vials, the vials were weighed again to determine the amount of plant tissue collected and then filled with purge-and-trap-grade methanol, to minimize volatilization losses during transport and to serve as the extraction solvent. The vials were again weighed to determine the amount of methanol added, capped, and shipped at $4 \pm 2\,^{\circ}C$ for analysis. Stem tissue samples were collected from the live oaks (*Quercus* spp.) by drilling a 6-millimeter (0.25-inch) hole directly into the stem, and directing the shaving into the pre-weighed vials using a funnel. For the plants that were not woody [(saw palmetto (*Serenoa repens*) and caster bean (*Ricinus communis*)], stem tissue was cut into small sections with stainless steel pruning shears and placed directly into the pre-weighed vials as described for the leaf samples. Roots samples were collected in a similar manner after digging a small pit at the base of each plant.

Trichloroethylene Samples at Hill Air Force Base

Leaf samples were collected in the same manner as previously described for the CCAS site, except that the pre-weighed vials contained 20 milliliters of a saturated sodium chloride solution acidified to pH 2 with phosphoric acid. This was done to facilitate the gas chromatographic analysis of the headspace following the procedure outlined in the U.S. Environmental Protection Agency SW-846 Method 5021 (http://www.epa.gov/epaoswer/hazwaste/test/5_series.htm) for soils. After samples were added, the vials were weighed again to determine the amount of plant tissue collected. The vials were capped and put on ice until analysis. Stem core samples, taken using a 12-millimeter (0.5-inch) increment borer (Forestry Suppliers, Inc.), were also placed directly into vials containing the acidified sodium chloride solution. Stem core samples were used at Hill AFB instead of drill shavings after it was determined through side-by-side sample collection and analysis that the core samples consistently yielded higher results by a factor of two (Lewis 2002). Root samples were to be collected only from trees in the seep location after being pulled from the ground as part of the "seep collection system improvement" process. However, only large woody roots, close to the tree trunks, were available for sampling because the fine roots were lost during the tree-removal process. Because these roots were not representative of the entire root structure, these samples were not analyzed.

Metabolite Samples

Samples of plant tissue (10 to 30 grams fresh mass) were collected for extraction and analysis of TCEt, TCAA, and DCAA. Tissue samples were placed directly in pre-weighed wide-mouth glass vials equipped with Teflon®-lined lids. After the plant-tissue samples were added, the vials were reweighed, capped, and shipped at $4 \pm 2\,°C$ for analysis.

Plant Tissue Extraction and Analysis Methods

Trichloroethylene at Cape Canaveral Air Station: Purge and Trap

Plant samples from CCAS were analyzed for TCE using a purge-and-trap gas chromatographic method adapted from U.S. Environmental Protection Agency SW-846 Methods 5030 and 8010B (1996). Upon arrival at the laboratory, the vials containing the plant samples and methanol were agitated for 24 hours in a rotary tumbler. A 250-microliter aliquot of the methanol extract was removed from the sealed vial with a syringe, introduced into a purge vessel, and diluted with 20 milliliters of deionized water. The water–methanol mixture was then purged with nitrogen for 12 minutes at 50 milliliters per minute and the TCE was concentrated on a Tenax trap. After a 20-second dry purge with nitrogen, the trap was thermally desorbed using a Dynatherm® desorber interfaced with a Shimadzu® GC14A gas chromatograph equipped with a 75 meter by 0.45 millimeter DB-VRX (2.5 micrometer film thickness) capillary column (J&W Scientific, Folsom, California) and electron capture detector. Column flow was 8 milliliters per minute of nitrogen and the column oven temperature was 50 °C isothermal. At these conditions, the retention time for TCE was 8.8 minutes. The column was periodically baked out at 240 °C between samples, if baseline drift was observed. The electron capture detector and injection port temperatures were set at 300 and 210 °C, respectively.

Trichloroethylene at Hill Air Force Base: Headspace

A headspace gas chromatographic method was used to determine TCE concentrations in the plant-tissue samples collected at Hill AFB. The headspace vials containing the plant tissue and modifier mixture were agitated for 24 hours in a rotary tumbler at 20 °C prior to analysis. This was followed by 12 hours of equilibration in a 40 °C water bath. A 10-microliter aliquot of the headspace was removed through the septa with a syringe and manually injected into a Shimadzu® GC14 gas chromatograph equipped with an electron capture detector and a 75 meter by 0.45 millimeter DB-VRX (2.5-micrometer film thickness) capillary column (J&W Scientific, Folsom, California). Column flow was 8 milliliters per minute of nitrogen and the column temperature was 90 °C isothermal. The detector and injection port temperatures were 300 and 210 °C, respectively.

Trichloroethylene Metabolites

Leaf-tissue samples were flash frozen in liquid nitrogen and macerated in a mortar. The "woody" tissues (stems and roots) were macerated using a coffee grinder. Approximately 5 grams of "processed" tissue were weighed into 60 milliliter Teflon® centrifuge tubes and combined with 15 milliliters of a 0.25 normal sodium hydroxide solution. The centrifuge tubes were shaken for 10 minutes on a reciprocating shaker and then centrifuged for 10 minutes at 9750 revolutions per minute. The supernatant was transferred from the Teflon® centrifuge tube to a glass centrifuge tube. This extraction procedure was performed a total of three times, each time combining the supernatant into the same glass centrifuge tube. The combined aqueous extracts were acidified to pH 1 with 50 percent sulfuric acid and, subsequently, extracted three additional times with 7 milliliters of methyl *tertiary*-butyl ether by shaking for 5 minutes followed by centrifugation for 5 minutes at 500 revolutions per minute. The supernatants from the triplicate methyl *tertiary*-butyl ether extractions were combined and brought to exactly 25 milliliters. The extract was dried over 2 grams of anhydrous sodium sulfate. A 2-milliliter aliquot of the dried extract was derivatized with diazomethane and analyzed by gas chromatography with electron capture detection using the same instrument and operational conditions as previously described for TCE. With these conditions, the retention times were 4 minutes for TCEt, 5.2 minutes for TCAA, and 4.4 minutes for DCAA. The column was periodically baked out at 240 °C between samples, if baseline drift was observed.

Plant-Transpiration Samples

Gas samples were collected to determine the potential flux of TCE from leaves to the atmosphere using the portable flow-through sampling apparatus illustrated in Figure 18-3. A glass chamber was placed over a representative section of each tree and sealed on the open end with closed-cell foam, latex silicon sealant, and electrical tape to produce a flexible yet tight seal around the stem and chamber. The chambers were fixed in the same location during

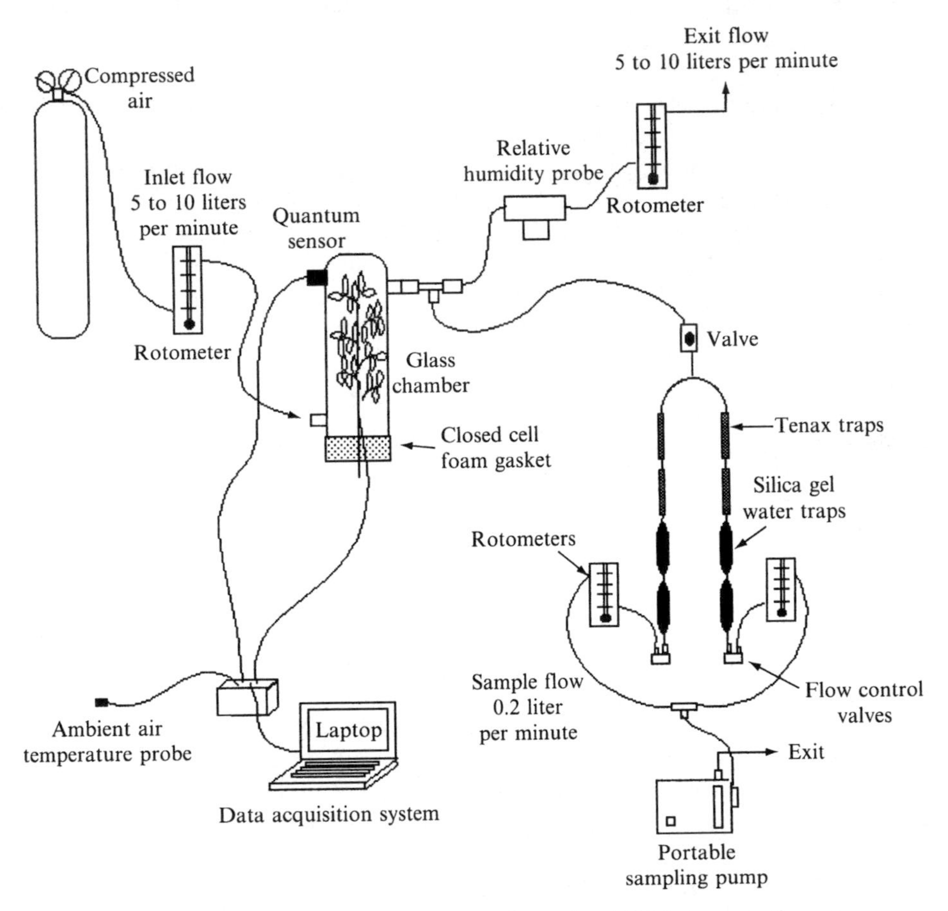

Figure 18-3 Schematic of chamber system used to collect TCE in transpiration gases.

replicate sample collection. Compressed breathing air containing 300 to 400 parts per million by volume of carbon dioxide (to maintain natural stomatal response) was used to purge the chambers of TCE and water vapor. The resulting slightly positive chamber pressure minimized the potential introduction of any TCE that might be in the ambient air surrounding the chamber (*i.e.*, TCE volatilizing directly from the soil surface). Using cylinders of compressed air also served to minimize external power requirements. However, because of the difficulties encountered in obtaining commercial cylinders of compressed air that were free of TCE during the sampling at CCAS, we filled our own small containers (20 liters at 120 pounds per square inch or 0.83 megapascal) using a laboratory compressor when sampling at Hill AFB. Each cylinder could then be easily checked for TCE contamination prior to use in the field and refilled if necessary.

All tubing and connections attached to the chamber were constructed of stainless steel to minimize sorption of TCE. Typically, samples were collected at 15- to 30-minute intervals at 100 to 200 milliliters per minute using portable sampling pumps. Portable sampling pumps were used to subsample the air leaving the chamber. Subsampling was necessary because relatively high flow rates (6 to 10 milliliters per minute) were used to minimize humidity increases within the chamber and to prevent the condensation of transpired water on the interior walls of the chamber. Tenax was used as the sorbent for the TCE traps because of high sorption capacity for volatile chlorinated organics and low affinity for water. Silica gel traps were used to determine the amount of water transpired. The volume of gas sample collected (calculated from the flow rate through the Tenax trap and the sampling time), based on preliminary modeling and reported breakthrough volumes, ranged from 3 to 6 liters. After sampling, Tenax traps were sealed with stainless steel caps, placed in bubble-pack envelopes, and shipped to the laboratory at Utah State University for analysis. Samples were shipped at ambient temperature to prevent moisture from condensing in the traps.

A minimum of three replicate samples were collected from each of the three plant species within the plume area at both sites. Samples were also collected from the control plants outside the plume area. Chamber blanks and ambient air samples were collected at each sample location.

The sampling manifold was designed to simultaneously collect duplicate trap samples (with breakthrough traps) during each sampling period. Samples were analyzed by gas chromatography with electron capture or mass spectrometry detection. The electron capture detection was about 10 to 50 times more sensitive than mass spectrometry, but more susceptible to misidentification and quantification of coeluting chromatographic peaks that could not be separated.

For gas chromatography with electron capture detection analysis, trap samples were introduced into a Shimadzu® GC14 gas chromatograph equipped with electron capture detector and capillary column (J&W Scientific DBVRX, 2.5 micrometer film thickness, 0.45 millimeter inside diameter by 75 meter length) using an Environchem® Model 710 thermal desorber. For gas chromatography with mass spectrometer analysis, a Tekmar® 6000 thermal desorber was used to desorb Tenax traps directly to a Hewlett-Packard® 6890 gas chromatograph, equipped with a 5973 mass spectrometer and capillary column (Hewlett-Packard® HP-624, 30 meter length by 0.25 millimeter inside diameter by 1.4 micrometer film thickness). An external standard approach was used to quantify the mass of TCE collected in each trap.

Water transpiration rates were determined with a portable balance by measuring the mass of condensed water that collected in the silica traps. Traps were weighed prior to and after being connected to the sample effluent stream. The weight of the water collected and the volume of effluent passing through the trap were used to calculate the transpiration rate. Transpiration measurements were used to determine the ratio of TCE to water transpired. This ratio, coupled with seasonal evapotranspiration rates, can be used to estimate the impact of vegetation on the flux of TCE to the atmosphere. Transpiration rates also indicate whether normal metabolic function is occurring.

Prior to going to the field, and between each sampling event, the interior chamber surfaces were rinsed with methanol. To evaluate potential sorption onto the chamber, several chamber rinses were analyzed for TCE.

Additional Samples

At CCAS, soil cores were obtained near each plant species to determine the root length and area density within the unsaturated soil column. This was not done at Hill AFB due to logistical problems. Groundwater samples were collected and analyzed for TCE and the metabolites previously mentioned. Surface flux measurements were also collected near each plant to quantify the amount of TCE volatilizing directly from the groundwater and soil surface.

RESULTS AND DISCUSSION

Direct pathways are uptake and metabolism, uptake and phytovolatilization, and uptake and accumulation. Trees may also impact the fate of TCE indirectly through enhanced rhizosphere degradation and by changing the subsurface environment through the removal of groundwater *via* transpiration. The sampling approach followed in this study was specifically designed to determine only if the direct pathways were occurring to a measurable extent.

Exposure of Vegetation to Trichloroethylene

High annual precipitation (127 centimeters per year or 50 inches per year, Mallander 1990) and shallow root distributions (70 percent in the top 60 centimeters or 2 feet, 90 percent in the top 120 centimeters or 4 feet) at the CCAS site (Doucette *et al.* 1998, Hayhurst 1998) suggest that groundwater use by existing vegetation is much less than that obtained by precipitation. This would reduce the exposure of the vegetation to TCE at this site. However, at Hill AFB, minimal precipitation during the growing season, lack of surface irrigation, and shallow groundwater (approximately 2.3 meters or 7.5 feet below ground surface) suggest that trees obtain a substantial portion of water needs from the contaminated groundwater. The expected high use of groundwater at Hill AFB should increase TCE uptake relative to CCAS. Root distribution was not examined at the Hill AFB OU2 site because cores could not be collected prior to tree removal. The fraction of groundwater used by mature trees is difficult to directly measure in a field setting, and measurement was not attempted at these sites. Stable isotope measurements (Dawson 1996) could be used to obtain this information if similar studies are performed in the future.

Roots may also be exposed to vapor phase TCE. While this pathway was not directly evaluated, measurable levels of TCE found in soil gas and surface flux samples suggest that this exposure route could be important and should be evaluated in future studies.

Groundwater Concentrations of Trichloroethylene

During the first sampling event at CCAS, groundwater-monitoring points were installed near each of the three test plants within the plume and near the live oak located east of the known extent of the TCE plume. Groundwater samples were collected for the analysis of TCE, DCAA, TCAA, and TCEt. Samples for TCE were analyzed using a purge-and-trap gas chromatography with mass spectrometry method as outlined in SW-846 Method 8260A (U.S. EPA 1996). For TCE metabolites, groundwater samples were acidified to a pH less than 1, extracted with methyl *tertiary*-butyl ether, derivatized with diazomethane and analyzed by gas chromatography with electron capture detection. Results from the groundwater sampling showed that the TCE concentrations varied spatially throughout the test area. The lowest concentration of TCE in groundwater was found near the live oak (*Quercus virginiana*) at an average of 0.5 ± 0.09 milligrams per liter (plus or minus one standard deviation) while the highest concentrations were found near the saw palmetto (*Serenoa repens*) (65 ± 26 milligrams per liter). By-products of reductive dechlorination, such as 1,1-dichloroethene (less than 1.1 to 1200 micrograms per liter), *cis*-1,2-dichloroethene (65 to 4800 micrograms per liter), *trans*-1,2-dichloroethene (less than 1.65 to 110 micrograms per liter), and vinyl chloride (less than 2 to 456 micrograms per liter), were also identified in the groundwater. Trichloroacetic acid (6 micrograms per liter) and TCEt (2 micrograms per liter) were also found at two separate monitoring points. However, TCEt was not found in the duplicate samples. Thus, the possibility of microbial degradation of TCE to the target metabolites in the groundwater is unlikely, but cannot be ruled out.

At Hill AFB, groundwater samples were collected from six different multiple level monitoring wells at the seep location, 3 weeks prior to the tree sampling. The samples were analyzed by purge-and-trap gas chromatography with an electron capture detector. Trichloroethylene concentrations ranged from 0.3 to 5.9 milligrams per liter within the seep location. No groundwater samples from wells at the canal or control locations were collected during the time that the trees were being sampled. However, historical groundwater data for OU2 indicate the following range of groundwater concentrations of TCE: seep location (1 to 10 milligrams per liter), canal location (0.1 to 1 milligram per liter), and control location (no detectable TCE).

Surface Emission Flux at Cape Canaveral Air Station and Hill Air Force Base

During the first sampling event at CCAS in June, 1997, measurable levels of TCE were found in all surface flux samples collected. However, as previously described, analysis of blanks and controls indicated that the source of TCE was the compressed breathing air used as the sweep gas for the flux measurements. Thus, no significant flux of TCE from the soil surface relative to blanks was observed. A second sampling event was carried out using the cleaner source of compressed air but again no statistical difference between samples and blanks was observed.

Because of the TCE contamination found in compressed breathing air during the CCAS sampling, pre-purified nitrogen was used as the sweep gas for the surface emission flux samples at the Hill AFB OU2 site. At the seep location, volatilization of TCE from the soil surface ranged from 28 to 750 milligrams per square meter per day while fluxes from undetetable to 9 milligrams per square meter per day were measured at the control location. As expected, the surface flux of TCE at the seep location decreased with increasing depth from surface to groundwater. The low levels of TCE found in several of the samples located at the control locations could indicate the presence of TCE contaminated groundwater outside the currently defined area of the plume, or analytical artifacts associated with inadequate Tenax-trap cleaning.

An additional factor to consider when evaluating the surface flux data is that the long-term surface emissions of TCE near plants could lead to the uptake or sorption of TCE vapor by leaves. While it is expected that this transfer would be small, the TCE tissue analysis cannot differentiate between uptake of atmospheric vapor *versus* translocation from the subsurface. However, ambient air samples collected during the sampling events showed no measurable concentrations of TCE.

TABLE 18-1 Maximum Concentrations (Milligrams per Kilogram of Dry Weight) of Trichloroethylene and Associated Metabolites in Plant Tissue Samples Collected at Cape Canaveral Air Station June and September 1997

Species	Tissue (month)	TCE[a]	TCEt[a]	TCAA[a]	DCAA[a]
Castor bean (*Ricinus communis*)	Root (June)	0.012(1/3)	<MDL[b](0/2)	<MDL (0/2)	<MDL (0/2)
	Stem (June)	<MDL (0/3)	< MDL (0/2)	0.207 (2/2)	< MDL (0/2)
	Stem (September)	NA[c]	0.054 (1/1)	< MDL (0/1)	2.57 (1/1)
	Leaf (June)	<MDL (0/3)	0.262 (2/2)	1.11 (2/2)	<MDL(0/2)
	Leaf (September)	NA	0.069 (1/2)	0.14 (2/2)	3.63 (2/2)
Saw palmetto (*Serenoa repens*)	Root (June)	0.048 (2/2)	<MDL (0/2)	0.131 (2/2)	<MDL (0/2)
	Stem (June)	0.002 (1/3)	<MDL (0/2)	0.394 (2/2)	<MDL (0/2)
	Stem (September)	NA	0.026 (1/1)	< MDL (0/1)	<MDL (0/1)
	Leaf (June)	<MDL (0/3)	0.086 (1/2)	0.125 (2/2)	<MDL (0/2)
	Leaf (September)	NA	< MDL (0/1)	< MDL (0/2)	0.48 (2/2)
Live oak (Quercus virginiana): plume	Root (June)	<MDL (0/2)	<MDL (0/2)	0.086 (2/2)	0.275 (2/2)
	Stem (June)	0.002 (2/3)	<MDL (0/2)	0.036 (1/2)	<MDL (0/2)
	Stem (September)	NA	0.058 (1/1)	<MDL (0/1)	0.602 (1/1)
	Leaf (June)	<MDL (0/3)	0.273 (4/6)	0.189 (3/3)	0.723 (1/3)
	Leaf (September)	NA	0.092 (1/1)	<MDL (0/1)	1.76 (1/1)
Live oak (Quercus virginiana): outside plume	Root (June)	<MDL (0/3)	<MDL (0/2)	0.063 (1/2)	<MDL (0/2)
	Stem (June)	<MDL (0/3)	<MDL(0/3)	0.038 (2/3)	<MDL (0/3)
	Stem (September)	NA	0.098 (1/1)	< MDL (0/1)	0.602 (1/1)
	Leaf (June)	<MDL (0/3)	0.295 (3/5)	0.132 (3/3)	<MDL (0/3)
	Leaf (September)	NA	< MDL (0/1)	0.399 (1/1)	1.17 (1/1)
Live oak (Quercus virginiana): control, Orlando, Florida	Leaf (September)	< MDL (0/1)	< MDL (0/1)	< MDL (0/1)	< MDL (0/1)

[a]Maximum concentration detected in dried tissue (number of samples greater than MDL[b] per total number of samples).
[b]MDL, Method detection limit for TCE = 0.0005 milligram per kilogram, TCEt = 0.02 milligram per kilogram, TCAA = 0.03 milligram per kilogram, and DCAA = 0.15 milligram per kilogram.
[c]Abbreviation used—NA: not available; (sample was either not collected or could not be analyzed).

Trichloroethylene and Metabolite Concentrations in Plant Tissue

Cape Canaveral Air Station

The results of the analysis of the plant tissue samples collected at CCAS in June and September 1997 are summarized in Table 18-1. Trichloroethylene, TCEt, TCAA, or DCAA were identified in all plant tissue types (root, stem, and leaf) and in all three species, including the on-site live oak (*Quercus virginiana*) believed to be outside the plume area. The highest concentrations of TCE (0.098 milligram per kilogram dry weight) were found in the trunk of the live oak (*Quercus virginiana*) that seemed to be outside the plume, while the highest concentrations of the metabolites (DCAA at 0.363 milligram per kilogram dry weight) were found in the leaf samples collected in September.

Detection of TCE and TCE metabolites in the live oak (*Quercus virginiana*) thought to be outside the plume area suggests exposure to TCE or an analytical problem. To further examine this finding, leaf samples were collected on July 21, 1997, from the two previously sampled on-site live oaks (*Quercus virginiana*) and a third live oak (*Quercus virginiana*) located off-site in Orlando, Florida. The off-site control showed no measurable levels of either TCE or the associated metabolites, while measurable levels of TCAA were found in most tissues of both of the on-site live oaks (*Quercus virginiana*) (0.036 to 0.399 milligram per kilogram dry weight). As previously suggested, accumulation of TCAA in the leaf tissue may be the result of previous exposure to TCE-contaminated groundwater, as live oak (*Quercus virginiana*) trees typically retain leaves for two growing seasons. No information is available regarding the stability of TCAA in leaf tissue, but some contaminants are stored in the vacuoles of the leaf. Another potential pathway that could account for the appearance of TCAA in the leaves is through the uptake and transformation of vapor phase TCE (Nzengung and Jeffers 2001) fluxing from the soil. However, as previously mentioned, ambient air samples collected at the site showed no detectable TCE and this pathway is unlikely to be important in most field situations.

Hill Air Force Base

At the Hill AFB OU2 site, the TCE concentrations ranged from 0.15 to 8.1 milligrams per kilogram on a dry weight basis in tissues from trees growing above the contaminant plume (Table 18-2). No TCE or metabolites were found in tissues from trees growing in the control location. Concentrations in the stems were an order of magnitude greater than that found in the leaves, and the poplar (*Populus* spp.) and willow (*Salix* spp.) trees had two to four times greater TCE concentration than the Russian olive (*Elaeagnaceae angustifolia*). Stem TCE concentrations correlated with the concentration of TCE in the groundwater below (*i.e.*, the higher the groundwater concentration the higher the stem core concentration).

TABLE 18-2 Average TCE and Metabolite Concentrations in Leaf and Stem Tissue of Poplar (*Populus* spp.), Russian Olive (*Elaeagnaceae angustifolia*), and Willow (*Salix* spp.) Collected at Hill Air Force Base Operating Unit 2 Seep and Control Locations, September 1999

Location	Tree species	Tissue	Milligrams per kilogram of dry weight			
			TCE[a]	TCEt[a]	TCAA[a]	DCAA[a]
Seep	Poplar (*Populus* spp.)	Leaves	0.364 ± 0.12	0.741 ± 0.35	0.0180 ± 0.011	MDL[b]
		Stem	8.14 ± 5.8	MDL	MDL	MDL
Control	Poplar (*Populus* spp.)	Leaves	MDL	MDL	MDL	MDL
		Stem	MDL	MDL	MDL	MDL
Seep	Russian olive (*Elaeagnacea angustifolia*)	Leaves	0.151 ± 0.05	MDL	0.390 ± 0.15	MDL
		Stem	1.94 ± 1.1	MDL	MDL	MDL
Control	Russian olive (*Elaeagnacea angustifolia*)	Leaves	MDL	MDL	MDL	MDL
		Stem	MDL	MDL	MDL	MDL
Seep	Willow (*Salix* spp.)	Leaves	0.339 ± 0.33	MDL	MDL	MDL
		Stem	5.11 ± 3.4	MDL	MDL	MDL
Control	Willow (*Salix* spp.)	Leaves	MDL	MDL	MDL	MDL
		Stem	MDL	MDL	MDL	MDL

[a]Average ± standard deviation of at least three replicate samples.
[b]MDL: method detection limits for TCE = 0.02 milligram per kilogram, TCEt = 0.05 milligram per kilogram, TCAA = 0.01 milligram per kilogram, and DCAA = 0.1 milligram per kilogram.

Metabolite concentrations in leaves ranged from below method-detection limits to 0.74 milligram per kilogram on a dry weight basis. No obvious trends were observed with regard to species or tissue type. The most prevalent metabolite was TCEt whereas DCAA was least prevalent. It is interesting to note, however, that metabolite concentrations found in the leaves collected from the seep and canal locations at Hill AFB were similar to those observed at CCAS (Doucette *et al.* 1998, Hayhurst 1998). This is despite the plant tissue concentration of TCE being significantly higher at Hill AFB. The reason for this is unclear.

Distribution of Trichloroethylene in Tree Stems at Hill Air Force Base Operable Unit 2

Vroblesky *et al.* (1999) reported that stem core concentrations of TCE decreased with increasing height for a single, bald cypress (*Taxodium distichum* L. Rich) tree sampled at the Department of Energy Savannah River Site in South Carolina. Those authors speculated that the decreasing concentration with increasing height might be due to the volatilization of TCE through the bark, transformation within the tree, or sorption to the plant tissues. At the OU2 seep location, the Russian olive (*Elaeagnaceae angustifolia*) appeared to follow a similar, but less pronounced trend, whereas the poplar (*Populus* spp.)

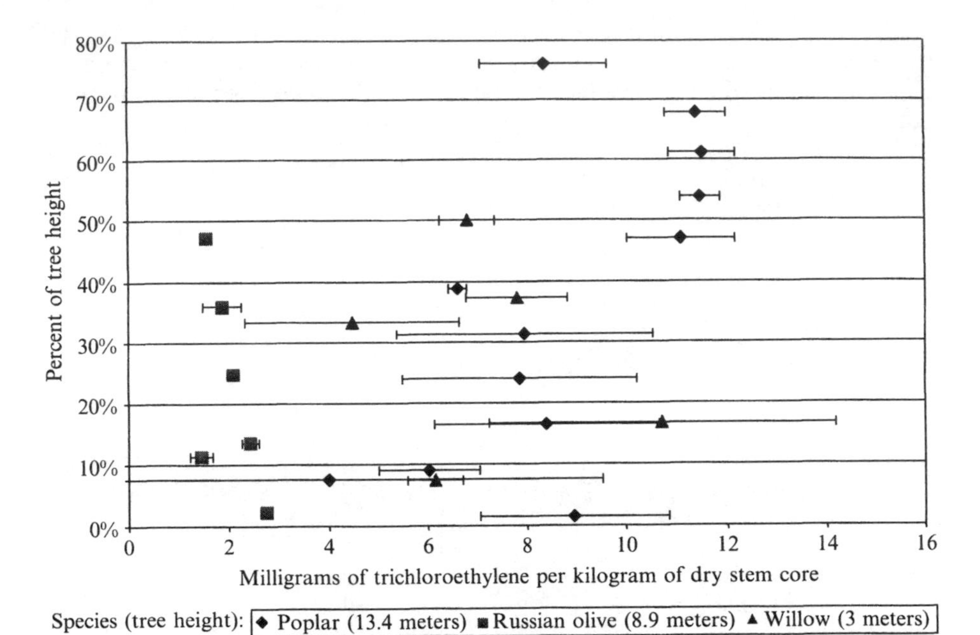

Figure 18-4 Vertical distribution of trichloroethylene in three tree species [poplar (*Populus* spp.), Russian Olive (*Eleagnaceae angustifolia*), and willow (*Salix* spp.)] at the seep location.

and willow (*Salix* spp.) trees showed no significant reduction in TCE concentration with an increase in height (Figure 18-4). In an ongoing project at another Hill AFB site, we monitored tree core concentrations of TCE in several cottonwood (*Populus deltoides*) and poplar (*Populus* spp.) trees, every month for a year. Concentrations measured in the winter months were relatively constant and only slightly less than those measured in the fall, suggesting that volatilization of TCE through the bark of a mature tree is probably not a significant loss mechanism at the Hill AFB site (Lewis 2002). The observation of uniform concentrations of TCE vertically within the trees at Hill AFB may indicate that rates of TCE uptake, sorption, and transformation have reached steady state.

Results of Transpiration Gas Sampling

At CCAS, no statistical difference between transpiration gas samples and apparatus blanks was found during both the June and September sampling events. However, at the Hill AFB OU2 site, significant differences between samples and apparatus blanks were observed for the seep and canal locations. Trichloroethylene was also found in some of the transpiration samples collected from the control trees, but at levels that were not significantly different from the apparatus blanks, and 10 to 100 times less than samples at the seep and

TABLE 18-3 Summary of Transpiration Data Collected from Three Tree Species at Hill Air Force Base, Operable Unit 2

	Poplar (*Populus* spp.)			Russian olive (*Elaeagnaceae angustifolia*)			Willow (*Salix* spp.)		
	Seep	Canal	Control	Seep	Canal	Control	Seep	Canal	Control
Number of samples	6	1	4	10	2	3	5	1	3
TCE trapped (nanograms)	105	54.8	MDL[a]	24.0	16.0	8.00	68.7	12.5	4.71
Total flow (liters per minute)	3.0	3.8	3.2	4.0	3.3	4.7	3.0	6.0	4.0
Subsample flow (liters per minute)	0.133	0.130	0.145	0.148	0.130	0.130	0.142	0.130	0.130
Sampling time (minute)	21	30	19	18	30	15	19	30	15
Water measured (grams)	0.057	0.150	0.110	0.086	0.205	0.100	0.110	0.080	0.047
TSC (milligrams of TCE per liter of transpired water)[b]	2.20	0.365	MDL	0.346	0.0803	0.0779	0.793	0.156	0.133

[a]MDL: method detection limit (1 nanogram of TCE on a trap).

[b]Water condensation in chamber during sampling may have caused transpiration stream concentrations (TSC) to be overestimated by a factor of 5.

canal locations. Table 18-3 summarizes the results of transpiration sampling at Hill AFB OU2, including the mass of TCE and water collected in each sampling, sampling rates, and transpiration stream concentrations (TSCs) for TCE. The TSC is the ratio of mass of TCE collected on the Tenax trap to the mass (or volume, 1 kilogram is 1 liter) of water collected on the desiccant. The TSCs are used for extrapolation of the mass flux of TCE *via* transpiration.

Trichloroethylene Transpired

The TSC values in Table 18-3 indicate the concentration of TCE transpired by the leaves in the sampling chambers at the seep and canal locations. The TSC values can be multiplied by a transpiration rate for the whole tree to predict the amount of TCE being phytovolatilized by each tree. This assumes that the TSCs are representative of all the leaves on the tree and are independent of the transpiration rate and the concentration of TCE in the groundwater.

One important consideration in obtaining accurate TSC values using a flow-through chamber is to maintain sufficient airflow through the chamber to prevent condensation of the transpired water. If water condenses in the chamber, the water flux will decrease, but the TCE flux will not be impacted, at least initially. Thus, water condensing in the chamber during the phytovolatilization measurements could result in TSC values that overestimate the ratio of TCE to water. It is important to note that some water vapor condensed in the chambers during these measurements even at flow rates of 3 to 6 liters per minute through the chamber. Subsequent field experiments performed at another Hill AFB site, show that water condensation could result in an overestimate of the TSC by as much as a factor of five (Zaugg *et al.* 2001). This potential overestimation should be considered when trying to estimate wholetree TCE efflux from the data in Table 18-3. To prevent condensation in current measurements, we have shifted to smaller chambers and slightly higher flow rates.

Scaling Transpiration and Trichloroethylene Efflux from Chambers to Entire Trees: Hill Air Force Base Example

The efflux of volatile compounds from whole trees is best extrapolated spatially and temporally from the ratio of TCE to water efflux as measured in the small chambers. This scaling requires the following:

1. The ratio of TCE to transpired water is the same in all leaves of the tree
2. The TCE is uniformly distributed within the root-zone
3. An estimate of the total transpiration rate of the tree

With access to ample water, and no competition from other trees, the daily transpiration rate of the largest trees in the seep location could be as high as 200 liters per day (Wullschleger *et al.* 1998). Using the highest measured TSC

(2.2 milligrams per liter) from Table 18-3, each large tree would be capable of transpiring 440 milligrams of TCE per day. Over a 120-day growing season, each tree would transpire 53 grams of TCE. A more realistic estimate of transpiration for a typical tree in this location is about 40 liters per day. Similarly, because of the condensation problem associated with the chamber sampling, the highest TSC value in Table 18-3 is probably high by a factor of nearly five. Thus, using more realistic estimates of TSC (0.44 milligram per liter) and transpiration rate (40 liters per day), the phytovolatilization removal rate would be about 2 milligrams of TCE each year per tree. Principles of scaling are discussed in more detail at (www.usu.edu/cpl).

Another removal mechanism for TCE is through leaf drop in the fall. However, the amount associated with leaf drop appears to be much less than removed by phytovolatilization. If the total TCE and TCE metabolite contaminant concentration in the leaf tissue is 1 to 2 milligrams per kilogram of dry tissue weight, and if the leaf production rate is 0.2 milligram per kilogram of dry tissue weight per year, the TCE removal rate would be 0.2 to 0.4 milligrams per square meter of ground area per year. Even using the low estimate for phytovolatilization (2 grams per tree per year), this is only 10 to 20 percent of the amount removed in phytovolatilization.

CONCLUSIONS

1. The identification of TCE, TCEt, TCAA, and DCAA in plants growing on TCE-contaminated groundwater at both sites establishes that the plant uptake and metabolism of TCE occurred in all the species sampled.
2. The concentrations of TCE in plant tissue were 10 to 100 times higher at Hill AFB than at CCAS, even though the groundwater concentrations below the vegetation were similar. We attributed the higher plant tissue TCE concentrations at Hill AFB to the higher fraction of contaminated groundwater used by the vegetation. This suggests that plants will have a greater impact on TCE attenuation at sites with lower rainfall during the growing season.
3. Trichloroethylene was found in transpiration gas samples collected from several trees located at the Hill AFB OU2 seep and canal locations, but no phytovolatilization was observed from any plants at the CCAS site. Estimates of TCE phytovolatilization by whole trees at Hill AFB range from 2 to 53 grams per tree per year.
4. The uniform vertical distribution of TCE in the trees, the presence of TCE in the transpiration gas, and the rough correlation with transpired water show that TCE is translocated through the trees. This suggests that uptake, sorption, and transformation within the tree are at steady state at this site. A steady state condition may be a requirement for significant phytovolatilization.

5. The significance of indigenous plants relative to other TCE-attenuation mechanisms was not specifically evaluated in this study. Additional studies addressing this issue are in progress.

Acknowledgment

The Air Force Center for Environmental Excellence (AFCEE) through Parsons Engineering Science, Inc., Denver, Colorado provided the primary funding for work at the CCAS field site and the Environmental Management Directorate at Hill AFB, UT for the work at Hill AFB OU2 site. The Utah Water Research Laboratory and the Utah Agricultural Experiment Station provided additional support. We would also like to thank the following individuals that were involved with sample collection, sample analysis, or data review: Bill Plaehn and Doug Downey (Parson Engineering Science, Inc.); Sam Taffinder (AFCEE, San Antonio, Texas); Robert Edwards (Booz, Allen and Hamilton, San Antonio, Texas); and Brady Orchard, Julie Chard, Scott Benson, Gus Koerner, and Ryan Dupont (Utah State University).

REFERENCES

Aitchison, E.W., S.L. Kelley, P.J.J. Alvarez, and J.L. Schnoor (2000) Phytoremediation of 1,4-dioxane by hybrid poplar trees. *Water Environ. Res.* **72** (3): 313–321.

Allen, R., L. Pereria, D. Raes, and M. Smith (1998) Crop evapotranspiration: guidelines for computing crop water requirements. FAO paper number 56. Food and Agriculture Organization, Rome, Italy.

Anderson, T.A. and B.T. Walton (1995) Comparative fate of ^{14}C trichloroethylene in the root zone of plants from a former solvent disposal site. *Environ. Toxicol. Chem.* **14**(12): 2041–2047.

Briggs, G.G., R.H. Bromilow, and A.A. Evans (1982) Relationships between lipophilicity and root uptake and translocation of non-ionised chemicals by barley. *Pestic. Sci.* **13**: 495–504.

Burken, J.G. and J.L. Schnoor (1998) Predictive relationships for uptake of organic contaminants by hybrid poplar trees. *Environ. Sci. Technol.* **32**: 3379–3385.

Camp, C.R., E.J. Sadler, and R.E. Yoder, eds. (1996) *Evapotranspiration and Irrigation Scheduling*. Proceedings of the American Society of Agricultural Engineers International Conference. Held at San Antonia, Texas, November 3–6. p. 1166.

Compton, H.R., D.M. Haroski, S.R. Hirsh, and J.G. Wrobel (1998) Pilot-scale use of trees to address VOC contamination. In: *Bioremediation and Phytoremediation: Chlorinated and Recalcitrant Compounds*. G.B. Wickramanyake and R.E. Hinchee eds. The First International Conference on Remediation of Chlorinated and Recalcitrant Compounds, Battelle Press, Columbus, Ohio. Held at Monterey, California, May 18–21. **C1–4**: 245–250.

Dawson, T.D. (1996) Determining water use by trees and forest from isotopic energy balance and transpirational analysis: the roles of tree size and hydraulic lift. *Tree Physiol.* **16**: 263–272.

Delrot, S., R. Atanassova, E. Gomés, and P. Coutos-Thévenot (2001) Plasma membrane transporters: machinery for uptake of organic solutes and stress resistance. *Plant Sci.* **161**(3): 391–414.

Doucette W.J., B. Bugbee, S. Hayhurst, W.A. Plaehn, D.C. Downey, S.A. Taffinder, and R. Edwards (1998) Phytoremediation of dissolved-phase trichloroethylene using mature vegetation. In: *Bioremediation and Phytoremediation: Chlorinated and Recalcitrant Compounds.* G.B. Wickramanyake and R.E. Hinchee, eds. The First International Conference on Remediation of Chlorinated and Recalcitrant Compounds, Battelle Press, Columbus, Ohio. Held at Monterey, California, May 18–21. **C1–4**: 251–256.

Gordon, M., N. Choe, J. Duffy, G. Ekuan, P. Heilman, I. Muiznieks, L. Newman, M. Ruszaj, B.B. Shurtleff, S. Strand, and J. Wilmoth (1997) Phytoremediation of trichloroethylene with hybrid poplars. In: *Phytoremediation of Soil and Water Contaminants.* E.L. Kruger, T.A. Anderson, and J.R. Coats, eds. American Chemical Society Symposium Series 664, Washington D.C., pp. 177–185.

Hayhurst, S.C. (1998) Evaluating the potential impact of existing vegetation on the fate of a trichloroethylene contaminated groundwater plume at Cape Canaveral Air Station (CCAS). M.S. thesis. Department of Civil and Environmental Engineering, Utah State University, Logan, Utah.

Hsu, F., R.L. Marxmiller, and A.Y.S. Young (1990) Study of root uptake and xylem translocation of cinmethylin and related compounds in detopped soybean and roots using a pressure chamber technique. *Plant Physiol.* **93**: 1573–1578.

Jones, S.A., R.W. Lee, and E.L. Kuniansky (1998) Phytoremediation of trichloroethylene (TCE) using cottonwood trees. In: *Bioremediation and Phytoremediation: Chlorinated and Recalcitrant Compounds.* G.B. Wickramanyake and R.E. Hinchee, eds. The First International Conference on Remediation of Chlorinated and Recalcitrant Compounds, Battelle Press, Columbus, Ohio. Held at Monterey, California, May 18–21. **C1–4**: 101–108.

Lewis, K. (2002) The relationship between tree-core and groundwater trichloroethylene concentrations for groundwater plume delineation. M.S. thesis. Department of Civil and Environmental Engineering, Utah State University, Logan, Utah.

Mallander, J.L. (1990) Climate of the Kennedy Space Center and vicinity. National Aeronautical and Space Administration Technical Memorandum No. 103498.

Marschner, H. (1995) *Mineral Nutrition of Higher Plants*, 2nd ed. Academic Press, New York.

McFarlane, J.C. (1995) Anatomy and physiology of plant conductive systems. In: *Plant Contamination: Modeling and Simulation of Organic Chemical Processes.* S. Trapp and J.C. McFarlane, eds. Lewis Publishers, Boca Raton, Florida, pp. 13–36.

McFarlane, J.C., T. Pfleeger, and J. Fletcher (1987) Transpiration effect on the uptake and distribution of bromacil, nitrobenzene, and phenol in soybean plants. *J. Environ. Qual.* **16**: 372–376.

Narayanan, M., L.C. Davis, and L.E. Erickson (1995) Fate of volatile chlorinated organic compounds in a laboratory chamber with alfalfa plants. *Environ. Sci. Technol.* **29**: 2437–2444.

Newman, L.A., S.E. Strand, N. Choe, J. Duffy, G. Ekuan, M. Ruszaj, B.B. Shurtleff, J. Wilmoth, and M.P. Gordon (1997) Uptake and biotransformation of trichloroethylene by hybrid poplars. *Environ. Sci. Technol.* **31**: 1062–1067.

Nilsen, E.T. and D.M. Orcutt (1996) *Physiology of Plants Under Stress.* John Wiley, New York.

Nzengung, V.A. and P. Jeffers (2001) Sequestration, phytoreduction, and phytooxidation of halogenated organic chemicals by aquatic and terrestrial plants. *Int. J. Phytoremed.* **3**(1): 13–40.

Orchard, B.J., W.J. Doucette, J.K. Chard, and B. Bugbee (2000a) A novel laboratory system for evaluating the fate of trichloroethylene in plants. *Environ. Toxicol. Chem.* **19**: 888–894.

Orchard, B.J., W.J. Doucette, J.K. Chard, and B. Bugbee (2000b) Uptake of trichloroethylene by hybrid poplar trees grown hydroponically in flow-through plant growth chambers. *Environ. Toxicol. Chem.* **19**: 895–903.

Russell, R.S. and V.M. Shorrocks (1959) The relationship between transpiration and the absorption of inorganic ions by intact plants. *J. Exp. Bot.* **10**: 301–316.

Schnabel, W.E., A.C. Dietz, J.G. Burken, J.L. Schnoor, and P.J. Alvarez (1997) Uptake and transformation of trichloroethylene by edible garden plants. *Water Res.* **4**: 816–824.

Schroll, R., B. Bierling, G. Cao, U. Dörfler, M. Lahaniati, T. Langenbach, I. Scheunert, and R. Winkler (1994) Uptake pathways of organic chemicals from soil by agricultural plants. *Chemosphere* **28**: 297–303.

Sicbaldi, F., G.A. Sacchi, M. Trevisan, and A.A.M. Del Re (1997) Root uptake and xylem translocations of pesticides from different chemical classes. *Pestic. Sci.* **50**: 111–119.

U.S. Environmental Protection Agency (U.S. EPA) (1996) Test methods for evaluating solid waste. SW-846, Revision 4, Washington, D.C.

Vroblesky, D.A., C.T. Nietch, and J.T. Morris (1999) Chlorinated ethenes from groundwater in tree trunks. *Environ. Sci. Technol.* **33**(3): 510–515.

Walton, B.T. and T.A. Anderson (1990) Microbial degradation of trichloroethylene in the rhizosphere: potential application to biological remediation of waste sites. *Appl. Environ. Microbiol.* **56**(4): 1012–1016.

Wullschleger, S., F. Meinzer, and R.A. Vertessy (1998) A review of whole-plant water use studies in trees. *Tree Physiol.* **18**: 499–512.

Zaugg, N.W., W.J. Doucette, B. Bugbee, K.L. Lewis, and C.J. Pajak (2001) Measurement and quantification of trichloroethylene phytovolatilization from phytoremediation systems: techniques and relationships useful in determining the fate of trichloroethylene in natural systems. Poster presented at the Air & Waste Management Association 94th Annual Conference & Exhibition. Held at Orlando, Florida, June 25–27

19

MULTIPLE-PROCESS ASSESSMENT FOR A CHLORINATED-SOLVENT PLUME

S. M. Eberts, G. J. Harvey, S. A. Jones, and S. W. Beckman

SUMMARY OF PRACTICAL IMPLICATIONS

A field project establishes that eastern cottonwood (*Populus deltoides*) trees partially attenuate low levels of trichloroethene (TCE) in groundwater. A multiple-process assessment, based on the first 5 years of performance data from a site developed using short-rotation-woody-crop-planting techniques, provides much of the science desired by the regulatory community. Trees planted where depth to water is 3.5 meters or less are estimated to reduce peak-growing-season outflow of contaminated groundwater by as much as 30 percent at full performance. Trees planted where depth to water is 3 meters or less delivered enough dissolved organic carbon to the aquifer to initiate microbial reductive dechlorination of the dissolved TCE. Hydraulic control was the primary *in situ* process for TCE-plume attenuation during the first 5 years of study. Enzyme transformation within the leaves may be the dominant attenuation process within the trees.

INTRODUCTION

Plants can withstand relatively high concentrations of organic chemicals without toxic effects while taking up and converting the chemicals to less toxic metabolites. Plant root dieback and exudation can build up enough organic carbon in the soil to stimulate biodegradation of organic chemicals in the rhizosphere—the soil that surrounds the roots (Schnoor 1997). The attenuation of organic chemicals, such as chlorinated solvents, by vegetation

Phytoremediation: Transformation and Control of Contaminants,
Edited by Steven C. McCutcheon and Jerald L. Schnoor.
ISBN 0-471-39435-1 (cloth)

involves multiple processes throughout the groundwater–soil–plant–atmosphere continuum and is the subject of ongoing investigation.

In 1996, a demonstration project was initiated by the U.S. Air Force at a site in Fort Worth, Texas, near Air Force Plant 4 (AFP 4) (Figure 19-1) to document the capability of eastern cottonwood (*Populus deltoides*) trees to attenuate low-level TCE contamination of groundwater. The ability to conduct long-term field-scale monitoring and evaluation at this site contributes to the understanding of multiple processes involved in phytoremediation technology, provides practical insights into field-scale applications, and provides much of the science desired by the regulatory community. The purpose of this chapter is to discuss what was learned from close observation of the groundwater-phytoremediation system during the first 5 years after planting. An in-depth description of the demonstration project, including an economic analysis, can be found in U.S. Environmental Protection Agency (2002).

Background

Air Force Plant 4 was built in 1942 to assemble aircraft for World War II. For decades, TCE was used in numerous industrial processes to clean metal parts before plating and painting of aircraft. The assembly of tens of thousands of aircraft and the use of TCE left a legacy of shallow groundwater contamination at AFP 4 and in the area immediately down gradient of the manufacturing facility.

Before development of the phytoremediation system near AFP 4, natural attenuation of TCE at the selected site was limited to dilution, dispersion, sorption, and possibly some volatilization. Some *in situ* biodegradation of the TCE plume may have occurred up gradient of the phytoremediation site (Rust Geotech 1996); however, site characterization and background sampling (1996) provided no evidence of *in situ* biodegradation of TCE across the site, which is 1.5 kilometers down gradient of the primary suspected source. Dissolved oxygen concentrations typically were greater than 5 milligrams per liter, total organic carbon generally was undetectable at a detection limit of 1.5 milligrams per liter, and ratios of TCE to daughter products remained relatively constant in the groundwater across the site during site characterization (Jacobs Engineering Group 1996).

To enable biodegradation of chlorinated solvents by means of reductive dechlorination, native or anthropogenic carbon must be available for use as an electron donor. Once dissolved oxygen is depleted, anaerobic microorganisms most often use available electron acceptors in the following order: nitrate, ferric iron oxyhydroxide, sulfate, and carbon dioxide. In the absence of dissolved oxygen and nitrate, chlorinated solvents compete with other electron acceptors and donors—especially sulfate and carbon dioxide. As a result, biodegradation of chlorinated solvents such as TCE is typically an electron-donor-limited process (Wiedemeier *et al.* 1996).

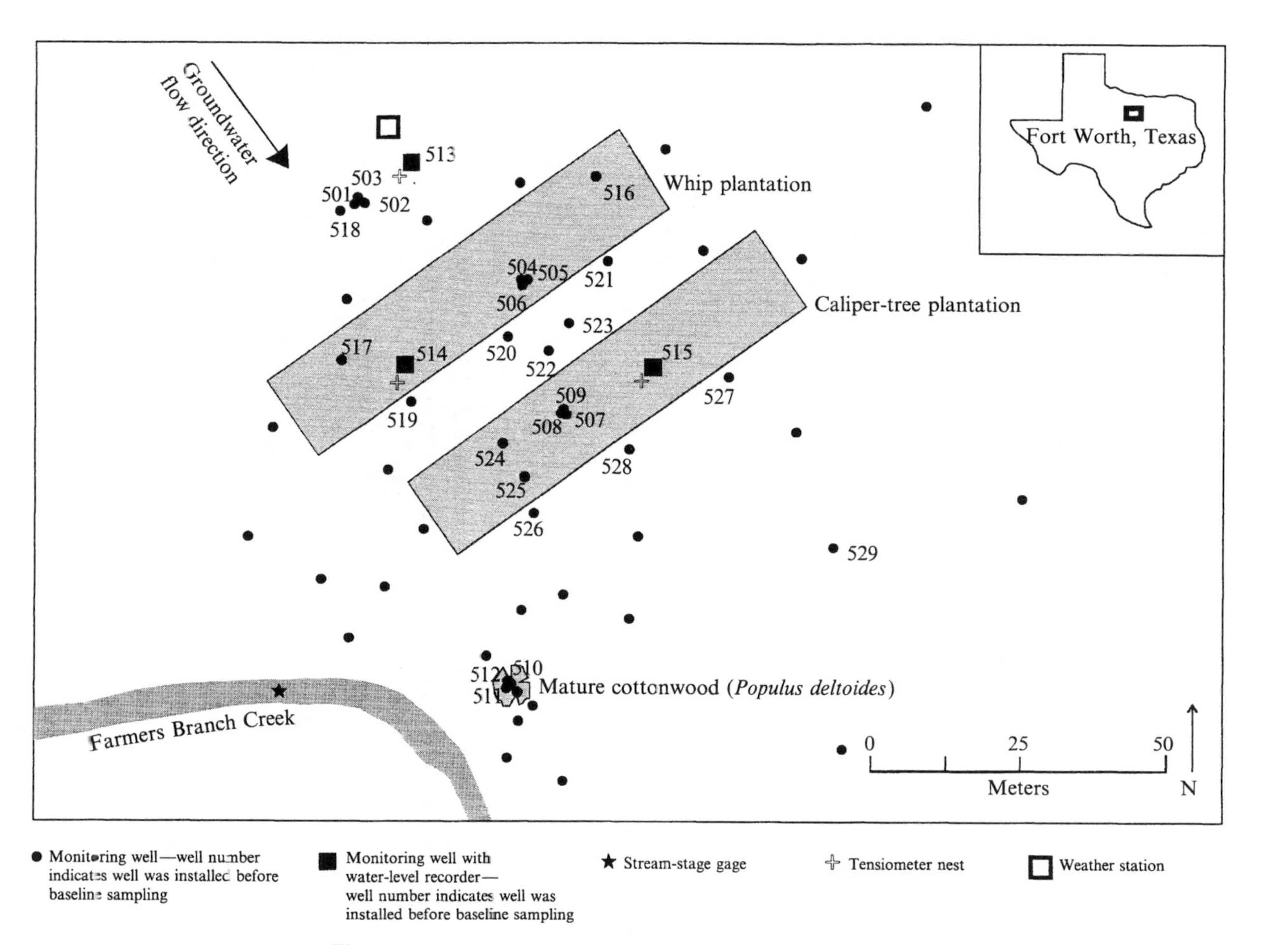

Figure 19-1 Layout of phytoremediation site in Fort Worth, Texas.

Conditions such as those at the site near AFP 4 that are characterized by dissolved oxygen concentrations greater than 1 milligram per liter, and with limited natural or anthropogenic carbon (electron donors) are referred to as Type-3 conditions. Under such aerobic conditions, reductive dechlorination of chlorinated solvents will not occur (Wiedemeier *et al.* 1996). (Type-1 and Type-2 conditions are characterized by a supply of anthropogenic or natural organic carbon, respectively, adequate for microbial reductive dechlorination.) Many chlorinated solvent plumes that are associated with military and industrial sites are in shallow aquifers that lack carbon sources and are well oxygenated from infiltrating precipitation—Type-3 conditions.

Eastern cottonwood (*Populus deltoides*) trees have the potential to add enough natural organic carbon to the groundwater at some chlorinated-solvent-contaminated sites to support *in situ* microbial reductive dechlorination of the contaminant; that is, to convert Type-3 to Type-2 conditions. In addition, phytoremediation by use of phreatophytes such as eastern cottonwood (*Populus deltoides*) has the potential to remove the contaminant mass through contaminant uptake at Type-3 sites.

Project Objectives

The overall objective of the project is to generate cost and performance data necessary to determine the applicability and limitations of phytoremediation technology for use at similar sites with similar contaminant profiles. A primary objective and six secondary project objectives were established to help guide investigations.

Reduction in Mass Flux

The primary objective of the project is to determine how effectively eastern cottonwood (*Populus deltoides*) trees reduce the mass flux of dissolved TCE within an aquifer across the down gradient end of a planted area. It was forecast that the physiological processes of the cottonwood (*Populus deltoides*) tree would reduce the TCE mass flux in the alluvial aquifer through a combination of hydraulic control of the contaminant plume and *in situ* biodegradation of the contaminant mass (natural pump-and-treat). Specifically, it was forecast that the tree roots would remove contaminated water from the aquifer and that TCE would be altered biologically within the trees or would be transpired and volatilized into the atmosphere. The trees also would promote microbial reductive dechlorination of dissolved TCE within the originally aerobic aquifer.

Elucidation of Biological, Hydrological, and Biogeochemical Processes

Secondary objectives were included in the project to elucidate the biological, hydrological, and biogeochemical processes that contribute to the effectiveness of phytoremediation of TCE-contaminated groundwater. Because phytoremediation can take years to become fully effective, much of the data associated with the secondary objectives were collected to prepare models to determine future performance. Secondary objectives include (1) documenting tree growth rates and root biomass, (2) analyzing tree transpiration rates to determine current and future water use, (3) analyzing hydrologic effects of tree transpiration on the contaminated aquifer, (4) analyzing contaminant uptake into tree organs, (5) evaluating subsurface oxidation–reduction processes in the groundwater, and (6) evaluating microbial contributions to reductive dechlorination.

Site Description

The site selected for the project is within a narrow finger (less than 100 meters in width) of a larger groundwater plume of TCE that originates up gradient of the site at AFP 4. Depth to groundwater at the time of site characterization ranged from 2.5 to 4 meters below the land surface. The saturated thickness of the alluvial aquifer, which is comprised of clayey sands and gravels, ranged from 0.5 to 1.5 meters. The porosity of the alluvium at the site, as determined in the laboratory, is 25 percent. Horizontal hydraulic conductivities for the aquifer, as determined from analysis of 11 slug tests, range from 1 meter per day (0.0012 centimeters per second) to 30 meters per day (0.035 centimeters per second) with a geometric mean of 6 meters per day (0.007 centimeters per second). The average hydraulic gradient across the site was just over 2 percent. The average groundwater velocity was approximately 0.5 meter per day. Groundwater discharges into Farmers Branch Creek down gradient of the site. A relatively impermeable, massively bedded shaley limestone underlies the aquifer. Concentrations of TCE in the alluvial aquifer ranged from 230 to 970 micrograms per liter and *cis*-1,2-dichloroethene (*c*DCE) concentrations ranged from 24 to 141 micrograms per liter. Groundwater was aerobic across the site at this time, most likely a result of oxygenated recharge from precipitation. The site is subhumid and receives approximately 80 centimeters of precipitation per year. Groundwater recharge from precipitation averages 6 centimeters per year. These site-characterization data suggest that roots of planted trees could potentially reach the water table, and that site conditions could benefit from processes that promote reductive dechlorination through the depletion of dissolved oxygen.

System Design

Two plantations of eastern cottonwood (*Populus deltoides*) trees (poplars) were established at the site during the spring of 1996. Eastern cottonwoods (*Populus deltoides*) were selected because of rapid growth, high transpiration rates, and direct uptake of water from the water table. Zhang *et al.* (1999) recently reported for poplar (*Populus* spp.) trees at a study area in southern England that between 15 and 60 percent of the transpired water came directly from the water table. The fact that eastern cottonwood (*Populus deltoides*) is native as far south and west as Texas also contributed to the decision to select these trees for the site (Southern Forest Nursery Management Cooperative, accessed December 19, 2000 at URL http://www.forestry.auburn.edu/sfnmc/class/cotton.html).

The two plantations at the site are rectangular (15 meters by 75 meters), designed so that the long sides are approximately perpendicular to the general direction of groundwater flow. This orientation was chosen so that the plantations serve as barriers to subsurface contaminant movement (Figure 19-1). The first plantation was planted with whips, which are sections of 1-year-old stems harvested from branches during the dormant season. The second plantation, 15 meters down gradient of the whips, was planted with 1-year-old seedlings of 2.5- to 3.8-centimeter caliper trunk diameter ("caliper" trees). Two sizes of trees were selected for this demonstration so that differences in the rate of growth, contaminant reduction, and costs could be compared for variations in the planting strategy. The whips were obtained from the Texas Forest Service in Alto, Texas, and include a mixture of the following cottonwood (*Populus deltoides*) clones: S7C1, S7C2, S7C4, S7C8, S7C13, S7C15, S7C20, S7C21, S13C15, S13C20, and KEN8. The caliper trees are the *Sioux Land* variety and were obtained from Gandy Nursery in Ben Wheeler, Texas.

The planting techniques used in this demonstration generally follow those established for short-rotation woody crops, which were developed for the production of bioenergy (heat, power, and transportation fuel) and fiber (pulp, paper, particle board, plywood, and other uses). This decision was made to facilitate scale-up of phytoremediation technology. A compilation of abstracts from a recent meeting of the International Poplar Commission (IPC 2000) provides an overview of current poplar (*Populus* spp.) and willow (*Salix* spp.) culture research topics (Isebrands and Richardson 2000).

Preparation for planting at the site involved trenching seven rows in each plantation to the depth of 1 meter, with a spacing of 2.4 meters between the rows. The whips, which were approximately 0.5 meters long, were placed 1.2 meters apart within the rows in the up gradient plantation, and the caliper trees were placed 2.4 meters apart within the rows in the down gradient plantation. Irrigation lines were placed within the trenches along with the vegetative material. After the trenches were backfilled, the whips extended 5 centimeters above land surface. The close spacing of the plantings, most notably the whips, was to allow tree growth and survival to conform to soil and aquifer heterogeneities. However, closely spaced cottonwoods (*Populus deltoides*) have been shown to have lower survival rates and smaller diameters than widely spaced cottonwoods (*Populus deltoides*; Krinard 1985). This planting strategy met the needs of this project, but may not have been adequate if the trees had been intended for biomass production. Recent research has shown that poplar spacing should provide approximately 6.5 square meters per tree if the mean tree diameter at harvest is targeted to be 15 centimeters, a size necessary for reasonable harvesting costs (DeBell and

Harrington 2000). Variations in TCE concentrations at the site were not expected to affect tree growth (Dietz and Schnoor 2001).

An agronomic assessment for macronutrients and micronutrients and the presence of hardpan at the site was done before planting. The need for fertilizer was determined from the soil characteristics that were identified, as well as from discussions with the Texas Forest Service, Tarrant County Agricultural Extension Service, and the Texas A&M Horticulture Department. A handful of slow-release Osmacote™ 14-14-14 fertilizer was applied around each whip and caliper tree. After planting was completed, fabric mulch and 10 centimeters of landscape mulch were placed along each of the planted rows to reduce weed competition. This placement of mulch was especially important for the newly planted whips.

The drip-irrigation system was required to supplement precipitation for the first two growing seasons. The trees were watered liberally during this time to encourage deep root development. Because the roots were expected to intercept percolating irrigation water (Licht and Madison 1994), irrigation was not considered to be an additional source of water to the aquifer.

Because phytoremediation processes require years to develop, the monitoring system was designed to measure small incremental changes in site conditions over time. The monitoring strategy for this project is more extensive than would be required for a typical phytoremediation system because of the desire to study multiple processes. The following monitoring stations were established:

1. Sixty-seven wells up gradient, within, down gradient, and surrounding the tree plantations
2. Continuous water-level recorders in three monitoring wells, including one up gradient of the tree plantations and two within the plantations
3. Nine tensiometers (in three nests) up gradient or within the tree plantations
4. A weather station to collect site-specific climate data
5. A stream gage on a creek adjacent to the site to record stream stage
6. Tree collars or tree probes installed periodically during the growing season to measure sap flow in selected trees

The location of the monitoring points with respect to the tree plantations is shown in Figure 19-1. Ten wells are not shown in Figure 19-1 because these are outside the area depicted. These wells were used to collect groundwater-level data in the surrounding area for use in calibrating a groundwater-flow model (presented in detail in the section "Hydrologic Effects of Tree Transpiration") that was used to simulate future system performance.

Eight of the sixty-seven wells were installed beneath six mature trees near the planted site; three of these wells were installed beneath a 19-year-old cottonwood (*Populus deltoides*) tree. Data were collected from these wells to provide early feedback on what may be achieved by the whip and caliper-tree plantations at the site. Transpiration rates for the mature cottonwood (*Populus deltoides*) tree were also estimated by use of tree probes.

MEASURE OF PERFORMANCE

Changes in the mass flux of TCE in the aquifer across the down gradient end of the planted area had to be calculated to address the primary project objective. To determine the mass flux of TCE at a given time, the volumetric flux of groundwater across the down gradient end of the site was multiplied by the average of the TCE concentrations in a row of wells immediately down gradient of the site. The volumetric flux of groundwater (Q) was calculated for each event (baseline, peak-growing season, and late growing season) according to Darcy's Law: $Q = -KiA$, where K is the hydraulic conductivity of the aquifer, i is the hydraulic gradient in the aquifer across the down gradient end of the planted area, and A is the cross-sectional area of the aquifer along the down gradient end of the planted area.

Three simplifying assumptions applied to these calculations

1. A single value of 6 meters per day (the geometric mean horizontal hydraulic conductivity determined for the study area) could be used for horizontal hydraulic conductivity in all calculations. Despite some actual variation in hydraulic conductivity along the down gradient end of the planted area, this assumption does not affect performance calculations because hydraulic conductivity is constant and cancels out when volumetric flux for selected events is compared to volumetric flux at baseline conditions. (November 1996 was used to represent baseline conditions in the aquifer because this time period is when the most comprehensive sets of water-level and groundwater-chemistry data were collected for the period before the tree roots reached the water table.)
2. The hydraulic gradient across the down gradient end of the planted area could be calculated at selected times using groundwater-elevation data from monitoring wells 522 and 529 (Figure 19-2). Well 522 is between the tree stands near the center of the planted area. Well 529 is down gradient and was outside the expected area of influence of the trees. These wells were chosen so as not to reflect increases in the hydraulic gradient across the up gradient end of the plantations. A corresponding potentiometric-surface map for each selected time was consulted to verify that changes in hydraulic gradient were caused by the influence of the trees rather than the slight changes in the direction of groundwater flow.
3. The thickness of the saturated zone at the selected times could be calculated from the average thickness of the aquifer in the monitoring wells immediately down gradient of the tree plantations (wells 526, 527, and 528) (Figure 19-2). The saturated thickness in each of these three wells was

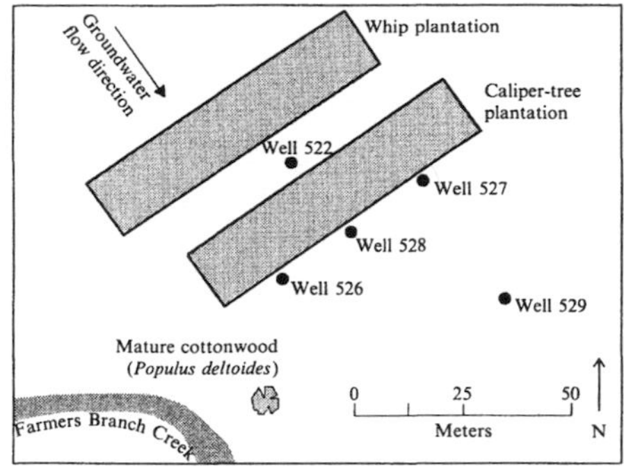

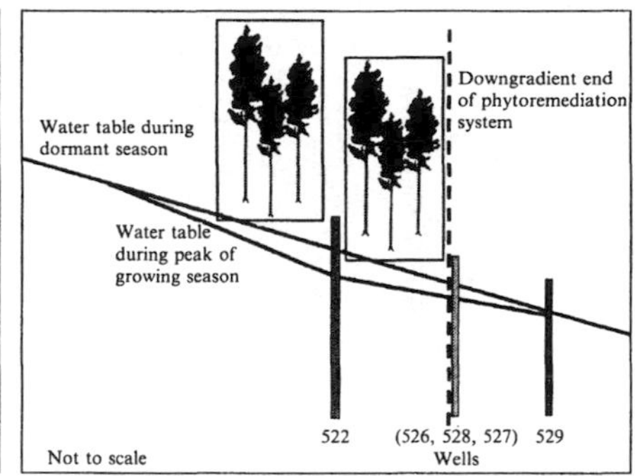

Figure 19-2 Wells used to monitor for changes in the volumetric flux of groundwater across the down gradient end of the phytoremediation system.

first normalized to wells in the surrounding area, to account for temporal changes in the saturated thickness of the aquifer unrelated to the planted trees. Specifically, the water-level data for these wells were adjusted by an amount equal to the difference between the water level at the selected time and the water level at baseline in wells outside the influence of the planted trees.

The mass flux (M_f) across the down gradient end of the planted area was subsequently calculated for the various events (baseline, peak-growing season, and late-growing season) according to $M_f = Q \cdot C$, where Q is the volumetric flux of groundwater and C is the average of the TCE concentrations in wells 526, 527, and 528 (Figure 19-2) for each event.

The following formula then was used to calculate the percentage change in the mass flux ($\Delta M_{f(\text{event } x)}$) of TCE at selected times that can be attributed to the planted trees: $\Delta M_{f(\text{event } x)} = (M_{f(\text{baseline})} - M_{f(\text{event } x)})(100)/M_{f(\text{baseline})}$, where event x is peak of the growing season (late June or beginning of July) 1997, 1998, 1999, or late in the growing season (end of September or beginning of October) 1997 or 1998.

OBSERVED CHANGES IN TRICHLOROETHENE MASS FLUX

The changes in the mass flux of TCE across the down gradient end of the site during the first three growing seasons are summarized in Table 19-1. Beyond the fourth growing season (1999), changes in the mass flux of TCE across the down gradient end of the planted area could not be calculated by use of the equations described in the previous section, because the cone of influence of the planted trees expanded to encompass all of the wells in the network that was established for the collection of baseline data (labeled wells in Figure 19-1). As a result, saturated thickness in the row of wells along the down

TABLE 19-1 Observed Changes in the Volumetric Flux of Groundwater and the Mass Flux of Trichloroethene (TCE) Across the Down Gradient End of the Planted Area, Fort Worth, Texas

Event	Hydraulic gradient across down gradient end of planted area (meter per meter)	Cross sectional area along down gradient end of planted area (square meters)	Volumetric flux of groundwater across down gradient end of planted area (cubic meters per day)	Change in volumetric flux of groundwater across down gradient end of planted area attributed to planted trees (percent)	Average TCE concentration in wells along down gradient end of planted area (micrograms per liter)	Mass flux of TCE across down gradient end of planted area (grams per day)	Change in mass flux of TCE across downgradient end of planted area attributed to planted trees (percent)
Baseline (1996)	0.0159	84	8.0	—	469	3.8	—
Peak 2nd season (1997)	0.0154	82	7.6	−5	535	4.1	8
Late 2nd season (1997)	0.0157	83	7.8	−2	—	—	—
Peak 3rd season (1998)	0.0143	82	7.0	−12	483	3.4	−11
Late 3rd season (1998)	0.0150	83	7.5	−6	473	3.5	−8
Peak 4th season (1999)	0.0153	81	7.4	−8	—	—	—

Note:– means that data is not available.

gradient end of the caliper-tree plantation could no longer be normalized to account for temporal changes in water levels unrelated to tree transpiration. Without this normalization of water-level data, the volumetric flux of groundwater—and, therefore, the mass flux of TCE for the sampling events—could not be compared to baseline conditions. Consequently, changes in the volumetric flux of groundwater across the down gradient end of the site (southeast) beyond the fourth growing season were simulated by use of the groundwater-flow model.

The data in Table 19-1 indicate that the TCE mass flux actually increased 8 percent during the peak of the second growing season, as compared to baseline conditions. The planted trees reduced outward flux of groundwater by 5 percent during the peak of the second season, but TCE concentrations in water samples from the row of wells immediately down gradient of the trees were higher, resulting in the increased TCE mass flux. These data suggest that mass flux of TCE out of the planted area during the peak of the second season could have been even greater without the hydraulic influence of the trees. The TCE mass flux during the third growing season was down 11 percent at the peak of the season and 8 percent near the end of the season, as compared to baseline conditions. Concentrations of TCE during the third season in the row of down gradient wells were similar to concentrations at baseline. Flux of groundwater out of the site during the peak of the fourth growing season was 8 percent less than at baseline. Variations in climatic conditions are the likely explanation for differences in outward flux of groundwater between the third and fourth seasons.

The relative contributions of hydraulic control and microbial reductive dechlorination as attenuation mechanisms within the aquifer in the early years of growth can be evaluated from these data. Specifically, the observed decrease in the volumetric flux of groundwater out of the site combined with no observed decrease in TCE concentration indicates that the primary mechanism for the reduction in mass flux during the first 3 years after planting was hydraulic control. Newman *et al.* (1999) also reported that the majority of TCE loss from an artificial aquifer planted with hybrid poplar (populus spp.) trees, 3 years after planting, resulted because of plant uptake of water. The largest observed reduction in hydraulic gradient at the site, 10 percent (from 0.0159 to 0.0143), occurred during the peak of the third growing season (June 1998). The maximum drawdown that could be attributed to the trees at this time was 10 centimeters and was observed between the whip and caliper-tree plantations. Although a drawdown cone could be mapped at the water table by the third growing season, a regional hydraulic gradient remained across the site. As a result, most of the contaminated groundwater continued to flow outward across the down gradient end of the planted area (Figure 19-3).

Preparation of the groundwater-flow model furthered understanding of the observed effects of tree transpiration on the aquifer. The model illustrates that the volume of water that was transpired directly from the aquifer during the third growing season was greater than the decrease in the groundwater

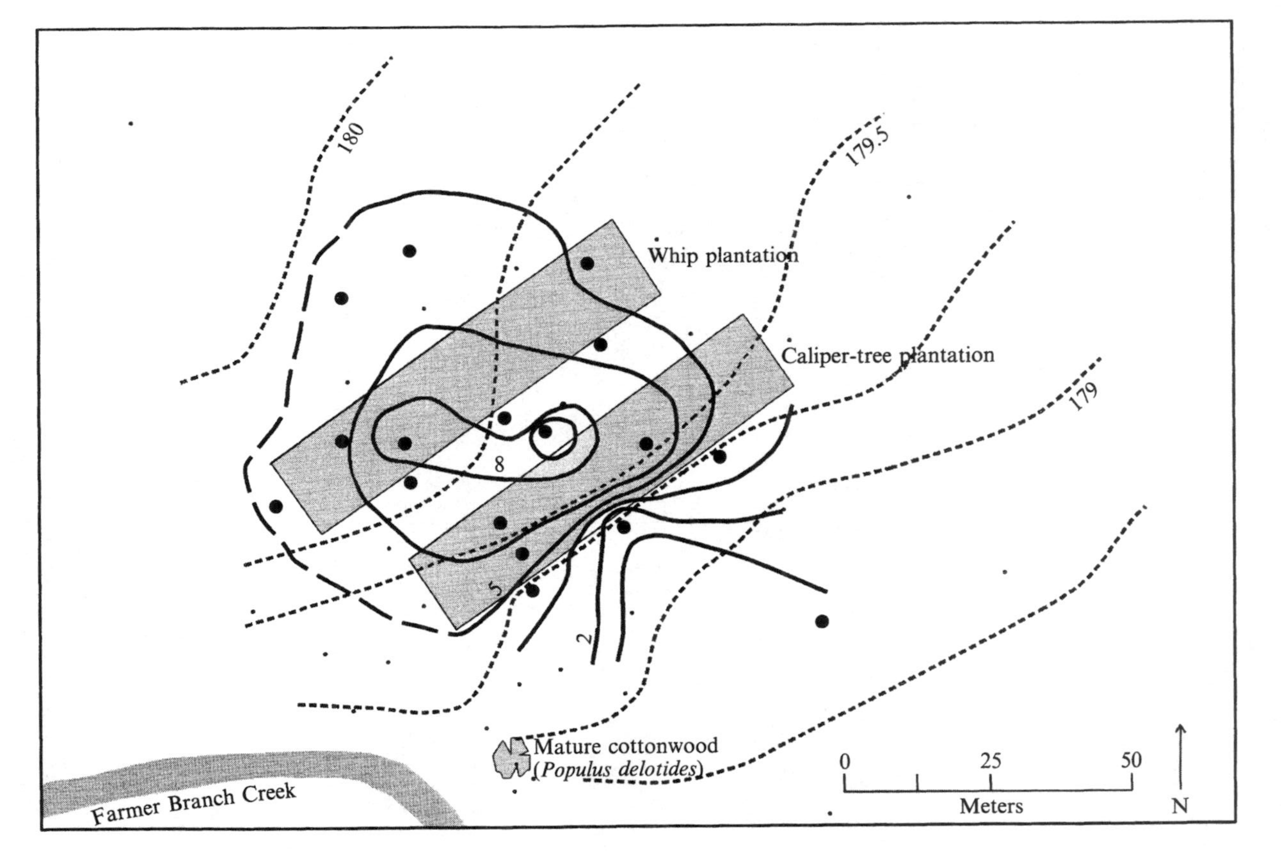

Figure 19-3 Drawdown at the water table that can be attributed to the planted trees, and water-table altitude, peak of third growing season (June 1998).

outflow that can be attributed to the trees. This is because the amount of groundwater inflow to the site increased because of an increased hydraulic gradient on the up gradient side of the drawdown cone created by the trees. The amount of contaminated groundwater removed from the aquifer by means of uptake during the peak of the third growing season was closer to 20 percent of the initial volumetric flux of water through the site, as opposed to the observed 12 percent decrease in outflow.

MEASURED AND MODELED BIOLOGICAL, HYDROLOGICAL, AND BIOGEOCHEMICAL PROCESSES

Scientists from numerous organizations and various disciplines conducted integrated investigations to address the secondary objectives, which were aimed at understanding processes associated with phytoremediation. Organizations that have been involved in elucidating phytoremediation processes at the site include Science Applications International Corporation, University of Georgia, U.S. Air Force, U.S. Environmental Protection Agency, U.S. Forest Service, and U.S. Geological Survey. Results of these investigations, along with lessons learned from the field over a 5-year period, are described in the following sections.

Tree Growth Rates and Root Biomass

The rate of tree growth (above and below ground) at the site was documented to determine the progression of the phytoremediation system over time. Evaluation of the growth rate of the planted trees over time was important because clonal differences in cottonwood (*Populus deltoides*) growth can take time to appear (Cooper and Ferguson 1979). Such documentation provides insight into the benefits and limitations of the planting strategy used in this demonstration, as well as the capability of the phytoremediation system to affect the groundwater hydrology of the site.

During the first three growing seasons, scientists with Science Applications International Corporation measured 52 whips and 51 caliper trees for trunk diameter, tree height, and canopy diameter several times. Tree roots in both plantations reached the water table at selected locations some time during the second growing season. The caliper trees had slightly more root biomass below 1.5 meters than the whips at this time (Ronald Hendrick, University of Georgia, written communication, 1998). Only 2 of the 52 whips and 3 of the 51 caliper trees failed to survive the first 3 years. Some of the trees in the plantations that had not been selected for measurement were stunted by beaver activity. Others were stunted or died as a result of monitoring well installation, which took place after planting and during the first growing season.

Trunk diameter increased to approximately 5 centimeters for the whips and 8 centimeters for the caliper trees during the first 3 years. Although the caliper trees were taller during the first growing season, the whips were able to approach the height of the caliper trees by the end of the third growing season (1998); at that time, average canopy diameter for the selected whips was 2.3 meters and for the caliper trees was 2.5 meters. More closely spaced than the caliper trees, the whips were nearer to achieving closed canopy at that time. Although light competition limits the maximum transpiration attainable for individual trees under a closed canopy, the maximum amount of water that can be transpired by the plantation as a whole is not limited by tree spacing and, typically, occurs once the canopy closes. By the end of the third growing season, performance data indicated no advantage to planting the more expensive caliper trees (8.00 U.S. dollars) as opposed to whips (0.20 U.S. dollars).

Notable differences in the whip- and caliper-tree plantations, however, showed up in subsequent growing seasons. By the fifth growing season (2000), much of the caliper-tree plantation had reached closed canopy. This result was not true for the whip plantation. In addition, the height of the caliper trees varied predictably with depth to groundwater as determined from measurements taken along an outer row of the plantation, whereas no relation between height and depth to groundwater was apparent in the whip plantation as determined from similar measurements (Figure 19-4). The 2000 growing season was characterized by extreme drought. This drought could have magnified effects of variations in soil structure on tree growth at the site (Hamblin 1985), differences in root biomass that may have resulted from variations in the planting strategy (whips versus caliper trees), and differences in the drought tolerance of the different clonal varieties that were planted (Tschaplinski *et al.* 1998, Khurana *et al.* 2000). Differences in the capability of the various trees to access groundwater may have been masked by irrigation in the early years after planting (Persson 1995). It is not known whether the predictable relation between tree growth and depth to groundwater in the

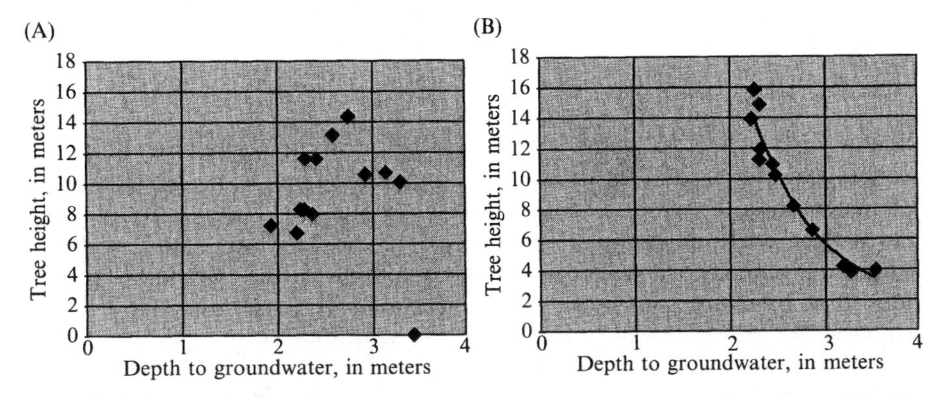

Figure 19-4 Tree height plotted against depth to groundwater, fifth growing season: (A) whip plantation, (B) caliper-tree plantation.

caliper-tree plantation, which is desirable for the purpose of phytoremediation system design, is because of the advantages provided by the larger initial root biomass or the monoclonal nature of the caliper-tree plantation, or both. Research on short-rotation woody crops, however, indicates that mixtures of vegetative material (such as the whip plantation) ensure greater plantation sustainability than do monocultures (such as the caliper-tree plantation) under elevated risk of disease (McCracken and Dawson 2000).

During the fifth growing season, trees growing where the water table dropped 3.5 meters below land surface exhibited signs of severe water stress. Maximum rooting depth for eastern cottonwood (*Populus deltoides*) has been reported as near 3.6 meters (Stone and Kalisz 1991). This observation may point to one limitation of the phytoremediation strategy at the site, which may be overcome by the use of more deeply rooting trees. An example of more deeply rooting trees that may warrant further study include—but should not be limited to—other poplar species (*Populus* spp.), willow (*Salix* spp.), mesquite (*Prosopis* spp.), eucalyptus (*Eucalyptus* spp.), and black locust (*Robinia* spp.) (Stone and Kalisz 1991). In addition, planting techniques that encourage the development of deeper roots could be used (Negri, Gatliff, *et al.* this book).

Tree Transpiration Rates

The amount of water the planted eastern cottonwood (*Populus deltoides*) trees can transpire throughout a lifecycle will determine, in part, the effectiveness of the phytoremediation system at the site. Transpired water can be derived from the near-surface soils and from the saturated zone (groundwater or aquifer). The amount of water transpired directly from the aquifer is of greatest interest in this investigation because the TCE and the daughter products at the site are dissolved in the groundwater.

Tree- and stand-level transpiration rates in the two age classes (whips and caliper trees) were quantified for the first five growing seasons. The climatic and physiological driving variables related to the observed patterns of transpiration were also investigated and are described by Vose *et al.* (this book).

Sap flow was measured on a statistical sampling of trees in the whip and caliper-tree plantations by use of sap flow gauges (Dynagage®, Dynamax Inc., Houston, TX) or thermal dissipation probes (TDP, Synamax, Inc., Houston, TX) during May, June, July, August, and October of the second and third growing seasons (1997 and 1998) and July of the fourth and fifth growing seasons (1999 and 2000, respectively). Sap flow also was measured by the use of thermal dissipation probes on six mature trees near the site in May, July, and September 1998, (Vose *et al.* 2000; James Vose, U.S. Forest Service, written communication, 2000).

Sap flow was less in individual trees within the whip plantation than in individual trees within the caliper-tree plantation for all months during the second growing season, except October. Mean total daily transpiration for trees in the whip plantation ranged from 9.2 kilograms of water per tree per day in June 1997 to 1.6 kilograms of water per tree per day in October 1997. Mean total daily transpiration for trees in the caliper-tree plantation ranged from 14.7 kilograms of water per tree per day in July 1997 to 0.92 kilogram of water per tree per day in October 1997. The average seasonal sap flow for trees in the whip plantation was half that of the trees in the caliper-tree plantation (0.34 kilogram of water per tree per hour *versus* 0.61 kilogram of water per tree per hour) during the second growing season. Preliminary estimates of stand-level transpiration were extrapolated from these transpiration rates by assuming that the amount of sap flow measured in the sample trees represented the population. The stand-level estimates indicated very little difference in the amount of water transpired from the whip and caliper-tree plantations during the second season, because the planting density of the whips was nearly twice that of the caliper trees. Stand-level sap flow was estimated at 15 600 kilograms per hectare per day for the whip plantation and 16 600 kilograms per hectare per day for the caliper-tree plantation. The associated estimates of growing-season transpiration for the second growing season were 23.5 centimeters and 25.5 centimeters for the whip and caliper-tree plantations, respectively (Vose *et al.* 2000).

The relative amount of water transpired by the whip and caliper-tree plantations changed appreciably by the fifth growing season. During July 2000, sap-flow rates for the selected trees in the whip plantation averaged 20 kilograms of water per tree per day, whereas sap-flow rates for the selected trees in the caliper-tree plantation averaged 80 kilograms of water per tree per day (James Vose, U.S. Forest Service, written communication, 2000). These data suggest that the caliper-tree plantation was appreciably outperforming the whip plantation in terms of water use by this time, even though there were twice as many trees in the whip plantation.

The sap-flow rate measured for the mature cottonwood (*Populus deltoides*) tree adjacent to the site was as high as 230 kilograms of water per day. This rate represents an upper limit for transpiration by a single tree at the site but most likely will not be attainable by individual trees within the plantations; individual trees within plantations generally do not reach the transpiration potential, as previously discussed (James Vose, U.S. Forest Service, written communication, 2000).

Because the plantations were not expected to reach maturity during the period of evaluation, a modeling approach was used to estimate future performance. Site-specific climate, sap flow, soil-moisture, and tree-root data were used to parametrize the model PROSPER (Goldstein *et al.* 1974), a mechanistic model of sap flow, to estimate stand-level transpiration for the phytoremediation system once the canopy closes. Results of the simulations vary according to assumptions made regarding future climatic conditions, soil moisture, and root-growth parameters. Simulated stand-level transpiration for closed canopy at the site is the same for the whip and caliper-tree plantations and ranges from 25 to 48 centimeters per growing season, depending on how the model was set up (Eberts *et al.* 1999).

The root biomass study was used to help determine the percentage of this simulated transpiration that may be derived from the saturated zone as opposed to the near-surface soils. Simulated transpiration of groundwater ranges from 12 to 28 centimeters per growing season at closed canopy, depending on model set up; this value is 48 to 58 percent of simulated total transpiration (Eberts *et al.* 1999).

Long-term, groundwater-level hydrographs from the site also were used to estimate transpiration directly from the aquifer. Diurnal fluctuations of the water table, which are attributed to transpiration from the aquifer, began to appear in the hydrographs of wells 514 and 515 (Figure 19-1) late in the second growing season. Notable diurnal fluctuations were recorded in the third growing season, although large data gaps resulted during the early part of the third season (because of problems with tree roots entangling the water level floats). By the fifth growing season, diurnal water-table fluctuations of more than 5 centimeters were recorded.

According to the method used to estimate transpiration of groundwater from groundwater-level hydrograph data (White 1932), the daily transpiration of ground water (q) can be calculated from $q = y(24r \pm s)$, where y is the specific yield of the aquifer, r is the slope of the hydrograph during the nightly recovery and is equal to the hourly groundwater inflow rate, and s is the daily change in the water table (Figure 19-5). Monthly transpiration of

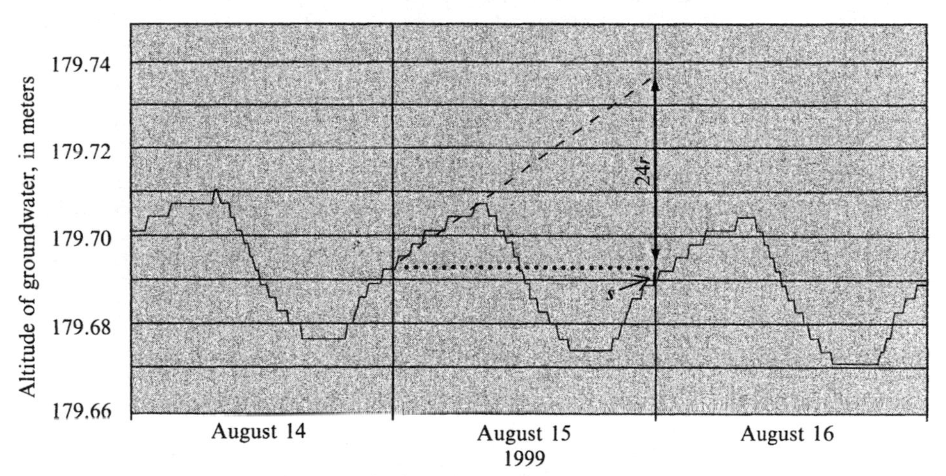

The daily transpiration of groundwater can be calculated from

$$q = y(24r \pm s)$$

where y is aquifer specific yield, r is the slope of the hydrograph during the nightly recovery and is equal to the hourly groundwater inflow rate, and s is the daily change in the water-table altitude

Figure 19-5 Method used to estimate transpiration of groundwater from a groundwater-level hydrograph (White 1932).

groundwater for the third, fourth, and fifth growing seasons was estimated by calculating the transpiration of groundwater for a single day at midmonth and multiplying the calculated rate by the number of days in the month. Growing-season estimates of groundwater transpiration were computed by summing the monthly estimates (Table 19-2). The aquifer-specific yields used in these calculations were estimated during calibration of the groundwater-flow model. The specific yield for the sand and gravel at well 514 in the whip plantation is 0.023, whereas the specific yield for the clay-rich material at well 515 in the caliper-tree plantation is 0.01.

The groundwater-level hydrographs indicate that transpiration from the contaminated aquifer increased from May until August during the fourth and fifth growing seasons. Daily transpiration of groundwater peaked sometime in August and then declined somewhat until the end of the growing season. Transpiration of groundwater at the end of each growing season was greater than that at the beginning. Peak transpiration of groundwater appears to occur later in the growing season than does peak total transpiration. This lag is most likely caused by the decrease in moisture in near-surface soils as the growing season progresses, which results in the trees transpiring a greater proportion of groundwater late in the season. Hall *et al.* (1996) and Zhang *et al.* (1999) also report an increase in the proportion of water transpired from groundwater as soil in the unsaturated zone dries out.

Computed transpiration of groundwater from sands and gravels near well 514 is 16.3 centimeters for the fourth growing season and 18.1 centimeters for the fifth growing season. Computed transpiration of ground water from clay-rich material near well 515 is 7.6 centimeters for the fourth growing season and 10.1 centimeters for the fifth growing season (Table 19-2). These values illustrate the effects of aquifer heterogeneities on the amount of water that is

TABLE 19-2 Transpiration of Groundwater That Can Be Attributed to the Planted Trees at the Site in Fort Worth, Texas

	Third season (1998)		Fourth season (1999)		Fifth season (2000)	
	Well 514	Well 515	Well 514	Well 515	Well 514	Well 515
May	—	—	0.4	0.3	1.3	1.2
June	—	—	1.1	0.3	2.2	1.6
July	—	—	1.3	0.6	2.6	1.6
August	2.2	0.9	4.2	1.6	4.8	2.2
September	1.2	0.3	3.4	1.6	3.5	1.9
October	0.6	0.3	3.4	1.6	3.7	1.6
November	NA	NA	2.5	1.6	NA	NA
Total	—	—	16.3	7.6	18.1	10.1

Abbreviation used— NA: data not applicable (end of growing season).
Note: Transpiration is in units of centimeters, — means data not available.

transpired from the aquifer. The position of the wells within the drawdown cone also may affect the amount of transpiration of groundwater that is reflected in the groundwater-level hydrographs.

The seasonal rates of transpired groundwater that were computed by means of the groundwater-level hydrographs are not directly comparable to the rates that were estimated by use of the sap-flow model. This is because the rates determined by means of the sap-flow model equal the volume of transpired groundwater divided by the planted area, whereas the rates determined by means of the groundwater-level hydrographs relate to the volume as distributed over the entire cone of influence of the trees, which is larger than the planted area. Nonetheless, the rates are similar. The average of the values determined from the hydrographs for the fifth growing season, when much of the planted area was approaching closed canopy (14 centimeters), lies between the growing season estimates for closed canopy determined by using the sap-flow model (12 to 28 centimeters).

Hydrologic Effects of Tree Transpiration

The groundwater-flow model was prepared using the code MODFLOW (McDonald and Harbaugh 1988) to help in understanding the observed effects of tree transpiration on the aquifer. The model also was used to simulate the effects of future transpiration of groundwater on the volumetric flux of water in the aquifer across the down gradient end of the planted area. All simulations are two-dimensional because vertical hydraulic gradients across the aquifer generally are less than 1 centimeter and no vertical patterns in TCE concentration are evident across the site.

The preparation, calibration, and validation of the groundwater-flow model are discussed briefly here to help illustrate the utility and the limitations of the model for simulating the effects of tree transpiration on the aquifer. The model grid was prepared to simulate an area of 0.5 square kilometers using 14-square-meter grid blocks so that the full effect of tree transpiration at the site, which is approximately 3500 square meters, could be evaluated without interference from the model boundaries. The up gradient model boundary is simulated as a specified-flux boundary; selected lateral model boundaries are simulated as no-flow boundaries and are coincident with groundwater flow lines. The remaining lateral model boundaries, as well as the down gradient boundary, are coincident with Farmers Branch Creek and are simulated as head-dependent flux boundaries. Fifty-eight water-level observations made during February 1998, which represents a time of near average water levels in the aquifer, and one estimate of average groundwater discharge to Farmers Branch Creek, which was determined from data in Rivers *et al.* (1996), were used to calibrate the model to steady-state aquifer conditions. A model-independent parameter estimation package (PEST, Watermark Computing 1994) was used to facilitate this calibration by estimating parameter values that minimized the difference between simulated and measured water levels and simulated and estimated groundwater discharge to the stream. Four parameters were estimated during the steady-state calibration: (1) horizontal hydraulic conductivity of the sand-and-gravel material at the site (8 meters per day), (2) horizontal hydraulic conductivity of the clay-rich material at the site (1 meter per day), (3) groundwater recharge from precipitation (6 centimeters per year), and (4) a conductance used to help simulate Farmers Branch Creek. The calibrated model matched the measured water levels and estimated groundwater discharge. The root mean square error associated with the water levels for this calibration was 0.4 meter; measured water levels had a range of approximately 15 meters across the study area. Simulated and estimated groundwater discharge differed by less than the error associated with the estimate.

Once the steady-state model was calibrated, a transient calibration was undertaken to estimate storage coefficients (specific yield) of the aquifer material so that transient simulations of the effects of tree transpiration on the aquifer could be made. Measurements of groundwater levels and the creek stage from an 11-day period associated with a single storm in December 1998 were used for the transient calibration. Specifically, recharge for this event and aquifer specific yield were estimated by entering into the model the change in the creek stage during the selected storm and then adjusting the model parameters until simulated and measured changes in water levels at the recorder wells matched within an acceptable range. Simulated changes in water levels were within 1.5 centimeters of measured water-level changes in all instances. (The total measured change in water levels for this 11-day event was 9 centimeters.) The estimated specific yield from the transient model calibration is 0.023 for the sand-and-gravel material and 0.01 for the more clay-rich material, as previously discussed. Groundwater recharge was estimated at 18 percent of total precipitation for this storm.

A final calibration step was undertaken to estimate average monthly groundwater recharge from precipitation for conditions before tree planting. This step was necessary so that the hydraulic effects of the eastern cottonwood (*Populus deltoides*) trees at the site could be evaluated over the course of an entire year. Nearly all groundwater recharge from precipitation occurs during the late autumn, winter, and early spring in this part of Texas; as a result, the annual groundwater recharge of 6 centimeters was distributed across the months of October through April in model simulations. The amount of recharge applied to each month was based on patterns of monthly precipitation recorded during a 50-year period at a meteorological station near the site. Yearly water-level fluctuations at the three wells with recorders, as well as data on groundwater discharge to Farmers Branch Creek reported in Rivers *et al.* (1996), were used to facilitate this final calibration step. In this calibrated model, the water table rises and falls over the course of one calendar year with essentially no change in storage within the aquifer at the end of the yearly cycle. Estimated groundwater recharge from precipitation is 18 percent of precipitation for the months of November through April, 5 percent of precipitation for the month of October, and 0 percent of precipitation for the months of May through September. This is the groundwater-flow model to which estimated or simulated transpiration of groundwater from the sap-flow model was added to determine the hydrologic effects of the phytoremediation system on the underlying alluvial aquifer at the site.

Transpiration of groundwater was simulated in the groundwater-flow model in all instances as wells with known pumping rates. Drawdown at the water table related to transpiration of groundwater was mapped for each simulation. In addition, a volumetric groundwater budget was computed for each simulation.

The calibrated groundwater-flow model was used first to simulate the hydrologic effects of transpiration of groundwater during the third growing season, so that simulated and observed drawdown could be compared to help test the validity of the models. When estimates for transpiration of groundwater for the third growing season were incorporated into the groundwater-flow model, the simulated drawdown at the center of the cone of influence was nearly 9 centimeters. The maximum observed drawdown at the water table for the same period was 10 centimeters.

The groundwater-flow model was then used to simulate the hydrologic effects of future transpiration on the underlying aquifer. Because the estimates of future transpiration of groundwater, which were derived from the sap-flow model, represent a range of possible climatic conditions, soil-water availability, and root-growth scenarios, there is a range of simulated drawdown and simulated reductions in the outflow of groundwater from the site for closed canopy conditions. Simulated drawdown at the water table at the peak of the growing season at closed canopy ranges from 12 to 25 centimeters at the center of the drawdown cone. The diameter across the simulated drawdown cones out to where drawdown is 3 centimeters or greater ranges from greater than 140 meters to greater than 210 meters (Figure 19-6). These simulations of future drawdown are associated with simulated decreases in the volumetric flux of groundwater across the down gradient end of the planted area, which range from 20 to 30 percent of the volumetric flux of water through the site before the trees were planted. The simulated volume of water transpired from the aquifer at closed canopy, however, ranges from 50 to 90 percent of the initial volumetric flux of groundwater at the site. The difference between the reduction in the volumetric outflow of groundwater and the volume of water transpired from the aquifer can be attributed to the combined increase in the hydraulic gradient on the up gradient side of the drawdown cone, which leads to an increase in the rate of groundwater inflow to the site, and the release of water from storage in the aquifer (Figure 19-7) (Eberts *et al.* 1999).

Model results indicate that a regional hydraulic gradient will remain across the planted area during future growing seasons. The volumetric flux of groundwater across the down gradient end of the planted area, however, will be notably reduced. Percentage reductions in the TCE mass flux because of

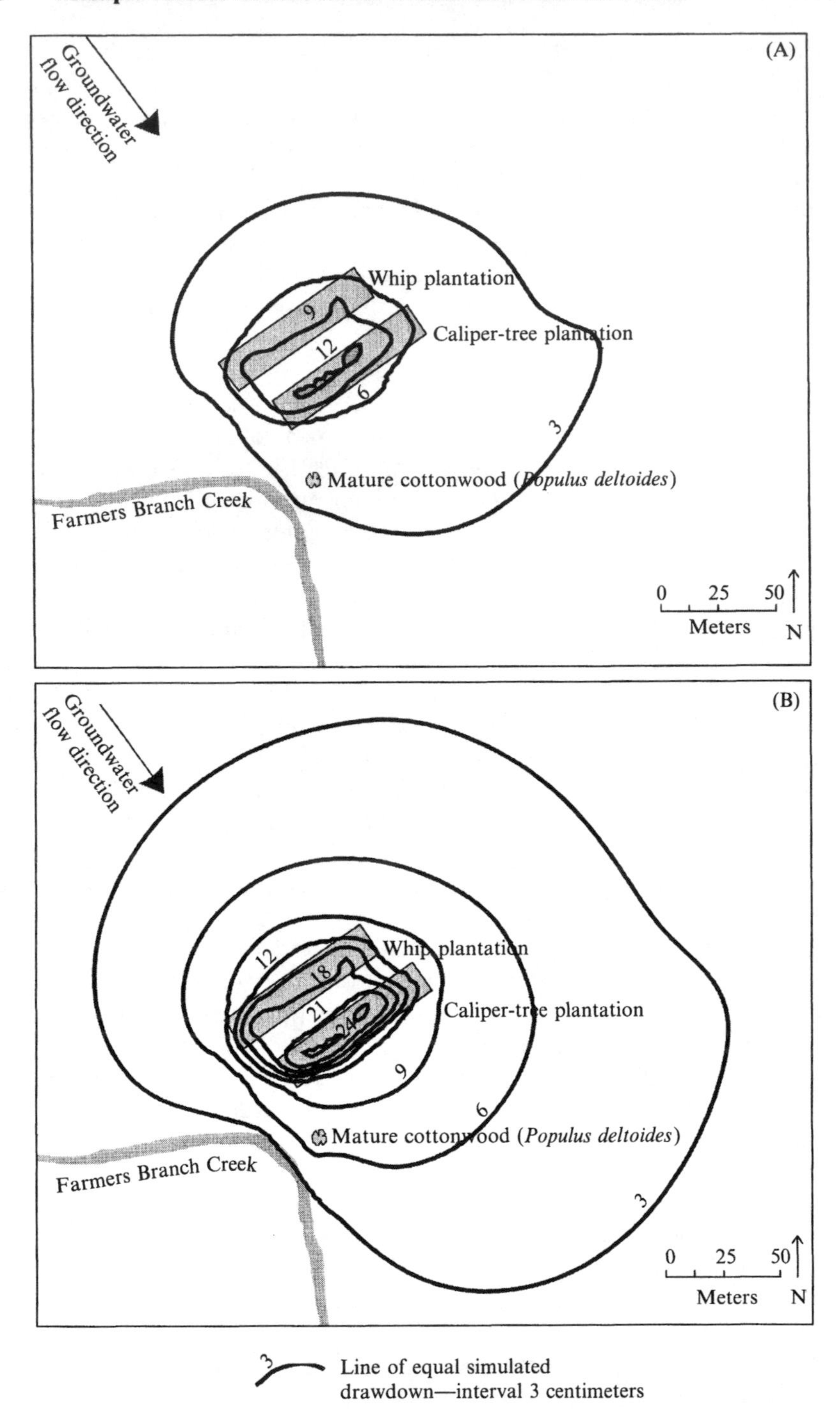

Figure 19-6 Simulated drawdown at the water table during peak of the growing season at closed canopy: (A) minimum, and (B) maximum.

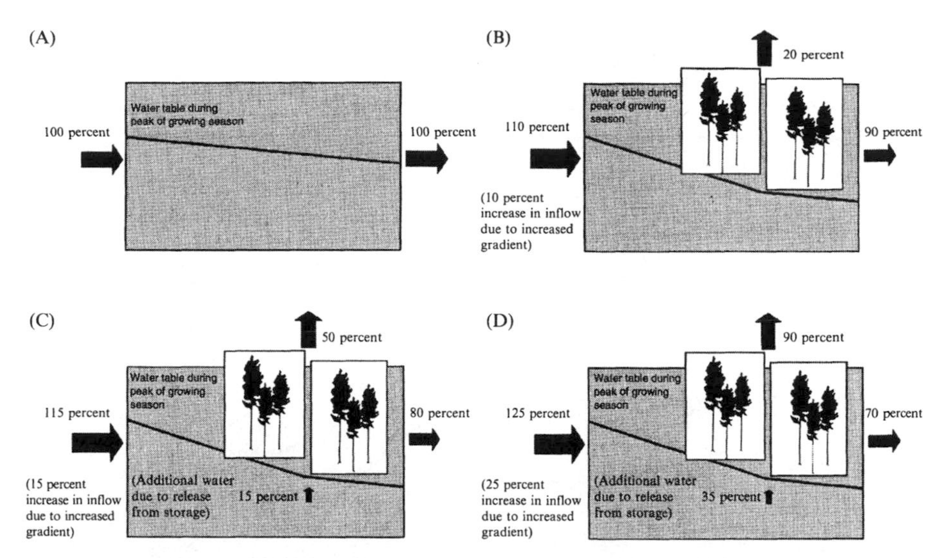

Figure 19-7 Simulated groundwater budget: (A) before treatment, (B) peak of third growing season, (C) peak of growing season at closed canopy, minimum predicted transpiration, and (D) peak of growing season at closed canopy, maximum predicted transpiration.

tree transpiration alone will be somewhat less than reductions in the volumetric flux of groundwater because membrane barriers at the root surface prevent TCE from being taken up at the same concentration as it occurs in the groundwater. The transpiration-stream concentration factor or fractional efficiency of uptake for TCE has been reported at 0.74 (Schnoor 1997). Model results also indicate no hydraulic control of the TCE plume during the dormant season (November through April).

A greater amount of hydraulic control may be possible if a larger area is planted, but the associated increase in groundwater inflow and release of water from storage in the aquifer will continue to limit hydraulic control of the contaminant plume. Full hydraulic control of the plume may not be desirable, however, because full control could result in an unacceptable decrease in flow in Farmers Branch Creek. This issue of an unacceptable decrease in flow in Farmers Branch Creek is particularly relevant because hydraulic control is not the only attenuation process associated with the phytoremediation system, and it is, therefore, unnecessary to rely upon hydraulic control for complete attenuation of the contaminant plume.

A map of observed drawdown at the water table near the end of the fifth growing season (September 2000), when much of the caliper-tree plantation had achieved closed canopy, can be used to help assess model results (Figure 19-8). The amount of drawdown that can be attributed to the planted trees in September 2000 was determined by comparing water levels measured in September 2000 with water levels measured in February 1998.

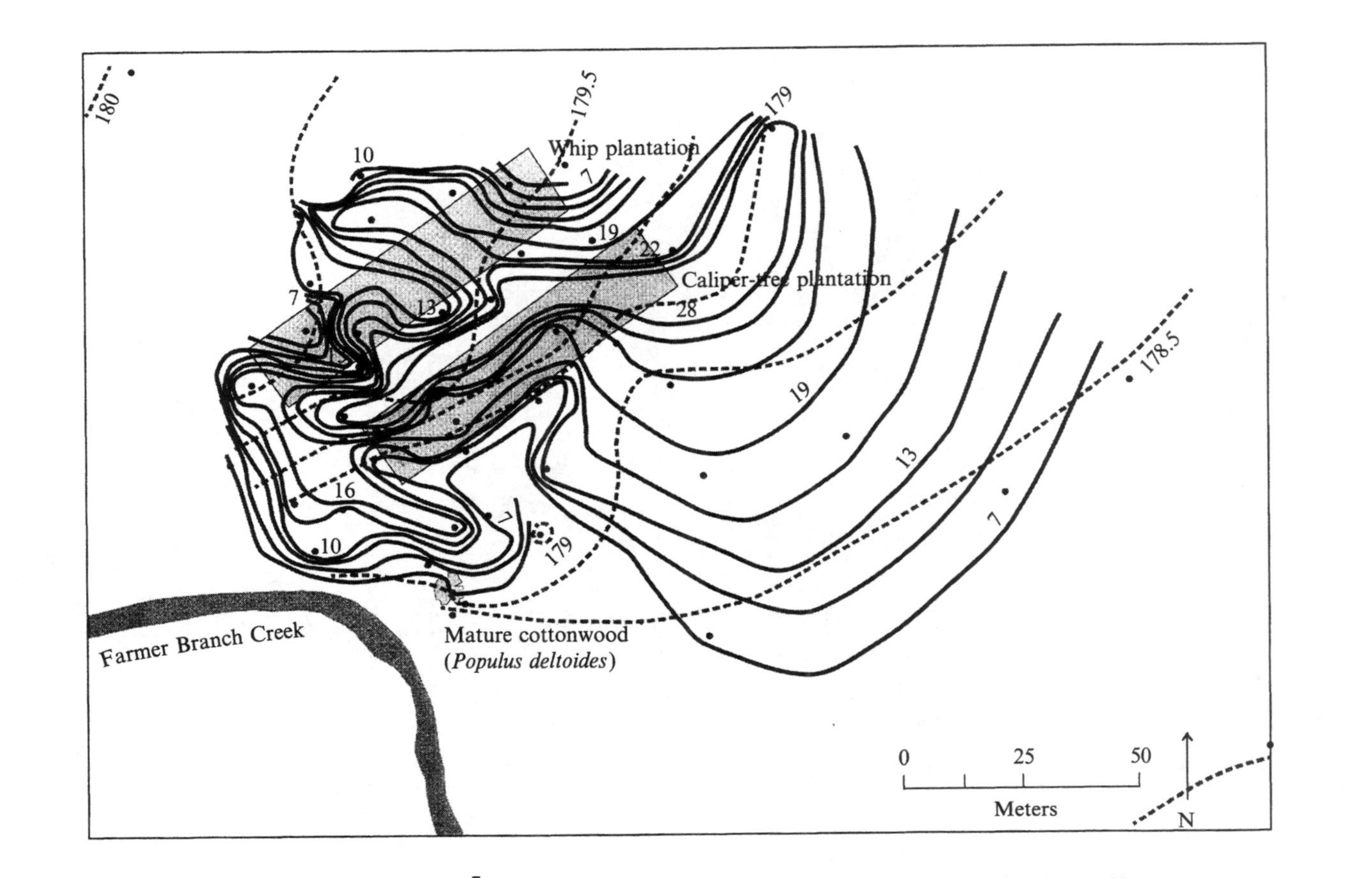

Figure 19-8 Drawdown at the water table that can be attributed to the planted trees, and water-table altitude, end of fifth growing season (September 2000).

February 1998 water levels were used for comparison because this was the largest available dormant-season data set and the groundwater-flow models indicated that there were no remnant effects of tree transpiration on the water table during the dormant season. At the end of the fifth growing season, the diameter across the observed drawdown cone out to where drawdown was 7 centimeters or greater was approximately 160 meters (Figure 19-8). The maximum observed drawdown at the center of the drawdown cone was near 29 centimeters. The maximum diameter across the simulated drawdown cones where drawdown is 7 centimeters or greater is 150 meters for closed canopy (Figure 19-6B). The maximum drawdown at the center of this simulated drawdown cone is nearly 25 centimeters. These results suggest that the groundwater-flow model can be used to draw conclusions about future effects of tree transpiration on the aquifer.

The drawdown cone that was observed during the fifth growing season, however, also illustrates some of the limitations of the sap flow and groundwater-flow models. Specifically, the observed drawdown cone has a more convoluted shape than the simulated drawdown cones. Although the observed drawdown cone may reflect some heterogeneities in the aquifer framework that are not simulated in the model, the differences in the shape of the observed and simulated drawdown cones beyond the first few growing seasons are more likely a result of variations in transpiration of groundwater across the planted area. The planted trees, especially in the whip plantation, have not grown or transpired groundwater uniformly across the plantations for the various reasons previously discussed. Not all of the variations in growth and transpiration of groundwater could be anticipated. As a result, these variations are not represented in the models. Rather, transpiration of groundwater was distributed evenly across the planted areas in the simulations. The trees within the plantations may eventually become more uniform in size, but future observations of groundwater levels will most likely be necessary to fully reveal the effects and benefits of the natural self-design of the system.

Contaminant Uptake into Tree Organs

The uptake of contaminated groundwater by the planted eastern cottonwood (*Populus deltoides*) trees results in the translocation of contaminants into various tree organs (roots, stems, and leaves). Translocated TCE may be metabolized within the trees or transpired through the stomata of the leaves (Newman *et al.* 1997). To assess the presence and magnitude of contaminant uptake and translocation of TCE at the site, samples of roots, stems, and leaves from five trees within each of the whip and caliper-tree plantations, as well as from the mature cottonwood (*Populus deltoides*) tree, were acquired and analyzed for volatile organic compounds, including TCE and the reductive microbial dechlorination products—*c*DCE and vinyl chloride (VC). Roots and leaves harvested during the growing season and leaves collected from the ground during the dormant season were analyzed only once for the oxidative daughter (metabolic) products trichloroethanol, trichloroacetic acid (TCAA), and dichloroacetic acid (DCAA). Leaves also were harvested and analyzed for dehalogenase activity to help determine the fate of TCE within the trees. In addition, bench-scale tests that utilized cuttings from the planted trees were done to further investigate the capability of the trees to take up and metabolize TCE.

Tissue samples collected for analysis of volatile organic compounds were maintained in frozen storage and were subjected to three successive freeze-thaw cycles to promote breakdown of the tissue before analysis by Standard Method 1624 (K.R. Chirgwin, ITS Environmental Laboratories, written communication, 1996). The volatile organic compound data are summarized in Table 19-3. The data indicate an increase over time in the number of trees that contained TCE and the reductive dechlorination product *c*DCE. Concentrations of TCE and *c*DCE were almost always higher in stem samples than in leaf samples. (Root samples were not collected during every sampling event.) Ratios of TCE/*c*DCE computed from average concentrations decreased with time in stem tissue (new growth) from both the whip and caliper-tree plantations. The ratio was 5.3 for trees in the whip plantation and 6.0 for trees in the caliper-tree plantation in October 1997, 3.1 and 4.5 in June 1998, and 2.4 and 2.8 in October 1998. By the end of the third growing season (1998), no notable differences were detected in the presence and concentration of volatile organic compounds in tree tissue from the whip and caliper-tree plantations. (Differences in tree growth and transpiration rates also had not shown up at this point in the development of the plantations but eventually appeared, as previously discussed.) Concentrations of TCE and *c*DCE generally were higher in the planted trees than in the mature cottonwood (*Populus deltoides*) trees, and the TCE/*c*DCE ratio was lower in the mature cottonwood (*Populus deltoides*) trees when both compounds were detected (0.64, October 1997 and 0.79, October 1998).

Trichloroethanol concentrations in leaves from two trees harvested during the third growing season ranged from 160 to 1060 micrograms per kilogram wet weight; root concentrations ranged from 210 to 640 micrograms per kilogram wet weight. Trichloroacetic acid concentrations in the harvested leaves ranged from 1300 to 2540 micrograms per kilogram wet weight. Trichloroacetic acid was not detected in the roots and DCAA was not detected in leaves or roots. Trichloroethanol concentrations in leaves collected from the ground during the third dormant season ranged from 250 to 620 micrograms per kilogram wet weight. Trichloroacetic acid concentrations ranged from 28 400 to 31 340 micrograms per kilogram wet weight and DCAA was not detected. Trichloroethanol and TCAA were not detected in leaves harvested from the mature cottonwood (*Populus deltoides*) tree during a single sampling event, whereas DCAA was 1210 micrograms per kilogram wet weight (Nzengung and Jeffers 2001).

Sapwood cores were collected with an increment borer from 23 mature trees that surround the site and were analyzed for the presence of TCE and *c*DCE (Don A. Vroblesky, U.S. Geological Survey, written communica-

TABLE 19-3 Average Concentration of Volatile Organic Compounds in Cottonwood (*Populus deltoides*) Tissue From Site in Fort Worth, Texas

Event	Analyte	Whip plantation			Caliper-tree plantation			Mature cottonwood (*Populus deltoides*)		
		Leaf	Stem	Root	Leaf	Stem	Root	Leaf	Stem	Root
October 1996	Trichloroethene	ND	26 (1)	ND	ND	ND	ND	—	ND	—
	Acrolein	ND	15.2 (3)	21.7 (3)	ND	7.0 (2)	9.1 (2)	—	ND	—
	Chloroform	ND	3.9 (1)	ND	ND	4.1 (1)	ND	—	ND	—
	Methylene chloride	ND	15 (2)	29 (3)	ND	10 (1)	ND	—	2.2	—
	cis-1,2-Dichloroethene	ND	ND	ND	ND	ND	ND	—	1.2	—
July 1997	Trichloroethene	ND	ND	—	ND	ND	—	ND	ND	—
	Acrolein	58.8 (5)	136 (3)	—	19 (1)	46.2 (5)	—	49	35	—
	Chloroform	ND	ND	—	0.73 (1)	ND	—	120	ND	—
	Methylene chloride	151 (5)	153 (3)	—	168 (5)	ND	—	ND	ND	—
	Toluene	0.73 (2)	ND	—	ND	ND	—	0.7	ND	—
	Tetrachloroethene	ND	ND	—	ND	71 (3)	—	ND	ND	—
October 1997	Trichloroethene	1.6 (2)	10.1 (3)	—	10.4 (3)	9.6 (3)	—	ND	6.4	—
	Acrolein	ND	20 (1)	—	ND	12.5 (4)	—	ND	ND	—
	Methylene chloride	8.3 (3)	6.6 (2)	—	ND	3.6 (5)	—	6.3	2.8	—
	cis-1,2-Dichloroethene	ND	1.9 (3)	—	ND	1.6 (3)	—	ND	10	—
	Toluene	ND	2.3 (3)	—	4.3 (2)	1.5 (1)	—	ND	ND	—
	Tetrachloroethene	ND	ND	—	ND	5.1 (2)	—	ND	ND	—
June 1998	Trichloroethene	ND	44 (1)	140 (1)	4.5 (2)	71 (1)	13 (1)	ND	13	—
	Acrolein	ND	ND	25 (1)	ND	ND	ND	ND	ND	—
	cis-1,2-Dichloroethene	ND	14 (1)	ND	ND	15.7 (3)	ND	ND	ND	—
	Toluene	1.4 (5)	2.3 (2)	1.1 (1)	1.1 (2)	2.0 (1)	0.91 (1)	ND	0.9	—
October 1998	Trichloroethene	ND	32.8 (5)	—	ND	24.6 (5)	—	ND	2.2	—
	Acrolein	ND	14.4 (5)	—	ND	ND	—	ND	ND	—
	cis-1,2-Dichloroethene	ND	13.5 (5)	—	ND	8.9 (4)	—	ND	2.8	—

Abbreviation used—ND: not detected, and —: means data not available, *Note*: Concentration is in units of micrograms per kilogram.
Number in parentheses represents the number of trees in which analyte was detected. Five trees from the whip plantation and five trees from the caliper-tree plantation were sampled for leaf and stem tissue during each sampling event. Fewer than five trees were sampled for root tissue.

tion, 1998). Sampled species included cottonwood (*Populus* spp.), oak (*Quercus* spp.), live oak (*Quercus virginiana*), cedar (*Juniperus virginiana*), willow (*Salix* spp.), hackberry (*Celtis* spp.), mesquite (*Prosopis glanulosa*), pecan (*Carya* spp.), elm (*Ulmus americana*), and one unidentified species. All cores were collected from a height of approximately 1.5 meters above the ground surface. Most of the trees that were sampled contained TCE and *c*DCE. In general, TCE concentrations detected within individual species decreased in the direction of decreasing TCE concentration in the groundwater, although TCE concentrations in the groundwater and in the tree tissue were not directly comparable. Differences in concentrations in these two media were expected because the trees transpire a combination of contaminated groundwater and uncontaminated water from infiltrating precipitation in the near-surface soil. No TCE was found in trees that grew in areas where the groundwater contained no TCE. Although most trees contained more TCE than *c*DCE, in areas where the depth to groundwater was approximately 1 meter or less, the trees contained substantially more *c*DCE than TCE. These data suggest the possibility that selected trees promote TCE dechlorination in areas where groundwater is shallow, or that tree-core data can be useful in locating areas of active reductive dechlorination in the groundwater. Analysis from one cottonwood (*Populus deltoides*) tree showed a substantially higher concentration of *c*DCE (238 nanomoles of gas per liter of core water) than TCE (3 nanomoles of gas per liter of core water) in leaves from a height of about 12 meters, whereas samples from the trunk of the tree contained more TCE than *c*DCE. Possible explanations for this observation include preferential volatilization of TCE or dechlorination of TCE within the tree (Don A. Vroblesky, U.S. Geological Survey, written communication, 1998).

Dehalogenase activity, which is believed to be responsible for reductive dechlorination of chlorinated solvents in plant tissues, was determined for leaf samples from seven trees that grew at or near the site. These trees included mature cedar (*Juniperus virginiana*), hackberry (*Celtis* spp.), oak (*Quercus* spp.), willow (*Salix* spp.), and mesquite (*Prosopis glanulosa*), as well as eastern cottonwood (*Populus deltoides*) from the whip and caliper-tree plantations. All leaf samples showed dehalogenase activity. Zero and pseudo first order disappearance rate constants for the leaf-enzyme transformation of TCE were determined; in general, the data adhered better to the first order model. The average and standard deviation for all seven rate constants is 0.049 ± 0.02 per hour [0.036 per hour for the planted eastern cottonwoods (*Populus deltoides*)]. This result corresponds to a half-life of 14.1 hours. The same mass (50 grams) was used in all experiments; as a result, there was no need to normalize the rate constants for plant mass for comparison. However, the averaged pseudo first order rate constant can be normalized for the plant mass for comparison to other systems (0.00098 per hour per gram). These kinetics are fast relative to other environmental transport and transformation processes for TCE, with the exception of volatilization (N. Lee Wolfe, U.S. Environmental Protection Agency, written communication, 1999).

Cuttings from eastern cottonwoods (*Populus deltoides*) at the site and willows (*Salix* spp.) from the surrounding area were harvested and used in bench-scale tests to further investigate the capability of the trees to phytotransform TCE, as well as tetrachloroethene (PCE). The cuttings were grown for 3 months in 25 percent of full strength Hoagland's solution before bioreactor experiments were conducted. The growth solution then was dosed with a saturated PCE or TCE aqueous solution. When the PCE concentration in solution decreased to 10 milligrams per liter, the experiments were terminated and the plants were sacrificed for analysis of PCE transformation products. The highest accumulation of parent compound (PCE) was in the roots and lower stems [cottonwood (*Populus deltoides*): 38 500 and 34 000 micrograms per kilogram of wet weight and willow (*Salix* spp.): 38 800 and 19 500 micrograms per kilogram of wet weight], which were submerged during the experiments. No PCE was detected in the leaves. Some TCE, which is a reductive dechlorination daughter product of PCE, was detected in the solution in experiments where the initial growth solution was dosed with PCE. Concentrations of TCE were lower in the roots and lower stems [cottonwood (*Populus deltoides*): 600 and 1000 micrograms per kilogram wet weight; willow (*Salix* spp.): not detected in roots or stems] than in solution in these experiments, and TCE was undetected in the leaves. The oxidative daughter products of PCE transformation (TCAA and DCAA) were detected in roots, lower stems, upper stems, and leaves [TCAA-cottonwood (*Populus deltoides*): 590, 1420, 6410, and 21 700 micrograms per kilogram wet weight; willow (*Salix* spp.): 1610, 110, 160, and 2 900 000 micrograms per kilogram wet weight] [DCAA-cottonwood (*Populus deltoides*): 220, 110, 880, and 1700 micrograms per kilogram wet weight; willow (*Salix* spp.): 2640, 820, 500, and 1 684 000 micrograms per kilogram wet weight]. The highest concentrations of these oxidative metabolites were detected in the leaves. The detection of both reductive and oxidative daughter products of PCE and TCE was used to conclude that more than one pathway may be involved in phytotransformation of PCE and TCE by cottonwood (*Populus deltoides*) and willow (*Salix* spp.) (Nzengung and Jeffers 2001). The root enzyme transformation of TCE in solution by willow (*Salix* spp.) was described by first order kinetics. The calculated half-life is 11 days (Valentine A. Nzengung, University of Georgia, written communication, 2000).

Transpired vapors were not sampled or analyzed from the trees at the site. Newman *et al.* (1999), however, reported transpiration of up to 9 percent of added TCE for the first 2 years in a controlled field study of the use of hybrid poplar (*Populus trichocarpa* × *Populus deltoides*) trees for the uptake and degradation of TCE. They reported no transpiration of TCE after 2 years. These observations are consistent with (1) the low concentrations of TCE in leaf samples measured during the first 3 years of the phytoremediation demonstration in Fort Worth, Texas, (2) the lower TCE/*c*DCE ratio in leaves compared to sapwood in the mature cottonwood (*Populus deltoides*) tree near the site, (3) the presence of dehalogenase activity and the fast kinetics of transformation of

TCE determined for leaves collected from the whip and caliper-tree plantations, (4) the occurrence of oxidative metabolites in the tissues, and (5) the lack of parent compounds detected in leaves associated with bioreactor experiments conducted on vegetative material from the site. This collective body of data suggests that transpiration of TCE is not the primary process associated with phytoremediation of TCE-contaminated groundwater at the site.

Subsurface Oxidation–Reduction Processes in Groundwater

Geochemical investigations were done to determine whether the trees planted at the site were capable of stimulating degradation of TCE in the groundwater below the root zone at the field scale. Previous research has demonstrated that soils in root zones of hybrid poplar (*Populus deltoides* × *Populus nigra* 'DN34' 'Imperial Corolina') trees are associated with enhanced numbers of beneficial bacteria when compared to soils in unplanted sites (Jordahl *et al.* 1997). In addition, such increases in bacteria, as well as increases in mycorrhizal fungi, soil organic carbon, and plant enzymes in the rhizosphere, encourage the degradation of organic chemicals in soils (Anderson and Walton 1992, Schnoor *et al.* 1995, Schnoor 1997). Controlled field trials have been used to demonstrate that hybrid poplar (*Populus trichocarpa* × *Populus deltoides*) trees can also remove TCE from flowing groundwater in a thin saturated (20-centimeter) zone at the bottom of test cells (Gordon *et al.* 1998). The extent to which planted trees can affect the geochemistry of a contaminated aquifer and the time required for changes to occur in a natural system, however, must be evaluated at the field scale. The site near AFP 4 provides an ideal setting for the study of geochemical changes in a TCE-contaminated aquifer that can be attributed to planted trees because native and anthropogenic carbon at the site were limited before the trees were planted (Type-3 conditions). Similar conditions did not exist at other sites selected for evaluation of phytoremediation of TCE-contaminated groundwater. As a result, observations from this evaluation cannot be directly compared to observations from the other sites. Specifically, the Aberdeen Proving Ground has a natural source of carbon in the groundwater (Type-2 conditions) (Hirsch *et al.* this book), and the Edward Sears site has an anthropogenic source of carbon (trimethylbenzene and xylenes) in the groundwater (Type-1 conditions) (U.S. Environmental Protection Agency 2001).

To assess changes in the geochemistry of the aquifer that can be attributed to the planted trees at the site near AFP 4, Science Applications International Corporation and U.S. Geological Survey scientists collected samples of groundwater up gradient, within, and down gradient of the planted trees, as well as beneath the mature cottonwood (*Populus deltoides*) tree, over a 5-year period. The samples were analyzed for volatile organic compounds (including TCE and *c*DCE), dissolved oxygen, ferrous and total iron, hydrogen sulfide, methane, carbon dioxide, dissolved hydrogen, total organic carbon, dissolved organic carbon, carbon-13 isotope values, alkalinity, and pH.

Conditions in the aquifer at the site were initially aerobic (dissolved oxygen greater than 5 milligrams per liter) (Jacobs Engineering Group, Inc. 1996). The concentration of dissolved oxygen beneath the mature cottonwood (*Populus deltoides*) tree adjacent to the planted site, however, was lower than 1 milligram per liter during baseline sampling in November 1996, and has generally remained lower than 1 milligram per liter over the course of the project. Although the average TCE/*c*DCE ratio in groundwater beneath and surrounding the planted area was 5 during baseline sampling, the TCE/*c*DCE ratio in groundwater from beneath the mature cottonwood (*Populus deltoides*) tree was an order of magnitude lower, at 0.5. In addition, total organic carbon was undetected at a detection limit of 1.5 milligrams per liter in groundwater from beneath the planted area during baseline sampling, whereas total organic carbon concentrations in groundwater beneath the mature cottonwood (*Populus deltoides*) tree ranged from 2.0 to 3.1 milligrams per liter. The total organic carbon in soil samples collected to define the baseline ranged from 630 to 2900 milligrams per kilogram beneath the planted area and was 5100 milligrams per kilogram beneath the mature cottonwood (*Populus deltoides*) tree. These data suggest that the mature cottonwood (*Populus deltoides*) tree may affect the chemistry of the soil and groundwater beneath, and provide early feedback on the biogeochemical processes that could eventually develop beneath the planted trees. Specifically, it was forecast that organic carbon is generated in the soil zone beneath the mature cottonwood (*Populus deltoides*) tree and is subsequently transported to the shallow groundwater as dissolved organic carbon (DOC) (Chapelle 2000). This increased DOC in the groundwater subsequently stimulates aerobic microbial activity that results in decreased dissolved oxygen levels in the aquifer beneath the tree, and results in local anaerobic conditions conducive to microbial reductive dechlorination. This forecast is consistent with observations of a poplar (*Populus italica*) vegetated riparian buffer strip in southern England. Specifically, Haycock and Pinay (1993) reported that poplars (*Populus italica*) in a riparian buffer strip contributed enough organic matter to the shallow groundwater during winter months to foster creation of anaerobic conditions in the aquifer and promote complete microbial denitrification of dissolved nitrate.

Lee *et al.* (2000) collected samples during the second and third years of the project to determine whether the eastern cottonwood (*Populus deltoides*) trees at the site were capable of creating anaerobic conditions that could facilitate reductive dechlorination of TCE in the groundwater according to the sequence $TCE + H_2 \rightarrow cDCE + HCl$. Specifically, terminal electron accepting processes in the aquifer were identified for the planted area, as well as for the area beneath the mature cottonwood (*Populus deltoides*). Because microbial-reductive dechlorination is driven by molecular hydrogen (H_2) and organic acids, which are waste products of fermentative bacteria that initiate oxidation of complex organic matter (Chapelle 2000), Lee *et al.* (2000) analyzed the groundwater for hydrogen (H_2) concentrations in addition to concentrations of the electron-acceptor oxygen and the reduced products ferrous iron, sulfide, and methane. Analyses for hydrogen (H_2) in natural groundwater are reported to be especially useful because hydrogen (H_2) is continuously cycled by microorganisms and is not transported by flowing groundwater in the same way that more stable compounds, such as ferrous iron, sulfide, and methane, are transported. As a result, hydrogen (H_2) concentrations, which increase in the following order—NO_3 reducing $<$ Mn(IV) reducing $<$ Fe(III) reducing $<$ SO_4 reducing $<$ methanogenic—are more representative of redox conditions near a well-screen than are concentrations of more stable compounds that can be transported from up gradient areas (Chapelle 2000). On the basis of multiple lines of evidence, Lee *et al.* (2000) concluded that highly reducing conditions are present in the aquifer beneath the mature cottonwood (*Populus deltoides*) tree. The terminal electron accepting process for this part of the aquifer was reported as being methanogenesis. The rest of the aquifer was reported as being in the state of oxygen reduction, except near well 514 in the whip plantation (Figure 19-1), where redox conditions in the aquifer were reported as approaching iron reducing.

Lee *et al.* (2000) also analyzed for DOC in the groundwater, as well as the acid, base, and neutral fractions of hydrophobic and hydrophilic organic compounds that comprise the DOC. [The analytical methods for DOC and DOC fractionation have a detection limit of 0.1 milligram per liter and are described in Leenheer and Huffman (1979).] The principal compound classes at the site in November 1997 (second dormant season) were hydrophobic neutral, hydrophobic acid, and hydrophilic acid. Hydrophilic acids (containing less than five carbon atoms) have been reported as labile or readily consumable by microorganisms. The highest measured concentrations of hydrophilic acid at the site were in groundwater from a well beneath the mature cottonwood (*Populus deltoides*) tree (0.6 milligram per liter) and from well 514 in the whip plantation (0.9 milligram per liter). These data indicate that leaching of litter and root systems in these areas could be producing hydrophilic acids that can be consumed by microorganisms, resulting in increased consumption of dissolved oxygen in the aquifer. During November 1997, when the wells at the site were sampled for analysis of DOC, the TCE/*c*DCE ratio in groundwater beneath the mature tree was 0.56; the ratio averaged 5 elsewhere in the aquifer, including the area near well 514 in the whip plantation. These ratios indicate that although DOC had increased in the aquifer locally beneath the whip plantation and redox conditions were approaching iron reducing, reductive dechlorination of TCE had not appreciably occurred anywhere beneath the planted trees by the second year of the project.

The TCE/*c*DCE ratios in groundwater reported by Lee *et al.* (2000) for November 1997 [beneath mature cottonwood (*Populus deltoides*): 0.56 and beneath whip and caliper-tree plantations: 5.0] are similar to the TCE/*c*DCE ratios computed from average concentrations in stem tissue for October 1997 [mature cottonwood (*Populus deltoides*): 0.64, whip plantation: 5.2, and caliper-tree plantation: 6.0]. The similarity in the TCE/*c*DCE ratios in groundwater and stem tissue indicates a relation between the chemistry of the groundwater and the chemistry of the stem tissue. By October 1998 (end of third growing season), the TCE/*c*DCE ratios computed from average concentrations in stem tissue from trees within the whip and caliper-tree plantations were 2.4 and 2.8, respectively. October 1998 was the last time that stem tissue (new growth) was sampled from the plantations. In September 2000 (end of fifth growing season), groundwater in two locations beneath the planted area had TCE/*c*DCE ratios as low as 2.5, whereas water from wells sampled just up gradient of the planted area still had TCE/*c*DCE ratios near 5. A map of TCE/*c*DCE ratios for September 2000 indicates that TCE/*c*DCE ratios in groundwater did not decrease in the general direction of flow at this time (Figure 19-9). Rather, the ratios appear to be related to depth to groundwater. Ratios of TCE/*c*DCE generally were lowest where depth to groundwater was less than 3 meters. The lack of systematic changes in groundwater chemistry along flow paths indicate that the groundwater was still in a state of geochemical transition at the end of five seasons.

Although not analyzed for hydrogen (H_2) or for the reduced products ferrous iron, sulfide, and methane during the September 2000 sampling event (end of fifth growing season), groundwater was analyzed for concentrations of dissolved oxygen and DOC, including concentrations of acid, base, and neutral hydrophobic and hydrophilic organic compounds. The groundwater beneath the mature cottonwood (*Populus deltoides*) had a dissolved oxygen concentration of 0.93 milligram per liter and a total DOC concentration of 1.7 milligrams per liter at this time. Groundwater from beneath the planted trees, where the TCE/*c*DCE ratio was 2.5 (well 514 within the whip plantation and well 515 within the caliper-tree plantation), had concentrations of dissolved oxygen between 1.6 and 1.7 milligrams per liter. The DOC concentration in water from these wells was 1.8 and 1.6 milligrams per liter, respectively. The average concentration of dissolved oxygen for all other wells at the end of the fifth growing season was 3.9 milligrams per liter, and the average DOC concentrations was 1.1 milligrams per liter. In general, concentrations of dissolved oxygen were lower where DOC concentrations were higher throughout the site at this time (Figure 19-10A). Wells 511 [beneath the mature cottonwood (*Populus deltoides*)], 514, and 515 were the only wells sampled where concentrations of hydrophilic acids were 0.4 milligram per liter or greater. Another notable relation was that between dissolved oxygen concentrations and TCE/*c*DCE ratios in groundwater from the site by the end of the fifth growing season; specifically, TCE/*c*DCE ratios in the groundwater decreased as concentrations of dissolved oxygen decreased (Figure 19-10B). These data suggest that by the end of the fifth growing season, there was a labile fraction of DOC derived from organic matter recently produced and leached into the shallow groundwater beneath parts of the plantations that were capable of stimulating some reductive dechlorination of TCE.

Field data collected during the fifth dormant season (January 2001) indicate that the trees were finally beginning to have a widespread effect on the

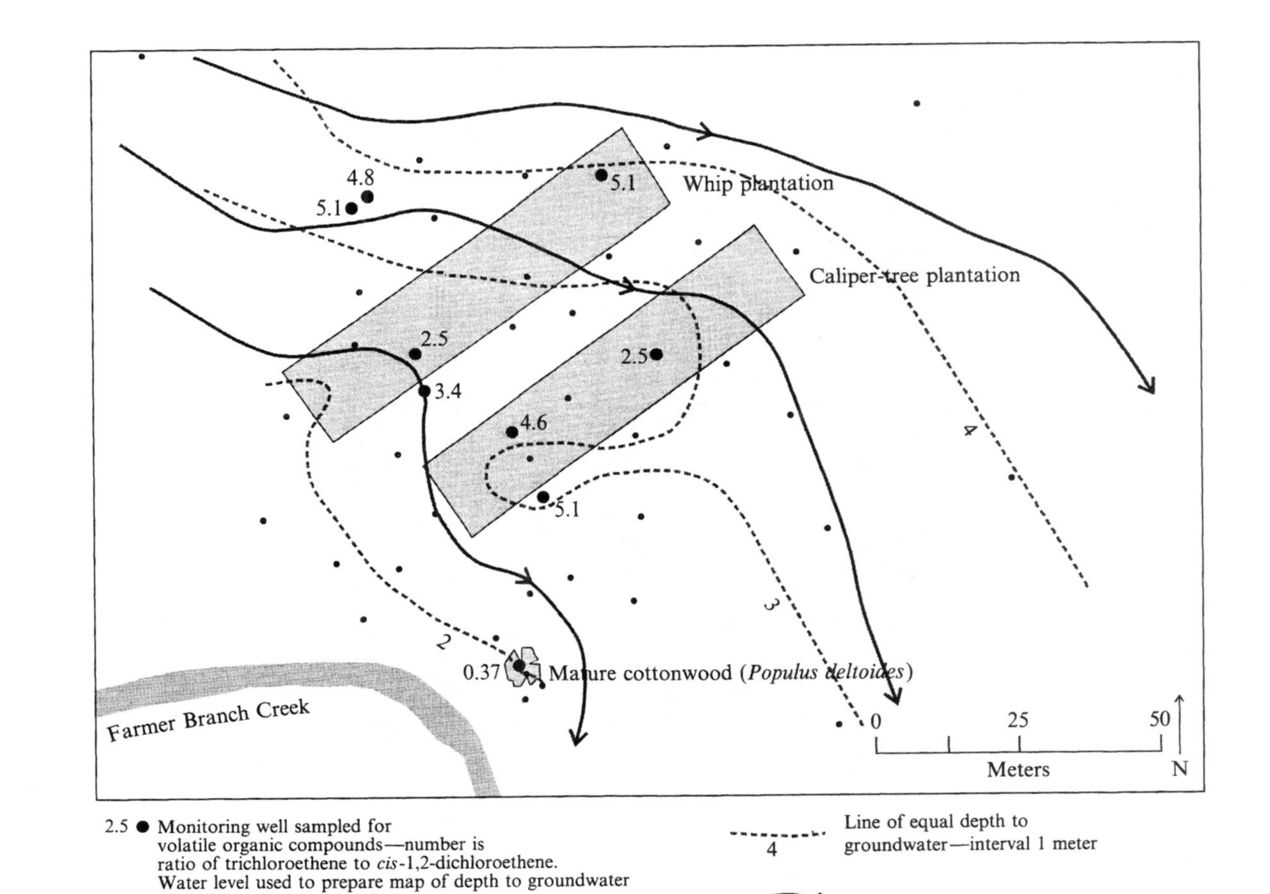

Figure 19-9 Ratio of trichloroethene to *cis*-1,2-dichloroethene, depth to groundwater, and groundwater flow paths at the end of fifth growing season (September 2000).

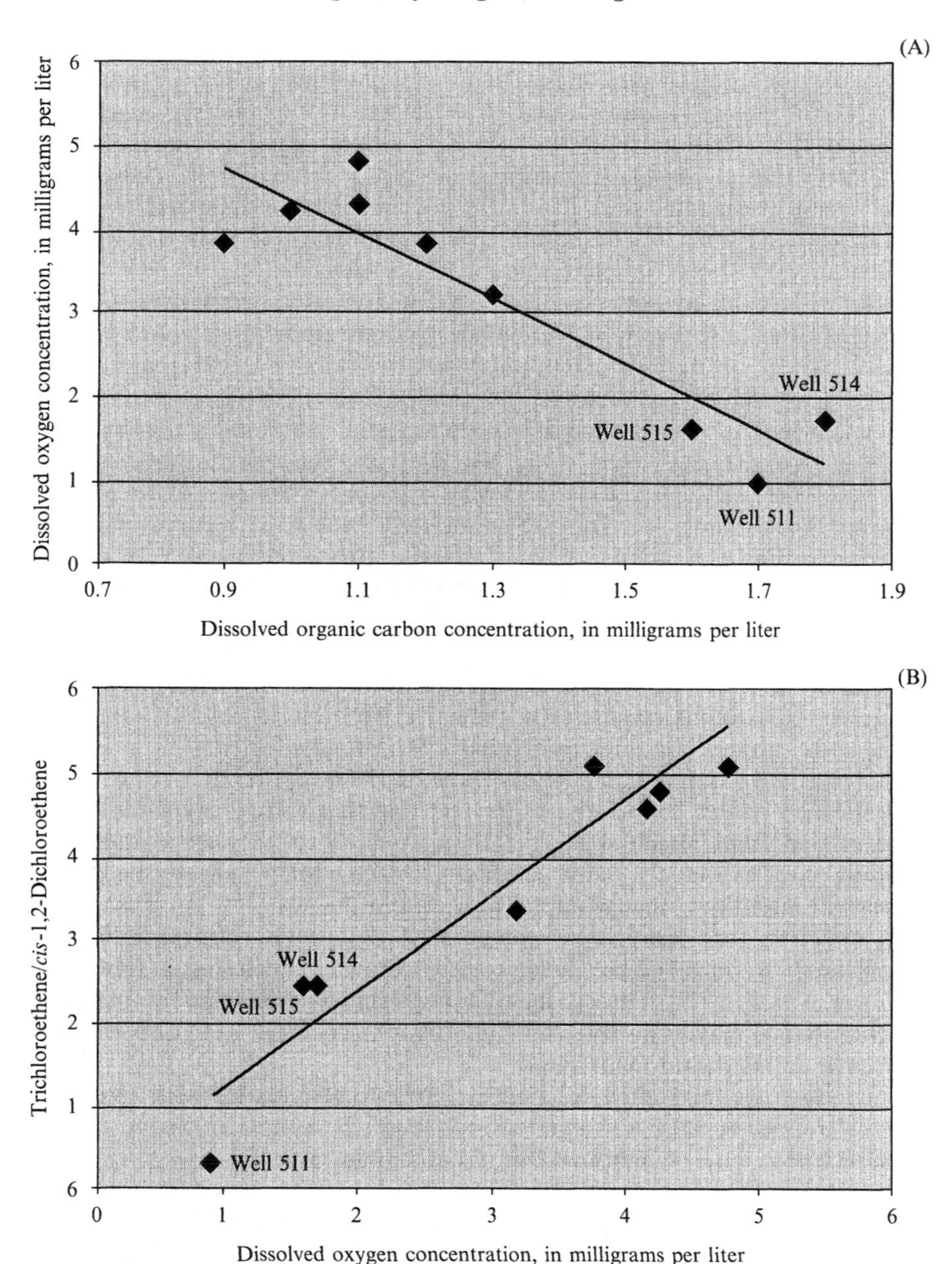

Figure 19.10 Relation between (A) dissolved oxygen concentration and dissolved organic carbon concentration, and (B) the ratio of trichloroethene to *cis*-1,2-dichloroethene and concentration of dissolved oxygen in groundwater sampled near the end of the fifth growing season (September 2000).

geochemistry of the groundwater. During the fifth dormant season, dissolved-oxygen concentrations were higher than 4.5 milligrams per liter in water from all up gradient wells and one well between the tree plantations (well 522), whereas concentrations were less than 3.5 milligrams per liter in water from all other wells at the site, including wells that were nearly 40 meters down gradient of the planted area. The mean concentration of dissolved oxygen in water from all wells, excluding the up gradient wells and well 522, was 1.76 milligrams per liter. The concentration of dissolved oxygen in six wells beneath or immediately down gradient of the planted trees was less than 1 milligram per liter. Ferrous iron and sulfide concentrations were elevated above baseline concentrations in some locations beneath and immediately down gradient of the tree plantations. Molecular hydrogen (H_2) concentrations indicated that nitrate-reducing, iron-reducing, and sulfate-reducing conditions were all present locally beneath the planted trees. Sulfate-reducing conditions prevailed beneath the caliper-tree plantation, which is the most down gradient plantation. Methane was detected in 11 wells within or immediately down gradient of the plantations. (Results of laboratory analyses of TCE and the associated daughter products had not been received at the time this manuscript was prepared.) These field data add to the body of evidence that the trees planted at the site can stimulate microbial activity, which results in the depletion of dissolved oxygen in the aquifer, and in the creation of local anaerobic conditions conducive to reductive microbial dechlorination. These data also support the conclusion that the groundwater was in a state of geochemical transition after 5 years. Hansen (1993) reports that soil carbon is significantly related (positively) to tree age and that there is a net addition of soil carbon from plantations older than about 6 to 12 years of age. This finding suggests that the redox conditions in the aquifer beneath the planted trees will most likely change appreciably during the next 5 or more years. It is not surprising that significant degradation of TCE in the rhizosphere was not reported for a controlled field study conducted over a much shorter time frame (Newman *et al.* 1999). Finally, these findings suggest that reductive microbial dechlorination may be an important attenuation process at a phytoremediation site even during the dormant season.

Although the data discussed thus far point to changes in redox conditions in the aquifer beneath the planted area over the course of years, there is evidence that redox conditions can also fluctuate over the course of weeks. Immediately after a groundwater recharge event in October 2000, concentrations of dissolved oxygen in water from all wells, except well 511 beneath the mature cottonwood (*Populus deltoides*), increased to greater than 3 milligrams per liter. Lee *et al.* (2000) also observed a temporary shift from anaerobic to aerobic conditions in parts of the planted area in June 1998. This shift between aerobic and anaerobic conditions may explain why there is no apparent accumulation of *c*DCE or VC in the aquifer beneath the whip and caliper-tree plantations, even where the TCE/*c*DCE ratio in the groundwater has decreased with time. Specifically, *c*DCE and VC may not accumulate because these compounds can biodegrade by means of aerobic oxidation (Chapelle 2000), and aerobic conditions still occurred, at least periodically, beneath all parts of the planted area by the end of the fifth growing season. In addition, *c*DCE and VC can undergo anaerobic oxidation, particularly under Fe(III)-reducing conditions (Chapelle 2000). Concentrations of dissolved oxygen in groundwater beneath the mature cottonwood (*Populus deltoides*) did not exceed 1 milligram per liter during the October 2000 recharge event, possibly because concentrations were initially lower beneath the mature tree than beneath the planted area. Concentrations of *c*DCE in groundwater beneath the mature cottonwood (*Populus deltoides*) are nearly twice as high as elsewhere in the aquifer, which is consistent with the persistent anaerobic conditions in this part of the aquifer. Some VC has been detected in the aquifer beneath the mature cottonwood (*Populus deltoides*). The production of VC, which is a known carcinogen, is not of concern at the site because the surrounding aquifer remains aerobic throughout the year and VC readily biodegrades in aerobic conditions.

Microbial Contributions to Reductive Dechlorination

In assessing the mechanisms and rates of biodegradation in an aquifer, it is best to examine the spatial distribution of the different microbial populations on the sediment and in the pore water, in addition to the concentrations and distribution of electron acceptors and reduced products in the groundwater. As a result, the U.S. Geological Survey did a reconnaissance study of microbial activity in soil and groundwater at the site in February and June 1998 (second dormant season and third growing season).

Sediment and groundwater samples were derived from cores collected from boreholes drilled specifically for microbial sampling, at sites within the plantations and at background sites outside the contaminant plume. It was considered important to use pore water from cores rather than water from wells to determine microbial communities within the groundwater because microbial communities within a well, on a well screen, and in the aquifer material immediately surrounding a well do not necessarily represent communities within the surrounding aquifer. When water flows through a well screen during sample collection, microorganisms growing on the screen can be picked up by the flowing water and incorporated into the water sample, even when the well has been purged sufficiently for the purpose of chemical analysis of the groundwater (E. Michael Godsy, U.S. Geological Survey, oral communication, 2000).

Microbial concentrations in the sediment and water samples at the site were determined using a five-tube most probable number analysis. Aerobes, fermenters, and denitrifiers were attached to the sediment and were in the water at two background sites in numbers consistent with results from other uncontaminated sites. The general populations within the whip and caliper-tree plantations were similar to those at the background sites, with the exception of increased numbers of anaerobic microorganisms. In addition, methanogenic bacteria were present in one core collected near well 514 within the whip plantation. The presence of methanogens suggests that reducing conditions were beginning to develop by the second season beneath the planted trees at this location (E. Michael Godsy, U.S. Geological Survey, written communication, 2001).

Sediment and groundwater samples collected from beneath the mature cottonwood (*Populus deltoides*) tree adjacent to the site had a greater number of microorganisms than at background sites for all physiological types. The relatively high numbers of microorganisms in the groundwater samples from beneath the mature cottonwood (*Populus deltoides*) are consistent with a microbial population that is active. The presence of hydrogen-oxidizing and acetate-fermenting methanogens at this location is also consistent with an anaerobic microbial population that is actively degrading organic compounds (E. Michael Godsy, U.S. Geological Survey, written communication, 2001).

Eleven water samples and two sediment samples were analyzed for low-molecular-weight hydrophilic acid compounds that could be used by microorganisms as a source of energy. Only the sediment sample from beneath the mature cottonwood (*Populus deltoides*) tree contained identifiable compounds, which include phenol, benzoic acid, and acetic acid. A cyclic hydrocarbon with side chain(s) that had most likely undergone partial biodegradation was also detected but could not be identified. The identified compounds are the most common intermediates observed in anaerobic ecosystems where complex organic compounds are undergoing biodegradation (Godsy *et al.* 1996) and are consistent with complex organic root exudates. These compounds may be serving as electron donors for the reductive dechlorination of TCE beneath the mature cottonwood (*Populus deltoides*) (E. Michael Godsy, U.S. Geological Survey, written communication, 2001).

DISCUSSION

Long-term (5-year) monitoring and evaluation at a site in Fort Worth, Texas, has enabled the performance of tree plantations to be documented at the field scale and has contributed to the understanding of multiple processes involved in phytoremediation of TCE-contaminated groundwater. The interoperability of data collected by scientists from numerous organizations and disciplines provides an insight into how various processes interrelate and adds to the strength of the individual investigations.

The data and observations from the 5-year project support the initial forecast that plantations of eastern cottonwood (*Populus deltoides*) trees can remove contaminated water from an aquifer and can then biologically alter TCE within the trees, or transpire and volatilize TCE to the atmosphere. In addition, the trees can promote *in situ* reductive microbial dechlorination of dissolved TCE within the aquifer.

The performance data have been synthesized and are discussed below in terms of implications for application of phytoremediation to other sites. In addition, insights into the multiple processes involved in phytoremediation, as well as the relative importance of individual processes at the field scale, are discussed.

Effects of the hydrogeologic setting on the performance of a groundwater phytoremediation system can be assessed from the demonstration data. Specifically, depth to water affects the capability of planted trees to root and take water from a contaminated aquifer. Depth to water ranged from 2.5 to 4 meters below land surface at the site at the time of planting. Most of the whips and 1-year-old seedlings ("caliper" trees) were able to root to the water table by the second growing season. The trees, however, did not thrive where the water table was greater than 3.5 meters below land surface. This limitation may not have been encountered if more deeply rooting trees had been planted. Trees that may warrant further study include—but should not be limited to—other poplar species (*Populus* spp.), willow (*Salix* spp.), mesquite (*Prosopis* spp.), eucalyptus (*Eucalyptus* spp.), and black locust (*Robinia* spp.). Deep-planting techniques also could have been used and have been used at other sites to encourage development of deeper roots (U.S. Environmental Protection Agency 2001; Negri, Gatliff, *et al.* this book). Such planting techniques, however, are more expensive than the methods developed for short-rotation woody crops.

A relation between depth to water and geochemical changes in the aquifer that can be attributed to the planted trees was also observed. After 5 years, geochemical changes in the aquifer were restricted to areas where depth to water was approximately 3 meters below land surface or less. It is not known whether this observation simply reflects groundwater that was still in a state of geochemical transition or if it is a physical limitation of the phytoremediation system. Observed geochemical changes in the shallowest parts of the aquifer appear to result from the introduction of DOC from the planted trees into the aquifer. Schiff *et al.* (1997) studied DOC in waters of forested watersheds and report that DOC in groundwater is largely from roots and leaching of litter. Generally this DOC is recalcitrant because of extensive recycling within the soil profile attributable to high rates of microbial activity and a high degree of contact with sorbing surfaces. The exception is where groundwater flow paths are shallow enough to interact with newly generated DOC before the microbially labile component has been fully consumed. Because a notable part of this labile DOC is from leaching of litter, the depth to which trees can appreciably alter the geochemistry of an underlying aquifer may be limited regardless of rooting depth. In areas where substantially more cDCE than TCE was detected in sapwood of mature trees near the site, depth to water generally was 1 meter or less (Don Vroblesky, U.S. Geological Survey, written communication, 1998). The depth to which the labile component of DOC beneath trees can be transported may warrant further study.

The hydrogeologic setting of the site also affected the development of the seasonal drawdown cone. Specifically, the observed drawdown cone commonly has a convoluted shape and reflects heterogeneities in the subsurface, as well as differences in the capability of various eastern cottonwood (*Populus deltoides*) clones to access groundwater. In general, a greater amount of drawdown was observed within sand and gravel as opposed to more clay-rich material. This finding suggests that the planted trees transpire more water from the most permeable parts of the aquifer where contaminant transport is likely to be greatest than from less permeable material.

Insights into timeframes necessary for a phytoremediation system to become effective can be derived from the demonstration data. By the peak of the third growing season, the trees had reduced the volumetric outflow of contaminated groundwater by 12 percent and the outward flux of contaminant mass by 11 percent. By the fifth growing season, the tree canopy was nearing closure in large sections of the plantations. As a result, the plantations were nearing maximum transpiration. Results of model simulations suggest that 48 to 58 percent of tree transpiration will be derived from the aquifer at closed canopy and the plantations may eventually decrease the volumetric outflow of contaminated groundwater by 20 to 30 percent during peak growing seasons.

It took nearly 5 years for the trees to deliver enough DOC to the underlying aerobic aquifer, so that microbial activity could create local anaerobic conditions in the groundwater and support reductive dechlorination of dissolved TCE. By the fifth dormant season, redox conditions beneath the planted trees, however, were not as reducing as the methanogenic conditions beneath the mature cottonwood (*Populus deltoides*) tree adjacent to the site. Rather, sulfate-reducing conditions prevailed beneath the caliper-tree plantation and locally beneath the whip plantation. The timing of these changes is consistent with published research indicating that 6 to 12 years are needed for a net addition of soil carbon beneath a poplar plantation (Hansen 1993). The observation that planting eastern cottonwood (*Populus deltoides*) trees may foster anaerobic conditions in an underlying aquifer implies that a similar approach may benefit sites contaminated with nitrate, as suggested by Haycock and Pinay (1993), as well as hexavalent chromium, uranium, and other halogenated hydrocarbons. On the other hand, the development of reducing conditions may mobilize metals and metalloids such as arsenic. Finally, data collected during this demonstration, as well as by Haycock and Pinay (1993), suggest that *in situ* microbial degradation of contaminants beneath a poplar plantation will not be limited to the growing season.

Various data sets highlight the importance of long-term monitoring for understanding the performance of a phytoremediation system. In the first 3 years of the project, no appreciable differences were found in the performance of the trees with respect to hydraulic control regardless of variations in the planting strategy (whips as opposed to caliper trees). After 5 years, however, individual caliper trees transpired nearly four times the water (80 kilograms per day) as trees planted as whips (20 kilograms per day). In addition, Nzengung and Jeffers (2001) detected the oxidative metabolites, trichloroethanol and TCAA (less than or equal to 1060 micrograms per kilogram of wet tissue weight and less than or equal to 2540 micrograms per kilogram of wet tissue weight, respectively), but not DCAA, in leaves harvested from the planted trees before the trees were 4 years old. On the other hand, trichloroethanol and TCAA were not detected in leaves harvested from the mature cottonwood (*Populus deltoides*) tree near the site, but DCAA was detected (1210 micrograms per kilogram of wet tissue weight). Nzengung and Jeffers (2001) concluded that stage of growth or age, species, and duration of exposure of the trees to dissolved TCE might affect the metabolic pathway within the trees.

The primary *in situ* process for attenuation of the TCE plume at the site during the first 5 years was hydraulic control. The relative importance of contaminant uptake (hydraulic control) and *in situ* reductive microbial dechlorination at full system performance has yet to be determined. The dominant attenuation process within the trees may be enzymatic transformation within the leaves. The similarity in the chemistry of groundwater and trunk and stem tissue from the site support this finding. Newman *et al.* (1999) also observed that leaf tissue may be more metabolically active than trunk and stem tissue. Shang *et al.* (2001), however, suggest that higher concentrations of oxidative metabolites in leaves than in roots may be caused by transport of metabolites to the leaves.

Finally, the techniques and methods used to implement this phytoremediation evaluation were developed for short-rotation woody crops and are generally in the public domain. This approach relies on the principles of self-design; subsurface heterogeneities are allowed to control tree growth and survival rates. The decision to use such an approach is consistent with the desire to facilitate cost-effective scale-up of phytoremediation technology and to foster technology transfer. The short-rotation-woody-crop planting strategy benefits from the extensive body of knowledge developed by the U.S. Departments of Agriculture and Energy for the bioenergy and fiber industries. Although some limitations were encountered, the phytoremediation system at the site overcame other limitations commonly associated with engineered groundwater cleanup technologies. Specifically, enhanced reductive microbial dechlorination of the dissolved TCE beneath the planted trees was observed in both low- and high-permeability aquifer material; many cleanup technologies cannot reach contaminants in low-permeability material. In summary, this field test illustrates a highly functional redundancy of processes associated with phytoremediation. Such redundancy is desirable in any remediation system.

Acknowledgments

The authors thank the U.S. Air Force Aeronautical Systems Center Environmental Safety and Health Division Engineering Directorate, the Department of Defense

Environmental Security Technology Certification Program (ESTCP), and the U.S. Environmental Protection Agency Superfund Innovative Technology Evaluation (SITE) Program for funding this evaluation. The authors extend a special thanks to Mr. David Lawrence (U.S. Air Force Aeronautical Systems Center Environmental Safety and Health Division Engineering Directorate), whose unwavering support enabled the phytoremediation system to be monitored for many years.

The use of business or trade names in this publication is for descriptive purposes only and does not imply endorsement by the U.S. Government.

REFERENCES

Anderson, T.A. and B.T. Walton (1992) Comparative plant uptake and microbial degradation of trichloroethylene in the rhiozospheres of five plant species—implications for bioremediation of contaminated surface soils. Publication No. 3809. Oak Ridge National Laboratory Environmental Sciences Division, Oak Ridge, Tennessee.

Chapelle, F.H. (2000) *Ground-Water Microbiology and Geochemistry*. John Wiley, New York.

Cooper, D.T. and B. Ferguson (1979) Avoid early selection for growth rate in cottonwood. In: *Proceedings of Southern Forestry Tree Improvement Conference 15*, Mississippi State University, Starkville, pp. 52–58.

DeBell, D.S. and C. Harrington (2000) Ten lessons from ten years of research in *Populus* production systems. In: 21st Session of the International Poplar Commission (IPC 2000), Poplar and Willow Culture, Meeting the Needs of Society and the Environment. J.G. Isebrands, and J. Richardson, compilers. General Technical Report NC-125. U.S. Department of Agriculture, Forest Service, North Central Research Station, St. Paul, Minnesota. Held in Vancouver, Washington, September 24–28. p. 44.

Dietz, A.C. and J.L. Schnoor (2001) Phytotoxicity of chlorinated aliphatics to hybrid poplar. *Environ. Toxicol. Chem.* **20**(2): 389–393.

Eberts, S.M., C.W. Schalk, J. Vose, and G.J. Harvey (1999) Hydrologic effects of cottonwood trees on a shallow aquifer containing trichloroethene. *Hydrol. Sci. Technol.* **15**(1–4): 115–121.

Godsy, E.M., D.F. Goerlitz, and D. Grbi-Gali (1996) Pathways of methanogenic biodegradation of creosote-derived aromatic compounds. In: U.S. Geological Survey Toxic Substances Hydrology Program-Proceedings of the Technical Meeting. D.W. Morganwalp and D.A. Aronson, eds. Water-Resources Investigations Report 94-4015, Vol. 2. U.S. Geological Survey. Held in Colorado Springs, Colorado, September 20–24, 1993. pp. 835–842.

Goldstein, R.A., J.B. Mankie, and R.J. Luxmoore (1974) Documentation of PROSPER a model of atmosphere-soil-plant water flow. EDFB-IBP-73/9, East Deciduous Forest Biome, U.S. Int. Biol. Prog. Oak Ridge National Laboratory, Oak Ridge, Tennessee. 75 pp.

Gordon, M., N. Choe, J. Duffy, G. Ekuan, P. Heilman, I. Muiznieks, M. Ruszaj, B.B. Shurtleff, S. Strand, J. Wilmoth, and L.A. Newman (1998) Phytoremediation of trichloroethylene with hybrid poplars. *Environ. Health Persp.* **106**(4): 1001–1004.

Hall, R.L., S.J. Allen, P.T.W. Rosier, D.M. Smith, M.G. Hodnett, J.M. Roberts, R. Hopkins, H.N. Davies, D.G. Kinniburgh, and D.C. Gooddy (1996) Hydrological effects of short rotation energy coppice. Energy Technology Support Unit Report ETSU B/W5/00275/REP. The Centre for Ecology and Hydrology, United Kingdom.

Hamblin, A.P. (1985) The influence of soil structure on water movement, crop root growth, and water uptake. *In.: Advances in Agronomy*. Academic Press, New York **38**: 95–157.

Hansen, E.A. (1993) Soil carbon sequestration beneath hybrid poplar plantations in the north central United States. *Biomass Bioenergy* **5**(6): 431–436.

Haycock, N.E. and G. Pinay (1993) Groundwater nitrate dynamics in grass and poplar vegetated riparian buffer strips during winter. *J. Environ. Qual.* **22**: 273–278.

Isebrands, J.G. and J. Richardson, compilers (2000) 21st Session of the International Poplar Commission (IPC 2000), Poplar and Willow Culture, Meeting the Needs of Society and the Environment. General Technical Report NC-125. U.S. Department of Agriculture, Forest Service, North Central Research Station, St. Paul, Minnesota. Held in Vancouver, Washington, September 24–28. 220 pp.

Jacobs Engineering Group (1996) Technology demonstration plan treatability study for plant-enhanced bioremediation of contaminated soil and groundwater—draft final. U.S. Air Force Aeronautical Systems Center, Air Force Plant 4, Fort Worth, Texas.

Jordahl, J.F., P.J. Alvarez, and J.L Schnoor (1997) Effect of hybrid poplar trees on microbial populations important to hazardous waste bioremediation. *Environ. Toxicol. Chem.* **16**(6): 1318–1381.

Khurana, D.K., S. Narljede, and M.B. Chandrashekhar (2000) Rooting behavior—an indicator of plantation success and growth in poplars in *Populus* production systems. In: 21st Session of the International Poplar Commission (IPC 2000), Poplar and Willow Culture, Meeting the Needs of Society and the Environment. J.G. Isebrands and J. Richardson, compilers. General Technical Report NC-125. U.S. Department of Agriculture, Forest Service, North Central Research Station, St. Paul, Minnesota. Held in Vancouver, Washington, September 24–28. p. 93.

Krinard, R.M. (1985) Cottonwood development through 19 years in a Nelder's Design. Research Note SO-322. U.S. Department of Agriculture, Forest Service, Southern Forest Experiment Station, New Orleans, Louisiana.

Lee, R.W., S.A. Jones, E.L. Kuniansky, G.J. Harvey, B. Sherwood-Lollar, and G.F. Slater (2000) Phreatophyte influence on reductive dechlorination in a shallow aquifer contaminated with trichloroethene (TCE). *Int. J. Phytoremed.* **2**(3): 193–211.

Leenheer, J.A. and E.W.D. Huffman, Jr. (1979) Analytical method for dissolved-organic carbon fractionation. Water-Resources Investigations Report 79–4. U.S. Geological Survey.

Licht, L.A. and M. Madison (1994) *Proceedings of the 87th meeting of the Air and Waste Management Association*, Cincinnati, Ohio.

McCracken, A.R. and W.M. Dawson (2000) Rust (*Melampsora epitea* var. *epitea*) on short-rotation coppice (SRC) willow. In: *Third Biennial Conference, Short-Rotation Woody Crops Operations Working Group*. State University of New York, Syracuse, New York, October 10–13, p. 111.

McDonald, M.G. and A.W. Harbaugh (1988) A modular three-dimensional finite-difference ground-water flow model. Techniques of Water-Resources Investigations of the U.S. Geological Survey, Book 6, Chapter A1.

Newman, L.A., S.E. Strand, N. Choe, J. Duffy, G. Ekuan, M. Ruszaj, B.B. Shurtleff, J. Wilmoth, P. Heilman, and M.P. Gordon (1997) Uptake and transformation of trichloroethylene by hybrid poplars. *Environ. Sci. Technol.* **31**(4): 1062–1067.

Newman, L.A., X. Wang, I.A. Muiznieks, G. Ekuan, M. Ruszaj, R. Cortellucci, D. Domroes, G. Karscig, T. Newman, R.S. Crampton, R.A. Hashmonay, M.G. Yost, P.E. Heilman, J. Duffy, M. Gordon, and S.E. Strand (1999) Remediation of trichloroethylene in an artificial aquifer with trees, a controlled field study. *Environ. Sci. Technol.* **33**(13): 2257–2265.

Nzengung, V.A. and P. Jeffers (2001) Sequestration, phytoreduction, and phytooxidation of halogenated organic chemicals by aquatic and terrestrial plants. *Int. J. Phytoremed.* **3**(1): 13–40.

Persson, G. (1995) Willow stand evapotranspiration simulated for Swedish soils. *Agric. Water Manag.* **28**: 271–293.

Rivers, G.A., E.T. Baker, and L.S. Coplin (1996) Geohydrologic units and water-level conditions in the Terrace Alluvial aquifer and Paluxy aquifer. May 1993 and February 1994, near Air Force Plant 4, Forth Worth, Texas. Water-Resources Investigations Report 96–4032. U.S. Geological Survey.

Rust Geotech (1996) Interim Air Force Plant 4 assessment of intrinsic bioremediation. Prepared for U.S. Department of the Air Force, Headquarters Aeronautical Systems Center. Contract No. DE-AC04-94AL96907. Wright Patterson Air Force Base, Ohio.

Schiff, S.L., R. Aravena, S.E. Trumbore, M.J. Hinton, R. Elgood, and P.J. Dillon (1997) Export of DOC from forested catchments on the Precambrian Shield of central Ontario: clues from ^{13}C and ^{14}C. *Biogeochemistry* **36**(1): 43–65.

Schnoor, J.L. (1997) Phytoremediation. Technology Evaluation Report TE-98-0. Ground-Water Remediation Technologies Analysis Center, Pittsburgh, Pennsylvania.

Schnoor, J.L., L.A. Licht, S.C. McCutcheon, N.L. Wolfe, and L.H. Carreira (1995) Phytoremediation of organic and nutrient contaminants. *Environ. Sci. Technol.* **29**(7): 318A–323A.

Shang, T.Q., S.L. Doty, A.M. Wilson, W.N. Howald, and M.P. Gordon (2001) Trichloroethylene oxidative metabolism in plants: the trichloroethanol pathway. *Phytochemistry* **58**: 1055–1065.

Stone, E.L. and P.J. Kalisz (1991) The maximum extent of tree roots. *Forest Ecol. Manag.* **46**: p. 77.

Tschaplinski, T.J., G.A. Tuskan, G.M. Gebre, and D.E. Todd (1998) Drought resistance of two hybrid *Populus* clones grown in a large-scale plantation. *Tree Physiol.* **18**: 653–658.

U.S. Environmental Protection Agency (2001) NATO/CCMS pilot study, evaluation of demonstrated and emerging technologies for the treatment of contaminated land and groundwater (phase III). 2000 Annual Report, Number 224, EPA 542-R-01-001, p. 85.

U.S. Environmental Protection Agency (2002) Short rotation woody crop groundwater treatment (SRWCGT) system. Innovative Technology Evaluation Report. U.S. Environmental Protection Agency, National Risk Management Research Laboratory, Cincinnati, Ohio.

Vose, J.M., W.T. Swank, G.J. Harvey, B.D. Clinton, and C. Sobek (2000) Leaf water relations and sapflow in eastern cottonwood (*Populus deltoides* Bartr.) trees planted for phytoremediation of a groundwater pollutant. *Int. J. Phytoremed.* **2**(1): 53–73.

Watermark Computing (1994) PEST, model-independent parameter estimation. Watermark Computing, Brisbane, Australia.

White, W.N. (1932) A method of estimating groundwater supplies based on discharge by plants and evaporation from soil, results of investigations in Escalante Valley, Utah. Water-Supply Paper 659–A. U.S. Geological Survey, pp. 1–105.

Wiedemeier, T.H., M.A. Swanson, D.E. Moutoux, E.K. Gordon, J.T. Wilson, B.H. Wilson, D.H. Kampbell, J.E. Hansen, P. Haas, and F.H. Chapelle (1996) Technical protocol for evaluating natural attenuation of chlorinated solvents in ground—water—draft revision 1. Air Force Center for Environmental Excellence, Technology Transfer Division, Brooks Air Force Base, San Antonio, Texas.

Zhang, H., J.I.L. Morison, and L.P. Simmonds (1999) Transpiration and water relations of poplar trees growing close to the water table. *Tree Physiol.* **19**: 563–573.

20

FIVE-YEAR PILOT STUDY: ABERDEEN PROVING GROUND, MARYLAND

S. R. Hirsh, H. R. Compton, D. H. Matey, J. G. Wrobel, and W. H. Schneider

SUMMARY OF PRACTICAL IMPLICATIONS

A phytoremediation demonstration at J-Field indicates that hybrid poplar (*Populus deltoides* × *Populus trichocarpa*) trees mitigate the migration of a groundwater plume of volatile organic compounds. Groundwater data reveal a cone of depression underlying the plantation, and sap-velocity data for the plantation indicate a maximum transpiration rate of approximately 7600 liters per day (2000 gallons per day). Tree-tissue and transpiration-gas data show that the trees capture contaminated groundwater. Volatile organic compounds are metabolized within the trees and possibly degraded within the rhizosphere. Low concentrations of volatile organic compounds and degradation products are transpired through the leaves and trunks but pose no threat to human or environmental receptors. A water budget analysis is being used to develop a transient groundwater-flow model to simulate seasonal hydraulic containment. These data will be used to assess the remedial benefits of planting additional trees. Geohydrologic conditions at J-Field are favorable for utilizing trees to intercept and treat contaminated groundwater and to supplement the natural biotic and abiotic degradation processes.

PHYTOREMEDIATION AND J-FIELD HYDROLOGY

Overview of Phytoremediation Study

Phytoremediation has been selected as a remedial technology for groundwater clean up at the J-Field Superfund site, Aberdeen Proving Ground, Maryland. The principal contaminants of concern at the site are chlorinated volatile organic compounds (VOCs), 1,1,2,2-tetrachloroethane (1,1,2,2-TeCA), and

Phytoremediation: Transformation and Control of Contaminants,
Edited by Steven C. McCutcheon and Jerald L. Schnoor.
ISBN 0-471-39435-1 (cloth)

trichloroethylene (TCE). The phytoremediation study was conducted under a joint program of the U.S. Army Directorate of Safety, Health, and Environment, the U.S. Environmental Protection Agency (U.S. EPA) Region III, and the U.S. EPA Environmental Response Team Center.

Several conventional groundwater remedial technologies have been studied or tested at J-Field during remedial investigation and feasibility studies (RI/FS). The success of these conventional technologies was significantly hindered by complex site conditions, including the presence of unexploded ordnance (UXO), a low-permeability aquifer, and a persistent VOC source fed by nonaqueous phase liquids. The conventional technologies that were evaluated included soil washing, soil-vapor extraction, groundwater pump-and-treat, and groundwater circulation wells.

In situ biotic and abiotic degradation processes have also been investigated. These studies demonstrate that VOC degradation actively occurs in the vicinity of a downgradient freshwater marsh. Despite the elevated groundwater contaminant concentrations, the risk to human health and the environment is low because of the lack of groundwater uses and the minimal impact on surface water (Yuen *et al.* 1998a, Pardue *et al.* 2000).

During spring 1996, a pilot-scale phytoremediation study was initiated at J-Field with the planting of 183 hybrid poplar (*Populus deltoids* × *Populus trichocarpa*) trees. The phytoremediation study evaluated the capacity of the trees to reduce dissolved-phase VOC mass loading to the marsh by providing hydraulic containment of the VOC plume. The study also evaluated the capacity of trees to provide dissolved-phase VOC mass reduction by transpiring and biotransforming VOCs and potentially enhancing *in situ* biodegradation by rhizosphere processes. This chapter summarizes the development and implementation of the 5-year phytoremediation remedial program, presents an overview of the comprehensive monitoring plan, and discusses preliminary conclusions on the remedial performance of the phytoremediation system.

Site Background

Aberdeen Proving Ground is located in Harford County, northeast of Baltimore, Maryland, along the western shore of the Chesapeake Bay. Over the years, military weapons testing and disposal activities at the J-Field testing area have caused extensive releases of chemicals to soil and groundwater. The Toxic Burning Pits at J-Field consist of several trenches where disposal activities occurred. Significant levels of VOCs have been detected in the groundwater at the Toxic Burning Pits area.

Groundwater Hydrology

Climate within the Chesapeake Bay region is temperate, with an annual precipitation of 114 centimeters per year (45 inches per year). Precipitation is fairly evenly distributed throughout the year, and Hughes (1995) estimated that between 1990 and 1991, approximately 38 percent of the precipitation infiltrated to recharge the surficial aquifer. Historic groundwater recharge estimates range from 33 to 58 centimeters (13 to 23 inches). Groundwater flow in the surficial aquifer is radial beneath the former Toxic Burning Pits, and eventually discharges into a nearby freshwater marsh. The marsh provides a prolific wetland habitat for many different species, including raccoons, river otter, white-tailed deer, and waterfowl. The marsh is dominated by reed grass (*Phragmites australis*) and is underlain by silt and peat. The abundant organic materials within the freshwater marsh provide an ideal environment for promoting the natural attenuation of contaminants by adsorption and biotic degradation (Yuen *et al.* 2000).

The source of the VOC plume, the Toxic Burning Pits, is on a local topographic high. This area contributes to groundwater recharge of the surficial aquifer. Seasonal variations in recharge result in 1.5-meter (4.9-foot) fluctuations of the water table that cause short-term shifts in hydraulic gradient and flow direction. The surficial aquifer is primarily low-permeability fine sand and clayey silt. These sediments are highly variable in texture and clay content, as evidenced by borehole geophysical testing methods, demonstrating that silt and clay lenses exist throughout the surficial aquifer (WESTON 2001). A representative cross section illustrating the high degree of heterogeneity of these sediments is shown in Figure 20-1. The heterogeneous nature of the sediments comprising the surficial aquifer contributes to complex hydrogeologic conditions.

Aquifer tests were conducted on the heterogeneous sediments of the surficial aquifer to characterize the hydraulic properties. The long-term aquifer tests indicate that the surficial aquifer is unable to sustain well yields greater than 3.8 liters per minute (1 gallon per minute). The highly heterogeneous

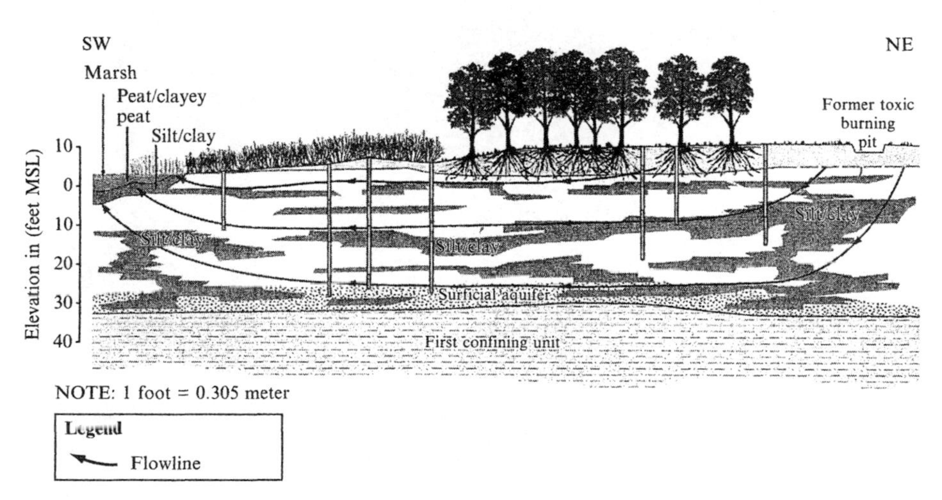

Figure 20-1 Representative cross section of the surficial aquifer along the southwest (SW) to northeast (NE) axis. The U.S. National Gesdetic Vertical Datun or mean sea level is abbreviated as MSL.

sediments exhibit relatively low hydraulic conductivities of 0.09 to 2.4 meters per day (0.3 to 8 feet per day) (Yuen *et al.* 1998b).

Contaminant Hydrology

The areas near the Toxic Burning Pits serve as a local groundwater recharge area for the surficial aquifer. Groundwater generally flows out from the recharge area towards the freshwater marsh. This flow pattern is reflected in the chemical distribution of TCE in the groundwater (Figure 20-2). This figure also illustrates the rapid decline in dissolved-phase concentrations (three orders of magnitude) from greater than 200 000 micrograms per liter to less than 200 micrograms per liter in the approximately 100 meters (330 feet) that the plume travels from the Toxic Burning Pits to the freshwater marsh. The VOC plume has reached the marsh areas on both the eastern and southwestern sides of the Toxic Burning Pits area, and the contaminants undergo significant biodegradation in the marsh before the groundwater discharges to surface water.

Biotic and abiotic degradation of VOCs readily occur as evidenced by (1) the reduction of VOC concentrations along the groundwater flow path, (2) the presence of biodegradation by-products such as vinyl chloride and ethane, and (3) the presence of methanogenic and sulfate reductive redox conditions that promote reductive dechlorination of TCE.

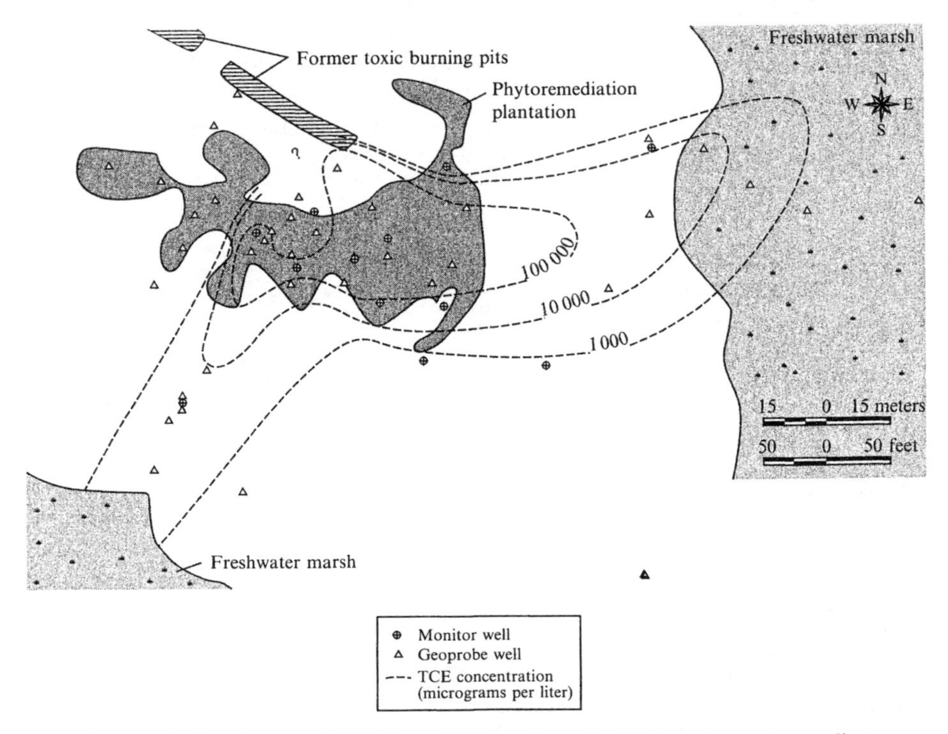

Figure 20-2 Distribution of TCE in surficial aquifer, July 2001 in micrograms per liter.

TABLE 20-1 Maximum Detected Concentrations of Contaminants of Concern at Toxic Burning Pits During July 2001 in Micrograms per Liter

Contaminant of concern	Groundwater concentration (micrograms per liter)
1,1,2,2-Tetrachloroethane (1,1,2,2-TeCA)	4 400 000
Trichloroethylene (TCE)	240 000
cis-1,2-Dichloroethylene (*c*-DCE)	280 000
Tetrachloroethylene (PCE)	11 000
trans-1,2-Dichloroethylene (*t*-DCE)	80 000
1,1,2-Trichloroethane (TCA)	52 000
Vinyl chloride	4200

At J-Field, the redox conditions of the groundwater underlying the upland area consist of a thin veneer of aerobic (oxic) groundwater overlying anaerobic (iron-reducing) groundwater that transitions to sulfate-reducing and ultimately methanogenic conditions as the groundwater flows into the marsh (Yuen *et al.* 2000). The results of groundwater sampling using dialysis samplers indicated that groundwater conditions are more oxidized within the rhizosphere (Pardue *et al.* 2000). The implication is that hybrid poplar (*Populus deltoides* × *Populus trichocarpa*) trees may be capable of enhancing the *in situ* biodegradation of the contaminants (*e.g.*, vinyl chloride) that are degradable under oxygen-reducing conditions.

A complicating factor at J-Field is the recent discovery of free-phase dense nonaqueous phase liquid (DNAPL) that underlies the Toxic Burning Pits. The DNAPL is suspected to exist as (1) a separate phase (mobile) liquid distributed in a thin veneer resting on silt and clay lenses, and (2) a residual phase (immobile) DNAPL entrapped in discontinuous, low permeability clay zones. Groundwater data collected between 1989 and 2001 indicate that the DNAPL continuously feeds the dissolved-phase VOC plume, as evidenced by steady concentrations, some of which are at or near the effective solubility of TCE and 1,1,2,2-TeCA. To characterize this source zone, a DNAPL investigation will be conducted to identify the three-dimensional subsurface heterogeneities and distribution of DNAPL. Table 20-1 presents a summary of the maximum VOC concentrations historically detected at J-Field.

PHYTOREMEDIATION PILOT STUDY DESIGN AND IMPLEMENTATION

Agronomic Samples

To assess the feasibility of phytoremediation as a remedial technology for the site, soil samples were collected and evaluated in relation to the suitability for plant growth following methodologies outlined in *Recommended Soils Testing*

Procedures for the Northeastern United States (Delaware Cooperative Extension 1995). The soil characteristics of J-Field support the use of hybrid poplar (*Populus deltoides* × *Populus trichocarpa*) trees, which are a suitable choice for phytoremediation because of the fast-growing, deep-rooting characteristics.

Tree-Planting Scheme and Tree Types

The phytoremediation system was designed in 1996. The area for planting the trees was selected in relation to the flow of contaminated groundwater and the wells that could be used in monitoring the system. An area of approximately 0.4 hectare (1 acre) was selected, and 183 hybrid poplar (*Populus deltoides* × *Populus trichocarpa*) trees were planted in a U-shaped configuration surrounding the Toxic Burning Pits. The configuration was designed to use the trees to intercept the VOC plume from migrating to the marsh. The trees were planted at 3-meter (10-feet) centers to permit vehicle and equipment access.

By 1998, several trees had died because of the stress induced by a drought, and replacement trees were planted the following year. Hybrid poplars (*Populus deltoides* × *Populus trichocarpa*) and two native tree species [tulip trees (*Liriodendron tulipifera*), a large, long-lived tree native to Aberdeen Proving Ground, and silver maple (*Acer saccharinum*)] were selected for planting. Using a diversity of native tree species lessens the plantation's susceptibility to mass wasting, an inherent weakness of monocultures. This approach follows the Executive Order (EO 13112 Invasive Species) that directs federal entities to restore disturbed sites with native species.

In the fall of 1998, several hybrid poplars (*Populus deltoides* × *Populus trichocarpa*) were excavated to assess the impacts of the plastic sleeves placed around the roots prior to planting. The plastic sleeves were used to encourage downward root growth and groundwater uptake. Excavation results revealed that the plastic sleeves and the high clay content of the soil restricted root growth to a limited area. Lateral and surface root growth were also restricted, which contributed to many of the trees blowing over during a tropical storm. Root depths were estimated to be 2 meters (7 feet) below ground surface. Because of the problems at the J-Field site, sleeves were not used in the second round of planting.

Based on observed root growth patterns, boreholes of various diameters and depths were used in the second round of planting to encourage more extensive root growth. In addition, when planting trees in tight soil formations, it is important to score, aerate, or otherwise loosen the soils surrounding the borehole. This procedure allows optimal root development and growth, and maximizes transpiration rates.

PHYTOREMEDIATION SYSTEM PERFORMANCE MONITORING

Monitoring was conducted to evaluate the performance of the phytoremediation plantation in achieving the goals of hydraulic containment of the

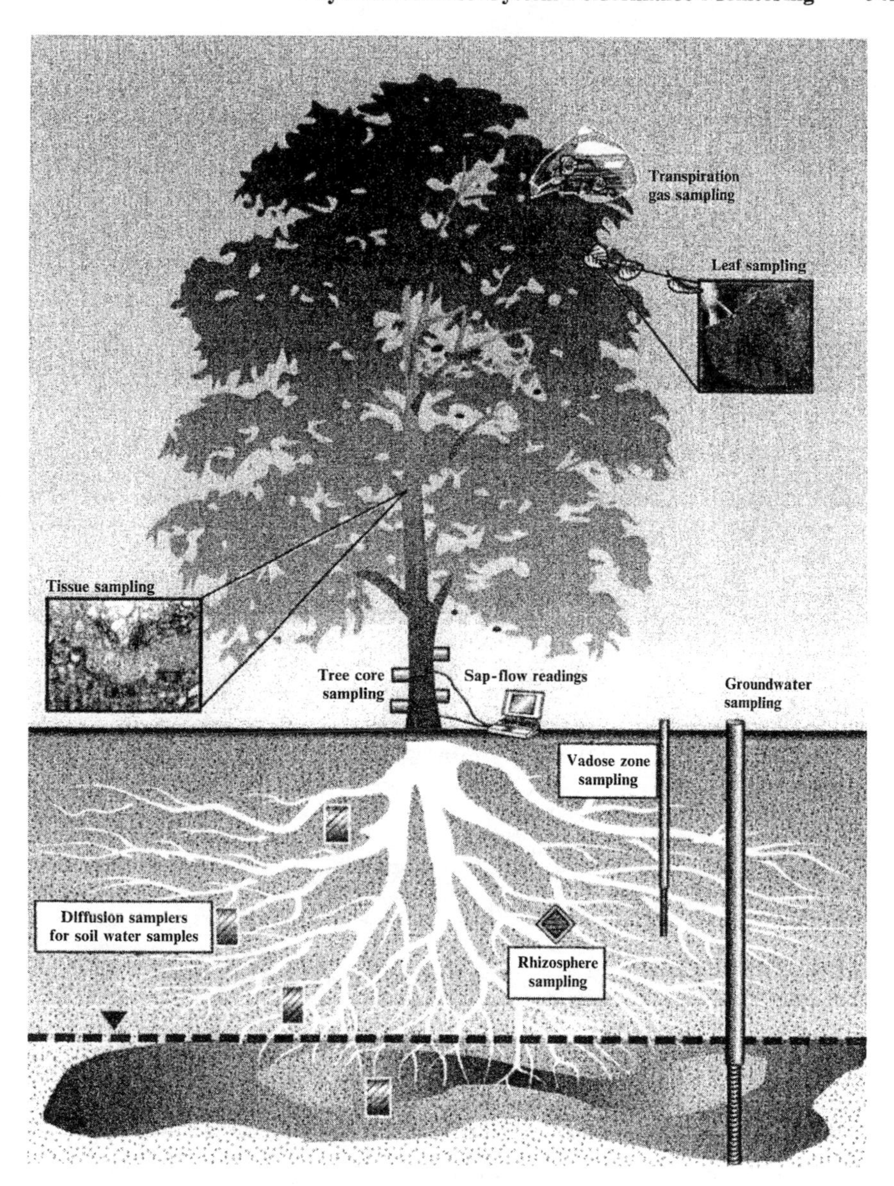

Figure 20-3 Locations of phytoremediation monitoring.

contaminated groundwater plume and mass removal of the contaminants. Figure 20-3 is a schematic that illustrates the phytoremediation sampling.

Tree Transpiration Rates

Transpiration rates for individual, mature trees were estimated by measuring sap flow using the Dynamax Inc., Dynagage® Flow 32 system. Sap flow, tree height, diameter-at-breast-height, and on-site weather conditions were examined seasonally over a 4-year period. Two methods were used to assess tree transpiration rates—an energy balance method and a thermal dissipation probe method. Dynagage® sensors were used to collect continuous sap flow rates for selected trees during seasonal monitoring periods from May 1997 to July 1999. Sap Velocity Thermal Dissipation Probes (TDP) were used in conjunction with the Dynagage® sensors during the July and October 1998 and July 1999 field sampling events. Details on sensor operations and methods are presented in Haroski *et al.* (2000).

Sap-flow and tree-growth data were used to calculate current and future transpiration rates. Local weather data were used to estimate the overall evapotranspiration potentital demand on a daily basis for the previous 3 years. The evapotranspiration potential from the plantation was compared with the sap flow data to generate a tree- and site-specific "crop index." Daily and yearly evapotranspiration potentials were calculated using on- and off-site weather data. Daily evapotranspiration was based on average temperature, total solar radiation, average humidity, and average wind speed. These data were used to predict an evapotranspiration potential index. The sum of evapotranspiration potential for all active days of the growing season was compared with sap flow or transpiration on a land area basis. The growing season evapotranspiration potential was compared with the total water consumed by the trees to determine the crop coefficient on an annual basis. The crop coefficient (*CC*) was computed from (*ETP*) using the following equation:

$$CC = \frac{\dfrac{\text{Sap flow for all active trees}}{\text{Land area}} \text{(liters per square meter)}}{ETP \text{ (millimeters)}} \tag{20-1}$$

The crop coefficient for the J-Field site was based on a 200-day period (April 15 to November 1) because transpiration is limited during the remainder of the year. The daily and annual crop coefficients were applied to test the limits of the long-term sap flow trend.

The sap flow results for this study were converted into water transpired per unit land area (as measured using a global positioning system) and expressed in millimeters per day. For ease of interpretation, the data are presented in liters and gallons per day. Results from this analysis indicated that peak transpiration would occur in approximately 10 to 15 years. In 1999, the trees were removing approximately 4130 liters per day (1090 gallons per day). At the end of 30 years, the trees are expected to remove approximately 7600 liters per day (2000 gallons per day), with individual trees removing approximately 49 liters per day (13 gallons per day) on average. Intercepted precipitation is included in these transpiration estimates.

Directly related to plant transpiration, the leaf area index (LAI) was calculated for use in an alternative method of estimating maximum transpiration rates. The LAI is defined as the quantity of leaf surface area per unit land area. Calculations and predictions of LAI can subsequently be used for estimating the time of canopy closure, when maximum stand-level transpiration occurs. The LAI at which canopy closure will most likely occur can be predicted by estimating yearly LAI (*e.g.*, by fitting a power equation trend to the data) and comparing results between years.

Leaf and stem area measurements were collected in July 1999 and compared with the trunk area to estimate the progression in LAI. Leaf area was calculated using a Dynamax leaf area meter (AM100). The leaf area meter is used to measure leaf area by scanning individual leaves and then summing all of the area measurements. Using these stem and leaf area calculations, an approximate leaf area to stem area ratio was computed and used to determine an approximate LAI.

The July 1999 average and maximum leaf and stem area ratios were used to calculate the average and maximum LAI for the site. Specifically, the leaf and stem ratios were multiplied by the stem area, yielding total leaf area that was then divided by the land area. This resulted in an average LAI of 2.6. The J-Field plantation is expected to reach canopy closure as the LAI approaches 7, based on trees planted on 3-meter (10-foot) centers. These calculations are based on only one data set, and further data collection may alter these predictions.

Predictions of LAI for a 30-year period were based on the growth data collected from 1997 to 1999. A power equation trend was best fit to predict the LAI over the first 10 years. This trend indicates rapid tree growth from 1996 to 1999, with no canopy closure occurring during this time. In years 8 to 11, however, almost complete canopy closure occurs. At this point, the LAI starts to increase more slowly and gradually flattens, increasing only slightly between years 17 to 30.

Monitoring of Groundwater Hydraulics

A comprehensive groundwater-monitoring program was implemented to define the capacity of the phytoremediation plantation to provide hydraulic containment. Water table elevation measurements have been continuously collected from 21 wells using automated data loggers, and manually collected on a monthly basis from 37 other wells.

The complex hydrology at J-Field requires careful interpretation of the groundwater data to account for confounding factors such as changing barometric pressure, tidal forcing, groundwater recharge, and delayed yield. The characteristics of the J-Field hydrologic system, specifically the low permeability aquifer, subsurface heterogeneities, and the proximity of the site to a freshwater marsh and tidal estuary, make evaluating these factors critical to accurately assess the effects of the plantation on the aquifer.

Preliminary examination of the data indicates that the aquifer response to these conditions is highly varied and depends upon groundwater recharge.

The water table fluctuates between the layers of sand and clayey silt. During the winter, groundwater recharge events create the peak water table elevation and the aquifer can become temporarily semiconfined (confined by a "leaky" unit through which recharge and discharge occur) as evidenced by an increase in barometric efficiency (change in hydraulic head *versus* change in atmospheric pressure). Beginning in mid-April, the hydrologic data illustrate diurnal water table fluctuations in response to the cyclic effects of evapotranspiration, with the declines occurring during the late afternoon, followed by aquifer recovery during the night. The monitoring data indicate that the water table declines approximately 4.6 centimeters (1.8 inches) in response to transpiration. A hydrograph for a monitor well located within the plantation (Figure 20-4) demonstrates the influence of the trees on the water table. During the seasonal lull in growth, the hydraulic gradients are downward and show that groundwater recharge is dominating the aquifer. As the growing season commences, a measurable reversal of the hydraulic gradients occurs, as groundwater flow is induced upwards towards the tree roots. The vertical shift is demonstrated by the vertical gradient reversal in response to transpiration. The aquifer response is observed to occur at depths of 7.6 meters (25 feet) below ground surface.

The effects of transpiration on the water table configuration are shown by the distinct differences in the groundwater elevation maps throughout the year. For example, during March, the water table is elevated because of in

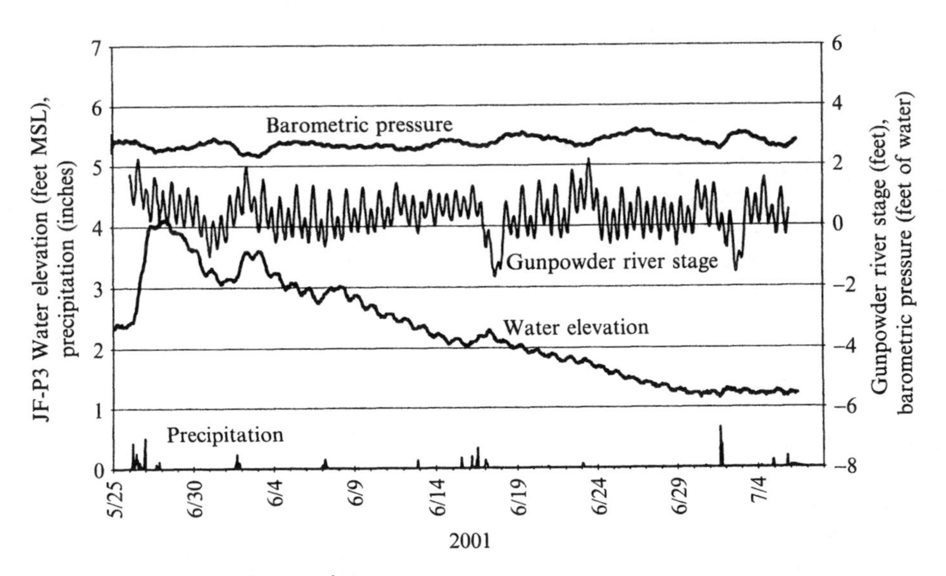

Figure 20-4 Hydrograph of Monitor Well JF-P3, Gunpowder River stage, barometric pressure, and precipitation. Mean sea level or the U.S. National Geodetic Vertical Datum is abbreviated as MSL.

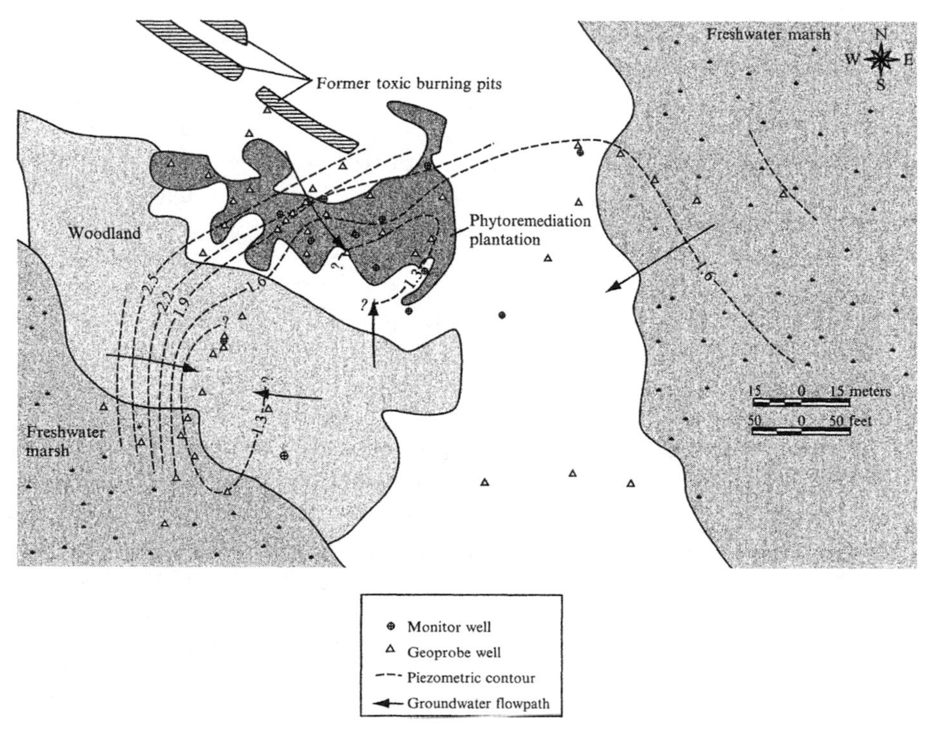

Figure 20-5 Water table elevations in feet (1 foot = 0.3048 meter), July 2001.

creased recharge, and groundwater flow is nearly radial toward the marsh with no evidence of diurnal cycles in response to transpiration.

During the summer, the aquifer experiences greater water stresses as a result of the combination of increased transpiration rates and reduced groundwater recharge, as shown by the discernible cone of depression that forms in the center of the plantation (Figure 20-5). Groundwater levels decline throughout the summer as aquifer storage is depleted, and eventually the water table declines below the freshwater marsh. This condition creates a temporary flow reversal (recharge boundary effect) that ultimately induces surface water infiltration back into the aquifer. The effects of barometric pressure and tidal loading on the water table were shown to be negligible during the growing season. Based on the hydraulic gradients in summer, the benefit of hydraulic containment of the VOC plume is significant because it reduces the migration rate of the plume and VOC mass loading to the marsh. Hydraulic containment is most evident in the region where groundwater flow is transitioning from a recharge zone (Toxic Burning Pits) to a discharge zone (freshwater marsh).

Monitoring of Groundwater Chemistry

Monitoring of contaminants in soil, groundwater, and surface water has been periodically conducted to assess water quality at J-Field. The monitoring has produced a chemical database suitable for characterizing the extent of VOC contamination and evaluating historic contaminant trends. An assessment of existing chemical data assists in establishing the fate and transport of contaminants and improves the phytoremediation design.

Groundwater sampling is being conducted using low-flow sampling methods. The samples have been analyzed for parameters that indicate the viability of natural attenuation processes to degrade VOCs using methods outlined in the U.S. EPA Technical Protocol (U.S. EPA 1998). Results of the monitored natural attenuation study are presented in Yuen *et al.* (1998a) and Yuen *et al.* (2000). Biotic and abiotic degradation of VOCs readily occur, as primary (parent) compounds 1,1,2,2-TeCA and tetrachloroethylene (PCE) were shown to degrade and, subsequently, to produce secondary (daughter) compounds such as dichloroethylene, vinyl chloride, and ethene.

The biodegradation of chlorinated solvents under the observed anaerobic conditions proceeds through reductive dechlorination or halorespiration. The reductive dechlorination process consists of VOCs being used by microorganisms as electron acceptors to support respiration. The process results in a chlorine atom being removed from the VOC and replaced with a hydrogen atom. The hydrogen for these reactions comes from the fermentation of other substrates or electron donors present in the aquifer. Transformations of VOCs can occur both chemically (abiotic) and biologically (biotic). Abiotic processes that occur most frequently are hydrolysis and dehydrochlorination (the elimination of hydrogen and chlorine molecules and the formation of a carbon double bond). Anaerobic biotic processes, generally, are reductions that involve either hydrogenolysis, the substitution of a hydrogen atom for chlorine, or dichloroelimination, where two adjacent chlorine atoms are removed, leaving a double bond between the carbon atoms.

The reductive dechlorination of chlorinated ethenes is primarily by hydrogenolysis and occurs sequentially from PCE and 1,1,2,2-TeCA to TCE to dichloroethylene to vinyl chloride. These reactions have also been reported in microcosm studies conducted from Aberdeen Proving Ground aquifer materials and observed in field-based studies (Lorah *et al.* 1998).

Monitoring of Rhizosphere

The potential for enhanced *in situ* biodegradation of VOCs in the rhizosphere was investigated using membrane dialysis samplers to collect high-resolution data from discrete vertical zones within the rhizosphere and aquifer areas. Samples were collected from both within and outside the phytoremediation plantation, and analyzed for the chemical parameters that would provide data to distinguish between gross microbial reactions within the aquifer and more discrete plant–microorganism interactions within the rhizosphere of the trees. These parameters include chlorinated aliphatic compounds, organic acids, chloroacetic acids, total organic carbon, chloride, and dissolved gases (ethane, ethene, methane, and carbon dioxide). Parameters measured also include potential root exudates such as (1) total dissolved organic carbon, (2) carbohydrate root exudates, and (3) degradation products of TCE in hybrid poplar (*Populus deltoides* × *Populus trichocarpa*) trees including trichloroacetic acid, dichloroacetic acid, and trichloroethanol.

Details of the sampling program and the study results are presented in Pardue *et al.* (2000). The study results indicated that reductive dechlorination continues to be the primary degradation pathway in groundwater and that areas within the rhizosphere are less reducing because of the introduction of carbon dioxide by the trees. Additional study is needed to evaluate the capacity for the more oxidized conditions to encourage the degradation of certain VOCs, like vinyl chloride. The sampling of root exudates indicates the rhizosphere does not contribute substantial increase of carbon to the groundwater. Analyses of samples collected for evaluating the specific degradation products for TCE in hybrid poplar (*Populus deltoides* × *Populus trichocarpa*) trees (trichloroacetic acid, dichloroacetic acid, and trichloroethanol) were inconclusive because membrane dialysis samplers did not collect a sufficient volume of groundwater for analysis.

Monitoring Tree Chemistry

Assessing the contribution of phytoremediation to the overall reduction of VOCs in groundwater is a critical component for evaluating remedial success. At J-Field, the investigation has focused on two predominant phytoremediation mechanisms: (1) phytodegradation–phytovolatilization, and (2) rhizodegradation.

The first mechanism involves the removal of VOC-contaminated groundwater by the trees and the subsequent biotransformation or volatilization of the contaminants. Excluding the contaminant mass that becomes adsorbed in the root zone before uptake into the tree, the transpired contaminants are biotransformed through various metabolic processes that promote and catalyze degradation of the VOCs. The contaminants are degraded in the tree with the breakdown products subsequently stored in the vacuole or incorporated into the tree tissues [Interstate Technology and Regulatory Council (ITRC) 2001]. The contaminants and biotransformation products can also be directly transpired by the trees and may be released to the atmosphere by volatilization through the trunk, stems, and leaves.

The contaminant mass that adsorbs in the root zone typically undergoes biodegradation in the rhizosphere. The proteins and enzymes that are produced and exuded by trees and soil organisms enhance this bioactivity. Root dieback and exudates such as carbohydrates and acids contain organic carbon that provides nutrients for the soil organisms, thereby enhancing biological activities. These processes were evaluated at J-Field to assess the efficiency of phytoremediation in removing VOCs from groundwater.

Monitoring of Tree Tissue

A primary concern of this investigation has been the accumulation of contaminants in tree tissue. Initial tissue sampling was conducted from 1997 to

2000 using a solvent extraction method that utilized hexane to partition contaminants from the tissue. The tissue samples consisted primarily of leaves and stems. Despite repeated efforts and numerous methods explored, no reliable extraction technique was identified that yielded consistent results for the tree tissue analysis. The sampling results based on the extraction method indicated that leaf tissue samples might not provide the most effective media for measuring parent compounds or the associated degradation products. The solvent extraction methods used to analyze VOCs in tree tissue were discontinued and replaced by two new methods in 2000.

The first replacement technique utilized sampling with an increment borer (Forestry Suppliers, Jackson, Mississippi) to collect a tree core. Biomass samples from the core were subsequently placed into pre-weighed 20-milliliter vials. Details of this technique are presented in Burken (2001). The vials were then weighed to determine the total mass of the collected sample. The headspace samples were injected into a HP5890 gas chromatograph (GC), with a dual electron capture detector-flame ionization detector. Following GC analysis, vials were stored until the cores could be placed in a 100 °C oven and dried to determine the initial core moisture and dry mass. Concentrations were then normalized using the initial weight of the biomass sample and experimentally determined partition coefficients. These partitioning coefficients were determined for air-biomass concentrations using fresh biomass samples from uncontaminated environments. These partitioning coefficients were 7.41 grams per liter for TCE, and 3.13 grams per liter for 1,1,2,2-TeCA.

The biomass samples were collected from tree cores to examine the efficiency of the hybrid poplars (*Populus deltoides* × *Populus trichocarpa*) for removing VOCs from groundwater and to assess the fate of these contaminants. Analyses of tree cores provide reliable results on whether trees remove TCE and other VOCs. In addition, tissue analysis may correlate with the associated groundwater concentrations (Vroblesky *et al.* 1999).

The first replacement sampling method based on core samples provided analytical results that were more consistent. Burken (2001) sampled the tree tissue and used the data to assess the degree of correlation between the distribution of VOCs in groundwater and the tree tissue. Groundwater and tree tissue samples were collected in tandem in both 2000 and 2001. Core samples were also taken at increasing heights above the ground from a live hybrid poplar (*Populus deltoides* × *Populus trichocarpa*) that was growing over the VOC plume. Following the collection of the tree cores, the root system of the tree was dug out of the ground, and core samples of the roots were collected. These core samples were analyzed for VOCs, and associated leaf and other tissue samples were collected and analyzed for VOCs and TCE degradation products (trichloroacetic acid, dichloroacetic acid, and trichloroethanol) using techniques reported by Newman *et al.* (1997). The analysis was completed to evaluate the potential fate of contaminants and to assess the efficiency of the trees to biotransform TCE.

The results of tree tissue sampling, using the tree coring method, indicated that groundwater contaminants are found throughout the tree. Burken (2001) demonstrated that the trees were removing VOCs from groundwater. The evidence included measurable TCE in the biomass samples that ranged between 0.1 and 75 micrograms per gram of plant by dry weight. Generally, the highest VOCs detected in the tree tissue corresponded to the groundwater source area near the DNAPL. However, the number of tree sampling locations compared to the number of groundwater monitoring wells was highly skewed, and this uneven sample distribution resulted in poor direct correlation between the tree and groundwater samples. To alleviate this problem, a more rigorous sampling program was implemented in July 2001 to improve the sampling distributions. The results will be reported in 2003. The tree coring method showed that the concentration of VOCs in the tree tissue decreased as the sampling elevation increased from the tree roots up to the tree branches. The decline in VOC concentrations is attributed to the tree volatilizing the VOCs through the trunk and to a lesser extent, the leaves. Transpiration gas data collected from the trunk and leaves provide direct evidence for this pathway, as discussed in the following section.

A second replacement tissue sampling technique was implemented in 2000 using solid-phase microextraction, a method that combines extraction, concentration, and sample introduction of organic compounds directly into a GC. A solid-phase microextraction unit consists of a length of fused silica fiber, coated with a polymer material, in some cases mixed with a solid adsorbent. The fiber is attached to a stainless steel plunger sheathed by a protective needle. Organic analytes in water and air samples can be collected directly on solid-phase microextraction fibers. Solid-phase microextraction can also be used to sample the headspace of solids and nonaqueous liquids. Sample workup is often the most time-consuming step in the analysis. However, solid-phase microextraction is quick and involves no solvents (J. Schneider, Argonne National Laboratory, 2001, personal communication to W. Schneider, Roy F. Weston, Inc.).

Plant tissue samples (leaves and corings) were collected for GC headspace analysis for trichloroacetic acid, TCE, and PCE. Solid-phase microextraction samples of the air transpired from the leaves and the trunk were collected for analysis of chloroform, TCE, and PCE. Preliminary data indicate that most of the transpiration of TCE and PCE is through the tree trunk. These data also demonstrate that biotransformation readily occurs. The results of these analyses will be reported in 2003.

Monitoring of Transpiration Gas

Two methods were used to sample the transpiration gas of leaves to assess whether the trees were removing contaminated groundwater. The first method involved placing a Tedlar® bag over several branches of the trees, sealing the bag around the stem, and adding compressed atmospheric air to the bag.

Temperature and humidity were monitored inside the bag. Based on high humidity levels and the presence of condensate, the bag apparatus was modified by the addition of a cold trap to remove water from the bag. Samples were then collected using Summa canisters approximately every half-hour. Tenax tubes, an on-site Viking SpectraTrak 620 gas chromatograph/mass spectrometer, and a Sciex® (PE Sciex Ltd., Ontario, Canada) model 6000E Trace Atmospheric Gas Analyzer were also used on-site for some sampling events. Summa canisters were chosen for use, as these provided the highest level of data quality and reproducibility.

Measurable concentrations of parent compounds 1,1,2,2-TeCA and TCE were found in only three trees using the Tedlar bag method. These trees were located in the area coincident with the highest groundwater contaminant concentrations. The logarithm of the octanol–water partitioning coefficient of 1,1,2,2-TeCA and TCE are similar, 2.39 and 2.38, respectively, in contrast to the vapor pressures of 6.5 *versus* 75 millimeters of mercury or solubilities of 2900 *versus* 1100 milligrams per liter. Hence, the relative concentrations in the transpiration gas over time are similar to those found in groundwater. This physiochemical condition is reflected in the consistent 1,1,2,2-TeCA to TCE ratio that is observed for both groundwater and transpiration gas. Given the disproportionately low number of trees with measurable parent compounds in the transpiration gas, the data indicate that volatilization through the leaves is not the primary fate of VOCs.

Monitoring Exposure Pathways

To address concerns that the trees were creating an exposure pathway for the groundwater contaminants, two sampling methods were employed to examine human and ecological risk. Air emissions sampling with an open-path Fourier transform infrared spectrometer was used to monitor ambient concentrations of the target compounds within the phytoremediation area on 31 July and 1 August 1997. Specific target compounds were 1,1,1-trichloroethane, 1,1,2,2-TeCA, TCE, PCE, and *cis*-dichloroethylene. The instrument had minimum detection limits ranging from 1.6 to 53 parts per million per meter of path length. The spectrometer was used in three areas on the site, and all monitoring events resulted in concentrations below the minimum detection limits for the target compounds, indicating that the contaminants released from the leaves, stems, and trunks are insignificant and do not pose a threat to human health or other ecological receptors.

Flux-chamber samples were collected during summer monitoring to assess air emissions from the shallow groundwater plume using methods outlined in Gussman and Halford (2000). Of the eight chambers sampled, TCE was detected in only three of the chambers, with concentrations ranging between 9.4 and 38 parts per billion by volume. The locations of the three chambers were not strictly correlated with the known shallow groundwater plume. The results indicated that the contaminants with higher vapor pressures, such as

TCE, are volatilizing through the vadose zone and contributing to measured mass emissions in the atmosphere.

In summary, the sampling program showed that there are measurable releases of VOCs (TCE and 1,1,2,2-TeCA) emitted from the plantation. However, as evidenced by the flux chamber sampling, TCE has been volatilizing and will continue to volatilize from the vadose zone. This evidence is supported by a previous soil gas survey conducted at J-Field. The absence of 1,1,2,2-TeCA in the flux chambers is attributed to the lower volatility. The open path spectrometer results indicate that VOCs released into the air through evaporation and transpiration do not approach a level that would cause risk to sensitive human or environmental receptors.

Monitoring Soil Health

To assess the health and functioning of the soil community, soil samples were taken for nematode identification and enumeration. Soil samples were collected from the base of selected trees in the plantation and from a reference location. Three soil samples were taken at approximately 0.15, 0.30, and 0.46 centimeters (6, 12, and 18 inches) from the base of the tree. Analyses were conducted to determine the total number and, based on esophageal morphology, to categorize the nematodes into bacterivore, fungivore, herbivore, predator, hatchling, and unknown. These data were compared to data previously collected from the site before the trees were planted.

The 1997 results revealed that total nematode abundance, specifically fungivore community abundance, had increased compared to samples taken prior to tree planting. Although the reasons for this increase are unknown, it is possible that the trees provide improved habitat for the nematodes or perhaps have mitigated the circumstances depressing the community. Samples from 1998 and 1999 showed decreases in total abundance from both 1997 and before trees were planted. Extreme drought conditions occurred during these sampling years, which may account for the decrease. Overall, nematode population analyses may be a useful parameter for evaluating soil community health at phytoremediation sites.

MODELING OF PHYTOREMEDIATION PERFORMANCE

Modeling Background

Several phases of groundwater modeling have been conducted to evaluate the hydrogeology and engineering design for groundwater remedial systems at J-Field. A three-dimensional numerical groundwater-flow model (MODFLOW, McDonald and Harbaugh 1988) was constructed to evaluate site data and improve the understanding of groundwater flow processes at J-Field (Hughes 1995). This model was later expanded to examine the effects of a proposed groundwater pump-and-treat system on the site hydrology (Quinn *et al.* 1996). After 1996, the site conceptual model continued to be refined with additional data, and the groundwater remedial action plan shifted toward an integrated engineered approach using technologies designed to provide dissolved-phase VOC mass removal. The model was updated again in late 1998 to support a feasibility study of the dissolved-phase VOC mass removal technologies (WESTON 2000a). This remedial approach did not account for DNAPLs that were not detected until 2001.

1998 Modeling Objectives

The main objectives of the 1998 modeling were as follows:

1. Update the existing groundwater model with the recent (post-1996) hydrogeochemical data
2. Evaluate the long-term effectiveness of four remedial technologies (groundwater extraction wells, groundwater circulation wells, phytoremediation, and *in situ* bioremediation) to provide dissolved-phase VOC mass removal
3. Evaluate the degree of hydraulic containment provided by the hybrid poplar (*Populus deltoides* × *Populus trichocarpa*) trees
4. Examine the costs and benefits of conducting dissolved-phase VOC mass removal in the source area *versus* natural degradation alone

Model Assumptions and Limitations

A significant factor in designing the model was the limited site-specific information on the transient processes that control groundwater recharge, vadose zone hydrology, *in situ* biodegradation, and transpiration. The groundwater model was designed to predict the effects of phytoremediation in conjunction with other remedial technologies. A challenging aspect of this design was the need to accurately represent these technologies in the model, while at the same time not exceeding either the model or data limitations. Based on the modeling objectives and considering the cost limitations, data gaps, and time constraints, a decision was made in early 1998 to refine the existing groundwater model. The refined model was designed to evaluate the effectiveness of the four proposed remedial technologies over a long-term operation schedule (30 years). To use the existing model given the constraints listed, the following assumptions were made:

1. The contributions of the vadose zone to the phytoremediation system (source of transpiration) are negligible
2. Steady-state conditions prevail over the duration of a 30-year analysis and short-term transient processes are negligible
3. The average annual transpiration and groundwater recharge rates are applicable to assess the effects of VOC uptake by the hybrid poplar (*Populus deltoides* × *Populus trichocarpa*) trees
4. *In situ* degradation of VOCs can be represented by first order degradation

Site Characterization Modeling

A three-dimensional geospatial model, based on extensive geological, geophysical, and chemical data collected at J-Field, was constructed using earth Vision® (Dynamic Graphics 1998). The model was used to characterize the aquifer framework such as the extent and thickness of strata that dictate plume movement. Contaminant distribution was modeled to demonstrate the lateral and vertical extent of the VOC plume and to quantify the VOC mass in the aquifer. Details of the methodologies used to develop the three-dimensional geospatial model are summarized in Schneider and Wrobel (1998). The geospatial data in earthVision was exported to Groundwater Vistas (Environmental Simulations, Inc. 2000) and used to update the existing groundwater model and to construct the contaminant transport model.

Groundwater Flow Modeling

Model Design

The existing MODFLOW model was updated with new hydrogeologic data. The model layers were developed based on the hydrostratigraphic framework in the geospatial model, and the model grid was telescoped further to refine the area of interest. The groundwater model was calibrated to mean groundwater elevations based on continuously measured water level data collected throughout 1998. Steady-state conditions were selected because the primary modeling objective was to predict the future long-term performance of the proposed integrated remedial technologies. Short-term seasonal variations were deemed insignificant over the course of 30 years. The groundwater model was calibrated using inverse regression techniques with the software UCODE (Poeter and Hill 1998). Details of the model design and model calibration are presented in Quinn *et al.* (1996) and WESTON (2000a).

The model was designed to incorporate the various remedial technologies being screened during the J-Field feasibility study (WESTON 2000b). One of the challenges of the model design was to find a suitable method for simulating the effects of the trees while accounting for other remedial technologies. For the modeling, it was assumed that all water withdrawn by the trees originated in the saturated zone. This assumption can be limiting in many cases, and the effect on the J-Field model is not well understood, although data indicate that the vadose zone is relatively thin [1 to 2 meters (3 to 7 feet)] and the soil moisture contents (average 0.10 in July) are relatively low.

The effects of hybrid poplar (*Populus deltoides* × *Populus trichocarpa*) trees were simulated with MODFLOW by using the module that simulates the effects of a well. A projected withdrawal rate for each tree was incorporated into the well module using thirty 1-year water stress periods. These stress periods are essentially an average rate of the groundwater withdrawal compiled for a given water-year cycle, based on the respective transpiration rates for the four seasons. Based on the crop index, these steady-state groundwater withdrawal rates were projected to increase as the plantation matures. The crop indices indicated that the plantation, upon reaching maturity, is expected to withdraw approximately 4200 liters per day (1100 gallons per day) over a water-year cycle (Gussman and Halford 2000).

In addition to hybrid poplar (*Populus deltoides* × *Populus trichocarpa*) trees, groundwater circulation wells were simulated in MODFLOW using the well module. The rates of extraction and re-injection applied to the model were 2 liters per minute (0.5 gallon per minute) as determined from aquifer tests, and these rates were held constant over the 30-year simulations.

Model Results

The groundwater model was used to examine the long-term, steady-state zone of hydraulic containment predicted once the hybrid poplar (*Populus deltoides* × *Populus trichocarpa*) trees reach maturity. Based on the maximum predicted transpiration rates [7600 liters per day (2000 gallons per day)], the groundwater model (MODFLOW-MODPATH) simulates a well-developed zone of capture generated by the hybrid poplar (*Populus deltoides* × *Populus trichocarpa*) trees that is centered at the southeastern extent of the plantation near well JF-P2. The transpiration effects are most evident in the aquifer zone where groundwater flow paths transition upward to discharge into the marsh. This zone is marked by a decrease in groundwater velocities. The predicted zones of influence decrease with depth as the trees provide the most influence in the upper 4.9 meters (16 feet) of the aquifer.

Contaminant Transport Modeling

Model Design

The primary modeling objective was to predict contaminant mass removal for remedial technologies screened during the feasibility study. During the modeling activities, mobile DNAPL had not yet been detected at J-Field, and predicting the comparative dissolved-phase mass removal rates for the integrated remedial technologies was the primary modeling objective. The fate and transport of 1,1,2,2-TeCA, TCE, and the respective transformation products were simulated using the contaminant transport model RT3D, which includes a multi-species, reactive transport module. This module uses first order, rate-limited kinetics to characterize the sequential degradation of VOCs.

The fate and transport modeling provided a method to compare the predicted dissolved-phase mass removal for the different remedial processes being screened at J-Field. Mass-removal estimates for the hybrid poplar (*Populus deltoides* × *Populus trichocarpa*) trees required the application of an uptake efficiency term, which has been described by Schnoor (1997). A detailed description of the model design that includes an explanation of the methods for differentiating mass removal between trees and wells within the framework of RT3D (Clement 1998) is presented in WESTON (2000a).

Model Results

The MODFLOW-RT3D model was used to evaluate the numerous configurations of the integrated remedial technologies. The model results indicated that poorly sustainable well yields limited efficient VOC dissolved-phase mass removal. With the groundwater circulation wells extracting only 2 liters per minute (0.5 gallon per minute), the simulated long-term VOC mass-removal rate was less than 23 kilograms per year (50 pounds per year). The field-scale treatability studies confirmed the inefficient VOC mass-removal capacity of the groundwater circulation wells due to the low permeability sediments at J-Field. The model results indicated that in comparison to natural remedial processes (*in situ* bioremediation and phytoremediation), the low-yielding groundwater circulation wells do not remove enough dissolved-phase VOC mass to significantly improve the groundwater quality at the point of discharge (freshwater marsh). Based on field evidence and modeling, the hybrid poplar (*Populus deltoides* × *Populus trichocarpa*) trees transpire enough groundwater during the spring and summer months to capture and take up contaminated groundwater. The model predicts that at the reported transpiration rates, the hybrid poplar (*Populus deltoides* × *Populus trichocarpa*) trees may remove up to 163 kilograms (360 pounds) of the dissolved-phase VOC mass after 30 years.

FINDINGS

Phytoremediation is an integral component of an alternative remedial action strategy developed for the J-Field site. Phytoremediation was accepted based on compelling monitoring data which demonstrated that natural processes (phytoremediation and *in situ* biodegradation) sufficiently reduce concentrations in a dissolved-phase VOC plume. The DNAPL source of the dissolved plume is being addressed under a separate action. J-Field possesses unique conditions that support a successful application of phytoremediation. These attributes include the remoteness of the site (restricted access and no groundwater users), low groundwater velocity, a VOC plume under steady-state conditions, and a freshwater marsh with anoxic sediments that are adsorbing and biodegrading VOCs, prior to the plume discharging to surface water.

After 5 years of phytoremediation performance monitoring at J-Field, the data demonstrate that the hybrid poplar (*Populus deltoides* × *Populus trichocarpa*) trees contribute to the remediation of a dissolved-phase VOC plume (as described in the following subsections). These data show that the hybrid poplar (*Populus deltoides* × *Populus trichocarpa*) trees seasonally transpire sufficient groundwater to influence groundwater flow directions and to provide hydraulic containment of the VOC plume. The data indicate that the trees transpire VOC-contaminated groundwater as evidenced by detectable concentrations of VOCs in tree tissue and transpiration gas samples. To date, the fate and transport pathway of transpired VOCs is not well characterized, but the current data indicate the VOCs are being biotransformed within the trees as well as directly transpired to the atmosphere. Because atmospheric monitoring indicates that the VOCs are photodegraded and diluted to barely detectable concentrations, the atmospheric pathway is deemed low risk to human health and the environment. The performance of the natural remedial processes will continue to be monitored to ensure that no unacceptable risks to human health and the environment occur because of the persistence of the dissolved-phase VOC plume.

Hydraulic Containment of the VOC Plume

That the trees transpire groundwater is evidenced by the following:

1. Sap-flow data indicated that in 1999, the 171 hybrid poplar (*Populus deltoides* × *Populus trichocarpa*) trees transpired 4130 liters per day (1090 gallons per day), and the trees are projected to transpire 7600 liters per day (2000 gallons per day) after 30 years
2. Groundwater data demonstrated that a discernible cone of depression forms beneath the hybrid poplar (*Populus deltoides* × *Populus trichocarpa*) trees during the summer as transpiration rates exceed the rates of release from aquifer storage
3. Groundwater data illustrated shifts in hydraulic gradients toward the trees both laterally and vertically [to depths of 7.6 meters (25 feet)]
4. Simulated groundwater flow conditions showed that the hybrid poplar (*Populus deltoides* × *Populus trichocarpa*) trees effectively capture the shallow component of the VOC plume centered near monitoring well JF-P2

Dissolved-Phase VOC Mass Removal

That the trees transpire contaminated groundwater is proven by the following:

1. Tree-tissue sampling indicated that the trees contain measurable concentrations of TCE and 1,1,2,2-TeCA. The TCE concentrations decreased as the sample height increased (from roots up to branches)
2. Leaf-tissue samples indicated that 1,1,2,2-TeCA and dichloroacetic acid were present
3. Transpiration-gas samples showed that the trees volatilize 1,1,2,2-TeCA and TCE
4. Transpiration samples demonstrated that the trees biotransform TCE as indicated by detectable concentrations of trichloroacetic acid
5. The tree tissue and transpiration gas data indicated that the reduction of TCE is attributable to multiple natural processes including adsorption to

roots and tree tissue, rhizosphere degradation, biotransformation by way of metabolism, and volatilization through the trunk and leaves

Hydrologic Aspects of J-Field

At J-Field, the performance of phytoremediation is influenced by the following hydrologic conditions:

1. Groundwater VOC concentrations will continue to remain relatively steady because DNAPLs continue to transfer contaminant mass to the dissolved-phase plume. Therefore, although the trees transpire and degrade VOCs, the attenuation of VOC groundwater concentrations will not occur until the DNAPL is either isolated or removed.
2. In spite of the DNAPL, the natural processes (phytoremediation and *in situ* bioremediation) continue to reduce VOC concentrations to near or below detection limits in surface water samples collected from the freshwater marsh. The highly reduced groundwater and low groundwater flow velocities provide an environment that supports biotic and abiotic degradation of VOCs, and the low permeability, organic-rich marsh sediments effectively retard contaminant migration.
3. The shallow, slow-moving groundwater system is ideal for using hybrid poplar (*Populus deltoides* × *Populus trichocarpa*) trees to hydraulically contain the VOC plume, and the trees enhance *in situ* biodegradation by increasing the groundwater residence times in the aquifer.

In summary, the data collected during the 5-year study indicate that phytoremediation satisfies the remedial objectives established for J-Field, and supported the decision to plant up to 600 additional trees in the fall of 2001 to improve the extent of hydraulic containment of the VOC plume.

Disclaimer

The U.S. Environmental Protection Agency has not reviewed this chapter and thus the work represents the views of the authors and not the Agency. Mention of trade names or commercial products does not constitute endorsement or recommendation for use.

REFERENCES

Burken, J.G. (2001) Uptake and volatilization of chlorinated solvents by poplars at field-scale. In: *Phytoremediation, Wetlands, and Sediments*. A. Leeson, E. Foote, K. Banks, and V.S. Magar, eds. The 6th International In Situ and On-Site Bioremediation Symposium. Battelle Press, Columbus, Ohio. Held in San Diego, California, June 4–7, **6**(5).

Clement, T.P. (1998) A modular computer code for simulating reactive multi-species transport in 3D groundwater aquifers. Prepared for the U.S. Department of Energy. Pacific Northwest Laboratories, Richland, Washington.

Delaware Cooperative Extension (1995) *Recommended Soil Testing Procedures for the Northeastern United States*. 2nd ed. Northeastern Regional Publication No. 493, Agricultural Experiment Stations of Connecticut, Delaware, Maine, Maryland, Massachusetts, New Hampshire, New Jersey, New York, Pennsylvania, Rhode Island, Vermont, and West Virginia. Revised December 15, 1995.

Dynamic Graphics (1998) earthVision geospatial modeling software. Alameda, California (www.dgi.com).

Environmental Simulations, Inc. (2000) Groundwater Vistas modeling software, V 2.52. Herndon, Virginia (www.groundwatermodels.com).

Gussman, C.D. and D.K. Halford (2000) J-Field phytoremediation project, field events and activities through July 31, 2000. Aberdeen Proving Ground, Edgewood, Maryland. Prepared by Lockheed Martin for U.S. Environmental Protection Agency, Emergency Response Team Center. EPA Contract No. 68-C99-223, EPA Work Assignment No. 0–034. Edison, New Jersey.

Haroski, D.M., S.R. Hirsh, J.G. Wrobel, and H.R. Compton (2000) Estimating hybrid poplar groundwater removal rates. Abstract. Second International Conference on Remediation of Chlorinated and Recalcitrant Compounds, Battelle Memorial Institute, Columbus, Ohio. Held in Moneterey, California, May 22–25.

Hughes, W.B. (1995) Groundwater flow and the possible effects of remedial actions at J-Field, Aberdeen Proving Ground, Maryland. Water-Resources Investigations Report 95–4075. U.S. Geological Survey, Baltimore, Maryland, p. 39.

Interstate Technology and Regulatory Council (IRTC) (2001) Phytotechnology technical and regulatory guidance document. Phytotechnologies Work Team. Washington, D.C. (http://www.itrcweb.org).

Lorah, M.M., L.D. Olsen, B.L. Smith, M.A. Johnson, and W.B. Fleck (1998) Natural attenuation of chlorinated volatile organic compounds in a freshwater tidal wetland, Aberdeen Proving Ground, Maryland. Water-Resources Investigations Report 97–4171. U.S. Geological Survey, Baltimore, Maryland.

McDonald, M.G. and A.W. Harbaugh (1988) A modular three-dimensional finite-difference ground-water flow model. Techniques of Water Resources Investigations, Book 6, Chapter A1. U.S. Geological Survey, Reston, Virginia.

Newman, L.A., S.E. Strand, N. Choe, J. Duffy, G. Ekuan, M. Ruszaj, B. Shurtleff, J. Wilmoth, P.E. Heilman, and M.P. Gordon (1997) Uptake and biotransformation of trichloroethylene by hybrid poplars. *Environ. Sci. Technol.* **31**(4): 1062–1067.

Pardue, J., W.A. Jackson, and L. Martino (2000) Dialysis sampler evaluation of rhizosphere phytoremediation processes in the poplar grove of J-Field, APG. Prepared by CFR Technical Services, Inc., Baton Rouge, Louisiana, and Argonne National Laboratory, Argonne, Illinois. Prepared for Directorate of Safety, Health and Environment, Aberdeen Proving Ground, Maryland.

Poeter, E.P. and M.C. Hill (1998) Documentation of UCODE, a computer code for universal inverse modeling. Water-Resources Investigations Report 98–4080. U.S. Geological Survey, Reston, Virginia.

Quinn, J.J., R.L. Johnson, T.L. Patton, and L.E. Martino (1996) An optimized groundwater extraction system for the Toxic Burning Pits area of J-Field, Aberdeen Proving Ground, Maryland. Prepared by Argonne National Laboratory, Argonne, Illinois. Prepared for Directorate of Safety, Health and Environment, Aberdeen Proving Ground, Maryland.

Roy F. Weston, Inc. (WESTON) (2000a) A three-dimensional groundwater flow and contaminant transport model to examine remedial alternatives at J-Field, Aberdeen Proving Ground, Maryland. Prepared by Roy F. Weston, Inc., West Chester, Pennsylvania for Directorate of Safety, Health and Environment, Aberdeen Proving Ground, Maryland.

Roy F. Weston, Inc. (WESTON) (2000b) Feasibility study report, J-Field surficial aquifer operable unit. Prepared by Roy F. Weston, Inc., West Chester, Pennsylvania for Directorate of Safety, Health and Environment, Aberdeen Proving Ground, Maryland.

Roy F. Weston, Inc. (WESTON) (2001) Technical impracticability evaluation for J-Field surficial aquifer. Prepared by Roy F. Weston, Inc., West Chester, Pennsylvania. for Directorate of Safety, Health and Environment, Aberdeen Proving Ground, Maryland.

Schneider, W. and J. Wrobel (1998) Interfacing 3D visualization technology with MODFLOW-RT3D and MODPATH to evaluate remedial alternatives. In: *Proceedings from MODFLOW98*. E. Poeter, C. Zheng, and M. Hill, eds. Colorado School of Mines, Golden, Colorado, October 4–8.

Schnoor, J.L. (1997) Phytoremediation. Technology Evaluation Report TE-98-01. Prepared for Groundwater Remediation Technology Remediation and Analysis Center, Pittsburgh, Pennsylvania.

U.S. Environmental Protection Agency (U.S. EPA) (1998) Technical protocol for evaluating natural attenuation of chlorinated solvents in groundwater. EPA/600/R-98/128. Office of Research and Development, Washington, D.C. (http://www.epa.gov/ada/download/reports/protocol.pdf)

Vroblesky, D.A., C.T. Neitsch, and J.T. Morris (1999) Chlorinated ethenes from groundwater in tree trunks. *Environ. Sci. Technol.* **33**(1): 510–515.

Yuen, C.R., J.J. Quinn, L. Martino, R.P. Biang, and T. Patton (1998a) Draft natural attenuation study of groundwater at Toxic Burning Pits area of concern at J-Field, Aberdeen Proving Ground, Maryland. Prepared by Argonne National Laboratory, Argonne, Illinois for Directorate of Safety, Health and Environment, Aberdeen Proving Ground, Maryland.

Yuen, C.R., L. Martino, Y.S. Chang, D. Dolak, R. VanLonkhuyzen, T. Patton, S. Prasad, J. Quinn, D. Rosenblatt, J. Vercellone, and Y.Y. Wang (1998b) Remedial investigation report for J-Field, Aberdeen Proving Ground, Maryland: vol.2—human health risk assessment. Final report. Prepared by Argonne National Laboratory, Argonne, Illinois for Directorate of Safety, Health and Environment, Aberdeen Proving Ground, Maryland.

Yuen, C.R., J. Quinn, L. Martino, and W. Davies (2000) Draft natural attenuation study of groundwater at Toxic Burning Pits area of concern at J-Field, Aberdeen Proving Ground, Maryland. Prepared by Argonne National Laboratory, Argonne, Illinois for Directorate of Safety, Health and Environment, Aberdeen Proving Ground, Maryland.

SECTION VI

MODELING, DESIGN, AND FIELD PILOT TESTING

21

MODELING AND DESIGN OF PHYTOREMEDIATION

L. C. Davis, L. E. Erickson, M. Narayanan, and Q. Zhang

SUMMARY OF PRACTICAL IMPLICATIONS

Modeling and design of effective treatment systems involving green plants is based on thermodynamics (phase equilibria) and transport (diffusion and advection) in the soil (including the vadose zone) and groundwater. Actively growing plants alter soil moisture and solute concentrations as water and solutes move from soil to plants. Organic contaminants may be biodegraded or transformed in bulk soil, in the rhizosphere, within the plant, and in the atmosphere (if volatile). Water management is a major issue in field settings to prevent contaminant plumes from moving off-site. A correctly designed plant system, alone or with other technologies, should contain the contaminant on-site, and result in transformation to less hazardous forms.

INTRODUCTION

Vascular plants have had eons over which to evolve metabolic responses to environmental chemicals, so it is not surprising that some plants show the ability to deal with recently synthesized xenobiotic chemicals. Much has been written about the capabilities of particular plants to transform, transfer, or sequester various chemicals (*cf.* U.S. EPA 1999, 2000). Yet a myriad of possibilities remains to be investigated. New metabolic pathways are still being discovered in plants, and genome sequencing indicates that a significant fraction of all the proteins encoded by plant genomes are of unknown or uncertain function (Plant Physiology 2000). This review focuses on a few examples of volatile organic compounds in the vadose zone. We have combined experimental and modeling approaches to formulate fundamental insight into phytoremediation processes to improve the design of phytoremediation (Davis *et al.* 1993, Shimp *et al.* 1993, Erickson *et al.* 1994). Better solutions to

Phytoremediation: Transformation and Control of Contaminants,
Edited by Steven C. McCutcheon and Jerald L. Schnoor.
ISBN 0-471-39435-1 (cloth)

many contamination problems await discovery, as phytoremediation (the engineered use of plants to enhance remediation) is a recent development.

Many organic contaminants have significant volatility at ambient temperatures (vapor pressure greater than 1 millimeter of mercury or 133 pascals). Hence these compounds are present in a gas phase as well as the liquid and sorbed phases of the vadose zone. This increases the complexity of detection and the range of mobilization pathways. Most of the design considerations and modeling described in this chapter are pertinent to most organic contaminants, but the extent of volatility and solubility introduces complexity that we address at appropriate points. Lindgren (2000) provides a useful table of contaminant characteristics with examples of materials, likely phase locations, and means of sampling for the vadose zone. All such classifications are arbitrary, and some important contaminants may fall on the borders of any particular cut off for soluble *versus* insoluble or volatile *versus* nonvolatile.

For the plant to play a significant direct role in biotransformation, the compounds of interest generally need to have at least modest water solubility (Lindgren 2000); otherwise the contaminants take a very long time to move enough of the material to the plant for transformation. Volatile but relatively insoluble compounds such as hexane have not been the subject of many investigations with plants. Harms *et al.* (this book) discuss two classes of nonvolatile, relatively insoluble compounds, polychlorinated biphenyls and polycyclic aromatic hydrocarbons, which are at the practical limits for plant biotransformation. Other chapters (Hutchinson *et al.* this book, Olson *et al.* this book) consider the role of rhizosphere organisms and processes that may be effective with the less volatile and less soluble compounds. We focus on volatile, soluble compounds, exploring how trichloroethylene (TCE), dichloromethane, and methyl *tertiary*-butyl ether (MTBE) enter and leave plants readily. Another chapter (Winnike-McMillan *et al.* this book) deals more broadly with MTBE and the potential for phytoremediation.

DESIGN CONSIDERATIONS FOR PHYTOREMEDIATION SYSTEMS

Engineering design involves the development of a detailed plan to accomplish one or more specific objectives, the agreed end points. Phytoremediation is but one means to achieve the desired cleanup levels for the vadose zone or groundwater [Schnoor 1997, Center for Waste Reduction Technologies 1999, Interstate Technology and Regulatory Council (ITRC) 1999, Kulakow *et al.* 1999, U.S. EPA 2000]. Other uses of plants are discussed elsewhere in this volume. Effective management for successful phytoremediation requires, as do all remediation strategies, implementation of sound principles (Quadrel and Lundgren 2000) to identify end points, and then developing and consistently following a plan.

The following list presents basic steps for a general phytoremediation design protocol:

1. Identify objectives and regulatory considerations
2. Characterize and evaluate contamination, risk, and environmental variables
3. Select and evaluate alternative phytoremediation processes and plant systems
4. Prepare the site and establish vegetation
5. Manage and monitor the vegetation, water, and contamination
6. Demobilize the site and dispose of residues as necessary

Steps 3 and 4 may require several iterations to obtain optimum outcomes. Because phytoremediation is an evolving field one, must consult current resources for particular applications. The ITRC (1999) decision tree is available (http://www.itrcweb.org) to assist those wanting to design and implement phytoremediation. Schnoor (1997), Kulakow *et al.* (1999), and authors of this book have also addressed some aspects of the design process. Comments on each of the six steps follow:

1. The environmental management objectives (end points) should be based on future land use, ecological risk issues, esthetics, acceptance by site managers and regulators, and required cleanup levels. The end points will be site specific and must be established through engagement of all stakeholders early in the process.
2. Site characterization is necessary to evaluate the extent of contamination, toxicity, and risk. Potential for movement of contaminants to receptor organisms, and associated risk, must be considered. Biodegradability and bioavailability of the contaminants, leaching, and plume control for water contamination must also be evaluated. In many settings, plants serve primarily to enhance or augment natural attenuation. A recent analysis of the potential for natural attenuation (National Research Council 2000) tabulated information available through 1998 (*cf*, pp. 87 to 90 and references cited therein). Healthy vegetation requires appropriate soil quality, climatic conditions, and site management. *The importance of adequate soil fertility is often underestimated, and almost all sites will benefit from added fertilizers if water is not limited.* Soil amendments, based on soil tests, are often required. The water holding capacity of the soil and specific nutritional needs of the chosen plant species should be considered. Water holding capacity and nutrient limitation are often a problem when working with highly disturbed sites where naturally developed topsoil has been removed, buried, or compacted. The presence or absence of spontaneous vegetation can often provide useful clues to the extent to which amendments and fertility enhancement may be necessary. Soil scientists should be consulted for each particular site early in the design process.
3. The central aspect of design is to select and then evaluate proposed solutions that meet both environmental objectives and regulatory consider-

ations. This chapter describes research results that have been used to formulate models to support this step of the design process. Modeling can often be used to indicate parameter sensitivity, showing how particular site conditions may affect performance. This is an inexpensive means to avoid some costly trial-and-error field studies. Each plant system must be designed for a particular site because site conditions affect the phytoremediation process. The method of planting should be included in the design, based on factors such as the desired rooting depth, ease of plant establishment, and heat, cold, drought, or flood tolerance of the chosen vegetation.

4. Once a remediation plan is developed, the soil and site preparations needed for planting and water management should proceed based on the appropriate time and means of planting. Specialists in horticulture, agronomy, or forestry must be consulted for any particular site. Failures of phytoremediation are usually the result of ignoring this part of the process.
5. The planted site must be managed and monitored. These activities include water management, periodic sampling to verify performance, pest management, and replanting if necessary. A recent report of the National Research Council (2000) discusses the critical need for thorough monitoring and evaluation for natural attenuation. In many instances, phytoremediation is a way to augment natural attenuation, so the enhancement requires equivalent monitoring. Most modeling results can be validated only by monitoring.
6. So far as we are aware, phytoremediation for organic materials has not required removal of residues. The situation is different for heavy metals and radionuclides, as discussed elsewhere (Schnoor 1997, U.S. EPA 2000).

Important media to which phytoremediation applications are made include contaminated groundwater, contaminated soil, and combinations of both. Phytoremediation design issues include some that occur in any remediation process; others are specific to plants. Issues that must be addressed include the following:

1. Contaminant treatability and rate of transformation
2. Plant uptake and metabolism of contaminant
3. Contaminant release to the atmosphere
4. Accumulation of contaminant or metabolites in plant biomass
5. Selection of climatically adapted cultivars with contaminant tolerance
6. Soil preparation and need for soil amendments
7. Competition of plants and microorganisms for nutrients
8. Oxygen profile in the soil
9. Irrigation and water management
10. Salinity of soil
11. Planting density
12. Areal extent of planting

13. Plant management (mowing, harvesting, and use of herbicides)
14. Pest management (beavers, deer, insects, diseases, and others)
15. Monitoring plan and access control
16. Time required for remediation
17. Public acceptance
18. Regulatory acceptance
19. Cost

Issues of public and regulatory acceptance are discussed in Marmiroli and McCutcheon (this book). Management issues are site specific. In the remaining portion of this chapter, we address the development of mathematical models to describe fate and transport in soil and plants, and show how these have implication for field implementation.

DEVELOPMENT OF MODELS

Physical Factors

Design of phytoremediation systems can often benefit from the use of modeling and simulation to predict performance. An upland plant comes into contact with three distinct phases—soil, water, and air. Within the soil, most plant processes depend on transfer of materials *via* water at the surface of the root. Volatile materials may pass directly from air to root and from root to air (Bromilow and Chamberlain 1995). Within the atmosphere, most materials leave or enter a plant through the gas phase, although some direct deposition onto the plant occurs in rainwater and aerosols, and plants exude some materials from tissues in nonvolatile forms. A very useful introduction to aspects of plant anatomy and physiology important to contaminant transfer may be found in Mc Farlane (1995). Jeffers and Liddy (this book) and Morikawa *et al.* (this book) discuss interactions of plants with air pollution; this chapter focuses on uptake from soil and water.

To develop a conceptual model of a plant and the interactions with the different phases, several investigators have taken a phase partitioning approach and have developed models for transfer of contaminants between phases *via* the plant. A useful review of earlier literature (Paterson *et al.* 1990) covers many studies for different plant species and different contaminants.

An early dynamic model for contaminant partitioning, including transfer from atmosphere to plant, was described by Schramm *et al.* (1987), while a second model was developed by Paterson and Mackay (1989, 1995). The PlantX model described by Trapp (1995) appears to be a general and powerful way to examine different phases interacting with plants. Matthies and Behrendt (1995) integrated the PlantX model into a larger modeling effort to consider low volatility pesticide fate in agricultural systems. In all of these modeling efforts, the plant is divided into a few simple compartments, such as roots, stems, and leaves. A limited number of processes are assumed to occur within each compartment.

The sections that follow describe systematic investigations of interactions between different plant compartments and the environment, using a combination of model simulations and testing with experimental results. Successful modeling requires gathering reliable parameter values for the model. Success is also critically dependent on the correct scientific basis. Our investigations aim to provide accurate parameters for transport processes of volatile contaminants in soil systems with growing plants. In these models, the scientific basis is that the contaminant is dissolved in the water phase in equilibrium with sorbed and gas phases in the soil, and that a nonaqueous phase liquid (NAPL) is not considered.

There are very few published experimental studies of readily soluble, volatile NAPL phases to address how plants might alter remediation of the contaminants. Zhang *et al.* (1997) investigated the fate of TCE as a NAPL in the saturated zone in a laboratory chamber with alfalfa (*Medicago sativa*). In these tests 20 milliliters (29.2 grams) of liquid TCE was injected directly into the soil at the inlet port of the system described in Table 21-1. Plants, established through the entire 1.1-meter length of the channel in this instance, enhanced vertical water flux and hence loss of TCE. Karthikeyan *et al.* (1999), and Karthikeyan *et al.* (2000) observed increased degradation of poorly soluble (approximately 2 milligrams per liter), slightly volatile (only 10 percent with a boiling point less than 190 °C) JP-8 jet fuel in planted systems, where the contaminant was introduced as NAPL.

TABLE 21-1 Experimental Conditions and Treatments for the Six-Channel System[a]

Location[b]	Contaminant	Concentration[c] (milligrams per liter)
Channels 1 and 2	Methyl *tertiary*-butyl ether	73
Channels 3 and 4	Dichloromethane	59
Channels 5 and 6	Trichloroethylene	45

[a]For all channels, dimensions are 110 centimeters long, 60 centimeters deep, and 10 centimeters wide; soil was alluvial silty sand (with less than 10 percent silt) collected near a closed landfill in Riley County, Kansas. The system was continuously illuminated with 12 cool white fluorescent lights (40 watts) at a height of 50 centimeters above the soil, giving a light intensity approximately 160 micromoles of protons or microeinsteins per square meter per second.

[b]For the experimental work of Davis *et al.* (2001), each channel had alfalfa (*Medicago sativa*) plants in the latter half of the channel. One liter per day of contaminant solution entered each channel 5 centimeters from the bottom of the channel at the unplanted end and the water table was maintained 10 centimeters from the bottom. The operation was continuous with steady state feed conditions.

For the experiment of Zhang *et al.* (1997), alfalfa (*Medicago sativa*) plants were present through the length of channels 1, 2, 3, 5, and 6. Channel 4 was unplanted. The water table was 35 centimeters from the bottom. A pulse of 20 milliliters of liquid trichloroethylene was introduced to channels 1, 2, 4, 5, and 6. Air sparging was applied to channels 1 and 6. Water supply was at a rate of 1 liter per day.

[c]The concentrations shown are daily averages for the input water.

Root–Soil Interaction

Root architecture and behavior are key variables for contaminant uptake from soil and water. Soil properties, particularly porosity, are also important. Much work has gone into understanding the ways in which plants are able to extract water from different kinds of soils (Shimp *et al.* 1993, Davis *et al.* 1998a). Patterns of root distribution have been examined for important crop species grown under both irrigated and nonirrigated conditions in different kinds of soils, beginning with the pioneering studies of Weaver (1926). Information is available for trees (Sprackling and Read 1979) and for native vegetation (Kutchera 1960). Now, sophisticated models enable prediction of plant water use, root growth, and productivity (Lynch *et al.* 1997) which allows simulation of efficient plant extraction systems. A field-scale model of this nature was described by Chang and Corapcioglu (1997) and applied to a hypothetical sandy loam soil contaminated with nonvolatile hydrocarbons (Corapcioglu *et al.* 1998). In the model, access to the contaminant depended mostly on root growth because the contaminant was taken to be insoluble and adsorbed to the soil. In such a case, plant characteristics, including rooting density and rhizosphere microbial populations, dominate the remediation process.

The models described in the following section, "Water and Contaminant Flux," based on the work of Tracy and Marino (1989), use experimentally derived characteristics of the plant root system of agricultural crops such as alfalfa (*Medicago sativa*) to model the pattern of root-water extraction. Understanding how plant roots support rhizosphere microbial populations and how root distribution and microbial populations together affect contaminant degradation is an area needing much more research. Shimp *et al.* (1993) reviewed the earlier literature, while Davis *et al.* (2002) reviewed work of the past decade.

Actively growing plants significantly alter soil moisture and solute distribution, transferring large quantities of water from soil to atmosphere, acting as solar-driven pumps, and simultaneously drawing solutes upward. The simplified scientific basis of most modeling to date (materials enter the plant only in the aqueous phase *via* roots and depart *via* leaves) may not be applicable for volatile contaminants. Increased capillary water flow brings soluble contaminants into the unsaturated zone, where volatile compounds may diffuse to the soil surface. Large losses of material to the air may occur during transport within the plant, independent of metabolism. Volatile contaminants may diffuse out of roots into the unsaturated soil, or from stems, twigs, and leaves to the atmosphere. Depending on the particular contaminant, soil porosity, and soil water content, such losses may be more important than losses of contaminant directly from deeper soil water to the atmosphere *via* the soil–air phase.

Water and Contaminant Flux

In field situations, water management is often a major issue because a contaminant is typically found dissolved in a plume of groundwater, migrating away from the site of initial contamination. Several chapters in this book discuss plume control efforts. Rates of surface runoff and infiltration determine how a dissolved contaminant moves from near the surface to deeper soils, potentially reaching an aquifer. Soil-water models should include both saturated and unsaturated flow, with the evapotranspiration contribution of plants altering the vertical transport component (Bodvarsson *et al.* 2000).

Tracy *et al.* (1993) described field scale modeling and simulation of plant influences on vertical fluxes for a dissolved contaminant. The simulation was based on a hypothetical substance with the solubility, microbial biodegradability, and sorption properties of benzene, but treated as nonvolatile. Initially, the contaminant was uniformly present in the upper 3 meters of a sandy soil, which had varied organic content and texture to reflect a real field situation. The soil was vegetated, with alfalfa (*Medicago sativa*) roots exponentially decreasing into the soil (Stanberry 1955). Different maximum rooting depths, respiration, and exudation rates were simulated. Plausible values for microbial biomass and an intrinsic degradation rate for the contaminant were assigned. For average precipitation and potential evapotranspiration conditions of northeast Kansas (both less than 1 meter per year), a significant portion of the original contaminant mass escaped to below 3 meters over a 6-month simulated growing season. Greater rooting depth, increased root exudation, and greater oxygen supply decreased, but did not prevent downward movement of the contaminant.

For a more realistic simulation of how varying water regimes might influence vertical contaminant migration, actual data for precipitation and potential evapotranspiration in 11 different years were used during the simulated growing period of 140 days (Tracy *et al.* 1994). The water table was set at 7 meters. The contaminant was introduced at a uniform concentration of 1 milligram per cubic meter in the top 3 meters of soil. Only 1 stimulated year of 11 showed an excess of precipitation over potential evapotranspiration per year. As anticipated, some contaminant migrated to below 4 meters in proportion to the precipitation to evapotranspiration ratio. During the 140-day simulation period, contaminant was largely depleted in the upper 1 to 2 meters of soil for all years. About half the contaminant remained at the end of the simulation period in any year, and much of that remaining fraction was below the main rooting depth of the plants. The distribution of contaminant over depth depended on the total precipitation deficit during the 140-day simulation. The contaminant remained closer to the surface, and hence more accessible to the roots and associated microorganisms when the deficit was greater. In reality, the fraction degraded depends on the soil microbial population, degradation rate, and initial contaminant concentration. In this instance, starting values were chosen so that the reference case gave only partial degradation within a single growing season.

One important conclusion from these simulations was that infiltration control is an important factor for effective remediation. Plants provide greatest benefit when the excess of evapotranspiration over infiltration is maximized.

The greater the evapotranspiration, the less likely the contaminant is to escape below the zone of effective phytoremediation. This case study clearly demonstrates the importance of limiting infiltration after establishment of vegetation.

Extension of the model developed by Tracy *et al.* (1993), Tracy *et al.* (1994) to two dimensions was described by Davis *et al.* (1993). In a simulation of atrazine, the contaminant was introduced in a groundwater plume upstream of a poplar (*Populus* spp.) tree plantation. Over the course of an 8-year simulation, there was a gradual increase of concentration of the contaminant within the planted area, while the fraction of atrazine passing beyond the planted area was markedly reduced. Such a simulation without microbial degradation of the contaminant was reasonable, because the transpiration stream concentration factor (TSCF) of atrazine is less than unity and atrazine is not volatile (vapor pressure (less than 10^{-9} atmospheres or less than 10^{-4} pascals) (Burken and Schnoor 1996). The TSCF is simply a ratio of the concentration in the water being transpired from a plant relative to that in the groundwater (Briggs *et al.* 1982). This factor is discussed latter in the section "Potential for Conaminant Uptake." Any material that is not degradable and which has a TSCF less than one will accumulate either in soil or in roots. Simple salts are the most common example of accumulating materials. If the material is degradable, the extent of potential accumulation will vary with the ratio of degradation to uptake rates.

A contaminant with the properties of benzene (but not volatile) was simulated (Tracy *et al.* 1992) for a groundwater plume passing through a planted buffer strip. Trees significantly enhanced the contaminant degradation both by supporting a larger microbial population and by drawing water upward. As anticipated, contaminant concentrations were higher at the down gradient edge in winter because of decreased evapotranspiration than in summer, when evapotranspiration exceeded infiltration. The fraction of the contaminant degraded during passage through the buffer strip depended on the rate constant chosen for the degradation reaction. This points out the need for good values of degradation rate constants in trying to predict behavior of contaminants.

Biodegradation in the Soil and Rhizosphere

Narayanan *et al.* (1995), Narayanan *et al.* (1998a, 1998b) modeled the fate of toluene in the subsurface below alfalfa (*Medicago sativa*). The model includes horizontal and vertical flow of water, evapotranspiration, fate of the contaminant in the soil and water phases, oxygen concentration in the water and gas phases, microbial numbers in the soil and water phases, and root exudate concentration (as a substrate for microbial growth). Gas phase diffusion of oxygen was important because biodegradation of toluene in the rhizosphere was oxygen-limited. Toluene in the saturated zone was not biodegraded. The simulations were consistent with experimental observations that showed no volatilization of toluene above a planted chamber, and little decrease of concentration in the groundwater flowing below the plants (Erickson *et al.* 1994, Narayanan *et al.* 1995). Similar experimental results were obtained with MTBE (Davis, unpublished observations), whereas phenol was completely degraded within the groundwater of our experimental system (Erickson *et al.* 1994).

The models of Narayanan *et al.* (1995), Narayanan *et al.* (1998a, 1998b) were used to investigate biodegradation of contaminants and loss to the atmosphere by gas phase diffusion (Narayanan *et al.* 1999a). Toluene appears to be biodegraded under aerobic conditions in the unsaturated zone, whereas TCE seems to disappear as a result of the upward transport to the atmosphere. Recent studies (Davis, unpublished observations) show that very little TCE leaves the alfalfa (*Medicago sativa*) plant above ground; instead TCE diffuses to the soil, from which the contaminant volatilizes. With a larger plant, some TCE moves upward in the plant and escapes from the stems and leaves rather than from the soil (Davis *et al.* 1999, Vroblesky *et al.* 1999).

Perhaps the greatest uncertainty in simulating degradation is the biological variability with environmental conditions. Complete degradation of TCE within the soil column even without plants has been documented (Davis 2000) but this result is atypical. Indigenous microorganisms often possess the capability to degrade contaminants (Fulthorpe *et al.* 1996), and may need an adaptation period for populations to increase enough to give readily measurable rates (Davis 2000). Root exudates enhance microbial populations in the rhizosphere and hence increase biodegradation rates (Erickson *et al.* 1994). Santharam *et al.* (1994) have modeled biodegradation in the root zone, documenting the beneficial effects of root exudates in soils contaminated with polycyclic aromatic hydrocarbons that adsorb strongly to soil. Burken and Schnoor (1996) experimentally showed the beneficial effect of added root exudates in atrazine degradation by indigenous organisms. Ames and Hoyle (1999) showed that added nutrients likely to be present in exudates (citrate and succinate) stimulated atrazine degradation in microcosms of soil obtained from a contaminated field site. Ames and Hoyle (1999) also found great heterogeneity within a single contaminated site, making difficult any *a priori* prediction of biodegradability for a specific contaminant at a specific site.

Estimation of Physical Parameters for use in Modeling

Phase Partitioning, Volatility, and Diffusion

The volatility and sorption of contaminants are of critical importance in the design of phytoremediation systems. Models are often simplified by assuming phase equilibrium at various steps. The dimensionless Henry's law constant provides the ratio, at equilibrium, for the concentration in a gas phase over an aqueous phase. However, mass balance accounting for all possible routes of contaminant removal from the soil or groundwater must consider dynamic processes. Diffusion coefficients in water and in air vary by several orders of magnitude. An "effective diffusion coefficient" for soil is a weighted average of these (Narayanan *et al.* 1999a). Bromilow and Chamberlain (1995) suggest that compounds with a dimensionless Henry's law constant greater than 10^{-4} move through soil primarily by diffusion in the air phase, so that soil water movement alone is not a reliable indicator of the fate. Plants also readily volatilize such compounds. Failure to detect a contaminant in leaves of a plant (Newman *et al.* 1999) does not necessarily mean that the compound was either rejected by the plant root or metabolized by the plant; the contaminant may simply have diffused out through the root, stem, twigs, and petioles (Zhang *et al.* 2001).

Modeling by Jury *et al.* (1990, 1992, both as corrected by Robinson 2000) has shown that volatility of contaminants strongly affects fate in soil. Highly volatile compounds such as benzene or dichloromethane are rapidly lost by gas phase diffusion from soils of moderate porosity unless confined within or below the saturated zone. Hydrocarbons like those found in gasoline move rapidly through soils with low to moderate moisture content, initially by sorption and wicking, then by diffusion and volatilization as the hydrocarbon phase evaporates (Gidda *et al.* 1999). In wet (nearly saturated) soils, gasoline moves much more slowly.

Washington (1996) provides a useful theoretical treatment of the partitioning behavior of volatile organic compounds for different climatic conditions. Measured values of the Henry's law constant are tabulated with the enthalpy of vaporization (required to estimate values of Henry's law constant over a range of temperatures according to the van't Hoff relationship). Washington (1996) found that both temperature and soil moisture controlled the fate of volatile organic compounds in unsaturated soil. He provided graphs that show the partitioning of a range of contaminants between gas, water, and soil phases. At higher temperatures and lower water content, a larger fraction of the volatile contaminant is in the gas phase, where the compound diffuses much more rapidly than in the aqueous phase. For many volatile compounds, diffusion through an aqueous phase may limit remediation using soil vapor extraction or air sparging. Evapotranspiration enhances gas phase diffusion and volatile contaminant loss to the atmosphere.

McCarthy and Johnson (1995) observed in a laboratory experiment that the relative water saturation of a soil had a major influence on vertical migration of TCE through the unsaturated zone. Because gas phase diffusivities are about four orders of magnitude larger than diffusivities in liquids, the effective diffusion coefficient depends strongly on the relative gas- to water-filled porosity of the soil. The TCE concentration decreased by three orders of magnitude over just 15 centimeters above the water table. Thus, volatilization to the atmosphere was limited by aqueous phase diffusion in the capillary fringe.

Narayanan *et al.* (1999a) provided both laboratory observation and modeling confirmation of these results for a two-dimensional planted system in which the contaminant entered horizontally from below and was constantly renewed with flowing groundwater. The most striking finding was that the concentration of TCE dropped sharply at the capillary fringe. Thus, the roots of a plant growing in the moist unsaturated vadose zone would be exposed to TCE concentrations far lower than that introduced with the groundwater. The experimentally measured concentration dropped from 300 milligrams per liter to less than 75 milligrams per liter over a vertical distance of less than 10 centimeters above the saturated zone. This was readily explicable by modeling the gas and aqueous phase movement of the contaminant. Most of the roots of an alfalfa (*Medicago sativa*) plant do not penetrate closer to the saturated soil than a few centimeters, experiencing a TCE concentration of 5 to 25 milligrams per liter despite the 300 milligrams per liter concentration entering the system with the groundwater. These data explain why plants can tolerate growing in a chamber with TCE concentrations greater than 300 milligrams per liter in the groundwater (Zhang *et al.* 1997) when toxicity of TCE is measurable at a five-to-ten-fold lower concentration (Ryu *et al.* 1996). Not only is TCE lost by transfer to the gas phase before contact with the root system, but TCE that enters the root system at depth from either the water or gas phase may diffuse out from the root as the contaminant is carried upward by the transpiration stream.

Diffusion and Advection Within the Plant

Plant roots and stems are generally impermeable to water and solutes carried in the transpiration stream. However, that impermeability is selective for water and polar substances and is caused by the hydrophobic nature of cell membranes and plant cell walls. The octanol–water partition coefficient K_{ow} indicates the equilibrium ratio of concentration in octanol, a surrogate for organic material, over an aqueous phase. Moderately nonpolar substances, with a logarithm of the octanol–water partition coefficient of about 1 to 4, pass through plant root tissue relatively easily (Briggs *et al.* 1982, Bromilow and Chamberlain 1995). This is assumed to be primarily because these nonpolar compounds dissolve in cell membranes relatively well.

Studies by Rusch and Rennenberg (1998) showed that black alder (*Alnus glutinosa*) trees might be major pathways for release to the atmosphere of methane and nitrous oxide from wet soils. Work by Hu *et al.* (1997) showed that the young roots of plants are very permeable to contaminants such as TCE. Further studies (Hu 1998, Hu *et al.* 1998a, 1998b, Davis *et al.* 1999) expanded on those observations to include woody tissues such as the stems of corkscrew willow (*Salix* spp.) and hybrid poplar (*Populus* spp.) trees. Zhang *et al.* (1999, 2001) have examined *in situ* transfer from water to air of contaminants such as TCE and MTBE with alfalfa (*Medicago sativa*).

While the diffusion coefficients of small molecules like TCE approach 10^{-5} square centimeters per second in pure water, the effective values within plant tissues are significantly smaller and vary with properties such as K_{ow} and the number of hydrogen bonds per molecule of solute (Stein 1967). However, these small values are not negligible. Our research shows that TCE moves more readily across plant stems than does MTBE, presumably because TCE is more lipophilic (Table 21-2) and is less retarded by the cell membranes (Briggs *et al.* 1982, Bromilow and Chamberlain 1995). With hybrid poplar (*Populus* spp.) and corkscrew willow (*Salix* spp.) stems, the effective diffusion coefficient for TCE was estimated to be 3×10^{-6} square centimeters per second (Davis *et al.* 1999).

Zhang *et al.* (2001) investigated alfalfa (*Medicago sativa*) stems which are of much smaller dimensions than those of trees, but which show diffusivities for MTBE near 10^{-7} square centimeters per second. For typical water use rates, half of the MTBE is lost from an alfalfa (*Medicago sativa*) stem within 15 centimeters from the ground surface, without ever reaching the leaves. For TCE, the loss is more marked, with little TCE detectable 10 centimeters above the ground surface (Davis, unpublished observations). The highly volatile dichloromethane (Table 21-2) is not detectable in the stems at all, being lost from the roots to the unsaturated soil presumably through diffusion (Davis, unpublished observations).

Vroblesky *et al.* (1999) studied mature bald cypress (*Taxodium distichum*) trees and detected TCE several meters aboveground when the root systems were in contact with contaminated water. Using smaller trees (1 to 2 centimeters diameter stems) in a controlled environment, Nietch *et al.* (1999) showed that TCE was transferred through bald cypress (*Taxodium distichum*) trees in proportion to transpiration during summer. Significant movement up the stem also occurred when the trees were leafless.

Orchard *et al.* (2000) used an elaborate plant growth chamber, trying to determine transfer of TCE from the roots to the shoots. They found a very small net transfer of TCE from the root chamber to the leaf chamber. The root chamber was maintained at a negative pressure relative to the leaf chamber, and downward flow of air probably extracted the TCE passing up through the stems. Downward gas flow would also explain the inability of the

TABLE 21-2 Physical Properties of Representative Volatile Contaminants

Compound	Henry's law constant (dimensionless)	log K_{ow}	log K_{oc}	Vapor pressure (millimeters of mercury at 25 °C)
Dichloromethane	0.1	1.25	0.94	350
Methyl *tertiary*-butyl ether	0.02	1.24	—	275
Trichloroethylene	0.38	2.53	2.03	75

Note: From calculations and references cited in Davis *et al.* (1998a), and Davis *et al.* (1998b) and Knox *et al.* (1993). The dimensionless Henry's law constant is the molar concentration in gas phase per molar concentration in aqueous phase. The log K_{ow} is the logarithm of the octanol–water partition coefficient. log K_{oc} is the logarithm of the organic carbon–water partition coefficient (cubic centimeters of solution per gram of wet plant weight).

investigators to establish anaerobic conditions in the root chamber, even when the chamber was continually purged with nitrogen (N_2) gas rather than air. A combination of diffusion and advection is known to provide oxygen (O_2) to the roots of wetland plants (Shimp *et al.* 1993) and TCE moves upward through bald cypress (*Taxodicum distichum*) when trees are dormant or dead (Nietch *et al.* 1999).

Observational Support for Modeling Results

We compared the vertical profiles of soil-water contamination for MTBE, TCE, and dichloromethane using the six-channel system described in Table 21-1. Each contaminant was introduced into the saturated zone of two channels, with plants in the latter half of each channel. For each compound the transition from contaminated to uncontaminated soil water was found closer to the surface for the planted half of the channel than for the unplanted half (Figure 21-1). This result is expected if plant roots are drawing more contaminated water to the surface. As previously observed during soil sampling (Davis, unpublished observations), and clearly shown here, the transition from greater to lesser contamination occurred deeper in the soil for TCE than

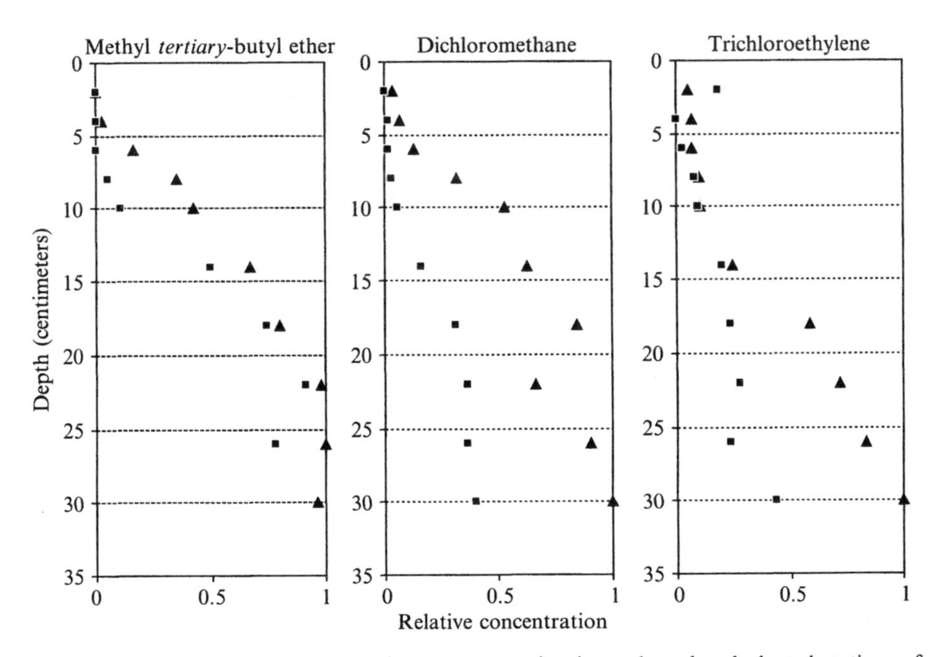

Figure 21-1 Relative soil-water contaminant concentration in unplanted and planted sections of soil. Sampling was done on the day of scheduled harvest when growth and water consumption were at maximum. Concentrations are expressed relative to the highest values observed in the planted half of the channel. Values for the planted area are shown as filled triangles and those for the unplanted area as filled squares. Results are shown for only one channel for each treatment.

for MTBE, consistent with the differences in the Henry's law constants and vapor pressures (Table 21-2). We have applied the model described by Narayanan *et al.* (1999a) to obtain an excellent fit to the profile shown in Figure 21-1, simply by using the appropriate Darcy water flux for each treatment condition and Henry's law constant for each compound (Davis *et al.* 2001).

Under the laboratory conditions of Table 21-1 with the water table 50 centimeters below the sandy soil surface, soil surface evaporation is about 2.5 millimeters of water per day and transpiration is up to three times that amount. In addition to drawing up the water more rapidly, the plant roots may transfer contaminant from deeper soil toward the surface by root uptake of contaminant in the deep soil, followed by diffusion from the roots and hydraulic lift of contaminated water (Negri, Gatliff *et al.* this book) into the shallower soil. Thus, plants may serve as a low cost method to pump out volatile contaminants. When there are microorganisms capable of degrading the contaminant in the rhizosphere, plants have the effect of promoting degradation, even if the plants do not metabolize the contaminant. This is discussed elsewhere for MTBE (Winnike-McMillan *et al.* this book).

Experimental Parameters for Field Situations

Plant Water Use

To make a major impact on remediation, plants must be in contact with the contaminants, alter the local soil environment, or draw contaminated water to the plant by transpiration. Modeling shows that plants may have greatest benefit when evapotranspiration exceeds infiltration (see the earlier section "Water and Contaminant Flux"). Generally, groundwater contaminants must move to the vadose zone for metabolism to occur, and deep-rooted plants are most effective at facilitating the necessary migration. Species that withdraw water from the capillary fringe and the groundwater aquifer are phreatophytes that include alfalfa (*Medicago sativa*), willow (*Salix* spp.), cottonwood (*Populus deltoides*), and salt cedar (*Tamarix* spp.). Depletion of soil water increases the depth of the vadose zone. All species, when actively growing, serve to reduce net infiltration, which will retard or prevent contaminant movement to the water table.

Davis *et al.* (1998a) discuss the variation of potential evapotranspiration with latitude and season, using well-established methods (Jensen and Heise 1963, Allen *et al.* 1998). Many different plant species may transpire a similar fraction of the potential evapotranspiration, but phreatophytes, which grow continually and withdraw water at the capillary fringe, will be most effective during hot, dry summers. Typically, temperate species use no more than 2 meters of water per year in a field setting, even when more is available. Most trees, and forbs such as alfalfa (*Medicago sativa*), that use the C-3 pathway of photosynthesis (Buchanan *et al.* 2000) use 500 to 800 grams of water for each gram of dry matter produced (Jensen and Heise 1963, Allen *et al.* 1998). For alfalfa (*Medicago sativa*), depending on climatic conditions, the peak biomass production is about 20 to 25 metric tons per hectare, with effective use of up to 1.6 meters of water per year (Stanberry 1955, Kansas Agricultural Experiment Station 1991). Some grasses, such as maize (*Zea mays*) and *Sorghum* spp., are more than twice as efficient in water use (Allen *et al.* 1998), and thus may be less water demanding. Higher yielding alfalfa (*Medicago sativa*) cultivars have been introduced over the past half-century, but many of these use no more water despite higher production (Kansas Agricultural Experiment Station 1991). *Eucalyptus* species in tropical settings may use up to 3 meters per year, an amount that is comparable to free water surface, evaporation (Davis *et al.* 1998a).

Potential for Contaminant Uptake

The potential maximum uptake can be calculated from (1) a known contaminant concentration in the groundwater, (2) the transpiration stream concentration factor (TSCF), and (3) the volume of water used. The evapotranspiration or water use can be estimated from plant biomass yield and water use efficiency (Allen *et al.* 1998). If the contaminant is not volatilized and not metabolized, the concentration in the plant may reach several hundred times that in the soil–water phase. Thus a contaminant may potentially be bioaccumulated in the food chain if the vegetation is a source of food for animals. When phytovolatilization is the removal process, the upward transport and release of the contaminant can be estimated from the product of the evapotranspiration rate of the soil water and the contaminant concentration in the soil water. These calculations are described in more detail in Davis *et al.* (1998a) and Narayanan *et al.* (1999b).

The TSCFs have been measured for a number of compounds (Briggs *et al.* 1982, Burken and Schnoor 1998, Davis *et al.* 1998b, Inoue *et al.* 1998) and may be predicted for others from the octanol–water partition coefficient (usually tabulated as the log K_{ow}). Different fits were obtained for different groups of compounds, but the data reasonably fit a bell-shaped curve with a maximum at log K_{ow} equal to 2 to 3 (Bromilow and Chamberlain 1995, Burken and Schnoor 1998, Burken this book). The maximum TSCF is about 0.75, indicating relatively little discrimination against the compound by the root system near the optimal K_{ow}.

A class of small polar but un-ionized molecules may pass through specialized aquaporins which normally serve to facilitate water and glycerol transfer (Buchanan *et al.* 2000). For instance, Aitchison *et al.* (2000) noted that the TSCF for dioxane is far greater than what would be predicted based on polarity. Likewise Makepeace *et al.* (1996) found that the TSCF for MTBE is much higher than predicted by the equation of Bromilow and Chamberlain (1995), which was developed with a diverse range of larger molecules.

Ionic compounds, having a small log K_{ow}, may depend on specific transport systems or electrogenic membrane crossing, driven by pH or ion gradients across the membrane, for uptake into the transpiration stream (Trapp 2000). Bromilow and Chamberlain (1995) indicate that the membranes of the endodermis may limit some polar compounds from the transpiration stream. Weak acids may accumulate if a significant fraction of the protonated neutral compound is present. Inoue *et al.* (1998) have discussed the mechanisms responsible for uptake of a series of small bases (amines), including both partitioning and ion-trapping effects. Proton removal to form the neutral species is important for effective amine uptake into the transpiration stream.

Nonpolar compounds, with log K_{ow} greater than 4, are scarcely taken up into the transpiration stream at all, as discussed in Trapp (1995) and Bromilow and Chamberlain (1995). The reasons for this are not well understood. Most larger hydrocarbon molecules are nonvolatile, move only with the aqueous phase in soil, and are strongly sorbed to the organic matter of the soil. Degradation of those compounds depends on rhizosphere effects (Hutchinson *et al.* this book, Olson *et al.* this book). Root architecture is a key variable for effective degradation, as mentioned earlier in the section "Root–Soil Interaction." Thus, although plants may metabolize such compounds *in vitro* (Harms *et al.* this book, Schwitzguebel and Vanek this book) metabolism is only a small *direct* contribution to remediation *in situ*.

Sorption and Translocation

Simulations of the transfer of less polar compounds (log K_{ow} greater than 2) through a plant requires accounting for the nonspecific sorption of a nonpolar contaminant in contact with plant organic matter, *e.g.*, cell walls, lignin, and membranes. Hu (1998) and Hu *et al.* (1998a, 1998b) measured the movement of TCE through sections of a poplar (*Populus* spp.) stem as a function of water flow. A sorption coefficient of 20 to 30 cubic centimeters of water per gram of dry plant biomass was necessary to explain the lag before contaminant was detected in outflow water. This sorption coefficient corresponds well to the reported K_{oc} value for TCE on organic matter in soil (Knox *et al.* 1993), as TCE does not bind to the cellulose fraction in the plant (Davis *et al.* 1998c). In the same experimental system, MTBE was scarcely sorbed to the plant at all. The derived sorption coefficient for MTBE was less than 0.1 cubic centimeters of water per gram of dry plant biomass, consistent with the much greater water solubility and polarity of MTBE (Table 21-2).

For strongly sorbing materials, monitoring the transfer through a plant from roots to leaves can produce erroneous values for the TSCF if the transfer rate is not approaching steady state. In the 1-day test of Makepeace (1993) and Makepeace *et al.* (1996), TSCFs for nonpolar contaminants such as ethylbenzene may be underestimated. By contrast, Burken and Schnoor (1998) allowed up to 10 days and noted up to 50 hours lag time before release from the plant for a nonpolar molecule such as ethylbenzene. Trapp (2000) modeled uptake and translocation into a growing plant. For strongly sorbing molecules, new growth of the roots constantly sorbs the contaminant prior to entry into the stem of the plant, resulting in a small TSCF. Thus, growing plants are not liable to bioaccumulate strongly nonpolar contaminants in the stems and foliage *via* water uptake. Deposition from the atmosphere may be a more important source (Paterson and McKay 1989, 1995, Riederer 1995).

Toxicity, Accumulation, or Metabolism Within a Plant

The likelihood of phytotoxicity can be estimated if the toxicity threshold (milligrams per gram of dry weight) of the substance in the plant is known, based on the conservative ratio of 1 kilogram water transpired for every gram of dry weight production by the plant (Allen *et al.* 1998). If the substance must be detoxified or sequestered within the plant, one can predict the likelihood of successful contaminant removal based on the energetics of the detoxification reactions, assuming that the necessary enzymes are present (Davis *et al.* 1998a), Davis *et al.* (2002). If a large fraction of total photosynthate (sugars) is required to sequester a contaminant, for instance as a glycoside, a plant will not grow well and hence contaminant accumulation may be small. Plants *per se* may not be very useful for cleaning up high concentrations of contaminants unless volatilization or mineralization of the contaminant occurs, or unless plants are harvested.

The potential for sequestration is specific to each contaminant and can occur through, for instance, storage in the vacuole or apoplast unmodified, or as a conjugate. Metabolism and detoxification depend on the presence of an enzymatic pathway for activation or reaction, as discussed elsewhere in this book. Noting the diverse structures and selectivity of herbicides and pesticides, few useful generalizations can be made about contaminant removal by plants; each compound must be tested individually. Even less is known about other organic chemicals. Extensive metabolism can occur with many different compounds at low levels. Transformation can usually be documented using isotopically labeled contaminants exposed to plants. The by-products or daughter products must be isolated or stable incorporation of the isotope must be shown. For almost any compound that can be metabolized by a plant, a host of microorganisms also may transform the chemical. Thus rhizosphere microbial degradation is a more general means by which plants enhance contaminant loss (Olson *et al.* 2001, Davis *et al.* 2002).

Volatilization from Plants

Narayanan *et al.* (1999b) estimated the impact of plant-dependent volatilization on air quality through transfer of contaminant from soil water to the atmosphere. Calculations followed U.S. EPA guidelines (U.S. EPA 1993) developed for estimating potential dispersion of air contamination from bio-venting sources. These guidelines are based on dispersed sources (100 meters across), being treated as point sources at a distance greater than 1 kilometer from the site. Note that any compound with a boiling point less than that of water and not sorbed or metabolized will be volatilized; if the boiling point is greater than that of water, the contaminant will accumulate to an extent which is a function of the boiling point.

Plants require one additional important consideration that differs from bioventing. When the contaminant must first dissolve in water and then enter the plants for phytovolatilization to occur, the amount of water vapor that a given volume of air can contain sets a definite limit for the concentration that will be observed in the atmosphere (Narayanan *et al.* 1999b, Winnike–McMillan *et al.* this book). At typical ambient temperatures of 21 °C, the water vapor concentration in the air (humidity) is 1 millimole per liter, compared to the approximately 55 moles per liter of liquid water. Thus, any contaminant transferred from an aqueous solution to air *via* the plant, whether by diffusion or transpiration, will necessarily be present at approximately 55 000-fold lower concentration in the air than in the deep soil water (expressed as a weight per unit volume). If the concentration within the soil water is known, maximum concentrations within the air aboveground can be predicted, independent of dispersion. (Note that this requires careful dimensional analysis because units commonly used to express concentrations in water and atmosphere differ.) Narayanan *et al.* (1999b) detail, for a number of compounds, the probability that regulatory limits will be exceeded within the confines of a contaminated site, or at some distance.

PLANT SELECTION FOR FIELD APPLICATIONS

The choice of plants depends very much on the objectives of the phytoremediation project, and may be quite different for plume control or a vegetative cap than for cleaning up petroleum contamination of unsaturated soil. Important considerations include the capability of the plants to establish and grow, the cost to plant and establish the vegetation, transpiration rates, and the potential to enhance the biodegradation rate and extent. The overall climatic adaptation and ecosystem restoration potential are also important in some settings.

As of 2003, no plant selection protocols had been formalized; but some useful compilations of candidate plants exist. First, other chapters in this book discuss the use of specific plants for specific contaminants and provide examples of successful uses of plants. Second, the U.S. EPA provides a convenient on-line listing (http://www.epa.gov/tio/download/remed/introphyto.pdf) of many species used at different sites to remediate a range of contaminants. This list contains proposed and successful as well as unsuccessful applications. The Phytoremediation Resource Guide (U.S. EPA 1999) includes abstracts of publications through early 1999, sorted by contaminant type, for the same studies considered in the cited website. Those may be examined for further information on how particular plants were applied at particular sites.

For common contaminants like TCE or MTBE there is no good evidence that particular plant species possess metabolic pathways not found in other species. Thus plant selection should be based primarily on suitable agronomic traits for a particular location. Plants such as alfalfa (*Medicago sativa*) that fix nitrogen may be advantageous where nitrogen is needed for biodegradation of organic contaminants. Climatic conditions such as seasonal rainfall pattern, length of growing season, and range of air temperatures must be considered in plant selection. In arid and semi-arid areas, species must be able to survive periods of drought and high temperature. In other areas where flooding is common, plants must be able to survive periods of soil saturation. Cold hardiness can be a problem with particular cultivars of alfalfa (*Medicago sativa*) or hybrid poplars (*Populus* spp.).

For erosion control, soil characteristics and topography must also be considered in choosing plants. Bulk density, soil texture, water retention, hydraulic conductivity, salinity, redox, pH, and phytotoxicity all affect the sustainability of vegetation. While soil amendments and preparation can be used to improve the water holding capacity and the availability of nutrients, selection of plant species or cultivars that are climatically adapted is important to control costs and obtain good establishment and growth. Guidance from local horticulture, agronomy, or forestry professionals is essential. Hardy perennial plants with extensive root systems that have long growing seasons and do not require extensive maintenance are best in many cases. The depth of the contamination will affect plant selection because the rooting depth of the potential plant species should be sufficient to accomplish the phytoremediation objective.

Available data on plant germination and growth in the contaminated soil should be reviewed, along with results of past phytoremediation investigations with the contaminant. Greenhouse or other pilot tests must be conducted in the absence of data on the ability of selected plants to germinate and grow in presence of the particular contaminant.

Different cultivars of the same species and different species of the same genus may differ in responses to climatic factors (Erickson *et al.* 2000b). Qiu *et al.* (1997) conducted a field study at the Union Carbide Seadrift Plant olefins production area in Texas that had clay soil (51 to 61 percent clay). When 12 different plant species were compared at the end of a 3-year study, Verde kleingrass (*Panicum coloratum*), of African origin, was far superior to the other plant species and the control without vegetation. Kleingrass (*Panicum coloratum*) was well adapted to the particular climatic conditions of the south Texas site and developed dense roots that were healthier and deeper than those of the other grasses. The concentrations of polycyclic aromatic hydrocarbons for the plots of the kleingrass (*Panicum coloratum*) were either below detection or approximately one or two orders of magnitude less than the control without vegetation and the other grass species (Qiu *et al.* 1997).

In some situations, native vegetation may be already providing benefits (Olson *et al.* 2001). However, many disturbed sites have only winter annuals, ephemerals, or weeds, which do not root into all of the contaminated soil and consume little of the annual precipitation. Planting phreatophytic and deep-rooted species may provide significant increases in water usage, bringing contaminants to the aerated vadose zone. In other settings with shallow contamination such as that described by Qiu *et al.* (1997), fibrous-rooted plants such as grasses may best promote rhizosphere degradation processes.

IRRIGATION AND WATER MANAGEMENT

Surface and subsurface water management are essential aspects of any phytoremediation plan. Sound water management plans are especially critical for vegetative caps and plume control. In this book, Harvey *et al.*; Hirsch *et al.*; Negri, Gatliff *et al.*; and Weaver *et al.* discuss plume-control. Design variables that are important for overall water management include transpiration characteristics, rooting depth, and area of planted vegetation. Surface runoff and infiltration control are vital for two reasons. First, percolation and mobilization of contaminants must be minimized in most designs. Second, infiltration of relatively clean water pushes plumes deeper into an aquifer and thus reduces potential to achieve plume control. Obviously, plants must be supplied sufficient water to survive drought. This may require irrigation if infiltration is restricted.

Where vegetative caps are applied, the rooting depth and the type of soil determine whether all of the rainfall can be contained or stored within the root zone. Table 21-3 gives estimated ranges of storage capacity for several different soils (Erickson *et al.* 2000a). A silty clay soil with an estimated storage capacity (field capacity minus permanent wilting point) of 14 to 19 percent water in a root zone of 3-meter depth can contain or store infiltration equal to 42 to 57 centimeters of water. This is significant part of the annual precipitation in many locations. Some plants have rooting depths that exceed

TABLE 21-3 Estimated Water Storage Capacity of Various Soils

Textural class	Storage (percent water)[a]
Silt loam	17 to 22
Loam	12 to 21
Silty clay	14 to 19
Sandy loam	4 to 16
Clay	11 to 15
Clay loam	5 to 10
Sand	2 to 5

[a]The available water storage capacity is defined as the difference between field capacity and the permanent wilting point of plants.

Note: Ranges are for data from Erickson *et al.* (2000a).

3 meters, and water use often ranges from 1 to 2 meters per year if sufficient water is available and nutrients are not limiting. A conventional soil cap of 1 meter will only store sufficient water for optimum transpiration over 1 to 3 weeks during summer conditions; so, only drought-tolerant species are useful with such shallow caps unless supplemental irrigation is planned.

The design of a vegetative cap is based on the water balance: storage is equal to precipitation minus runoff minus evapotranspiration. Storage is the soil water of the root zone in excess of that present at the wilting point of the vegetation. For a site with a given depth of roots in soil of known porosity, water storage as a function of time may be forecast using past precipitation data and climatic conditions (Tracy *et al.* 1994). In a conservative design, the maximum transient storage should be less than about 70 percent of the maximum storage capacity, based on the data in Table 21-3. To keep the vegetation alive over the design life or specified growing season, soil moisture should exceed the wilting point water content.

For interception and control of a plume, the water balance also requires that plume flow in plus infiltration is equal to evapotranspiration plus plume flow out. The goal here is to have no plume flow off-site. The capability to manage surface water to reduce infiltration on the site and to reduce the quantity of groundwater flowing into the contaminated area by reducing infiltration at nearby locations will vary from site to site. Successful design to intercept and control a plume requires sufficient vegetation to establish a cone of influence, so that evapotranspiration is the only water loss from the plume of contaminated water. Under some conditions where the plume is small, the water table may be lowered substantially, allowing aerobic conditions for biodegradation at a greater depth. Besides control of the plume, the contaminant concentration must be reduced to an acceptable level over the lifetime of the project. The amount of water transpired also depends on plant health, availability of water, temperature, and humidity. Potential annual evapotranspiration and water use by plants both vary with climatic conditions and thus with geographical location (Jensen and Heise 1963, Allen *et al.* 1998, Davis *et al.* 1998a). Vose and Harvey (this book) discuss measuring transpiration at the field scale.

Irrigation may be needed to establish vegetation; however, in most applications where a plume is present the goal is for the plants to have deep roots such that the plume provides the water for plant growth. When contaminated water is found at a depth which exceeds the ability of the vegetation to bring it to the surface, or is below the reach of deep planting approaches (Negri, Gatliff *et al.* this book), mechanical pumps can be used to lift the contaminated water to the roots of the plants. In some applications, drip irrigation within the rooting depth of the plants is recommended to avoid contaminant exposure at the soil surface. Solar plant pumping can be augmented with mechanical pumping to achieve complete plume control. Water from mechanical pumping can be piped to a location where it can be used for irrigation or treated by another technique.

TIME REQUIREMENTS

Phytoremediation for organic compounds may be selected when the time available for restoration or stabilization is 3 or more years. It is less likely to be useful for short times because perennials require at least a season to establish. Olson *et al.* (2001) have documented a natural phytoremediation process at a former organic sludge disposal site where very large decreases of contaminant level are associated with vegetation development. The properly designed vegetated site may still be used for other purposes while remediation goes on, provided that both the remediation and the site use plan are integrated into the design (Hong *et al.* 2001, Trapp and Christiansen this book). Thus a site may not need to be "out of use" for the duration of the remediation.

Under aerobic conditions, considerable progress is often observed during the first year of a 2-year period when biodegradation of petroleum compounds in soil is the objective (Karthikeyan *et al.* 1999, Banks *et al.* 2000, Erickson *et al.* 2000b). The results of Karthikeyan *et al.* (2001), shown in Figure 21-2, indicate that most of the biodegradation of petroleum hydrocarbons took place during the first year when contaminated sediments from washing vehicles were applied to land. In other work shown in Table 21-4, biodegradation

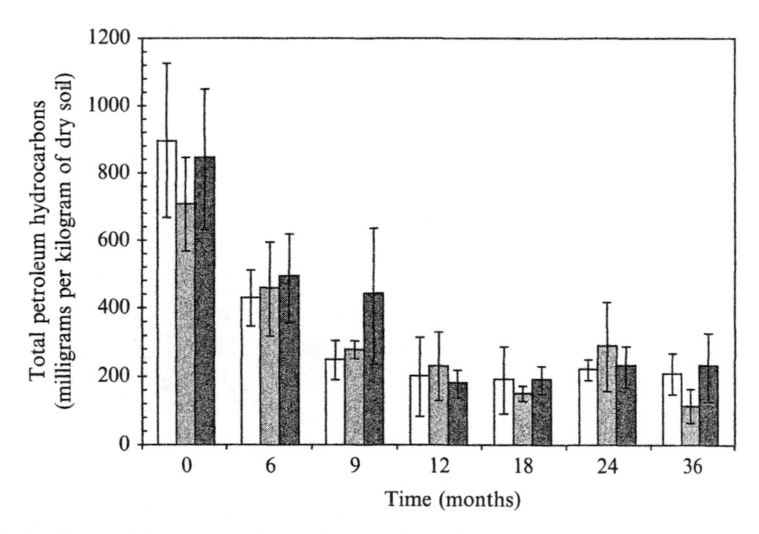

Figure 21-2 Effect of time on total petroleum hydrocarbon concentration, Fort Riley field trial with contaminated sediments from washing vehicles. Three treatments consisted of: (1) grass mixture [tall fescue (*Festuca arundinacea*) plus western wheatgrass (*Agropyron smithii*)]; (2) legume plus grass [sweet clover (*Melilotus officinalis*), red clover (*Trifolium pratense*), birds-foot trefoil (*Lotus corniculatus*), tall fescue, (*Festuca arundinacea*)]; and (3) unplanted control. Quadruplicate plots of a randomized complete block design were analyzed; variation is indicated by standard deviation error bars. From left to right in each group, the open bars represent measurements on the grassed plots, the middle bars represent measurements from the plots planted with legumes, and the rightmost bars represent measurements from the unplanted plots. All plots were fertilized before planting. Few legumes survived beyond a year.

TABLE 21-4 Disappearance of Total Petroleum Hydrocarbons of Jet Fuel JP-8 in the Presence of Alfalfa (*Medicago sativa*)

Date[a]	Sample depth (centimeters)	Channel 1	Channel 2
March	0 to 10	215 ± 59[b]	89 ± 68 (2 UD)[c]
	10 to 20	1249 ± 355	198 ± 22
	20 to 30	405 ± 118	148 ± 99
May	0 to 10	76 ± 45 (2 UD)	50 (3 UD)
	10 to 20	733 ± 109	128 ± 70 (1 UD)
	20 to 30	107 ± 59	76 ± 46 (2 UD)

[a]Initial contamination in December gave a calculated average of 1700 milligrams of jet fuel per kilogram of dry soil from JP-8 injected *via* cannula, continuously during withdrawal between 35 and 15 centimeters depth using 3 milliliters per location in each of 20 locations.
[b]Total petroleum hydrocarbon concentration (milligrams of jet fuel per kilogram of dry soil) was measured in triplicate samples drawn from composites of 8 locations sampled for the indicated depth. Data are for the front rows of a two-channel planted system after 3 months and 5 months. Channels were 10 centimeters wide, 90 centimeters long, 35 centimeters deep. Channel 1 was irrigated by maintaining a saturated zone about 25 to 30 centimeters below the soil surface whereas Channel 2 was irrigated using controlled drip irrigation 20 centimeters below the soil surface. Data from Karthikeyan *et al.* (1999). Composite samples in each channel at 8 months gave only undetectable levels (Karthikeyan unpublished observations).
[c]Abbreviation used UD: undetectable. (The detection limit of the gas chromatograph was 100 milligrams of jet fuel per kilogram of dry soil; values below the detection limit were assigned a value of 50 milligrams of jet fuel per kilogram of dry soil. The number in parentheses denotes the number of undetectable values recorded.)

of JP-8 jet fuel (boiling point 159 to 263 °C) was found to be almost complete after 5 months when drip irrigation was used in a laboratory study (Karthikeyan *et al.* 1999). Mathematical models that include biodegradation, plant uptake, and gas phase diffusion to the atmosphere (Narayanan *et al.* 1998a, 1998b) may be used to estimate remediation time requirements for contaminants in a vadose zone. In cases where all the water moving in a plume can be transpired by the plants, the time to accomplish this can be estimated, based on water usage rates and plume volume.

When a separate hydrocarbon phase is present as a nonaqueous phase liquid or solid hydrocarbons, the time for biodegradation and biostabilization may be much longer. Santharam *et al.* (1994) successfully modeled experimental results for polycyclic aromatic hydrocarbon in the rhizosphere. The degradation of jet fuel introduced as a nonaqueous phase liquid in a laboratory system has been modeled (Karthikeyan 2001). The time required to restore a closed refinery site or sludge basin may be 10 years or more (Olson *et al.* 2001). One must have a clearly defined and attainable end point for success (normally expressed as milligrams of contaminant per kilogram of dry soil) to say when the process is complete.

POTENTIAL FOR INTEGRATION OF PHYTOREMEDIATION WITH OTHER TECHNOLOGIES

There are many opportunities to use treatment trains that include more than one technology. *In situ* bioremediation and natural attenuation are often implicit in the design of a phytoremediation process, because rhizosphere effects may be dominant (Davis *et al.* 2002). Air sparging or bioventing to provide oxygen to microorganisms can enhance aerobic biodegradation and enhance movement of volatile contaminants, while rhizosphere organisms degrade the contaminants. Soil removal for portions of a spill where very toxic concentrations are located can be followed by phytoremediation to treat the remaining contamination at the site. Composting or land-farming may be used to reduce the toxicity of contaminated material, followed by incorporation into soil where vegetation is established to provide a final treatment or treatment of residuals after closure (Schwab and Banks 1994).

Design processes to support phytoremediation, particularly to couple the approach with other treatment options, are not yet well developed. All models are approximations and those that are available for phytoremediation are based on limited scientific information. As emphasized earlier, choice of plants may be highly specific for sites and contaminants. Experienced professionals who understand the individual elements of the phytoremediation process should be present in the design team.

Acknowledgments

This is contribution # 01-110-B of the Kansas Agricultural Experiment Station. This research was partially supported by the U.S. EPA under assistance agreements R-815709, R-819653, R-825549, and R-825550 to the Great Plains-Rocky Mountain Hazardous Substance Research Center for Regions 7 and 8. It has not been submitted to the U.S. EPA for peer review and therefore may not necessarily reflect the views of the agency and no official endorsement should be inferred. The U.S. Department of Energy, Office of Restoration and Waste Management, Office of Technology Development, U.S. Department of Defense, and the Center for Hazardous Substance Research also provided partial support.

REFERENCES

Aitchison, E.W., S.L. Kelley, P.J.J. Alvarez, and J.L. Schnoor (2000) Phytoremediation of 1,4-dioxane by hybrid poplar trees. *Water Environ. Res.* **72**: 313–321.

Allen, R.G., L.S. Periera, D. Raes, and M. Smith (1998) Crop evapotranspiration: guidelines for computing crop water requirements. FAO Irrigation and Drainage Paper # 56. Food and Agriculture Organization of the United Nations, Rome, Italy.

Ames, R.A. and B.L. Hoyle (1999) Biodegradation and mineralization of atrazine in shallow subsurface sediments from Illinois. *J. Environ. Qual.* **28**: 1674–1681.

Banks, M.K., R.S. Govindaraju, A.P. Schwab, and P. Kulakow (2000) Field demonstration. In: *Phytoremediation of Hydrocarbon Contaminated Soil.* S. Fiorenza, C.L. Oubre, and C.H. Ward, eds. Lewis Publishers, Boca Raton, Florida, pp. 3–88.

Bodvarsson, G., S. Finsterle, H.H. Liu, C.M. Oldenburg, K. Preuss, E. Sonnenthal, and Y.-S. Wu (2000) Flow and transport modeling of the vadose zone. In: *Vadose Zone Science and Technology Solutions.* B.B. Looney and R.W. Falta, eds. Battelle Press, Columbus, Ohio, pp. 591–827.

Briggs, G.G., R.H. Bromilow, and A.A. Evans (1982) Relations between lipophilicity and root uptake and translocation on non-ionized chemicals by barley. *Pestic. Sci.* **13**: 495–504.

Bromilow, R.H. and K. Chamberlain (1995) Principles governing uptake and transport of chemicals. In: *Plant Contamination: Modeling and Simulation of Organic Chemical Processes.* S. Trapp and J.C. Mc Farlane, eds. Lewis Publishers, Boca Raton, Florida, pp. 37–68.

Buchanan, B.B., W. Gruissem, and R.L. Jones (2000) *Biochemistry and Molecular Biology of Plants.* American Society of Plant Physiologists, Rockville, Maryland.

Burken, J.G. and J.L. Schnoor (1996) Phytoremediation: plant uptake of atrazine and role of root exudates. *J. Environ. Eng.* **122**: 958–963.

Burken, J.G. and J.L. Schnoor (1998) Predictive relationships for uptake of organic contaminants by hybrid poplar trees. *Environ. Sci. Technol.* **32**: 3379–3385.

Center for Waste Reduction Technologies (1999) Guidelines for successful phytoremediation. American Institute of Chemical Engineers, New York.

Chang, Y-Y. and M.Y. Corapcioglu (1997) Effect of roots on water flow in unsaturated soils. *J. Irrig. Drain. Eng.* **123**: 202–209.

Corapcioglu, M.Y., R.L. Rhykerd, C.L. Munster, M.C. Drew, K. Sung, and Y-Y. Chang (1998) Modeling phytoremediation of land contaminated by hydrocarbons. In: *Bioremediation and Phytoremediation; International Conference on Remediation of Chlorinated and Recalcitrant Compounds.* G.B. Wickramanayake and R.E. Hinchee, eds. Battelle Press, Columbus, Ohio, pp. 239–244.

Davis, L.C. (2000) Long-term monitoring to observe natural attenuation in the laboratory. In: *Proceedings of the 2000 Conference on Hazardous Waste Research.* L.E. Erickson and M. M Rankin, eds. Kansas State University, Manhattan, Kansas, pp. 105–110 (http://www.engg.ksu.edu/HSRC/).

Davis, L.C., L.E. Erickson, E. Lee, J.F. Shimp, and J.C. Tracy (1993) Effects of plants on the bioremediation of contaminated soil and ground water. *Environ. Prog.* **12**: 67–75.

Davis, L.C., M.K. Banks, A.P. Schwab, Muralidharan Narayanan, L.E. Erickson, and J.C. Tracy (1998a) Plant-based bioremediation. In: *Bioremediation: Principles and Practice, Vol. 2. Biodegradation Technology Developments.* S.K. Sikdar and R.L. Irvine, eds. Technomic Publ., Lancaster, Pennsylvania, pp. 183–219.

Davis, L.C., S. Vanderhoof, J. Dana, K. Selk, K. Smith, B. Goplen, and L.E. Erickson (1998b) Chlorinated solvent movement through plants monitored by Fourier transform infrared (FT-IR) spectroscopy. *J. Hazard. Subs. Res.* **1**(4): 1–26 (http://www.engg.ksu.edu/HSRC/).

Davis, L.C., S. Vanderhoof, and D. Lupher (1998c) Sorption of trichloroethylene (TCE) and other halogenated aliphatics on plant materials. In: *Proceedings of the*

1998 Conference on Hazardous Waste Research. L.E. Erickson and M.M Rankin, eds. Kansas State University, Manhattan, Kansas, pp. 319–325 (http://www.engg.ksu.edu/HSRC/).

Davis, L.C., D. Lupher, J. Hu, and L.E. Erickson (1999) Transport of trichlorethylene through living plant tissues. In: *Proceedings of the 1999 Conference on Hazardous Waste Research.* L.E. Erickson and M. M Rankin, eds. Kansas State University, Manhattan, Kansas, pp. 203–209 (http://www.engg.ksu.edu/HSRC/).

Davis, L.C., C. Pitzer, S. Castro, and L.E. Erickson (2001) Henry's constant, Darcy's law, and contaminant loss. In: *Proceedings of the 2001 Conference on Hazardous Waste Research.* L.E. Erickson and M. M Rankin, eds. Kansas State University, Manhattan, Kansas, pp. 2–15 (http://www.engg.ksu.edu/HSRC/).

Davis, L.C., S. Castro, Q. Zhang, and L.E. Erickson (2002) Benefits of vegetation for soils with organic contaminants. *Crit. Rev. Plant Sci.* **21**: 457–491.

Erickson, L.E., M.K. Banks, L.C. Davis, A.P. Schwab, M. Narayanan, K. Reilly, and J.C. Tracy (1994) Using vegetation to enhance in situ bioremediation. *Environ. Prog.* **13**: 226–231.

Erickson, L.E., L.C. Davis, and P.A. Kulakow (2000a) Phytoremediation. In: *Vadose Zone Science and Technology Solutions.* B.B. Looney and R.W. Falta, eds. Battelle Press, Columbus, Ohio, pp. 1090–1097.

Erickson, L.E., P.A. Kulakow, and L.C. Davis (2000b) Phytoremediation of petroleum contaminated soil. In: *Vadose Zone Science and Technology Solutions.* B.B. Looney and R.W. Falta, eds. Battelle Press, Columbus, Ohio, pp. 1234–1237.

Fulthorpe, R.R., A.N. Rhodes, and J.M. Tiedje (1996) Pristine soils mineralize 3-chlorobenzoate and 2,4-dichlorophenoxyacetate via different microbial populations. *Appl. Environ. Microbiol.* **62**: 1159–1166.

Gidda, T., W.H. Stiver, and R.G. Zytner (1999) Passive volatilization of gasoline in unsaturated soils. *J. Contam. Hydrol.* **39**: 137–159.

Hong, M.S., W.F. Farmayan, I.J. Dortch, C.Y. Chiang, S.K. McMillan, and J.L. Schnoor, (2001) Phytoremediation of MTBE from a groundwater plume. *Environ. Sci. Technol.* **35**: 1231–1239.

Hu, J. (1998) Study on fate and transport of contaminants in plants. M.S. thesis. Department of Chemical Engineering, Kansas State University, Manhattan, Kansas.

Hu, J., M. Narayanan, L.C. Davis, and L.E. Erickson (1997) Modeling root uptake and transport of trichloroethylene. In: *Proceedings of the 27th Annual Biochemical Engineering Symposium.* Colorado State University, Fort Collins, Colorado, pp. 113–121.

Hu, J., L.C. Davis, and L.E. Erickson (1998a) Modeling contaminant transport in plants. In: *Proceedings of the 1998 Annual Conference on Hazardous Waste Research.* L.E. Erickson and M.M. Rankin, eds. Kansas State University, Manhattan, Kansas, pp. 285–295 (http://www.engg.ksu.edu/HSRC).

Hu, J., L.C. Davis, and L.E. Erickson (1998b) Uptake of trichloroethylene by plants. In: *Proceedings of the 28th Annual Biochemical Engineering Symposium.* P.J. Reilley, ed. Iowa State University, pp. 93–100.

Inoue, J., K. Chamberlain, and R.H. Bromilow (1998) Physicochemical factors affecting the uptake by roots and translocation to shoots of amine bases in barley. *Pestic. Sci.* **54**: 8–21.

Interstate Technology and Regulatory Council (ITRC) (1999) Phytoremediation decision tree. Washington, D.C. (http://www.itrcweb.org).

Jensen, M.J. and H.R. Heise (1963) Estimating evapotranspiration from solar radiation. *J. Irrig. Drain. Eng. Div. Am. Soc. Civil Eng.* **89**: 15–41.

Jury, W.A., D. Russo, G. Streile, and H. El Abd (1990) Evaluation of volatilization by organic chemicals residing below the soil surface. *Water Resour. Res.* **26**: 13–20.

Jury, W.A., D. Russo, G. Streile, and H. El Abd (1992) Correction to evaluation of volatilization by organic chemicals residing below the soil surface. *Water Resour. Res.* **28**: 607–608.

Kansas Agricultural Experiment Station (1991) 1991 Kansas alfalfa performance tests. Report of Progress # 650, pp. 12–14.

Karthikeyan, R. (2001) Experimental and modeling studies on fate and transport of petroleum contaminants in soils with plants. Ph.D. dissertation. Biological and Agricultural Engineering, Kansas State University, Manhattan, Kansas.

Karthikeyan, R., L.C. Davis, K.R. Mankin, L.E. Erickson, and P.A. Kulakow (1999) Biodegradation of jet fuel (JP-8) in the presence of vegetation. In: *Proceedings of the 1999 Conference on Hazardous Waste Research.* L.E. Erickson and M.M. Rankin, eds. Kansas State University, Manhattan, Kansas, pp. 243–256 (http://www.engg.ksu.edu/HSRC/).

Karthikeyan, R., L.E. Erickson, L.C. Davis, and K.R. Mankin (2000) Fate and transport of jet fuel in soils with plants: experiments and modeling. In: *Proceedings of the 9th International Conference of the Integrated Petroleum Environmental Consortium.* K. Sublette, ed. University of Tulsa, Tulsa, Oklahoma. 21 pp., 5 tables, and 13 figures.

Karthikeyan, R., P. Kulakow, L.E. Erickson, and B.A. Leven (2001) Vegetated treatment of vehicle wash sediments: a field demonstration. In: *Proceedings of the 30th Annual Biochemical Engineering Symposium.* D.S. Kompala, ed. University of Colorado, Boulder, pp. 79–85.

Knox, R.C., D.A. Sabatini, and L.W. Canter (1993) *Subsurface Transport and Fate Processes.* Lewis Publishers, Boca Raton, Florida, pp. 393–404.

Kulakow, P., L.E. Erickson, and L. Jackson (1999) Plant system technologies for environmental management in the petroleum industry. Chevron Research and Technology Company, Richmond, California.

Kutchera, L. (1960) *Wurzelatlas Mitteleuropischer Ackerenkrauter und Kulturpflanzen.* DLG-Verlag-Gmbh, Frankfurt.

Lindgren, E. (2000) Performance monitoring. In: *Vadose Zone Science and Technology Solutions.* B.B. Looney and R.W. Falta, eds. Battelle Press, Columbus, Ohio, pp. 511–563.

Lynch, J.P., K.L. Nielsen, R.D. Davis, and A.G. Jablokow (1997) SimRoot: modelling and visualization of root systems. *Plant Soil* **188**: 139–151.

Makepeace, V.D. (1993) Using extractive Fourier transform infrared (FT-IR) spectroscopy to monitor bioremediation of gasoline components. M.S. thesis. Department of Chemistry, Kansas State University, Manhattan, Kansas.

Makepeace, V.D., L.C. Davis, J. Dana, K. Selk, K. Smith, R.M. Hammaker, W.G. Fateley, and L.E. Erickson (1996) Measuring contaminant flux through plants by Fourier transform infrared (FT-IR) spectrometry. In: *Proceedings of the HSRC and*

WERC Joint Conference on the Environment. Albuquerque, New Mexico, pp. 577–582.

Matthies, M. and H. Behrendt (1995) Dynamics of leaching, uptake and translocation: the simulation model network atmosphere–plant–soil (SNAPS). In: *Plant Contamination: Modeling and Simulation of Organic Chemical Processes*. S. Trapp and J.C. Mc Farlane, eds. Lewis Publishers, Boca Raton, Florida, pp. 215–243.

McCarthy, K.A. and R.L. Johnson (1995) Measurement of trichloroethylene diffusion as a function of moisture content in sections of gravity-drained soil columns. *J. Environ. Qual.* **24**: 49–55.

Mc Farlane, J.C. (1995) Anatomy and physiology of plant conductive systems. In: *Plant Contamination: Modeling and Simulation of Organic Chemical Processes*. S. Trapp and J.C. Mc Farlane, eds. Lewis Publishers, Boca Raton, Florida, pp. 13–34.

Narayanan, M., L.C. Davis, J.C. Tracy, L.E. Erickson, and R.M. Green (1995) Experimental and modeling studies of the fate of organic contaminants in the presence of alfalfa plants. *J. Hazard. Mater.* **41**: 229–249.

Narayanan, M., J.C. Tracy, L.C. Davis, and L.E. Erickson (1998a) Modeling the fate of toluene in a chamber with alfalfa plants 1. theory and modeling concepts. *J. Hazard. Subs. Res.* **1**(5a) (http://www.engg.ksu.edu/HSRC/).

Narayanan, M., L.C. Davis, J.C. Tracy, and L.E. Erickson (1998b) Modeling the fate of toluene in a chamber with alfalfa plants 2. numerical results and comparison study. *J. Hazard. Subs. Res.* **1**(5b) (http://www.engg.ksu.edu/HSRC/).

Narayanan, M., N.K. Russell, L.C. Davis, and L.E. Erickson (1999a) Fate and transport of trichloroethylene in a chamber with alfalfa plants. *Int. J. Phytoremed.* **1**: 387–411.

Narayanan, M., L.E. Erickson, and L.C. Davis (1999b) Simple plant-based design strategies for volatile organic pollutants. *Environ. Prog.* **18**: 231–242.

National Research Council (2000) *Natural Attenuation for Groundwater Remediation*. National Academy Press, Washington, D.C.

Newman L.A., X. Wang, I.A. Muiznieks, G. Ekuan, M. Ruszaj, R. Cortellucci, D. Domroes, G. Karscig, T. Newman, R.S. Crampton, R.A. Hashmonary, M.G. Yost, P.E. Heilman, J. Duffy, M.P. Gordon, and S.E. Strand (1999) Remediation of trichloroethylene in an artificial aquifer with trees: a controlled field study. *Environ. Sci. Technol.* **33**(13): 2257–2265.

Nietch, C.T., J.T. Morris, and D.A. Vroblesky (1999) Biophysical mechanisms of trichloroethene uptake and loss in baldcypress growing in shallow contaminated groundwater. *Environ. Sci. Technol.* **33**: 2899–2904.

Olson, P.E., J.S. Fletcher, and P.R. Philp (2001) Natural attentuation/phytoremediation in the vadose zone of a former industrial sludge basin. *Environ. Sci. Pollut. Res.* **8**: 243–249.

Orchard, B.J., W.J. Doucette, J.K. Chard, and B. Bugbee (2000) Uptake of trichloroethylene by hybrid poplar trees grown hydroponically in flow-through plant growth chambers. *Environ. Toxicol. Chem.* **19**: 895–903

Paterson, S. and D. Mackay (1989) Modeling the uptake and distribution of organic chemicals in plants. In: *Intermedia Pollutant Transport: Modeling and Field Measurements*. D.T. Allen, Y. Cohen, and I.R. Kaplan, eds. Plenum Press, New York, pp. 269–281.

Paterson, S. and D. Mackay (1995) Interpreting chemical partitioning in soil-plant-air systems with a fugacity model. In: *Plant Contamination: Modeling and Simulation of Organic Chemical Processes*. S. Trapp and J.C. Mc Farlane, eds. Lewis Publishers, Boca Raton, Florida, pp. 191–214.

Paterson, S., D. Mackay, D. Tam, and W.Y. Shiu (1990) Uptake of organic chemicals by plants: a review of processes, correlations and models. *Chemosphere* **21**: 297–331.

Plant Physiology, special issue (2000) Arabidopsis genome: a milestone in plant biology. *Plant Physiol.* **124**: 1449–1854.

Qiu, X., T.W. Leland, S.I. Shah, D.L. Sorenson, and E.W. Kendall (1997) Field study: Grass remediation of clay soil contaminated with polycyclic aromatic hydrocarbons. In: *Phytoremediation of Soil and Water Contaminants*. E.L. Kruger, T.A. Anderson, and J.R. Coats, eds. ACS Symposium Series 664, American Chemical Society, Washington, D.C., pp. 186–189.

Quadrel, M. and R. Lundgren (2000) Managing an effective vadose zone project. In: *Vadose Zone Science and Technology Solutions*. B.B. Looney and R.W. Falta, eds. Battelle Press, Columbus, Ohio, pp. 61–130.

Riederer, M. (1995) Partitioning and transport of organic chemicals between the atmospheric environment and leaves. In: *Plant Contamination: Modeling and Simulation of Organic Chemical Processes*. S. Trapp and J.C. Mc Farlane, eds. Lewis Publishers, Boca Raton, Florida, pp. 153–190.

Robinson, N.I. (2000) Cumulative losses of buried volatile chemicals. *Water Resour. Res.* **36**: 801–806.

Rusch, H. and H. Rennenberg (1998) Black alder (*Alnus glutinosa*) trees mediate methane and nitrous oxide emission from the soil to the atmosphere. *Plant Soil* **201**: 1–7.

Ryu, S.B., L.C. Davis, J. Dana, K. Selk, and L.E. Erickson (1996) Evaluation of toxicity of trichloroethylene for sterile-grown plants. In: *Proceedings of the HSRC and WERC Joint Conference on the Environment*. Albuquerque, New Mexico, pp. 583–588 (http://www.engg.ksu.edu/HSRC/).

Santharam, S.K., L.E. Erickson, and L.T. Fan (1994) Modeling the fate of polynuclear aromatic hydrocarbons in the rhizosphere. In: *Proceedings of the 9th Annual Conference on Hazardous Waste Remediation*. L.E. Erickson, D.L. Tillison, S.C. Grant, and J.P. McDonald, eds. Kansas State University, Manhattan, Kansas, pp. 333–350.

Schnoor, J.L. (1997) Phytoremediation. Technology Evaluation Report TE-97-01. Ground-Water Remediation Technologies Analysis Center, Pittsburgh, Pennsylvania, 38 pp.

Schramm, K.-W., A. Reischl, and O. Hutzinger (1987) UNITTree: a multimedia compartment model to estimate the fate of lipophilic compounds in plants. *Chemosphere* **16**: 2653–2663.

Schwab, A.P. and M.K. Banks (1994) Biologically mediated dissipation of polyaromatic hydrocarbons in the root zone. In: *Bioremediation Through Rhizosphere Technology*. T.A. Anderson and J. Coats, eds. American Chemical Society Symposium Series 563, pp 132–141.

Shimp, J.F., J.C. Tracy, L.C. Davis, E. Lee, W. Huang, L.E. Erickson, and J.L. Schnoor (1993) Beneficial effects of plants in remediation of soil and groundwater contaminated with organic materials. *Crit. Rev. Environ. Sci.* **23**: 41–77.

Sprackling, J.A. and R.A. Read (1979) Tree root systems in eastern Nebraska. Nebraska Conservation Bulletin 37. Conservation and Survey Division, Institute of Agriculture and Natural Resources, University of Nebraska, Lincoln, Nebraska, 73 pp.

Stanberry, C.O. (1955) Irrigation practices for the production of alfalfa. *Yearbook of Agriculture*. U.S. Department of Agriculture, Washington, D.C., pp. 435–443.

Stein, W.D. (1967) *The Movement of Molecules Across Cell Membranes*. Academic Press, New York.

Tracy, J.C. and M.A. Marino (1989) Solute movement through root–soil environment, *J. Irrig. Drain. Eng.* **115**: 608–625.

Tracy, J.C., L.E. Erickson, J. Shimp, and L.C. Davis (1992) Modeling the beneficial effects of vegetation in the management of landfill leachates. In: *Proceedings of the Air and Waste Management Association 86th Annual Meeting*, Paper No. 92-27.03.

Tracy, J.C., L.E. Erickson, and L.C. Davis (1993) Rate limited degradation of hazardous organic contaminants in the root zone of a soil. In: *Proceedings of the Air and Waste Management Association 86th Annual Meeting*, Paper No. 93-WA-89.02.

Tracy, J.C., H. Ramireddy, L.E. Erickson, and L.C. Davis (1994) Effects of climatological variability on the performance of vegetative systems in remediating contaminated soil. In: *Proceedings of the Air and Waste Management Association 87th Annual Meeting*, Paper No. 94-WA-86.01.

Trapp, S. (1995) Model for uptake of xenobiotics into plants. In: *Plant Contamination: Modeling and Simulation of Organic Chemical Processes*. S. Trapp and J.C. Mc Farlane, eds. Lewis Publishers, Boca Raton, Florida, pp. 107–151.

Trapp, S. (2000) Modeling uptake into roots and subsequent translocation of neutral and ionizable organic compounds. *Pestic. Manag. Sci.* **56**: 767–788.

U.S. EPA (Environmental Protection Agency) (1993) Estimation of air impacts for bioventing systems used at Superfund sites. Air/Superfund National Technical Guidance Series. EPA-451/R-93-003, Washington, D.C.

U.S. EPA (Environmental Protection Agency) (1999) Phytoremediation resource guide. Office of Solid Waste and Emergency Response, Technology Innovation Office, EPA542-B-99-003, Washington, D.C. (http://www.epa.gov/tio).

U.S. EPA (Environmental Protection Agency) (2000) Introduction to phytoremediation. National Risk Management Research Laboratory, Office of Research and Development, EPA/600/R-99/107, Cincinnati, Ohio.

Vroblesky, D.A., C.T. Nietch, and J.T. Morris (1999) Chlorinated ethenes from groundwater in tree trunks. *Environ. Sci. Technol.* **33**: 510–515.

Washington, J.W. (1996) Gas partitioning of dissolved volatile organic compounds in the vadose zone: principles, temperature effects and literature review. *Ground Water* **34**: 709–718.

Weaver, J.E. (1926) *Root Development of Field Crops*, McGraw-Hill, New York

Zhang, Q. (1999) Phytoremediation of methyl *tert*-butyl ether (MTBE) in groundwater: experimental and modeling studies. Ph.D. dissertation. Department of Chemical Engineering, Kansas State University, Manhattan, Kansas.

Zhang, Q., L.E. Erickson, and L.C. Davis (1997) Effect of air sparging on fate and transport of trichloroethylene in chambers with alfalfa plants. In: *Proceedings of the*

1997 Annual Conference on Hazardous Waste Research. L.E. Erickson, M.M. Rankin, S.C. Grant, and J.P. McDonald, eds. pp. 564–573 (http://www.engg.ksu.edu/HSRC/).

Zhang, Q., L.C. Davis, and L.E. Erickson (1999) An experimental study of phytoremediation of methyl *tert*-butyl ether (MTBE) in groundwater. In: *Proceedings of the 1999 Conference on Hazardous Waste Research*. L.E. Erickson and M.M. Rankin, eds. Kansas State University, Manhattan, Kansas, pp. 227–242 (http://www.engg.ksu.edu/HSRC/).

Zhang, Q., L.C. Davis, and L.E. Erickson (2001) Transport of methyl *tert*-butyl ether (MTBE) through alfalfa plants. *Environ. Sci. Technol.* **35**: 725–731.

22

HYDROLOGIC FEASIBILITY ASSESSMENT AND DESIGN IN PHYTOREMEDIATION

J. W. Weaver, S. C. McCutcheon, and S. L. Hutchinson

SUMMARY OF PRACTICAL IMPLICATIONS

Several design considerations for phytoremediation are based on the capability of trees to remove contaminated water from aquifers. These capabilities depend upon the amount of water that can be withdrawn, the vertical distribution of contaminants, and flow patterns in the aquifer. For each of these, an example field site is given where methods were applied to quantify the characteristics. These include a model of forest evaporation for estimating the balance between precipitation and evapotranspiration, a simple model of recharge-driven plume diving, and the use of groundwater flow modeling for assessing various features of the groundwater flow.

ASSESSMENT AND CHARACTERIZATION OF PHYTOCONTAINMENT

To clean up a contaminated site, characterizing the type and extent of pollutants at the site and developing a conceptual remediation plan are essential. A proficient plan covers the expected effectiveness of the remedial technology, projection of the time required to achieve cleanup, and cost. The best way of assessing these factors is to use models to determine the capability of the existing or projected vegetation to contribute to the control and cleanup of the site. No other method has proven applicable and useful for projecting future scenarios, especially for selection and design.

Phytohydraulic containment (Hong *et al.* 2001) of contaminated groundwater depends on the evapotranspiration rate exceeding the flow rate of groundwater from the water table down to the lowest extent of the contaminated plume. Phytodegradation and phytotransformation require that the

Phytoremediation: Transformation and Control of Contaminants,
Edited by Steven C. McCutcheon and Jerald L. Schnoor.
ISBN 0-471-39435-1 (cloth)

contaminants be in contact with the plants, rhizosphere microorganisms, and any associated exogenous enzymes, all of which can degrade or transform pollutants. Because roots grow from the surface downward to the water table and remain in the capillary zone in most cases, the main factor influencing the effectiveness of phytoremediation is the ability of the plants to draw contaminated water to the roots. Two critical elements are the rate and amount of water withdrawn directly from the aquifer (*versus* soil water from precipitation) and the vertical distribution of contaminants. In other cases to be illustrated in this chapter, plant control of infiltration will limit leaching from contaminated soil. Thus, the local water balance is the vital basis for the assessment and design of phytoremediation to control and remove soil and shallow groundwater contaminants.

The physiologies of vegetation at a site, local hydrology and geology, and the chemical characteristics of the contaminants control the leaching and uptake of contaminated water. The capabilities to transpire large amounts of water, to put roots into the capillary or phreatic zones, and to tolerate stress from other plants and wildlife, contaminants, disease, and occasional lack of nutrients and water are the vital plant attributes for phytocontainment and treatment. The local hydrology involves meteorology (precipitation, wind speed and direction, and air temperature); interception; surface conditions that control runoff *versus* infiltration; percolation through the vadose zone; quantity of flow and direction; and change in storage in the surficial aquifer. The local soil stratigraphy, porosity and hydraulic conductivity of different layers, and hydraulic gradient are the important hydrogeologic factors. The dominant chemical characteristic is the extent of sorption and uptake.

The art of using vadose and groundwater models and measurements to assess and design phytohydraulic control and treatment is progressing rapidly. Hong *et al.* (2001) used modeling to guide the field design of a methyl *tertiary*-butyl ether (MTBE) phytoremediation application in Houston, Texas. They simulated the saturated and unsaturated zones and established that deep-planted (approximately 5 meters or 15 feet) hybrid poplar (*Populus deltoides* × *Populus nigra* 'DN-34' 'Imperial Carolina') trees could control this shallow methyl *tertiary*-butyl ether plume after full canopy development. Field studies are underway to confirm the projections. Quinn *et al.* (2001) estimated the effects of hybrid poplars (*Populus trichocarpa* × *Populus deltoides*) planted to a depth greater than 8 meters (25 feet) into a confined aquifer at Argonne National Laboratory near Chicago, Illinois. Simulations with the U.S. Geological Survey standard groundwater model MODFLOW and the associated particle tracking code indicate that a high degree of containment should be achieved after tree canopy closure. Extensive parts of the trichloroethylene contaminated confined aquifer are expected to be dewatered, or nearly so, on a seasonal basis. Eberts *et al.* (this book) and Hirsch *et al.* (this book) along with this chapter, complete a summary of the pioneering work with vadose and phreatic zone modeling to assess and design phytocontainment and treatment systems based on evapotranspiration. Landmeyer

(2001) reviews some of the measurement techniques that may be useful in phytocontainment and treatment assessments. Ferro *et al.* (2001) reviews measurement and assessment techniques (not involving modeling) applied at a petroleum hydrocarbon contaminated site in Ogden, Utah.

This chapter presents three field examples that illustrate important design considerations for the application of phytoremediation to shallow aquifer contamination. The first case presents a method for estimating the effective evapotranspiration rate for a site in Joliet, Illinois. The second case illustrates the need for vertical characterization of plumes in general by introducing a clear-cut example of a landscape feature that causes plume diving. Hydrogeologic features that cause plumes to migrate deeper into aquifers work against phytoremediation because of the increased evapotranspiration required to control the plume. The second case study also reveals that conventional monitoring approaches that use long-screened wells have a great potential for missing plumes or underestimating the length. The third case explores a proposed phytoremediation system at a site in Florida where a contaminant plume dived upon emerging from under a paved area. Methods to control the vertical location of the plume and potential application of phytoremediation are assessed using the U.S. Geological Survey modular groundwater flow model, MODFLOW (McDonald and Harbaugh 1988).

AVERAGED TRANSPORT THROUGH THE VADOSE ZONE

Joliet Army Ammunition Plant

Trinitrotoluene (TNT) and hexahydro-1,3,5-trinitro-1,3,5-triazine (RDX) plumes were found in the shallow aquifer at the Joliet Army Ammunition Plant. Figure 22-1 shows the area in a 1:7200 scale aerial photograph of the site. Of primary interest was a forested area along the banks of Prairie Creek. In this area of the site, there is a possibility either of naturally occurring phytoremediation by existing trees or of enhancing phytoremediation by planting new trees. Either option depends upon the capability of the trees to remove contaminants from the shallow aquifer. The objective of the work at the Joliet plant was to assess the potential for net withdrawal of water from the aquifer.

The photograph reveals some of the irregular nature of the forest cover in this area. A significant fraction of the forested area consists of a lower density of forest cover where lower evapotranspiration rates would be expected. A vegetation survey conducted at the site showed that the lower density forests contained a number of dead or dying trees. Thus, the role of the existing vegetation in phytoremediation of the contaminant plume depends on the influence of the forest cover and the capability to continue extracting water in the future over this section of the groundwater flow path. Because of the relatively low quality of the existing vegetation, replanting with young

Figure 22-1 Aerial photograph (1:7200) of the Joliet Army Ammunition Plant site showing Prairie Creek and the forested area overlying the contaminated aquifer. Note that the side of the photograph is oriented north and south.

trees [hybrid poplars (*Populus* spp.) or willows (*Salix* spp.)] or native prairie grasses might provide a more sustainable treatment system.

Evapotranspiration Estimates

Several equations have been developed for estimating evapotranspiration. The most applicable for this analysis is the Penman-Monteith equation

(Jensen *et al.* 1990). This equation is based on a site energy budget. The budget is based on (1) calculation of all the short- and long-wave energy available for evaporation at the plant canopy, and (2) water vapor diffusing out of the leaves, controlled by the surface or stomatal resistance, and then out into the atmosphere, controlled by aerodynamic resistance (Shuttleworth 1993). The equation for evapotranspiration ET in millimeters per day is as follows

$$ET = \frac{1}{\lambda}\left[\frac{\Delta A + \rho_a c_p \frac{D}{r_a}}{\Delta + \gamma\left(1 + \frac{r_s}{r_a}\right)}\right] \tag{22-1}$$

where λ is the latent heat of vaporization equal to $2.501 + 0.002361\,T_w$ in million joules per kilogram, in which T_w is the water temperature at the interface with the air; Δ is the gradient of the saturated vapor pressure in kilopascals per °C; A is the available energy in million joules per square meter per day; ρ_a is the density of moist air in kilograms per cubic meter equal to $3.486P_a/(275 + T_a)$, in which P_a is atmospheric pressure in kilopascals and T_a is air temperature in °C; c_p is the specific heat of moist air equal to 1.013 kilojoules per kilogram per °C; D is the vapor pressure deficit in kilopascals at a reference height; r_a is the aerodynamic resistance in seconds per meter; γ is the psychrometric constant in kilopascals per °C equal to $c_p P_a/(1000\varepsilon\lambda)$, in which ε is the ratio of the molecular weight of water vapor to that of dry air or 0.622; and r_s is the surface resistance of the land cover in seconds per meter that also includes the effect of vegetation. [See Shuttleworth (1993) for a description of the reference grass or crop method to estimate relative resistance of a different vegetative land cover.] Most of these variables are strongly related to the meteorological conditions of the site, affirming that the feasibility and success of phytocontainment is very site specific.

A practical approach (Shuttleworth 1993) for estimating evapotranspiration, based on the energy balance and aerodynamic effects, uses generalized long-term climatic data to approximate evapotranspiration. For the evaluation at the Joliet site, existing climatic data were used to estimate evapotranspiration. In the U.S., the Solar and Meteorological Observation Network, SAMSON (U.S. Department of Commerce 1993), provides hourly climate observations recorded for a 30-year period (1961 to 1990) from 239 stations located around the country. The data from SAMSON include (1) hourly precipitation (rainfall and snowfall) amounts, (2) air temperature, (3) percent cloud cover, (4) solar radiation, and (5) relative humidity. These data provide the means to estimate averaged hourly, daily, monthly, or annual evaporation amounts for locations in all parts of the U.S. For Joliet, the nearest SAMSON station is located at Chicago. The complete hourly SAMSON file for this station consists of approximately 38 megabytes of data. Practical use of these data in Equation (22-1) requires post processing to generate, in this case, monthly rainfall and climatic estimates from the 28-year-long record for the Chicago station.

Tables 22-1 and 22-2 give a summary of the processed SAMSON data for the Chicago station. These values were used in Equation (22-1) to generate the forest evapotranspiration estimates given in Table 22-3. The net extraction, defined here as the potential evapotranspiration less the rainfall, gives an estimate of the amount of water that the forested area near Prairie Creek can remove from the aquifer. For the months of December through March,

TABLE 22-1 Monthly Storm Statistics Determined from the SAMSON Database for the Chicago Station

Month	Average interarrival time (days)	Average depth (millimeters)	Average storm duration (hours)	Average number of storms
January	1.70	2	0.133	18.0
February	1.59	2	0.123	16.3
March	1.55	3	0.140	19.3
April	1.39	4	0.128	20.3
May	1.69	5	0.113	17.1
June	3.07	6	0.099	15.9
July	2.21	7	0.100	14.2
August	2.13	8	0.105	14.5
September	1.99	6	0.106	14.9
October	2.07	4	0.125	14.5
November	1.72	4	0.143	17.7
December	1.55	3	0.144	19.3

TABLE 22-2 Monthly Temperature and Relative Humidity Statistics Determined from the SAMSON Database for the Chicago Station

Month	Average temperature (°C)	Minimum temperature (°C)	Maximum temperature (°C)	Average dew point temperature (°C)	Average relative humidity
January	−5.83	−32.2	17.8	−10.19	0.72
February	−3.33	−25.0	21.7	−7.86	0.71
March	2.82	−22.2	30.6	−2.49	0.70
April	9.21	−13.3	32.8	2.17	0.65
May	15.16	−4.4	33.9	7.70	0.64
June	20.47	3.3	39.4	13.10	0.66
July	23.12	5.6	38.9	16.53	0.68
August	22.14	5.0	37.8	16.17	0.71
September	18.01	−1.1	36.1	12.10	0.71
October	11.59	−7.2	32.2	5.35	0.69
November	4.67	−17.2	24.4	−0.19	0.73
December	−2.66	−31.7	21.7	−6.74	0.76

TABLE 22-3 Monthly Estimated Evapotranspiration, Rainfall, and Net Extraction Estimates Determined from the SAMSON Database for the Chicago Station and the Forest Evapotranspiration Equation (22-1)

	Estimated evapotranspiration			Net extraction	
Month	(millimeters per day)	(millimeters)	Monthly precipitation (millimeters)	(millimeters)	(gallons per acre)
January	4.87	151	42.7		
February	6.21	174	34.7		
March	9.88	306	66.6		
April	11.34	352	92.3	259	2.8×10^5
May	12.22	379	86.4	293	3.1×10^5
June	14.80	444	95.7	348	3.7×10^5
July	13.85	429	99.0	330	3.5×10^5
August	12.27	380	118	262	2.8×10^5
September	10.42	313	90.6	222	2.4×10^5
October	8.62	267	63.1	204	2.1×10^5
November	5.75	173	77.7	95	0.4×10^5
December	5.50	171	65.2		

the net extraction is assumed to be zero because average temperatures were less than 4.4 °C (40 °F). Thus, in the warmer months of the year, the estimated evapotranspiration exceeds the amount of water available from rainfall. Net extraction ranges from 222 to 348 millimeters per month. These correspond to 2.4×10^5 to 3.7×10^5 gallons per acre per month.

The estimated amount of water evaporated and transpired in excess of the rainfall amount was assumed to be drawn from the aquifer. This analysis assumed, however, that the trees can draw water from the aquifer without hydraulic limitation. The depth to the aquifer and the low hydraulic conductivity of the aquifer materials limit the actual rate of withdrawal. Another limitation was the variable density of vegetation in the forest. Dead or dying trees reduce the amount of water extracted from the aquifer. This analysis suggested that in the absence of site-specific evapotranspiration measurements, which would quantify how the existing vegetation actually draws water from the aquifer, there is the potential for the trees to create a net withdrawal of water from the Joliet surficial aquifer during the summer months. Nevertheless, additional site investigation and performance monitoring of any phytoremediation remedy is necessary. These feasibility and application procedures require collection of site-specific climatic and hydrogeologic data to verify and adjust the forecasts from Equation (22-1). Sap-flow measurements and sampling of tissues from trees are necessary to provide verification of remediation efficacy.

VERTICAL CHARACTERIZATION OF CONTAMINANT DISTRIBUTIONS

Typical Characteristics of Surficial Aquifers

With a positive potential for withdrawal of water from an aquifer, there are at least two important additional considerations. First is the vertical location of contaminants in the aquifer. Before being able to assess the prospects for removal, accurate delineation of contamination is required. The following example describes a case where accurate vertical delineation was used to reveal, in addition to the distribution of contaminants, landscape features that were responsible for diving of the plume (Weaver *et al.* 1996). Because the contaminants move with the water, the location of aquifer recharge is also an important consideration for phytoremediation to be successful. This case indicates clearly that phytoremediation, which is dependent upon the capability of trees to remove water from the top of an aquifer, requires an understanding of the vertical distribution of the contaminants. The second consideration is the depth to which water is removed from the surficial aquifer. At each site, these recharge and discharge zones are important determinants of the groundwater flow. The importance of characterizing recharge, evapotranspiration, and discharge is illustrated in a second example where phytoremediation was considered to remediate a chlorinated solvent plume. See the section "Feasibility Investigation for Phytohydraulic Containment and Treatment."

Vertical dimensions that are small relative to the length of the flow system characterize many aquifers. In order to better visualize the stratigraphy or vertical distribution of contaminants, however, commonly drawings are vertically exaggerated, which visually exaggerates the importance of vertical variations. Aquifers and some contaminant plumes can extend over distances of kilometers at least, whereas the thickness is of the order of meters. This results in ratios of length to thickness on the order of 100:1. For analysis of the flow at water tables or in surficial aquifers, the approach taken by Dupuit and Forchheimer (Bear 1972) recognizes the two orders of magnitude difference in scale. The approach also uses the assumptions that the head in the aquifer is depth-independent (vertical potentials) and that the discharge is proportional to the slope of the water table (Darcy's equation).

Two more observations are significant. First, because the hydraulic conductivity of a confining layer is orders-of-magnitude lower than that of an aquifer, flow in the confining layer tends to be almost vertical (Charbeneau 1999). Thus, because of the typical scale, the flow in the aquifer is nearly horizontal as a consequence of vertical potentials and nearly vertical in confining units as a consequence of permeability distributions. Conceptually, the three-dimensional character of flow has been reduced to aquifers with planar flow and confining units with vertical flow. The second important observation concerns the apparent dilution of contaminant plumes that is commonly characterized by aquifer dispersivity. In tabulating dispersivity, Gelhar *et al.*

(1992) stated that vertical dispersivities in aquifers were not often measured, but that available data suggested that the values of vertical dispersivity were two orders-of-magnitude lower than longitudinal dispersivity. Therefore, the dilution of plumes over the thickness of the aquifer is likely to be minimal when compared to longitudinal dispersion. The conventional explanation for the difference in dispersivities is that sediments resulting from depositional systems have a preferential orientation—that of the water that deposited the sediments. The practical implication is that transport of contaminants perpendicular to main direction of deposition is minimized.

The conclusions drawn from these observations are that water entering the aquifer vertically is turned so that flow is nearly horizontal. This follows from the Dupuit observations. As recharge accretes along the length of the flow path, more and more water is added to the aquifer. Because of continuity and limited vertical dispersivity, this water is added to the top of the aquifer and contaminant plumes appear to "dive" or move deeper into the aquifer. Likewise, when removed from a surficial aquifer due to evapotranspiration, water is removed from the top and a contaminant plume would appear to rise back to the water table. Such flow characteristics are similar in concept to the classic work of Toth (1963) but consistent with the scaling observations of Dupuit and Forchheimer.

At some sites, unlined drainage ditches, leaking water mains and sewer pipes, irrigation, and the flow pattern in the aquifer can determine the vertical distribution of contaminants. Where recharge is the likely cause of diving, the amount of water that infiltrates into the area above the plume, and the amount that this recharge contributes to flow in the aquifer determine where and how much diving will take place.

Plume Diving: East Patchogue, New York

A gasoline release at an East Patchogue, New York, underground storage tank (*i.e.*, UST) facility created large benzene, toluene, ethylbenzene and xylene (BTEX), and methyl *tertiary*-butyl ether (MTBE) plumes. The plumes were detected because the contaminants showed up in a private water supply well, located 1200 meters (4000 feet) down gradient from the source. The well screen was about 15 meters (50 feet) below the water table, where much of the methyl *tertiary*-butyl ether mass was located. The site investigation started at this point and proceeded up gradient to identify the source.

Because of the importance of the aquifer for drinking water supply, the State of New York undertook an extensive investigation of the site that included vertical characterization of the plumes. Multilevel samplers with 15-centimeter (6-inch) screens at 1.5-meter (5-foot) intervals were used. A resulting vertical section through the plume showed that BTEX and methyl *tertiary*-butyl ether tended to dive into the aquifer with distance from the source (Figure 22-2). Furthermore, a significant amount of diving occurred as the benzene and xylene plumes passed under a gravel pit. At East Patchogue, the maximum depth of the hydrocarbon plumes reached 7.62 meters (25 feet) below the water table.

By studying the well logs and performing a detailed hydraulic characterization of the aquifer with a borehole flowmeter, vertical migration controlled by stratigraphy was ruled out because the hydraulic conductivities varied by less than a factor of two over the aquifer. Weaver *et al.* (2001) present an Internet calculator for plume diving in simple aquifer systems (www.epa.gov/athens/learn2model) that is based on a mass balance and Dupuit-Forchheimer flows in aquifers (Strack 1984). The calculator reproduces the upper bound on the contaminant distribution by accounting for enhanced recharge from the gravel pit (Figure 22-2). Because of the lack of stratigraphic evidence for preferential flow paths and the ability of the model to reproduce the observed plume diving, recharge-driven plume diving is the likely reason for the observed vertical distribution of the contaminants at East Patchague (Weaver and Haas 2001).

This example sheds light not only on how recharge pushes the plume downward, but also what happens when water discharges from aquifers. Where water comes up at discharge points, so will the contaminants: along wetlands, streams, rivers, lakes, or the ocean. The ocean is the expected

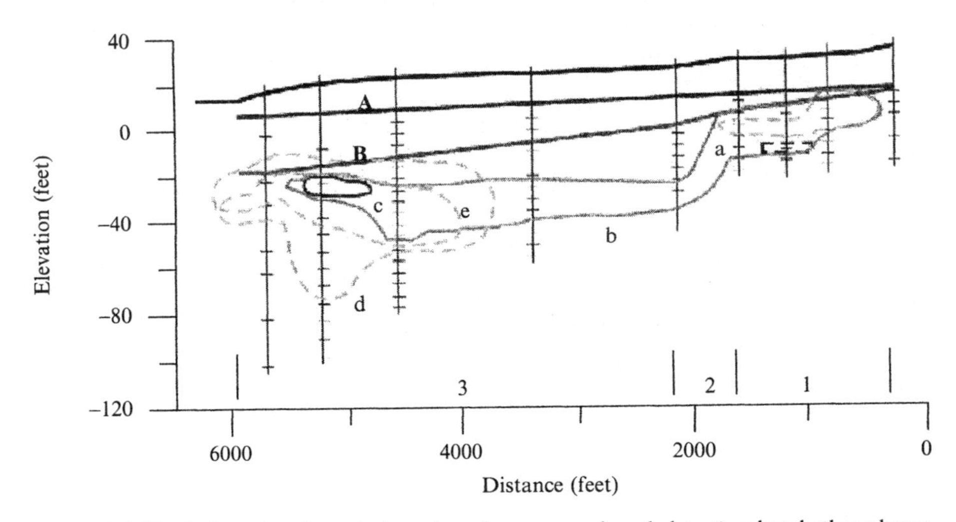

Figure 22-2 Vertical section through the xylene, benzene, and methyl *tertiary*-butyl ether plumes at East Patchogue, New York. A leaking tank was located at the right-hand edge of the drawing and the flow is to the left. The aquifer was divided into three recharge zones labeled 1, 2, and 3 below the plumes. In zones 1 and 3, the recharge was 559 millimeters per year (22 inches per year), whereas zone 2 recharge dominated by a gravel pit was 1120 millimeters per year (44 inches per year). The water table is marked "A" and the predicted top of the plumes by "B" (see www.epa.gov/athens/onsite for how the top of the plumes was derived). The plume labels are xylene: a—concentrations of 1000 micrograms per liter; benzene: b—100 micrograms per liter and c—1000 micrograms per liter; and methyl *tertiary*-butyl ether: d—100 micrograms per liter and e—1000 micrograms per liter. Note that 1 foot is equal to 0.3048 meter.

destination of the methyl *tertiary*-butyl ether plume at East Patchogue, where the groundwater discharges into Great South Bay, adjacent to the southern shore of Long Island. The groundwater and contaminants move upward as the plume approaches the discharge point at the bottom of the Bay.

Despite the low vertical dispersivity, contaminant distributions still, of course, vary over the vertical. This was seen in the East Patchogue plume where the maximum concentration at each sample location was higher than the average taken over the entire plume thickness. With conventional sampling techniques, there is effective averaging over the length of the well screens used. With no knowledge of the underlying vertical contaminant distribution, the well screen might be placed anywhere relative to the contaminant distribution. This could easily lead to well screens that sample part of the contaminant distribution and part clean water. The resulting concentrations will be between undetectable and the maximum for the location (Weaver *et al.* 2001; see the average borehole concentration calculator that is a part of the OnSite system at www.epa.gov/athens/onsite for a specific example). Therefore, the same hydrologic characterization of a plume determines the feasibility for phytohydraulic containment and treatment, and also cost effectively dictates when and where the cleanup should be monitored.

FEASIBILITY INVESTIGATION FOR PHYTOHYDRAULIC CONTAINMENT AND TREATMENT

The Orlando Naval Training Center is one of thousands of U.S. Department of Defense facilities contaminated with chlorinated solvents. The plume is shallow and within the rooting depth of proven phytoremediation approaches using tree planting. The Training Center has been closed and the Navy is cleaning up and preparing the site to be turned over to the City of Orlando. As such the Naval Southern Command in Charleston, South Carolina has pilot tested a number of innovative remediation methods. One of the methods investigated was the use of phytoremediation to cost effectively clean up the plume. For this purpose, the EPA National Exposure Research Laboratory in Athens, Georgia was requested by the Navy to establish guidelines for determining the feasibility of applying phytocontainment and treatment and to apply those guidelines at the Orlando Naval Training Center.

The initial phase involved (1) designating appropriate characterization data for an assessment, (2) selecting the appropriate flow and transport models from a multimedia context, (3) vegetation surveys for indigenous phytoremediation potential (conducted by Navy forester), (4) pioneering sap-flow measurements of existing trees and extrapolation to stand-level estimates of existing evapotranspiration for assessments, (5) projecting improved evapotranspiration from replanting, and (6) assessing chlorinated solvent degradation rates in various media by vegetation and microorganisms. This section reports on the selection and application of appropriate models for assessment of the feasibility of phytocontainment and treatment. Existing stand assessments of evapotranspiration and projections were conducted by the U.S. Forest Service Coweeta Hydrologic Laboratory (Vose and Elliot 2000) based on sap-flow measurements and extrapolation and projection methods as described in Vose *et al.* (this book). Nzengung (2000) collected vegetation from the site, soil and aquifer cores, and wetland and lake sediments to determine the level of ongoing attenuation. These laboratory incubations provide expected rates and kinetics of degradation and transformation of tetrachloroethylene and trichloroethylene that contaminate the site.

This section covers all other important components of a feasibility investigation—selecting and setting up a groundwater model, and parametrizing that model to simulate the effects of existing and projected vegetation, and runoff and infiltration control. Depth of the plume capture is used as the criteria for projecting the likelihood of using vegetation to capture and treat the plume. As a result, these projections are conservative because the transformation and degradation resulting from contaminant reactions in lake and wetland sediments and from plant enzymatic processes have not been fully assessed to date.

The Orlando Naval Training Center in Florida has a plume of chlorinated solvents that has resulted in detection of significant amounts of vinyl chloride in adjacent Lake Druid shown in Figure 22-3. Lake Druid is also bordered by

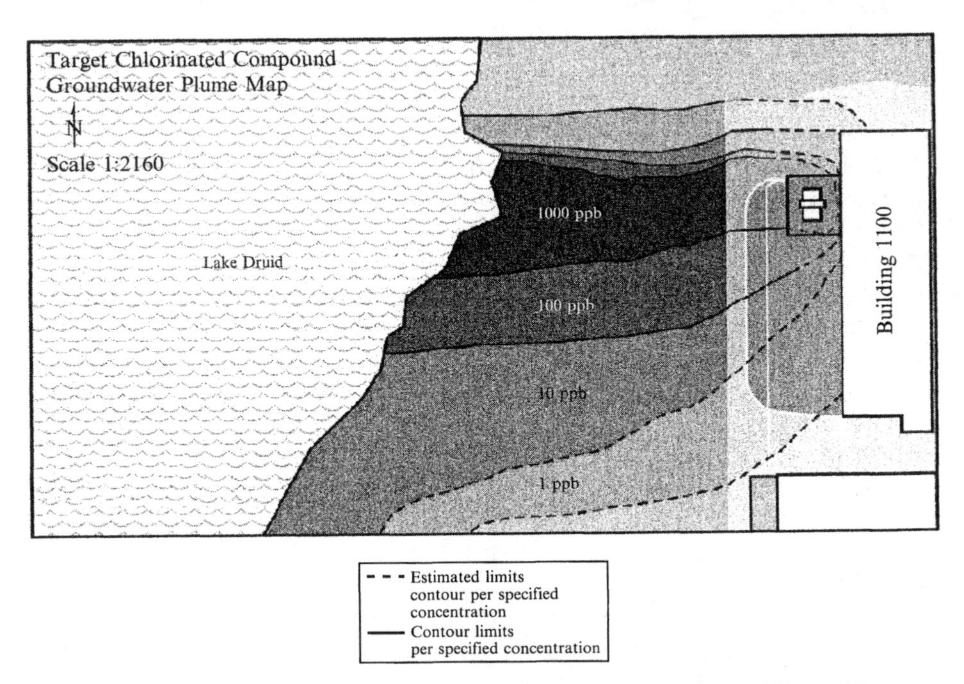

Figure 22-3 Site map of the Orlando Naval Training Center showing the Building 1100 source area and the shoreline of Lake Druid. The light shading surrounding Building 1100 delineates impervious parking lots and roads. Note that ppb is the concentration in micrograms per liter.

a number of residences and represents a valuable esthetic and recreational asset to adjoining residents.

Site Assessment and Apparent Plume Diving

A groundwater plume of chlorinated ethenes originated at the site of a laundry and dry cleaning facility. Prior to the closing in 1994, hazardous materials and water contaminated with chlorinated solvents were released to the groundwater from the surge tank beside Building 1100, and from spills of tetrachloroethylene and trichloroethylene in the facility. The contaminated groundwater plume flowed from the site of the former laundry underneath and into the nearby Lake Druid. The site plan shows that the source area is contained within an area of very high impervious cover. This consists of the buildings, paved driveways, and parking lots. Immediately down gradient from the parking lot is an unlined drainage ditch. The ditch that receives runoff from the parking lot and buildings on the site runs perpendicular to and crosses the plume that emerges from beneath the pavement. Beyond a narrow grassy area lies a densely wooded area that borders Lake Druid, where the plume discharges. Located nearly over the top of the plume is a creek that feeds the lake. Much of the wooded area is a seepage wetland.

Figure 22-4 shows the distribution of the chlorinated ethene contaminants taken over a vertical section of the aquifer spanning the distance from the former laundry to the lakeshore. The data were obtained using a direct push probe operated on a 0.6-meter (2-foot) interval. These results show that the

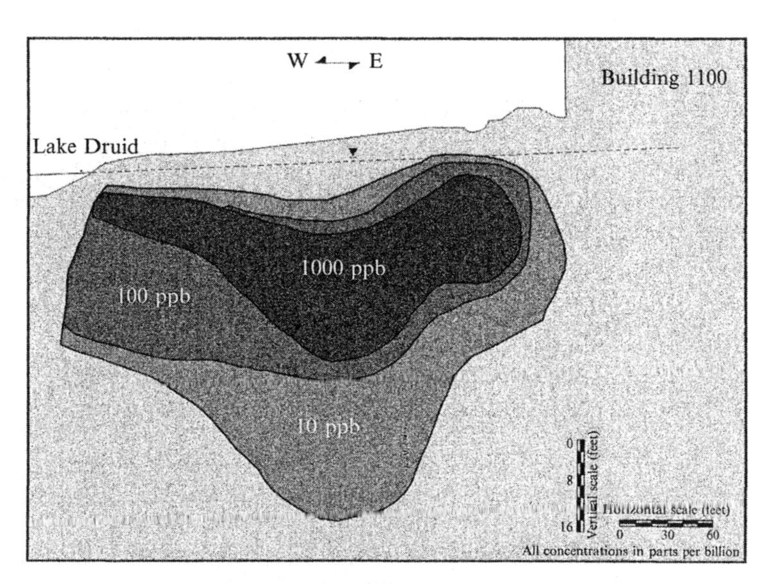

Figure 22-4 Vertical section of chlorinated ethene plume at Orlando Naval Training Center. Lake Druid is on the left and the contaminant source on the right. Note that 1 foot = 0.3048 meter and ppb is the concentration in micrograms per liter.

concentrations are low in the upper part of the aquifer, once the plume is down gradient of the ditch. In these locations the depth of uncontaminated water having undetectable amounts of chlorinated ethenes ranges up to 2.4 meters (7 feet) below the water table. The horizontal distance spanned by the diving portion of the plume is approximately 67 meters (220 feet). Thus the plume moves deeper into the aquifer upon emergence from the paved area and upon passing under the unlined ditch.

Model Application for Orlando Naval Training Center

The effects of recharge and discharge on the groundwater flow at Building 1100 were simulated with two models. First, the simple screening model for plume diving was applied (Weaver *et al.* 2001). The model assumes that flow occurs between two points with fixed heads, the aquifer is a single uniform layer, and the flow system can be described by a set of one-dimensional segments. These assumptions are satisfied for a screening analysis of the Orlando Naval Training Center plume, although other details of the flow system require the use of a numerical model. This consideration led to the application of MODFLOW with particle tracking for a more detailed evaluation.

The screening-level plume diving calculation allows determination of the vertical displacement of the plume generated by the supply and withdrawal of water from the aquifer. The simple model was used to assess the amount of diving that would occur as the plume passed under the unlined ditch. With an estimated average annual rainfall of 1400 milliliters (55 inches) at Orlando, the amount of water that could run off into the ditch is very large. Estimates based on the area of the pavement surrounding the source ranged up to the equivalent of 38 000 millimeters per year (1500 inches per year) concentrated in the ditch. The actual amount of infiltration through the ditch was not determined for this study. However, the leakage through the ditch would be a function of the transient depth of water, soil properties, soil antecedent moisture content, and the vegetation in the ditch. The online calculator (www.epa.gov/athens/onsite) was used to determine how much water would be required to infiltrate through the ditch to cause the observed diving of the plume. The purpose of this exercise was to determine if the observed plume diving could be attributed to reasonable recharge amounts ranging from the rainfall rate of 1400 milliliters (55 inches) to the estimated runoff amount of 38 000 millimeters per year (1500 inches per year). Figure 22-5 shows the results from infiltration amounts of 1400 milliliters per year (55 inches per year) and 2800 millimeters per year (110 inches per year). Thus, if the ditch recharges double the average annual precipitation to the aquifer from the impervious area, this amount can account for the observed plume diving. This amount of water 2800 millimeters per year (110 inches per year) is less than 10 percent of the maximum estimated flow in the ditch and thus forms a plausible explanation for the vertical distribution of the contamination observed at the site. In the upper portion of the aquifer, there is no evidence for stratigraphic

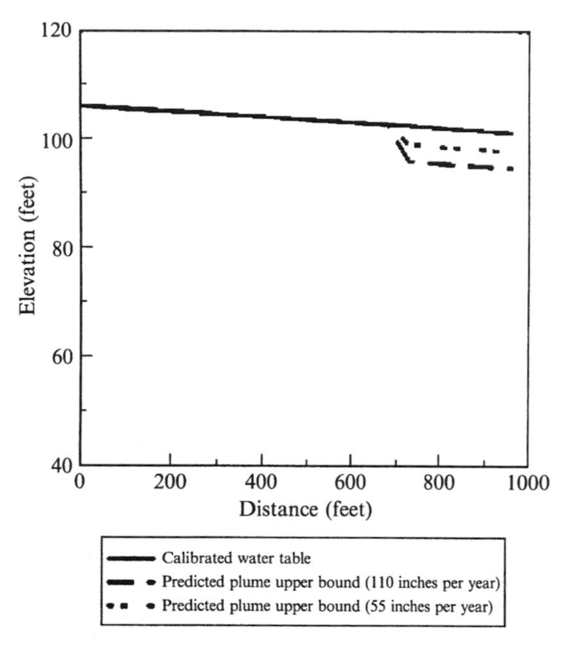

Figure 22-5 Predicted plume diving using the OnSite plume diving calculator (www.epa.gov/athens/onsite) and two estimates of infiltration through the unlined ditch. Note that 1 foot is 0.3048 meter and 1 inch is 25.4 millimeters.

preferential flow paths and density-driven plume diving is not evident because the plume dives only after emerging from below the pavement.

Numerical Modeling

The U.S. Geological Survey modular groundwater flow (MODFLOW) code with particle tracking was used to perform more detailed simulations of the site. These simulations provided a more precise determination of the vertical and horizontal distribution of flow and a tool with which to evaluate the various design alternatives. Some of the parameters used in the model were taken from a previous modeling study (Halford 1998) that was not focused on the plume and Lake Druid. Halford (1998) calibrated the model to regional flow conditions. Calibrated parameters for hydraulic conductivity, porosity, and aquifer and layer thicknesses from Halford (1998) were used in the current model. Boundary conditions were taken to match observed water levels collected around the site and consistent with the large-scale flows simulated by Halford (1998). The need for site-scale water levels constrained the modeled flow domain to the area immediately surrounding Lake Druid. For an adequate assessment of the potential effects of vegetation, the area around Lake Druid and Building 1100 was modeled in more detail, both horizontally and vertically. To provide sufficient vertical delineation of the plume, the model simulations were based on 12 layers to adequately resolve the lake bathymetry. The refinements provided sufficient resolution to explore the factors that control the vertical movement of the contaminants moving from Building 1100 to Lake Druid.

The amount of transpiration from either the existing vegetation or by any proposed plantings was difficult to estimate. A generalized estimate of 1300 millimeters per year (51 inches per year) was available from the Florida Agricultural Extension Service for central Florida. An estimate of 500 millimeters per year (20 inches per year) was determined for nearby orange groves (Summer 1996). Finally, a site-specific estimate of 200 to 300 millimeters per year (8 to 12 inches per year) was made by Vose and Elliot (2000) for the existing mature trees over the down gradient end of the plume. The estimate was made by measuring the sap-flow rate in nine trees in spring, summer, and fall. The estimate included only mature trees and did not account for the understory vegetation. Vose and Elliot (2000) believed that this might not be too severe a limitation because the understory may be cycling shallow soil moisture, and they cite studies that show that understory transpiration can offset reduced overstory transpiration in open stands. The potential evapotranspiration at full stocking of the stand was 600 to 900 millimeters per year (24 to 35 inches per year), a higher value because of the relatively sparse distribution of the existing mature trees (also see Vose *et al.* this book). With the estimated precipitation of 1400 millimeters per year (55 inches per year) in central Florida, the estimated evapotranspiration ranges from 14 to 93 percent of that value. The lower values reflected the site-specific measurements.

These values suggested that in any case there was no strong driving force for removal of water from the aquifer by the vegetation. The generalized or regional estimate of evapotranspiration of 1300 millimeters per year (51 inches per year) came close to balancing precipitation. With the site-specific estimate of 200 to 300 millimeters per year (8 to 12 inches per year) accounting for some of the precipitation, the remainder would evaporate, contribute to soil moisture, or run off.

The MODFLOW results reproduced the presumed flow pattern in the cross section of the aquifer (Figure 22-6). The simulated downward flow just down gradient of the paved area accurately reflects the observed plume (Figure 22-4). The 12-layer representation of the aquifer allows adequate representation of the bathymetry of the lake in order to properly resolve the pattern of groundwater flow to the bottom of the lake. The discharge pattern was consistent with measurements of the attenuated plume emerging into the lake (data not shown in this chapter). The significance of these simulations was that the recharge and discharge patterns of the aquifer alone, can explain the vertical distribution of the contaminants (Figure 22-4). Furthermore, the use of 12 thin layers to represent the surficial aquifer is unusual (see for example Halford 1998) but necessary to represent a plume in this hydrogeologic setting.

These observations lead to the following strategies for phytoremediation at the site. First, the prospects for the trees transpiring sufficient water to pull

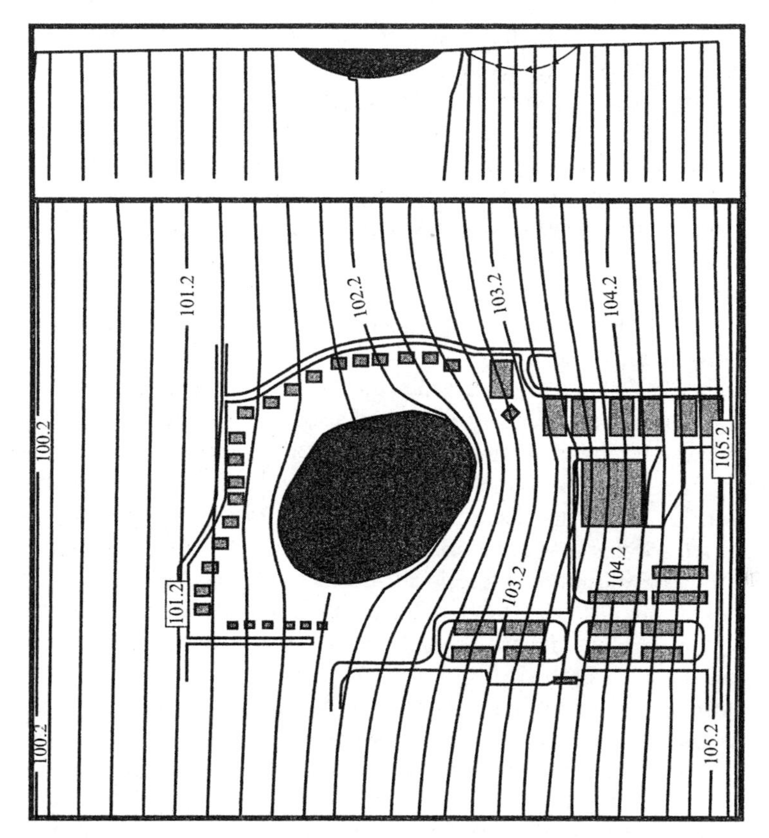

Figure 22-6 MODFLOW simulation of part of the Orlando, Florida U.S. Naval Training Center showing the effects of infiltration through the unlined ditch and upwelling at the shoreline. The upper panel represents a vertical section through the source and Lake Druid. The nearly vertical lines represent constant head contours, which correspond to the contours (labeled in feet above mean sea level; note that 1 foot is 0.3048 meter) on the plan view shown in the lower panel. The arrows in the upper panel represent a flow path from the unlined ditch to a discharge point along the shoreline. The scale is approximately 1:8280. The sides of the lower panel are vertically oriented north (top) to south (bottom).

the contaminant plume up to the root zone for treatment is too unlikely for phytoremediation alone to be considered as an adequate cleanup remedy. This follows from the fact that the existing vegetation, extracting at least of 200 millimeters per year (8 inches per year), has not contained the existing contaminant plume.

Because the transpiration of trees alone cannot contain the plume, clearly the combined effects of replanted trees and natural attenuation in the seepage wetland and lake sediments needed to be investigated next. Nzengung (2000) found significant potential for microbial and enzymatic degradation of chlorinated solvents in the wetland and lake sediments, thus requiring a refined hydraulic analysis. The refined analysis established that enhanced evapotranspiration would pull some of the plume upward into the seepage wetland and probably achieve better attenuation. This next investigation established more contact with a wider range of shallow rooted aquatic and wetland plants at the lakeshore without threatening the health of the existing wetland (Medina *et al.* this book). Nevertheless, the quantification of attenuation rates (Nzengung 2000) was deemed too uncertain to forecast whether this enhanced natural attenuation would be sufficient without a field pilot investigation. These pilot investigations are currently (2003) underway. Because forecasting was not possible for the enhanced natural attenuation option, other supplemental design scenarios were also investigated. These included diversion of the surface runoff (from the impervious parking lots and roofs) from the unlined ditch to alleviate some plume diving, and planting more trees through the parking lot cover and over the source to contain the plume. Because of the ecological sensitivity of the seepage wetlands, an investigation of alternative planting was useful to avoid any temporary disruption of the wetland with supplemental planting if possible, especially to avoid planting alien species that are of regulatory concern to the State of Florida.

Simulations of eliminating the localized recharge to the aquifer at the edge of the parking lot established that most but not all of the downward movement of plume below the water table would be eliminated. However, some infiltration through the down gradient tree and shrub cover would continue because rainfall in this humid area exceeds potential evapotranspiration (Vose *et al.* this book). In addition, this sandy aquifer has high conductivity that limits contact of the plume with roots, wetland soils, and lake sediments, even under the best of conditions.

Finally, if trees were planted over the source zone, the simulations demonstrate that transpiration could remove water from the upper part of the aquifer despite the hydraulic conductivity. With the proper selection of trees that transform and degrade chlorinated solvents, some of the contaminated water could be treated, leaving less for natural attenuation. This could have the effect of removing contaminants from the aquifer, before the plume enters wetland soil and lake sediments. The MODFLOW results show that upward gradients would be established through the upper 7 meters (23 feet) of the aquifer if the evapotranspiration rate was 800 millimeters per year and through the upper 10 meters (30 feet) of the aquifer if the evapotranspiration was 1300 millimeters per year. These amounts exceed the site specific measured values of 200 to 300 millimeters per year (Vose and Elliot 1999) because the observed values were net values that integrated the effect of the trees working against infiltration. The higher values used in the model were selected because recharge was assumed to be zero due to the presence of the pavement. Forcing the trees to draw water from the aquifer exclusively over the source area maximized the impact on the contaminated portion of the aquifer by maximizing the withdrawal of water from the source zone. Overall, runoff diversion, tree planting over the source, and enhanced natural attenuation in the seepage wetland and nearer the lakeshore have the potential to manage contamination at this site but the uncertainty in enhanced attenuation requires a field pilot test to be sure this is the best remedy for this particular plume.

FEASIBILITY AND DESIGN IMPLICATIONS

Design of phytoremediation systems requires delineation of contaminant plumes and the parameters governing flow and transport in shallow aquifers. Because trees remove water and contaminants from the aquifer, estimates of evapotranspiration are needed to design phytocontainment systems. For the Joliet Army Ammunition Plant, the estimates were based on a forest evaporation model (Shuttleworth 1993) using the Penman-Monteith equation (Jensen *et al.* 1990) with climate data from the U.S. SAMSON database. From these readily available data, the feasibility of phytoremediation can be assessed.

Vertical delineation of contaminant plumes is best accomplished with short-screened or point sampling that eliminates (1) mixing of clean and contaminated water caused by screen placement that is only partly contained within the plume, and (2) averaging of concentration over long well screens. Both of these sampling effects or artifacts cause contaminant concentrations to be underestimated. Further, if the horizontal and vertical boundaries of the plume are not accurately defined, then transport processes are not correctly determined and phytoremediation potential can not be accurately assessed.

Sampling at the Orlando Naval Training Center revealed that the chlorinated solvent plume was depressed once the plume emerged from below the paved parking lot located adjacent to the source. Because of the presence of an unlined ditch adjacent to the parking lot, the plume dived due to localized recharge from the bottom of the unlined ditch. This plume characteristic was quantified with two models. These models included a simplified analytical solution of flow and a three dimensional numerical model of the site (MODFLOW) to simulate groundwater flow. Like the field data, the numerical model needed to be sufficiently resolved in the vertical to simulate the vertical distribution of flow and the bathymetry of the lake. Thus, a greater-than-usual number of layers were required for this phytocontainment assessment.

Both of these models showed that with reasonable rates of recharge and evapotranspiration and using some regional scale MODFLOW parameters, the observed plume diving could be accounted for just from the hydraulics of flow in the aquifer. Thus, the application of phytoremediation to this site was established as critically dependent upon the localized pattern of flow in the aquifer. The analysis of the Naval Training Center site shows that the localized plume characteristics must be known to plan the locations for tree planting or design other remedial options for the site.

At the Naval Training Center, observations and simulations show that the contaminated water discharges to Lake Druid. Thus, the shoreline vegetation has the potential to intercept and treat the contaminated aquifer water. This interception of the groundwater flow is dependent upon evapotranspiration and all other characteristics of flow in the aquifer. Other elements of a remedy at this site include plume control by eliminating runoff from the parking lot to the unlined ditch. This would have the effect of preventing some of the downward migration of the contaminants. The depth of contamination at the source, however, is approximately 10 meters (34 feet). The depth of contamination is due to the location of the sump and the denser-than-water nature of the contaminants. If trees were planted and successfully grown at the source, then some of the contaminated water would migrate upward. The effectiveness of this source control depends upon the evapotranspiration rate that could be achieved by the trees and by the continued prevention of recharge in the parking lot.

Overall, runoff diversion, source phytocontainment, and phyto-enhancement of natural attenuation may be sufficient to remediate this site over a long period (until the nonaqueous phase is completely dissolved). However, due to the uncertainty in extrapolating treatment and attinuation kinetics, field pilot studies are necessary. These pilot investigations are underway at the Orlando site and other source removal remedies have been tested.

Acknowledgments

Dan Reible and John Quinn provided insightful and through peer reviews used in the evaluation of this chapter. Jerry Schnoor accepted a revised version of the chapter for this book. Support for the work came from the U.S. Environmental Protection Agency Office of Research and Development Contaminated Sites Program and a cooperative interagency research agreement with the U.S. Navy Southern Command. Cliff Casey of the U.S. Navy was instrumental in conceiving of and pursuing the work at the Orlando Naval Training Center. This chapter has been reviewed in accordance with the U.S. Environmental Protection Agency peer and administrative review policies and approved for publication. Mention of trade names or commercial products does not constitute endorsement or recommendation for use.

REFERENCES

Bear, J. (1972) *Dynamics of Fluids in Porous Media*. Dover, New York.

Charbeneau, R.J. (1999) *Groundwater Hydraulics and Pollutant Transport*. Prentice Hall, Upper Saddle River, New Jersey, pp. 593.

Ferro, A., J. Chard, R. Kjelgren, B. Chard, D. Turner, and T. Montague (2001) Groundwater capture using hybrid poplar trees: evaluation of a system in Ogden, Utah. *Inter. J. Phytoremed* **3**(1): 87–104.

Gelhar, L.W., C. Welty, and K.R. Rehfeldt (1992) A critical review of data on field scale dispersion in aquifers. *Water Resour. Res.* **28**(7): 1955–1974.

Halford, K.J. (1998) Assessment of the potential effects of phytoremediation on ground-water flow around area C at Orlando Naval Training Center, Florida. U.S. Geological Survey Water Investigations Report 98–4110.

Hong, M.S., W.F. Farmayan, I.J. Dortch, C.Y. Chiang, S.K. McMillan, and J.L. Schnoor (2001) Phytoremediation of MTBE from a groundwater plume. *Environ. Sci. Technol.* **35**(6): 1231–1239.

Jensen, M.E., R.D. Burman, R.G. Allen, eds. (1990) *Evapotranspiration and Irrigation Water Requirements*. Manual No. 70. American Society of Civil Engineers, New York.

Landmeyer, J.E. (2001) Monitoring the effect of poplar trees on petroleum-hydrocarbon and chlorinated-solvent contaminated ground water. *Int. J. Phytoremed.* **3**(1): 61–85.

McDonald, M.G. and A.W. Harbaugh (1988) A modular three-dimensional finite-difference ground-water flow model. U.S. Geological Survey Techniques of Water Resources Investigations Book 6, Chapter A1.

Nzengung, V.A. (2000) Data on phytotransformation and natural attenuation processes for the degradation of perchloroethene (PCE) and trichloroethene (TCE): data specific for the Orlando Naval Training Center (NTC), Orlando, Florida, task 1 to 4 (separate reports). Contract report. Contract #: UKC019 (QT-RT-99-002122), U.S. Environmental Protection Agency National Exposure Research Laboratory, Athens, Georgia.

Pennington, J.C., D.W. Harrelson, M. Zakikhani, D. Gunnison, J. Clarke, C.J. McGrath, H. Fredrickson, J.H. May, E. Perkins, C.A. Hayes, and D. Ringelberg (1998) Feasibility of using natural attenuation as a remedial alternative for explosives-contaminated groundwater at site L1. Joliet Army Ammunition Plant, Joliet, Illinois. U.S. Army Corps of Engineers, Waterways Experiment Station Technical Report EL-98-8, Vicksburg, Mississippi.

Quinn, J., M.C. Negri, R.R. Hinchman, L.P. Moos, and J.B. Wozinak (2001) Predicting the effect of deep-rooted hybrid poplars on the groundwater flow system at a large-scale phytoremediation site. *Int. J. Phytoremed.* **3**(1): 41–60.

Shuttleworth, J. (1993) Evaporation. In: *Handbook of Hydrology*. D.R. Maidment, ed. McGraw-Hill, New York, pp. 4.1–4.53.

Strack (1984) Three-dimensional streamlines in Dupuit-Forcheimer flow. *Water Resour. Res.* **20**(7): 812–822.

Summer, D.M. (1996) Evapotranspiration from successional vegetation in a deforested area of the Lake Wales Ridge, Florida, U.S. Geological Survey Water-Resources Investigations Report 96-4244, 37 pp.

Toth, J., (1963) A theoretical analysis of groundwater flow in small drainage basins. *J. Geophys. Res.* **68**: 4375–4387.

U.S. Department of Commerce (1993) Solar and Meterological Surface Observation Network 1961–1990. National Climatic Data Center (http://www.ncdc.noaa.gov).

Vogel, T., K. Huang, R. Zhang, M. Th. van Genuchten (1996) The HYDRUS code for simulating one-dimensional water flow, solute transport, and heat movement in variably-saturated media. U.S. Salinity Laboratory, Agricultural Research Service, U.S. Department of Agriculture. Research Report 140, Riverside, California.

Vose, J.M. and K.J. Elliot (1999) Preliminary estimates of transpiration at the Naval Training Center, Orlando, Florida: results from phase I. Report by the U.S. Department of Agriculture Forest Service Coweta Hydrologic Laboratory, Otto, North Carolina.

Weaver, J.W., C.B. Sosik, and J.W. Washington (2001) OnSite: the online site assessment tool. U.S. Environmental Protection Agency National Exposure Research Laboratory. Athens, Georgia. (www.epa.gov/athens/onsite) (In review).

Weaver, J.W., J.T. Haas, and J.T. Wilson (1996) Analysis of the gasoline spill at East Patchogue, New York. In: *Procedings of Non-Aqueous Phase Liquids (NAPLs) in Sub-surface Environment: Assessment and Remeidation.* L. Reddi, ed. American Society of Civil Engineers, Washington, D.C., November 12–14, pp. 707–718.

23

WASTE MANAGEMENT USING TREES: WASTEWATER, LEACHATE, AND GROUNDWATER IRRIGATION

J. L. Jordahl, M. F. Madison, J. K. Smesrud, H. M. Emond, and M. Q. Motte

SUMMARY OF PRACTICAL IMPLICATIONS

Phytoirrigation provides a relatively inexpensive means of moving impaired water to a planted area or forest for treatment, greatly expanding the ways in which phytoremediation can be used. Irrigation systems can be used to apply water at the land surface or below to meet the requirements of treatment and for regulatory and public acceptance. Irrigation-system designs based on trees are particularly advantageous because of the high water use, deep rooting, and low operations and maintenance costs of tree systems. This chapter introduces the rationale for using trees with irrigation to manage contaminated water, identifies key limitations, describes example projects, and provides a general design guidance.

GLOSSARY

Contaminated water: Any water requiring additional treatment for the intended use, including municipal or industrial wastewater, reclaimed water, landfill leachate, or contaminated groundwater. Contaminated water is the global term used in this chapter to refer to any water that is not potable and that may be treated for other uses with a tree-based phytoirrigation system.

Phytoremediation: Transformation and Control of Contaminants,
Edited by Steven C. McCutcheon and Jerald L. Schnoor.
ISBN 0-471-39435-1 (cloth)

Pan evaporation: The loss of water by evaporation from the open water surface of a pan under very specific conditions.

Phytoirrigation: The use of irrigation and trees to beneficially reuse, reclaim, or treat wastewater, reclaimed water, landfill leachate, or contaminated groundwater. Phytoirrigation could be used with many different plant species, but for the purposes of this chapter, the term refers only to tree-based applications.

TREES AND PHYTOREMEDIATION

Tree-based treatment systems can be used to manage wastewater and contaminated groundwater economically while meeting environmental protection standards and providing other benefits. Trees live for many years and have relatively low maintenance requirements after establishment compared to mechanical treatment systems. Trees can significantly improve site esthetics and provide a marketable product at harvest. Some trees can root deeply and take up water and nutrients at high rates, and some are capable of transforming or mineralizing certain organic compounds.

Trees are often the plants of choice for phytoremediation because of growth characteristics, physiology, and potential secondary economic benefits. Examples of species being used or investigated are listed in Table 23-1.

Remediation systems must often operate for many years to be successful. Unlike many herbaceous species, most woody species provide remedial benefits for decades without replanting. Hybrid poplars (*Populus* spp.),

TABLE 23-1 Tree Species Used in Phytoremediation

Species	Water source or contaminant	Reference
poplar (*Populus* spp.)	Municipal wastewater, landfill leachate, chlorinated solvents, methyl *tertiary*-butyl ether (MTBE)	Madison *et al.* (2000), Newman *et al.* (1999)
willow (*Salix* spp.)	Municipal wastewater and landfill leachate	Hasselgren (1999a, 1999b, 1999c)
Eucalyptus spp. and salt cedar (*Tamarix* spp.)	Hydraulic control, arsenic	Tossell *et al.* (1998)
red maple (*Acer rubrum*)	Landfill leachate	Shrive *et al.* (1994)
Monterey pine (*Pinus radiata* D. Don)	Municipal wastewater	Barton *et al.* (1999)
red mulberry (*Morus rubra*)	Polycyclic aromatic hydrocarbons (PAHs)	Hegde and Fletcher (1996)
milo (*Thespesia populnea*) and kiawe (*Prosopis pallida*)	Petroleum hydrocarbons	Sun *et al.* (2000)

capable of extremely fast growth rates, are relatively long-lived. Native cottonwood trees, such as *Populus trichocarpa*, may live more than 200 years (DeBell 1990). Moreover, once trees are established, maintenance requirements (such as mowing and applying pesticides and herbicides) are simpler and less costly than for mechanical treatment systems. A secondary benefit of trees is an economic return based on a harvestable product. For treatment of continuous sources of contaminated water, harvesting trees for paper, plywood, solid wood products, or biofuels can provide an economic return that can, at least partially, defray the costs of operation (CH2M HILL 2000). In many European applications, wood products are the primary reason for using trees, and waste utilization is secondary. In Sweden, a bioenergy program based on willows (*Salix* spp.) was initiated in the 1970s after the general oil crisis. Almost 1 percent of the Swedish arable land (approximately 18 000 hectares) supports biofuel production, with specifically selected willow (*Salix* spp.) clones (Hasselgren 1999a).

The visual structure and diversity that trees provide can also significantly improve site esthetics, an issue of considerable importance to many industries and utilities. Furthermore, many contaminated sites, such as brownfields, are located in highly visible urban areas. Trees screen these distressed sites.

Among the many tree species being used for phytoremediation, Salicaceae (*Populus* spp. and *Salix* spp.) are often the most appropriate. Trees in this family are phreatophytes, meaning that these trees are capable of rooting deeply to transpire groundwater. Salicaceae are found on several continents over a wide range of climatic conditions, are easy to establish from stem cuttings, grow rapidly, and produce a tremendous amount of biomass. Hybrid poplars (*Populus* spp.) are the fastest growing trees in North America (Heilman *et al.* 1995). These trees take up nutrients and transpire at comparatively high rates (Pallardy and Kozlowski 1981, Heilman *et al.* 1996), minimizing land area requirements. Poplars (*Populus* spp.) appears to have degradative capabilities for chlorinated solvents (Newman *et al.* 1997, Shang *et al.* this book). Because knowledge of Salicaceae genetics and culture is extensive, with literally hundreds of hybrid clones available, designers can select varieties with specific characteristics best matched to climatic, edaphic (related to the soil), and treatment requirements.

APPLICATIONS

The production of trees with contaminated irrigation water, or "phytoirrigation," represents one way to transform waste into a useful resource at a reasonable cost. Water application methods for phytoirrigation include drip, micro-spray, sprinkler, and furrow irrigation. If carefully managed, phytoirrigation may be possible even on steeply sloping land without adverse impacts on surface water quality (Speir *et al.* 1999). Hydraulic loading of contaminated water may be quite different from routine agricultural applications if specific constituents rather than the agronomic water uptake rate dictate loading.

Wastewater Reuse

Municipal and industrial wastewaters are being successfully reused to grow Poplars (*Populus* spp.) and other trees (see Table 23-1) for fuel, fiber, and construction materials. This section provides examples of how wastewater has been applied to poplar (*Populus* spp.), willow (*Salix* spp.), and pine (*Pinus radiata*) trees.

Municipal Wastewater Reuse at Woodburn, Oregon

The municipal wastewater reuse system operated by the City of Woodburn Wastewater Treatment Plant in Oregon is the first in the U.S. The 34-hectare hybrid poplar (*Populus* spp.) plantation was designed and installed as the first phase of a full-scale system to beneficially reuse wastewater treatment plant effluent (CH2M HILL 1999a, 1999b, Madison *et al.* 2000). The City chose phytoirrigation of the treated wastewater as a cost-effective and environmentally attractive method of reducing summer discharges to the Pudding River. As an important secondary benefit, the site is designed to grow large-diameter trees that will be sold for wood products to partially defray installation and operating costs.

The high rate of water and nutrient uptake by hybrid poplars (*Populus* spp.) in Oregon in June, July, August, and September coincides with the period of restricted stream discharge. During July and August, the discharge permit issued to the City of Woodburn restricts ammonia concentrations and the temperature of the treated water based on a total maximum daily load (TMDL) assessment. As a result, the Woodburn treatment plant pumps 5034 cubic meters per day to the poplar (*Populus* spp.) plantation instead of discharging to the Pudding River. In the final development phase, the phytoirrigation flow is expected to increase to 11 400 cubic meters per day on 120 hectares.

Before the 34-hectare system was installed in 1999, a 3-hectare demonstration site was monitored for 4 years (1995 to 1998). Loading rates for both the 3-hectare and 34-hectare plantations are shown in Table 23-2. Nutrient concentrations in the wastewater applied over the period 1995 to 1999 ranged from 3.8 to 18.8 milligrams of total phosphorus (P) per liter, 4.33 to 17.5 milligrams of NH_3-N per liter, and 5.8 to 21.6 milligrams of NO_3+NO_2-N per liter (Monthly City of Woodburn Wastewater Treatment Plant Monitoring Report for 1995 through 1999). In 2002, the third year of growth, the design hydraulic loading for the 34-hectare site was scheduled to be 1750 millimeters, and the design nitrogen-loading rate was targeted to be 292 kilograms of N per hectare available for uptake by the trees. The treated municipal wastewater is applied July through September (CH2M HILL 1999b).

TABLE 23-2 Annual Volume of Wastewater and Nutrients Applied at Woodburn, Oregon for 3-Hectare Demonstration and 34-Hectare Full Scale System (Phase I)

Parameter	1995	1996	1997	1998	1999[a]	2000
Total irrigation (millimeters per unit area)	762	889	533	632	271	508
Total nitrogen (kilograms of N per hectare)	67	213	120	161	57	247
Total phosphorus (kilograms of P per hectare)	22	36	30	40	16	No data

[a]Full scale application beginning in 1999.

Swedish Biomass Production Using Wastewater

A 5-year pilot study on municipal secondary effluent irrigation of willow (*Salix* spp.) plantations in Sweden was completed in 1997. Hasselgren (1999b) concluded that the three commercial willow (*Salix viminalis*) clones (ORM, ULV, and RAPP) tested showed the highest annual stem growth, 4 to 12 tons of dry matter per hectare, using an average wastewater application rate of 6 millimeters per day during the growth period May to October. Tree growth was three to five times higher compared to control stands without effluent irrigation. Annual nutrient application rates, using six fixed wastewater application rates between 2 and 12 millimeters per day during the 5-year period, averaged 52 to 313 kilograms of N per hectare, 9 to 53 kilograms of P per hectare, and 56 to 334 kilograms of K per hectare. Average annual nutrient uptake in the fully established willows (*Salix* spp.) was 45 to 57 kilograms of N per hectare, 9 to 10 kilograms of P per hectare and 25 to 28 kilograms of K per hectare measured in the stems. The content of wastewater nitrogen and phosphorus at an average irrigation rate of 2 millimeters per day corresponded well with the nutrient need of the willow (*Salix viminalis*) stands. Although maximum growth occurred at about three times this application rate, the increased growth did not result in greater nutrient extraction. A sustainable system should not be loaded by nutrients in excess of tree requirements, as this will result in loss of nitrogen to the air, groundwater, or surface water plus accumulation of phosphorus in the soil profile. The removal of nutrients and organic material in the pilot system, on a mass balance basis to 1 meter below the ground surface, was 82 to 93 percent of total nitrogen, 90 to 97 percent of total phosphorus and 74 to 82 percent of biochemical oxygen demand.

Full-scale wastewater irrigation of willow (*Salix* spp.) has been in operation since 1997 and 1998 at three medium-sized treatment plants in southern Sweden. The Swedish environmental authorities require pretreatment of the wastewater for odor and pathogen control, and application of wastewater below the canopy and near the soil surface to control aerosol formation and drift. For the first few years, the authorities also required risk assessments of the possible transmission of pathogens to the surroundings. Two more willow (*Salix* spp.) irrigation facilities, both approximately 60 hectares in size, began operations in 2001. One of these uses secondary effluent while the other irrigates with nitrogen-rich liquid from the sludge dewatering process.

Spray Irrigation of New Zealand Pine Trees

Spray irrigation has been used since 1991 to apply municipal wastewater onto a Monterey pine (*Pinus radiata*) forest in New Zealand (Barton *et al.* 1999). An average of 71 millimeters of effluent was applied each week using sprinkler irrigation, with annual loadings of 406 kilograms of N per hectare and 98 kilograms of P per hectare. The resulting annual stream loading was within target levels of 24.5 milligrams of N per year and 2.5 milligrams of P per year (Tomer *et al.* 2000).

Landfill Leachate Reuse

Leachate is produced as water infiltrates through landfill waste. A primary constituent of leachate that determines the site loading rate is often nitrogen, a plant macronutrient. Typical municipal landfill leachate may also contain low levels of heavy metals and organic contaminants.

Treatment of landfill leachate is another application well suited for phytoirrigation. In Sweden, about 35 landfills use leachate to irrigate on-site trees both natural coniferous forests and specific plantations of fast-growing deciduous trees (Hasselgren *et al.* 1999c). The first facility in Sweden, consisting of irrigation of willow (*Salix* spp.) plantations with methanogenic landfill leachate, was established in 1985. High-salinity, high-ammonia leachate is also being used to irrigate willow (*Salix* spp.) in the United Kingdom (Riddell-Black *et al.* 2000). Studies at a landfill near Hamilton, Ontario, Canada, showed that with proper management, leachate could be used to successfully irrigate both red maple (*Acer rubrum* L.) and hybrid poplar (*Populus nigra* × *Populus maximowiczii*) (Shrive *et al.* 1994). Leachate irrigation studies are also being conducted with both willow (*Salix* spp.) and Poplar (*Populus* spp.) in Minnesota (Rog and Isebrands 2000).

Depending on the water quality, leachate may have to be diluted or partially treated prior to irrigation so that agronomic rates, toxicity levels, and regulatory limits are not exceeded. Without care, some leachate irrigation systems can lead to reduced tree performance or failure. Shrive and McBride (1995) found that red maple (*Acer rubrum*) saplings subjected to a flood irrigation regime that saturated the root zone resulted in a significant decrease in net photosynthesis. Avoiding even short-term saturation with leachate should help to minimize plant stress.

The Riverbend Landfill near McMinnville, Oregon, U.S. (Figure 23-1) is the site of a successful demonstration that shows that hybrid poplar (*Populus* spp.) trees can reuse leachate.This landfill has been beneficially using leachate as a source of irrigation water and nutrients for hybrid poplar (*Populus* spp.) production since 1992.

Figure 23-1 Riverbend Landfill (operated by Waste Management) leachate reuse system showing landfill cap, leachate storage lagoon, and 6.9-hectare hybrid poplar (*Populus* spp.) tree plantation.

TABLE 23-3 Riverbend Landfill Application Rates and Average Soil Pore Water Concentrations: 1993 to 1999

Year	1993	1994	1995	1996	1997	1998	1999
Irrigation rate (millimeters)	230	480	810	440	420	620	460
Total applied nitrogen (kilograms of N per hectare)	58	273	522	320	293	366	340
1.2-meter lysimeter (milligrams of NO_3-N per liter)	–	–	–	5.4	9.3	14.1	16.4
1.8-meter lysimeter (milligrams of NO_3-N per liter)	–	–	–	10	10.7	11.1	10.8
2.4-meter lysimeter (milligrams of NO_3-N per liter)	23	14.4	6.4	7.1	7.8	9.2	8.9

The primary constituent removed from the Riverbend leachate is ammonia nitrogen. Performance data for this system are shown in Table 23-3. From the time the tree plantation reached full canopy in 1995 through the 1999 irrigation season, the average annual nitrate–nitrogen concentrations in the soil pore water below the effective root zone were reduced to below 10 milligrams of N per liter, the U.S. drinking water standard. Average nitrogen concentrations in the leachate have been approximately 100 milligrams of N per liter. The total annual nitrogen loading to the tree plantation has stabilized at slightly more than 330 kilograms of N per hectare. The plantation is managed with deficit irrigation, meaning that the trees could take additional water if nitrogen loading was not limiting.

Irrigation of Contaminated Groundwater

One of the most rapidly growing areas of phytoremediation is plume interception. This is the use of trees to treat shallow contaminated groundwater containing such chemicals as dissolved-phase chlorinated solvents; benzene, toluene, ethylbenzene, and xylenes (BTEX); and nutrients such as nitrate. This application depends on tree root withdrawal from the water table, or from the capillary fringe to control or treat the contaminants. Plume interception may be constrained by land use above the aquifer, depth to groundwater, or the presence of soil layers that present physical or chemical barriers to deep rooting. Deep planting (Negri, Gatliff *et al.* this book) and mechanical pumping and irrigating with this water or "phytoirrigation" provide a means of overcoming these site limitations. Phytoirrigation of loblolly pine (*Pinus taeda*) and cottonwood (*Populus trichocarpa*) is being investigated at the U.S. Savannah River Site for removal and degradation of chlorinated solvents in a groundwater too deep for conventional plume interception approaches (Hitchcock *et al.* 2002, L. Newman, Savannah River Ecology Laboratory, Aiken, South Carolina, personal communication 2002). Phytoirrigation provides an alternative or supplement to such conventional treatment systems as air stripping and granular activated carbon absorption used to remediate extracted groundwater. However, applying water containing certain organics or metals to uncontaminated soil through irrigation could result in an increased volume of soil requiring treatment if the phytoremediation system is not effective. The design and associated regulatory review criteria must ensure that contaminants are unlikely to spread into clean media.

Another example of phytoirrigation of contaminated groundwater is underway at Beale Air Force Base in California (U.S.). At this site, a combination of trees, irrigation, and a containment wall are being used to control and treat a plume containing chlorinated solvents, primarily trichloroethylene. A backup granular activated carbon treatment system is available for times of the year when the phytoirrigation system cannot provide adequate hydraulic control. In this application, evapotranspiration of the trees and grasses provide an inward hydraulic gradient across the containment walls. A 600 meter barrier wall, keyed into the underlying bedrock, provides a low-permeability boundary around the contaminated area to reduce the quantity of water that moves through the wastes. Groundwater within the barrier wall that is not taken up directly by roots is collected by a subsurface drainage system and is used to irrigate the vegetation.

The tree plantation consists of an open stand of Fremont cottonwood *Populus fremontii* (trees) interspersed with interior live oak (*Quercus wislizenii*). Evapotranspiration is dominated by Fremont cottonwood (*Populus fremontii*) and live oak (*Quercus wislizenii*) during the summer, and by live oak (*Quercus wislizenii*) and native grasses during the winter. A complex plant community with a number of other species was also established as a part of the important goal of ecosystem restoration. These other species include a

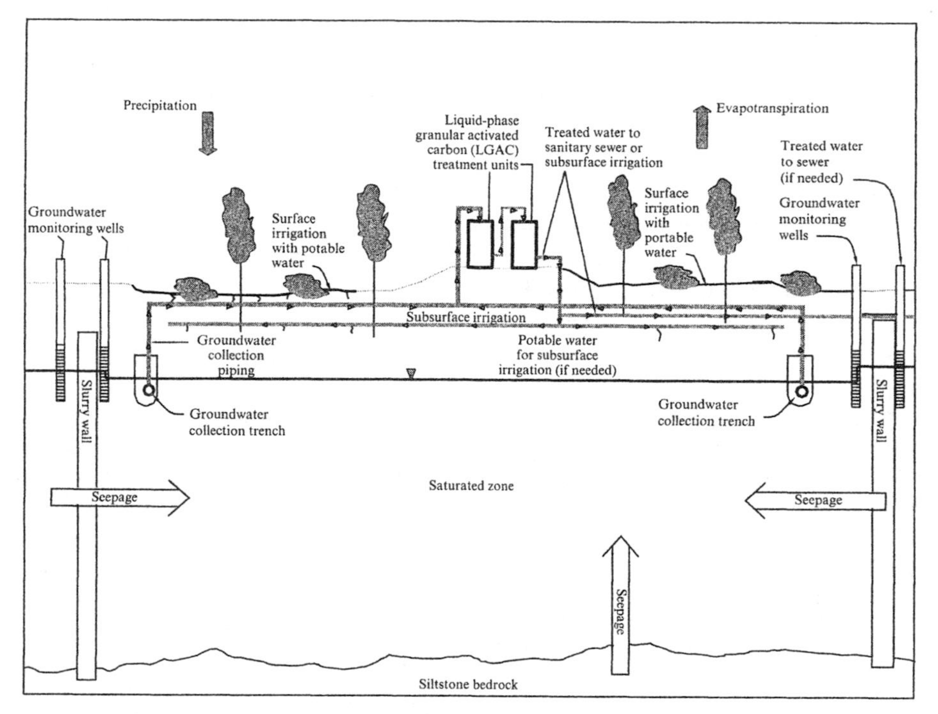

Figure 23-2 Design of irrigated phytoremediation system within bentonite barrier wall for Beale Air Force Base site.

number of grasses, legumes, forbs, and shrubs such as deergrass (*Muhlenbergia rigens*), meadow barley (*Hordeum brachyantherum*), clustered field sedge (*Carex praegracilis*), and narrow-leaved willow (*Salix exigua*). Figure 23-2 shows a cross section of the Beale application.

DESIGN COMPONENTS

General guidance on design and implementation of many types of phytoremediation systems is increasingly available (CH2M HILL 1999, ITRC 2001). Extensive design guidance on land treatment of municipal wastewater can be found in U.S. EPA (1981). New Zealand and Australia have recently published on the Internet, guidelines for wastewater irrigation. Design issues of special importance for phytoirrigation include project planning, waste stream characterization, site suitability for tree planting and growth, tree system design, irrigation system design, site preparation, and soil moisture monitoring.

Phytoirrigation design is complex and site specific. An approach based on first principles is the most likely to succeed. Loading rates, for example, are often determined based on several technical factors such as local climate, soil properties, and irrigation water properties, which must be integrated with regulatory requirements and decisions made as part of the public involvement process. Consequently, this chapter does not include extensive "rules of thumb" and other simple numerical design criteria.

Project Planning

The critical first steps in properly executing a phytoirrigation project are to determine and prioritize the objectives, and then to prepare a plan for design and installation. Criteria for measuring success and determining the project end points must be established in the project plan and negotiated as conditions in the permit. This helps to ensure that the project is not saddled with unrealistic expectations.

Site project planning may include mitigation measures and environmental enhancements. The plan for a phytoirrigation project can recommend a logical progression from a distressed brownfield to a site with lush vegetation in transition to a reclaimed, green field site with a wide range of potential uses. A partially remediated site with a surface condition safe for public access can become a community asset if the site is managed as a park or recreation area while the deeper contamination is further treated by the plant system (see for example Trapp and Christiansen this book). Similarly, a site growing trees can produce marketable wood as a natural resource to spur the local economy and offset a portion of the operational costs.

Remediation projects generally involve risk and environmental impacts. Often risk management becomes a primary design criterion that influences all major decisions in the design process. The risk of human exposure to volatile subsurface contaminants may dictate that trees are planted as whips in dibble holes rather than as rooted stock in trenches, that the irrigation system be built entirely on the surface without trenched-in pipelines, and that automated controls are used to minimize operator time on-site. Cleanup plans for complex sites should be an extension of the site risk assessment, as long as the assessment includes ecological health.

Characterization of Waste Stream Quality

The quantity, quality, and variability of the waste stream must be defined in the early stages of design to ensure plant health, and to establish loading rates, acreage, storage, and fertilizer requirements. The nature of the water determines the specific properties that should be measured. Salinity and pH data are required for any water to be used for tree irrigation. Nutrient-rich waters, such as wastewater and leachate, require data on nitrogen and phosphorus as well as standard wastewater parameters such as total suspended solids, alkalinity, and biochemical oxygen demand. The amount and nature of the solids influence the selection of the appropriate irrigation system and the extent of preliminary filtration required to prevent the obstruction of small orifices in the irrigation system. Suspended solids can also plug conducting soil pores. Waters with a high sodium adsorption ratio can disperse soil particles (Levy *et al.* 1999), adversely affecting soil hydraulic properties and system performance. If the water is high in dissolved salts, then sodium, calcium, magnesium, and chloride concentrations should also be monitored. Irrigation water temperature data are required if significantly different from ambient. A heavy-metals analysis is required for landfill leachate and many industrial wastewaters. Xenobiotic organic compounds are equally important to define, and may be the focus of some applications. Concentrations of the principal constituents and known degradation products should be measured to predict remedial performance, assess risks to human health and the environment, and to avoid application of phytotoxic concentrations.

Groundwater to be used in phytoirrigation is often dominated by a few specific constituents of concern. A common error is to narrowly focus on these few, while ignoring phytotoxic levels of nontarget or unregulated parameters such as salinity. The variability of parameters must also be known to determine the potential for abrupt increases in concentration and the potential impact on tree health. Table 23-4 provides the basic parameters used to characterize waste streams.

TABLE 23-4 Waste Stream Characterization

Parameter	Comments
Total dissolved solids and electrical conductivity	Determines plant selection, potential for groundwater degradation, and leaching fraction required
5-Day biochemical oxygen demand	Can limit loading rates for food processing wastewaters
Total organic carbon	Affects chlorine demand for irrigation system maintenance
Total suspended solids	Can plug soil pores and irrigation system orifices
pH and alkalinity	Especially important with high salinity wastewaters
Nitrogen (total Kjeldahl N, NH_3-N, NO_3-N, NO_2-N)	Complete characterization needed to estimate fate where nitrogen loading will be limiting
Phosphorus	Phosphorus loading may be limiting for some systems; determines fertilization needs
Calcium, magnesium, sodium, chloride	Especially important with saline waters—risk of soil dispersion and ion toxicity (Na and Cl)
Heavy metals (zinc, lead, cadmium, copper, and others)	Determines the feasibility of treating leachate, contaminated groundwater, and industrial wastewater
Volatile organic compounds, semivolatile organic compounds, and degradation products	Determines the feasibility of treating leachate and contaminated groundwater

Site Suitability for Trees

Thorough site characterization is necessary to prevent unacceptable system performance, damaged infrastructure, or even personal injury. Soil characteristics, site hydrogeology, and existing utilities must be well understood before installation.

Soil Characterization

Tree-based projects require a thorough understanding of the rooting medium. Published soil maps, if available, often provide excellent preliminary data, but site-specific investigations are required before installation of phytoirrigation systems. Physical and chemical impediments such as soil horizons restrictive to tree survival and growth must be identified. Rooting depth is required information for water balance models used in irrigation system design and scheduling, and is determined by a combination of inherent plant characteristics, site hydrology, and soil conditions such as restrictive layers. Field infiltration tests should be conducted to avoid ponding and runoff of applied water. The results of these tests are used to select appropriate hydraulic loading rates and irrigation application equipment. Laboratory analysis is required to determine soil fertility, salinity, and soil water parameters such as particle size distribution and water holding capacity. Water holding capacity, the amount of soil water retained between field capacity and the wilting point (soil water potential is between about −0.033 and −1.5 megapascals for many soils), is one of the most important soil characteristics affecting irrigation scheduling decisions.

Hydrogeological Characterization

A hydrogeological investigation should reveal the presence of water-producing aquifers, seasonal high water tables, and the location of water supply wells. Integration of this information reveals the potential risks of phytoirrigation systems to groundwater, helps define appropriate adjustments to system design to mitigate these risks, and helps to determine groundwater monitoring needs.

Utilities

Existing utilities can be damaged in the process of installation or operation. Obvious risks include excavation damage to pipelines or telecommunication cables. Less obvious risks include plugging of drainage systems such as tiles and sewers by tree roots, damage that may not be apparent for years. Phreatophytic trees such as poplar (*Populus* spp.) and willow (*Salix* spp.) are noted for this trait. Expanding tree roots are capable of breaking power cables and concrete structures. Appropriate setbacks must be incorporated into plantation layout to prevent such damage.

Tree Plantation Design

The design of a tree plantation requires a clear understanding of overall project objectives. Design components to meet those objectives include the selection of appropriate tree species and varieties, type of plant material, tree spacing, understory species, hydraulic and nutrient loading, soil amendments, site preparation, and climatic considerations such as the length of the growing season, potential evapotranspiration, and precipitation.

System Objectives

The objectives of phytoirrigation may include beneficial reuse of nutrients such as nitrogen and phosphorus, transformation or accumulation of a contaminant such as trichloroethylene, or reductions in contaminated water volume through evapotranspiration. Secondary objectives may include eventual harvest of wood products, creation of green space, improved site esthetics, or ecosystem restoration.

Species and Variety Selection

One of the most important design considerations for phytoirrigation is choosing appropriate plant species and varieties. Selected trees must be capable of achieving the desired treatment objective and be adapted to the irrigation water, soils, and climate of the site. Typically, achieving a high evapotranspiration rate is an important goal. Critical site conditions for plant selection include water chemistry (Table 23-4), and adaptation to pests and diseases of the area. Any factor that compromises tree health and growth will reduce performance. For example, hybrid poplar (*Populus* spp.) clones that include either *Populus trichocarpa* or *Populus maximowiczii* parents are quite susceptible to *Septoria* canker if used in the U.S. Midwest (Heilman 1995). Certain varieties may result in a more valuable final wood product because of straighter stems or better paper processing properties. Significant differences in damage from voles (*Microtus* spp.) have been observed among hybrid poplar (*Populus* spp.) tree species at phytoremediation sites in Oregon. Salt tolerance is a very important selection criterion, as differences between species and varieties can be significant (Banuelos *et al.* 1999, Shannon *et al.* 1999).

Once the species have been selected, the best approach is to obtain plants from nurseries located as close as possible to the site (Dickman and Isebrands 1999). Local nurseries will generally be able to recommend and provide varieties adapted to the local area. Use of alien plants may be unacceptable to local community groups and sometimes to regulators. Use of native or naturalized plants will generally ensure some resistance to local pests and diseases, but may not afford the greatest remedial efficiency.

Tree species may be pilot tested through greenhouse studies or demonstration projects in the field. Field tests are preferred because greenhouse tests may not always be representative of important field conditions. Growth rates and evidence of plant stress, such as chlorosis, should be used to evaluate the suitability of candidate species.

For the full-scale Woodburn site, hybrid poplar (*Populus trichocarpa* × *Populus deltoides*) clones were selected for the yield potential and adaptation to the Willamette Valley, Oregon. Insufficient data were available to select clones best adapted for utilization of municipal wastewater, and growth potential was assumed to be proportional to water and nutrient uptake. For New Zealand wastewater applications, *Eucalyptus* spp. performs better than pine (*Pinus radiata*) (Nicholas *et al.* 2000). For the Beale site, use of native species for ecosystem restoration was an important concern; therefore, the native Fremout cottonwood (*Populus fremontii*) was selected rather than a hybrid poplar (*Populus* spp.).

When using Salicaceae, a number of different forms of plant material may be used. These include use of stem cuttings, whips, poles, and bare root or potted trees. Use of larger or rooted plants result in more rapid establishment and reduce weed competition, but tree purchase costs and planting costs are much higher than for smaller trees or cuttings, whips, and poles. Longer whips and poles are commonly used for deep planting applications, but for phytoirrigation, the main advantage of taller plant material is less shading by weeds during establishment. Economics, especially planting costs, drive most larger installations (greater than 5 hectares) toward short stem cuttings.

Tree Spacing

A clear definition of project objectives and priorities helps to determine the most appropriate tree spacing. For example, enough space must be left between tree rows to allow planned maintenance activities such as mowing or spraying. The influence of tree spacing on hydraulic capacity of these systems is often overemphasized. Once the tree plantation forms a complete canopy, spacing has little effect on evapotranspiration rate or nutrient requirements. The impact of spacing on hydraulic and nutrient loading is primarily a concern during the establishment phase. Establishing dense initial plantings with the intention of thinning may provide small increases in early capacity, but thinning operations are often neglected and the resulting mature tree stands are excessively dense. For hybrid poplar (*Populus* spp.) plantations with harvest as an important objective, 3.7- by 3.7-meter spacing is becoming increasingly standard in the U.S. Pacific Northwest. This spacing allows flexibile commercial harvest of poplars (*Populus* spp.) for pulp wood or solid wood markets. Water use requirements of tree stands with an open or narrow structure that never truly develops a canopy are difficult to predict and may require more in-depth modeling to estimate hydraulic loading limits.

At the Woodburn site, trees were planted on a 4- by 2-meter spacing. The 4-meter spacing between rows was used to allow access by biosolids application trucks, and the 2-meter spacing between trees within the row was determined to be the most appropriate compromise between early system capacity and marketable trees at harvest. The site is designed to be thinned to a 4 by 4-meter spacing after the third year.

At the Beale site, trees were planted in a 4.6-meter diamond-shaped pattern with positions within the rows staggered. This spacing allowed high tree density, but also sufficient space for growth of species important for ecosystem restoration, including understory grasses, legumes, forbs, and shrubs.

The Swedish *Salix* program for biofuel production has standardized tree spacing. Spacing to maximize biofuel production is much closer than that used for most phytoremediation applications in the U.S. A double-row concept is used with 0.75 meter between the two rows in the double-row and a distance of 1.50 meters between double-rows (Hasselgren 1999b). The harvesting equipment covers one double-row at a time. The distance between plants in the rows is 0.3 to 0.4 meter, depending on which willow (*Salix* spp.) clone is used. Normally 15 000 to 17 000 25-centimeter cuttings are planted on each hectare.

Understory

Commercial nurseries and plantations typically keep the soil between trees completely bare to minimize competition for light, water, and nutrients. Grassy or weedy areas around trees can also encourage problematic tree pests such as voles (*Microtus* spp.) (Heilman 1995). Sometimes, when trees are small, grasses are maintained or planted to provide increased water and nutrient utilization thus reducing the land area required for phytoirrigation. Understory seedings may also provide some control of problematic weeds. In the tree establishment years, some understory seedings may need to be mowed periodically to reduce pest cover and to prevent interference with irrigation. Understory seedings often diminish as the trees develop a complete canopy. At the Woodburn site, water and nutrient utilization objectives led to selecting a grass understory for the area between tree rows. This cover was a combination of existing perennial ryegrass (*Lolium perenne*) for some fields and new seedings of dwarf perennial ryegrass and hard fescue (*Festuca ovina*) for other fields.

Seasonal and Regional Considerations

Unlike most mechanical systems, phytoremediation systems are strongly affected by the season of the year and annual senescence for many species. Knowledge about the effect of changing seasons on system performance is limited, especially for non-nutrient contaminants such as trichloroethylene. Large storage lagoons are used at the Woodburn and Riverbend (see Figure 23.1) sites to balance water flow with evapotranspiration demands. A range of plant species is used at the Beale site to provide some evapotranspiration capacity year-round.

Nutrient Loading Considerations

For irrigation with wastewater, landfill leachate, and some nutrient-rich groundwater, nutrient loading may control application rates rather than hydraulic loading. Nitrogen concentration typically limits the volume that can be applied with these waters. Nitrogen applied to plantations may be assimilated as plant tissue, carried away in runoff, immobilized as soil organic matter, volatilized as ammonia, leached as nitrate or ammonium to groundwater, or denitrified. All of these processes should be carefully evaluated during design to avoid adverse environmental impacts, including groundwater pollution, eutrophication of surface waters, and global warming. Numerous general references are available to guide nitrogen loading calculations (U.S. EPA 1981, Follet *et al.* 1991). General experience should be supplemented by information from local extension services and land grant universities to obtain the best possible estimate of nitrogen fate.

Estimates for the volatile loss of ammonia from sprinklers vary widely. For example, according to Meisinger and Randall (1991), 5 to 30 percent losses of applied nitrogen due to ammonia volatilization can be expected during the application of liquid urea and ammonium nitrate. For sprinkler-irrigated wastewater, Broadbent and Reisenauer (1985) suggest that volatile losses are less than 20 percent of the total nitrogen applied. The U.S. EPA (1981) suggests a 10 percent loss of total applied nitrogen for soils with low cation exchange capacity and a pH above 7.8. For general design purposes, the U.S. EPA (1981) recommends including the effects of ammonia volatilization in a 15 to 25 percent denitrification loss estimate.

At the Riverbend site, approximately 337 kilograms of nitrogen per hectare have been applied for years without deep percolation of excess nitrate. Data are not sufficient at this site to develop a mass balance for nitrogen, but presumably immobilization and denitrification in the soil are major sinks. At the Woodburn site, soil testing is conducted annually to track forms and amounts of nitrogen to adjust loading rates as needed.

Site Preparation

The objectives of site preparation are to minimize vegetative competition for the trees, apply and incorporate soil amendments, and provide low bulk density favorable for root development. Soil amendments may include organic matter, fertilizer, and amendments to adjust soil pH. Fertilization and pH adjustment requirements can be determined through standard agricultural soil tests. Large applications of organic amendments are often required for many industrial sites to improve the physical and chemical conditions of the soil. All soil amendments should be sufficiently mixed into the soil.

The center of the tree rows should be tilled deeply or "ripped" to a depth of at least 50 centimeters. Ripping when subsoil conditions are dry will provide the best possible soil-shattering action to reduce compaction and promote rapid tree root growth. Best results are obtained by subsequently tilling the soil thoroughly by rototilling a 1.2-meter strip to a depth of at least 13 centimeters centered on the tree rows. Multiple passes will probably be required. Tillage is most effective on moist soils. At Riverbend, for example, irrigation with river water was carried out prior to rototilling to facilitate effective tillage of very dry soils. A post-emergence herbicide, such as glyphosate, should be combined with a residual herbicide and applied to the tilled strips before planting. This herbicide combination aids control of any existing vegetation in the short term as well as germinating grass and broadleaf weed seeds for 2 to 3 months following application. Thereafter, grass competition in the tree rows can be controlled with selective herbicide applications. Herbicide options for postemergence broadleaf weed control in hybrid poplar (*Populus* spp.) plantations are limited.

Irrigation Design for Tree Systems

An effective phytoirrigation design must adequately address the project objectives, site hydrology, and water requirement for the trees. The first step in designing the irrigation system is to prepare a water balance that defines the site hydrology. Once the tree water requirements are determined from the water balance results, an irrigation system concept can be developed. The formulation of irrigation management approaches must also be integral to the selection of irrigation system components, in order to achieve a final design that meets all the site objectives.

Water Balance

A water balance is a critical tool for evaluating the hydrologic performance of a phytoirrigation system. Used to identify all flows of water to and from the system, the water balance allows for the estimation of irrigation requirements and deep percolation potential. A water balance used for project planning purposes is typically performed using historical precipitation data and historical reference evapotranspiration estimates from a nearby meteorological station (*e.g.*, see Weaver *et al.* this book), measured or estimated soil properties, and other data regarding plant water uptake. For operational purposes, a water balance is used to adjust irrigation schedules and to evaluate the hydrologic processes affecting the phytoremediation system.

For project planning purposes, the water balance should be estimated on a monthly basis for the median year and 1-in-10-year (or 1-in-20-year) wet and dry years, at a minimum. Irrigation requirements during the first few years of tree growth are lower than during later years. Consequently, monthly water balances are often developed separately for the first few years of tree growth. Depending upon the site objectives and required precision, daily water balance modeling may also be required. For example, sites that are managed to ensure that deep percolation does not contaminate groundwater will require higher precision forecasting.

The water balance within a tree plantation root zone for a given period of time (*e.g.*, hour, day, month, or year) is described in terms of consistent units of volume or unit depth as

$$\Delta S = I + P - R - ET - D \qquad (23\text{-}1)$$

where ΔS is the change in soil water storage, I is irrigation, P is precipitation, R is runoff, ET is evapotranspiration, and D is the water volume that percolates below the root zone. The water balance component most difficult to measure is the deep percolation below the root zone. Consequently, this component is most often estimated as a residual of the water balance.

The change in soil water storage can be measured directly with soil monitoring devices placed at several depths throughout the root zone. Using measurements from these devices, the total soil water storage within the root zone can be estimated by integrating soil water content with depth. To estimate the maximum long-term soil water storage within the root zone, the soil water holding capacity is often used with the simpler water balance models. Estimates of soil water holding capacity are widely published in soil surveys and are thus readily used in simple water balance calculations.

Irrigation application volumes are measured with flow meters installed in the irrigation delivery system, with rain gages in the field for sprinkler or micro-spray applications, or by recording the period of irrigation when constant flow devices (*i.e.*, pressure-compensating nozzles or emitters) are used. The ultimate volume or unit depth of irrigation water used in the root zone water balance must account for only that amount of water that reaches the soil surface. This amount excludes water lost to wind drift off of the cropped area and evaporation of the irrigation spray in the air. Application systems do not apply water uniformly to target crops. To account for this nonuniformity, water is generally over-applied in some areas of the field so that the areas receiving lesser amounts receive at least some minimum application. The required amount of over-application can be estimated with statistical knowledge of the nonuniformity of irrigation across a field and the area requiring some specified minimum application amount (Hart and Reynolds 1965). For highly uniform systems, like a well-designed drip or micro-spray system, an average application depth can be used in the water balance. Where significant nonuniformity is encountered, separate water balances may need to be performed for areas receiving the minimum and maximum irrigation amounts.

For most U.S. phytoremediation sites, precipitation data are available from local meteorological stations. It is important, however, to be sure that the climate at the station selected is representative of the climate at the phytoremediation site. Where this is not the case, precipitation can be easily measured at the site by manual or automated methods.

Evapotranspiration is difficult to measure directly under field conditions, but can be closely estimated by a number of methods (Jensen *et al.* 1990). The most direct method of measuring evapotranspiration is with large weighing-lysimeters. However, such lysimeters are very costly and must be very large to obtain a representative sample within a tree plantation. Indirect estimation methods include energy balance approaches such as the Bowen ratio and Penman methods, the mass transfer approach with eddy flux correlation, the pan evaporation method, and thermal dissipation sap-flow measurement (Vose *et al.* this book). The most widely used methods in agricultural systems are the energy balance and pan evaporation methods. With these latter two methods, evaporation of free water from a small reservoir or micro-meteorological measurements are used to estimate the evapotranspiration of a reference crop [grass or alfalfa (*Medicago sativa*)] under standard conditions. The most reliable estimation method for reference evapotranspiration, regardless of climatic conditions, is the modified Penman method (Jensen *et al.* 1990). The FAO–Penman-Monteith method is one form of the modified Penman equation. The FAO–Penman-Monteith method has been accepted by the Food and Agriculture Organization (FAO) of the United Nations, the International Commission for Irrigation and Drainage, and the World Meteorological Organization as the best reference evapotranspiration estimation approach at this time (Allen *et al.* 1998). The Penman methods employ measurements of solar radiation, wind speed, air temperature, and relative humidity to characterize the evaporation demand of the climate (see Weaver *et al.* this book). Measurements of these parameters are taken either from local meteorological stations or collected at the site.

Empirically based coefficients are then used to relate the reference crop or stand evapotranspiration to the evapotranspiration of the tree plantation of interest as

$$ET_c = K_c ET_r \tag{23-2}$$

where ET_c is the tree stand evapotranspiration, K_c is the crop or stand coefficient, and ET_r is the reference stand evapotranspiration. Use of Equation (23-2) assumes that the tree plantation is disease free and is not nutrient or water stressed. Water stress can be avoided with appropriate irrigation, but when unavoidable or even desirable as part of the site water management approach, K_c corrections can be applied (Allen *et al.* 1998). Direct application of Equation (23-2) also assumes that the tree plantation maintains a uniform, closed plant canopy over an area greater than 3 to 5 hectares (Allen *et al.* 1998) with a perimeter to total planted area ratio less than about 400 meters per hectare (530 feet per acre). For many plume interception applications, only a few rows of trees may be planted, creating a plantation with predominant edge effects. These conditions can be accounted for using an approximate K_c value for border crops

$$K_c = \min\left(1.2 + \frac{F_r H}{W} \text{ or } 2.5\right) \tag{23-3}$$

where F_r is a stomatal resistance correction factor, H is the mean vertical height of the tree canopy, and W is the width of the tree canopy (Allen *et al.* 1998). The K_c value of 2.5 translates to an approximate upper limit on the water use per unit area of any tree crop.

In the initial year of planting, water use is the least because the trees are relatively small. Cover vegetation between rows of trees can be used to enhance tree stand evapotranspiration before canopy closure. An approximate K_c value for intercropping that is corrected for the combined water use of overlapping vegetation can be used

$$K_{c\text{-intercrop}} = \frac{f_1 H_1 K_{c1} + f_2 H_2 K_{c2}}{f_1 H_1 + f_2 H_2} \tag{23-4}$$

where f_1 and f_2 are the areas of crop canopies 1 and 2 per unit land area, H_1 and H_2 are the plant canopy heights of crops 1 and 2, and K_{c1} and K_{c2} are the stand coefficients for crops 1 and 2 (Allen *et al.* 1998). Water use will increase as trees continue to grow in subsequent years until the tree stand canopy closes. Although tree growth continues after canopy closure, the boundary layer characteristics between the atmosphere and the canopy, which controls vapor transfer, change little after canopy closure. Consequently, tree stem density has little to no influence on tree stand water use after canopy closure.

The initial tree planting density and the rate of tree growth influences the number of seasons required to attain a closed canopy. Water-use measurements under drip irrigation in eastern Oregon (Gochis and Cuenca 2000) provided stand coefficient values for hybrid poplar (*Populus* spp.) in the first through third year of growth, and projected likely values for fourth year coefficients. Use of the fourth year coefficients provides a reasonably conservative design assumption for long-term steady-state water demands. Oregon sites with good soils, irrigation, and fertilization have demonstrated the validity of this assumption. However, it is not unusual to require from 4 to 5 years to reach steady-state water demand in many regions.

Runoff is often difficult to measure except at a small plot scale. As a result, field-scale runoff is most often estimated by modeling site hydrology using the curve number approach (SCS 1985, 1986), or by using general rainfall–runoff relations (Dastane 1974).

Deep percolation is usually estimated as the residual of the soil water balance. This water balance component can be directly measured with a variety of buried lysimeters (Wilson *et al.* 1995, Stephens 1996); however, instrumentation costs often are beyond the reach of most project budgets.

Water Balance Model Selection

The hydrologic models available to assess field-scale soil-water balances vary widely. These models include HYDRUS, UNSATH, LEACHM, HELP, and tipping-bucket models (Allen *et al.* 1998). Selection of the appropriate model for the specific application is often a difficult task. Each model has specific strengths and limitations that must be carefully weighed in the context of project objectives.

Irrigation

Current irrigation technologies for phytoirrigation include drip irrigation, micro-spray irrigation, and to a far lesser degree, sprinkler and flood irrigation. Drip and micro-spray irrigation are generally favored over other methods because of the potential for delivering irrigation water uniformly and efficiently, thus reducing runoff and undesirable deep percolation below the root zone.

Benefits of Irrigation

One of the common objectives of phytoirrigation is to use wastewater beneficially to eliminate or reduce the need to discharge the water into streams, lakes, and estuaries. Some contaminated sites require management of storm water or groundwater from a pump-and-treat system. Although stormwater and groundwater may contain only very low levels of contamination, costly treatment and permission from regulatory agencies may still be required before discharge is allowed to public waterways. Phytoirrigation can be cost-effective, and in many climatic zones, can result in zero discharge during the growing season. Combined with wastewater storage, zero discharge through phytoirrigation may be achievable year-round in some climates.

Trees that are properly irrigated exhibit more predictable growth and remediation benefits than nonirrigated trees in nearly every climatic condition. A well-managed irrigation system can compensate for the variability of natural precipitation while maintaining available water holding capacity to minimize site runoff or deep percolation. A significant benefit of phytoirrigation is that controlled water applications can also precisely place nutrients in the desired location in the soil profile. A metering pump and a chemical injection system can supplement nutrients in the irrigation water to encourage optimal tree and microbial growth to degrade contaminants. Nutrient injection into the irrigation water (fertigation) allows the operator to apply a widely variable combination of nutrients, as needed, in response to plant tissue or soil sampling analysis. The water-soluble nutrients applied with fertigation are more predictably available at the time of application than nutrients applied as dry, surface-broadcast fertilizers. The cost and potential disruption of nutrient addition by fertigation is minimal compared to dry chemical applications that require vehicle traffic on the site. Sites with multiple level subsurface and surface irrigation systems can even apply nutrients to one level of the root zone and not to others to manipulate root development. Adding a hydroponic nutrient solution to the irrigation water can result in successful establishment of trees at sites that might otherwise be considered too infertile. Small, frequent irrigations with nutrient-enriched water may be the single most important component of tree management at the most difficult sites—those with very rapidly draining soils. Small frequent fertigations maintain a more constant water and nutrient condition in the root zone than any combination of surface-applied, dry-chemical fertilizers and precipitation. Small, frequent irrigations also help to minimize the plant stress caused by salinity.

Irrigation and subsurface drainage systems can also be used to enhance oxygen transport into the subsurface. Sites with deep, subsurface irrigation can have air injected at low pressure through the irrigation delivery system between irrigation events. Buried drains can also be used to improve aeration of the root zone with passive barometric pumping of air into the drains resulting from subtle pressure differences over hills or other surface irregularities. Increasing the oxygen content deeper in the soil profile improves the potential for roots and microbial populations to expand to greater depths.

The irrigation system also can be designed and operated to either restrict root growth to a relatively shallow depth or to encourage roots to grow to the maximum extent possible, depending upon the site soil and tree physiology. The desired rooting depth should be considered when designing the irrigation system and the irrigation schedule. Distributing the water uniformly over the entire soil surface in small, frequent irrigations with sprinklers can apply all of the water that the trees will use without wetting the soil deeper than about a meter. Most roots will grow only deep enough to collect the water the trees require, and the readily-available water will result in the maximum above-ground biomass. By contrast, a drip irrigation system that has one high-flow-rate emitter located at the base of each tree can apply all of the water that the tree needs in a relatively infrequent pulse of water. This pulse will move as a vertical column of water from the surface to the maximum potential depth of rooting. The frequency and duration of the pulsing will depend upon a number of factors, including soil type and tree transpiration rates. Both of these irrigation systems can apply the same amount of water and grow approximately the same tree mass, but will result in dramatically different rooting depths. Most phytoirrigation sites require an irrigation system that is somewhere between these two extremes, which results in the use of micro-spray systems or multiple drip emitters for each tree. Some sites have special needs, such as shallow-rooted shrubs and forbs for surface cover, and deep-rooting trees to enhance phytoremediation at the maximum possible depth. These sites require hybrid irrigation systems that combine surface and subsurface irrigation; these systems may also have deep, subsurface irrigation with venting or air injection. In regions prone to high winds, deeper-rooted trees are necessary to minimize lodging (trees being blown down).

Selection of an Irrigation Application System

The selection of an appropriate irrigation application system will depend upon a number of factors, including phytoremediation objectives, topography, soil conditions, water and power availability, and cost. The objective of some phytoirrigation systems may be to beneficially reuse wastewater, whereas for others the objective may be to sustain plant growth with a limited supply of water. When high water use is important, micro-spray and sprinkler methods are favored because more water is required to satisfy both evaporation and wind drift losses, and crop demands. Where wind drift onto neighboring property must be avoided, aerial applications may not be feasible. Drip irrigation is usually the favored option when human contact with the water is a concern or low water use is required.

Preferential flow can rapidly transport contaminants down beyond root-zone influence. Selection of an appropriate irrigation delivery system reduces the potential for preferential flow (Selker 1996). In cracking clays and other soils with large macropores, saturated flow may largely bypass the root zone in preferential flow paths. In dry coarse soils, preferential flow may funnel infiltrating water into concentrated "fingers" of flow. In general, water flow under preferential flow conditions is very difficult to predict. Problems resulting from preferential flow include inefficient irrigation delivery to the root zone, inefficient salt leaching, and faster rates of contaminant travel to groundwater. Knowledge of the potential preferential flow processes at a site can be used to select irrigation equipment. Low-rate irrigation, including drip and micro-spray, may be used on cracking and macroporous soils to deliver water under unsaturated flow conditions because water will not flow down cracks and macropores unless the surrounding soil is saturated. For coarse soils subject to fingered flow, high-rate irrigation, including sprinkler irrigation, may be used to encourage saturated flow conditions, which overwhelm the physical causes of fingered flow, resulting in a more uniform wetting front.

Drip Irrigation

Drip irrigation equipment generally consists of small orifice water emitters located at a selected spacing (*e.g.*, 60 centimeters) along a 1- to 3-centimeter polyethylene tube. Pressure compensating emitters are especially useful for sites with uneven topography or with long drip tube lengths, and can be selected to produce nearly uniform flows with line pressures from 70 to 415 kilopascals. A few specialized emitters are suitable for subsurface installation, either because the design mechanically precludes root intrusion, or the emitter plastic is impregnated with a herbicide that retards root intrusion. Most common drip emitters will become clogged with roots if installed within the active root zone.

The soil-wetting pattern of a drip emitter changes with flow rate, duration of wetting, spacing, and soil type. For surface drip, the wetted volume of soil is roughly spherical with a flattened top and an extended bottom. For subsurface drip, the shape is more spherical, which reduces the potential for surface ponding. Analytical approximations of soil-water content, or pressure as a function of radial distance from a drip emitter, can be used to estimate the wetting pattern shape and extent (Warrick 1974). This information is useful when selecting the appropriate drip emitter flow rates and spacings that will work with the site soils.

Drip irrigation requires a relatively clean supply of water, free of suspended particles large enough to clog the emitters. Commonly, a self-cleaning filter (screen, disc, or sand media) that is equivalent to the filtration of a 160-mesh (approximately 100 micrometers) screen is used to remove particles from the irrigation water.

The most common use of drip irrigation for tree phytoremediation sites involves surface drip tubing located along tree rows with closely spaced emitters (1 to 2 meters) that create a continuous band of wetting along the tube. The tubes are laid along every row of trees and may only wet half to three-fourths of the total site area because the spaces between tree rows may not be affected. Emitter flow rate and spacing can be adjusted to control the percentage of the soil volume that is wetted.

The most significant benefits of drip irrigation include the following:

1. Precise water application
2. Low pressure requirements with flow control available at every emitter
3. Very low flow rates that allow the largest field sizes and smallest irrigation volumes
4. Least potential for aerosol drift or public contact and vandalism if the system is subsurface
5. Well suited for fertigation and injection of some pesticides
6. Irrigation application efficiency greater than 90 percent, with the least evaporation losses of any irrigation delivery system
7. Lowest-cost permanent system, except for flood irrigation; about 2500 U.S. dollars per hectare (year 2000 U.S. dollars) for surface drip irrigation: the in-field application system of tubing with emitters

The disadvantages of drip systems include the need for filtration, requirement of specialized design and installation experience for sophisticated systems, and the inability to visually observe the performance of a buried system.

Micro-Spray Irrigation

Micro-spray irrigation, distinguished from sprinkler irrigation by much lower flow rates, is more like a drip system in many aspects. Similar to drip emitters, micro-spray emitters are supplied water by polyethylene tube. The micro-spray emitters usually consist of a small orifice that sprays water against a splash plate to create fine droplets that are distributed in a circular pattern. Some spray heads have a spinning splash plate; some have a rotating stream deflector that distributes a single rotating stream over a larger diameter circle than that obtained from a stationary splash plate. Many of these devices are available with pressure-compensating, flow-control orifices and perform well over a pressure range of about 170 to 340 kilopascals. The flow rates of micro-spray emitters range from about 3 to 230 liters per hour, and the spray patterns range from about 3 to 15 meters in diameter. Using the smaller sprayers on tree sites allows irrigation in the tree rows without spreading water between rows much further than would occur with drip irrigation. The largest micro-spray emitters can be used to provide a blanket application of water to the irrigated surface and to irrigate two or more rows of trees per emitter, reducing the number of emitters and amount of tubing that is needed. Irrigation efficiencies with micro-spray systems vary almost as much as the type of emitters, but are often between 75 and 85 percent.

The wastewater phytoirrigation system at Woodburn, Oregon, utilizes a pressure-compensating, rotatory micro-spray irrigation system to grow 34 hectares of hybrid poplar (*Populus* spp.) trees. The tree farm has rows 4 meters apart and uses one row of micro-spray emitters on every two rows of trees. The longest supply tubes are 245 meters long, 25 millimeters in diameter, and operate at 240 to 300 kilopascals. Each field is about 3 hectares in size and requires 1300 liters per minute to irrigate. Micro-spray irrigation was selected for the Woodburn site in order to only require coarse screening of the secondary-treated sewage, and to provide full coverage of the site surface. Biosolids are applied between the tree rows and are "incorporated" by breaking up and transporting the solids on the soil surface with irrigation water. The treatment plant injects chlorine into the wastewater for disinfection, and with about 15 milligrams of chlorine per liter residual in the irrigation water, there appears to be a secondary benefit of insect and fungal disease control. The irrigation system is automated to irrigate each field for 1 to 4 hours per day to meet the irrigation demand. The sophisticated automation is integrated with the wastewater treatment plant control computer and communicates with the remote terminals in the field through fiber optic cable.

The most significant benefits of micro-spray irrigation include the following:

1. Relatively large orifices allow use of poorer quality water
2. Full surface area coverage is possible with flat trajectory spray that has low aerosol drift potential
3. Surface supply tubes and the application system can be removed for harvest and replaced after replanting
4. The field supply tubes and application system cost about 5000 U.S. dollars per hectare (year 2000 U.S. dollars), which is the lowest of all permanent spray or sprinkler systems

A disadvantage is that micro-spray systems require a large number of support stakes and spray heads, which are hard to mow around in grass fields. Also, some off-site wind drift must always be considered.

Monitoring and Control

Irrigation control can be either manual or automated, and typically a system will have a combination of the two. Turning the system on and off for the season will usually be done manually to save costs and ease operations. Other operations, such as the daily irrigation of a field or automatic flushing of drip irrigation lines at the end of an irrigation event, are best handled with automated controls.

Automated controls range in degree of sophistication. A supervisory control and data acquisition system, such as the one installed at Woodburn, can control and monitor many functions, including operating pumps, filters, flow meters, automated valves, meteorological sensors, and soil moisture monitoring sensors. The control and monitoring system for phytoirrigation will often be an integrated part of a larger site treatment system. Programmable logic controllers can also offer automation as stand-alone units without a supervisory control system.

A simpler automated system, such as off-the-shelf, multiple station irrigation timers, can open and close valves to irrigation lines on a schedule, as long as the water supply is continuous. This type of system is usually adequate when the irrigation system stands alone and is not part of a more elaborate system. More elaborate systems include supplemental wastewater treatment, or pumping contaminated groundwater, as is done at the Beale phytoremediation site.

Monitoring

To verify that a phytoirrigation system is functioning as designed, several types of monitoring programs are essential. Tracking of contaminant concentrations and water fluxes in the vadose zone is always required, and is described in more detail elsewhere (Wilson *et al.* 1995, Stephens 1996). Agronomic monitoring programs are needed to detect any major environmental stresses, and usually include a strategy for visual monitoring of plant health, insects and pathogens, animal damage, and weed pressure. Additional information on monitoring phytoremediation systems may be found in standard references (CH2M HILL 1999, ITRC 2001). Several parameters that are especially important for phytoirrigation are described in this section as follows: soil moisture monitoring to meet both the water requirements of the trees and to assess the potential for contaminant transport, irrigation water quality monitoring to assure continued plant health and appropriate constituent loading, and meteorological monitoring that affects irrigation scheduling.

Soil Moisture

Soil moisture monitoring is critical in irrigated systems to ensure that plants are not stressed from either the lack of water or the presence of surplus water. In addition, soil moisture must be monitored in conjunction with the quality of water in the soil pores to assess contaminant transport.

Typically, soil moisture monitoring is done at least weekly to keep up with changes resulting from climatic conditions. Sensors should be installed in nested groups so that soil moisture can be determined at a minimum of 30- and 60-centimeter depths, and sometimes at a 90-centimeter depth, for irrigation scheduling and maintenance of optimal conditions for tree growth. If deep percolation of constituents is a concern, soil moisture sensors should be installed at greater depths, extending to below the root zone. Ideally, moisture sensors should be installed at the same depths as any suction lysimeters that are installed. Soil moisture should be monitored in at least three locations, or more, depending on site size and the heterogeneity of field conditions.

Many kinds of soil moisture monitoring equipment are available, ranging in complexity from simple sensors that are manually read in the field to sensors that require data downloading directly to a computer, to sensors with automated translation of data to a supervisory control and data acquisition system as was designed for the site at Beale Air Force Base. At the Beale site, the control system monitored soil moisture sensors and groundwater level sensors, to determine whether to irrigate or whether water should be pumped from the well collection system. If the site is too dry, and the groundwater level is not high enough for pumping, irrigation can be temporarily supplemented with freshwater to satisfy the needs of the plants.

One of the most reliable, low-cost soil moisture monitors is a tensiometer. A tensiometer measures the soil water pressure by equilibration of water in the instrument with water held in the soil surrounding the instrument tip. When installed in nests (*i.e.*, in the same location at different depths), tensiometers can be used to determine soil water pressure gradients and hence, the direction of the soil water flux. The drawback is that the instrument must be regularly serviced in the field to maintain water in the instrument, and the gage is subject to damage from freezing. Tensiometers are used at both the Riverbend and Woodburn sites to provide low-cost soil moisture data.

Time-domain reflectometry (TDR) is widely used to monitor volumetric soil water content (*i.e.*, in units of water volume divided by soil volume). Three nests of TDR sensors are used to monitor the soil moisture profile to a depth of 2.4 meters at the Riverbend site. Application of TDR in saline soils (soil solution electrical conductivity greater than 0.2 siemens per meter or 2 dS/m) must be carefully evaluated because measurements show a shift in the slope of the water content (Nadler *et al.* 1999) and an attenuation of the electrical wave-form, which reduces the accuracy of the readings. This problem can be mitigated for TDR measurements in saline soils by using epoxy coated TDR probes; however, this is a special application not commonly available through commercial vendors. Frequency domain reflectometry is increasingly used in place of TDR. Frequency domain reflectometry is very similar to TDR, in that the dielectric constant of bulk soil and water around the probe is estimated and related to the volumetric water content in the soil. Frequency domain reflectometry is less costly than TDR and uses simple oscillating amplifiers that change frequency in response to wave travel time down the wave guides, which is affected by the dielectric constant of the surrounding soil medium.

Heat-dissipation sensors used for soil moisture monitoring are comprised of a porous material that surrounds a heating element and a thermal detector. These sensors are typically used to measure soil water matric potential in the range of −10 to −1000 kilopascals and are insensitive to changes in soil salinity. Due to manufacturing variability, each sensor must be individually calibrated in a laboratory, which usually costs more than the intial cost of the sensor. Stephens (1996) also reported that these instruments became inoperative after 2 years of operation in the field, raising doubts about suitability for long-term monitoring.

Irrigation Water Quality and Meteorological Conditions

Data on water quality are often collected as part of normal operations by the facilities providing the water. Otherwise, water quality can be sampled during conveyance to the field or during actual application. For all applications but freshwater irrigation, water quality is typically sampled monthly, or whenever there is a projected change. Table 23-4 lists typical parameters that characterize waste streams.

The proper functioning of an irrigation system requires information on the ambient climate. The amount of precipitation reaching the site has implications for the irrigation schedule. Inadequate natural moisture can mandate an increase in irrigation to the site, while an excess of precipitation can necessitate a reduction in the amount of irrigation being applied. This is important both from the point of view of maximizing plant health, as well as reducing any chance of deep percolation from the site.

Typically, climatic monitoring is done with a simple weather station, which consists of at least a rain gauge and a thermometer. Depending on the degree of sophistication required, other factors can be recorded such as relative humidity, wind speed, and solar radiation.

Other Design Considerations

Resolving technical issues of system design is only a partial solution. Public acceptance, siting and permitting issues, and ecological considerations are also critical components of the project. Documenting secondary benefits of tree system installation, such as carbon sequestration, may also be increasingly important.

For some projects, consideration of other community benefits and amenities beyond site remediation is very important. Drip irrigation greatly diminishes odors, and may only be required around site margins to achieve the required level of public acceptance. The visual screening that mature trees provide can greatly improve the esthetics of treatment facilities. Many contaminated sites are in communities with limited green space. Creation of parks and open spaces, groves, and forest buffers around facilities can allow phytoirrigation systems to provide multiple benefits. These benefits are attainable only if human health and ecosystem risk are rigorously addressed, and community input is sought and seriously considered throughout project design and implementation.

To minimize pumping and conveyance costs, the tree plantation should be placed as close as possible to the source of the water that requires treatment. Generally, the public also prefers that these systems be placed as close as possible to related facilities to minimize the total area impacted.

Phytoremediation systems can enhance or diminish habitat. Increasing habitat is not necessarily a goal for all projects. For some systems, an attractive nuisance may be created, increasing the risk of exposing wildlife to chemical constituents of concern (CH2M HILL 1999). This risk is especially important if contaminant uptake, accumulation without degradation, harvest, and disposal is the waste management approach selected. Assessing the positive or negative impacts on the ecosystem habitat requires knowledge and integration of remediation mechanisms, impacted wildlife species, and surrounding habitats. Habitat enhancements, such as created wetlands, can be constructed on the margins or within tree-based treatment systems to provide increased habitat diversity.

Reforestation has been seen for some time as an important tool to mitigate elevated concentrations of greenhouse gases in the atmosphere. As the carbon trading market develops, production of woody biomass in phytoirrigation systems may produce marketable credits for sequestered carbon. Such credits may provide another source of revenue to offset the costs of phytoremediation systems.

MAINTENANCE AND MANAGEMENT OF KEY RISK FACTORS

Once well-established, phytoirrigation using trees generally requires relatively low levels of maintenance, but a number of issues should be considered in the planning and design phases. Replanting may be required to achieve the target tree density. Thinning and pruning may be required to develop a high quality tree if harvest is a project objective, and to maintain irrigation application uniformity if branches interfere with sprinkler irrigation. When trees are planted with a grass cover (as was done at Woodburn and Riverbend), grass growth will be fairly vigorous until the tree canopy closes and shades the grass. Regular mowing of the area between tree rows may be required.

Weeds, insects, diseases, and animals can reduce the performance of the tree plantation. Weeds are the single greatest threat to most tree systems in the first year. A combination of herbicides, mowing, and possibly some manual weeding is required.

Tree-based phytoremediation systems may require special management, and may not be suitable for all sites. A summary of significant siting and operations and maintenance issues is provided in Table 23-5. One example of a proactive approach is the Swedish willow (*Salix* spp.) plantations which consist of carefully selected plant material resistant to frost, rust, and insects as a result of comprehensive plant breeding programs started 15 years ago (Larrson *et al.* 1998).

COSTS

The costs associated with a phytoremediation site are capital costs and operations and maintenance costs. Capital costs include the initial and post-harvest investment in installed materials and equipment. Specifically, the costs include land purchase (if necessary), site preparation and soil amendments, construction and installation of equipment, tree purchase, and tree planting. Operations and maintenance costs, which recur annually, are usually projected for the defined life span of the system. Operations and maintenance costs typically include those for equipment, including that of irrigation pump stations, water distribution and application systems, tree monitoring systems, plantation management, site maintenance, power, harvest, nutrients, soil amendments, pesticides, and replanting.

The costs of tree-based phytoremediation can be offset by potential revenue from trees. However, because the market for tree products fluctuates significantly based on supply and demand, these revenues tend to be omitted from the cost evaluation of a system. Because remediation is the primary goal of these systems, any return from the system that can reduce net costs is usually thought of as an extra benefit.

The following cost estimates apply to phytoirrigation systems ranging in size from approximately 0.4 to 40 hectares. These estimates are based on tree spacing on a grid pattern averaging about 3 to 4.6 meters square, and a permanent drip or micro-spray irrigation system. Costs for a subsurface drainage system and sump pump are not included. These cost estimates must be taken with the *caveat* that each site has specific considerations and requirements, which will dictate a unique set of costs.

For irrigated hybrid poplar (*Populus* spp.) tree applications established during the late 1990s in the U.S., per hectare capital costs have ranged from a low of around 12 350 to more than 247 000 U.S. dollars. This wide range of costs reflects the varied infrastructure that may be required, such as pumping systems, access roads, and utilities. The range also reflects the economies of scale that can be realized.

TABLE 23-5 Summary of Key Siting and Operations and Maintenance Issues

Siting or operations and maintenance issue	Management tools
Phytotoxicity of target or nontarget constituents in irrigation water	Dilution Plant selection—herbaceous plant species may be more tolerant than trees
Phreatophytic tree roots [*e.g.*, poplar (*Populus* spp.) and willow (*Salix* spp.)] can enter and restrict the flow of subsurface drains, sewers, and water pipes. Expanding roots can break buried power and communication cables and even small diameter pressure pipelines	Careful mapping of site infrastructure before implementation and relocate as needed Nonperforated drains near trees Root barrier fabrics to prevent root contact with buried facilities
Even fast-growing trees provide little remedial benefits in the year of establishment	Intercropping with herbaceous species Use larger acreage Alternate technologies
Weed pressure, especially in establishment year	Till several times during the year before planting Appropriate herbicides Weed barrier fabrics around tree stem for small plantations
Animal damage: beaver (*Castor* spp.), elk (*Cervus elephus*), porcupines (*Erethizon dorsatum*), voles (*Microtus* spp.), deer (*Odocoileus* spp.), pocket gophers (*Thomomys* spp.), and others	Knowledge of and adaption to the most likely pests Vigilant monitoring Timely implementation of control measures Fencing
Insects and disease, especially for poplars (*Populus* spp.)	Only use trees adapated to local conditions and pests Monitoring and timely implementation of control measures
Large land area required	Site selection—May be key limitation on some sites
Severe weather (wind, ice, flooding, and unseasonable frost)	Only use trees adapted to local conditions Site selection Provide excess capacity Backup treatment systems

Annual operations and maintenance costs also vary over a wide range, based on the intensity of maintenance required by the site, the level of sophistication of operations, and the level of monitoring needed for the system.

Excluding environmental monitoring, annual operations and maintenance costs for a phytoremediation site can range in the neighborhood of 7400 to 25 000 U.S. dollars per hectare. Environmental monitoring can impose additional costs of 25 000 to 124 000 U.S. dollars per hectare or more, particularly on smaller sites that require extensive monitoring and reporting.

Acknowledgments

The authors would like to thank Frank Tiwari, Dean Morrison, and Frank Sinclair of the City of Woodburn, Oregon; Mike O'Brien of Beale Air Force Base; and Dan Wilson and George Duvendack of Waste Management's Riverbend Landfill for their cooperation in the development of this chapter and their vision and commitment to the systems described herein. Important information on applications in Sweden were provided by Kenth Hasslegren of SWECO VBB VIAK AB.

REFERENCES

Allen, R.G., L.S. Pereira, D. Raes, and M. Smith (1998) Crop evapotranspiration: guidelines for computing crop water requirements. Irrigation and Drainage Paper No. 56. Food and Agriculture Organization of the United Nations, Rome, Italy.

Bañuelos, G.S., M.C. Shannon, H. Ajwa, J.H. Draper, J. Jordahl, and L. Licht (1999) Phytoextraction and accumulation of boron and selenium by poplar (*Populus*) hybrid clones. *Int. J. Phytoremed.* **1**(1): 81–96.

Barton, L., C.D.A. McLay, L.A. Schipper, and C.T. Smith. (1999) Denitrification rates in a wastewater irrigated forest soil in New Zealand. *J. Environ. Qual.* **28**: 2008–2014.

Broadbent, F.E. and H.M. Reisenauer (1985) Fate of wastewater constituents in soil and groundwater: nitrogen and phosphorus. In: *Irrigation with Reclaimed Municipal Wastewater – A Guidance Manual.* G.S. Pettygrove and T. Asano, eds. Lewis Publishers, Chelsea, Michigan.

CH2M HILL (2000) Biosolids land application farm site feasibility and conceptual design. Prepared for Metropolitan Wastewater Commission in Association with Mater Engineering, Portland, Oregon.

CH2M HILL (1999a) (1999) Operations and monitoring: City of Woodburn reclaimed water and biosolids reuse system. Prepared for City of Woodburn, Oregon. Portland, Oregon.

CH2M HILL (1999b) Reclaimed water reuse management plan for the Woodburn WWTP poplar plantation. Prepared for City of Woodburn, Oregon. Portland, Oregon.

CH2M HILL (1999c) Guidance for successful phytoremediation. Prepared by CH2M HILL for the Center for Waste Reduction Technologies (CWRT) and the American Institute of Chemical Engineers New York.

Dastane, N.B. (1974) Effective rainfall in irrigated agriculture. Irrigation and Drainage Paper No. 25. Food and Agriculture Organization of the United Nations, Rome, Italy.

DeBell, D.S. (1990) *Populus trichocarpa* Torr. & Gray: black cottonwood. In: *Silvics of North America: Hardwoods. Vol. 2.* R.M. Burns and B.H. Honkala, eds. U.S. Department of Agriculture Agricultural Handbook 654, pp. 570–576.

Dickman, D.I. and J.G. Isebrands (1999) *Caveat emptor*: nursery professionals should use only those poplar clones that have proved to be hardy and pest- and disease-resistant. *American Nurseryman* **1**: 60–65.

Follet, R.F., D.R. Keeney, and R.M. Cruse (1991) *Managing Nitrogen for Groundwater Quality and Farm Profitability*. Soil Science Society of America, Madison, Wisconsin.

Gochis, D.J. and R.H. Cuenca (2000) Plant water use and crop curves for hybrid poplars. *J. Irrig. Drain Eng.* **126**(4): 206–214.

Hart, W.E. and W.N. Reynolds (1965) Analytical design of sprinkler systems. *Trans. Am. Soc. Agric. Eng.* **8**(1): 83–85, 89.

Hasselgren, K. (1999a) Sewage sludge recycling in short-rotation energy forestry–a pilot study. *Waste Manage. Res.* **17**: 251–262.

Hasselgren, K. (1999b) Irrigation of short-rotation energy forestry with secondary wastewater effluent. The Swedish Water and Wastewater Works Association Report No.1999–5, Stockholm, Sweden (In Swedish, English summary).

Hasselgren, K., D. Roy, and K.M. Persson (1999c) Leachate treatment in soil–plant systems. *RVF Nytt.* **5**: 23–26 (In Swedish by the Swedish Association of Waste Management, Malmo, Sweden).

Hegde, R.S. and J.S. Fletcher (1996) Influence of growth stage and season on the release of root phenolics by mulberry as related to development of phytoremediation. *Chemosphere* **2**: 2471–2479.

Heilman, P.E., T.M. Hinckley, D.A. Roberts, and R. Ceulemans (1996) Production Physiology. In: *Biology of Populus*. vol. 1. R.F. Stettler, H.D. Bradshaw, Jr., P.E. Heilman, and T.M. Hinckley, eds. National Research Council of Canada, Ottawa, pp. 459–489.

Heilman, P.E., R.F. Stettler, D.P. Hanley, and R.W. Carkner (1995) High yield hybrid poplar plantations in the Pacific Northwest. Pacific Northwest Extension Bulletin 356, Washington State University Cooperative Extension, Pullman.

Hitchcock, H.R., C.D. Barton, and L. Newman (2002) Implementing ecological engineering at the Savannah River Site through phytoremediation. In: *Ecological Engineering: Implementing the Profession.* Am. Ecological Eng. Soc. Held in Burlington, Vermont, April 28–30.

Interstate Technology and Regulatory Council (ITRC) (2001) Phytotechnology technical and regulatory guidance document. Phytotechnologies Team. Washington, D.C. (http://www.itrcweb.org).

Jensen, M.E., R.D. Burman, and R.G. Allen, eds. (1990) *Evapotranspiration and Irrigation Water Requirements.* Manual on Engineering Practice No. 70, American Society of Civil Engineers, New York.

Larsson, S., G. Melin, and H. Rosenqvist (1998) Commercial harvest of willow wood chips in Sweden. In: *Biomass for Energy and Industry*. Proc. 10th European Conference and Technology Exhibition, C.A.R.M.E.N., Rimpar, Germany. Held in Wurzburg, Germany, June 8–11, pp. 200–203.

Levy, G.J., A. Rosenthal, J. Tarchitzky, I. Shainberg, and Y. Chen (1999) Soil hydraulic conductivity changes caused by irrigation with reclaimed wastewater. *J. Environ. Qual.* **28**: 1658–1664.

Madison, M.F., F. Sinclair, and M.R. Khan (2000) Wastewater irrigation of poplars minimizes surface water degradation. Paper presented at the American Society of Agricultural Engineers 4th Decennial National Irrigation Symposium. Held in Phoenix, Arizona, November 14–16.

Meisinger, J.J. and G.W. Randall (1991) Estimating nitrogen budgets for soil-crop systems. In: *Managing Nitrogen for Groundwater Quality and Farm Profitability*. R.F. Follet, D.R. Keeney, and R.M. Cruse, eds. Soil Science Society of America, Madison, Wisconsin.

Nadler, A., A. Gamliel, and I. Peretz (1999) Practical aspects of salinity effects on TDR-measured water content: a field study. *Soil Sci. Soc. Am. J.* **63**: 1070–1076.

Newman, L., S. Strand, J. Duffy, G. Ekuan, M. Raszaj, B. Shurtleff, J. Wilmoth, P. Heilman, and M. Gordon (1997) Uptake and biotransformation of trichloroethylene by hybrid poplars. *Environ. Sci. Technol.* **31**: 1062–1067.

Newman, L.A., X. Wang, I.A. Muiznieks, G. Ekuan, M. Ruszaj, R. Cortellucci, D. Domroes, G. Karscig, T. Newman, R.S. Crampton, R.A. Hashmonay, M.G. Yost, P.E. Heilman, J. Duffy, M.P. Gordon, and S.E. Strand. (1999). Remediation of trichloroethylene in an artificial aquifer with trees: a controlled field study. *Environ. Sci. Technol.* **33**: 2257–2265.

Nicholas, I.D., J.M. Carnus, and G.R. Oliver (2000) Comparative performance of tree species in New Zealand wastewater irrigation systems. In: *The Forest Alternative: Principles and Practice of Residuals Use*. Symposium Proceedings. Held in Seattle, Washington, July 14–16, 1997.

Pallardy, S.G. and T.T. Kozlowski (1981) Water relations of *Populus* clones. *Ecology* **62**: 159–169.

Riddell-Black, D., R. Marshall, and R. Fergeson (2000) *Salix* as a means of cost-effective, onsite management of landfill leachate. In: 21st Session of the International Poplar Commission (IPC 2000) Poplar and Willow Culture: Meeting the Needs of Society and the Environment. U.S. Forest Service General Technical Report NC-215. St. Paul, Minnesota.

Rog, C. and J.G. Isebrands (2000) Growth and contaminant uptake by hybrid poplars and willows in response to application of municipal landfill leachate. In: 21st Session of the International Poplar Commission (IPC 2000) Poplar and Willow Culture: Meeting the Needs of Society and the Environment. U.S. Forest Service General Technical Report NC-215. St. Paul, Minnesota.

Selker, J.S. (1996) Applying preferential flow concepts to horticultural water management. Salt Management Workshop Proceedings. *HortTech.* **6**(2): 107–110.

Shannon, M.C., G.S. Banuelos, J.H. Draper, H. Ajwa., J. Jordahl, and L. Licht (1999) Tolerance of hybrid poplar trees (*Populus*) irrigated with varied levels of salt, selenium, and boron. *Int. J. Phytoremed.* **1**(3): 273–288.

Shrive, S.C. and R.A. McBride (1995) Physiologic responses of red maple saplings to sub-irrigation with an untreated municipal landfill leachate. *Waste Manage. Res.* **13**: 219–239.

Shrive, S.C., R.A. McBride, and A.M. Gordon (1994) Photosynthetic and growth responses of two broad-leaf tree species to irrigation with municipal landfill leachate. *J. Environ. Qual.* **23**: 534–542.

Soil Conservation Service (SCS) (1985) National Engineering Handbook, Section 4, Hydrology, U.S. Department of Agriculture, Washington, D.C.

Soil Conservation Service (SCS) (1986) Urban Hydrology for Small Watersheds, Technical Release 55, U.S. Department of Agriculture, Washington, D.C.

Speir, T.W., A.P. van Schaik, H.A. Kettles, K.W. Vincent, and D.J. Campbell (1999) Soil and stream water impacts of sewage effluent irrigation onto steeply sloping land. *J. Environ. Qual.* **28**: 1105–1114.

Stephens, D.B. (1996) *Vadose Zone Hydrology*. Lewis Publishers, Boca Raton, Florida.

Sun, W.H., W.J. Lo, and P.K. Jones (2000) Evaluation of coastal trees for the phytoremediation of petroleum contaminated soils in Hawaii. In: *Proceedings of SoilRem 2000: International Conference of Soil Remediation*, Y.M. Luo, S.P. McGrath, Z.H. Cao, F.J. Zhao, Y.X. Chen, and J.M. Xu, eds. Hangzhou, China, pp. 100–104.

Tomer, M.D., T.H. Charleson, C.T. Smith, L. Barton, A.J. Thorn, and G.J.H.P. Gielen (2000) Evaluation of treatment performance and processes after six years of wastewater application at Whakarewarewa Forest, New Zealand. In: *The Forest Alternative: Principles and Practice of Residuals Use*. Symposium Proceedings. Held in Seattle, Washington, July 14–16, 1997.

Tossell, R.W., K. Binard, L. Sangines-Uriarte, M.T. Rafferty, and N.P. Morris (1998) Evaluation of tamarisk and eucalyptus transpiration for the application of phytoremediation. In: *Bioremediation and Phytoremediation: Chlorinated and Recalcitrant Compounds*. G.B. Wickramanyake and R.E. Hinchee, eds. Battelle Press, Columbus, Ohio. Held in Monterey, California, May 18–21, **C1-4**: 257–262.

U.S. Environmental Protection Agency (U.S. EPA) (1981) Process design manual for land treatment of municipal wastewater. Report EPA 625/1–81–013, Washington, D. C.

Warrick, A.W. (1974) Time-dependent linearized infiltration. I. point sources. *Soil Sci. Soc. Am. J.* **38**: 383–386.

Wilson, L.G., L.G. Everett, and S.J. Cullen (1995) *Handbook of Vadose Zone Characterization and Monitoring*. Lewis Publishers, Boca Raton, Florida.

24

SALT TOLERANT PLANTS TO CONCENTRATE SALINE WASTE STREAMS

M. C. Negri, R. R. Hinchman, and T. L. Settle

SUMMARY OF PRACTICAL APPLICATIONS

Produced water is a salty brine generated during the production of oil and gas. Brine disposal is typically achieved by reinjection in deep wells, and the cost is highly dependent on volume. We investigated, evaluated, and developed the principal components of a cost-effective, engineered bioreactor system that will, through natural plant processes of transpiration and salt tolerance, reduce the volume of produced water. After a screening of plant effectiveness at the greenhouse scale, the system was tested at the pilot scale at a gas and oil field in Oklahoma. Saltwater cordgrass (*Spartina alterniflora*) and great bulrush (*Scirpus validus*) were chosen to treat the wastewater with chloride levels between 30 000 and 60 000 milligrams per liter in a two-compartment bioreactor. These tests confirmed greenhouse data and demonstrated volume reductions of 75 percent in less than 5 days. Compared to evaporation from an open surface, plants achieved 30 percent more volume reduction in the same time. Cost savings between 36 and 70 percent are feasible under full field conditions.

INTRODUCTION AND BACKGROUND

Water that is brought to the surface during natural gas and oil production from gas wells is called "produced water." This water originates from the same geologic formation as the gas or oil, which includes ancient marine deposits, as well as coal beds of various geologic ages. The constituents of produced water reflect these geochemical conditions. Elevated concentrations of dissolved salt, often greater than seawater, is perhaps the most common characteristic. Other constituents can include bicarbonates, carbonates, sulfates, hydrocarbons, and other organic compounds, trace elements, and heavy metals.

Phytoremediation: Transformation and Control of Contaminants,
Edited by Steven C. McCutcheon and Jerald L. Schnoor.
ISBN 0-471-39435-1 (cloth)

Produced water disposal is subject to Environmental Protection Agency (EPA) regulations (U.S. EPA 1986) and the costs of current treatment and disposal methods are becoming unacceptably high. With ever more stringent discharge regulations for saline contaminants, particularly in the arid west, the only viable alternative is often the reinjection of the salty brines into old oil and gas wells. There are two conceptual routes to improve the treatment of saline wastewater. The first is to remove the salt contaminant and make the quality of the water acceptable for discharge or reuse (desalination). The second is to reduce the volume of the water and concentrate the salt (dewatering), thus decreasing the volume of the wastewater to be treated.

The cost of disposal by well injection is directly related to the volume of water disposed. In comparison, the salinity of the water disposed is a relatively minor, maintenance related cost factor. Therefore, disposal costs can be lowered if the volume of produced water can be decreased, even if the salinity of the water to be disposed increases.

The volume of produced water to be disposed by well injection is often decreased by using solar evaporation ponds, a solution that requires a significant amount of space and time. In this chapter, we will illustrate a novel approach to enhance wastewater reduction through plant transpiration. This approach can result in significant savings to the oil and gas industry by reducing the requirements of time and space to achieve volume reduction.

Water constitutes 70 to 90 percent of the total plant weight, and is essential for plant nutrition and growth. The uninterrupted uptake of water by the roots, translocation to other plant tissues, and evaporation through leaf stomata (transpiration) is fundamental to the life of a majority of plants. Evapotranspiration is the loss of water by plant transpiration plus that by direct evaporation from an open water surface. Plants differ by the amount of biomass generated per unit of water used. In crops, this ratio describes plant water-use efficiency. In the usual agricultural setting, a crop is efficient when the plants can produce a large amount of biomass utilizing very little water. In situations where plants are grown to reduce wastewater volume, an "efficient" plant is one that has an elevated consumption of water per unit of biomass. We adopted this reversed concept of efficiency in the development of a waste-treatment method that used green plants to reduce the volume of produced waters.

To develop biological treatment systems for saline produced waters, halophytes are a logical choice to take advantage of the natural capabilities of these plants to grow and be highly productive at elevated external salinity levels. Those plants that occupy highly saline habitats along seacoasts, in salt and brackish marshes, in estuaries, and in dry, saline interior regions have already undergone strong selection for the capability to tolerate salt. This large pool of plants provides numerous candidates for use in engineered, plant-based systems designed to treat saline wastewater. Very exhaustive reviews of halophyte plants and the associated uses are given by Aronson (1989), Chokur-Allah *et al.* (1997), and Yensen (1999). Although some experimentation using halophytes have been conducted for the removal of salt from soil (Yensen *et al.* 1999), very limited work has been conducted for the treatment of brine except by the authors of this chapter.

Basically, higher plants have two mechanisms for tolerating elevated external salt concentrations, and these are fundamentally different. Some halophytes accumulate salt to high levels in tissues, whereas others exclude the salt (Sutcliffe 1962). Some plants that accumulate salt exude it through special salt glands located on the leaf surfaces where salt crystals often accumulate until washed off by rain. A capability to exclude salts does not come without a price. The compensatory processes that must occur in the tissues to maintain osmotic balance in the absence of high internal salt concentrations consume energy and substrates that might otherwise have gone into growth. Thus, it is not surprising to observe a decrease in growth and plant stature in these species as the salt concentration in the growth medium increases. Furthermore, there is a limit to salt concentration beyond which even halophytes cannot grow or live.

Plant Screening at the Greenhouse Scale

A plant mass balance of the relative abundance of water (approximately 90 percent of plant biomass) and of salt (1 to 3 percent) and the uptake (liters of water per day *versus* grams of salt per day) from the waste stream favors "dewatering," which includes salt exclusion or uptake and exudation rather than "desalination" (using salt-accumulating plants). On the basis of available literature, a number of halophytes were selected for testing in the greenhouse, to compare effectiveness at decreasing the volume of saline produced waters.

Halophytes were selected based on the degree of salt tolerance, transpiration rates, biomass production, nutrient requirements, ease of propagation, and possibilities for beneficial reuse of biomass. These salt excluders and nonexcluders included saltwater cordgrass (*Spartina alterniflora*), coastal dropseed (*Sporobolus virginicus*), perennial glasswort (*Salicornia virginica*), sawgrass (*Cladium jamaicense*), vermilion cordgrass (*Spartina alterniflora* var. *vermilion*), and great bulrush (*Scirpus validus*). These plants were tested hydroponically at the bench scale with three initial salt concentrations (0, 1.5, and 3 percent). Details of the materials and methods are reported by Negri *et al.* (1997). As a consequence of the methods used, evapotranspiration was measured over an essentially continuous range of salt concentrations, up to the salt tolerance limit of the species. To an extent, this regime mimics the normal salt fluctuations to which salt marsh plants are exposed as a result of tidal changes, drying, and the dilution by rainfall of seawater in saturated wetland soil. As the tolerance limit approached, the plants went through stages characterized by a slowing or cessation of growth that included severe wilting, drying up of the leaves, and eventually, death of the plant.

A summary of evapotranspiration rates for all six plant species tested is given in Figure 24-1. Observations during several experimental runs indicate that under a variety of conditions mature plants were more effective at

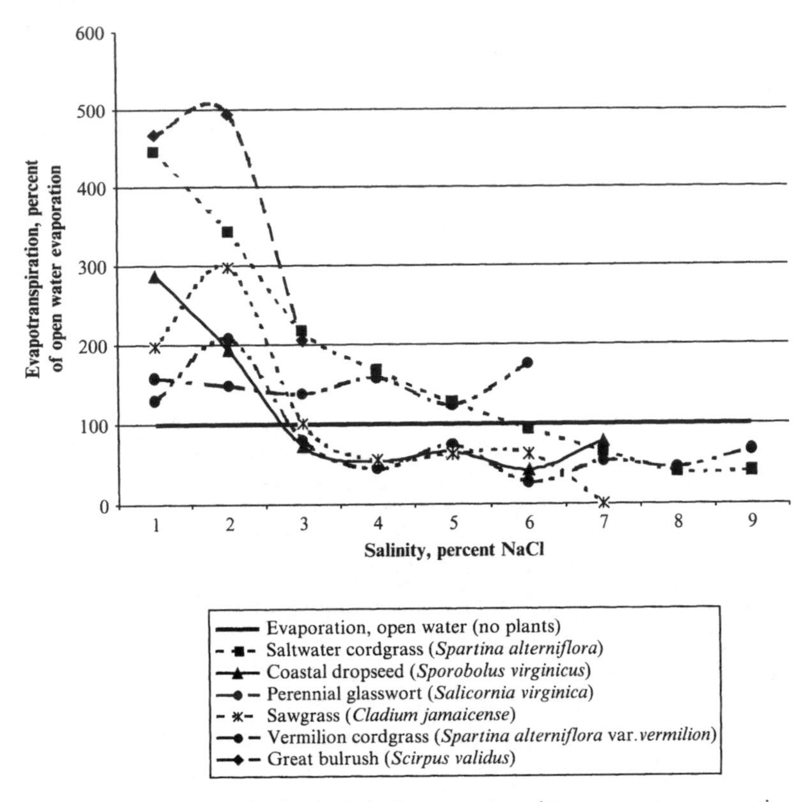

Figure 24-1 Evapotranspiration by halophytes compared to open water evaporation.

maintaining large evapotranspiration rates than an equal biomass of young plants. A review of performance, characteristics, and adaptability to a bio-reactor follows for each plant.

Saltwater Cordgrass (Spartina alterniflora)

Saltwater cordgrass (*Spartina alterniflora*) attains a large height (up to 1.5 meters) in natural habitats. The growth habit of the multiple, long, flat leaves provides a large surface area of transpiring tissue. Considered an excellent component in our bioreactor system, this plant maintained a mean evapo-transpiration rate per container that was larger than the open water evapo-ration rate up to a salt concentration of almost 6 percent. Saltwater cordgrass (*Spartina alterniflora*) is, therefore, appropriate to treat higher salinity waste-water. Winter dormancy could, however, slow both growth and evapotran-spiration rates, as described in the following paragraph for Vermilion cordgrass (*Spartina alterniflora* var. *vermilion*).

Vermilion cordgrass (*Sportina alterniflora* var. *vermilion*) is a variety of soltwater cordgrass (*Spartina alterniflora*), which can grow considerably taller and more robustly than the standard cordgrass discussed earlier. However, experimental evapotranspiration rates recorded both on a total and on a fresh weight basis were considerably lower than the smaller cordgrass, probably because of partial winter dormancy. Even though the mean total evapotranspiration rates were relatively small, these rates were consistently greater than the open water evaporation rates (up to salt levels over 5 percent in the nutrient solution) in which these plants were grown.

Coastal Dropseed (Sporobolus virginicus)

On a fresh weight basis, coastal dropseed (*Sporobolus virginicus*) has the largest evapotranspiration rate of any halophyte tested to date. Coastal dropseed (*Sporobolus virginicus*) is also very salt tolerant, but the natural growth habit is small (up to 30 to 40 centimeters in height) compared to saltwater cordgrass (*Spartina alter niflora*). Thus, total biomass production was smaller, which resulted in smaller total evapotranspiration rates for an equivalent size container.

Perennial Glasswort (Salicornia virginica)

Perennial glasswort (*Salicornia virginica*) is a true halophyte, which is very salt tolerant and takes up salt to relatively high tissue concentrations. Perennial glasswort (*Salicornia virginica*) has a relatively slow growth rate and the natural growth habit is small (shorter than 50 centimeters in height) compared to the grasses tested. Because of these factors, total evapotranspiration rates were the smallest of the species tested to date. However, perennial glasswort (*Salicornia virginica*) maintained total evapotranspiration rates greater than open water up to a salt concentration of more than 2 percent in growth medium.

Sawgrass (Cladium jamaicense)

Sawgrass (*Cladium jamaicense*) is a sedge of medium stature that attains a height of 1 to 3 meters in a natural habitat. In these tests, sawgrass (*Cladium jamaicense*) showed good salt tolerance and acceptable total evapotranspiration rates. Growing in nutrient solutions with up to 3 percent salt, the maximum total evapotranspiration rates were relatively large and the mean rate was considerably greater than the open water rate. Sawgrass (*Cladium jamaicense*) would be a good candidate species for the treatment of moderate salinity wastewater. Perennial grasswort (*Salicornia virginica*) and sawgrass (*Cladium jamaicense*) both had greater evapotranspiration rates under saline conditions than in a nutrient solution without salt. This is a common response in salt tolerant plants, and is an example of the tendency to grow and function best under the conditions to which the plants are adapted, even when these conditions are more harsh compared to those for which these plants may be occasionally exposed (*e.g.*, salt dilution by tidal flushing or rainfall in the halophyte habitat) (Long and Mason 1983).

Great Bulrush (Scirpus validus)

Great bulrush (*Scirpus validus*) exhibited very large evapotranspiration rates and growth (biomass production) at salinities up to 3 percent. The available data indicates that these rates may drop significantly at higher salt concentrations. Total evapotranspiration rates were the largest for any plant tested. Great bulrush (*Scirpus validus*) also had the largest positive difference from the mean total evapotranspiration rates compared to open water—up to 3 percent salt concentration. The large rates are also maintained when calculated on a per gram fresh weight basis. These results indicate that great bulrush (*Scirpus validus*) would be an ideal candidate for the treatment of lower salinity wastewater.

Development of a Bioreactor Model and Field Verification at a Pilot-Scale

The results from the greenhouse studies were used to develop a conceptual model of a "dewatering" or salt exclusion bioreactor and to predict effectiveness at reducing the volume of produced water. Such a bioreactor, depicted in Figure 24-2, was conceived and developed as two connected wastewater processing compartments operating much like two adjacent rice paddies, each containing different and possibly multiple plant species. The wastewater flows into the first compartment and remains until the water attains (*via* evaporation and plant transpiration) a higher salinity level and a concomitant reduction in volume. The salinity level of each compartment is determined by the tolerance of the plants in that compartment. The water must be replaced by incoming lower salinity wastewater before the concentrated salt begins to have adverse, elevated stress effects on the plants in that compartment. To do

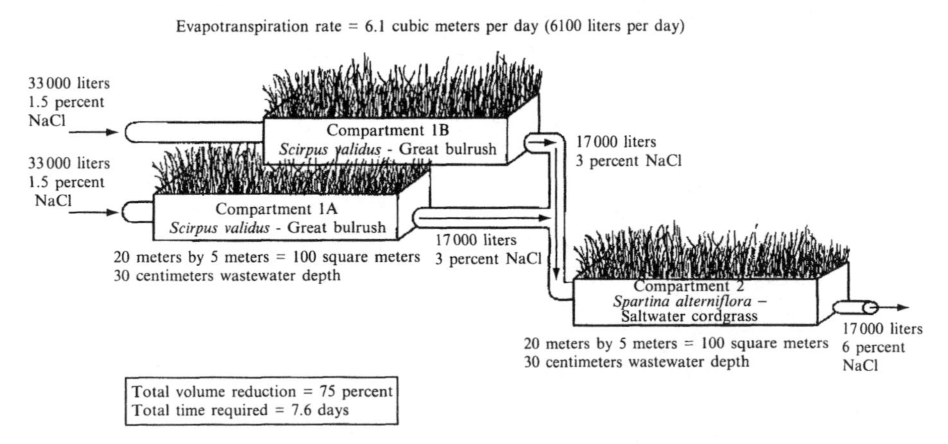

Figure 24-2 Plant bioreactor model.

this, the wastewater (reduced in volume and greater in salinity) is moved by gravity to the next compartment, which contains plants of a higher salt tolerance than the previous upstream compartment and the process is repeated. A measurement of conductivity of the water in the first compartment determines when water should be moved to the second one.

Among the candidate plant species studied in the greenhouse tests, great bulrush (*Scirpus validus*) was chosen to grow in the first compartment, due to the greater evapotranspiration rates and smaller salt tolerance. Saltwater cordgrass (*Spartina alterniflora*) was chosen to grow in the second compartment, which would receive wastewater at a higher salinity level.

Using a conservative value of 40 percent of the maximum evapotranspiration rate for the plants selected, the bioreactor model was designed to treat 67 cubic meters of wastewater over a total surface of 300 square meters (200 square meters for the first stage and 100 square meters for the second) at a depth of 30 centimeters. A volume reduction of 75 percent was projected to occur in 7.6 days.

Based on the developed concept, an experimental pilot-scale bioreactor was then assembled and successfully tested at two oil and gas fields in Oklahoma (Settle *et al.* 1998) (Figure 24-3). The units operated on 100 percent produced water and were monitored for volume reduction, salinity, and contaminants. The batch processors were made of two separate flow-through troughs filled with pea gravel, each with a capacity of approximately 0.7 cubic meter, that processed wastewater of increasing salinity. The plant species used included: great bulrush (*Scirpus validus*), saltwater cordgrass (*Spartina alterniflora*), and Vermilion cordgrass (*Spartina alterniflora* var. *vermilion*). The two sites, about 55 kilometers (35 miles) apart, had a

Figure 24-3 Pilot bioreactor operated at an oil/gas well (Devon Energy) in Oklahoma.

produced water production of 1 and 20 barrels per day (42 gallons or 0.16 cubic meter, and 840 gallons or 3.2 cubic meters), respectively.

The set of tests demonstrated salt tolerance, transpiration rates, and the advantages of halophyte evapotranspiration *versus* open water evaporation. In both cases, within a few days of transpiration the saltwater cordgrass (*Spartina alterniflora*) leaves were covered with salt crystals. In the first field test, conducted from May to October 1995, 30 000 milligrams of chloride per liter (estimated total dissolved solids somewhat greater than 50 000 milligrams per liter) produced water was used. Although some problems were encountered, during both field tests, a constant 75 percent reduction in produced water volume was achieved in only 4 days using this system.

In the second field test, conducted from May to October 1996, produced water with 60 000 milligrams chloride per liter (estimated total dissolved solids somewhat greater than 100 000 milligrams per liter) was used. Parallel troughs were set up to allow for continuous treatment of the incoming produced waters. In this test, halophytes consistently reduced the volume of produced water by 30 percent more than the open water control (no plants).

Halophyte leaves from this second field test were harvested without removing salt crystals adhering to the leaves. Under the advice of the Natural Resources Conservation Service in Baton Rouge, Louisiana, both the incoming produced waters and the salt covered leaves were tested for toxic contaminants and no significant levels were found. Following this reassurance, salt covered halophyte leaves were then given to a local rancher who used them for cattle feed. Subsequently, cattle were allowed to graze on the halophytes.

The cost and related savings of the disposal of produced water at the second site were calculated (Settle *et al.* 1998). The typical cost to transport and dispose of the produced water at this site was 1.20 U.S. dollars per barrel (7.50 U.S. dollars per cubic meter) of produced water. As transportation and disposal cost are reduced proportionally with the volume of the waste, a cost saving of 0.9 U.S. dollar per barrel (5.60 U.S. dollars per cubic meter) was projected, partly offset by the cost of plants, testing equipment, materials, and data acquisition. For the pilot adopted for these tests, the wetland plant and other costs were approximately 0.46 U.S. dollar per barrel (2.88 U.S. dollars per cubic meter) of produced water treated, but further decreases in these offset costs are anticipated as reuse of plants and equipment, management costs, and crop revenues become increasingly favorable, especially in dry locales that often experience shortages of forage. Cost savings between 36 and 70 percent are considered within reach under full field conditions.

FINDINGS

Greenhouse studies have shown that the primary factors that determine the amount of water removed from a closed system *via* evapotranspiration are the total aerial plant biomass of transpiring tissue and the surface area. Therefore, large, robust species of grasses or grasslike plants (sedges and rushes) best match the profile of the "ideal plant" for volume reduction. Those sedges and rushes exhibiting moderate to high salt tolerance (halophytes native to salt or brackish marshes, or coastal environments) would be appropriate for treating saline produced water. Of the species tested to date, those best matching the "ideal plant" profile were saltwater cordgrass (*Spartina alterniflora*), sawgrass (*Cladium jamaicense*), and great bulrush (*Scirpus validus*). Winter dormancy of candidate bioreactor plants is a potential problem that requires additional untreated water holding capacity or research on different plant species. Winter dormancy could slow both growth and evapotranspiration rates of certain plants, particularly in the northern latitudes of the U.S.

Recent developments have shown that phytoremediation is also a viable alternative for a number of other oil and gas field related constituents including heavy metals and petroleum hydrocarbons. In light of these findings, volume reduction through evapotranspiration could be accompanied by the reduction of hydrocarbon contaminants by degradation mediated by the extensive fine-root systems of produced water-transpiring halophytes.

The salt tolerant plants used in the pilot test have definite potential in the treatment of produced water as a cheaper alternative to solar evaporation ponds. Potential savings of 36 to 70 percent were calculated from pilot tests of 1 to 20 barrels (0.16 to 3.2 cubic meters) of incoming produced water per day; the tests reflected mere volume reduction and did not consider additional advantages gained by reducing treatment times. Extrapolation of the batch and pilot scale data to large-scale field conditions depends on a number of factors and will require careful consideration of all pertinent variables. These factors include choice of equipment and lining of the field installation, spacing treatment wetlands or bioreactors to maximize transpirates, the length of growing seasons, frequency of harvesting, effect of mixing species, winter dormancy, and specific climatic considerations such as rainfall and relative humidity. Full field scaling up would be the essential step to optimize design selections and costs.

Acknowledgments

This project was funded by the Gas Technology Institute, Des Plaines, Illinois, and by Devon Energy Corporation, Oklahoma City, Oklahoma. This chapter has been created by the University of Chicago as operator of Argonne National Laboratory under Contract No. W-31-109-ENG-38 with the U.S. Department of Energy. The

REFERENCES

Aronson, J.A. (1989) HALOPH—a data base of salt tolerant plants of the world. Office of Arid Lands Studies, The University of Arizona, Tucson, Arizona.

Chokur-Allah, R., C.V. Malcolm, and A. Hamdy (1997) *Halophyte and Biosaline Agriculture*. Marcel Dekker, New York.

Long, S.P. and C.F. Mason (1983) *Saltmarsh Ecology*. Blackie-Chapman and Hall, New York.

Negri, M.C., R.R. Hinchman, and J. Mollock (1997) Biotreatment of produced waters for volume reduction and contaminant removal. In: *Proceedings of the 4th International Petroleum Environmental Conference*. The Integrated Petroleum Environmental Consortium (IPEC), University of Tulsa, Oklahoma. Held in San Antonio, Texas, Sept. 9–12.

Settle, T., G.N. Mollock, R.R. Hinchman, and M.C. Negri (1998) Engineering the use of green plants to reduce produced-water disposal volume. SPE paper 39763, presented at the 1998 Society of Petroleum Engineers Permian Basin Oil and Gas Recovery Conference, Held in Midland, Texas, March 5–7.

Sutcliffe, J.F. (1962) *Mineral Salts Absorption in Plants*. Pergamon Press, New York.

U.S. Environmental Protection Agency (U.S. EPA) (1986) Wastes from the exploration, development, and production of crude oil. Natural Gas and Geothermal Energy, Technical Report EPA/530-SW-86-051. Office of Solid Waste and Emergency Response, Washington, D.C.

Yensen, N.P. (1999) *Salt-Tolerant Plants and Their Uses*. NyPa Publishing Company, Tucson, Arizona.

Yensen, N.P., R.R. Hinchman, M.C. Negri, G.N. Mollock, T. Settle, C.S. Keiffer, D.J. Carty, B. Rogers, R. Martin, R. Erickson, and R.H. Adams (1999) Using halophytes to manage oilfield saltwater: disposal by irrigation/evapotranspiration and remediation of spills. In: *Proceedings of the 6th Annual International Petroleum Conference*. The Integrated Petroleum Environmental Consortium (IPEC), The University of Tulsa, Oklahoma. Held in Houston, Texas, Nov.

SECTION VII

LATEST ADVANCES

25

METABOLISM AND GENETICS OF ATMOSPHERIC NITROGEN DIOXIDE CONTROL USING POLLUTANT-PHILIC PLANTS

H. Morikawa, M. Takahashi, and Y. Kawamura

SUMMARY OF PRACTICAL IMPLICATIONS

This chapter describes five topics on phytoremediation of atmospheric nitrogen dioxide (NO_2), a major air pollutant. These include the following: (1) screening of 217 plant taxa for high capability to assimilate the nitrogen (N) from nitrogen dioxide (more than a 600-fold variation in the capability was discovered); (2) manipulation of genes involved in nitrate metabolism to eventually produce transgenic pollutant-philic plants; (3) identification of denitrification in plants to develop gas-to-gas conversions in plants to transform nitrogen dioxide to gaseous nitrogen (N_2); (4) experimentation involving novel aspects of the metabolic fate of nitrogen dioxide in plants, which show nitrogen compounds that are not recoverable by Kjeldahl digestion; and (5) prospects for phytoremediation of air pollution with the invention of "green walls" that cover the vertical area of buildings and highways in urban areas with nitrogen dioxide-philic plants.

GLOSSARY

Antisense RNA: Artificial, single-stranded ribonucleic acid (RNA) molecule that is complementary in sequence to all or part of a molecule of messenger RNA and hybridizes (forms a molecular hybrid), thus interfering with a specific RNA molecule.

Phytoremediation: Transformation and Control of Contaminants,
Edited by Steven C. McCutcheon and Jerald L. Schnoor.
ISBN 0-471-39435-1 (cloth)

Apoplast: Continuous, (nonliving) cell wall and xylem phase outside the plasma membrane.

Atomic percentage of ^{15}N (atom percent ^{15}N): Corresponds to [(the number of ^{15}N molecules) × 100]/(the total number of ^{15}N plus ^{14}N molecules).

Chloroplast: Intracellular organelle that contains a green pigment (chlorophyll), in which photosynthesis and amino acid synthesis occur.

Reductase: Enzyme that catalyzes the reduction of a substrate (the name of which precedes reductase, *e.g.*, nitrite reductase) by adding hydrogen or electrons (with or without addition of protons).

Symplast: Space consisting of all the plant cell tissues connected by plasmodesmata (which connect the cytoplasm of one cell with another) that is surrounded by a plasma membrane.

Synthetase: Enzyme that catalyzes a reaction producing a substance (the name of which precedes synthetase) by concomitant breaking of a diphosphate bond in a nucleoside triphosphate such as adenosine triphosphate (ATP).

SOLUTION TO POLLUTION

In order to mitigate environmental pollutants and to establish sustainable living on a global scale, we need to address at least three equally important issues. These issues include changing life styles to stop or decrease the emission of pollutants, developing technologies to avoid or greatly decrease emissions, and decontaminating pollutants that are already in the environment.

Gaseous pollutants such as nitrogen oxides (NO_x), once emitted, disperse into the atmosphere rapidly. Centralized treatment of these scattered pollutants is energy intensive (equivalent to the entropy), and costly. By contrast, plants are driven by solar energy, self-reproducing, and concentrate and detoxify pollutants. Therefore, plants should be very useful to clean up dispersed, ambient pollutants (Figure 25-1).

In fact, natural and planted vegetation are an efficient sink for various air pollutants including nitrogen oxides (Yunus *et al.* 1996), carbon dioxide (CO_2) (Pacala *et al.* 2001), and polycyclic aromatic hydrocarbons (Simonich and Hites 1994, 1995). Accordingly, several investigations proposed that plants should be utilized to reduce pollutant concentrations in the atmosphere (Hill 1971, Okano *et al.* 1988, Simonich and Hites 1994, 1995, Weber *et al.* 1995).

With the eventual aim of providing a solution to global air-pollution problems from plant science and plant biotechnology, we have been investigating the production of a novel "nitrogen dioxide-philic plant" that can grow with atmospheric nitrogen dioxide (NO_2) as the sole nitrogen (N) source (Kamada *et al.* 1992, Morikawa *et al.* 1992). To create such plants, understanding the biochemical and molecular biological fundamentals of the metabolism of nitrogen dioxide in plants is essential. Genetic engineering of plants based on the genomics and proteomics (study of genes and proteins) is one of the challenges.

Figure 25-1 Green wall with nitrogen dioxide-philic plants to clean up polluted air in an urban area.

SCREENING OF NATURALLY OCCURRING NITROGEN DIOXIDE-PHILIC PLANTS

The air in polluted urban areas in Japan contains about 0.05 parts per million of nitrogen dioxide, similar or even higher amounts of nitrogen oxide (NO), and 0.1 parts per million of nonmethane hydrocarbons (http://www.env.go.jp/en/index.html). Methane and other volatile organic hydrocarbons photochemically produce a hydroxyl radical (highly reactive molecule containing a single unpaired electron), which reacts with nitrogen oxides producing photooxidative ozone. Nitrogen dioxide and these oxidants are very toxic gases in the lower atmosphere (troposphere) for animals, humans, and plants.

Road transport, the major anthropogenic source in most developed countries, produced in 1984 as much as 75 percent of the nitrogen oxides in some metropolitan cities, and the percentage has continued to rise with increased vehicular volume. In many developing countries as well, petrol-fueled motor vehicles are the principal source of nitrogen oxides (Yunus *et al.* 1996).

Plants take up nitrogen dioxide (Zeevaart 1974, Lee and Schwartz 1981, Rowland *et al.* 1985, Ramge *et al.* 1993, Larcher 1995) and incorporate the nitrogen into organic nitrogenous compounds (Durmishidze and Nutsubidze 1976, Rogers *et al.* 1979, Yoneyama and Sasakawa 1979, Kaji *et al.* 1980, Wellburn 1990). Little is known, however, about whether this activity varies among plant species and families apart from the study by Durmishidze and Nutsubidze (1976).

We therefore studied the assimilation of nitrogen dioxide in 217 taxa of the higher plants, including 50 wild herbaceous plants (42 genera of 15 families) collected from roadsides, 60 cultivated herbaceous plants (55 genera of 30 families), and 107 cultivated woody plants (74 genera of 45 families). The 217 taxa were analyzed for response to fumigation with nitrogen dioxide as described elsewhere (Morikawa *et al.* 1998a). Briefly, the plants were fumigated with 4 ± 0.1 parts per million $^{15}NO_2$ (51.6 atom percent ^{15}N) for 8 hours at 22 °C under fluorescent lights (70 micromoles of photons per square meter per second), with a 70 percent relative humidity, and a carbon dioxide concentration at atmospheric levels (0.03 to 0.04 percent). Leaves were harvested from fumigated plants, washed with distilled water, and dried at 80 °C for 3 days. Total reduced nitrogen in the fine powder of the plants was determined by the Kjeldahl method (Morikawa *et al.* 1998a). The ^{15}N contents in samples were determined by mass spectrometry.

Two parameters were determined. One was the content of reduced nitrogen derived from nitrogen dioxide in the leaves of fumigated plants (milligrams of N per gram of dry weight). This parameter is a measure of the assimilation of nitrogen dioxide by plants during 8 hours of fumigation. The other was the percentage of reduced nitrogen derived from nitrogen dioxide in the total reduced nitrogen, which corresponds to the plant affinity for nitrogen dioxide. This is a measure of the utilization efficiency of nitrogen dioxide in the nitrogen metabolism of plants.

The amount of reduced nitrogen derived from nitrogen dioxide differed by a factor of 657 between the highest [Australian manna or ribbon gum (*Eucalyptus viminalis*), 6.57] and lowest [blushing bride (*Tillandsia ionantha*) and *Tillandsia caput-medusae*, 0.01] in the 217 taxa. The affinity for nitrogen dioxide differed about 80-fold between these two species, manna gum (*Eucalyptus viminalis*) (12.5 percent) and blushing bride (*Tillandsia ionantha*) (0.16 percent).

Table 25-1 shows the distribution of taxa having reduced nitrogen derived from nitrogen dioxide and affinity for nitrogen dioxide among the 217 taxa and in some selected families. Of the 217 taxa analyzed, only 16 taxa had reduced nitrogen derived from nitrogen dioxide of greater than 3 milligrams of N per gram of dry weight. Similarly, only nine taxa had greater than 10 percent affinity for nitrogen dioxide. Compositae, Myrtaceae, Solanaceae, and Salicaceae had a few taxa with high reduced nitrogen derived from nitrogen dioxide, and high affinity for nitrogen dioxide. None of the plants examined from the families Gramineae, Theaceae, and Rosaceae showed high reduced nitrogen derived from nitrogen dioxide or high affinity for nitrogen dioxide.

Figure 25-2 shows the affinity for nitrogen dioxide of each of the 217 plants plotted against the content of reduced nitrogen derived from nitrogen dioxide. The relationship between these two parameters is somewhat complex. Data were fitted to a third order polynomial regression (solid line) as an aid to the eye. Nine species involving six trees and three herbs had high affinities for nitrogen dioxide as shown in Figure 25-2 (see also Table 25-1). Five species had high reduced nitrogen derived from nitrogen dioxide; northern

TABLE 25-1 Number of Taxa Having Reduced Nitrogen Derived from and Affinity for Nitrogen Dioxide among 217 Taxa and Selected Families

	Reduced N from NO_2 (milligrams of N per gram dry weight)						NO_2 affinity (percent)		
	>5	4 to 5	3 to 4	2 to 3	<2	Total	>10	5 to 10	<5
All taxa	5	4	7	30	171	217	9	61	147
Compositae	2[a]	0	2	10	16	30	1[b]	15	14
Gramineae	0	0	0	1	12	13	0	0	13
Myrtaceae	1[c]	2	0	2	5	10	2[d]	6	2
Theaceae	0	0	0	1	8	9	0	2	7
Rosaceae	0	0	1	1	6	8	0	2	6
Solanaceae	1[e]	0	0	1	0	2	1[e]	1	0
Salicaceae	1[f]	0	1	0	1	3	1[f]	2	0

[a]Fireweed (*Erechtites hieracifolia*) and red flower ragleaf (*Crassocephalum crepidioides*).
[b]Fireweed (*Erechtites hieracifolia*).
[c]Manna gum (*Eucalyptus viminalis*).
[d]Manna gum (*Eucalyptus viminalis*) and *Eucalyptus cinerea*.
[e]Tobacco (*Nicotiana tabacum*).
[f]Black poplar (*Populus nigra*).

Japanese magnolia (*Magnolia kobus*) (4.92 milligrams of N per gram of dry weight), manna gum (*Eucalyptus viminalis*) (6.57 milligrams of N per gram dry weight), black poplar (*Populus nigra*) (5.14 milligrams of N per gram of dry weight), tobacco (*Nicotiana tabacum*) (5.72 milligrams of N per gram of dry weight), and fireweed (*Erechtites hieracifolia*) (5.72 milligrams of N per gram of dry weight).

These plants are "nitrogen dioxide-philic" because the nitrogen in nitrogen dioxide seems to have an important role in nitrogen metabolism. Plants of these types are suited for use as vegetation in the roadside green zones and parks to reduce the atmospheric concentration of nitrogen dioxide.

The amount of reduced nitrogen derived from nitrogen dioxide varied within a plant family; about 20-fold in the Compositae (30 taxa), 8-fold in the Gramineae (13 taxa), 5-fold in the Myrtaceae (10 taxa), and 62-fold in the Theaceae (9 taxa). Variation in the affinity for nitrogen dioxide within a family was less pronounced than the amount of reduced nitrogen derived from nitrogen dioxide; about 2, 5, 4, and 21-fold for Compositae, Gramineae, Myrtaceae, and Theaceae, respectively. The reduced nitrogen derived from nitrogen dioxide also varied among the individuals in a species, being 26-fold in goldenrod (*Solidago altissima*) (19 individuals), 21-fold in garden sorrel (*Rumex acetosella*) (12 individuals), 18-fold in red flower ragleaf (*Crassocephalum crepidioides*) (8 individuals), 12-fold in tobacco (*Nicotiana tabacum*) (14 individuals), 11-fold in plantin (*Plantago lanceolata*) (4 individuals), 12-fold in northern Japanese magnolia (*Magnolia kobus*) (3 individuals), 11-fold in Japanese spindle (*Euonymus japonicus*) (3 individuals), and 7-fold in *Prunus cerasoides* (10 individuals). These results imply that the further screening of

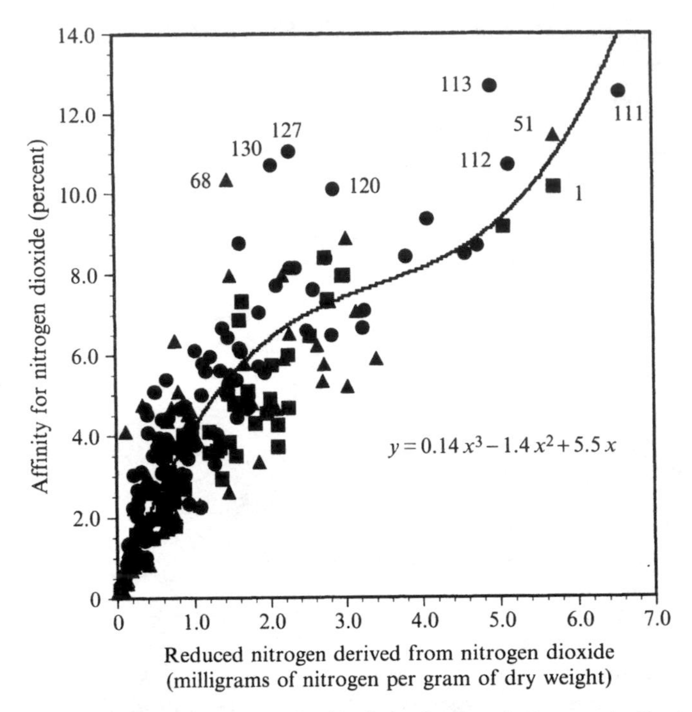

Figure 25-2 The percentage of nitrogen dioxide-derived reduced nitrogen in the total reduced nitrogen, defined as the "NO_2-affinity," plotted against the content of nitrogen dioxide–derived reduced nitrogen (milligrams of nitrogen per gram of dry weight) in 217 individual taxa. (■) Wild herbaceous plants collected from the roadside (50 taxa), (▲) cultivated herbaceous plants (60 taxa), and (●) cultivated woody plants (107 taxa). Nine species that had NO_2-affinities greater than 10 percent were indicated by the numbers (113) northern Japanese magnolia (*Magnolia kobus*) (12.7 percent), (111) manna gum (*Eucalyptus viminalis*) (12.5 percent), (51) tobacco (*Nicotiana tabacum*) (11.4 percent), (127) jasmine (*Gardenia jasminoides*) (11.1 percent), (112) black poplar (*Populus nigra*) (10.7 percent), (130) *Eucalyptus cinerea* (10.7 percent), (68) borage (*Borago officinalis*) (10.4 percent), (1) fireweed (*Erechtites hieracifolia*) (10.1 percent), and (120) Chinese tallow tree (*Sapium sebiferum*) (10.1 percent). The solid line is a third order polynomial regression used as a visual aid. The correlation coefficient (r) for this fit is 0. 87.

plants at the individual level in a species or at the species level in a genus for high nitrogen dioxide assimilation, done along the lines of this study, is imperative to optimize selection of plant lines for phytoremediation of air pollution. Details of the results for the 217 taxa were published previously (Morikawa *et al.* 1998a) and updated at http://www.mls.sci.hiroshima-u.ac.jp/mpb/Home.html.

Clearly, the capability to assimilate nitrogen dioxide and the affinity for this air pollutant are enormously diversified among naturally occurring plants. The assimilation of nitrogen dioxide in plants involves complex processes. Nitrogen dioxide molecules are first taken up through stomata (microscopic openings in the leaves) into leaf tissues, transported through the cell walls and plasma membrane into the leaf cells, and metabolized.

The nitrogen dioxide, after uptake into the stomatal cavity, reacts with water to yield nitric acid (HNO_3), nitrous acid (HNO_2), and nitric oxide (NO), although the quantitative details are only poorly understood (Zeevaart 1974, Rowland *et al.* 1985). Lee and Schwartz (1981) proposed that equal amounts of nitrate (NO_3^-) and nitrite (NO_2^-) are formed from nitrogen dioxide at the cell-wall surface and both are transported inside the cell across the plasma membrane. Ramge *et al.* (1993) proposed that reducing reagents such as ascorbate in the cell wall reduce nitrogen dioxide to a nitrite ion, and that the ion is transported into the cell. Rowland *et al.* (1985) determined that nitric acid will ionize to yield nitrate ions and protons at the pH of cell wall water but that a comparable volume of nitrous acid will be present as a mixture of nitrite ions, protons, and undissociated acid. Consequently for plant metabolism, these gases are in five different chemical forms, all of which can pass through cell membranes, but only two of which (nitrate and protons) are normally present in appreciable quantities inside the cell. The overall efficiency to take up these different chemical species through the stomata, cell walls, and plasma membrane into cells is a vital factor that determines the capability of plants to assimilate nitrogen dioxide, and thereby explains the variation among plant taxa. Involvement of the cuticle layer (waxy layer covering the surfaces of the leaves) in nitrogen dioxide uptake into leaf tissues is also postulated (Larcher 1995), at least in certain plant species, which makes the transport process even more complex.

The variation of metabolic activities among plant species, if any, is another important factor that affects the variation in capabilities of plants to assimilate nitrogen dioxide. In fact, genetic engineering of plants with genes involved in nitrate metabolism will improve nitrogen dioxide assimilation (see section "Feasibility of Transgenic Nitrogen Dioxide-Philic Plants"). Also, the metabolism of nitrogen dioxide in plants is more complicated than anticipated so far, and substantial amounts of nitrogen dioxide are converted into unknown nitrogen compounds that are not recoverable by the Kjeldahl method (see section "Metabolic Fate of the Nitrogen Dioxide in Plants"). Wellburn (1990) reported that nitrogen dioxide inhibits photosynthesis in plants. Therefore, differences among plant species in the resistance to the toxic effects of nitrogen oxides, if any, may possibly be another cause for the differences found in the reduced nitrogen derived from nitrogen dioxide, and the affinity for nitrogen dioxide among the plant species studied.

Based on an integrated bioinformatics analysis using DNA (deoxyribonucleic acid) microarrays, proteomics, and databases of known metabolic interaction, Ideker *et al.* (2001) have identified 997 genes that are involved in the yeast galactose-utilization pathway. A similar approach to the metabolic control of nitrogen dioxide in plant leaves is needed to understand the genetic and biochemical causes for the observed diversity in the assimilation of nitrogen dioxide among plant species. This information will be very useful to engineer nitrogen dioxide-philic plants.

FEASIBILITY OF TRANSGENIC NITROGEN DIOXIDE-PHILIC PLANTS

Most nitrogen dioxide molecules taken up into plants are converted, either chemically or enzymatically, to nitrate and nitrite, assimilated to ammonia, and incorporated into amino acids and other organic nitrogenous compounds. The conversion occurs through a primary nitrate assimilation pathway (Zeevaart 1976, Yoneyama and Sasakawa 1979, Kaji *et al.* 1980, Rowland *et al.* 1985, Wellburn 1990, Morikawa *et al.* 1992, 1998b, Ramge *et al.* 1993) as shown in

$$\begin{array}{l} \qquad\quad NO_2\ (+4) \\ \quad\ \swarrow \qquad \searrow \\ NO_3^-\ (+5) \rightarrow NO_2^-\ (+3) \rightarrow NH_3\ (-3) \rightarrow \\ \qquad \text{Glutamine}\ (-3) \rightarrow \text{Glutamate}\ (-3) \\ \qquad\qquad\qquad\qquad \rightarrow \text{Other organic N compounds}\ (-3) \end{array} \tag{25-1}$$

The numbers in parentheses in Equation (25-1) are the formal oxidation states of nitrogen in the respective compounds. In the primary nitrate pathway, nitrate is converted to nitrite by a two-electron reduction catalyzed by the first enzyme, nitrate reductase. The second enzyme, nitrite reductase, catalyzes the six-electron reduction to convert nitrite to ammonia. Ammonia thus formed is incorporated into glutamic acid to form glutamine by glutamine synthetase, the third enzyme.

We thought these three enzymes may play key roles in the assimilation of nitrogen dioxide, and therefore studied transgenic plants in which these enzymes are genetically modified. Using a gene engineering approach, we first constructed "expression vectors" for the gene of each of these enzymes. Each vector consisted of three DNA fragments for each gene: promoter, terminator, and complementary DNA (cDNA). The former two fragments (about 1000 base pairs each) are essential to allow vigorous synthesis of messenger RNA and thereby synthesis of the enzyme protein corresponding to the last fragment. We employed a promoter originated from the genome of a cauliflower mosaic virus (a potent plant virus), and a terminator originated from a bacterium *Agrobacterium tumefaciens* (a potent plant parasite in soil). The last fragment of the vector, which bears the code for each enzyme, is placed between the promoter and terminator. The expression vector was then inserted into a "DNA propagation vector" or plasmid, which allows vigorous propagation of the genetic element in bacterial cells (normally *Escherichia coli*). The resulting chimeric (foreign-gene-bearing) plasmids were designated pBSL16, pSNIRH, and pAGS2, which bear the cDNA of the nitrate reductase gene from tobacco (*Nicotiana tabacum*) (Vincentz and Caboche 1991), that of nitrite reductase gene isolated from spinach (*Spinacia oleracea*) (Back *et al.* 1988), and that of glutamine synthetase gene from *Arabidopsis thaliana*, respectively. The pAGS2 contained a 1545 base-pair long cDNA that covers the whole polypeptide code for the *Arabidopsis thaliana* glutamine synthetase (Morikawa *et al.* unpublished).

For the selection of transformants, we used another plasmid vector designated pCH (Goto *et al.* 1993), which bears a gene (Gritz and Davies 1983) that confers resistance to hygromycin B, an antibiotic. The pSNIRH contains this expression vector but the other two do not.

Each plasmid was separately introduced, alone or together with plasmid pCH, into root sections of *Arabidopsis thaliana* by particle bombardment as described in Takahashi and Morikawa (1996). Transgenic plants were selected for hygromycin resistance, and the presence of the introduced cDNA sequences in the *Arabidopsis thaliana* genome was confirmed by the polymerase chain reaction (PCR) method using gene-specific primers (Takahashi *et al.* 2001). The first to third generation plants bearing chimeric genes of tobacco (*Nicotiana tobocum*) nitrate reductase cDNA, spinach (*Spinacia oleracea*) nitrite reductase cDNA, and *Arabidopsis* (*thaliana*) glutamine synthetase cDNA were analyzed further.

The nitrate reductase activity was determined by the method of Wray and Fido (1990) using the reduced form of nicotinamide adenine dinucleotide (NADH) as an electron donor. A typical value for wild type plants was 5.48 ± 0.80 nanomoles of NO_2^- per minute per milligram of protein (average of five samples ± standard deviation), whereas activities of nine transgenic plant lines varied from about 0.96- to 1.8-fold compared to the control values, depending on the individual lines of transgenic plants (Figure 25-3A).

The nitrite reductase activity was determined according to the method of Wray and Fido (1990) using methyl viologen as an electron donor. Typical nitrite reductase activity of the control plants was 150.6 ± 30.7 nanomoles of NO_2^- per minute per milligram of protein (average of five samples ± standard deviation), whereas activities of twelve transgenic plant lines varied from about 0.8- to 1.8-fold *versus* the control values, depending on the individual transformant lines (Figure 25-3B). Of the twelve nitrite reductase transformant lines studied, four had highly significant (probability of error less than 1 percent) high nitrite reductase activities compared to the wild-type control. One had 180 percent higher nitrite reductase activity (Figure 25-3B).

The glutamine synthetase (transferase) activity was determined according to Rhodes *et al.* (1975), with one unit of glutamine synthetase activity being defined as 1 micromole of γ-glutamylhydroxamate (GHA) formed per minute per milligram of protein. The typical value for wild type plants was 4.18 ± 0.03 micromole of γ-GHA per minute per milligram of protein (average of three samples ± standard deviation), whereas activities of eight transgenic plant lines varied from about 1.1- to 1.5-fold *versus* the control values, depending on the individual transformant lines (Figure 25-3C).

Nitrogen dioxide assimilation in our nitrite reductase transgenic *Arabidopsis thaliana* plants was analyzed using ^{15}N-labeled nitrogen dioxide and mass spectrometry as described elsewhere (Morikawa *et al.* 1998a, Moribawa *et al.* 1998b; see also the section "Screening of Naturally Occurring Nitrogen Dioxide-Philic Plants"). Typical values for the contents of reduced nitrogen derived from nitrogen dioxide were 1.18 ± 0.08 milligrams of N per gram of dry weight (average of five samples $\pm$ standard deviation) for the control plants. Neither the nitrate reductase nor glutamine synthetase transformants showed a significant increase in the content of reduced nitrogen derived from nitrogen dioxide (see Figure 25-3A and C). Of the twelve nitrite reductase transformant lines studied, three had significantly (probability of error less than 1 percent) high levels of the reduced nitrogen derived from nitrogen dioxide compared to the control. Two of the three had 140 percent higher reduced nitrogen derived from nitrogen dioxide (see Figure 25-3B). A 40 percent increase in the capability of these transgenic plants to assimilate nitrogen dioxide is statistically significant and may provide a basis for further improvement of the nitrite reductase gene by bioengineering. The potential to overexpress (*i.e.*, increased production) nitrite reductase could lead to a new phytoremediation strategy.

The flux control coefficient is a measure of the effect of change in a single enzyme activity on the flux (Kacser and Porteous 1987, Stitt and Sonnewald 1995). The coefficients for nitrate reductase, nitrite reductase, and glutamine synthetase were estimated as reported elsewhere (Runquist and Kruger 1999). The nitrite reductase flux control coefficient for nitrogen dioxide assimilation was estimated to be about 0.4. A similar value of the flux control coefficient of nitrite reductase in nitrogen dioxide assimilation was obtained using transgenic tobacco (*Nicotiana tabacum*) plants (Vaucheret *et al.* 1992) which contain antisense nitrite reductase cDNA, and hence much reduced nitrite reductase activity. The flux control coefficients of nitrate reductase and glutamine synthetase were much smaller than this value (−0.01 and −0.1). Together, these findings indicate that nitrite reductase is a controlling enzyme in nitrogen dioxide assimilation by plants.

In transgenic *Arabidopsis thaliana* plants bearing chimeric spinach (*Spinacia oleracea*) nitrite reductase cDNA (Takahashi *et al.* 2001), three positive correlations were found for nitrite reductase gene expression. These correlations were between the total nitrite reductase messenger RNA (determined by a competitive PCR method) and total nitrite reductase protein (determined by western blot) contents (correlation coefficient equal to 0.74), between the total nitrite reductase protein and nitrite reductase activity (correlation coefficient equal to 0.71), and between nitrite reductase activity and content of reduced nitrogen derived from nitrogen dioxide (correlation coefficient equal to 0.65). Of twelve transgenic lines, four had significantly (probability of error less than 1 percent) high levels of nitrite reductase activity compared to the wild type control, and three had significantly (probability of error less than 1 percent) high reduced nitrogen derived from nitrogen dioxide. Each of the latter three had one to two copies of spinach (*Spinacia oleracea*) nitrite reductase cDNA per haploid genome (minimum set of genome DNAs). These results clearly indicate that increases in nitrite reductase activities can enhance the nitrogen

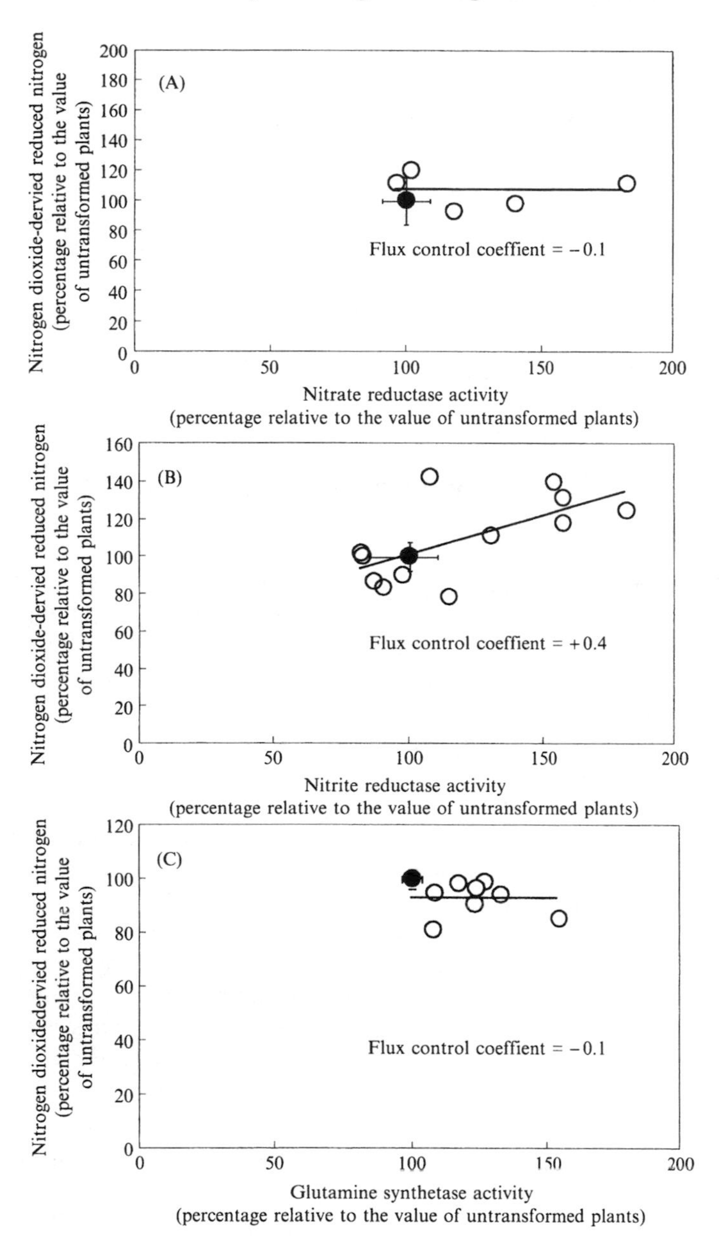

Figure 25-3 Reduced nitrogen derived from nitrogen dioxide plotted against the enzyme activities of nitrate reductase (A), nitrite reductase (B), and glutamine synthetase (C) in transgenic *Arabidopsis thaliana* plants. Transgenic plant lines containing chimeric genes of (A) tobacco (*Nicotiana tabacum*) nitrate reductase, (B) spinach (*Spinacia oleracea*) nitrite reductase, and (C) *Arabidopsis* (*thaliana*) glutamine synthetase were fumigated with 4 parts per million of $^{15}NO_2$ for 8 hours. The leaves were then harvested and analyzed for the reduced nitrogen by the Kjeldahl method followed by mass spectrometry. The enzyme activities were determined as described in the text. Each value (*open circles*) of transgenic plant lines corresponds to the percentage relative to the control (*closed circle*) of untransformed *Arabidopsis thaliana* plants.

dioxide assimilation capability of plants. This is further support and encouragement for the eventual goal of producing a novel "nitrogen dioxide-philic plant" that can grow with atmospheric nitrogen dioxide as the sole nitrogen source (Kamada *et al.* 1992, Morikawa *et al.* 1992).

Denitrification in Transgenic Plants to Transform Nitrogen Dioxide Gas Directly to Dinitrogen Gas

Denitrification is the process by which nitrate or nitrite ions in water or soil are reduced to a gaseous form of nitrogen, generally nitrogen gas (N_2) or nitrous oxide (N_2O) (Zumft 1997) as shown in Equation (25-2). The numbers in parentheses in the following equation are the formal oxidation states of nitrogen in the respective compounds

$$NO_3^-(+5) \rightarrow NO_2^-(+3) \rightarrow NO(+2) \rightarrow N_2O(+1) \rightarrow N_2(0) \qquad (25\text{-}2)$$

This process has an important role in the global nitrogen cycle (Williams *et al.* 1992, Conrad 1996), through which as much as 2×10^{11} kilograms of nitrogen is metabolized per year. Denitrification was long thought to be a characteristic of prokaryotes (Zumft 1997) and some fungi (Shoun *et al.* 1992). However, Vanecko and Varner (1955), Harper (1981), Mulvaney and Hageman (1984), Dean and Harper (1986), and Klepper (1990) reported some evidence indicating that plants have denitrifying capability. For example, detached wheat (*Triticum aestivum*) leaves that take up nitrite ions were reported to emit nitrogen gas (N_2) (Vanecko and Varner 1955). Nevertheless, the plant materials used in these experiments were not grown aseptically, and therefore the possibility that microorganisms participated in the denitrification could not be excluded.

We have shown that genetically engineered tobacco (*Nicotiana tabacum*) plants successfully convert nitrate and nitrite into nitrous oxide (Goshima *et al.* 1999). In these transgenic tobacco (*Nicotiana tabacum*) plants, the activity of nitrite reductase, a key enzyme in the nitrate-assimilation pathway, is largely reduced (less than 5 percent of the wild type) by the expression of nitrite reductase cDNA in an antisense orientation (Vaucheret *et al.* 1992). These tobacco (*Nicotiana tabacum*) plants were grown aseptically, and therefore the involvement of microorganisms in the emission of nitrous oxide in this experiment can be excluded.

The emission of nitrous oxide in this transgenic tobacco (*Nicotiana tabacum*) is largely inhibited by treatment with tungstate, a potent inhibitor for nitrate reductase (Goshima *et al.* 1999). The emission of nitrous oxide from the engineered tobacco (*Nicotiana tabacum*) was repressed to 26 percent by treatment with 150 micromoles of tungstate, in which 74 percent of the nitrate reductase activity was eliminated (Figure 25-4).

In addition, the emission of nitrous oxide appeared to be highly dependent on the nitrite level in plant cells. The emission of nitrous oxide from engineered tobacco (*Nicotiana tabacum*) plants was increased 25-fold when nitrite

(A)

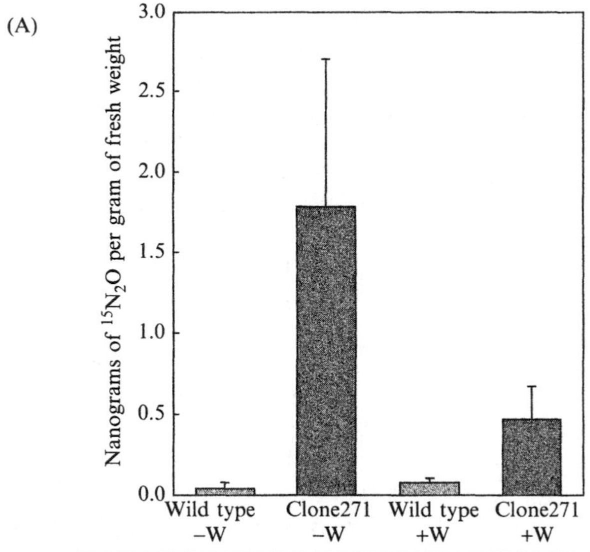

(B)

Plants	Nitrate recuctase activity (nanomoles per minute per milligram of protein)
Wild type, −W	6.43 ± 1.71
Wild type, +W	1.57 ± 0.27
Clone 271, −W	12.61 ± 1.07
Clone 271, +W	2.04 ± 0.32

Figure 25-4 Emission of nitrous oxide from transgenic tobacco (*Nicotiana tabacum*) plants and inhibition by tungstate treatment. (A) Emission of nitrous oxide from clone 271 or wild-type plants that were grown on $^{15}NO_3^-$-containing medium supplemented with or without tungstate for 2 weeks. These plants were transferred to the same medium in which ^{15}N was replaced by ^{14}N and grown for another 2 weeks. The ^{15}N-labeled nitrous oxide emitted from the plants during the second 2 weeks was analyzed by gas chromatography–mass spectrometry. The values +W and −W correspond to the result for plants that were incubated with and without tungstate, respectively. Bars give the mean value ± one standard deviation of four independent experiments. (B) Nitrate reductase activities of clone 271 or wild type plants that were grown on NO_3^--containing medium supplemented with (+W) or without (−W) tungstate for 2 weeks. Each value is the mean of results from four samples with the standard deviation.

concentration in the culture medium was increased from 0.5 to 2 millimoles. No emission of nitrous oxide from tobacco (*Nicotiana tabacum*) plants was found when grown on ammonia labeled with ^{15}N, indicative that NH_4^+-nitrogen was not metabolized to nitrous oxide in these tobacco (*Nicotiana tabacum*) plants.

Therefore, we deduced the following mechanism for the emission of nitrous oxide in the engineered tobacco (*Nicotiana tabacum*)

$$\begin{array}{l} NO_3^- \rightarrow NO_2^- \rightarrow NH_4^+ \rightarrow \text{glutamine} \rightarrow \text{amino acids (nitrate assimilation)} \\ \qquad\quad \downarrow \\ \qquad\quad ? \rightarrow N_2O \end{array} \qquad (25\text{-}3)$$

Consistent with this is the fact that the nitrous oxide emission by clone 271 is highly heat-labile and light dependent, indicative that probably nitrous oxide is emitted by an enzymatic process in plants (Goshima and Morikawa unpublished results). Based on the analyses of the amount of ^{15}N derived from nitrate incorporated into the plant tissue and the amount of ^{15}N derived from emitted nitrous oxide by gas chromatography–mass spectrometry, it was found that slightly more than 0.01 percent of the nitrate-nitrogen incorporated into plants is converted to the nitrous oxide-nitrogen (Goshima *et al.* 1999).

Wildt *et al.* (1997) reported that various higher plants such as sunflower (*Helianthus annuus*), tobacco (*Nicotiana tabacum*), corn (*Zea mays*), soybean (*Glycine max*), spruce (*Picea* spp.), sugarcane (*Saccharum officinarum*), rape (*Brassica napus*), and spinach (*Spinacia oleracea*), emit nitric oxide. They estimated that the emission of nitric oxide by plants on a global scale to be 2.3×10^{11} grams of nitrogen per year. Very recently, Smart and Bloom (2001) have reported that wheat (*Triticum aestivum* L. 'Veery 10') leaves emit significant amounts of nitrous oxide, and that nitrous oxide production in leaves occurred during photoassimilation of nitrite ions in the chloroplast. Smart and Bloom (2001) also estimated that production of nitrous oxide by plants could account for as much as 12 percent of the global emission of nitrous oxide. Taken together, these results suggest that the denitrification process is not limited to bacteria or fungi, but plants also appear to denitrify nitrate and nitrite.

We have discovered that even untransformed wild-type tobacco (*Nicotiana tabacum*) plants as well as the engineered tobacco (*Nicotiana tabacum*) plants emit nitrous oxide when fumigated with nitrogen dioxide (Morikawa and Hakata, unpublished results). Nitrous oxide is converted to nitrogen gas in a single step by a nitrous oxide reductase encoded by the *nos Z* gene (Zumft 1997) in bacterial nitrifiers, such as *Pseudomonas stutzeri* (see Equation 25-2). Introduction of the *nos Z* gene to plants and subsequent expression may confer the denitrifying capability to convert nitrogen dioxide gas to nitrogen gas. Engineering of such gas-to-gas converting transgenic plants for the roofs and walls of buildings, and highway corridors is an intriguing strategy (see the section "Prospects for Phytoremediation of Air Pollution" at the end of this chapter).

METABOLIC FATE OF THE NITROGEN DIOXIDE IN PLANTS

Because the nitrogen from nitrogen dioxide taken up into plant cells was shown to be converted to organic nitrogen (amino groups of amino acids, Yoneyama and Sasakawa 1979), it has been assumed that nitrogen dioxide is metabolized through the primary nitrate pathway (see Equation 25-1). However, Arimura *et al.* (1998) discovered that the metabolic fate of the nitrogen of nitrogen dioxide is somewhat different from that of nitrate in plants. The total nitrogen derived from nitrogen dioxide taken up by plants was not equal to the sum of organic nitrogen as measured by the Kjeldahl procedure plus inorganic nitrogen, and the difference was designated as unidentified nitrogen. In the case of *Arabidopsis thaliana* leaves fumigated with 4 parts per million ^{15}N-labeled nitrogen dioxide for 8 hours, about 30 percent of the total nitrogen was unidentified, together with about 70 percent of the reduced Kjeldahl nitrogen and 1 to 2 percent of the inorganic nitrogen.

Table 25-2 summarizes the content of unidentified nitrogen in the leaves of three plant species including *Arabidopsis thaliana*. All of the plants that were fumigated with nitrogen dioxide, as described in the footnotes of Table 25-2, had the unidentified nitrogen. Conceivably, the formation of unidentified compounds from nitrogen dioxide is a common phenomenon in plants.

It has been known that the nitrogen of nitro compounds and nitrosamines is not recoverable by the Kjeldahl method when no additional reagents such as reducing agents were added to the digestion mixture (*e.g.*, Bradstreet 1965). More recently, we have determined putative nitro compounds in the leaves of *Arabidopsis thaliana* fumigated with 4.0 ± 0.4 parts per million $^{15}NO_2$ (51.6 atomic percent of ^{15}N) for 4 hours. From fumigated leaves, a cationic fraction, which contained a major part of the unidentified nitrogen (Morikawa and Arimura unpublished results), was prepared by using cation exchange resin, and subjected to Kjeldahl digestion. After the distillation of ammonia, the residue was analyzed for ^{15}N by a mass spectrometer. From this analysis, the putative nitro compounds were estimated to be 7.2 percent of the total unidentified nitrogen of *Arabidopsis thaliana* leaves.

The chemistry of nitrogen oxides makes clear that these compounds include or are easily transformed to reactive species under physiological conditions to form various nitro- and nitroso-compounds in biological systems (Suzuki *et al.* 1991, (Suzuki *et al.*) 1993, Beckman 1996, Nonoyama *et al.* 1999). Nitrogen dioxide is a strong nitrating reagent in the presence of ozone or oxygen gas at room temperature or even below 0 °C (Suzuki *et al.* 1991, Suzuki 1993). Nitric oxide is formed from nitrogen oxide by reaction with water (Zeevaart 1974, Rowland *et al.* 1985). In the presence of superoxide, nitric oxide forms peroxynitrite ($ONOO^-$), which is a strong nitrating agent (Beckman 1996). Goldstein *et al.* (2000) and Sawa *et al.* (2000) reported that tyrosine residues on

TABLE 25-2 Unidentified Nitrogen in Plant Leaves Fumigated with $^{15}NO_2$

Plant species	Unidentified nitrogen (percent of total NO_2-N)[a]
Arabidopsis thaliana	30.9 ± 10.0[b], 21.4 ± 9.5[c]
Manna gum (*Eucalyptus viminalis*)	19.0 ± 3.5[d]
Tobacco (*Nicotiana tabacum*)	18.8 ± 4.7[b]

[a]Values are the means of results from three to five independent experiments plus the standard deviation.
[b]Plants were fumigated with 4 parts per million $^{15}NO_2$ for 4 hours.
[c]Plants were fumigated with 0.1 parts per million $^{15}NO_2$ for 4 hours.
[d]Plants were fumigated with 4 parts per million $^{15}NO_2$ for 8 hours.

proteins could be nitrated by peroxynitrites formed from nitric oxide and superoxide in animal cells

Because the pK_a (pH at which equal amounts of the ionized and protonated forms of nitrous acid exist) of nitrous acid is rather high (around 3.4 to 3.6), substantial amounts of nitrite ions are protonated in plant cells (pH 5 to 7) to form nitrous acids (Beckman 1996). Therefore, when plants are fumigated with nitrogen dioxide, the intracellular level of nitrite ions is increased and hence that of nitrous acids. Nitrous acids are in reversible equilibrium with nitrosonium ions and hydroxide ions (Beckman 1996). Nitrosonium ions are strong nitrosating agents to form carcinogenic nitrosamines with amines. Because plant cells contain high amounts of amines in both the apoplast and symplast, nitrosoamines may be good candidates for the compound bearing unidentified nitrogen.

Nitrous acid is easily broken into a hydroxyl ion and nitric oxide, and the latter molecule may be released in the solution (Beckman 1996). It is possible that nitric oxide may be released by a similar mechanism in plant cells and reacts with the superoxide there to produce peroxynitrite ($ONOO^-$). It is also possible that tyrosine residues in plant cells can be similarly nitrated by the resulting peroxynitrite, as in animal cells (Goldstein *et al.* 2000, Sawa *et al.* 2000).

Stamler and Hausladen (1998) first reported that high levels of nitric oxide exert a nitrosative stress on humans, parallel to oxidative stress exerted by high levels of active oxygen such as superoxide. Liu *et al.* (2001) reported that high levels of cellular nitric oxide induce nitrosation of the thiol groups (*S*-nitrosation) of proteins and, thereby, inhibit signal transduction and enzyme activities. Compounds containing *S*-nitroso substituents also are good candidates for the compounds bearing unidentified nitrogen.

Taken together, the unidentified nitrogen bearing compounds such as nitro and nitroso compounds can be formed from nitrogen dioxide in plant cells. Hence, these compounds are a novel and integral metabolite in the nitrogen metabolism in plants. We are now investigating structures of these compounds.

The compounds such as nitroaromatics and nitrosamines are sometimes carcinogenic and harmful to humans, other animals, and plants (Beckman 1996). In addition, nitration and *S*-nitrosation of proteins, if any, are harmful to biological systems (Stamler and Hausladen 1998, Goldstein *et al.* 2000, Sawa *et al.* 2000, Liu *et al.* 2001). Our preliminary cytotoxicity test suggested that these unidentified nitrogen-bearing compounds are toxic to the seedlings of *Arabidopsis thaliana*.

The 19th-century Kjeldahl method is still the sole method available for the analysis of reduced nitrogen in biological systems. Our finding is that a substantial amount of nitrogen dioxide (up to 40 percent) is converted to Kjeldahl-unrecoverable, and thus unidentified compounds indicate that development of a new analytical technology is needed to quantify the nitrogenous metabolites in plants. The required analytical advances should provide a detailed pathway analysis of nitrogen metabolism in certain plants. In addition, understanding the mechanisms of the formation and degradation of these metabolites and the mode of action in plants and animals is pivotal not only in the remediation of air pollution, but also in basic nitrogen metabolism.

Prospects for Phytoremediation of Air Pollution Using Nitrogen Dioxide-Philic "Green Walls"

de Nevers (1995) calculated that to pump out the polluted air of the Los Angeles basin every day at a speed of 10 meters per second to somewhere far in the dessert requires the construction of a huge pipe having about a 3000-meter diameter. Such a large pipe is far beyond current structural engineering capabilities. Clearly, a less expensive, low energy, prevention and dispersed treatment approach is necessary. One of the best tools to mitigate atmospheric pollutants is plants driven by solar energy that self-reproduce and spread in the environment.

It is intriguing to calculate how plants mitigate atmospheric nitrogen dioxide when the walls of buildings and of highway corridors and roofs are covered with plants. Genetically engineered perennial plants that can convert nitrogen dioxide gas to nitrogen gas will be the best plant material. However, because such plants are not available yet, we choose, in the following calculation, petunia (*Petunia* × *hybrida*) that has not been genetically engineered. This plant has a rather high capability to assimilate nitrogen dioxide among the 217 taxa we studied (Morikawa *et al.* 1998a). The reduced nitrogen derived from nitrogen dioxide was 2.72 milligrams of N per gram of dry weight when fumigated at 4 parts per million NO_2 for 8 hours. Based on the assumption that the capability of petunia (*Petunia* × *hybrida*) to assimilate nitrogen dioxide changes linearly with the concentration of nitrogen dioxide (Rogers *et al.* 1979), the assimilation of nitrogen dioxide is estimated to be 11.6 nanograms of NO_2 per minute per gram fresh weight at 50 parts per billion NO_2. We also assume (1) a planted density of 300 plants (or approximately 2 kilograms in fresh weight) per square meter, each 0.2 meter high; (2) nitrogen dioxide concentrations of 50 parts per billion, 500 parts per billion, or 5000 parts per billion, depending on the type of urban area; and (3) plants metabolically active 12 hours per day for 365 days a year. Therefore, the annual assimilation by petunia (*Petunia* × *hybrida*) plants per square meter is

$$\begin{aligned}&(1.44\,\text{milligrams of } NO_2 \text{ per hour}) \times 12 \times 365 \text{ days} =\\ &\qquad 6.31 \text{ grams of } NO_2 \text{ per year per square meter}\end{aligned} \tag{25-4}$$

We forecast the novel planting of petunia (*Petunia* × *hybrida*) walls on the following three type of spaces in Japan:

1. A novel space available for vegetation is on the roof and building surfaces, which is estimated by the Organization for Landscape and Urban

Greenery Technology Development of Japan (http://www.greentech.or.jp/). In this estimation, the spaces necessary for building maintenance, construction, and other use are deducted. This area is estimated to be 46 389 hectares in the eleven major cities in Japan. The average concentration of nitrogen dioxide in this area is not known, and therefore is tentatively assumed to be 50 parts per billion.

2. The wall surface area of highways (where nitrogen dioxide concentration is tentatively assumed to be 500 parts per billion) in Japan is estimated to be 10 584 hectares (http://www.mlit.go.jp/).
3. The wall surface area of the city highway corridors in Tokyo and Osaka (where nitrogen dioxide concentration is tentatively assumed to be 5000 parts per billion) is estimated to be 960 hectares (http://www.mlit.go.jp/).

The data for such surface areas in other major cities in Japan are not available to us at present, and therefore the data are not included in the estimates that follow.

In total, petunias (*Petunia* × *hybrida*) on these surfaces can remove 1.57×10^4 tons of NO_2 per year. This roughly corresponds to 3.1 percent of the annual emission of nitrogen dioxide by cars in these areas in Japan where the total annual emission is about 500 000 tons of nitrogen dioxide (http://www.env.go.jp/). When we can increase the ability of plants ten times by using genetic engineering, such petunia (*Petunia* × *hybrida*) vegetation will clean up more than 30 percent (*i.e.*, 3.1 percent × 10) of the current annual emission of nitrogen dioxide by cars in Japan. Therefore, we believe that the creation of "green walls" that cover the walls of buildings and highway corridors in urban areas and rooftops with nitrogen dioxide-philic plants is potentially a partial solution to the severe air pollution occurring in many large cities. Basic and applied studies for such projects are very important subjects for future research on phytoremediation in this new century.

Acknowledgments

The pBSL16 was a gift from Dr. Kyoko Ogawa, Japan Women's University, Tokyo, Japan. The cDNA of the spinach nitrite reductase gene was a gift from Dr. S. Rothstein of the University of Guelph, Canada. This work was supported in part by the program Research for the Future (JSPS-RFTF96L00604) and by Grants-in-Aid for Scientific Research (nos. 10294513, 11740446, and 13556002) from the Japan Society for the Promotion of Science. Financial supports of Grants-in-Aid for Scientific Research on Priority Areas (nos. 05266213, 11151223, and 12025224) from the Ministry of Education, Science, Culture and Sports, Japan are also greatly appreciated.

REFERENCES

Arimura, G., M. Takahashi, N. Goshima, and H. Morikawa (1998) Metabolic fate of nitrogen dioxide nitrogen differs from that of nitrate nitrogen in plant leaves.

Abstract, *Annual Meeting of the American Society of Plant Physiologists*. Held in Madison, Wisconsin, June 27–July 1, p. 59.

Back, E., W. Burkhart, M. Moyer, L. Privalle, and S. Rothstein (1988) Isolation of cDNA clones coding for spinach nitrite reductase: complete sequence and nitrate induction. *Mol. Gen. Genet.* **212**: 20–26.

Beckman, J.S. (1996) The physiological and pathological chemistry of nitric oxide. In: *Nitric Oxide. Principles and Actions*. J. Lancaster, Jr., ed. Academic Press, New York, pp. 1–82.

Bradstreet, R.B. (1965) *The Kjeldahl Method*. Academic Press, New York, pp. 9–88.

Conrad, R. (1996) Soil microorganisms as controllers of atmospheric trace gases (H_2, CO, CH_4, OCS, nitrous oxide, and NO). *Microbiol. Rev.* **60**: 609–640.

Dean, J.V. and J.E. Harper (1986) Nitric oxide and nitrous oxide production by soybean and winged bean during the *in vivo* nitrate reductase assay. *Plant Physiol.* **82**: 718–723.

de Nevers, N. (1995) *Air Pollution Control Engineering*. McGraw-Hill, Singapore, pp. 4–5.

Durmishidze, S.V. and N.N. Nutsubidze (1976) Absorption and conversion of nitrogen dioxide by higher plants. *Doklad. Biochem.* **227**: 104–107.

Goldstein, S., G. Czapski, J. Lind, and G. Merényi (2000) Tyrosine nitration by simultaneous generation of NO and O_2 under physiological conditions. *J. Biol. Chem.* **275**: 3031–3036.

Goshima, N., T. Mukai, M. Suemori, M. Takahashi, M. Caboche, and H. Morikawa (1999) Emission of nitrous oxide (N_2O) from transgenic tobacco expressing antisense nitrite reductase mRNA. *Plant J.* **19**: 75–80.

Goto, F., S. Toki, and H. Uchimiya (1993) Inheritance of a co-transferred foreign gene in the progenies of transgenic rice plants. *Transgenic Res.* **2**: 300–305.

Gritz, L. and J. Davies (1983) Plasmid-encoded hygromycin B resistance: the sequence of hygromycin B phosphotransferase gene and its expression in *Escherichia coli* and *Saccharomyces cerevisiae*. *Gene* **25**: 179–188.

Harper, J.E. (1981) Evolution of nitrogen oxide(s) during *in vivo* nitrate reductase assay of soybean leaves. *Plant Physiol.* **68**: 1488–1493.

Hill, A.C. (1971) Vegetation: a sink for atmospheric pollutants. *J. Air Pollut. Contr. Assoc.* **21**: 341–346.

Ideker, T., V. Thorsson, J.A. Ranish, R. Christmas, J. Buhler, J.K. Eng, R. Bumgarner, D.R. Goodlett, R. Aebersold, and L. Hood (2001) Integrated genomic and proteomic analyses of a systematically perturbed metabolic network. *Science* **292**: 929–934.

Kacser, H. and J.W. Porteous (1987) Control of metabolism: what do we have to measure? *Trends Biochem. Sci.* **12**: 5–14.

Kaji, M., T. Yoneyama, T. Totsuka, and H. Iwaki (1980) Absorption of atmospheric nitrogen dioxide by plants and soils VI. transformation of nitrogen dioxide in the leaves and transfer of the nitrogen through the plants. In: Studies on Effects of Air Pollutants on Plants and Mechanisms of Phytotoxicity. Research Report No. 11 National Institute for Environmental Studies, Japan, pp. 51–58.

Kamada, M., A. Higaki, Y. Jin, M. Ayabe, M. Seki, T. Sawasaki, S. Ida, G. Toyohara, K. Irifune, and H. Morikawa (1992) Transgenic "air-pollutant-philic plants" produced by particle bombardment. In: *Research in Photosynthesis Vol IV*. N. Murata, ed. Kluwer Academic Publishers, Dordrecht, The Netherlands, pp. 83–86.

Klepper, L. (1990) Comparison between NOx evolution mechanisms of wild-type and nr1 mutant soybean leaves. *Plant Physiol.* **93**: 26–32.

Larcher, W. (1995) *Physiological Plant Ecology*. Springer-Verlag, Berlin, Germany, p. 421.

Lee, Y.-N. and S.E. Schwartz (1981) Reaction kinetics of nitrogen dioxide with liquid water at low partial pressure. *J. Phys. Chem.* **85**: 840–848.

Liu, L., A. Hausladen, M. Zeng, L. Que, J. Heitman, and J.S. Stamler (2001) A metabolic enzyme for *S*-nitrosothiol conserved from bacteria to humans. *Nature* **410**: 410–494.

Morikawa, H., A. Higaki, M. Nohno, M. Kamada, M. Nakata, G. Toyohara, K. Fujita, and K. Irifune (1992) "Air-pollutant-philic plants" from nature. In: *Research in Photosynthesis Vol. IV*. N. Murata, ed. Kluwer Academic Publishers, Dordrecht, The Netherlands, pp. 79–82.

Morikawa, H., A. Higaki, M. Nohno, M. Takahashi, M. Kamada, M. Nakata, G. Toyohara, Y. Okamura, K. Matsui, S. Kitani, K. Fujita, K. Irifune, and N. Goshima (1998a) More than a 600–fold variation in nitrogen dioxide assimilation among 217 plant taxa. *Plant Cell Environ.* **21**: 180–190.

Morikawa, H., M. Takahashi, and K. Irifune (1998b) Molecular mechanism of the metabolism of nitrogen dioxide as an alternative fertilizer in plants. In: *Stress Responses of Photosynthetic Organisms*. K. Satoh and N. Murata, eds. Elsevier, Amsterdam, The Netherlands, pp. 227–237.

Mulvaney, C.S. and R.H. Hageman (1984) Acetaldehyde oxime, a product formed during the *in vivo* nitrate reductase assay of soybean leaves. *Plant Physiol.* **76**: 118–124.

Nonoyama, N., K. Chiba, K. Hisatome, H. Suzuki, and F. Shintani (1999) Nitration and hydroxylation of substituted phenols by peroxynitrite: kinetic feature and an alternative mechanistic view. *Tetrahedron Lett.* **40**: 6933–6937.

Okano, K., T. Machida, and T. Totsuka (1988) Absorption of atmospheric NO_2 by several herbaceous species: estimation by the ^{15}N dilution method. *New Phytol.* **109**: 203–210.

Pacala, S.W., G.C. Hurtt, D. Baker, P. Peylin, R.A. Houghton, R.A. Birdsey, L. Heath, G.T. Sundquist, R.F. Stallard, P. Ciais, P. Moorcroft, J.P. Caspersen, E. Shevliakova, B. Moore, G. Kohlmaier, E. Holland, M. Gloor, M.E. Harmon, S.M. Fan, J.L. Sarmiento, C.L. Goodale, D. Schimel, and C.B. Field (2001) Consistent land-and atmosphere-based U.S. carbon sink estimates. *Science* **292**: 2316–2320.

Ramge, P., F.-W. Badeck, M. Plöchl, and G.H. Kohlmaier (1993) Apoplastic antioxidants as decisive elimination factors within the uptake process of nitrogen dioxide into leaf tissues. *New Phytol.* **125**: 771–785.

Rhodes, D., G.A. Rendon, and G.R. Stewart (1975) The control of glutamine synthetase level in *Lemna minor* L. *Planta* **125**: 201–211.

Rogers, H.H., J.C. Campbell, and R.J. Volk (1979) Nitrogen-15 dioxide uptake and incorporation by *Phaseolus vulgaris* (L.). *Science* **206**: 333–335.

Rowland, A., A.J.S. Murray, and A.R. Wellburn (1985) Oxides of nitrogen and their impact upon vegetation. *Rev. Environ. Health* **5**: 295–342.

Runquist, M. and N.J. Kruger (1999) Control of gluconeogenesis by isocitrate lyase in endosperm of germinating castor bean seedlings. *Plant J.* **19**: 423–431.

Sawa, T., T. Akaike, and H. Maeda (2000) Tyrosine nitration by peroxynitrite formed from nitric oxide and superoxide generated by xanthine oxidase. *J. Biol. Chem.* **275**: 32467–32474.

Shoun, H., D.H. Kim, H. Uchiyama, and J. Sugiyama (1992) Denitrification by fungi. *FEMS Microbiol. Lett.* **73**: 277–281.

Simonich, S.L. and R.L. Hites (1994) Importance of vegetation in removing polycyclic aromatic hydrocarbons from the atmosphere. *Nature* **370**: 49–51.

Simonich, S.L. and R.A. Hites (1995) Organic pollutant accumulation in vegetation. *Environ. Sci. Technol.* **29**: 2905–2914.

Smart, D.R. and A.J. Bloom (2001) Wheat leaves emit nitrous oxide during nitrate assimilation. *Proc. Nat. Acad. Sci. USA* **98**: 7875–7878.

Stamler, J.S. and A. Hausladen (1998) Oxidative modifications in nitrosative stress. *Nat. Sruct. Biol.* **5**: 247–249.

Stitt, M. and U. Sonnewald (1995) Regulation of metabolism in transgenic plants. In: *Annual Review of Plant Physiology and Plant Molecular Biology* Annual Review, Palo alto, California. **46**: 341–368.

Suzuki, H., T. Marushima, K. Shimizu, and K. Tsukamoto (1991) The ozone-mediated reaction of nonactivated arenes with nitrogen oxides. *Chem. Lett.* Jan.: 817–818.

Suzuki, H., T. Marushima, I. Kozai, and T. Mori (1993) Ozone-mediated nitration of alkylbenzens and related compounds with nitrogen dioxide. *J. Chem. Soc., Perkin Trans. 1: Organic and Bio–Organic Chemistry* (14): 1591–1597.

Takahashi, M. and H. Morikawa (1996) High frequency stable transformation of *Arabidopsis thaliana* by particle bombardment. *J. Plant Res.* **109**: 331–334.

Takahashi, M., Y. Sasaki, S. Ida, and H. Morikawa (2001) Nitrite reductase gene enrichment improves assimilation of NO_2 in *Arabidopsis*. *Plant Physiol.* **126**: 731–741.

Vanecko, S. and J.E. Varner (1955) Studies on nitrite metabolism in higher plants. *Plant Physiol.* **30**: 388–390.

Vaucheret, H., J. Kronenberger, A. Lepingle, F. Vilaine, J.-P. Boutin, and M. Caboche (1992) Inhibition of tobacco nitrite reductase activity by expression of antisense RNA. *Plant J.* **2**: 559–569.

Vincentz, M. and M. Caboche (1991) Constitutive expression of nitrate reductase allows normal growth and development of *Nicotiana plumbaginifolia* plants. *EMBO J.* **10**: 1027–1035.

Weber, P., S. Nubbaum, J. Fuhrer, H. Gfeller, U. Schlungger, P. Brunold, and H. Rennenberg (1995) Uptake of atmospheric $^{15}NO_2$ and its incorporation into free amino acids in wheat (*Triticum aestivum*). *Physiol. Plant* **94**: 71–77.

Wellburn, A.R. (1990) Why are atmospheric oxides of nitrogen usually phytotoxic and not alternative fertilizers? *New Phytol.* **115**: 395–429.

Wildt, J., D. Kley, A. Rockel, P. Rockel, and H.J. Segschneider (1997) Emission of NO from several higher plant species. *J. Geophys. Res.* **102**: 5919–5927.

Williams, E.J., G.L. Hutchinson, and F.C. Fehsenfeld (1992) NOx and N_2O emissions from soil. *Global Biogeochem. Cycles* **6**(4): 351–388.

Wray, J.L. and R.J. Fido (1990) Nitrate reductase and nitrite reductase. In: *Methods in Plant Biochemistry Vol. 3*. P.M. Dey and J.B. Harborne, eds. Academic Press, London, pp. 241–256.

Yoneyama, T. and H. Sasakawa (1979) Transformation of atmospheric nitrogen dioxide absorbed in spinach leaves. *Plant Cell Physiol* **20**: 263–266.

Yunus, M., N. Singh, and M. Iqbal (1996) Global status of air pollution: an overview. In: *Plant Response to Air Pollution*. M. Yunus and M. Iqbal, eds. John Wiley, New York, pp. 1–34.

Zeevaart, A.J. (1974) Induction of nitrate reductase by NO_2. *Acta. Bot. Neerl.* **23**: 345–346.

Zeevaart, A.J. (1976) Some effects of fumigating plants for short periods with NO_2. *Environ. Pollut.* **11**: 97–108.

Zumft, W.G. (1997) Cell biology and molecular basis of denitrification. *Microbiol. Mol. Biol. Rev.* **61**: 533–616.

26

TREATMENT OF ATMOSPHERIC HALOGENATED HYDROCARBONS BY PLANTS AND FUNGI

P. M. Jeffers and C. D. Liddy

SUMMARY OF PRACTICAL IMPLICATIONS

Plants and fungi are included among the sources and sinks of small halogenated hydrocarbons in the troposphere. Some of these anthropogenic halocarbons may be important participants in stratospheric ozone depletion and in global warming. Thus, quantitative details of the generation and destruction of these halocarbons are required for modeling assessments that may determine whether restrictions on manufacture and use of these compounds are warranted. Plants and fungi appear to be a non-negligible additional sink for atmospheric methyl bromide (CH_3Br), but probably are not significant sinks for any other atmospheric halocarbons of interest.

CHLOROFLUOROCARBON IN THE ATMOSPHERE

The troposphere contains trace amounts of numerous halocarbons ranging in complexity from the monohalomethanes to airborne polychlorinated biphenyl (PCB) congeners. The sources of these compounds vary as widely as do the chemicals. The entire range of chlorofluorocarbon compounds is strictly anthropogenic. The easily discernable presence of these chemicals in global air samples indicates our naive, carefree past handling practices. The Freons™ are the primary source of chlorine atoms in the stratosphere and thus are the major culprits responsible for stratospheric ozone depletion. However, bromine atoms also catalyze ozone destruction with an efficiency estimated at 50

Phytoremediation: Transformation and Control of Contaminants,
Edited by Steven C. McCutcheon and Jerald L. Schnoor.
ISBN 0-471-39435-1 (cloth)

times that of chlorine atoms (Wofsy *et al.* 1975, Yung *et al.* 1980, Solomon *et al.* 1992). Thus, methyl bromide (CH_3Br) becomes part of the atmospheric ozone cycle because this widely used agricultural fumigant is the probable major source of bromine atoms in the stratosphere.

SOURCES OF ATMOSPHERIC HALOCARBONS

The methyl halides present a fascinating story. Methyl chloride (CH_3Cl) often appears on the athletic field in a spray can to instantly chill a sprained ankle or knee. Methyl bromide degreases wool, extracts oils from nuts, or removes caffeine from coffee beans, although the much wider use is as an agricultural fumigant. Methyl iodide (CH_3I) is a standard methylating agent in synthetic organic chemistry. In addition, there are many natural sources of the methyl halides. Biomass burning liberates half as much methyl bromide as does the agricultural and industrial use. The oceans are a major source, although also a major sink of methyl halides. Lee-Taylor and Holland (2000) recently estimated that fungal decomposition of woody litter may account for significant amounts of methyl bromide, and Gan *et al.* (1998) demonstrated that species of the mustard genus, *Brassica*, are prodigious *in vivo* producers of methyl bromide, in proportion to the amount of bromide ion in the soil.

The two studies mentioned above were based on work by Harper (1985) who observed certain fungi to produce methyl halides, especially methyl chloride, from the halide ion, and by Wuosmaa and Hager (1990) who isolated a methyl chloride transferase enzyme from wood-rotting fungi and from marine red algae. Muramatsu and Yoshida (1994) discovered methyl iodide volatilized by rice (*Oryza sativa*) and oat (*Avena sativa*) plants grown on soils containing about 2 milligrams of I^- per liter per kilogram of dry soil and suggested that soil microorganisms or some action of the roots methylated the iodide. Saini *et al.* (1995) found that leaf disks of 87 diverse taxonomic species produced measurable amounts of methyl halides when floated *in vitro* on solutions containing high concentrations of halide ions. Further work by this group (Attieh *et al.* 1995) led to isolation of a methyl transferase that, in conjunction with the cofactors adenosyl methionine and adenosyl homocysteine, accounted for the production of methyl halides and methanethiol in cabbages (*Brassica oleracea*).

Three very recent papers have suggested additional natural and potentially large sources of the methyl halides

1. Ocean surface air-sampling (Yokouchi *et al.* 2000) indicates that methyl chloride, the most abundant halocarbon in the atmosphere, is emitted in large amounts by terrestrial plants found in warm coastal land, particularly from tropical islands

2. Rhew *et al.* (2000) specifically found and quantified methyl chloride and methyl bromide emissions from coastal salt marshes, estimating that these ecosystems might produce as much as 10 percent of the total atmospheric halomethanes
3. Finally, Keppler *et al.* (2000) discovered and demonstrated an abiotic production mechanism active in soils and sediments, and requiring only oxidizable organic matter such as guaiacol, halide ion, and Fe(III) as an electron sink

Montzka *et al.* (1999) and Butler (2000) have reviewed the sources, concentrations, trends, and relative importance of atmospheric halocarbons as ozone depleting agents.

Previous chapters of this book have discussed the transport of trichloroethene from aquifers to the atmosphere by the processes of transpiration. The "standard" air-stripping remediation process also transfers trichloroethene from groundwater to the atmosphere. Atmospheric trichloroethene is not generally considered a major problem because the concentration is negligibly low and the lifetime is short as a result of the reaction with hydroxyl radicals in the troposphere. In groundwater, trichloroethene has a hydrolysis lifetime estimated to be many centuries (Jeffers *et al.* 1989). On the other hand, although methyl bromide has a tropospheric concentration of only 10 parts per trillion by volume and a relatively short atmospheric lifetime, the contribution to stratospheric ozone depletion is undisputed.

In addition to the numerous natural sources mentioned, the anthropogenic sources of methyl halides are significant. Coupled with the myriad uses of these compounds, the ease of manufacture and the low boiling points, especially of methyl chloride (−24.2 °C) and methyl bromide (3.6 °C), assure finding these pollutants in the atmosphere.

The relative importance on ozone depletion is a function of the true lifetime of anthropogenic methyl bromide in the atmosphere. A short lifetime implies a fast natural cycle in the troposphere, and very little entry of commercially produced methyl bromide into the stratosphere. Conversely, a long lifetime would indicate a slow natural cycle and greater human impact. The 1997 "best estimate" was a lifetime of 0.7 year (Methyl Bromide Global Coalition/NASA 1997). In addition to methyl bromide contributing to ozone depletion, the presence of halogenated hydrocarbons also poses other possible threats to the environment. All of the halocarbons are strongly infrared-active, and thus contribute to global warming. For these and other similar reasons, a consideration of the potential of phytoremediation for halocarbons seems appropriate.

Plants are clearly a significant source of atmospheric methyl iodide, methyl bromide, and methyl chloride. Plants and some fungi may be significant sinks for these trace gases and thus, at the same time, may provide a route for enriching soil in the associated halides. Degradation of larger monohaloalkanes appears sufficiently slow as to conclude that plants probably serve as no measurable sink. Our estimate is that global degradation of methyl bromide by plants may be as large as 2 gigagrams per year, an additional 1 percent of the known tropospheric sinks. Thus, plants do have potential for the removal and treatment of at least some volatile halogenated hydrocarbons that pose a danger to our environment.

TRANSFORMATIONS OF HEXACHLOROETHANE AND PERBROMOETHENE

Under "environmental conditions" of 25 °C and pH 7, methyl bromide has a hydrolysis half-life of about 20 days, but trichloroethene and chloropicrin show absolutely no reactivity. However, anaerobic sediments and aquatic plants effect relatively rapid dehalogenation of a number of compounds. Specifically, anaerobic sediments rapidly reduce hexachloroethane to perchloroethene, and perbromoethene to tribromoethene (Jafvert and Wolfe 1987).

When a sprig of the aquatic plant parrot feather (*Myriophyllum aquaticum*) was added to solutions of methyl bromide, trichloroethene, and chloropicrin, all three compounds underwent first order concentration loss with half-lives of about 20 hours. In addition, dichloronitromethane was an intermediate of the chloropicrin transformation, displaying a concentration that rose then fell, clearly indicating a sequential reduction process (Jeffers, unpublished). These observations led, early in 1996, to some simple vapor-phase experiments with various plants and tree leaves contacting gaseous hexachloroethane or perbromoethene.

EXPERIMENTAL APPROACH

We have compiled and presented below a set of related experiments involving the treatment of atmospheric halocarbons by plants and fungi. Each experiment is numbered and results are presented in corresponding order in the section "Results and Discussion."

1. Hexachloroethane is a solid at room temperature, with a vapor pressure of about 1 torr (133 pascals). We ran experiments with hexachloroethane concentrations ranging from saturated vapor (some solid hexachloroethane in the reaction vessel) to slightly under 1 part per million by volume. Plants contacting the hexachloroethane vapor included parrot feather (*Myriophyllum aquaticum*), wood sorrel (*Oxalis corniculata* L.), a common garden weed water oak (*Quercus nigra* L.), flowering dogwood (*Cornus florida* L.), and silver maple (*Acer saccharinum* L.). Early experiments were run in 250 milliliter Erlenmeyer flasks with foil-covered rubber stoppers that were cracked open to withdraw gas samples. Later experiments utilized 17 milliliter Pyrex™ screw-cap test tubes with Teflon®-faced septa. An inner aluminum foil seal

cut loss of hexachloroethane and perchloroethene to negligible levels. Some experimental vials were repeatedly sampled, and some experiments had one tube sacrificed for each experimental point. Analysis for hexachloroethane and perchloroethene was by gas chromatography with mass spectroscopy used to confirm reactant and product identities. An electron capture detector allowed for quantitative measurements, with calibration by injection of dilute methanol solutions of the pure compounds.

Early qualitative experiments tested solid hexachloroethane and solid perbromoethene with parrot feather (*Myriophyllum aquaticum*). In attempts to minimize possible contributions by microorganisms, the parrot feather (*Myriophyllum aquaticum*) was pre-treated with dilute hypochlorite solution for one experiment, and in a second experiment three 17 milliliter test tubes containing 20 sorrel (*Oxalis corniculata* L.) leaves each were irradiated with 40, 60, and 80 kilorads from a [^{60}Co] source before the perbromoethene reactant gas was introduced. After a fresh set of sorrel (*Oxalis corniculata* L.) leaves had been in contact with 34 parts per million by volume hexachloroethane for 3 hours, these leaves were exposed to a fresh airflow for 1 minute, placed in a clean, foil-sealed tube, and the subsequent evolution of hexachloroethane and perchloroethene was measured for 73 hours.

2. An extensive series of investigations, begun early in 1996, focused on atmospheric methyl bromide production and reactive removal from air by numerous plants, plant materials, and fungi (including rusts). In the earliest experiments, Jeffers and Wolfe (1996b) subjected the stems and feathery leaves of parrot feather (*Myriophyllum aquaticum*), an entire swamp or yellow iris (*Iris pseudocorus*) plant, and drained samples of *Draparnaldia*, a filamentous freshwater algae to methyl bromide at 5 to 200 parts per million by volume in air. The concentration was monitored using gas chromatography for up to 114 hours.

3. Following the demonstration described above, a total of 100 grams of entire plants of wood sorrel (*Oxalis corniculata* L.) with rinsed roots was placed in five 50 milliliter beakers each containing 25 milliliters of deionized water, and stationed evenly along a 1 by 0.1 by 0.1 meter glass reactor. Spiking and sampling were done through Teflon®-faced septa waxed over 5 millimeter holes spaced evenly along the top plate. Methyl bromide was added to the air at an initial concentration of 5 parts per million by volume. A matching control reactor was prepared and treated in exactly the same manner, except that no plants were placed in the beakers of water. The plant reactor was re-spiked at 89 hours, and again at 148 hours; the methyl bromide concentrations in both reactors were monitored for a total of 196 hours.

4. Extensions and refinements of the early experiments utilized plant leaf samples in sets of 20 milliliter autosampler vials with crimp-sealed foil-faced Teflon® septa (Jeffers *et al.* 1998). For these experiments, control vials spiked and sealed in the same way as test vials but containing no plant

material were regularly prepared for each experiment. These control vials were analyzed at the conclusion of an experiment, thus covering the total experimental period. The test vials included 39 different plant species: 9 herbaceous, 12 coniferous, and 18 deciduous. Similar experiments were performed with samples of local woodland soil taken from just beneath the topmost leafy cover.

5. Two 1 gram sections of spinach (*Spinacia oleracea*) root were boiled for about 2 minutes and were placed in 20 milliliter autosampler vials with crimp-sealed foil-faced septa, spiked with methyl bromide, and later analyzed, as were two 1 gram sections of spinach (*Spinacia oleracea*) root that had not been boiled.

6. Three 36-square centimeter spinach (*Spinacia oleracea*) leaf samples in 58 milliliter vials were flooded 10 times at half-hour intervals with 10 parts per million by volume of methyl bromide. Three matched leaf samples were not treated with methyl bromide. All the samples were ground, digested in hydrogen peroxide (H_2O_2), evaporated to dryness, the residue was extracted with deionized water, filtered, extracted with methylene dichloride (CH_2Cl_2) and the aqueous volume was analyzed for inorganic bromide by ion chromatography.

Radiolabeled methyl bromide ($^{14}CH_3Br$) was added to 10 crimp-cap, 20 milliliter septum vials, each containing a 15-square centimeter section of spinach (*Spinacia oleracea*) leaf. Ten control flasks were prepared similarly, but with nonlabeled methyl bromide. The control flasks were tested for reaction extent of the methyl bromide and for production of methane or carbon dioxide. Vials from the radiolabeled set were extracted after 4 hours and after 14 days with acetonitrile, dilute acid, or dilute base, and the solutions were checked for activity by scintillation counting. Leaf material from the remaining vials was combusted and the carbon dioxide (CO_2) absorbed into a sodium hydroxide (NaOH) solution that was checked by scintillation counting for radioactivity.

7. Spinach (*Spinacia oleracea*), cabbage (*Brassica oleracea*), and cauliflower (*Brassica oleracea botrytis*) plants were grown on soil augmented by adding 100 milligrams of Br^- (150 milligrams KBr) to 1 kilogram of dry soil. Matched plants were grown on the same soil with no added bromide. Air samples were taken from beakers that had been placed over the plants for several hours. Leaves were taken from these plants, placed in 20 milliliter septum vials for several hours, and then sampled for production of methyl bromide. In related experiments, some "lawn mushrooms" were watered with a dilute potassium bromide solution with a set of similar mushrooms watered with tap water containing no added bromide, again to test for production of methyl bromide.

8. Five species of mushrooms were grown from spores under sterile conditions in 40-milliliter vials about one-third filled with growth medium, and when the tissue had developed to a disk of about 10 millimeters in diameter

by 2 millimeters in height, the vials were spiked with methyl bromide and Freon® 12. Methyl bromide concentrations were monitored for 50 hours.

Sections of the caps of several mushrooms (species undetermined) were placed in vials and spiked with methyl bromide. These vials were then sampled and analyzed by a gas chromatograph equiped with a mass spectrometer (GC/MS) over a 50-hour period. These two experiments involving fungal degradation of halocarbons led to related experiments involving plant species afflicted by fungal growths. Loblolly pine (*Pinus taeda*) needles covered with black fungal rust were tested as were leaves from Norway maple (*Acer Platanoides*) trees in the Cortland, New York area, suffering from a widespread black disk leaf fungus.

9. We have recently extended our degradation experiments to additional monohaloalkanes. Our most recent experiments have utilized a Hewlett-Packard GC/MS model 5989A, operating in single ion mode set for the major fragment ion of the reactant of interest. Stock mixtures of a halocarbon reactant and of Freon® 12 (CCl_2F_2 added as an internal standard) in room air were prepared in 2 liter flasks such that a several milliliter aliquot added by syringe to a 43 milliliter Teflon®-faced septum vial gave a starting reactant concentration of typically 10 parts per million by volume. Sets of 43 milliliter screw-cap septum vials were prepared with measured cut-leaf sections or entire leaves. The vials were spiked with the reaction mix by syringe past the just-open cap. Then the cap was tightened and the vial was agitated so that the leaf would affect thorough mixing. Analysis of 50 to 100 microliter samples withdrawn from the vials by gas-tight syringe covered time spans of several hours to several days, depending on the reactivity of the halocarbon–leaf combination.

RESULTS AND DISCUSSION

Experimental Results

1. In all the preliminary experiments with hexachloroethane, an essentially linear growth of the perchloroethene product peak was observed with time. The same rate of perchloroethene liberation was measured with the chemically disinfected parrot feather (*Myriophyllum aquaticum*) and with a non-treated control. Tetrabromoethene was observed as a gaseous product of leaves exposed to perbromoethene. The same rate of tetrabromoethene production was observed with [^{60}Co]-irradiated sorrel (*Oxalis corniculata* L.) leaves as with untreated leaves.

As shown in Figure 26 1, with 20 sorrel (*Oxalis corniculata* L.) leaves (about 1 square centimeter each) or with a 20 square centimeters oak (*Quercus nigra* L.) leaf section in 17 milliliter tubes dosed with 33 parts per million by volume of hexachloroethane, between 90 and 99 percent of the hexachloroethane disappeared within the first hour. A slow, linear evolution of perchloroethene was observed over the next 24 hours. However, only 5 percent [oak (*Quercus nigra* L.)] to 15 percent [sorrel (*Oxalis corniculata* L.)] reactive transformation to gaseous perchloroethene occurred within 24 hours. (Note that the product concentrations shown in Figure 26-1 have been magnified by a factor of 10 for emphasis.) The hexachloroethane loss can almost certainly be ascribed to adsorption or partitioning into the leaf. However, perchloroethene may also be produced by the plant and be adsorptively held, so the gaseous perchloroethene observed may be only a fraction of the total reactive output for the plant. In addition, because transformation of perbromoethene to tetrabromoethene was observed, perchloroethene may also undergo further transformation within the plants, leading to a smaller observed perchloroethene production rate than the actual rate of hexachloroethane transformation. Hexachloroethane adsorption was plant-specific, with parrot feather (*Myriophyllum aquaticum*) and oak (*Quercus nigra* L.) absorbing less than dogwood (*Cornia florida*), maple (*Acer saccharinum*), and sorrel (*Oxalis corniculata* L.).

The hexachloroethane adsorption was strong and was not rapidly reversible. Five sorrel (*Oxalis corniculata* L.) leaves were exposed to 34 parts per million by volume of hexachloroethane for 3 hours and were then exposed to fresh air for 1 minute. Within 7 minutes after the leaves were returned to the sealed test

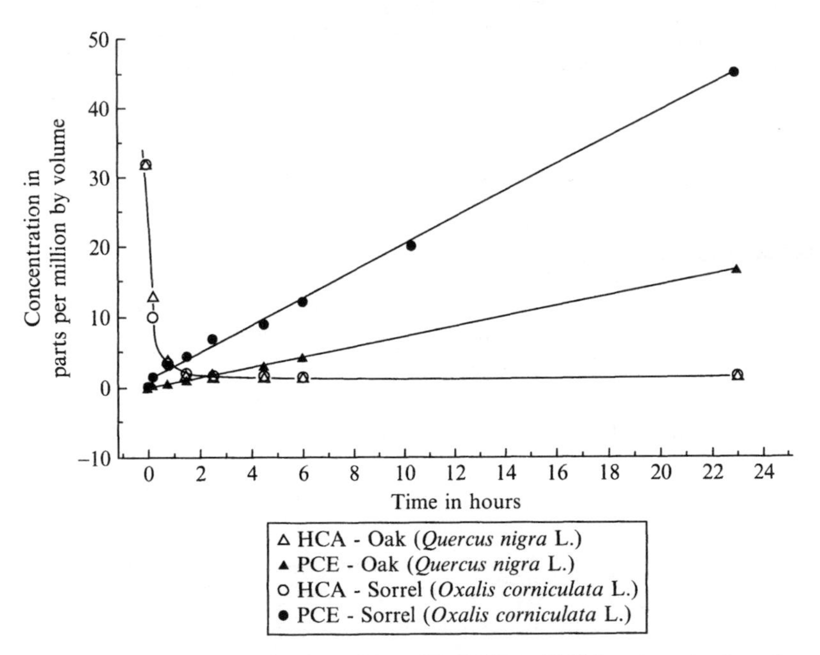

Figure 26-1 Absorption and reactive loss of hexachloroethane (HCA) and production of gaseous perchloroethene (PCE) by oak (*Quercus nigra* L.) and sorrel (*Oxalis corniculata* L.) leaves. The figure magnifies the product concentration 10 times to emphasize the linear product evolution.

tube, the hexachloroethane reached a maximum concentration of 1 part per million by volume, after which it decreased to a final value smaller by a factor of 50, whereas perchloroethene increased throughout the experiment.

A total of 28 different experiments representing different configurations of containers, plants, plant parts, reactants, concentrations, and sampling and analysis protocols all indicated that hexachloroethane and perbromoethene sustained reactive transformation. The product appearance rate scaled linearly with leaf area and with the amount of reactant absorbed. Cross-species rates probably differed by less than a factor of 60, for the species of plants tested. Several experiments in this series with carbon tetrachloride and with 1,1,1-trichloroethane indicated a much slower reaction, if any; therefore, the halocarbon–plant reactivity is a function of both plant species and of halocarbon structure.

2. The 2 grams (wet weight) of parrot feather (*Myriophyllum aquaticum*) removed methyl bromide by an apparently first-order process from an initial 5 parts per million by volume in a 500 milliliter flask with a half-life of 35 hours over a 70-hour period. The air contacting a 2 grams (wet weight) of parrot feather (*Myriophyllum aquaticum*) sample dosed with 200 parts per million by volume of methyl bromide was checked for methane production with none observed up to 145 hours. The gas chromatograph with a flame ionization detector system was capable of detecting ambient air methane about 1 part per million by volume. Ions chromatography did indicate bromide ions in ground plant tissue extracted with water that quantitatively correlated to the methyl bromide delivered to the flask.

Methyl bromide concentration followed first-order decay and decreased by a factor of 20 over 114 hours with the *Iris pseudocorus* plant, from an initial concentration of 20 parts per million by volume. Less than 30 percent loss of methyl bromide occurred from a control flask over this time.

The algae *Drapernaldia* spp. also removed methyl bromide from the headspace of the flask, but these experiments were complicated by the presence of numerous microorganisms, clearly seen through a microscope at a magnification of 200 times. A further complication appeared to be the air-to-water transport process that was required, even with drained but very wet algae samples. No attempts were made to unscramble these complications (Jeffers and Wolfe 1996b).

3. During the 196 hours, methyl bromide in the control reactor decreased by about a factor of 10, ascribed to leakage, adsorption to the Apiezon™ black wax sealing bead, or solution in the water. The wood sorrel (*Oxalis corniculata* L.)-filled reactor displayed nearly first order decay from 5 to about 0.005 parts per million by volume following each of the three methyl bromide additions, at a loss rate approximately six times that of the control system.

4. All 9 herbaceous, 12 coniferous, and 18 deciduous plants tested removed methyl bromide from air by apparent first order kinetics. Figure 26-2 shows a typical first order plot of methyl bromide reaction with spinach

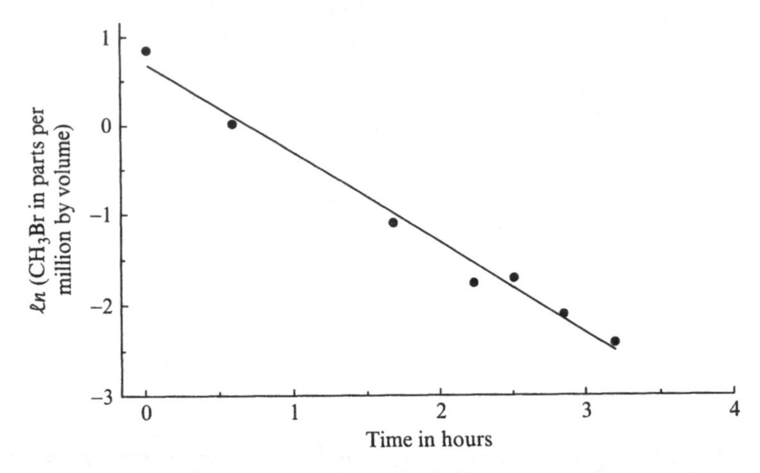

Figure 26-2 First-order loss of methyl bromide with spinach (*Spinacia oleracea*) leaves. This experiment utilized a 7 centimeters by 1 centimeter leaf section of spinach (*Spinacia oleracea*) in a 43-milliliter vial, initially spiked with 10 parts per million by volume of methyl bromide. First-order loss of methyl bromide is observed here for nearly five half-lives.

(*Spinacia oleracea*). The list of thirty-nine different leaves and rate constants for methyl bromide reactive removal (Table 26-1) shows a range from 0.2 red leaf lettuce (*Lactuca* spp.,) to 28.0 spinach (*Spinacia oleracea*) in units millimeters per hour, with a mean of about 0.10 for herbaceous plants and deciduous trees, and about 0.30 for conifers. Further, the control vials generally indicated no significant decrease in methyl bromide by adsorption or leakage. Rate measurements with *Magnolia grandiflora* leaves held at 4 °C and 23 °C yielded an Arrhenius activation energy of 27 kilojoules per mole, whereas the hydrolysis of methyl bromide and the Cl^-/Br^- aqueous ion exchange reaction have activation energies of 103.5 and 105.7 kilojoules per mole, respectively (Jeffers and Wolfe 1996c). These observations implicate the enzyme-catalyzed processes that typically provide very low energy barrier pathways for biological reactions.

TABLE 26-1 Removal Rate Data for Methyl Bromide by Plant Leaves

Plant	Area (square centimeters)	Mass (grams)	Initial concentration of methyl bromide (parts per million by volume)	Time (hours)	Rate constant (millimeters per hour)
Herbaceous					
Clover (*Trifolium* spp.)	10	0.12	10	1 to 23	0.38
Red leaf lettuce (*Lactuca* spp.)	7	0.15	10	17	0.20
Tot soi	7	0.16	10	6	0.76
Arugula (*Eruca vesicaria*)	7	0.14	10	17	0.98

Corn (*Zea mays*) leaf	7	0.10	10	41	0.38
Broccoli (*Brassica* spp.)	7	0.50	10	1 to 64	1.38
Cabbage (*Brassica oleracea*)	7	0.45	10	10	2.00
Spinach (*Spinacia oleracea*)	7	0.15	10	0 to 3.5	28.0
Centipede grass (*Eremochloa ophiuroides*)	45	2.3	10	4.7	2.66
Deciduous					
Cherry (*Prunus* spp.)	66	1.34	1.0	5.75	0.36
White birch (*Betula papyrifera*)	7	0.10	1.0	42.5	2.08
American beech (*Fagus grandifolia*)	7	0.04	1.0	41.4	0.334
Schumardii oak (*Quercus schumardii*)	7	0.08	1.0	40.8	0.62
Post oak (*Quercus stellata*)	7	0.10	1.0	41.0	0.884
Red oak (*Quercus rubra*)	7	0.09	1.0	40.5	1.43
Water oak (*Quercus phellos*)	7	0.12	1.0	49.4	0.802
White oak (*Quercus alba*)	7	0.10	1.0	50.4	1.462
Red maple (*Acer rubrum*)	7	0.13	1.0	40.7	0.79
Sugar maple (*Acer saccharum*)	7	0.06	1.0	40.5	0.168
Silver maple (*Acer saccharinum*)	7	0.10	1.0	50.3	0.748
Pecan (*Carya illindinensis*)	7	0.11	1.0	48.7	0.552
Crab apple (*Pyrus angustifolia*)	7.5	0.16	1.0	48.9	1.136
Black walnut (*Juglans nigra*)	7	0.11	1.0	49.5	0.208
Sycamore (*Platanus occidentalis*)	7	0.13	1.0	49.9	1.506
Tulip poplar (*Liriodendron tulipifera*)	7	0.12	1.0	50.1	0.19
Ginkgo (*Ginkgo biloba*)	15	0.25	1.0	18.5	0.734
Magnolia grandiflora	25	1.00	10	18	1.902
Coniferous					
Fir (*Abies firma*)	7	0.84	1.0	43.1	3.4
White fir (*Abies concolor*)	17.1	0.82	1.0	43.0	0.854
Red cedar (*Thuja plicata*)	10	0.40	1.0	42.6	1.01
American arborvitae (*Thuja occidentalis*)	7	0.66	1.0	42.4	1.6
Cypress (*Taxodium distichum*)	7	0.65	1.0	41.9	3.14
Slash pine (*Pinus elliotii*)	4.6	0.24	1.0	41.5	1.138
Yellow pine (*Pinus echinata*)	6	0.50	1.0	49	2.16
Loblolly pine (*Pinus taeda*)	14	1.87	10	4.6	3.92
Short leaf pine (*Pinus echinata*)	9	0.50	1.0	2.1	5.22
Blue spruce (*Picea pugens*)	9	0.64	1.0	1.9	11.54
Norway spruce (*Picea abies*)	4.7	0.61	1.0	1.0	8.94
Cedar (*Juniperus virginiana*)	10	0.55	1.0	1.1	11.98

Other recent experiments have indicated that all green plants and plant materials as diverse as spinach (*Spinacia oleracea*) root and wheat (*Triticum aestivum*) berries remove methyl bromide from air within the concentration range of 200 to 0.1 parts per million by volume by an apparently first-order process. Several experiments with spinach (*Spinacia oleracea*) and maple (*Acer saccharum*) have shown linear first order loss kinetics to below ambient methyl bromide values of 0.01 parts per million by volume.

5. The boiled Spinach (*Spinacia oleracea*) roots showed no degradative activity, whereas the fresh roots removed methyl bromide with a half-life under 1 hour. Re-spiking of these samples gave duplicate results. The white spinach (*Spinacia oleracea*) root degraded methyl bromide as effectively as the green leaf implying that the degradation reaction is not directly related to photosynthesis. Flushing the vials with argon prior to methyl bromide addition had no effect on the removal rate, nor was the removal altered by storing the vials in the dark. Although all parts of the spinach (*Spinacia oleracea*) plant were active, *Magnolia grandiflora* flowers neither adsorbed nor degraded methyl bromide.

Loss rates from samples of local woodland soil samples and comparable masses of plant leaves yielded soil rate constants of the same order as those reported by Shorter *et al.* (1995). Rate constant values for the leaves were smaller by a factor of 2 to 10.

6. Radiolabeled methyl bromide degraded by spinach (*Spinacia oleracea*) leaves produced no $^{14}CO_2$ or $^{14}CH_4$. About half of the [^{14}C] was recovered in either acetonitrile or aqueous acid-base extraction of ground plant tissue. This recovery is not unexpected if the methyl group is incorporated into a relatively small biomolecule. Combustion of the plant material led to 100 percent recovery of [^{14}C] as radiolabeled $^{14}CO_2$.

Ion chromatography in the ground and extracted blank spinach (*Spinacia oleracea*) leaf samples gave 5.1, 2.9, and 4.0 area counts for bromide ion, compared to 36.8, 70.9, and 41.0 area counts found with the methyl bromide treated leaves, equivalent to 20 to 50 percent of the total bromine added as methyl bromide. This recovery seems reasonable in view of the processing involved with these experiments and the relatively short time allowed for reaction between successive methyl bromide addition to the test spinach (*Spinacia oleracea*) leaves.

7. Methyl bromide was observed in inverted beakers covering each cabbage (*Brassica oleracea*) and cauliflower (*Brassica oleracea botrytis*) plant growing on bromide-doped soil, but no methyl bromide was found in air samples from plants grown on soil that contained no bromide, nor was any methyl bromide observed with any spinach (*Spinacia oleracea*) plant grown on either bromide augmented or nonaugmented soil. Leaves taken from these plants and placed in septum vials for several hours exhibited equivalent results; cabbage (*Brassica oleracea*) and cauliflower (*Brassica oleracea botrytis*) leaves clipped from plants growing on bromide-containing soil rapidly produced easily measurable amounts of methyl bromide, but no methyl bromide was found with any spinach (*Spinacia oleracea*) leaves or with cabbage (*Brassica oleracea*) and cauliflower (*Brassica oleracea botrytis*) leaves that had grown on bromide-free soil. The cabbage (*Brassica oleracea*) and cauliflower (*Brassica oleracea botrytis*)

plants were sampled for 14 months with unchanging results. Leaves from plants grown on nonsalted soil were placed in a beaker containing dilute potassium bromide. Again, cabbage (*Brassica oleracea*) and cauliflower (*Brassica oleracea botrytis*) leaves produced methyl bromide, within hours of contacting the salt solution, but spinach (*Spinacia oleracea*) leaves treated in the same fashion produced no methyl bromide.

Within a day, the bromide-fed lawn mushrooms (two unidentified species), placed in capped vials, produced methyl bromide in easily measured amounts, but none was observed with the control mushrooms of the same species that had not been watered with a potassium bromide solution. These experiments confirmed the findings of Gan *et al.* (1998), and indicate a much lower or zero capacity of spinach (*Spinacia oleracea*) to biosynthesize methyl bromide, in comparison with the cabbage family (*Brassica*). Thus, we found no apparent correlation between plants that produce methyl bromide and plants that degrade the compound. Spinach (*Spinacia oleracea*), one of the best degraders, is a nonproducer. The cabbage family (*Brassica*), a group of prodigious producers, has limited degradation. These observations imply that different enzyme systems are involved in the two processes. Literature reports on grape (*Vitaceae* spp.) fumigated with methyl bromide (Liyanage *et al.* 1993) and on human erythrocytes exposed to methyl bromide (Hallier *et al.* 1990) suggest that glutathione may be involved in the degradation pathway.

8. The various experiments we performed with mushrooms have some important implications. Several mushrooms randomly picked from the lawn and garden, when placed in vials and spiked with methyl bromide yielded the highest removal rates we have measured. Experiments with five species of mushrooms grown on sterile medium represent our best evidence that systems other than microorganisms have developed enzyme systems that are capable of degrading methyl bromide. Figure 26-3 shows the rather complex dynamic concentrations observed when methyl bromide was injected into the headspace of the mushroom-sterile growth medium vials. Loss of methyl bromide in the blank vials is ascribed to absorption and diffusion into the gelatinous medium. Loss with the *Pholiota squarrosa* tissue was essentially the same as was observed with the blank, implying that this species of mushroom does not degrade methyl bromide. The other four mushroom species degraded methyl bromide at varying rates, as we have observed with various leafy plants. In these mushroom-tissue growth medium vials, the fate of methyl bromide is controlled by a combination of reactive loss and adsorption into the medium, so disappearance is neither linear nor logarithmic, as Figure 26-3 shows. The rapid degradation of methyl bromide by mushrooms led to further investigations involving plants with fungal growths. Loblolly pine (*Pinus taeda*) needles covered with a black rust removed methyl bromide a factor of two faster than unafflicted needles. Leaves from Norway maple (*Acer platanoides*) trees with a black disk leaf fungus degrade methyl bromide considerably more rapidly than do healthy leaves.

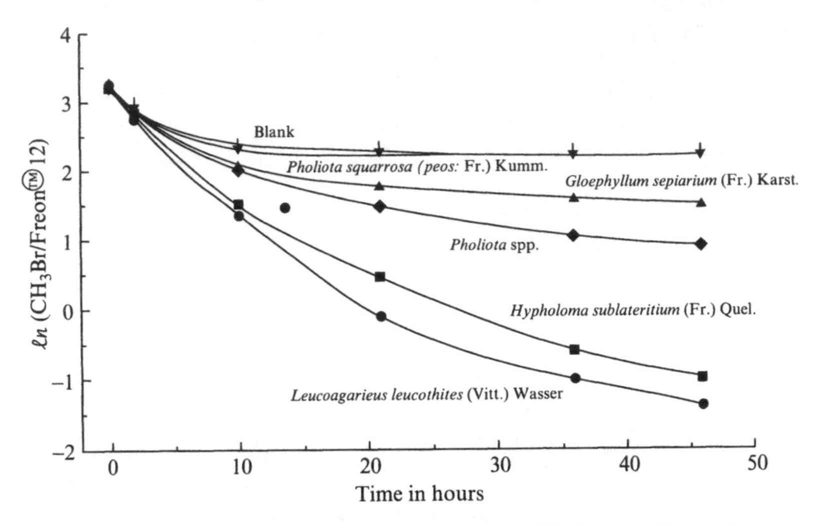

Figure 26-3 Absorption and reactive loss of methyl bromide (CH_3Br) with five mushroom species grown from spores on sterile medium. Initial methyl bromide concentration was 10 parts per million by volume.

9. In experiments with leaf samples in vials that were spiked with alkyl mono-halide-Freon® mixtures, with a few exceptions, the Freon® concentration remained constant, whereas the halocarbon concentration decreased by apparent first order kinetics, and apparent first order rate constants were deduced from a logarithmic concentration–time plot, as shown in Figure 26-2. With spinach (*Spinacia oleracea*) as the test plant, methyl bromide and methyl iodide degrade at about the same rate. Relative to methyl bromide, rates are: methyl chloride—0.02; ethyl iodide—0.05, ethyl bromide—0.005; and *n*-propyl iodide—0.1. The relative rates of methyl iodide degradation by cabbage (*Brassica oleracea*), sugar maple (*Acer saccharum*), and spinach (*Spinacia oleracea*) are very similar to the relative rates for these leaves with methyl bromide. It is interesting to note that methyl bromide and ethyl bromide undergo neutral hydrolysis with almost identical rate constants (Jeffers and Wolfe 1996a), thus the factor of 200 favoring methyl bromide over ethyl bromide in reactive loss with spinach (*Spinacia oleracea*) leaves is further evidence that the loss process is enzyme moderated.

Quantitative Relations

There is little evidence that methyl bromide partitions strongly to plants. The reactive removal correlates with leaf area, and is apparently first order over a very wide gas-phase concentration range. The situation with hexachloroethane and perchloroethene is quite different; rapid, extensive partitioning to plants was observed. We then assume that for methyl bromide removal by plants, the removal rate is proportional to collisions of methyl bromide molecules with the plant surface

$$\frac{-\mathrm{d}N}{\mathrm{d}t} = kZ_s A \tag{26-1}$$

where N is the number of methyl bromide molecules; Z_s is the kinetic theory wall collision number, that is, the number of collisions per unit area of a surface per second; k is the proportionality constant, in this case, the "true" dimensionless reaction rate constant; and A is the total active plant surface area. From kinetic theory

$$Z_s = 0.25\frac{(N)}{V}u \tag{26-2}$$

where $u = (8\kappa T/\pi m)^{1/2}$ = average molecular speed and V is the volume. In this relation, κ is the Boltzmann constant, T is the absolute temperature, and m is the molecular mass. Dividing Equation (26-1) by the volume of the container, V

$$\frac{-\mathrm{d}(N/V)}{\mathrm{d}t} = k(1/4)(N/V)u(A/V) \quad \text{or} \quad \frac{-\mathrm{d}C}{\mathrm{d}t} = k^*(A/V)C = k_{\mathrm{obs}}C \tag{26-3}$$

The final form of Equation (26-3) is a typical first order rate equation with $k_{\mathrm{obs}} = k^*(A/V)$. To compare experiments utilizing different areas of plant and different volume containers, the quantity that is fundamental is the k^*, obtained from the observed first order rate constant by multiplying by the volume to area ratio. Note that the k_{obs} has "typical" units for a first order process of reciprocal time, but the "true" rate constant will have units of (volume divided by area) per time, reducing to length per time.

Note also that the analysis presented above is for foliage of a given area in a closed container of given volume, and assumes that reaction is slow compared with diffusive mixing. A more detailed treatment would consider diffusion and boundary layer constraints, although the k^* values we have deduced and report in Table 26-1 should represent relative rates appropriately for different plants.

The rate of methyl bromide removal by spinach (*Spinacia oleracea*) leaves can be impressive, with half-lives on the order of several minutes with large area to volume ratios. Our experiments have not found an upper limit of the apparently first-order reaction, so a phytoreactor for treatment of off-gases from a manufacturing or fumigation process could be very efficient, and the spinach (*Spinacia oleracea*) grown would be a fine source of bromide in the diet. Beyond methyl bromide and methyl iodide, plants may have some utility in remediation of atmospheric halocarbons, but reactive removal of most of these compounds is not likely to be rapid, useful, or environmentally important.

ISSUES

There are possible policy implications of the methyl bromide—plant connections. Because of the capability to rapidly degrade methyl bromide, plants and fungi may provide a major sink for the removal and treatment of at least several xenobiotic volatile halogenated hydrocarbons that pose a threat to our environment. Further, plants and fungi clearly are major participants in cycling bromine from inorganic bromide in soils and the oceans, to methyl bromide in the atmosphere, and, through plant accumulation and decay, to bromide ion in soils. Yet, depending on the fluxes of these processes, human production and use may cause very little fluctuation in the natural sequences that determine the atmospheric budget of methyl bromide.

Acknowledgments

P.M. Jeffers was a National Research Council Senior Fellow in 1996, visiting in Dr. N.L. Wolfe's lab at the U.S. EPA research facility in Athens, Georgia. During this stay, the happy coincidence of homogeneous hydrolysis studies (Jeffers) and aquatic plant/sediment/aqueous halocarbon transformation (Wolfe and others in his research group) led to the first investigations of plant and atmospheric halocarbon interactions. The National Research Council and Lee Wolfe are gratefully acknowledged. Recent financial support of this research effort by the donors of The Petroleum Research Fund, administered by the American Chemical Society, is acknowledged. Cortland undergraduate students who served as research assistants include Christina D. Liddy, Kimberly F. Davis, Robert Swarthout, Jeanine Prendergast, and Alyssa Huebner.

REFERENCES

Attieh, J.M., A.D. Hanson, and H.S. Saini (1995) Purification and characterization of a novel methyltransferase responsible for biosynthesis of halomethanes and methanethiol in *Brassica oleracea*. *J. Biol. Chem.* **270**: 9250–9257.

Butler, J.H (2000) Better budgets for methyl halides? *Nature* **403**: 260–261.

Gan, J., S.R. Yates, H.D. Ohy, and J.J. Sims (1998) Production of methyl bromide by terrestrial higher plants. *Geophys. Res. Lett.* **25**: 3595–3598.

Hallier, S., S. Deutschmann, C. Reichel, H.M. Bolt, and H. Peter (1990) A comparative investigation of the metabolism of methyl bromide and methyl iodide in human erythrocytes. *Int. Arch. Occup. Environ. Health* **62**: 221–225.

Harper, D.B. (1985) Halomethane from halide ion—a highly efficient fungal conversion of environmental significance. *Nature* **315**: 55–57.

Jafvert, C.T. and N.L. Wolfe (1987) Reduction of selected halogenated ethanes in anaerobic sediments. *Environ. Toxicol. Chem.* **4**: 827–831.

Jeffers, P.M. and N.L. Wolfe (1996a) Hydrolysis of halogenated hydrocarbons, Chapter 4. American Chemical Society Symposium Series 652. *Fumigants: Environmental Fate, Exposure, and Analysis.* J.N. Seiber, ed. Washington, D.C., pp. 32–41.

Jeffers, P.M. and N.L. Wolfe (1996b) Degradation of methyl bromide by green plants, Chapter 6. American Chemical Society Symposium Series 652. *Fumigants: Environmental Fate, Exposure, and Analysis.* J.N. Seiber, ed. Washington, D.C., pp. 53–59.

Jeffers, P.M. and N.L. Wolfe (1996c) On the degradation of methyl bromide in seawater. *Geophys. Res. Lett.* **23**: 1773–1776.

Jeffers, P.M., L. Ward, L. Woytowitch, and N.L. Wolfe (1989) Homogeneous hydrolysis rate constants. *Environ. Sci. Technol.* **23**: 965–968.

Jeffers, P.M., N.L. Wolfe, and V. Nzengung (1998) Green plants: a terrestrial sink for atmospheric $CH_3 B_r$. *Geophys. Res. Lett.* **25**(1): 43–46.

Keppler, F., R. Eiden, V. Niedan, J. Pracht, and H.J. Scholer (2000) Halocarbons produced by natural oxidation processes during degradation of organic matter. *Nature* **403**: 298–301.

Lee-Taylor, J.M. and E.A. Holland (2000) Litter decomposition as a potential natural source of methyl bromide. *J. Geophys. Res.* **105**: 8857–8865.

Liyanage, C., D.A. Luvisi, and D.O. Adams (1993) The glutathione content of grape berries is reduced by fumigation with methyl bromide. *Am. J. Enol. Viticult.* **44**: 8–12.

Methyl Bromide Global Coalition/National Aeronautics and Space Administration (NASA) (1997) Summary. Methyl Bromide State of The Science Workshop, Monterey, California.

Montzka, S.A., J.H. Butler, J.W. Elkins, T.M. Thompson, A.D. Clark, and L.T. Lock (1999) Present and future trends in the atmospheric burden of ozone-depleting halogens. *Nature* **398**: 690–694.

Muramatsu, Y. and S. Yoshida (1994) Volatilization of methyl iodide from the soil—plant system. *Atmos. Environ.* **29**: 21–25.

Rhew, R.C., B.R. Miller, and R.F. Weiss (2000) Natural methyl bromide and methyl chloride emissions from coastal salt marshes. *Nature* **403**: 292–295.

Saini, J.M., J.M. Attieh, and A.D. Hanson (1995) Biosynthesis of halomethanes and methanethiol by higher plants *via* a novel methyltransferase reaction. *Plant Cell Environ.* **18**: 1027–1033.

Shorter, J.H., C.E. Kolb, P.M. Krill, R.A. Kerwin, R.W. Talbot, M.E. Hines, and R.C. Harriss (1995) Rapid degradation of atmospheric methyl bromide in soils. *Nature* **377**: 717–719.

Solomon, S., M. Mills, L.E. Heidt, W.H. Pollock, and A.F. Tuck (1992) On the evaluation of ozone depletion potentials. *J. Geophys. Res.* **97**: 825–842.

Wofsy, S.C., M.B. McElroy, and Y.L. Yung (1975) The chemistry of atmospheric bromine. *Geophys. Res. Lett.* **2**: 215–218.

Wuosmaa, A.M. and L.P. Hager (1990) Methyl chloride transferase: a carbocation route for biosynthesis of halometabolites. *Science* **240**: 160–162.

Yokouchi, Y., Y. Noijiri, L.A. Barrie, D. Toom-Sauntry, T. Machida, Y. Inuzuka, H. Akimoto, H.-J. Li, Y. Fujinuma, and S. Aoki (2000) A strong source of methyl chloride to the atmosphere from tropical coastal land. *Nature* **403**: 295–298.

Yung, Y.L., J.P. Pinto, R.T. Watson, and S.P. Sander (1980) Atmospheric bromine and ozone perturbations in the lower stratosphere. *J. Atmos. Sci.* **37**: 339–353.

27

PHYTOREMEDIATION OF METHYL *TERTIARY*-BUTYL ETHER

S. K. Winnike-McMillan, Q. Zhang, L. C. Davis, L. E. Erickson, and J. L. Schnoor

SUMMARY OF PRACTICAL IMPLICATIONS

The capability of plants to remediate methyl *tertiary*-butyl ether (MTBE) at contaminated sites is promising to the point that at least one field test is underway and others anticipated. Both hybrid poplar (*Populus deltoides* × *Populus nigra* 'DN34' 'Imperial Carolina') cuttings and alfalfa (*Medicago sativa*) facilitate the removal of MTBE from saturated soil in the laboratory. The primary removal of MTBE is by soil volatilization facilitated by plant transpiration. Very little of the contaminant is metabolized or accumulated in laboratory studies. The MTBE that volatilizes and the small amount that diffuses from stems or transpires through leaves do not represent a risk in the atmosphere due to the rapid dispersion and degradation. Despite the different behavior of MTBE in groundwater, plants, and the atmosphere, concurrent phytoremediation of gasoline spills containing MTBE and other contaminants is expected to be very useful in the future.

INTRODUCTION

Methyl *tertiary*-butyl ether was first developed as a fuel oxygenate in the late 1970s to enhance the octane of gasoline and to improve air quality. In 1990, the U.S. Clean Air Act Amendments required fuel oxygenates to be added to gasoline in an effort to reduce atmospheric concentrations of carbon monoxide and ozone in some urban areas. The use of MTBE increased 16-fold from 1984 to 1993, to 24 billion pounds (approximately 10×10^6 metric tons). By 1998, MTBE was the fourth most common chemical product in the U.S.

Phytoremediation: Transformation and Control of Contaminants,
Edited by Steven C. McCutcheon and Jerald L. Schnoor.
ISBN 0-471-39435-1 (cloth)

(Johnson *et al.* 2000). Oxygenates are added to over 30 percent of the gasoline used in the U.S., and MTBE is the most commonly used of these additives (Squillace *et al.*, 1996). After formulation, MTBE constitutes up to 15 percent of gasoline by weight, giving rise to a daily usage of more than 10 million gallons (3.8×10^7 liters) in the U.S. (Johnson *et al.* 2000). However, continued use of MTBE is being questioned because of reports of adverse health effects, uncertainty about the efficacy of air pollution control, and the potential for contamination of groundwater (Anonymous 1997, Mehlman 1998). Several U.S. states have required that MTBE be phased out over time. A recent proposal was that the U.S. President use the authority under the Toxic Substances Control Act to ban MTBE across the U.S., but production continued in late 2000 at a rate of more than 11 million gallons or nearly 40 million liters per day.

As a part of the U.S. Geological Survey National Water Quality Assessment program from 1993 to 1994, 60 volatile organic compounds were analyzed for occurrence in groundwaters and surface waters. Of these 60 organic compounds, MTBE was the second most frequently detected chemical in samples of shallow groundwater from U.S. urban areas (Zogorski *et al.* 1998). Very little was detected in agricultural areas or deep groundwater wells. Surveys have reported finding MTBE in 5 to 10 percent of U.S. potable groundwater supplies (Stuckey 2000). Based on current distribution of MTBE sources and well pumping rates, Johnson *et al.* (2000) assessed the likelihood that more wells will become contaminated by MTBE. Although new and much more extensive databases are needed, the projected number of U.S. wells likely to experience contamination is in the thousands due to a large fraction of all community water supply wells lying within 1 kilometer of leaking underground storage tanks.

Methyl *tertiary*-butyl ether contamination of soil and groundwater can be separated into two categories: point sources, such as leaking underground storage tanks, and nonpoint sources, including infiltration and diffusion from urban air into shallow groundwater. General urban air contamination arises from fugitive emissions at gasoline stations and from gasoline-powered engines. Point sources are typically known quantifiable oxygenated gasoline spills onto the ground or leaks from underground storage tanks. Contamination by MTBE commonly occurs in conjunction with BTEX gasoline constituents (benzene, toluene, ethyl benzene, and xylenes). Gasoline from leaking underground storage tanks is the most frequent source of significant groundwater pollution with as much as one million gallons (approximately 4 million liters) of gasoline released into the ground each year (Hartley and Englande 1992), implying MTBE releases of hundreds of tons. It is estimated that out of three million underground storage tanks in use in the U.S., as many as 500 000 may be leaking petroleum into the ground (Hartley and Englande 1992). Actual leakage may be much higher than this implied level of 2 gallons per tank per year (8 liters per tank per year).

Properties of Methyl *Tertiary*-Butyl Ether

Methyl *tertiary*-butyl ether is an aliphatic ether with a molecular weight of 88.15 grams per mole (Figure 27-1) (Squillace *et al.* 1997). This compound is

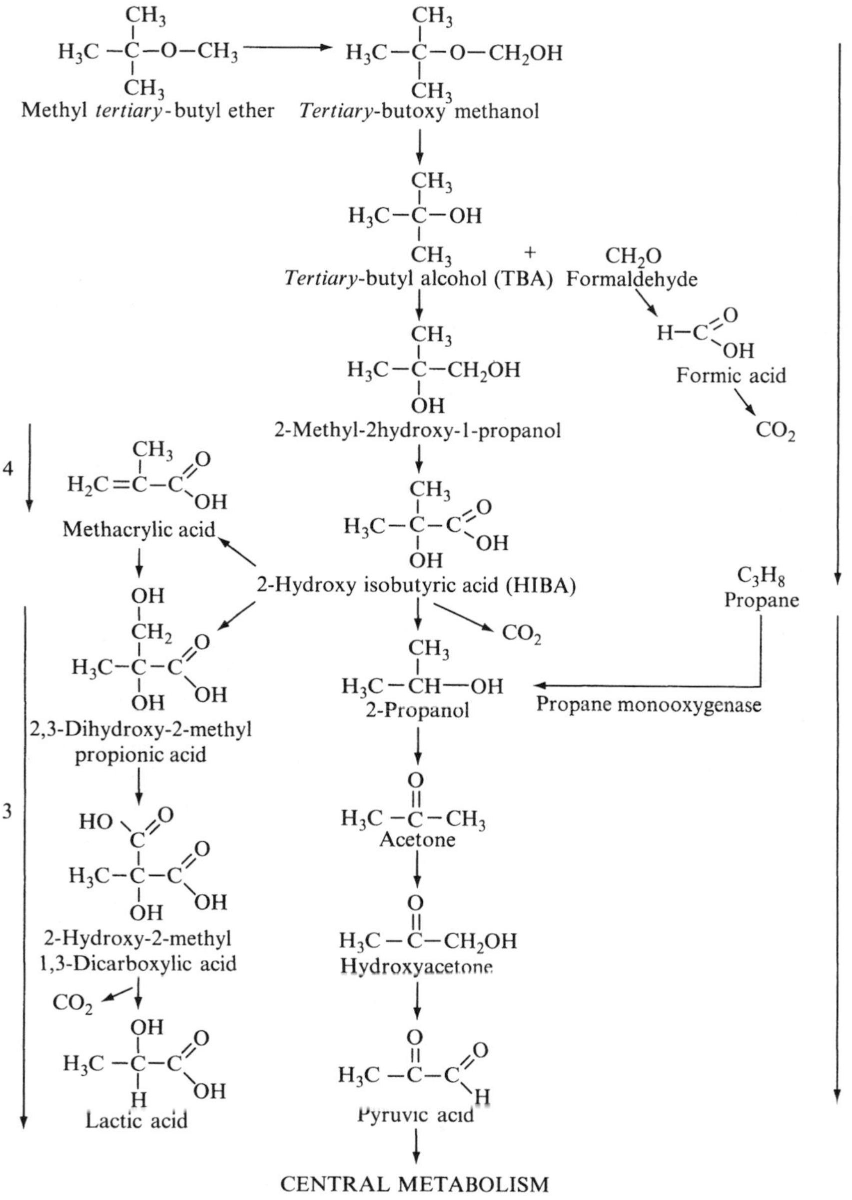

Methyl *tertiary*-butyl ether
Tertiary-butoxy methanol
CH_3 + CH_2O
Tertiary-butyl alcohol (TBA) Formaldehyde
Formic acid
CO_2
2-Methyl-2hydroxy-1-propanol
Methacrylic acid
2-Hydroxy isobutyric acid (HIBA)
C_3H_8
Propane
CO_2
2-Propanol
Propane monooxygenase
2,3-Dihydroxy-2-methyl propionic acid
Acetone
2-Hydroxy-2-methyl 1,3-Dicarboxylic acid
Hydroxyacetone
CO_2
Lactic acid
Pyruvic acid
CENTRAL METABOLISM

Figure 27-1 Methyl *tertiary*-butyl ether degradation pathway proposed by the investigation of propane oxidizing bacteria. Adapted from Steffan *et al.* (1997).

volatile, colorless, and flammable (Caprino and Togna 1998). Because chemical structure influences many of the properties such as solubility, volatilization potential, and resistance to microbial degradation, it is useful to compare the physical and chemical properties of MTBE with those of the BTEX compounds with which it usually occurs in the environment (Table 27-1).

There is good evidence for microbial degradation of MTBE by several processes, as well as field data showing natural attenuation. Several processes for MTBE have been identified.

TABLE 27-1 Chemical and Physical Properties of Gasoline, Methyl *Tertiary*-Butyl Ether, Benzene, and Toluene

Property	Gasoline	Methyl *tertiary*-butyl ether	Benzene	Toluene
Molecular weight, grams per mole	≈ 100	88.15	78.11	92.13
Specific gravity	0.72 0.74	0.744 0.7404 to 0.7578	0.88 0.8765	0.8669
Boiling temperature, °C		53.6 to 55.2	80.1	110.6
Water solubility, milligrams per liter	100 to 200	43 000 to 54 300	1780 1791	534.8
Vapor pressure at 25 °C, millimeters mercury		245 to 251	95.19	28.4
Log K_{ow}		1.20 0.94 to 1.30	2.13 1.56 to 2.15	2.73 2.11 to 2.80
Henry's law constant, atmospheres-cubic meters per gram mole		5.87×10^{-4} 1.4×10^{-3} 3×10^{-3} 5.28×10^{-4}	5.43×10^{-3}	5.94×10^{-3}
Dimensionless Henry's law constant (milligrams per liter of air) per (milligrams per liter of water)		0.0240 0.0572 0.123 0.0216 mean: 0.0564	0.222	0.243
Log K_{oc}		1.05 $K_{oc} \approx 11$ cubic centimeters per gram	1.1 to 2.5 median = 1.9 $K_{oc} \approx 80$ cubic centimeters per gram	1.56 to 2.25

K_{ow} is the dimensionless octanol–water partitioning coefficient.
K_{oc} is organic carbon normalized sorption coefficient.
Log is the base 10 logarithm.
Note: One atmosphere is 1.0133×10^{-5} pascals in the Henry's law constant and 1 millimeter of mercury of vapor pressure is 133.32 pascals. Adapted from Squillace *et al.* (1997) and references cited therein.

Anaerobic Microbial Transformation

Mormile *et al.* (1994) tested both pure and mixed cultures to anaerobically degrade MTBE. Two acetogens, *Acetobacterium woodii* and *Eubacterium limosum*, known to degrade phenyl methyl ethers, showed no MTBE transformation. However, the microbial cells consumed other gaseous substrates from the headspace of the incubation system and produced formate and acetate with no apparent toxicity caused by MTBE (Mormile *et al.* 1994). The capability of mixed cultures to anaerobically transform MTBE and *tertiary*-butyl alcohol was investigated under denitrifying and methanogenic conditions, with rates of transformation increasing with nutrients added (Yeh and Novak 1994). Methyl *tertiary*-butyl ether was shown to degrade in the soil taken from only one site out of three and only when nutrients were added. The degradation rates were much slower than those of the *tertiary*-butyl alcohol and complete degradation took approximately 270 days. More recently Landmeyer *et al.* (1998) and Finneran and Lovley (2001) observed anaerobic degradation of MTBE under Fe(III) reducing conditions and Bradley *et al.* (2001) observed MTBE biodegradation under nitrate-reducing conditions

Aerobic Transformation in Mixed Cultures

Aerobic microbial transformation of MTBE was first successfully shown by Salanitro *et al.* (1994) using a mixed culture isolated from the sludge of an industrial chemical biotreatment operation. This culture, consisting of organisms from *Corynebacterium*, *Pseudomonades*, and *Achromobacters* was able to utilize MTBE at 2 milligrams per liter as a sole carbon source although none of the organisms were shown to grow alone on MTBE. Approximately 80 percent of the [^{14}C]-MTBE was mineralized to $^{14}CO_2$ or incorporated into new microbial cells after an incubation period of approximately 8 hours (Salanitro *et al.* 1994). *Tertiary*-butyl alcohol was observed as an intermediate. With 20 milligrams of MTBE per liter, little or no [^{14}C]-MTBE was metabolized to $^{14}CO_2$ or cellular material, suggesting that MTBE may be inhibitory to aerobes at high concentrations (Salanitro *et al.* 1994).

A consortium of aerobes from refinery activated sludge was also effective in transforming MTBE (Cowan and Park 1996), as was an aquifer consortium studied by Borden *et al.* (1997). Bradley *et al.* (1999) and Landmeyer *et al.* (2001) observed rapid metabolism of [^{14}C]-labeled MTBE using naturally occurring streambed microorganisms under mixed aerobic and anaerobic conditions in microcosms. Eweis *et al.* (1998) examined a compost-based biofilter microbial community and found MTBE degradation.

Aerobic Cometabolism

A fungus of the genus *Graphium* has been shown capable of cometabolizing MTBE with butane (Hardison *et al.* 1997), and Garnier *et al.* (1999) observed cometabolism with pentane by *Pseudomonas.* Steffan *et al.* (1997) reported indigenous propane-oxidizing bacteria isolated from contaminated and uncontaminated soils that could degrade MTBE. Isolate ENV425 that degrades MTBE was identified by fatty acid analysis to be most closely related to members of the genus *Nocardia.* All isolates were first grown with propane, then fed concentrations of MTBE in the range of 20 to 100 milligrams per liter, which was used as a sole carbon and energy source (Steffan *et al.* 1997). The cells were fed 20 milligrams per liter metabolized MTBE at a rate of at least 3.8 nanomoles per minute per milligram of cells without a significant lag period. Complete degradation occurred after approximately 4 hours, with *tertiary*-butyl alcohol accumulating nearly stoichiometrically, and with evolution of $^{14}CO_2$ from the methyl group. The microorganisms were capable of transforming *tertiary*-butyl alcohol at a rate of 1.8 nanomoles per minute per milligram of cells, about half that of MTBE. Following 10 hours of transformation of *tertiary*-butyl alcohol, approximately 70 percent of the radiolabeled compound was still present in solution, suggesting the formation of water-soluble *tertiary*-butyl alcohol metabolites. Continued incubation showed 60 percent conversion to $^{14}CO_2$ after 24 hours and greater than 90 percent conversion after several days (Steffan *et al.* 1997). Figure 27-1 is a pathway for MTBE degradation proposed by Steffan *et al.* (1997).

Aerobic Transformation in Pure Cultures Lacking Other Substrates and Prior Induction

Four pure cultures of diverse genera have been shown capable of transforming at least a fraction of MTBE in a defined medium (Mo *et al.* 1997). The three strains investigated by Mo *et al.* (1997) represented *Methylobacterium, Rhodococcus,* and *Arthrobacter*. Also, a proteobacteria strain was derived from the biofilter examined by Eweis *et al.* (1998).

Potential Biotreatments for Methyl *Tertiary*-Butyl Ether

Salanitro *et al.* (2000) have successfully used bioaugmentation and oxygenation with a consortium designated MC-100 to enhance degradation of MTBE within an aquifer. Landmeyer *et al.* (2001) successfully used oxygenation only to enhance biodegradation of an MTBE-contaminated aquifer. Successful treatment was achieved in laboratory studies through cometabolism in an upflow fluidized-bed bioreactor (Stringfellow and Oh 2002) and in a vapor phase treatment using a trickling filter (Fortin and Deshusses 1999). In the latter case, contaminated soil from two different locations was used as an inoculum. Prince (2000) has reviewed most of the earlier literature on microbial MTBE degradation. Despite several instances of successful metabolism, MTBE appears to transform only slowly under typical field spill conditions, resulting in a need for additional cost-effective treatment methods.

A potentially useful alternative is phytoremediation. Plants have the benefit of absorbing large quantities of subsurface water by solar driven pumping (transpiration) and the roots support diverse microbial populations (Shimp *et al.* 1993, Davis *et al.* 1998a, Narayanan *et al.* 1999). Deep-rooted poplar (*Populus* spp.) trees have been studied in several contexts but there is little information on the capability to transpire or metabolize MTBE. Davis *et al.* (1998b) showed by use of Fourier transformed infrared spectroscopy that MTBE passes freely through hybrid poplars (*Populus deltoides* × *Populus nigra* 'DN34' 'Imperial Carolina') in 1- to 2-day experiments although no effort was made to determine metabolic processes that might occur within the plant. Landmeyer *et al.* (2000) detected MTBE in cores of trees above a gasoline-contaminated aquifer but no effort was made to track the movement through the trees.

Rubin and Ramaswami (2001) have also investigated the uptake and phytoremediation of MTBE using hybrid poplar (*Populus deltoides* × *Populus nigra* 'DN34', & IC) cuttings. They reported transpiration of MTBE through plants with no evidence of MTBE transformation in the poplar (*Populus* spp.) saplings (Ramaswami and Rubin 2001).

UPTAKE OF METHYL *TERTIARY*-BUTYL ETHER BY HYBRID *POPULUS* IN HYDROPONIC SOLUTION

Experimental Setup

To determine the potential for hybrid poplars (*Populus deltoides* × *Populus nigra* 'DN34' 'Imperial Carolina') to take up, metabolize, and translocate MTBE, a hydroponic experiment was performed, measuring the capability of hybrid poplar (*Populus deltoides* × *Populus nigra* 'DN34' 'Imperial Carolina') cuttings to remove [^{14}C]-MTBE from aqueous solution. Uptake of [^{14}C]-radiolabeled MTBE by hybrid poplar (*Populus deltoides* × *Populus nigra* 'DN34' 'Imperial Carolina') cuttings was investigated in a manner similar to methods pioneered by Burken (1993) and Burken and Schnoor (1996). A schematic of the reactor design is provided in Figure 27-2. However, in the first experiment, there was no attempt made to collect gases volatilizing from plant leaves. The study included varying concentrations of MTBE in solution to determine any toxic response. Hydroponic studies used one-quarter-strength Hoagland's inorganic nutrient solution, fed to the root zone of batch reactors containing hybrid poplar (*Populus deltoides* × *Populus nigra* 'DN34' 'Imperial Carolina') cuttings. Eight-inch (200-millimeter) long cuttings (0.75 to 1 inches or 19 to 25 millimeters in diameter) were affixed with pre-drilled screw caps and pre-drilled Teflon®-lined septa. Teflon® tape was wrapped around the stem and acrylic caulk was used to seal the caps and septa to the cutting. Hybrid poplar (*Populus deltoides* × *Populus nigra* 'DN34' 'Imperial Carolina') cuttings were first rooted in hydroponic solution until a vascular root system appeared,

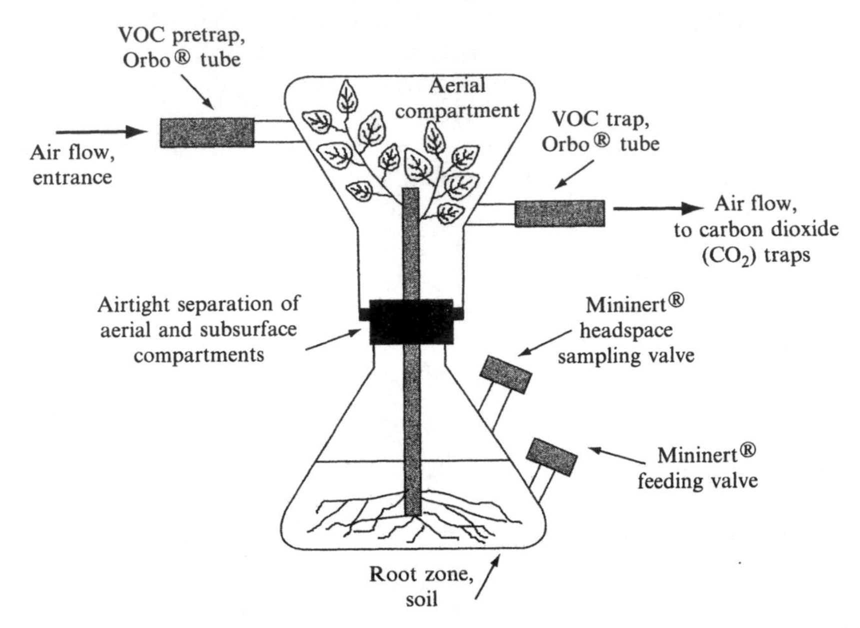

Figure 27-2 Schematic of the reactor design used to investigate hybrid poplar (*Populus deltoides* × *Populus nigra* 'DN34' 'Imperial Carolina') cuttings growing in soil.

then placed in 1-liter bioreactors containing 400 milliliters of nutrient solution. The bioreactors consisted of a 1-liter Erlenmeyer flask modified by attaching a sampling port to the bottom and top of each flask. The cuttings were maintained in a laboratory growth chamber at 28 °C under artificial grow lights that provided a photon flux at the leaf surface of 100 to 160 micromoles of photons per square meter per second, the full photosynthetically active range.

Controls were included in triplicate to distinguish the uptake of MTBE by hybrid poplar (*Populus deltoides* × *Populus nigra* 'DN34' 'Imperial Carolina') cuttings from losses out of the system. First, a capped control with only nutrient solution and [^{14}C]-MTBE was included to determine the contribution of biodegradation possibly occurring inside the hydroponic solution and sorption to the glass. To determine loss of compound through leaks in the system, controls were used with solid glass rods in place of a tree cutting. Finally, two sets of planted cuttings were cut off just above the cap to create decapitated controls, which were necessary to observe the effect of an excised tree on [^{14}C]-MTBE removals. Two sizes of cuttings were used; the first was approximately the same diameter as the glass rod and the second was similar to the intact tree reactors.

After being sealed, all reactors were spiked with a combination of [^{14}C]-radiolabeled MTBE ("hot," *i.e.*, radioactive), 6 microcuries per reactor, and pure MTBE ("cold," *i.e.*, nonradioactive) to a total aqueous concentration of 10.18 milligrams per liter, (creating a cold:hot MTBE ratio of 12 037:1). Concentrations of cold MTBE were then increased to 100, 1000, and 10 000 milligrams per liter. Reactors were prepared for all concentrations in triplicate. Samples were collected daily to monitor the disappearance of [^{14}C]-MTBE from hydroponic solution. Transpiration was measured gravimetrically each day, and replacement nutrient solution was added through the lower Mininert® port. Prior to adding nutrient solution, all headspace gas surrounding the root system was removed through a series of traps to capture any MTBE that was present in the headspace above the hydroponic solution. The air was pumped out through an activated carbon trap capturing pure compound and other organic metabolites (Orbo® tube 32, large, Supelco), and a trap that allowed air to bubble through 1 molar sodium hydroxide (NaOH) to capture any carbon dioxide (CO_2). To extract the trapped MTBE, the activated carbon was placed into 4-milliliter glass vials and submerged in 2 milliliters of methanol (Fisher) for 24 to 48 hours. Samples of 100 microliters were taken and injected into 15 milliliters of Scintiverse® for counting on a liquid scintillation counter. One milliliter samples were taken from the sodium hydroxide (NaOH) in the carbon dioxide (CO_2) traps and injected into an Ultima Gold scintillation cocktail (Packard) for scintillation counting.

After 10 days, total [^{14}C]-uptake was quantified in the roots, lower stem, upper stem, leaves, and petioles of the hybrid poplar (*Populus deltoides* × *Populus nigra* 'DN34' 'Imperial Carolina') cuttings *via* oxidation to $^{14}CO_2$ in a R.J. Harvey Bio-Oxidizer, and subsequently by scintillation counting. Oxidation efficiency was determined to be 92 percent and was calibrated with pre- and post-analysis efficiency testing.

Results from Hydroponic Solutions

Based on recorded observations of visual symptoms, biomass, and transpiration rate, MTBE was found not toxic to hybrid poplar (*Populus deltoides* × *Populus nigra* 'DN34' 'Imperial Carolina') cuttings at concentrations as high as 1000 milligrams of MTBE per liter. However, the trees dosed with 10 000 milligrams of MTBE per liter died after only a few days. Transpiration rate correlated well with biomass measurements and was a good measure of growth and vitality. Hybrid poplar (*Populus deltoides* × *Populus nigra* 'DN34' 'Imperial Carolina') cuttings were healthy and grew well in all experiments at concentrations as high as 1000 milligrams of MTBE per liter.

Figure 27-3 displays removal of [^{14}C]-MTBE from solution by the full and excised cuttings as well as the controls. The data for the full ($n = 5$) and excised ($n = 3$) cuttings are shown as an average with the error bars representing one standard deviation. Uptake of MTBE occurred immediately in all planted reactors and continued throughout the investigation. For the five planted reactors, 66.98 ± 6.09 percent of the total MTBE was removed from hydroponic solution at the end of the 10-day experiment.

Methyl *tertiary*-butyl ether removal after 10 days for the excised cuttings was 42.97 ± 3.32 percent. The data from the toxicity portion of the experiment

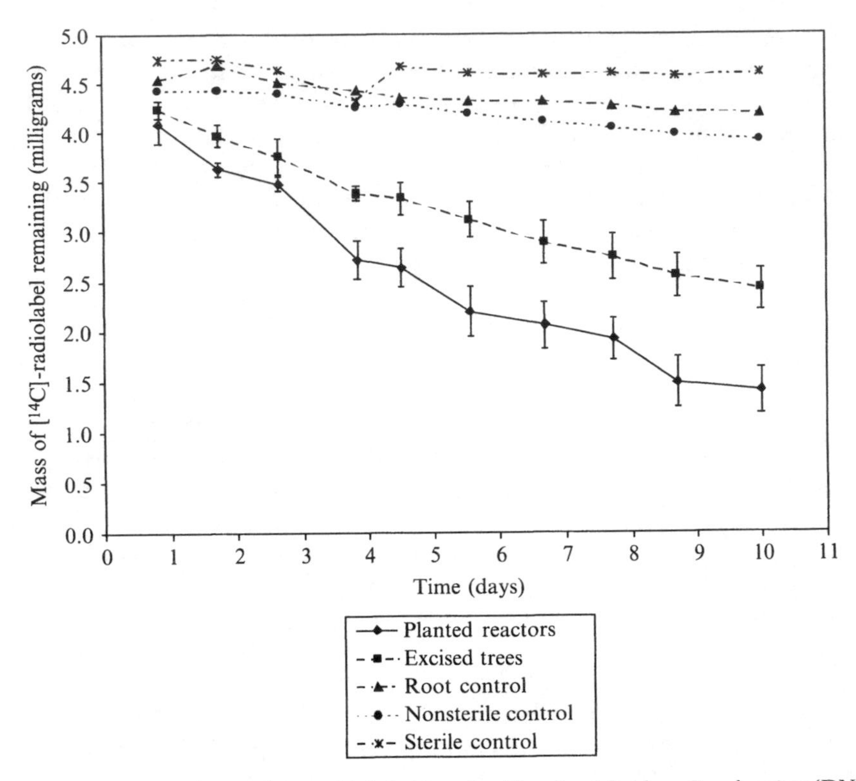

Figure 27-3 Hydroponic experiment with hybrid poplar (*Populus deltoides* × *Populus nigra* 'DN34' 'Imperial Carolina') cuttings: mass of methyl *tertiary*-butyl ether remaining in aqueous solution as determined by liquid scintillation counting. At time = 0, 4.3 milligrams of methyl *tertiary*-butyl ether was added for an initial concentration of 10.2 milligrams per liter in solution. Error bars on the data for full trees ($n = 5$) and excised trees ($n = 3$) represent one standard deviation.

yielded a comparable result for the cuttings dosed with 10 000 milligrams per liter—51.52 ± 2.29 percent was removed from solution after 11 days. The removal rate was similar for the excised cuttings (k = 0.060 ± 0.002 per day and the correlation coefficient $r^2 = 0.994$), and for the cuttings that died at 10 000 milligrams of MTBE per liter (k = 0.067 per day) as evidenced by the first order rate coefficients. The reproducibility of the removal rates and percent removed from solution are a strong indication that the cause is attributable to movement of MTBE through the dead stem as opposed to leaks from the enclosed system. Excised cuttings are apparently capable of volatilizing MTBE (and very little water) from solution to the atmosphere.

After 10 days, the experiment was terminated and the cuttings were sacrificed. The various components were bio-oxidized and the radioactivity measured with the liquid scintillation counter to determine the amount of radiolabeled compound in the plant tissue. Figure 27-4 summarizes the radiolabel recoveries for the full cutting and displays the overall mass balance.

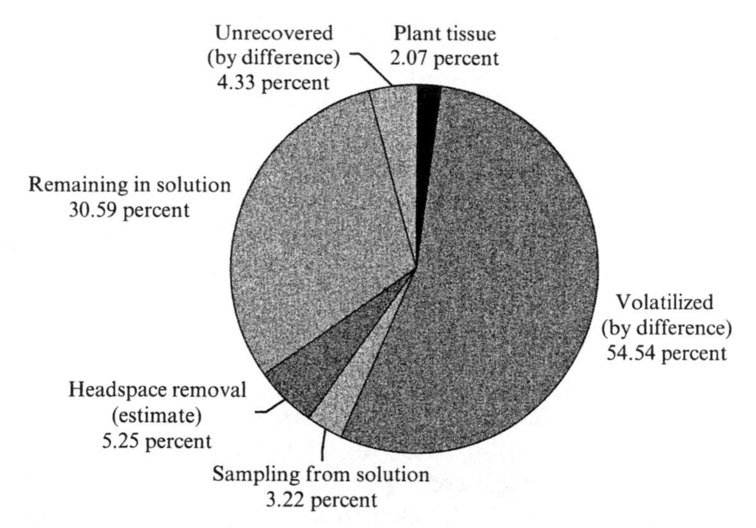

Figure 27-4 Hydroponic experiment: graph depicting [^{14}C]-radiolabel distribution and recovery results for the reactors containing hybrid poplar (*Populus deltoides* × *Populus nigra* 'DN34' 'Imperial Carolina') cuttings as determined by liquid scintillation counting at the termination of the 10-day investigation. Results are given as a percent of the total [^{14}C]-MTBE initial spike (6 microcuries) added to reactors at time = 0. The amount of radiolabel that was reported as volatilized was determined as the difference in the amount remaining in solution at the termination of the experiment between the unsterilized and planted reactors. The amount reported as unrecovered was the difference between the unsterilized and sterile reactors.

The amount of MTBE volatilized through plants was found to be 54.5 percent. This was estimated by the difference in the amount of radiolabel remaining in solution between the planted cuttings and the sterile controls. This estimate assumes that any removal from the excised tree reactors was due to uptake and not to leaks (Winnike 1998). However, leaks of MTBE through the septums and seals can be substantial, and leaks have been shown to rival transpiration of MTBE through leaves in such experiments (Hong *et al.* 2001).

The average [^{14}C]-radiolabel remaining in solution was 33.02 ± 6.09 percent. This is approximately twice as much as that found in the toxicity investigations in the same reactors at the same concentration. This value was expected to be similar due to the similar experimental conditions and duration. The reason for this difference lies in the relationship between cumulative transpiration and uptake of MTBE. A comparison of average transpiration rates shows that the cuttings in the toxicity experiment transpired more (32.3 ± 11.24 milliliters per day) than the cuttings in the hydroponic experiment (26.5 ± 10.0 milliliters per day). Because the leafy portion was exposed to the atmosphere, the amount of radiolabeled MTBE transpired through the leaves was not measured directly.

UPTAKE EFFICIENCY AND TOXICITY

Methyl *tertiary*-butyl ether was taken up by the plants and translocated up the vascular system of the plant in the transpiration stream. The efficiency of uptake of MTBE from hydroponic solution by hybrid poplar (*Populus deltoides* × *Populus nigra* 'DN34' 'Imperial Carolina') cuttings was estimated by the change in MTBE mass in the solution and the volume fraction of water transpired each day. Because of the high transpiration rates of all cuttings, it was necessary to add nutrient solution to the reactors daily, diluting the initial concentration of MTBE. This resulted in a continual decrease in concentration and mass of MTBE in the root compartment. For chemicals that are readily taken up and not transformed in the roots, the efficiency of uptake is identical to the dimensionless transpiration stream concentration factor (*TSCF*). The *TSCF* is defined as the chemical concentration in the xylem water divided by the chemical concentration in the external solution (Briggs *et al.* 1982). The TSCF was estimated to be in the range of 0.5 to 0.8 for MTBE for hybrid poplar (*Populus deltoides* × *Populus nigra* 'DN34' 'Imperial Carolina') cuttings based on

$$U = (TSCF)\,(C)\,(T) \tag{27-1}$$

in which U is the rate of mass uptake (milligrams per day), C is the concentration in solution (milligrams per milliliter), and T is the rate of transpiration (milliliters per day). The concentrations used to estimate *TSCF* were based on initial experimental concentrations of 10, 100, and 1000 milligrams per liter. The *TSCF* value was not calculated for the toxic concentration of 10 000 milligrams per liter.

UPTAKE OF METHYL *TERTIARY*-BUTYL ETHER BY HYBRID *POPULUS* PLANTED IN SOIL

Experimental Setup

To assess the capability of hybrid poplar (*Populus deltoides* × *Populus nigra* 'DN34' 'Imperial Carolina') cuttings to remove MTBE from soil, an Iowa agricultural silt-loam soil (Nodeway-Ely series) was taken from an uncontaminated field site with a fraction organic matter of 0.025 (Burken 1993). The soil was air-dried, pulverized, and passed through a 2-millimeter sieve. The cuttings were allowed to root hydroponically for 2 weeks until a full vascular root system developed. Four cuttings were transplanted to individual reactors containing 500 grams of air-dried soil, and 180 milliliters of nutrient solution was added intermittently to the flask to achieve uniform soil moisture content (Figure 27-2). The reactors were initially at 100 percent saturation and acclimated for 7 days. During this time, the trees transpired and the soil–water saturation level was reduced 70 to 90 percent.

After this acclimation, the leafy portion of the cutting was enclosed by attaching an inverted 1-liter Erlenmeyer flask modified with an inlet port for air flow near the top and an outlet near the bottom. The perimeter of the open end was sealed with acrylic caulk to fit snugly around the screw cap. Parafilm® tape was used to seal this joint, and a Mininert® valve was placed on the top sampling port of the bottom flask to create a fully encapsulated system. Reactors were treated with doses of 3.95 milligrams of cold MTBE and 7.1 microcuries of [^{14}C]-MTBE mixed with the feed water (resulting in a ratio of cold:hot MTBE of 10 069:1). Air was purified and pumped through the aerial portion of the reactor at 1.0 to 3.0 liters per minute. The air inlet and outlet ports of the Erlenmeyer flask were fitted with activated carbon traps to capture [^{14}C] transpired through the plants (two traps were used in series on the outlet port to ensure capture). The scrubbed air then passed through traps to capture any $^{14}CO_2$ that was mineralized by soil microorganisms or by the plant in the aerial portion of the reactor.

Results with Hybrid *Populus* in Soil

The primary goal of this test was to quantify volatilization of [^{14}C]-MTBE from leaf tissue. By enclosing the leafy portion of the cutting, a total mass balance was achieved. The total recovery was lower than expected (67.4 ± 6.3 percent). However, this test provided valuable information with regard to the major pathways for MTBE removal. The unrecovered portion seemed to be partially the result of leaks from the enclosed reactor vessel (Figure 27-2). The four planted cuttings remained healthy throughout the test, which was terminated after 33 days because the leaves grew too large and were pressing into the sides of the aerial compartment. The longer duration of this experiment (compared to the prior experiment in hydroponic solution) was possible because of slower growth of the cuttings in the soil. Transpiration was monitored by gravimetric analysis.

The rate of transpiration varied widely throughout the test because of excess humidity in the aerial compartment. As humidity increased to high levels, transpiration decreased dramatically due to aerodynamic resistance. In addition, the average transpiration rates for all reactors were significantly lower than under hydroponic conditions as a result of the decreased water availability in the soil. Although [^{14}C]-MTBE uptake was not quantified daily, the fluctuation in transpiration rates very likely influenced the rate of MTBE uptake from the soil water. The full set of transpiration data is provided in Winnike (1998).

Radiolabel recoveries on a percent mass basis are presented in Figure 27-5. Because only liquid scintillation counting was utilized, it was uncertain whether the radiolabel recovered was [^{14}C]-MTBE or a [^{14}C]-metabolite. Only

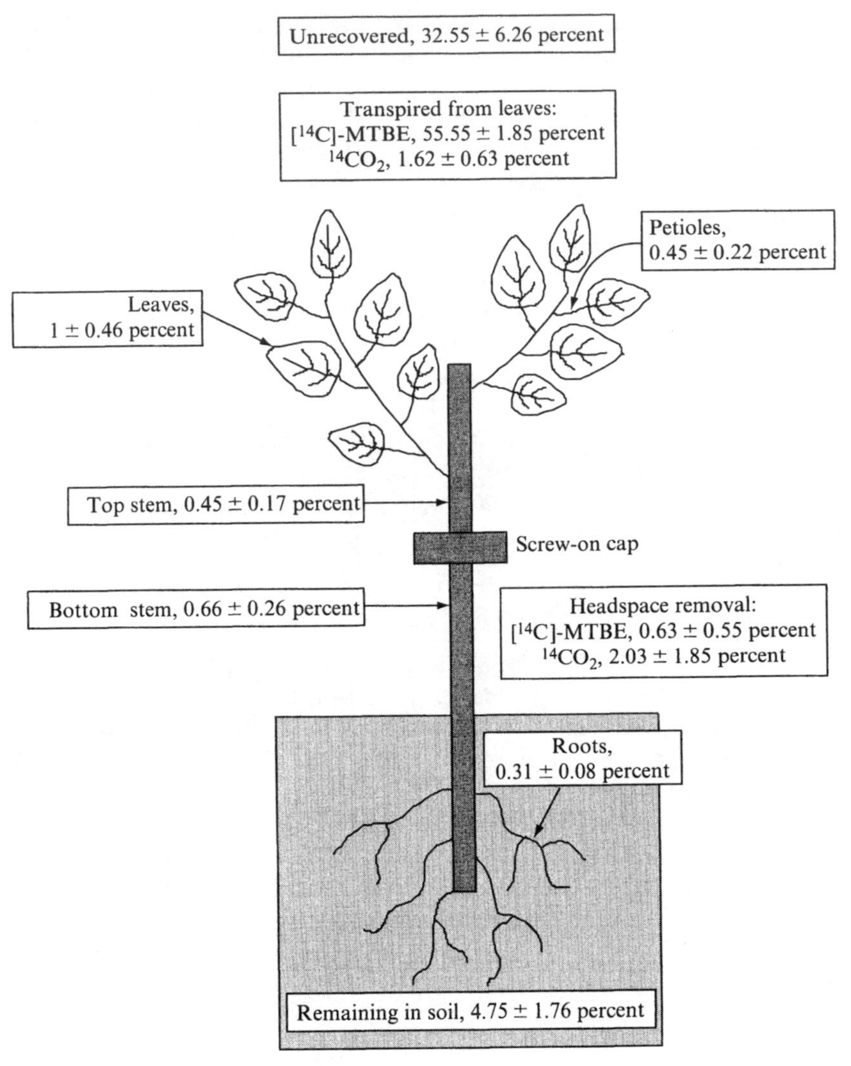

Figure 27-5 Schematic and mass balance of methyl *tertiary*-butyl ether uptake by hybrid poplars (*Populus deltoides* × *Populus nigra* 'DN34' 'Imperial Carolina') in soil reactor.

a small percentage of the initial [^{14}C]-radiolabel remained in the soil (4.75 ± 1.76 percent), indicating that 33 days were adequate to remove a majority of the MTBE from the soil. In subsequent tests, the vast majority of the radiolabel exists as parent compound, MTBE, in hybrid poplar (*Populus deltoides* × *Populus nigra* 'DN34' 'Imperial Carolina') tissues.

The [^{14}C]-radiolabel that was transpired constituted the most significant flux of MTBE movement (55.55 ± 1.85 percent). Assumedly the recovered radiolabel from the activated carbon traps was [^{14}C]-MTBE (subsequent analyses in the laboratory indicated that MTBE was predominant over other potential metabolites such as *tertiary*-butyl alcohol. The air also bubbled through a sodium hydroxide (NaOH) trap to capture $^{14}CO_2$ from the mineralization of MTBE. A small percent of the initial MTBE input was recovered as $^{14}CO_2$ (1.62 ± 0.63 percent), which indicates that either hybrid poplar (*Populus deltoides* × *Populus nigra* 'DN34' 'Imperial Carolina') cuttings or microorganisms in the soil were able to mineralize a portion of the MTBE to carbon dioxide (CO_2).

The headspace in the root compartment was sampled and analyzed for both [^{14}C] and $^{14}CO_2$ prior to watering and at the end of the test. This amount was shown to be fairly significant with recoveries of 0.63 ± 0.55 percent and 2.03 ± 1.85 percent for [^{14}C] and $^{14}CO_2$, respectively. The observed range of 0.41 to 4.50 percent demonstrates the high variability in $^{14}CO_2$ recovery. The amount of [^{14}C] remaining in plant tissue was approximately 3 percent. On the basis of total plant mass, the leaves contained 14.72 ± 4.02 milligrams of MTBE per kilogram, the petioles 11.10 ± 3.86 milligrams of MTBE per kilogram, and roots ± 3.76 milligrams of MTBE per kilogram. The leaves contained the largest total percentage (34.27 ± 9.55 percent) and the roots accounted for the smallest (11.14 ± 2.12 percent).

CHANNEL TESTS WITH ALFALFA

Tests were carried out in a 6-channel system (Zhang *et al.* 1998). Table 27-2 provides information on the experimental system and the tests that were conducted. The results are described fully elsewhere (Zhang *et al.* 1998, Zhang 1999, Zhang *et al.* 1999, 2000, 2001). Table 27-3 and Figure 27-6 show that vegetation affects the upward movement of MTBE from the saturated zone to the soil surface. Channel 4 has a much smaller fraction of MTBE and water moving to the soil surface. The recovery of MTBE was larger for Channel 4, which was barren of vegetation and unseeded with microbial cultures capable of biodegrading MTBE. As shown in Table 27-3, the measured loss of MTBE to the atmosphere at the soil surface was much greater than that which entered the plant and then passed into the atmosphere. The fraction of water that was lost as vapor to the atmosphere was considerably larger than the measured fraction of MTBE that entered the atmosphere (Zhang *et al.* 1999). The total recoveries of MTBE in the vegetated channels were less than 100 percent for all of the tests. The incomplete recovery of MTBE was probably due to enhanced rhizosphere biodegradation in the soil. Microcosm experiments with soils from the channel system were conducted, and a disappearance rate for MTBE of up to 5 milligrams per kilogram of soil per day was observed.

TABLE 27-2 Experimental Conditions and Treatments in Each Channel for the Six-Channel Experimental System

	Planted with 10 alfalfa (*Medicago sativa*) plants	Air sparged at 0.24 liter per channel per day	Additional bacteria (2 liters of suspension added on June 20, 1998)
Channel 1	Yes	Yes	#33 (*Rhodococcus*)
Channel 2	Yes	No	#33 (*Rhodococcus*)
Channel 3	Yes	No	No
Channel 4	No	No	No
Channel 5	Yes	No	#41 (*Arthrobacter*)
Channel 6	Yes	Yes	#41 (*Arthrobacter*)

Note: Channel dimensions were 1.1 meters long, 0.65 meter deep, and 0.10 meter wide. The soil was alluvial silty sand soil (with less than 10 percent silt) collected near a landfill in Riley County, Kansas. The system was illuminated with 12 cool white fluorescent lights (40 watts) at a height of 0.50 meter above the soil; light intensity was approximately 160 micromoles of photons or micro-einsteins per square meter per second. The channels were fed with an MTBE-water solution of 0.84 millimole per liter at 1 per day from June 30, 1998 to September 20, 1998 (83-day duration). Distilled water was fed from September 21 to December 25, 1998 (96-day duration) and the channels fed with an MTBE-water solution of 0.84 millimole per liter at 1 per day from Feb. 6, 1999 until the data in Figures 28-7 and 28-8 were collected during the summer of 1999. Strain #33 (*Rhodococcus*) and Strain #41 (*Arthrobacter*) were obtained from C.F. Kulpa (Mo *et al.* 1997, Zhang *et al.* 1999).

TABLE 27-3 Approximate Mass Balance Data for Methyl *Tertiary*-Butyl Ether (MTBE) for Each of the Six Channels

Channel #	1[a]	2[a]	3	4	5[b]	6[b]
Total water added (liters)	186	192	191	185	199	197
Evapotranspired water (liters)	81	116	109	37[c]	106	108
Fraction of water lost as vapor	0.44	0.60	0.57	0.20	0.53	0.55
MTBE added (millimoles)	68.1	69.1	68.5	69.1	68.8	71.7
MTBE from groundwater (millimoles)[d]	42.4	34.5	37.6	63.0	44.7	41.2
MTBE from gas (millimoles)[e]	15.3	17.0	21.8	12.0	19.1	14.0
Total MTBE recovered (millimoles)	57.7	51.5	59.4	75.0	63.8	55.2
MTBE groundwater recovery (fraction)	0.62	0.50	0.55	0.91	0.65	0.57
MTBE soil gas recovery (fraction)	0.22	0.25	0.32	0.17	0.28	0.20
Estimated average alfalfa (*Medicago sativa*) uptake of MTBE (fraction)	0.015	0.029	0.024	0.0	0.032	0.035
Total Recovery of MTBE (fraction)	0.87	0.78	0.89	1.08	0.96	0.81

[a]Strain #33 (*Rhodococcus*) was added.
[b]Strain #41(*Arthrobacter*) was added.
[c]There is only evaporation of water in the unplanted channel.
[d]"From groundwater" is the MTBE mass recovered from the groundwater effluent.
[e]"From gas" is the MTBE mass recovered from the soil surface based on gas flux measurements at the soil surface (Zhang *et al.* 1998).

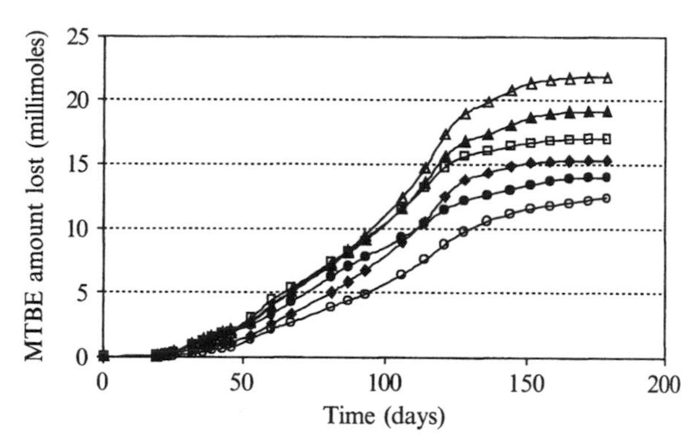

◆ channel 1, ● channel 2, ▲ channel 3, □ channel 4, ○ channel 5, and △ channel 6

Figure 27-6 Cumulative loss of methyl *tertiary*-butyl ether to the atmosphere through the soil surface for each of six channels. Tests conducted from June 30 to December 25, 1998.

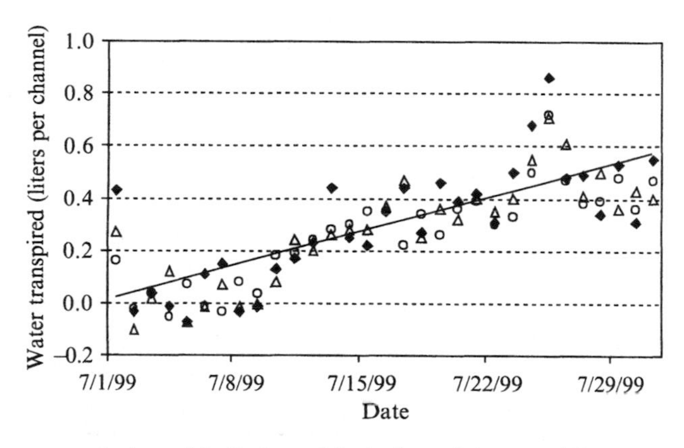

◆ channel 1, ○ channel 5, △ channel 6, —trend line

Figure 27-7 Water transpired daily by alfalfa (*Medicago sativa*) in three channels during July 1999; plants were cut back to 60 millimeters on July 1.

The water transpired daily by the plants increased from the time the alfalfa (*Medicago sativa*) was cut until the next harvest as shown in Figure 27-7. The transpiration rate was on average 5 millimeters per day (0.5 cubic centimeter per square centimeter per day) at the end of July 1999. This value was in agreement with predicted transpiration rates (Davis *et al.* 1998a). Figure 27-8 provides information on the concentration of MTBE in the plant water as a function of plant height. Experiments conducted with plant stems show that MTBE diffuses through the stem surface into the atmosphere (Zhang *et al.* 2001). The estimated values of the diffusivity for MTBE range from 8×10^{-8} to 8×10^{-7} square centimeters per second for alfalfa (*Medicago sativa*) plant stems (Zhang *et al.* 2001).

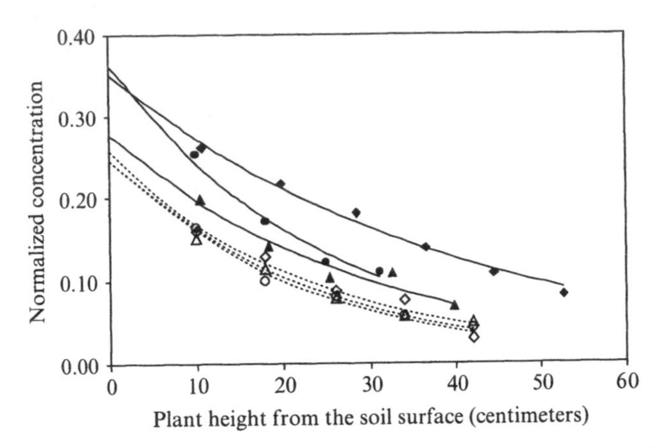

◆ plant #1, ● plant #2, ▲ plant #3, ◇ plant #4, ○ plant # 5, and △ plant #6

Figure 27-8 Methyl *tertiary*-butyl ether dimensionless concentration in plant water as a function of alfalfa (*Medicago sativa*) stem position from the soil surface. Dimensionless feed concentration of methyl *tertiary*-butyl ether is 1 which corresponds to 0.84 millimoles per liter.

FATE OF METHYL *TERTIARY*-BUTYL ETHER IN THE ATMOSPHERE

Site-specific determinations are necessary to ensure that the amount of MTBE transferred to the atmosphere does not exceed air quality standards. This analysis takes into account the total amount of compound volatized per unit time and the mixing zones in the atmosphere. This can be compared to the large quantities of MTBE introduced into the atmosphere because of gasoline pumping at refueling stations and engine operation. The MTBE introduced to the atmosphere caused by phytoremediation is significantly less.

The MTBE released to the atmosphere is rapidly dissipated. The published information on the degradation half-life of MTBE in the atmosphere indicates that MTBE readily degrades chemically because of the hydroxyl radicals (OH^-) present. Squillace *et al.* (1997) have reviewed the fate of MTBE in the atmosphere and report that the half-life of MTBE can be as short as 3 days at 25 °C.

The following method to estimate the concentration of MTBE being released through vegetation is based on mass balance considerations. If the MTBE flows upward into the unsaturated zone together with the water drawn up by transpiration, the air required to hold the associated water vapor is the critical control volume. For example, consider the case in which there is 88 milligrams of MTBE per liter in the groundwater, and the water and MTBE move into the atmosphere together at 25 °C where the vapor pressure of water is 3173 pascals (23.8 millimeters of mercury) or a mole fraction of 0.0313 at saturation in air with 1×10^5 pascals (1 atmosphere) pressure. A liter of water contains 55.6 gram moles which requires 43 434 liters of air to dissolve the water under saturation conditions at 25 °C and 1×10^5 pascals (1 atmosphere) of pressure. The 88 milligrams of MTBE per liter is 1 millimole per liter in water, but only 0.001/1776 = 0.56 parts per million by volume in air. Thus, the concentration of MTBE in air is diluted because of the water vapor dissipation process that partially controls the rate of evapotranspiration. Because (1) MTBE concentrations in groundwater are often much smaller than 88 milligrams per liter, (2) the air is usually not saturated with water vapor, and (3) the ambient air contains some water vapor, the values of MTBE in the atmosphere as a result of phytoremediation are very small. This is especially true compared to values at a refueling station where MTBE vapors in equilibrium with the liquid fuel may be as high as 50 000 parts per million by volume.

FIELD STUDIES

The basic processes observed at the laboratory scale (*i.e.* biodegradation, volatilization, and transpiration), also take place at a field scale. Salnitro *et al.* (2000) observed aerobic biodegradation in the field with and without bioaugmentation using MTBE-degrading microorganisms. Wilson *et al.* (2000) reported biodegradation under anaerobic conditions. Hong *et al.* (2001) demonstrated the beneficial contributions of vegetation with respect to plume control. Landmeyer *et al.* (2000) reported the uptake of MTBE into vegetation.

Phytoremediation is an effective technology for remediation of fuel-contaminated sites as shown elsewhere in this book. Based on the results of laboratory and field studies reviewed in this chapter, vegetation increases the rate of evapotranspiration and the upward movement of MTBE into the unsaturated zone and the atmosphere. The beneficial effects of vegetation for MTBE are similar to those for BTEX compounds in fuel-contaminated soils; however, MTBE tends to persist longer in the atmosphere compared to BTEX compounds because MTBE is more resistant to photochemical degradation. Furthermore, MTBE is more mobile in soil and groundwater, and MTBE is not degraded readily in the root zone. Where these different properties of BTEX and MTBE are taken into account, simultaneous phytoremediation of shallow groundwater plumes from fuel spills should be feasible at many sites (see Weaver *et al.* this book).

CONCLUSIONS

Methyl *tertiary*-butyl ether is a compound of concern because of possible carcinogenic effects, persistence in soil and groundwater, and potential resistance to degradation by microorganisms. The widespread occurrence of this compound in groundwater is relatively recent, and conventional treatment is either ineffective or expensive. Phytoremediation provides an inexpensive, esthetically pleasing, and effective alternative for treating numerous organic contaminants. These investigations show that hybrid poplar (*Populus deltoides* ×

Populus nigra 'DN34' 'Imperial Carolina') cuttings and alfalfa (*Medicago sativa*) are capable of cleaning up water and soil contaminated with MTBE. Specific conclusions are as follows:

- Hybrid poplar (*Populus deltoides* × *Populus nigra* 'DN34' 'Imperial Carolina') cuttings tolerate concentrations of MTBE of at least 1000 milligrams per liter. The concentrations observed in the environment near a source of MTBE contamination can be as high as 200 milligrams per liter and hybrid poplars (*Populus deltoides* × *Populus nigra* 'DN34' 'Imperial Carolina') could effectively take up MTBE at this concentration without apparent toxic effects.
- Small hybrid poplar (*Populus deltoides* × *Populus nigra* 'DN34' 'Imperial Carolina') cuttings are effective at removing MTBE from hydroponic solution. Tests demonstrated that cuttings dosed with 10, 100, and 1000 milligrams per liter were able to achieve an MTBE removal of 30.6 percent remaining in solution within 10 days from 400-milliliter reactors.
- Small hybrid poplar (*Populus deltoides* × *Populus nigra* 'DN34' 'Imperial Carolina') cuttings and alfalfa (*Medicago sativa*) are effective at removing MTBE from soil. Hybrid poplar (*Populus deltoides* × *Populus nigra* 'DN34' 'Imperial Carolina') cuttings removed 95.2 ± 1.8 percent within 33 days when MTBE was initially at a soil water concentration of 22 milligrams per liter (soil concentration of 7 milligrams per kilogram). The evapotranspiration rate and amount of MTBE volatilized are considerably larger in alfalfa (*Medicago sativa*) vegetated soil compared to soil without vegetation. The evapotranspiration rate is small after alfalfa (*Medicago sativa*) plants are cut back; the rate increases nearly linearly with time as the plants increase in size over 29 days.
- The primary removal mechanism for MTBE was volatilization into the soil gas and atmosphere. When soil volatilization was not possible, significant amounts also transpired through hybrid poplar (*Populus deltoides* × *Populus nigra* 'DN34' 'Imperial Carolina') leaves and stem. Shown to move with water into the plant (*TSCF* of 0.5 to 0.8), MTBE also volatilized through the stem of excised small hybrid poplar (*Populus deltoides* × *Populus nigra* 'DN34' 'Imperial Carolina') plants and alfalfa (*Medicago sativa*) stems. Removal occurred at a slower rate than for actively transpiring hybrid poplar (*Populus deltoides* × *Populus nigra* 'DN34' 'Imperial Carolina') and alfalfa (*Medicago sativa*) plants. However, the volatilization is typically expected to be very small, especially compared to the MTBE volatilized at gasoline refueling stations. The small amount of MTBE that does volatilize into the atmosphere disperses and degrades photochemically with a half-life on the order of days.

This research provides evidence that phytoremediation could be a potential alternative for the remediation of MTBE contaminated sites.

Acknowledgments

We thank the U.S. Environmental Protection Agency Hazardous Substances Research Center, Great Plains and Rocky Mountain Region, and Equilon Enterprises LLC for financial support of this chapter. We benefited from collaboration and discussion with Marjorie Hong, Walter Farmayan, and Lee Newman.

REFERENCES

Anonymous (1997) Fueling the debate. *Environ. Health Persp.* **105**(10): 1042–1043.

Borden, R.C., R.A. Daniel, L.E. LeBrun, and C.W. Davis (1997) Intrinsic biodegradation of MTBE and BTEX in a gasoline-contaminated aquifer. *Water Resour. Res.* **33**: 1105–1115.

Bradley, P.M., J.E. Landmeyer, and F.H. Chappelle (1999) Aerobic mineralization of MTBE and *tertiary*-butyl alcohol by streambed microorganisms. *Environ. Sci. Technol.* **33**: 1877–1879.

Bradley, P.M., F.H. Chapelle, and J.E. Landmeyer (2001) Methyl *t*-butyl ether mineralization in surface-water sediment microcosms under denitrifying conditions. *Appl. Environ. Microbiol.* **67**(4): 1975–1978.

Briggs, G.G., R.H. Bromilow, and A.A. Evans (1982) Relationships between lipophicity and root uptake and translocation of non-ionized chemicals by barley. *Pestic. Sci.* **13**: 495–504.

Burken, J.G. (1993) Vegatative uptake by *Populus* spp. and mineralization of atrazine in variable soil types. Masters thesis. University of Iowa, Iowa City.

Burken, J.G. and J.L. Schnoor (1996) Phytoremediation: plant uptake of atrazine and role of root exudates. *J. Environ. Eng.* **122**(11): 958–963.

Caprino, L. and G.I. Togna (1998) Potential health effects of gasoline and its constituents: a review of current literature (1990–1997) on toxicological data. *Environ. Health Persp.* **106**(3): 115–125.

Cowan, R.M. and K. Park (1996) Biodegradation of the gasoline oxygenates MTBE, ETBE, TAME, TBE and TAA by aerobic mixed cultures. *Hazard. Ind. Wastes* **28**: 523–530.

Davis, L.C., M.K. Banks, A.P. Schwab, M. Narayanan, L.E. Erickson, and J.C. Tracy, (1998a) Plant-based remediation. In: *Bioremediation: Principles and Practice. Vol. 2, Biodegradation Technology Developments*. S.K. Sikdar and R.L. Irvine, eds. Technomic, Lancaster, Pennsylvania, pp. 183–219.

Davis, L.C., S. Vanderhoof, J. Dana, K. Selk, K. Smith, B. Goplen, and L.E. Erickson (1998b) Movement of chlorinated solvents and other volatile organics through plants monitored by Fourier transform infrared (FT-IR) spectrometry. *J. Hazard. Sub. Res.* 4–1: 1–26.

Eweis, J.B., E.D. Schroeder, D.P.Y. Chang, and K.M. Scow (1998) Biodegradation of MTBE in a pilot-scale biofilter. In: *Natural Attenuation: Chlorinated and Recalcitrant Compounds*. G.B. Wickrmanayake and R.E. Hinchee, eds. Battelle Press, Columbus, Ohio, pp. 341–346.

Finneran, K.T. and D.R. Lovley (2001) Anaerobic degradation of methyl *tert*-butyl ether (MTBE) and *tert*-butyl ether (TBE). *Environ. Sci. Technol.* **35**: 1785–1797.

Fortin, N.Y. and M.A. Deshusses (1999) Treatment of methyl *tert*-butyl ether vapors in biotrickling filters. 1. reactor startup, steady-state performance and culture characteristics. *Environ. Sci. Technol.* **33**: 2980–2986.

Garnier, P.M., R. Auria, C. Auger, and S. Revah (1999) Cometabolic biodegradation of methyl *t*-butyl ether by *Pseudomonas aeruginosa* grown on pentane. *Appl. Microbiol. Biotechnol.* **51**(4): 498–503.

Gomez, M.M., C.O. Abernathy, and J.T. Du (1997) Drinking water health advisory for methyl *tertiary*-butyl ether. American Chemical Society Division of Environmental Chemistry Preprints of Papers, Washington, D.C., pp. 370–372.

Hardison, L.K., S.S. Curry, L.M. Ciuffetti, and M.R. Hyman (1997) Metabolism of diethyl ether and cometabolism of methyl *tert*-butyl ether by a filamentous fungus, a *Graphium* sp. *Appl. Environ. Microbiol.* **63**: 3059–3067.

Hartley, W.R. and A.J. Englande, Jr. (1992) Health risk assessment of the migration of unleaded gasoline – a model for petroleum products. *Water Sci. Technol.* **25**: 65–73.

Hong, M.S., W.F. Farmayan, I.J. Dortch, C.Y. Chiang, S.K. McMillan, and J.L. Schnoor (2001) Phytoremediation of MTBE from a groundwater plume. *Environ. Sci. Technol.* **35**: 1231–1239.

Johnson, R., J. Pankow, D. Bender, C. Price, and J. Zogorski (2000) MTBE: To what extent will past releases contaminate community water supply wells. *Environ. Sci. Technol.* **34**: A210–A217.

Landmeyer, J.E., F.H. Chapelle, R.M. Bradley, J.F. Pankow, C.D. Church, and P.G. Tratnyek, (1998) Fate of MTBE relative to benzene in a gasoline-contaminated aquifer (1993–1998). *Ground Water Mon. Rem.* **18**: 93–102.

Landmeyer, J.E., D.A. Vroblesky, and P.M. Bradley (2000) MTBE and BTEX in trees above gasoline-contaminated ground water. In: *Case Studies in the Remediation of Chlorinated and Recalcitrant Compounds*. G.B. Wickramanayake, A.R. Gavaskar, J.T. Gibbs, and J. L. Means, eds. Battelle Press, Columbus, Ohio, pp. 17–24.

Landmeyer, J.E., F.H. Chapelle, H.E. Herlong, and P.M. Bradley (2001) Methyl *tert*-butyl ether biodegradation by indigenous aquifer microorganisms under natural and artificial oxic conditions. *Environ. Sci. Technol.* **35**: 1118–1126.

Mehlman, M.A. (1998) Editorial: pollution by gasoline containing hazardous methyl *tertiary*-butyl ether (MTBE). *Arch. Environ. Health* **53**(4): 245–246.

Mo, K., C.O. Lora, A.E. Wanken, M. Javanmardian, X. Yang, and C.F. Kulpa (1997) Biodegradation of methyl *t*-butyl ether by pure bacterial cultures. *Appl. Microbiol. Biotechnol.* **47**: 67–72.

Mormile, M.R., S. Liu, and J.M. Suflita (1994) Anaerobic biodegradation of gasoline oxygenates: extrapolation of information to multiple sites and redox conditions. *Environ. Sci. Technol.* **28**: 1727–1732.

Narayanan, M., L.E. Erickson, and L.C. Davis (1999) Simple plant-based design strategies for volatile organic pollutants. *Environ. Progress* **18**: 231–242.

Prince, R.C. (2000) Biodegradation of methyl *tertiary*-butyl ether (MTBE) and other fuel oxygenates. *Crit. Rev. Microbiol.* **26**: 163–178.

Ramaswami A. and E. Rubin (2001) Measuring phytoremediation parameters for volatile organic compounds: focus on MTBE. *Pract. Periodical Hazard. Toxic Radioact. Waste Manag.* **5**: 123–129.

Rubin E. and A. Ramaswami (2001) The potential for phytoremediation of MTBE. *Water Res.* **35**: 1348–1353.

Salanitro, J.P., L.A. Diaz, M.P. Williams, and H.L. Wisniewski (1994) Isolation of a bacterial culture that degrades methyl *t*-butyl ether. *Appl. Environ. Microbiol.* **60**: 2593–2596.

Salanitro, J.P., P.C. Johnson, G.E. Spinnler, P.M. Maner, H.L. Wisniewski, and C. Bruce (2000) Field-scale demonstration through aquifer bioaugmentation and oxygenation. *Environ. Sci. Technol.* **34**: 4152–4162.

Schnoor, J.L., L.A. Licht, S.C. McCutcheon, N.L. Wolfe, and L.H. Carreira (1995) Phytoremediation of organic and nutrient contaminants. *Environ. Sci. Technol.* **29**(7): 318–323.

Shimp, J.F., J.C. Tracy, L.C. Davis, E. Lee, W. Huang, L.E. Erickson, and J.L. Schnoor (1993) Beneficial effects of plants in the remediation of soil and groundwater contaminated with organic materials. *Crit. Rev. Environ. Sci. Technol.* **23**(1): 41–77.

Squillace, P.J., J.S. Zogorski, W.G. Wilber, and C.V. Price (1996) Preliminary assessment of the occurrence and possible sources of MTBE in groundwater in the United States, 1993–1994. *Environ. Sci. Technol.* **30**: 1721–1730.

Squillace, P.J., J.F. Pankow, N.E. Korte, and J.S. Zogorski (1997) Review of the environmental behavior and fate of methyl *tert*-butyl ether. *Environ. Toxicol. Chem.* **16**(9): 1836–1844.

Steffan, R.J., K. McClay, S. Vainberg, C.W. Condee, and D. Zhang (1997) Biodegradation of the gasoline oxygenates methyl *tert*-butyl ether, ethyl *tert*-butyl ether, and *tert*-amyl methyl ether by propane-oxidizing bacteria. *Appl. Environ. Microbiol.* **63**: 4216–4222.

Stringfellow, W.T. and K.C. Oh (2002). Initiation of MTBE biotreatment in fluidized-bed bioreactors. *J. Environ. Eng.* **128**(9): 852–861.

Stuckey, H.T. (2000) The benefits and problems associated with MTBE. *Environ. Protect.* **11**(7): 49–52.

Wilson, J.T., J.S. Cho, B.H. Wilson, and J.A. Vardy (2000) Natural attenuation of MTBE in the subsurface under methanogenic conditions. EPA/600/R-00/006. National Risk Management Research Laboratory, U.S. Environmental Protection Agency, Cincinnati, Ohio. (http://www.epa.gov/ORD/NRMRL/mtbe/MTBE report.pdf).

Winnike, S.K. (1998) Phytoremediation of methyl *tert*-butyl ether (MTBE) by hybrid poplar trees. Masters thesis. University of Iowa, Iowa City.

Yeh, C.K. and J.T. Novak (1994) Anaerobic biodegradation of gasoline oxygenates in soils. *Water Environ. Res.* **66**: 752–774.

Zhang, Q. (1999) Phytoremediation of methyl *tert*-butyl ether (MTBE) in groundwater–experimental and modeling studies. Ph.D. dissertation. Kansas State University, Manhattan, Kansas.

Zhang, Q., L.C. Davis, and L.E. Erickson (1998) Using vegetation to treat methyl *tert*-butyl ether contaminated groundwater. In: *Proceedings of the 1998 Conference on Hazardous Waste Research–Bridging Gaps in Technology and Culture*. U.S. Environmental Protection Agency Hazardous Substance Research Center: Great Plains/Rocky Mountains, Kansas State University, Manhattan, Kansas. Held in Snowbird, Utah, May 18–21. pp. 262–272 (http://www.engg.ksu.edu/HSRC).

Zhang, Q., L.C. Davis, and L.E. Erickson (1999) An experimental study of phytoremediation of methyl *tert*-butyl ether (MTBE) in groundwater. *J. Hazard. Sub. Res.* **2**: 4–1 to 4–19. (http://www.engg.ksu.edu/HSRC).

Zhang, Q., L.C. Davis, and L.E. Erickson (2000) Plant uptake of methyl *tert*-butyl ether (MTBE) from groundwater. In: *Environmental and Pipeline Engineering 2000*. R.Y. Surampalli, ed. American Society of Civil Engineers, Reston, Virginia, pp. 166–173.

Zhang, Q., L.C. Davis, and L.E. Erickson (2001) Transport of methyl *tert*-butyl ether (MTBE) through alfalfa plants. *Environ. Sci. Technol.* **35**: 725–731.

Zogorski, J.S., G.C. Delzer, D.A. Bender, P.J. Squillace, T.J. Lopes, A.L. Baehr, P.E. Stackelberg, J.E. Landmeyer, C.J. Boughton, M.S. Lico, J.F. Pankow, R.L. Johnson, and N.R. Thomson (1998) MTBE: summary of findings and research by the U.S. Geological Survey. In: *1998 Annual Conference of the American Water Works Association*, Denver, Colorado.

28

PHYTOREMEDIATION OF CYANIDE-POLLUTED SOILS

S. A. J. Trapp and H. Christiansen

SUMMARY OF PRACTICAL APPLICATIONS

A series of laboratory and field investigations establish that the phytoremediation of cyanide wastes in soil is a feasible option and should be considered for cleanup of abandoned gasworks. Cyanide laced soils are a worldwide problem. During the 19th and the 20th century, gas manufacturing works in towns and cities provided street and home lighting. Cyanide waste was a typical by-product and frequently dumped near the production facilities. The contaminated soils contain some easily liberatable cyanide, which can be deadly, but mostly complexed cyanide that is far less toxic to animals and plants. Trees, all of which naturally produce cyanide, seem ideal to take up and mineralize cyanide to benign products and to reduce or prevent infiltration of cyanide to groundwater during cleanup. A risk assessment and preliminary results at the former Holte gasworks in Denmark demonstrate that even secondary uses for recreation are safe during remediation.

GLOSSARY

Cyanide compounds: Chemicals that have a -CN group, both organic and inorganic, including the very toxic cyan hydrogen, HCN. Free cyanide is the sum of cyan hydrogen [HCN] and cyanide anions [CN^-]; easily liberatable cyanides are cyanide compounds which easily liberate free cyanide, *e.g.*, potassium cyanide (KCN); complexed cyanides are stable complexes, such as Prussian blue ($Fe_4[Fe(CN)_6]_3$); and total cyanide is the sum of all. Concentrations of cyanide compounds are frequently expressed in milligrams of CN; this means that 1 milligram of KCN equals 0.4 milligrams of CN, and 1 milligram of $Fe_4[Fe(CN)_6]_3$ equals 0.545 milligrams of CN.

Dry weight: Weight of a sample after drying, usually at 105 °C.

Phytoremediation: Transformation and Control of Contaminants,
Edited by Steven C. McCutcheon and Jerald L. Schnoor.
ISBN 0-471-39435-1 (cloth)

Fresh weight: Weight of a plant sample without drying.
Limit of determination: Amount or concentration of chemical that can be analytically distinguished from zero with a given method.
Transfer factor: Ratio between the concentrations in plant tissue at harvest and the concentration in the medium (soil or solution) in which the plant was growing. The transfer factor is identical to the bioconcentration factor for roots. For leaves, the transfer factor indicates that uptake is from soil, whereas the bioconcentration factor is based on uptake from air. The transfer factor does not necessarily represent chemical equilibrium or steady state, but depends on the time of harvest or sampling.
Wet weight: See fresh weight.

INTRODUCTION

Old gasworks and coal coking sites are sources of cyanides in the environment. There are about 3000 sites each in the U.S. and in Great Britain, 1000 sites in Germany (Mansfeldt 2000), and 150 sites in Denmark.

The gas from coking plants contained Prussic acid (0.5 to 1 liters per cubic meters), which was precipitated by iron oxides (often bog iron). Typically, for 1000 cubic meters of gas, 8 to 21 kilograms of ore was used. After gas cleaning, the residue contained high amounts of sulfur (40 to 50 percent) and cyanide (1 to greater than 2 percent by mass), mainly in the form of iron complexes, such as Prussian blue, $Fe_4[Fe(CN)_6]_3$. Other components are ferricyanide ($FeCN_6^{3-}$) and thiocyanates (*e.g.*, HSCN), but rarely free cyanide (HCN or CN^-) (Kjeldsen 1999). These waste products were frequently used as fill material, often directly on the gasworks ground and close to waterworks. Cyanides are not the only problem; polycyclic aromatic hydrocarbons and other organic pollutants are also frequently found on these sites (Shifrin *et al.* 1996).

Phytoremediation combined with pump-and-treat has been used since December 2000 at the former gasworks site at Holte, commune Søllerød, Denmark. At this site, aromatic and polycyclic hydrocarbons and, in particular, cyanides are being cleaned up. The site is next to the waterworks of the town. Figure 28-1 shows the former gasworks.

Several investigations and reviews deal with the behavior and fate of cyanides in soil and groundwater (Smith and Mudder 1991, Shifrin *et al.* 1996, Gosh *et al.* 1999, Kjeldsen 1999). A great deal less is known about the uptake of cyanides into plants, the phytotoxicity, and the degradation kinetics. This paper presents a review of cyanide fate and toxicity and an evaluation of the potential of phytoremediation for degrading and thereby reducing the toxicity of cyanides.

Figure 28-1 The former gasworks at Holte. Reproduced with permission from the Byhistorisk Arkiv, Søllerød Museum, Denmark.

CHEMISTRY AND TOXICITY OF CYANIDE COMPOUNDS

Chemistry of Cyanide Compounds in Soil and Groundwater

The chemical behavior of cyanide compounds in soil, groundwater, and vegetation is very complex because many processes act simultaneously. The processes include precipitation, dissolution, sorption, complexation, acid-base reaction, volatilization, transformation, abiotic and biotic degradation, and biological formation. Table 28-1 lists the relevant cyanide species.

The major part of the cyanides in wastes from manufactured gas plants is iron-complexed, *e.g.*, as solid or dissolved Prussian blue. From dissolved Prussian blue, ferrocyanide forms by the reaction (K denotes the dimensionless equilibrium constant of the reaction, Kjeldsen 1999)

$$Fe_4[Fe(CN)_6]_3(s) \leftrightarrow 4Fe^{3+} + 3Fe(CN)_6^{4-} \quad (\log K = -66.5) \tag{28-1}$$

TABLE 28-1 Important Cyanide Species (Oppenheimer 1928a, 1928b, Kjeldsen 1999)

Substance	Synonym	Chemical formula	Properties
Cyan hydrogen	Prussic acid, hydrocyanic acid	HCN	Weak acid
Ferrocyanide	Hexacyanoferratc(II)	$[Fe(CN)_6]^4$	Fe(II)-complex
Ferricyanide	Hexacyanoferrate(III)	$[Fe(CN)_6]^{3-}$	Fe(III)-complex
Ferric ferrocyanide	Prussian blue	$Fe_4[Fe(CN)_6]_3$	Fe(II)-Fe(III)-complex
Cyanate	Cyanic acid	HOCN	Acid
Thiocyanate	Rhodan hydrogen, rhodic acid, sulfocyanic acid	HSCN	Acid

By redox reaction, ferricyanide forms

$$Fe(CN)_6^{3-} + e^- \leftrightarrow Fe(CN)_6^{4-} \quad (\log K = -6.0) \tag{28-2}$$

Both iron complexes may release the cyanide anion

$$Fe^{3+} 6CN^- \leftrightarrow Fe(CN)_6^{3-} \quad (\log K = 43.9) \tag{28-3}$$

$$Fe^{2+} + 6CN^- \leftrightarrow Fe(CN)_6^{4-} \quad (\log K = 36.9) \tag{28-4}$$

The neutral form of the cyanide anion is cyan hydrogen (HCN):

$$H^+ + CN^- \leftrightarrow HCN \quad (pK_a = 9.2) \tag{28-5}$$

Prussic acid is volatile with a vapor pressure of 88.8 kilopascals at 20 °C; the dimensionless Henry's law constant K_{AW} is 0.00158. Prussic acid is both water and lipid soluble, with a water solubility of 1 kilogram of HCN per liter at 20 °C and a $\log K_{ow}$ of –0.25 (BGVV 2000), in which $\log K_{ow}$ is the dimensionless octanol–water partition coefficient.

The dissolution and dissociation of cyanide species depends largely on the pH and redox potential of the medium. In a solution with 1000 milligrams of $Fe_4[Fe(CN)_6]_3$ per liter of Prussian blue and at pH 6.7, 2.8 milligrams of CN per liter of easily liberatable cyanide was measured.

Toxicity of Cyanides

Human toxicity

The toxicity of prussic acid is well known and has been the topic of many crime stories, *e.g.*, Christie (1945) and Kesselring (1941). The cyan hydrogen (HCN) that is not dissociated is, perhaps due to the faster membrane permeability, more toxic than the corresponding anion (CN^-). Prussic acid is one of the fastest acting poisons known to date. The mode of action is a reaction with the iron-containing cytochrome oxidase in the mitochondria (Faust 1994). Almost immediate suffocation is the consequence. Cyanide is not accumulated and is readily metabolized. "There is no evidence that chronic exposure to free cyanide results in teratogenic, mutagenic, or carcinogenic effects" (Smith and Mudder 1991).

Iron-complexed cyanide, on the other hand, is far less toxic (Table 28-2). Ferrocyanide is a supplement in winter throwing salts. Potassium hexacyanoferrate(II) is added to red wine to give a more attractive color (Faculty for Chemistry Bielefeld 2000). Prussian blue has been used to whiten yellow sugar types (Oppenheimer 1928a), to reduce the sorption of radioactive caesium, and as an antidote against thallium. The therapeutic dose is 250 milligrams of $Fe_4[Fe(CN)_6]_3$ per kilogram of body weight per day (Shifrin *et al.* 1996).

TABLE 28-2 Human Toxicity of Cyanides (Shifrin *et al.* 1996, U.S. EPA 1990)

Substance	Chemical formula	LD_{50}[a] (milligrams per kilogram of bodyweight)	RfD[b] (milligrams per kilogram of bodyweight per day)
Prussic acid	HCN	0.5 to 3.5	0.02
Prussian blue	$Fe_4[Fe(CN)_6]_3$	Greater than 5000	32

[a] LD_{50} is the dosage where 50 percent of exposed persons die.
[b] RfD is subchronic reference dose.

The toxicity of iron-complexed cyanides is related to the release of free cyanide, *e.g.*, under ultraviolet light. Levels below 1 milligram of CN per liter of iron-complexed cyanide have been shown to be lethal in closed test chambers when exposed to direct or diffuse sunlight. The lethal concentration (LC_{50}) of ferrocyanide to 50 percent of a juvenile rainbow trout population held 96 hours in covered aquariums decreased from 752 milligrams of CN per liter of total cyanide in the dark to 33 milligrams of CN per liter under 680 Lux light intensity for 18 hours per day. The corresponding LC_{50} values for ferricyanide are 1210 milligrams of CN per liter and 10.8 milligrams of CN per liter (Smith and Mudder 1991).

Phytotoxicity

Before the invention of the ammonia synthesis process by Haber and Bosch in 1913, cyanide compounds, mainly cyanamides, were used as fertilizing herbicides (Fuller 1984). Calcium cyanamide (Ca=N-CN) is still commercially used (Börner 1995). At Holte in Denmark, local farmers sought cyanide wastes from the gasworks (probably as a herbicide or as a fertilizer). Although these wastes had been used in the 19th and early 20th century agriculture, only a few studies deal with uptake and phytotoxicity of cyanides.

Fuller (1984) studied the use of sodium cyanide (NaCN) and potassium cyanide (KCN) as fertilizer (a degradation product is ammonia). Bean (*Phaseolus vulgaris*) plants showed strongly reduced growth when 50 to 100 milligrams of NaCN were added per kilogram of soil (Wallace *et al.* 1977). At the same time, metals such as aluminum, sodium, and iron, and also heavy metals (zinc, copper, nickel, and cobalt) were taken up faster. The transfer factors into plants of $[^{14}C]$ for leaves, stems, and shoots was 0.63, 0.8, and 0.68 (related to dry weight) with 50 milligrams of ^{14}CN per kilogram of soil and 0.58, 1.41, and 0.93, respectively, with 100 milligrams of ^{14}CN per kilogram of soil.

Several plants produce cyanides to ward off herbivores and fungal attack. Concentrations above 3 grams of HCN per kilogram of tissue can be formed in some plants. During several metabolic reactions, plants are confronted with cyanide as a by-product, *e.g.*, during the ethylene synthesis of mature tissue. Therefore, plants had to evolve effective detoxifying strategies. One of these strategies is the use of rhodanese (thiosulfate-thiotransferase) which converts cyanide (CN^-) into the less toxic thiocyanate (SCN^-). Another detoxifying enzyme is β-cyanoalanine synthease, which connects free cyanide and cysteine to cyanoalanine. The final metabolite is asparagine, an essential amino acid (Manning 1988). Additionally, mitochondria of all plants investigated so far possess a cyanide-resistant alternative oxidase system, which is, however, less efficient (Parlar and Angerhöfer 1995, Shugaev 1999).

We found trees growing on old gasworks sites without any visible signs of toxicity. At the gasworks site at Holte, several willow species (*Salix* spp.) and well-developed black elder (*Sambucus nigra*) bushes were found growing directly on cyanide-polluted soil.

MATERIALS AND METHODS

We conducted laboratory tests to supplement limited information on phytotoxicity of free and complexed cyanide and to define the degradation kinetics, with the goal of establishing phytoremediation as an *in-situ* treatment method for cyanide-polluted soils. The tests were performed with potassium cyanide (KCN) and iron-complexed cyanide (Prussian blue) in nutrient solutions, and with cyanide-polluted soil samples from the manufactured-gas production site at Holte, Denmark.

Chemical Analysis

The concentrations of cyanides in solution, soil, and plant material were analyzed by the standardized method ISO/DIS 11262 "Soil Quality—Determination of Cyanide" (draft 1998). The analysis distinguishes three fractions: easily liberatable cyanide, complexed cyanide, and total cyanide. The method is very similar, but not identical to the U.S. EPA Method 9010 (Smith and Mudder 1991). Additionally, free cyanide was measured by direct photometric analysis.

Free Cyanide

Free cyanide is directly measured photometrically. Prussic acid reacts with chloramine-T during formation of cyanogen chloride. Cyanogen chloride reacts with pyridine-4-carboxylic acid and 1,3-dimethylbarbituric acid to form a colored complex. Free cyanide is the sum of cyan hydrogen [HCN] and cyanide [CN^-].

Easily Liberatable Cyanide

Cyanide is released as hydrogen cyanide (HCN) from a sample by reflux distillation at pH 4 for 1 hour and collected in a sodium hydroxide scrubber solution, followed by the same colorimetric determination used for free cyanide. "Easily liberatable" cyanide (similar to weak-acid dissociable cyanide) includes free cyanide plus compounds and complexes that release cyan hydrogen (HCN) or cyanide (CN^-) at pH 4, *e.g.*, alkali salts such as potassium cyanide (KCN), and copper, zinc, cadmium, and nickel complexes.

Complexed Cyanide

After addition of 20 milliliters of orthophosphoric acid, the solution remaining from the determination of easily liberatable cyanide is boiled for 2 hours. The released cyanide is collected in a new sodium hydroxide scrubber solution to determine the complexed cyanide. This fraction contains iron, manganese, and gold complexes and some organic forms.

Total Cyanide

Total cyanide is the sum of easily liberatable and complexed cyanide. It is measured as the complexed cyanide, but without prior distillation of easily liberatable cyanide. The total cyanide analysis will recover all cyanide species except cobalt complexes and some organic forms.

Soil

The concentrations in soil were analyzed at the beginning and at the end of the experiments. Ten grams of soil (wet weight) was homogenized and put in a distillation flask. One hundred milliliters of distilled water, 10 milliliters of zinc sulfate, and 50 milliliters of buffer (potassium hydrogen phthalate at pH 4) was added. The solution was boiled for 1 hour to distill cyan hydrogen (HCN) into a sodium hydroxide trap. Then, the sodium hydroxide was taken from the trap and analyzed photometrically for free cyanide. Afterward, 2 milliliters of tin(II)chloride solution, 10 milliliters of copper(II)-solution, and 20 milliliters of orthophosphoric acid (85 percent on a mass-per-mass basis) were added to the remaining solution. After 2 hours distillation of the mixture, the complexed fraction was determined from the collection in a new sodium hydroxide trap.

Nutrient Solution

The concentrations of free, easily liberatable, complexed, and total cyanide in the nutrient solution were determined by collecting a 10 to 20 milliliter sample. In experiments with potassium cyanide, cyanide was directly analyzed photometrically, because we assumed that all the cyanide was present as cyan hydrogen (HCN). Occasionally, easily liberatable cyanide was determined to confirm this.

Plant Material

Leaves and roots were harvested, weighed, and immediately transferred to water-filled flasks, which were carefully closed and stored at 2 °C in the dark until distillation. Small pieces of the stem were cut off with rose scissors and stored in the same way. Plant tissue from the controls was also analyzed for cyanide.

Analytical Performance

The analysis is very sensitive but time-consuming. Lower detection limits were reported to be 0.002 ± 0.001 milligram of CN per liter (Smith and Mudder 1991). We found the limit of determination (blank plus three standard deviations) to be 0.04 milligram of CN per kilogram or milligrams of CN per liter for a 10-gram sample or 10-milliliter solution, respectively, which is 0.4 micrograms of CN absolute. For the plant material analysis, only 1 to 4 grams could be used, giving a limit of determination for plant matter of 0.1 to 0.4 milligram of CN per kilogram. When a peak was quantifiable, the result was presented, even when the value was less than the limit of determination. The time-consuming analysis of cyanides allowed only a few replicates and standard deviations could be determined only in some cases.

Analytical interferences are most important for the complexed and the total cyanide determination. During boiling under strongly acidified conditions, nitrite and thiocyanate (HSCN) may convert to cyanide: 0.2 milligram of HSCN per liter may contribute as much as 0.01 milligram of CN per liter. Some other compounds, such as carbonates and sulfur compounds, may also interfere with the analysis (Smith and Mudder 1991).

Experimental Setup for Phytotoxicity Testing

A phytotoxicity test was especially designed for short-term acute toxicity of compounds in solution or the toxicity of polluted soil to trees (Trapp *et al.* 2000). The test uses growth and transpiration as parameters. Photosynthesis and transpiration, and therefore growth and transpiration, are closely related. Cuttings of basket willows (*Salix viminalis*) and balsam poplars (*Populus trichocarpa*) were rooted for up to 6 weeks in buckets containing tap water under natural sunlight. Trees were then transferred into 500 milliliter Erlenmeyer flasks with about 400 milliliters of nutrient solution and sealed with a cork stopper (with a hole for the stem).

In a climate-controlled chamber at constant temperature and humidity (24.5 ± 0.5 °C and 45 ± 5 percent relative humidity) with artificial, continuous light, the loss of weight after 24 hours was measured. The difference in weight was almost exclusively due to transpiration. Then, the plants were transferred into flasks with soil or a spiked solution. Controls were transferred into unpolluted soil or remained in nutrient solution. The weight was measured every 24 hours. The weight loss, compared to initial loss, was one toxicity criteria. The test can be run for any length of time, because weighing does not injure the plants. Usually, between 72 and 144 hours are sufficient to measure a pronounced effect, depending on the compounds tested. Figure 28-2 shows the setup of the system in the laboratory.

The toxicity of polluted soils was measured in a similar way. Approximately 500 grams of polluted soil (and unpolluted soil for the controls) was placed into the Erlenmeyer flasks, and 150 milliliters of nutrient solution was

Figure 28-2 Toxicity test system with willows (*Salix viminalis*) and poplars (*Populus trichocarpa*) in soil and in nutrient solution.

added. The mixture was shaken immediately before placing the cutting. The roots were covered with soil by gently shaking the Erlenmeyer flasks, then the flasks were sealed, and the test proceeded as with trees in nutrient solution.

Quantification of Toxic Effects

To compare the toxic effect on cuttings with different initial transpiration (before the toxicant was added), the weight loss was expressed as relative transpiration. The transpiration was normalized with respect to the initial transpiration (to eliminate the necessity of finding cuttings with similar initial transpiration) and with respect to the transpiration of uncontaminated control cuttings (to include the effect of normal growth of the cuttings during the test). The mean normalized relative transpiration (*NRT*) was calculated by

$$NRT(C, t) = \frac{\frac{1}{n} \cdot \sum_{i=1}^{n} \frac{T_i(C, t)}{T_i(C, 0)}}{\frac{1}{m} \cdot \sum_{j=1}^{m} \frac{T_j(0, t)}{T_j(0, 0)}} \tag{28-6}$$

where C is concentration (milligrams per liter), t is time period (hours), T is absolute transpiration (grams per hour), i is replicate 1, 2, ... , n, and j is control 1, 2, ... , m.

Growth rate and water use efficiency were also used as parameters to quantify toxic effects. Growth rate was determined by weighing the plants (fresh weight) before and after the experiments. The water use efficiency of biomass production (*WUE*) was determined by dividing growth in grams by transpiration in liters. Reduced water use efficiency is a sign of sublethal stress (Larcher 1995).

RESULTS

Basket Willows (*Salix viminalis*) in Potassium Cyanide Solution

The purpose was to see whether plants are able to use cyanide as a nitrogen source, and what levels can be tolerated. Controls were established with and without nitrogen. All exposed plants had cyanide as the only nitrogen source. Potassium cyanide dissociates in aqueous solution and releases cyan hydrogen (HCN). Based on the speciation model MINTEQ (U.S. EPA 1991), 99.1 percent of all cyanide in the nutrient solution was present as cyan hydrogen (HCN). The correlation between the potassium cyanide dosage and the normalized relative transpiration (*NRT*) was very good (Figure 28-3). Basket willows (*Salix viminalis*) in a 5 milligrams of KCN per liter solution showed a reduction of transpiration of more than 50 percent after 72 hours and died after approximately 3 weeks. Trees exposed to 20 or 50 milligrams of KCN per liter quickly dried out and were probably dead after less than 1 week. Plants with 1 milligram of KCN per liter in the nutrient solution grew better than all other experimental variants at the end of the test (statistically significant at a probability of 0.05). The lack of nitrogen in the nutrient solution of

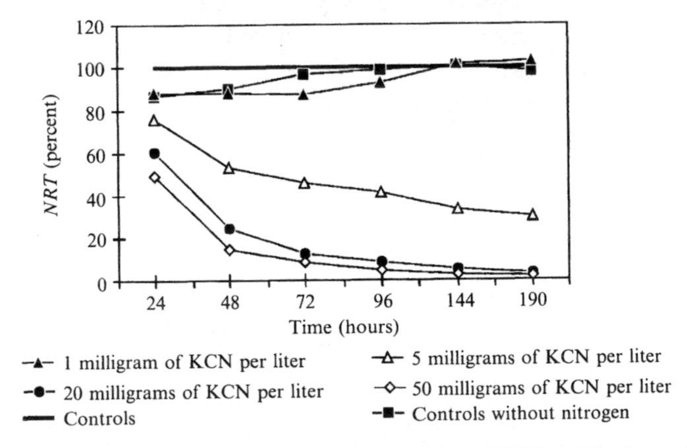

Figure 28-3 Time *versus* normalized relative transpiration (*NRT*) of basket willows (*Salix viminalis*) in potassium cyanide solution; four replicates for test solutions and five replicates for controls.

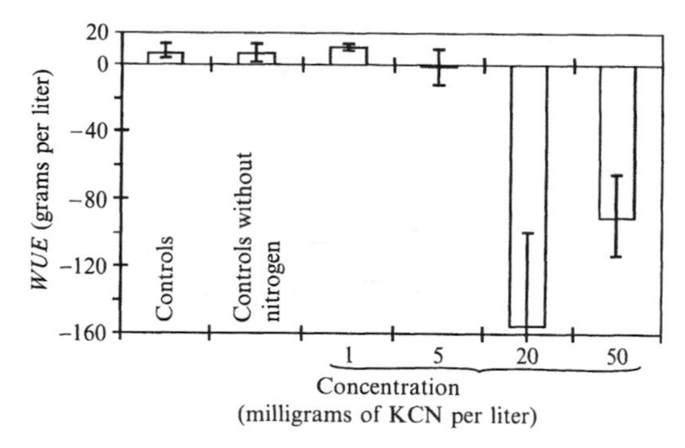

Figure 28-4 Water use efficiency (*WUE*, growth in grams of biomass per liter of transpired water) of basket willows (*Salix viminalis*) in potassium cyanide solutions.

one control decreased the transpiration initially but had no significant effect on growth. With 5 milligrams of KCN per liter or more, growth and water use efficiency (growth per volume of transpired water) were negative (Figure 28-4).

A statistical program using weighted nonlinear regression was selected to calculate effect concentrations (*EC*) and confidence intervals. The program uses a lognormal distribution to describe toxicity data and uses inverse estimation, taking into account the covariance within the control response for calculation of confidence intervals (Andersen 1994). Plants with no intact leaves still had a relative transpiration between 5 and 20 percent, which means that the EC_{50} value, a 50 percent reduction of relative transpiration, denotes severe damage. The EC_{10}, often interpreted as *LOEC* (lowest-observed-effect-concentration) was 0.76 milligram of KCN per liter (0.3 milligram of CN per liter) (95 percent confidence interval from 0.45 to 1.27 milligrams of KCN per liter) for $t = 72$ hours. The EC_{50} was 4.47 milligrams of KCN per liter (95 percent confidence interval from 3.51 to 5.69 milligrams of KCN per liter). The EC_{50} of free cyanide for various fish species has been reported between 0.05 and 0.2 milligram of CN per liter in a dynamic 96-hour test (UNEP 2000). This indicates that basket willow (*Salix viminalis*) trees can tolerate much higher concentrations of free cyanide than fish.

Balsam Poplars (*Populus trichocarpa*) in Prussian Blue

The results of the toxicity test with balsam poplars (*Populus trichocarpa*) in Prussian blue are shown in Figures 28-5 and 28-6. Controls without nitrogen showed increased transpiration. This might be a sign of stress, or a normal reaction in order to gain more nutrients. Plants with 10 milligrams of $Fe_4[Fe(CN)_6]_3$ per liter as the only nitrogen source did not show this increase. No conclusion could be drawn as to whether plants could use ferric ferrocyanide as a nitrogen source, or whether transpiration was simply inhibited. Higher doses clearly had a inhibitory effect, though not exactly correlated to the dose. At the highest dose, 2500 milligrams of $Fe_4[Fe(CN)_6]_3$ per liter, the

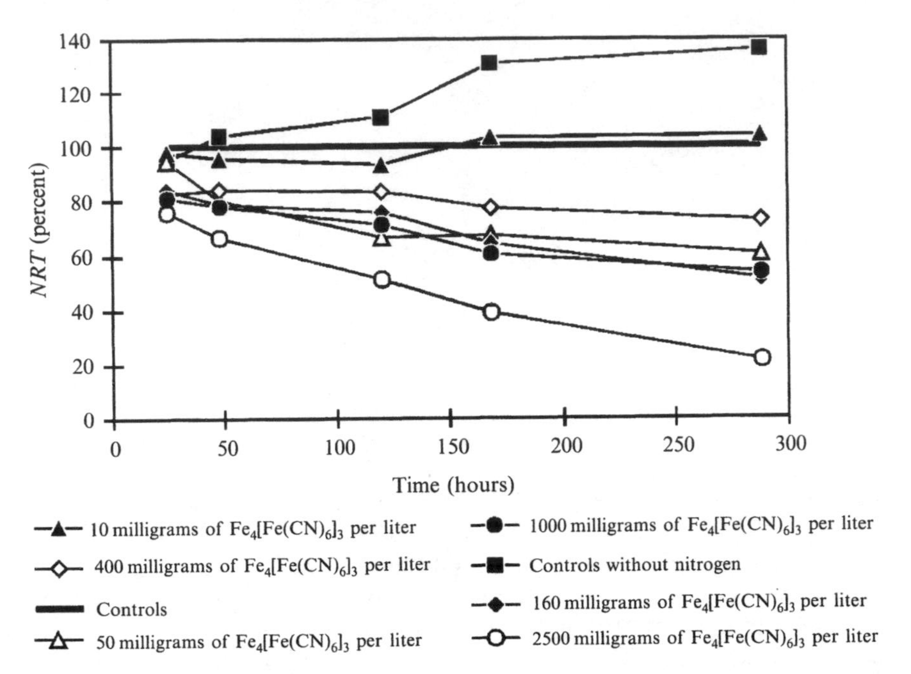

Figure 28-5 Time *versus* normalized relative transpiration (*NRT*) of balsam poplars (*Populus trichocarpa*) in Prussian blue solution; three replicates for all test solutions and five replicates for controls.

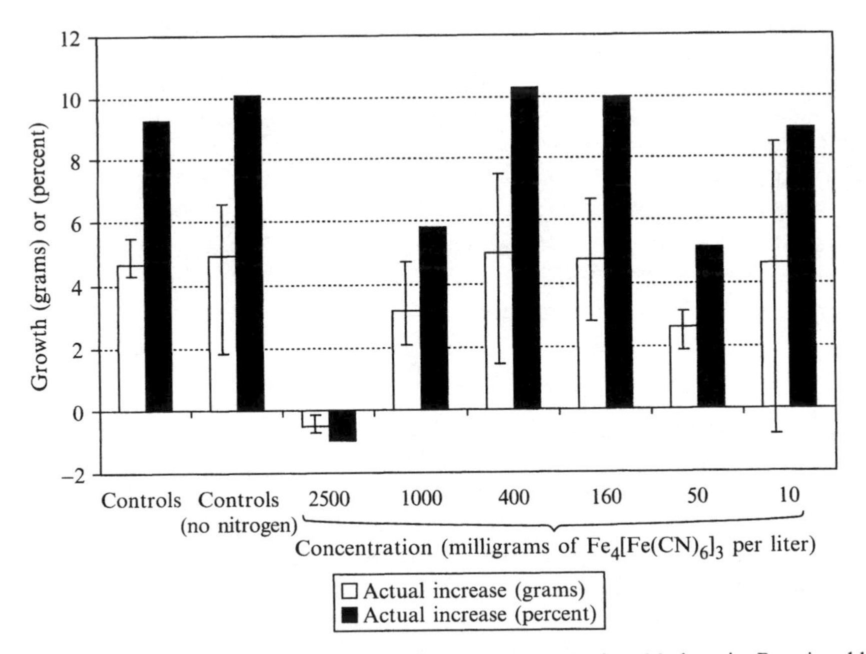

Figure 28-6 Growth of balsam poplars (*Populus trichocarpa*) after 16 days in Prussian blue solutions.

plants had 80 percent reduced transpiration. All plants survived Prussian blue, at all concentrations. The growth showed large variations within each group, and results for all but the highest dose did not differ significantly. At the highest dose, growth for all trees was negative (Figure 28-6). Some old leaves had been dropped, and new leaves were much smaller (about one fifth of the leaf area) than those of the plants exposed to lower doses, but of a deeper green color.

Basket Willows (*Salix viminalis*) Grown in Gasworks Soil

Soil samples were collected from the gasworks at Holte, Denmark. The area is a former river valley with good soil. The measured cyanide concentrations are given in Table 28-3. A blue-colored lump of soil from the Holte gasworks site had the highest concentrations. Control soils were taken from the campus of the Technical University of Denmark at Lyngby, Denmark.

Soils 1 and 3 with a total cyanide concentration of 42 to 45 milligrams of CN per kilogram of dry weight and about 452 milligrams of CN per kilogram of dry weight, respectively, had a limited acute toxicity. The transpiration was slightly reduced (Figure 28-7), but the growth of the basket willows (*Salix viminalis*) was good (Figure 28-8). Soil 2 with about 330 milligrams of CN per kilogram of dry weight of total cyanide was a poorer substrate for basket willow (*Salix viminalis*) growth. This soil had a very low pH (Table 28-3). Basket willows (*Salix viminalis*) planted in the blue lump of soil, with the very high cyanide content, died quickly.

Generally, basket willows (*Salix viminalis*) tolerated higher doses of easily liberatable cyanide in these soils than from the solution. Probably, measured easily liberatable cyanide is not all cyan hydrogen (HCN), or it is not completely bioavailable in soil; Smith and Mudder (1991) reported sorption of cyan hydrogen (HCN) to organic matter in soil.

TABLE 28-3 Cyanide Concentrations of Soil Samples from the Gasworks at Holte

	Concentration in milligrams as CN per kilogram of dry weight						
	Complexed, replicates		Easily liberatable, replicates		Total, replicates		Soil pH mean (standard deviation of 3 replicates)
Sample	1	2	1	2	1	2	
Soil 1	28.9	40.6	1.1	2.3	41.8	44.6	7.1 (0.19)
Soil 2	248.4	284.5	16.2	17.9	327.3	331.6	3.7 (0.06)
Soil 3	402.4	510.6	15.4	43.4	452.0	452.4	4.8 (0.09)
"Blue lump"	7775	15 385	281.3	256.6	8056	15 652	ND[a]

[a]Abbreviation used—ND: not determined.

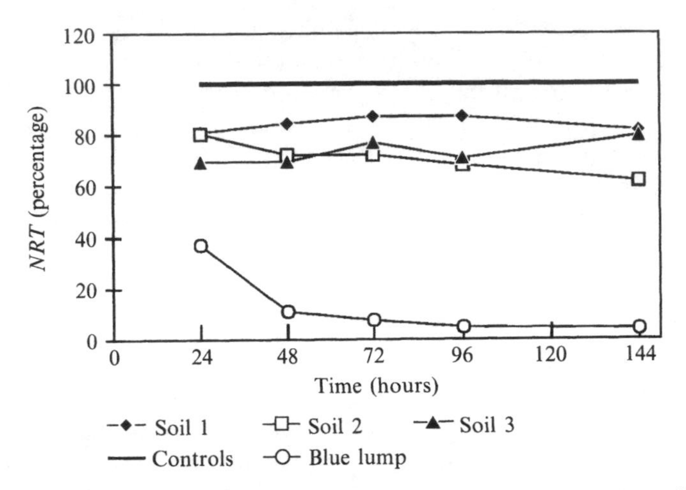

Figure 28-7 Time *versus* normalized relative transpiration (*NRT*) of basket willows (*Salix viminalis*) grown in Holte Gasworks soils.

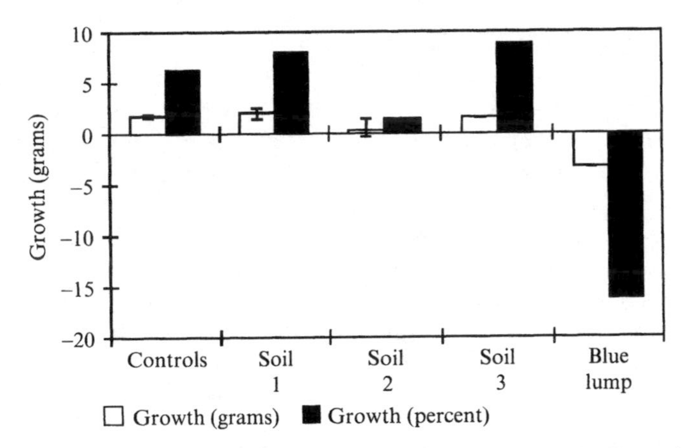

Figure 28-8 Growth of basket willows (*Salix viminalis*) grown in Holte Gasworks soils.

Growth of Vegetation at the Holte Gasworks Site

The site where the soil sample 1 had been taken had a vegetation cover in secondary succession (natural development of vegetation cover). Together with Ole Kusk, biologist at the Technical University of Denmark, a species list was made (Table 28-4). Plants were sampled from a strip of approximately 3 by 8 meters. Indicator values of Ellenberg (1979) for Central European plant species are given when available. The scale ranges from 1 to 9—the higher the value, the more the plant prefers the parameter. The designation × means insensitivity to the parameter. The plants growing on the site prefer direct sunlight, are indifferent to temperature (on an island in Scandinavia), and tend to prefer

TABLE 28-4 Species List with Indicator Values of Ellenberg (1979)

Botanical name	Light	Temperature	Humidity	Nitrogen	Abundance[a]
Maple (*Acer pseudoplatanus*)	4	×	6	7	
Mugwort (*Artemisia vulgaris*)	7	×	6	8	+
Birch (*Betula* spp.)	7	×	×	×	
Strawberry (*Fragaria vesca*)	7	×	5	6	–
Mayweed (*Matricaria maritima*)					–
Poppy (*Papaver dubium*)	6	6	4	5	–
Plantain (*Plantago major*)	8	×	5	6	–
Blackberry (*Rubus fruticosus*)	×	×	×	×	–
Raspberry (*Rubus idaeus*)	7	×	5	8	+
Dock (*Rumex obtusifolius*)	7	5	6	9	+
Black elder (*Sambucus nigra*)	7	5	5	9	++
Figwort (*Scrophularia nodosa*)	4	5	6	7	–
Thistle (*Sonchus arvensis*)	7	5	5	×	–
Dandelion (*Taraxacum officinale*)	7	×	5	7	+
Great nettle (*Urtica dioica*)	×	×	6	8	++
Elm (*Ulmus glabra*)	4	5	7	7	
Mean	6.5	5.2	5.4	7.2	
Number of species insensitive to the parameter	2	9	2	3	

[a]++ means dominant; + means frequent; and – is rare.

Note: Because the site is too small, abundance values for large species like trees are not meaningful and thus are not reported.

humidity as evidenced by an averaged index greater than five. The dominant plants are adapted to high nitrogen content. In particular, black elder (*Sambucus nigra*) and great nettle (*Urtica dioica*) occurred very frequently.

Uptake of Cyanide into Plants

The cyanide concentrations in soil and hydroponic solution were measured before and after the toxicity experiment, and the content in the plants only at the end of the exposure. Almost all of the cyanide vanished from the potassium cyanide solutions, in which the trees rooted for 8 days. Concentrations above the determination limit were found only in the solution with the highest potassium cyanide dose. Details are given in the following section "Willows (*Salix* spp.) in Potassium Cyanide."

Willows (*Salix* spp.) *in Potassium Cyanide*

Concentrations in hydroponic solutions and willow (*Salix* spp.) tissues are shown in Table 28-5. The concentrations in plant tissue listed in Table 28-5

TABLE 28-5 Concentrations in Hydroponic Solution and Willow (*Salix* spp.) Tissue at the Beginning and the End of Exposure to Potassium Cyanide

Initial solution in milligrams per liter as:		Final solution in milligrams of CN per liter		Tissue concentrations in milligrams of CN per kilogram of fresh weight					
				Roots		Leaves		Stems	
KCN	CN	Total	Easily liberatable	Total	Easily liberatable	Total	Easily liberatable	Total	Easily liberatable
1	0.4	0.003	0.0007	< blank	< blank	0.23	0.028	0.06	0.016
5	2	0.01	0.0032	1.87	1.45	0.16	0.05	0.14	< blank
20	8	0.014	0.0049	26.4	23.7	2.9	2.37	0.125	0.025
50	20	0.057	0.0183	49.6	32.2	9.7	6.1	4.7	3.3

< blank means that the value is below the values of unexposed control plants (blanks).

have been corrected by subtracting the cyanide content measured in the control plants, because cyanide is also produced naturally in the plant metabolism or may be formed during the analyses. The blank values for roots were 0.84 milligram of CN per kilogram of fresh weight of easily liberatable cyanide (three measurements with a standard deviation of 0.73 milligram of CN per kilogram of fresh weight) and 0.51 milligram of CN per kilogram of fresh weight of complexed cyanide (four measurements with a standard deviation of 0.38 milligram of CN per kilogram of fresh weight). For stems, the blank values were 0.10 milligram of CN per kilogram of fresh weight of easily liberatable cyanide (four measurements with a standard deviation of 0.12 milligram of CN per kilogram of fresh weight) and 0.12 milligram of CN per kilogram of fresh weight of complexed cyanide (six measurements with a standard deviation of 0.16 milligram of CN per kilogram of fresh weight). For leaves, the blank values were 0.35 milligram of CN per kilogram of fresh weight of easily liberatable cyanide (six measurements with a standard deviation of 0.22 milligram of CN per kilogram of fresh weight) and 0.41 milligram of CN per kilogram of fresh weight of complexed cyanide (three measurements with a standard deviation of 0.21 milligram of CN per kilogram of fresh weight). Cyanide in solution after 8 days is mainly complexed. Concentrations in roots increase with the concentrations in the external solution and are mainly easily liberatable cyanide. Cyanide concentrations in leaves and stems are smaller than in the roots, and mainly complexed at lower doses, but easily liberatable at the higher (deadly) doses.

Poplars (Populus trichocarpa) in Prussian Blue

The uptake of cyanide from a prussian blue solution can principally occur in at least six forms which are as follows:

1. As colloidal or dissolved Prussian blue $Fe_4[Fe(CN)_6]_3$
2. As $Fe(CN)_6^{3-}$ or as $Fe(CN)_6^{4-}$
3. As HCN or as CN^-

Contrary to the potassium cyanide experiment, cyanide is mainly present in complexed form in poplar (*Populus trichocarpa*) leaves and stem (roots were not analyzed) (Table 28-6). This indicates that complexed cyanides are actually taken up and translocated, but to a lesser extent than free cyanide.

Willows (Salix spp.) in Gasworks Soil

Only one laboratory plant grown in the gasworks soil was analyzed (from soil 2). The exposure period was 9 days. The concentration of easily liberatable cyanide was higher in the willow (*Salix* spp.) roots, than in the soil. Complexed cyanide dominated in the leaves (Table 28-7).

TABLE 28-6 Nominal and Measured Concentrations in Hydroponic Solutions and Poplar (*Populus* spp.) Tissues Before and After Prussian Blue Exposure

Nominal in milligrams per liter as:		Concentration in hydroponic solutions in milligrams of CN per liter				Tissue concentrations in milligrams of CN per kilogram of fresh weight			
		Initial measured		Final measured		Stems		Leaves	
$Fe_4[Fe(CN)_6]_3$	CN	Total	Easily liberatable	Total	Easily liberatable	Easily liberatable	Complexed	Easily liberatable	Complexed
10	5.45	4.7	0.22	8.8[a]	5.2[a]	0.25	0.19	0.40	1.66
50	27.25	23.2	0.14	21.2	0.64	0.11	0.27	1.08	0.42
160	87.2	74.1	0.44	96.9	0.72	0.27	0.75	0.53	2.28
400	218	185.9	2.0	180.7	1.53	1.65	1.88	2.74	4.36
1000	545	463	2.8	591.9	3.1	2.75	5.97	3.08	9.54
2500	1362.5	1157.4	6.9	1007.4	8.77	5.65	16.48	3.76	14.50

[a]Measured value is questionable.

TABLE 28-7 Cyanide Concentration in Soil (Milligrams of CN per Kilogram of Dry Weight) and Willow (*Salix* spp.) (Milligrams of CN per Kilogram of Fresh Weight) Grown in Holte, Denmarks Gasworks Soil 2

	Total, replicates		Easily liberatable, replicates		Complexed, replicates	
	1	2	1	2	1	2
Soil, initial	327.3	331.6	16.2	17.9	248.4	284.5
Soil, final	294.8	303.3	20.0	ND	252.9	ND
Leaves	ND	ND	1.2	ND	4.9	ND
Roots	ND	ND	42.1	ND	14.3	ND

Abbreviation used—ND not determined.

Uptake of Cyanides into Vegetation at the Holte Gasworks Site

From two sites on the former gasworks at Holte, plant and soil samples were taken. Controls of black elder (*Sambucus nigra*) were taken from the campus of the Technical University of Denmark in Lyngby, controls of willows (*Salix* spp.) were grown in the laboratory. The material was analyzed for cyanides. Table 28-8 shows results for black elder (*Sambucus nigra*) and willow (*Salix* spp.). Site 1 was identical with the strip where the vegetation was determined (Table 28-4). The site 1 soil had a neutral to slightly alkaline pH (7.5). Site 2, about 100 meters away, had acidic soil (pH 4.8).

Soils from sites 1 and 2 showed a similar cyanide level (Table 28-8). The investigated black elder (*Sambucus nigra*) roots had a diameter of about 10

TABLE 28-8 Concentrations of Cyanides in Soils (Milligrams of CN per Kilogram of Dry Weight) and Black Elder (*Sambucus nigra*) and Willow (*Salix* spp.) (both in Milligrams of CN per Kilogram of Fresh Weight) Growing at the Holte Gasworks Compared to Control Plants

Sample	Site	Easily liberatable	Complexed
Soil	1	90	800
Elder roots	1	4.4	11.2
Black elderberries (*Sambucus nigra*)	1	0.6	0.2
Black elderberries (*Sambucus nigra*) from controls	University campus	0.2	0.2
Willow (*Salix* spp.) leaves	1	0.4	3.6
Soil	2	95	932
Willow (*Salix* spp.) roots	2	199	411
Willow (*Salix* spp.) leaves	2	0.3	2.6
Willow (*Salix* spp.) roots from controls	Laboratory	0.7	0.84
Willow (*Salix* spp.) leaves from controls	Laboratory	0.4	1.9

millimeters. These roots were directly growing into and through soil 1, which had 890 milligrams of total cyanide per kilogram. The roots had elevated levels of cyanide. The ratio of easily liberatable cyanide concentrations between the root and the soil was 0.05. For complexed cyanide, the ratio was only 0.01. In elderberries (*Sambucus nigra*), free cyanide content was almost as low as in unexposed controls. Complexed cyanide is equal in both samples. Cyanide in the controls may be formed during boiling in acid, but also by secondary metabolism of plants. Black elder (*Sambucus nigra*) is known to be a cyanide-producing plant (Frohne and Jensen 1985).

There was no difference in easy liberatable cyanide content between leaves of unexposed controls and leaves of willows (*Salix* spp.) growing on the cyanide-polluted soils 1 and 2. The willow (*Salix* spp.) roots analyzed were fine roots and had a very high content of both easily liberatable and complex cyanide. It is hard to imagine that the 199 milligrams of CN per kilogram of fresh root that were liberated from willow (*Salix* spp.) roots through cooking during the analysis were present in free form (HCN and CN^-) in the roots without killing the plant. We do not know whether β-cyanoalanine, the detoxified metabolite of cyan hydrogen (HCN), is part of the easily liberatable cyanide fraction.

There is an important difference between willow (*Salix* spp.) trees grown in the laboratory and outdoors; with the small basket willows (*Salix viminalis* spp.) in the laboratory (about 0.4 meter length), a measurable transfer of cyanides from hydroponic solution or soil into leaves was found (Tables 28-5 to 28-7). With the much taller trees outdoors (samples were taken at a height between 1.5 and 3 meters), the transfer into leaves was very small or not detectable. This is remarkable because the content of easily liberatable cyanide in roots of outdoor plants was 5 times higher than in the laboratory plants (Table 28-7), and even 29 times higher for complexed cyanide.

Loss of Cyanide During the Experiments

It was not possible to establish a mass balance for the cyanide in these experiments, as the volatilization of cyan hydrogen (HCN) was not accounted for in the experimental setup. Still, some results are interesting.

Controls

In a control experiment with 2 milligrams of KCN per liter (nominal 0.8 milligram of CN per liter) without plants, the initial measured concentration was 0.81 milligram of CN per liter of free cyanide. After 3 days in the dark, 0.804 milligram of CN per liter or 99.3 percent was recovered (five replicates with a standard deviation of 0.033 milligram of CN per liter).

Willows (Salix spp.) in Potassium Cyanide Solutions

During the 8-day experiments of willows (*Salix* spp.) growing in potassium cyanide solutions, the majority of applied cyanide had vanished (Table 28-5).

In all experiments, less than 1 percent of the initially applied cyanide was recovered from the hydroponic solutions. In plants, only about 1 percent of the applied amount was found. The percentages recovered from the plants were smaller at higher doses (0.25 percent at 50 milligrams of KCN per liter). This can be explained by the reduced transpiration of the plants that were exposed to higher doses of potassium cyanide (KCN).

Willows (Salix spp.) in Gasworks Soil

The loss of total cyanide from the soil during the 9-day period due to willow (*Salix* spp.) was 9.2 percent. Given the analytical uncertainties, no conclusions should be drawn from this result.

Poplars (Populus trichocarpa) in Prussian Blue

At the end of the 16-day Prussian blue experiments, the cyanide concentration occasionally increased because water was taken up faster than Prussian blue (Table 28-6). Mass losses in 16 days were between 30.5 and 42 percent, except for the smallest concentration, where an increase of mass was measured (probably an error). Prussian blue has a low water solubility (0.25 milligrams of $Fe_4[Fe(CN)_6]_3$ per liter at pH 7; Smith and Mudder 1991) and is mainly present in colloidal form. The mass loss could be due to settling of these colloids. The loss was actually highest at the highest dose, and precipitated particles were found at the bottoms of these flasks. The loss was therefore probably not caused by poplar (*Populus trichocarpa*) uptake or degradation.

MASS BALANCE OF FREE CYANIDE

The unexplained loss of cyanide in the toxicity experiments with potassium cyanide prompted further investigations of the fate of cyanide in the system. Possible removal processes are volatilization, complexation, storage in the trees, and metabolism by either microorganisms or the trees.

Initial Experiments to Isolate Microbial Effects

We were unsuccessful in isolating the silver willow (*Salix alba*) trees from bacteria and fungi. Autoclaving damages trees. The use of antibiotics resulted in the outbreak of a fungal disease. The use of antibiotics combined with an agent against fungi (carbenicillin at 300 milligrams per liter, streptomycin at 100 milligrams per liter, natamycin at 50 milligrams per liter kills fungi) or conserving agents (benzoic acid) seriously affected the trees; the leaves turned yellow or white and were dropped. Therefore, we could not determine by experiments whether metabolism of cyanides is by the trees exclusively or assisted by the associated microorganisms. Initial experiments were made to investigate the metabolic capacity of microorganisms in the system.

A sublethal dosage (for the trees) of 2 milligrams of KCN per liter was added to Erlenmeyer flasks for the following experimental setups:

1. Autoclaved nutrient solution and equipment
2. Unsterilized nutrient solution
3. Nutrient solution, in which willows (*Salix alba*) had rooted for 1 week
4. Nutrient solution and an autoclaved willow (*Salix alba*) stem with roots (2 hours at 125 °C)
5. Nutrient solution and a dried-out willow (*Salix alba*) stem without roots
6. A cell suspension of ground living willow (*Salix alba*) leaves in nutrient solution
7. A living willow (*Salix alba*) tree rooting in a nutrient solution

Experiment 2 was conducted under septic conditions, but without inoculation of bacteria. In experiment 3, potassium cyanide was added directly into an Erlenmeyer flask containing nutrient solution in which a living willow (*Salix alba*) had rooted for 1 week. The idea was to see if root exudates or root-associated bacteria have any influence on degradation of cyanide. In experiment 4, 5, and 7, willow (*Salix alba*) stems or trees were transferred into the Erlenmeyer flasks with a cyanide spiked nutrient solution. The flasks were carefully closed with drilled cork stoppers and sealed with Teflon™ tape. Then the flasks were placed in a desiccator to determine loss into the gas phase (see the section "Results of Initial Experiments").

The light source was a 60 watt plant light from FDB, Albertslund, Denmark, with a 450 Lumen light stream, placed 0.3 meters from the desiccator. After 24 hours of continuous light exposure, samples were analyzed for free cyanide by direct photometry in two to four replicates. Some were also analyzed for easily liberatable, complexed, and total cyanide. Some of the measured values were below the limit of determination. Quantifiable amounts were nonetheless given, although these may be uncertain.

Results of the Initial Experiments

The difference among the seven treatments is clear (Table 28-9 and Figure 28-9). Free cyanide was quickly eliminated from all Erlenmeyer flasks containing living willow (*Salix alba*) tissue (experiments 6 and 7). This occurred for the intact trees as well as for the leaf cell suspension. Because ethylene is produced in mature leaves (Larcher 1995), the detoxification of the cyanide by-product should be expected in leaf tissue. Maybe this explains the very effective elimination of cyanide by leaf cell suspensions. Some removal was caused by complexation. Even the dried stem was active, with greater than 80 percent removal of cyanide. From all willow (*Salix alba*)-free samples, the majority of cyanide could be recovered. In the gas traps, only 1 percent of the applied amount of cyanide was found.

TABLE 28-9 Results of the Initial Experiments to Assess the Cyanide Mass Balance

No.	Experiment	Initial free milligrams of CN per liter (easy/complexed/total)	Final milligrams of CN per liter free (easy/complexed/ total)
1	Autoclaved nutrient solution	0.631 ± 0.045 (NA[a])	0.604; $s = 0.0028$ (NA[a])
2	Unsterilized nutrient solution	0.427 ± 0.054 (NA[a])	0.410; $s = 0.056$ (NA[a])
3	Inoculated, by willow (*Salix alba*) roots	0.246 ± 0.003 (NA[a])	0.243; $s = 0.0012$ (NA[a])
3	Inoculated by willow (*Salix alba*) roots, replicate in nutrient solution	0.458 ± 0.002 (NA[a])	0.410 ± 0.01 (NA[a])
4	Autoclaved stems and roots	0.394 ± 0.003 (0.344/0.044/0.388)	0.278 ± 0.028 (0.134/0.072/0.205)
5	Dried willow (*Salix alba*) stem	0.394 ± 0.003 (0.344/0.044/0.388)	0.075 ± 0.001 (0.0575/0.0063/0.064)
6	Willow (*Salix alba*) leaf cell suspension	0.460 ± 0.003 (NA[a])	< det[b] (NA[a])
7	Living willow (*Salix alba*) tree	0.394 ± 0.003 (0.344/0.044/0.388)	< det[b] (0.02/0.005/0.025)

Abbreviation used—[a]NA: not analyzed.
[b]< det is no quantifiable peak.

Note: For duplicate sample analysis ($n = 2$), $\pm$ the minimum and maximum is given; for ($n = 4$), the standard deviation s is given.

Fate of Potassium Cyanide in the Soil-Tree-Air System

Because the quick loss of free cyanide from the Erlenmeyer flasks with willows (*Salix* spp.) was repeatedly confirmed, we were very curious to find out what role the trees played and why. The presence of willows (*Salix* spp.) was obviously a prerequisite for the removal of cyanide. But was free cyanide really that quickly metabolized, or was the cyanide simply complexed? Could the cyanide escape *via* the cork stoppers or the stems into the gas phase, or was cyanide stored in the plants?

In the follow-up experiments, three Erlenmeyer flasks with basket willows (*Salix viminalis*) were sealed in a desiccator to capture gaseous cyanide (Figure 28-10). Air was blown through the desiccator with a flow of 0.535 liter per minute (0.53 to 0.55 liter per minute, three measurements). A gasometer was placed at the inlet. At the outlet, two 1-molar sodium hydroxide traps, identical to those used in the analysis of total cyanide, trapped escaping cyan hydrogen (HCN). The background content of cyan hydrogen (HCN) in the air was measured for 12 hours, but none was found. The gasometer was placed at the outlet for 1 day. Measured air flows were identical to those at the inlet.

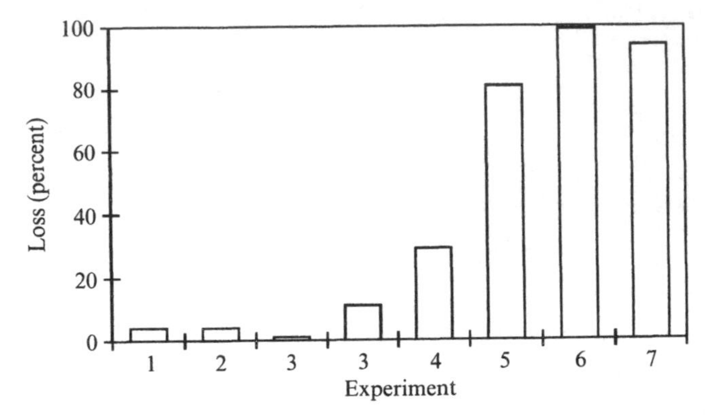

Figure 28-9 Loss in the initial experiments calculated from measurement of free cyanide (except experiment 7).

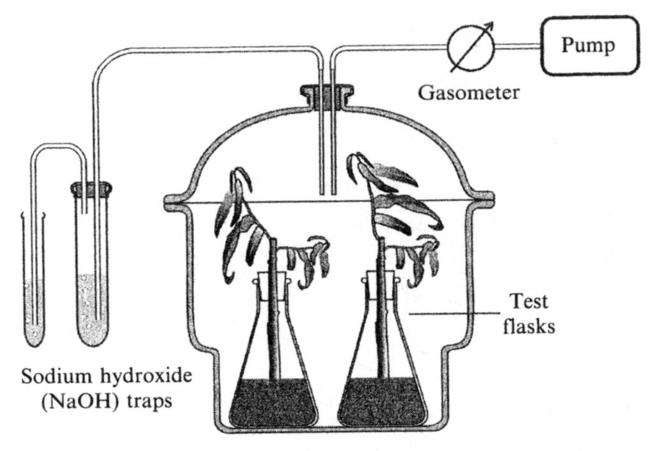

Figure 28-10 Experimental system for the determination of the mass balance of cyanides in basket willow (*Salix viminalis*) tree cuttings.

After exposure to 24 hours of light, one of the plants was replaced by a new one in a fresh potassium cyanide (KCN) solution. The light was switched off and, after 24 hours, the second plant was replaced. Then followed a 24 hours light period, then a 72 hour dark (weekend) period with one new plant, and a final light period. In summary, two plants had been 24 hours in the light, one 24 hours in the dark, and two in 48 hours of light and 96 hours of darkness, but one in a final 24 hours of light. The temperature difference in the desiccator between the light and dark periods was less than 2 °C.

The cyanide content of the samples was determined as usual (*i.e.*, easily liberatable, complexed, and total), except for the gas traps, where cyan hydrogen (HCN) was measured directly as free cyanide. The efficiency of the traps was determined by comparing amounts in both.

Loss from Solution

In these experiments, both easily liberatable and complexed cyanide were measured and used for the mass balance (Table 28-10). The recovered amounts are higher than previously because a considerable fraction of cyanide in solution was present in complexed form at the end of the experiment (up to 77 percent of the remaining cyanide in solution, which corresponds with up to 15.9 percent of the initially applied cyanide). The highest amount of cyanide (79 percent of the initial cyanide) was recovered from the hydroponic solution of the plant that was 24 hours in the dark. The highest elimination of cyanide from the solution occurred with the second plant in light, the test with the smallest dosage (less than 1 percent of the initial cyanide remained in solution).

The fact that cyanide vanished more quickly in the presence of willows (*Salix viminalis*), and faster during light periods, suggests uptake in the transpiration stream and translocation upwards, out of the Erlenmeyer flasks. Except for the second experiment in light, the correlation between transpiration and loss of cyanide is very high (*i.e.*, the correlation coefficient r was 0.99 based on four measurements). On the other hand, the plants transpired only between 1.3 percent (dark period) and 10.6 percent (24 hours light) of the nutrient solution (250 milliliters) per day. So if translocation upwards was the main elimination process, the cyanide concentration in the xylem would have had to be at least six times that in the external hydroponic solution. We are not aware of any physico–chemical process that leads to this high an accumulation. The ion trap mechanism, which accumulates weak acids, does not apply. First, the pH in solution (pH 6.7) was higher than in the xylem

TABLE 28-10 Potassium Cyanide Mass Balance Experiments: Basket Willow (*Salix viminalis*) Transpiration, Concentration, and Mass of Cyanide in Solution

Experimental conditions		Initial free cyanide		Final cyanide mass		Loss
Exposure to light	Transpiration (grams of water)	Concentration (milligrams of CN per liter)	Mass (milligrams of CN)	Easily liberatable (milligrams of CN)	Complexed (milligrams of CN)	Loss of cyanide (percent)
24 hours light (1)	26.5	0.484	0.121	0.026	0.017	64.2
24 hours light (2)	8.7	0.316	0.079	0.0006	< det[a]	> 98.8
24 hours dark	3.3	0.484	0.121	0.077	0.019	20.7
6 days (a)[b]	33.6	0.484	0.121	0.013	< det[a]	> 89
6 days (b)[c]	40.1	0.484	0.121	0.0012	0.004	95.7

[a] < det means no quantifiable peak.

[b] 6 days (a) was 24 hours light, 24 hours dark, 24 hours light, and 72 hours dark.

[c] 6 days (b) was 24 hours dark, 24 hours light, 72 hours dark, and 24 hours light.

Note: easily liberatable cyanide includes free cyanide.

(pH 5.5) and second, the pK_a (9.2) of cyan hydrogen (HCN) is too high for the process to occur (Trapp 2000).

Volatilization

A fraction of free cyanide that was translocated upwards to the basket willow (*Salix viminalis*) leaves, volatilized. The measured contents of the gas traps are shown in Table 28-11. In total, 0.56 milligram of CN was added to all flasks. Only 8.3 micrograms or 1.5 percent of the applied cyanide was found in the gas traps (after correction for the efficiency of the traps). Therefore, volatilization was not the dominant elimination process. Despite this, the gas phase concentration followed the expected light–dark cycle (Figure 28-11).

Concentrations in Plant Tissue

The majority of loss of free cyanide from the hydroponic solutions was neither due to complexation nor to volatilization. The remaining processes were

TABLE 28-11 Amounts of Cyanide in the Gas Traps, Trapping Efficiency, and Air Flux

Conditions	In micrograms as CN		Trapping efficiency (percent)[a]	Total air volume (cubic meters)
	Trap 1	Trap 2		
24 hours light (1)	0.92	0.64	51.6[b]	0.77
24 hours dark	1.24	0.16	98.3	0.77
24 hours light (2)	1.7	0.16	99.1	0.77
72 hours dark	0.5	0.14	92.2	2.31
24 hours light (3)	0.62	0.32	73.4	0.77

[a] 1 – (amount in trap 2 / amount in trap 1)^2.
[b] Occasional gas escape observed.

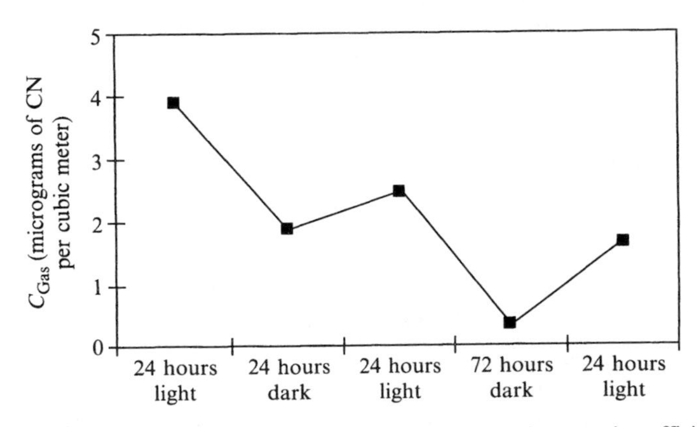

Figure 28-11 Calculated gas phase concentration C_{Gas} (corrected for trapping efficiency) due to cyanide volatilization from basket willow (*Salix viminalis*) cuttings.

TABLE 28-12 Measured Cyanide Concentrations and Masses in Basket Willow (*Salix viminalis*) at the End of the Experiment

Light exposure to different plant organs	Fresh weight (grams)	Total cyanide (milli-grams as CN per kilogram of fresh weight)	Total cyanide (milli-grams of CN)	Easily liberatable (milligrams of CN)	Complexed (milligrams of CN)	Total; easily liberatable; and complexed (percentage of initially applied cyanide)
24 hours light (1)						
Roots	2.9	1.14	0.0033	0.0024	0.001	2.7; 1.9; 0.8
Stem	55.5	0.09	0.005	0.003	0.002	4.1; 2.4; 1.7
Leaves	4.0	0.30	0.0012	0.0006	0.0007	0.9; 0.5; 0.4
24 hours light (2)						
Roots	3.0	0.23	0.0007	0.00015	0.00055	3.2; 0.2; 3.0
Stem	40.6	0.003	0.00013	0.00013	< det[a]	0.1; 0.1; ND[b]
Leaves	3.5	< det[a]	< det[a]	< det[a]	< det[a]	ND
24 hours dark						
Roots	4.2	1.38	0.006	0.005	0.0007	5.0; 4.18; 0.12
Stem	58.6	0.11	0.006	0.0036	0.0027	5.1; 2.9; 2.2
Leaves	3.9	0.26	0.001	0.00035	0.00065	0.8; 0.3 ; 0.5
6 days[c]						
Roots	2.7	0.13	0.00035	0.0003	0.00005	0.3; 0.2; 0.1
Stem	33.7	0.06	0.002	0.0013	0.0007	1.6; 1.0; 0.6
Leaves	2.5	0.04	0.0001	0.0001	< det[a]	0.1; 0.1; nd
6 days[d]						
Roots	3.5	0.17	0.0006	0.0003	0.0003	0.5; 0.25; 0.25
Stem	48.3	0.008	0.00038	0.00021	0.00017	0.3; 0.17; 0.14
Leaves	3.2	0.22	0.0007	0.00015	0.00055	0.6; 0.1; 0.5

[a]< det means no quantifiable peak.
[b]Abbreviation used—ND: no data.
[c]Final period was dark.
[d]Final period was light.

Note: See note for Table 28-10 to describe light exposure conditions.

accumulation and metabolism in basket willow (*Salix viminalis*) cuttings. Table 28-12 gives the amounts recovered from various plant organs. The concentrations in roots seem to be independent of the transpiration or the light conditions. Higher cyanide concentrations were measured in leaves after 6 days exposure when the final period was light. The stems (samples were taken from the lower stem) generally had the smallest concentrations. The majority of the missing cyanide was not found in the plants.

Transfer into Basket Willows (*Salix viminalis*)

Transfer factors were related to the initial and the final concentration in the solution (Figures 28-12 and 28-13). Because initially only free cyanide was present and the cyanide taken up may form complexes, transfer factors were related to the total cyanide concentration. In case of peaks that could not be quantified, concentrations were assumed to be zero in calculating transfer factors.

The transfer into plants is higher when related to the final rather than the initial concentrations in the solution, in particular for the 6-day exposure

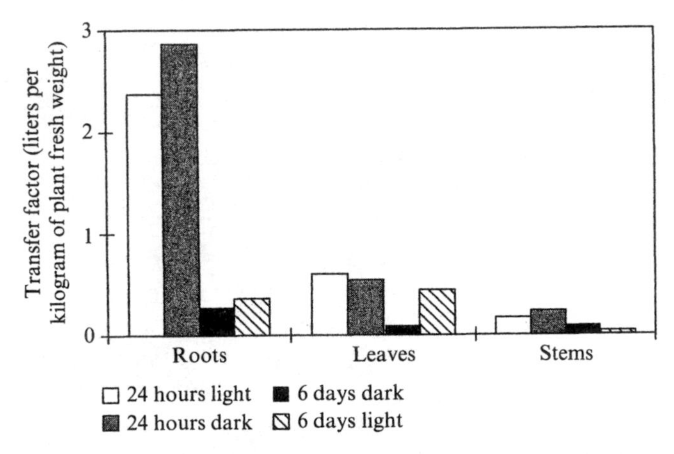

Figure 28-12 Transfer factors for total plant cyanide *versus* initial solution concentration [milligrams of CN per kilogram of fresh weight of basket willow (*Salix viminalis*)] divided by milligrams of CN per liter solution.

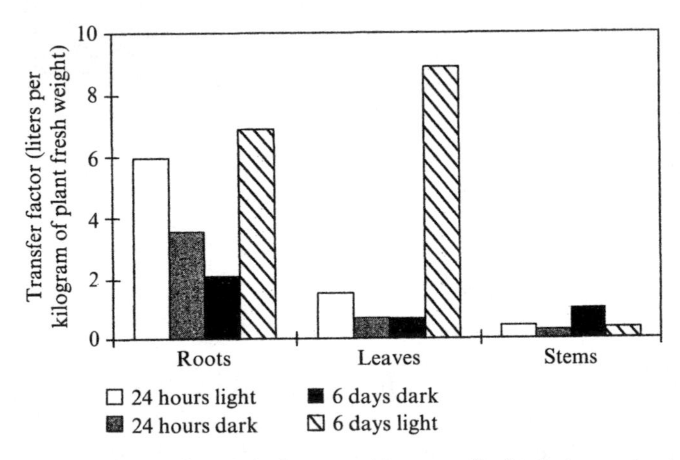

Figure 28-13 Transfer factors for total plant cyanide *versus* final solution concentration [milligrams of CN per kilogram of fresh weight of basket willow (*Salix viminalis*)] divided by milligrams of CN per liter solution.

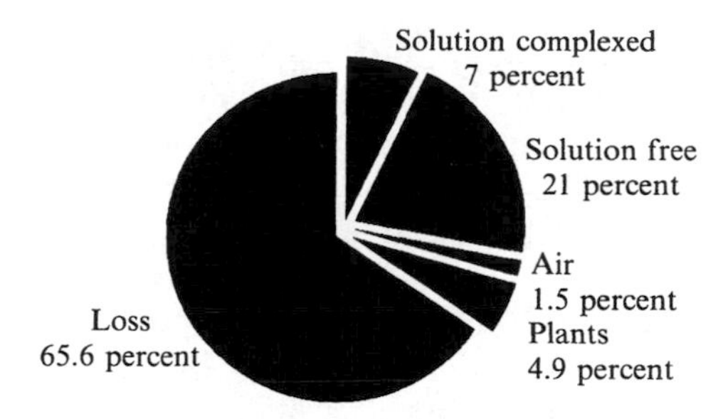

Figure 28-14 Fate of added free cyanide in the mass balance investigations of basket willow (*Salix viminalis*).

period. This is, of course, due to the elimination of cyanide in the Erlenmeyer flasks during the exposure period. The transfer into leaves is higher for the light periods.

In total, less than 35 percent of the applied 0.56 milligram of CN was recovered, and less than 5 percent thereof in plant tissue (Figure 28-14). The highest fraction was free cyanide remaining in the solution (mainly for the cuttings maintained 24 hours in darkness). Our conclusion is that cyanides are most likely degraded, and certainly faster in light.

RISK ASSESSMENT

The uptake of cyanide into plants might lead to a risk that has not been addressed in current risk assessments (Smith and Mudder 1991, Shifrin *et al.* 1996, Gosh *et al.* 1999).

Risk of Eating Exposed Leaves and Fruits

The reference dose for chronic exposure of cyan hydrogen is 0.02 milligram of CN per kilogram of body weight per day (Table 28-2; U.S. EPA 1990). For an 8-year-old boy with a 20 kilogram body weight, a 0.4 milligram as CN of free cyanide daily dosage would be acceptable. The lethal dose (50 percent mortality) is between 0.5 and 3.5 milligrams times a 20 kilogram body weight, which is 10 to 70 milligrams as CN of cyan hydrogen (HCN). Assuming that the toxicity of complexed cyanide in leaves is negligible, then the following risks can be calculated.

Willow (*Salix* spp.) leaves grown on the contaminated soil with 800 milligrams as CN per kilogram of dry weight of complexed and 90 milligrams as CN per kilogram dry weight of easily liberatable cyanide showed the same level of cyanide as unexposed controls (Table 28-8). As a result, there is no increased risk from cyanide transfer into leaves. Elderberries (*Sambucus* spp.) from the contaminated soil had a measured concentration of 0.6 milligram CN per kilogram of fresh weight of easily liberatable cyanide, compared to

0.2 milligram as CN per kilogram of fresh weight in controls. In the worst case, this can be assumed to be all cyan hydrogen. The acceptable daily dose of 0.4 milligram of cyanide would then be found in 667 grams of elderberries (*Sambucus* spp.). The 50 percent risk of a lethal dosage [10 milligrams of cyan hydrogen (HCN)] begins at a consumption of 16.7 kilograms of berries. Cyanide in fruits therefore provides a higher risk for poisoning, both because of higher concentrations and higher possible consumption. For example, homemade elderberry wine with berries from the gasworks site at Holte might be lethal when consumed in high amounts (compare Kesselring 1941).

Animals Feeding From the Trees

Animals such as caterpillars probably get all necessary food from trees. Therefore, the intake, related to body weight, is expected to be much higher, compared to human consumption. During July 2000, aphids were found on trees that were just rooted. We conducted an ecotoxicology test. Willows (*Salix* spp.) with aphids were transferred into a nutrient solution with 2 milligrams of KCN per liter. This is a dosage that the trees tolerate without obvious effects. The aphids on the trees survived with no difference to the controls. After 4 weeks, the test was terminated.

We concluded that cyanide doses tolerated by willow (*Salix* spp.) trees are also tolerated by these parasites, or that the trees metabolized the cyanide before it reached the aphids. However, we cannot predict the effects on other animal species. Nevertheless, an accumulation of free cyanide in the food web does not seem likely. Free cyanide is either detoxified or reacts immediately, *e.g.*, with the iron of the cytochrome oxidase (Faust 1994). The reaction product, an iron-complexed cyanide, is less toxic than free cyanide.

Volatilization from Leaves and Gaseous Exposure

Inhalation of prussic acid can kill humans almost immediately. The lethal concentration is 270 milligrams of HCN per cubic meter for immediate death and 110 to 135 milligrams of HCN per cubic meter for half an hour exposure (Faust 1994). Concentrations between 0.11 and 0.99 milligrams of HCN per cubic meter have no effect on sensitive persons (Shifrin *et al.* 1996).

In the previous experiments, the highest measured gas phase concentration of free cyanide was 3.92 micrograms of CN per cubic meter. This is about 2800 times below the acceptable workplace concentration in Germany (11 milligrams of CN per cubic meter, BGVV 2000) and still below the concentration that affects sensitive persons. If potassium cyanide concentrations in hydroponic solutions had been five times higher, higher gas phase concentrations might be expected, but the dosage would be deadly to the trees. Therefore, the risk of cyanide volatilization from leaves can be neglected because the trees would die before detrimental air concentrations occur.

Shifrin *et al.* (1996) calculated direct volatilization from soil containing 5 milligrams as CN per kilogram of free cyanide and found the cyan hydrogen (HCN) concentration in air to be 0.077 milligrams of CN per cubicmeter, which is also below the hazardous dose. But in a recent risk assessment from the Netherlands, volatilization of cyan hydrogen and inhalation of ambient air is seen as one of the major risks from historical soil contamination with cyanides (Köster 2001). Levels of 1 to 2 milligrams of CN per cubic meter of soil gas were found at a former gasworks site with a total cyanide concentration of 90 grams of CN per kilogram of soil (Köster 2001).

Direct Soil Uptake and Ingestion of Roots

The concentration of easily liberatable cyanide in gasworks soil was up to 95 milligrams of CN per kilogram. This means that the ingestion of more than 4 grams of soil per day may lead to toxic effects; the ingestion of little more than 100 grams of soil may be deadly. Moreover, blue-colored stones such as the "blue lump" (Table 28-3) might attract playing children. This lump had an average concentration of about 270 milligrams as CN per kilogram of easily liberatable cyanide. We concluded that the risk of cyanide poisoning by direct soil uptake is very high. That was the reason why the Holte gasworks site was carefully covered, as described in the section "Phytoremediation of Cyanide-Polluted Soils at Holte."

The thick roots of black elder (*Sambacus nigra*) had an easily liberatable cyanide concentration as 4.4 milligrams as CN per kilogram. That means that less than 100 grams per day of these (or similar) roots can be ingested without the risk of poisoning. More than 2.3 kilograms of these roots could be deadly.

Groundwater

According to Gosh *et al.* (1999), easily liberatable cyanide (weak-acid dissociable cyanide) is degraded during groundwater passage, whereas iron-complexed cyanide is stable. The gasworks at Holte were close to the waterworks. Drinking water wells at the waterworks are monitored for cyanide to exclude any risk. If the iron-complexed cyanide resides in the vadose zone, as at the gasworks site at Holte, cyanide leaching to groundwater will most likely be reduced by planting trees, because iron-complexed cyanides are nonvolatile, and the transport downwards will be lessened by transpiration once the trees mature.

PHYTOREMEDIATION OF CYANIDE-POLLUTED SOILS AT HOLTE

The experiments were initiated to investigate the feasibility of phytoremediation at the abandoned Holte gasworks site (Søllerød commune, Denmark). The following conclusions were drawn. Total cyanide concentrations in soils of up to 1000 milligrams as CN per kilogram, mainly iron-complexed, were tolerated by willows (*Salix* spp.) and black elder (*Sambucus nigra*). Transpiration of trees will reduce infiltration of water, and thus minimize the leaching of contaminants to groundwater. Sublethal doses of cyanide are metabolized quickly inside plants. Ingestion of leaves from trees that grow in cyanide-contaminated soil is not a risk to human health (except, of course, for naturally toxic leaves), because the trees die before accumulating toxic amounts of cyanide. Ingestion of fruits or products from fruits could present a risk. Ingestion of soil and roots is a high risk, and measures had to be taken to avoid that at the Holte site.

Phytoremediation was considered to be a potentially successful remediation method for the Holte gasworks (6000 square meters). About 2500 fast growing poplar (1000 *Populus trichocarpa* × *Populus maximoviczii*, 500 *Populus canescens*, and 1000 *Populus* × *robusta*) trees were planted. The whole area was covered with FiberTex®, which was covered with 100 millimeters of wood chips. These measures were needed to avoid ingestion of soil, *e.g.*, by playing children, and the growth of herbs and weeds that are potentially used for food or as medicinal plants. A monitoring of the cyanide concentrations in the poplar (*Populus* spp.) leaves will be carried out to give added assurance of safety during remediation. In the year 2001, a mean of 0.35 milligram as CN per kilogram of dry weight of total cyanide (four measurements with a standard deviation was 0.18 milligram as CN per kilogram of dry weight) was found in the poplar (*Populus* spp.) leaves (L.C. Larsen, personal communication). In order to control leaching of pollutants into the aquifer, water is pumped from wells when necessary and purified or discharged into the sewer. During and after the remediation, the site, which is situated in the center of the town Holte, is being used as a public park (Figure 28-15).

Figure 28-15 Phytoremediation of the former gasworks site at Holte (June 2001). Photo by S. Trapp.

Acknowledgments

The authors thank Jens C. Tjell and Hans Mosbæk, Lars C. Larsen, K. Ole Kusk, Kim C. Zambrano, Ulrich Karlson, anonymous reviewers, Steven C. McCutcheon, Ines Koch, Morten Larsen, Birte Breil, and Worasiri Jaiplord Pedersen. Thanks to Hedeselskabet A/S, Roskilde, Denmark, for financial support.

REFERENCES

Andersen, H. (1994) Statistical methods for evaluation of toxicity of wastewater (Danish). M.Sc. thesis 7/94. Institute for Mathematical Statistic and Operation Analysis, Technical University of Denmark.

BGVV (2000) Chemis-CIV substance data for cyan hydrogen (German). (http://www.bfr.bund.de).

Börner, H. (1995) *Unkrautbekämpfung*. Gustav Fischer, Jena, Germany.

Christie, A. (1945) *Sparkling Cyanide*. Collins, London, United Kingdom.

Ellenberg, H. (1979) *Zeigerwerte der Gefässpflanzen Mitteleuropas* (Indicator values of central European vascular plants). Erich Goltze KG, Göttingen, Germany.

Faculty for Chemistry Bielefeld (2000) Exemplarische Zusatzstoffe in Lebensmitteln (Examples of food additions). (http://dc2.uni-bielefeld.de/dc2/wsu-bclm/kap_05.htm).

Faust, R.A. (1994) Toxicity summary for cyanide. Report. Oak Ridge Laboratory, Tennessee. (http://risk.lsd.ornl.gov/tox/profiles/cyanide_f_V1.shtml).

Frohne, D. and U. Jensen (1985) *Systematik des Pflanzenreiches unter besonderer Berücksichtigung chemischer Merkmale und pflanzlicher Drogen*. G. Fischer, Stuttgart, Germany.

Fuller, W.H. (1984) Cyanides in the environment with particular attention to the soil. In: *Conference on Cyanides and the Environment*. Geotechnical Engineering Program, Colorado State University, Fort Collins, pp. 19–46.

Gosh, R.S., D.A. Dzombak, R.G. Luthy, and D.V. Nakles (1999) Subsurface fate and transport of cyanide species at a manufactured-gas plant site. *Water Environ. Res.* **71**: 1205–1216.

Kesselring, J. (1941) *Arsenic and Old Lace*. Theater play first shown on Broadway, New York.

Kjeldsen, P. (1999) Behaviour of cyanides in soil and groundwater: a review. *Water, Air Soil Pollut.* **115**: 279–307.

Köster, H.W. (2001) Risk assessment of historical soil contamination with cyanides; origin, potential human exposure and evaluation of intervention values. RIVM report 711 701 019, Bilthoven, The Netherlands.

Larcher, W. (1995) *Physiological Plant Ecology*. Springer, Berlin, Germany.

Manning, K. (1988) Detoxification of cyanide by plants and hormone action. In: *Cyanide Compounds in Biology*. Ciba Foundation. John Wiley, Chichester, United Kingdom.

Mansfeldt, T. (2000) Ehemalige Gaswerk-und Zechen-Kokereistandorte in Nordrhein-Westfalen. *UWSF-Z. Umweltchem. Ökotox.* **12**: 122–123.

Oppenheimer, C. (1928a) Anorganische Chemie. In: *Kurzes Lehrbuch der Chemie in Natur und Wirtschaft*. C. Oppenheimer and J. Matula, eds. Band I. 2nd ed. Georg Thieme, Leipzig, Germany.

Oppenheimer, C. (1928b) Organische Chemie. In: *Kurzes Lehrbuch der Chemie in Natur und Wirtschaft* C. Oppenheimer and J. Matula, eds. Band II. 2nd ed. Georg Thieme, Leipzig.

Parlar, H. and D. Angerhöfer (1995) *Chemische Ökotoxikologie*. Springer, Berlin, Germany.

Shifrin, N.S., B.D. Beck, T.D. Gauthier, S.D. Chapnick, and G. Goodman (1996) Chemistry, toxicology, and human health risk of cyanide compounds in soil at former manufactured gas plant sites. *Regul. Toxicol. Pharmacol.* **23**: 106–116.

Shugaev, A.G. (1999) Alternative cyanide-resistant oxidase in plant mitochondria: structure, regulation of activity, and presumable physiological role. *Russ. J. Plant Physiol.* **46**: 262–273.

Smith, A. and T. Mudder (1991) *The Chemistry and Treatment of Cyanide Waste*. Mining Journal Book Ltd., London, United Kingdom.

Trapp, S. (2000) Modeling uptake into roots and subsequent translocation of neutral and ionisable organic compounds. *Pest Manage. Sci. (formerly Pestic. Sci.)* **56**: 767–778.

Trapp, S., K.C. Zambrano, K.O. Kusk, and U. Karlson (2000) A phytotoxicity test using transpiration of willows. *Arch. Environ. Contam. Toxicol.* **39** (2): 154–160.

UNEP/OCHA United Nations Environmental Program and Office for the Coordination of Humanitarian Affairs (2000) Cyanide spill at Baia Mare, Romania. Report. (http://www.natural-resources.org).

U.S. Environmental Protection Agency (U.S. EPA) (1990) Iris, Integrated risk information system; cyanide, free. Consulted January 2002 at (http://www.epa.gov/iris/subst/0031.htm).

U.S. Environmental Protection Agency (U.S. EPA) (1991) MINTEQA2/PRODEFA2, a geochemical assessment model for environmental systems: version 3.0. Report EPA/600/3–91/021. Athens, Georgia,USA. Available from the US-EPA web site January 2002 at (http://www.epa.gov/ceampubl/softwd_t.htm).

Wallace, A., J.W. Cha, and R.T. Mueller (1977) Cyanide effects on the transport of trace metals in plants. *Comm. Soil Sci. Plant Anal.* **8**: 709–712.

29

PHYTOREMEDIATION OF PERCHLORATE

V. A. Nzengung and S. C. McCutcheon

SUMMARY OF PRACTICAL IMPLICATIONS

Perchlorate is taken up from contaminated waters by a variety of plants under a wide range of conditions similar to those observed at contaminated sites. Several but not all terrestrial and wetland plants mineralize perchlorate to chloride and oxygen. Some plants accumulate perchlorate to more than 0.1 percent by weight. Some halophytes and phreatophytes adapted to the U.S. southwest (location of many sites of perchlorate contamination) do well in degrading perchlorate or spurring bacterial mineralization. Some trees spur the acclimation of microorganisms, and these bacteria mineralize perchlorate faster than plants. Chloride, nitrate, and other ions competitively affect perchlorate uptake and mineralization.

CONCERNS ABOUT PERCHLORATE AND THE LIMITED WASTE MANAGEMENT OPTIONS

The chemical industry produces perchlorate salts on a large scale for a wide range of applications. Ammonium perchlorate is used in the production of solid rocket fuel, explosives, pyrotechnics, and blasting formulations (Ataway and Smith 1993). Perchlorate salts are also used on a large scale as a component of automobile air bag inflators. Other industrial or commercial applications of perchlorate salts include use in nuclear reactors and electronic tubes; as additives in lubricating oils; in the tanning and finishing of leathers; as a mordant for fabrics and dyes; in electroplating, aluminum refining, and rubber manufacture; and in the production of paints and enamels (Siddiqui *et al.* 1998). The approximate percentage of end use is 92 percent as an oxidizer, 7 percent as an explosive, and 1 percent for all other uses.

The current perchlorate disposal inventory has 25×10^6 kilograms (55 million pounds). Over the next 8 to 10 years, this amount is expected to

Phytoremediation: Transformation and Control of Contaminants,
Edited by Steven C. McCutcheon and Jerald L. Schnoor.
ISBN 0-471-39435-1 (cloth)

increase to 74×10^6 kilograms (164 million pounds) (Phillips Laboratory 1997). Due to a short shelf life, perchlorate must be removed from inventory periodically and replaced with a fresh supply. The accepted method for removal of solid propellant from rocket motors is high-pressure water washout. This method generates wash water containing elevated concentrations of ammonium perchlorate. Although ammonium perchlorate can be recovered from these aqueous solutions, the process is costly. Because of these practices, perchlorate is often found in surface water and groundwater associated with aerospace development, testing, or manufacture (less than 0.002 to 400 milligrams of ClO_4^- per liter, AWWARF 1998 and unpublished field observations by Nzengung). Wastewaters generated from aerospace manufacturing contain perchlorate concentrations on the order of 1000 milligrams of ClO_4^- per liter (Herman and Frankenberger 1998). See Betts (2000) for a map of reported sites with perchlorate contamination in the U.S.

The U.S. Safe Drinking Water Act regulations currently do not cover perchlorate. However, the U.S. Environmental Protection Agency (EPA) has indicated that perchlorate may be a candidate for regulation in the near future. In the interim, the California Department of Health Services (2002) and agencies in other states have established provisional action levels for drinking water from 4 to 18 micrograms of ClO_4^- per liter. A new reference dose is under consideration by the EPA, based on recently concluded toxicology studies (Renner 1999). The primary human health concern related to perchlorate is that exposure can prevent the uptake of iodine by the thyroid gland (Wolff 1998), leading to concerns about thyroid hormone production and carcinogenic, neurodevelopmental, developmental, reproductive, and immunotoxic effects (Wallace *et al.* 1998, Renner 1999).

A number of physical and chemical processes to treat perchlorate-contaminated sites have been developed in recent years. Some of these processes (*i.e.*, granular activated carbon filtration, ion exchange, chemical reduction, and ultraviolet irradiation) have been tested with limited success. Activated carbon adsorption used for potable and wastewater treatment is ineffective in treating perchlorate-contaminated waters due to poor sorption of perchlorate anions to carbon (AWWARF 1998). Ion exchange is potentially useful for wastewater cleanup but is not widely used because of high costs (Glass 1998). Other advanced processes for the removal of perchlorate anions, such as reverse osmosis and nanofiltration were applied successfully; however, these processes are expensive to implement for large volumes of contaminated waters (Herman and Frankenberger 1998, Wallace *et al.* 1998).

A biological process developed by the Research Laboratory at Tyndall Air Force Base for treatment of wastewaters contaminated with perchlorate consists of two steps, *i.e.*, an anaerobic reactor followed by an aerobic reactor (Wallace *et al.* 1998). Development of this process has progressed from laboratory and bench scale pilot testing, to field applications. This microbial process reduces perchlorate to ubiquitous chloride and oxygen. The process was applied successfully to remove perchlorate from water containing 9000 milligrams of ClO_4^- per liter to below 0.5 milligram of ClO_4^- per liter (Rikken *et al.* 1996, van Ginkle *et al.* 1996). However, a continuous supply of labile carbon is required to sustain the microbial transformation (Wallace *et al.* 1998). Miller and Logan (2000 and references therein) isolated perchlorate-reducing hydrogen-oxidizing autotrophic bacteria, which they used in fixed-film bioreactors to successfully degrade perchlorate in the micrograms per liter and milligrams per liter concentration ranges. The fixed-film process removed perchlorate from water under hydrogen-oxidizing conditions at a high rate and provided a more reliable source carbon; however, pilot-scale tests are needed to establish practicality. In summary, no technology is currently available for the treatment of large volumes of water, sediment, or soil contaminated with perchlorate. Consequently, the development of efficient and low-cost strategies for the remediation of perchlorate-contaminated sites are of immense interest.

Phytoremediation, the use of green plants to cleanse soil and water contaminated with organic or inorganic pollutants, has shown great potential as an alternative strategy for the remediation of low levels of contamination. Plant-based technologies provide an attractive remediation strategy, because complete transformation of the target compound to innocuous products can be achieved. This technology can be implemented for on-site and *in situ* cleanup of some large volumes and expansive areas of contaminated soils or waters (including some groundwaters) without excavation or pretreatment.

Loehr *et al.* (1998) reviewed the state-of-the-science in biological degradation of perchlorate, which is quite limited kinetically, perhaps because this compound does not occur naturally in too many soils that are low in deposited salts. Coats *et al.* (1999) found that only six species of microorganisms used perchlorate degradation to provide energy, but that dissimilatory perchlorate reduction is ubiquitous and more metabolically diverse than originally understood. Although specific reductive enzymes different from the nitrite reductase may facilitate microbial degradation, plant degradation seems to be an artifact of nitrite assimilation and reduction (van Wijk and Hutchinson 1995). Therefore, while some plants may mineralize perchlorate using the the nitrite reductase, more important benefits of using trees may be in providing a carbon source *via* root abandonment and exudation, providing biofilm substrate in the rhizosphere, and removing competing nitrate ions to facilitate microbial degradation of perchlorate.

One of the first checks for the applicability of phytoremediation involves investigating phytotoxicity. Early agronomists found that the phytotoxicity of perchlorate varied with the concentration, type of plant, soil conditions, and lighting in which the plants were grown. Cook (1937) found that the lethal dose of aqueous perchlorate to four different plants [stinkweed, wild mustard, wild oats, and lamb's-quarters (*Chenopodium album*)] was approximately 2500 milligrams per liter, which is about one to two orders of magnitude higher than the concentrations used in this study and found at contaminated sites (AWWRAF 1998). Lauffs (1902) discovered homeostasis in that small concentrations of perchlorate had a stimulating influence on plant growth.

Meanwhile, Aberg (1948) observed that 50 to 995 milligrams of ClO_4^- per liter (0.5 to 10 millimoles per liter) of potassium perchlorate caused growth distortions to leaves of 5-day-old wheat (*Triticum aestivam*). When used for the spraying of plants grown in soil, Cook (1937) observed that perchlorate was considerably less toxic than chlorate. The reason seems to be that perchlorate has a higher activation energy than chlorate, and thus chlorate more readily transforms into chlorite, thought to be highly toxic to plants (van Wijk and Hutchinson 1995). Perchlorate exposures of up to 300 milligrams of ClO_4^- per liter to 6-month-old eastern cottonwood (*Populus deltoides*) and willow (*Salix nigra* and *Salix caroliniana*), and 1-year-old *Eucalyptus cinerea* and loblolly pine (*Pinus taeda*) reported in Nzengung *et al.* (1999a), produced no obvious toxicity. However, Susarla *et al.* (1999) observed that some halophytes [pickleweed (*Allenrolfea occidentalis*) and perennial glasswort (*Salicornia virginica*)] actually grew better in the presence of perchlorate. Clearly, additional phytotoxicity work is warranted.

Only Tollenaar and Martin (1972) seemed to address phytotoxicity to perchlorate in soils or fertilizers. They found that the 0.1 to 0.3 percent levels of $KClO_4$ in Chilean nitrate caused leaf rugosity in the soybean (*Glycine max* [L.] Merr. 'Amsoy') when applied at levels of 300 to 900 kilograms of $NaNO_3$ per hectare. This translated into 0.3 to 3 kilogram of $KClO_4$ per hectare.

EXPERIMENTAL OBJECTIVES

Plant Selection Criteria

Wetland and terrestrial plants were tested to identify species capable of removing perchlorate from water. Studies were conducted to study the rate and mechanisms of plant-mediated transformation of perchlorate. The specific objectives were

1. To evaluate the capability of selected plant species to decontaminate water containing perchlorate
2. To determine the fate of perchlorate taken up by plants (*e.g.*, distribution in plant tissue and accumulation *versus.* transformation)
3. To identify the phytoremediation mechanisms
4. Study different influences on phytoremediation of perchlorate

These objectives led to the investigation of the following plants:

1. Reference plants previously used in phytoremediation investigations
2. Several phreatophytes for exploring potential tree-pumping strategies involving high evapotranspiration rates, rapid growth, and extensive and deep rooting to enhance rhizodegradation in shallow groundwater

3. Tarragon (*Artemisia dracuculus sativa*) and spinach (*Spinacia oleracea*) as rich sources of different enzymatic activities

Table 29-1 lists the plants tested. See Nzengung *et al.* (1999a) and Nzengung *et al.* (1999b) on the preparation and testing of the phreatophytic trees.

TABLE 29-1 Different Plants, Algae, Microbial Mats, and Bioreactor Systems Used in Long-Term Perchlorate Degradation Studies

Plant name	Type of bioreactor or container
16-month-old eastern cottonwood (*Populus deltoides*) from Carswell Air Force Base, Fort Worth, Texas	1. 2-liter sand and hydroponic bioreactors 2. Sand (bottom) plus soil (above) planted with one tree and holding 10 liters of water
16-month-old black willow (*Salix nigra*) from Carswell Air Force Base, Fort Worth, Texas	1. 2-liter sand and hydroponic bioreactor 2. Sand (bottom) plus soil (above) planted with one tree and holding 10 liters of water
16-month-old Willow (*Salix caroliniana*) from a wetland at the University of Central Florida, Tampa, Florida	1. 2-liter hydroponic bioreactor 2. Sand (bottom) plus potting soil (above) planted with one tree and holding 10 liters of water 3. Ebb-and-flow bioreactor
1-year-old *Eucalyptus cinerea* from commercial nursery, Athens, Georgia	1. 2-liter sand and hydroponic bioreactors
1-year-old loblolly pine (*Pinus taeda*) from the University of Georgia Botany Department, Athens	1. Planted in sand and soil mixtures
Parrot feather (*Myriophyllum aquaticum*)	1. 40-liter glass aquarium with freshwater lake sediment at the bottom 2. 108-liter glass tank with 25 percent Hoagland solution
Selected wetland plants	1. Constructed wetland containing 12 liters of water and four types of plant species [Cattails (*Typha latifolia*), Canadian waterweed (*Elodea canadensis*), duckweed (*Spirodela polyrhiza*), and green algae (spp. unidentified)]
Tarragon (*Artemisia dracunculus sativa*) from commercial nursery, Athens, Georgia	1. 20-milliliter serum bottles with the pureed whole plants plus deionized water
Fresh spinach (*Spinacia oleracea*) from Athens, Georgia grocery	1. 20-milliliter serum bottles with the pureed whole plants plus deionized water
Mixed-species microbial mats	1. 60-milliliter serum bottles 2. 5- and 50-liter bioreactors
Freshwater green algae	1. 50-liter rectangular reactor

Note: Source of the microbial mats was as described in O'Niell (1999) and O' Niell *et al.* (1998).

Extraction of Perchlorate from Plant Matter

The concentration of perchlorate in different plant tissues was measured to determine if perchlorate was accumulated or phytodegraded in the organs of dosed plants. At the termination of the phytoremediation tests, representative plants were removed from the media, rinsed with deionized water, and sectioned into roots, lower stem, upper stem, branches, and leaves. The wet weights of the whole plant and of the sectioned tissues were measured before extraction. Each sectioned tissue (3 to 5 grams) was extracted several times by blending for 30 to 60 minutes with a 100 milliliter volume of deionized water or 1 millimole per liter of sodium hydroxide (pH11) aqueous solution, sonicated for 30 minutes, boiled for 30 minutes, and cooled to room temperature. Boiling the extracts precipitated the proteins and also increased the fraction of perchlorate extracted from the plant tissues by approximately 10 percent. The remaining solid biomass was separated from solution by centrifugation followed by filtration through 0.45- and 0.2-micrometer Gelman Acrodisk ion membranes (Fisher Scientific, Fairlawn, New Jersey). Using a large mass of plant tissues and dilution of the boiled extracts with deionized water minimized interferences from extracted plant organic acids during ion chromatographic analysis. The number of extractions needed to ensure complete removal of the extractable perchlorate ions and by-products depended on the plant fraction and type of plant. However, no more than three extractions were needed for complete recovery of the extractable perchlorate and the associated metabolites from plant organs (Nzengung *et al.* 1999a).

For quality assurance and quality control, two equal masses of the same bulk plant samples were extracted, one half with deionized water (resistance of 18 megasiemens across a centimeter distance and the other half with a solution of 1 milligram of ClO_4^- per liter. Control plants were used to further verify the method. This extraction method was further validated by extraction and analysis of $^{36}ClO_4^-$ and [^{36}Cl] activity in samples of roots, stems, branches, and leaves of plants used in greenhouse experiments. This information was used for mass balance determinations, from which recoveries of greater than 92 percent were calculated. In a subset of plants, the concentration of perchlorate in the leaves was monitored for several days or weeks following the complete removal of perchlorate added to the reactors. Leaves (3 grams) harvested from a cross section of branches on each tree were later extracted and analyzed for perchlorate and the associated by-products.

RHIZO AND PHYTODEGRADATION OF PERCHLORATE

A series of longer-term, bioreactor tests lasting one or more months were conducted to determine the primary phytoremediation mechanisms of perchlorate removal. These included tests with tree cuttings, wetland plants, algae, and microbial mats in saturated sand and hydroponic solutions conducted in the laboratory and in greenhouses as listed in Table 29-1. Perchlorate concentrations of 10 to 300 milligrams of ClO_4^- per liter (Nzengung *et al.* 1999a, Nzengung *et al.* 1999b, Nzengung and Wang 2000, O'Niell *et al.* 2000) in the range measured in groundwater at most field sites were successfully degraded to below the ion chromatographic method detection limit of 0.002 milligram of ClO_4^- per liter using rooted cuttings of trees, wetland plants, algae, and microbial mats. Willows (*Salix* spp.) were identified as the best overall candidates for phytoremediation of perchlorate in permanently saturated conditions; these trees were easily maintained and progressively perform better than the other trees under stressful experimental conditions (Nzengung *et al.* 1999a, Nzengung *et al.* 1999b).

Phytodegradation and rhizodegradation were the two primary phytoremediation mechanisms identified by these tests (Nzengung *et al.* 1999a, Nzengung *et al.* 1999b). *Eucalyptus cinerea*, willow (*Salix* spp.), and eastern cottonwood (*Populus deltoides*) trees grown in perchlorate-dosed media for the first time showed that an initial slow uptake of perchlorate through the transpiration stream was followed by a more rapid rate of removal, which was due to microbial degradation in the root zone (rhizodegradation). The results shown in Figure 29-1 confirm that the disappearance rate of perchlorate from hydroponic solutions dosed to achieve 10 and 122 milligrams of ClO_4^- per liter initially follows pseudo first order kinetics, which changes to more rapid zero-order rhizodegradation after many weeks. In selected tests, plants were dosed in multiple cycles following removal of each dose (Figure 29-2). Dosing after microbial acclimation resulting in perchlorate removal from solution by more rapid zero-order kinetics.

Planted sand and hydroponic bioreactors dosed for the first time with perchlorate followed three distinct reaction phases (Figures 29-1 and 29-2). Initially, perchlorate concentrations decreased, which was accompanied by a linear increase in the volume of water transpired by each tree. This phase was attributed to perchlorate uptake into the plant with the transpiration stream. The initial phase was followed by a progressive increase in water uptake without any significant loss of perchlorate from solution. It appeared that the perchlorate taken up by the tree and transported to the leaves approached a toxic level, or that the chlorite and chloride resulting from degradation becomes toxic. The willow (*Salix nigra*) and other trees seemed to develop defense mechanisms to resist further perchlorate uptake or perhaps anion uptake in general. Figure 29-2 also indicated that declining perchlorate concentrations and new doses of perchlorate might have caused a change in the transpiration rate as a defense mechanism. The last phase was characterized by a very rapid removal of perchlorate described by zero-order kinetics and not influenced by the water uptake rate of the tree, although the transpiration rate of the willow (*Salix nigra*) tree increased as the concentration of perchlorate droped rapidly. In addition, the presence of higher nitrate concentrations in the rhizosphere delayed the onset of the rapid kinetic phase due to rhizodegradation (Nzengung and Wang 2000). Doddema *et al.* (1978) found that the

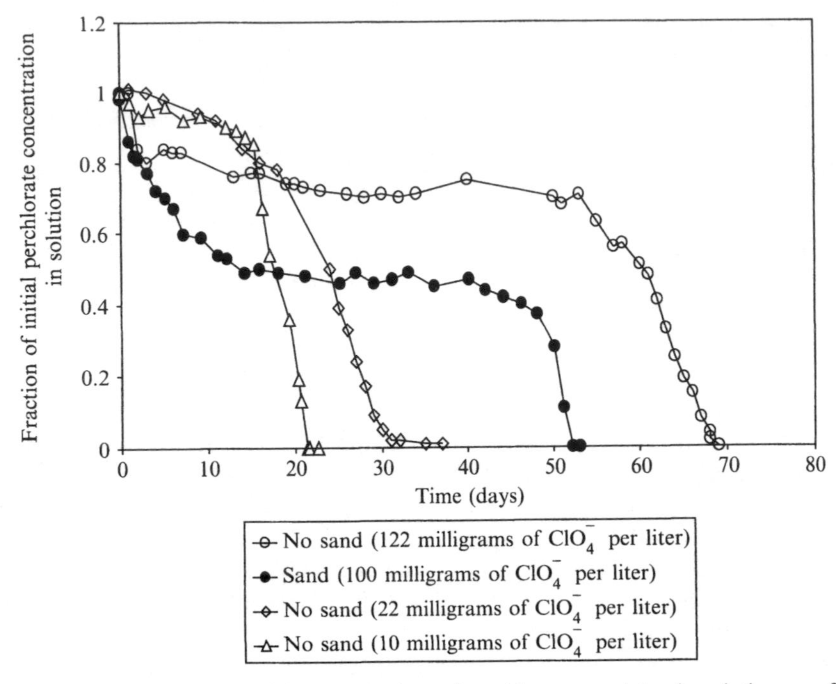

Figure 29-1 Fraction of the initial concentrations of perchlorate remaining in solution as a function of time for pH 5.6 and nitrate concentrations of less than 200 milligrams of NO_3^- per liter (less than 3.2 millimoles per liter). The data represent the first dose of perchlorate added to willow (*Salix* spp.) planted bioreactors. Reproduced by permission from Nzengung *et al.* (1999a). Copyright 1999 by the American Chemical Society.

uptake and assimilation of nitrate could result in a shift in the metabolic state that inhibits the uptake of chlorate by some plants.

Rhizodegradation of Perchlorate

Indirect and direct approaches demonstrated the importance of microbial degradation of perchlorate in the root zone of plants (rhizodegradation). The observed rapid decrease in the concentration of perchlorate ions following a relatively slow initial removal phase, shown in Figures 29-1 and 29-2, was attributed to rhizodegradation. The initially slow kinetic phase (lag phase) corresponded to the time needed for the biostimulation of perchlorate-degrading microorganisms in the root zone. This lag phase varied with the concentration of perchlorate in the rhizosphere, the nitrogen source, and the fraction root mass of the plant (Nzengung and Wang 2000). Generally, rhizodegradation was initiated more rapidly under the following conditions: low nitrate relative to perchlorate concentration, high dissolved organic carbon (electron source), and low perchlorate concentration in the root zone. In contrast to experiments conducted at higher perchlorate concentrations of

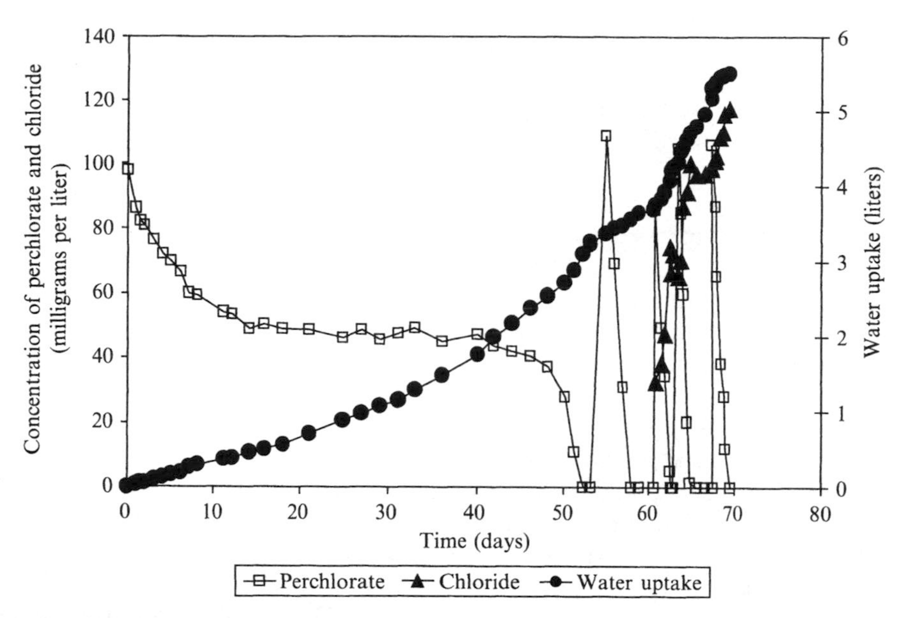

Figure 29-2 Perchlorate loss and cumulative water uptake as a function of time in sand bioreactors dosed five times with approximately 100 milligrams of ClO_4^- per liter of Hoagland solution. Also shown is the increasing chloride concentration in the willow (*Salix nigra*) root zone from rhizodegradation of multiple doses of perchlorate. Reproduced with permission from Nzengung *et al.* (1999a). Copyright 1999 by the American Chemical Society.

greater than 100 milligrams of ClO_4^- per liter, rhizodegradation was initiated in a shorter time (20 to 30 days) at lower perchlorate concentrations of 10 to 30 milligrams of ClO_4^- per liter (Figure 29-1).

In ebb-and-flow bioreactors where aerobic conditions were maintained in the rhizosphere throughout the study, the removal of perchlorate from solution was very slow and no evidence of rhizodegradation was observed (O'Niell *et al.* 2000). The ebb-and-flow bioreactor consists of an upper hydroponic tank (20 liters) containing six willow (*Salix caroliniana*) trees and a lower reservoir (60 liters) containing the bulk of the contaminated water and nutrients for the trees. The willow (*Salix caroliniana*) trees were established in the system for 3 months prior to dosing once with sodium perchlorate at 165 milligrams of ClO_4^- per liter. The ebb-and-flow bioreactor continuously circulated perchlorate-contaminated water through the rhizosphere.

Figure 29-2 shows that initial exposure of willows (*Salix nigra*) to perchlorate-dosed nutrient (Hoagland) solution resulted in a more rapid removal of perchlorate from the growth medium. The rapid kinetics observed after the second, third, fourth, and fifth dose of perchlorate added to the bioreactors (Figure 29-2) suggested that the predominant reaction mechanism was rhizodegradation. Each dose of perchlorate was degraded to below the method detection limit concentration of 0.002 milligram of ClO_4^- per liter before the reactors were dosed again. The rate of reduction of subsequent doses of perchlorate added to the media increased by several orders of magnitude and was described by zero-order kinetics. For example, the rate of degradation of the second dose of perchlorate added to the willow (*Salix nigra*) bioreactor was much faster than that of the initial spike. Specifically, 109 milligrams of ClO_4^- per liter was degraded to below the method detection limit of 0.002 milligrams of ClO_4^- per liter in less than 72 hours. When this reactor was dosed for a third time with an initial concentration of 86 milligrams of ClO_4^- per liter, the perchlorate was degraded to a point below the method detection limit in 43 hours. Figure 29-2 showed that after dosing each bioreactor for a total of five times, the reaction rate did not decrease. The averaged zero order rate constant for this experiment was 2.35 ± 0.37 milligrams of ClO_4^- per liter per hour, average water uptake rate was 92.3 ± 10.7 milliliters per day, and the average chloride mass balance was estimated to be 92.1 ± 6.7 percent. The nearly stoichiometric degradation of perchlorate to chloride in the rhizosphere suggests that there was no accumulation of potential intermediates (Nzengung *et al.* 1999a).

Radiolabeled [^{36}Cl]-perchlorate tests provided additional direct evidence that the rapid removal of perchlorate from the rhizosphere of plants was a result of rhizodegradation. Figures 29-3 showed that the concentration of perchlorate decreased to undetectable levels (0.002 milligram of ClO_4^- per liter) in the rhizosphere whereas the total [^{36}Cl] activity remained approximately the same. If the removal of perchlorate from solution was due to uptake and degradation in the plant (phytodegradation), the [^{36}Cl] activity and concentration of perchlorate measured in solution should decrease proportionately. In experiments where diluted Miracle-Gro® was used as the growth media, the mass balance based on the [^{36}Cl] activity in solution was better than 95 percent. However, a lower recovery (80 percent) of [^{36}Cl] activity was observed if a nitrate nitrogen source (Hoagland solution) rather than a urea and ammonium nitrogen source (Miracle-Gro®) was used as the growth solution. The relatively poor mass balance obtained in experiments with Hoagland solution was attributed to the greater uptake of perchlorate by plants in the presence of nitrate. In addition, the rate of rhizodegradation of perchlorate was faster for trees grown in Miracle-Gro® than those grown in Hoagland solution (Figure 29-3).

The total inorganic and total organic carbon measurements in the growth solution taken at the beginning and end of experiments showed an increase in total inorganic carbon and a small decrease in total organic carbon. For example, in one experiment the total inorganic carbon increased from 2.3 milligrams per liter at the beginning of the experiment to 79.5 milligrams per liter, while total organic carbon decreased from 87.3 to 61.8 milligrams per liter. We believe that the role of plants and other photoautotrophic organisms in the degradation of perchlorate is to provide natural

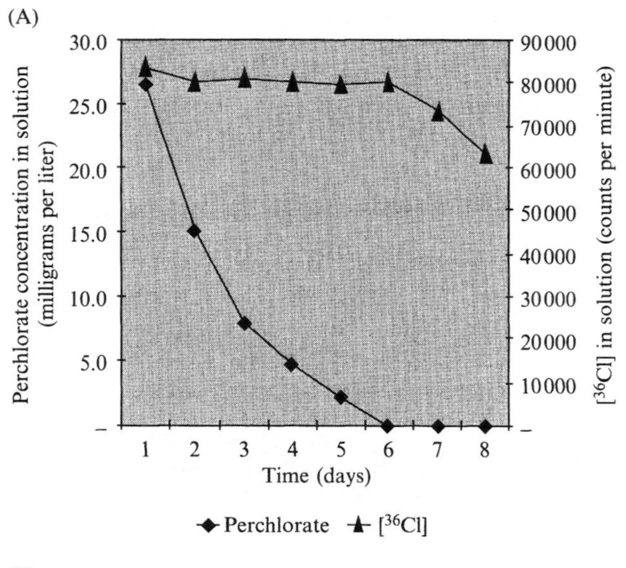

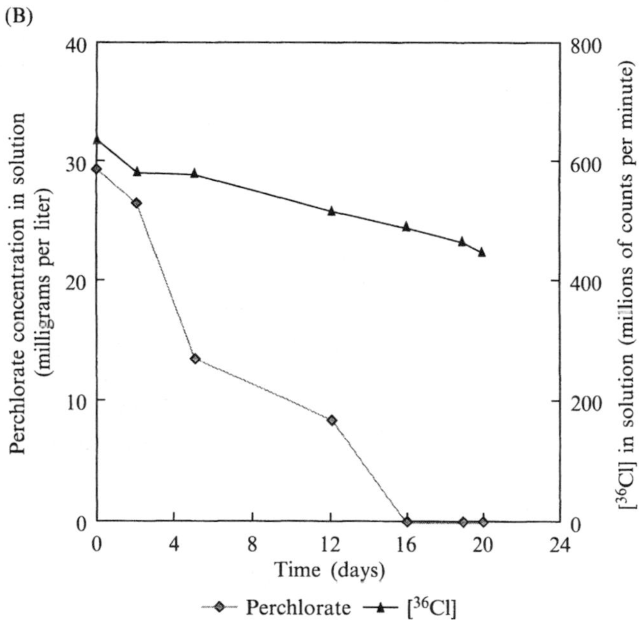

Figure 29-3 Evidence of rhizodegradation of perchlorate using [^{36}Cl]-perchlorate as a tracer in (A) Miracle-Gro® and (B) Hoagland solutions in which willow (*Salix caroliniana*) trees were growing. A 1-milliliter dose containing 147 700 becquerels (3.993 microcuries) was added to each 2 liter bioreactor. The [^{36}Cl] activity was assayed with a Beckman 5801 liquid scintillation counter.

sources of organic carbon, such as acetate, sugars, and alcohols, which were oxidized by rhizosphere bacteria to inorganic carbon and electrons used in the mineralization of perchlorate. If acetate is the source of reducing equivalent for the reduction of perchlorate to chloride, the following reaction can be written

$$CH_3COO^- + ClO_4^- \rightarrow 2HCO_3^- + H^+ + Cl^- \quad (29\text{-}1)$$

Rikken *et al.* (1996) proposed a three-step mechanism of perchlorate reduction by anaerobic bacteria, in which chlorate and chlorite are intermediate products and bicarbonate and chloride are end products

$$ClO_4^- \rightarrow ClO_3^- \rightarrow ClO_2^- \rightarrow Cl^- \quad (29.2)$$

The biodegradation rate of perchlorate in planted bioreactors and those in unplanted controls packed with 75 millimeters sand at the bottom and 225 millimeters of potting soil (Majestic brand from an Athens, Georgia nursery, Dondero 2001) at the top was similar. The pore space held approximately 10 liters of water. Six-month-old willows (*Salix nigra*) and cottonwood (*Populus deltoides*) trees and 1-year-old loblolly pine (*Pinus taeda*) trees were planted. The controls contained the same amount of sand, soil, water, and perchlorate, but without a tree. Each tree was dosed multiple times to achieve 100 to 300 milligrams of ClO_4^- per liter after the previous dose was removed to below the ion chromatographic method detection level of 0.002 milligram of ClO_4^- per liter. The concentration in the leaves was measured during and after complete removal of perchlorate from solution. The potting soil was very rich in nutrients that could sustain high microbial activity. Thus, the role of the trees as a potential nutrient source was minimized. This finding provided additional evidence that natural attenuation of perchlorate in groundwater is limited by the absence of suitable carbon or electron sources.

Figure 29-4 shows directly that raw growth solution taken from around the roots of willow (*Salix nigra*) trees that had been used in phytoremediation of perchlorate showed perchlorate biodegradation activity whereas the same root zone solution that was filtered using a 0.45 micrometer membrane filter, boiled for 2 hours, or autoclaved showed no activity. Equal volumes (50 milliliters) of filtered, boiled, or autoclaved and untreated media were each placed in three sterilized serum bottles, dosed immediately with 10 milligrams of ClO_4^- per liter, and sealed. The vials were continuously mixed on a shaker until sacrificed for analysis. Bacteria were then isolated that showed perchlorate degradation activity. Each isolate was grown in mineral salt media dosed to achieve 200 milligrams of ClO_4^- per liter. Carbon as an electron source was added in the form of 200 to 400 milligrams per liter of acetate. These findings clearly suggest that microorganisms associated with the root zone mediate the rhizodegradation of perchlorate. Nzengung and Wang (2000) obtained additional direct evidence of rhizodegradation of perchlorate by demonstrating

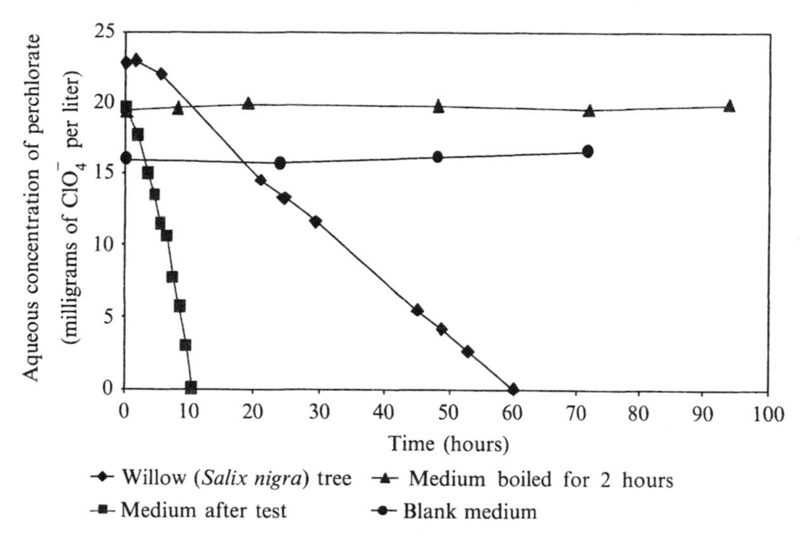

Figure 29-4 Degradation of perchlorate in growth media from root zone of trees used in phytodegradation studies. Reproduced by permission from Nzengung *et al.* (1999b). Copyright. Battelle Press.

that one of the multiple species of bacteria isolated from the root zone of willow (*Salix nigra*) trees degrade perchlorate to chloride. The isolated perchlorate-degrading bacteria utilized acetate as an electron and a carbon source. These bacteria have only been tentatively identified and full characterization is in progress. The practical benefit of studying rhizosphere bacteria that metabolize perchlorate is that rhizodegradation can be initiated and enhanced, and uptake and the less effective phytodegradation can be minimized in food crops at impacted sites. Additionally, such an approach should minimize unanticipated changes to the natural ecosystem resulting from the introduction of foreign plant species.

Phytodegradation of Perchlorate

Three different approaches were followed to obtain evidence that confirms phytodegradation of perchlorate. These include (1) monitoring changes in concentration of perchlorate in tissues of plants previously used in greenhouse experiments (Figures 29-5 and 29-6); (2) monitoring changes in concentration of perchlorate in young, old, and dead leaves of trees grown on perchlorate-contaminated medium in laboratory tests and at perchlorate contaminated field sites (Figure 29-7); and (3) degradation of perchlorate using pureed edible plants and crude extracts of those plants (Figure 29-8). Comparing the results obtained from these three different studies have led us to conclude that phytodegradation is a much slower process than rhizodegradation.

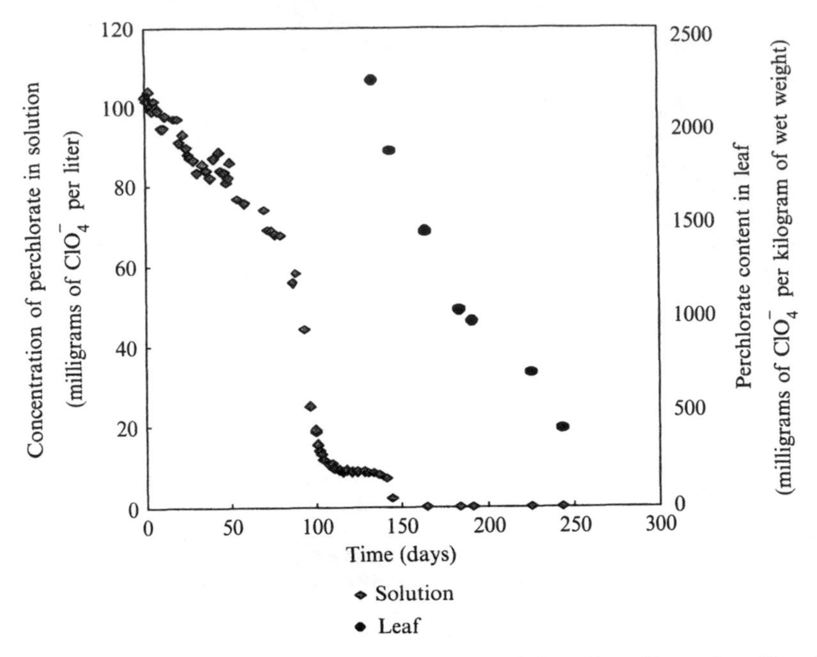

Figure 29-5 Evidence of rhizodegradation and phytodegradation of perchlorate by willow (*Salix caroliniana*) trees. Experiments were conducted in hydroponic bioreactors dosed to achieve 112 milligrams of ClO_4^- per liter and the nutrient solution was 25 percent of the full strength Hoagland solution.

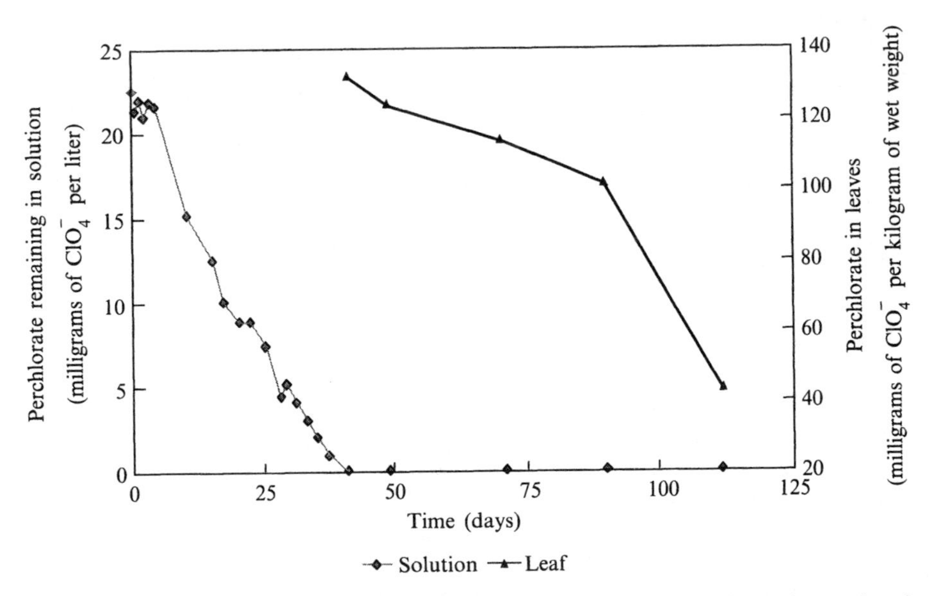

Figure 29-6 Phytodegradation of the fraction of 22 milligrams of ClO_4^- per liter taken up into in leaves of willow (*Salix caroliniana*) trees.

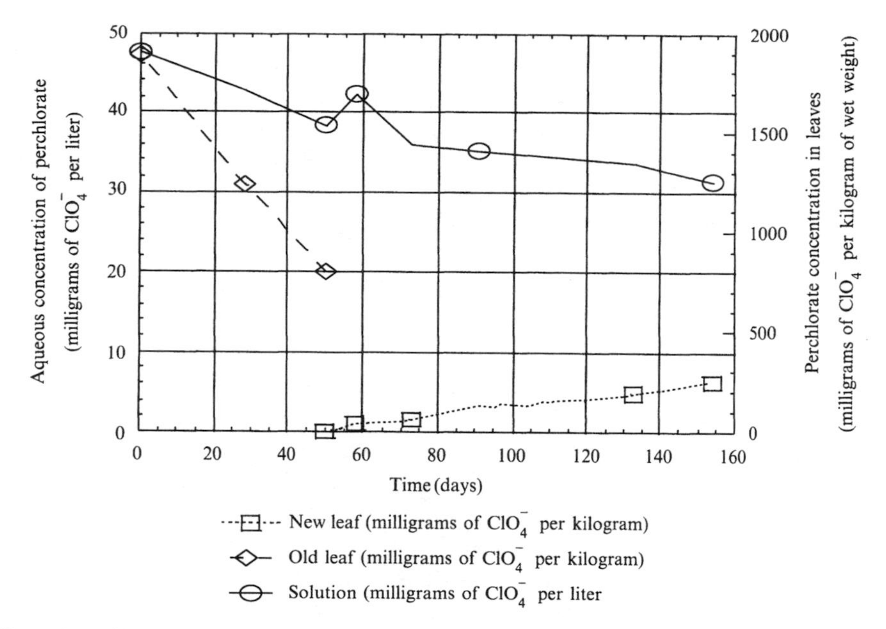

Figure 29-7 Phytodegradation of perchlorate in old willow (*Salix caroliniana*) leaves and uptake and temporal accumulation of perchlorate in new willow (*Salix caroliniana*) leaves.

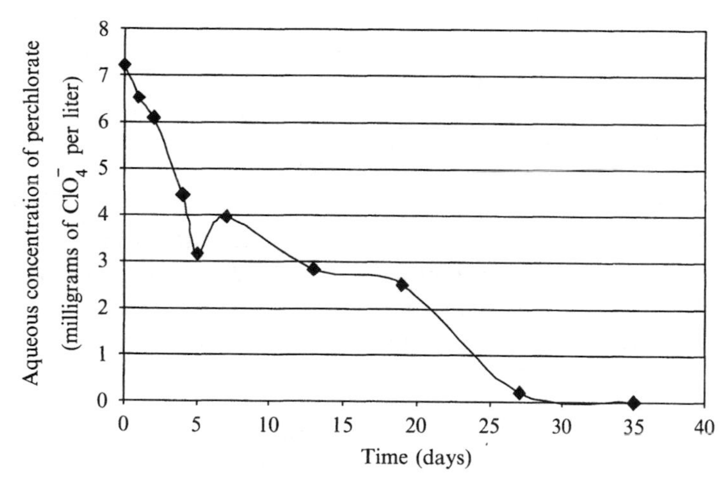

Figure 29-8 Phytodegradation of perchlorate in suspensions of pureed whole French tarragon (*Artemisia dracunculus sativa*) plants. Reproduced by permission from Nzengung and Wang (2000). Copyright. Kluwer Academic/Plenum Publisher.

Nzengung *et al.* (1999b) determined that about 85 percent of the perchlorate taken up by woody plants is temporarily stored in the leaves, and 15 percent distributed among the roots, stem, and branches. Laboratory and field data indicated that the amount of perchlorate ion in the leaves of trees grown on perchlorate-contaminated media was determined by how long the tree had been growing in perchlorate-contaminated media, when the leaves were harvested during the growing season, the concentration of perchlorate in the rhizosphere, and the water uptake rate of the tree. Willow (*Salix caroliniana*) trees grown hydroponically in 112 milligrams of ClO_4^- per liter for about 120 days contained a maximum of 2200 milligrams of ClO_4^- per kilogram of wet weight of extractable perchlorate in the leaves (Figure 29-5). Meanwhile, willow (*Salix nigra*) trees grown in 100 milligrams of ClO_4^- per liter for 26 days contained 850 milligrams of ClO_4^- per kilogram of wet weight and the trees grown on 22 milligrams of ClO_4^- per liter contained 130 milligrams of ClO_4^- per kilogram of wet weight. For trees dosed five times consecutively with 100 milligrams of ClO_4^- per liter, the extracted fresh leaves contained 261 milligrams of ClO_4^- per kilogram of wet weight; on the same trees, leaves showing senile characteristics contained 755 milligrams of ClO_4^- per kilogram of wet weight, whereas the chloride concentrations in the same leaves were 226 and 803 milligrams of Cl^- per kilogram of wet weight, respectively. When the concentration of perchlorate in the solution decreased to undetectable levels, the concentration of perchlorate in the leaves of these trees started to slowly decrease (Figures 29-5 and 29-6). The smaller-than-predicted concentration extracted from leaves of willow (*Salix nigra*) trees used in long-term experiments suggested that perchlorate did not just accumulate in the leaves but was slowly transformed by the plant deoxygenase or reducing enzymes. The results presented in Figures 29-5, 29-6, and 29-7 confirm that the perchlorate taken up into the leaves increased to a maximum before slowly decreasing to below detection levels after perchlorate concentrations drop below detection limits in solution.

The concentration of perchlorate in leaves harvested from pine (*Pinus palustris*), sweet gum (*Liquidambar styraciflua*), and willow (*Salix* spp.) trees growing in perchlorate-contaminated water at the former Longhorn Army Ammunition Plant site in Texas was measured in the winter, spring, and summer months to verify if perchlorate is recycled by plants. The dead leaves collected from under sweet gum (*Liquidambar styraciflua*) and willow (*Salix* spp.) trees in the winter months (January and February) contained no detectable concentration of perchlorate. Leaves harvested in spring contained lower levels of perchlorate than leaves collected in the summer months. Pine (*Pinus palustris*) needles harvested in the winter months contained higher concentrations of perchlorate than those harvested in spring and summer. These results suggested seasonal variations in perchlorate temporarily accumulated in leaves of plants growing over contaminated sites.

Further confirmation of phytodegradation was obtained from experiments in which perchlorate was degraded with crude extracts of French tarragon (*Artemisia dracunculus sativa*) and spinach (*Spinacia oleracea*), and the minced plants (Figure 29-8, Table 29-2). One gram of the pureed plant was added to 20-milliliter serum bottles, filled with deionized water, and dosed with perchlorate to obtain an initial solution concentration of 7 to 8 milligrams of ClO_4^-

per liter. The crude extracts of French tarragon (*Artemisia dracunculus sativa*) and spinach (*Spinacia oleracea*) were prepared by extracting 34 grams of the pureed plant with 0.5 liter of deionized water. Different dilutions of the raw extract were prepared (Table 29-2), dosed with about 23 milligrams of ClO_4^- per liter, and incubated for 11 days. For the microbial mats, 1 gram of the wet mat was weighed into 60 milliliter vials, and the headspace filled with 50 milliliter of deionized water and dosed with perchlorate to obtain a solution concentration of 10 milligrams of ClO_4^- per liter. The vials were mixed continuously on a rotary shaker and sacrificed for analysis at predetermined intervals. Nine samples were used in each of the replicate experiments. Controls contained only deionized water dosed with the same perchlorate concentration as the samples. The pellets were separated from solution by centrifugation, and 1 milliliter of the liquid-phase was diluted and analyzed for perchlorate.

The activity observed in the crude extracts of the herbs did not increase proportionately with the concentration of the extract added to the reaction solution, as one would expect (Table 29-2). The degradation of perchlorate by the crude extract and minced herbs provided additional evidence that the reactions in the plant tissues were enzymatically catalyzed. Possible transformation products of perchlorate such as chlorate, chlorite, and hypochlorite were not detected in the plant extracts. These results suggested that although perchlorate may temporarily accumulate in leaves of plants grown on perchlorate-contaminated soils and water, over the long-term the perchlorate was degraded, and the potential for recycling and continued exposure to perchlorate was minimized. Also, decaying leaves that fall to the ground should be a suitable source of carbon and electrons needed for microbial biodegradation of perchlorate in soils.

Effect of Nitrate Concentration and Nitrogen Source

Two sets of tests verified the competition of nitrate (NO_3^-) during phytoremediation of perchlorate, including the rhizodegradation of perchlorate. First,

TABLE 29-2 Percent Perchlorate Degraded in Different Concentrations of Crude Extracts of French Tarragon (*Artemisia dracunculus sativa*) and Spinach (*Spinacia oleracea*) after 11 Days of Exposure

	Volume of crude extract in solution (milliliters)							
Percent perchlorate degraded by extracts of:	0 (control)	1	2	3	5	10	15	45
French tarragon (*Artemisia dracunculus sativa*)	2.9	3.8	2.1	41.5	100	11.9	18.3	38.1
Spinach (*Spinacia oleracea*)	0.0	0.0	0.3	28.1	100	100	100	0.0

Note: The initial concentration of perchlorate was 23.5 milligrams of ClO_4^- per liter in 20 milliliters of aqueous solution of the extract (Nzengung and Wang 2003).

willow (*Salix* spp.) trees were grown hydroponically (see Table 29-1) in diluted Hoagland solution supplemented with sodium nitrate to achieve three desired concentrations in the range of less than 100, 100 to 300, and greater than 400 milligrams of NO_3^- per liter, or approximately 10, 25, and 50 percent of a full strength Hoagland solution, respectively. To obtain a growth solution free of nitrate, Miracle-Gro,® an ammonium and urea nitrogen source diluted to 500 and 2000 milligrams per liter, was substituted in some experiments. Each set of reactors using the Hoagland solution was dosed multiple times with approximately 100 milligrams of ClO_4^- per liter. The solutions with Miracle-Gro® or Hoagland solution was also dosed with perchlorate to obtain approximately 22 milligrams of ClO_4^- per liter. The chloride, nitrate, and acetate concentrations and the pH of the growth solutions were measured during the course of each test. Controls consisted of willow (*Salix* spp.) trees grown in perchlorate-free media and dosed bioreactors with no plants.

In Figure 29-9, the fastest perchlorate degradation kinetics were observed when nitrate was not present and decreased with increasing nitrate concentration. First- and zero-order kinetics described the utilization rates of nitrate and perchlorate, respectively, at the lowest nitrate concentration. The average perchlorate utilization rates were 23.6 ± 3.7 and 2.3 ± 0.2 micromoles per liter per hour (2.35 ± 0.37 and 0.23 ± 0.02 milligrams of ClO_4^- per liter per hour) for black willows (*Salix nigra*) grown in sand bioreactors with the nitrate concentration maintained at less than 100 and 100 to 300 milligrams of NO_3^- per liter, respectively. Nzengung and Wang (2000) attributed the slow

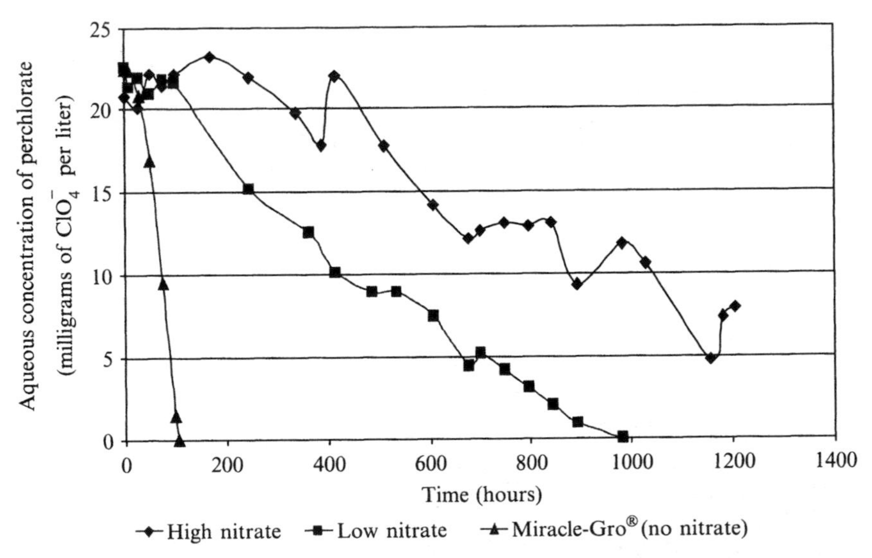

Figure 29-9 Effect of nitrate concentration on phytoremediation of perchlorate by black willow (*Salix nigra*) trees. Reproduced by permission from Nzengung and Wang (2000). Copyright. Kluwer Academic / Plenum Publisher.

reduction rates of perchlorate at high nitrate concentrations to competing reactions in which both anions were utilized as electron acceptors. This was independently confirmed in similar hydroponic studies in which willows (*Salix caroliniana and Salix nigra*) were grown on a nutrient solution with only ammonium and urea as the nitrogen sources (see Figure 29-3). Thus, the type of nitrogen source may have an important influence on phytoremediation of perchlorate.

Second, four different sets of batch vials were prepared and used to verify the effect of nitrate (NO_3^-) and acetate on rhizodegradation of perchlorate in root zone solutions taken from willow (*Salix* spp.) bioreactors. A total of 400 milliliters of growth solution was withdrawn from the rhizosphere of the willow (*Salix* spp.) reactor after perchlorate concentrations were completely degraded below the ion chromatography detection limit of 0.002 milligrams of ClO_4^- per liter. The root zone solution (50 milliliters) was transferred into each of eight serum vials and amended with different ions. Sodium nitrate was added to a pair of the vials to obtain a concentration of 200 milligrams of NO_3^- per liter. Acetate was added to a second pair of vials to obtain an initial concentration of 400 milligrams per liter, and to the third pair of vials, both nitrate (NO_3^-) and acetate were added to obtain initial concentrations of 200 and 400 milligrams per liter, respectively. The remaining pair of vials that contained the unamended media was used as controls. Other controls consisted of deionized water only or deionized water with the same concentrations of nitrate, acetate, or both compounds. The samples and controls were each dosed with perchlorate to obtain an initial solution concentration of 100 milligrams of ClO_4^- per liter, sealed with serum wrap, and incubated at ambient temperature. One milliliter of root zone solution was withdrawn every 24 hours for analysis.

Nzengung *et al.* (1999a) showed that the root zone growth media withdrawn from the rhizosphere of willow (*Salix nigra*) trees also degraded perchlorate to chloride. However, no perchlorate loss or chloride formation was observed in portions of the same growth media supplemented with sodium nitrate ($NaNO_3$) to achieve 200 milligrams of NO_3^- per liter. These results offer further confirmation that high nitrate activity in the rhizosphere may hinder the reduction of perchlorate as a result of competition for electrons between nitrate and perchlorate, or may inhibit bacterial growth at high nitrate concentrations.

The same growth media amended with acetate or both acetate and nitrate did not inhibit the reduction of perchlorate to chloride. This suggests that a high concentration of carbon or electrons may reverse the inhibitory effects observed at high nitrate activity in the root zone. Acetate is one of many plant exudates (Schnoor *et al.* 1995) that the rhizosphere bacteria could have used as a substrate for perchlorate reduction. The acetate concentration measured in the willow (*Salix nigra*) rhizosphere ranged between 124 and 254 milligrams of per liter. Acetate and other organic compounds exuded into the reactor by the willow (*Salix nigra*) trees may stimulate the growth of bacterial consortia that utilized nitrate and perchlorate as terminal electron acceptors. Because perchlorate was degraded in the unamended media in the absence of willow (*Salix nigra*) roots but not in the filtered, boiled, and autoclaved samples, we concluded that microorganisms associated with the root zone mediated the rhizodegradation reactions.

FINDINGS AND RECOMMENDATIONS

The use of vascular plants for phytoremediation of contaminated soil and water showed that some plant species play an important role in causing rhizodegradation or taking up and transforming perchlorate. Perchlorate uptake and degradation was dependent on (1) the plant species, (2) the perchlorate concentration, (3) the support media *versus* hydroponics, and (4) the presence or absence of competing anions.

The effectiveness of phytoremediation of perchlorate-contaminated environments may depend on the concentration of competing terminal electron acceptors, such as nitrate, and the the type of nitrogen supplied to the plants. Woody plants grown on ammonium and urea removed perchlorate from solution at a higher rate and had lower perchlorate concentrations in the leaves than plants grown on a nutrient solution with nitrate as the nitrogen source. The initial or prolonged exposure of plants to perchlorate-dosed media stimulates the growth of perchlorate-degrading microorganisms in the rhizosphere. Two phytoprocesses identified as important in the remediation of perchlorate-contaminated water are (1) uptake and phytodegradation of perchlorate in the tree branches and leafs, and (2) rhizodegradation. Generally, a pseudo first order model described phytodegradation kinetic data within the range of 10 to 100 milligrams of ClO_4^- per liter, and after acclimation rhizodegradation followed zero-order kinetics. Because perchlorate does not volatilize from water an appropriate perchlorate remediation scheme may involve an intensively cultivated plantation of phreatophytic trees irrigated with contaminated water, or a flow through constructed wetland planted with selected plants.

The multitude of influential factors identified in these experiments and unexplored factors of concern necessitate additional research to develop approaches for field application of vascular plants for phytoremediation. Based on the results of these experiments and ecological knowledge of the species evaluated, the following species should be investigated further for phytoremediation of perchlorate. Sweet gum (*Liquidambar styraciflua*), eastern cottonwood (*Populus deltoides*), and black willow (*Salix nigra*) are recommended for on-site and *in situ* phytoremediation of contaminated soils in uplands, including areas with shallow groundwater accessible to plant roots, and if production of biomass for harvest is of interest. For on-site or *in situ* phytoremediation of contaminated areas that are inundated, or for wetlands created for phytoremediation, additional investigation on the use of parrot feather

(*Myriophyllum aquaticum*), cattail (*Typha latifolia*) duckweed (*Spirodela polyrhiza*), and microbial mats is recommended. Finally, extracts from tarragon (*Artemis dracunculus sativa*) may be useful for injection into mechanized flow-through reactors or systems where groundwater is extracted, exposed to phytoremediation plants, then re-injected into the aquifer, or for similar flow-through systems for contaminated surface water. A similar process using horseradish (*Armoracia lopathifolia*) to treat phenol-contaminated wastewaters was found to be feasible by Dec and Bollag (1994). An ecologically sound approach would be to screen indigenous plants and identify suitable species for phytodegradation of perchlorate if none of the species identified in these tests are identified in a vegetation survey of the site.

Acknowledgment

Cooperative interagency research agreements with the U.S. Air Force Restoration Division at Wright Patterson Air Force Base and the Human Systems Centre at Brooks Air Force Base supported this perchlorate work along with two other grants directly to the University of Georgia from the Restoration Division. Greg Harvey served as the Restoration Division project manager. Anna Dondera and Dr. Chuhua Wang of the University of Georgia were instrumental in collecting the data summarized in this chapter. Sridhar Susarla with the University of Georgia provided comments on the literature review and context for the chapter. This paper has been reviewed in accordance with the U.S. Environmental Protection Agency peer and administrative review policies and approved for publication. Mention of trade names or commercial products does not constitute endorsement or recommendation for use.

REFERENCES

Aberg, B. (1948) On the mechanism of the toxic action of chlorates and some related substances upon young wheat plants. *The Annals of the Royal Agricultural College of Sweden* **15**: 37–107.

Ataway, H. and M. Smith (1993) Reduction of perchlorate by an anaerobic enrichment culture. *J. Ind. Microbiol.* **12**: 408.

AWWARF (1998) Perchlorate issue group presentations. (http://www.awwarf.com/newprojects/percsum.html).

Betts, K.S. (2000) Accelerating perchlorate detection. *Environ. Sci. Technol.* **34**(11): 245A–246A.

California Department of Health Services (2002) Perchlorate in California drinking water (http://www.dhs.cahwnet.gov/ps/ddwem/chemicals/perchl/perchlindex. htm).

Coats, J.D., U. Michaelidou, R.C. Bruce, S.M. O'Connor, J.N. Crespi, and L.A. Achenbach (1999) Ubiquity and diversity of dissimilatory (per)chlorate-reducing bacteria. *Appl. Environ. Microbiol.* **65**(12): 5234–5241.

Cook, W.H. (1937) Chemical weed killers. I. relative toxicity of various chemicals to four annual weeds. *Can. J. Res.* **15**: 299–323.

Dec, J. and J.-M. Bollag (1994). Use of plant material for the decontamination of water polluted with phenols. *Biotechnol. Bioeng.* **44**: 1132–1139.

Doddema, H., J.J. Hofstra, and W.J. Feenstra (1978) Uptake of nitrate by mutants of *Arabidopsis thaliana*, disturbed in uptake or reduction of nitrate. 1. effect of nitrogen source during growth on uptake of nitrate and chlorate. *Physiol. Plant* **43**: 343.

Dondero, A.C. (2001) Phytoremediation of perchlorate under greenhouse and natural conditions. M. S. thesis. Department of Geology, University of Georgia, Athens.

Glass, D.J. (1998) The 1998 United States market for phytoremediation (http://www.channel 1.com/dglassassoc).

Herman, D.C. and W.T. Frankenberger, Jr. (1998) Microbial-mediated reduction of perchlorate in groundwater. *J. Environ. Qual.* **27**: 750–754.

Lauffs, A. (1902) Uber einige physiologische Wirkungen des Perchlorats auf die Pflanze. Diss Konigsberg. 29 pp.

Loehr, R.C., L.E. Katz, and D.R. Opdyke (1998) Fate and transport of ammonium perchlorate in the subsurface. Report to the Brooks Air Force Base HSC/XRE support contractor, Environmental and Water Resources Engineering Program, University of Texas, Austin.

Miller, J.P. and B.E. Logan (2000) Sustained perchlorate degradation in autotrophic, gas-phase, packed-bed bioreactor. *Environ. Sci. Technol.* **34**: 3018–3022.

Nzengung, V.A. and C. Wang (2000) Influences on phytoremediation of perchlorate contaminated water. In: *Perchlorate in the Environment*. E. Urbansky, ed. American Chemical Society (ACS) Special Symposium Series, Kluwer Academic/Plenum Publishers, New York, Chap. 21, pp. 219–229.

Nzengung, V.A. (2003) Phytoremediation of perchlorate–contaminated soils and water. Report to U.S. Air Force Restoration Division, Wright Patterson Air Force Base, Dayton, Ohio, February.

Nzengung, V.A., C. Wang, and G. Harvey (1999a) Plant-mediated transformation of perchlorate into chloride. *Environ. Sci. Technol.* **33**(9): 1470–1478.

Nzengung, V.A., C. Wang, G. Harvey, S.C. McCutcheon, and N.L. Wolfe (1999b) Phytoremediation of perchlorate contaminated water: laboratory studies. In: *Phytoremediation and Innovative Strategies for Specialized Remedial Applications*. A Lesson and B.C. Alleman, eds. Fifth International In Situ and On-Site Bioremediation Symposium. Battelle Press, Columbus, Ohio, held at San Diego, California, April 19–22, pp. 239–244.

O'Niell, W. (1999) Biosorption and transformation of tetrachloroethylene and trichloroethylene using mixed-species microbial mats. Ph.D. dissertation, Department of Geology, University of Georgia, Athens.

O'Niell, W.L., V.A. Nzengung, J.E. Noakes, J. Bender, and P. Phillips (1998) Biosorption and transformation of tetrachloroethylene and trichloroethylene using mixed species microbial mats. *J. Hazard. Sub. Res.* **1**(3): 203–226.

O'Niell, W., V.A. Nzengung, and A. Adesida (2000) Treatment of perchlorate contaminated water in microbial mat, algae, and ebb-and-flow hydroponic bioreactors. In: *Case Studies in the Remediation of Chlorinated and Recalcitrant Compounds*. G.B. Wickramanayake, A.R. Gavaskar, J.T. Gibbs, and J.L. Means, eds. Second International Conf. Remediation of Chlorinated and Recalcitrant Compounds. Battelle Press, Columbus, Ohio, held in Monterey, California, May 22–25. **C2**(7): 101–106.

Phillips Laboratory (1997) Reclaimed ammonium perchlorate characterization for rocket propellants (http://www.plk.af.mil).

Renner, R. (1999) EPA draft almost doubles safe dose of perchlorate in water. *Environ. Sci. Technol.* **33**(5): 110A–111A.

Rikken, G.B., A.G.M. Kroon, and C.G. van Ginkel (1996) Transformation of (per)chlorate into chloride by a newly isolated bacterium: reduction and dismutation. *Appl. Microbiol. Biotechnol.* **45**(3): 420–426.

Schnoor, J.L., L.A. Licht, S.C. McCutcheon, N.L. Wolfe, and L.H. Carreira (1995) Phytoremediation of organic and nutrient contaminants. pilot studies are demonstrating the promise and limitations of using vegetation for remediating hazardous wastes in soils and sediments. *Environ. Sci. Technol.* **29**: 308A–323A.

Siddiqui, M., M.W. LeChevallier, J. Ban, T. Phillips, and J. Pivinski (1998) Occurrence of perchlorate and methyl tertiary butyl ether (MTBE) in groundwater of the American water system. American Water Works Service Company, Vorhees, New Jersey, September 30.

Susarla, S., S.T. Bacchus, S.C. McCutcheon, and N.L. Wolfe (1999) Potential species for phytoremediation of perchlorate. U.S. Environmental Protection Agency Report EPA/600/R-99/069. Athens, Georgia.

Tollenaar, H. and C. Martin (1972) Perchlorate in Chilean nitrate as the cause of leaf rugosity in soybean plants in Chile. *Phytopathology* **62**: 164–166.

van Ginkel, C.G., G.B. Rikken, A.G.M. Kroon, and S.W.M. Kengen (1996) Purification and characterization of chlorite dismutase: a novel oxygen-generating enzyme. *Arch. Microbiol.* **166**: 321–326.

van Wijk, D.J. and T.H. Hutchinson (1995) The ecotoxicity of chlorate to aquatic organisms: a critical review. *Ecotoxicol. Environ. Safe.* **32**: 244–253.

Wallace, W., S. Beshear, D. Williams, S. Hospadar, and M. Owens (1998) Perchlorate reduction by a mixed culture in an up-flow anaerobic fixed bed reactor. *J. Ind. Microbiol. Biotechnol.* **20**: 126–131.

Wolff, J. (1998) Perchlorate and the thyroid gland. *Pharmacol. Rev.* **50**(1): 89–105.

30

DATABASES AND PROTOCOL FOR PLANT AND MICROORGANISM SELECTION: HYDROCARBONS AND METALS

T. C. McIntyre

SUMMARY OF PRACTICAL IMPLICATIONS

Information on plants and other organisms that have the potential for phytoremediation of contaminants aids in the planning and establishment of removal and restoration projects for contaminated sites. Environment Canada is compiling databases on plants, bacteria, fungi, and other organisms that remediate petroleum hydrocarbons and accumulate metals and metalloids. The databases PHYTOPET (for PHYTOremediation of PETroleum hydrocarbon contaminants) and PHYTOREM (for PHYTOREmediation of Metals) are user-friendly, searchable, and summarize all the known literature. Environment Canada has also drafted a Protocol, "Environmental Technology Verification (ETV) Protocol for Metal Contaminants." The program provides suppliers of specified environmental technologies and services that meet the required criteria, with the right to use the licensed ETV verification mark.

PHYTOREMEDIATION APPLICATIONS

Phytoremediation has good potential as a flexible and cost-effective means to clean up contaminated sites. Two of the most important applications are the use of plants and other organisms to degrade petroleum hydrocarbons and accumulate metals and metalloids (hereafter the term metals covers both).

Phytoremediation: Transformation and Control of Contaminants,
Edited by Steven C. McCutcheon and Jerald L. Schnoor.
ISBN 0-471-39435-1 (cloth)

Plants and other organisms extract and degrade contaminants using a variety of processes such as:

1. Phytoextraction—the capability of terrestrial plants, algae, and plant tissues to sequester toxic elements, especially metals, by uptake or biosorption. Contaminated organisms can be harvested for disposal or extraction (biomining when at commercial levels) (Wehrheim and Wettern 1994, Brown *et al.* 1995, Kim *et al.* 1995, Wang *et al.* 1995, Wilson and Edyvean 1995, Gardea-Torresdey *et al.* 1996a).
2. Rhizofiltration—the use of plant roots to absorb and sometimes precipitate contaminants from polluted waters (*e.g.*, Dushenkov *et al.* 1995).
3. Phytostabilization—the use tolerant plants to stabilize contaminants by reducing bioavailability.
4. Phytodegradation—the use of plants and associated microorganisms to degrade organic pollutants.
5. Phytovolatilization—the use of plants to volatilize pollutants.

Information on useful phytoremediation species is limited and scattered in the literature. Environment Canada has organized information into two databases called PHYTOPET and PHYTOREM. The PHYTOPET database is a compilation and global inventory of organisms, with a focus on plants, for phytoremediation of petroleum hydrocarbons. PHYTOREM is a similar database for metals and metalloids.

Plants and other organisms have a remarkable capacity to accumulate contaminants, *e.g.*, the common sunflower (*Helianthus annuus*) (Salt *et al.* 1995), Indian mustard (*Brassica juncea*) (Watanabe 1997), alpine pennycress (*Thlaspi caerulescens*) (Baker *et al.* 1994, Comis 1995), and brake fern (*Pteris vittata*) (Ma *et al.* 2001). Some of the highest accumulation of metals occurs in the New Caledonia tree (*Sebertia acuminata*) (Brown 1995) and the Australian shrub (*Hybanthus floribundus*) (reported by Streit and Stumm 1993 as first reported by Harborne 1988). The purpose of this chapter is to describe the compilation of data for plants and other organisms that degrade hydrocarbons, or accumulate, hyperaccumulate, or tolerate metals.

PLANTS THAT TOLERATE AND SPUR DEGRADATION OF TOTAL PETROLEUM HYDROCARBONS

The database PHYTOPET specializes in plant species that spur the rhizodegradation of petroleum hydrocarbons in terrestrial and aquatic environments, including wetlands. The plants (both monocots and dicots), bacteria, protozoa, and fungi covered by the database show potential for phytoremediation of a wide range of hydrocarbons—from crude oil to polyaromatic hydrocarbons, such as chrysene and benzo(*a*)pyrene (Hutchinson *et al.* this book, Olson *et al.* this book). In the database, special attention has been paid to the Canadian plants from the oil-producing regions in the Western Prairie and Boreal ecozones for reclamation of oil-contaminated sites. The relationships between plants, microorganisms, and other species that can phytoremediate similar chemicals or mixtures of contaminants (metals, pesticides, and hydrocarbons) have also been considered.

The database uses many of the same fields as the PHYTOREM database for plants (exchanging information on the metal elements for information on the hydrocarbons). In addition, the database compiles information on plant-associated microorganisms. The information includes concentrations of the contaminant before and after treatment, length of treatment period, soil characteristics, age of the plant when exposed to the contaminant, and requirements for phytoremediation. Data in PHYTOPET also include tolerance to salinity and Western Canadian occurrence.

Many native grasses show promise to phytoremediate hydrocarbons [e.g., *Agropyron* spp., gramas (*Bouteloua* spp.), and buffalo grasses (*Buchloe* spp.)]. Poplar (*Populus* spp.) trees also are potentially useful. Cultivated plants with phytoremediation potential include carrot (*Daucus carota*), red fescue (*Festuca rubra*), alfalfa (*Medicago sativa*), and ryegrass (*Lolium* spp.).

PLANTS THAT ACCUMULATE METALS

Scope and Organization of the PHYTOREM Database

The global inventory of PHYTOREM covers both terrestrial and aquatic plants and other organisms (such as bryophytes, lichen, fungi, algae, and bacteria) that have potential value for phytoremediation of metals. Species included in the database have an ability or potential to tolerate, accumulate, or hyperaccumulate specific metals, or be useful as sorbents for metals.

An intensive literature search was conducted for the database, for accumulation of or tolerance to the 19 selected elements listed in Table 30-1. Reviews on metal accumulation provided global data (Atri 1983, Baker and Brooks 1989, Shaw 1990, Baker *et al.* 1991, Roberts and Proctor 1992, Markert 1993, Frankenberger and Losi in Skipper and Turco 1995, Reeves *et al.* 1996). Although a large number of species are known to grow over serpentine substrates, only those specifically shown to be hyperaccumulators or to accumulate substantial levels of heavy metals are included in the database.

TABLE 30-1 Metals and Metalloids Scanned for the Development of the PHYTOREM Database

Aluminum (Al)	Cobalt (Co)	Mercury (Hg)	Radium (Ra)
Arsenic (As)	Copper (Cu)	Molybdenum (Mo)	Strontium (Sr)
Beryllium (Be)	Chromium (Cr)	Nickel (Ni)	Uranium (U)
Cadmium (Cd)	Lead (Pb)	Palladium (Pd)	Zinc (Zn)
Cesium (Cs)	Manganese (Mn)	Platinum (Pt)	

The database contains information on organisms that are both hyperaccumulators and accumulators (Table 30-2). The term hyperaccumulator was originally coined for plants with greater than 1000 milligrams per kilogram (0.1 percent) nickel in dried aboveground tissues (Brooks *et al.* 1977, Reeves 1992). This definition has been broadened to other elements and all parts of plants. Values for accumulation were based on either literature, or a value of 100 or 200 milligrams per kilogram of dry weight was assigned where none has been specified.

Information in the database includes geographical origin of the organisms and habitat characteristics; taxonomy; environmental effects, health effects, and uses of the species; cultivation practices; sources of material or species studied; weedy or other significant relatives; and mode of action (*i.e.*, tolerance and accumulation). See Table 30-3 for details of criteria and fields. The database is worldwide in scope, but focuses on species from Canada, the U.S., and high-altitude species from warmer regions.

The database, compiled using *Microsoft Access 97*® software, allows for easy access by many database formats and programs and can be updated as needed. The database consists of three organism and two reference tables (Figure 30-1A). The organism tables are (1) vascular plants; (2) algae, lichens, fungi, and mosses; and (3) bacteria. The main table of reference includes all citations that relate to concentration levels of elements accumulated or tolerated, and the second is a table of general botanical and reference works consulted. The database as designed can be searched by queries (Figure 30-1B) and includes three data entry forms (botany, metal, and plant) and reporting formats for organisms and reference records.

Data sets can be quickly retrieved by applying appropriate search criteria, and viewed in tabular format. An example of a data screen for a species retrieved through the filtering process is given in Figure 30-2.

The database has specific search criteria, some of which are as follows:

1. To access the elements radium and uranium, use the codes Rd and Ur respectively. Searches for accumulators or hyperaccumulators, however, use commonly used short forms for these elements. For example, the character code RaA will bring up accumulation data for radium.
2. To retrieve all records of species accumulating (A), hyperaccumulating (H), tolerating (T), or precipitating (P) an element, use the symbol for the element surrounded by asterisk in the elem_action field, *e.g.*, "*Pb*" as the query or filter criterion. To retrieve species accumulating or hyperaccumulating a particular element, the search criterion should be a combination of the element and the desired accumulation code, *e.g.*, "*PbH*" for lead hyperaccumulators.
3. There is no separate field to distinguish coniferous species; however, species can be retrieved by using the family Pinaceae or by retrieving all records with a growth form of TR for tree.

TABLE 30-2 Values and Sources for Hyperaccumulation and Accumulation Criteria for Inclusion in the PHYTOREM Database

Element	Hyperaccumulation criteria (milligrams per kilogram of dry weight)	Plants, where appropriate, which were the basis of criteria	References for data	Accumulation criteria (milligrams per kilogram of dry weight)	Number of records in the database	Species with highest recorded value {origin} [value (milligrams per kilogram of dry weight)]
Al	1000	Barley (*Hordeum vulgare*), horse bean (*Vicia faba*)	Grauer and Horst (1990)	100	25	Hairy goldenrod (*Solidago hispida*) {Canada} [6820]
As	1000		Porter and Petersen (1975)	100	4	Colonial bent (*Agrostis tenuis* = *Agrostis capillaris*) {cultivation} [2000]
Cd	100		Brown *et al.* (1995)	10	37	Eel grass (*Vallisneria spiralis*) {India} [6242]
Co	1000		Baker and Brooks (1989)	100	27	*Haumaniastrum robertii* {Africa} [10 200]
Cr	1000	General review	Baker and Brooks (1989)	100	35	Alfalfa (*Medicago sativa*) {cultivation} [7700]
		Dicoma niccolifera, Sutera fodina	Wild (1974), Brooks and Yang (1984)			
		Alfalfa (*Medicago sativa*)	Tiemann *et al.* (1997)			
		Smooth Water hyssop (*Bacopa monnieri*) (aquatic species)	Gupta *et al.* (1994)			
		Azolla spp.	Priel (1995)			
		Duckweed (*Spirodela polyrhiza*), kariba weed (*Salvinia molesta*)	Srivastav *et al.* (1994)			
		Water lettuce (*Pistia stratiotes*)	Sen *et al* (1987)			
Cs	?			?	1	Sunflower (*Helianthus annuus*) {cultivation} [high absorption rate]

(*continues*)

TABLE 30-2 (*Continued*)

Element	Hyperaccumulation criteria (milligrams per kilogram of dry weight)	Plants, where appropriate, which were the basis of criteria	References for data	Accumulation criteria (milligrams per kilogram of dry weight)	Number of records in the database	Species with highest recorded value {origin} [value (milligrams per kilogram of dry weight)]
Cu	1000		Baker and Brooks (1989)	100	67	Creosote bush (*Larrea tridentata*) {U.S.} [23 700 biosorption]
Hg	1000		Atri (1983), Baker and Brooks (1989)	100	35	Water lettuce (*Pistia stratiotes*) {pantropical} [1100]
Mn	10 000		Baker and Walker (1990)	200	28	*Macademia neurophylla* {New Caledonia} [51 800]
Mo	1500	Alpine pennycress (*Thlaspi caerulescens*)		?	1	Alpine pennycrest (*Thlaspi caerulescens*) {Europe} [1500–1800]
Ni	1000		Brooks *et al.* (1977) Reeves (1992)	100	372	*Psychotria douarrei* {New Caledonia} [47 500] Shrub violet (*Hybanthus floribundus*) {Australia} [2% total ash content]
Pb	1000		Baker and Brooks (1989)	200	79	Indian mustard (*Brassica juncea*) {cultivation} [26 200]
Sr	?			?	1	Sunflower (*Helianthus annuus*) {cultivation} [high absorption Rate]
Ur	?			?	3	Sunflower (*Helianthus annuus*) {cultivation} [>15 000]
Zn	10 000		Baker and Brooks (1989)	500	48	Alpine pennycrest (*Thlaspi caerulescens*) {Europe} [52 000]

? indicates that criteria have not been defined for hyperaccumulation or accumulation of these elements.

Note: No reports were found for accumulation of beryllium, palladium, platinum, and radium.

TABLE 30-3 Fields, Description, and Access for PHYTOREM Database

Name	Description [and explanatory comments]	Plants	Bacteria	Algae	Access
Entry_seq	Automatic counter field to record sequence of data entry [allows multiple records taken from a single summary document to be grouped together for purposes of review and data management]	✔	✔	✔	
Type_org	Vascular plant = VP, bacteria = BA, algae = AG, lichens = LI, fungi = FU, and bryophytes = BR [field allows tables of different organisms to be combined and provides the ability to distinguish records by type of organism]	✔	✔	✔	
Growth_form	For plants only: fern = FE, graminoid = GR, herb = HE, shrub = SH, succulent = SU, vine = V, and tree = T	✔			
Sci_name	Complete scientific name with authorities	✔	✔	✔	✔
Synonym	Common synonym	✔		✔	
Com_name	Common English name(s)	✔	✔	✔	
Cv_strain	Cultivar or strain name, or code; also transgenic variants	✔	✔	✔	
Family	Taxonomic category: family	✔		✔	✔
Order	Taxonomic category: order	✔		✔	✔
Subclass	Taxonomic category: subclass	✔		✔	✔
Class	Taxonomic category: class	✔		✔	✔
Duration	For plants only: A(nnual), B(iennial), A/B, P(erennial), A/P, and B/P	✔			
Origin	Country or region of origin of plants on which report(s) is based [Canadian and U.S. species or those of other continents or countries can be sorted]	✔	✔	✔	✔
World_range	World range of the species	✔			

(*continues*)

TABLE 30-3 (*Continued*)

Name	Description [and explanatory comments]	Plants	Bacteria	Algae	Access
Primary_habitat	T = terrestrial, A = aquatic, T/A = terrestrial and aquatic [provides the ability to distinguish between terrestrial, aquatic, marsh, and wetland species (T/ A)]	✔		✔	✔
Hab_descr	Habitat description [allows for sorting to distinguish plants from serpentine soils and to distinguish cultivated species (crop plants)]	✔			
Namer_occ	North American occurrence: N(ative); E(xotic); X (not present); ? (status unknown) [allows North American species to be categorized]	✔			
Namer_spp	Indication of whether other species in the genus are present in North America	✔			
Sig_relatives	Other species of significance in the genus	✔			
Cult_wild	For plants: crop plant = C, horticultural species = H, wild = W. For algae and other: cultured = C or wild = W	✔		✔	✔
Cult_info	Propagation and/or test studies, or experimental conditions	✔		✔	
Cult_source	Source of cultivated material as indicated in publication	✔	✔	✔	
Impact_attributes	UN = unknown, WD = weed, NX = noxious properties, HY = hybridizes, and PH = disease and insect pest host	✔			
Impact_description	Documentation of potential impact [information on invasive potential of an exotic weed]	✔			
Uses	General uses of plant and citations [medicinal or edible]	✔			
Bio_notes	For plants: notes on such topics as pollination and dispersal mechanism. For bacteria: notes on important and interesting aspects of biology.	✔	✔		
Gen_notes	For plants and others: toxicity of metal to the plant or other organisms and other pertinent information and citations. For bacteria: general notes on procedures of the study or results and citations.	✔	✔	✔	

Elem_conc_ref	Citation and concentration of elements with organ of storage	✔	✔	✔	✔
Elem_form_ref	Element form if specifically indicated [*e.g.*, Cu(II)]	✔	✔	✔	
Elem_chel_ref	Element, chelate used and citation	✔	✔	✔	
Elem_action	Hyperaccumulate = H, accumulate = A, tolerate = T, rhizosphere concentration = R, and precipitate = P [*e.g.*, PbH is lead hyperaccumulation]	✔	✔	✔	✔
Storage_sites	For plants: sites where element concentrations were measured [*e.g.*, root, shoot, or leaf]. For bacteria: sites where element measured [*e.g.*, cell wall].	✔	✔		
Biofiltr_use	Use for biofiltration: yes (Y) or no (N) [used to distinguish species useful for such a function as indicated in the publication]	✔	✔	✔	
Biofiltr_ref	References to the biofiltration publications [generally the same as the primary reference]	✔	✔	✔	
Tolerance_info	Information on conditions under which tolerance occurs	✔		✔	
Summary_ref	Reference to summary papers where several to many species are listed	✔	✔	✔	✔
Primary_ref	Primary literature reference [obtained from summary reference if summary document was used]	✔	✔	✔	✔

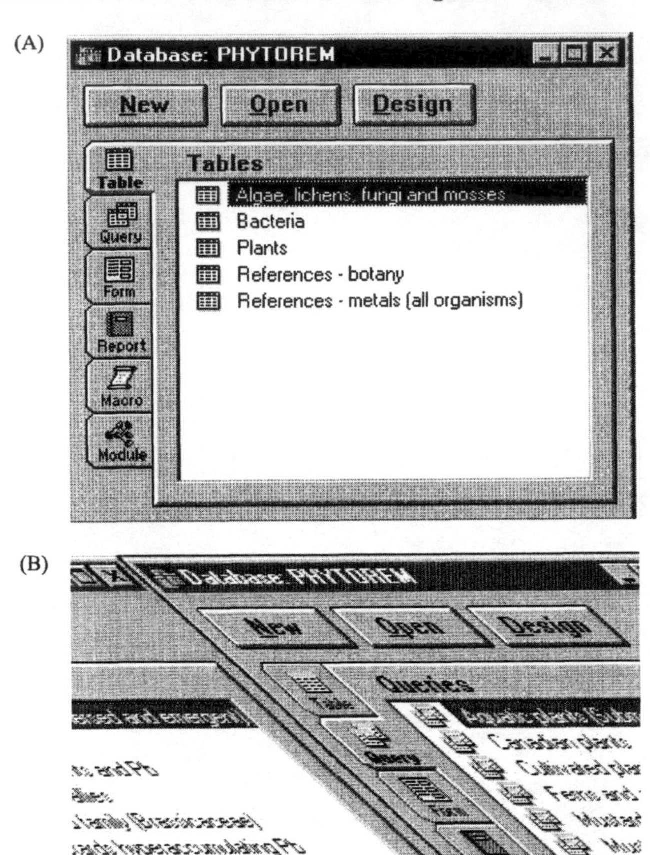

Figure 30-1 PHYTOREM database windows showing contents from the tables (A) and query (B) menus.

4. Species are reported as rhizofiltration agents for heavy metals. Studies dealing with nutrient culture experiments, although not aimed at identifying such agents, have been recorded with a Y for "yes" in the field biofiltr_use, because of the potential for such use.

Summary of the Database Contents

The database provides an extensive, though preliminary, collection of records on phytoremediation potential. At present, the database contains records of

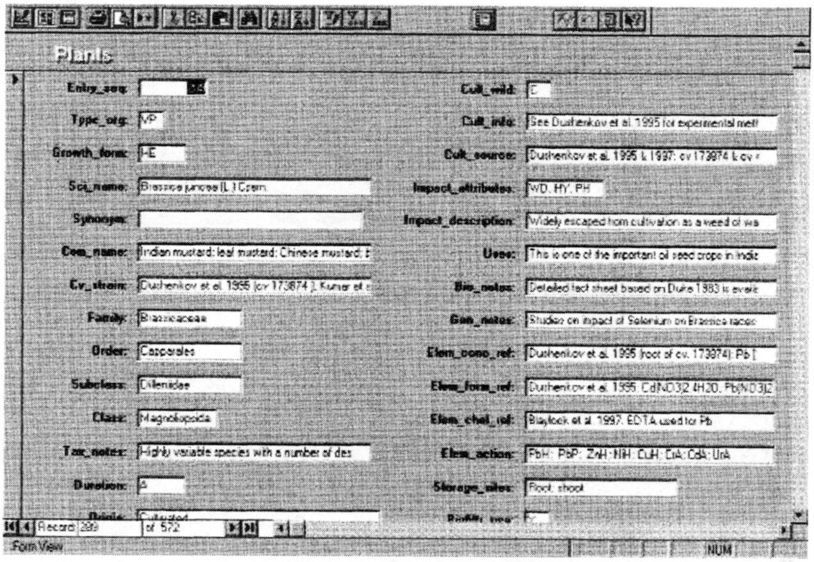

Figure 30-2 Example of a plant report form from a search of the PHYTOREM database.

775 species spread over 76 families, 39 orders, 9 subclasses, and 2 classes (dicots and monocots). Data come from 39 countries of origin (based on names used in the publications).

The greatest proportion (465) of species recorded in the plant table accumulate, hyperaccumulate, or tolerate a single metal of the 19 elements surveyed. Some species take up two elements (66) or three (25), with a balance of 15 species capable of accumulating four or more elements. Table 30-4 summarizes the species in the latter category with elements and degree of accumulation.

The highest number of records (260) was found for nickel followed by lead and copper (Figure 30-3). No records were found of species accumulating beryllium, palladium, platinum, or radium. Species with the highest concentration of the elements recorded in the database (listed in the right-hand column of Table 30-2) provide some interesting possibilities for phytoremediation, including several commonly cultivated species such as sunflower (*Helianthus annuus*) and alfalfa (*Medicago sativa*).

Data Update Recommendations

Updates that could be considered include (1) information from a more extensive literature search, (2) fields for data on storage sites (*e.g.*, roots), (3) mechanisms of tolerance or accumulation of metals by plants, (4) species that remediate radionuclides, (5) plants that remediate organic and other inorganic compounds and stabilize contaminated sites, and (6) expansion of data for

TABLE 30-4 Plants Capable of Accumulating Four or More Metals

Scientific name	Common name	Origin and characteristics	Elements and degree of accumulation[a]
Azolla filiculoides	Water fern	Africa, floating	CuA, NiA, PbA, and MnA
Bacopa monnieri	Water hyssop	India, emergent species	HgA, CuH, CrH, PbA, and CdH
Eichhornia crassipes	Water hyacinth	Pantropical/subtropical, troublesome weed	CdH, CrA, ZnA, HgH, PbH, and CuA
Hydrilla verticillata	Hydrilla	Southern Asia but introduced and spreading as a troublesome weed in the warmer states of the USA	CdH, CrA, HgH, and PbH
Lemna minor	Duckweed	Native to North America and widespread	PbH, CdH, CuH, and ZnA
Pistia stratiotes	Water lettuce	Pantropical and native to southern U.S. but an aquatic weed	CuT, CdT, HgH, and CrH
Salvinia molesta	Water fern	India	CrH, NiH, PbH, and ZnA
Spirodela polyrhiza	Giant duckweed	Native to North America	CdH, NiH, CrH, PbH, and ZnA
Vallisneria americana	Tape grass	Native to Europe and North Africa but widely cultivated in the aquarium trade	CuH, CdH, CrA, and PbH
Brassica juncea	Indian mustard	Cultivated	PbH, PbP, ZnH, NiH, CuH, CrA, CdA, and UrA
Helianthus annuus	Sunflower	Cultivated	PbH, UrH, SrH, CsH, CrA, CdA, CuA, MnA, NiA, and ZnA
Agrostis castellana	Bent grass	Portugal	AsH, PbA, ZnA, MnA, and AlA
Thlaspi caerulescens	Alpine pennycress	Europe	ZnH, CdH, CoH, CuH, NiH, PbH, and CrA
Athyrium yokoscense	Fern	Japan	CuH, CdA, ZnH, and PbH

[a]H: hyperaccumulator; A: accumulator; P: precipitator; and T: tolerant.

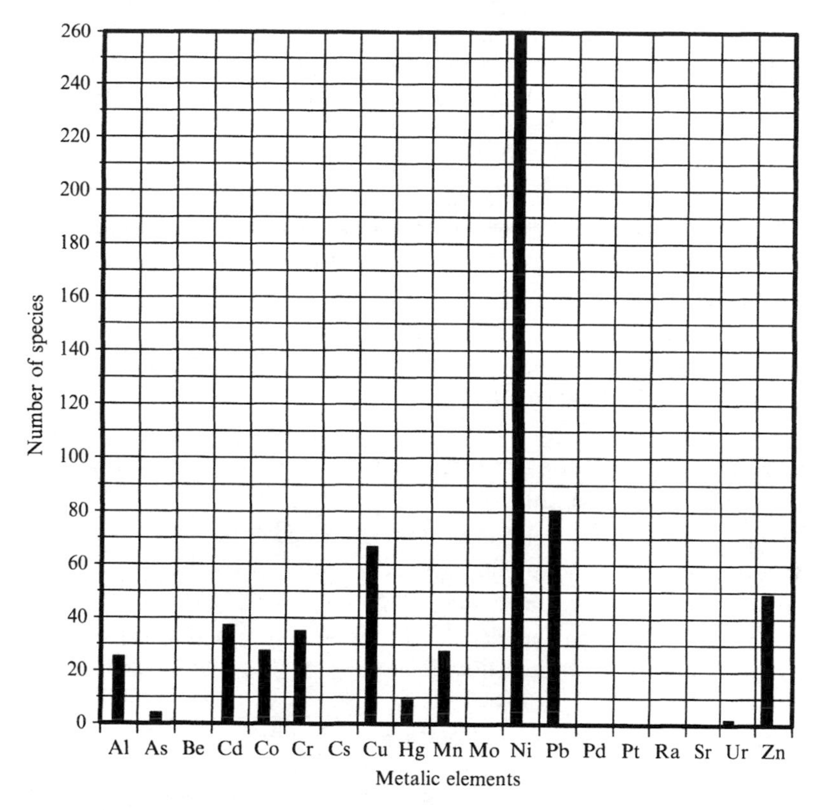

Figure 30-3 Number of species records for the 19 elements scanned.

algae, in particular those with a high absorptive capacity. In addition, the database could also provide pictures of plants and sites.

INORGANIC PHYTOREMEDIATION VERIFICATION PROTOCOL

Environment Canada is formulating a protocol to establish the limits of applicability for the phytoremediation of metals and metalloids in contaminated soil and water that will aid economic adoption and sustainable use. The pollutants (no organic contaminants to this point) are arsenic (As), cadmium (Cd), chromium (Cr), copper (Cu), lead (Pb), mercury (Hg), nickel (Ni), selenium (Se), silver (Ag), and zinc (Zn). Use of the verification protocol will allow the use of a Canadian Environmental Technology Verification symbol in remediation products. The applicant will be required to establish that a particular application has been tested (and under what conditions) (*e.g.*, proof of principle at the bench, greenhouse, plot, or full-scale testing). Only applications that involve phytoextraction, rhizofiltration, phytovolatilization, phytostabilization, and combinations of these physiological processes will be certified initially. All phytoremediation applications (or technologies) will be evaluated based on the following criteria and information:

1. Total amount, concentration, and speciation of elements in contaminated soil, rhizosphere soil water, water below the root zone, groundwater, wastewater, roots, shoots, and air (where phytovolatilization is a factor) during and after treatment
2. Amount and timing of supplements added to treated soils and waters as related to mobilization or immobilization of contaminants, including co-contaminants, and expected effects
3. Seepage or infiltration volumes from treatment wetlands or soil plots and concentrations of contaminants and by-products for the range of expected moisture conditions
4. Range of meteorological conditions for effective phytoremediation application including, but not limited to, seasonal precipitation, air temperatures, humidity, wind speed and direction, and solar radiation
5. Time required to achieve cleanup or risk-reduction standards for various levels of initial contamination in soil or water, plants selected, soil or sediment conditions, and weather
6. Soil, sediment, or water quality during and after the treatment, including texture, mineralogy, moisture content, cation-exchange capacity for soils, organic carbon and nutrient content, pH, and redox potential for all media
7. List of plants and description of communities and any rooting depth, growing habits, and foliage requirements for effective cleanup
8. Unusual ecological conditions to avoid including endangered and threatened plants and wildlife, weeds, invasive species, bioaccumulation (in the absence of netting, fencing, motion detectors, or alarms), and wetland changes (*e.g.*, drying up of seepage wetlands)
9. Risk-reduction or cleanup levels achieved in soil, groundwater, and air for different land uses
10. Residual management options (*e.g.*, land filling in hazardous waste landfills, composting, and pyrolysis) and any risks during handling, transport, and disposal of residuals
11. Closure plans and options for vegetative cover

These criteria will be focused on specific issues for terrestrial and aquatic remediation. Criteria and information requirements will be weighted, depending on the application. This protocol is different from that used in the U.S. (RTDF 2000, ITRC 2001); however, the differences are expected to diminish over time with continued collaborative projects between Environment Canada and the U.S. Environmental Protection Agency.

CURRENT AND FUTURE EXPECTATIONS

Plant-based remediation and restoration applications are expected to receive increasing attention in Canada and elsewhere as options for removal of the more recalcitrant inorganic and organic constituents at contaminated sites. These Environment Canada databases and protocol could provide valuable assistance in the pre-selection of organisms, help clarify the regulatory responsibilities of proponents in securing necessary approvals prior to the use of specific organisms, and provide valuable insight into genetic materials to adapt or modify representative Canadian plant species to the climatic circumstances unique to specific sites. Future initiatives include (1) database expansion to include one for radioactive materials (PHYTORAD); (2) a Canadian network to identify, store, protect, propagate, and possibly distribute germplasm of candidate plant cultivars; (3) protocols that outline responsible provisions; (4) methods to recover inorganic contaminants from contaminated biomass; (5) a complete overview of the regulatory issues inherent in the use of plants as remediation and restoration agents; (6) definition of bioavailability and the role of plants in natural attenuation; and (7) establishment of mechanisms to limit phytoremediation site access by herbivores and omnivores.

Acknowledgments and Disclaimer

Acknowledgements
The author is extremely grateful for the opportunity to realize the conceptual development of PHYTOREM and PHYTOPET databases through the funding of the Canadian Biotechnology Strategy and the Panel on Energy Research and Development. Data were compiled and provided by Dr. Erich Haber of National Botanical Services in Canada for PHYTOREM and by Dr. J.J. Germida, Dr. R.E. Farrell, and Dr. C.M. Frick of the University of Saskatchewan for PHYTOPET. Environment Canada headquarter's library staff, Jean-François Belanger and Marie Jetten and the Canadian Museum of Nature librarian Mireille Boudreau were invaluable. Finally the author extends an incredible debt of gratitude to Dr. Marcos Alvarez of Environment Canada for his editorial support. Copies of the PHYTOREM and PHYTOPET databases are available on CD-ROM from the author at terry.mcintyre@ec.gc.ca.

Disclaimer
Environment Canada does not endorse the use of, nor does it attempt to determine the merits of any specific technology or technology provider through publication in this document, nor does Environment Canada assume any liabilities with respect to the use of any information, apparatus, method, or process discussed in this document. Mention of trade names or commercial products does not constitute endorsement or recommendation for use. Further, the views expressed in the document remain solely those of the authors.

REFERENCES

Atri, F.R. (1983) *Schwermetalle und Wasserpflanzen*. Gustav Fischer Verlag, Stuttgart, Germany.

Azadpour, A. and J.E. Matthews (1996) Remediation of metal-contaminated sites using plants. *Remediation* Summer: 1–18.

Baker, A.J.M. and R.R. Brooks (1989) Terrestrial higher plants which hyperaccumulate metallic elements—a review of their distribution, ecology and phytochemistry. *Biorecovery* **1**: 81–126.

Baker, A.J.M. and P.L. Walker (1990) Ecophysiology of metal uptake by tolerant plants. In: *Heavy Metal Tolerance in Plants: Evolutionary Aspects*. A.J. Shaw, ed. CRC Press, Boca Raton, Florida, pp. 155–177.

Baker, A.J.M., J. Proctor, and R.D. Reeves (1991) The vegetation of ultramafic (serpentine) soils. In: *Proceedings of the First International Conference on Serpentine Ecology*. A.J.M. Baker, J. Proctor, and R.D. Reeves, eds. Intercept Ltd., Andover, United Kingdom. Held at University of California, Davis, June 19–22.

Baker, A.J.M., R.D. Reeves, and A.S.M. Hajar (1994) Heavy metal accumulation and tolerance in British populations of the metallophyte *Thlaspi caerulescens* J. & C. Presl (Brassicaceae) *New Phytol.* **127**: 61–68.

Black, H. (1995) Absorbing possibilities: phytoremediation. *Environ. Health Persp.* **103**: 1106–1108.

Brooks, R.R. and X.H. Yang (1984) Elemental levels and relationships in the endemic serpentine flora of the Great Dyke, Zimbabwe, and their significance as controlling factors for this flora. *Taxon* **33**: 392–399.

Brooks, R.R., J. Lee, R.D. Reeves, and T. Jaffré (1977) Detection of nickeliferous rocks by analysis of herbarium specimens of indicator plants. *J. Geochem. Explor.* **7**: 49–77.

Brown, K.S. (1995) The green clean. *BioScience* **45**: 579–582.

Brown, S.L., R.L. Chaney, J.S. Angle, and A.J.M. Baker (1995) Zinc and cadmium uptake by hyperaccumulator *Thlaspi caerulescens* and metal tolerant *Silene vulgaris* grown on sludge-amended soils. *Environ. Sci. Technol.* **29**: 1581–1585.

Coghlan, A. (1997) Plants gobble up waste explosives. *New Scientist* 1 February: p. 19.

Comis, D. (1995) Green remediation: using plants to clean the soil. *J. Soil Water Conserv.* May–June: 184–187.

Cunningham, S.D. and W.R Berti (1993) Remediation of contaminated soils with green plants: an overview. *In Vitro Cell. Dev. Biol.* **29P**: 207–212.

Cunningham, S.D. and C.R. Lee (1995) Phytoremediation: plant-based remediation of contaminated soils and sediments. In: *Bioremediation: Science and Applications*. H.D. Skipper and R.F. Turco, eds. Soil Science of America Special Publication **43**: 145–156.

Dushenkov, V., N.P.B.A. Kumar, H. Motto, and I. Raskin (1995) Rhizofiltration: the use of plants to remove heavy metals from aqueous streams. *Environ. Sci. Technol.* **29**: 1239–1245.

Frankenberger, W.T., Jr. and M.E. Losi (1995) Applications of bioremediation in the cleanup of heavy metals and metalloids. In: *Bioremediation: Science and Applica-*

tions. H.D. Skipper and R.F. Turco, eds. Soil Science of America Special Publication **43**: 173–210.

Gardea-Torresdey, J.L., J. Bibb, K.J. Tiemann, J.H. Gonzalez, and J. Arenas (1996a) Adsorption of copper ions from solution by heavy metal stressed *Larrea tridentata* (creosote bush) biomass. In: *Proceedings of the HSRC/WERC Joint Conference on the Environment*. L.E. Erickson, D.L. Tillison, S.C. Grant, and J.P. McDonald, eds. Great Plains/Rocky Mountain Hazardous Substance Research Center, Kansas State University, Manhattan. Held in Albuquerque, New Mexico, May 21–23. (http://www.engg.ksu.edu/HSRC/ 96proceed/gardea1.html).

Gardea-Torresdey, J.L., L. Tang, and J.M. Salvador (1996b) Copper adsorption by esterified and unesterified fractions of Sphagnum peat moss and its different humic substances. *J. Hazard. Mat.* **48**: 191–206.

Grauer, U.E. and W.J. Horst (1990) Effect of pH and nitrogen source on aluminium tolerance of rye (*Secale cereale* L.) and yellow lupin (*Lupinus luteus* L.). *Plant Soil* **127**: 13–21.

Gupta, M., S. Sinha, and P. Chandra (1994) Uptake and toxicity of metals in *Scirpus lacustris* L. and *Bacopa monnieri* L. *J. Environ. Sci. Health* **A29**: 2185–2202.

Harborne, J.B. (1988) *Introduction to Ecological Biochemistry*. 3rd ed. Academic Press, London, 356 pp.

Hartman, W.J., Jr. (1975) An evaluation of land treatment of municipal wastewater and physical siting of facility installations, U.S. Department of the Army, Washington, D.C.

Interstate Technology and Regulatory Council (ITRC) (2001) Phytotechnology technical and regulatory guidance document. Phytotechnology Team. Washington, D.C. (http://www.itrcweb.org).

Kim, I. (1996) Harnessing the green clean. *Chem. Eng*. December: 39–41.

Kim, Y.H., Y.J. Yoo, and H.Y. Lee (1995) Characteristics of lead adsorption. *Biotechnol. Lett*. **17**: 345–350.

Ma, L.Q., K.M. Komar, and C. Tu (2001) A fern that accumulates arsenic. *Nature* **409**: p. 579.

Markert, B. (1993) *Plants as Biomonitors: Indicators for Heavy Metals in the Terrestrial Environment*. VCH Publishing Group (Wiley-VCH Verlag), New York.

Moffat, A.S. (1995) Plants proving their worth in toxic metal cleanup. *Science* **269**: 302–303.

Parry, J. (1995) Plants absorb heavy metals: nature provides a low-cost alternative to chemical extraction or immobilization. *Pollut. Eng*. February: 40–41.

Porter, E.K. and P.J. Peterson (1975) Arsenic accumulation by plants on mine waste (United Kingdom). *Sci. Total Environ*. **4**: 365–371.

Priel, A. (1995) Purification of industrial wastewater with the Azolla fern. *World Water Environ. Eng*. **18**: p. 13.

Reeves, R.D. (1992) The hyperaccumulation of nickel by serpentine plants. In: *The Vegetation of Ultramafic (Serpentine) Soils*. A.J.M. Baker, J. Proctor, and R.D. Reeves, eds. Intercept Ltd., Andover, U.K., pp. 253–277.

Reeves, R.D., A.J.M. Baker, A. Borhidi, and R. Berazaín (1996) Nickel-accumulating plants from the ancient serpentine soils of Cuba. *New Phytol*. **133**: 217–224.

Remediation Technology Demonstration Forum (RTDF) (2000) A database of existing phytoremediation field sites (http://www.rtdf.org/public/phyto/siteprof/usersearch/phyto_list.cfm).

Roberts, B.A. and J. Proctor, eds. (1992) *The Ecology of Areas with Serpentinized Rocks: a World View*. Kluwer Academic Publishers, Dordrecht, The Netherlands.

Rouhi, A.M. (1997) Plants to the rescue: crops with big appetites for certain elements could be key players in soil cleanup. *Chem. Eng. News* **75**(2): 21–23.

Salt, D.E., M. Blaylock, N.P.B.A. Kumar, V. Dushenkov, B.D. Ensley, I. Chet, and I. Raskin (1995) Phytoremediation: a novel strategy for the removal of toxic metals from the environment using plants. *Biotechnology* **13**: 468–474.

Sen, A.K., N.G. Mondal, and S. Mandal (1987) Studies of uptake and toxic effects of Cr (VI) on *Pistia stratioides*. *Water Sci. Technol.* **19**: 119–127.

Shaw, A.J. ed. (1990) *Heavy Metal Tolerance in Plants: Evolutionary Aspects*. CRC Press, Boca Raton, Florida.

Srivastav, R.K., S.K. Gupta, K.D.P. Nigam, and P. Vasudevan (1994) Treatment of chromium and nickel in wastewater by using aquatic plants. *Water Res.* **28**: 1631–1638.

Streit, B. and W. Stumm (1993) Chemical properties of metals and the process of bioaccumulation in terrestrial plants. In: *Plants as Biomonitors: Indicators for Heavy Metals in the Terrestrial Environment*. B. Markert, ed. VCH Publishing Group (Wiley-VCH Verlag), New York, pp. 31–62.

Tiemann, K.J., J.L. Gardea-Torresdey, G. Gamez, O. Rodriquez, and S. Sias (1997) Study of the ligands involved in metal binding to alfalfa biomass. Abstract 9. In: Abstracts From The 1997 Conference on Hazardous Waste Research, L.E. Erickson, M.M. Rankin, S.C. Grant, and J.P. McDonald, eds. Great Plains/Rocky Mountain Hazardous Substance Research Center, Kansas State University, Manhattan, Kansas. Held in Kansas City, Missouri, May 20–22, (http://www.engg.ksu.edu/HSRC/ 97abstracts/doc9.html).

Wang, T.C., J.C. Weissman, G. Ramesh, R. Varadarajan, and J.R. Benemann (1995) Bioremoval of toxic elements with aquatic plants and algae. In: *Bioremediation of Inorganics*. R.E. Hinchee, J.L. Means, and D.R. Burris, eds. Third International *In Situ* and On-Site Bioreclamation Symposium. Battelle Press, Columbus, Ohio. Held in San Diego, California, April 24–27, pp. 65–69.

Watanabe, M.E. (1997) Phytoremediation on the brink of commercialization. *Environ. Sci. Technol.* **31**: 182–186.

Wehrheim, B. and M. Wettern (1994) Biosorption of cadmium, copper and lead by isolated mother cell walls and whole cells of *Chlorella fusca*. *Appl. Microbio. Biotechnol.* **41**: 725–728.

Wild, H. (1974) Indigenous plants and chromium in Rhodesia. *Kirkia* **9**: 233–241.

Wilson, M.W. and R.G. Edyvean (1995) Biosorption for the removal of heavy metals from industrial wastewaters. In: *Biodeterioration and Biodegradation* 9. A. Bousher and M. Chandra, eds. Proceedings of the 9th International Biodeterioration and Biodegradation Symposium. Leeds, U.K., Sept. 1993, pp. 47–51.

31

FIELD EVALUATIONS OF PHYTOTECHNOLOGIES

S. A. Rock

SUMMARY OF PRACTICAL IMPLICATIONS

Phytotechnologies are a set of innovative cleanup methods for contaminated soils and water that are slow to develop, low in energy requirements, and to a certain degree self-maintaining. With only a limited performance record, the advantages and weaknesses are difficult to define presently. There is a certain and growing place in the remediator's toolbox for innovative plant-based systems, but each application must be considered carefully. While various phytotechnologies are at different stages of public and regulatory acceptance, most regulators still require local field trials to determine if a particular technology will be effective on a specific site.

This chapter summarizes the field trials and evaluations that the U.S. Environmental Protection Agency (U.S. EPA) National Risk Management Research Laboratory has observed and analyzed. Various programs sponsored these evaluations to encourage the development and appropriate use of innovative technology. Several of the evaluations were pilot tests to determine if the proposed phytotechnology would be appropriate for a particular site. These evaluations and field tests provide the framework or protocol for future demonstrations, help determine whether a technology may be applicable for any site, and set reasonable expectations for the technologies. This chapter is not meant as guidance, or to interpret federal and state laws, or to set precedents or policies.

INTRODUCTION

Whether any technology can and should be applied to any given site is based on the site conditions, the cleanup goals, and relevant regulations. Plant-based technologies are sensitive to climate, elevation, and soil quality, as well as to the properties of the contaminants and media. These considerations

Phytoremediation: Transformation and Control of Contaminants,
Edited by Steven C. McCutcheon and Jerald L. Schnoor.
ISBN 0-471-39435-1 (cloth)

must be clearly defined and understood by site owners, contractors, regulators, and stakeholders before any remediation technology can be implemented, and especially so for an innovative phytotechnology.

In long-term remediation projects, many opportunities arise to gather data for later analysis. This matching of remediation and research is sometimes serendipitous, sometimes deliberate. Because some phytotechnologies take a long time to establish a trend, some regulatory agencies allow full-scale evaluations before final regulatory acceptance. The information garnered from these long-term, site-specific field evaluations can thus be used to achieve similar performance on sites with comparable climate, soil, and contaminant conditions.

The same evaluation and selection standards apply to phytotechnologies as any remediation technique; namely, can the technology meet the relevant cleanup standards? How do we know? Despite the application of the same criteria, both federal and state agencies lack specific phytotechnology selection and evaluation guidance. There has simply not been enough evidence from enough locations to justify widespread acceptance or denial of any technique.

In an attempt to help state and federal regulators evaluate the likelihood of success for a given phytotechnology proposed for use on a specific site, the U.S. EPA Office of Research and Development sponsors research and conducts field evaluations of innovative technologies. The results of such evaluations are often applicable to conditions at similar sites, the eventual goal being to gain a nationwide understanding of the opportunities and limitations of the various phytotechnologies, leading to swifter and surer regulatory decisions.

SUPERFUND INNOVATIVE TECHNOLOGY EVALUATION PROGRAM

The U.S. EPA Office of Solid Waste and Emergency Response and Office of Research and Development established the Superfund Innovative Technology Evaluation (SITE) Program in response to the Superfund Amendments and Reauthorization Act of 1986. The primary purpose of the SITE Program is to promote the development, evaluation or demonstration, and use of innovative technologies to clean up Superfund sites across the country. Once accepted into the SITE Program, one or more innovative technologies are chosen for evaluation based on the potential applicability of each technology to remediate the site. Developers of the technology are responsible for operating innovative systems on the site and for paying the costs of the evaluation. The U.S. EPA is responsible for project planning, sampling and analysis, quality assurance and quality control, preparing reports, disseminating information, and transporting and disposing of treated waste materials.

McCormick and Baxter: Pentachlorophenol and Polycyclic Aromatic Hydrocarbons in Soil

The first SITE phytoremediation project took place at McCormick and Baxter, a former wood treatment site in Portland, Oregon. The wood preservatives pentachlorophenol and creosote with polycyclic aromatic hydrocarbons contaminated shallow soil. A perennial grass which grows well in the Northwest was tested for survival and contaminant degradation in both greenhouse and field plots. Phytokinetics, Inc., of Logan, Utah, conducted this evaluation. This technology was accepted into the Emerging Technology Program in 1994. The study consisted of two phases: the first phase was a 180-day greenhouse study; the second phase took place at the site. Perennial ryegrass (*Lolium perenne)* was used for both phases.

For both phases, contaminant removal for planted, unplanted, and undisturbed soils were compared. Planted soils caused faster removal of pentachlorophenol. From an initial concentration of 162 milligrams per kilogram, the 60-day results showed that an undisturbed soil remains largely unchanged (148 milligrams per kilogram). The nutrient-supplemented, but unplanted, soil (microbial degradation or land farming) lowered concentrations to 91 milligrams per kilogram, while the planted soil dropped to 62 milligrams per kilogram. Polycyclicaromatic hydrocarbon concentrations showed a similar trend (Ferro *et al.* 1999). By the sixth month, the planted soil tested below the pentachlorophenol regulatory level of 50 milligrams per kilogram for this site. At the end of the evaluation in the eleventh month, neither of the unplanted controls had reached 50 milligrams per kilogram. A regulatory agency evaluating this technology for a similar site could approve such a proposed treatment based on this evidence, although treatability tests would be prudent.

Carswell: Trichloroethene in Shallow Groundwater

The first SITE phytotechnology project for groundwater control took place at the Carswell Golf Club at the U.S. Naval Air Station, Fort Worth (Texas), which is adjacent to Air Force Plant 4. An aircraft assembly building is the suspected source of trichloroethene at the site. In April 1996, approximately 660 trees were planted in two plots at the site in an effort to intercept, contain, and degrade the trichloroethene plume in the groundwater, as described by Eberts *et al.* (this book).

Constructed in 1942, Plant 4 currently produces F-16 aircraft, radar units, and various aircraft and missile components. Since 1953, Plant 4 has produced B-36, B-58, and F-111 aircraft. Historically, the manufacturing processes at Plant 4 have generated an estimated 5000 to 5400 metric tons (5500 to 6000 tons) of waste per year, including waste solvents, oils, fuels, paint residues, and miscellaneous spent chemicals. Throughout most of the history of Plant 4, the waste oil, solvents, and fuels were disposed of at on-site landfills, or were burned in fire-training exercises. Plant 4 is on the National

Priorities List and is being cleaned up in accordance with the Comprehensive Environmental Response, Compensation, and Liability Act as amended by the Superfund Amendments and Reauthorization Act.

The reduction in groundwater volumetric flux was approximately 12 percent by the peak of the third growing season and was not considered to significantly control the down gradient migration of contaminants. The maximum water level drawdown at the center of the system during the peak of the third growing season was approximately 0.1 meters (0.3 feet). Transpiration and rooting extent data support the water level data. To date, up to approximately one third of the plume has been controlled by this treatment. Modeling performed by the U.S. Geological Survey (Eberts *et al.* this book) indicates that maximum transpiration may occur around 2007 and will result in an approximate by 30 percent decrease in the groundwater volumetric flux out of the site. The increase in the groundwater gradient caused by the zone of depression under the trees and the seasonal effect of tree dormancy makes complete plume control difficult (U.S. EPA 2001a).

While these observed and simulated reductions in groundwater flux are positive, the reductions may not satisfy a regulatory containment requirement. The system effectiveness may be increased with the addition of more trees. The system at Carswell is a low-cost, easy to implement and maintain remedial system that is suited for a long-term contaminant reduction strategy. The system produces virtually no process residuals and requires minimal maintenance. The system has increased in effectiveness over the years.

Ogden: Organic Contaminants in Soil and Groundwater

The SITE Program accepted this evaluation at the former Chevron Terminal site, Ogden, Utah, in 1995. The objective was to assess the capability of higher plants to reduce the concentration of petroleum hydrocarbons in near-surface soils, to modify the groundwater gradient, and reduce petroleum hydrocarbons in the saturated zone. Alfalfa (*Medicago sativa*) and fescue (*Festuca* spp.) were used for soil remediation, while hybrid poplar (*Populus deltoides* × *Populus nigra* 'DN34' 'Imperial Carolina') and juniper (*Juniperus* spp.) trees were investigated to treat the saturated groundwater zone. Preliminary soil phytotoxicity tests were conducted over a range of contaminant concentrations to select plants that would tolerate those contaminants at the concentration levels present at this site. Greenhouse treatability studies were used to select the most successful varieties.

A triple row of hybrid poplars (*Populus deltoides* × *Populus nigra* 'DN34' 'Imperial Carolina') was planted in the path of the plume, perpendicular to the direction of groundwater flow. Groundwater wells were used to measure the contaminant concentration and aquifer level over the 1997 and 1998 growing seasons. Ferro *et al.* (2001) published these data and data for 1999, and analyzed the information.

The evaluation determined the following:

1. Total petroleum hydrocarbon concentrations in the soil plots planted with alfalfa (*Medicago sativa*) and fescue (*Festuca* spp.) were reduced by 30 percent annually
2. A depression of the water table could not be detected (Ferro *et al.* 2001)

As at Carswell, the trees at Ogden did not capture the entire plume. However, the trees did take up some contaminated groundwater and removed soil contaminants.

Magic Marker and Fort Dix: Phytoextraction of Lead

Phytotech, Inc. developed a process that combines soil amendments (ethylenediaminetetraacetic acid or EDTA) and selected metal-accumulating crops to extract and accumulate lead and other metals from shallow soils. The SITE Program evaluated the process between May 1997 and November 1998 at the former Magic Marker factory in Trenton, New Jersey. In June 1999, Edenspace Systems Corporation of Reston, Virginia, purchased Phytotech, Inc.

The 3-hectare (7-acre) Magic Marker site, an urban brownfield, was occupied by a lead–acid battery manufacturing company from 1947 to 1979. At the site, previous investigations identified lead in the surface soils above the residential regulatory limit of 400 milligrams per kilogram of dry soil.

There were two primary objectives for this SITE project. The first was to determine with a 90 percent level of confidence that the average concentration of lead accumulated in the aboveground plant tissue in each crop harvested from the treatment plot would be greater than or equal to 200 milligrams of lead per kilogram on a dry weight basis. The second objective was the achievement of a minimum 15 percent reduction in soil lead at sample locations within the treatment plot where the baseline concentrations exceeded the state residential regulatory standard of 400 milligrams per kilogram of dry soil (U.S. EPA 2001b).

The project evaluated multiple crops grown in a 9.1 by 17 meter (30 by 57 feet) treatment plot over two growing seasons. Two crops of Indian mustard (*Brassica juncea*) grew and were harvested over the spring and summer of 1997. One crop of sunflower (*Helianthus annus*) grew in the summer of 1998. A 9.1 by 12 meter (30 by 40 feet) control plot was also established to evaluate a no-treatment scenario. Harvested plant tissue samples were collected from the treatment plot for evaluating the amount of lead uptake in each crop. Soil samples were collected prior to the first crop planting and immediately following harvest of the third crop. These samples were for evaluating the change in lead contaminant concentrations in the top 0.15 meters (6 inches) of soil.

Results of the evaluation (Table 31-1) indicated that lead accumulated in the aboveground plant tissue. Samples from the first and second crops of

TABLE 31-1 Summary of Magic Marker Crop Growth and Plant Tissue Data

	Crop 1	Crop 2	Crop 3
Crop name	Indian mustard (*Brassica juncea*)	Indian mustard (*Brassica juncea*)	Sunflower (*Helianthus annus*)
Tissue sampling date	July 27, 1997	Sept. 18, 1997	Aug. 26, 1998
Crop growth data			
Growing period	June 7 to July 27	July 28 to Sept 18	June 20 to August 26
Growing period length	50 days	52 days	67 days
Temperature[a]	22 °C (71 °F)	21 °C (70 °F)	23 °C (74 °F)
Regional precipitation[a]	79 millimeters (3.1 inches)	124 millimeters (4.9 inches)	66 millimeters (2.6 inches)
Amendment dosage[b]	Initial	Optimized	Optimized
Plant tissue data			
Aboveground tissue lead concentration[c]	830	2300	400
Root tissue lead concentration[c]	—[d]	89	290
Aboveground tissue mass (wet)[e]	86.5 kilograms	164 kilograms	141 kilograms
Aboveground tissue mass (dry)[e]	10.5 kilograms	12.2 kilograms	34.4 kilograms
Total lead extracted from soil[f]	8.7 grams	28 grams	14 grams

[a]Estimated average based on NOAA historical data for the state of New Jersey.
[b]The amendment mixture and dosage was changed after crop 1. Note that the amendments and ratios used are proprietary.
[c]Test method was U.S. EPA SW-846 3050/6010. All values have been, expressed in milligrams of lead per kilogram of plant tissue, corrected for moisture content to express the concentration on a dry weight basis, and rounded to two significant digits.
[d]No sample collected.
[e]Values are rounded to three significant digits. A total of five samples were collected for crops 1 and 2, and ten samples were collected for crop 3. Each sample represented the biomass harvested from 0.84 square meter (1 square yard). Wet mass values are the average sample weight times the treatment plot area (159 square meters or 190 square yards). Dry mass values were then determined using moisture content data.
[f]Aboveground tissue concentration times aboveground tissue mass. Values are rounded to two significant digits.

Indian mustard (*Brassica juncea*) contained 830 and 2300 milligrams of lead per kilogram, respectively, on a dry weight basis. Lead in the aboveground plant tissues of the single crop of sunflowers (*Helianthus annus*) was measured at an average concentration of 400 milligrams of lead per kilogram of plant dry weight. All three of these average values exceeded the project objective of achieving a minimum lead uptake of 200 milligrams per kilogram. Based on estimates of plant biomass from each planting, the total mass of lead removed from the treatment plot by all three crops was approximately 51 grams.

There was disparity between the amount of lead measured in shoots and the lead reduction percent measured in the soil from the treatment plot. Plant uptake alone did not account for the magnitude of reductions encountered in the soil. Assuming that the conditions responsible for the observed 24 percent decrease in soil lead concentration in the control plot were also operative in the treatment plot, then the reductions attributable to phytoextraction may be on the order of 4 percent (the difference between 28 and 24 percent). This level of reduction is consistent with the mass of lead removed using the plant uptake data.

To follow the evaluation at the Magic Marker site, a further field test was conducted at the Small Arms Firing Range 24 of Fort Dix in New Jersey. The objective was to clean up lead-contaminated soil to achieve the New Jersey Department of Environmental Protection residential standard of 400 milligrams per kilogram or a Toxicity Characteristics Leaching Procedure (TCLP) concentration of 5 milligrams per liter or less for each sample collected.

The phytoextraction test was conducted in a 0.506 hectare (1.25 acre) *ex situ* lined treatment cell. The cell was equipped with the following:

1. A drainage system to collect excess water from irrigation
2. A lined precipitation basin
3. A setup to recirculate the water through the irrigation system as needed

The investigation included three cropping cycles, followed by collection of soil samples for lead analysis. The average plant uptake on a dry weight basis of the three crops was

1. Indian mustard (*Brassica juncea*): 1400 milligrams of lead per kilogram (June 2000)
2. Sunflower (*Helianthus annus*): 1750 milligrams of lead per kilogram (August 2000)
3. Mixed grasses: 4400 milligrams of lead per kilogram (October 2000)

All three crops were subject to above average rainfall and were also irrigated with leachate, containing both lead and EDTA. Rain filled the drainage basin, requiring water tanks to store the excess beyond what could be used for irrigation. At the end of the third crop, the 420 cubic meters (110 000 gallons) of excess water were tested and found to contain 160 milligrams of lead per liter.

The initial average soil lead concentration was 515 milligrams per kilogram, and 290 milligrams per kilogram remained in the soil at the conclusion of the evaluation. An amount of lead equivalent to 28 milligrams per kilogram in the soil was contained in the water drained from the cell, and a maximum amount equivalent to 20 milligrams per kilogram was drawn into the plant material. The remaining final 177 milligrams per kilogram of soil lead apparently removed from the soil, remain unaccounted. With a total lead concentration below 400 milligrams per kilogram, a lead TCLP less than 5 milligrams per liter, a soil pH greater than 6, and no increase in groundwater lead due to the protective liner, the developer could claim that all project goals were met.

Though project goals were technically met, it is difficult to justify calling this project a successful demonstration of phytoextraction. Even by an optimistic estimate, three plant crops removed only the equivalent of 20 milligrams of lead per kilogram of soil. The average initial lead in soil was 515 ± 36 milligrams per kilogram and the final average concentration was 290 ± 67 milligrams per kilogram. The amount of lead removed by the plants (20 milligrams per kilogram) was less than the uncertainty due to soil inhomogeneity. It was also less than the amount of lead remaining in the drained water (28 milligrams per kilogram) and much less than the unaccounted for amount (177 milligrams per kilogram). While it is unknown where the unaccounted lead went, it is clear that not much went into the plants. It must be concluded that phytoextraction, as conducted at Range 24, was ineffective in removing lead.

This evaluation showed that it is extremely difficult to control field-scale phytoextraction when additives chelate the lead. Further, laboratory studies are necessary to see if the use of chelating agents can be optimized at a lower dosage, consistent with the rate of plant uptake of chelated lead and infiltration. Therefore, uncontrolled field evaluations should not be conducted until laboratory optimization or bench-scale proof of principle is obtained.

The use of chelates to mobilize metals is inherently risky. At Fort Dix, the high level of lead in the drainage water was directly related to the intentional mobilization of metals which, without a liner, would have resulted in an uncontrolled release to surface or groundwater. Regulators should be wary of this technology as applied at Fort Dix if applied without a liner (U.S. EPA 2001b). The expense of lined, *ex situ* phytoextraction may not be cost competitive with soil washing and other remediation techniques.

Argonne National Laboratory-East: TreeWells®, Trichloroethene, and Tritium

The 317/319 Area at Argonne National Laboratory-East (approximately 2 hectares or 5 acres) contains several sites used in the past to dispose of solid and liquid waste from various laboratory activities. Because of these past activities, volatile organic compounds and tritium have been released into the groundwater at depths approximately of 6 to 9 meters (20 to 30 feet) and are moving off-site. The Department of Energy Accelerated Site Technology Deployment Program and EM-40 funded Argonne to deploy a phytoremediation system to supplement the existing pump-and-treat system.

In the 317 French drain area and in three distinct areas down gradient, approximately 800 hybrid poplars (*Populus trichocarpa* × *Populus deltoides*) and willows (*Salix* spp.) were planted in the spring of 1999 at varying, predetermined depths as an engineered plantation. The objective was to provide hydraulic control and contaminant uptake, transformation, and transpiration from the contaminated groundwater (Negri, Gatliff *et al.* this book). Control plots were set up both in the planting areas and at another Argonne-East location to determine tree-induced effects in contrast with general changes induced by weather and time. Monitoring wells, moisture sensors, and sap flow meters were also installed to document effectiveness. Groundwater, air, soil, tree tissue, and transpired vapor will be collected and analyzed at regular intervals to determine contaminant fate (SAIC 1999).

Applied Natural Sciences, Inc. is the contractor who provided the concept, provided the trees, and did the fieldwork to install the remediation system at Argonne-East. Edward Gatliff of Applied Natural Sciences is the holder of the patents assigned to the TreeMediation® and TreeWell® technologies applied in this evaluation.

The U.S. EPA joined this project in 1999, just before planting, in the role of independent evaluator. The final SITE sampling was due to occur after the growing season of 2001, with an analysis and final report to follow.

Jones Island: Phytoremediation of Dredged Sediments

The U.S. Army Corps of Engineers maintains approximately 140 harbors and channels for commercial and recreational users around the Great Lakes. Because of the migration of sediments, annual dredging of approximately 2×10^6 to 4×10^6 cubic meters (3 to 5 million cubic yards) is required to maintain these navigable waterways. In 1967, the Corps began investigating environmentally sound alternatives to the open water disposal of dredged material. It was during this time that the concept of a confined disposal facility was first conceived and implemented. There are now 45 confined disposal facilities in the Great Lakes region. Of these 45 facilities, 28 remain operational and 17 are full. Six of the 28 operational facilities are now nearing design capacity (*i.e.*, greater than or equal to 85 percent full). Because the construction of replacement confined disposal facilities is cost-prohibitive, Corps policy now encourages the development of beneficial uses for dredged material.

The Jones Island confined disposal facility in Milwaukee, Wisconsin, has quantifiable amounts of polycyclic aromatic hydrocarbons, polychlorinated biphenyls, and metals. Typically, the concentration of these contaminants is low (barely exceeding solid waste criteria), but high enough to restrict management options. As with other high-volume, low-concentration wastes, disposal alternatives are extremely limited. Phytotransformation is a potential management alternative. A cooperative partnership between the Corps and the Milwaukee Port Authority has the long-term goal of developing a cost-effective treatment and disposal option for handling and managing dredged material, and eventually creating a marketable product. The project joined the SITE Program in 2000.

A confined disposal facility naturally supports dense vegetation. The dominant vegetation is reed canary grass (*Phalaris arundinacea*), sandbar willow (*Salix interior*), and tall nettle (*Urtica procera*).

The dredged materials generally consist of brown to black silt. The concentration of target pollutants varies across the site. Polycyclic aromatic hydrocarbon concentrations range from undetectable to 120 milligrams per kilogram; polychlorinated biphenyl concentrations range from undetectable to 4 milligrams per kilogram; and diesel range organics from 5 to 1300 milligrams per kilogram. Depth profiles show higher diesel range organics levels in the 0.6 to 1.2 meters (2 to 4 feet) interval than the 0 to 0.3 meter (0 to 1 foot) interval.

Four treatments are being evaluated in field plots. The first treatment plot is a rotation of clover (*Trifolium* spp.) and corn (*Zea mays*). The second treatment plot uses willow (*Salix* spp.) cuttings; a third plot allows natural vegetation to take root and grow. The fourth plot is an unplanted control. The final soil concentration sampling will occur following the second growing season in 2002. Summary reports will be available after project completion (SAIC 2001).

The primary objective of this SITE project is to determine whether the planned treatments can attain residual soil contaminant levels for critical analytes, polycyclic aromatic hydrocarbons, polychlorinated biphenyls, and diesel range organics, consistent with current Wisconsin Department of Natural Resources requirements. No official standards exist for the beneficial use of dredged sediments; therefore, the treatment goals for this evaluation are derived from the most appropriate and relevant promulgated rules currently available as follows:

1. Polycyclic aromatic hydrocarbons. The Wisconsin Department of Natural Resources has suggested that the most appropriate standard is NR 538, "Beneficial Use of Industrial Byproducts" of the Wisconsin Administrative Code. This document lists the numerical criteria for regulated polycyclic aromatic hydrocarbon compounds.
2. Polychlorinated biphenyls. The Toxic Substances Control Act [40 CFR 761.61(a)(4)] for dredged material set the level for no further action. Dredged materials with less than 1 milligram of polychlorinated biphenyls per liter are exempt from notification (§538.14), environmental monitoring (§538.20), and property owner notification (§538.22).
3. Diesel range organics. The state-required cleanup levels are set forth in NR 720.09(4) of the Wisconsin Administrative Code. The generic diesel range organic cleanup levels are based on hydraulic conductivity of the soils at the site. For site soils that exhibit a hydraulic conductivity greater than 1×10^{-6} centimeters per second, diesel range organics must be 100 milligrams per kilogram or less. For site soils that exhibit conductivity less than 1×10^{-6} centimeters per second, a cleanup level of 250 milligrams per kilogram or less is required.

In this case, the Corps has a specific problem and enough time to conduct a careful evaluation on-site. The Corps will use the results of this evaluation to determine if this phytotechnology should be applied full scale at this site. If the scale-up is successful, a protocol may result to implement similar treatments on other confined disposal facilities in the Great Lakes.

Cooperation and planning meetings between the Corps, the State of Wisconsin, and U.S. EPA helped clarify the goals and set very specific criteria for evaluation. This project illustrates how regulatory agencies can work with regulated entities to evaluate if an innovative treatment will be sufficiently protective.

REMEDIATION TECHNOLOGY AND DEVELOPMENT FORUM

Field demonstrations are being undertaken cooperatively by a public and private partnership known as the Remediation Technologies Development Forum (RTDF). Initiated by the U.S. EPA Office of Research and Development and the Technology Innovation Office, the RTDF encourages collaboration among industry members and with various federal, state, and local agencies. Among the RTDF teams is one called Phytoremediation. The phyto action team consists of the following three subgroups:

1. The total petroleum hydrocarbon in soil group
2. The alternative covers group
3. The chlorinated solvents group

The first two groups have fieldwork underway.

Total Petroleum Hydrocarbon Work Group

The total petroleum hydrocarbon group is open to any site owner, manager, or investigator who assures site access and regulatory acceptance for the duration of the project, agrees to follow the consensus-defined protocol, and provides the in-kind resources needed to successfully complete the project. The group has designed a common field study plan or protocol to develop transferable data. From an initial eight sites, the use of the protocol has grown to applications at thirteen, including two sites in Canada that recently joined the effort. Others outside the group, *e.g.*, the Jones Island project, described in the previous section "Jones Island: Phytoremediation of Dredged Sediments," are applying the protocol for soil and sediment remediation. The full protocol is available at www.RTDF.org/phyto/phytodoc.htm with the first annual report that describes initial sampling and planting on many of the sites.

The purpose of the field study protocol is to determine the efficacy of agricultural and other plants for degradation of aged-petroleum hydrocarbons in soil at multiple locations and under varied climatic conditions. The minimum plot size is 6 by 6 meters (20 by 20 feet). Each treatment requires at least four replications. The statistical design is a randomized, complete block. The design requires one composite sample each year per plot for soil and one for plant tissue samples from eight random locations in the plot. The treatments for each site include at least a planting optimized for local conditions as follows:

1. Grass, mix of species, including trees
2. An unplanted and unfertilized control, kept weed-free
3. A standard mixture of 10 to 15 percent annual or perennial rye; 20 to 25 percent legumes including alfalfa (*Medicago sativa*), clover (*Trifolium* spp.), and bird's-foot trefoil (*Lotus corniculatus*); and 60 to 70 percent fescue (*Festuca* spp. varieties chosen for local conditions).

Each site uses common test methods and, as much as possible, common laboratories (Table 31-2). Each site was tested for agronomic conditions such as pH, salinity, and available nutrients, and tested annually for the following contaminant concentrations: polycyclic aromatic hydrocarbons using gas chromatography, total petroleum hydrocarbons using dichloromethane as

TABLE 31-2 Summary of Analyses Planned for each Remediation Technology Development Forum (RTDF) Sampling Event

Sampling event	Matrix	Analysis	Purpose
T = initial	Soil	PAH, TPH, agronomic needs, biomarkers, and TPHCWG protocol	Determine variability and agronomic amendments
0	Soil	PAH, TPH, microbial, biomarkers, and TPHCWG protocol	Establish baseline
1	Soil	PAH, TPH, vegetative cover (species percent), root density, and biomarkers	Assess first season results
2	Soil	PAH, TPH, vegetative cover, root density, and biomarkers	Assess second season results
3	Soil, plant	PAH, TPH, microbial, biomarkers, TPHCWG protocol, vegetative cover, root density, and plant uptake	Assess third season results and assess plant uptake

Note: The following abbreviations are used — PAH: polycyclic aromatic hydrocarbons; TPH: total petroleum hydrocarbons; and TPHCWG: a TPH contaminant working group.

a solvent, and total petroleum hydrocarbon fractions using the total petroleum hydrocarbon working group method. Biomarkers provide additional information about the microbial processes (http://www.rtdf.org/public/phyto/protocol/protocol99.htm).

Pooling resources and sharing experiences have allowed working group members to build a robust, standardized protocol, providing results that give confidence to corporate and regulatory bodies alike. As results come in from dissimilar sites across the continent using similar procedures, more data to determine the rate and extent of remediation on any given site will be gathered (Table 31-3). Such evidence is one of the prime determinants for regulatory acceptance of an innovative technology on a given site. Some RTDF members expect to use phytodegradation on multiple sites if it proves effective (Kulakow 2000).

Alternative Cover Assessment Program

The Alternative Cover Assessment Program (ACAP) was launched in 1998 to address a growing interest in the use of alternative landfill covers based on evapotranspiration. Under ACAP, test facilities have been constructed at landfill sites across the country that do not pollute local groundwater. At several sites, alternative covers are being evaluated side-by-side with traditional covers. The resulting performance and economics data will be used to develop cost-effective and environmentally sound covers for the landfill sites, guidelines for alternative cover designs throughout the country, and improved numerical models to be used by design and regulatory engineers.

The ACAP sites have cover designs that emphasize evapotranspiration. These sites utilize plants to return water from the soil to the atmosphere during the growing season. This requires a cover with soil water storage that minimizes year-round drainage.

To date, the design of most landfill covers in the U.S. has been based on criteria developed by U.S. EPA for use in closing either Resource Conservation and Recovery Act Subtitle C (hazardous waste) landfills or Subtitle D (municipal solid waste) landfills. Under Subtitle D of the Resource Conservation and Recovery Act, the U.S. EPA provides recommended minimum criteria for covers placed on municipal solid waste landfills. The recommended covers fall into four categories (Table 31-4) based on the type of liner underlying the waste at each site. The criteria are primarily based on preventing infiltration of water through the cover and into the waste.

The Resource Conservation and Recovery Act stipulates that a regulatory agency may approve an alternative if the final cover design achieves an infiltration reduction equivalent to that of the recommended design. No design guidance exists for alternate cover designs that use equivalent performance standards.

TABLE 31-3 Summary of Total Petroleum Hydrocarbons in Soil at Remediation Technology Demonstration Forum (RTDF) Field Sites as of June 2001

		Date of sampling events						
		Time 0		Time 1				
	Date planted	A	B	A	B	Time 2	Time 3	Microbial analysis
Site A, California	12/1998	12/1998		6/1999		6/2000	6/2001	No
Site B, Ohio	4/1999	6/1999	11/1999	11/1999	12/2000	10/2001	11/2002	No
Site C, Alaska	6/1999	6/1999		9/1999		9/2000	11/2001	Yes
Site D, Alaska	9/1998	6/1998		7/1999	9/1999	9/2000	11/2001	Yes
Site E, Alaska	9/1998	8/1998		6/1999	11/1999	9/2000	11/2001	Yes
Site F, New York	6/1999	6/1999		11/1999		11/2000	11/2001	No
Site G, Kansas	10/1999	10/1999		10/2000		10/2001	11/2002	Yes
Site H, Rhode Island	5/2001	5/2001					11/2003	No
Site I, Missouri	11/2000	4/2001					11/2003	Yes
Site J, Arkansas	10/1999	10/1999		7/2000	5/2001		11/2002	Yes
Site K, Indiana	5/1999	5/1999		5/2000		11/2000	11/2001	Yes
Site L, Saskatchewan, Canada	10/2001	9/2001					11/2004	Yes
Site M, Alberta, Canada	10/2001	9/2001					11/2004	Yes

TABLE 31-4 Recommended Regulatory Landfill Cover Designs

Category	Existing liner	Cover requirement
A	No liner	1. 18-centimeter (6-inch) erosion layer 2. 46-centimeter (18-inch) barrier layer with $K_{sat} < 10^{-5}$ centimeters per second or K_{sat} of underlying soils, whichever is smaller
B	Soil liner with a saturated hydraulic conductivity, $K_{sat} < 10^{-6}$ centimeters per second	1. 18-centimeter (6-inch) erosion layer 2. 46-centimeter (18-inch) barrier layer with $K_{sat} < 10^{-6}$ centimeters per second
C	Soil liner with $K_{sat} < 10^{-7}$ centimeters per second	1. 18-centimeter (6-inch) erosion layer 2. 46-centimeter (18-inch) barrier layer with $K_{sat} < 10^{-7}$ centimeters per second
D	Composite liner (soil layer having a $K_{sat} < 10^{-7}$ centimeters per second overlain by geomembrane)	1. 18-centimeter (6-inch) erosion layer 2. geomembrane 3. 46-centimeter (18-inch) barrier layer with $K_{sat} < 10^{-5}$ centimeters per second

K_{sat} is saturated hydraulic conductivity in centimeters per second.

The ACAP has the following three phases:

1. Phase 1. Review of previous data collection efforts and applicable numerical models
2. Phase 2. Design and construction of a network of alternative cover testing facilities
3. Phase 3. Monitoing of operations for 5 years and analyzing field results

The ACAP team completed phase 1 in September 1999. The design and construction segments of Phase 2 were completed in November 2000. Under phase 2, test facilities for evaluating alternative covers have been constructed at twelve landfills. Table 31-5 lists all the landfills participating in the ACAP in chronological order from June 1999 to November 2000. The U.S. EPA initiated Phase 3 activities in 2001.

The basic components of alternative covers at the participating sites are vegetation and soil. Different communities of trees, shrubs, and grasses are incorporated, depending on the local soil and climate. The cover soils chosen are available locally and the vegetation is generally native to the area. A typical demonstration site contains at least one test section consisting of a large-scale, pan-type lysimeter to monitor percolation through the tested covers. Of the twelve sites participating in ACAP, nine involved the construction of pan lysimeters. Seven of these sites include a prescriptive Resource

TABLE 31-5 Installations Involved in the Alternative Cover Assessment Program (ACAP)

Name of landfill facility and location	Owner	Test sections constructed	Prescriptive test sections	Alternative test sections	Average annual precipitation in millimeters per year (inches per year)	Vegetation on test sections
Altamont Landfill and Resource Recovery Facility, Livermore, California	Waste Management, Inc.	2	1	1	343 (13.5)	Local grasses
Bluestem Landfill Site No.2, Marion, Iowa	Bluestem Solid Waste Agency	3	2	1	925 (36.4)	Hybrid poplars (*Populus* spp.)
Center Hill Landfill, Cincinnati, Ohio	City of Cincinnati	Leachate collection system	Not applicable	Not applicable		Hybrid poplars (*Populus* spp.)
Uranium Mill Tailings Repository, Monticello, Utah	U.S. Department of Energy	3.04-hectare (7.5-acre) collection area	Not applicable	Not applicable	381 (15)	Mixture of native and adapted plants
Douglas County Landfill, Bennington, Nebraska	Waste Management of Nebraska	3	1	2	711 (28)	Local warm and cool seasonal grasses
Finley Buttes Regional Landfill, Boardman, Oregon	Waste Connections, Inc.	3	1	2	221 (8.7)	Mixture of crested grasses
Green II Landfill, Logan, Ohio	Consortium of responsible parties	Leachate collection system	Not applicable	Not applicable		Hybrid poplars (*Populus* spp.)

Kiefer Landfill, Sloughhouse, California	Sacramento County Public Works Agency	2	0	2	437 (17.2)	Native grasses and trees
Lake County Landfill, Polson, Montana	Lake County Solid Waste Management District	2	1	1	345 (13.6)	Native perennial grasses, forbs, and shrubs
Lewis and Clark County Landfill, Helena, Montana	Lewis and Clark County	1	0	1	305 (12.0)	Native perennial grasses, forbs, and shrubs
Marine Corps Logistics Base Landfill, Albany, Georgia	U.S. Marine Corps	2	1	1	1280 (50.4)	Hybrid poplars (*Populus* spp.)
Monterey Peninsula Landfill, Marina, California	Monterey Regional Waste Management District	2	1	1	411 (16.2)	Mixture of perennial grasses

Conservation and Recovery Act cover test section to facilitate a side-by-side comparison between the prescriptive cover and the alternative covers. The remaining three sites involve equipment to monitor the performance of the alternative cover designs at full scale.

Over 5 years, the U.S. EPA will monitor and record the climatological conditions (*e.g.*, rainfall, snowfall, air temperature, solar radiation, and humidity) at all sites plus soil parameters (moisture content, moisture potential, and temperature) at ten of the sites. Annual reports will discuss each site and the performance of the alternative covers. Most of the ACAP sites were constructed to evaluate a proposed alternative design for full-scale application in the next 3 to 5 years. State and regional regulators participated with the site owners and the cover design team in drafting the evaluation criteria for each test facility.

The data collected through ACAP may also lead to the development of new computer models for designing and evaluating future landfill covers by providing a reliable data set that includes many climates, soil types, and cover designs. Equivalence is a difficult standard to achieve, especially when disparate covers achieve the same ends. For example, conventional covers are designed to exclude all moisture, while evapotranspiration covers store water for release during the growing season. Many laws and regulations allow some flexibility for regulators to accept technology that was not available at the time the regulation was drafted, while others are less flexible. The ACAP data should help establish how to measure performance equivalence (U.S. EPA 2001c).

INTERSTATE TECHNOLOGY AND REGULATORY COUNCIL

Established in 1995, the Interstate Technology and Regulatory Council (ITRC) is a state-led, national coalition of personnel from the regulatory and technology programs of more than 38 states, 3 federal agencies, and tribal, public, and industry stakeholders. The organization is devoted to reducing barriers and speeding interstate deployment of better, more cost-effective, and innovative environmental technologies.

Various tools have been developed and services provided by the ITRC to accomplish this goal. The ITRC Technical and Regulatory Guidance Documents, each of which deals with a specific type of technology, enable more thorough, faster reviews by state agencies of permit applications, site investigations, and remediation plans for full-scale deployment. Use of these documents by states in regulatory reviews also fosters greater consistency in technical requirements among states, and results in reduced fragmentation of markets for technologies caused by differing state requirements. The ITRC offers several 2-hour teleconference and internet courses, and 2-day training sessions called "Introduction to Phytotechnologies." More information about the ITRC and the documents and training schedule can be found on the Internet at http://www.itrcweb.org.

The ITRC Phytoremediation work team recently completed the "Phytotechnology Technical and Regulatory Guidance Document" (http://www.itrcweb.org). The intent of this document is to determine the potential effectiveness of a phytotechnology at a given site. The document complements existing introductory documents. The technical and regulatory document incorporates a previous document called the "Phytoremediation Decision Tree," and provides decision-making scenarios for soil, groundwater, and sediments. Each decision tree also includes basic information. In addition to the decision trees, a brief overview of various types of phytoremediation and of stakeholder concerns is included. A glossary of terms used in the phytoremediation field is also included as a resource for the user.

The ITRC documents and training courses help educate regulators and those who prepare technology proposals for regulators. While the ITRC does not claim to offer answers to every concern, all the relevant questions are disclosed so that proposers can offer answers with the initial proposal to use innovative technologies.

REGULATORY ACCEPTANCE

Regulation requires environmental cleanup and, in many cases, provides criteria to determine when the cleanup has been completed. Knowing which laws are driving the remediation tells a great deal about what options are acceptable. Some states have laws that seemingly prohibit long-term cleanups such as most phytotechnologies, whereas other states have exemptions and incentives for the use of innovative treatment. One example of regulatory flexibility may occur when contaminated groundwater is used to irrigate the plants used in a phytotechnology. If contaminated groundwater is pumped to the surface and then applied to plants, some state regulations regarding surface discharge or reinjection of contaminated groundwater might be applicable. Some states have exceptions to bans on reinjection based on the treatment that can be forecast for irrigation of the right vegetation (U.S. EPA 2000).

Based on the limited experience of phytotechnologies in field applications, three categories of treatments are most likely to be accepted. First, the partial interception of groundwater containing chlorinated solvents, total petroleum hydrocarbons, or methyl *tertiary*-butyl ether occurred at Carswell, Ogden, and other sites. In addition, Thompson *et al.* (this book) establish that created wetlands can control and treat explosives in groundwater. Second, landfill covers based on evapotranspiration are ready for sites without a current groundwater concern. Finally, phytodegradation of low molecular weight organic contaminants in soil are possible in places like the McCormick and Baxter site and the RTDF total petroleum hydrocarbon sites.

Innovative treatment technologies are more acceptable to regulatory agencies when existing containment systems are in place. Therefore, phytotechnologies may be successfully implemented as one stage of a treatment train, as in the case of a tree plantation situated on a groundwater plume, to transpire the water and degrade the contaminants as a supplement to an existing pump-and-treat system. Such use could result in cost savings if the trees decrease the operating and maintenance costs of the other treatment technologies in the train.

REFERENCES

Ferro, A.M., S.A. Rock, J. Kennedy, J.J. Herrick, and D.L. Turner (1999) Phytoremediation of soils contaminated with wood preservatives: greenhouse and field evaluations. *Int. J. Phytoremed.* **1**(3): 289–306.

Ferro, A., J. Chard, R. Kjelgren, B. Chard, D. Turner, and T. Montague (2001) Groundwater capture using hybrid poplar trees: evaluation of a system in Ogden, Utah. *Int. J. Phytoremed.* **3**(1): 87–104.

Kulakow, P. (2000) The 2000 annual report of the RTDF phytoremediation action team – TPH subgroup: cooperative field trials. U.S. Environmental Protection Agency, Washington, D.C. (http://www.rtdf.org/public/phytodoc.htm).

SAIC (1999) Argonne National Laboratory quality assurance project plan, Scientific Applications International Corporation. Contract 68-C5-0036. Report to U.S. Environmental Protection Agency. National Risk Management Research Laboratory, Cincinnati, Ohio.

SAIC (2001) Jones Island quality assurance project plan, Scientific Applications International Corporation. Contract 68-C5-0036. Report to U.S. Environmental Protection Agency, National Risk Management Research Laboratory, Cincinnati, Ohio.

U.S. EPA (2000) Introduction to phytoremediation. U.S. Environmental Protection Agency, Report EPA/600/R-99/107. National Risk Management Research Laboratory, Cincinnati, Ohio (http://www.epa.gov/clariton/clhtml/pubtitle.html).

U.S. EPA (2001a) Phytoremediation at Carswell Air Force Plant 4—short rotation woody crop groundwater treatment system. Innovative Technology Evaluation Report, U.S. Environmental Protection Agency, draft. National Risk Management Research Laboratory, Cincinnati, Ohio.

U.S. EPA (2001b) Phytoextraction of lead in soils at Magic Marker and Ft. Dix. Innovative Technology Evaluation Report, U.S. Environmental Protection Agency, draft. National Risk Management Research Laboratory, Cincinnati, Ohio.

U.S. EPA (2001c) Alternative Cover Assessment Program. Phase II report, U.S. Environmental Protection Agency, draft. National Risk Management Research Laboratory, Cincinnati, Ohio.

Index of Names of Plants

[Common names are those used by authors of the different chapters and are not standardized to any one country, though North American usage is prevalent. Not all scientific synonyms were available for indexing.]

Grasses, legumes, rushes, and sedges

Other forbs

Trees, shrubs, and vines

Index

[Note on use: due to time constraints and length of the compiled material in this book not all terms could be fully indexed and linked; thus some general references to additional chapters are used occasionally. Not all chemical names, especially infrequently used pesticide names, could be indexed and thus the reader should see general classes of compounds like herbicides and pesticides. See the plant name index preceding this index.]

ENVIRONMENTAL SCIENCE AND TECHNOLOGY

A Wiley-Interscience Series of Texts and Monographs

Edited by JERALD L. SCHNOOR, *University of Iowa*
ALEXANDER ZEHNDER, *Swiss Federal Institute for Water Resources and Water Pollution Control*

PHYSIOCHEMICAL PROCESSES FOR WATER QUALITY CONTROL
Walter J. Weber. Jr., Editor
pH AND pION CONTROL IN PROCESS AND WASTE STREAMS
F. G. Shinskey
AQUATIC POLLUTION: An Introductory Text
Edward A. Laws
INDOOR AIR POLLUTION: Characterization, Prediction, and Control
Richard A. Wadden and Peter A. Scheff
PRINCIPLES OF ANIMAL EXTRAPOLATION
Edward J. Calabrese
SYSTEMS ECOLOGY: An Introduction
Howard T. Odum
INTEGRATED MANAGEMENT OF INSECT PESTS OF POME AND STONE FRUITS
B. A. Croft and S. C. Hoyt, Editors
WATER RESOURCES: Distribution, Use and Management
John R. Mather
ECOGENETICS: Genetic Variation in Susceptibility to Environmental Agents
Edwards J. Calabrese
GROUNDWATER POLLUTION MICROBIOLOGY
Gabriel Bitton and Charles P. Gerba, Editors
CHEMISTRY AND ECOTOXICOLOGY OF POLLUTION
Des W. Connell and Gregory J. Miller
SALINITY TOLERANCE IN PLANTS: Strategies for Crop Improvement
Richard C. Staples and Gary H. Toenniessen, Editors
ECOLOGY, IMPACT ASSESSMENT, AND ENVIRONMENTAL PLANNING
Walter E. Westman
CHEMICAL PROCESSES IN LAKES
Werner Stumm, Editor
INTEGRATED PEST MANAGEMENT IN PINE-BARK BEETLE ECOSYSTEMS
William E. Waters, Ronald W. Stark, and David L. Wood, Editors
PALEOCLIMATE ANALYSIS AND MODELING
Alan D. Hecht, Editor
BLACK CARBON IN THE ENVIRONMENT: Properties and Distribution
E. D. Goldberg
GROUND WATER QUALITY
C. H. Ward, W. Giger, and P. L. McCarty, Editors
TOXIC SUSCEPTIBILITY: Male/Female Differences
Edward J. Calabrese
ENERGY AND RESOURCE QUALITY: The Ecology of the Economic Process
Charles A. S. Hall, Cutler J. Cleveland, and Robert Kaufmann
AGE AND SUSCEPTIBILITY TO TOXIC SUBSTANCES
Edward J. Calabrese
ECOLOGICAL THEORY AND INTEGRATED PEST MANAGEMENT PRACTICE
Marcos Kogan, Editor

ENVIRONMENTAL SCIENCE AND TECHNOLOGY
List of Titles (*Continued*)

AQUATIC SURFACE CHEMISTRY: Chemical Processes at the Particle Water Interface
Werner Stumm, Editor
RADON AND ITS DECAY PRODUCTS IN INDOOR AIR
William W. Nazaroff and Anthony V. Nero, Jr., Editors
PLANT STRESS–INSECT INTERACTIONS
E. A. Heinrichs, Editor
INTEGRATED PEST MANAGEMENT SYSTEMS AND COTTON PRODUCTION
Ray Frisbie, Kamal El-Zik, and L. Ted Wilson, Editors
ECOLOGICAL ENGINEERING: An Introduction to Ecotechnology
William J. Mitsch and Sven Erik Jørgensen, Editors
ARTHROPOD BIOLOGICAL CONTROL AGENTS AND PESTICIDES
Brian A. Croft
AQUATIC CHEMICAL KINETICS: Reaction Rates of Processes in Natural Waters
Werner Stumm, Editor
GENERAL ENERGETICS: Energy in the Biosphere and Civilization
Vaclav Smil
FATE OF PESTICIDES AND CHEMICALS IN THE ENVIRONMENT
J. L. Schnoor, Editor
ENVIRONMENTAL ENGINEERING AND SANITATION, Fourth Edition
Joseph A. Salvato
TOXIC SUBSTANCES IN THE ENVIRONMENT
B. Magnus Francis
CLIMATE-BIOSPHERE INTERACTIONS
Richard G. Zepp, Editor
AQUATIC CHEMISTRY: Chemical Equilibria and Rates in Natural Waters, Third Edition
Werner Stumm and James J. Morgan
PROCESS DYNAMICS IN ENVIRONMENTAL SYSTEMS
Walter J. Weber, Jr. and Francis A DiGiano
ENVIRONMENTAL CHEMODYNAMICS: Movement of Chemicals in Air, Water, and Soil, Second Edition
Louis J. Thibodeaux
ENVIRONMENTAL MODELING: Fate and Transport of Pollutants in Water, Air, and Soil
Jerald L. Schnoor
TRANSPORT MODELING FOR ENVIRONMENTAL ENGINEERS AND SCIENTISTS
Mark M. Clark
ECOTOXICOLOGY: Ecological Fundamentals, Chemical Exposure, and Biological Effects
Gerrit Schüürmann and Bernd Markert, Editors
FORMULA HANDBOOK FOR ENVIRONMENTAL ENGINEERS AND SCIENTISTS
Gabriel Britton
SEDIMENT FLUX MODELING
Dominic M. DiToro
PHYTOREMEDIATION: Transformation and Control of Contaminants
Steven C. McCutcheon and Jerald L. Schnoor, Editors